IHS™ Jane's®
All the World's Aircraft
Unmanned

2013-2014

Editor: Martin Streetly

ISBN 978 0 7106 3040 7 - All the World's Aircraft Development & Production
ISBN 978 0 7106 3041 4 - All the World's Aircraft In Service
ISBN 978 0 7106 3043 8 - All the World's Aircraft Unmanned
ISBN 978 0 7106 3077 3 - All the World's Aircraft Full Set

© 2013 IHS. All rights reserved.
No part of this publication may be reproduced or transmitted, in any form or by any means, electronic, mechanical, photocopying, recording or otherwise, or be stored in any retrieval system of any nature, without prior written permission of IHS Global Limited. Applications for written permission should be directed to Christopher Bridge.

Any views or opinions expressed by contributors and third parties are personal to them and do not represent the views or opinions of IHS Global Limited, its affiliates or staff.

Disclaimer of liability
Whilst every effort has been made to ensure the quality and accuracy of the information contained in this publication at the time of going to press, IHS Global Limited, its affiliates, their officers, employees and agents assume no responsibility as to the accuracy or completeness of and, to the extent permitted by law, shall not be liable for any errors or omissions or any loss, damage or expense incurred by reliance on information or any statement contained in this publication.

Advertisement
Advertisers are solely responsible for the content of the advertising material which they submit to us and for ensuring that the material complies with applicable laws. IHS Global Limited is not responsible for any error, omission or inaccuracy in any advertisement. IHS Global Limited will not be liable for any damages arising from any use of products or services or any actions or omissions taken in reliance on information or any statement contained in advertising material. Inclusion of any advertisement is not intended to endorse any views expressed, nor products or services offered, nor the organisations sponsoring the advertisement.

Third party details and websites
Any third party details and websites are given for information and reference purposes only and IHS Global Limited does not control, approve or endorse these third parties or third party websites. Further, IHS Global Limited does not control or guarantee the accuracy, relevance, availability, timeliness or completeness of the information contained on any third party website. Inclusion of any third party details or websites is not intended to reflect their importance, nor is it intended to endorse any views expressed, products or services offered, nor the companies or organisations in question. You access any third party websites solely at your own risk.

Use of data
The company and personal data stated in any directory or database may be used for the limited purpose of enquiring about the products and services of the companies listed who have given permission for their data to be used for this purpose only. You may use the data only to the extent, and in such a manner, as is necessary for the authorised purpose. You must comply with the Data Protection Act 1998 and all other applicable data protection and privacy laws and regulations. In particular, you must not use the data (i) for any unlawful, harmful or offensive purpose; (ii) as a source for any kind of marketing or promotion activity; or (iii) for the purposes of compiling, confirming or amending your own database, directory or mailing list.

Trade Marks
IHS and Jane's are trade marks of IHS Global Limited.

This book was produced using FSC® certified paper
Printed and bound in the UK by Polestar Wheatons

Contents

Alphabetical list of Advertisers .... [4]

Users' Charter .... [12]

Executive Overview .... [7]

Acknowledgements .... [13]

Glossary .... [14]

How to use .... [20]

UNMANNED AERIAL VEHICLES 1

AERIAL TARGETS 293

CONTROL AND COMMUNICATIONS 369

LAUNCH AND RECOVERY SYSTEMS 399

Contractors 415

Alphabetical index 427

Manufacturers' index 433

Alphabetical list of advertisers

Schiebel Elektronische Geraete
Gesellschaft MBH, Margaretenstrasse 112, A-1050, Vienna, Austria ....Bookmark and [5]

IHS Users' Charter

This publication is brought to you by IHS, a global company drawing on more than 100 years of history and an unrivalled reputation for impartiality, accuracy and authority.

Our collection and output of information and images is not dictated by any political or commercial affiliation. Our reportage is undertaken without fear of, or favour from, any government, alliance, state or corporation.

We publish information that is collected overtly from unclassified sources, although much could be regarded as extremely sensitive or not publicly accessible.

Our validation and analysis aims to eradicate misinformation or disinformation as well as factual errors; our objective is always to produce the most accurate and authoritative data.

In the event of any significant inaccuracies, we undertake to draw these to the readers' attention to preserve the highly valued relationship of trust and credibility with our customers worldwide.

If you believe that these policies have been breached by this title, or would like a copy of IHS's Code of Conduct for its editorial teams, you are invited to contact the Group Publishing Director.

www.ihs.com

SCHIEBEL
CAMCOPTER® S-100
UNMANNED AIR SYSTEM

www.schiebel.net

24/7 ready for your mission

EDITORIAL AND ADMINISTRATION

Managing Director: Michael Dell, e-mail: michael.dell@ihs.com
Group Publishing Director: Sean Howe, e-mail: sean.howe@ihs.com
Director EMEA Editing and Design: Sara Morgan, e-mail: sara.morgan@ihs.com
Director IHS Jane's Reference and Data Transformation: Chris Bridge, e-mail: chris.bridge@ihs.com
Product Manager Defence Equipment & Technology: Emma Cussell, e-mail: emma.cussell@ihs.com
Compiler/Editor: Welcomes information and comments from users who should send material to:

Research and Information Services
IHS Jane's, IHS Global Limited, Sentinel House, 163 Brighton Road, Coulsdon, Surrey CR5 2YH
e-mail: yearbook@ihs.com

SALES OFFICES

Europe/Middle East/Africa/Asia Pacific
Tel: (+44 0) 13 44 32 83 00 Fax: (+44 0) 13 44 32 80 05
e-mail: customer.support@ihs.com

North/Central/South America
Tel: Customer care to 1–800–IHS-CARE or 1–800–447–2273
e-mail: customercare@ihs.com

ADVERTISEMENT SALES OFFICES

UNITED KINGDOM
IHS Jane's, IHS Global Limited
Sentinel House, 163 Brighton Road,
Coulsdon, Surrey CR5 2YH, UK
Tel: (+44 20) 32 53 22 89 Fax: (+44 20) 32 53 21 03
e-mail: defadsales@ihs.com

Janine Boxall, Global Advertising Sales Director
Tel: (+44 20) 32 53 22 95 Fax: See UK
e-mail: janine.boxall@ihs.com

Richard West, Senior Key Accounts Manager
Tel: (+44 20) 32 53 22 92 Fax: See UK
e-mail: richard.west@ihs.com

Carly Litchfield, Advertising Sales Manager
Tel: (+44 20) 32 53 22 91 Fax: See UK
e-mail: carly.litchfield@ihs.com

UNITED STATES
IHS Jane's, IHS Global Inc.
110 N Royal Street, Suite 200,
Alexandria, Virginia 22314, US
Tel: (+1 703) 683 37 00 Fax: (+1 703) 836 55 37
e-mail: defadsales@ihs.com

Janet Berta, US Advertising Sales Director,
Tel: (+1 703) 236 24 10 Fax: (+1 703) 836 55 37
e-mail: janet.berta@ihs.com

Drucie DeVries, South and Southeast USA
Tel: (+1 703) 836 24 46 Fax: (+1 703) 836 55 37
e-mail: drucie.devries@ihs.com

Dave Dreyer, Northeastern USA
Tel: (+1 703) 438 78 38 Fax: (+1 703) 836 55 27
e-mail: dave.dreyer@ihs.com

Janet Murphy, Central USA
Tel: (+1 703) 836 31 39 Fax: (+1 703) 836 55 37
e-mail: janet.murphy@ihs.com

Richard L Ayer, Western USA and National Accounts
127 Avenida del Mar, Suite 2A, San Clemente, California 92672, US
Tel: (+1 949) 366 84 55 Fax: (+1 949) 366 92 89
e-mail: ayercomm@earthlink.com

REST OF THE WORLD
Australia: *Richard West* (UK Office)
Benelux: *Carly Litchfield* (UK Office)
Brazil: *Drucie DeVries* (USA Office)
Canada: *Janet Murphy* (USA Office)
Eastern Europe (excl. Poland): MCW Media & Consulting Wehrstedt

Dr Uwe H Wehrstedt
Hagenbreite 9, D-06463 Ermsleben, Germany
Tel: (+49 03) 47 43/620 90 Fax: (+49 03) 47 43/620 91
e-mail: info@Wehrstedt.org

Germany and Austria: *MCW Media & Consulting Wehrstedt* (see Eastern Europe)
Greece: *Carly Litchfield* (UK Office)
Hong Kong: *Carly Litchfield* (UK Office)
India: *Carly Litchfield* (UK Office)
Israel: *Oreet International Media*
15 Kinneret Street, IL-51201 Bene Berak, Israel
Tel: (+972 3) 570 65 27 Fax: (+972 3) 570 65 27
e-mail: admin@oreet-marcom.com
Defence: Liat Heiblum
e-mail: liat_h@oreet-marcom.com

Italy and Switzerland: *Ediconsult Internazionale Srl*
Piazza Fontane Marose 3, I-16123 Genoa, Italy
Tel: (+39 010) 58 36 84 Fax: (+39 010) 56 65 78
e-mail: genova@ediconsult.com

Japan: *Carly Litchfield* (UK Office)
Middle East: *Carly Litchfield* (UK Office)
Pakistan: *Carly Litchfield* (UK Office)
Poland: *Carly Litchfield* (UK Office)
Russia: *Anatoly Tomashevich*
4-154, Teplichnyi Pereulok, Moscow, Russia, 123298
Tel/Fax: (+7 495) 942 04 65
e-mail: to-anatoly@tochka.ru

Scandinavia: *Falsten Partnership*
23, Walsingham Road, Hove, East Sussex BN41 2XA, UK
Tel: (+44 1273) 77 10 20 Fax: (+ 44 1273) 77 00 70
e-mail: sales@falsten.com

Singapore: *Richard West* (UK Office)
South Africa: *Richard West* (UK Office)
Spain: *Carly Litchfield* (UK Office)

ADVERTISING COPY
Sally Eason (UK Office)
Tel: (+44 20) 32 53 22 69 Fax: (+44 20) 87 00 38 59/37 44
e-mail: sally.eason@ihs.com

For North America, South America and Caribbean only:
Tel: (+1 703) 68 33 700 Fax: (+1 703) 83 65 5 37
e-mail: us.ads@ihs.com

Executive Overview

All So Much Hot Air?

The February 2013 cancellation of Northrop Grumman's Long Endurance Multi-Intelligence Vehicle (LEMV) appears to have ended America's brief love affair with remotely piloted/optionally manned airships as a means of providing medium- to high-altitude persistent surveillance. Always at the exotic end of the UAS technology spectrum, the demise of the LEMV programme brings the total of United States (US) unmanned military airship efforts that have been terminated over the last two years to six. While the looming spectre of sequestration has had an obvious impact on the LEMV effort, the remaining five appear to have succumbed primarily to technical and managerial problems. With this in mind it is instructive to look in some detail at these various programmes in order to understand what has gone wrong and whether or not the technology has a future.

In alphabetic order, Lockheed Martin Mission Systems and Sensors' (LM MS2) High-Altitude Long Endurance Demonstrator (HALE-D) airship is the first of our half dozen casualties. Forming part of the US Army's High Altitude Airship (HAA) programme, the 73.2 m (240 ft) long by 21.3 m (70 ft) diameter HALE-D had a hull volume of 14,158 m³ (499,986 cu ft) and was designed to stay aloft for more than 15 days at a station keeping altitude of 18,288 m (60,000 ft). As such, the AV was intended to act as a recoverable and reusable demonstrator for an operational HAA that would be able to reach 19,812 m (65,000 ft) and fly sorties of more than 30 days at a time. Again, HALE-D was able to carry an approximately 20.8 kg (46 lb) payload (made-up of a prototype configuration of Thales Communications' 30 to 512 MHz band Multi-channel Multi-band Airborne Radio (MMAR), an ITT Exelis high-resolution EO system and an L-3 Communications Mini Common DataLink (MCDL) unit), was powered by a pair of 2 kW electric propulsion motors and utilised an array of thin-film solar cells (rated at 15 kW) and re-chargeable lithium polymer batteries (offering a 40 kW/h storage capacity) as its power source.

As *IHS Jane's* understands the programme, the HALE-D effort began during 2003 under the auspices of the US Missile Defense Agency and in 2008 was subsequently transferred to the oversight of the US Army's Space and Missile Defense Command. As of October 2008, HALE-D was expected to make its maiden flight during August 2009, with the actual event taking place approximately two years later on 27 July 2011. Here, HALE-D number one rose from the Akron Airdock in Ohio at 05.47 local time and ascended to an altitude of approximately 9,754 m (32,000 ft) before a detected helium level anomaly resulted in the flight being terminated approximately three

Lockheed Martin's HALE-D technology demonstrator made its maiden flight from Akron, Ohio during the early hours of 27 July 2011 *(Lockheed Martin)*

hours into its planned 14 day mission. Thereafter, the AV was brought down at a "predetermined location" in southwestern Pennsylvania, an event that is understood to have resulted in the destruction of the vehicle's envelope and solar cell array together with fire damage to its payload. In a somewhat upbeat, considering the fact that the HALE-D prototype had been all but destroyed, media release, Lockheed Martin's then Vice President, Ship and Aviation Systems Dan Schulz characterised the flight as having not reaching its target altitude but as having demonstrated "a variety of advanced technologies including launch and control of the airship, [its] communications links, [its] unique propulsion system, solar array electricity generation, [a] remote piloting communications and control capability, in-flight operations and controlled vehicle recovery to a remote unpopulated area".

By October 2012 and the publication of the US Government Accountability Office's *Defense Acquisitions: Future Aerostat and Airship Investment Decisions Drive Oversight and Coordination Needs* (GAO-13-81) the HALE-D programme office was reporting that the effort no longer had "current" funding with which to continue the HALE-D demonstration. (It is worth noting here that the above report is must read for anybody interested in aerostat and remotely piloted airship technology.) Elsewhere within the US Army's HAA effort, mention should be made of the HiSentinel sub-programme that made use of a novel remotely piloted airship design developed by the Southwest Research Institute (SwRI). Here, the HiSentinel V was designed for launch from remote sites and became airborne in a flacid state, with complete hull inflation being achieved by the expansion of its helium lifting agent at altitude. The US Army acquired one example each of SwRI's solar-powered HiSentinel 50 and HiSentinel 80 designs, with the former reaching an altitude of 20,239 m (66,400 ft) for three minutes during a June 2008 test flight and the latter suffering a propulsion system failure eight hours into a planned 24 hour mission that was flown in December 2010. As of October 2012 (and according to GAO-13-81 and like HALE-D), continued demonstration of HiSentinel in the HAA context was no longer being funded.

While both HALE-D and HiSentinel actually flew, Lockheed Martin's Integrated Sensor Is Structure (ISIS) AV never even got off the ground. Launched late in US Fiscal Year (FY) 2004, the 137 m (450 ft) long by 46 m (151 ft) diameter ISIS airship was

An artist's impression of the proposed HAA sensor platform that shows to advantage the type's planned dorsal array of thin-film solar cells *(Lockheed Martin)*

[8] Executive Overview

A computer-generated impression of an on-station ISIS persistent surveillance airship *(Lockheed Martin)* 1365359

being developed under contract to the US Defense Advanced Project Agency (DARPA). It was intended to demonstrate the feasibility of an operational, stratospheric, airship-based, autonomous, unmanned sensor system that would be capable of providing persistent surveillance over periods measured in years. The system would be equipped with a lightweight, dual-band (Ultra High Frequency (UHF - 300 MHz to 3 GHz) and X-band [8 to 12.5 GHz]) Active Electronically Scanned Array (AESA) air-to-air and air-to-ground radar. So equipped, the fully developed capability would be able to track "advanced" cruise missiles and dismounted enemy combatants at ranges of up to 324 n miles (600 km; 373 miles) and 162 n miles (300 km; 186 miles) respectively. A 6,000 m² (64,583 sq ft - 600 m² (6,458 sq ft) in the demonstrator) AESA being used to achieve the required performance. Here, the AESA was to be integrated into its host's structure (hence the ISIS acronym) and was to rely on aperture size rather than brute power to achieve the specified range requirements. Key research areas within the ISIS programme were identified as including the development of the necessary ultra-lightweight antennas, antenna calibration technologies, power systems, station keeping approaches and the airship configuration required to support "extremely large antennas".

Looking at some of these in more detail, the ISIS radar (which was under development by Raytheon Space and Airborne Systems) took the form of a dual-band UHF/X-band sensor that would be capable of simultaneous Air- and Ground Moving Target Indication (AMTI/GMTI) functionality; was designed to operate at an altitude of between 19,812 m (65,000 ft) and 21,336 m (70,000 ft) and which made use of flexible AESA panels that were to be bonded to the hull of a 150 to 300 m (492 to 984 ft) long airship. Again, the full-scale architecture was estimated to have been likely to have used as many as four million low-power transceiver modules, with the technology used being derived from that developed for "low-cost" cellular telephones. Again, Raytheon envisaged the AESA panels used as being approximately one centimetre thick and as requiring a bonding agent that was capable of withstanding an ambient temperature of −80° C (−112°F). Such an environment (combined with the sensor's minimised power requirement) was expected to eliminate the need for the sort of cooling system usually associated with high performance MTI radars, while the use of an aperture size rather than output power approach to performance achievement facilitated the use of solar-regenerative power (with fuel cells instead of batteries) for both the radar and the airship's station keeping and other sub-systems.

Overall, the ISIS payload was likely to have represented more than 30% of the overall host airship's mass (as against no more than 3% in a conventional aerostat-sensor combination) and the host platform was intended to be able to reach sustained and sprint airspeeds of 60 kt (111 km/h; 70 mph) and 100 kt (185 km/h; 115 mph) respectively. In terms of sensor performance, the ISIS radar's GMTI mode was designed to detect dismounts across its entire line-of-sight, have target detection/location and tracking ranges of 600 and 300 km (373 and 186 miles, both at a 3° grazing angle) respectively and to provide "LSRS [Littoral Surveillance Radar System] - like resolution". Again, ISIS GMTI functionality would provide wide-area foliage penetration with "Joint STARS [Joint Surveillance Target Attack Radar System] precision across an extremely large operational area". In AMTI mode, the ISIS sensor was designed for "the theoretical limit at the radar horizon"(effectively, 600 km (373 miles) from the cited operational altitude) and combined the search, track and fire-control functions in a "single platform". In terms of deployment, an operational ISIS platform would be launched from a site (or sites) within the continental US (which would also be home to a "permanent"ISIS ground station/s); would be capable of "global deployment" within 10 days and would have a 10 year service life. As such, the capability would have the potential to replace a range of existing surveillance platforms including the E-2 airborne early warning and control aircraft, the E-3 Airborne Warning And Control System (AWACS) and the E-8 Joint STARS.

Looking at the ISIS demonstrator's programmatics in some detail, Phase 1 (a feasibility study that included an objective system design, the establishment of technology requirements and an analysis of sensor options) is understood to have been launched during US FY2004 and to have been completed by the end of US FY2005. Subsequently, a programme conceptual design review is reported to have been undertaken at the mid-point of US FY2006, an event that was followed by Phase 2 of the programme. Here, the emphasis was on system studies, core technology development and continuation of the architecture's objective design. As part of this, 8 August 2006 saw Raytheon Space and Airborne Systems announce that DARPA had awarded it, a then year, USD8 million contract with regard to the development of a suitable dual-band AESA for use in the ISIS airship and the means by which such an array would be bonded to the host vessel's hull or "other structure[s]". A month later, 26 September 2006 saw Lockheed Martin's Akron, Ohio business unit report that it had been awarded an approximately then year USD10 million, two year long contract with regard to the continued development of "advanced material technology and next-generation hull material[s] for [use in the ISIS] stratospheric [airship]".

The Phase 2 ISIS effort culminated in a mid-US FY2008 technology readiness review, while Phase 3 appears to have been launched at the end of the first quarter of USFY2009 and was intended to include system design demonstration, "large scale integration"activities, "high fidelity flight test simulation", the construction of a one third scale technology

A computer graphic showing the disposition of the lifting gas envelopes and the dual-band AESA within a generic ISIS stratospheric surveillance airship. Note also the station keeping 'motor' on the starboard side of the vehicle's nose section *(Lockheed Martin)* 1299108

demonstrator and flight testing of that demonstrator. Here, a key event occurred on 27 April 2009 when Lockheed Martin's Palmdale, California-based Advanced Development Programs business unit (the corporation's famed 'Skunk Works ') was awarded a then year USD100 million increment (from an existing then year USD400 million cost plus fixed fee contract) with regard to Phase 3 of the ISIS programme. At the time of the announcement, work on the effort was to be undertaken at facilities in El Segundo, California (37% workshare); Palmdale, California, (33% workshare); Denver, Colorado, (14% workshare); Akron, Ohio, (3 % workshare); Litchfield Park, Arizona, (3% workshare); Mesa, Arizona, (3% workshare); Frederica, Delaware, (2% workshare); Huntsville, Alabama, (2% workshare); Monrovia, California, (2% workshare) and Sunnyvale, California, (1% workshare) and the programme was scheduled for completion by the end of March 2013.

The program's contracting activity was DARPA (Arlington, Virginia) and in an associated release, Lockheed Martin characterised the work as involving itself as systems integrator and Raytheon as radar developer. In terms of the effort involved, the release went on to describe the award as centring on an "industry team" designing, building, testing and flight demonstrating a one third scale ISIS airship equipped with a Raytheon developed "low-power density radar". Again, the flight test element of the effort would involve autonomous AV functionality and a 90 day on-station operating period, with the total "key technologies" demonstration being envisaged as being completed within a 12 month period. ISIS Phase 3 is noted as having been "procured under a limited source competition with two bids solicited and two bids received". Here (and aside from Lockheed Martin), IHS Jane's understands that both Northrop Grumman and a consortium of Boeing and the United Kingdom (UK) contractor Hybrid Air Vehicles Ltd (HAV) were interested in the evolving ISIS programme. Which of the two was the second bidder in the Phase 3 competition remains unclear.

As of May 2009, DARPA was reporting that the ISIS hull material had achieved:
- An areal (of, pertaining to, or the nature of an area) density of 90.6 g/m^2 (against a required value of <=100 g/m^2)
- A matrix glass transition temperature of –101°C (–150°F, against a specified figure of <=–90°C [–130°F])
- A fibre strength-to-weight ratio of 1,274 kN.m/kg (against a required value of >=1,000 kN.m/kg)
- Retention of better than 85% fibre strength after a 'life' of 22 years (against a specified figure of >85% strength after five years).

For its part, the evolving AESA had demonstrated a 1.8 kg/m^2 areal density (against a required value of <=2 kg/m^2), a power consumption on-receive of 4.7 W/m^2 (against a specified figure of <=5.0 W/m^2) and viable array-to-hull bonding. Again, the architecture's transceiver modules were reported to have achieved a figure of merit of 1.1 × 10^4 W-2 (against a required value of >=1 × 10^4 W-2) and a mean time to failure value of >1.98 × 10^6 hours (against a specified figure of >10^6 hours). This latter value represented "demonstrated" Technology Readiness Level 5. Last but not least, the ISIS regenerative power sub-system was rated at 779 W-h/kg against a required demonstrated value of 400 W-h/kg.

As of the given date (May 2009), the Phase 3 ISIS Preliminary Design Review (PDR) was scheduled for the early part of US FY2010, with its Critical Design Review (CDR) to follow in early US FY2011. Thereafter, the described demonstrator flight test programme was to take place circa the third quarter of US FY2012. Moving forward, 25 January 2011 saw US contractor ATK announce that Lockheed Martin had selected it to provide a Thermal Control Sub-system (TCS) for the ISIS programme. In an associated release, ATK went on to note that the ISIS TCS contract involved it in the design, development, construction and testing of hardware that would be capable of performing heat acquisition, transport and rejection for two "complete thermal systems". Again, the baseline design for both systems was described as comprising a pumped, two-phase, fluid loop that interfaced with "large, lightweight"radiator panels attached to the vehicle's power bay. Seven months later, August 2011 saw usually reliable sources suggesting that the ISIS demonstrator would now make its maiden flight from Lockheed Martin's Akron, Ohio facility during "late 2013" and would be deployed over the Florida Keys for an up to three month long trial period. Thereafter, the capability would be "placed at the disposal" of US Southern and Northern Commands for approximately nine months and the demonstrator vehicle would be followed by a full-scale operational system at sometime after 2015 subject to the availability of USAF funding. Again, pushing back the ISIS demonstrator's first flight from 2012 to 2013 was said to be due to the need to complete "key"radar component bench testing prior to the procurement of "materials" with which to build the demonstrator's sensor.

By the time GAO-13-81 came to be published in October 2012, the report's authors were characterising the ISIS demonstrator programme as having "experienced technical challenges stemming from [sub-system] development and radar [antenna] panel manufacturing" issues as a consequence of which, DARPA had "temporarily" delayed airframe development activities during the early part of 2012 in order to focus on radar risk reduction activities. During the ensuing hiatus, the ISIS demonstrator team was expected to formulate an airship risk reduction plan and to conduct "limited airship activities". Thereafter, DARPA was expected to "reassess the future plan for [the programme] with the [USAF]". With the benefit of hindsight (and the looming effects of sequestration), this assessment seems somewhat optimistic and in this writer's opinion (and despite the technological interest of the proposed ISIS radar technology) it seems unlikely that ISIS will be resurrected in its current form any time soon.

While HALE-D, HiSentinel and ISIS were intended to demonstrate the viability of persistent stratospheric ISR, the next entrants (MAV6's M1400-I 'Blue Devil 2' and Northrop Grumman's LEMV) in this sorry tale of woe were both intended to provide optionally manned persistent surveillance tailored to the needs of ongoing operations in Southwest Asia. Of the two, the oddly named MAV6 describes its M1400-I 'Blue Devil 2' architecture as being an "airship-based C4ISR [Command, Control, Communications, Computers, Intelligence, Surveillance and Reconnaissance] aerial fusion node and weapons system platform [that integrates] multiple distributed and local sensors with onboard processing [of acquired data] into a common operating picture". Elsewhere, America's Joint Improvised Explosive Device Defeat Organization (JIEDDO - noted as being one of the system's sponsors alongside the USAF and the US Army) characterised 'Blue Devil II' as being a "unique, developmental, integrated, multi-intelligence, auto-tipping and cueing C-IED [Counter Improvised Explosive Device] airborne ISR capability" that integrated "the highest resolution wide field-of-view EO sensor with high-definition cameras and signals intelligence geo-location sensors", with imagery being sent "simultaneously" to "a tactical operations [centre] and remote video terminals in real-time".

MAV6 describes its optionally-manned M1400-I 'Blue Devil 2I' AV as having been constructed in association with TCOM. Again, the M1400-I airship's design was conventional and incorporated four cruciform steering/stabilising surfaces arranged around its aft end. Motive power was to be provided by Rolls-Royce or Honeywell diesel engines (mounted in a ventral "power car") that were to drive port and starboard turbine propulsors, with platform manoeuvre being aided by a rear-mounted manoeuvre engine. Ram air turbines (for electrical power generation) were mounted on the sides of the AV's envelope. In terms of onboard mission systems, M1400-I payload options have been listed as comprising communications, GMTI radar and processing pallets; a wide area EO/IR imaging capability; a SAR radar and multiple high-definition motion video sensors, with the whole being housed primarily in a 7.0 (L) × 3.1 (W) × 2.1 (H) m (23 × 10 × 7 ft) "payload car" that had a volume of 456 m^3 (16,104 cu ft) and was suspended from the bottom of the airship's envelope ahead of a power car. Overall available payload volume is given as having

Executive Overview

A schematic showing connectivity and sensor provision aboard the USAF's proposed 'Blue Devil 2' optionally-manned, long-endurance, multi-intelligence platform *(USAF)* 1365300

been 538 m³ (19,000 cu ft), with individual payload modules being up/down loadable in "less than four hours".

In more detail, unconfirmed sources suggest that the Sierra Nevada Corporation's 'Gorgon Stare' EO ISR sensor and/or the BAE Systems Autonomous Real-time Ground Ubiquitous Surveillance Imaging System (ARGUS-IS) were considered for the wide area surveillance provision, with Northrop Grumman's GMTI/SAR Vehicle And Dismout Exploitation Radar (VADER) being a candidate for the AV's radar system. Other mission systems/capabilities that have associated with the 'Blue Devil 2' architecture include; the 'Pennant Race' Signals Intelligence (SIGINT) payload; unspecified "receivers for ground sensors"; a Ku-band (12.5 to 18 GHz) satellite communications system; a USAF Tactical Targeting Network Technology (TTNT) application; an unmanned ground system relay; a Cubic Defense Tactical Common DataLink (TCDL) application; an AOptix Technologies high-bandwidth laser downlink; and provision for communications with the L-3 Communications Remotely Operated Video Enhanced Receiver (ROVER) architecture. Elsewhere, US contractor Rockwell Collins is known to have been selected to supply the 'Blue Devil II' programme's AV flight control systems, radios and ad hoc real-time communications and ground control station "leveraging"capabilities.

With regard to system functionality, MAV6 described a generic 'Blue Devil II' based "ISR constellation" as comprising four M1400-I airships and as:

- offering "24/7" persistence, inboard data processing ("more information with less bandwidth ") and actionable warfighter information "in 15 seconds or less"
- providing automatic sensor-to-sensor cueing amongst an arrangement of 40 plus sensors (10 or more per M1400-1 platform) and a "streamlined PED [Processing, Exploitation and Dissemination] footprint"that would include a "consolidated Multi-Mission Ground Station (MMGS)" that would be capable of controlling "multiple UAV missions"
- being simultaneously capable of "near real-time dissemination of fused ISR [information] ", full motion video sensor control hand-off and long-term forensic data storage".

Focusing specifically on the 'Blue Devil II's' onboard processing capability, one source suggests that the architecture was designed to incorporate a "near" super computer package that would have been the equivalent in power to "2,000 single-core servers ", was designed to handle up to "300 terabytes [of data] per hour" and would have been able to group data to meet specific end-user requirements. Last but not least, the M1400-I was designed to be flown autonomously, under remote control or by an onboard pilot. Here, the pilot's cockpit appeared to have been located at the front end of the type's payload car and is said to have mimicked "that [installed in the airship's] ground control station". In terms of physical and performance data, the 'Blue Devil 2' airship had a volume of 37,000 m³ (1,306,644 cu ft), an overall length of 112.7 m (370 ft), a payload weight of between 1,134 kg (2,500 lb) and 3,402 kg (7,500 lb) depending on the duration of the specific missio. The airship would also have a maximum cruising speed, operating altitude and endurance of 80 kt (148 km/h; 92 mph), 6,096 m (20,000 ft) and between 72 and 216 hours (environmental conditions and payload dependent) respectively.

Always an ambitious programme, 'Blue Devil 2' was launched in October 2010 in support of "multiple validated CENTCOM [US Central Command] urgent needs" and was intended to contribute directly to "force protection and [C-IED] missions" for coalition forces operating in the region. The architecture's downfall probably stems in part from its US FY2010 transfer from US Army to USAF oversight, with the latter service's Materiel Command (Wright-Patterson AFB, Ohio) being known to have awarded MAV6 a then year USD86.2 million "follow-on" 'Blue Devil 2' contract during March 2011. Two months later (May 2011), reliable sources were suggesting that the 'Blue Devil 2's' hull was scheduled for completion in July 2011, with the AV's maiden flight to follow during the following September 2011 prior to deployment to Afghanistan (for operational evaluation) during February 2012. Again (and as of December 2010), the US Navy's MZ-3A airship was being identified as a pilot trainer and sensor testbed for the operational 'Blue Devil 2' capability. Eight months later (August 2011), usually reliable sources were reporting MAV6 as still being scheduled to deliver the prototype 'Blue Devil 2' during February 2012 and to have had an operational "vision" for the platform that saw it as a "central node in the [USAF's] constellation of surveillance and intelligence aircraft". Here, air vehicles such as the MQ-9A and RQ-4 UASs, the LEMV hybrid airship (see following) and the E-8C Joint STARS manned aircraft would transmit data (via line-of-sight links) to the 'Blue Devil 2' platform which, in turn, would use its onboard servers to process the data and to filter out non-useful information.

Elsewhere in the programme, 17 August 2011 saw US contractor Rockwell Collins announce that the USAF had awarded it a then year USD86.2 million contract with regard to the supply of flight control, control and communications equipment and capabilities for the 'Blue Devil 2' airship, with a second then year USD12,054,022 cost plus fixed fee award going to L-3 Communications Integrated Systems (Greenville, Texas) to cover provision of integration and test support for the M1400-I AV. In the latter case, the USAF's Aeronautical Systems Group at Wright-Patterson AFB was the effort's contracting activity, while the Rockwell Collins award was specified as including the AV's "flight control system, vehicle control system and radios" together with a "real-time, ad-hoc communications capability" (sourced by Rockwell Collins' networking solutions business) and "ground control station leveraging capabilities" (with the work to be carried out by the contractor's simulation and training solutions arm). Moving forward, unconfirmed sources report TCOM as having inflated the M1400-I AV's envelope "with air"at its Elizabeth City, North Carolina facility during early September 2011, a process that is said to have been followed by the replacement of the "air" with helium during the first week of the following October.

Intimations of 'Blue Devil 2's' nemesis began to appear during the early part of 2012 when media reports began to surface suggesting that the programme was in trouble. Here, it was suggested that programme delays, cost overruns and technical difficulties (including problems with the AV's envelope, the rigidity and weight of its stabilising fins and the platform's flight control software) were of increasing concern to the USAF who, it was said, had not been totally committed to the programme from the start. Eventually, May 2012 saw the USAF issue a 'Blue Devil 2' stop-work order citing MAV6's perceived inability to bring the project to a conclusion within the existing time and cost constraints. At this point, unconfirmed sources were suggesting that the programme was 12% over budget and eight months behind schedule. Again, MAV6 is said to have unsuccesfully attempted to interest the USN in completing the effort and by August 2012, a company spokesman confirmed that the stop work notice had been received and noted that the M1400-I airship had been placed in storage "awaiting further government direction" This said, the contractor continued to keep faith with the concept, with the spokesman reporting the creation of a spin-off, MAV6 airship specific business unit that

IHS Jane's All the World's Aircraft: Unmanned 2013-2014 © 2013 IHS

would continue to promote the M1400 and the ongoing development of a "conventional UAV" variant of the 'Blue Devil 2' payload.

While designed to do nearly the same job in the same theatre-of-operations in the same time scale, Northrop Grumman's LEMV initially looked a safer bet than 'Blue Devil 2' by virtue of its having the backing of a larger, more established contractor and single service (the US Army) sponsorship. Taking the form of a hybrid, non-rigid, optionally manned airship that could be configured for either heavy lift or persistent ISR, LEMV was built around UK contractor HAV's 92.1 m (302 ft) long HAV-304 AV. As such, the AV was powered by a quartet of Thielert/Centurion heavy-fuel engines that generated forward and vectored thrust, with the latter being used for take-off, landings and ground manoeuvres. Again, lift was provided by both the AV's envelope (60%) and its aerodynamic shape (40%) and the vehicle is noted as having been able to take-off within roughly three times its own length using a "clear" but not necessarily improved launch area. Recovery was "similar to that of a traditional airship", with flight control being by means of an onboard pilot, a pilot located in a ground control station or an autonomous flight control system. Baseline ISR payloads for the vehicle included four EO sensor turrets (two × L-3 Wescam MX-15HDi and two × L-3 Wescam MX-20D equipments), a GMTI radar (possibly Northrop Grumman's VADER system) and a SIGINT package together with an onboard communications capability, with all the equipment being drawn from a range that was already "in-theatre" in southwest Asia. Other potential ISR-related payloads included a blue force tracker and an electronic attack package and the LEMV as a whole was "recoverable and re-usable ", was designed to accept plug and play payloads and was intended to augment existing ISR platforms with "additional capabilities".

A more detailed look at the LEMV ISR specifics included (as set out in a May 2009 Statement Of Objectives - SOO):

- Un-tethered and unmanned functionality
- An optional manning capability for self-deployment during operations over the continental US
- The ability to deploy and sustain operations at a nominal cruise operating altitude of 6,096 m (20,000 ft) above Mean Sea Level (MSL)
- A minimum payload capacity of 1,134 kg (2,500 lb)
- The ability to fly missions that included weather avoidance in both national and international airspace
- Incorporation of a recoverable payload with interfaces to facilitate destruction or zeroing out if captured by hostile forces
- The ability to forward deploy and to support extended operations from austere forward operating locations
- Provision of an environmental enclosure (or enclosures) for the payload which would be considered as part of the payload and incorporated the necessary sub-system/s required to maintain the required environmental conditions
- The ability to fly a 2,172 n mile (4,023 km; 2,500 mile) round-trip mission
- Use of a modular payload bay (gondola) that incorporated plug and play interfaces and which would be able to support payloads of up to 2,268 kg (5,000 lb) at altitudes "as low as" 3,048 m (10,000 ft) if required
- The ability to deliver at least 16 kW of electrical power to the payload
- A dash speed of 80 kt (148 km/h; 92 mph)
- The ability to remain on-station for three continuous weeks at a designated location or locations and within a 19 n mile (3.5 km; 2.2 mile) radius circle of a given on-station point 50% of the time, within a 40 n mile (75 km; 47 mile) radius 75% of the time and within a 81 n mile (150 km; 93 mile) radius 95% of the time
- A defined growth path to a payload weight of 3,175 kg (7,000 lb), electrical power provision of 73 kW and a mission duration of at least one month during which, the vehicle would be able to remain within a 1.1 n mile (2.0 km; 1.2 miles) radius circle of a given on-station point for 50% of the time and within a 27 n mile (50 km; 31 mile) radius 95% of the time
- The ability to maintain an average cruising speed of 20 kt (37 km/h; 23 mph) true airspeed at the nominal cruising altitude and at capacity payload weight for the duration of the flight
- Provision of a global, autonomous, joint-interoperable C2 system that would operate from a fixed command centre in the US, with control being by means of available satellite communications and terrestrial and/or aerial capabilities.

A display model showing the general layout of the LEMV hybrid, non-rigid surveillance airship *(IHS/Patrick Allen)*

Executive Overview

Other C2 requirements included tactical transportability, utilisation of existing equipment, operation and health monitoring every 10 to 60 seconds, flight and navigation capabilities sufficient to achieve the previously noted station keeping metrics, a complete way-point/route/trip navigation capability, autopilot provision and secure and encrypted communication links incorporation of a redundant, independent, rapid deflation device sub-system.

Elsewhere within the SOO, LEMV candidate ISR missions were listed as including the theatre support, civil application, homeland security and "space mission" roles. Here (and under the theatre support banner), specifics included horizontal and upward viewing optical and radar surveillance, broadband data relay (beacon, sensor, laser communications, C2 and intelligence information being cited), the carriage of intelligence sensors and utility tasks such as flight test support, experimental equipment use and the testing of new airship sub-systems. Civil applications were listed as wireless telecommunications (transmit/receive, radio relay and range extension), support for US FAA communications (range-extension relay and as a transponder beacon adjunct) and participation in US Bureau of Land Management, US National Oceanic and Atmospheric Administration and US Environmental Protection Agency survey work. For their part, homeland security applications were listed as real-time, multi-band, persistent area surveillance (horizontal and downward viewing domains); border patrol; counter-terrorism, counter-narcotics smuggling and communications linking and relay. Last but not least, cited "space mission" areas included Space Force "enhancement" and "application"; Command, Control, Computers and Communications (C4); ISR; communications support; position/navigation augmentation and weapons targeting.

Within the LEMV consortium put together by Northrop Grumman, HAV acted as the AV's design authority and provide the vehicle's payload and fuel bays, propulsers (and their associated ducts), skeleton structural elements, fins and control surfaces while US contractor Warwick Mills was responsible for prototyping and verification tasks. For its part, US contractor ILC Dover undertook fabric layout and vehicle inflation, while AAI provided the US Army's family of One System ground control and data reception architectures with which, LEMV was to be compatible. Elsewhere, US contractors Aurora Flight Systems, Quantum Research International, Rockwell Collins and SAIC were also involved in the programme.

Looking back, the LEMV effort grew out of the US DoD's Persistent Elevated Reconnaissance Surveillance and Intelligence Unmanned System (PERSIUS) programme. Here, the PERSIUS effort was intended to provide a joint service hybrid unmanned aircraft technology demonstrator with an overall length of 76.2 m (250 ft), an envelope volume of 27,609 m^3 (975,004 cu ft), an endurance of 21 days (at an altitude of 6,096 m [20,000 ft]) and a payload capacity of 1,134 kg (2,500 lb). Within this context, Lockheed Martin is known to have flown its P791 hybrid buoyant craft (bearing the US civil registration N791LM) for the first time on 31 January 2006, with the P791 apparently being intended to be the prototype for an operational PERSIUS AV that would be delivered during December 2009. In the event, 2009 saw PERSIUS being superseded by LEMV.

A general LEMV description was set out in the cited SOO that was first issued on 22 April 2009. A month later (May 2009), a US Army Research, Development, Test and Evaluation (RTD & E) budget item justification document identified the LEMV as being part of Project 978 which covered both "space control"and the LEMV. Associated funding requests included then year USD195,000 during US FY2009 to cover a LEMV related small business innovative research/small business technology transfer programme, followed by an omnibus then year USD80 million sum to cover activity during US FY2010. Within this global figure, USD3 million was earmarked for LEMV matériel development management, acquisition planning, testing and initial demonstration planning, with a further USD1 million being devoted to definition of the LEMV system's architectural requirements and co-ordination with its

LEMV AV number one made its maiden flight on 7 August 2012 *(US Army)*

combat developer with regard to a concept-of-operations. Here, specifics were to include co-ordination with other services on technology development, system engineering, concept trade studies, programme risk identification and the development of acquisition documentation. Elsewhere, USD4 million was requested for a range of risk reduction activities, with the remaining USD72 million being allocated to LEMV contract award, design and fabrication initiation and PDR and CDR activities. Actual LEMV acquisition was to take the form of an Other Transaction Authority (OTA) procurement that would be designed to increase participation from non-traditional DoD contractors. Again, the acquisition was to be pursued "rapidly" with all developmental and operational testing to be completed within 18 months of the award.

On 28 August 2009, the original LEMV SOO was amended to reflect its logistical support for consortia forums to establish LEMV-related ISR technology consortia. At this time, US contractors the AAI Corporation (Hunt Valley, Maryland), Carolina Unmanned Vehicles Inc (Raleigh, North Carolina), Matrix International Inc (McLean, Virginia) and UAV Communications Inc (Newport News, Virginia) had all registered as LEMV "interested vendors". Moving forward, 29 December 2009 saw the publication of a LEMV OTA schemata that if consummated, would launch a five year long technology demonstration that would include fabrication of the air vehicle, integration of its payload and ancillary systems, test and programme support. Performance testing was to be completed within 18 months of contract award after which, the AV was to be deployed to Afghanistan for the remainder of the OTA term. As of the given date (and subject to US Congressional approval notification), release of an OTA RFP was planned for 29 January 2010.

Moving forward again, 14 June 2010 saw Northrop Grumman announce that it had been awarded a then year USD517 million OTA contract with regard to the supply of up to three × LEMV systems. Under the terms of the deal, LEMV numbers two and three were options, with LEMV number one being scheduled to be inflated 10 months after contract signature, to make its first flight no later than 13 months after signature, to enter flight testing (at the Yuma Proving Ground in Arizona) 16 months after signature and to be deployed to southwest Asia no later than 18 months after signature. Here, LEMV number one would self-deploy under the control of an onboard pilot. Moving forward, 2 September 2010 saw L-3 Wescam announce that Quantum Research International had ordered two MX-15HDi and two MX-20D EO sensor turrets for installation aboard LEMV AV number one. In an associated release, L-3 Wescam reported that the planned four turret LEMV configuration would "enable both general surveillance and precise targeting missions on four separate geographical locations simultaneously from a sustained altitude of 20,000 ft [6,096 m] MSL". Again, each of the turrets being supplied was to be equipped with 1,080 pixel imaging cameras; would offer "multiple [high definition] feeds streaming from the cameras within each turret" and would make use of technology that would "generate bolder colours, greater contrasts, increased image sharpness and more pixels on target".

At the time of the award announcement, delivery of the four turrets was to begin during 2010 and was to be completed by January 2011. Two months later, 4 November 2010 saw Northrop Grumman announced that the LEMV programme team had completed the architecture's System Readiness Review, its Initial Baseline Review and its PDR in "less than four months". In an accompanying media release Northrop Grumman itemised the consortium building LEMV as comprising the AAI Corporation, HAV, ILC Dover, SAIC, Warwick Mills and a "team of technology leaders from 18 US states and three countries" in addition to itself. In terms of programmatics, the release further noted that first air vehicle inflation was scheduled for "next spring"(2011), first flight for "mid-next summer" (again, 2011) and acceptance testing for December 2011.

As 2011 began, Northrop Grumman announced on 1 February that it had completed the LEMV architecture's CDR. In an accompanying release the contractor went on to note that in addition to acting as LEMV prime, it was "lead" on system integration and "flight and ground control operations"and that it was now expecting LEMV AV number one to have made its maiden flight by "mid-to-late summer" 2011. Again, a US Army Joint Military Utility Assessment "in an operational environment" was pencilled in for "early 2012". As of late summer 2011, usually reliable sources were reporting Northrop Grumman as having inflated the first of 19 sections within the prototype ISR LEMV's hull during June 2011 and as "being on track for first flight by late July or August [2011]". In the event, LEMV AV made its maiden flight from Joint Base McGuire-Dix-Lakehurst, New Jersey on 7 August 2012. Lasting 90 minutes, this first sortie was flown with a safety pilot aboard and is reported to have demonstrated the AV's airworthiness (including safe launch and recovery) and to have verified its flight control system and "system level performance". According to the US Army, all the flight's objectives were met. In an accompanying release, the service further reported LEMV as being capable of staying aloft for "more than" 21 days at altitudes of "greater than" 6,706 m (22,000 ft); as having a radius-of-action of 1,738 n miles (3,219 km; 2,000 miles); as being able to carry a 1,247 kg (2,749 lb) payload; as offering 16 kW of payload electrical power; as being runway independent and as having a fuel consumption value that was "10 times less than [that of] comparable capabilities ".

While the programme's media releases remained upbeat, the fact that LEMV number one had first taken to the air at a time when it was supposed to have been delivered to Afghanistan for operational evaluation did not auger well for the future. Equally, the fact that it did not fly again during 2012 was bad news and bearing in mind the increasingly fraught financial situation the US military was finding itself in, it is perhaps not surprising that February 2013 saw the US Army announce that it was pulling out of the programme. Here, a stop-work statement noted that LEMV had been "initially designed to support operational needs in Afghanistan in Spring 2012" and went on to state that it obviously would not be providing the required capability in "the time frame required". Accordingly, and also due to "technical and performance challenges and the limitations imposed by constrained resources", the "Army had determined to discontinue the LEMV development effort".

The last of the 'sorry six' was another high-altitude surveillance demonstrator that never got off the ground. Sponsored by the USN and making use of Global Near Space Services Star Light "two-stage, saucer-shaped, lighter-than-air, unmanned communications and surveillance vehicle", this US FY2010 science and technology demonstration effort platform was intended to operate at altitudes of between 19,812 m and 25,908 m (65,000 and 85,000 ft) and to create an AV that could remain on-station for between 30 and 120 days depending on the time of year and the prevailing wind conditions. In programmatic terms, the Star Light demonstrator programme involved one AV and started and finished during US FY2010 due to "insufficient funding".

Taken together, the six described programmes involved seven AVs (of which four flew and one was lost in a crash landing) at a cost of USD1,041.5 million (2012 values) over the period US FY2007 and US FY2012. Again, all six programmes are currently either terminated or are in limbo and in no case has an operational system emerged from their collective development efforts. In trying to understand why these six programmes have failed, the present author identifies a toxic brew made up of over ambitious technology targets and time frames, changing operational requirements, a lack of unambiguous support for the various efforts and the large shadow of sequestration as being the chief culprit. In terms of over ambition, both 'Blue Devil 2' and LEMV were developed against very tight deadlines which in the event proved impossible to meet. Digging down, 'Blue Devil 2' looked on paper to be a relatively easy project that involved modifying a commercially available airship design, integrating available and well-understood payloads and scaling-up an existing flight control system.

As recounted previously, all of these assumptions proved wrong, with the situation not being helped by MAV6 being a start-up company and the 'Blue Devil 2' programme switching sponsors services halfway through and ending up under the oversight of a service (the USAF) that appears not to have been fully committed to seeing the programme through to the bitter end. Add to this a rapidly receding operational requirement that is persistent ISR over Afghanistan and vulnerability to being seen as a 'nice to have' rather than a 'must have' in a time of near financial crisis; it is perhaps surprising that 'Blue Devil 2' got as far as it did. Most of the above is also true for LEMV where despite having the backing of one of America's largest defence contractors in Northrop Grumman, development of the novel hybrid AV that was at the centre of the effort proved more difficult and time consuming than at first thought. Equally (and from a European perspective), the cancellation of LEMV might just have a whiff of the same 'not invented here' mind set that scuppered Airbus's ambitions with regard to the USAF's future in-flight refuelling tanker programme.

Looking at the strategic systems discussed (HALE-D, HiSentinel, ISIS and Star Light), all four attempted to push forward the technological envelope in order to attain the required altitudes and endurances while at the same time utilising primarily solar energy to provide power for their onboard payloads and station-keeping equipment. While none of the above are show stoppers in their own right, it looks as though trying to combine them was a step too far. Again, none of the four systems seems to have had a clear enough goal other than a general interest in a stratospheric, ultra-long endurance surveillance rather than a specific requirement for, say, an anti-ballistic missile detection platform. Mission 'go-away' also affected the 'tactical' 'Blue Devil 2' and LEMV applications, with the withdrawal from southwest Asia effectively ending the need for large, slow and potentially vulnerable platforms that can only really operate in non-contested airspace.

If the foregoing appears unremittingly gloomy, it should be said that MAV6, SwRI and HAV all continue to promote remotely piloted airships for use in military applications. Again (and in response to the recommendations outlined in GAO-13-81), the US DoD is on record as stating that while it has "no intent for robust short term investment [in the technology]", it is "committed to staying abreast of developments in the airship community, identifying potential airship solutions to capability gaps, monitoring commercial activity and engaging in [future] collaborative efforts when appropriate". Accordingly (and assuming relatively rapid adjustment to post-sequestration realities), it looks as though the US is not completely out of the remotely piloted airship market despite the failure of the 'first generation' programmes outlined here. In terms of future possibilities, it seems to this observer that the genre does have a future (particularly in the areas of heavy lift and persistent surveillance in support of 'civil powers') but one which probably will not see the US at the fore front any time soon.

Acknowledgements

The Editor would like to thank all those UAV manufacturers around the world for supplying and updating entry data; his *IHS Jane's* colleagues (in alphabetic order) Michael J Gething, Robert Hewson, Paul Jackson, Kenneth Munson (the originator and driving force behind IHS Jane's UAV reference coverage) and Huw Williams as well as all those involved in the physical and electronic publication of *IHS Jane's All The World's Aircraft: Unmanned* (JAWAU). Here, the Editor would like to recognise the work of Edward Hordley (*JAWAU's* Content Editor), Sarah Loughlin (Senior Information Services Researcher) together with that of IHS Jane's Image Services and Pre-Press teams. To all, a large and heartfelt 'thank you'.

Martin Streetly
February 2013

Editor's Biography

Martin Streetly

Martin Streetly is a full-time author and analyst who specialises in all aspects of airborne Intelligence, Surveillance, and Reconnaissance (ISR) and the history, technology and application of defence electronics (with a particularly emphasis on Electronic Warfare). He is Editor of *IHS Jane's All The World's Aircraft: Unmanned* (JAWAU) and a Contributing Editor on *IHS Jane's C4ISR and Mission Systems: Air*. Prior to their absorption into the *IHS Jane's C4ISR and Mission Systems* portfolio, Martin was the Editor of the *IHS Jane's Radar and Electronic Warfare Systems* yearbook and the Compiler/Editor of the *IHS Jane's Electronic Mission Aircraft* product. Elsewhere, he has also acted as a Contributing Editor on *JAWAU's* predecessor *IHS Jane's Unmanned Aerial Vehicles and Targets*.

Over the past 30 years, Martin has been a regular contributor to various IHS Jane's publications such as *IHS Jane's Defence Weekly* and *IHS Jane's International Defence Review* as well as a variety of other international defence and technology publications. This includes the *Journal of Electronic Defense* (acting as the magazine's European Editor for 12 years up to March 2001), *The Knowles Report, Defence Helicopter, Asian Military Review, Digital Battlespace, Microwave Journal* (as the publication's European Correspondent), *Flight International, Defence, Naval Forces, Military Simulation & Training*, the *NAVINT* naval intelligence newsletter and the Unmanned Vehicle Systems annual; *UAS Global Perspective* annual. Over time, he has appeared on the UK's Channel 4 news programme, the BBC and the Discovery Channel. During the 1991 Gulf War, he worked with the UK's Independent Television News Ltd and a range of international newspapers (including the *New York Times* and the *Jerusalem Post*) and has been invited to lecture on EW technology by industry, NATO, the Government of the United Arab Emirates and the Association of Old Crows.

Over and above the described body of work, Martin has published four books on the history and technology of airborne EW, the details of which are as follows:

Confound and Destroy: 100 Group and The Bomber Support Campaign Macdonald & Jane's Publishers Ltd, London, 1978 and Jane's Publishing Ltd, London 1985.

World Electronic Warfare Aircraft Jane's Publishing Ltd, London 1983 and 1984.

The Aircraft of 100 Group Robert Hale Ltd, London, 1984.

Airborne Electronic Warfare: History, Techniques and Tactics Jane's Publishing Ltd, London, 1988.

Of these, *Confound and Destroy* is considered by many as being the definitive study of the birth of airborne EW in the UK, while over 5,000 copies of the two editions of *World Electronic Warfare Aircraft* have been sold worldwide. Most recently, Martin has been working on a history of airborne SIGINT that has the working title of *On Watch*.

IHS Jane's Defence Equipment & Technology

In 2012 IHS released a new generation of products across the Defence Equipment & Technology portfolio.

The new portfolio is designed to provide our customers with the highest quality defence reference content, grouped by subject and aligned to fit with their workflows, underpinned by new structured content and overlaid with powerful new search, manipulation and navigation tools.

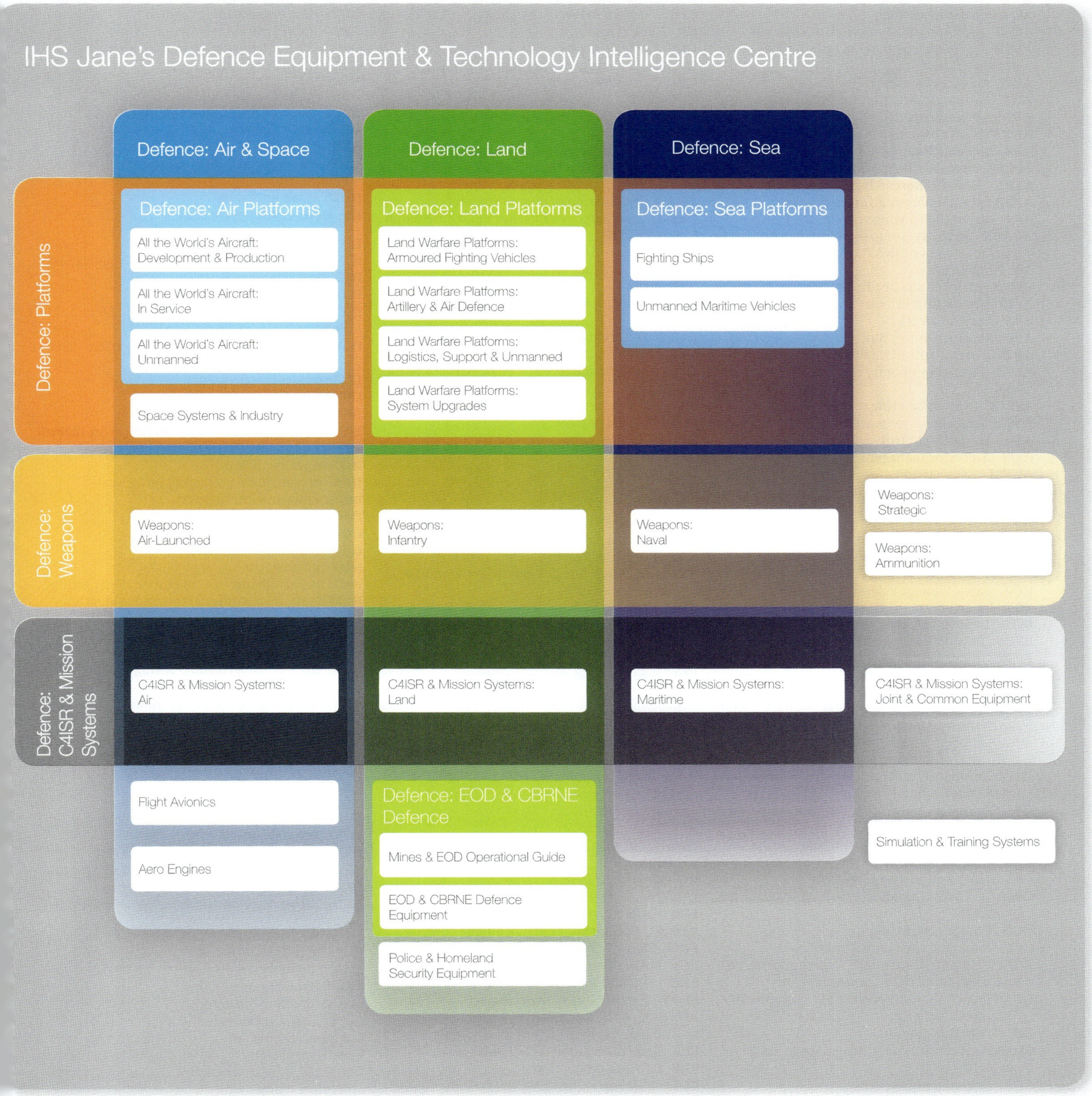

IHS Jane's Defence Equipment & Technology Intelligence Centre

	Defence: Air & Space	Defence: Land	Defence: Sea	
Defence: Platforms	**Defence: Air Platforms** All the World's Aircraft: Development & Production All the World's Aircraft: In Service All the World's Aircraft: Unmanned Space Systems & Industry	**Defence: Land Platforms** Land Warfare Platforms: Armoured Fighting Vehicles Land Warfare Platforms: Artillery & Air Defence Land Warfare Platforms: Logistics, Support & Unmanned Land Warfare Platforms: System Upgrades	**Defence: Sea Platforms** Fighting Ships Unmanned Maritime Vehicles	
Defence: Weapons	Weapons: Air-Launched	Weapons: Infantry	Weapons: Naval	Weapons: Strategic Weapons: Ammunition
Defence: C4ISR & Mission Systems	C4ISR & Mission Systems: Air	C4ISR & Mission Systems: Land	C4ISR & Mission Systems: Maritime	C4ISR & Mission Systems: Joint & Common Equipment
	Flight Avionics Aero Engines	**Defence: EOD & CBRNE Defence** Mines & EOD Operational Guide EOD & CBRNE Defence Equipment Police & Homeland Security Equipment		Simulation & Training Systems

Whether you're a researcher, an analyst, a planner, a strategist or a trainer, whether you're in the military, industry or academia, IHS will help you to understand the defence landscape and support you in your critical intelligence processes.

Contact us or visit
ihs.com/janesnextgen to find out more.

Glossary of Aerospace Terms

The following 'across the board' abbreviations and acronyms are used in this publication; others with more individual meanings are explained in the text as they occur.

A Ampère
AA Anti-aircraft
AAA Anti-aircraft artillery
AAM Air-to-air missile
AC Alternating current
ACAS Airborne collision avoidance system
ACTD Advanced concept technology demonstration
ADC Air data computer
ADF Automatic direction finder/finding
ADS Air data sensor
ADT Air data terminal
AFB Air Force Base (US)
AFCS Automatic flight control system
AFRL Air Force Research Laboratory (US)
AFRP Aramid fibre-reinforced plastics
AFV Armoured fighting vehicle
AGARD Advisory Group for Aerospace Research and Development (NATO)
Ah Ampère hour
AHRS Attitude and heading reference system
Al-Li Aluminium-lithium
AoA Angle of attack
APU Auxiliary power unit
ARINC Aeronautical Radio Inc (US company)
ARPA Former temporary title of DARPA (which see)
ASROC Anti-submarine rocket
ASW Anti-submarine warfare
ATC Air traffic control
ATM Air traffic management
AUVSI Association for Unmanned Vehicle Systems International (US)
AUW All-up weight
AV Air vehicle
Avgas Aviation gasoline
AVO Air vehicle operator

BDA Battle damage assessment
BER Bit error rate
BITE Built-in test equipment
BLOS Beyond line of sight
BPSK Bi-phase shift keying
BVR Beyond visual range
BW Bandwidth
BWB Bundesamt für Wehrtechnik und Beschaffung (Germany)

C2 Command and control
C3I Command, control, communications and intelligence
C4 Command, control, communications and computers
C of A Certificate of Airworthiness
CAD Computer-aided design
CAS Calibrated airspeed
CASA Civil Aviation Safety Authority (Australia)
CCD Charge coupled device
CCI Command and control interface
CCIR Comité Consultatif International des Radiocommunications (France)
CD (1) Circular dispersion; (2) chrominance difference
CDL Common datalink
CDMQ Commercially developed, military qualified
CDR Critical design review
CEP Circular error probability
CFAR Constant false alarm rate
CFRP Carbon fibre-reinforced plastics
CG Centre of gravity
CIA Central Intelligence Agency (US)
CKD Component knocked down
CMOS Complementary metal oxide semiconductor
CMT Cadmium mercury telluride (CdHgTe)

COA Certificate of Authorisation (US)
comint Communications intelligence
CONOPS Concept of operations
CONUS Continental United States
COTS Commercial off-the-shelf
CR Close range
CRT Cathode ray tube
CTOL Conventional take-off and landing
CW Continuous wave

DARO Defense Airborne Reconnaissance Office (US)
DARPA Defense Advanced Research Projects Agency (US)
dB Decibel
DEAD Destruction of enemy air defences
DERA Defence Evaluation and Research Agency (UK)
DF Direction-finding
DGA Direction Générale de l'Armement (France)
DGAC Direction Générale de l'Aviation Civile (France)
DGPS Differential GPS
DLI Datalink interface
DND Department of National Defence (Canada)
DoD Department of Defense (US)
DoF Degrees of freedom
DPCM Digital pulse code modulation

EAS Equivalent airspeed
ECA Experimental Certificate of Airworthiness (US)
ECCM Electronic counter-countermeasures
ECM Electronic countermeasures
ECR Electronic combat reconnaissance
EEPROM Electronically erasable programmable read-only memory
EHF Extra high frequency
EISA Extended industry standard architecture
ELF Extremely low frequency
elint Electronic intelligence
ELT Emergency locator transmitter
EMD Engineering and manufacturing development
EMI Electromagnetic interference
EMP Electromagnetic pulse
EO Electro-optical
EOD Explosive ordnance disposal
EPLRS Enhanced position location and reporting system
ERP Effective radiated power
ESM Electronic support (or surveillance) measures
EW Electronic warfare

FAA Federal Aviation Administration (US)
FADEC Full authority digital engine control
FBW Fly-by-wire
FCS Future Combat System (US Army)
FLIR Forward-looking infra-red
FLOT Forward line of own troops
FM Frequency modulation
FoV Field of view
FPA Focal plane array
FRP Full rate production
FSAT Full-scale aerial target
FSED Full-scale engineering development
FSK Frequency shift keying
FSRWT Full-scale rotary-wing target
FY Financial year

GCI Ground controlled intercept
GCS Ground control station (or system)
GDT Ground data terminal
GEN Generation

GFE Government-furnished equipment
GFRP Glass fibre-reinforced plastics
GLCM Ground-launched cruise missile
GMTI Ground moving target indicator
GOTS Government off-the-shelf
GPS Global positioning system
GPWS Ground proximity warning system
GSE Ground support equipment

HAE High-altitude endurance
HALE High-altitude, long endurance
HF High frequency
HFE Heavy-fuel engine
HMMWV High-mobility multipurpose wheeled vehicle (US)
HUD Head-up display
Hz Hertz (cycles per second)

IAS Indicated airspeed
ICAO International Civil Aviation Organisation
IDF Israel Defence Force
IED Improvised explosive device
IEEE Institute of Electrical and Electronic Engineers
IEWS Intelligence, electronic warfare and sensors
IF Intermediate frequency
IFF Identification, friend or foe
IFOR Implementation Force (NATO)
IFR (1) Instrument flight rules; (2) in-flight refuelling
IGE In ground effect
IIRS Imagery interpretability rating scale
imint Imagery intelligence
IMU Inertial measurement unit
INS Inertial navigation system
InSb Indium antimonide
I/O Input/output
IOC Initial operating (or operational) capability
IOT&E Initial operational test and evaluation
IR Infrared
IR&D Internal research and development
IRLS Infra-red linescan
IRST Infra-red search and tracking
ISA International standard atmosphere
ISAF International Security Assistance Force (UN)
ISR Intelligence, surveillance and reconnaissance
ISTAR Intelligence, surveillance, targeting, acquisition and reconnaissance

JAA Joint Aviation Authorities (Europe)
JATO Jet-assisted take-off
JPO Joint Project Office (US)
JSIPS Joint Services Imagery Processing System
JSOW Joint Stand-Off Weapon
JTIDS Joint Tactical Information Distribution System

lb st Pounds static thrust
LCD Liquid crystal display
LCS Littoral Combat Ship
LiSO2 Lithium disulphide
LLTV Low-light television
LO Low observables
LOS Line of sight
LPC (1) Linear predictive coding; (2) low-pressure compressor
LR Long range
LRF Laser range-finder
LRIP Low-rate initial production
LRU Line-replaceable unit
LTA Lighter than air

MAE Medium-altitude endurance
MALE Medium-altitude, long endurance
MANPADS Man-portable air defence system

masint Measurements and signatures intelligence
MAV Micro air vehicle
MDI Miss-distance indicator
MEMS Micro-electromechanical system
MER Multiple ejector rack
MFD Multifunction display
MIL-STD Military standard(s) (US)
MLRS Multiple Launch Rocket System
MMI Man-machine interface
MMW Millimetre wave
MoD Ministry of Defence
Mogas Motor (automobile) gasoline
MoU Memorandum of Understanding
MOUT Military operations in urban terrain
MPCS Mission planning and control station (or system)
MPO Mission payload operator
MPS Mission planning system
MR Medium range
MRE Medium-range endurance
MTBF Mean time between failures
MTCR Missile Technology Control Regime
MTI Moving target indicator
MTOW Maximum take-off weight
MTTR (1) Multitarget tracking radar; (2) mean time to repair

NACA National Advisory Committee for Aeronautics (US)
NAS (1) Naval Air Station (US); (2) national airspace (US)
NASA National Aeronautics and Space Administration (US)
NATMC NATO Air Traffic Management Committee
NATO North Atlantic Treaty Organisation
NBC Nuclear, biological and chemical (warfare)
NCW Network-centric warfare
NEC Network-enabled capability
Ni/Cd Nickel/cadmium
NIIRS National imagery interpretability rating scale (US)
NLOS Non-line of sight
NOLO No onboard live operator (US Navy)
NTSC National Television Standards Committee (US)
NULLO Not utilising live local operator (US Air Force)

OAV Organic air vehicle
OEF Operation 'Enduring Freedom'
OEI One engine inoperative
OEM Original equipment manufacturer
OGE Out of ground effect
OIF Operation 'Iraqi Freedom'
OLOS Out of line of sight
OPA Optionally piloted aircraft
OPAV Optionally piloted air vehicle
OPV (1) Optionally piloted vehicle; (2) offshore patrol vessel
OTH Over the horizon

PAL (1) Phase alternation line; (2) programmable array logic
PCI Personal computer interface
PCM Pulse code modulation
PDR Preliminary design review

PIM (1) Position of intended movement; (2) previously intended movement
PIP Product improvement programme
PLRS Position location and reporting system
POC Proof of concept
PPC Pulse position coded
PPI Planned position indicator
PRF Pulse repetition frequency
PRI Pulse repetition interval
PtSi Platinum silicide
PWM Pulsewidth modulation

QPSK Quadrature phase shift keyed

R&D Research and development
RAAF Royal Australian Air Force
RAM Random access memory
RAN Royal Australian Navy
RAST (1) Recovery, assist, secure and traverse (helicopter); (2) radar-augmented subtarget
RATO Rocket-assisted take-off
RCO Remote-control operator
RCS Radar cross-section
RDT&E Research, development, test and evaluation
RF Radio frequency
RFA Rectangular format array
RFI Request for information
RFP Request for proposals
RISC Reduced instruction set computer
RMS (1) Reconnaissance management system; (2) root mean squared
ROA Remotely operated aircraft
RON Research octane number
RPA (1) Remotely piloted aircraft; (2) rotorcraft pilot's associate
RPH Remotely piloted helicopter
rpm Revolutions per minute
RPV Remotely piloted vehicle
RSTA Reconnaissance, surveillance and target acquisition
R/T Receiver/transmitter
RTS (1) Remote tracking station; (2) request to send
RVT Remote video terminal
RWR Radar warning receiver

SAM Surface-to-air missile
SAR (1) Synthetic aperture radar; (2) search and rescue
satcom Satellite communications
SBIR Small business innovative research (US contract type)
SCSI (1) Small computer system interface; (2) single card serial Interface
SEAD Suppression of enemy air defences
sfc Specific fuel consumption
SFDR Spurious free dynamic range
SFOR Stabilisation Force (NATO)
SHORAD Short-range air defence
shp Shaft horsepower
Sigint Signals intelligence
SINCGARS Single channel ground and airborne radio system
S/L Sea level
SPIRIT (Trojan) Special Purpose Integrated Remote Intelligence Terminal
SPRITE Signal processing in the element

SR Short range
SSB Single sideband
STANAG Standardisation NATO Agreement
STOL Short take-off and landing

Tacan Tactical air navigation
TAS True airspeed
TBD To be determined
TBO Time between overhauls
TCAS Traffic collision and avoidance system
TCDL Tactical common datalink
TCS Tactical control station (or system)
TED Transferred electron device
TER Triple ejector rack
T/FDOA Time/frequency difference of arrival
TFT Thin film transistor
TICM Thermal imaging common modules
TMD Theatre missile defence
TMT Telemetry
T-O Take-off
TTL Transistor/transistor logic
TUAV Tactical unmanned aerial vehicle
TV Television
TWT Travelling wave tube

UAS Unmanned (or uninhabited) aircraft system
UAV Unmanned (or uninhabited) aerial vehicle
UCAR Unmanned (or uninhabited) combat armed rotorcraft
UCARS UAV common automated recovery system (US)
UCAV Unmanned (or uninhabited) combat air vehicle
UCS Universal control station (NATO)
UHF Ultra-high frequency
UN United Nations
UNSA Uninhabited naval strike aircraft
UOR Urgent operational requirement
USAF United States Air Force
USD Unmanned (or uninhabited) surveillance drone (NATO)
USMC United States Marine Corps
USN United States Navy
UTCS Universal target control station
UTM Universal Transverse Mercator
UTV Unmanned (or uninhabited) target vehicle
UV Ultra-violet

VCR Video cassette recorder
VDU Video (or visual) display unit
VFR Visual flight rules
V/H Velocity/height (ratio)
VHF Very high frequency
VHS Very high speed
VLA (1) Very light aircraft; (2) very large array
VLAR Vertical launch and recovery
VLF Very low frequency
VLSI Very large scale integration
VME Virtual memory environment
VOR VHF omnidirectional radio range
VTOL Vertical take-off and landing
VTR Video tape recorder

WAAS Wide area augmentation system
WAS Wide area search

Conversion Factors

Conversion factors used in *IHS Jane's All the World's Aircraft: Unmanned* are as follows:

From	To	Multiply by	From	To	Multiply by
acres	ha	0.404686	lb/hp	kg/kW	0.60864
cc	cu in	0.06102	lb/h/hp	g/h/kW	608.29
cu ft	m³	0.0283168	lb/h/hp	µg/J	169.0
cu ft	litres/dm³	28.3392	lb/h/lb st	mg/Ns	28.325
cu in	litres/dm³	0.0164	lb/lb st	kg/kN	101.972
cu in	cc	16.387	lb-s	kN-s	0.0044489
cv	hp	0.98632	lb/sq ft	kg/m²	4.88243
cv	kW	0.7355	lb st	kN	0.0044483
ft	m	0.3048	litres/dm³	cu ft	0.035287
ft/min	mph	0.011364	litres/dm³	cu in	60.9756
ft/min	m/s	0.00508	litres/dm³	Imp gallons	0.219975
g	oz	0.03527	litres/dm³	US gallons	0.264177
g/h/kW	lb/h/hp	0.001644	m	ft	3.28084
ha	acres	2.471053	m²	sq ft	10.7639
hp	cv	1.01387	m³	cu ft	35.3147
hp	kW	0.7457	m/s	ft/min	196.8504
Imp gallons	litres/dm³	4.54596	mg/Ns	lb/h/lb st	0.0353
Imp gallons	US gallons	1.20095	µg/J	lb/h/hp	0.00592
in	mm	25.4	miles, mph	km, km/h	1.609344
kg	lb	2.20462	miles, mph	n miles, kt	0.86898
kg/h/kW	lb/h/hp	2.95644	mm	in	0.03937
kg/kN	lb/lb st	0.009807	n miles, kt	km, km/h	1.852
kg/kW	lb/hp	1.643	n miles, kt	miles, mph	1.15078
kg/m²	lb/sq ft	0.204816	oz	g	28.3495
km, km/h	miles, mph	0.621371	sq ft	m²	0.092903
km, km/h	n miles, kt	0.5399568	sq miles	sq n miles	0.7553062
km²	sq miles	0.3861	sq miles	km²	2.58999
km²	sq n miles	0.2915533	sq n miles	km²	3.4299045
kN	lb st	224.80455	sq n miles	sq miles	1.3239663
kN-s	lb-s	224.77	US gallons	Imp gallons	0.83267
kW	hp	1.341	US gallons	litres/dm³	3.785411
lb	kg	0.453592			

Radio and radar bands and frequencies. Radio and radar performance can be referred to either by wavelength (measured in mm, cm, m or km) or by frequency (measured in MHz or GHz). IHS Jane's house style, as used in this publication, is to use the frequency band, with the frequency range, where known, in parentheses: for example, G-band (5.25 to 5.85 GHz). However, some originators' diagrams used herein may show use of wavelength bands. These can be understood using the accompanying table.

Band designation	Abbreviation	Frequency	Wavelength
General			
Extremely Low Frequency	ELF	30 Hz – 3 kHz	10,000 – 100 km
Very Low Frequency	VLF	3 – 30 kHz	100 – 10 km
Low Frequency	LF	30 – 300 kHz	10 – 1 km
Medium Frequency	MF	300 kHz – 3 MHz	1 km – 100 m
High Frequency	HF	3 – 30 MHz	100 – 10 m
Very High Frequency	VHF	30 – 300 MHz	10 – 1 m
Ultra High Frequency	UHF	300 MHz – 3 GHz	1 m – 10 cm
Super High Frequency	SHF	3 – 30 GHz	10 – 1 cm
Extremely High Frequency	EHF	30 – 300 GHz	1 cm – 1 mm
NATO Radar and Electronic Warfare			
A-band		0 – 250 MHz	100 m – 120 cm [1]
B-band		250 – 500 MHz	120 – 60 cm
C-band		500 MHz – 1 GHz	60 – 30 cm
D-band		1 – 2 GHz	30 – 15 cm
E-band		2 – 3 GHz	15 – 10 cm
F-band		3 – 4 GHz	10 – 7.5 cm
G-band		4 – 6 GHz	7.5 – 5 cm
H-band		6 – 8 GHz	5 – 3.75 cm
I-band		8 – 10 GHz	3.75 – 3 cm
J-band		10 – 20 GHz	3 – 1.5 cm
K-band		20 – 40 GHz	1.5 – 0.75 cm
L-band		40 – 60 GHz	0.75 – 0.5 cm
M-band		60 – 100 GHz	0.5 – 0.3 cm

Band designation	Abbreviation	Frequency	Wavelength
Radar and Satellite Communications			
P-band		230 MHz – 1 GHz	120 – 30 cm [1]
L-band		1 – 2 GHz	30 – 15 cm
S-band		2 – 4 GHz	15 – 7.5 cm
C-band		4 – 8 GHz	7.5 – 3.75 cm
X-band		8 – 12.5 GHz	3.75 – 2.5 cm [1]
Ku-band		12.5 – 18 GHz	2.5 – 1.6 cm [1]
K-band		18 – 26.5 GHz	1.6 – 1.1 cm [1]
Ka-band		26.5 – 40 GHz	1.1 – 0.75 cm [1]
mm-band		40 – 100 GHz	0.75 – 0.3 cm

[1] *Approximate value*

Professional Organisations

International

AUVSI
Association for Unmanned Vehicle Systems International
2700 South Quincy Street, Suite 400, Arlington, Virginia 22206, United States
Tel: (+1 703) 845 96 71
Fax: (+1 703) 845 96 79
email: info@auvsi.org
Web: www.auvsi.org
Formed: 1972 (as National Association for Remotely Piloted Vehicles)
Membership: 250 corporate and academic and more than 6,000 individual members from government, military, industry and academia internationally. Represents 1,500 organisations in 50 countries.

UCARE
UAVs: Concerted Actions for Regulations
Contact details as for UVS International except:
Web: www.ucare-network.org

UVS International
86 rue Michel Ange, F-75016 Paris, France
Tel: (+33 1) 46 51 88 65
Fax: (+33 1) 46 51 05 22
e-mail: info@uvs-international.org
Web (1): www.uvs-international.org
Web (2): www.uvs-info.com
Formed: 16 May 1997 (as Euro UVS; present name adopted February 2004)
Membership: 252 corporate, military and institutional members from 37 countries and 10 international organisations worldwide; 108 honorary members from 23 countries and eight international organisations.

National

AESiNT
Asociación Española de Sistemas No Tripulados
c/o Logstar Aviación, C/Fray Francisco 27, E-01007 Vitoria-Gasteiz, Spain
Tel/Fax: (+34 917) 47 82 71
e-mail: jmliquete@logstar.es
Formed: Early 2004
Membership: No information received.

JUAV
Japan Unmanned Aerial Vehicles Association
c/o Fuji Heavy Industries Ltd, 1-1-11 Yonan, Utsunomiya, Tochigi 320-8564, Japan
Tel: (+81 28) 684 70 60
Fax: (+81 28) 684 70 71
e-mail: kitagaway@uae.subaru-fhi.co.jp
Web: www.juav.org (in Japanese only)
Formed: September 2004
Membership: Fuji, Hirobo, Hitachi, Kawada, Kawasaki, Mitsubishi Heavy Industries, Mitsubishi Electric, Sky Remote, Yamaha and Yanmar.

Korea UVS Association
Korea Unmanned Vehicle Systems Association
c/o Korea Aerospace Research Institute, PO Box 113, No. 45 Eoeung-dong, Yu-Sung, 305600 Taejon, South Korea
Tel: (+82 42) 860 23 52
Fax: (+82 42) 860 20 06
Web: www.korea-uvs.org (in Korean only)
Formed: 29 August 2003
Membership: Over 30 Korean corporate, military and institutional members.

UAV DACH
UAV DACH
c/o Von Bothmer, Am Kölnkreuz 17, D-53340 Meckenheim, Germany
Tel/Fax: (+49 2225) 83 95 29
e-mail: info@uavdach.org
Web: www.uavdach.org (in German only)
Formed: 2000
Membership: German-speaking working group; name derived from names of original member countries: Germany (D), Austria (A) and Switzerland (CH); now also from Netherlands. Current 12 members comprise Autoflug, Diehl BGT Defence, DLR, EADS Deutschland, EMT, ESG, IABG, Rheinmetall Defence Electronics and Stemme UMS from Germany; Schiebel Elektronische from Austria; RUAG Aerospace from Switzerland; and ADSE from Netherlands (September 2006).

UAVS
Unmanned Aerial Vehicle Systems Association
The Granary, 1 Waverly Lane, Farnham, Surrey GU9 8BB, United Kingdom
Tel: (+44 1252) 73 25 77
Fax: (+44 1252) 73 25 01
e-mail (1): secretary@uavsuk.org
e-mail (2): secretary@uavsuk.co.uk
e-mail (3): secretary@uavsuk.com
Web (1): www.uavs.org
Web (2): www.uavsuk.co.uk
Web (3): www.uavsuk.com
Formed: November 1998
Membership: More than 30 corporate and other members from UAV systems industry and academia in the UK (March 2007).

UNITE
UAV National Industry Team
c/o Am Tech, 499 Seaport Court, Suite 100, Redwood City, California 94063, United States
Tel: (+1 650) 569 38 38
Fax: (+1 650) 569 38 39
e-mail: scarmona@amtech-usa.org
Formed: 2002
Membership: AeroVironment, Aurora Flight Sciences, Boeing, General Atomics — Aeronautical Systems, Lockheed Martin and Northrop Grumman. (According to a 7 February 2007 announcement, UNITE was intending to disband later in that year.)

UVS Canada
Unmanned Vehicle Systems Canada
PO Box 81005, Ottawa, Ontario K1P 1B10, Canada
Tel: (+1 613) 845 01 45
Fax: (+1 613) 248 49 32
e-mail: admin@uvscanada.org
Web: www.uvscanada.org
Formed: November 2003
Membership: Includes some 300 individual and corporate members from industry, government and academia in Canada, the US and Asia.

UVS New Zealand Australia
Unmanned Vehicle Systems New Zealand and Australia Association Inc, 17 Arwen Place, PO Box 58642, East Tamaki, Greenmount, Auckland 1730, New Zealand
Tel: (+64 9) 273 63 07
Fax: (+64 9) 273 63 08
e-mail: info@uvs-nza.com
Web: www.uvs-nza.com
Formed: Date unknown
Membership: No information received.

TERMS AND CONDITIONS FOR THE SALE OF HARDCOPY PRODUCTS

All orders for the sale of hardcopy Products are subject to the following terms and conditions:

1. **DEFINITIONS**

    "Client" means the person, firm or company or any other entity that purchases the Products from IHS.

    "Delivery Point" where applicable, means the location as defined in the Order Confirmation where delivery of the Products is deemed to take place.

    "Directory Products" means IHS's proprietary database or any part thereof, including without limitation, details of particular company/organisation, key personnel, financial/statistical information, products/services description, organisational structure and any other information pertaining to such company(s)/organisation(s) operating in various industrial sectors.

    "Fees" means the money due and owing to IHS for Products supplied including any order processing charge and as set forth in the Order Confirmation. Fees are exclusive of taxes, which will be charged separately to the Client.

    "Products" means any publication, database, supplied to the Client in physical or electronic media, more specifically mentioned in the Order Confirmation. Products include Directory Products.

    "Order Confirmation" includes the order form or confirmation email or any other document which IHS sends to the Client to confirm that IHS has accepted the Client's order and which identifies the name of the Client, Product(s) being supplied, period of supply, delivery information, media of supply, Fees and any terms or conditions unique to the particular Product to be supplied hereunder.

2. Client will pay IHS the Fees as set forth in the Order Confirmation within 30 days from the date of the invoice. Any payments not received by IHS when due will be considered past due, and IHS may choose to accrue interest at the rate of five percent (5%) above the European Central Bank "Marginal lending facility" rate. Client has no right of set-off. Client will pay all the value-added, sales, use, import duties, customs or other taxes where applicable to the purchase of Products. IHS may request payment of the Fees before shipping the Products.

3. IHS grants to Client a nonexclusive, nontransferable license to use the Products for its internal business use only. Client may not copy, distribute, republish, transfer, sell, license, lease, give, disseminate in any form (including within its original cover), assign (whether directly or indirectly, by operation of law or otherwise), transmit, scan, publish on a network, or otherwise reproduce, disclose or make available to others, store in any retrieval system of any nature, create a database or create derivative works from the Product or any portion thereof, except as specifically authorized herein. Any information related to third party company and/or personal data included in the Directory Product(s), may be used by Client for the limited purpose of enquiring about the products and services of the companies/organisations listed therein and who have given permission for their data to be used for this purpose only. Client must comply with the UK Data Protection Act and all other applicable data protection and privacy laws and regulations. In particular, Client must not use such data (i) for any unlawful, harmful or offensive purpose; (ii) as a source for any kind of marketing or promotion activity; or (iii) for the purposes of compiling, confirming or amending its own database, directory or mailing list.

4. Client must not remove any proprietary legends or markings, including copyright notices, or any IHS-specific markings on the Products. Client acknowledges that all data, material and information contained in the Products are and will remain the copyright property and confidential information of IHS or any third party and are protected and that no rights in any of the data, material and information are transferred to Client. Client will take any and all actions that may reasonably be required by IHS to protect such proprietary rights as owned by IHS or any third party. Any unauthorised use may give rise to IHS bringing proceedings for copyright and/or database right infringement against the Client claiming an injunction, damages and costs.

5. Any dates specified in the Order Confirmation for delivery of the Products are intended to be an estimated time for delivery only and shall not be of the essence. IHS shall not be liable for any delay in the delivery of the Products. Unless otherwise agreed by the parties, packing and carriage charges are not included in the Fees and will be charged separately. The Products will be despatched and delivered to the Delivery Point as per Client's preferred method of delivery and as agreed by IHS. If special arrangements are required, then IHS reserves the right to additional charges. Except as provided hereunder, for all Products supplied hereunder, delivery is deemed to occur and risk of loss passes upon despatch of Products by IHS.

6. If for any reason IHS is unable to deliver the Products on time due to Client's failure to provide appropriate instructions, documents or authorisations etc; (i) any risk in the Products will pass to the Client; (ii) the Products will be deemed to have been delivered; and (iii) IHS may store the Products until delivery, whereupon the Client will be liable for all related costs and expenses.

7. Except as otherwise required by law, Client will not be entitled to object or to return or reject the Products or any part thereof unless the Products are damaged in transit. IHS's sole obligation and Clients' exclusive remedy for any claim with respect to such damaged Products will be to replace the damaged Products without any charge. No returns will be accepted by IHS without prior agreement and a returns number issued by IHS to accompany the Products to be returned. All return shipments are at the Client's risk and expense.

8. The possession and usage rights of the Products in accordance with clause 3 above will not pass to Client until IHS has received in full all sums due to it in respect of: (i) Fees; and (ii) all other sums which are or which become due to IHS from Client on any account. Until such rights have passed to Client, the Client will: (i) hold the Products in a fiduciary capacity; (ii) store the Products (at no cost to IHS) in such a way that they remain readily identifiable as IHS property; (iii) not destroy, deface or obscure any identifying mark or packaging on or relating to the Products; and (iv) maintain the Products in satisfactory condition and keep them insured on IHS' behalf for their full price against all risks to the reasonable satisfaction of IHS.

9. The quantity of any consignment of Products as recorded by IHS on despatch from IHS' place of business shall be conclusive evidence of the quantity received by the Client on delivery unless Client can provide conclusive evidence proving otherwise. IHS shall not be liable for any non-delivery of the Products (even if caused by IHS' negligence) unless Client provides conformed claims to IHS of the non-delivery. Any such conformed claim for non-receipt of the Products must be made in writing, quoting the account and Order Confirmation number to the IHS' Customer Service Department, within thirty (30) days of the estimated date of delivery as stated in the Order Confirmation.

10. The Products supplied herein are provided "AS IS" and "AS AVAILABLE". IHS does not warrant the completeness or accuracy of the data, material, third party advertisements or information as contained in the Product or that it will satisfy Client's requirements. IHS disclaims all other express or implied warranties, conditions and other terms, whether statutory, arising from course of dealing, or otherwise, including without limitation terms as to quality, merchantability, fitness for a particular purpose and noninfringement. To the extent permitted by law, IHS shall not be liable for any errors or omissions or any loss, damage or expense incurred by reliance on information, third party advertisements or any statement contained in the Products. Client assumes all risk in using the results of the Product(s).

11. If the Products supplied hereunder are subscription based, except as otherwise provided herein the period of supply will run for one calendar year from the start date as specified in the Order Confirmation and the Fees will cover the costs of supply of all issues of the Product published in that year. If Client attempts to cancel the Product subscription anytime during such period; (i) the Fees payable for that year will be invoiced by IHS in full; or (ii) where Client has already paid the Fees in advance, any Fees relating to the remaining period shall be forfeited. In addition to other rights and subject to the provisions of this clause, IHS in its sole discretion may discontinue the supply the Products in the event Client commits breach of any of the provision of these terms and conditions.

12. In the event of breach of any of the provision of these terms and conditions by IHS, IHS' total aggregate liability for any damages/losses incurred by the Client arising out of such breach shall not exceed at any time the Fees paid for the Product which is the subject matter of the claim. In no event shall IHS be liable for any indirect, special or consequential damages of any kind or nature whatsoever suffered by the Client including, without limitation, lost profits or any other economic loss arising out of or related to the subject matter of these terms and conditions. However, nothing in these terms and conditions shall limit or exclude IHS' liability for (i) death or personal injury caused by its negligence; (ii) fraud or fraudulent misrepresentation; or (iii) any breach of compelling consumer protection or other laws.

13. Client represents and warrants that it will not directly or indirectly engage in any acts that would constitute a violation of United States laws or regulations governing the export of United States products and technology.

14. The parties will comply with all applicable country laws relating to anti-corruption and anti-bribery, including the US Foreign Corrupt Practices Act and the UK Bribery Act. The parties represent and affirm that no bribes or corrupt actions have or will be offered, given, received or performed in relation to the procurement or performance of these terms and conditions. For the purposes of this clause, "bribes or corrupt actions" means any payment, gift, or gratuity, whether in cash or kind, intended to obtain or retain an advantage, or any other action deemed to be corrupt under the applicable country laws.

15. All Products supplied herein are subject to these terms and conditions only, to the exclusion of any other terms which would otherwise be implied by trade, custom, practice or course of dealing. Nothing contained in any Client-issued purchase order, Clients' acknowledgement, Clients' terms and conditions or invoice will in any way modify or add any additional terms to these terms and conditions. IHS reserves the right to amend these terms and conditions from time to time.

16. These terms and conditions and any dispute or claim arising out of or in connection with them or their subject matter shall be governed by and construed in accordance with the laws of England and Wales and shall be subject to the exclusive jurisdiction of the English Courts.

How to use

This title is designed to describe both the airborne and the ground-based elements of the principal Unmanned Aerial Vehicle (UAV) and aerial target systems known to be currently under development, in production and/or in service worldwide. It does so within a common entry structure (see below), to enable informed comparison between systems of a comparable nature.

Structure

The title is divided into four product sections, the first two of which are devoted to the air vehicles of each system. The next two deal with major subsystems. All product entries are listed in alphabetical order of country of origin, and within each country in alphabetical order of manufacturer or other prime contractor. The title is completed by an alphabetical listing of manufacturers' addresses and contact details, and separate indexes, by manufacturer and equipment name, for the product entries.

Unmanned Aerial Vehicles: The major section, this covers UAVs of every size and type, both fixed-wing and rotary-wing. Lighter than air systems are also included.

Aerial targets: This section covers all kinds of aerial target, including ballistic and towed targets. Where a manufacturer may offer a UAV variant of a standard target, this is also noted.

Control and communications: Coverage in this section ranges from dedicated, customised complete ground control stations to individual elements thereof such as remote video terminals and datalinks.

Launch and recovery systems: Covers systems known to be in general use. However, it should be noted that many UAV and target manufacturers also produce such systems tailored to their own products but not adaptable for other use.

Record Structure

Air vehicle (UAV and target) entries are described under the following main headings:

Type: Gives a quick-reference summary of each aircraft's primary purpose.

Development: Gives the history of development up to the time of publication, or entry into service if earlier.

Description: This is subdivided under the following subheadings:

Airframe: Gives a brief description of the air vehicle's general configuration and construction.

Mission payloads: Describes the main sensor or weapon payloads installed in, or capable of carriage by, each air vehicle.

Guidance and control: Details the method(s) by which the air vehicle is commanded and controlled throughout its mission. (*Editor's note:* Readers should be wary of the term 'autonomous', which tends to be over-used in manufacturers' product literature as though it were synonymous with 'automatic'. Strictly, 'autonomous' should imply that the air vehicle is capable of making its own decisions without human input. Where the word appears in this volume, it has been taken in editorial good faith, but the editor is not always in a position to determine its strict accuracy.)

Transportation: Is intended primarily to give an idea of the portability of the system.

System composition: Gives, wherever possible, an overall picture of the number of air and ground vehicles, personnel and support equipment making up a standard or typical fielded system.

Launch: Details the method, with alternatives where applicable, by which the air vehicle is despatched for a mission.

Recovery: The means, also with alternatives if applicable, by which the air vehicle is retrieved at the end of a mission.

Power plant: Aircraft power plant (including fuel capacity).

Variants: Where more than one major version of a particular type exists, the principal differences — and any alternative names or designations — are detailed under this heading.

Specifications: Are given in standardised form under the five main headings of Dimensions, External; Dimensions, Internal; Areas; Weights and Loadings and Performance to the fullest extent of the information available.

Readers should note that weights given by manufacturers are, unfortunately, not to a consistent standard. Thus, aircraft empty weight may or may not include such items as onboard (non-payload) avionics; some payload figures may include fuel; maximum take-off or launching weight may include the weight of a launch booster; and so on. These factors will be clarified wherever such details have been provided.

All speeds are assumed to be TAS (true airspeed) unless otherwise indicated. Maximum level speeds are rounded down, stalling speeds rounded up. Altitudes are rounded to the nearest 5 m or 20 ft.

Status: Gives the current known operational status of each system or programme, indicating whether it is in service, in production, under development or evaluation. Also includes any other known information considered to be of interest.

Customers: These are identified either by individual name, country or region, to the extent that this information is openly available. Specific force or unit allocations are also given, if known. However, it will be appreciated that, for reasons of defence security and/or customer confidentiality, this information cannot always be given.

Contractor: Gives the name and basic location of each product's manufacturer; full addresses and contact numbers appear in the Manufacturers' Addresses list.

Subsystem entries, due to the diverse nature of the products concerned, are described under a more basic set of headings, as follows:

Type, **Development** and **Contractor** are as above.

Status contains customer details, where known.

Specifications groups the basic technical information under similar headings for size, weight and 'performance', although data listed under the last-named necessarily vary with the nature of the product.

Description All other information, including variants where applicable, is grouped under this catch-all heading.

Images

At least one photograph is included for each equipment wherever possible. Line drawings and graphics are also provided in some cases. Images are annotated with a seven-digit number which uniquely identifies them in the *IHS Jane's* image database.

UNMANNED AERIAL VEHICLES

Argentina

CANA Guardian

Type: Multirole tactical UAV.

Development: Main objective of the Argentine Navy's two-year, USD164,000 Guardian programme is to develop a system to improve command and control of its surveillance and reconnaissance capability in combined or specific naval operations. It is also expected to provide secondary value by increasing UAV operating experience, capacity for manufacturing in composites, and eventual dual use for military and civil scientific applications. By August 2007, the four-month Phase 1 (platform design and construction and prototype flight test) had been completed. Phase 2 (visual flight with telemetry; autopilot incorporation; autonomous navigation and distance command; image, data and video set-up; real-time data transmission) and Phase 3 (GCS completion and sensor integration) were in progress at that time, with Phase 4 (launch and recovery system evaluation) and Phase 5 (series production) to follow.

Description: *Airframe:* Typical pod and twin boom configuration with untapered high wings and tailplane, sweptback fins and rudders and pusher engine. Fixed tricycle landing gear. Composites construction.

Mission payloads: Intended payloads include a gyrostabilised pan-tilt-zoom video camera, EO/IR camera or FLIR, with onboard video and real-time, LOS data transmission.

Guidance and control: Initial (2007) flight testing limited to runway operation, within visual control range and with ceiling limited to 1,000 m (3,280 ft). Projected operational characteristics include automatic operation, re-programmable in flight, and eventual autonomous operation.

Launch: Conventional wheeled take-off initially; rocket-assisted or catapult launch projected.

Recovery: Conventional wheeled landing initially; net and arrester hook systems to be explored.

Power plant: One (unidentified) 150 cc, 12.7 kW (17 hp) two-cylinder two-stroke engine; two-blade pusher propeller.

Guardian

Dimensions, External	
Wings, wing span	5.00 m (16 ft 4¾ in)
Weights and Loadings	
Weight	
Weight empty	44 kg (97 lb)
Max T-O weight	77 kg (169 lb)
Payload, Max payload	30 kg (66 lb)
Performance	
Altitude	
Service ceiling, projected	3,000 m (9,840 ft)
Speed	
Cruising speed, max	65 kt (120 km/h; 75 mph)
Radius of operation, projected	27 n miles (50 km; 31 miles)
Power plant	1 × piston engine

Status: Under development in 2007-08. Potential customers are Argentine Navy (shipboard use) and Marine Corps (from land vehicles). Flight testing continuing 2009.

Contractor: Comando de Aviación Naval Argentina (Argentine Navy) Buenos Aires.

Prototype CANA Guardian tactical UAV *(Argentine Navy)* 1295111

EA Lipán M3

Type: Brigade-level tactical UAV.

Development: The Argentine Army, through its Dirección de Investigación, Desarrollo y Producción (Research, Development and Production Department), is believed to be the first agency in the country to undertake development of unmanned aircraft, having begun the Lipán programme in 1996. It is now in production and service.

Lipán M3 brigade-level surveillance UAV *(Argentine Army)* 1295110

The Lipán M3 in flight *(Argentine Army)* 1295109

Description: *Airframe:* High-mounted wings, with compound taper on leading-edges; pod fuselage and twin tailbooms, with inverted V tail surfaces. Fixed tricycle landing gear.

Mission payloads: Two-axis, gyrostabilised pan/tilt/zoom ventral turret for CCD daylight TV camera or lightweight FLIR. Real-time imagery downlink; Two 4 to 8 mm Varifocal camera, one in nose for steering, one in port side of nose controlling artificial horizon.

Guidance and control: GCS comprises two militarised computers, plus a utility rack with video processing and telemetry equipment, installed in three high-impact transportation cases. Manual control during take-off and landing; remainder of mission autonomous, using MicroPilot autopilot and 1,000-waypoint GPS navigation.

System composition: Three air vehicles and payloads, GCS, transport trailer, fuel containers, generator, amplifier and ground (runway) support equipment.

Launch: Conventional, automatic, wheeled take-off.

Recovery: Conventional, automatic, wheeled landing.

Power plant: One 13.3 kW (17.8 hp) piston engine (type not stated); two-blade pusher propeller. Fuel capacity 16 litres (4.2 US gallons; 3.5 Imp gallons).

Variants: *Lipán M1:* Technology demonstration prototype; in development 1996-98. Conventional twin-tailboom tail unit. Manual (remote piloting) control within visual radius; CCD video sensor; 10 kg (22.0 lb) payload capacity; maximum T-O weight 40 kg (88.2 lb); 3-hour endurance.

Lipán M2: Operational evaluation prototype; in development 1999-2002. Redesigned wing, with winglets; enlarged (but still conventional) twin-tail unit. Manual control; radius increased to 22 n miles (40 km; 25 miles); CCD video sensor; 15 kg (33.1 lb) payload capacity; maximum T-O weight 50 kg (110 lb); 3-hour endurance.

Lipán M2-B: Modified M2, developed 2003-07. Winglets deleted, and tail unit modified to inverted V configuration, similar to that of US RQ-7 Shadow 200. Pre-programmed, autonomous control. Sensor, weights and radius/endurance as for M2.

Lipán M3: Operational version, to which main description applies; developed 2006-07. Configuration as for M2-B. Re-programmable in flight.; CCD video and FLIR sensors; increased payload and MTO weights; extended endurance.

Lipán XM4: Further development planned for 2008, incorporating automatic take-off and landing, ultrasonic altimeter and an onboard energy generator.

Lipán M3

Dimensions, External	
Overall	
length	3.43 m (11 ft 3 in)
height	1.14 m (3 ft 9 in)
Wings, wing span	4.38 m (14 ft 4½ in)
Weights and Loadings	
Weight, Max T-O weight	60 kg (132 lb)
Payload, Max payload	20 kg (44 lb)
Performance	
Altitude, Operating altitude	2,000 m (6,560 ft)
Speed	
Max level speed	91 kt (169 km/h; 105 mph)
Loitering speed	43 kt (80 km/h; 49 mph)
Radius of operation	22 n miles (40 km; 25 miles)
Endurance	5 hr
Power plant	1 × piston engine

Status: Delivery of first system to Destacamento de Inteligencia de Combate 601 (601 Combat Intelligence Detachment) of the Argentine Army, December 2007. Second system due in 2008, third in 2009 and fourth was scheduled for 2010.

Contractor: Ejército Argentino (Argentine Army)
Buenos Aires.

Nostromo Cabure

Type: Tactical mini-UAV.

Development: Designed for RSTA, force and infrastructure protection, convoy security, BDA and law enforcement applications. Public debut at Sinprode defence exhibition in Buenos Aires in September 2006. In July 2007, Simrad Optronics of Norway acquired 30 per cent of Nostromo Defensa. An agreement signed with Innocon of Israel in 2011 could make Naviator GCS and flight control computer available for this system.

Description: *Airframe*: High-mounted wings, with flaps; taper and dihedral on outer panels. Pod and boom fuselage; T tail. Pusher engine. No landing gear. Composites construction (glass fibre, Kevlar and carbon fibre).

Mission payloads: Up to three fixed CCD cameras (daylight or IR); or gimbal-stabilised micro EO sensor.

Guidance and control: Pre-programmed from sturdy laptop GCS with integrated video monitor, simple interface and advanced graphical interface for mission planning and control. Mavionics or MicroPilot autopilot.; GPS/INS manual, assisted or automatic navigation. Instead of conventional waypoint navigation, 'spline' navigation concept enables aircraft to fly a specific trajectory by marking a line on a tactical map on the GCS monitor screen. Video (2.4 GHz) and telemetry (900 MHz) datalinks; digital video downlink, with encryption.

Transportation: Cabure can be dismantled and carried in a soldier's backpack.

Launch: Can be launched by hand, from a take-off trolley or by bungee catapult.

Recovery: Deep stall to belly landing. Avionics package includes a data fusion algorithm and 17 highly coupled Kalman filters and a calibrated inertial sensor that allows landings to be controlled and improves operational safety in heavy crosswinds. At customer's option, can be fitted with a small parachute for emergency recoveries.

Powerplant: Battery-powered electric motor (300 to 450 W, depending upon mission requirements); two-blade pusher propeller.

Cabure	
Dimensions, External	
Overall, length	1.20 m (3 ft 11¼ in)
Wings, wing span	1.70 m (5 ft 7 in)
Weights and Loadings	
Weight, Max launch weight	3.5 kg (7.00 lb)
Performance	
Altitude, Service ceiling	2,000 m (6,560 ft) (est)
Speed	
Cruising speed	49 kt (91 km/h; 56 mph)
Loitering speed	32 kt (59 km/h; 37 mph)

Radius of operation	5.4 n miles (10 km; 6 miles)
Endurance, typical, 350 W engine	1 hr 12 min
Power plant	1 × electric motor

Status: Awarded a four-month law enforcement and homeland security demonstration contract by Argentine government in September 2007, understood to be for evaluation by Argentine Air Force and Marine Corps. One system ordered by US company for demonstration in law enforcement applications; several Cabure 2 to France, Germany and Spain for R&D; one system to an Argentinian company for forest fire monitoring. Some 28 systems were reported on order.

Contractor: Nostromo Defensa SA
Córdoba.

Nostromo Yarara

Type: Small tactical UAV.

Development: Nostromo began researching and developing UAVs and aerial targets in 2000. Design of Yarara thought to have begun in late 2005 or early 2006, and first system delivered by August 2006 (said to be first export of a production UAV from South America). Public debut at Fuerza Aérea Argentina air show on 10th of that month. Applications seen as ISR, target acquisition, force and infrastructure protection, BDA, border patrol and law enforcement. First flight with heavy fuel motor August 2009. In early 2011 Nostromo Defensa announced an agreement with Innocon of Israel, which could make Naviator GCS and flight control systems available for Yarara and future air vehicles developed by the two companies.

Description: *Airframe*: Shoulder-mounted wings with tapered leading-edges and tips; pod and boom fuselage; T tail unit; pusher engine. Conventional three-axis control surfaces. Fixed tricycle landing gear. Composites construction.

Mission payloads: Fixed, nose-mounted CCD TV camera for situation awareness. Steerable IAI Tamam MicroPOP EO suite; or EO, IR or multispectral camera on non-stabilised gimbal.

Guidance and control: Fully autonomous, pre-programmed or remotely operated. Sturdy laptop GCS with integrated video monitor, simple interface and advanced graphical interface for mission planning and control. MicroPilot or Mavionics autopilot; GPS/INS; manual, assisted or automatic 'spline' navigation similar to that of Cabure (which see). Video (2.4 GHz) and telemetry (900 MHz) datalinks; optional digital video link. GCS can control up to three air vehicles at a time.

Transportation: System packs into four containers and is transportable by HMMWV or similar vehicle.

System composition: Typical system comprises three air vehicles, payloads, GCS, batteries, engine starter and flight simulator.

Launch: Conventional, automatic or manually controlled, wheeled take-off.

Recovery: Conventional, automatic or manually controlled, wheeled landing.

The Cabure hand-launched mini-UAV *(Nostromo)* 1290210

Three air vehicles make up a typical Yarara system *(Nostromo)* 1132986

Cabure on display in UK, September 2007 *(IHS/Patrick Allen)* 1328763

One of the Yararas sold to the US DoD *(J M Barragan)* 1290213

Power plant: Initial production aircraft have either a 3.7 to 6.0 kW (5 to 8 hp) two-stroke or 3.0 to 4.5 kW (4 to 6 hp) four-stroke, locally produced, piston engine with two-blade pusher propeller. From 2009 a Cubewano Sonic 35 3.7 kW (5 hp) four-stroke heavy-fuel Wankel engine was tested on Yarara using JP8 fuel; other fuels compatible.

Yarara

Dimensions, External	
Overall, length	2.47 m (8 ft 1¼ in)
Wings, wing span	3.98 m (13 ft 0¾ in)
Areas	
Wings, Gross wing area	1.41 m² (15.2 sq ft)
Weights and Loadings	
Weight	
Weight empty	15.5 kg (34 lb)
Max T-O weight	30 kg (66 lb)
Payload, Max payload	5.0 kg (11.00 lb)
Performance	
T-O	
T-O run	60 m (197 ft)
Altitude, Service ceiling	3,000 m (9,840 ft) (est)
Speed	
Max level speed	79 kt (146 km/h; 91 mph)
Cruising speed	62 kt (115 km/h; 71 mph)
Stalling speed, power on	25 kt (47 km/h; 29 mph)
Radius of operation	27 n miles (50 km; 31 miles)
Endurance, 5.5 hp two-stroke	6 hr
Landing	
Landing run	40 m (132 ft)
Power plant	1 × piston engine

Status: In production and service. Two systems delivered to an unidentified US customer in 2006; one system to an Argentine oil company in 2007; one 'Yarara-C' system for US Department of Defense reported in late 2007. By 2008 12 air vehicles had been produced against orders for 14.

Contractor: Nostromo Defensa SA
Córdoba.

Armenia

MoD Krunk-25

Type: Short-range tactical UAV.

Development: Developed by Armenian Air Force Institute; name translates as Crane. First reported in June 2011 and seen publicly for first time, in Independence Day parade, three months later.

Description: *Airframe:* High-mounted wings; pod and twin tailboom configuration. Fixed tricycle landing gear. Composites construction.

Krunk-25 taking part in Independence Day parade, 21 September 2011 *(Armenian MoD)* 1440874

Mission payloads: Video camera and stills camera in stabilised ventral turret.

Guidance and control: Pre-programmed via autopilot. GCS crew of three.

Launch: Conventional wheeled take-off.

Recovery: Conventional wheeled landing.

Power plant: One piston engine (type and rating not known); two-blade propeller.

Krunk-25

Weights and Loadings	
Weight, Max T-O weight	60 kg (132 lb)
Performance	
Altitude	
Service ceiling	
operational	3,500 m (11,480 ft)
max	4,500 m (14,760 ft)
Speed	
Cruising speed, max	81 kt (150 km/h; 93 mph)
Endurance	5 hr

Status: Apparently in service in 2011.

Contractor: Armenian Ministry of Defence
Yerevan.

Australia

Aerosonde Aerosonde

Type: Long-range/endurance UAV.

Development: Development of the Aerosonde stems from a 1991 design study, followed by concept demonstration trials (part-funded by the US Office of Naval Research) between 1992 and 1994. The aircraft made its first fully autonomous flight, including take-off and landing, on 22 September 1997, at Trout Lake, Washington, USA. Applications include surveillance, reconnaissance, meteorological experiments and environmental monitoring.

By mid-1998, approximately 30 **Mk 1** Aerosondes had been built. In August 1998, an Aerosonde air vehicle completed the first transatlantic flight ever achieved by a UAV. In 26 hours and 45 minutes, it flew the 1,765 n miles (3,270 km; 2,032 miles) separating Newfoundland, Canada, and the Outer Hebrides, UK, under autonomous guidance by autopilot and GPS. Only 4 kg (8.8 lb) of its 5 kg (11.0 lb) fuel load was consumed.

The original aircraft has been continually upgraded, the **Mk 2** introduced in 2000 being followed by the **Mk 3** in early 2001 and by the lighter-weight **Mk 3.2** in August 2005. Current version is the **Mk 4**, introduced in March 2006, which sports winglets at the wingtips and offers 30 per cent more in payload volume and 20 per cent more by weight, as well as an improved endurance. Aerosonde flight hours now exceed 5,000, including a flight into Hurricane Noel in November 2007. The aircraft operated from NASA's Wallops Flight Facility in Virginia and spent more than 17 hours in the air, including approximately 7.5 hours navigating the hurricane's eye and eye wall. This project was a partnership between AAI Corporation, NASA and the National Oceanic and Atmospheric Administration (NOAA).

In June 2006, Aerosonde Pty Ltd was acquired by AAI Corporation, which is an operating unit of Textron Systems.

Description: *Airframe:* Small fuselage pod nacelle with rear-mounted engine and high-mounted wing; twin tailbooms with inverted V tail unit. Construction mainly of composites and balsa wood. No landing gear.

Aerosonde car-top launch *(Aerosonde)* 1195209

The current, wingleted Aerosonde Mk 4 *(Aerosonde)* 1195208

Unmanned aerial vehicles > Australia > **Aerosonde Aerosonde – BAE Systems Kingfisher**

Mounting an Aerosonde for car-top launch *(Aerosonde)* 1122597

Aerosonde in flight *(Aerosonde)* 1195372

Status: In production and service. By late 2007, Aerosonde had produced more than 150 air vehicles, which had been flown with a wide range of payloads under varied extreme climatic conditions. Trials using Aerosonde vehicles in 2009 included an EW payload for the Defence Science and Technology Organisation in Australia. Four Aerosondes were deployed on environmental research in the Antarctic in 2009.

Customers: Aerosondes are used to provide customer-specified services such as surveillance and military research and environmental monitoring as well as meteorology and icing research. Aerosondes have been operated on behalf of NASA; US Office of Naval Research (ONR), National Oceanic and Atmospheric Administration (NOAA), Department of Energy (DoE), National Science Foundation, National Centre for Atmospheric Research, National Weather Service (NWS), Barrow Arctic Science Consortium, US Army Cold Regions Research and Engineering Laboratory, University of Colorado and Georgia Institute of Technology; Australian Defence Force (Army) and Bureau of Meteorology; Taiwan Central Weather Bureau; Japanese and South Korean Meteorological Research Institutes; University of Hokkaido; and others.

Contractor: Aerosonde Pty Ltd
Notting Hill, Victoria.

BAE Systems Kingfisher

Type: UAV technology demonstrator.

Development: BAE Systems Australia collaborated with the University of Sydney's Australian Centre for Field Robotics (ACFR) in 1999, in a three-year R & D programme known as ANSER (Autonomous Navigating and Sensing Experimental Research), to develop and demonstrate technologies to enhance network-centric warfare and autonomous UAV operations. Known as SLAM (Simultaneous Localisation And Map-building), these technologies were aimed at guidance of, and DDR (decentralised data fusion) from, multiple air vehicles. Potential applications include tactical and combat UAVs, reconnaissance, surveillance, avionics architecture and systems integration. The early air

Mission payloads: EO/IR payloads, as well as more than a dozen scientific payloads, have been fielded, including various radio sondes and other meteorological monitoring instruments; sensors and cameras; remote sensing equipment; laser scanner; radiometers; and radar altimeter. Onboard electrical power available for payloads was increased to 75 W in the Mk 4, with a separate 40 W available for the guidance system.

Guidance and control: Laptop computer GCS. Fully autonomous en-route mission profiles, with GPS or DGPS navigation. Controlled by two microprocessors (one for flight, one for payloads). Loral/Conic CRI-400 series 9600 band half-duplex UHF modem. Up/down command links share a single 25 kHz channel. Satcom links available. Integrated with the AAI One System Ground Control Station, which allows operation of multiple air vehicles and dissemination of video data.

System composition: One or more air vehicles, payloads, single GCS and ground crew of three.

Launch: From car roof-rack or by catapult, autonomously or under operator control.

Recovery: Belly landing, autonomously or under operator control.

Power plant: One 1.3 kW (1.75 hp), 24 cc Aerosonde H single-cylinder piston engine with fuel injection; two-blade fixed-pitch pusher propeller.

Fuel capacity (premium unleaded), 7 litres (1.85 US gallons; 1.54 Imp gallons).

Aerosonde Mk 4

Dimensions, External	
Overall	
length	1.70 m (5 ft 7 in)
height	0.60 m (1 ft 11½ in)
Fuselage	
height	0.19 m (7½ in)
width	0.18 m (7 in)
Wings, wing span	2.90 m (9 ft 6¼ in)
Engines, propeller diameter	0.51 m (1 ft 8 in)
Areas	
Wings, Gross wing area	0.55 m² (5.9 sq ft)
Weights and Loadings	
Weight	
Weight empty, typical	9.5 kg (20.00 lb)
Max launch weight	15.0 kg (33 lb)
Fuel weight, Max fuel weight	1.0 kg (2.00 lb)
Payload, Max payload	5.5 kg (12.00 lb)
Performance	
Climb, Rate of climb	150 m/min (492 ft/min)
Altitude	
Operating altitude at intermediate weight	6,095 m (20,000 ft)
Speed	
Max level speed	62 kt (115 km/h; 71 mph)
Cruising speed	49 kt (91 km/h; 56 mph)
Loitering speed	49 kt (91 km/h; 56 mph)
Range	
maximum payload	300 n miles (555 km; 345 miles) (est)
still-air range, no payload, no reserves	1,800 n miles (3,333 km; 2,071 miles) (est)
Endurance	
max payload	5 hr (est)
no payload, no reserves	30 hr (est)
Power plant	piston engine

Kingfisher Mk 1 technology demonstrator *(Paul Jackson)* 1122586

Preparing a Kingfisher Mk 1 for flight *(BAE Systems)* 1122587

Kingfisher Mk 1 engine detail *(Paul Jackson)* 1122588

Kingfisher UAV Mk 2 *(BAE Systems Australia)* 1353874

Brumby Mk 3 ANSER air vehicle *(ACFR)* 1196995

Kingfisher UAV Mk 2 *(BAE Systems Australia)* 1353876

vehicle testbeds used in the programme were named Brumby; developed versions have the name Kingfisher.

Description: *Airframe: Brumby Mk 3:* Mid-mounted, clipped-delta wings with two-segment trailing-edge elevons; twin inset fins and rudders; pusher engine; fixed tricycle landing gear. Construction of composites and aluminium alloy. Rectangular canards.

Kingfisher 1: Inherits fuselage (minus canards), power plant and landing gear of Brumby Mk 3, allied to new, constant chord, unswept, high-lift wing and conventional twin-boom tail assembly.

Kingfisher 2: Wings of Mk 1, adjustable C of G to allow nose-mounted payloads, new tail and landing gear, with improved engine and new box-section fuselage having solid nose and rectangular ventral cutout for sensor turrets.

Mission payloads: Modular, interchangeable payloads (any two from range of TV camera, laser range-finder or scanning MMW radar). Mounted in removable nose section of Brumby Mk 3/Kingfisher 1; in ventral sensor turret on Kingfisher 2.

Guidance and control: BAE Systems GCS, incorporating ISR management system (latter co-developed with CDL Systems, Canada). Air vehicle is autonomous during take-off and landing phases; fully autonomous with INS/GPS navigation, guidance and control during normal flight. UHF command datalink; S-band air-to-ground and air-to-air datalink; preprogrammed and in-flight navigation waypoint uplink. Electrical power supply from 1 kW generator, draws 125 W for flight automation. Fully decentralised, multiple-sensor, single- and multiple-platform, multiple-target tactical picture compilation (tracking); SLAM tactical picture compilation and navigation without GPS.

System composition: Two air vehicles, each carrying up to four terrain and inertial sensors; mission planning and control station; air-to-ground and air-to-air communications links; ground support equipment.

Launch: Conventional wheeled take-off.

Recovery: Conventional wheeled landing.

Power plant: Brumby Mk 3 and Kingfisher 1: One 11.9 kW (16 hp) Desert Aircraft DA-150 flat-twin engine; two-blade pusher propeller.C: One 28.3 kW (38 hp) UEL AR741 engine; two-blade pusher propeller

Kingfisher 2: One 28.3 kW (38 hp) UEL AR741 engine; two-blade pusher propeller

Variants: *Brumby Mk 1:* Developed by students at the University of Sydney; first flight November 1997. Subsequently adopted as ANSER programme prototype and used for communications relay. No longer operated

Brumby Mk 2: Single-payload, intermediate airframe design, built by the ANSER team for sensor testing. First flight April 2000. No longer operated

Brumby Mk 3: Multipayload ANSER vehicle; bulged, transparent forward fuselage and rectangular, mid-mounted canard surfaces; autonomous flight control system. First flight early 2001. Developed as a research platform for operation in harsh environments, and to demonstrate low-cost rapid prototyping for UAV airframe design. No longer operated

Kingfisher 1: Further refinement of Brumby Mk 3. Test-flown at Kingaroy, Queensland, in May 2005. No longer operated.

Kingfisher 2: Developed version, of revised configuration. Substantial increases in payload mass and volume and ease of fitment.

Brumby Mk 3, Kingfisher Mk 1, Kingfisher Mk 2

Dimensions, External	
Overall length	
Brumby Mk 3	2.19 m (7 ft 2¼ in)
Kingfisher Mk 1	3.73 m (12 ft 2¾ in)
Kingfisher Mk 2	4.20 m (13 ft 9¼ in)
height, Brumby Mk 3	0.70 m (2 ft 3½ in)
Wings wing span	
Brumby Mk 3	2.82 m (9 ft 3 in)
Kingfisher Mk 1	4.23 m (13 ft 10½ in)
Kingfisher Mk 2	4.13 m (13 ft 6½ in)
Areas Wings Gross wing area	
Brumby Mk 3	1.63 m² (17.5 sq ft)
Kingfisher Mk 1, Kingfisher Mk 2	2.67 m² (28.7 sq ft)
Weights and Loadings Weight	
Weight empty, Brumby Mk 3	25 kg (55 lb)
Operating weight, empty	
Brumby Mk 3	32 kg (70 lb)
autonomous, Brumby Mk 3	97 kg (213 lb)
Max T-O weight	
Brumby Mk 3	45 kg (99 lb)
Kingfisher Mk 1	60 kg (132 lb)
Kingfisher Mk 2	121 kg (266 lb)
Payload Max payload	
Brumby Mk 3	7 kg (15.00 lb)
Kingfisher Mk 1	12 kg (26 lb)
Kingfisher Mk 2	22 kg (48 lb)
Performance Altitude	
Service ceiling, Brumby Mk 3	1,370 m (4,500 ft)
Speed	
Max level speed	100 kt (185 km/h; 115 mph)
Cruising speed	
min, Brumby Mk 3	65 kt (120 km/h; 75 mph)
max, Brumby Mk 3	75 kt (139 km/h; 86 mph)
Endurance	
Brumby Mk 3	2 hr
Kingfisher Mk 1, Kingfisher Mk 2	3 hr

Status: In August 2002, successful completion was announced of a series of flight tests involving multiple UAVs (two Brumby Mk 3s) in the performance of fully decentralised picture compilation. They were remotely piloted over terrain at the Sydney University test range at Marulan, New South Wales, where they sensed a number of artificial ground features, processed the data on board and built up a real-time picture of ground activity without requiring use of a central processing facility. The ANSER team aimed to create a four-UAV testbed facility to develop and demonstrate decentralised data fusion technologies further at BAE Systems Australia's facility at West Sale airfield in south-eastern Victoria, where airspace expansion is underway. Development of the Kingfishers was continuing in 2010.

Contractor: BAE Systems Australia
Richmond, Victoria.

Cyber Technology CyberEye II

Type: MALE surveillance UAV.

Description: *Airframe:* Shoulder-mounted wings with downturned tips; pod and boom fuselage; pylon-mounted dorsal pod with pusher engine; V tail. Fixed tricycle landing gear. All-composites construction.

Mission payloads: Multiple options, including gimbal-mounted Sony daytime pan-tilt-zoom video camera with ×26 zoom (standard fit); daytime video with ×36 zoom; or 60 Hz infra-red camera. Max video range 16 n miles (30 km; 18.5 miles). Real-time, encrypted, analogue or digital video downlink, with a range of up to 32 n miles (60 km; 37 miles). Engine-driven generator provides 12 V and 6 V DC power for payload and onboard systems. Back-up storage batteries for emergency power.

Guidance and control: Fully autonomous and reprogrammable in flight, from UAV Navigation GCS03 ground station with DGPS navigation and AP04 autopilot. GCS can control up to six air vehicles simultaneously

The V-tailed CyberEye II has a 10-hour endurance *(Cyber Technology)* 1395204

Rear view of the CyberEye II *(Cyber Technology)* 1395254

through a single encrypted, frequency-hopping datalink with a range of more than 54 n miles (100 km; 62 miles). Two-man operation.

Transportation: Air vehicle and GCS fit into a 3 m × 80 cm (9.8 × 2.6 ft) case (total weight less than 150 kg; 330 lb).

Launch: Conventional, automatic wheeled take-off.

Recovery: Conventional, automatic wheeled landing. Parachute for emergency recovery.

Power plant: One 7.5 kW (10 hp) two-cylinder two-stroke engine, driving a three-blade carbon fibre pusher propeller.

CyberEye II

Dimensions, External	
Overall, length	2.80 m (9 ft 2¼ in)
Wings, wing span	4.50 m (14 ft 9¼ in)
Engines, propeller diameter	0.61 m (2 ft 0 in)
Weights and Loadings	
Weight, Max T-O weight	80 kg (176 lb)
Payload, Max payload	20 kg (44 lb)
Performance	
Altitude	
Operating altitude, typical	305 m (1,000 ft)
Service ceiling	3,050 m (10,000 ft)
Speed	
Max level speed	86 kt (159 km/h; 99 mph)
Cruising speed, max	54 kt (100 km/h; 62 mph)
Stalling speed	38 kt (71 km/h; 44 mph)
Radius of operation	
datalink, normal	54 n miles (100 km; 62 miles)
datalink, with additional tracking	81 n miles (150 km; 93 miles)
Endurance	10 hr
Power plant	1 × piston engine

Status: In production. Said to be currently operational with 'various' defence forces; customers include Royal Thai Air Force, which has purchased 'several' platforms.

Contractor: Cyber Technology (WA) Pty Ltd
Bibra Lake, Western Australia.

Cyber Technology CyberQuad

Type: Close-range VTOL mini-UAV.

Development: Cyber Technology was formed in February 2006. Its other UAV or aerial target products include the CyberEye II, CyberWraith and CyBird. Development of the CyberQuad, with a view to use in urban situations unsuited to other types of UAV (inside buildings, for example), began in late 2008.

Description: *Airframe:* Quad-rotor platform comprising four fan ducts, each encircling a two-blade rotor; enclosed central area houses batteries, payload and flight control avionics. Opposing pairs of rotors are counter-rotating. Vehicle is supported on a four-hoop landing gear.

Mission payloads: Low-light CCD video camera in Mini, with interchangeable lenses; Maxi has a high-definition video camera with ×12 optical zoom and image stabiliser. Both versions can carry a 1.2, 2.4 or 5.8 GHz video downlink transmitter. Four separate sensor packages available for AirGuard, depending on mission.

Guidance and control: Fully or semi-autonomous operation, with GPS-based waypoint navigation, via nine-channel radio command uplink in hand-held controller. Laptop GCS with touch-screen display optional. Manoeuvrability controlled by differential operation of ducted rotors.

Mechanical simplicity and low noise are keynotes of the CyberQuad design *(Cyber Technology)* 1395255

Attitude and heading stabilised by inertial sensors and high-frequency speed controllers.

Transportation: System is easily transportable in a robust carrying case.

Launch: Automatic vertical take-off.

Recovery: Automatic vertical landing.

Power plant: Battery-powered brushless electric motors, each driving a two-blade ducted rotor. Battery duration approx 30 minutes.

Variants: *Mini:* Baseline version.

Maxi: Slightly larger version, with additional battery pack(s) for enhanced payload/endurance capability.

AirGuard: Customised version, with sensor packages for air quality measurement, industrial or chemical emissions monitoring or WMD detection.

CyberQuad Mini, CyberQuad Maxi

Dimensions, External	
Overall	
length	0.53 m (1 ft 8¾ in)
height, excl landing gear	0.16 m (6¼ in)
width	0.53 m (1 ft 8¾ in)
Weights and Loadings	
Weight	
Max T-O weight	
CyberQuad Mini	2 kg (4.00 lb)
CyberQuad Maxi	3 kg (6.00 lb)
Payload	
Max payload	
CyberQuad Mini	0.5 kg (1.00 lb)
CyberQuad Maxi	0.8 kg (1.00 lb)
Performance	
Climb	
Rate of climb	
max, at S/L, CyberQuad Mini	300 m/min (984 ft/min)
max, at S/L, CyberQuad Maxi	600 m/min (1,968 ft/min)
Altitude, Operating altitude	1,000 m (3,280 ft)
Speed	
Max level speed	
CyberQuad Mini	27 kt (50 km/h; 31 mph)
CyberQuad Maxi	32 kt (59 km/h; 37 mph)
Radius of operation	0.5 n miles (km; miles)
Endurance	
CyberQuad Mini	25 min
CyberQuad Maxi	35 min

Status: As of mid-2010, CyberQuad was understood to be in service with, or under evaluation by, several Australian and overseas agencies.

Contractor: Cyber Technology (WA) Pty Ltd
Bibra Lake, Western Australia.

Cyber Technology CyberWraith

Type: Multipurpose mini-UAV.

Development: The CyberWraith can double as either a UAV or as a reusable target for weapon testing and operational training.

Description: *Airframe:* Mid-mounted 'semi-delta' wings; cylindrical fuselage; twin sweptback fins and rudders attached to wing trailing-edges. No landing gear. All-composites construction.

Mission payloads: As specified by customers.

Guidance and control: Fully autonomous via UAV Navigation GCS03 ground station and AP04 autopilot.

Launch: Automatic launch by catapult.

Recovery: Automatic, to belly landing.

Power plant: One 0.14 kN (31 lb st) unidentified turbojet. Fuel capacity 20 litres (5.3 US gallons; 4.4 Imp gallons).

The twin-tailed, jet-powered CyberWraith UAV and target
(Cyber Technology) 1395256

CyberWraith

Dimensions, External	
Overall, length	2.50 m (8 ft 2½ in)
Wings, wing span	2.50 m (8 ft 2½ in)
Weights and Loadings	
Payload, Max payload	6 kg (13.00 lb)
Performance	
Speed	
Max level speed, IAS	190 kt (352 km/h; 219 mph)
Stalling speed, IAS	70 kt (130 km/h; 81 mph)
Power plant	1 × turbojet

Status: Understood to be in production and/or in service; customer(s) not identified.
Contractor: Cyber Technology (WA) Pty Ltd
Bibra Lake, Western Australia.

V-TOL Aerospace Warrigal

Type: Tactical mini-UAV.
Development: From conception, the Warrigal (name of a mystical bird) was developed by V-TOL Aerospace in conjunction with S & B Model Aircraft to offer law enforcement agencies a low-cost, reliable and versatile tactical mini-UAV capability. Designed to gain national airspace regulatory approval over urban terrain, the airframe is manufactured with public safety and impact countermeasures (system survivability) as the key objectives. Its targeting and surveillance capabilities are also designed to ensure mission and operational reliability and versatility for the end-user by providing real-time video or near-real-time high-resolution imagery for day or night, urban or rural situational awareness.

Description: *Airframe:* Mid-mounted wings with tapered leading-edge; bullet-shaped fuselage; sweptback fin; no landing gear. Composites construction (impact optimised, energy-absorbing extruded polypropylene).

Mission payloads: Retractable, stabilised day or night camera (see under Variants below). Options include recording, targeting and tracking software.

Guidance and control: Camera-centric 'point and fly' operation, with autonomous or semi-autonomous waypoint programming. Typical GCS set-up is manpack or vehicle set-up with laptop, video capture card/LCD DV evidence recorder, encrypted frequency-hopping tactical data/video link with portable omni or directional antennas and power supplies. A day/night, all-weather, 'sense and avoid' capability is also planned.

Launch: By hand or bungee catapult.

Recovery: Deep stall landing (in urban operations); elsewhere (where space permits) by traditional belly landing.

Power plant: One Hyperion 3025-6 brushless, 40 to 60 A electric motor, powered by lithium-polymer batteries; two-blade pusher propeller.

Variants: *Warrigal Tactical:* Designed to carry pan/tilt or pan/tilt/zoom CCD TV or miniature thermal camera for real-time surveillance to more than 1.6 n miles (3 km; 1.9 miles); video optionally extendable to more than 3.2 n miles (6 km; 3.7 miles).

Warrigal Commercial: Designed to carry a fixed high-resolution digital camera to capture near-real-time images to create geo-referenced small-area or short-range linear high-resolution mosaic picture maps.

The Warrigal electric-powered tactical mini-UAV
(V-TOL Aerospace) 1290290

Warrigal Tactical, Warrigal Commercial

Dimensions, External	
Overall, length	1.19 m (3 ft 10¾ in)
Wings, wing span	1.52 m (4 ft 11¾ in)
Weights and Loadings	
Weight	
Weight empty	4.0 kg (8.00 lb)
Max launch weight	5.0 kg (11.00 lb)
Performance	
Altitude	
Operating altitude	
P/T and hi-res	60 m to 120 m (200 ft to 400 ft)
P/T/Z	150 m (500 ft) (est)
Speed	
Max level speed	60 kt (111 km/h; 69 mph)
Cruising speed	30 kt (56 km/h; 35 mph)
Radius of operation	
standard, control link	3.2 n miles (5 km; 3 miles)
optional upgrade, control link	27 n miles (50 km; 31 miles)
Endurance	40 min
Power plant	1 × electric motor

Status: Tactical version available, commercial version under development mid-2008. Two further Warrigal prototypes built by late 2008. In October 2009 V-TOL Aerospace, with partners, launched a programme to obtain certification of Warrigal, the declared objective being the first approval of a UAV by a national regulator to operate over populous areas day and night.
Contractor: V-TOL Aerospace Pty Ltd
Rocklea, Queensland.

Austria

Schiebel Camcopter 5.1

Type: Unmanned helicopter.
Development: The Camcopter 5.1 can be used for: ground and aerial surveillance; target acquisition and designation; minefield and surface ordnance survey; day/night traffic; border patrol and observation; NBC survey; and environmental monitoring. Small size and stealth characteristics minimise the risk of detection. System upgraded from original model 5.0 to 5.1 standard in early 1998 with more powerful engine, wheeled landing gear and reconfigured mobile GCS.

Upgraded 5.1 Mk 2 announced in early 2000; has 28.3 kW (38 hp) UEL AR 741 rotary engine, increasing hover ceiling OGE to 3,810 m (12,500 ft); 900 W power generation capability; and, eventually, satcom datalink. Superseded by Camcopter S-100.

Description: *Airframe:* Modular design, with faceted fuselage shell; built of robust, high-strength, lightweight materials. Two-blade main rotor with Bell/Hiller type stabiliser; two-blade tail rotor. Twin-skid (5.0) or three-wheel (5.1) landing gear. Air vehicle is man-portable, and can be transported in a light utility vehicle such as a pick-up truck.

Mission payloads: Nose-mounted colour CCD camera, with real-time C-band video downlink, for piloting. Additional pan-and-tilt-mounted, gyrostabilised sensors such as thermal imagers, radars, laser range-finder/designators and NBC detectors can be carried, up to 60 × 33 × 30 cm (23.6 × 13 × 11.8 in) in size. Real-time C-band, bidirectional datalink for data transfer, analysis and recording. Data transceiver

Camcopter at Webster Field, Maryland, July 2003 *(Kenneth Munson)* 0558397

Unmanned aerial vehicles > Austria > Schiebel Camcopter 5.1 – Schiebel Camcopter S-100

Camcopter 5.1 without body fairing *(Schiebel)* 0044417

receives control commands, and transmits navigation and operational data continuously to the GCS. Payload mounting base has RS-232 interface.

Guidance and control: GCS consists of a Mission Control Unit (MCU) with C-band spread system up/down datalink and a customised Sensor Control Unit (SCU), both housed in shock-resistant containers. The UAV has INS/DGPS navigation, and can be preprogrammed to fly preselected routes, or operated under manual control. In automatic mode, operator can observe flight at MCU and flight control follows a preprogrammed mission plan; in manual mode, operator directs and observes flight from the MCU, which is linked to the onboard CCD pilot camera. In both modes, Air vehicle is computer-stabilised in flight and can hover automatically in one position.

Mission control computer generates geographically based mission data, displaying the air vehicle's location, programmed route, waypoints and mission data on-screen on a digital map. Pilot control sub-unit uses a control stick and panel to select flight mode (VTOL, automatic, manual or homing). Flight path can be observed via video downlink and monitored on an instrument panel.

System composition: Air vehicle; GCS, including a pilot navigation and flight control sub-unit; customised SCU.

Launch: Conventional helicopter take-off.

Recovery: Conventional helicopter landing.

Power plant: One 16.4 kW (22 hp) Limbach L 275 E air-cooled two-stroke engine with remote start. Fuel capacity 13.5 litres (3.6 US gallons; 3.0 Imp gallons), expandable to 30.5 litres (8.1 US gallons; 6.7 Imp gallons).

Camcopter 5.1

Dimensions, External	
Overall, height	0.80 m (2 ft 7½ in)
Fuselage	
length	2.50 m (8 ft 2½ in)
width, max	0.82 m (2 ft 8¼ in)
Rotors, rotor diameter	3.09 m (10 ft 1¾ in)
Tail rotor, tail rotor diameter	0.50 m (1 ft 7¾ in)
Weights and Loadings	
Weight	
Weight empty	43 kg (94 lb)
Max T-O weight	68 kg (149 lb)
Fuel weight	
Max fuel weight, standard	17.4 kg (38 lb)
Payload	
Max payload, incl fuel	25 kg (55 lb)
Performance	
Climb	
Rate of climb, max, at S/L	180 m/min (590 ft/min)
Altitude	
Service ceiling	3,000 m (9,840 ft)
Hovering ceiling, OGE	1,700 m (5,575 ft)
Speed	
Cruising speed, max	49 kt (91 km/h; 56 mph)
Radius of operation	
datalink, standard	5.4 n miles (10 km; 6 miles)
datalink, enhanced	54 n miles (100 km; 62 miles)
Endurance, max	6 hr

Status: Production complete. Three systems purchased by US Army for humanitarian demining; three air vehicles purchased by Thales, France, for payload demonstrations; selected by German Army for AAMIS airborne minefield surveillance technology programme. Active demonstration programme with US Air Force, Army, Marine Corps and Border Patrol; shipboard demonstration aboard US Coast Guard cutter *Valiant* in 2000 with Joint Interagency Task Force for counter-drug and search and rescue applications. Demonstrated to UK MoD in January 2000. Successor system is Camcopter S-100.

Customers: Production complete. Austria; Egyptian Navy (two 5.1 Mk 2 systems); German Army; Thales (France, three systems); US Army (three systems). Also evaluated by British Army (Royal Artillery) and US Naval Surface Warfare Center.

Contractor: Schiebel Elektronische Geräte GmbH
Vienna.

Schiebel Camcopter S-100

Type: VTOL UAV system.

Development: The Camcopter S-100 is a versatile, fully autonomous VTOL tactical UAV system, developed to provide a balance between advanced capabilities and operation in tactical environments. Mission applications include terrestrial and naval surveillance. For maritime applications it can, like its Camcopter 5.1 predecessor, land on ships equipped with a helicopter deck without the use of additional landing aids. The S-100 made its first flight in the second quarter of 2004.

A two-year joint development programme between Schiebel and the UAE Air Force's UAV Research and Technology Centre resulted in a derivative known as **Al-Saber** (Watchkeeper), of which nine prototypes were built. These were delivered in October 2005 for in-country acceptance trials that were completed in Abu Dhabi in March 2006. Delivery of 40 production systems (80 air vehicles) began in 2006. A teaming agreement with the Boeing Company covering the pursuit of marketing and support opportunities for Camcopter S-100 was signed in August 2009. Underslung payload capability, Remotely Operated Video Enhanced Receiver (ROVER) surveillance capability and extra fuel was introduced from 2010.

Description: *Airframe:* Two-blade main rotor; streamlined pod and boom fuselage. The latter is a carbon fibre monocoque for superior strength-to-weight ratio. Dorsal tailfin, with T tailplane, plus underfin. Three-point landing gear, comprising twin single main legs and reinforced tip of underfin.

Mission payloads: The S-100 has capacity for a wide range of payload/endurance combinations, which can be mounted in or on one of the two payload bays, an internal auxiliary electronics bay, or fuselage-side hardpoints, with real-time telemetry and dual video downlinks. The primary payload bay, located directly beneath the main rotor shaft, is capable of mounting payloads weighing up to 50 kg (110 lb). The side hardpoints are capable of carrying loads of up to 10 kg (22 lb). The secondary bay, located in the nose section, is also capable of mounting payloads weighing up to 10 kg. The auxiliary electronics bay is intended for additional instrumentation such as IFF, ACAS or other electronic equipment.

Standard sensor is an IAI Tamam POP 200 stabilised and gimballed EO/IR unit, combining an InSb focal plane array thermal imager and a colour CCD TV camera. Other suitable payloads could include lidar, integrated spotlights, loudspeakers, rope/net-dropping containers, a multispectral imager and ground-penetrating or synthetic aperture radar.

An S-100 was demonstrated in UAE colours for the first time at the 2011 IDEX trade exhibition *(IHS/Patrick Allen)* 1436439

A head-on view of an S-100 equipped with an L-3 Wescam MX-10 EO payload *(Schiebel)* 1436478

Schiebel Camcopter S-100 < Austria < Unmanned aerial vehicles 9

An in-flight view of an S-100 fitted with an L-3 Wescam MX-10 EO payload *(Schiebel)* 1436479

Schiebel demonstrated the S-100 at Berlin's ILA Airshow during 2010 *(Schiebel)* 1418210

An in-flight view of an S-100 at the LIMA 2011 tradeshow in Malaysia *(Schiebel)* 1365369

An example of EO/IR imagery generated by an S-100 during the Berlin 2010 air show *(Schiebel)* 1418211

An in-flight view of an S-100 at the 2011 Paris Airshow *(Schiebel)* 1365370

During sea trials, the S-100 has been tested aboard 14 different types of naval vessel *(Schiebel)* 1409709

An S-100 photographed during sea trials aboard the Italian Navy frigate *Bersagliere (Schiebel)* 1365371

An S-100 VTOL UAV caught in the hover *(Schiebel)* 1377377

The ROVER system has been integrated with Camcopter, enabling field commanders to see the same imagery as the UAV operator. The information is displayed on a small monitor or laptop carried in a backpack by ground units.

An underslung payload capability of up to 80 kg (176 lb) was being offered from 2010 for re-supply missions.

In January 2011 the L-3 Wescam MX-10 EO/IR payload was flown on the S-100 for the first time at Wiener Neustad, Austria. This payload was also trialled on a UAE S-100 at the Abu Dhabi Autonomous System Investments company test range in early 2011.

Guidance and control: The S-100 can complete its entire mission automatically, from take-off to landing, controlled by a triple-redundant flight computer based on proven flight control methods and algorithms and using a C-band command uplink. The operator can take over manually at any point during the flight and can add, delete and modify waypoints at any time via an interactive user interface.

In manual mode, the operator gives directional commands via a simple joystick; directional commands are then interpreted by the flight control computer to provide control surface inputs appropriate to the desired manoeuvre, while assuring that the air vehicle remains within its

Unmanned aerial vehicles > Austria > Schiebel Camcopter S-100

An S-100 at sea *(Schiebel)* 1414032

An S-100 photographed during sea trials *(Schiebel)* 1414034

During October 2008, the S-100 was trialled aboard the French Navy frigate *Montcalm*, with the effort involving both automatic launch and recovery *(Schiebel)* 1294757

During September 2007, an S-100 UAV was test flown from an Indian Navy Offshore Patrol Vessel operating in the Arabian Sea *(Schiebel)* 1343557

A general view of some of the components that make-up the S-100 system *(IHS)* 1419978

performance limitations. In both modes, the air vehicle is stabilised automatically. The flight control system is said to provide a positional accuracy of ±1 m (3.3 ft).

The control station is scalable, ranging from operation from two laptop computers (one for mission planning and control, the other for payload control and imagery exploitation) to larger, integrated suites. The mission planning and control workstation displays air vehicle position and status information in real time on a user-friendly, aviation-style instrument panel, together with integrated checklists and failure procedures. Mission planning and preparation is done using Geographical Information System (GIS) data, and the entire mission can be viewed and rehearsed within a 3D synthetic environment to which other layers of GIS data (such as threat zones, no-fly zones and other intelligence information) can be added for greater situational awareness.

The payload control workstation allows the payload operator to control the payload while also having access to mission planning information. Video viewing, capture and recording are also available.

The command system has been designed to allow control of multiple air vehicles via either a stand-alone network or through integration into existing tactical networks. Air vehicle electronics, including datalinks, are designed to be completely modular, allowing for upgrades or the fitting of customer-specific components.

Within Austria, Schiebel's demonstrators fly with an experimental civil registration (OE-VXW) and operate under a Permission to Fly issued by AustroControl.

Transportation: The complete system can be deployed in two tactical 4 × 4 trucks.
System composition: Two air vehicles, payloads, control station and support equipment (including pilot control unit).
Launch: Conventional helicopter take-off.
Recovery: Conventional helicopter landing.
Power plant: One 41.0 kW (55 hp at 7,100 rpm) Diamond rotary engine with dual pump-fed fuel injection. Internal fuel capacity (Avgas 100LL) 57 litres (15.1 US gallons; 12.5 Imp gallons). An external tank can be carried to extend endurance to up to 10 hours. A heavy fuel engine was scheduled to become available during 2013.

Camcopter S-100

Dimensions, External	
Fuselage	
length	3.11 m (10 ft 2½ in)
height, max.	1.12 m (3 ft 8 in)
width, max.	1.24 m (4 ft 0¾ in)
Rotors, rotor diameter	3.40 m (11 ft 1¾ in)
Tail rotor, tail rotor diameter	0.66 m (2 ft 2 in)
Weights and Loadings	
Weight	
Weight empty	110 kg (242 lb)
Max T-O weight	200 kg (440 lb)
Payload	
Max payload	
useful, (sensor + fuel)	100 kg (220 lb)
sensor	50 kg (110 lb)
Performance	
Altitude, Service ceiling	5,485 m (18,000 ft)
Speed	
Never-exceed speed	120 kt (222 km/h; 138 mph)
Cruising speed, best endurance	55 kt (102 km/h; 63 mph)
Endurance	
with 34 kg (75 lb) payload	6 hr
with external fuel tank	up to 10 hr

Status: Delivery of prototype/pre-series systems began in October 2005. Delivery of the Al-Saber version for UAE began in June 2006.

In September 2006, the company's OE-VXW demonstrator completed a series of nine routine night flights along a 70 km (43.5 mile) section of the Austria/Slovakia border on behalf of the Austrian Ministry of the Interior, using the thermal imager of its POP 200 EO/IR sensor. Additional equipment for these flights included a Mode C transponder, enabling ATC centres at the nearby Vienna Schwechat and Bratislava International airports to track each flight.

Schiebel and Diehl of Germany in June 2007 extended an agreement to market the Camcopter S-100 in Germany as a shipboard reconnaissance system on K130 corvettes.

Camcopter S-100 received a Permit to Fly under EASA regulations in June 2007. Permit to Fly is issued to EU-registered aircraft which do not have a type certificate, but can fly under EASA-approved conditions.

Demonstrations of the S-100 to the Indian Navy were completed in October 2007 from an Offshore Patrol Vessel (OPV) in the Arabian Sea. In 2008 a series of trials and demonstrations were conducted from a Pakistan Navy Type 21 frigate, from the French Navy frigate *Montcalm*, on a Spanish patrol vessel of the Guardia Civil and from German K130 corvettes. More than 500 take-offs and landings have been logged from six different classes of naval vessel.

In June 2009 the S-100 was flown at the Paris Le Bourget Airshow, the first UAV to be displayed this way at Le Bourget. It was demonstrated at Paris again in 2011. Similar firsts were claimed in June 2010 at ILA Berlin and in February 2011 at IDEX Abu Dhabi.

In August 2009 a teaming agreement was signed between Schiebel and the Boeing Company covering the marketing of the S-100 in the United States and other territories. In October 2009 Schiebel announced a sub-contract to provide a Camcopter S-100 for the US DoD Yellow Jacket programme intended to examine the feasibility of using a UAV to detect IEDs. The payload in this case, which is designed by CenTauri Solutions, was a high definition electro-optical sensor together with an unintentional electromagnetic emissions detector. An S-100 was flown at Fort Bragg in late 2009 in a demonstration of a psychological warfare mission, carrying an American Technology Corporation loudspeaker, a leaflet drop capability and an IAI POP300 EO/IR camera. In 2010 the Camcopter S-100 completed 150 hours of trials with the French Army and Navy under a leasing contract with the DGA. The tests demonstrated maritime surveillance from a coastal base and support of troops in an urban environment. The payload was the Thales Optronics Agile 2 EO/IR sensor.

During the November 2010 G-20 Heads of State meeting in Seoul an S-100 was deployed on surveillance duties as part of the security screen, the only UAV to do so. Also in November 2010, the S-100 was demonstrated to military and government agencies in Spain.

Two Camcopter S-100 systems were delivered to Jordan in February 2011 for operation by the Jordan Armed Forces Reconnaissance Squadron, following a contract awarded in July 2010 by the King Abdullah Design and Development Bureau. These vehicles are equipped with the L-3 Wescam MX-10 payload.

A partnership with French shipbuilder DCNS was announced in 2011 for deploying Camcopter S-100 on the newly-built Gowind-class Offshore Patrol Vessel *L'Adroit* for trials in late 2011. The ship has been outfitted during the build process to carry S-100, said to be first vessel to be so equipped for a UAV.

Also in 2011, the S-100 was deployed in Malaysia by the national oil company Petronas for pipeline inspection duties. It has been flown on power line monitoring demonstrations in Austria.

In April 2012, the S-100 was flown from the ITS *Bersagliere* (a Soldati-class frigate) and 'successfully carried out a number of missions for observers from the Italian Navy'.

During July 2012, the S-100 played an 'integral role' in the European Defence Agency's Intelligent Control of Adversary Radio communications (ICAR) effort. Here, an 'essential aspect' of the project was the prevention of the wireless triggering of improvised explosive devices.

Customers: As of late 2012, Schiebel had contracts for more than 140 Camcopter S-100s from three export customers, comprising sales and leases of systems.

Contractor: Schiebel Elektronische Geräte GmbH
Vienna.

Belgium

VITO Mercator

Type: HALE research UAV.

Development: Mercator is the airborne component in a Belgian research programme named Pegasus, first proposed in 2000 as a means of providing up-to-date remote sensing data on the Earth's environment. Funding by the Flemish government was approved in June 2004, with bids invited six months later for a suitable unmanned platform and its ground control station (GCS). This resulted in VITO issuing a EUR11 million contract in June 2005 to Verhaert Space for a solar-powered aircraft able to perform as a 'stratellite', flying at altitudes between 14 and 20 km for extended periods of up to eight months. Verhaert was to provide the GCS, integrate subsystems and manage the flight test programme; QinetiQ, which later acquired a 90 per cent controlling interest in Verhaert, was subcontracted to supply the air vehicle.

On 6 June 2006, an initial test flight was made by a scaled-down (less than 10 kg; 22 lb gross weight) proof-of-concept aircraft named Mercator Low, at the Pampa Range in Helchteren. Since then, Mercator development has continued in partnership with that of QinetiQ's similar Zephyr programme, resulting in preparations for a Mercator 1 maiden flight in late 2010.

Description: *Airframe:* Constant-chord wings, with dihedral on outer panels; rectangular tail surfaces. 'Fuselage' is a single aluminium alloy boom to which wings and tail unit are attached. MEDUSA pod is of aluminium and titanium within a carbon fibre shell. No landing gear.

Mission payloads: Mercator 1 has a dedicated payload module named MEDUSA, an acronym derived from Monitoring Equipment and Devices for Unmanned Systems at [high] Altitude, which is funded by the European Space Agency (ESA). This is attached at the front of the boom 'fuselage', and contains a mission sensor, a GPS receiver and communications equipment. Over time, this will be used to trial various sensors, the first of which will be a wide-swath digital colour camera to undertake photogrammetric and disaster monitoring missions. The camera has pan capability, near real-time data transmission, and is expected to achieve an imagery level of 30 cm ground resolution from 18 km altitude. An infra-red camera and lidar are among other sensors expected to be installed later.

Guidance and control: Autonomous flight profile, with capability for in-flight re-programming. Mobile GCS, with satcom S-band two-way datalink and GPS navigation.

Launch: Runway or balloon launch.

Recovery: Runway landing.

Computer-generated image of Mercator 1 in flight *(QinetiQ)* 1395277

Mercator Low scale demonstrator, exhibited at RAF Waddington in July 2006. Note the different tail configuration *(Paul Jackson)* 1151597

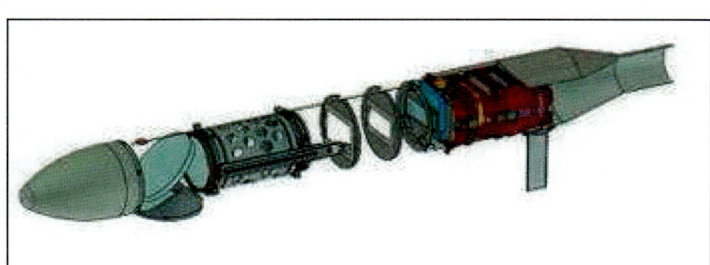

Cutaway of Mercator 1's MEDUSA sensor pod *(VITO)* 1395279

The full-size Mercator 1 *(VITO)*

Power plant: Two electric motors in wing-mounted nacelles, each driving a two-blade propeller. Powered by solar energy during daylight and by storage batteries at night.

Mercator

Dimensions, External	
Overall	
length, MEDUSA pod	1.00 m (3 ft 3¼ in)
Fuselage, width	0.12 m (4¾ in)
Wings, wing span	18.00 m (59 ft 0¾ in)
Weights and Loadings	
Weight, Max T-O weight	32 kg (70 lb)
Payload, Max payload	2.6 kg (5.00 lb)
Performance	
Altitude	
Operating altitude	
by day	18,000 m (59,060 ft)
by night	15,000 m (49,220 ft)
Speed, Cruising speed	39 kt (72 km/h; 45 mph)
Radius of operation, datalink	91 n miles (168 km; 104 miles)
Power plant	2 × electric motor

Status: The Belgian Ministry of Mobility issued Mercator 1 a Permit to Fly on 2 September 2010, and the aircraft and its GCS were installed at Bertrix air base later that month. As of late October 2010, it was awaiting suitable weather conditions in order to make a maiden flight of around 24 hours' duration. First flight with the camera-carrying MEDUSA pod, which will aim to last for three days and two nights, is planned for mid-2011.

Contractor: VITO NV
Mol.

Brazil

Avibras Falcao

Type: Medium-range tactical UAV.

Development: The Falcão is the third and most recent VANT-RE member of the Brazilian minitsry of defence's Veiculo Aéreo Não Tripulado programme, a mockup being displayed at the LAAD exhibition in Rio de Janeiro in April 2011.

Description: *Airframe:* Low-mounted constant-chord wings; circular-section fuselage; conventional tail surfaces; rear-mounted engine. Fixed tricycle landing gear. Composites (CFRP/GFRP) construction.

Mission payloads: EO sensor with laser rangefinder; FLIR; SAR or SAR/GMTI radar. Satcom datalink.

Guidance and control: Flight Technologies autopilot. Automatic take-off and landing based on DGPS and radar altimeter. Control and navigation system based on flight tests in 2007-08 using a small Brazilian Air Force Acauã UAV, in a joint programme with Brazilian Army and Navy R & D bodies and the CTA.

System composition: Three air vehicles; truck-mounted GCS, datalink terminal and transporter. Launcher truck optional.

Launch: Conventional, automatic wheeled take-off. Option to launch from truck-mounted rail.

Recovery: Conventional, automatic wheeled landing.

Power plant: One unidentified piston engine, driving a three-blade pusher propeller.

Falcao

Dimensions, External	
Overall, length	5.90 m (19 ft 4¼ in)
Wings, wing span	10.80 m (35 ft 5¼ in)
Weights and Loadings	
Weight, Max T-O weight	630 kg (1,388 lb)
Payload, Max payload	150 kg (330 lb)
Performance	
T-O	
T-O run	600 m (1,969 ft)
Altitude, Service ceiling	4,570 m (15,000 ft)
Speed	
Cruising speed, max	97 kt (180 km/h; 112 mph)
Radius of operation	
LOS	81 n miles (150 km; 93 miles)
with satcom	1,349 n miles (2,498 km; 1,552 miles)
Endurance	15 hr
Landing	
Landing run	600 m (1,969 ft)
Power plant	1 × piston engine

Status: Under development.
Contractor: Avibras Indústria Aeroespacial SA
São Paulo.

Falcão mockup at LAAD show in April 2011 *(Robert Hewson)*

Falcão rear view *(Robert Hewson)*

Flight Technologies Horus 100

Type: Close-range tactical mini-UAV.

Development: The Horus 100 is one of three current systems involved in the Brazilian MoD's VANT programme (Veiculo Aéreo Não Tripulado). The design process is understood to have started in 2007, to fulfil the Mini-VANT requirement for a portable, short-range tactical system usable at platoon, company or battalion level. It utilises a Flight Technologies (formerly Flight Solutions) air vehicle known as the **FS-02 AvantVision**.

Description: *Airframe:* Wing pylon-mounted above simple pod and boom fuselage. Dihedral on outer panels. No landing gear. All-composites construction.

Mission payloads: Dual-sensor (EO/IR) payload for day and night operation. Real-time imagery downlink.

Guidance and control: Autonomous, with real-time telemetry and datalink. Suitcase-size ground control unit, operable by two people.

Transportation: Man-portable; dismantled air vehicles and ground control unit can be carried in two backpacks.

System composition: Three air vehicles, three individual reception units, and command and control unit. Set-up time 3 minutes.

Launch: Hand-launched.

Recovery: Belly landing.

Power plant: One electric motor (type and rating not stated); two-blade propeller.

Horus 100 hand-launched mini-UAV *(Flight Technologies)*

FS-01 Watchdog in its original configuration *(Flight Technologies)*

FS-02

Dimensions, External	
Overall, length	1.90 m (6 ft 2¾ in)
Wings, wing span	2.10 m (6 ft 10¾ in)
Weights and Loadings	
Weight, Max launch weight	5.8 kg (12.00 lb)
Payload, Max payload	0.7 kg (1.00 lb)
Performance	
Altitude	
Operating altitude, typical	500 m (1,640 ft)
Service ceiling	1,525 m (5,000 ft)
Radius of operation	6.5 n miles (12 km; 7 miles)
Endurance	1 hr
Power plant	1 × electric motor

Status: Development continuing in 2013.
Contractor: Flight Technologies
São José dos Campos.

Flight Technologies Horus 200

Type: Medium-range tactical UAV.
Development: The Horus 200 is the medium-range 'VANT-Tactic' member of the Brazilian MoD's VANT (Veiculo Aéreo Não Tripulado) programme of UAS development, for use at regiment, brigade or division level. The air vehicle, named **FS-01 Watchdog**, was unveiled by Flight Technologies (then known as Flight Solutions) in April 2007 at the Latin American Aerospace and Defence (LAAD) exhibition in Rio de Janeiro, at which time it was already undergoing flight testing. It was reportedly derived from a 2005 design by the University of Minas Gerais.

Description: *Airframe:* Typical pod and twin tailboom configuration; high-mounted, high aspect ratio wings, with dihedral; pusher engine. Fixed tricycle landing gear. Composites construction.
  Mission payloads: Dual-sensor (EO/IR) turret for day and night operation; options include SAR or elint package. Real-time imagery and data downlinks.
  Guidance and control: Autonomous, with option to revert to operator control.
  System composition: Three air vehicles in trailer, three individual reception units, truck-mounted GCS, C2I shelter, plus launch/ground support vehicle. Three-person operation. Set-up time ten minutes.
  Launch: From truck-mounted ramp.
  Recovery: Conventional wheeled landing.
  Power plant: One flat-twin piston engine (type and rating not revealed); two-blade pusher propeller.

Horus 200 model at LAAD 2011 exhibition *(Robert Hewson)*

FS-01

Dimensions, External	
Overall, length	2.80 m (9 ft 2¼ in)
Wings, wing span	4.07 m (13 ft 4¼ in)
Weights and Loadings	
Weight, Max launch weight	70 kg (154 lb)
Payload, Max payload	30 kg (66 lb)
Performance	
Altitude	
Operating altitude, typical	1,500 m (4,920 ft)
Service ceiling	6,095 m (20,000 ft)
Speed	
Cruising speed, max	103 kt (191 km/h; 119 mph)
Radius of operation, mission	81 n miles (150 km; 93 miles) (est)
Endurance	3 hr
Landing	
Landing run	100 m (329 ft)
Power plant	1 × piston engine

Status: One three-aircraft system was ordered by the Brazilian Army in early 2008 at a reported contract value of USD770,000. These prototypes (also known as VANT VT15) were delivered in 2010; evaluation was continuing in 2013.
Contractor: Flight Technologies
São José dos Campos.

Bulgaria

Aviotechnica Sokol

Type: Close-range surveillance mini-UAV.
Development: Revealed in March 2008. Aviotechnica refers to 'series 1, 1M, 2 and 3', but has not provided explanatory details. Sokol is designed for close-range, low-altitude surveillance.
Description: *Airframe:* Constant-chord, unswept shoulder-mounted wings and low-set tailplane; conventional three-axis control surfaces; box-section fuselage. All-composites construction, with carbon fibre load-bearing elements. Fixed tricycle landing gear.
  Mission payloads: Video camera
  Guidance and control: Remote control via 10-channel MC-24 portable radio and onboard receiver with digital servos.
  Launch: Conventional wheeled take-off.
  Recovery: Conventional wheeled landing.
  Power plant: One 3.0 kW (4.0 hp) single-cylinder two-stroke engine, driving a two-blade tractor propeller.

Sokol UAV *(Aviotechnica)*

Sokol

Dimensions, External
Overall
 length.. 1.90 m (6 ft 2¾ in)
 height... 0.65 m (2 ft 1½ in)
Wings, wing span... 2.50 m (8 ft 2½ in)
Weights and Loadings
Weight, Max launch weight................................ 16.0 kg (35 lb)
Payload, Max payload... 4.0 kg (8.00 lb)
Performance
Altitude, Service ceiling....................................... 1,200 m (3,940 ft)
Speed, Max level speed..................................... 75 kt (139 km/h; 86 mph)
Radius of operation, mission.............................. 2.7 n miles (5 km; 3 miles)
Endurance, max.. 2 hr

Status: New in 2008. Production status not revealed.
Contractor: Aviotechnica S.p. Ltd
 Plovdiv.

Aviotechnica Yastreb-2S

Type: Recoverable airborne jammer UAV.

Development: Variant of Yastreb-2M aerial target (which see); development started in 1993. Flight testing began in mid-1994, and initial trials were said to be highly effective, demonstrating the ability to suppress communications within a range of 5.4 n miles (10 km; 6.2 miles) while flying above 1,000 m (3,280 ft). It is the air vehicle component of a UAV system known as Glarus. The aerial target Yastreb was being offered with a rotary engine in 2008 to increase speed and ceiling, an upgrade that could be applied to Yastreb-2S.

Description: *Airframe:* Tapered mid-wing monoplane with circular-section fuselage, V tail and two rectangular ventral fins. NACA 64A series wing aerofoil section. Steel tube fuselage frame with duralumin and composites skin; wings and tail surfaces have metal spars. Ailerons located at approximately mid-span of wings; fixed tab on each outer wing trailing-edge; tail surfaces have inset ruddervators. No landing gear, but underfuselage reinforced to absorb landing impact.

Mission payloads: Signal augmenters of target version are replaced by an AJ-045A radio jammer (*smutitel*, the S in the air vehicle designation) which is an airborne version of the standard Bulgarian land forces' shell-delivered R045/046 Starshel, manufactured by the Samel '90 defence electronics company. The standard Starshel operates in five frequency subranges in the HF/VHF (18 to 104 MHz) waveband; the AJ-045A is designed to jam all kinds of radio equipment in the 27 to 90 MHz frequency ranges. The Yastreb-2S can jam receivers with an antenna diameter of up to 4 m (13.1 ft) that are within 5.4 n miles (10 km; 6.2 miles) of the UAV.

Guidance and control: Three-channel autopilot controls engine throttle, ailerons and ruddervators, and stabilises air vehicle airspeed, height and course. Flight control can be by direct remote piloting from the ground, or automatic via the autopilot with provision for overriding by the ground controller. Telemetry downlink data are displayed in analogue and digital form on a GCS TV monitor. Jamming missions can be either preprogrammed or remotely controlled from the GCS.

Launch: By small booster rocket motor from truck-mounted zero-length launcher.

Recovery: Parachute recovery to landing on underfuselage airbags. Deployment of parachute is automatic in the event of a control link failure.

Power plant: One 12.3 kW (16.5 hp) DB-250 two-stroke engine; two-blade propeller.

Yastreb-2S

Dimensions, External
Overall
 length.. 2.85 m (9 ft 4¼ in)
 width... 0.533 m (1 ft 9 in)
Fuselage
 height, max... 0.28 m (11 in)
Wings, wing span... 3.52 m (11 ft 6½ in)
Weights and Loadings
Weight, Max launch weight................................ 66 kg (145 lb)
Payload, Max payload... 4.5 kg (9.00 lb)
Performance
Altitude, Operating altitude................................. 500 m to 1,800 m (1,640 ft to 5,905 ft)
Speed
 Max level speed.. 97 kt (180 km/h; 112 mph)
 Cruising speed.. 86 kt (159 km/h; 99 mph)
Radius of operation... 27 n miles (50 km; 31 miles) (est)
Endurance... 1 hr 30 min

Status: Current status unknown.
Customers: Bulgarian Army.
Contractor: Aviotechnica S.p. Ltd
 Plovdiv.

Canada

Aeryon Scout™

Type: Micro VTOL surveillance UAV.

Development: Aeryon's Scout™ is an 'easy-to-operate' VTOL surveillance micro UAV that was first identified by *IHS Jane's* during 2012 and is described as being suitable for military, public safety and industrial and commercial applications. As such, the less than 1.4 kg (3.0 lb) Scout™ AV can fly in sustained winds of up to 27 kts (50 km/h; 31 mph) and gusts of up to 43 kts (80 km/h; 50 mph); can operate within a temperature range of −30°C to +50°C (−22°F to +122°F) and has an endurance and LOS range of 25 minutes and 1.6 n miles (3.0 km; 1.9 miles) respectively. Identified Scout™ operational usage during 2012 included wildfire monitoring (the University of Alaska Fairbanks, US Forestry Service and USAF Combustion and Atmospheric Dynamics Research Experiment - Rx-CADRE), oil pipeline inspection (BP Alaska in the Prudhoe Bay, Alaska region) and crop and livestock monitoring (Isis Geomatics).

Description: *Airframe:* The Scout™ micro UAV comprises a circular central body that supports four two-bladed rotor arms, a ventral quick-change payload module, what is presumed to be a dorsal communications module and four take-off/landing feet.

Mission payloads: The Scout™ micro UAV can accommodate the Photo35™ (3-axis, stabilised, high-resolution stills imager), VideoZoom10X™ (video with ×10 optical zoom) and Thermal FLIR (thermal IR video) hot-swappable payload modules together with custom payloads that are designed for specialised imaging and other sensing applications.

Guidance and control: Autonomous capabilities with point-and-click navigation and camera controls. Fly-safe features and automated flight planning software are included to reduce operator workload.

Transportation: Carry case or backpack, with the AV components being designed for snap fit assembly.

System composition: AV and a control tablet identified.

Launch: Conventional vertical take-off.

Recovery: Conventional vertical landing.

Status: Most recently, Aeryon Laboratories Inc was continuing to promote the Scout™ VTOL micro UAV.

Contractor: Aeryon Laboratories Inc,
 Waterloo, Ontario, Canada.

A general view of the elements that make up the Scout™ VTOL micro UAV system *(Aeryon Laboratories)*

Draganfly Draganflyer X series

Type: VTOL mini-UAV.

Development: Draganfly Innovations was founded in 1998, initially producing small radio-controlled flying machines for the hobby market. The Draganflyer X6 is understood to have been the company's entry into the UAV marketplace.

Description: *Airframe:* Aluminium frame, with central module containing avionics and battery; three or four radiating arms (see *Variants*), each with two- or four-blade rotor unit at tip, with carbon fibre blades. Twin-skid detachable landing gear.

Mission payloads: Payloads suspended beneath central module. Over time, identified options have included a Panasonic 12.1 MP digital still/video camera; Tau Micro or FLIR Systems Photon 640 IR camera; Watech 902 low-light video camera; or a micro analogue camera. Video transmitted via 5.8 GHz downlink; 2.4 GHz two-way telemetry datalink.

A general view of the four-arm, four-rotor Draganflyer X4 *(Draganfly)* 1356261

The Draganflyer X6 has three arms and six rotors *(Draganfly)* 1356262

A Draganflyer X6 AV with an alternative payload *(Draganfly)* 1356263

Guidance and control: Remotely piloted. The architecture's ground station comprises PC-based main unit (with an embedded telemetry transceiver and a video receiver), a portable (suitcase-size) video display unit and a hand-held AV controller. Elsewhere, such AVs feature built-in payload stabilisation and the X6 configuration is further noted as having a GPS position hold capability.

Transportation: With skid gear removed, the X6 AV's rotor arms can be folded to allow it to be transported in a 14 cm (5.5 in) diameter tube (the X8 configuration is also foldable for transport). Again, all versions can be carried in a backpack.

Launch: Conventional helicopter take-off.

Recovery: Conventional helicopter landing.

Power plant: Rotors on each arm each of which, is driven by a separate brushless electric motor. Power is provided by a lithium-polymer battery carried in the AV's central module.

Variants: *Draganflyer X4:* The Draganflyer X4 is the smallest version in the Draganflyer X range and is understood to have been introduced in mid-2009. As such, it features four rotor arms that each carry a single two-blade rotor (here, opposing pairs rotate in opposite directions). Identified X4 sub-variants comprise the **X4/E4** (optimised for educational and research applications), the **X4-ES** (optimised for law enforcement and fire service applications) and the **X4-P** (featuring enhanced endurance and the ability to carry an up to 0.7 kg payload).

Draganflyer X6: Launched in August 2008. Draganflyer X6 configurations feature three rotor arms, with each arm carrying a pair of contra-rotating rotors. Again, Draganflyer X6 is said to have double the payload capacity of the X4 variant. Within the X6 family, the **X6-ES** sub-variant is optimised for law enforcement and fire service applications.

Draganflyer X8: The Draganflyer X8 is described as being a larger, four-arm version of X6, with each arm being equipped with a pair of contra-rotating rotors. Again, the AV utilises a carbon fibre airframe and is said to be capable of carrying double the payload of the X6 configuration. The Draganflyer X8 was unveiled in September 2009.

Draganflyer X6, Draganflyer X4, Draganflyer X8

Dimensions, External
Overall
 length
  Draganflyer X4 .... 0.645 m (2 ft 1½ in)
  Draganflyer X6 .... 0.91 m (2 ft 11¾ in)
  Draganflyer X8 .... 0.87 m (2 ft 10¼ in)
  folded, Draganflyer X6 .... 0.68 m (2 ft 2¾ in)
  folded (with rotors and landing
   gear), Draganflyer X8 .... 0.70 m (2 ft 3½ in)
 height
  Draganflyer X4 .... 0.21 m (8¼ in)
  Draganflyer X6 .... 0.254 m (10 in)
  Draganflyer X8 .... 0.32 m (1 ft 0½ in)
 width
  Draganflyer X4 .... 0.645 m (2 ft 1½ in)
  Draganflyer X6 .... 0.91 m (2 ft 11¾ in)
  folded, Draganflyer X6 .... 0.30 m (11¾ in)
  Draganflyer X8 .... 0.87 m (2 ft 10¼ in)
  folded (with rotors and landing
   gear), Draganflyer X8 .... 0.35 m (1 ft 1¾ in)

Weights and Loadings
Weight
 Weight empty
  incl battery, Draganflyer X4 .... 0.68 kg (1.00 lb)
  incl battery, Draganflyer X6 .... 1.0 kg (2.00 lb)
  Draganflyer X8 .... 1.7 kg (3.00 lb)
 Max T-O weight
  Draganflyer X4 .... 0.98 kg (2.00 lb)
  Draganflyer X6 .... 1.50 kg (3.00 lb)
  Draganflyer X8 .... 2.7 kg (5.00 lb)
Payload
 Max payload
  Draganflyer X4 .... 0.25 kg (0.00 lb)
  Draganflyer X6 .... 0.50 kg (1.00 lb)
  Draganflyer X8 .... 0.80 kg (1.00 lb)

Performance
Climb
 Rate of climb
  max, at S/L .... 420 m/min (1,378 ft/min)
  max, Draganflyer X8 .... 120 m/min (393 ft/min)
Altitude
 Service ceiling .... 2,440 m (8,000 ft)
 ASL, Draganflyer X8 .... 2,438 m (8,000 ft)
Speed
 Max level speed
  Draganflyer X6 .... 27 kt (50 km/h; 31 mph)
  approximate, Draganflyer X8 .... 27 kt (50 km/h; 31 mph)
Endurance
 no payload .... 20 min
 approximate, no payload,
  Draganflyer X8 .... 20 min
Power plant, Draganflyer X8 .... 8 × electric motor

Status: Most recently, Dragonfly Innovations Inc has continued to promote the Draganflyer X4/E4, X4-ES, X4-P, X6, X6-ES and X-8 VTOL mini-UAVs. As of early 2013, Draganfly was reporting its X series customer base as including six × American and Canadian law enforcement agencies, 12 × industrial concerns (with Honeywell International, the NEC Corporation and Quanser being identified as operating X6 sub-variants), four × film and aerial photographic companies and 65 × universities and educational establishments.

Contractor: Dragonfly Innovations Inc
Saskatoon, Saskatchewan.

MicroPilot CropCam

Type: Crop control surveillance UAV.

Development: MicroPilot, a maker of UAV autopilots, produced a low-cost UAV called **MP-Vision**, initially as a trainer and demonstrator for its products. It developed a specialised version of this, the CropCam, for precision photography of agricultural and forest areas to aid farmers and conservationists to monitor the health of their domains. The company formed a CropCam division to market and support the system.

Description: *Airframe:* Typical sailplane configuration of high aspect ratio wings, with winglets, slim fuselage and T-tail. No landing gear. Glass fibre construction.

Mission payloads: Standard sensor is a Pentax Digital Optio camera; a version with a thermal imager is reportedly under development.

Guidance and control: Fully automatic radio control (72 MHz hobby band) from launch to landing, by means of MicroPilot MP2128 autopilot, Trimble GPS navigation and MicroPilot Horizon GCS.

Launch: Hand, bungee or catapult launch.

Recovery: Deep stall to belly landing, or parachute recovery.

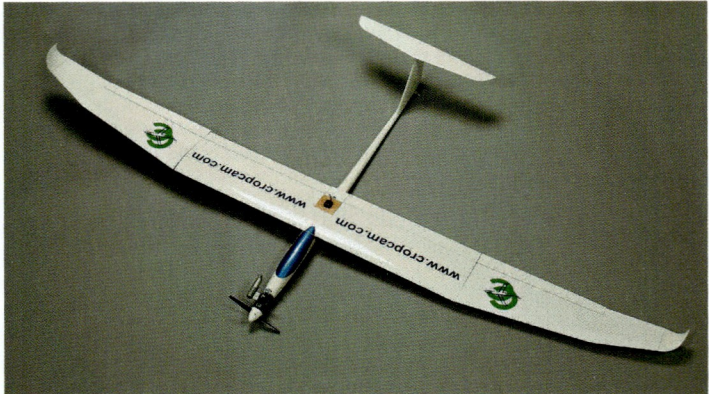

CropCam is based on the airframe of a hobby-market glider (MicroPilot)

CropCam flying in the UK's U-MAP programme in August 2008 (QinetiQ)

Power plant: One 2.5 cm³ (0.15 cu in) Axi brushless electric motor, powered by lithium-polymer batteries; two-blade tractor propeller. Glow-plug engine/liquid fuel version available optionally.

CropCam

Dimensions, External	
Overall, length	1.22 m (4 ft 0 in)
Wings, wing span	2.44 m (8 ft 0 in)
Weights and Loadings	
Weight, Max launch weight	2.7 kg (5.00 lb)
Payload, Max payload	0.45 kg (0.00 lb)
Performance	
Altitude, Operating altitude	120 m to 670 m (400 ft to 2,200 ft)
Speed	
Max level speed	52 kt (96 km/h; 60 mph)
Cruising speed	30 kt (56 km/h; 35 mph) (est)
Radius of operation, LOS, mission	0.9 n miles (1 km; 1 miles)
Endurance	20 min
Power plant	1 × electric motor

Status: Over time, production of CropCam has exceeded 250 units, with customers in Canada and exports to at least 18 countries including Australia, India, Malaysia, Russia, the UK and US. In the UK, the CropCam was in use with QinetiQ as systems integrator, as part of the Welsh Assembly's U-MAP (UAVs for Managing Agricultural Practice) programme run by Aberystwyth University. The Cropcam system has also been used for wildlife surveys in Canada and Finland. Most recently, MicroPilot has continued to promote both the MP-Vision and CropCam AVs, with the CropCam entity continuing to market the CropCam UAV system.

Contractor: MicroPilot
Stony Mountain, Manitoba.

MMIST CQ-10A SnowGoose

Type: Cargo, communications and ISR UAV.
Development: *Note: Where trademark is not added the UAV in question is the US Military version.*

The SnowGoose™ multi-role UAV has its origins in the US Special Operations Command's (USSOCOM) Wind Supported Aerial Delivery System (WSADS) aerial leaflet delivery programme. Here, the air-launched WSADS was to offer precise delivery of leaflets and reduce the risk to aircrew and transport aircraft. MMIST's SnowGoose™ (a version of the company's Sherpa™ guided parafoil system) was chosen to fulfil the requirement after initial trials and was subsequently formally designated as the CQ-10A SnowGoose. In conjunction with the US Army Unmanned Aerial Vehicle Systems (UAVS) office, an ACTD programme called Air Launched Extended Range Transporter (ALERT) was established with the goal of equipping and demonstrating the baseline WSADS with multiple, existing, non-developmental payloads such as EO/IR cameras, communications relays, meteorological sensors and dropsondes. As such, ALERT was part of the Expendable UAV (XUAV) ACTD managed by the USN's Naval Air Systems Command (NAVAIR) for USSOCOM. In addition to payload demonstrations, ALERT included upgrades to the basic air vehicle including the introduction of a SATCOM datalink (for command and control of the AV and its payloads), an IFF transponder, navigation lights and a feed into the Blue Force Tracking (BFT) system. A multi-purpose payload interface unit was also added to the basic configuration to allow up to 1 kW of vehicle electrical power to be shared with various payloads. All ALERT upgrades subsequently incorporated into the CQ-10A SnowGoose FRP configuration. Elsewhere, the CQ-10A has been used to deliver a range of supply types (including vaxipacks, trauma kits, water, petrol (gasoline) and blood products) and has been used as an emplacement vehicle for unattended ground sensors. Other significant programme events have included:

June 2003 A CQ-10A Milestone C LRIP decision was reached during June 2003.

2004 A military assessment of the CQ-10A SnowGoose was conducted as part of the US Army's 2004 Quartermaster Liquid Logistics Exercise (QLLEX) at Fort Hunter Liggett, California, with a reserve unit from the Service's 383rd Quartermasters Battalion operating the architecture after a three week operator's training course. During the effort, multiple missions (including night-time recoveries) were flown each day, with 'mission highlights' including a multi-mission demonstration sortie that combined simultaneous communications relay and EO/IR surveillance with a subsequent re-supply effort that delivered a cargo to within 50 m (165 ft) of its intended recipients. Subsequent to the exercise,

Ground launch of a SnowGoose™ AV from an HMMWV (MMIST)

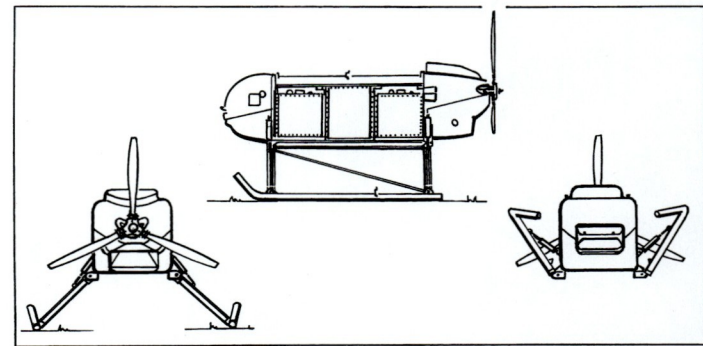

Rear, side and front (folded) elevations of the SnowGoose™ AV's central fuselage module (IHS/John W Wood)

A SnowGoose™ AV being air launched from a C-130 transport aircraft *(MMIST)* 1143574

A general view of the SnowGoose™ central fuselage module *(IHS/Patrick Allen)* 1025392

A SnowGoose™ AV photograph of the moment of lift-off from a HMMWV launch vehicle *(MMIST)* 1143577

An in-flight view of a SnowGoose™ AV delivering a 45.4 kg (100 lb) payload *(MMIST)* 1143573

SnowGoose is said to have been highly rated for its communications relay, re-supply and ISR capabilities.

August 2004 *IHS Jane's* sources report MMIST as having received a CQ-10A FRP contract for 200 × AVs (to be delivered over a five year period) during August 2004.

2005 As of 2005, *IHS Jane's* sources were reporting MMIST as having been considering a CQ-10B configuration that would incorporate a third wing kit (to facilitate both self-launch without reliance on ground equipment and cargo extraction) and would have more than four times the range of the CQ-10A.

February 2005 *IHS Jane's* sources report the first production CQ-10A operational set as having been fielded by the US Army's 3rd Psychological Operations Battalion during February 2005. As of the given date, CQ-10A production and fielding was said to be proceeding at a minimum rate of two × AVs per month.

Description: *Airframe:* The SnowGoose™ AV incorporates a central fuselage module (containing fuel, a payload, the vehicle's propulsion system and its guidance unit), an air or ground launch parafoil and a foldable skid landing gear.

Mission payloads: The SnowGoose™ AV incorporates six cargo bays (three each side) that can be used to carry modular fuel bins, cargo bins or fixed electronic payloads, with the whole facilitating an 'easy' trade-off between payload capacity and endurance over the AV's full flight envelope. Each standard cargo bin is suitable for dispensing up to 45.4 kg (100 lb) of medical supplies, food, water, leaflets, ammunition, fuel, tools or spare/replacement parts. Here, High-Altitude Low-Opening (HALO) parachutes provide low-observability delivery of 45.4 kg (100 lb) bundles with 'autonomous precision'. As of 2005, payloads flown had included an EO/IR camera, a LOS communications relay, a meteorological sensing unit, a wind sonde dispenser, a security loudspeaker, a high-capacity (2 Mbps) SATCOM link (possibly) and Frequency Modulated (FM) and TV broadcast equipment (possibly).

Guidance and control: The SnowGoose™ AV features a fully autonomous guidance, navigation and control system based on MMIST's Sherpa™ aerial delivery system's (see separate entry) parachute control unit. System autonomy includes waypoint navigation, avoidance areas, air launch, landing and cargo delivery execution (based on in-flight, real-time wind measurements). An airborne guidance unit (AGU) performs all navigation and control functions and the AV's flight plan is programmed using an industry-standard laptop computer with map underlay. Upload to the AGU takes place before launch or via the SATCOM datalink. A remote manual override is available (using a LOS RF datalink) to allow the operator to interrupt an autonomous mission to fly and land the system, or to deploy cargo manually.

Launch: The SnowGoose™ AV can be air-launched from a C-17, C-130 or C-141 transport aircraft, or ground-launched from a HMMWV, a flatbed truck or a logistics trailer. Four AVs can be deployed in-flight from a single C-130.

Recovery: The SnowGoose™ AV performs fully autonomous landings over a wide variety of unprepared surfaces. A four-person recovery team can retrieve the AV with an unmodified HMMWV.

Power plant: One 84.4 kW (113.3 hp) Rotax 914 UL turbocharged flat-four engine, driving a three-blade composites pusher propeller. A number of Diesel engine options have been integrated into the SnowGoose™ AV and were being evaluated in ground and flight tests during 2005.

CQ-10A

Dimensions, External	
Overall	
length	2.90 m (9 ft 6¼ in)
height	
to top of fuselage	1.50 m (4 ft 11 in)
over vertical propeller blade	2.24 m (7 ft 4¼ in)
Fuselage	
height	0.76 m (2 ft 6 in)
width	0.71 m (2 ft 4 in)
Dimensions, Internal	
Payload bay, length	1.60 m (5 ft 3 in)
Weights and Loadings	
Weight	
Weight empty	340 kg (749 lb)
Max launch weight	609 kg (1,342 lb)
Fuel weight, Max fuel weight	19 kg (41 lb)
Payload	
Max payload	
per cargo bin	45.4 kg (100 lb)
total	249 kg (548 lb)
Performance	
Altitude	
Operating altitude, min	60 m (200 ft)
Service ceiling	
self-propelled, air launch	4,570 m (15,000 ft) (est)
self-propelled, ground launch	5,485 m (18,000 ft) (est)
Speed, Never-exceed speed	32 kt (59 km/h; 36 mph)
Range, max, zero wind conditions	486 n miles (900 km; 559 miles)
Endurance	
with 45.4 kg (100 lb) cargo, air launch	15 hr (est)
with 45.4 kg (100 lb) cargo, ground launch	17 hr (est)
with 90 kg (200 lb) cargo, air launch	12 hr (est)
with 90 kg (200 lb) cargo, ground launch	14 hr (est)
Power plant	1 × piston engine

Status: Most recently, Mist Mobility Integrated Systems Technology has continued to promote the SnowGoose™ cargo, communications and ISR UAV and notes that the CQ-10A configuration has been 'fielded and deployed' by the US Army and is a US military Program of Record.

Customers: US Army Special Operations Command (up to 200 air vehicles).

Contractor: Mist Mobility Integrated Systems Technology Inc
Nepean, Ontario.

MMIST Sherpa

Type: Gliding UAV.

Development: The Sherpa™ series is a family of COTS guided parachute delivery systems that incorporate a ram air parachute, a Parachute Control Unit (PCU), a hand-held remote controller, a mission planning programme and ancillaries. As such, Sherpa™ provides an aerial supply chain that can precisely delivery of cargo in hostile situations such as civil strife, search and rescue missions or in times of military necessity where expeditious resupply is an operational requirement. Again, the Sherpa™ architecture has formed a prime component of the US Army's Joint Precision Airdrop System (JPADS) ACT initiative. The systems has also been used within the US Army's Precision and Extended Glide Airdrop-L Systems (PEGASYS) and the US Air Force Precision Air Delivery System (PADS), with both programmes being managed by the US Army Soldier Center in Natick, Massachusetts. Of these, JPADS was designed to provide a seamless and flexible system that could deliver equipment resupply capabilities worldwide within 24 hours; was payload independent; could be launched from any type of aerial delivery configured aircraft from altitudes of up to 7,620 m (25,000 ft) and could land at a predetermined target point within a 100 m (330 ft) CEP.

In addition to the cited JPADS and PEGASYS programmes, both Sherpa™ and the related MMIST CQ-10A SnowGoose (see separate entry) have been included in US Integrated Logistics Aerial Resupply (ILAR) demonstrations, which have been designed to integrate and synchronise airland, airdrop and sling-load operations with other components of the distribution process, with the ultimate goal of providing a full range of aerial delivery support/services that takes advantage of joint intermodal enablers and is fully supportive of special operations missions. In the ILAR context, both systems have completed successful and safe airdrops of multi-class configured loads and commodity/capability modules with landing accuracies of 75 m (246 ft) or less. Other programme features have included use of the Sherpa™ Provider 1200 configuration by the US Army's 1st Air Delivery Platoon/Combat Service Support Battalion/1st Force Service Support Group to re-supply USMC units operating in Iraq's Al Anbar province during 2005 (with an average landing error of 69 m (226 ft) from the targetted impact point) and the use of Sherpa™ precision aerial delivery systems by 18 customers worldwide during 2011.

A Sherpa™ Provider gliding UAV aerial delivery system photographed at the point of impact with the ground *(MMIST)*
1175857

Description: *Airframe*: The generic Sherpa™ airframe configuration comprises a ram air parachute and an aluminium roll cage assembly that are operable in ground wind speeds of up to 30 kt (56 km/h; 35 mph) at elevations of up to 915 m (3,000 ft); in wind speeds of up to 150 kt (278 km/h; 173 mph, Sherpa™ Ranger configuration) at an altitude of 7,620 m (25,000 ft) and in conditions of up to 100 per cent humidity. Again, Sherpa™ systems are splashproof and water resistant (but not fully waterproof, due to the nature and functions of such components as their servo-motors and battery sections) and can be returned to service after immersion in shallow water after factory rehabilitation (including cleaning, drying, re-assembly and test).

Mission payloads: Soldier teams have used the Sherpa™ in training demonstrations and training exercises for autonomous delivery of vaxipaks, trauma kits, water, fuel and fragile payloads such as blood. Post-delivery testing confirmed that blood products were delivered undamaged. Payload weight ranges for the various Sherpa™ configurations are given in *Specifications*.

Guidance and control: Sherpa™ series equipments are pre-programmed before launch and offer manual override and in-flight landing point re-programming facilities via an associated (and optional) hand controller. Here, the unit is fitted with a proportional joystick and has a beacon mode that sends an encoded GPS position (in the form of a one-time burst signal) to the AV so that it can automatically switch to a new landing point target. As well as autonomous GPS guidance and/or manual control, the Sherpa™ Provider series guidance system includes a dynamic wind sensing facility (for automatic flight-path redefinition), programmable High Altitude High Opening (HAHO) and High Altitude Low Opening (HALO) mission profiles, waypoints and 'no-fly' zones and optional plug-in Selective Availability Anti-Spoofing Module (SAASM) GPS/INS guidance. IHS Jane's sources suggest that a 'well-trained' operator can execute a manual landing within 20 m (66 ft) of a designated target point.

System composition: The Sherpa™ Provider 1200 system used in Iraq consisted of a commercial laptop computer, an airborne guidance unit, an 83.6 m² (900 sq ft) ram air canopy (designated as the MMIST 1200 canopy) and shipping and accessory containers. Identified accessories included batteries for the parachute control unit (24 V 5 Ah); a hand-held controller (12 V 2.5 Ah); a mission planner cable; a GPS repeater; a toolkit; antennas and battery recharger.

Launch: Upon leaving the cargo aircraft, the Sherpa™ configuration deploys a parachute, then flies the cargo autonomously to the pre-programmed landing point. A small drogue chute is used to stabilise the parafoil and its cargo during free-fall and to extract the main canopy when it deploys. IHS Jane's sources describe the Sherpa™ architecture as being 'easily' mounted on standard aerial cargo systems and as being compatible with a range of fixed-wing aircraft and helicopters including the C-17, C-123, C-130, C-141, CH-47 and CH-53 types. Again, dispatch range is dependent upon predominant winds during the flight and is calculated by the mission planner once all required data has been entered.

Recovery: Sherpa™ systema are designed to withstand normal landing impacts with little or no post-flight maintenance. As such, they can be recovered without the use of pyrotechnics or other consumables and can be readied for re-use within 45 to 90 minutes. The post-recovery process includes checking the system, recharging its batteries, packing the canopy and drogue, setting the drogue ring and dispatch and programming the new mission's parameters.

Power plant: None.

Variants: The Sherpa™ system is available in **Provider 600**, **Provider 1200**, **Provider 2200**, **Provider 10,000** and **Ranger** configurations, with the different formats being defined by their maximum payload in pounds (see *Specifications*).

Sherpa™ Provider 600, Sherpa™ Provider 1200, Sherpa™ Provider 2200, Sherpa™ Provider 10,000, Sherpa™ Ranger

Weights and Loadings	
Payload	
Max payload	
Sherpa™ Provider 600	272 kg (600 lb)
Sherpa™ Provider 1200	544 kg (1,200 lb)
Sherpa™ Provider 2200	998 kg (2,200 lb)
Sherpa™ Provider 10,000	4,536 kg (10,000 lb)
Sherpa™ Ranger	318 kg (700 lb)

Status: Most recently, Mist Mobility Integrated Systems Technology Inc has continued to promote the Sherpa™ Provider 600, Sherpa™ Provider 1200, Sherpa™ Provider 2200, Sherpa™ Provider 10,000 and Sherpa™ Ranger GPS guided aerial delivery systems.
Customers: In 2011, 18 customers around the world (including US Army; Portugal, other (unspecified) NATO member countries and Singapore) had procured Sherpa™ gliding UAV aerial delivery systems.
Contractor: Mist Mobility Integrated Systems Technology Inc
Nepean, Ontario.

China

ALIT CH-91

Type: Tactical surveillance UAV.
Development: IHS Jane's first became aware of the Aerospace Long-March International Trade Company (ALIT) CH-91 tactical surveillance UAV during the early part of 2013 at which time, the type was being noted as being in production. Roles envisaged for the type included ISR, precision targeting and artillery fire correction, battle damage assessment, Geographic Information System (GIS) data collection, wildfire and pipeline monitoring, meteorological measurement and emergency communications establishment. As of the given date, ALIT was also being reported as developing a CH-91 configuration that would be equipped with a conventional tricycle undercarriage. Again, the CH-91 can be deployed within 60 minutes, with tear-down time being given as 20 minutes.
Description: *Airframe:* The CH-91 features a 'boxy' body that carries shoulder-mounted wings (with 'winglets', control surfaces at mid span and supporting twin tail booms that support an inverted V-shaped horizontal surface) and a rear-mounted pusher engine that is equipped with a two-bladed propeller. Again, the AV is fitted with a twin skid landing gear.
Mission payloads: Assumed to be mission appropriate.
Guidance and control: No details available.
Launch: The CH-91 is launched by means of either a bungee system or a rocket booster.
Recovery: The CH-91 is recovered by means of an arrester net or a parachute, with the landing skids absorbing the shock of impacting the Earth's surface.
Power plant: The CH-91 is powered by a 16.4 kW (22 hp) twin cylinder petrol engine driving a two-bladed pusher propeller.

A general view of what is almost certainly a mock-up of the CH-92 ISR and air-to-surface strike UAV mounted on a tricycle undercarriage *(ALIT)* 1455373

reported as being under test and scheduled for production during 2014. CH-92 can be deployed in less than 60 minutes, with tear-down being billed as taking less than 40 minutes.
Description: *Airframe:* The CH-92 features a streamlined fuselage (with a slightly enlarged nose section) that supports mid-mounted wings and V-shaped vertical tail surfaces. Again, the type features a two-bladed pusher propeller at the rear end of its fuselage. What is thought to be a mock-up has been photographed with a tricycle undercarriage (see *Recovery*).
Mission payloads: Assumed to be mission appropriate.
Guidance and control: No details available.
Launch: Bungee system, 'conventional' launch from a moving truck or by means of a tricycle undercarriage.
Recovery: By means of parachute, landing skid or tricycle undercarriage.
Power plant: One × 37.25 kW (50 hp) two-cylinder petrol engine driving a two-bladed pusher propeller.

CH-91

Dimensions, External	
Overall	
length	3.5 m (11 ft 5¾ in)
height	1.4 m (4 ft 7 in)
Wings, wing span	4.4 m (14 ft 5¼ in)
Weights and Loadings	
Weight, Max T-O weight	110 kg (242 lb)
Payload, Max payload	20 kg (44 lb)
Performance	
Altitude	
Operating altitude	
min	1,980 m (6,500 ft)
max	2,500 m (8,200 ft)
Service ceiling	4,480 m (14,700 ft)
Speed	
Max level speed	97 kt (180 km/h; 112 mph)
Cruising speed, average	73 kt (135 km/h; 84 mph)
Radius of operation	81 n miles (150 km; 93 miles)
Power plant	1 × piston engine

Status: Most recently, ALIT has continued to promote the CH-91 tactical surveillance UAV.
Contractor: Aerospace Long-March International Trade Company Ltd, Beijing.

The CH-91 tactical surveillance UAV *(ALIT)* 1455372

ALIT CH-92

Type: ISR and air-to-surface strike UAV.
Development: IHS Jane's first became aware of the Aerospace Long-March International Trade Company Ltd (ALIT) CH-92 ISR and air-to-surface strike UAV during 2013 at which time, the AV was being

CH-92

Dimensions, External	
Overall	
length	4.1 m (13 ft 5½ in)
height	1.6 m (5 ft 3 in)
Wings, wing span	9.0 m (29 ft 6¼ in)
Weights and Loadings	
Weight, Max T-O weight	300 kg (661 lb)
Payload, Max payload	60 kg (132 lb)
Performance	
Altitude	
Operating altitude	
lower limit	1,980 m (6,500 ft)
upper limit	2,500 m (8,200 ft)
Service ceiling	5,975 m (19,600 ft)
Speed	
Max level speed	103 kt (190 km/h; 118 mph)
Cruising speed, up to	86 kt (160 km/h; 99 mph)
Radius of operation	135 n miles (250 km; 155 miles)
Endurance	10 hr
Power plant	1 × piston engine

Status: Most recently, ALIT has continued to promote the CH-92 ISR and air-to-surface strike UAV.
Contractor: Aerospace Long-March International Trade Company Ltd, Beijing.

ALIT CH-901

Type: ISR, BDA, meteorological survey and precision attack UAV.
Development: IHS Jane's first became aware of the Aerospace Long-March International Trade Company (ALIT) CH-901 tube-launched UAV during 2013 at which time, ALIT was describing the AV as having been designed primarily to provide Special Forces with ISR, BDA and precision attack capabilities. Again, ALIT characterises CH-901 as also being applicable to meteorological survey and as of the given date, was anticipating the start of CH-901 production during 2014. When being used as a strike weapon, CH-901 has a 5 m (16.4 ft) CEP accuracy, with the value rising to 50 m (164.1 ft) CEP when the AV is being employed for targeting.

Unmanned aerial vehicles > China > ALIT CH-901 – AVIC Whirlwind Scout

A mock-up of the CH-901 ISR, BDA, meteorological survey and precision attack UAV *(ALIT)*   1455374

Description: *Airframe:* CH-901 features a cylindrical body that houses deployable wings, horizontal and vertical tail surfaces and two-bladed pusher propeller. The AV's wings are mounted above its fuselage and feature slight leading edge sweep on their outer sections. Again (and when used in the ISR/BDA/meteorological survey roles), the CH-901 incorporates a belly skip for recovery.

   Mission payloads: Assumed to be mission specific and including meteorological wind speed, temperature and humidity sensors and an unspecified type of warhead for use in the precision attack role.

   Guidance and control: No details available.

   Transportation: CH-901 is transported in its launch tube.

   Launch: Catapult launch from the AV's transportation tube.

   Recovery: Either by parachute or landing skid (ISR/BDA/metereological survey roles).

   Power plant:

CH-901

Dimensions, External	
Overall	
length	1.5 m (4 ft 11 in)
height	0.6 m (1 ft 11½ in)
Wings, wing span	2.0 m (6 ft 6¾ in)
Weights and Loadings	
Weight, Max T-O weight	9 kg (19.00 lb)
Payload, Max payload	2 kg (4.00 lb)
Performance	
Altitude	
Operating altitude	
min	10 m (32 ft)
max	1,495 m (4,900 ft)
Speed	
Max level speed	81 kt (150 km/h; 93 mph)
Cruising speed, average	51 kt (95 km/h; 59 mph)
Radius of operation	8.1 n miles (15 km; 9 miles)
Power plant	1 × piston engine

Status: Most recently, ALIT has continued to promote the CH-901 ISR, BDA, meteorological survey and precision attack UAV.

Contractor: Aerospace Long-March International Trade Company Ltd, Beijing.

AVIC Night Eagle

Type: Short-range tactical UAV.

Development: The Night Eagle was displayed at 8th Airshow China in Zhuhai, November 2010, but no development history was revealed. A superficial similarity to the Australian/US Aerosonde air vehicle may be noted.

Description: *Airframe:* Cigar-shape fuselage nacelle with rear-mounted engine; constant-chord high wings; twin tailbooms, supporting inverted V tail surfaces. Twin-skid landing gear.

   Mission payloads: Undernose ISR sensor turret.

   Guidance and control: No details disclosed.

   Launch: By rocket booster.

   Recovery: Parachute recovery.

   Power plant: Single piston engine (type and rating not disclosed), driving a two-blade pusher propeller.

Night Eagle model on display in late 2010 *(Robert Hewson)*   1395285

Night Eagle

Weights and Loadings	
Weight, Max launch weight	95 kg (209 lb)
Performance	
Altitude, Service ceiling	3,500 m (11,480 ft)
Speed	
Max level speed	91 kt (169 km/h; 105 mph)
Cruising speed, max	81 kt (150 km/h; 93 mph)
Radius of operation, mission	64 n miles (118 km; 73 miles)
Endurance, max	3 hr
Power plant	1 × piston engine

Status: Stated in November 2010 to be in service with PLA units for battlefield surveillance, targeting, artillery fire correction and BDA.

Contractor: AVIC Aviation Techniques Company Beijing.

AVIC Whirlwind Scout

Type: VTOL micro UAV.

Development: In the size, weight and performance class of the US Honeywell T-Hawk, which it closely resembles, this small UAV was first seen at Airshow China, Zhuhai, in November 2010. In operation, AVIC claims it is inaudible at a stand-off range of 125 m (410 ft), can detect a human being at 560 m (1,837 ft) and identify him at 70 m (230 ft). It was also declared to be stable in gusty conditions.

Description: *Airframe:* Barrel-shaped ducted fan, with four control vanes in fan slipstream; four curved 'landing legs'.

   Mission payloads: Unspecified ISR sensor(s).

   Guidance and control: Pre-programmed, with 100-waypoint GPS/INS navigation. GCS has capacity for storing up to 4 hours' worth of downlinked data.

   Transportation: Man-portable.

   System composition: Air vehicle, GCS, simulator and support equipment.

   Launch: Vertical take-off.

   Recovery: Vertical landing.

   Power plant: Single turbocharged piston engine (rating not disclosed), driving a ducted fan.

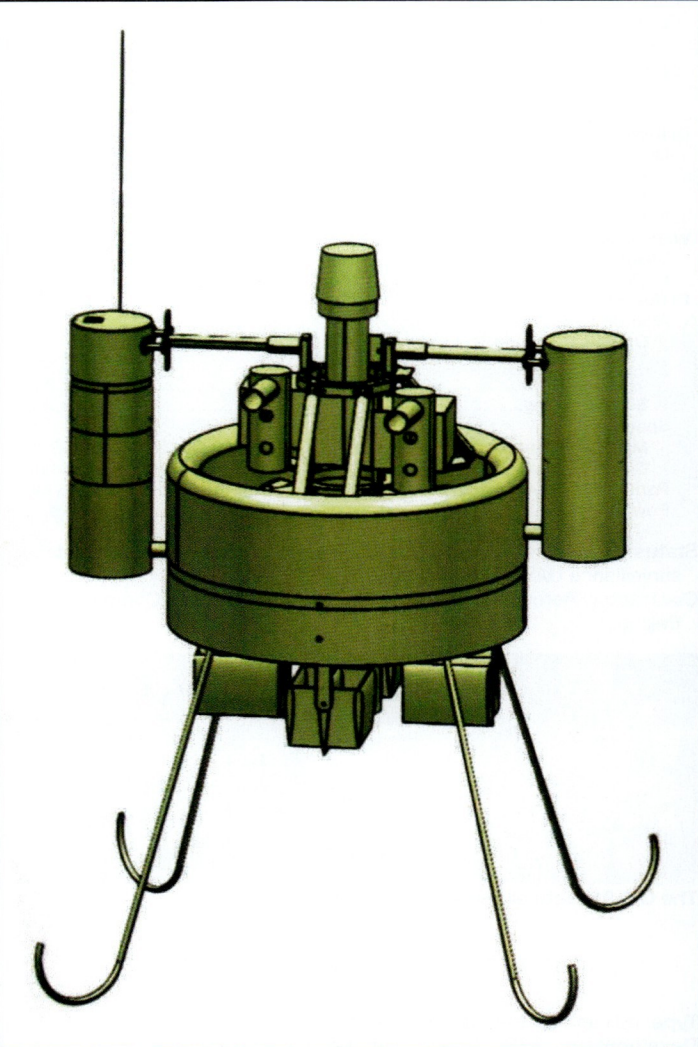

AVIC Defence's 'perch and stare' Whirlwind Scout *(Robert Hewson)*   1395288

Whirlwind Scout

Weights and Loadings
Weight, Max T-O weight .... 8.0 kg (17.00 lb)
Performance
Altitude, Operating altitude .... 3,000 m (9,840 ft)
Speed
   Max level speed .... 48.5 kt (90 km/h; 56 mph)
   Cruising speed .... 32 kt (59 km/h; 37 mph)
Radius of operation, datalink .... 5.4 n miles (10 km; 6 miles) (est)
Endurance .... 20 min (est)
Power plant .... 1 × piston engine

Status: Under development in 2010-11.
Contractor: AVIC Aviation Technology Company
  Beijing.

BUAA BZK-005

Type: Multirole MALE UAV.

Development: This UAV was first noted in an AVIC promotional video at the 2006 Airshow China in Zhuhai. Development is thought to have been started by BUAA in 2005, reportedly in conjunction with Hongdu Aircraft Industries Corporation (HAIC). It was first seen 'in the flesh' in October 2009, when a photograph of two aircraft, said to have been taken at an airfield near Beijing, appeared on the Chinese internet. Pending the emergence of firmer evidence, the following description should be regarded as provisional.

Description: *Airframe*: Slightly swept, low-mounted wings; streamlined fuselage nacelle with rear-mounted engine; twin-boom tail unit with sweptback, outward-canted fins and rudders. Retractable tricycle landing gear. Believed to be of composites construction.

  Mission payloads: EO/IR turret under nose, with real-time data transmission.

  Guidance and control: Bulged upper lobe of nose section indicates probable presence of satcom antenna

  Launch: Conventional wheeled take-off.

  Recovery: Conventional wheeled landing.

  Power plant: One rear-mounted piston engine, driving a three-blade pusher propeller.
(estimated)

2006 video grab of the BZK-005 MALE air vehicle *(AVIC)*

BZK-005

Dimensions, External
Wings, wing span .... 18 m (59 ft 0¾ in)
Weights and Loadings
Weight, Max T-O weight .... 1,250 kg (2,755 lb)
Payload, Max payload .... 150 kg (330 lb)
Performance
T-O
   T-O run .... 600 m (1,969 ft)
Altitude, Service ceiling .... 8,000 m (26,240 ft)
Speed, Cruising speed .... 118 kt (219 km/h; 136 mph) (est)
Endurance, max .... 40 hr
Landing
   Landing run .... 500 m (1,641 ft)
Power plant .... 1 × piston engine

Status: Unknown as of 2012, although one usually knowledgeable source has described the system as being "in service with PLA Department of Chief of Staff".
Contractor: Beijing University of Aeronautics and Astronautics
  Beijing.

BUAA Chang Hong

Type: High-altitude air-launched multipurpose UAV.

Development: The Chang Hong (Long Rainbow) is based on the US Teledyne Ryan Model 147H (AQM-34N), several of which were shot down over mainland China before the banning of reconnaissance overflights in 1972. Its development by the Beijing Institute (now University) of Aeronautics and Astronautics, under the original designation WZ-5, began when the Institute was tasked in 1969 with developing a high-altitude drone for daylight photographic reconnaissance. Two prototypes were built and flown in 1972, followed by two more in 1976. Government technical certification was received in February 1980, and the Chang Hong entered service for both training and tactical reconnaissance in the following year. The reverse-engineered airframe has no visible external differences from the American RPV and the Chinese Wopen 11 (WP11) power plant is based on the American vehicle's 8.54 kN (1,920 lb st) Teledyne CAE J69-T-41 turbojet.

Description: *Airframe*: Based on that of Northrop Grumman BQM-34A Firebee aerial target. Fuselage is subdivided into a radar (tracking) compartment, camera compartment, fuel cells, engine bay, avionics bay and parachute compartment.

  Mission payloads: The original payload was an optical camera, which could be rotated to any one of five positions to photograph to either side of the aircraft as well as directly beneath the flight path. Currently able to carry appropriate and more modern sensors for reconnaissance, atmospheric sampling, geological survey or target drone missions.

  Guidance and control: Preprogrammed flight profiles, aircraft climbing automatically to operating altitude after release to follow a preset flight plan. Altitude, airspeed, flight time and range/endurance are controlled by the programme. The latest version, shown in model form at Airshow China in November 2000, has a digital flight control and management system, including a new inertial navigation system with embedded GPS.

  System composition: Y-8E drone control aircraft, six UAVs, Mi-8 recovery helicopter, two mobile GDTs, three truck-mounted GCS shelters, crew bus and 15 personnel.

First live sighting of the BZK-005, October 2009

Shaanxi Y-8E drone carrier with two Chang Hong UAVs underwing 0505113

BUAA Chang Hong high-altitude reconnaissance UAV, developed from the US AQM-34N 0505112

Display model of Chang Hong in latest version *(Photo Link)* 0100808

Chang Hong on servicing trolley *(Robert Hewson)* 0105851

   Launch: Air-launched. Original carrier aircraft were modified Tu-4 bombers, but a dedicated version of the Shaanxi Y-8 transport, the Y-8E, is now in use as a drone carrier for the Chang Hong.
   Recovery: On arrival at recovery site, the aircraft deploys a parachute (automatically or on operator's command) for helicopter mid-air (MARS) recovery.
   Power plant: One 8.34 kN (1,874 lb st) BUAA WP11 turbojet.

Chang Hong

Dimensions, External	
Overall, length	8.97 m (29 ft 5¼ in)
Wings, wing span	9.76 m (32 ft 0¼ in)
Weights and Loadings	
Weight, Max launch weight	1,700 kg (3,747 lb)
Payload	
Max payload, mission	65 kg (143 lb)
Performance	
Altitude, Operating altitude	17,500 m (57,415 ft)
Speed, Max level speed	432 kt (800 km/h; 497 mph) at 17,500 m (57,415 ft)
Range	1,350 n miles (2,500 km; 1,553 miles)
Endurance	3 hr
Power plant	
8.34 kN (1,874 lb st) BUAA WP11	1 × turbojet

Status: Production complete. Remains in service. Used for geological survey and scientific research purposes such as atmospheric sampling, as well as for military reconnaissance and target drone roles. Was offered for export in 2000, no foreign sales subsequently reported.
   Customers: Chinese armed forces and civil agencies.
Contractor: Developed by Beijing University of Aeronautics and Astronautics
   Beijing.

CAC HALE UAV

Type: Multirole HALE UAV.
Development: A model of this conceptual HALE design was shown at Airshow China, Zhuhai, in October/November 2006. No details were disclosed. Configuration is closer to the US Global Hawk than GAIC's Soar Dragon (Xianglong) design, with which one may suppose it is in contention to fulfil a future PLA Air Force requirement. In late 2008 images emerged from China which appeared to show a UAV of this configuration taxi-ing at Chengdu.
Description: *Airframe:* Low-mounted, tapered, high aspect ratio wings; V tail surfaces; dorsally mounted pod for turbojet or turbofan engine; retractable tricycle landing gear.
   Mission payloads: No details given.
   Guidance and control: No details given. Bulbous front fuselage suggests accommodation for satcom antenna.
   Launch: Conventional wheeled take-off.
   Recovery: Conventional wheeled landing.
Status: In conceptual design stage in late 2006. The air vehicle may have emerged in 2008 with flight tests starting in 2009.
Contractor: Chengdu Aircraft Industry (Group) Company
   Chengdu, Sichuan.

Model of CAC conceptual HALE UAV displayed in late 2006 *(Robert Hewson)* 1151590

CAC Tianyi

Type: Short-range tactical UAV.
Development: The Tianyi (Sky Wing) UAV was seen publicly for the first time at Airshow China, Zhuhai, in October/November 2006. Examples have been built and flown.
Description: *Airframe:* Shoulder-mounted wings with tapered leading-edges; pod and twin tailboom configuration, with sweptback fins and low-set tailplane; pusher engine; fixed tricycle landing gear.
   Mission payloads: No details revealed, but EO and/or IR sensors assumed.
   Guidance and control: No details revealed.
   Launch: Conventional wheeled take-off.
   Recovery: Conventional wheeled landing.
   Power plant: One piston engine (details not stated); two-blade pusher propeller.

Tianyi

Dimensions, External	
Overall	
length	2.465 m (8 ft 1 in)
height	0.97 m (3 ft 2¼ in)
Wings, wing span	4.18 m (13 ft 8½ in)

The Chengdu Tianyi tactical UAV *(Robert Hewson)* 1151589

Tianyi

Areas	
Wings, Gross wing area	1.574 m² (16.9 sq ft)
Weights and Loadings	
Weight, Max launch weight	80 kg (176 lb)
Payload, Max payload	20 kg (44 lb)
Performance	
Altitude, Service ceiling	3,000 m (9,840 ft)
Speed	
Max level speed	97 kt (180 km/h; 112 mph)
Cruising speed, max	81 kt (150 km/h; 93 mph)
Radius of operation	54 n miles (100 km; 62 miles)
Endurance	3 hr
Power plant	1 × piston engine

Status: Not stated, but possibly in production from late 2006. No subsequent report of entry into service found.

Contractor: Chengdu Aircraft Industry (Group) Company
Chengdu, Sichuan.

CAC Wing-Loong

Type: Surveillance MALE UAV.

Development: According to contemporary sources, development began in May 2005, a prototype flying for the first time in October 2007. In early references, the UAV was also referred to by the alternative name **Yilong**.

Description: *Airframe:* Mid-mounted, tapered, high aspect ratio wings; V tail surfaces; rear-mounted engine and pusher propeller. Retractable tricycle landing gear. Aluminium fuselage; composites radome.

Mission payloads: Include day/night EO/IR and laser designator, ECM and small underwing air-to-surface missiles.

Guidance and control: Stated to be fully autonomous, but no details released.

Launch: Conventional, automatic wheeled take-off.

Recovery: Conventional, automatic wheeled landing.

Power plant: One 74.6 kW (100 hp) turbocharged piston engine, driving a three-blade pusher propeller.

Yilong

Dimensions, External	
Overall	
length	9.055 m (29 ft 8½ in)
height	2.775 m (9 ft 1¼ in)

Wing-Loong pictured for the first time 1365909

Exhibition model of the Yilong/Wing-Loong 1356254

Computer image of Yilong/Wing-Loong carrying underwing missiles
1356255

Fuselage	
height	1.09 m (3 ft 7 in)
width	1.065 m (3 ft 6 in)
Wings, wing span	14.00 m (45 ft 11¼ in)
Weights and Loadings	
Weight, Max T-O weight	1,100 kg (2,425 lb)
Fuel weight, Max fuel weight	300 kg (661 lb)
Payload	
Max payload	200 kg (440 lb)
useful (fuel + payload)	350 kg (771 lb)
Performance	
T-O	
T-O run	800 m (2,625 ft)
Altitude, Service ceiling	5,000 m (16,400 ft)
Speed, Max level speed	151 kt (280 km/h; 174 mph)
Range, max	2,160 n miles (4,000 km; 2,485 miles)
Endurance, max	20 hr
Landing	
Landing run	600 m (1,969 ft)
Power plant	1 × piston engine

Status: Reported to have completed performance and payload trials in October 2008. Development and/or initial production evidently continuing through 2010, when official data first released publicly.

Contractor: Chengdu Aircraft Industry (Group) Company
Chengdu, Sichuan.

CASC CH-3

Type: Armed tactical UAV.

Development: The CH-3 is said to have been developed by the 11th institute of the China Aerospace Science and Technology Corporation (CASC). It was first seen at Airshow China, Zhuhai in November 2008. As of late 2009, no information on the development history had been revealed. The aircraft has also been referred to in some Chinese reports as the **Rainbow-3**.

Description: *Airframe:* Rear-mounted, cranked, mid-wings with endplate fins and rudders, canard foreplanes, pod fuselage with pusher engine. Fixed tricycle landing gear.

Mission payloads: Understood to comprise EO/IR turret and laser designator. Max payload quoted as 60 kg (132 lb), but has been displayed with a pair of 45 kg (99 lb) CASC AR-1 air-to-surface missiles underwing.

Guidance and control: No details yet known.

Launch: Conventional wheeled take-off.

Recovery: Conventional wheeled landing.

Power plant: One piston engine (type and rating not revealed), driving a three-blade pusher propeller.

CH-3

Weights and Loadings	
Weight, Max T-O weight	640 kg (1,410 lb)
Performance	
Altitude, Service ceiling	5,000 m (16,400 ft)
Range	1,295 n miles (2,398 km; 1,490 miles)
Endurance	12 hr
Power plant	1 × piston engine

Status: In development and/or production. According to the Chinese media in October 2009, the CH-3 was then the subject of a recent export order, although the customer country was not identified.

Contractor: China Aerospace Science and Technology Corporation
Beijing.

CH-3 on show in 2008 with two underwing AR-1 missiles
(Robert Hewson) 1296388

CASIC LT series

Type: Micro-UAV.

Development: China's first efforts in the field of micro-UAVs were seen publicly at the November 2002 China Airshow, when three MAVs of different sizes but similar general configuration (here referred to as A, B and C for convenience) were displayed. The smallest (A) was reported to have the Chinese name *Zhangzhongbao*, translating as "Treasure in the

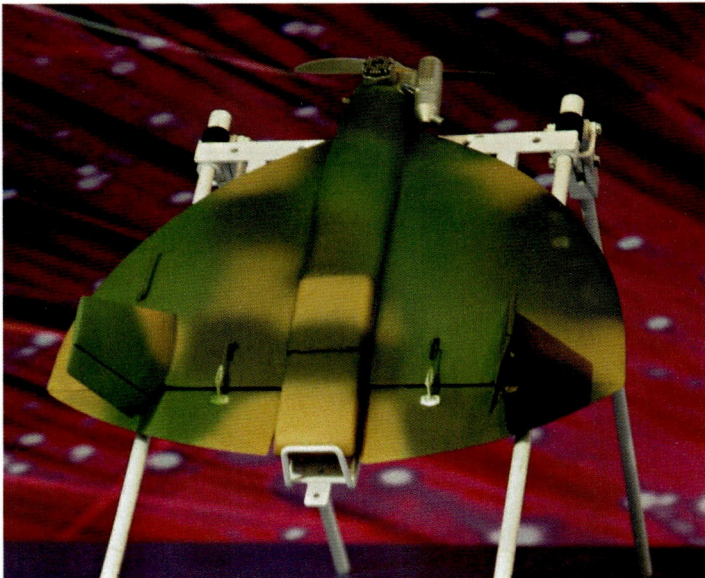
CASIC LT series MAV displayed in November 2006 (Robert Hewson)

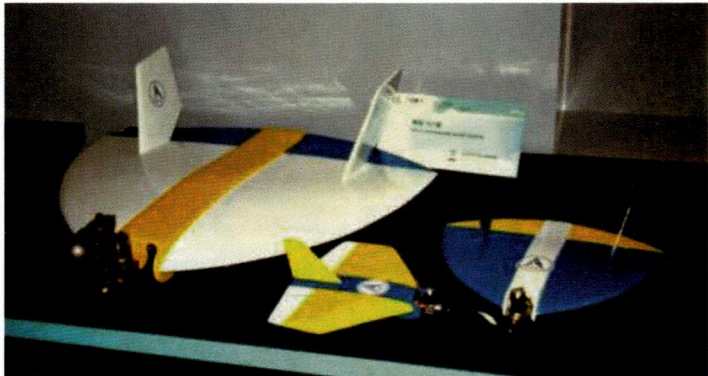
The trio of MAV prototypes first shown in 2002 (Robert Hewson)

Palm". All three had been developed from 1999 by the Beijing Institute of Aerodynamics (BIA), a subsidiary of China Aerospace and Technology Centre (CASC), and were said to then be awaiting a go-ahead for series production, having completed testing three months earlier; limited specification data were available at that time. No reappearance occurred in 2004, but in November 2006 an MAV substantially similar to types B and C, in camouflage finish, was displayed as a CASIC product and labelled as 'LT series'. No further information or data were given. Appearance can be seen in the accompanying illustrations. No further information released at the 2008 Airshow.

Mission payloads: Prototypes each fitted with video camera allowing live transmission of imagery.
Guidance and control: Understood to be preprogrammed, with facility for operator input during flight. Navigation and piloting systems reported to include gyroscopes and miniature GPS receivers.
Launch: Hand or bungee launch assumed.
Recovery: Belly landing assumed.
Power plant: Appear to be model-aircraft type (possibly glow-plug) engines; two-blade tractor propeller.

MAV A, MAV B, MAV C

Dimensions, External	
Wings	
wing span	
MAV A	0.22 m (8¾ in)
MAV B	0.30 m (11¾ in)
MAV C	0.60 m (1 ft 11½ in)
Performance	
Altitude	
Service ceiling	
MAV A, MAV B	800 m (2,620 ft)
MAV C	1,000 m (3,280 ft)
Speed	
Max level speed	
MAV A, MAV B	54 kt (100 km/h; 62 mph)
MAV C	108 kt (200 km/h; 124 mph)
Cruising speed, MAV A, MAV B	38 kt (70 km/h; 44 mph)
Radius of operation	
MAV A, MAV B	5.4 n miles (10 km; 6 miles)
MAV C	19.5 n miles (36 km; 22 miles)
Endurance	
MAV A	20 min
MAV B	30 min
MAV C	40 min

Status: May now be in production and/or service.
Contractor: China Aerospace Science and Industry Corporation (CASIC) Beijing.

CASIC WJ-600

Type: Armed reconnaissance UAV

Development: The WJ-600 is one of many previously unseen Chinese UAVs shown in model form at the 2010 Airshow China in Zhuhai, most of them with a frustrating lack of further information. The Primary role for the WJ-600 was said to be medium-altitude surveillance, combined with the ability for long-range targeting and post-strike assessment.

Description: *Airframe:* High aspect ratio high wings, with single hardpoint each side near the root; circular-section fuselage, slightly bulged at the nose and having a dorsal air intake aft of the wing; conventional tail surfaces. No landing gear shown on the exhibited model.
Mission payloads: Internal sensor payloads may include EO, SAR, EW or communications relay systems. Weapons displayed with the November 2010 model included air-to-surface missiles designated KD2 and TB1, and a ZD1 laser-guided bomb.
Guidance and control: Not disclosed.
Launch: Vehicle-launched (details not disclosed).
Recovery: Not disclosed.
Power plant: One turbojet or turbofan engine (type and rating not revealed).

WJ-600

Weights and Loadings	
Payload, Max payload	130 kg (286 lb)
Performance	
Altitude, Service ceiling	10,000 m (32,800 ft) (est)
Speed	
Cruising speed, max	324 kt (600 km/h; 373 mph)
Endurance	6 hr (est)
Power plant	1 × turbojet

Status: Unconfirmed. In November 2010, *IHS Jane's* was told that the system was "in service and operational". However, other reporters were apparently told that it was "in the early stages of flight testing".
Contractor: China Aerospace Science and Industry Corporation (CASIC) Beijing.

WJ-600 model, shown with two of its potential small weapon payloads (Robert Hewson)

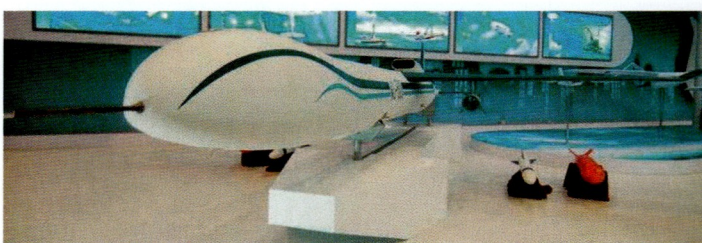
Frontal view of the CASIC WJ-600

CATIC U8

Type: Helicopter UAV.

Development: Limited information regarding this small VTOL UAV was available on the Chinese exhibit at the 2010 Singapore Air Show, CATIC literature for a U8E export version describing it as 'multirole', for various civil, military and parapublic applications including surveillance, anti-terrorist roles and crop-spraying.

As usual with CATIC promotions, the aircraft's actual manufacturer was not identified. The U8E does, however, bear very close resemblance to a somewhat smaller design known as the **CUH**, which had figured several years earlier in the literature of a company called Jiangxi CL Light Helicopter Corporation, a joint venture set up in Jingdezhen by a Chinese American and the nearby China Helicopter Research and Development Institute. The CUH had a 3.4 m (11 ft 1.9 in) rotor, 75 kg (165 lb) take-off weight with a 23 kg (50.7 lb) payload, and was powered by an 11.2 kW (15 hp) Hirth F36 piston engine. It seems highly probable that the U8 is the current outcome of that association. Earlier reports suggest that it flew for the first time in 2007.

U8E poster displayed at Singapore Air Show, February 2010
(IHS/Michael J Gething) 1395201

Brochure illustration of the U8 in military camouflage, February 2010
(CATIC) 1356300

Description: *Airframe:* Two-blade main and tail rotors; pod and boom fuselage; swept fins above and below tailboom, plus tailplane with small auxiliary fins at tips. Fixed twin-skid landing gear.
  Mission payloads: Various, depending upon mission; none specifically listed. EO sensor turret under nose.
  Guidance and control: Information not provided.
  Launch: Conventional helicopter take-off.
  Recovery: Conventional helicopter landing.
  Power plant: One piston or turboshaft engine (type and rating not stated).
(U8E)

U8E

Dimensions, External	
Overall, height	1.47 m (4 ft 9¾ in)
Fuselage	
length, tail rotor turning	3.74 m (12 ft 3¼ in)
Skids, skid track	1.00 m (3 ft 3¼ in)
Rotors, rotor diameter	3.86 m (12 ft 8 in)
Weights and Loadings	
Weight, Max T-O weight	220 kg (485 lb)
Payload, Max payload	40 kg (88 lb)
Performance	
Altitude, Service ceiling	3,500 m (11,480 ft)
Speed	
Max level speed	81 kt (150 km/h; 93 mph)
Cruising speed	65 kt (120 km/h; 75 mph)
Radius of operation	40 n miles (74 km; 46 miles)
Endurance	4 hr
Power plant, optional	1 × piston engine
	1 × turboshaft

Status: Apparently developed and reportedly available. The Xinhua news agency reported in July 2011 that the U8 had successfully completed its first series of test flights.
Contractor: China National Aero-Technology Import & Export Corporation Beijing.

GAIC Harrier Hawk

Type: General purpose UAV.
Development: The Harrier Hawk was first shown, in model form, at the November 2010 Airshow China in Zhuhai and was then subsequently displayed again at the Dubai Airshow a year later, though no evidence had been seen by then of actual manufacture or flight testing. Some design

Harrier Hawk model at Dubai Airshow in November 2011
(Paul Jackson) 1395376

Harrier Hawk design exhibits many characteristics of a stretched Evektor EuroStar *(Paul Jackson)* 1395377

features suggest possible evolution from Czech Evektor EuroStar lightplane, one of which had been flight tested as a UAV in China several years earlier. AVIC has described it as having been "developed on the basis of maturely existing UAV systems".
Description: *Airframe:* Low-wing monoplane with upturned tips; tapered fuselage; slightly swept fin and rudder. Retractable tricycle landing gear. Mainly composites construction.
  Mission payloads: Typical, but unspecified, reconnaissance and surveillance sensors for civil and/or military applications; or communications relay. Real-time imagery downlink.
  Guidance and control: No details revealed.
  Launch: Conventional wheeled take-off.
  Recovery: Conventional wheeled landing.
  Power plant: One 73.5 kW (98.6 hp) Rotax 912S flat-four engine; three-blade propeller. Fuel capacity 220 litres (58.1 US gallons; 48.4 Imp gallons).

Harrier Hawk

Dimensions, External	
Overall	
length	7.00 m (22 ft 11½ in)
height	2.60 m (8 ft 6¼ in)
Wings, wing span	11.30 m (37 ft 1 in)
Weights and Loadings	
Weight, Max T-O weight	700 kg (1,543 lb)
Payload, Max payload	100 kg (220 lb)
Performance	
Altitude	
Operating altitude	
lower	500 m (1,640 ft)
upper	7,000 m (22,960 ft)
Service ceiling	7,500 m (24,600 ft)
Speed	
Max level speed	124 kt (230 km/h; 143 mph)
Cruising speed	97 kt (180 km/h; 112 mph)
Endurance	
with 60 kg (132 lb) payload	16 hr
Power plant	1 × piston engine

Status: Promotion continued in late 2011, perhaps in the hope of attracting export orders. As of February 2013, no evidence had been found of any domestic use.
Contractor: Guizhou Aviation Industry Group Guiyang, Guizhou.

GAIC Xianglong

Type: Multirole HALE UAV.
Development: Debut, in model form, at Airshow China in Zhuhai, October/November 2006, inevitably inviting comparison with US Global Hawk. 'CADI' logo on model indicates design input from Chengdu Aircraft Design and Research Institute. Intended applications said to include reconnaissance, surveillance, other intelligence-gathering, BDA and relay. Its English name is Soar Dragon.
Description: *Airframe:* Low-mounted, narrow, sweptback wings with winglets at tips; additional winglets at approximately two-thirds span, extremities of which are attached to forward-swept horizontal surfaces which meet upper fuselage at base of tall, sweptback fin and rudder. Retractable tricycle landing gear.
  Mission payloads: Surveillance and targeting sensors probable.

2006 model of Guizhou's Soar Dragon candidate for a future multirole HALE system *(Robert Hewson)* 1151591

A Xianglong prototype undergoing ground testing, possibly at Chengdu, in 2011 1395352

Guidance and control: No details revealed. Satcom antenna in bulged nose.
Launch: Conventional wheeled take-off.
Recovery: Conventional wheeled landing.
Power plant: One turbofan engine (type and rating not stated), with dorsal intake.
(provisional)

Soar Dragon

Dimensions, External	
Overall	
length	14.00 m (45 ft 11¼ in)
height	5.40 m (17 ft 8½ in)
Wings, wing span	23.00 m (75 ft 5½ in)
Weights and Loadings	
Weight, Max launch weight	7,500 kg (16,534 lb)
Payload, Max payload	650 kg (1,433 lb)
Performance	
Altitude, Operating altitude	18,000 m (59,050 ft)
Speed, Cruising speed	405 kt (750 km/h; 466 mph)
Range	3,780 n miles (7,000 km; 4,349 miles)
Power plant	1 × turbofan

Status: Stated on debut, late 2006, to be in conceptual design stage. Film unofficially released in 2008 appeared to show an air vehicle of this configuration undergoing ground tests; clearer images emerged in mid-2011. An unofficial report from China has given the maiden flight as 7 November 2009.
Contractor: Guizhou Aviation Industry Group
Guiyang, Guizhou.

GAIC WZ-2000

Type: Jet-powered surveillance UAV.
Development: First shown in model form at Airshow China, Zhuhai, November 2000; labelled as WZ-9 (*wu zhen:* unmanned reconnaissance) at opening of show, but retitled WZ-2000 a few days later. A prototype in this configuration was reported to have flown by that time. However, by the time of the November 2002 Airshow China, the displayed model revealed an almost total redesign, the mid-mounted delta wings having been replaced by tapered, sweptback surfaces, the twin intakes by a

Three-quarter front aspect of the sweptwing WZ-2000 *(Robert Hewson)* 0576818

Revised configuration of the WZ-2000 as seen in late 2002 *(Robert Hewson)* 0528764

2000 display model of the WZ-2000 *(Photo Link)* 0100802

single orifice, and the fuselage shape and length also being considerably different.
Description: *Airframe:* Originally seen with bullet-shaped fuselage, twin dorsal air intakes, mid-mounted delta wings and twin outward-canted fins and rudders. Late 2002 configuration showed low-mounted, tapered, sweptback wings, box-section lengthened fuselage and single dorsal intake.
Mission payloads: No specific information, but EO, IR and/or SAR payloads would normally be expected, and elint-gathering sensors may also be envisaged. Real-time datalinks likely; domed nose, similar to that on such US types as Predator and Global Hawk, suggests probable accommodation for satcom antenna.
Guidance and control: No specific information, but expected to be preprogrammed and to incorporate GPS navigation.
Launch: No information.
Recovery: No information.
Power plant: Two 3.00 kN (674.5 lb st) turbojets.

WZ-2000

Dimensions, External	
Overall, length	4.50 m (14 ft 9¼ in)
Wings, wing span	3.00 m (9 ft 10 in)
Weights and Loadings	
Payload, Max payload	50 kg (110 lb)
Performance	
Speed, Max level speed	459 kt (850 km/h; 528 mph)

Status: Possibly a technology demonstrator. Development evidently continuing in 2002–03. An unidentified Guizhou UAV reported to have made its maiden flight at Shuanyang on 12 December 2003, is assumed to have been the WZ-2000. The Xinhua news agency has also reported that on 15 August 2005 China's "first high-end and multifunctional remote sensing" UAV had made a successful test flight at Anshun, equipped with a remote sensing system developed jointly by GAIC and Beijing University. No subsequent reports have been found.
Contractor: Guizhou Aviation Industry Group
Guizhou.

Weifang Tianxiang V750

Type: Helicopter UAV.
Development: Basis of this Chinese UAV is the US-designed Brantly B-2B two-seat helicopter, production rights to which were acquired in 2007 by Qingdao Haili Helicopter Manufacturing Co, which completed the first Chinese-assembled example in 2009. Others in the consortium developing the unmanned version include China Electronic Technology

Maiden flight of the Brantly-based V750, 5 May 2011
(Chinese Internet) 1438327

Institute (CETI) and Xian Flight Automatic Control Research Institute. On 7 May 2011, Chinese state media announced the maiden flight of the V750, which had taken place two days earlier.

Expected applications include reconnaissance/surveillance, battle damage assessment, monitoring of powerlines and forest fires, coastal patrol, marine or mountain search and rescue, anti-smuggling operations, scientific missions and traffic control. Features include deck landing capability, and shipboard use is thought to be one of the most likely early uses, possibly from the PLA Navy's Type 056 corvettes.

Description: *Airframe:* Three-blade main and two-blade tail rotors. Constant-taper, roughly circular-section fuselage; small fixed tailplane each side of tailcone. Fixed twin-skid landing gear. Mainly metal construction.

Mission payloads: Various, depending upon mission.

Guidance and control: Autonomous or remotely controlled, with modes switchable during flight. As of May 2011, company spokesman indicated that the INS and radio navigation systems still required further testing.

Launch: Vertical take-off.
Recovery: Vertical landing.
Power plant: One 134 kW (180 hp), vertically mounted Lycoming IVO-360-A1A flat-four engine.

V750

Dimensions, External	
Overall	
length, rotors turning	8.53 m (27 ft 11¾ in)
height	2.11 m (6 ft 11 in)
Skids, skid track	1.73 m (5 ft 8 in)
Rotors, rotor diameter	7.24 m (23 ft 9 in)
Tail rotor, tail rotor diameter	1.30 m (4 ft 3¼ in)
Weights and Loadings	
Weight, Max launch weight	757 kg (1,668 lb)
Payload, Max payload	80 kg (176 lb) (est)
Performance	
Altitude, Service ceiling	3,000 m (9,840 ft)
Speed, Max level speed	87 kt (161 km/h; 100 mph) at S/L
Range, max	270 n miles (500 km; 310 miles)
Radius of operation, mission	81 n miles (150 km; 93 miles)
Endurance, max	4 hr (est)

Status: Development continued through 2011 and 2012.
Contractor: Weifang Tianxiang Aviation Industry Co Ltd
Weifang, Shandong.

Xian ASN-15

Type: Lightweight, low-cost reconnaissance and surveillance UAV.
Development: Public debut at Airshow China, Zhuhai, November 2000.
Description: *Airframe:* Parasol-wing monoplane with dihedral on outer panels; slender pod and boom fuselage; T tail. No landing gear.
Mission payloads: CCD camera with real-time video downlink, or film camera.
Guidance and control: LOS radio control.

Launching the man-portable ASN-15 0527014

System composition: Three air vehicles; GCS; remote-control transmitter; video receiver; VCR; monitor; LED display.
Launch: Hand or rail launch.
Recovery: Belly skid landing or parachute recovery.
Power plant: One small single-cylinder piston engine, mounted above wing centre-section; two-blade propeller.

ASN-15

Dimensions, External	
Overall, length	1.80 m (5 ft 10¾ in)
Wings, wing span	3.00 m (9 ft 10 in)
Weights and Loadings	
Weight, Max launch weight	6.5 kg (14.00 lb)
Performance	
Altitude	
Operating altitude	
min	50 m (165 ft)
max	500 m (1,640 ft)
Speed, Max level speed	48 kt (89 km/h; 55 mph) at S/L
Radius of operation, LOS	5.4 n miles (10 km; 6 miles)
Endurance	1 hr
Power plant	1 × piston engine

Status: Believed to be in production and service with the PLA.
Customers: Chinese armed forces.
Contractor: Xian ASN Technical Group
Xian, Shaanxi.

Xian ASN-104 and ASN-105B

Type: Reconnaissance and surveillance UAVs.
Development: China's Pilotless Vehicle Research Institute, on the campus of the Northwestern Polytechnical University, has been involved in the development of UAVs for more than 30 years. More than a dozen individual types have been developed, at least seven of which have received government approval. Some have been produced in small numbers, including two types of which 33 examples were exported. Development of the D-4 began in March 1980, originally as a low-altitude, low-speed UAV for civil aerial survey applications. First flight was made in November 1982, government technical certification was granted in December 1983, and production started in late 1985.

The Xian ASN Technical Group, created in 1992, has a workforce of more than 400.

The hand-launched Xian ASN-15 *(Photo Link)* 0100804

ASN-105B on display at Zhuhai, November 2000 *(Photo Link)* 0100805

D-4 RD reconnaissance UAV 0505114

ASN-105B on zero-length launcher 0001597

Description: *Airframe:* Mid-wing monoplane, with tapered outer wings; wings and tailplane detachable for transportation and storage. Honeycomb sandwich and GFRP construction. Twin underfuselage landing skids.

Mission payloads: Usual D-4 RD sensors are a 100 mm (3.9 in) aerial photogrammetry camera (frame size 18 x 18 cm; 7.1 x 7.1 in), and a CCD video camera, with real-time video downlink transmitter, or a single infra-red linescanner. Wingtip equipment pods can also be fitted. An onboard generator driven by the piston engine provides electrical power for the UAV's avionics and sensor equipment.

The ASN-104/-105B can provide real-time reconnaissance and surveillance for up to two hours, payloads consisting of an 18 x 18 cm (7.1 x 7.1 in) panoramic camera and an LLTV camera with zoom lens; the latter can cover an area of 1,700 km² (656.4 sq miles) during a typical mission. Provision is made for CCD TV camera, airborne video recorder or IR linescanner alternative payloads.

Guidance and control: Remote-control uplink; video and telemetry downlinks; or can be flown autonomously by preprogramming the onboard analogue autopilot. GPS and GLONASS navigation.

System composition: ASN-104/-105B system comprises six air vehicles, main and secondary GCSs in two mobile command shelters, and two other shelters for photo processing and imagery interpretation, plus ground crew of six to eight persons.

Launch: From lightweight zero-length launcher, assisted by reusable underfuselage solid rocket booster that is jettisoned after take-off.

Recovery: By parachute, deployed from a dorsal compartment near the tail; lands on under-fuselage skids which have oleo-pneumatic shock-absorption to absorb landing impact.

Power plant: One 22.4 kW (30 hp) Xian HS-510 four-cylinder two-stroke engine; two-blade propeller.

Variants: *D-4 RD:* Initial production version, originally used mainly for civil tasks (large-scale aerophotogrammetry, geophysical survey, aerial mapping and remote sensing). Early production rate was 15 per year, but this was subsequently increased for supplies also to Chinese armed forces, with whom roles include front-line reconnaissance and electronic jamming.

ASN-104: Increased-capability, shorter range development of D-4 RD.

ASN-105B: As ASN-104 except for increased control range.

D-4 RD, ASN-104, ASN-105B

Dimensions, External	
Overall length	
ASN-104	3.32 m (10 ft 10¾ in)
ASN-105B	3.75 m (12 ft 3¾ in)
height	
excl skids, ASN-104	0.93 m (3 ft 0½ in)
excl skids, ASN-105B	1.40 m (4 ft 7 in)
Wings	
wing span	
ASN-104	4.30 m (14 ft 1¼ in)
ASN-105B	5.00 m (16 ft 4¾ in)
Weights and Loadings	
Weight	
Max launch weight	
ASN-104	140 kg (308 lb)
ASN-105B	170 kg (374 lb)
Payload	
Max payload	
ASN-104	30 kg (66 lb)
ASN-105B	40 kg (88 lb)
Performance	
Altitude	
Operating altitude	
D-4 RD Dozor-2	100 m to 3,000 m (320 ft to 9,840 ft)
ASN-104, ASN-105B	100 m to 3,200 m (320 ft to 10,500 ft)
Service ceiling, ASN-105B	6,000 m (19,680 ft)
Speed	
Max level speed	
D-4 RD Dozor-2	111 kt (206 km/h; 128 mph)
ASN-105B	108 kt (200 km/h; 124 mph)
Cruising speed, ASN-104, ASN-105B	81 kt (150 km/h; 93 mph)
Radius of operation	
D-4 RD Dozor-2	54 n miles (100 km; 62 miles)
ASN-104	32 n miles (59 km; 36 miles)
ASN-105B	81 n miles (150 km; 93 miles)
Endurance	
ASN-104	2 hr
ASN-105B	7 hr
Power plant	1 × piston engine

Status: In production and service.

Customers: Chinese armed forces.

Contractor: Xian ASN Technical Group
Xian, Shaanxi.

Xian ASN-206

Type: Short-range multirole UAV.

Development: The ASN-206 made its public debut at China Air Show in November 1996. Military and civil applications include day and night reconnaissance, battlefield surveillance, artillery target location and adjustment, border and traffic patrol, NBC detection, atmospheric sampling, aerial photography and survey, and disaster monitoring.

Description: *Airframe:* Tapered, high-mounted wings; pod fuselage; twin tailbooms with twin fins and rudders, bridged by mid-mounted tailplane with elevator. Twin underfuselage landing skids.

Mission payloads: Frame or panoramic film camera; black and white or colour TV camera; LLTV; IR imager; laser altimeter; target location/artillery adjustment equipment; or other according to mission. Airborne video recorder; real-time data downlink.

Guidance and control: Radio-command flight control and management system, with GPS navigation. Telemetry uplink and imagery downlink.

System composition: Air vehicle(s) and launch truck; six other truck-mounted shelters for command and control, mobile control station, information processing, power supply, maintenance and transportation.

Launch: By booster rocket from zero-length launcher.

Recovery: Parachute recovery.

Power plant: One 37.3 kW (50 hp) SAEC (Zhuzhou) HS-700 four-cylinder two-stroke engine; two-blade wooden pusher propeller.

The ASN-206 on display at Zhuhai in 1996 *(Kenneth Munson)* 0001598

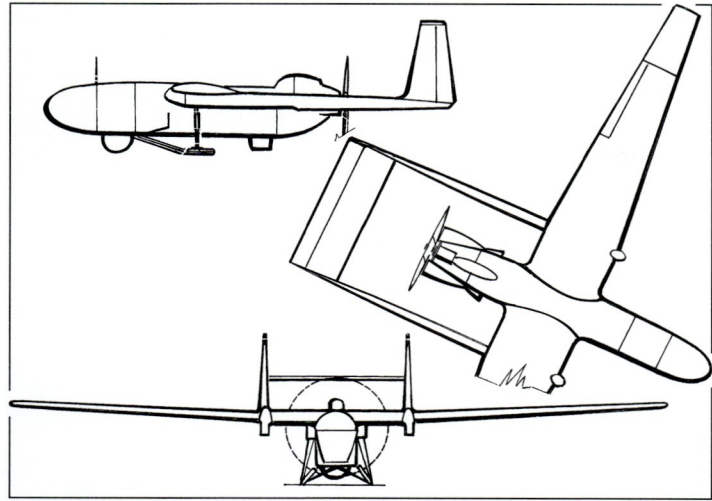

ASN-206 general arrangement *(IHS/John W Wood)*

ASN-206

Dimensions, External
- Overall
  - length..3.80 m (12 ft 5½ in)
  - height..1.40 m (4 ft 7 in)
- Wings, wing span...6.00 m (19 ft 8¼ in)
- Tailplane, tailplane span...................................1.30 m (4 ft 3¼ in)

Weights and Loadings
- Weight, Max launch weight...............................222 kg (489 lb)
- Payload, Max payload..50 kg (110 lb)

Performance
- Altitude, Service ceiling....................................5,000 m (16,400 ft) (est)
- Speed, Max level speed...................................113 kt (209 km/h; 130 mph)
- Range...81 n miles (150 km; 93 miles)
- Endurance
  - typical...4 hr
  - max..8 hr

Status: In service.
Customers: Chinese armed forces.
Contractor: Xian ASN Technical Group
Xian, Shaanxi.

Xian ASN-207

Type: Medium-range multirole UAV.
Development: The ASN-207 was revealed in 2002. It appears to be an enlarged development of the ASN-206 (which see), with enhanced capability.
Description: *Airframe:* Tapered, high-mounted wings; pod fuselage; twin tailbooms with twin fins and rudders, bridged by mid-mounted tailplane with elevator. No landing gear.
Mission payloads: MOSP (which see) combined EO/IR/laser range-finder/designator in retractable ventral turret; can also be equipped with SAR/MTI, FLIR or comint, elint, sigint, jamming or decoy payloads. Imagery and telemetry datalinks may be assumed.
Guidance and control: Autonomous operation, with INS, global navigation satellite system (pylon-mounted dorsal antenna) and air data computer for accurate positioning. Can be flown singly as mission aircraft or in two-aircraft mode with one acting as relay.
System composition: Air vehicles; truck-mounted GCS, MCS and GDT; one or more portable video terminals.
Launch: Assumed to be by booster rocket(s) from zero-length launcher.
Recovery: Assumed to have parachute or parafoil recovery.
Power plant: One piston engine (type and rating not stated); pusher propeller. See under Weights for fuel details.

ASN-207

Dimensions, External
- Overall
  - length..6.00 m (19 ft 8¼ in)
  - height..2.10 m (6 ft 10¾ in)
- Wings, wing span...9.30 m (30 ft 6¼ in)

Weights and Loadings
- Weight
  - Weight empty..250 kg (551 lb)
  - Max launch weight
    - relay aircraft...410 kg (903 lb)
    - mission aircraft..480 kg (1,058 lb)
- Fuel weight, Max fuel weight............................130 kg (286 lb)
- Payload
  - Max payload
    - relay aircraft..30 kg (66 lb)
    - mission aircraft..100 kg (220 lb)

Performance
- Altitude
  - Service ceiling
    - relay aircraft...8,000 m (26,240 ft)
    - mission aircraft..6,000 m (19,680 ft)
- Speed, Max level speed...................................97 kt (180 km/h; 112 mph)
- Radius of operation
  - relay aircraft at 4,000 m (13,120 ft).............108 n miles (200 km; 124 miles)
  - mission aircraft at 2,000 m (6,560 ft)..........324 n miles (600 km; 372 miles)
- Endurance..16 hr
- Power plant..1 × piston engine

Status: Market-ready; may already be in production and/or service. At the Zhuhai exhibition in 2010, Xian ASN claimed to have two types of UAV in service with the PLA and others in production.
Contractor: Xian ASN Technical Group
Xian, Shaanxi.

Xian ASN-209

Type: Multirole tactical UAV.
Development: The Medium-Altitude, Medium-Endurance (MAME) ASN-209 is the most recent extrapolation of this Group's ASN-206/207 series of tactical UAVs.
Description: *Airframe:* Tapered, high-mounted wings; pod fuselage with rear-mounted engine; twin tailbooms with twin fins and rudders, bridged by mid-mounted tailplane with elevator. Twin underfuselage landing skids.
Mission payloads: Military payloads can include GMTI radar, ground target designators, elint, EW and communications relay. For civil applications, quoted payloads include EO, SAR or specialised equipment for forest fire prevention, atmospheric and climate monitoring and cloud seeding.
Guidance and control: Probably autonomous, but details not disclosed. LOS and BLOS datalinks.
System composition: Air vehicle, payload(s), datalinks, truck-mounted GCS and transporter/launcher vehicle.

ASN-207 surveillance and relay UAV

ASN-209 model at Singapore Air Show, February 2010 *(Paul Jackson)*

ASN-209 in flight, showing landing skids extended *(CATIC)*

Launch: By booster rocket from truck-mounted rail launcher.
Recovery: Skid landing or parachute recovery.
Power plant: One unidentified piston engine (type and rating not stated), driving a two-blade pusher propeller.

ASN-209

Dimensions, External	
Overall	
length	4.275 m (14 ft 0¼ in)
height	1.54 m (5 ft 0¾ in)
Wings, wing span	7.50 m (24 ft 7¼ in)
Weights and Loadings	
Weight, Max launch weight	320 kg (705 lb)
Payload, Max payload	50 kg (110 lb)
Performance	
Altitude, Service ceiling	5,000 m (16,400 ft)
Speed	
Max level speed	97 kt (180 km/h; 112 mph)
Cruising speed	76 kt (141 km/h; 87 mph) (est)
Radius of operation, LOS, mission	54 n miles (100 km; 62 miles)
Endurance, max	10 hr
Power plant	1 × piston engine

Status: Being promoted for both civil and military applications in 2010.
Contractor: Xian ASN Technical Group
  Xian, Shaanxi.

Xian ASN-211

Type: Close-range tactical UAV.
Development: At Airshow China in November 2010, the Xian ASN group exhibited numerous UAVs in a wide variety of configurations, including the ASN-211 ornithopter design, although available details were sparse. It evidently aims to address a need for a close-range reconnaissance/surveillance system for use by front-line troops.
Description: *Airframe:* Box-shape fuselage with tapered rear end; single fin and rudder. High-mounted flapping wings. No landing gear. Composites construction.
  Mission payloads: EO and/or IR sensor assumed. There also appears to be a window for a piloting camera in the nose.
  Guidance and control: Electric-powered biomimetic imitation of insect flight, including short and/or vertical take-off and landing and ability to hover over a target.
  Transportation: Man-portable.
  Launch: Hand launch.
  Recovery: Belly landing.
  Power plant: Electric motor (rating not revealed).

ASN-211

Dimensions, External	
Wings, wing span	0.80 m (2 ft 7½ in)
Weights and Loadings	
Weight, Max T-O weight	0.220 kg (0.00 lb)
Performance	
Altitude, Service ceiling	200 m (660 ft)
Speed, Max level speed	19 kt (35 km/h; 22 mph)
Power plant	1 × electric motor

Status: In development as of late 2010; possibly continuing in 2013.
Contractor: Xian ASN Technical Group
  Xian, Shaanxi.

The flapping-wing ASN-211 on display in Zhuhai in November 2010 *(Xinhua)* 1395383

Xian ASN-213

Type: Technology demonstrator.
Development: This unorthodox UAV was shown for the first time at Airshow China in Zhuhai in November 2010, though with very little accompanying detail. Its main feature of interest is the wing, whose configuration can be changed to suit different phases of its mission profile: fully spread for take-off and landing; or folded for attack or rapid climb. Company

ASN-213 with wings fully spread 1395284

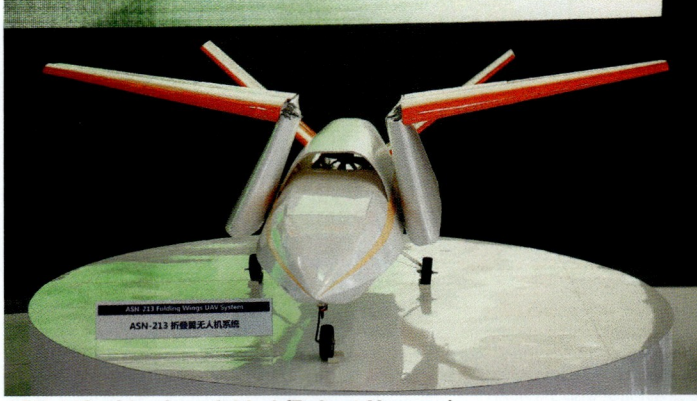

ASN-213 with wings folded *(Robert Hewson)* 1395283

spokesmen did not reveal whether the aircraft had been test-flown at that time.
Description: *Airframe:* Wings are hinged at the roots and about one-third span, folding upwards. The bulbous fuselage has a rear-mounted engine and V tail surfaces. Fixed tricycle landing gear.
  Mission payloads: Not stated, but presumably would consist mainly of test measuring equipment.
  Guidance and control: All the ASN Group's UAV designs are understood to be operated from a common integrated ground station. This would presumably be a current version of that developed earlier for its ASN-104/105 tactical unmanned systems.
  Launch: Conventional wheeled take-off.
  Recovery: Conventional wheeled landing.
  Power plant: One piston engine (rating unknown), driving a two-blade pusher propeller.

ASN-213

Dimensions, External	
Overall	
length	1.50 m (4 ft 11 in)
height	0.63 m (2 ft 0¾ in)
Wings	
wing span	
spread	2.01 m (6 ft 7¼ in)
folded	1.42 m (4 ft 8 in) (est)
Areas	
Wings, Gross wing area	0.67 m² (7.2 sq ft)
Weights and Loadings	
Weight, Max T-O weight	5.0 kg (11.00 lb)
Payload, Max payload	1.0 kg (2.00 lb)
Performance	
Power plant	1 × piston engine

Status: Undergoing development in 2010-11, unknown as of 2013.
Contractor: Xian ASN Technical Group
  Xian, Shaanxi.

Xian ASN-216

Type: Close-range tactical mini-UAV.
Development: Unveiled at Airshow China in November 2010.
Description: *Airframe:* High-wing configuration; pod and boom fuselage with rear-mounted engine and T tail. Fixed tricycle landing gear.
  Mission payloads: Undernose ISR sensor turret.
  Guidance and control: No details disclosed, but assumed to be generally as for other ASN Group UAVs; refer to entry for Xian ASN-104 and ASN-105B GCS.
  Launch: Wheeled take-off or catapult launch.
  Recovery: Wheeled landing or parachute recovery.
  Power plant: Single piston engine (type and rating not stated), driving a two-blade pusher propeller.

ASN-216 on show at Zhuhai in November 2010 *(Robert Hewson)*

ASN-216

Dimensions, External	
Overall, length	2.40 m (7 ft 10½ in)
Wings, wing span	3.75 m (12 ft 3¾ in)
Weights and Loadings	
Weight	
Max T-O weight	20.0 kg (44 lb)
Max launch weight	20.0 kg (44 lb)
Payload, Max payload	6.0 kg (13.00 lb)
Performance	
Altitude, Service ceiling	4,500 m (14,760 ft)
Speed	
Max level speed	75 kt (139 km/h; 86 mph)
Cruising speed, max	49 kt (91 km/h; 56 mph)
Loitering speed	32 kt (59 km/h; 37 mph)
Radius of operation	
mission, control limits	21.5 n miles (39 km; 24 miles)
Endurance, max	6 hr
Power plant	1 × piston engine

Status: Not disclosed.
Contractor: Xian ASN Technical Group
Xian, Shaanxi.

Xian ASN-217

Type: Close-range mini-UAV.
Development: Unveiled at Airshow China in November 2010.
Description: *Airframe:* Parasol wing with dihedral on outer panels; tapered, cylindrical fuselage; slightly-swept fin and rudder and low-set tailplane. No landing gear.
  Mission payloads: Small stills or TV camera.
  Guidance and control: Not disclosed.
  Transportation: Man-portable.
  Launch: By hand or from catapult.
  Recovery: Belly landing or parachute recovery.
  Power plant: Electric motor (rating not disclosed), driving a two-blade propeller.

ASN-217

Dimensions, External	
Overall, length	1.70 m (5 ft 7 in)
Wings, wing span	2.70 m (8 ft 10¼ in)
Weights and Loadings	
Weight, Max launch weight	5.5 kg (12.00 lb)
Payload, Max payload	1.5 kg (3.00 lb)
Performance	
Altitude, Service ceiling	4,000 m (13,120 ft)
Speed	
Max level speed	64 kt (119 km/h; 74 mph)
Cruising speed, max	38 kt (70 km/h; 44 mph)
Loitering speed	22 kt (41 km/h; 25 mph)
Radius of operation	
mission, control limit	10.8 n miles (20 km; 12 miles)
Endurance, max	1 hr 30 min
Power plant	1 × electric motor

Status: Presentation in late 2010 implied that this system was in use in civil applications such as disaster relief, weather monitoring, aerial mapping, search and rescue, and powerline/pipeline inspection.
Contractor: Xian ASN Technical Group
Xian, Shaanxi.

Xian ASN-229A

Type: Armed reconnaissance MALE UAV.
Development: Unveiled at the November 2010 air show in Zhuhai, the ASN-229A is the latest and so far the largest in Xian's growing family of twin-boom, pusher-engined tactical UAVs. Applications are quoted as including day or night reconnaissance, battlefield surveillance, target location and artillery fire adjustment.
Description: *Airframe:* High-mounted, high aspect ratio wings with taper on outer panels; elongated fuselage nacelle with rear-mounted engine; twin-boom twin-tail empennage. Skid landing gear.
  Mission payloads: Undernose ISR sensor turret. Hardpoint beneath each wing for small precision-guided weapon.
  Guidance and control: No details disclosed, but assumed to be generally as for other ASN Group UAVs; refer to entry for Xian ASN-104 and ASN-105B GCS. Satcom datalink.
  Launch: By booster rocket(s).
  Recovery: Parachute recovery.
  Power plant: Single piston engine (rating not disclosed), driving a two-blade pusher propeller.

ASN-229A

Dimensions, External	
Overall, length	5.50 m (18 ft 0½ in)
Wings, wing span	11.00 m (36 ft 1 in)
Weights and Loadings	
Weight, Max launch weight	800 kg (1,763 lb)
Payload, Max payload	100 kg (220 lb)
Performance	
Altitude	
Operating altitude	8,000 m (26,240 ft)
Service ceiling	10,000 m (32,800 ft) (est)
Speed	
Cruising speed, max	97 kt (180 km/h; 112 mph)
Radius of operation	
mission, with satellite relay	1,079 n miles (1,998 km; 1,241 miles)
Endurance, max	20 hr
Power plant	1 × piston engine

Status: Under development in late 2010; forecast for possible service entry by end of 2011.
Contractor: Xian ASN Technical Group
Xian, Shaanxi.

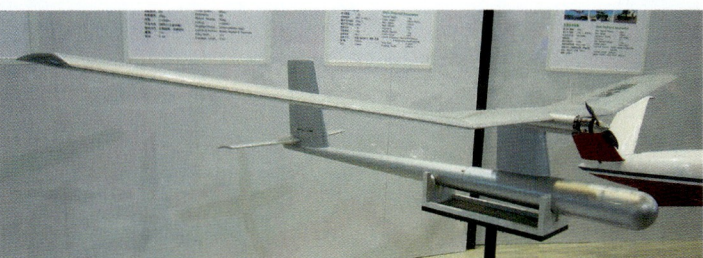
The hand-launched, electric powered ASN-217 *(Robert Hewson)*

2010 display model of the ASN-229A *(Robert Hewson)*

Czech Republic

Track System Heros

Type: Helicopter UAV.

Development: Development of the Heros, in collaboration with other Czech companies, began in 2007. It made its public debut at a NATO Day exhibition in Ostrava in September 2008, and was due to begin flight testing shortly afterwards. It is intended for both military and civil applications.

As of early 2009, the company also had a fixed-wing UAV, named RoboStar, under development.

Description: *Airframe:* Twin coaxial and counter-rotating three-blade rotors. Pod and boom fuselage; twin fins and rudders. Twin-skid landing gear.

Mission payloads: As appropriate for civil missions such as power and pipeline inspection, border patrol, aerial photography, or detection of hazardous chemicals, pollution or explosives. Military versions could be equipped for ISTAR, communications jamming or search and rescue operations. Telemetry downlink has a range of 80 km (50 miles).

Guidance and control: Fully autonomous or operator-controlled modes, based on a Tracking System proprietary flight control system. Air vehicle is also equipped with an 'advanced' anti-collision system said to be capable of recognising objects as small as 10 cm (4 in).

Launch: Conventional helicopter vertical take-off.

Recovery: Conventional helicopter vertical landing.

Power plant: Two 37.0 kW (49.6 hp) Rotax 503 UL two-cylinder inline engines. Fuel capacity 114 litres (30.1 US gallons; 25.1 Imp gallons).

Heros

Dimensions, External	
Overall, height	2.08 m (6 ft 10 in)
Fuselage	
length	4.675 m (15 ft 4 in)
width, max	1.775 m (5 ft 10 in)
Weights and Loadings	
Weight	
Weight empty	290 kg (639 lb)
Max T-O weight	465 kg (1,025 lb)
Payload, Max payload	120 kg (264 lb)
Performance	
Speed	
Max level speed	97 kt (180 km/h; 112 mph)
Cruising speed	75 kt (139 km/h; 86 mph)
Radius of operation, mission	135 n miles (250 km; 155 miles)
Endurance	
max, in hover	2 hr 30 min
max, in forward flight	4 hr

Status: Under development in 2008-09, unknown as of end of 2012.

Contractor: Track System a.s.
Hradec Králové.

VTUL Manta

Type: Tactical surveillance UAV.

Development: Development of the Manta, part-funded by the Czech Ministry of Industry and Trade, was started in 2006 by VTÚL a PVO engineers in collaboration with the Aveko company and the Czech Technical University in Prague, and is said to have been completed in March 2009. It is seen as a potential successor to the in-service Sojka III system, but also for non-military applications. Main differences from Sojka are the fitment of a fixed landing gear, and a composites airframe that allows carriage of almost as much payload within a considerably lower take-off weight, with correspondingly enhanced performance. The complete system including the Manta air vehicle has the name **Mamok**.

Description: *Airframe:* High-wing monoplane with upward-curving wingtips/winglets; Pod and twin tailboom configuration; pusher engine. Fixed tricycle landing gear. Composites construction (carbon fibre and Kevlar).

Mission payloads: MicroView stabilised EO sensor.

Guidance and control: Fully or semi-automatic control system, with Neuron RS autopilot.

Launch: Conventional wheeled take-off or pneumatic catapult launch for military version; civil version can be modified to launch from atop a moving car.

Recovery: Conventional wheeled landing. Provision for parachute for rough-terrain recoveries.

Power plant: One ZDZ 210 two stroke twin cylinder piston engine, driving a two-blade pusher propeller.

Manta

Dimensions, External	
Overall, length	2.47 m (8 ft 1¼ in)
Wings, wing span	4.37 m (14 ft 4 in)
Weights and Loadings	
Weight, Max T-O weight	62 kg (136 lb)
Payload, Max payload	10.0 kg (22 lb)
Performance	
Altitude, Service ceiling	3,000 m (9,840 ft)
Speed, Max level speed	108 kt (200 km/h; 124 mph)
Radius of operation	32 n miles (59 km; 36 miles)
Endurance	4 hr (est)
Power plant	1 × piston engine

Status: Development completed in 2009; production orders awaited.

Contractor: Air Force Research Institute (VTÚL a PVO) Prague.

Heros camouflaged to represent its military applications *(Track System)* 1290342

Heros in civil display paint scheme *(Track System)* 1290343

Manta in flight *(Air Force Research Insitute)* 1416316

Manta air vehicle, part of the Mamok UAS *(VTÚL)* 1356258

VTUL Optoelektron 1

Type: Close-range surveillance mini-UAV.
Development: Unveiled in mid-2009.
Description: *Airframe:* Constant-chord high wings; pod and boom fuselage; T tail; pusher engine. No landing gear. Composites construction.
  Mission payloads: Nose-mounted TV camera with real-time video downlink.
  Guidance and control: Neuron RS4 autopilot. Apart from minor software changes, this UAV uses essentially the same flight and mission control avionics as the Institute's larger Manta UAV. GCS embodies digital map display and a flight simulator for crew training.
  Transportation: System is man-portable in two backpacks. Set-up time is 5 minutes.
  Launch: Hand or bungee-assisted launch.
  Recovery: Belly skid landing or parachute recovery.
  Power plant: Electric motor (rating not stated), powered by an 11 Ah battery and driving a two-blade pusher propeller.

Optoelektron 1

Dimensions, External
  Overall, length .... 1.20 m (3 ft 11¼ in)
  Wings, wing span .... 2.40 m (7 ft 10½ in)
Weights and Loadings
  Weight, Max launch weight .... 5.0 kg (11.00 lb)
Performance
  Altitude, Service ceiling .... 2,000 m (6,560 ft)
  Speed
    Max level speed .... 54 kt (100 km/h; 62 mph)
    Cruising speed, max .... 43 kt (80 km/h; 49 mph)
    Loitering speed .... 38 kt (70 km/h; 44 mph)
  Radius of operation, mission .... 2.7 n miles (5 km; 3 miles)
  Endurance, max .... 1 hr
  Power plant .... 1 × electric motor

Status: Under development in 2010, unknown as of end of 2012.
Contractor: Air Force Research Institute (VTÚL a PVO) Prague.

Optoelektron 1 (Air Force Research Institute) 1416322

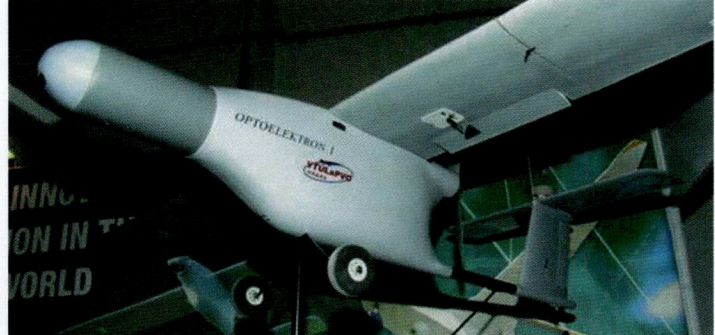
The Optoelektron 1 hand-launched mini-UAV (VTÚL) 1356251

VTÚL Sojka III

Type: Tactical reconnaissance UAV.
Development: The Sojka (Jay) system was developed for tactical reconnaissance at ranges of up to 54 n miles (100 km; 62 miles) from the GCS, at altitudes from 200 to 2,000 m (660 to 6,560 ft), with real-time transmission of optical and telemetric data. It evolved from numerous prototypes and development models designated E 50 (target, tested from 1986 onwards) and E 80 (target/reconnaissance, 1990 onwards) by Czech Air Force Research Institute. A marketing alliance with Hungary (Army Institute for Military Technology and AviaTronic) finished in 1995.

Final development and flight testing took place in 1993 and early 1994. The system underwent operational evaluation with the Czech Army in 1995, took part in domestic air defence exercises in 1996, and became fully operational in 1998. An upgraded Sojka III/TVM version entered service in 2001, with improved performance resulting mainly from the installation of a UEL rotary engine of slightly greater power than the original flat twin. Other improvements have included a modular payload capability, a more effective (digital) datalink, increased endurance and down-sizing of the ground control and launcher elements.

Sojka III preflight inspection (LOM PRAHA) 1416328

Sojka III/ML air vehicle on launch rail (IHS/Patrick Allen) 1182528

Ground control relay station truck for the Sojka III/ML (IHS/Patrick Allen) 1182534

GCS workstation (LOM PRAHA Air Force Research Institute) 1416332

Unmanned aerial vehicles > Czech Republic > VTÚL Sojka III

Workstations inside the Sojka ground relay truck *(IHS/Patrick Allen)* 1182535

Sojka III/TVM being launched *(VTÚL)* 1151535

Sojka III/TVM reconnaissance UAV 0068483

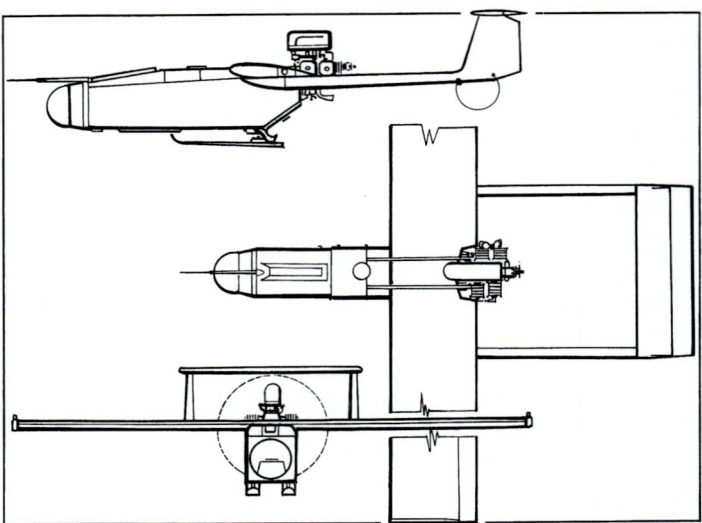
The Czech-designed Sojka III surveillance UAV *(IHS/John W Wood)* 0518066

Description: *Airframe:* High-wing monoplane with pusher engine and twin-boom tail unit. Construction of glass and carbon fibre composites; sandwich wing structure. No landing gear.

Mission payloads: Standard reconnaissance payload is a set of daylight colour CCD TV cameras installed in a container. Embedded fixed cameras or single axis stabilised platform with two camera or Microview turret. Pilot's camera in the nose.

Guidance and control: UAV control is either semi- or fully automatic. The flight plan can be preprogrammed before take-off or during flight. There may be up to four variants of flight plan in autopilot memory and the system can switch between them in flight. The UAV can return to operator on command. An onboard safety system stops the power plant and deploys a parachute if any malfunction in a flight control system occurs. The flight can also be cancelled at any moment by the operator's single pushbutton command. Navigation is via GPS and there is a real-time datalink between the UAV and its GCS, allowing the ground crew to monitor onboard EO sensors and air vehicle position (displayed on a digital map). Main flight data are displayed on monitors at the GCS.

System composition: Three or four air vehicles, one GCS, one rail launcher with booster rockets, one ZSK/RE air-conditioned transport container and one off-road recovery vehicle. Containerised GCS, launcher and transport container are attached to field trucks by ISO-1D standard fasteners. Sojka III was developed to MIL standards. Operating crew of six; set-up time less than 45 minutes.

Launch: By reusable solid-fuel rocket booster from 14 m (50 ft), truck-mounted foldable ramp with 20° elevation.

Recovery: Underfuselage skids for belly landing on flat areas of grass, clay, sand or concrete. For more difficult terrain or emergency use, a recovery parachute is employed.

Power plant: Initial version: One 22.0 kW (29.5 hp) ÚVMV M 115 two-cylinder two-stroke engine; two-blade fixed-pitch wooden pusher propeller. Fuel capacity 19 litres (5.0 US gallons; 4.2 Imp gallons)

Modernised version: One 28.3 kW (38 hp) UEL AR 741 rotary engine.

Variants: *Sojka III/TVM:* Standard reconnaissance version; *as described.*

Sojka III/ML: As for TVM version, but with added mobile relay station.

Sojka III/TVM, Sojka III/ML

Dimensions, External	
Overall	
length	3.78 m (12 ft 4¾ in)
height	1.08 m (3 ft 6½ in)
Fuselage, length	2.10 m (6 ft 10¾ in)
Wings, wing span	4.50 m (14 ft 9¼ in)
Dimensions, Internal	
Payload bay, volume	0.075 m³ (2.6 cu ft)
Weights and Loadings	
Weight	
Max launch weight	
Sojka III/TVM	180 kg (396 lb) (est)
Sojka III/ML	145 kg (319 lb)
Fuel weight, Max fuel weight	12 kg (26 lb)
Payload, Max payload	20 kg (44 lb)
Performance	
Climb	
Rate of climb	
max, at S/L, Sojka III/TVM	540 m/min (1,771 ft/min)
Altitude	
Operating altitude	
min	50 m (160 ft)
max, Sojka III/TVM	2,000 m (6,560 ft)
max, Sojka III/ML	4,000 m (13,120 ft)
Speed	
Max level speed	
Sojka III/TVM	111 kt (206 km/h; 128 mph)
Sojka III/ML	113 kt (209 km/h; 130 mph)
Loitering speed	65 kt (120 km/h; 75 mph)
Radius of operation	
mission, Sojka III/TVM	54 n miles (100 km; 62 miles)
mission, Sojka III/ML	108 n miles (200 km; 124 miles)
Endurance	
Sojka III/TVM	2 hr
Sojka III/ML	4 hr (est)

Status: Remains in service with no plans announced to replace this system.

Customers: Czech Army (106 Reconnaissance Battalion in Prostejov, No 345 UAV Squadron). One system in service, with nine air vehicles.

Contractor: Air Force Research Institute (VTÚL a PVO) Prague.

Denmark

Sky-Watch Huginn X1

Type: VTOL mini UAV.

Development: *IHS Jane's* became aware of the Sky-Watch Huginn X1 VTOL mini UAV during 2012. Huginn X1's manufacturer Sky-Watch was established during 2009 and specialises in the development and production of drones and associated control technology.

Description: *Airframe:* The Huginn (mythological Raven) X1 mini UAV comprises a centre body that is equipped with a forward-facing payload mounting, a cruciform arrangement of four deployable electric motors (each driving a two-bladed rotor, with the four rotors arranged in counter-rotating pairs) and a dorsal communications antenna. Each of the electric motor housings incorporates a ventral take-off/landing skid. The Huginn X1 can operate in wind speeds of up to 8 m/s (26 ft/s), a temperature range of –7°C to +40°C (19°F to 104°F) and a relative humidity of 98 per cent.

Mission payloads: A 640 × 480 resolution FLIR Systems Quark uncooled thermal imaging camera core (17-micron pixels and in-flight calibration). A dual camera mount is an option.

Guidance and control: Autonomous onboard navigation system (including automatic battery status warning (for safe return home); automatic return to home following datalink loss; GPS (0.5-1.5 m (1.6-4.9 ft) accuracy), heading and pressure height hold; real-time waypoints and sonar altimeter (for low-altitude flight) functionality). AV equipped with solid state 3-axis Micro Electro-Mechanical System (MEMS) gyros and accelerometers, 3-axis magnetometers, barometric pressure and external/internal temperature sensors, a sonar altimeter and a GPS receiver. One × 5.8 GHz analogue video transmitter (baseline) and an 868 or 900 MHz datalink (1.1 n miles (2.0 km; 1.2 miles) LOS range). A supporting ground station and flight controller are described in a separate entry.

Transportation: 3.8 kg (8.4 lb), 55.0 × 43.0 × 21.5 cm (21.7 × 16.9 × 8.5 in), water and dust tight carrying case. Suitable for carriage by car (automobile) or as hand luggage aboard an airliner.

System composition: One × AV, one × ground station (with associated software package), one × battery charger, one × hand controller, four × spare rotors, four × custom batteries and one × carrying case.

Launch: Automatic vertical take-off.

Recovery: Automatic vertical landing.

Power plant: Four × 710 rpm/V brushless electric motors (450 g (0.002 lb) re-chargeable (75 minutes from full discharge) lithium polymer battery).

Huginn X1

Dimensions, External	
Overall	
length	
AV folded for carriage	0.45 m (1 ft 5¾ in)
AV deployed	0.50 m (1 ft 7¾ in)
height	
AV folded for carriage	0.15 m (6 in)
AV deployed	0.22 m (8¾ in)
width	
AV folded for carriage	0.14 m (5½ in)
AV deployed	0.50 m (1 ft 7¾ in)
Rotors, rotor diameter	0.26 m (10¼ in)
Weights and Loadings	
Weight	
Weight empty	0.94 kg (2.00 lb)
Performance	
Altitude	
Service ceiling, up to	3,050 m (10,000 ft)
Speed, Max level speed	19 kt (36 km/h; 22 mph)
Endurance, up to (payload dependent)	25 min
Power plant	4 × electric motor

Status: As of February 2013, Sky-Watch has continued to promote the Huginn X1 VTOL mini UAV.

Contractor: Sky-Watch, Støvring, Denmark.

The Huginn X1 VTOL mini UAV *(David Oliver)* 1356696

Finland

Patria MASS

Type: Multirole short-range mini-UAV.

Development: Development of the MASS (Modular Airborne Sensor System), Patria's first venture into the UAV arena, is understood to have begun in 2005. Manufacturer's flight testing has been completed.

Description: *Airframe:* High-wing monoplane, with box-section fuselage and V tail surfaces. Aerodynamic control by flaps and ruddervators. No landing gear. Construction is of expanded polystyrene, and aircraft dismantles into eight components, which can be push-fit assembled without the use of tools.

Mission payloads: Modular payloads can include a stabilised EO camera (optionally accompanied by a chemical or nuclear radiation emission detector); a stabilised IR camera; pan-and-tilt cameras controllable from the GCS; or a nuclear (optionally biological) pollution sampler. A fixed 'see and avoid' camera is installed for all configurations, and the aircraft has onboard sensor data recording capability.

Guidance and control: Can be operated fully autonomously, requiring no piloting skills. Air vehicle and GCS are operable by one person.

Transportation: Complete system weighs less than 40 kg (88 lb), is portable by two persons, and can be carried in a single backpack or transportation case.

System composition: One to three air vehicles, payloads, communications suite, laptop GCS, telescopic antenna mast, launching equipment, batteries and spares. System deployment, turnaround and packing can be accomplished in less than 8 minutes.

Launch: Bungee cord launch.

Patria's MASS mini UAV *(IHS/Patrick Allen)* 1329505

The very simple and lightweight MASS *(IHS/Patrick Allen)* 1182331

Recovery: Belly landing.
Power plant: Battery-powered electric motor, driving a two-blade pusher propeller.

MASS

Dimensions, External	
Overall, length	1.05 m (3 ft 5¼ in)
Wings, wing span	1.50 m (4 ft 11 in)
Weights and Loadings	
Weight	
Weight empty	3.0 kg (6.00 lb)
Payload, Max payload	0.5 kg (1.00 lb)
Performance	
Altitude, Operating altitude	50 m to 150 m (160 ft to 500 ft)
Speed	
Max level speed	65 kt (120 km/h; 75 mph)
Cruising speed, optimum	32 kt (59 km/h; 37 mph)
Endurance	1 hr 15 min
Power plant	1 × electric motor

Status: No sales reported by mid-2010. Patria delivered a two-aircraft MASS system to the Finnish Defence Forces for evaluation starting in October 2007, part of a programme set to test surveillance solutions for replacement of anti-personnel mines. The test system includes a ground control station, a digital data link and spares. Over 500 missions have been completed, including field trials outside Finland, by 2009.
Contractor: Patria Aerostructures Oy
Tampere.

France

Aeroart Aelius

Type: Amphibious UAV.
Development: This highly unorthodox craft has received little media prominence, even in the French press, since the concept was first announced at the Paris Air Show in June 2005. The concept is that of an unmanned vehicle capable of operation above, on and under water, essentially by employing a wing that can morph into a hydrofoil for surface manoeuvres and retract for submarine use. Aeroart was formed in October 2006 to proceed with its development and testing.
Test flights took place at the Centre d'Essais en Vol (CEV) at Cazaux, beginning in May 2007.
Description: Airframe: Single-step hull, with dorsal engine intake; sweptback fin and rudder. Sweptback (possibly all-moving) foreplanes. Non-swept, mid-mounted main wing at rear, outer panels of which have skid tips and hinge downwards to form main units of hydrofoil landing gear, third element of which is a retractable undernose skid. To submerge, nose unit retracts and wing panels/skids fold back and retract under inner wing to minimise drag. Construction of composites (85 per cent) and aluminium alloy.
Mission payloads: No details revealed.
Guidance and control: Details not reported.
Launch: Conventional take-off from water, or from wheeled trolley on land.
Recovery: Conventional landing on water.
Power plant: One turbojet (type and rating not specified) for normal flight; unidentified turbine for underwater propulsion.
Variants: POC prototype: This scaled-down demonstrator is reported as having a 2.30 m (7 ft 6.6 in) wing span, 20 kg (44.1 lb) gross weight and endurance of 30 minutes. It was used in 2007-08 for initial tests to validate flight characteristics of the basic aircraft configuration.
Development prototype: Joined test programme in 2008.
H250: Prospective production version, 2009.

Aelius, H250

Dimensions, External	
Overall	
length	
Aelius	3.00 m (9 ft 10 in)
H250	4.00 m (13 ft 1½ in)
Wings	
wing span	
Aelius	5.00 m (16 ft 4¾ in)
H250	6.00 m (19 ft 8¼ in)
Areas	
Wings	
Gross wing area, Aelius	3.00 m² (32.3 sq ft)
Weights and Loadings	
Weight	
Max T-O weight	
Aelius	75 kg (165 lb)
H250	125 kg (275 lb)
Payload	
Max payload	
Aelius	25 kg (55 lb)
H250	55 kg (121 lb)
Performance	
Altitude	
Service ceiling, Aelius	5,000 m (16,400 ft)
Speed	
Max level speed	
airborne, Aelius	87 kt (161 km/h; 100 mph)
airborne, H250	97 kt (180 km/h; 112 mph)
under water, Aelius	6 kt (11 km/h; 7 mph)
Cruising speed	
airborne	68 kt (126 km/h; 78 mph)
under water	3 kt (6 km/h; 3 mph)
Radius of operation	
airborne, H250	216 n miles (400 km; 248 miles)
under water, Aelius, H250	21.5 n miles (39 km; 24 miles)
Endurance	
airborne, Aelius	6 hr
airborne, H250	18 hr
Power plant	1 × turbojet

Status: Development apparently completed. In 2009, company website was promoting a slightly larger and more capable version under the designation H250.
Contractor: Aeroart SAS
Mérignac.

Aeroart Aves

Type: Multirole tactical UAV.
Development: Start of development of Aves was announced by Aeroart in early February 2009, at which time preliminary design was said to have been completed.
Description: Airframe: Dihedral wings mounted atop a simple tubular open frame 'fuselage'; twin tailbooms supporting inward-canted fins and rudders bridged by a common tailplane. Fixed tricycle landing gear.
Mission payloads: Interchangeable payload packages fit within rectangular rack beneath wing centre-section, and can include 'most observation types', high-definition camera and/or a retractile ISR/exploration pod, physical or chemical sampler pods, or optional additional fuel. Aves can also be used as an 'aerial taxi' to transport and/or air-drop high-value or time-sensitive loads, or rescue equipment.
Guidance and control: No details announced.
Launch: Conventional wheeled take-off.
Recovery: Conventional wheeled landing.
Power plant: One piston engine (type and rating not specified); two-blade pusher propeller.

Aves

Dimensions, External	
Overall, length	1.80 m (5 ft 10¾ in)
Wings, wing span	2.60 m (8 ft 6¼ in)
Dimensions, Internal	
Payload bay, volume	0.025 m³ (0.9 cu ft)
Weights and Loadings	
Weight, Max launch weight	24 kg (52 lb)
Payload, Max payload	13 kg (28 lb)
Performance	
Speed	
Cruising speed, max	59 kt (109 km/h; 68 mph)
Radius of operation, mission	22 n miles (40 km; 25 miles)
Endurance	
with 3 kg (6.6 lb) payload	12 hr
with max payload	2 hr
Power plant	1 × piston engine

Status: Under development. Aeroart expected to fly a prototype during 2009.
Contractor: Aeroart SAS
Mérignac.

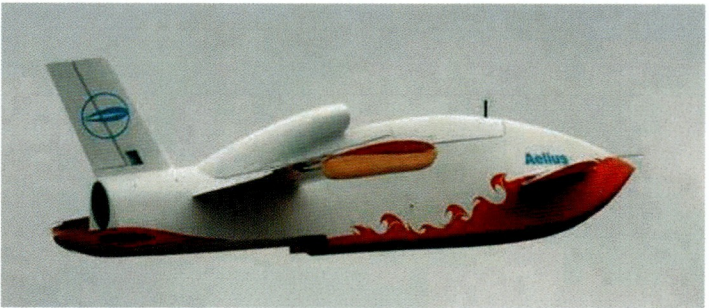

Aelius prototype during an early test flight at the CEV *(Aeroart)*

Aeroart Featherlite

Type: Close-range observation and training UAV.

Development: Based on experience gained with the experimental Aelius and Seagnos UAVs, the Featherlite (originally named Plume) is thought to be Aeroart's first commercially marketed design. It was unveiled at the Paris Air Show in June 2009 and is aimed at such civil applications as security, firefighting, mapping, research and operator training.

Description: *Airframe:* Simple high-wing design with upward-curved tips, slender fuselage and conventional tail unit. No landing gear. Construction mainly of foam plastics.
  Mission payloads: In standardised, mission-specific modules, including real-time high-definition video.
  Guidance and control: Includes automatic launch and recovery.
  Transportation: Man-portable.
  Launch: Hand or catapult launch.
  Recovery: Belly landing.
  Power plant: One electric motor (type and rating not specified); two-blade tractor propeller.

Featherlite

Dimensions, External	
Overall, length	1.20 m (3 ft 11¼ in)
Wings, wing span	1.90 m (6 ft 2¾ in)
Weights and Loadings	
Weight, Max launch weight	1.5 kg (3.00 lb)
Payload, Max payload	0.25 kg (0.00 lb)
Performance	
Speed	
Cruising speed, max	22 kt (41 km/h; 25 mph)
Radius of operation, mission	5.4 n miles (10 km; 6 miles)
Endurance, max	1 hr 30 min
Power plant	1 × electric motor

Status: Being promoted from 2009.
Contractor: Aeroart SAS
  Mérignac.

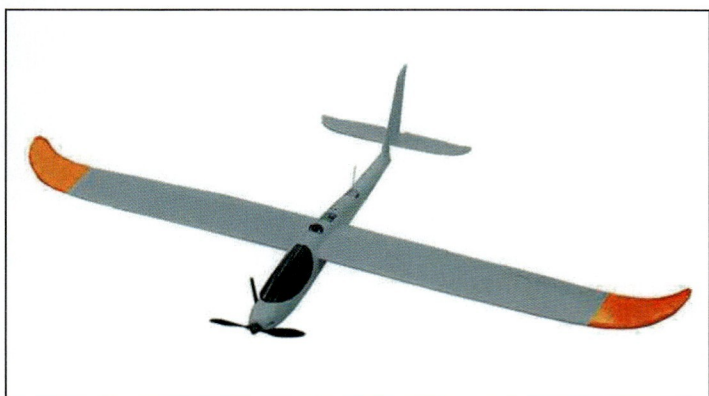

The ultra-lightweight, low-cost Featherlite *(Aeroart)*   1356273

Aeroart Seagnos 80

Type: Maritime surveillance UAV.

Development: The twin-hulled Seagnos was developed in 2008 as a type suitable for patrolling territorial waters in such roles as fishery surveillance and environmental monitoring.

Description: *Airframe:* Twin single-step hulls design, each with separate foreplane and tail unit; pylon-mounted parasol wing bearing twin engines. Payload bay between hulls.
  Mission payloads: Real-time video or high-resolution camera.
  Guidance and control: No details announced.
  Launch: Conventional water take-off.
  Recovery: Conventional water landing.
  Power plant: Two 4.2 kW (5.6 hp) piston engines; three-blade tractor propellers.

The prototype waterborne Seagnos during early testing *(Aeroart)*   1356226

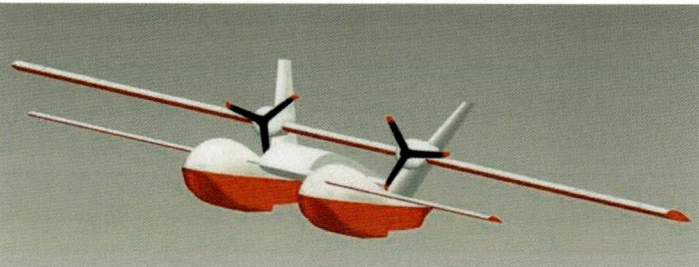

Computer impression of the Seagnos 80 *(Aeroart)*   1356274

Seagnos 80

Dimensions, External	
Overall, length	2.10 m (6 ft 10¾ in)
Wings, wing span	4.70 m (15 ft 5 in)
Weights and Loadings	
Weight, Max launch weight	40 kg (88 lb)
Payload, Max payload	12 kg (26 lb)
Performance	
Altitude, Service ceiling	4,000 m (13,120 ft)
Speed	
Max level speed	77 kt (143 km/h; 89 mph)
Cruising speed	59 kt (109 km/h; 68 mph)
Radius of operation	32 n miles (59 km; 36 miles)
Endurance, max	6 hr
Power plant	2 × piston engine

Status: In 2008, Aeroart announced expressions of civil and military interest from four prospective customers, two French and two American, with plans to begin marketing the Seagnos in 2009. As of late 2009 however, no evidence had been found to indicate if this had begun.

Contractor: Aeroart SAS
  Mérignac.

Alcore Biodrone

Type: Man-portable lightweight UAV.

Development: Described by Alcore as a biomimetic (biological imitation) UAV, based on study of raptors and other birds with long flight endurance; intended for use on close-range surveillance and as an aerosonde.

Description: *Airframe:* High-wing monoplane with twin tailbooms and inverted V tail unit; pusher engine; no landing gear. Composites construction (carbon fibre, Kevlar and epoxy).
  Mission payloads: IR linescan or colour CCD camera with real-time video downlink.
  Guidance and control: GCS comprises one or two portable computers. Remotely controlled via S-band datalink; GPS navigation. Operated by crew of two.
  Launch: Hand or rail launch.
  Recovery: Skid landing.
  Power plant: One 1.2 kW brushless electric motor, powered by lithium polymer (LiPo) batteries; pusher propeller with two folding blades.

Biodrone

Dimensions, External	
Overall	
length	1.80 m (5 ft 10¾ in)
height	0.30 m (11¾ in)
Fuselage, length	1.00 m (3 ft 3¼ in)
Wings, wing span	3.40 m (11 ft 1¾ in)
Weights and Loadings	
Weight	
Weight empty	6.0 kg (13.00 lb)
Max launch weight, hand launch	9.0 kg (19.00 lb)
Payload, Max payload	3.0 kg (6.00 lb)
Performance	
Altitude, Service ceiling	300 m (980 ft)
Speed	
Max level speed	70 kt (130 km/h; 81 mph)
Cruising speed	32 kt (59 km/h; 37 mph)
Loitering speed	19 kt (35 km/h; 22 mph)

Biodrone on its 3 m rail launcher   0081747

Unmanned aerial vehicles > France > Alcore Biodrone – Alcore Maya

Biodrone in flight *(Alcore)* 1170753

Biodrone

Radius of operation	10.8 n miles (20 km; 12 miles)
Endurance	2 hr
Power plant	1 × electric motor

Status: Available and in service.
   Customers: France (MoD, DGA and DSP/STTC).
Contractor: Alcore Technologies SA
   Cergy-Pontoise.

Alcore Easycopter

Type: Aerial photography helicopter UAV.
Development: Developed for civilian aerial video and digital photography applications.
Description: Airframe: Construction of carbon composites with polyurethane drive belts.
   Mission payloads: Digital or daylight video camera.
   Guidance and control: Remote control with GPS navigation and autostabilisation; COTS datalink with radius of at least 0.5 n mile (1 km; 0.6 mile). Portable computer GCS with three scrolled windows (map, dashboard and video monitor). Power provided by 12 V lithium polymer (LiPo) batteries.
   Launch: Vertical take-off.
   Recovery: Vertical landing.
   Power plant: One 180 W (0.24 hp) brushless electric motor.

Easycopter

Dimensions, External	
Overall, height	0.65 m (2 ft 1½ in)
Rotors, rotor diameter	0.65 m (2 ft 1½ in)
Weights and Loadings	
Weight, Max T-O weight	1.6 kg (3.00 lb)
Performance	
Endurance	15 min
Power plant	1 × electric motor

Status: In production or available off-the-shelf. The company provides complete air vehicles such as Easycopter to other specialised integrators.
Contractor: Alcore Technologies SA
   Cergy-Pontoise.

The Easycopter VTOL UAV *(Alcore)* 1170754

Alcore Futura

Type: Tactical reconnaissance and surveillance UAV, aerial target or UCAV.
Development: Began in 2000 as the basis for a family of multirole applications.
Description: Airframe: Shoulder-mounted, clipped-delta wings, with elevons; single large vertical fin. Composites construction. No landing gear.

Futura UAV *(Alcore)* 1170755

Futura UCAV *(Alcore)* 1170756

   Mission payloads: IR linescan or colour CCD camera with real-time video downlink; as UCAV, can mount a fragmentation warhead in enlarged bulbous nose.
   Guidance and control: Portable (microcomputer) GCS. Remote control with GPS navigation and S-band datalink.
   Transportation: Entire system can be carried by a single all-terrain vehicle, and is air-transportable.
   Launch: Catapult launch.
   Recovery: Skid landing.
   Power plant: One 0.22 kN (50 lb st) unidentified turbojet.

Futura

Dimensions, External	
Overall	
length	2.00 m (6 ft 6¾ in)
height	0.60 m (1 ft 11½ in)
Wings, wing span	2.00 m (6 ft 6¾ in)
Weights and Loadings	
Weight	
Weight empty	20 kg (44 lb)
Max launch weight	70 kg (154 lb)
Payload, Max payload	10 kg (22 lb)
Performance	
Altitude, Service ceiling	300 m (985 ft)
Speed	
Max level speed	194 kt (359 km/h; 223 mph)
Cruising speed	162 kt (300 km/h; 186 mph)
Loitering speed	70 kt (130 km/h; 81 mph)
Range, max	189 n miles (350 km; 217 miles)
Radius of operation, mission	27 n miles (50 km; 31 miles)
Endurance	1 hr 10 min
Power plant	1 × turbojet

Status: Reported to be in production by Alcore in 2011, and by licensees. Production numbers undisclosed.
   Customers: Not disclosed.
Contractor: Alcore Technologies SA
   Cergy-Pontoise.

Alcore Maya

Type: VTOL mini-UAV.
Development: Development of the Maya began in early 2002 as a second-generation UAV following the company's fixed-wing Azimut, Biodrone and Epsilon series. A tail-sitting VTOL design was chosen in

Maya as shown at Eurosatory exhibition in June 2004
(IHS/Patrick Allen) 1066613

preference to a conventional helicopter layout, this being regarded as offering greater platform stability in gusty conditions. A 30 cm (11.8 in), 1.5 kg (3.3 lb) demonstration prototype flew for the first time in April 2002. Carrying a 100 g (3.5 oz) autopilot and miniature GPS transceiver, it achieved vertical and horizontal speeds of 23 kt (43 km/h; 27 mph) and 87.5 kt (162 km/h; 101 mph) respectively and was entered later that year in a DGA competition overseen by Onera. Although unsuccessful on that occasion, further refinement continued, and the Maya was shown publicly at the Eurosatory defence exhibition in Paris in June 2004.

Description: **Airframe:** Four-blade shrouded rotor and four X-configuration tailfins, each with movable control surface; conical centrebody; four-leg non-retractable landing gear. Composites construction.

Mission payloads: CCD drive by roll/heading.

Guidance and control: Pre-programmed or LOS remote control from portable GCS via radio datalink; GPS navigation.

Launch: Conventional 'tail-sitter' vertical take-off.

Recovery: Conventional vertical landing.

Power plant: One 0.6 kW (8 hp) rotary piston engine. Suitable also for electric propulsion.

Maya

Dimensions, External	
Overall, height	0.34 m (1 ft 1½ in)
Weights and Loadings	
Weight	
Weight empty	2.0 kg (4.00 lb)
Max T-O weight	2.5 kg (5.00 lb)
Payload, Max payload	0.5 kg (1.00 lb)
Performance	
Climb	
Rate of climb, max, at S/L	300 m/min (984 ft/min)
Altitude	
Operating altitude	50 m (160 ft)
max	1,000 m (3,280 ft)
Speed	
Max level speed	58 kt (107 km/h; 67 mph)
Cruising speed, normal	29 kt (54 km/h; 33 mph)
Radius of operation, mission	15 n miles (27 km; 17 miles)
Power plant	1 × piston engine

Status: Developed and available. Although no details have emerged of its production and/or service status, the company was continuing to promote the Maya system via its website from 2010.

Contractor: Alcore Technologies SA
Cergy-Pontoise.

Bertin HoverEye

Type: Close-range VTOL mini-UAV.

Development: Bertin studies of mini-UAV designs began in 1999, the first realisation being a 30 cm (11.8 in), 1.5 kg (3.3 lb) spherical vehicle known as SmartBall which made its maiden flight in the third or fourth quarter of 2001. This proved to be somewhat vulnerable to wind gusts, but backing

HoverEye during FELIN trials in October 2007 *(DGA)* 1169327

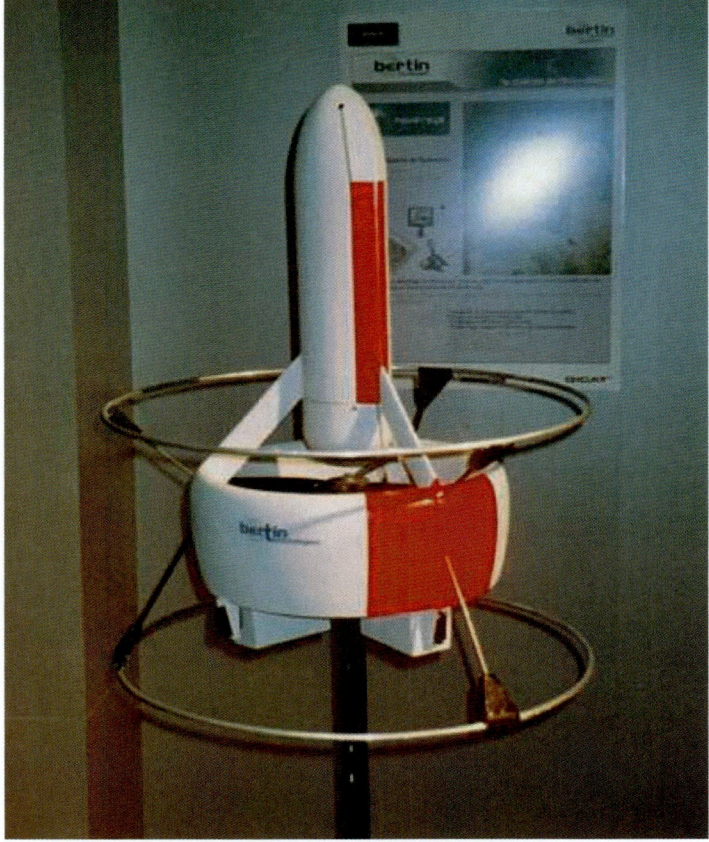

HoverEye prototype test vehicle *(Kenneth Munson)* 1290211

for further development was received from the DGA in mid-2003, leading to debut of the more stable HoverEye in 2004. Gradual refinement of the design has continued, and it was earmarked as a constituent element, together with UGVs, in the French Army's FELIN future soldier system for dismounted close combat and intelligence in urban warfare.

Description: *Airframe:* Twin counter-rotating rotors turning within an outer duct; central pod 'fuselage' houses mission payload, avionics and power plant. Glass fibre and carbon fibre construction.

Mission payloads: CMOS colour TV day camera or uncooled IR night camera. Spread spectrum datalink. Alternatives could include biochemical or IED detectors or communications equipment.

Guidance and control: Automatic or semi-automatic from miniaturised ground control unit. GPS/INS navigation.

Transportation: Complete system, including GCS, is transportable in a backpack.

Launch: Vertical take-off, followed by tilting for forward flight.

Recovery: Vertical landing.

Power plant: One 1 kW electric motor.

HoverEye

Dimensions, External	
Overall, height	0.60 m (1 ft 11½ in)
Weights and Loadings	
Weight, Max T-O weight	3.2 kg (7.00 lb)
Payload, Max payload	0.2 kg (0.00 lb)
Performance	
Altitude, Operating altitude	200 m (660 ft) (est)
Speed	
Max level speed	54 kt (100 km/h; 62 mph)
Loitering speed	5.5 kt (10 km/h; 6 mph)
Radius of operation	
LOS	1.1 n miles (2 km; 1 miles)
without LOS	0.5 n miles (km; miles)
Endurance	36 min (est)
Power plant	1 × electric motor

Status: Marketed since 2005 in collaboration with Sagem Défense Sécurité, whose Odin system is also based on HoverEye. Interaction with unmanned land systems was demonstrated at Mourmelon in October 2007 as part of the DGA's Fantassin à Equipements et Liaisons Intégrées (FELIN) programme. Operational trials were set to continue during 2008. The system continued to be promoted by both companies in 2010.

Contractor: Bertin Technologies
Montigny-le-Bretonneux.

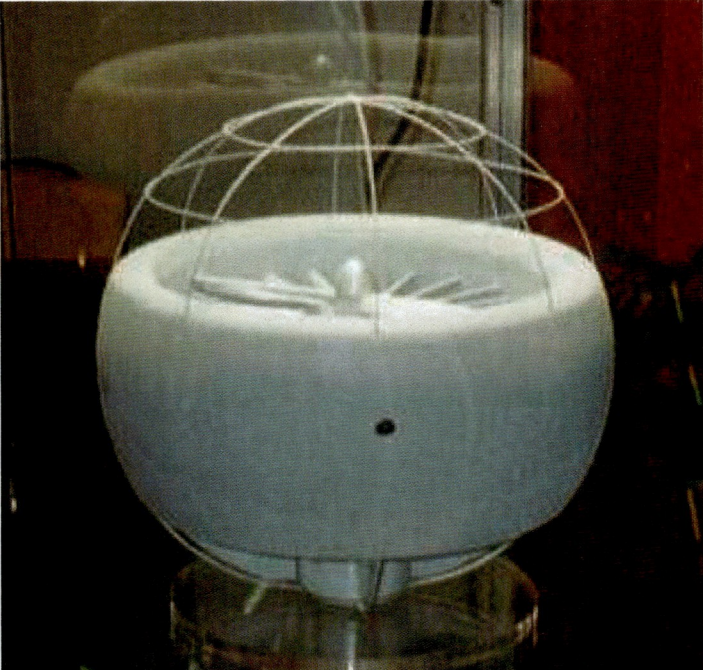
The original Bertin SmartBall *(Kenneth Munson)* 1290212

Cassidian (EADS DC) Eagle (Harfang)

Type: Medium-altitude, long-endurance strategic intelligence-gathering UAV.

Development: Unveiled at the Paris Air Show in June 1999, the Eagle is based on the airframe of the IAI Heron. It was initiated by the then Aerospatiale Matra to satisfy a French Air Force MALE requirement known as SIDM *(Système Interimaire de Drone MALE)*. Invitations to tender for this requirement were issued by the French DGA in 2001, and the Eagle 1 was subsequently ordered for the French Air Force. Military name is Harfang.

Description: *Airframe:* Manufactured by IAI, and based on that company's Heron (which see), but with large bulbous nose section to house satcom antenna. High-mounted, high-aspect ratio wings, equipped with anti-icing system and bearing twin fins and rudders on slender tailbooms; large central fuselage nacelle; pusher engine. Retractable tricycle landing gear.

Mission payloads: Wide range of sensors possible, according to mission. Basic configuration comprises an IAI Tamam MOSP EO/IR payload, a tactical communications relay and, for search and rescue operations, a Sarsat distress beacon location system. For other missions, can be equipped with a laser illuminator (target designation), an Elta EL/M-2055 synthetic aperture radar (all-weather reconnaissance) or elint/comint packages.

Guidance and control: Eagle uses LOS and satcom command links and datalinks, both to control the UAV and to transmit images and data (at 8 Mbytes/s), at ranges compatible with civil and military communications

HoverEye current configuration *(IHS/Patrick Allen)* 1181881

Eagle 1 at Paris Air Show, June 2003, with nose panel removed to reveal satcom antenna *(E R Hooton)* 0558420

Harfang F-SDAU/1021 at Bagram airfield Afghanistan, December 2009 *(USAF)* 1416243

Cassidian (EADS DC) Eagle (Harfang) < France < Unmanned aerial vehicles

Tail insignia of F-SDAU. *(USAF)* 1416244

Eagle 1 flying in civil-controlled airspace at Asian Aerospace, Singapore, in February 2004 *(Paul Jackson)* 0573599

satellite coverage areas. Autoland function incorporates DGPS, radio altimeter and back-up laser system for all-weather operation.

System composition: Three air vehicles; secure satellite datalink system for real-time data transmission; GCS has mission control centre and operator stations for AV and payload control and data exploitation and processing.

Launch: Conventional runway take-off, using IAI Malat automatic T-O and landing system.

Recovery: Conventional and automatic runway landing.

Power plant: *Eagle 1:* One 84.6 kW (113.4 hp) Rotax 914 F turbocharged engine, driving a two-blade pusher propeller.

Eagle 2: One 895 kw (1,200 shp) Pratt & Whitney Canada PT6A turboprop.

Variants: *Eagle 0:* Prototype/testbed. Features include anti-icing system, remote video terminal, radio relay and advanced GCS. Radar and MOSP payloads. Test flown in 1998 with Rotax 914 piston engine. Operated only within LOS, with remote radio control T-O and landing; Tadiran Spectralink C-band (4 GHz) datalink (range 135 n miles; 250 km; 155 miles). In June 2002, Eagle 0 was evaluated by Sweden's FMV in a series of five flights as part of a programme to develop a network-centric data collection and dissemination system. Flying from Kiruna airbase near the Arctic Circle, the UAV was equipped with a Swedish Space Corporation VDL Mode 4 IFF, EO/IR and SAR sensors, a civil aviation transponder and atmospheric research instrumentation. Real-time imagery from detected and identified targets was transmitted simultaneously by secure datalink to GCSs in Kiruna and Stockholm.

Eagle 1: Initial production version. First flight, in Israel, 2 June 2003. Automatic T-O and landing (ATOL, using augmented DGPS, OPATS laser tracker and radar altimeter); nose-mounted Ku-band satcom antenna; integrated avionics (radio and IFF); underwing hardpoints for sensor pods, auxiliary fuel tanks or other stores; and modular GCS. Principal missions are imint, target acquisition and designation, and communications relay.

An Eagle 1 configured for CSAR took part in a joint service exercise ('Desert Rescue XI') at NAS Fallon, Nevada, USA, in mid-2003. Selected in September 2003 by UK MoD, to whom one aircraft (ZJ989) delivered for MALE aspect of Joint UAV Experimentation Programme (JUEP). From 7 to 21 October 2003, this aircraft flew 12 missions (73 flight hours) during a JUEP exercise at the British Army Training Unit in Suffield, Alberta, Canada, equipped with a high-definition SAR, an EO/IR dual-sensor and a radio relay.

Eagle 1 in Royal Air Force markings for UK trials in Canada 0583303

Eagle MPR: Projected maritime patrol radar version of Eagle 1 for coastal and offshore missions. No recent reports of this version.

Eagle 2: Larger follow-on version with Pratt & Whitney Canada PT6A turboprop, enhanced altitude and speed capability. Greater capacity and scope for additional payloads such as sigint, SAR/MTI, ESM, environmental monitoring and cartography. Netherlands joined development programme on 1 December 2003. Not built.

Eagle 1, Eagle 2

Dimensions, External	
Overall length	
Eagle 1	9.30 m (30 ft 6¼ in)
Eagle 2	13.00 m (42 ft 7¾ in)
Wings wing span	
Eagle 1	16.60 m (54 ft 5½ in)
Eagle 2	26.00 m (85 ft 3½ in)
Weights and Loadings	
Weight	
Weight empty, Eagle 1	657 kg (1,448 lb)
Max T-O weight	
Eagle 1	1,250 kg (2,755 lb)
Eagle 2	3,900 kg (8,598 lb)
Fuel weight	
Max fuel weight, Eagle 1	250 kg (551 lb)
Payload	
Max payload	
Eagle 1	250 kg (551 lb)
Eagle 2	500 kg (1,102 lb)
Performance	
T-O	
T-O run	
Eagle 1	600 m (1,969 ft)
Eagle 2	1,000 m (3,281 ft) (est)
Altitude	
Service ceiling	
Eagle 1	7,620 m (25,000 ft)
Eagle 2	13,715 m (45,000 ft) (est)
Speed	
Max level speed	
Eagle 1	112 kt (207 km/h; 129 mph)
Eagle 2	230 kt (426 km/h; 265 mph) (est)
Range	
Eagle 1	918 n miles (1,700 km; 1,056 miles)
Eagle 2	1,565 n miles (2,898 km; 1,801 miles) (est)
Endurance	
Eagle 1	24 hr
Eagle 2	24 hr (est)
Time on station at 550 n miles (1,018 km; 632 miles),	
Eagle 1	12 hr
at 800 n miles (1,481 km; 920 miles),	
Eagle 2	12 hr (est)
Landing	
Landing run	
Eagle 1	600 m (1,969 ft)
Eagle 2	1,000 m (3,281 ft) (est)
Power plant	1 × turboprop

Status: *Eagle 1:* French Air Force order for 24 air vehicles and 25 GCSs. First system originally due for delivery to CEV Istres (later CEV Cazaux, then CEAM Mont-de-Marsan) during second half of 2004, thence to French Air Force base at Cognac in third quarter of 2005. However, programme reportedly affected by cost escalation and need to find alternatives for some US-embargoed equipment, and delivery to CEV not accomplished until August 2005. From December 2007 a team from the DGA and the French Air Force carried out verification the system consisting of three aerial vehicles, payloads, the laser designator, two ground stations, the line-of-sight (LOS) and satellite data link system. Flight acceptance operations were concluded in June 2008 with the system being transferred into the ownership of the French Ministry of Defence and named Harfang. Deployed to Afghanistan February 2009 with first sortie on 17 February. Operating unit is Escadron d'Expérimentation Drone (EED) 01.330, home base Mont-de-Marsan. Air vehicles are F-SDAU/1021; F-SDAY/1022 and F-SDAZ/1023. Deployed over Lourdes in 2008 during Papal visit, and over Deauville during G8 summit in May 2011, for security purposes.

Eagle 2: Feasibility stage began 2002 and continuing to mid-2006; projected development stage was 2007–10; production 2009–12 and service entry 2012 or 2013.

Customers: Eagle 1 ordered for French Air Force.
Contractor: Cassidian (an EADS company).
Vélizy-Villacoublay, France.

Cassidian (EADS DC) Tracker

Type: Close-range surveillance mini-UAV.
Development: Developed in co-operation with, and manufactured by, Survey-Copter; public debut October 2003. In 2006, Survey-Copter was promoting a smaller version under the designation DVF-2000.
Description: *Airframe:* Straight-winged, twin-fuselage configuration with double-T tail unit. No landing gear. Manufactured by Survey-Copter.
 Mission payloads: Gyrostabilised Survey-Copter two-axis colour CCD camera with ×20 zoom is standard; IR camera optional. Real-time imagery downlink.
 Guidance and control: Fully autonomous weControl GPS-based flight guidance and navigation system, using two laptop computers.
 Transportation: Man-portable in backpack.
 System composition: Two air vehicles, interchangeable gyrostabilised CCD and IR payloads, datalink and portable, multifunction GCS. Entire system can be stowed in a backpack. Operating crew of two.
 Launch: Hand-launched.
 Recovery: Belly landing under automatic control.
 Power plant: Two electric motors, each driving a two-blade folding propeller.
Variants: *Tracker:* EADS version; *as described.*
 DVF-2000: Smaller Survey-Copter version. Wing span 2.50 m (8 ft 2.4 in), max launching weight less than 7 kg (15.4 lb), range 2.7 n miles (5 km; 3.1 miles), endurance 1 h 30 min; other details as for Tracker.

Tracker

Dimensions, External	
Overall, length	1.40 m (4 ft 7 in)
Wings, wing span	3.60 m (11 ft 9¾ in)
Weights and Loadings	
Weight, Max launch weight	7.5 kg (16.00 lb)
Payload, Max payload	1.8 kg (3.00 lb)

Performance

Altitude	
Operating altitude, typical	300 m (980 ft) (est)
Service ceiling	2,000 m (6,560 ft) (est)
Speed, Cruising speed	32 kt (59 km/h; 37 mph)
Range, datalink	5.4 n miles (10 km; 6 miles)
Endurance	2 hr (est)
Power plant	2 × electric motor

Status: Under development in 2003–04, including one demonstration flight in 2003 for UK Joint UAV Experimentation Programme (JUEP). Awarded French DGA contract in January 2005 for 25 systems to fulfil French Army DRAC (*Drone de Reconnaissance Au Contact*) requirement. This represented the first tranche, with 35 additional systems ordered in 2008, en route to eventual total of 160 systems over a five-year period, with overall contract value of EUR30 million. Operator training was completed in 2008 and the systems became operational with the French Army in Afghanistan in 2010. Deliveries were scheduled to continue until 2013. Several systems have been delivered to undisclosed export customers.
Contractor: Cassidian (an EADS company).
Vélizy-Villacoublay, France.

Flying Robots FR101

Type: Multirole UAV.
Development: Formed in June 2004, Flying Robots introduced the FR101 at the Paris Air Show a year later, followed by two additional variants at the 2007 event. All are aimed at a wide range of potential civil applications. Typical examples include patrol and surveillance of international borders, power and pipelines, road and maritime traffic; monitoring of fire, pollution, radiation and other hazards; atmospheric research and weather forecasting; geological survey; and broadband communications. Unusually for a UAV, the FR101 design also allows for the onboard carriage of a human pilot.
Description: *Airframe:* Parafoil wing, from which is suspended a stainless steel and foam sandwich fuselage supporting short twin tailbooms and twin fins and rudders. Pusher engine with ducted propeller. Fixed tricycle landing gear.
 Mission payloads: Can accommodate a variety of day and night cameras, gyrostabilised turrets, thermal sensors, communications relay or scientific equipment. S-band (2.4 GHz) video downlink. Can also air-drop or soft-land cargo loads.
 Guidance and control: Fully automated, with three flight modes: remote controlled, onboard joystick, or GPS-based with unlimited programmed waypoints and real-time override. Dual VHF and UHF transceivers for command uplink and telemetry downlink. Dual GPS

Tracker hand-launched, man-portable mini-UAV *(IHS/Patrick Allen)*

Tracker nose detail *(IHS/Patrick Allen)*

FR101 Téléporteur delivering supplies for ASF in Africa *(Flying Robots)*

Flying Robots FR101 – Fly-n-Sense Project 360 < France < Unmanned aerial vehicles

The parafoil-winged FR101 multirole UAV *(Flying Robots)* 1290329

receivers. Fixed-base or mobile GCS can control several sensors simultaneously.

Transportation: Transportable by truck. Set-up time, by two-person team, 15 minutes.
Launch: Conventional, automatic wheeled take-off.
Recovery: Conventional, automatic, wheeled landing. Parafoil allows glide to soft landing if engine fails; pre-programmed procedure for recovery if command signal is lost. Ballistic parachute for other emergency situations.
Power plant: *FR101:* One 73.5 kW (98.6 hp) Rotax 912 ULS flat-four piston engine, driving a two-blade, ducted pusher propeller. Standard fuel capacity 160 litres (US gallons; Imp gallons). Provision for auxiliary fuel tanks.
FR50: One 3W-157B2TS piston engine (rating not stated).
Variants: *FR50:* Smaller, close-range version of FR101, introduced in 2007.
Téléporteur: Dedicated version to deliver emergency equipment and supplies for disaster relief and other humanitarian missions. Designed to provide an unmanned airborne shuttle link between locations up to 300 km (186 miles) apart, especially in remote or hazardous areas, with pre-programmed return to starting point.

FR101, FR50

Dimensions, External
  Fuselage
    length, FR101 .... 3.20 m (10 ft 6 in)
    height, FR101 .... 2.10 m (6 ft 10¾ in)
    width, FR101 .... 1.46 m (4 ft 9½ in)
  Wings
    wing span, FR101 .... 14.40 m (47 ft 3 in)
  Engines
    propeller diameter, FR101 .... 1.66 m (5 ft 5¼ in)
Dimensions, Internal
  Payload bay
    length
      min, FR101 .... 0.92 m (3 ft 0¼ in)
      max, FR101 .... 1.40 m (4 ft 7 in)
    width, FR101 .... 1.00 m (3 ft 3¼ in)
    depth, FR101 .... 0.87 m (2 ft 10¼ in)
    volume, FR101 .... 1.00 m³ (35 cu ft)
Areas
  Wings, Gross wing area .... 42.0 m² (452.1 sq ft)
Weights and Loadings
  Weight
    Weight empty
      FR101 .... 250 kg (551 lb)
      FR50 .... 40 kg (88 lb)
    Max T-O weight
      FR101 .... 600 kg (1,322 lb)
      FR50 .... 70 kg (154 lb)
  Payload
    Max payload
      FR101 .... 250 kg (551 lb)
      FR50 .... 20 kg (44 lb)
Performance
  T-O
    T-O run
      FR101 .... 80 m (263 ft) (est)
      FR50 .... 20 m (66 ft)
  Altitude
    Service ceiling
      FR101 .... 4,000 m (13,120 ft)
      FR50 .... 3,000 m (9,840 ft)
  Speed
    Max level speed
      FR101 .... 38 kt (70 km/h; 44 mph)
      FR50 .... 32 kt (59 km/h; 37 mph)
    Cruising speed
      FR101 .... 32 kt (59 km/h; 37 mph)
      FR50 .... 19 kt (35 km/h; 22 mph)
  Range
    BLOS, FR101 .... 648 n miles (1,200 km; 745 miles)
    BLOS, FR50 .... 108 n miles (200 km; 124 miles)
  Radius of operation
    datalink, FR101 .... 81 n miles (150 km; 93 miles)
    datalink, FR50 .... 43 n miles (79 km; 49 miles)
  Endurance
    FR101 .... 30 hr (est)
    FR50 .... 8 hr
  Landing
    Landing run
      FR101 .... 20 m (66 ft)
      FR50 .... 15 m (50 ft)
  Power plant .... piston engine

Status: In production and in service. FR101 has been sold in 'several' countries (details not given). Téléporteur customers in 2008 included Aviation Sans Frontières (ASF).
Contractor: Flying Robots SA
Illkirch Graffenstaden.

Fly-n-Sense Project 360

Type: VTOL mini-UAV.
Development: This new UAV was unveiled at the Eurosatory defence exhibition in Paris in June 2010, though with minimal specifications data. Its name indicates ability to secure 360-degree imaging coverage.
Description: *Airframe:* Tubular metal structure with two-blade main rotor and twin-skid landing gear.
  Mission payloads: Stabilised, high-resolution video camera or thermal imager.
  Guidance and control: Remotely piloted; C-band communications links.
Launch: Vertical take-off.
Recovery: Automatic vertical landing.
Power plant: One 3 kw (4 shp) miniature turboshaft.

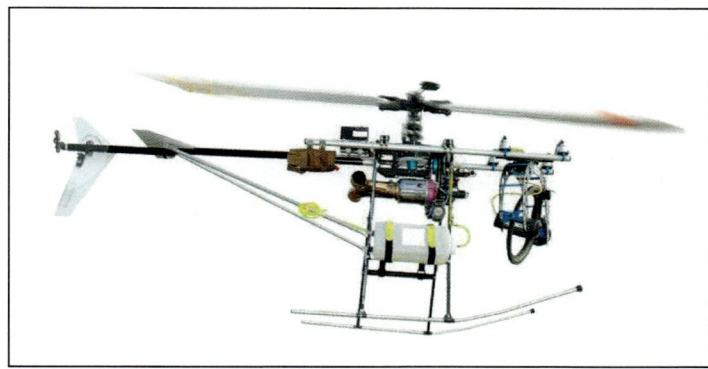

Project 360 in side elevation *(Fly-n-Sense)* 1395276

A Project 360 prototype on public debut in June 2010 *(IHS/Patrick Allen)* 1418794

Project 360

Dimensions, External	
Rotors, rotor diameter	2.50 m (8 ft 2½ in)
Weights and Loadings	
Payload, Max payload	20 kg (44 lb)
Performance	
Endurance, max	30 min
Power plant	1 × turboshaft

Status: Under development in 2010; planned to become available in 2011.
Contractor: SAS Fly-n-Sense
Mérignac.

Fly-n-Sense Scancopter

Type: Reconnaissance and surveillance VTOL micro-UAVs.
Development: The Scancopter series offers two solutions for the conduct of close-range ISR operations.
Description: *Airframe: CB 350:* Domed central module containing payload, avionics and batteries, from which protrude three arms, each supporting a pair of coaxial, counter-rotating three-blade rotors.
CB 750: Module for payload, avionics and batteries is at centre of a trio of circular ducts, within each of which rotate a pair of coaxial, counter-rotating two-blade rotors.
Mission payloads: Micro camera in CB 350. Payloads for CB 750 can include video and thermal cameras, chemical and acoustic sensors. Real-time data downlinks.
Guidance and control: Automatic, via ENAC Paparazzi autopilot with waypoint navigation; or remotely piloted via radio command uplink. Laptop GCS, with radio modem and radio-control transmitter.
Transportation: Man-portable.
Launch: Vertical take-off.
Recovery: Vertical landing.
Power plant: Battery-powered electric motor(s); ratings not quoted.
Variants: *Scancopter CB 350:* Designed specifically for missions in otherwise inaccessible areas, including building interiors.
Scancopter CB 750: Designed for missions in urban areas such as threat detection, site security and environmental issues.

Scancopter CB 350, Scancopter CB 750

Dimensions, External	
Overall	
width	
Scancopter CB 350	0.75 m (2 ft 5½ in)
Scancopter CB 750	0.35 m (1 ft 1¾ in)
Weights and Loadings	
Weight	
Max T-O weight	
Scancopter CB 350	0.6 kg (1.00 lb)
Scancopter CB 750	2.0 kg (4.00 lb)
Payload	
Max payload, Scancopter CB 750	0.5 kg (1.00 lb)
Performance	
Speed	
Cruising speed, Scancopter CB 750	58 kt (107 km/h; 67 mph)
Endurance	
max, Scancopter CB 350	15 min
max, Scancopter CB 750	30 min
Power plant	1 × electric motor

Status: Developed and available; promotion began in 2010.
Contractor: SAS Fly-n-Sense
Mérignac.

Fly-n-Sense Seeker FNS 900

Type: Multirole micro UAV.
Development: Fly-n-Sense was formed in July 2008 and introduced its first UAVs during the first half of 2009. The Seeker is designed for security, industrial, environmental and research applications in the civil sector.
Description: *Airframe:* Mid-mounted sweptback wings with endplate fins; bullet-shaped fuselage with sweptback central fin; no horizontal tail surfaces. No landing gear.
Mission payloads: Various visual, chemical, acoustic or other sensors, according to mission. Real-time imagery and data downlink.
Guidance and control: Fully automatic or remotely piloted flight, with ENAC Paparazzi autopilot and GPS waypoint navigation. Laptop GCS, radio modem and radio-control transmitter.
Transportation: Man-portable.
Launch: Hand launch.
Recovery: Belly landing.
Power plant: Battery-powered electric motor (rating not quoted), driving a two-blade propeller.

Seeker

Dimensions, External	
Overall, length	0.75 m (2 ft 5½ in)
Wings, wing span	0.90 m (2 ft 11½ in)
Weights and Loadings	
Weight, Max launch weight	2.0 kg (4.00 lb)
Payload, Max payload	0.5 kg (1.00 lb)
Performance	
Speed	
Max level speed	62 kt (115 km/h; 71 mph)
Cruising speed	39 kt (72 km/h; 45 mph)
Endurance	1 hr (est)
Power plant	1 × electric motor

Status: Developed and available; promotion began in 2010. Production and service status not known.
Contractor: SAS Fly-n-Sense
Mérignac.

Scancopter CB 750 *(Fly-n-Sense)*

Scancopter CB 350 *(IHS/Patrick Allen)*

Seeker FNS 900 exhibited at Eurosatory in June 2010 *(IHS/Patrick Allen)*

HELIPSE HE-190

Type: VTOL UAV system

Description: The HE-190 is fitted with battery level indicators.

The helicopter is delivered complete and tested with motorisation. It is equipped with:
- 6 servos (4 for the swash plate, 1 throttle, 1 tail rotor);
- gyrometer for tail rotor;
- automatic run regulator;
- wiring professional;
- system integrating the energy sources, electronics switches, double power Lipo 3 cells 5 mAh (necessary to supply the autopilot and the ground/air communication).

All the fixing points are designed to receive all payload types.

The HE-190 is equipped with an automatic flight controller, the NAV 4.

Airframe: Rotor head using the Bell/Hiller with double shock absorbers, dihedron of 1 degrees. Principal drive by notched belt, secondary drive by helical gears, helix angle 17 degrees. Tail rotor drive by double notched belts. Tail rotor with automatic alignment system on shock absorber.

Launch: Vertical take-off, followed by tilting for forward flight.

Recovery: Vertical landing.

Power plant: Engine OS 160 (2.9 kW/4 hp) using 1.8 to 2 litres an hour (depending on type of flight) Muffled tuned pipe

HE-190	
Dimensions, External	
Overall	
length	2.25 m (7 ft 4½ in)
height	0.70 m (2 ft 3½ in)
Rotors, rotor diameter	1.90 m (6 ft 2¾ in)
Tail rotor, tail rotor diameter	0.36 m (1 ft 2¼ in)
Weights and Loadings	
Payload, Max payload	7 kg (15.00 lb) (est)

Status: Used in the Japanese Fukushima nuclear disaster in 2011.

Contractor: HELIPSE
La Couronne

HE-190 remotely operated helicopter *(HELIPSE)* 1295701

The HELIPSE unmanned helicopter (HE-190) *(HELIPSE)* 1295702

Infotron IT 180-5

Type: VTOL UAV.

Development: Evolving from an earlier design that had a spherical body with both sets of rotors on top, the IT 180-5 flew for the first time in 2004, making its public debut at the 2005 Paris Air Show. It was officially launched for civil applications at Le Bourget in June 2007, and has subsequently entered production and service.

Description: ***Airframe:*** Circular central body, sandwiched between two three-blade, coaxial and counter-rotating rotors. Tripod landing gear.

The IT 180-5 VTOL UAV *(IHS/Patrick Allen)* 1326201

Operator's control box for the IT 180-5 *(IHS/Patrick Allen)* 1173654

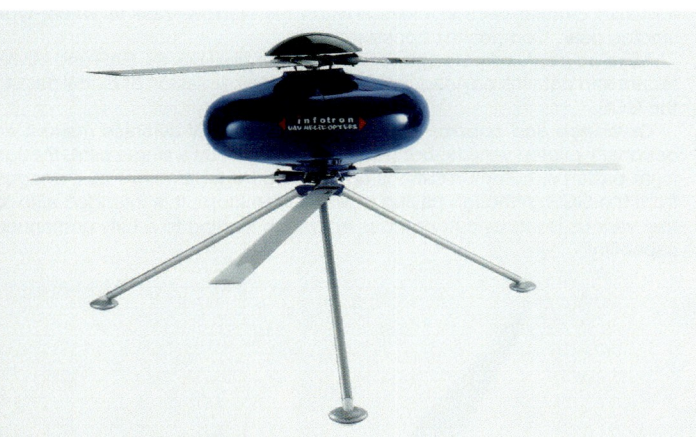

Overall view of the IT 180-5 *(Infotron)* 1209127

Mission payloads: Integrated multiple sensors include video (gyrostabilised or fixed) or high-resolution digital stills camera, thermal imager or other sensors to customer's requirements, for such applications as civil and military security, industrial and environmental inspection and surveillance. Payloads are suspended below the UAV on a boom or cradle. Imagery can be stored on board or transmitted to the GCS by real-time downlink.

Guidance and control: Can be pre-programmed for fully automatic flight, including take-off and landing phases; or operator controlled; GPS navigation and positioning. Embedded software controls altitude and sends accurate instructions to the air vehicle via a secure S-band (2.4 GHz) double radio link. One-man operation, using laptop ground control unit.

Transportation: Man-portable.

Launch: Automatic vertical take-off.

Recovery: Automatic vertical landing.

Power plant: *IT 180-5 TH:* One 26 cm³ two-stroke Diesel engine.
IT 180-5 EL: One DC brushless electric motor.
Variants: *IT 180-5 TH:* Piston-engined ('thermal') version, for optimum range and endurance.
IT 180-5 EL: Electric-powered version, trading range and endurance for near-silent operation.

IT 180-5 EL, IT 180-5 TH

Weights and Loadings	
Weight	
Weight empty	10.0 kg (22 lb)
Max launch weight	15.0 kg (33 lb)
Payload	
Max payload, incl fuel	5.0 kg (11.00 lb)
Performance	
Altitude, Service ceiling	3,000 m (9,840 ft)
Speed, Max level speed	48 kt (89 km/h; 55 mph)
Radius of operation	
datalink, mission	2.7 n miles (5 km; 3 miles) (est)
Endurance	
IT 180-5 TH	1 hr 30 min
IT 180-5 EL	30 min
Power plant, IT 180-5 EL	1 × electric motor

Status: In production and in service. In France, has been used by such major entities as SNCF (for viaduct inspection) and RTE (for installing high-voltage power lines). Other unusual uses have included controlled triggering of avalanches by air-dropping a small explosive charge. Infotron provides services with the system as well as direct sales.
Contractor: Infotron
Massy.

Sagem/Onera Busard

Type: Optionally piloted systems testbed.
Development: Launched in 2005, the Busard programme is a five-year collaborative venture between Sagem, the French defence research agency Onera, the French Air Force Academy and the CEV flight test centre, using an adapted (S10-VTX) variant of the German Stemme S10 motor glider (F-COSP), acquired in late 2004, as the platform. Although its name conveniently translates as 'Harrier' it was actually coined as an acronym of *Banc Ultra-léger pour Systèmes Aéroportés de Recherche sur les Drones*. Its purpose is to test UAV payloads and other technologies such as 'sense and avoid' that will lead to the safe use of unmanned aircraft in controlled civil airspace. Sagem Défense Sécurité was the prime contractor, with the TNO (the Netherlands), the ONERA (France) and ELSPESA (Spain) as co-contractors. Payloads for these tests were installed in fuselage bays and/or twin inboard equipment pods under the wings.
Description: *Airframe:* Three-part, shoulder-mounted wings (outer panels foldable); circular-section fuselage with T tail. Narrow-track tailwheel-type landing gear. Composites construction.
Mission payloads: Designed to carry a range of EO, IR, SAR and MMW radars and datalink payloads, with real-time transmission of digital data to the GCS.
Guidance and control: Busard's complete UAV avionics make it an optionally piloted vehicle, but it is currently flown by a single pilot, though flight patterns, communications and payload operation are all managed from the GCS. Although having a pilot 'in the loop', it is intended also to test various levels of autonomous operation leading to a fully unmanned capability.

Launch: Conventional wheeled take-off.
Recovery: Conventional wheeled landing.
Power plant: One 84.6 kW (113.4 hp) Rotax 914 F turbocharged flat-four engine, buried in centre-fuselage, with extension shaft to a two-blade propeller in retractable nosecone.
Fuel capacity 90 litres (23.8 US gallons; 19.8 Imp gallons) standard, 120 litres (31.7 US gallons; 26.4 Imp gallons) optional.
(standard S10-VT)

Busard

Dimensions, External	
Overall	
length	8.42 m (27 ft 7½ in)
height	1.80 m (5 ft 10¾ in)
Wings, wing span	23.00 m (75 ft 5½ in)
Areas	
Wings, Gross wing area	18.70 m² (201.3 sq ft)
Weights and Loadings	
Weight	
Weight empty	660 kg (1,455 lb)
Max T-O weight	850 kg (1,873 lb)
Performance	
Altitude, Service ceiling	9,145 m (30,000 ft)
Speed	
Cruising speed, max	134 kt (248 km/h; 154 mph) at 3,048 m (10,000 ft)
Stalling speed, flaps down	42 kt (78 km/h; 49 mph)
Radius of operation	
mission, standard fuel	348 n miles (644 km; 400 miles)
mission, optional max fuel	464 n miles (859 km; 534 miles)

Status: Experimental flying continuing in 2008. Eventual unmanned configuration is planned to achieve an average endurance of 10 hours, or 20 hours at 7,620 m (25,000 ft) altitude and 108 n miles (200 km; 124 miles) from base.
Contractor: Sagem Défense Sécurité
Paris.

Sagem Patroller

Type: MALE UAV.
Development: Continuing the development begun in the Sagem/ONERA Busard demonstrator programme, the Patroller utilises a Stemme S15 motor glider as the platform aircraft. A prototype made its first flight on 10 June 2009 at the Kemijarvi test ground in Finland, making its public debut at the Paris Air Show a few days later. First phase of flight tests (eight sorties) completed on 30 June 2009. Further tests followed at Cergy-Pontoise airfield near Paris in 2010.
Description: *Airframe:* Three-part shoulder-mounted wings, with Schempp-Hirth airbrakes and winglets; circular-section fuselage with T tail. Tricycle landing gear. Composites (CFRP) construction.
Mission payloads: Prototype equipped with Sagem EuroFlir 410 EO/IR sensor in gyrostabilised undernose turret, plus an OHB synthetic aperture radar in underwing pod, a Ku-band LOS link and a low data rate

Patroller on an early test flight at Kemijarvi *(Sagem)*

F-COSP, the French-registered Busard, with camera turret in the port underwing pod *(Paul Jackson)*

Inside the Patroller GCS *(IHS/Patrick Allen)*

Patroller lined up for its maiden flight in Finland, 10 June 2009 *(Sagem)*

Patroller on show at Le Bourget, June 2009 *(IHS/Patrick Allen)*

satcom video downlink. Other potential payloads include maritime radar or communications relay equipment. Underwing fuel tanks test flown during 2010.

Guidance and control: Patroller is operated from the same ground control station as the company's SDTI Sperwer tactical UAV, with which it shares a large percentage of the avionics, including the automatic flight control system.

Launch: Conventional, automatic wheeled take-off.

Recovery: Conventional, automatic wheeled landing.

Power plant: One 84.6 kW (113.4 hp) Rotax 914 F turbocharged flat-four engine, buried in centre-fuselage with extension shaft to a three-blade constant-speed tractor propeller.

Patroller

Dimensions, External	
Overall	
length	8.50 m (27 ft 10¾ in)
height	2.45 m (8 ft 0½ in)
Wings, wing span	18.00 m (59 ft 0¾ in)
Areas	
Wings, Gross wing area	17.86 m² (192.2 sq ft)
Weights and Loadings	
Weight, Max T-O weight	1,050 kg (2,314 lb)
Payload, Max payload	250 kg (551 lb)
Performance	
T-O	
T-O run	250 m (821 ft)
Altitude, Service ceiling	7,620 m (25,000 ft)
Speed	
Max level speed	162 kt (300 km/h; 186 mph)
Loitering speed	51 kt (94 km/h; 59 mph)
Endurance, max	30 hr (est)
Landing	
Landing run	250 m (821 ft)

Status: Under development starting in 2009. Flight testing continuing at Istres in southern France in 2010.

Contractor: Sagem Défense Sécurité
Paris.

Sagem Sperwer A and Ugglan

Type: Surveillance and target acquisition UAV.

Development: Sperwer (Sparrowhawk) is derived from the Sagem Crecerelle, but has a redesigned airframe. Development began in the early/mid-1990s.

Description: *Airframe:* Low/mid-mounted cropped-delta wings, with elevons; box-section fuselage; twin outward-canted fins and rudders. Wings de-iced by hot air from engine exhaust.

Mission payloads: Sagem IRIS dual EO/IR sensor payload in retractable turret (OLOSP in Danish, Dutch and Swedish aircraft), plus piloting camera in nose. Greek aircraft also have communications relay package, extending mission radius to 108 n miles (200 km; 124 miles). Sperwer can also be equipped with SAR, ESM/elint package, communications jammer or atmospheric data-gathering payload.

Guidance and control: Take-off and landing under ground operator control; remainder of mission can be preprogrammed or remotely controlled via UHF uplink. INS/GPS navigation; Ku-band (15 GHz)

Sperwer *(Sagem)*

Sperwer in flight *(GAM STAT-Valence)*

Unmanned aerial vehicles > France > Sagem Sperwer A and Ugglan

Canadian personnel preparing a Sperwer for flight in Afghanistan *(Canadian DND)* 0577488

Sperwer in Royal Netherlands Army insignia *(Kenneth Munson)* 0114404

Sperwer leaves launcher *(Sagem)* 1353805

Sperwer SDTI in French Army insignia *(IHS/Patrick Allen)* 1024049

Ugglan (owl): Modified version for Swedish Army. Maximum launching weight 320 kg (705 lb), including 75 kg (165 lb) payload; maximum speed 119 kt (220 km/h; 137 mph); operating height range 300 to 3,000 m (985 to 9,840 ft); mission radius 38 n miles (70 km; 43.5 miles).

Sperwer A

Dimensions, External	
Overall	
length	3.50 m (11 ft 5¾ in)
height	1.10 m (3 ft 7¼ in)
Wings, wing span	4.20 m (13 ft 9¼ in)
Weights and Loadings	
Weight	
Weight empty	212 kg (467 lb)
Max launch weight	350 kg (771 lb)
Payload, Max payload	50 kg (110 lb)
Performance	
Altitude, Service ceiling	4,570 m (15,000 ft)
Speed	
Max level speed	127 kt (235 km/h; 146 mph)
Cruising speed	94 kt (174 km/h; 108 mph)
Loitering speed	90 kt (167 km/h; 104 mph) (est)
Range, datalink	108 n miles (200 km; 124 miles)
Radius of operation, mission	40 n miles (74 km; 46 miles)
Endurance	6 hr (est)

Status: Production complete. In service with Royal Netherlands Army, Canadian Army, French Army and Greek Army (Sperwer); and Swedish Army (Ugglan). Total of 14 systems ordered by March 2006, at which time 88 air vehicles delivered. Two additional systems ordered by Greece in second quarter of 2006.

A Sagem/Tenix Defence/Saab Aerospace team offered a mix of Sperwer and Aerosonde to meet the Australian Defence Force's Joint Project 129 requirement for a tactical UAV to enter service in about 2007, but was unsuccessful.

Customers:

Canada: Canada became sixth customer country in 2003, following CAD33.8 million contract with Oerlikon-Contraves Inc that September for one system (four air vehicles). These deployed to Kabul, Afghanistan, at end of October 2003 with Army's 2nd Regiment of Royal Canadian Horse Artillery 'Operation Athena' support for International Security Assistance Force (ISAF). Two attrition replacement air vehicles were acquired in 2005 for a further CAD4.17 million, this pair being of similar standard to the French SDTI version. Some 107 training and operational missions were flown between November 2003 and July 2004 for the loss of two air vehicles, (21 November 2003 and 21 January 2004). The system was returned to Canada in August 2004. In December 2005, Canada was reported to be negotiating the purchase of a further five air vehicles. Ten were received from Denmark in 2006. The Canadian Armed Forces estimated that the system would need to be replaced by 2009, when deployments of Heron were scheduled to start in Afghanistan.

Denmark: Danish order was for two systems (12 air vehicles), delivered in 2001 and intended for use in southern Iraq. Danish company Terma manufactured and integrated various subsystems; UAVs were maintained for the Danish Army by the Royal Danish Air Force's Air Materiel Command, but Danish Army Sperwer operations were suspended in 2005 without being deployed operationally. Ten surviving air vehicles sold to Canada in 2006.

France: Two systems (18 air vehicles, four ground stations, two launchers and other equipment) ordered by French Army under EUR54 million contract in August 2001, for delivery to two sections of the 61st Artillery Regiment at Chaumont-Semoutier as replacement for Crecerelle. Meets SDTI requirement (*Système de Drone Tactique Interimaire*), and completed SDTI acceptance flight tests in January 2004; deliveries began later that year.

Greece: Two systems, including GCSs and a relay station for real-time data transmission, ordered by Greece in July 2002; contract value EUR35.7 million (USD34.8 million). First system delivered in late 2004, and second in early 2005; entered Greek Army service mid-2005. Two additional systems ordered in May or June 2006.

datalink. Target co-ordinates are transferred directly to an artillery network data terminal integrated in the GCS for target designation and real-time fire adjustment. Intelligence gathered is relayed to GCS, processed and analysed in real time, and transmitted to end-users via artillery or intelligence networks. Data are exchanged in accordance with NATO Adat-P3 format for interoperability with other C^3I systems.

System composition: *Sperwer:* Up to eight air vehicles and payloads, plus seven 4-tonne class trucks (three for air vehicle transportation and recovery, one each for GCS, GDT, launch catapult and maintenance unit) and two trailers for power generators and spare parts.

Ugglan: Three air vehicles; GCS and GDT integrated in small, tracked armoured vehicle; GDT antenna can be deployed up to 0.5 n mile (1 km; 0.6 mile) away from the GCS, to which it is linked by fibre optic cable.

Launch: Automatic, by pneumatic catapult.

Recovery: Parachute and triple airbag recovery system.

Power plant: One 48.5 kW (65 hp) Rotax 586 two-cylinder two-stroke engine; four-blade pusher propeller. Fuel capacity 120 litres (31.7 US gallons; 26.4 Imp gallons).

Variants: ***Sperwer A:*** Ordered for Royal Netherlands Army (launch customer); first flight (in France) in second half of 1996. Since ordered also by Canadian, Danish, French, Greek and Swedish armies. Following emergence of Sperwer B derivative, is now officially referred to as Sperwer A. Awarded certification by Netherlands MoD (extendable to other countries) in December 2000; cleared by French DGA to operate in ATC-managed national airspace. *Description applies to this version.*

Sperwer B: Improved derivative; *described separately.*

CU-161: Designation of Sperwers supplied to Canadian Army.

Tårnfalken (tower falcon): Name of Sperwers supplied to (but no longer used by) Danish Army.

Netherlands: Royal Netherlands Army order, placed in November 1995 and valued then at approximately USD81.6 million), was for four Sperwer systems (34 air vehicles and payloads). Opeval conducted at Mourmelon, France, in September 1999; certification by RNLAF resulted in temporary C of A December 2000; final qualification review December 2001; type certificate January 2002. Operator is No 101 RPV Batterie.

Sweden: Swedish Army ordered three platoon-level Ugglan systems (nine air vehicles) in June 1997; deliveries began in June 1999. Deployed at Vidsel, Älvdalen, Karvalen and Krak. Equipped for surveillance, BDA, mine detection and decoy roles. Operator training by simulator installed in RSwAF Tp 61 (Bulldog) aircraft.

Contractor: Sagem Défense Sécurité
Paris.

Sagem Sperwer B

Type: Tactical surveillance UAV.

Development: Two new variants of the Sperwer were unveiled at the June 2001 Paris Air Show: an HV (*haute vitesse* or high velocity) jet-powered variant for battlefield surveillance, target designation or EW roles, then envisaged as a potential replacement for France's CL-289 Piver system; and an LE long-endurance version for surveillance and/or communications relay, which had first flown at Mourmelon in December 2001. Brief details of these have appeared in earlier editions. The HV project has since been abandoned, but an extended capacity LE, first shown in armed configuration at the Eurosatory exhibition in June 2002, had become further modified (with assistance from Dassault) and redesignated as Sperwer B by the time of the same show in June 2004. It is reported as having flown for the first time in February 2004, and offers twice the payload and endurance of the original Sperwer, which is now officially referred to retrospectively as Sperwer A.

A one-year contract was awarded to Sagem in February 2006 to begin development work on a future generation UAV for the French Army, and the company has confirmed that it plans to propose the Sperwer B for this requirement. Sperwer B carrying Israeli Rafael Spike LR air-to-surface missiles was displayed at Paris in 2005. The initial development contract for Sperwer B was scheduled for completion in 2007 and it was reported that a further development contract would follow in 2007/08.

Description: *Airframe:* Uses the basic Sperwer fuselage, but has extended-span, strengthened, clipped double-delta wings with marked outboard dihedral, and moderately swept canard surfaces mid-mounted on a lengthened nose. Construction is generally similar to that of Sperwer A.

Mission payloads: Potential payloads include SAR/GMTI, EO/IR sensor, laser target designator, radar jammer, elint, comint or communications relay. Ku-band (15 GHz) datalink. Mode 3C transponder/IFF. In mid-2006, Sagem was also offering a medium-speed, redundant satellite llink on the Sperwer B, extending its BLOS range to 'several hundred' kilometres.

Maximum payload of the armed version is 100 kg (220 lb), of which more than 60 kg (132 lb) can be in the form of external weapons, mounted on Rafaut lightweight underwing pylons. The aircraft has been tested with TDA 'intelligent' rockets and Rafael Spike LR air-launched missiles.

Guidance and control: Autonomous/preprogrammed flight, with provision for operator override. Same common GCS, GDT, digital avionics and datalinks as Sperwer A; VHF relay with ATC. GCS can control two air vehicles simultaneously.

Launch: Fully automated catapult launch, using Sperwer/Crecerelle common launcher.

Recovery: Fully automated landing.

Sperwer B launch from an MC2555LLR at the Robonic Arctic Test UAV Flight Centre (RATUFC) at Kemijarvi, Finland *(Robonic)* 1151537

Sperwer B

Dimensions, External	
Overall, length	3.50 m (11 ft 5¾ in)
Wings, wing span	6.80 m (22 ft 3¾ in)
Weights and Loadings	
Weight, Max launch weight	350 kg (771 lb) (est)
Payload	
Max payload	
internal	35 kg (77 lb)
external	65 kg (143 lb) (est)
Performance	
Altitude, Service ceiling	6,095 m (20,000 ft)
Speed, Max level speed	81 kt (150 km/h; 93 mph)
Radius of operation, datalink	108 n miles (200 km; 124 miles)
Endurance	12 hr

Status: Development continuing in 2008. On 9 June 2006, a series of endurance flights began at the Robonic Arctic Test UAV Flight Centre (RATUFC) at Kemijarvi, Finland. Sagem also has a DGA (French procurement agency) contract for an armed configuration feasibility study.

Contractor: Sagem Défense Sécurité
Paris.

Survey-Copter Bicopt CH

Type: VTOL mini-UAV.

Development: Introduced in 2010.

Description: *Airframe:* Two-blade main and three-blade tail rotors; pod and boom fuselage; twin engines. Tall, twin-skid landing gear.

Mission payloads: Two- or three-axis, gyro-stabilised gimbal for 'plug and pay' payloads that can include T150 Visair (combined daylight + IR), T130VI, T130 IR or company's own Survey-6XL cameras. Analogue downlink for real-time imagery transmission; digital telemetry downlink.

Guidance and control: Pre-programmed and automatic via wePilot GCS-03 flight control system with AP-1000 autopilot; or under manual control with high-level speed commands. Autopilot has a built-in auto-rotation feature. GCS comprises two suitcase-size PCs and control units. Joysticks are used to fly aircraft manually and to control gimbal and cameras. GCS is coupled with datalink for real-time air vehicle tracking. System operable by a crew of two.

Launch: Automatic vertical take-off.

Recovery: Automatic vertical landing.

Power plant: Two piston engines (type and rating not stated), with optional electrical self-starting. Fuel capacity 3 litres (0.8 US gallon; 0.7 Imp gallon).

Rafael Spike missile under the wing of a Sperwer B *(IHS/Patrick Allen)* 1144931

Sperwer B on display in September 2005 *(IHS/Patrick Allen)* 1150089

Bicopt CH

Dimensions, External	
Fuselage, length	1.95 m (6 ft 4¾ in)
Rotors, rotor diameter	2.20 m (7 ft 2½ in)
Weights and Loadings	
Weight, Max T-O weight	30 kg (66 lb)
Payload, Max payload	5 kg (11.00 lb)

Unmanned aerial vehicles > France > Survey-Copter Bicopt CH – Survey-Copter Copter 1B

A Bicopt CH equipped with the T150 Visair dual-sensor turret *(Survey-Copter)* 1421307

Bicopt CH

Performance	
Altitude	
Operating altitude, typical	150 m (490 ft)
Service ceiling	2,500 m (8,200 ft)
Speed, Max level speed	0 kt (0.015 km/h; 0 mph)
Radius of operation	
mission, standard	5.4 n miles (10 km; 6 miles)
mission, optional	13.5 n miles (25 km; 15 miles)
Endurance	
standard	1 hr 15 min
optional	2 hr
Power plant	2 × piston engine

Status: Developed.
Contractor: Survey-Copter
  Pierrelatte.

Survey-Copter Blimp 37M

Type: Unmanned airship.
Development: The Blimp 37M is understood to have made its maiden flight in March 2007. Applications are said to include photography and photogrammetry; reconnaissance and surveillance of urban areas, borders and other sensitive sites; pollution monitoring; and public safety missions.
Description: *Airframe:* Helium-filled non-rigid envelope. Four rectangular tailfins, indexed in X configuration.
  Mission payloads: Daylight TV and IR cameras.
  Guidance and control: Manual guidance and tracking via digital datalink.

Blimp 37M *(Survey-Copter)* 1345391

Launch: Conventional airship take-off.
Recovery: Conventional airship landing.
Power plant: Electrical motor(s) (details not provided).

Specifications

Dimensions	
Envelope: Length	9.00 m (29 ft 6.3 in)
Max diameter	3.00 m (9 ft 10.1 in)
Weights	
Max payload	10.0 kg (22.0 lb)
Performance	
Ceiling	150 m (490 ft)
Endurance	1 h (battery pack life, 5 minutes to recharge battery pack)

Status: Production and/or service status not stated, but being promoted during 2012.
Contractor: Survey-Copter
  Pierrelatte.

Survey-Copter Copter City

Type: VTOL mini-UAV.
Development: Introduced in 2010. Details are as described for the Survey-Copter Bicopt CH, except as noted below.
Description: *Mission payloads:* Standard payload is a T180 Visair (daylight + IR) camera turret, mounted on a two-axis gimbal.
  Power plant: Two electric motors (type and rating not stated).

Copter City

Dimensions, External	
Fuselage, length	1.60 m (5 ft 3 in)
Rotors, rotor diameter	1.80 m (5 ft 10¾ in)
Weights and Loadings	
Weight, Max T-O weight	12.750 kg (28 lb)
Payload, Max payload	2 kg (4.00 lb)
Performance	
Altitude, Service ceiling	1,500 m (4,920 ft)
Radius of operation	5.4 n miles (10 km; 6 miles)
Endurance	
normal flight	40 min
hovering only	35 min
Power plant	2 × electric motor

Status: Developed and available. Being promoted in 2012.
Contractor: Survey-Copter
  Pierrelatte.

A Survey-Copter Copter City VTOL mini-UAV equipped with a T series Visair dual camera sensor *(Survey-Copter)* 1499243

Survey-Copter Copter 1B

Type: Aerial observation VTOL mini-UAV.
Development: Developed initially for close-range aerial photography; more capable later versions are for industrial and military applications. Public debut at UV '98 in Paris, May 1998.
Description: *Airframe:* Typical pod and boom configuration, with two-blade main and tail rotors; sweptback upper and lower fins at end of tailboom; skid landing gear. Sensors gyrostabilised in all axes and mounted to provide 360° pan and 95° tilt freedom of movement.
  Mission payloads: Can include CCD, zoom, IR or thermal imaging cameras; other options include pollution, temperature, radioactivity or mine detection sensors. Real-time video downlink via 5.7 GHz HF transmitter. In mid-1998, the Copter 1B was upgraded by substituting a stabilised video frame and being given improved altitude, range and endurance performance. Specific imager types associated with the Copter 1B are the T 130 V1, T-130 IR-Sf, SURVEY 2, SURVEY 3 and COOL PIX equipments.

An in-flight view of the Survey-Copter Copter 1B aerial observation VTOL mini-UAV *(Survey-Copter)* 1377461

A ventral view of the Survey-Copter Copter 1B aerial observation VTOL mini-UAV that emphasises the type's payload installation *(Survey-Copter)* 1113527

Guidance and control: Fully autonomous, with waypoint navigation. Guidance by gyroscope and orientation sensors; tracking by GPS with moving map display. Two-way datalink provides information on speed, altitude, position, engine temperature and rotor speed.
Launch: Conventional, automatic, helicopter take-off.
Recovery: Conventional, automatic, helicopter landing.
Power plant: One 26 cc single-cylinder two-stroke.

Copter 1B

Dimensions, External	
Overall, length	1.95 m (6 ft 4¾ in)
Rotors, rotor diameter	1.80 m (5 ft 10¾ in)
Weights and Loadings	
Weight, Max T-O weight	15 kg (33 lb)
Payload, Max payload	5 kg (11.00 lb)
Performance	
Altitude, Service ceiling	2,500 m (8,200 ft)
Speed, Max level speed	22 kt (40 km/h; 25 mph)
Range	5.4 n miles (10 km; 6 miles)
Endurance	45 min

Status: As of end-2012, Survey-Copter was continuing to promote the Copter 1B aerial observation VTOL mini-UAV.
Contractor: Survey-Copter
Pierrelatte.

Survey-Copter DVF 2000
Type: Close-range surveillance mini-UAV.
Development: Arising from its work in manufacturing airframes for the EADS DC Tracker, Survey-Copter began developing this broadly similar platform in 2006.
Description: *Airframe:* Straight-winged, twin-fuselage configuration with double-T tail unit and central engine and sensor pod; dihedral on outer wings. Triple landing skids under nose and at front of tailbooms. Composites construction.
Mission payloads: Daylight TV or IR camera standard, on two- or three-axis, gyrostabilised, nose-mounted, gimballed turret, with real-time imagery downlink. Alternatives can include monospectral or multispectral sensors, low-light cameras, laser designation and telemetry systems, or others such as radar or scanner at customer's request.
Guidance and control: Autopilot-based pre-programmed flight profiles, with provision to override for assisted manual mode. Suitcase-sized GCS
Launch: Hand or catapult launch.
Recovery: Automatic return to skid landing.
Power plant: Two electric motors (rating not stated), driving a single two-blade pusher propeller.

DVF 2000 with alternative 'plug-in' payload modules *(Survey-Copter)* 1345388

The hand- or catapult-launched DVF 2000 mini-UAV *(Survey-Copter)* 1345387

DVF 2000

Dimensions, External	
Overall, height	0.30 m (11¾ in)
Fuselage, length	1.20 m (3 ft 11¼ in)
Wings, wing span	3.00 m (9 ft 10 in)
Weights and Loadings	
Weight, Max launch weight	7.8 kg (17.00 lb)
Payload, Max payload	1.0 kg (2.00 lb)
Performance	
Altitude, Service ceiling	2,500 m (8,200 ft)
Speed, Cruising speed	32 kt (59 km/h; 37 mph)
Radius of operation, mission	2.7 n miles (5 km; 3 miles)
Endurance	1 hr 30 min
Power plant	2 × electric motor

Status: Total of 15 reported in service for aerial survey and video inspection with customers in France, Germany, Sweden and the Middle East in 2009.
Contractor: Survey-Copter
Pierrelatte.

Thales Spy Arrow
Type: Close-range surveillance mini-UAV.
Development: Unveiled at Eurosatory defence exhibition in Paris, June 2008. Configuration at that time was of aircraft with double-delta wings with curved tips and twin vertical tailfins. Subsequently changed to more simplified plain delta with blunt tips and single fin, improving flight stability in gust conditions and tolerance to multiple hard landings.
Description: *Airframe:* Low-mounted delta wings with elevons; single swept fin and rudder; rear-mounted engine with two-blade pusher propeller. No landing gear. Polystyrene foam construction.
Mission payloads: Standard payload is a stabilised, COTS daylight video camera with real-time video (2.4 GHz) and telemetry (869 MHz) downlinks. Options include an IR camera and atmospheric, chemical or biological sensors.
Guidance and control: Pre-programmed or manual operation, with autopilot and GPS navigation, from laptop PC ground control unit connected to a small antenna box. One-person operation; no specialised

Production Spy Arrow as shown at Le Bourget, June 2009 *(IHS/Patrick Allen)* 1380493

Unmanned aerial vehicles > France > Thales Spy Arrow – WorkFly EyesFly MV5B1

Spy Arrow hand launch *(Thales/Bernard Rousseau)* 1356242

Spy Arrow complete system *(Thales/Bernard Rousseau)* 1356243

piloting skills required. Can be flown in wind speeds of up to 19 kt (36 km/h; 22 mph).

Transportation: Man-portable.
System composition: Thought to comprise four air vehicles per system.
Launch: Hand launch under automatic electronic flight control is standard; catapult launch optional.
Recovery: Automatic to belly landing.
Power plant: Electric motor (rating not stated), driving a two-blade pusher propeller.

Spy Arrow

Dimensions, External	
Overall	
length	0.525 m (1 ft 8¾ in)
height	0.185 m (7¼ in)
Wings, wing span	0.67 m (2 ft 2½ in)
Weights and Loadings	
Weight, Max launch weight	0.5 kg (1.00 lb)
Payload, Max payload	0.07 kg (0.00 lb)
Performance	
Altitude, Operating altitude	30 m to 335 m (100 ft to 1,100 ft)
Speed	
Max level speed	54 kt (100 km/h; 62 mph)
Cruising speed	27 kt (50 km/h; 31 mph)
Radius of operation, LOS, mission	1.6 n miles (3 km; 1 miles) (est)
Endurance, max	30 min
Power plant	1 × electric motor

Status: Prototype system supplied to Direction Générale des Armements (DGA) in second quarter of 2008 for field trials by French special forces. Two other systems to French Army for operational evaluation in Afghanistan. Order for 'a small number' of production systems was reported in early 2009; first of these was due for delivery in December 2009. Reported under test in Afghanistan in 2011.

Contractor: Thales France
Elancourt.

WorkFly EyesFly MV5B1

Type: VTOL mini-UAV.
Development: WorkFly has been perfecting its 'flying saucer' unmanned system for several years. This latest version was unveiled at the Eurosatory defence exhibition in June 2010. It is conceived specifically for civil applications such as site inspection and environmental monitoring, and can be operated with the permission of the site owner without need for certification.
Description: *Airframe:* Central module housing rotor head, payload and other avionics, surrounded by doughnut-shaped cage and supported on hooped landing gear.
Mission payloads: EyesFly is a gyrostabilised platform for a video or stills camera, or other sensor specified by a customer.
Guidance and control: Automatic flight profiles, with GPS navigation. EyesFly operates on an Axon cable tether which controls its flight zone and working height. Up to six 30-minute sorties can be flown from a single battery charge.
Transportation: Man-portable.
Launch: Automatic vertical take-off.
Recovery: Automatic vertical landing.
Power plant: Battery-powered electric motor (rating not stated), driving a three-blade rotor.

EyesFly

Dimensions, External	
Overall	
height	0.60 m (1 ft 11½ in)
width	1.20 m (3 ft 11¼ in)
Weights and Loadings	
Weight, Max T-O weight	7.0 kg (15.00 lb)
Performance	
Altitude	
Operating altitude, max	150 m (500 ft)
Endurance, max	3 hr
Power plant	1 × electric motor

Status: Promoted in 2010.
Contractor: WorkFly
Neuilly-sur-Marne.

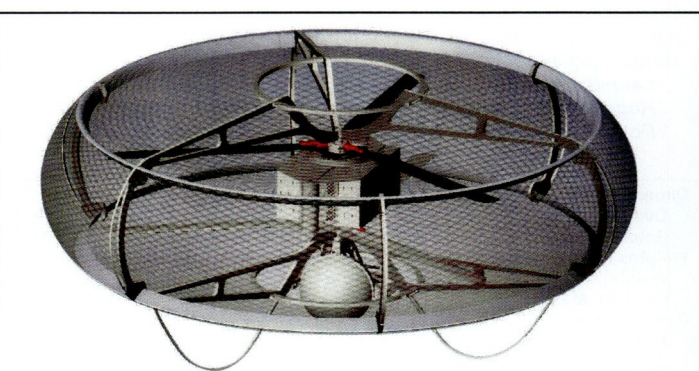

General appearance of the EyesFly MV5B1 *(WorkFly)* 1395244

EyesFly MV5B1 on show at Eurosatory, June 2010 *(IHS/Patrick Allen)* 1418827

Germany

AirRobot AR 100-B

Type: Surveillance micro-UAV.

Development: The AirRobot series has been developed as inexpensive, near-silent systems for reconnaissance, surveillance, intelligence, search and rescue, documentation and site inspection applications. It can provide airborne imaging or data transmission while hovering or flying, or from the ground while perching, and can be managed by one operator without the need for specialised piloting skills.

Description: *Airframe:* Central module contains control avionics and datalinks, with payload suspended beneath. Four booms extend at right angles from base of module, each with a two-blade rotor at their outer end. Entire apparatus is protected by an outer ring, braced to rotor booms, to avoid damage to rotors in the event of collision with an obstacle. Fixed landing gear comprises four half-hoops anchored to undersides of rotor booms.

Mission payloads: Four standard payloads available: (1) daylight colour duo video camera (wide/narrow FoV, switchable lens); (2) dawn/low light level black and white video camera; (3) 10 MP digital colour stills camera; or (4) 7 to 14 micron thermal imaging camera. All are self-contained and sealed, can be interchanged in less than 1 minute, and have a stepless mechanism allowing remotely controlled tilt of up to 100°. Real-time data control and transmission by digital HF radio link or fibre-optic cable. Analogue or digital video and telemetry downlinks. Since 2010 the payload has an antishake mechanism to facilitate stable zoom images.

Guidance and control: GCS with twin joysticks and sunlight-readable TFT for live video. Fully autonomous flight and 'hover and stare' operation using GPS or optical position lock; latter maintains UAV height and direction even in conditions where a GPS signal is not accessible. Air vehicle attitude, altitude and directional control are also stabilised electronically, backed up by a combination of gyroscopic, barometric and magnetic sensors. It can move in all directions, including sideways and backwards, and can operate in wind speeds of up to 15 kt (28 km/h; 17 mph).

Transportation: Man-portable in storm case or backpack.

System composition: One or more air vehicles, payloads and ground control unit.

Launch: Automatic vertical take-off.

Recovery: Automatic vertical landing. Can self-land autonomously when battery limit is reached or communication link is lost.

Power plant: Four brushless and gearless electric motors, each powered by a 16.8 V, 3.55 A, snap-in lithium RC battery and driving a two-blade rotor at 2,000 rpm.

AirRobot AR 100 quad-rotor micro UAV *(IHS/Patrick Allen)* 1173370

AirRobot joystick control unit *(IHS/Patrick Allen)* 1173378

AirRobot AR 100 in flight *(IHS/Patrick Allen)* 1321990

AR 100 centre body and surveillance sensor *(IHS/Patrick Allen)* 1173369

Variants: *AR 70:* Same payload capacity as AR 100, but has a three-arm rotor system similar to that of AR 150 and smaller ring diameter of 70 cm (2 ft 3.6 in). Rotor arms can be folded and locked in position for transportation. Expected to become available in fourth quarter of 2009.

AR 100-B: Pproduction version in 2009. *Description applies to this version.*

AR 150: Enlarged version, with 1.0 kg (2.2 lb) payload capacity. Three-arm rotor system, each with two two-blade coaxial and contra-rotating rotors; ring diameter 150 cm (4 ft 11.1 in). Prototype testing 2008-09, release expected February 2010.

AR 100-B

Weights and Loadings	
Weight, Max T-O weight	1.3 kg (2.00 lb)
Payload, Max payload	0.2 kg (0.00 lb)
Performance	
Altitude	
Operating altitude, typical	140 m (460 ft)
Service ceiling	1,000 m (3,280 ft)
Speed, Max level speed	15 kt (28 km/h; 17 mph)
Radius of operation	
analogue signal	0.54 n miles (1 km; miles)
digital signal	0.81 n miles (1 km; miles)
Endurance	28 min (est)
Power plant	4 × electric motor

Status: AR 100-B in production and service. Used by police, army, fire and security services. Other versions under development, as described above.

Contractor: AirRobot GmbH & Co KG
Arnsberg.

Cassidian Barracuda

Type: Experimental UCAV/URAV.

Development: The Barracuda UCAV/URAV demonstrator was publicly revealed during May 2006 and has been co-developed with Airbus Military (formerly EADS-CASA) in Spain. Programme management has been in the hands of Cassidian Air Systems (formerly the Military Air Systems business of EADS Defence and Security). In terms of purpose, Cassidian describes Barracuda as being a technology testbed that incorporates a modular structure and a flexible configuration in order to facilitate the testing of a wide variety of systems and flight profiles.

Description: Airframe: Moderately sweptback, mid-mounted wings; basically elliptical-section fuselage with hard chine along longitudinal datum; dorsal intake for turbofan engine; sweptback horizontal tail and angular, outward-canted twin fins and rudders. Aerodynamic control surfaces are actuated electromechanically; retractable tricycle landing gear has hydraulic actuation and nosewheel steering, with electrically operated carbon brakes. Carbon fibre composite construction, with the work being carried out at facilities in Augsburg and Manching (Germany) and at Getafe (Spain, wings).

Mission payloads: The Barracuda demonstrator can accommodate EO/IR imaging, laser target designation, emitter location and SAR radar equipments in its payload bay, with *IHS Jane's* sources suggesting the possibility or external stores carriage.

In in-flight view of Barracuda AV number two *(Cassidian)*

Barracuda AV number one photographed during its maiden flight on 2 April 2006 *(Cassidian)*

A display model of the Barracuda demonstrator that shows to advantage the AV's general configuration *(IHS Jane's/Patric Allen)*

Barracuda AV number two photographed during the June-July 2012 Agile UAV-NCE demonstration programme *(Cassidian)*

Guidance and control: Pre-programmed and autonomous, based on triplex flight control and navigation unit. Broad-band LOS and BLOS datalinks, secure and jamming-resistant crypto-links, MIDS/Link 16 communications and satcom. A dual KAM-500 data acquisition unit installation is used to monitor the vehicle, its sense and avoid capability, its auto taxi system and its onboard avionic buses amongst other elements. Again, the Barracuda's avionic fit is billed as being an open and modular architecture that is capable of accommodating a wide range of sensors and datalinks.

Launch: Conventional wheeled take-off.
Recovery: Conventional wheeled landing.
Power plant: One 14.19 kN (3,190 lb st) Pratt & Whitney Canada JT15D-5C non-afterburning turbofan. See under Weights for fuel details.

Barracuda

Dimensions, External	
Overall, length	8.25 m (27 ft 0¾ in)
Wings, wing span	7.22 m (23 ft 8¼ in)
Weights and Loadings	
Weight	
Weight empty	2,300 kg (5,070 lb)
Max T-O weight	3,250 kg (7,165 lb)
Fuel weight, Max fuel weight	650 kg (1,433 lb)
Payload, Max payload	300 kg (661 lb)
Performance	
Power plant	1 × turbofan

Status: Programme launched in January 2003, passing preliminary and critical design reviews in April and November of that year. Wings delivered from Spain in November 2004; first captive-carry flight, on a Dornier 228, December 2004; rolled out at Augsburg 1 March 2005; second captive-carry flight at Manching in April 2005, and third at San Javier air base in Spain December 2005. Taxi tests at Manching began 6 January 2006; 20-minute first flight, at San Javier air base, took place on 2 April 2006.

Initial phase of flight test programme completed by May 2006. The air vehicle crashed during later tests off the Spanish coast. A second prototype (99+91) was built, flying for the first time in Goose Bay, Canada, in July 2009. Four flights were made before the aircraft was shipped to Germany to await further tests. A further series of four flights were completed at Goose Bay in 2010. The second airframe was identical to the first vehicle, but the software and datalink arrangements were revised. In more detail, the then EADS noted that Barracuda number two would be all electric, with the hydraulic gear retraction and brakes fitted to the first Barracuda replaced by electrical systems from Liebherr and Dunlop respectively.

Moving forward, Cassidian undertook a serial of five Barracuda sorties during June and July 2012, with the effort being designed to demonstrate operations using multiple networked UAVs and the autonomous distribution of roles between UAVs in complex mission scenarios. Carried out at Goose Bay in Canada, the serial was designated as the Agile UAV in a Network Centric Environment (Agile UAV-NCE) programme and involved both the surviving Barracuda AV and a Learjet business aircraft that was modified to act as a surrogate UAV. During the trial, the two AVs flew missions during which, each AV had its own role and was autonomously co-ordinated and synchronised with its partner. Role distribution was pre-defined, with co-ordination between the AVs being largely automatic. The effort further demonstrated the capability to upload new waypoints and mission segments during individual sorties. For its part, the Barracuda flew completely autonomously using pre-programmed flight profiles that included auto-taxiing. Ground station monitoring of both UAVs was in a safety context only and Cassidian claims that Agile UAV-NCE represented the first co-ordinated trial of two jet-powered UAVs where the AVs were tasked with differing roles and engaged different targets.

Contractor: Cassidian, Friedrichshafen.

Cassidian (EADS DS) Talarion

Type: HALE UAV.

Development: Previously known simply as 'Advanced UAV', this European UAS originated as an ISTAR successor to the aborted Euromale programme. It was named Talarion (after the winged sandals of Hermes in Greek mythology) in May 2009, shortly before a full-size mockup was unveiled at the Paris Air Show. At that time, few specification details had been released and a development contract was still awaited, but this was expected to be awarded in the near future.

Acting through the German BWB (Federal Office of Defence Technology and Procurement), the defence ministries of France, Germany and Spain awarded EADS a EUR60 million, 15-month risk-reduction study contract for a high-performance UAS in December 2007, and this was undertaken by the Military Air Systems business unit of EADS Defence and Security. Its stated objective was "to consolidate the capability requirements for Surveillance and Reconnaissance (SR) and Fast Reconnaissance (FR), and to propose appropriate technical solutions". To this end, EADS proposed a modular approach and a two-airframe solution: a long-span, long-stay SR version for medium to

Full-size Talarion SR mockup at the Paris Air Show, June 2009 *(IHS/Patrick Allen)*

high altitude operations and a short-wing FR version for use at lower levels. Separately, within the study framework, Thales (France), Indra (Spain) and EADS DS's Defence Electronics unit reviewed aspects of the radar payload and the future joint development and integration of a SAR for the UAV.

Successful conclusion of the risk-reduction study was announced on 28 May 2009, enabling details to be revealed during the Paris Air Show the following month. EADS Defense and Security division became Cassidian from September 2010. Cassidian state Talarion is intended "to be the first unmanned aerial system that will operate in civil airspace". The twin-engine configuration may be intended to satisfy future airworthiness requirements.

Description: *Airframe:* Shoulder-mounted wings (long- or short-span, according to version); bulged over-nose radome on SR version, housing satcom antenna and other electronics; twin jet engines in rear-mounted nacelles; swept fin and low-set tailplane. Retractable tricycle landing gear. Composites construction.

Mission payloads: Ventral bay amidships for SAR/GMTI or maritime radar module. forward of this, an EO/IR/laser rangefinder-designator

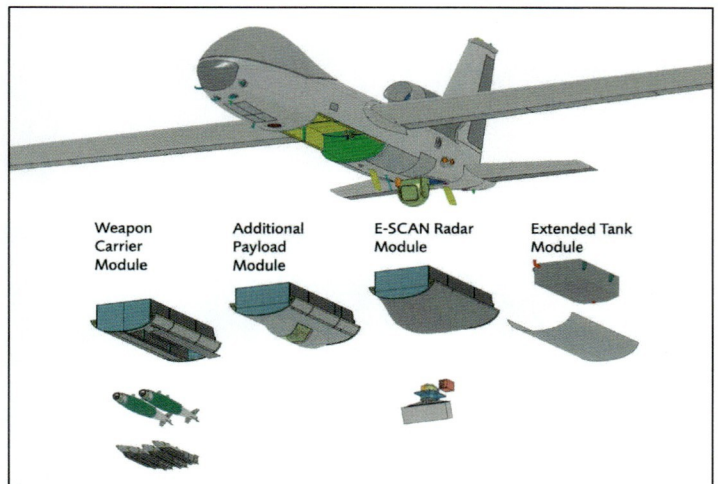
The Talarion UAV structure showing its payload bay and four payload module options *(EADS)*

Full-size mock-up of the EADS Talarion UAV (front three-quarter close-up) under development for France, Germany and Spain *(EADS Defence and Security/Military Air Systems)*

EADS 'Advanced UAV' graphic illustrating both the long-span SR and (below) short-span FR versions of the Talarion *(EADS)*

sensor turret (retractable on SR version). Also reported to have other side- and rear-mounted electronic surveillance sensors. Specific equipments not yet identified.

Guidance and control: No details announced.
System composition: Proposal is based on three air vehicles and one ground segment per system.
Launch: Conventional, automatic wheeled take-off.
Recovery: Conventional, automatic wheeled landing.
Power plant: Two gas turbine engines, dorsally mounted on shoulders of rear fuselage. As of July 2009, preferred choice was reported to be the 8.45 kN (1,900 lb st) Williams FJ33-5A turbofan.

Talarion

Dimensions, External	
Wings, wing span	27.90 m (91 ft 6½ in)
Weights and Loadings	
Weight, Max T-O weight	7,000 kg (15,432 lb)
Payload, Max payload	680 kg (1,499 lb) (est)
Performance	
Altitude, Service ceiling	14,000 m (45,940 ft)
Speed, Max level speed	200 kt (370 km/h; 230 mph)
Endurance, max	20 hr (est)
Power plant	2 × turbofan

Status: EADS' original proposal, valued at development costs of EUR1.5 billion in 2009, was based on 15 three-aircraft systems: six each for France and Germany plus three for Spain. This was awaiting development approval by the three governments in 2011. In May 2011 Turkey joined the programme, with TAI Turkish Aerospace Industries signing an MoU as an industrial partner. First flight of the Talarion prototype scheduled for 2014.

Contractor: Cassidian (an EADS company).
Munich, Germany.

EMT Aladin

Type: Close-range battlefield mini-UAV.
Development: Designed to meet requirement announced by German Army in May 2000; first of two prototypes flown in mid-2000; selected as BWB competition winner in July 2000 and development contract awarded in March 2001. Intended for reconnaissance and target location and identification.
Description: Airframe: Parasol-mounted wings, outer panels of which have dihedral and upturned tips; slender boom fuselage; T tail unit. Mission payload pod is pylon-mounted beneath fuselage. No landing gear.
Mission payloads: Choice of four colour daylight video cameras with different angles of view, or an IR sensor, with C-band real-time video/telemetry datalink.
Guidance and control: Autonomous or manual control via digital autopilot, with automatic terrain avoidance. Miniaturised, 17 kg (37.5 lb) man-portable (backpackable) mission planning and flight control station, with 2-D or 3-D digital map display and image evaluation and storage facilities. Flight path can be updated during mission via UHF command uplink.
Transportation: Complete system is transportable in a number of vehicles, including the standard Fennek 4 × 4 off-road vehicle; air vehicle can be dismantled and stowed in a 62 × 22 × 47 cm (24.4 × 8.7 × 18.5 in) box.
System composition: Two or three air vehicles, one GCS and two operators. Set-up time approximately five minutes.
Launch: Hand- or bungee-launched.
Recovery: Autonomous, deep-stall belly landing at preselected location.
Power plant: Battery-powered electric motor; two-blade propeller.

Aladin hand launch and ground control

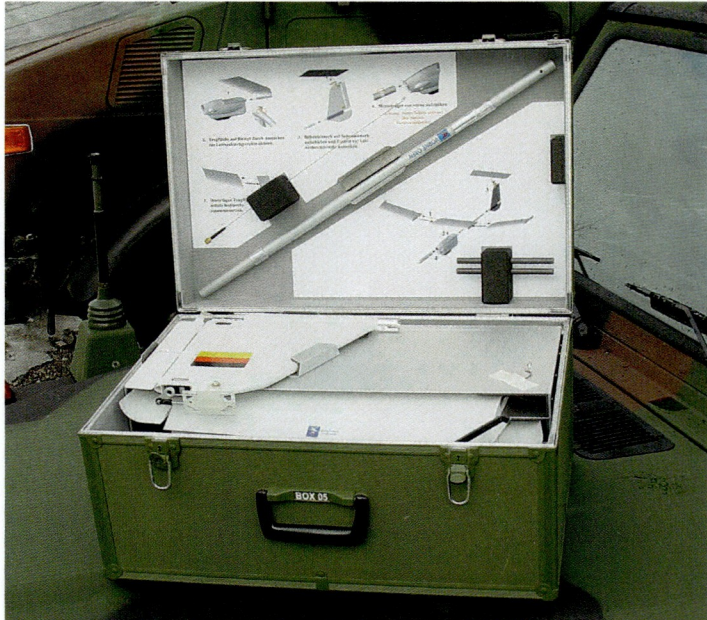

Aladin packed in its box *(EMT)*

The complete Aladin system

EMT Aladin – EMT LUNA < Germany < Unmanned aerial vehicles

Aladin in flight *(EMT)* 0580985

Aladin

Dimensions, External
- Overall
  - length .... 1.50 m (4 ft 11 in)
  - height .... 0.32 m (1 ft 0½ in)
- Wings, wing span .... 1.46 m (4 ft 9½ in)

Weights and Loadings
- Weight, Max launch weight .... 3.0 kg (6.00 lb) (est)

Performance
- Altitude, Operating altitude .... 30 m to 200 m (100 ft to 660 ft)
- Radius of operation
  - datalink, mission .... 2.7 n miles (5 km; 3 miles)
- Endurance, standard batteries .... 1 hr (est)
- Power plant .... 1 × electric motor

Status: In service. Six systems ordered in July 2002 and delivered to armoured reconnaissance units of German Army in March 2003, for experimental deployment with contingent of International Security Assistance Force (ISAF) in Afghanistan, where they flew 71 sorties for the loss of five aircraft.

By April 2005, EMT had received a contract increasing the total German Army order to 121 systems, some equipped for night reconnaissance, with deliveries starting in August 2005. By April 2010 more than 6500 missions had been flown by the German Army.

Five systems were supplied to the Royal Netherlands Army, these formed part of that country's contribution to ISAF in Afghanistan in 2007. The Norwegian Army ordered one Aladin system for evaluation in September 2006 trials at Setermoen by its Norwegian Battle Lab and Experimentation (NOBLE) unit.

Customers: German Army; Netherlands Army (now withdrawn); Norwegian Armed Forces (evaluation), German Federal Police.

Contractor: EMT Ingenieurgesellschaft
Penzberg.

EMT FanCopter

Type: Urban reconnaissance and surveillance mini-UAV.
Development: Began in first half of 2004. Version with a much simplified structure emerged 2006.
Description: *Airframe:* Two coaxial main rotors, three outrigged steering rotors.
Mission payloads: Stabilised pan-tilt payload. Daylight video camera, low-light video camera, video zoom camera, high-resolution stills camera

FanCopter in working environment *(EMT)* 1416236

or thermal imager. Optical flow sensor with ground view camera. Further payloads in development.
Guidance and control: Miniaturised, portable control station, with datalink for real-time control and video transmission.
Launch: Autonomous vertical take-off.
Recovery: Autonomous vertical landing.
Power plant: Battery-powered brushless electric motors.

FanCopter

Dimensions, External
- Rotors, rotor diameter .... 0.60 m (1 ft 11½ in) (est)

Weights and Loadings
- Weight, Max T-O weight .... 1.5 kg (3.00 lb) (est)

Performance
- Radius of operation .... 0.8099 n miles (1 km; miles) (est)
- Endurance
  - observation time, perch and stare .... 3 hr (est)

Status: Under development in 2004-06, in production from 2007. Two pre-production systems delivered to the German Armed Forces in 2006-2007. A production order for 19 systems was placed in October 2008. Orders to South Africa (2007) and USA followed (2009).

Contractor: EMT Ingenieurgesellschaft
Penzberg.

EMT LUNA

Type: Short-range battlefield RSTA UAV.
Development: The LUNA takes its name from the German Army's *Luftgestützte Unbemannte Nahaufklärungs Ausstattung* (airborne unmanned close reconnaissance system) programme for a close-range, brigade-level system for introduction at the beginning of the 21st century.

Seven manufacturing teams (four German, two French and one from the UK) took part in a hardware demonstration at the Wildflecken range in Germany in August 1996; an eighth (the TechMent Midget RPG) was demonstrated in Sweden. The BWB (German MoD) issued RFPs in March 1997 with the original intention of down-selecting two of these systems for

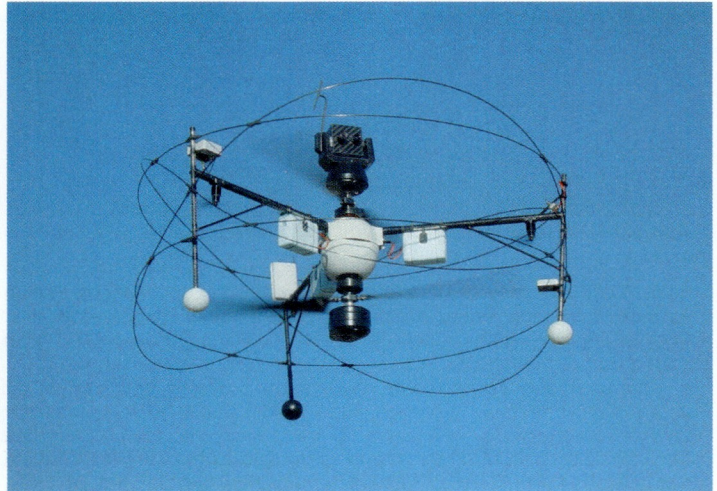
EMT FanCopter 2010 configuration *(EMT)* 1416237

Detail of engine and folded propeller blades *(Paul Jackson)* 0024430

Unmanned aerial vehicles > Germany > **EMT LUNA**

LUNA GCS during exercises in Norway, mounted on an armoured, air-transportable DURO 6 × 6 all-terrain vehicle *(EMT)* 0580993

German Army LUNA in service near Kabul, 2003 *(EMT)* 0563300

LUNA payload bay with daylight colour video and IR cameras *(EMT)* 0054211

Demonstrating LUNA's man-portability *(EMT)* 0580996

Readying LUNA for a launch near Kabul in 2003 *(EMT)* 0580988

The short-range LUNA in flight *(EMT)* 0580989

A pair of LUNAs stowed for transportation *(EMT)* 0580990

a final fly-off. However, in October 1997 it named EMT as the sole development contract winner.

The LUNA X-2000 prototype, which first flew in September 1996, is essentially an unmanned motor glider, able to turn off its engine for silent operation near and over a target, restarting it once the UAV is out of aural range. It made its public debut at the Berlin Air Show in May 1998.

It is the only UAV in use in Germany with Cat. II certification (authorised to fly in civil airspace over thinly populated areas).

One LUNA was employed in late 2003 as testbed for the EADS Dornier MiSAR miniature radar.

Description: *Airframe:* High-aspect ratio shoulder-wing monoplane; cruciform tail unit with dependent auxiliary fins and rudders; dorsally mounted pusher engine. Glass fibre epoxy composites construction.

Mission payloads: Daylight CCD colour TV camera, with zoom, , 1.7 kg (3.75 lb) thermal imager in ventral bay; nose-mounted colour camera for piloting. EADS MiSAR (miniature SAR) successfully tested on board a LUNA in early 2004. Relay module for using LUNA as relay station for reconnaissance drones. Real-time, eight-channel imagery downlink. Optional onboard imagery storage. Onboard electrical power 200 W from battery-backed generator.

Guidance and control: Preprogrammed powered or gliding flight and/or radio-command guidance via autopilot; DGPS navigation and/or datalink autotracking, with back-up dead-reckoning. Real-time data transmission. HF (5 MHz) command uplink and eight-channel UHF data downlink; G-band (5 GHz) tracking antenna in vehicle and at GCS. Automatic recontacting after contact loss. Return to base function.

EMT LUNA – Mavionics Carolo < Germany < Unmanned aerial vehicles

EADS MiSAR radar on the LUNA *(EADS)* 0580855

System composition: Ten air vehicles, two launch catapults and two vehicle-mounted GCS; operating crew of 14. Turnaround time less than 15 minutes.

Launch: From 9 m (29.5 ft) rail by EMT bungee catapult, foldable for transport.

Recovery: Mobile net recovery is standard to allow landing in difficult or mine-infested terrain, alternative parachute recovery system.

Power plant: One 6.0 kW (8.0 hp) two-cylinder two-stroke engine with restart capability; three-blade pusher propeller with folding blades. Fuel capacity 3 litres (0.8 US gallon; 0.7 Imp gallon).

LUNA

Dimensions, External	
Overall	
length	2.28 m (7 ft 5¾ in)
height	0.78 m (2 ft 6¾ in)
Wings, wing span	4.17 m (13 ft 8¼ in)
Engines, propeller diameter	0.56 m (1 ft 10 in)
Dimensions, Internal	
Payload bay, volume	0.01 m³ (0.4 cu ft)
Weights and Loadings	
Weight	
Weight empty	20 kg (44 lb)
Max launch weight	40 kg (88 lb) (est)
Payload, Max payload	5 kg (11.00 lb)
Performance	
Climb	
Rate of climb, max, at S/L	300 m/min (984 ft/min)
Altitude	
Operating altitude, max	500 m (1,640 ft)
Service ceiling	3,000 m (9,840 ft)
Speed	
Max level speed, IAS	70 kt (130 km/h; 81 mph)
Cruising speed, normal, IAS	38 kt (70 km/h; 44 mph)
Loitering speed, IAS	26 kt (48 km/h; 30 mph)
Radius of operation	
datalink	54 n miles (100 km; 62 miles) (est)
OFF-LINE	54 n miles (100 km; 62 miles) (est)
Endurance	5 hr (est)
Best glide ratio	18

Status: Ten development aircraft ordered in October 1997, including four for end-user's trials programme. Some of these were deployed to Prizren, Kosovo, on 27 March 2000 and from 3 April flew 176 missions during a two-month period of trials. Further orders since then. By May 2004, LUNAs with KFOR had flown more than 600 sorties in Kosovo and Macedonia. A detachment of two systems (20 air vehicles) was deployed in April 2003 with the International Security Assistance Force (ISAF) in Afghanistan, where it is still in operation. More than 6,000 operational flights have been logged in the Balkans and Afghanistan by early 2010. Four additional LUNA systems were ordered by Germany in July 2009 for delivery by 2010. This contract covered 40 air vehicles, eight ground control stations, eight launchers and eight net landing systems All systems are carried by mine protected vehicles.

An order from the Pakistan Army for three systems was reported in March 2006.

Customers: German Army; Pakistan Army.

Contractor: EMT Ingenieurgesellschaft
Penzberg.

EMT Museco

Type: Helicopter UAV.

Development: Revealed in late 2008. Designed for land and sea-based, day and night operations including reconnaissance, surveillance, radio relay or rescue mission support.

Impression of the EMT Museco *(EMT)* 1290316

Description: *Airframe:* Utilises the Swiss UAV Neo S-300 (which see) as the air vehicle.

Mission payloads: Standard sensor is a three-axis stabilised, swivelling, modular EO platform with ×26 zoom. Options include high-resolution IR camera; digital stills camera; meteorological sensors; gas and particle samplers; radio relay; minesweeping sensors; and radioactive contamination sensors. Sensor module equipped with gyro, rate sensors, magnetic compass, air data sensors and accelerometers. Duplex microwave datalink (transmission range more than 100 km; 62 miles), with directional antennas in aircraft and GCS.

Guidance and control: Modular, compact GCS. Integrated mission planning and flight control system utilises the operationally proven GCS avionics of the EMT LUNA fixed-wing UAS, having several protected workstations equipped with 'virtual cockpit' aircraft control; high-resolution colour monitors for real-time image evaluation and 3-D mission planning; and a wide range of available software for data evaluation, depending upon customer's requirements. GCS also permits repeated replays of entire flight mission for post-flight evaluation, simulation and training. It can be easily integrated into both ships and land vehicles. Navigation, autopilot and aircraft system management are all fully digital.

Launch: Conventional helicopter take-off.

Recovery: Conventional helicopter landing.

Power plant: One 14 kw (18.8 shp) unidentified heavy-fuel turboshaft.

Museco

Dimensions, External	
Overall	
length	2.75 m (9 ft 0¼ in)
height	0.95 m (3 ft 1½ in)
Fuselage	
width, max	0.56 m (1 ft 10 in)
Skids, skid track	0.85 m (2 ft 9½ in)
Rotors, rotor diameter	3.00 m (9 ft 10 in)
Tail rotor, tail rotor diameter	0.65 m (2 ft 1½ in)
Weights and Loadings	
Weight, Max T-O weight	75.0 kg (165 lb)
Payload, Max payload	25.0 kg (55 lb)
Performance	
Endurance	2 hr (est)
Power plant	1 × turboshaft

Status: Under development in 2008-09, unknown as of end of 2012.

Contractor: EMT Ingenieurgesellschaft
Penzberg.

Mavionics Carolo

Type: Research mini-UAV.

Development: The Institute of Aerospace Systems at the Technical University of Braunschweig has, over time, developed a number of micro- and mini-UAV prototypes, including the Carolo family. Mavionics was formed in 2004 as a spin-off business unit to market selected designs commercially, utilising its own autopilot and AV flight control systems. Earlier Carolo models, of various design configurations, included the C40, P50, P70, P200 and T140.

As of 2010, the versions being promoted were the T200 and P330.

Description: *Airframe:* As described under Variants below. Composites construction.

Mission payloads: T200: Meteorological sensor, developed by Braunschweig University Institute of Aerospace Systems, for wind vector, temperature and humidity research within lower atmosphere. Standard 868 MHz (optionally 900 MHz) telemetry downlink.

P330: Standard payload is a digital photo camera.

Guidance and control: Autonomous or remotely controlled. Mavionics MINC autopilot system, with GPS/INS navigation. Desktop or laptop PC with MAVCDesk GCS software

System composition: Air vehicle, datalink, GCS software and remote control (last-named for manual flight and as back-up during automatic mode).

Launch: Bungee catapult or hand launch (manual control standard, automatic optional).

Recovery: Belly landing (manual control standard, automatic optional).

Power plant: T200: Two wing-mounted brushless electric motors, powered by lithium-polymer batteries; two-blade tractor propellers.

The twin-engined Carolo T200 with a nose-mounted meteorological sensor *(Mavionics)* 1290334

General view of the MD4-200 configuration *(IHS/Patrick Allen)* 1326911

P330: One nose-mounted brushless electric motor; two-blade tractor propeller.

Variants: Carolo T200: Twin-engined, high-wing monoplane; rectangular-section fuselage; T tail. No landing gear. Also available in single-engined form as P200.

Carolo P330: Single-engined, high-wing sailplane/motor glider configuration, with T tail. No landing gear.

Carolo T200, Carolo P330

Dimensions, External	
Overall	
length, Carolo T200	1.80 m (5 ft 10¾ in)
Wings	
wing span	
Carolo T200	2.00 m (6 ft 6¾ in)
Carolo P330	3.30 m (10 ft 10 in)
Weights and Loadings	
Weight	
Max launch weight	
Carolo T200	5.6 kg (12.00 lb)
Carolo P330	5.0 kg (11.00 lb)
Payload	
Max payload	
Carolo T200	1.0 kg (2.00 lb)
Carolo P330	0.4 kg (0.00 lb)
Performance	
Altitude	
Operating altitude, Carolo T200	700 m (2,300 ft)
Speed	
Cruising speed	
normal, Carolo T200	39 kt (72 km/h; 45 mph)
normal, Carolo P330	54 kt (100 km/h; 62 mph)
Radius of operation	
mission, Carolo T200	1.3 n miles (2 km; 1 miles)
Endurance	
max, Carolo T200	45 min (est)
max, Carolo P330	1 hr
Power plant	
Carolo T200	2 × electric motor
Carolo P330	1 × electric motor

Status: Developed and available. Carolo T200 use has included deployment with the British Antarctic Survey from December 2006 to January 2008. It has also been used in programmes with clients in Norway and Spain.

Contractor: Mavionics GmbH
Braunschweig.

MD4-200 in flight *(IHS/Patrick Allen)* 1326530

Microdrones MD4 series

Type: VTOL micro-UAV.

Description: Airframe: Central, hemispherical module of moulded carbon fibre housing propulsion unit, payload and avionics. From this radiate four arms (two of which are optionally foldable), each supporting a two-blade rotor. Twin-skid landing gear.

Mission payloads: Standard sensors include black and white camera, colour video camera or 12 MP digital photo camera. MD4-200 has 2.4 GHz encrypted imagery downlink and 35 MHz telemetry datalink, and can carry day camera, thermal imager or other sensor such as jammer or air sampler. MD4-1000 utilises Coded Orthogonal Frequency Division Multiplexing (COFDM) wideband digital communications for both telemetry and downlink and can carry sensor alternatives in addition to camera.

Guidance and control: Autonomous or remotely operated flight profiles. Position holding using GPS in MD4-200, or DGPS/INS in MD4-1000. Former can 'perch' tethered, latter untethered. Ground station includes video receiver, directional antenna (optional) and 12 or 230 V power supply, housed in a strong, air- and water-tight wheeled case with fold-down handles. Optional additions are notebook PC to record missions and copy data to CD or DVD; and videoglasses for flights beyond line of sight.

Transportation: Man-portable.
Launch: Vertical take-off.
Recovery: Vertical landing.

The larger MD4-1000, with MD4-200 in the foreground *(Microdrones)* 1356220

Power plant: Four 250 W brushless electric motors, each powered by a 14.8 V, 2,300 mAh lithium-polymer battery and driving a two-blade carbon fibre rotor.

Variants: *MD4-200:* Current baseline production version.
MD4-1000: Much-enlarged version.

MD4-200, MD4-1000

Dimensions, External	
Rotors	
distance between rotor centres	
MD4-200 Seeker 400	0.7 m (2 ft 3½ in) (est)
rotor diameter	
each, MD4-200 Seeker 400	0.37 m (1 ft 2½ in)
Weights and Loadings	
Weight	
Weight empty	
MD4-200 Seeker 400	0.90 kg (1.00 lb)
MD4-1000	3.8 kg (8.00 lb)
Max T-O weight	
MD4-200 Seeker 400	1.0 kg (2.00 lb)
Payload	
Max payload	
MD4-200 Seeker 400	0.25 kg (0.00 lb)
MD4-1000	1.2 kg (2.00 lb)
Performance	
Altitude	
Operating altitude	
typical, MD4-200 Seeker 400	150 m (490 ft) (est)
Service ceiling	
MD4-200 Seeker 400	3,000 m (9,840 ft) (est)
Radius of operation	
LOS, mission, MD4-200 Seeker 400	0.5 n miles (km; miles) (est)
NLOS, mission, MD4-1000	2.7 n miles (5 km; 3 miles) (est)
Endurance	
max, MD4-200 Seeker 400	30 min
max, MD4-1000	1 hr
Power plant	4 × electric motor

Status: MD4-200 in production from April 2006. In service with worldwide customers, including Merseyside Police, British Transport Police and West Midlands Fire Service in the UK.

Contractor: Microdrones GmbH
Siegen.

Rheinmetall KZO

Type: RSTA and BDA tactical UAV.

Development: The German Army's KZO (*Kleinfluggerät Zielortung*: small air vehicle for target location) is a derivative of the former MBB (later STN Atlas) Tucan (Toucan) series of experimental UAVs. It was previously known as Brevel.

Developed originally to support long-range artillery systems such as the PzH 2000 self-propelled howitzer, KZO is designed for brigade level, real-time, day and night observation of enemy forces and post-strike BDA at distances of up to 81 n miles (150 km; 93 miles).

Description: *Airframe:* Small, stealthy low-wing monoplane; no horizontal tail surfaces; pusher engine. Wings fold for container storage and transportation, and incorporate hot-air anti-icing of the leading-edge flaps. Construction of composites. No landing gear.

Mission payloads: Nose configured for installation of various modular payloads. Known examples include Carl Zeiss OPHELIOS 8 to 12 micron stabilised day/night (FLIR) sensor or synthetic aperture radar with ×8 zoom; optional playback recorder for deferred in-flight transmission, allowing data storage of up to 10 minutes of video footage when real-time transmission is not possible. Sensors and real-time data transmission links are highly resistant to jamming, and are capable of transmitting up to 54 n miles (100 km; 62 miles) through dense jamming, and up to 81 n miles (150 km; 93 miles) in more favourable conditions. Provision for laser range-finder/designator.

Guidance and control: GCS in 4.6 m (15 ft) long, NBC- and EMP-protected shelter with C^3I links to Adler terminal; three workstations, computer-aided for mission planning, flight monitoring and image/target evaluation. KZO flies preprogrammed flight and mission profiles, including automated launch and recovery sequences, but flight path can be altered from GCS during mission; control of KZO can also be handed over to another GCS during mission. Although equipped with a GPS

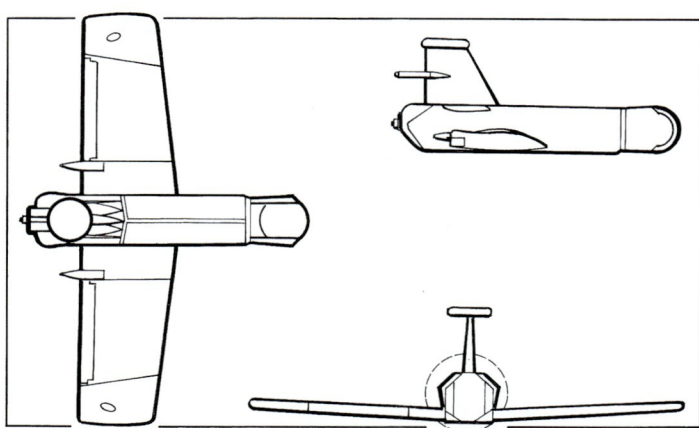
KZO general arrangement *(IHS/John W Wood)*   0518061

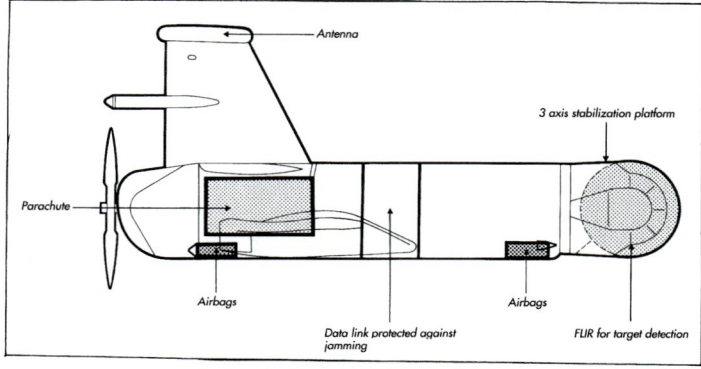
KZO internal features   0518062

Impression of the KZO in flight *(Rheinmetall)*   1122541

KZO deliveries to the Bundeswehr began in November 2005 *(Rheinmetall)*   1122542

KZO in German Army service *(Rheinmetall)*   1122516

Rheinmetall KZO

KZO parachute/airbag recovery 0518227

Recovery: Parachute and airbag recovery system. Required recovery area 200 × 200 m (660 × 660 ft).
Power plant: One 23.9 kW (32 hp) Fichtel & Sachs two-cylinder two-stroke engine; two-blade pusher propeller.
Variants: KZO: Version for the German Army. *Detailed description applies to this version.*
ECM version: Prospective version for jamming of VHF and UHF communications in the 20 to 500 MHz frequency bands.
ESM version: Prospective stand-off ESM and sigint version for communications and radar systems.
Prospector: Name announced in November 2004 for version of KZO bid for US Army Future Combat System (FCS) Class III UAV requirement, in a teaming arrangement with Teledyne Brown Engineering (TBE); USD3.7 million, 10-month Phase 1 (requirements assessment and risk reduction) contract from Boeing to TBE as one of three Class III finalists announced 25 August 2005.. Tail unit modified to inverted V configuration, said to improve lift and endurance; more fuel in extended fuselage; increased wing span; payload increased to 113 kg (250 lb). This FCS requirement was deferred during 2007.

KZO

Dimensions, External	
Overall	
length	2.26 m (7 ft 5 in)
height	0.96 m (3 ft 1¾ in)
Fuselage	
height	0.36 m (1 ft 2¼ in)
width	0.36 m (1 ft 2¼ in)
Wings, wing span	3.42 m (11 ft 2¾ in)
Weights and Loadings	
Weight, Max launch weight	161 kg (354 lb)
Payload, Max payload	35 kg (77 lb)
Performance	
Altitude	
Operating altitude	300 m to 3,500 m (980 ft to 11,480 ft)
Service ceiling	4,000 m (13,120 ft)
Speed	
Max level speed	118 kt (219 km/h; 136 mph)
Cruising speed, operating	81 kt (150 km/h; 93 mph)
Loitering speed	65 kt (120 km/h; 75 mph)
Radius of operation, datalink	81 n miles (150 km; 93 miles)
Endurance	3 hr 30 min

system, KZO normally navigates by means of position measuring and automatic comparison of the aerial view with the pre-programmed digital map in the GCS, enabling it to avoid the susceptibility of GPS to disruption and ECM. EADS Defence Electronics jamming-resistant Ku-band single command datalink transmits via fibre optic cable to GDT; separate TV and telemetry downlinks. GCS receives and processes downlinked imagery and transmits information to command headquarters. Two air vehicles, in different mission modes, can be controlled simultaneously.
System composition: Rapidly deployable mobile batteries comprising 10 air vehicles per system, plus vehicle-mounted GCS; truck-mounted GDT; launch vehicle; recovery vehicle; maintenance and fuelling vehicle. Set-up time less than 30 minutes.
Launch: By jettisonable booster rocket directly from a container mounted on a flatbed military truck. Required launch area 100 × 100 m (330 × 330 ft). Can be launched in winds of up to 29 kt (54 km/h; 34 mph), regardless of wind direction. Can also be launched by catapult.

Status: Flight testing of definitive prototypes began in October 1994 at the German Army test centre in Meppen; payload flight trials started in mid-1996; operator training began in June 1997. A five-aircraft system was delivered for operational evaluation in April 1998; nearly 200 flights had been made by mid-1999. The first production contract, then valued at DM600 million, was awarded on 19 June 1998, calling for six KZO systems (60 air vehicles, 12 GCSs and accompanying ground support equipment). Deliveries of production systems began on 28 November 2005 and were scheduled for completion by the end of 2007.
As of early 2006, the first system was being used by training units and the second was awaiting deployment to Afghanistan. KZO was deployed in Afghanistan during 2009.
Customers: German Army (KZO). First system divided between artillery school at Idar/Oberstein and technical school at Aachen; second system equipped Artillery Battalion 71.
KZO is being promoted in Poland, with Rheinmetall teaming with BUMAR Group's PIT to offer KZO to the Polish Armed Forces.
Contractor: Rheinmetall Defence Electronics GmbH
Bremen.

Greece

EADS 3 Sigma Nearchos

Type: Multirole UAV.
Development: The Nearchos was designed for medium-range, medium-endurance, high-payload capacity, and navigational targeting accuracy. Potential applications include aerial reconnaissance, battlefield surveillance, ESM/ECM, target acquisition, BDA, communications data relay, traffic surveillance, forest fire detection, boundary and forestry patrol, geological and oceanographic applications, and monitoring of pollution and natural disaster situations.
Description: Airframe: Pod and twin tailboom configuration with high-mounted, non-swept wings; pusher engine; fixed tricycle landing gear.
Mission payloads: Typical payloads can include a stabilised modular platform, low-resolution TV camera and ECM/ESM equipment. Optional other payloads include a high-resolution TV camera, high-speed cameras, FLIR, video recorder, thermal imaging system, and a laser range-finder and targeting system.
Guidance and control: The Eniochos ground navigation control (GNC) is capable of operating up to four air vehicles simultaneously, each one autonomously or each linked to the others. Each air vehicle is monitored constantly and controlled from the GCS; data are downlinked continuously by all air vehicles.

Nearchos multirole UAV (EADS 3 Sigma) 1136389

Launch: Conventional wheeled take-off or catapult.
Recovery: Conventional wheeled landing; parachute recovery optional. The parachute recovery system is energised by the onboard flight termination system or by a preprogrammed or operator command.
Power plant: One 28.3 kW (38 hp) UEL AR 741 rotary engine, driving a pusher propeller. Fuel capacity up to 100 litres (26.4 US gallons; 22.0 Imp gallons).

Nearchos

Dimensions, External
Overall
  length..3.95 m (12 ft 11½ in)
  height
    excl landing gear..0.52 m (1 ft 8½ in)
    incl landing gear..1.15 m (3 ft 9¼ in)
Fuselage, width...0.43 m (1 ft 5 in)
Wings, wing span...5.10 m (16 ft 8¾ in)
Areas
Wings, Gross wing area..2.95 m² (31.8 sq ft)
Weights and Loadings
Weight
  Weight empty...60 kg (132 lb)
  Max launch weight
    catapult launch..150 kg (330 lb)
    wheeled T-O..190 kg (418 lb)
Performance
Altitude, Service ceiling...7,000 m (22,965 ft)
Speed, Max level speed...119 kt (220 km/h; 137 mph)

Status: In service, notably as a research platform in programmes co-funded by the Greek Ministry of Development and the General Secretariat of Research and Technology. Also, and in collaboration with the Technical University of Crete and the National Technical University of Athens, Nearchos has been used as the platform for the development of a collision avoidance system and a fire detection system for UAVs.

Contractor: EADS 3 Sigma SA
Chania, Crete.

Nearchos landing (EADS 3 Sigma)

India

ADE Imperial Eagle

Type: Close-range mini-UAV.
Development: The Imperial Eagle made its public debut at Aero India, February 2011. Its development history was not disclosed.
Description: *Airframe:* Shoulder-wing design, with outer panel dihedral; pod and boom fuselage, with dorsally mounted pusher engine; conventional tail unit. No landing gear.
  Mission payloads: Nose-mounted daylight colour TV camera with ×10 optical and ×4 digital zoom; or Miricle 307 KS uncooled thermal imager with 39° field of view. S-band telemetry and imagery downlink.
  Guidance and control: Fully automatic via autopilot and UHF command uplink; GPS-based navigation and tracking.
  Transportation: Man-portable.
  Launch: Hand launch.
  Recovery: Soft belly landing.
  Power plant: Battery-powered electric motor (rating not stated); two-blade pusher propeller.

Imperial Eagle

Dimensions, External
Overall, length..1.20 m (3 ft 11¼ in)
Wings, wing span...1.60 m (5 ft 3 in)
Weights and Loadings
Weight, Max launch weight...2.3 kg (5.00 lb)
Performance
Altitude, Operating altitude.........................30 m to 3,000 m (100 ft to 9,840 ft)
Speed
  Max level speed...48 kt (89 km/h; 55 mph)
  Loitering speed...22 kt (41 km/h; 25 mph)
Radius of operation, mission.............................5.4 n miles (10 km; 6 miles)
Endurance, max...1 hr
Power plant..1 × electric motor

Status: Being promoted as of end 2011.
Contractor: Aeronautical Development Establishment
Bangalore.

The hand-launched Imperial Eagle SUAS (Robert Hewson)

ADE/NAL Slybird

Type: Close-range mini-UAV.
Development: The Slybird system is a joint development by the ADE and India's National Aerospace Laboratories. It was first shown publicly at the Aero India exhibition in Bangalore in February 2011. Successful autonomous demonstrations were said to have been flown at that time.

Slybird on view at Aero India in February 2011 (Robert Hewson)

Description: *Airframe:* Parasol-wing design, with dihedral on outer panels. Pod and boom fuselage, with dorsally mounted pusher engine; T tail. No landing gear. Composites construction.
  Mission payloads: Various and interchangeable, including EO, IR, bio-chemical and metrological sensors.
  Guidance and control: Fully automatic, with GPS/INS navigation. GCS features real-time control of hardware and software, enabling fail-safe flight operation and user-friendly interfacing for route planning, choice of operational modes, payload control and target localisation. ADE-developed multi-channel (C-, L- and S-bands) datalink system.
  Transportation: Man-portable in robust carrying cases. Set-up time less than 5 minutes.
  Launch: Hand launch.
  Recovery: Belly landing.
  Power plant: One 300 W (0.4 hp) brushless electric motor; two-blade pusher propeller.

Slybird

Dimensions, External
Overall
  length..1.20 m (3 ft 11¼ in)
  height..0.30 m (11¾ in)
Wings, wing span...1.55 m (5 ft 1 in)
Weights and Loadings
Weight, Max launch weight...2.38 kg (5.00 lb)
Payload, Max payload..0.3 kg (0.00 lb)
Performance
Climb
  Rate of climb, max, at S/L..............................180 m/min (590 ft/min)
Altitude, Service ceiling..4,570 m (15,000 ft)
Speed
  Max level speed...54 kt (100 km/h; 62 mph)
  Cruising speed...29 kt (54 km/h; 33 mph)
  Stalling speed..17 kt (32 km/h; 20 mph)
Radius of operation, mission.............................5.4 n miles (10 km; 6 miles)
Endurance, max...1 hr
Power plant..1 × electric motor

Status: Being promoted in 2011.
Contractor: Aeronautical Development Establishment
Bangalore.

ADE Nishant

Type: Short-range tactical UAV.
Development: *IHS Jane's* sources report development of the Nishant (Dawn) tactical UAV as having begun during the early 1990s and as having built on the Bangalore-based Aeronautical Development Establishment's

A display model of the Nishant short-range tactical UAV *(Paul Jackson)* 0527027

A ground view of the Nishant AV mounted on its hydraulic/pneumatic launch catapult. This particular AV is not fitted with landing 'feet' attached to the forward ends of its tail booms *(IHS/Patrick Allen)*
1311396

(ADE) experience with the Kapothaka 'mini remotely piloted vehicle' testbed (130 kg (287 lb) all up weight, 90 minute endurance and a payload made up of TV and panoramic cameras). Key programme events include:

January 1995 The first of three Nishant prototypes is reported to have made its maiden flight.

December 1996 The Nishant UAV is reported to have been displayed publicly for the first time at the 1995 Aero India trade show.

Mid-2004 *IHS Jane's* sources were reporting the Indian Army as conducting 'end-user trials' of the Nishant UAV.

June 2005 The Nishant UAV is understood to have made its 100th test flight during June 2005.

30 September 2005 The Indian Government is reported to have given its approval for production of three Nishant systems (one firm, two as options and involving 12 AVs). Here, the work was to be undertaken by Hindustan Aeronautics Ltd and was valued at then year INR15 million.

3 May 2007 India's then Minister of State for Defence A K Antony is reported to have characterised the Nishant system as being in 'limited series production'.

December 2010 As of the given date, India's Defence Research and Development Organisation (DRDO) was reporting that a Nishant variant with a wheeled undercarriage was under development; that the integration of colour video and SAR radar payloads into the architecture was 'in progress' and that the Indian Army had placed an order for four Nishant AVs and their associated ground systems after 'successful user evaluation trials'.

February 2011 As of the given date, *IHS Jane's* sources were reporting the first four Nishant AVs as being 'in the process of delivery' to the Indian Army, with a further eight scheduled for delivery during the period 2013-2014.

Nishant is designed for use in the day/night battlefield reconnaissance, general surveillance, target tracking and localisation and artillery fire correction roles and readers should also note that there are unconfirmed reports of Taneja Aerospace producing 14 Nishant AVs for an 'operational evaluation programme'.

Description: *Airframe:* Box-shaped fuselage with an aerodynamically shaped nose section, provision for a ventral payload, a dorsal digital datalink antenna and a rear-mounted engine driving a two-bladed pusher propeller. High mounted wings at mid fuselage and carrying port and starboard tail booms. Twin vertical tail surfaces with a horizontal tail surface mounted between the rear ends of the tail booms. When not fitted with a wheeled undercarriage, the Nishant airframe features six landing 'feet' attached to the four corners of its lower fuselage and at the forward ends of its tail booms. *IHS Jane's* sources describe the Nishant airframe as being built using carbon fibre composites and the AV makes use of conventional ailerons, elevator and rudders.

Mission payloads: ADE developed Gimballed Payload Assembly (GPA) that is equipped with daylight TV and FLIR sensors (Elbit (Elop) equipment according to *IHS Jane's* sources) and offers detection ranges of 0.8-1.3 n miles (1.5-2.5 km; 0.9-1.6 miles), 2.2-2.7 n miles (4-5 km; 2.5-3.1 miles) and 5.4-6.5 n miles (10-12 km; 6.2-7.5 miles) against human beings, trucks and 'buildings' respectively. Other payload options developed or under consideration are reported to include a colour video camera, a '35 mm miniature panoramic camera', a SIGINT package and a SAR radar.

Guidance and control: Ground-mobile GCS mounted on a Tata eight wheeled truck chassis that has provision for an AV controller, a mission commander and a payload operator; that features an electronic map display, a mission planning and validation capability, semi-automatic AV checkout and an AV pilot training simulator and which provides AV/payload command and control and tracking. An associated antenna vehicle (mounted on the same Tata eight wheeled chassis as the Nishant GCS) incorporates a jam-resistant datalink (possibly operating at L- (1 to 2 GHz) band), a ground data terminal and a single-axis tracker. The architecture's antenna vehicle is remotely controlled from the GCS via a 300 m (985 ft) fibre-optic cable.

System composition: A Nishant system comprises four AVs, a truck-mounted GCS, an antenna vehicle, a truck-mounted launcher, an avionics preparation/maintenance vehicle, a mechanical maintenance vehicle, an AV transportation vehicle and multiple power supply vehicles. *IHS Jane's* sources note individual Nishant architectures as being supported by a ground crew of 10.

Launch: By mobile hydraulic/pneumatic catapult (mounted on a Tata eight wheeled truck chassis) that is capable of launching an AV with an all up weight of 375 kg (827 lb) at a speed of 42-45 m/s (138-148 ft/s). Conventional wheeled take-off (on a fixed tricycle undercarriage) as an alternative.

Recovery: Parachute and twin airbag recovery to belly landing. Conventional wheeled landing (on a fixed tricycle undercarriage) as an alternative.

Power plant: One 38.0 kW (51 hp) UEL AR 801R rotary engine; two-blade pusher propeller.

Nishant

Dimensions, External	
Overall	
length	4.63 m (15 ft 2¼ in)
height	1.183 m (3 ft 10½ in)
Wings, wing span	6.64 m (21 ft 9½ in)
Weights and Loadings	
Weight	
Weight empty	252 kg (555 lb)
Max launch weight	375 kg (826 lb)
Payload, Max payload	45 kg (99 lb)
Performance	
Altitude, Service ceiling	3,600 m (11,820 ft)
Speed	
Max level speed	100 kt (185 km/h; 115 mph)
Cruising speed	
maximum	81 kt (150 km/h; 93 mph)
minimum	68 kt (126 km/h; 78 mph)
Stalling speed	59 kt (110 km/h; 68 mph)
Radius of operation	
datalink, (payload)	54 n miles (100 km; 62 miles)
command link	86 n miles (159 km; 99 miles)
Endurance	4 hr 30 min

Status: As of end 2012, examples of the Nishant short-range tactical UAV are reported to have been procured by the Indian Army and the type was continuing to be promoted by India's DRDO.

Contractor: Aeronautical Development Establishment Bangalore.

ADE Rustom

Type: Family of tactical and MALE UAVs.

Development: Rustom is a development programme for an indigenous family of tactical (Rustom-1) and MALE systems, with the latter offering similar capabilities to those of the Heron AVs currently in use by the Indian Air Force and Navy. Rustom AVs are intended for use by all three of Indian armed forces together with agencies such as the Coast Guard and Police. Of the two, the Rustom-1 is derived from the Rutan Long-EZ based Light Canard Research Aircraft and is reported to have made its official maiden flight on 16 October 2010. A full-size Rustom MALE mockup was displayed at the 2009 Aero India trade show. Other key programme events include:

8 April 2012 The Rustom-1 demonstrator 'successfully' completed its 14th test sortie during which, it reached an altitude of approximately 3,505 m (11,500 ft); flew at a speed of more than 76 kt (140 km/h; 87 mph); achieved a range of 27 n miles (50 km; 31 miles); used its lean mixture control system at altitude and followed a series of pre-programmed waypoints.

Description: *Airframe:* **Rustom-1:** Canard configuration with rectangular fore planes and slightly swept back wings with vertical end plates. Fixed

The mockup of the Rustom MALE UAV that was displayed at the February 2009 Aero India trade show *(Robert Hewson)*

tricycle undercarriage. Two-bladed pusher propeller. Dorsal antenna array. Provision for a nose-mounted payload.

Rustom MALE: High-mounted, high aspect ratio wings; mainly rectangular section fuselage with bulbous nose; very tall tailfin with T tailplane. Retractable tricycle landing gear. All-composites (GFRP) construction.

Mission payloads: **Rustom-1:** Gimballed EO payload reportedly demonstrated.

Rustom MALE: To include (reportedly Israel-sourced) EO and IR sensors, SAR and maritime patrol radars, communications relay, elint and comint packages. Design incorporates provision for underwing stores hardpoints.

Guidance and control: Rustom-1 and MALE : Pre-programmed; dual-redundant flight control system; GPS navigation; satcom and datalink. GCS configuration derived from that of ADE Nishant tactical UAV.

Launch: **Rustom-1 and MALE:** Conventional wheeled take-off.
Recovery: **Rustom-1 and MALE:** Conventional wheeled landing.
Power plant: **Rustom MALE:** Two 84.5 kW (113.3 hp) Rotax 914 F turbocharged flat-four engines, each driving a three-blade tractor propeller. See under Weights for fuel details.

All data refers to Rustom MALE

Rustom

Dimensions, External	
Overall, length	14 m (45 ft 11¼ in) (est)
Wings, wing span	20 m (65 ft 7½ in) (est)
Weights and Loadings	
Weight, Max T-O weight	1,800 kg (3,968 lb)
Fuel weight, Max fuel weight	1,000 kg (2,204 lb)
Payload, Max payload	350 kg (771 lb)
Performance	
Altitude	
Operating altitude	9,145 m (30,000 ft)
Service ceiling	10,670 m (35,000 ft)
Speed	
Max level speed	121 kt (224 km/h; 139 mph)
Cruising speed, max	94 kt (174 km/h; 108 mph)
Loitering speed	67 kt (124 km/h; 77 mph)
Radius of operation	
LOS, mission	135 n miles (250 km; 155 miles)
with relay	189 n miles (350 km; 217 miles)
Endurance, max, on station	24 hr (est)

Status: As of October 2012, Rustom-1 and Rustom MALE development was continuing.
Contractor: Aeronautical Development Establishment Bangalore.

The Rustom-1 demonstrator made its 14th test flight on 8 April 2012 *(DRDO)*

ADRDE Akashdeep

Type: Non-rigid surveillance aerostat.
Development: Although not unveiled until the completion of flight trials in December 2010, development of the Akashdeep (Punjabi for Sky Light) is said to have begun in 2006 by the Aerial Delivery Research and Development Establishment (ADRDE), a component of India's Defence Research and Development Organisation (DRDO). The system architecture is said to be suitable for military, paramilitary and civilian applications, including disaster management. Bharat Electronics was among several industrial and other DRDO contributors to the programme.
Description: *Airframe:* Conventional ovoid, helium-filled envelope with three tapered tail surfaces in inverted Y configuration.

This view of the 2009 Rustom MALE UAV mockup shows it fitted with ventral EO and radar sensor housings together with a dorsal satellite communications antenna and COMINT antennas at the ends of its horizontal tail surfaces *(Robert Hewson)*

Detail of Akashdeep EO sensor turret and comint antenna array *(DRDO/ADRDE)*

Akashdeep aerostat at rest during December 2010 flight test programme (DRDO/ADRDE)

Mission payloads: Day/night EO imaging sensor (360° FoV) and comint payloads (the former from a Nishant UAV) were installed for December 2010 development trials. Surveillance radar and elint payloads are also said to have been developed in India for this application. A data downlink is incorporated in the aerostat's tether cable.

Guidance and control: Ground-based uplink command and control of payload operation.

Launch and Recovery: By hydraulic winch.

Specifications

Dimensions	
Length overall	9.80 m (32 ft 2 in)
Envelope volume	2,000 m³ (70,630 cu ft)
Performance	
Operating altitude	1,000 m (3,280 ft)
Sensor range	10.8 to 54 n miles (20 to 100 km; 12.4 to 62 miles)

Status: As of December 2010, delivery of one system to the Indian Army was reportedly expected to take place "in the near future". Later, the system performed an aerial security role during the February 2011 Aero India show in Bangalore.

Development is continuing, according to reports that the ADRDE has plans to field an improved system capable of operating at 5,000 m (16,400 ft) altitude and providing a sensor range of up to 135 n miles (250 km; 155 miles).

Contractor: Aerial Delivery Research and Development Establishment Agra.

Aurora Integrated Systems Altius Mk II

Type: Medium-range surveillance mini-UAV.

Development: Aurora was formed in 2010 as a new start-up company, with funding support from the Tata Group and the Indian Ministry of Science and Technology. Other UAV programmes in 2011 were the Sky Dot and Urban View (which see). The company also produces a range of tethered aerostats, the Skyview 50, 100HD and 200.

Description: Airframe: Traditional high-wing, tractor-engined monoplane. Fixed tricycle landing gear.

Mission payloads: Altius is integrated with a stabilised dual-sensor gimbal, enabling missions in light and weather conditions. Sensor suite is modular; interchangeable payload options include day TV camera with zoom and 22.5° FoV; a less than 6 micron FLIR with multiple FoV; a laser rangefinder with more than 4 km range and 5 metre CEP accuracy; a laser spotter/designator; and an analogue or digital video downlink.

Guidance and control: Fully autonomous and mission-programmable. Two-person GCS (air vehicle and payload operators), with functionally interchangeable consoles, allowing interactive mission planning, retasking, playback and operator mission training. Digital, jamming-resistant telemetry command and datalinks.

System composition: Air vehicle(s), a vehicle-mounted GCS and three operators. Set-up time less than 1 hour.

Launch: Conventional, automatic wheeled take-off.

Recovery: Conventional, automatic wheeled landing.

Power plant: One two-cylinder two-stroke engine (rating not stated); two-blade propeller. Fuel capacity 24 to 40 litres (6.3 to 10.6 US gallons; 5.3 to 8.8 Imp gallons).

Altius Mk II

Dimensions, External	
Overall, length	3.20 m (10 ft 6 in)
Wings, wing span	5.50 m (18 ft 0½ in)
Weights and Loadings	
Weight	
Weight empty	35 kg (77 lb)
Max T-O weight	80 kg (176 lb)
Payload, Max payload	20 kg (44 lb)
Performance	
T-O	
T-O run, unpaved runway	100 m (329 ft) (est)
Climb	
Rate of climb, max, at S/L	150 m/min (492 ft/min) (est)
Altitude	
Operating altitude	
min	455 m (1,500 ft)
max	2,440 m (8,000 ft)
Service ceiling	3,660 m (12,000 ft)
Speed	
Max level speed	86 kt (159 km/h; 99 mph) (est)
Cruising speed	59 kt (109 km/h; 68 mph)
Radius of operation, datalink	189 n miles (350 km; 217 miles)

Status: Being promoted as of end 2012.
Contractor: Aurora Integrated Systems Pvt Ltd Bangalore.

Aurora Integrated Systems Sky Dot

Type:

Development: Revealed at Aero India show in February 2011. Optimised for covert and urban terrain operations, crowd monitoring and disaster assistance.

Description: Airframe: Sweptback flying-wing design, with endplate fins; pusher engine. No landing gear.

Mission payloads: Daylight TV camera capable of detecting a man at 150 m (492 ft) range or vehicle at 250 m (820 ft); or LLTV camera with corresponding detection ranges of 70 m (230 ft) and 150 m.

Guidance and control: Fully autonomous, from common GCS as described for company's Altius Mk II.

Transportation: One-person-portable in hand-held bag. Complete system weighs less than 5 kg (11 lb).

Launch: Hand launch.

Recovery: Belly landing.

Power plant: Electric motor (rating not stated), powered by lithium-polymer batteries; two-blade pusher propeller with folding blades.

Sky Dot hand-launched mini-UAV *(AIS)* 1395302

Urban View hand launch *(AIS)* 1395303

Sky Dot

Dimensions, External	
Overall, length	0.40 m (1 ft 3¾ in)
Wings, wing span	0.80 m (2 ft 7½ in)
Weights and Loadings	
Weight, Max launch weight	0.5 kg (1.00 lb) (est)
Performance	
Climb	
Rate of climb, max, at S/L	150 m/min (492 ft/min) (est)
Altitude	
Operating altitude	
min	60 m (200 ft)
max	185 m (600 ft)
Service ceiling	3,050 m (10,000 ft)
Speed	
Max level speed	43 kt (80 km/h; 49 mph)
Cruising speed	30 kt (56 km/h; 35 mph) (est)
Radius of operation, mission	2.7 n miles (5 km; 3 miles)
Endurance, max	40 min
Power plant	1 × electric motor

Status: Being promoted in 2011.
Contractor: Aurora Integrated Systems Pvt Ltd
Bangalore.

Aurora Integrated Systems Urban View

Type: Close-range surveillance UAV.
Development: Revealed at Aero India show in February 2011.
Description: *Airframe:* Slightly sweptback mid-mounted wings; single fin; no horizontal tail surfaces; pusher engine. No landing gear. Airframe can be assembled without tools.

Mission payloads: Three payload options are offered:
(1) Dual EO cameras (front- and side-looking) with optical zoom, lock-on targeting and tracking capability;
(2) Retractable two-axis (azimuth and elevation) stabilised gimbal for EO camera with lock-on targeting and tracking capability;
(3) Forward-looking IR camera.

Full duplex, jamming-resistant digital datalink. Analogue video downlink standard; or, optionally, a digital downlink with encryption facility.

Guidance and control: Fully autonomous, with pre-set search patterns, using AIS common GCS as described for Altius Mk II UAV.
Transportation: One/two-soldier man-portable in backpack. Total system weight 10 to 15 kg (22 to 33 lb). Set-up time less than 5 minutes.
Launch: Hand launch.
Recovery: Belly landing.
Power plant: One electric motor (rating not stated), powered by lithium-polymer batteries; two-blade propeller with folding blades.

Urban View

Dimensions, External	
Overall, length	0.80 m (2 ft 7½ in)
Wings, wing span	1.50 m (4 ft 11 in)
Weights and Loadings	
Weight, Max launch weight	2.0 kg (4.00 lb) (est)
Performance	
Climb	
Rate of climb, max, at S/L	240 m/min (787 ft/min) (est)
Altitude	
Operating altitude	90 m to 305 m (300 ft to 1,000 ft)
Service ceiling	4,570 m (15,000 ft)
Speed, Max level speed	32 kt (59 km/h; 37 mph) (est)
Radius of operation	8.1 n miles (15 km; 9 miles) (est)
Endurance	1 hr (est)
Power plant	1 × electric motor

Status: Being promoted as of end 2012. System is stated to have been extensively tested in rural, urban and marine environments.
Contractor: Aurora Integrated Systems Pvt Ltd
Bangalore.

Kadet Defence Systems FireBee

Type: Close-range battlefield surveillance UAV.
Development: Development of the FireBee started in January 2010. It is designed for 'over the hill' gathering of real-time visual intelligence, with all maintenance at field level. Potential applications include front-line surveillance, BDA, law enforcement, border patrol, search and rescue and environmental monitoring.
Description: *Airframe:* Sweptback, tail-less flying wing configuration; no landing gear. Carbon fibre and Kevlar construction.
Mission payloads: Central bay for interchangeable modular payloads. Options include fixed EO camera with ×10 zoom; fixed thermal imager; or stabilised pan-and-tilt camera gimbal with option for automatic object tracking, object metadata and interchangeable (EO/thermal imager) nosecone. Analogue video link standard; digital secure video link optional.
Guidance and control: Spread spectrum frequency-hopping command and control. Automatic target detection and tracking. Communication and video range is dictated by GCS configuration, and is extendable.
Transportation: Complete system can be carried in two 20 kg (44.1 lb) backpacks.
Launch: Automatic launch by bungee catapult.

Urban View at Aero India in February 2011 *(Robert Hewson)* 1395304

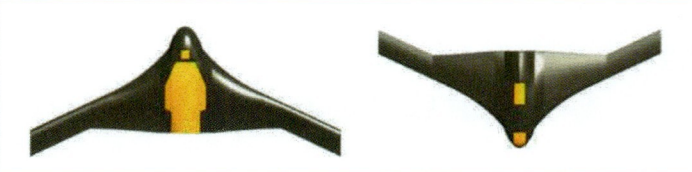

General arrangement views of the FireBee air vehicle
(Kadet Defence Systems) 1395225

Recovery: Automatic recovery by parachute.
Power plant: Battery-powered electric motor (rating not stated); pusher propeller.

FireBee

Dimensions, External	
Wings, wing span	2.44 m (8 ft 0 in)
Weights and Loadings	
Weight, Max launch weight	7.0 kg (15.00 lb)
Payload, Max payload	1.0 kg (2.00 lb)
Performance	
Radius of operation, LOS, mission	5.4 n miles (10 km; 6 miles)
Endurance, max	1 hr (est)
Power plant	1 × electric motor

Status: Under development in 2010.
Contractor: Kadet Defence Systems
Kolkata.

Kadet Defence Systems Trogon

Type: Close-range reconnaissance and surveillance mini-UAV.
Development: Design and development of the Trogon began in August 2007, with the first flight being made on 10 November that year. Maiden flight of the production version took place on 1 December 2007. Two versions, (R and U), are available.
Description: *Airframe:* Blended wing/body modular design, with endplate fins. No landing gear. All-composite construction.
Mission payloads: Fixed-view, forward-looking CCD day camera with real-time imagery downlink.
Guidance and control: Trogon R is operated manually under UHF radio control. Autonomous Trogon U has a three-axis GPS-based autopilot with a long-range datalink and portable GCS.
Transportation: Man-portable and backpackable; easily transportable by air, land and sea. Set-up time is 30 minutes.
System composition: A typically fielded Trogon R system consists of three air vehicles, one hand-held ground controller, a bungee launcher, two personnel and a mini LCD screen. In addition, Trogon U has a portable GCS and a long-range data uplink.
Launch: Operator controlled from bungee launcher.
Recovery: Operator controlled to belly landing or net recovery. Optional 'return home' capability.
Power plant: One 0.75 kW Li/Po battery-powered brushless electric motor; two-blade pusher propeller.

Trogon

Dimensions, External	
Overall	
length	0.46 m (1 ft 6 in)
height	0.10 m (4 in)
Wings, wing span	1.22 m (4 ft 0 in)
Engines, propeller diameter	0.28 m (11 in)
Weights and Loadings	
Weight	
Weight empty	1.0 kg (2.00 lb)
Max launch weight	2.25 kg (4.00 lb)
Payload, Max payload	0.5 kg (1.00 lb)
Performance	
Altitude, Operating altitude	100 m to 1,000 m (320 ft to 3,280 ft)
Speed	
Max level speed	70 kt (130 km/h; 81 mph)
Cruising speed, max	40 kt (74 km/h; 46 mph)
Loitering speed	20 kt (37 km/h; 23 mph)
Radius of operation	
datalink, mission, Trogon R	2.7 n miles (5 km; 3 miles)
datalink, mission, Trogon U	5.4 n miles (10 km; 6 miles)
Endurance, max	45 min
Power plant	1 × electric motor

Status: In December 2007, the Trogon system was being evaluated by Special Forces, Northern Command, Indian Army.
Contractor: Kadet Defence Systems
Kolkata.

The Trogon flying-wing UAV *(Kadet Defence Systems)* 1290244

MKU Erasmus

Type: Medium-range tactical UAV.
Development: The Erasmus design made its debut, together with the company's smaller TERP close-range system, in February 2008.
Description: *Airframe:* Constant-chord, shoulder-mounted wings; pod fuselage; twin booms supporting twin fins and rudders with enclosed tailplane and elevators; pusher engine. Fixed tricycle landing gear. Composites construction.
Mission payloads: Dual gimballed EO and thermal imagingg sensors. Image exploitation, targeting and tracking software.
Guidance and control: As described for TERP. Datalink encryption optional.
Transportation: Complete system can be transported in two strengthened cases. Set-up time 45 minutes.
Launch: Conventional wheeled take-off.
Recovery: Conventional wheeled landing.
Power plant: One 75 cm³ two-cylinder piston engine, driving a two-blade pusher propeller.

Erasmus

Dimensions, External	
Fuselage, length	1.52 m (4 ft 11¾ in)
Wings, wing span	3.20 m (10 ft 6 in)
Performance	
T-O	
T-O run	150 m (493 ft)
Altitude	
Operating altitude	1,000 m to 3,000 m (3,280 ft to 9,840 ft)
Service ceiling	4,000 m (13,120 ft)
Speed	
Max level speed	86 kt (159 km/h; 99 mph) (est)
Cruising speed, max	54 kt (100 km/h; 62 mph)
Loitering speed	45 kt (83 km/h; 52 mph)
Radius of operation, LOS, datalink	81 n miles (150 km; 93 miles)
Endurance	5 hr (est)
Landing	
Landing run	150 m (493 ft)
Power plant	1 × piston engine

Status: No orders reported by April 2009. Marketing was continuing at that time.
Contractor: MKU Pvt Ltd
New Delhi.

Erasmus tactical MALE UAV *(IHS/Patrick Allen)* 1376642

MKU TERP

Type: Close-range tactical UAV.
Development: The TERP (Tactical Electrical Reconnaissance Probe) is designed for military, paramilitary and civil applications. It was launched, together with the company's larger Erasmus, in February 2008.
Description: *Airframe:* High-mounted wings with dihedral outer panels; pod and boom fuselage; T tail; pusher engine. All-composites construction.
Mission payloads: Standard payloads are an LLTV camera with ×40 zoom or a thermal imaging camera, mounted on a gimbal which retracts into the fuselage and is enclosed by 'beetle-wing' doors.

TERP at IDEX 2009 *(IHS/Patrick Allen)*

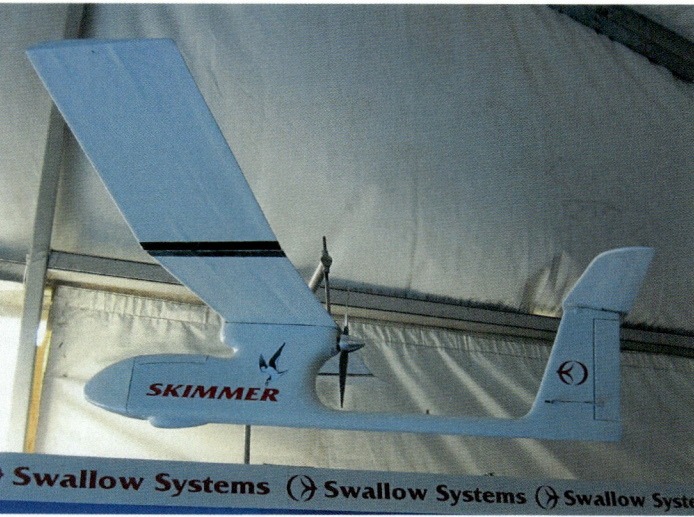
Skimmer on display at Aero India, February 2011 *(Robert Hewson)*

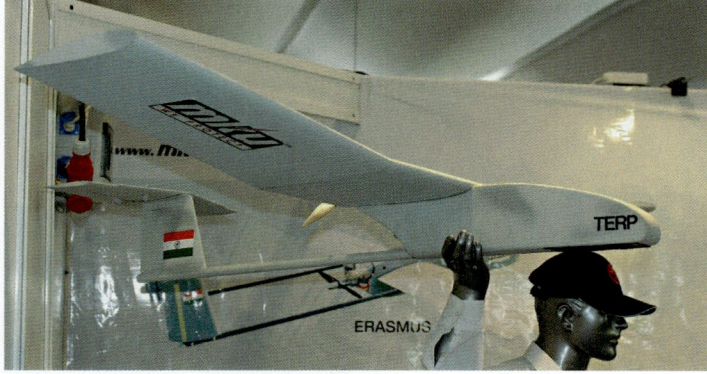
TERP at ILA Berlin 2008 *(Paul Jackson)*

Guidance and control: Laptop GCS, with touch screen and moving map overlay; can control multiple air vehicles. GPS-based navigation (up to 1,000 waypoints). L-, S- or C-band datalinks available. TERP can be flown in autonomous, targeting or manual mode, with en-route updating.
Transportation: Man-portable; can be dismantled and carried in two backpacks. It can be re-assembled, without tools, in less than 10 minutes.
Launch: Hand launch standard; bungee catapult optional.
Recovery: Deep stall to belly landing standard; net recovery optional.
Power plant: One brushless electric motor, driving a two-blade pusher propeller.

TERP

Dimensions, External	
Overall, length	1.32 m (4 ft 4 in)
Wings, wing span	1.83 m (6 ft 0 in)
Performance	
Altitude	
Operating altitude	90 m to 1,220 m (300 ft to 4,000 ft)
Service ceiling	5,945 m (19,500 ft)
Speed	
Max level speed	75 kt (139 km/h; 86 mph)
Cruising speed	30 kt (56 km/h; 35 mph)
Radius of operation, LOS, mission	8.1 n miles (15 km; 9 miles)
Endurance, max	1 hr 30 min (est)
Power plant	1 × electric motor

Status: Marketing and promotion continuing.
Contractor: MKU Pvt Ltd
New Delhi.

Swallow Systems Skimmer

Type: Close-range mini-UAV.
Development: Said by the company to be the outcome of more than six years' research and evaluation through some 300 hours of flight testing, the Skimmer was unveiled publicly at Aero India in February 2011. It is optimised for military RSTA roles or such civil applications as reconnaissance of forest fires or other natural calamities, wildlife monitoring, area mapping and perimeter surveillance.
Description: *Airframe:* Pod and boom fuselage, atop which are mounted engine nacelle and high wing with outer panel dihedral; T tail. No landing gear. Composites construction (glass fibre, carbon fibre and Kevlar).
Mission payloads: Forward-looking, daylight colour TV camera as standard. Available options include a switchable (forward/side-looking) daylight colour TV camera, a standard daylight camera with ×10 zoom; LLTV camera or an uncooled thermal imager. Real-time video downlink.

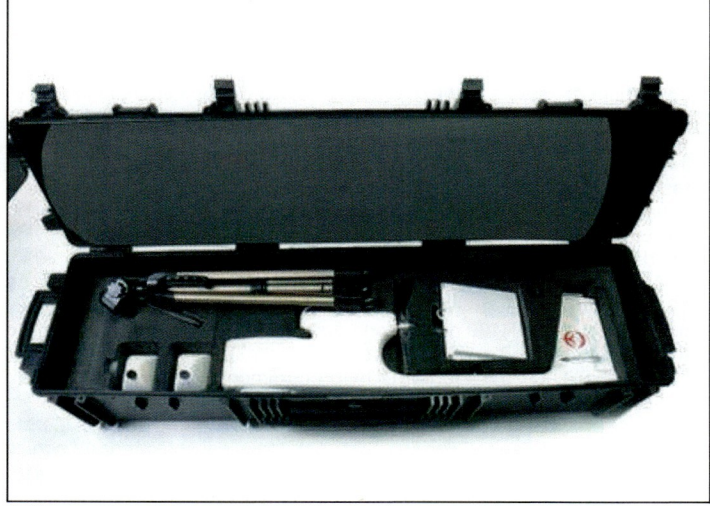
A dismantled Skimmer in its carrying case *(Swallow Systems)*

Guidance and control: Two-person operation via an FHSS jamming-resistant command link, with telemetry data superimposed on the video. The system supports pre-programmed navigation (up to 500 GPS waypoints), with manual override and a 'Return home' capability if the command link is lost. The GCS is capable of 4 hours continuous operation on fully charged battery, which is rechargeable by AC power/car adapter
Transportation: Man-portable in weatherproof carrying cases. Set-up time 7 minutes.
System composition: Air vehicle(s), carrying case, GCS case, Panasonic Toughbook CF-10 touch-screen notebook PC and antenna tripod.
Launch: Hand launch.
Recovery: Belly landing.
Power plant: Electric motor (rating not stated), powered by single rechargeable lithium battery; two-blade pusher propeller.

Skimmer

Dimensions, External	
Overall, length	1.22 m (4 ft 0 in)
Wings, wing span	1.68 m (5 ft 6¼ in)
Weights and Loadings	
Weight, Max launch weight	2.5 kg (5.00 lb)
Performance	
Altitude	
Operating altitude	300 m (985 ft) (est)
Service ceiling	2,745 m (9,000 ft)
Speed	
Max level speed	51 kt (94 km/h; 59 mph)
Stalling speed	14 kt (26 km/h; 17 mph)
Radius of operation, mission	2.7 n miles (5 km; 3 miles)
Endurance, max	1 hr
Power plant	1 × electric motor

Status: Being promoted in 2012. Company reported "serious enquiries" by possible customers and stated a manufacturing capacity to produce between 10 and 15 systems per month.
Contractor: Swallow Systems Pvt Ltd UAV Solutions
Tal Halol.

International

Dassault Neuron

Type: Experimental UCAV.

Development: With participation from Aviation Design, Dassault began developing an AVE (*Aéronef de Validation Expérimentale*) named Petit Duc (Little Owl) as a subscale UCAV technology demonstrator. This aircraft, AVE-D, which flew for the first time on 18 July 2000, was intended as the first stage of a three-stage project named LogiDuc (Logistics to Demonstrate a UCAV). A second machine, designated AVE-C, which differed from its predecessor in having no fins, made its maiden flight in early June 2003. Intended follow-on models were: Moyen Duc (Long-Eared Owl), a multirole tactical version; and Grand Duc (Great Horned Owl), a full-size operational UCAV carrying internally stored weapons.

In June 2003, the DGA announced the commitment of EUR300 million (USD348 million) to the Grand Duc programme and in December of that year Saab Aerostructures signed an MoU to become a principal partner in the programme. Belgium (Sabca), Greece (Hellenic Aerospace Industries), Italy (Alenia Aeronautica), Spain (EADS CASA) and Switzerland (RUAG Aerospace) subsequently joined and the programme became a European international programme under the new name Neuron. Belgium withdrew its participation in March 2005 but is thought likely to rejoin in due course. Dassault remains the prime contractor, with a 50 per cent share of the programme, the remaining 50 per cent being shared among the other partners.

Saab responsibilities include the fuselage, fuel system, avionics and some general design and flight testing. EADS CASA's share is the wing, GCS and datalink integration. Alenia is designing the 'smart' integral weapons bay, electrical power and air data systems, and is also involved in general design and some flight testing. RUAG undertakes wind tunnel testing and weapon interfaces, while HAI produces the rear fuselage and exhaust system. General concept, final assembly and major flight testing are undertaken by Dassault. Thales has been subcontracted to provide the datalinks; other major suppliers had yet to be identified by early 2006.

A full-size Neuron mockup was unveiled at the Paris Air Show on 13 June 2005. As of early 2006, programme cost estimates had reportedly risen to nearer EUR400 million, and first flight was then targeted for the third quarter of 2010 rather than first quarter 2009 as originally envisaged. In February 2006 the French DGA procurement agency launched the project. During 2008 RUAG completed wind tunnel tests of ground effect characteristics and bird strike tests on the wing leading edge. A nozzle made by HAI was scheduled to undergo mechanical and integration trials with the Adour engine starting in January 2009. An inlet demonstrator system was produced by Dassault in 2008. By 2009 the first flight date

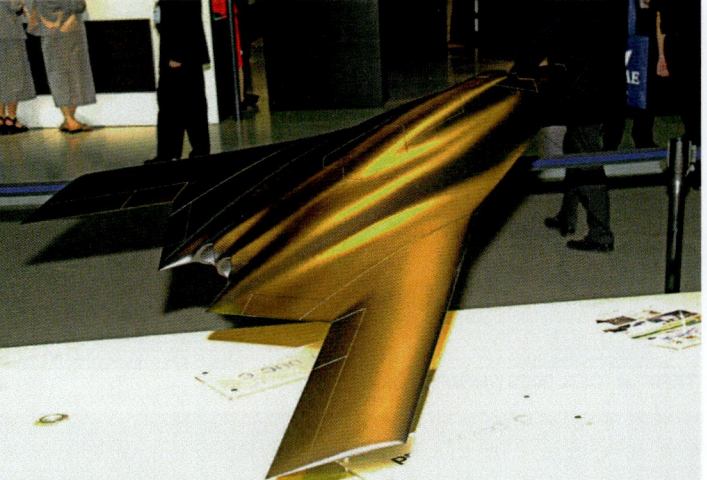

Model of the tailless AVE-C *(IHS/Patrick Allen)* 1024091

Dassault AVE-D first test vehicle *(Dassault)* 0073576

was estimated to be the end of 2011 and flight tests were to run for 18 months in France, Sweden and Italy. The fuselage was handed over to Dassault by Saab Aerospace in January 2011 with assembly of the complete air vehicle expected to take nine months.

Description: *Airframe:* All-composites, stealthy, flying wing design, with 'W' trailing-edge, dorsal engine air intake, retractable tricycle landing gear.

Mission payloads: Modular weapons or sensors in internal weapons bay. Expected to include laser-guided bombs of up to 500 kg size.

Guidance and control: To be configured for network-centric warfare (NCW) interface. NATO-compliant (STANAG 7085) BLOS datalinks by Thales Land and Joint Systems: high-rate for imagery, video, radar and control/communications; low-rate for transmission security.

Launch: Wheeled take-off.

Recovery: Wheeled landing.

Power plant: One Rolls-Royce/Turbomeca Adour non-afterburning turbofan.

(provisional)

Neuron mockup at Paris Air Show, June 2005 *(IHS/Patrick Allen)* 1144425

Neuron fuselage during handover ceremony at Saab in early 2011 *(Saab Aerospace)* 1436485

Neuron

Dimensions, External	
Overall, length	9.50 m (31 ft 2 in)
Wings, wing span	12.50 m (41 ft 0¼ in)
Weights and Loadings	
Weight, Max T-O weight	6,000 kg (13,227 lb)
Performance	
Altitude, Service ceiling	10,670 m (35,000 ft)
Speed, Max level Mach number	0.7
Endurance	12 hr (est)
Power plant	1 × turbofan

Status: Development continuing.

Contractor: Dassault Aviation
St Cloud, France.

Northrop Grumman/IAI MQ-5 and RQ-5 Hunter

Type: Short-range ISTAR UAV.

Development: Hunter (formerly known as JIMPACS: Joint Improved Multimission Payload Aerial surveillance, Combat Survivable) is based on an IAI air vehicle originally known as Impact, and was designed with specific operational goals to meet US short-range (UAV-SR, later renamed Joint Tactical UAV) requirements. The first air vehicle made its initial flight, in Israel, on 30 September 1990, and was delivered to the

Northrop Grumman/IAI MQ-5 and RQ-5 Hunter

The MQ-5B Hunter making its first flight *(Northrop Grumman)* 1122565

RQ-5A Hunter in flight *(Northrop Grumman)* 1122563

Rocket-assisted launch of an RQ-5A *(Northrop Grumman)* 1122562

The MQ-5C E-Hunter *(Northrop Grumman)* 1122564

USA in December 1990. The Avionics and Surveillance Group of TRW (absorbed into Northrop Grumman in late 2002) and IAI Malat jointly submitted proposals for test and evaluation of the system, which was performed between December 1990 and March 1992. First relay flight took place in July 1991.

Hunter was selected as winner of the UAV-SR competition in June 1992 and, in February 1993, the TRW/IAI team received a USD169 million LRIP (low-rate initial production) contract to produce seven systems (average eight air vehicles per system). The original BQM-155A designation was changed to RQ-5A in 1999. The first production Hunter was flown on 22 February 1994. The first system was handed over to the DoD on 14 April 1995, and all seven (total of 72 air vehicles, including attrition replacements) had been delivered by the end of September 1995. At that time, Hunters had flown well in excess of 3,300 flight hours. Shipboard compatibility was demonstrated in 1993 from USS *Essex*, in preparation for a potential US Navy and Marine Corps version, but this was later abandoned.

In November 1995, the US Joint Requirements Oversight Council (JROC) recommended terminating the Hunter programme, mainly on the grounds that sufficient funding would not be available for both Hunter and the close-range Tactical UAV (TUAV). (In the event, the Hunter outlived the original TUAV selectee, the Alliant RQ-6A Outrider.) The JROC recommendation was accepted by DARO in January 1996. Test flying resumed in December 1995 after a grounding due to a number of accidents during that year; modifications included fitment of redesigned ailerons.

Since that time, Hunter has continued to evolve, both as a result of extensive use in developing UAV CONOPS in general, and in its continuing use as a testbed for later and more versatile mission payloads. As of January 2003, pending introduction of the RQ-7A Shadow 200, it remained the only fully operational UAV in the US Army's inventory. On 22 July 2002, a Hunter completed the first phase of an autoland programme, making two take-offs, four touch-and-gos and one landing under the automatic control of an SNC UCARS system.

A weapons development programme - the US Army's first on an operational UAV - also began to take shape in 2002 when, on 23 September, simulated weapon drops were made at Fort Huachuca to demonstrate the feasibility of arming the UAV. This was followed by actual weapon firings a few weeks later at the White Sands Missile Range in New Mexico. On 9 October 2002, two Northrop Grumman Bat (Brilliant Anti-Tank) acoustic and IR-guided submunitions, each fitted with a flight data recorder, selected two moving targets (a BMP infantry fighting vehicle and a T-72 main battle tank) from a group of eight and impacted them both. Two days later a pair of warhead-armed standard production Bats were similarly launched, and again both found their targets. The missiles were fired from Bat UAV ejection tubes (BUETs) developed by Systima. For these trials, the Hunter's wing span was increased in order to locate the pylon-mounted launch tubes outboard of the main landing gear, incidentally increasing the UAV's ceiling by around 610 m (2,000 ft) as an additional benefit. Following the demonstration, a fast-track programme was launched to fit 78 Bats with BUETs and upgrade six Hunters to deliver them. These Hunters were among a force of 16 deployed to Kuwait in January 2003 for Operation 'Iraqi Freedom', which were given extra fuel in an extended wing centre-section, boosting their endurance by 50 per cent. An improved version of Bat known as Viper Strike, using a semi-active laser seeker to locate its target, was developed by Northrop Grumman for use against point targets in congested urban areas; this was similarly trialled at the end of March 2003 and received US Army approval a year later, resulting in the armed version receiving the designation MQ-5B.

Description: *Airframe:* Robust pod-and-twin-tailboom high-wing monoplane, built of low-observable composites. One tractor and one pusher engine to improve single-engine survivability. Fixed tricycle landing gear.

Mission payloads: Basic sensor payloads are the IAI Tamam MOSP combined TV/FLIR sensor and an Elta C-band airborne data relay system. A modular building block approach enables these sensors to be replaced by alternative packages. An onboard power supply of 1.1 kW is available for payloads.

Payloads demonstrated during and since 1996 have included VHF/UHF communications relay between two GCSs 65 n miles (120 km; 74.5 miles) apart; SAR/MTI; L-3 Communications TCDL; laser designator/range-finder for Hellfire missiles; radar jammer; lightweight comint; sigint; and communications jammer. Trials with Northrop Grumman's TUAVR radar took place between 8 and 14 May 2003.

In early 2000, TRW equipped US Army Hunters with a chemical threat detection system called Safeguard, which combines an IRLS with a thermally stabilised FFT infra-red spectrometer for cloud particle analysis. A new Elta third-generation FLIR and 770 mm spotter for the daylight camera were part of a planned 2001 retrofit on the Task Force Hunter aircraft deployed to Macedonia. French Army laser designator trials took place at the CEV and CEAM during May and June 1999. In January 2006, Northrop Grumman announced a successful series of flights with a new Adaptive Joint Intelligence payload on the RQ-5A, enabling it to share multiple types of communication (communications relay, sigint or EW) simultaneously. Tactical common data link was fitted to an MQ-5B by the end of 2009, increasing data transfer rate, doubling communications range, and enabling additional payload capabilities.

In addition to the basic E-O/IR payload, the MQ-5B and C carry a laser designator; the MQ-5C can also carry a SAR.

Weapons development began in 2002, as described under Development heading above. MQ-5B capable of carrying Viper Strike laser-guided missiles or Textron BLU-108 submunitions underwing.

Guidance and control: Preprogrammed or remotely controlled. The RQ-5A advanced control and mission planning system includes two IAI GCS-3000 ground control stations, a mission planning station with remote video terminals, a miniaturised control unit and microwave band datalinks (two Ku-band uplinks and two downlinks). Control from the cockpit of an AH-64 Apache helicopter has been demonstrated. Total systems operation is readily attainable with only minimum field user training. An onboard power supply of 3 kW is available for avionics. The MQ-5B is controlled by the US Army's 'One System' GCS, first of which was delivered to the 224th MI Battalion in October 2005; this will be introduced retrospectively for the RQ-5A as well.

One aircraft was taken out of store in late 1997 and fitted with a Litton (now Northrop Grumman) LR-100 radar warning receiver for SEAD trials with the USAF's UAV Battle Lab in January 1998. In tests at the Melrose Bombing Range near Cannon AFB, New Mexico, the Hunter UAV was used in conjunction with two Block 50 F-16 fighters equipped with an improved data modem (IDM) datalink, the UAV locating target radars and transmitting their position directly to the fighters' pilots.

Continuing the development of this concept, in a programme called AMUST-D (Airborne Manned/Unmanned System Technology - Demonstration), Boeing's Phantom Works received a USD1.2 million,

RQ-5A Hunter with underwing BUETs (Bat UAV ejection tubes) *(Kenneth Munson)* 0528601

Hunter II prototype *(Northrop Grumman)* 1122573

Arrester hook deployed for runway landing 0518064

Bat-armed Hunter taking off at White Sands Missile Range *(Northrop Grumman)* 0529701

22-month contract in early 1999 to develop and flight test communications protocols between piloted aircraft and UAVs. The first such pairing, in August 2000, was between the Hunter UAV and the AH-64D Apache Longbow attack helicopter, for which TRW developed the interface.

In September 2009 Northrop Grumman equipped a Hunter with an automatic takeoff and landing system, and it demonstrated launch and recovery at Fort Huachuca, Arizona.

Ground support for Belgian B-Hunters is handled by Eagle consortium comprising Sonaca (50 per cent share), Thales Belgium (25 per cent) and the Malat and Elta Divisions of IAI (25 per cent).

System composition: Six (standard) to eight air vehicles, payloads and ADTs; two trailer-mounted GDTs; two or three GCS/mission planning stations in HMMWV-mounted shelters; one launch and recovery shelter (also HMMWV mounted); one launch and recovery terminal; one trailer-mounted RATO launcher; four remote video terminals; five mobile (towed) power units; one 10 kW generator; one power interface unit; plus ground support and training equipment.

Launch: Conventional wheeled take-off; can operate from unprepared strips. Rocket-assisted take-off optional. Automatic take-off standard on B-Hunter, MQ-5B, MQ-5C and Hunter II; demonstrated on RQ-5A.

Recovery: Conventional wheeled landing using retractable hook and arrester cable. Automatic (laser-guided) landing system on B-Hunter; UCARS automatic landing for MQ-5B, MQ-5C and Hunter II, demonstrated on RQ-5A. Parachute for emergency recovery.

Power plant: *RQ-5A:* Two 50.7 kW (68 hp) Moto Guzzi two-cylinder four-stroke engines initially, one at front and one at rear of fuselage nacelle; two-blade wooden propellers (one tractor, one pusher). Fuel capacity 189 litres (50 US gallons; 41.6 Imp gallons).
MQ-5B and Hunter II: Heavy-fuel engines. Increased fuel capacity in larger wings.
MQ-5C: Mogas or heavy-fuel engines. Increased fuel capacity in larger wings.

Variants: *RQ-5A Hunter:* Standard (initial LRIP) version. Fielded improvements include 'wet' extended centre wing (WECW), providing 50 per cent increase in endurance, and weaponisation. Modification of 11 to MQ-5B standard was under way in 2006. *Detailed description applies to this version except where indicated.*

B-Hunter: Version of RQ-5A ordered by Belgium under USD71.4 million contract announced on 10 December 1998 for three systems, each with six air vehicles and two GCSs. Improvements, based on US operating experience, include an IAI Malat fully automatic take-off and landing system, advanced avionics and a modernised GCS. Replacement for Belgian Army's Epervier, which was withdrawn in 1999 after some 22 years' service. Prototype lost in crash in Israel in August 2000, but Belgian deliveries (first three aircraft) began early 2001. Assembled in Belgium by Eagle consortium headed by Sonaca.

MQ-5B Hunter: US Army multimission upgrade of RQ-5A. Increased wing span and fuel load, greater endurance and higher operating altitudes than RQ-5A; weapon-carrying capability; dual-redundant avionics, including Northrop Grumman LN-251 GPS/INS; heavy-fuel engine; automatic take-off and landing. First flight 8 July 2005; remained airborne for more than 21 hours in endurance test on 4/5 January 2006. 18 on order as of early 2006; service entry second quarter 2006. In late 2009 an MQ-5B deployed to Afghanistan was equipped with tactical common data link.

MQ-5C E-Hunter: Enhanced endurance/range version, combining Hunter fuselage and avionics with the extended-span laminar flow wet wings, tailbooms and tail unit of the IAI Heron (which see) and retractable mainwheels. Increased payload, higher ceiling and greater endurance. Offered to US Army as conversion kit; could be installed on standard Hunter in the field in less than 3 hours. First flight (in Israel) July 1995; maiden flight in US, after further development, 17 March 2005.

Hunter II: Medium-altitude endurance version, based on Hunter fuselage combined with longer wing and tailbooms of IAI Heron air vehicle. Company-funded initial flight tests, between 27 December 2004 and 12 January 2005, were announced by Northrop Grumman in the latter month as having been successfully completed, with further testing planned through the first quarter of that year. Second RQ-5A converted, followed by 29 April 2005 order to Aurora Flight Sciences for third aircraft, this powered by a heavy-fuel engine. Stated to 'feature a software architecture that can easily accommodate new payloads and data handling requirements; state-of-the-art avionics; a weapons capability and communications subsystem that will allow it to share data seamlessly with current battlefield networks'. Semi-finalist in US Army extended range multipurpose (ERMP) competition to find Hunter replacement, but unsuccessful.

RQ-5A, MQ-5B, MQ-5C, Hunter II

Dimensions, External
Overall
  length
    RQ-5A, MQ-5B .... 7.01 m (23 ft 0 in)
    MQ-5C .... 7.47 m (24 ft 6 in)
    Hunter II .... 9.27 m (30 ft 5 in)
  height
    RQ-5A, MQ-5B .... 1.65 m (5 ft 5 in)
    Hunter II .... 2.29 m (7 ft 6¼ in)
Wings
  wing span
    RQ-5A .... 8.84 m (29 ft 0 in)
    MQ-5B .... 10.44 m (34 ft 3 in)
    MQ-5C, Hunter II .... 16.61 m (54 ft 6 in)

Dimensions, Internal
Payload bay
  volume, RQ-5A .... 282.5 m³ (9,976 cu ft)

Weights and Loadings
Weight
  Weight empty, RQ-5A .... 540 kg (1,190 lb)
Max T-O weight
  RQ-5A .... 726 kg (1,600 lb)
  MQ-5B .... 885 kg (1,951 lb)
  MQ-5C .... 998 kg (2,200 lb)
  Hunter II .... 1,496 kg (3,298 lb)
Fuel weight
  Max fuel weight, RQ-5A .... 136 kg (299 lb)
  Fuel with max payload
    internal, RQ-5A .... 178 kg (392 lb)

RQ-5A, MQ-5B, MQ-5C, Hunter II

Payload	
Max payload	
internal, RQ-5A	113 kg (249 lb)
internal, Hunter II	136 kg (299 lb)
external, MQ-5B, MQ-5C	118 kg (260 lb)
external, Hunter II	317.5 kg (699 lb)
Payload with max fuel	
MQ-5B	227 kg (500 lb)
MQ-5C	304 kg (670 lb)
Performance	
T-O	
T-O run, at S/L, RQ-5A	200 m (657 ft)
Climb	
Rate of climb, max, at S/L	232 m/min (761 ft/min)
Altitude	
Service ceiling	
RQ-5A	4,570 m (15,000 ft)
MQ-5B	6,095 m (20,000 ft) (est)
MQ-5C	7,620 m (25,000 ft) (est)
Hunter II	8,535 m (28,000 ft)
Speed	
Max level speed	
RQ-5A	110 kt (204 km/h; 127 mph)
MQ-5B, MQ-5C	120 kt (222 km/h; 138 mph)
Hunter II	160 kt (296 km/h; 184 mph)
Cruising speed, max	80 kt (148 km/h; 92 mph)
Loitering speed	60 kt (111 km/h; 69 mph)
Radius of operation	108 n miles (200 km; 124 miles) (est)
Endurance	
at 4,575 m (15,000 ft) at 100 n miles (185 km; 115 miles) from base, RQ-5A	8 hr
max, RQ-5A	12 hr
max, MQ-5B	21 hr (est)
max, MQ-5C, Hunter II	30 hr
Power plant, RQ-5A	2 × piston engine

Status: All seven US LRIP systems delivered by September 1995. Plans for eventual 50 systems (US Army 24, Navy 18, Marine Corps five, plus three for training) curtailed by termination of programme in early 1996. One complete and one partial system, plus a number of spare air vehicles, were kept in service for development work and operator training; remainder placed in store, from which 12 were deployed in March 1999 for operations in the Kosovo theatre. Some, or all, of these had been upgraded by fitment of a laser designator. Orders for MQ-5B, as of early 2006, were for 18 air vehicles; as of mid-2006, a total of 40 was said to be required. In November 2008 the US Army ordered 12 more MQ-5B air vehicles, six ground stations and other related equipment. Total Hunter flight hours had exceeded 80,000, including more than 53,000 hours of combat operation, by the end of 2009. The year 2010 marked the eleventh year of deployed US Army service for the MQ-5.

Customers: Belgium: Belgian Army deliveries (three aircraft) began in early 2001, with initial trials at Elsenborn in March and at Koksijde later; remaining deliveries from September 2001 and continuing into 2002. Equip No. 80 UAV Squadron (so renamed from former two platoons of the 80th *Artillerie Batterie d'Observation et de Surveillance*) at Elsenborn; IOC achieved 1 July 2004. Deployed to Tuzla, Bosnia, from 22 June 2005 on behalf of EUFOR peacekeeping force for four-month tour, ending on 30 October 2005.

France: French MoD ordered a four air vehicle Hunter system and two GCSs from IAI in February 1996 for special operation by the Direction du Renseignement Militaire. This was delivered to the CEAM at Mont-de-Marsan on 20 January 1998 and equipped EE 01.330 'Adour' at Creil. An upgrade package, including an improved sensor payload, was offered to the French Army by EADS in 2002. French Hunters were deployed to the Democratic Republic of the Congo in mid-2003. They were put up for sale in mid-2005.

Israel: Israeli Hunters (local name *Cachlileet*: Magpie) equip No. 200 Squadron at Palmachim.

United States: LRIP systems 1 and 2 to US Army 1995 (C and D Companies, 304 Military Intelligence Battalion, Fort Huachuca, Arizona); others to White Sands Missile Range, Yuma Proving Grounds and elsewhere for continuing test and evaluation. Operated in 1996 by A Company, 504 Military Intelligence Brigade (Aerial Exploitation) at Fort Hood, Texas, and D Company, 304 MI Battalion of 111 MI Brigade at Fort Huachuca, Arizona. System at Fort Hood used for exercise participation, payload development, and US Army operational concepts (CONOPS) evaluation (plus US Navy CONOPS in 1997). A partial system at Fort Huachuca is used primarily for training.

First US Army unit to receive MQ-5B (18 ordered) is 224th MI Battalion at Fort Bragg, which began training in January 2006 following delivery of first 'One System' GCS in October 2005. It deployed to Iraq in April 2006, concurrent with conversion of the 15th MI Battalion; third unit, from October 2006, is the 1st MI Battalion.

US Department of Homeland Security used two Hunters for US/Mexico border patrol for a few months from November 2004, replacing Hermes 450s used in similar role earlier that year.

Contractor: Northrop Grumman Unmanned Systems
San Diego, California, United States. Israel Aircraft Industries
Tel-Aviv, Israel.

Iran

HESA Ababil

Type: Multirole RPV or UAV.

Development: HESA established a subsidiary named Qods Aviation Industries in 1984 to specialise in RPV development. The Ababil programme began at Qods (Arabic for 'Jerusalem') in 1986 and the first deliveries were made in 1993. Since then a number of variants have been developed, and at an Iranian air show in late 2002 the Ababil was presented as a HESA, rather than Qods, product.

Description: *Airframe:* Mainly cylindrical fuselage with ogival nosecone, large sweptback vertical fin (twin, smaller fins on Ababil-T) and pusher engine; swept wings at rear, mounted to fuselage underside; swept metal construction foreplanes on top of fuselage near nose. Metal construction (all-composites on Ababil-T).

Mission payloads: Radar and IR (flare) augmentation devices and acoustic MDI in Ababil-B; small camera (for navigation?) in Ababil II; TV camera with real-time imagery downlink in Ababil-S, plus onboard digital processor; Ababil-T has daylight TV sensor and HE warhead.

Guidance and control: Basic Shahid Noroozi Iranian-developed remote control system incorporates altitude hold, radar tracking and telemetry, with flight data displayed and stored on computer. Alternative autopilot system for longer ranges displays radar navigation and flight data. Ababil-S can be preprogrammed via mission computer and has multichannel digital communications links; both -S and -T versions have GPS navigation.

System composition: Two air vehicles; GCS; launcher; ground support vehicles.

Launch: From Benz 911 truck-mounted pneumatic launcher designed and manufactured by HESA; can also be zero-length launched with booster rocket assistance. Launchable from land or ship's deck.

Recovery: By belly skid landing or HESA (Qods) cruciform parachute.

Power plant: One 18.6 kW (25 hp) Meggitt WAE 342 Hurricane two-cylinder two-stroke engine standard, driving two-blade pusher propeller. One 22.4 kW (30 hp) P 73 Wankel-type rotary engine optional. Fuel capacity 17 litres (4.5 US gallons; 3.7 Imp gallons).

Variants: *Ababil-B:* Initial production version, in service from 1993. Used mainly as aerial target for Iranian Army air defence units.

Ababil II: Close-range UAV; reportedly first flown in October 1997, but not revealed until March 1999. Improved flight control system; may have been prototype for Ababil-S.

Ababil-B aerial target version *(Qods)* 0044402

HESA Ababil-S surveillance UAV *(Robert Hewson)* 0528961

An Ababil-T departing its launcher *(Robert Hewson)*

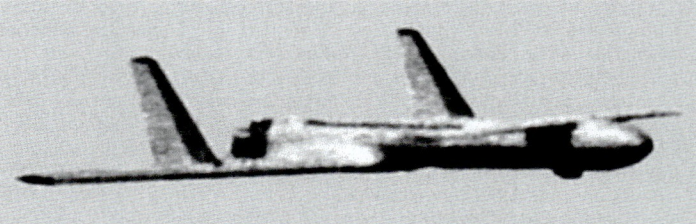
The warhead-carrying Ababil-T attack UAV *(HESA)*

Wreckage of the Mirsad-1 shot down by an ISAF F-16 on 7 August 2006 *(IDF)*

Tailfin of the wrecked Mirsad-1 shows insignia of Islamic Resistance, the armed wing of Hizbullah *(IDF)*

Ababil-S: Medium-range reconnaissance and surveillance UAV version, announced early 2000.

Ababil-T: Short/medium-range attack UAV, with 45 kg (100 lb) HE warhead; distinguishable by twin-tailed configuration. Can engage both fixed and mobile targets. Hizbullah name **Mirsad-1**.

Ababil-B, Ababil II, Ababil-S, Ababil-T

Dimensions, External	
Overall	
length, Ababil II	2.88 m (9 ft 5½ in)
height, Ababil II	0.91 m (2 ft 11¾ in)
Fuselage	
width, Ababil II	0.25 m (9¾ in)
Foreplanes	
foreplane span, Ababil II	1.69 m (5 ft 6½ in)
Wings	
wing span, Ababil II	3.25 m (10 ft 8 in)
Engines	
propeller diameter, Ababil II	0.61 m (2 ft 0 in)
Areas	
Wings	
Gross wing area, Ababil II	1.76 m² (18.9 sq ft)
Weights and Loadings	
Weight	
Max launch weight, Ababil II	83 kg (182 lb)
Payload	
Max payload, Ababil II	40 kg (88 lb)
Performance	
Altitude	
Service ceiling	
Ababil II	3,300 m (10,820 ft)
Ababil-S	4,265 m (14,000 ft)
Speed	
Launch speed, Ababil II	32.92 m/s (108 ft/s)
Max level speed	
Ababil II	200 kt (370 km/h; 230 mph)
Ababil-S	162 kt (300 km/h; 186 mph)
Radius of operation	
mission, Ababil-B	16.2 n miles (30 km; 18 miles)
mission, Ababil II	65 n miles (120 km; 74 miles)
mission, Ababil-S	81 n miles (150 km; 93 miles)
LOS, mission, Ababil-T	27 n miles (50 km; 31 miles)
mission, GPS, Ababil-T	81 n miles (150 km; 93 miles) (est)
Endurance, Ababil-B	1 hr 30 min

Status: In production and service. A newspaper report from Tehran said 58 Ababil air vehicles were planned for production in March 2006-March 2007. A UAV shot down by US fighters over Iraq in 25 February 2009 was reported to have been an Iranian-operated Ababil.

Customers: Iranian armed forces. Exports are understood to have included eight Ababil-T/Mirsad-1s in about October 2004 to the armed wing of the Lebanese organisation Hizbullah. Incursions by these into Israeli airspace in November 2004, April 2005 and August 2006 have resulted in the loss of at least two, on 7 November 2004 and 7 August 2006.

Contractor: Iran Aircraft Manufacturing Industries (HESA), Esfahan.

HESA Karrar

Type: Armed UAV.

Development: This previously unseen UAV was unveiled on 22 August 2010 at the Malek-e-Ashtar University of Technology in the presence of Iranian President Mahmoud Ahmadinejad and Defence Minister Brigadier General Ahmad Vahidi.

Preliminary estimates place the Karrar (Farsi for 'Striker') in roughly the same size and weight class as a trio of well-known aerial targets: South Africa's Denel Skua (which it most closely resembles), the US MQM-107 Streaker, and India's ADE Lakshya. A Lakshya clone named Raa'd (Thunder) is used by Iran; there is also some evidence that 'unidentified third parties' received some Skua technical data via one of that target's export customers. It may also be commented that, apart from the black-painted example at the unveiling, other photographs released at the time by Iranian news agencies show the Karrar in the all-over red paint finish normally associated with target aircraft.

Power plant of the Karrar has not been positively identified, but the domestically produced Tolloue 5 turbojet or the French Microturbo TR 60-5 (which powers the Lakshya) are thought to be the most likely candidates. Iranian officials are quoted as saying that the aircraft has both attack and surveillance capabilities, though there is no visible evidence of any onboard EO/IR sensors or satcom navigation capability. However, suggestions that the Karrar is solely a 'one-way' attack drone are belied by the presence of a recovery parachute compartment aft of the dorsal air intake.

Description: *Airframe:* Low wing, clipped-delta configuration, with cylindrical, blunt-nosed fuselage having dorsal intake for engine; twin arrowhead-shape endplate tailfins. No landing gear.

Mission payloads: Reported weapon payloads include a single 227 kg (500 lb) precision-guided 'MK 82 type' bomb on the centreline, or two underwing stations for 113 kg (250 lb) bombs, Kosar anti-shipping missiles (Iranian variant of the Chinese C-701), or Nasr-1 short-range cruise missiles.

Guidance and control: Karrar is estimated to be preprogrammed for GPS navigation to a BLOS target waypoint, with manual control for weapon release or emergency override.

Launch: Ground launch from zero-length ramp by booster rocket. Also claimed to be capable of air launch.

Recovery: Parachute recovery.

Power plant: Single turbojet (see above), with likely rating of about 4.2 to 4.4 kN (950 to 1,000 lb st).

Karrar

Dimensions, External	
Overall, length	4 m (13 ft 1½ in)
Wings, wing span	2.5 m (8 ft 2½ in)
Weights and Loadings	
Weight, Max launch weight	700 kg (1,543 lb) (est)
Payload, Max payload	227 kg (500 lb)
Performance	
Speed, Max level speed	486 kt (900 km/h; 559 mph)
Radius of operation	270 n miles (500 km; 310 miles) (est)
Power plant	1 × turbojet

Status: In production; service status (as of end 2011) not established with certainty. Unveiling may mark conclusion of development phase and readiness for serial production. There is also speculation that it may be an interim stage towards development of a true cruise missile.

Contractor: Iran Aircraft Manufacturing Industries (HESA), Esfahan.

Karrar at its official unveiling in August 2010, wearing "Bomber Jet" titles along its nose section (FARS News Agency) 1395258

Karrar air vehicles undergoing final assembly (FARS News Agency) 1395259

Karrar with underfuselage bomb (FARS News Agency) 1395260

Karrar makes a rocket-assisted ground launch (FARS News Agency) 1395261

Qods Mohadjer

Type: Short-range multirole UAV family.

Development: The Mohadjer (also transliterated as Mohajer) was apparently developed during the 1980–88 Iran-Iraq war, being operated by the Pasdaran (Iranian militia) during the later stages of that conflict — in an attack role with underwing RPG-7 anti-tank rockets, according to one account. Its name is Arabic for 'migrant'. The UAV first came to public notice at an armed forces display in Iran in September 1999, at which time

Rocket-assisted rail launch of a Mohadjer 3 (IHS/H Farmehr) 0048921

The Mohadjer tactical UAV, probably a Mk 3 or 4 1151626

the latest version, Mohadjer 4, was reported to have entered production in late 1997.

Description: *Airframe:* Mid-mounted, untapered wings; bullet-shaped fuselage; twin tailbooms; twin fins and rudders bridged by horizontal tail surface. Construction substantially of composites. Retractable belly skids or wheeled landing gear, according to version.

Mission payloads: Vary according to version and armed forces' requirements, but designed for reconnaissance, surveillance, artillery fire support, ECM and communications relay. Mohadjer 2 and 3 have frame or video (monochrome or colour) cameras; Mohadjer 4 has IR camera and is reportedly also capable of communications relay and EW roles. All three versions have an onboard digital processor and can downlink sensor imagery.

Guidance and control: Hyarat 3 guidance and control system. Normally preprogrammed via mission computer and radio uplink. Waypoints and other mission profiles can be updated during flight in Mohadjer 4/Hodhod, which has GPS navigation.

Launch: Rail launched from land or on board ship by PL3 pneumatic catapult (Mohadjer 2), booster rocket (Mohadjer 3), or by conventional wheeled take-off. Rail can be truck-mounted or wheeled for independent mobility.

Recovery: Parachute recovery, skid or wheeled landing.

Power plant: Single 18.6 kW (25 hp) flat-twin piston engine (type not known) in Mohadjer 2 and 3; 28.3 kW (38 hp) unidentified engine in Mohadjer 4; two-blade pusher propeller in each case.

Variants: *Mohadjer 2:* Baseline reconnaissance and surveillance version; skid landing gear only.

Mohadjer 3: Also known as **Dorna** (bluebird). More capable than Mohadjer 2, but lacks GPS of Mohadjer 4. Choice of skid or wheel landing gear.

Mohadjer 4: Also known as **Hodhod** (a hooded bird). Most recent and most capable version; can be used for communications relay; also said to have 'impressive' ECM capability; equipped with GPS.

Mohadjer 2, Mohadjer 3, Mohadjer 4

Dimensions, External	
Overall	
length, Mohadjer 2	2.91 m (9 ft 6½ in)
height, Mohadjer 2	1.03 m (3 ft 4½ in)
Wings	
wing span, Mohadjer 2	3.80 m (12 ft 5½ in)
Tailplane	
tailplane span, Mohadjer 2	0.915 m (3 ft 0 in)
Weights and Loadings	
Weight	
Weight empty, Mohadjer 2	70 kg (154 lb)
Max launch weight	
Mohadjer 2	85 kg (187 lb)
Mohadjer 4	175 kg (385 lb)
Performance	
Altitude	
Service ceiling	
Mohadjer 2	3,355 m (11,000 ft)
Mohadjer 4	5,485 m (18,000 ft)
Speed, Max level speed	108 kt (200 km/h; 124 mph)
Radius of operation	
mission, Mohadjer 2	27 n miles (50 km; 31 miles)
mission, Mohadjer 3	54 n miles (100 km; 62 miles)
mission, Mohadjer 4	81 n miles (150 km; 93 miles)
Endurance	
Mohadjer 2	1 hr 30 min
Mohadjer 4	7 hr
Power plant, Mohadjer 2, Mohadjer 3	1 × piston engine

Status: In service; production believed to be complete. A Mohadjer may have been shot down over Iraq on 19 April 2001. A Mohadjer was displayed in the Tehran National Armed Forces Day parade in March 2008.

Customers: Iranian armed forces. Mohadjer 4 reported in use by Iranian Border Guards to detect illegal drug trafficking.

Contractor: Qods Aviation Industries
Tehran.

Model of the Mohadjer 4 *(Robert Hewson)*

Qods Mohadjer 2 UAV *(Robert Hewson)*

Drive-by of a pair of Mohadjers during an Iranian Army display in 2006

Qods Talash

Type: Close-range operator training UAV and target.

Development: Talash 1 (endeavour), the first unmanned aircraft produced by Qods, was developed to train ground operators in remote piloting techniques. Talash 2 was designed as a basic training target for AA gunnery systems.

A longer nose and raked wingtips identify Talash 2

Talash 1 has equi-tapered wings

Description: *Airframe:* Fairly basic 'model aeroplane' configuration high-wing monoplane with fixed mainwheels and a tailskid (Talash 1). Talash 2 has retractable belly skid to absorb landing impact if parachute recovery system fails. Composites construction.

Mission payloads: No specific details known.

Guidance and control: Radio command and control within visual range.

Launch: Conventional wheeled take-off (Talash 1). Rail launch by pneumatic (compressed nitrogen) catapult or from zero-length launcher by booster rocket (Talash 2).

Recovery: Conventional wheeled landing (Talash 1). Qods cruciform parachute recovery system, or by belly skid landing (Talash 2).

Power plant: One 30 cc single-cylinder two-stroke engine (type and rating not known); two-blade tractor propeller.

Variants: *Talash 1:* Initial version.

Talash 2: Upgraded version. Modified airframe, higher speed, improved handling and increased control range. Alternative designation **Hadaf 3000** (Target 3000)

Talash 1, Talash 2

Dimensions, External	
Overall	
length	
Talash 1	1.70 m (5 ft 7 in)
Talash 2	1.90 m (6 ft 2¾ in)
Wings	
wing span	
Talash 1	2.64 m (8 ft 8 in)
Talash 2	2.10 m (6 ft 10¾ in)
Weights and Loadings	
Weight	
Max T-O weight	
Talash 1	12 kg (26 lb)
Talash 2	11 kg (24 lb)
Performance	
Altitude, Service ceiling	2,745 m (9,000 ft)
Speed	
Max level speed	
Talash 1	48 kt (89 km/h; 55 mph)
Talash 2	65 kt (120 km/h; 75 mph)
Radius of operation, radio control	0.2699 n miles (km; miles)
Endurance	30 min
Power plant	1 × piston engine

Status: In production and service.

Customers: Iranian armed forces.

Contractor: Qods Aviation Industries
Tehran.

Israel

Aeronautics Aerolight, Aerosky and Aerostar

Type: Tactical UAVs.

Development: Aeronautics Defence Systems (ADS) specialises in comprehensive unmanned solutions for the military and civilian market. It has a workforce of more than 300, and has established an Unmanned Vehicles Academy (UVA) to support its customers with services and training. Strategic partners include Israel Military Industries, General Dynamics (US) and Motorola; is the owner of Italian engine company Zanzottera Technologies and CommTact of Israel (datalinks); and is part owner of the Israeli companies BVR (simulators) and Controp (optical payloads).

The Aerostar TUAV family was introduced in 2000.

Description: *Airframe:* High-wing monoplanes with pod fuselage, pusher engine and single (Aerolight) or twin (Aerosky, Aerostar) tailbooms and T-tail unit. Fixed tricycle landing gear. Composites construction, in-house by ADS.

Mission payloads: All off-the-shelf, to customer's requirements. Pan-tilt-zoom optical or stabilised Controp EO/IR camera in Aerolight; stabilised, gimbal-mounted day/night EO sensor standard in Aerosky and Aerostar. CommTact UAV datalink.

Guidance and control: Aeronautics generic mission control station (MCS) enables a user to perform various real-time missions with unmanned air, surface or ground vehicles. The system consists of advanced software and user-friendly displays for comprehensive mission management, simulation, briefing and debriefing. The MCS can be provided in various configurations, and can be installed in shelters, on board vehicles or inside headquarters. A compact and lightweight remote payload control station (RPCS) (also referred to as remote video terminal (RVT Level 3, as classified by NATO STANAG 4586) is used for receiving real-time video images directly from the unmanned platform, by units in the vicinity of the mission area, independently of the MCS. The RPCS allows the user full control of the payload, and can be used by ground forces, special forces, police units and others. Aeronautics' own Unmanned Miniature Applications System (UMAS) is an advanced, multi-application command and control, navigation and communications system which features 'plug and play' interface to and between any system, sensor or network and enables system commonality and interoperability.

System composition: Three (Aerolight), or three to six (Aerosky, Aerostar) air vehicles, with payloads; ground control MCS, RPCS and UMAS; logistics support.

Launch: By catapult or conventional wheeled take-off.

Recovery: Parafoil recovery or conventional wheeled landing.

Power plant: Aerolight: One 4.5–8.2 kW (6–11 hp) piston engine (type unspecified); two-blade pusher propeller.

Aeronautics Aerostar tactical UAV *(NMSU)*

ADS Aerolight small UAV *(Aeronautics)*

Aeronautics Aerostar tactical UAV *(NMSU)*

Unmanned aerial vehicles > Israel > Aeronautics Aerolight, Aerosky and Aerostar

Aerosky: One 11.2–14.9 kW (15–20 hp) Zanzottera piston engine; two- or three-blade pusher propeller.
Aerostar: One 28.3 kW (38 hp) Zanzottera piston engine; two-blade pusher propeller.
Variants: *Aerolight:* Close-range (OTH) reconnaissance and surveillance version; also suitable as UAV operator training platform.
Aerosky: Close-range version for reconnaissance, surveillance, target acquisition and designation; also suitable for operator training.
Aerostar: Short/medium-range ISTAR version.

The close-range Aerosky *(IHS/Kenneth Munson)*

Multisensor Aerostar with Rafael ESM. Irkut has acquired a similar system *(IHS/Kenneth Munson)*

Aeronautics Aerostar *(Aeronautics)*

Aeronautics Aerostar tactical UAV *(Aeronautics)*

Aerolight, Aerosky, Aerostar

Dimensions, External
  Overall
    length
      Aerolight .... 2.57 m (8 ft 5¼ in)
      Aerostar .... 4.50 m (14 ft 9¼ in)
    height
      Aerolight .... 0.84 m (2 ft 9 in)
      Aerostar .... 1.22 m (4 ft 0 in)
  Wings
    wing span
      Aerolight .... 4.06 m (13 ft 3¾ in)
      Aerosky .... 4.50 m (14 ft 9¼ in)
      Aerostar .... 6.50 m (21 ft 4 in)
Weights and Loadings
  Weight
    Max T-O weight
      Aerolight .... 36.3 kg (80 lb)
      Aerosky .... 70.3 kg (154 lb)
      Aerostar .... 200 kg (440 lb)
  Payload
    Max payload
      Aerolight .... 8.0 kg (17.00 lb)
      Aerosky .... 18.1 kg (39 lb)
      Aerostar .... 49.9 kg (110 lb)
Performance
  Altitude
    Service ceiling
      Aerolight .... 3,050 m (10,000 ft)
      Aerosky .... 4,570 m (15,000 ft)
      Aerostar .... 5,485 m (18,000 ft)
  Speed
    Max level speed
      Aerolight .... 80 kt (148 km/h; 92 mph)
      Aerostar .... 110 kt (204 km/h; 127 mph)
    Loitering speed
      Aerolight .... 50 kt (93 km/h; 58 mph)
      Aerostar .... 60 kt (111 km/h; 69 mph)
  Radius of operation
    datalink, Aerolight .... 27 n miles (50 km; 31 miles)
    datalink, Aerosky .... 54 n miles (100 km; 62 miles)
    datalink, Aerostar .... 108 n miles (200 km; 124 miles) (est)
  Endurance
    Aerolight .... 5 hr
    Aerosky .... 5 hr (est)
    Aerostar .... 14 hr

Status: Aerolight and Aerostar are in production, and in service "on four continents"; Aerosky status uncertain.

Aerolight: This UAV is the main operator trainer for the Israel Air Force UAV training school. One aircraft took part on 31 March 2003 in a test with the US Navy's 'Hairy Buffalo' NP-3C Orion network-centric warfare testbed, when the latter became the service's first fixed-wing platform to demonstrate Level IV successful airborne control of a UAV during flight. The test, using one control station on the ground and one aboard the NP-3C, was part of a programme to assess the utility of an adjunct UAV role for the US Navy's future Multimission Maritime Aircraft (MMA) programme.

Aerostar: This version was being operated for the IDF/AF by the manufacturer in 2003, on surveillance and anti-smuggling missions over the Gaza Strip. Another, displayed at the 2003 Paris Air Show, was equipped with a new sensor package combining a standard optical payload with Rafael Top-Scan ESM. The system purchased by Russian manufacturer Irkut in 2003 is similar to this latter version.

From August 2003, Aerostars operated by company personnel were used for round-the-clock protection of about 100 US-owned oil platforms, scattered off the Angolan coast and licensed to drill by the national oil company Sonangol. Since then, they have been employed on daily patrols of the Sonangol oilfields, and by January 2005 the Aerostars in this project had accumulated more than 6,000 flight hours.

From September 2004, Aerostars produced, managed and handled by General Dynamics Ordnance and Tactical Systems of St Petersburg, Florida, have provided training support to the US Naval Strike and Air Warfare Center in Fallon, Nevada.

Poland ordered two Aerostar systems (eight air vehicles) together with GCS and support in 2010 and the Dutch Armed Forces leased Aerostar systems from Israel for deployment in Afghanistan 2009-2010.

In early 2006, Aeronautics and GD were jointly seeking FAA certification for the Aerostar. Trials on behalf of the Israeli Highway Police were announced in February 2006. As of August 2006, Aerostar was also being used by New Mexico State University (NMSU) to test a 'sense and avoid' autonomous anti-collision system destined for installation on the GA-ASI Predator UAV.

Customers: *Aerolight:* Israel Air Force; US Navy; other (unidentified) customers.
Aerostar: Angola (Sonangol oil company); Israel Defence Force/Air Force; Russian Federation (Irkut); Nigeria (USD260 million March 2006 contract for three systems, for Coast Guard use); US Navy; New Mexico State University; Poland two systems (USD30 million February 2010).

Contractor: Aeronautics Defense Systems Ltd
Yavne.

Aeronautics Dominator II

Type: Multirole MALE UAV.

Development: The Aeronautics Dominator II UAV (designated as the Dominator XP in export configuration) is based on the airframe of the Diamond Aircraft Industries DA42 NG Twin Turbo general aviation aircraft. Key programme events include:

23 July 2009 The prototype Dominator II UAV (Israeli civil registration 4X-UAG) made its maiden flight.

October 2010 Israeli media sources were reporting Aeronautics as being close to completing the second prototype Dominator II and to have test flown AV number one with an increased fuel capacity that had enabled it to extend its endurance to 24 hours (against a planned target of 28 hours). Elsewhere, 2010 also saw Aeronautics and US contractor Boeing sign a Dominator II marketing memorandum of understanding under which, Boeing was to be responsible for initial marketing of the AV in Australia, Austria, NATO member states and Switzerland.

February 2011 Israeli media sources were reporting Aeronautics as having launched series production of the Dominator II and to have been in negotiation with "more than one" potential customer for the system.

January 2012 Israeli media sources were reporting the Israeli Ministry of Defence as having blocked the sale of Dominator XP AVs to Abu Dhabi for undisclosed reasons.

Description: *Airframe:* All-composites design: low-wing monoplane with twin wing-mounted engines, tapered rear fuselage and T tail. Retractable tricycle landing gear.

Mission payloads: Prototypes fitted with Rafael Recce-U EO/IR in undernose turret. Production versions could accommodate various optical, satcom or EW payloads.

Guidance and control: Dominator II operates under the control of the same UMAS automated flight control system as the company's other UAVs.

Launch: Conventional and automatic wheeled take-off.
Recovery: Conventional and automatic wheeled landing.
Power plant: Two 123.5 kW (165.6 hp) Austro Engine AE 300 turbo-Diesel engines; three-blade tractor propellers.

Dominator II

Dimensions, External	
Overall length	8.53 m (27 ft 11¾ in)
height	2.44 m (8 ft 0 in)
Wings, wing span	13.41 m (44 ft 0 in)
Areas	
Wings, Gross wing area	16.28 m² (175.2 sq ft)
Weights and Loadings	
Weight, Max T-O weight	1,780 kg (3,924 lb)
Payload, Max payload	400 kg (881 lb)
Performance	
Altitude, Service ceiling	9,145 m (30,000 ft)
Speed	
Max level speed	180 kt (333 km/h; 207 mph)
Loitering speed	80 kt (148 km/h; 92 mph)
Radius of operation, mission	162 n miles (300 km; 186 miles)
Endurance, max	28 hr

Status: As of September 2012, the Dominator II/XP programme was understood to be live, with Aeronautics being credited with the production of two prototypes and at least eight production AVs (latter figure unconfirmed). Sources further report that the Turkish Army has leased two Dominator II AV for border surveillance duties, with Aeronautics providing the launch and recovery personnel and Turkey the system operators. As of September 2012, such a deal had not been confirmed.

Contractor: Aeronautics Ltd
Yavne.

Dominator II maiden flight, 23 July 2009 *(Aeronautics)* 1363383

Civil-registered Dominator II prototype *(Aeronautics)* 1356250

Aeronautics Orbiter

Type: Close-range surveillance mini-UAV.

Development: Introduced late 2004. Intended for military (battalion or company level) and homeland security missions.

Description: *Airframe:* Sweptback, flying-wing configuration, with winglets; minimal, tapered, circular-section fuselage with transparent nosecap; no landing gear. Can be quickly assembled and dismantled.

Mission payloads: Controp D-STAMP (day) or L-STAMP (LLTV) sensor or high-resolution, stabilised colour CCD camera. Night sensor optional. Real-time data transmission.

Guidance and control: By personal ground control station (PGCS); or by Aeronautics UMAS, as described for the company's Aerostar family. Avionics and datalink described as 'advanced'; GPS/INS navigation. Payload LOS and guided targets flight. Designed for single-operator use after minimal training.

Transportation: Man-portable. Entire system is fully fieldable in two backpacks.

System composition: Air vehicle(s), hand-held PGCS and datalink unit.

Launch: Can be launched by hand, bungee cord or catapult.
Recovery: Automatic parachute recovery.
Power plant: Brushless electric motor; two-blade pusher propeller.

Orbiter in service with the Polish Armed Forces *(Aeronautics)* 1369051

Orbiter backpacks, PGCS, air vehicle and launcher *(Aeronautics)* 1194764

Aeronautics Orbiter mini-UAV *(Aeronautics)* 1140352

80 Unmanned aerial vehicles > Israel > **Aeronautics Orbiter** – **Aeronautics Picador**

Orbiter recovers by parachute *(Aeronautics)* 1369054

The Orbiter can be hand-, bungee- or catapult-launched *(Aeronautics)* 1132054

Launched from a Land Rover *(Aeronautics)* 1369050

Orbiter air vehicle in flight *(Aeronautics)* 1194765

```
Performance
 Altitude
  Operating altitude
   min .................................................................................. 150 m (500 ft)
   max ................................................................................ 610 m (2,000 ft)
   Service ceiling ........................................................... 3,050 m (10,000 ft)
 Speed, Max level speed ............................................. 75 kt (139 km/h; 86 mph)
 Radius of operation, LOS, datalink ........................ 8.1 n miles (15 km; 9 miles)
 Endurance ............................................................................................. 2 hr
 Power plant ............................................................................. 1 × electric motor
```

Status: Operational in Afghanistan and Chad, according to the manufacturer.

Customers: Polish Army. Three systems ordered by Polish Ministry of National Defence for its Special Forces in November 2005; estimated contract value USD500,000; contract for purchase of a further six systems signed in July 2007 at a cost of USD3 million. Another un-named European NATO country has bought Orbiter. The Irish Defence Forces bought two systems with six air vehicles in 2007 and have deployed on peace monitoring operations. Two CIS member states are also customers.

Contractor: Aeronautics Defense Systems Ltd
 Yavne.

Aeronautics Picador

Type: Helicopter UAV.

Development: The Picador is Aeronautics' first VTOL UAV and is based on the H2S two-seat kit-built helicopter produced by Dynali SA Hélicoptères of Belgium, of which ADS is part-owner. When it was unveiled at the Paris Air Show in June 2009, some flights of a half-scale model had been conducted. Flight control systems, navigation and communication with GCS were all demonstrated with scaled models. A full-scale Picador, which was expected to appear in late 2009, has yet to emerge.

Description: *Airframe:* Two-blade main and eight-blade shrouded tail rotor. Stainless steel fuselage frame; composites blades on both rotors. Fixed tailplane with endplate fins. Cockpit/cabin of manned version replaced by new flat-plate, faceted fuselage. Twin-skid landing gear.

 Mission payloads: Can be equipped with a variety of payloads including EO/IR; laser pointer or designator; maritime or SAR radar; communications relay; elint or sigint packages; or other sensors to customer's requirements.

```
                    Orbiter
Dimensions, External
 Overall
  length ............................................................................... 1.00 m (3 ft 3¼ in)
  height ............................................................................... 0.12 m (4¾ in)
 Wings, wing span ................................................................. 2.20 m (7 ft 2½ in)
Weights and Loadings
 Weight, Max landing weight .................................................... 6.5 kg (14.00 lb)
 Payload, Max payload ............................................................. 1.2 kg (2.00 lb)
```

Picador general appearance *(Aeronautics)* 1356227

Guidance and control: All subsystems controlled by Aeronautics' own UMAS digital flight control system (MTBF 30,000 hours). Autonomous mission programming and GPS/IMU navigation. Commtact advanced multichannel datalink system (C-, L- and S-bands and UHF). Option for remote video terminals: Level 2 displays video and data, Level 3 adds direct payload control.
Launch: Automatic vertical take-off.
Recovery: Automatic vertical landing.
Power plant: One 118 kW (158 hp) Subaru EJ25 flat-four engine.

Picador

Dimensions, External	
Overall	
length	6.58 m (21 ft 7 in)
height	2.58 m (8 ft 5½ in)
Fuselage, width	2.00 m (6 ft 6¾ in)
Rotors, rotor diameter	7.22 m (23 ft 8¼ in)
Tail rotor, tail rotor diameter	0.84 m (2 ft 9 in)
Weights and Loadings	
Weight, Max T-O weight	720 kg (1,587 lb)
Payload, Max payload	180 kg (396 lb)
Performance	
Altitude, Service ceiling	3,660 m (12,000 ft)
Speed, Max level speed	110 kt (204 km/h; 127 mph)
Radius of operation	108 n miles (200 km; 124 miles)
Endurance	8 hr
Power plant	1 × piston engine

Status: Picador is described as a "development project under continuous process" by Aeronautics. No full-scale prototype had appeared by 2011.
Contractor: Aeronautics Defense Systems Ltd
Yavne.

AeroTactiX Skyzer 100

Type: Reconnaissance and surveillance mini-UAV.
Development: Revealed in 2010, and offered as a low-noise, easy to use platform for military, paramilitary and civil missions such as urban warfare, law enforcement, traffic control and air pollution monitoring.
Description: *Airframe:* Mid-mounted, sweptback wings, with winglets; pod fuselage; twin tailbooms, bridged by single horizontal surface; pusher engine. No landing gear. Composites construction.
Mission payloads: High-resolution daytime CCD TV camera; FLIR sensor for night use; or gimballed or fixed stills camera. Real-time imagery downlink; high-speed telemetry transceiver.
Guidance and control: Fully autonomous from laptop GCS in Pelican case; INS/GPS navigation; integral airspeed sensor. 'Return home' capability at end of mission or if command link is lost.
Transportation: Man-portable in single backpack.
Launch: Hand or catapult launch.
Recovery: Belly landing.

The Skyzer 100 swept-wing mini-UAV *(AeroTactiX)* 1395246

Power plant: Brushless electric motor (rating not stated), powered by lithium-polymer batteries; two-blade pusher propeller.

Skyzer 100

Dimensions, External	
Overall, length	1.40 m (4 ft 7 in)
Wings, wing span	2.80 m (9 ft 2¼ in)
Weights and Loadings	
Weight, Max launch weight	6.0 kg (13.00 lb)
Performance	
Altitude	
Operating altitude	90 m to 915 m (300 ft to 3,000 ft)
Service ceiling	2,440 m (8,000 ft)
Speed	
Max level speed	50 kt (93 km/h; 58 mph)
Loitering speed	25 kt (46 km/h; 29 mph)
Radius of operation	
normal	7 n miles (13 km; 8 miles)
with high-gain directional antenna	26 n miles (48 km; 29 miles)
Endurance	1 hr 30 min
Power plant	1 × electric motor

Status: Was promoted in 2010.
Contractor: AeroTactiX Ltd
Petah-Tikva.

BlueBird Blueye

Type: Small tactical UAV.
Development: BlueBird developed the Blueye for both military and civil applications, with particular emphasis on providing photogrammetric coverage of areas of interest. Potential military roles include border patrol, improvised explosive device (IED) detection and convoy escort. Civilian missions could include environmental resource monitoring and pipeline inspection.
Description: *Airframe:* Hump-backed, box-section fuselage with V tail surfaces; parafoil in lieu of fixed wings. Fixed, four-wheel landing gear.
Mission payloads: Blueye carries a photogrammetric system that provides digital orthophotos — aerial photographs that are equivalent to a map of the same scale, thereby enabling exact measurement of distances on the Earth's surface. By comparing two orthophotos taken at intervals, it can also 'flag up' any changes noted between one image and the other, relaying the information to the GCS via datalink in near real time. Geographical co-ordinates can be overlaid on the orthophoto. The system manages the entire process automatically. In addition, elevation data can be derived from the process to create a visualisation of the area from which a 3D stereoscopic model can be built.
Alternative payload options include day (TV) or night (IR) stabilised sensors.

The Blueye UAV complete with its parafoil 'wing' *(BlueBird)* 1290318

Blueye main body *(BlueBird)* 1290319

Guidance and control: Pre-programmed throughout, including take-off and landing; GPS-based navigation, with provision for real-time manual override. GCS crew of two.
Transportation: Transportable in Jeep or small van.
Launch: Automatic wheeled take-off.
Recovery: Automatic wheeled landing.
Power plant: One 12.7 kW (17 hp), 150 cm³ piston engine, driving a two-blade tractor propeller.

Blueye

Dimensions, External	
Fuselage	
length	1.70 m (5 ft 7 in)
height	1.05 m (3 ft 5¼ in)
width	1.40 m (4 ft 7 in)
Wings	
wing span, parafoil	4.50 m (14 ft 9¼ in)
Weights and Loadings	
Weight, Max launch weight	60.0 kg (132 lb)
Payload, Max payload	15.0 kg (33 lb)
Performance	
Altitude, Service ceiling	2,000 m (6,560 ft)
Speed	
Cruising speed, max	35 kt (65 km/h; 40 mph)
Loitering speed	22 kt (41 km/h; 25 mph)
Radius of operation	27 n miles (50 km; 31 miles)
Endurance	8 hr (est)
Power plant	1 × piston engine

Status: Believed to be in service.
Contractor: BlueBird Aero Systems Ltd
Kadima.

BlueBird Boomerang

Type: Close-range tactical mini-UAV.
Development: In development from about 2004.
Description: *Airframe:* All-swept flying wing configuration, with endplate fins and rudders and minimal fuselage. No landing gear. All-composites construction.
Mission payloads: Standard payload is a gyrostabilised, high-resolution daylight TV camera with ×10 optical zoom. IR camera for night observation optional. RS-232 datalink.
Guidance and control: Fully automatic, including launch and recovery.
Transportation: Man-portable in backpack.
Launch: Automatic launch by catapult.
Recovery: Automatic landing by parachute and airbag.
Power plant: Brushless electric motor (rating not stated), driving a two-blade pusher propeller.

Boomerang

Dimensions, External	
Overall, length	1.10 m (3 ft 7¼ in)
Wings, wing span	2.75 m (9 ft 0¼ in)
Weights and Loadings	
Weight, Max launch weight	9.0 kg (19.00 lb)
Payload, Max payload	2.5 kg (5.00 lb)
Performance	
Altitude, Service ceiling	3,000 m (9,840 ft)
Speed	
Cruising speed, max	59 kt (109 km/h; 68 mph)
Loitering speed	32 kt (59 km/h; 37 mph)
Radius of operation, mission	11 n miles (20 km; 12 miles)
Endurance, max	2 hr 30 min
Power plant	1 × electric motor

Status: Production and service status uncertain. Marketing continuing in 2008-09.
Contractor: BlueBird Aero Systems Ltd
Kadima.

The Boomerang flying wing mini-UAV *(BlueBird)* 1290320

BlueBird MicroB

Type: Close-range tactical mini-UAV.
Development: Latest addition to BlueBird product range. Development thought to have begun in about 2006.
Description: *Airframe:* All-swept flying wing configuration, with endplate fins and minimal fuselage. No landing gear. Composites construction.
Mission payloads: EO/IR or other multisensor package in interchangeable nose module. Real-time imagery downlink.
Guidance and control: Pre-programmed and automatic, or will follow camera LOS when in 'camera guide' mode. Provision for operator override. Portable GCS.
Transportation: Man-portable.
Launch: Hand or automatic catapult launch.
Recovery: Automatic recovery by parachute.
Power plant: Brushless electric motor (rating not stated), driving a two-blade pusher propeller.

MicroB

Dimensions, External	
Overall, length	0.60 m (1 ft 11½ in)
Wings, wing span	0.90 m (2 ft 11½ in)
Weights and Loadings	
Weight, Max launch weight	1.0 kg (2.00 lb)
Payload, Max payload	0.45 kg (0.00 lb)

MicroB in flight *(BlueBird)* 1290322

MicroB on its rail launcher *(BlueBird)* 1290321

The MicroB tactical mini-UAV *(IHS/Patrick Allen)* 1329416

MicroB

Performance
Speed
  Cruising speed, max..49 kt (91 km/h; 56 mph)
  Loitering speed..24 kt (44 km/h; 28 mph)
Radius of operation..5.4 n miles (10 km; 6 miles)
Endurance..1 hr (est)
Power plant..1 × electric motor

Status: Production and service status uncertain.
Contractor: BlueBird Aero Systems Ltd
Kadima.

BlueBird SpyLite

Type: Short-range surveillance UAV.

Development: Previously known as SkyLite B, under 2005 teaming arrangement with Rafael; renamed when this partnership ended in 2008.

Description: *Airframe:* Tubular fuselage with nose-mounted sensor; V-configuration tailfins. No landing gear.
  Mission payloads: Three-axis stabilised payloads for surveillance; target acquisition, identification and designation; BDA; and law enforcement. Analogue or digital datalink with real-time video transmission.
  Guidance and control: Backpack- or vehicle-mounted laptop GCS. Fully autonomous operation, with moving target tracking and tracker. Two-person crew.
  Transportation: Man-portable in two backpacks.
  System composition: Three air vehicles, payloads, GCS, launcher and ground/field support equipment.
  Launch: Automatically by catapult from zero-length launcher.
  Recovery: Automatic parachute recovery. Turnaround time less than 10 minutes.
  Power plant: Brushless electric motor (rating not stated), driving a two-blade pusher propeller.

SpyLite

Dimensions, External	
Overall, length	1.15 m (3 ft 9¼ in)
Fuselage, height	0.12 m (4¾ in)
Wings, wing span	2.40 m (7 ft 10½ in)
Engines, propeller diameter	3.20 m (10 ft 6 in)
Weights and Loadings	
Weight, Max launch weight	6.3 kg (13.00 lb)
Payload, Max payload	1.3 kg (2.00 lb)
Performance	
Altitude	
Operating altitude, max	11,000 m (36,080 ft)
Service ceiling	3,000 m (9,840 ft)

Air vehicle as depicted under its former name of SkyLite B *(Rafael)* 1116921

Speed
  Max level speed...54 kt (100 km/h; 62 mph)
  Cruising speed..40 kt (74 km/h; 46 mph)
  Loitering speed...32 kt (59 km/h; 37 mph)
Radius of operation
  mission, analogue link...8.1 n miles (15 km; 9 miles)
  mission, digital link...18.9 n miles (35 km; 21 miles)
Endurance, max...4 hr
Power plant..1 × electric motor

Status: In production and in service.
Contractor: BlueBird Aero Systems Ltd
Kadima.

Elbit Systems Hermes 90

Type: Short-range tactical UAV.

Development: According to *Jane's* sources, the Hermes 90's ancestry dates back to a mid-1990s design by IAI known as Eye View, rights to which were acquired by BAE Systems in 2005 and transmuted into that company's Skylynx I and II, of which the latter competed (unsuccessfully) in a US Marine Corps Tier II competition in 2006. Ownership then evidently reverted to Israel (Elbit), whose essentially similar (though slightly different dimensionally) version was initially designated Skylark II-LE (for long endurance), despite being of a different configuration to the company's existing Skylark II. Under that title, it was bid as the platform by a General Dynamics-led team in a renewed US Navy/Marine Corps STUAS Tier II competition in 2008. Part of this evaluation included shipboard launch and recovery tests in the Mediterranean in September/October 2008. Subsequently, the Skylark II-LE was rebranded as a member of Elbit's Hermes family, first appearing under this title at the Australian International Airshow in Avalon in March 2009.

Description: *Airframe:* High-mounted wings; pod fuselage with rear-mounted pusher engine, above which is a pylon supporting a single tailboom carrying inverted V tail surfaces. Fixed three-skid landing gear (nose and two main).
  Mission payloads: Standard payload is an Elbit/Elop Micro-CoMPASS turret containing a daylight colour TV camera and MWIR (mid-wave infra-red) thermal imager (both with continuous zoom), plus a laser target illuminator. Optional alternatives include an EW (sigint/comint) or communications relay (multiband tactical radios) suite.
  Guidance and control: Utilises same universal (UGCS) or dual (DGCS) ground stations as other members of the Hermes family. Fully autonomous flight; LOS and BLOS datalinks. GCS crew of two.
  Launch: Catapult launch. Automatic mode optional.
  Recovery: Skid landing. Automatic mode optional.
  Power plant: Not revealed, but thought likely to be 28.3 kW (38 hp) UEL AR741 rotary engine; two-blade pusher propeller.

Hermes 90

Dimensions, External	
Wings, wing span	5.00 m (16 ft 4¾ in)
Weights and Loadings	
Weight, Max launch weight	90 kg (198 lb) (est)
Payload, Max payload	30 kg (66 lb)

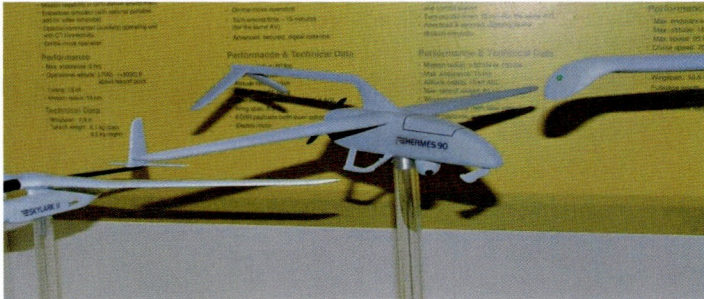

Hermes 90 on display at Australian International Airshow, March 2009 *(IHS/Damian Kemp)* 1308858

The mid-1990s IAI Eye View, from which the Hermes 90 airframe is derived. *(IAI)* 0518228

Hermes 90

Performance
- Altitude, Service ceiling..4,570 m (15,000 ft)
- Radius of operation................................81 n miles (150 km; 93 miles)
- Endurance...15 hr (est)

Status: Being promoted in 2013.
Contractor: Elbit Systems Ltd
  Haifa.

Elbit Systems Hermes 450

Type: Long-endurance tactical UAV.
Development: The Hermes 450 was initially powered by a pair of UEL AR 741 rotary engines, each of 28.3 kW (38 hp). By 1997, the design had been refined into an improved version (known originally as the 450S), having modified outer wings and a single, but higher-powered, tail-mounted rotary. In this form, it was ordered into production for the IDF in June of that year.
Description: Airframe: High-aspect ratio high-wing monoplane with turned-down tips and single tail-mounted engine, V tail unit (included angle 100°) and non-retractable tricycle landing gear. Construction is of composites.
  Mission payloads: Any customer-specified payloads within air vehicle's capacity. Typical sensors include SAR/GMTI, Controp DSP-1 day/night EO or Elop CoMPASS IV multisensor, laser designator, sigint, elint, comint or communications relay. The electrical power available for payloads is 1.6 kW.
  Guidance and control: Fully autonomous flight with in-flight redirection capability; GPS navigation, advanced dual computers, dual datalinks and redundant electrical and avionics systems. It is piloted only during the take-off and landing phases (Optional DGPS automatic T-O and landing available). Modularity and maintainability have been designed into the aircraft from the outset. There is also provision for the integration of an OLOS wide-band digital satellite link to extend the operational radius and obviate dependence upon a separate (relay) aircraft. The air vehicle was displayed at the 2001 Paris Air Show with BAE Systems satcom antenna in a dorsal pod; flight testing of this configuration was completed in 2005.
  System composition: Four to six air vehicles. The Elbit GCS includes a truck-mounted S-280 shelter with two or three 48 cm (19 in) consoles (mission control and payload control), displays, computers, recorders and hard-copy printer, a trailer-mounted GDT, a remote video terminal, tactical communications, a flightline tester/loader and an external power supply (generator and emergency batteries). The system is manned by a ground crew of two.
  Launch: Wheeled (optionally automatic) take-off is standard, but a catapult launch ability was demonstrated in April 2005 using a Robonic MC 2055L launcher, as part of the Elbit entry for the Australian Army's JP 129 tactical UAV competition.
  Recovery: Wheeled (optionally automatic) landing, with arrester cable.
  Power plant: One 38.8 kW (52 hp) UEL AR 801 rotary engine; two-blade pusher propeller. See under Weights for 450 fuel details; 450 LE has additional 50 litres (13.2 US gallons; 11.0 Imp gallons) in two underwing tanks.
Variants: Hermes 450: Standard version. *Description applies to this version except where indicated.*

Elbit 450 in Brazilian colours *(Elbit Systems Ltd)*

Hermes 450 LE at Paris Air Show, June 2003 *(IHS/Kenneth Munson)*

Singapore Air Force Hermes 450 displayed in 2010 *(IHS/Patrick Allen)*

Hermes 450 overflying Arizona in 2004 for US Customs and Border Protection *(US CBP/Gerald L Nino)*

British Army Hermes 450 ZK510 in hangar, Basra, 2008 *(IHS/Patrick Allen)*

Hermes 450, in service with the Israel Defence Forces

Hermes 450 mobile shelter, GCS consoles and typical mission display *(Elbit)* 0528603

Hermes 450 is the first UAV in Israel to receive certification by civil aviation authorities *(Elbit Systems)* 1290196

Hermes 450 LE: Longer-range, twin-payload version which was unveiled at the Paris Air Show in June 2003. Aft-mounted payload bay and two permanent wing-mounted fuel tanks.

WK 450: Designation of version under development for UK Thales Watchkeeper programme and described separately under that heading.

Hermes 450, Hermes 450 LE

Dimensions, External
 Overall
  length.. 6.10 m (20 ft 0¼ in)
  height.. 2.37 m (7 ft 9¼ in)
 Fuselage, width.. 0.52 m (1 ft 8½ in)
 Wings
  wing span... 10.51 m (34 ft 5¾ in)
  wing chord, constant.. 0.69 m (2 ft 3¼ in)
 Tailplane, tailplane span... 2.95 m (9 ft 8¼ in)
 Wheels, wheel track.. 1.45 m (4 ft 9 in)
Dimensions, Internal
 Payload bay
  volume, Hermes 450.. 0.30 m³ (10.6 cu ft)
Areas
 Wings, Gross wing area... 6.90 m² (74.3 sq ft)
Weights and Loadings
 Weight
  Weight empty, Hermes 450.................................. 200 kg (440 lb)
  Max T-O weight, Hermes 450............................... 450 kg (992 lb)
 Fuel weight
  Max fuel weight
   usable, Hermes 450... 105 kg (231 lb)
 Payload
  Max payload
   Hermes 450... 150 kg (330 lb)
   with satcom antenna, Hermes
   450.. 120 kg (264 lb)
Performance
 T-O
  T-O run... 350 m (1,149 ft)
 Climb
  Rate of climb, max, at S/L................................... 274 m/min (900 ft/min)
 Altitude
  Operating altitude, max...................................... 5,485 m (18,000 ft)
 Speed
  Max level speed.. 95 kt (176 km/h; 109 mph)
  Cruising speed.. 70 kt (130 km/h; 81 mph)
  Stalling speed... 42 kt (78 km/h; 49 mph)
 Radius of operation, Hermes 450......................... 108 n miles (200 km; 124 miles)
 Endurance
  Hermes 450.. 20 hr
  Hermes 450 LE... 30 hr

Status: Production of the Hermes 450 began in 1997 and it first entered service in 1999. Further IDF orders were reportedly placed in mid-2000 and mid-2004, the latter being for delivery over a period of three years. No reliable production figures are known but a WK 450 Watchkeeper aircraft seen in 2005 bore constructor's number 103. Elbit stated that the Hermes 460 had accumulated more than 65,000 flying hours by May 2007. In the same month Hermes 450 became the first UAV to receive civil certification in Israel (4X-USY). The Hermes 450 has also been in operational service with the British Army in Afghanistan and Iraq since 2007, conducting intensive operations in both theatres where it flew more than 10,000 hours in a 12 month period, with typical missions lasting 14 hours.

Customers: Israel Defence Forces (No 200 Squadron at Palmachim; possibly also No 155 (Hatserim) and No 146 (Ramon). Botswana Defence Force ordered one system in 2001 and Georgia acquired a number of systems, two of which have been shot down. Croatia was reported as a customer in late 2006.

One or more systems were delivered in mid-2003 to the Joint UAV Test and Evaluation Center at NAS Fallon, Nevada, under a contract awarded to the Elbit subsidiary EFW (Elbit Fort Worth). The systems were used to provide information on joint interoperability questions, including missions in support of US Navy carrier air wing training.

In June 2004, it was announced that EFW had leased a two-aircraft system to the US Department of Homeland Security, Customs and Border Protection to supplement manned aircraft patrolling the Arizona/Mexico border for various applications, including illegal immigration and drug smuggling. This pilot scheme was the first non-military use of UAVs for US border protection. Starting in June 2004, these aircraft flew nearly 480 hours of patrol and are reported as having prevented 11 drug-smuggling attempts — resulting in seizure of over 227 kg (500 lb) of marijuana — and more than 780 attempts at illegal entry to the US.

An order for Hermes 450 was placed by Singapore in 2007.

The system was procured for the British Army under an Urgent Operational Requirement named Project Lydian, operated by 32 Regiment Royal Artillery. Accurate numbers were not disclosed, but it is thought that at least ten air vehicles entered service from 2007 under an initial USD110 million contract, with one vehicle lost in Iraq in 2008.

An un-named European country ordered Hermes 450 in July 2008 under a USD20 million contract for vehicles and ground stations, with a delivery date given as 2009.

A contract covering an undisclosed number of Hermes 450 for "a country in the Americas" was announced in September 2008. In December 2010 Brazil selected the Hermes 450 following a demonstration of the system by Elbit's Brazilian subsidiary Aeroeletronica. Two air vehicles and a GCS are to be supplied initially under a contract finalised in 2011.

Contractor: Elbit Systems Ltd
Haifa.

Elbit Systems Hermes 900

Type: Tactical MALE UAV.

Development: Unveiled 11 June 2007, shortly before presentation at Paris Air Show. System is fully compatible with missions and support infrastructure of the smaller Hermes 450 and retains all the capabilities of that system while featuring larger, multi-payload configurations, higher flight altitude and greater endurance.

Description: *Airframe:* Mid-mounted wings with full-span flaps. Mainly cylindrical fuselage with pronounced bulge above nose; rear-mounted engine; V tail. Fully retractable tricycle landing gear. Composites construction.

Mission payloads: System supports a wide variety of missions and specialised applications, payloads for which can include EO, IR imaging laser rangefinder and laser designator, SAR/GMTI radar, comint D/F, elint and EW. Aircraft is also equipped with ATC radio, radio relay, IFF transponder and built-in autonomous emergency procedures.

Guidance and control: Elbit Systems' Universal Ground Control Station (UGCS, which see) can, with two GDTs, control two air vehicles simultaneously, each air vehicle and its payloads being operated by a single operator. System maintains secure and redundant LOS datalink and redundant BLOS satcom.

System composition:
Launch: Automatic wheeled take-off.

Hermes 900 during its first flight during which the landing gear was not retracted *(Elbit Systems)*

Current Hermes 1500, with long-span wings and ventral radome

Hermes 900 makes its first flight, 9 December 2009 *(Elbit Systems)*

Recovery: Automatic wheeled landing.
Power plant: One 78.3 kW (105 hp) Rotax 914 F turbocharged flat-four engine, driving a pusher propeller.

Hermes 900

Dimensions, External	
Wings, wing span	15.00 m (49 ft 2½ in)
Weights and Loadings	
Weight, Max T-O weight	970 kg (2,138 lb)
Payload, Max payload	300 kg (661 lb)

Status: First flight was made on 9 December 2009, with series production in begun in 2010. In May 2010 an undisclosed number of Hermes 900 systems was ordered by the IDF.
Contractor: Elbit Systems Ltd
Haifa.

Elbit Systems Hermes 1500

Type: Medium-altitude, long-endurance UAV.
Development: Corps to command level MALE UAV, geared for heavy payloads and long endurance. Revealed at Paris Air Show, June 1997, superseding Hermes 750 shown two years earlier. Prototype rolled out in early 1998 and made a 55 minute first flight on 22 May. The original 10 m (32 ft 9.7 in) span wing was subsequently extended to 15 m (49 ft 2.6 in), first flight in this form taking place in early 2002; further extension to 18 m has been made since then. A maritime surveillance version was displayed at the Paris Air Show in June 2005, having flown for the first time earlier that year. Development has been slow by Elbit standards.
Description: *Airframe:* High-aspect ratio high-wing monoplane; twin underwing podded engines with tractor propellers; high-aspect ratio V tail unit; fully retractable tricycle landing gear, with mainwheel brakes. Underfin originally fitted but later deleted. Composites construction, similar to that of Hermes 450.
Mission payloads: Multiple payload capability. Intended primarily for surveillance and reconnaissance. Adaptable to visint (EO/IR staring, scanning or long-range oblique), SAR/GMTI or ISAR radar, communications relay, and 'virtually any' sigint, comint or elint payload, tailored to customer requirements. Up to 9 kW electrical power available.
Guidance and control: Fully autonomous flight with in-flight redirection capability. GPS navigation, advanced dual computers, dual datalinks, and redundant electrics, avionics and engines. Piloted during T-O and landing phases only, but automatic (ATOL) system available optionally. Wideband digital radio LOS and/or satellite communications. GCS is common with others of the Hermes family.
Launch: Wheeled runway take-off.
Recovery: Wheeled runway landing.

Hermes 1500 in maritime configuration at the 2005 Paris Air Show *(IHS/Patrick Allen)*

This view emphasises the high aspect ratio of the Hermes 1500 wings

Power plant: Two 73.5 kW (98.6 hp) Rotax 914 F turbocharged flat-four engines; two-blade propellers. Fuel tanks in wings. Two pylon-mounted underwing auxiliary fuel tanks, each approx 150 litres (39.6 US gallons; 33.0 Imp gallons), on maritime version.

Hermes 1500

Dimensions, External	
Overall, length	9.40 m (30 ft 10 in)
Fuselage	
height, max	0.70 m (2 ft 3½ in)
width, max	0.90 m (2 ft 11½ in)
Wings, wing span	18.00 m (59 ft 0¾ in)
Dimensions, Internal	
Payload bay, volume	2 m³ (71 cu ft)
Weights and Loadings	
Weight	
Max T-O weight	
normal	1,650 kg (3,637 lb)
maritime version	1,750 kg (3,858 lb)
Payload	
Max payload	
normal	300 kg (661 lb)
maritime version	400 kg (881 lb)
Performance	
Climb	
Rate of climb, max, at S/L	275 m/min (902 ft/min)
Altitude, Service ceiling	10,060 m (33,000 ft)
Speed	
Max level speed	130 kt (241 km/h; 150 mph) at S/L
Cruising speed, max	120 kt (222 km/h; 138 mph)
Loitering speed	80 kt (148 km/h; 92 mph)

Hermes 1500

Radius of operation	
LOS, mission	108 n miles (200 km; 124 miles)
OLOS, mission	540 n miles (1,000 km; 621 miles) (est)
Endurance	
max, normal	26 hr
max, maritime version	50 hr (est)

Status: Development was reported to be continuing in 2008, possibly with the emphasis on maritime patrol. Listed as a active programme by Elbit in 2009.

Contractor: Elbit Systems Ltd
Haifa.

Elbit Systems Skylark I

Type: Close-range tactical mini-UAV.

Development: Unveiled at the Paris Air Show in June 2003, the Skylark I is a man-portable system designed for unit-level military or paramilitary use in such roles as 'over the hill' reconnaissance, artillery direction, perimeter security, border and coastal surveillance and law enforcement. The initial version entered Israeli service in March 2005. Towards the end of that year, Elbit announced "significant improvements" in a version which it has designated **Skylark IV**, which it then demonstrated to the Polish MoD. Skylark I LE (Long Endurance) has a greater span and curved wingtips.

Description: *Airframe:* Narrow tubular fuselage with underslung, pylon-mounted payload pod; angular tail surfaces; high-aspect ratio wings, with 13° dihedral on outer panels. Construction, mostly of composites, by Elbit subsidiary Aero-Design & Development Ltd. Quick-connect, snap-together module assembly. No landing gear.

Mission payloads: Interchangeable, fully gimballed and stabilised, daylight colour CCD TV camera with ×10 zoom, FLIR or other, in underfuselage pod. Real-time continuous downlink of video and telemetry data within LOS, reportedly by Tadiran Spectralink StarLink datalink. A new thermal imaging payload, developed by Elbit and weighing only 700 to 800 g (25 to 28 oz), was being evaluated by the IDF in early 2006.

Guidance and control: Miniature ground control unit (MGCU), weighing about 10 kg (22 lb). Fully autonomous (preprogrammed) flight, with GPS positioning; if datalink signal is lost, UAV automatically gains altitude and returns to preset landing area. Sensor payload is commanded and controlled from suitcase-sized portable tactical computer. Encrypted UHF uplink and D/E-band telemetry/video downlink initially.

System composition: Three air vehicles, three (normally two TV and one FLIR) to five payloads and MGCU, contained in two backpacks. Total system weight approximately 40 kg (88 lb) initially for Skylark IV, but expected to reduce to about 30 kg (66 lb) with further development and change from analogue to digital avionics. Operating crew of two.

Launch: Hand- or bungee-launched.

Recovery: Preprogrammed automatic recovery by deep stall to ventral airbag landing. Has demonstrated ability to land routinely within 5 m (15 ft) of operator.

Power plant: Battery-powered electric motor, driving a two-blade propeller.

The hand-launched Skylark I *(Elbit)* 1116829

Skylark I and its MGCU *(IHS/Patrick Allen)* 1149510

Preparing a Skylark I for hand launch *(IHS/Patrick Allen)* 1149516

The Elbit Skylark I mini-UAV *(Elbit)* 0561217

Skylark I LE, identified by curved wingtips *(Elbit)* 1308652

Unmanned aerial vehicles > Israel > Elbit Systems Skylark I – EMIT Dragonfly 2000

Skylark I LE *(Elbit)* 1409835

Skylark IV, Skylark I LE

Dimensions, External	
Overall	
length, Skylark IV	2.20 m (7 ft 2½ in)
Wings	
wing span	
Skylark IV	2.00 m (6 ft 6¾ in)
Skylark I LE	2.90 m (9 ft 6¼ in)
Tailplane	
tailplane span, Skylark IV	0.70 m (2 ft 3½ in)
Engines	
propeller diameter, Skylark IV	0.30 m (11¾ in)
Dimensions, Internal	
Payload bay	
length, Skylark IV	0.72 m (2 ft 4¼ in)
width, Skylark IV	0.10 m (4 in)
Weights and Loadings	
Weight	
Max launch weight	
TV, Skylark IV	4.5 kg (9.00 lb)
FLIR, Skylark IV	5.0 kg (11.00 lb)
Skylark I LE	6.3 kg (13.00 lb)
Performance	
Altitude	
Operating altitude	
min, normal, Skylark IV	300 m (985 ft)
max, normal, Skylark IV	455 m (1,500 ft)
Service ceiling, Skylark IV	5,000 m (16,400 ft)
Speed	
Max level speed, Skylark IV	60 kt (111 km/h; 69 mph)
Cruising speed, Skylark IV	35 kt (65 km/h; 40 mph)
Endurance	
Skylark IV	1 hr 30 min
Skylark I LE	3 hr
Power plant	1 × electric motor

Status: In production and in service. Four Australian Army systems deployed to Iraq late 2005 for operations in Al Muthanna province; remaining two systems retained in Australia for training and preparation purposes.

Customers: Israel Defence Force Ground Forces Command (LRIP of possibly 10 systems ordered 2004, delivered from March 2005; Skylark IV orders placed late 2005; a further in January 2009 order worth USD40 million for Skylark I LE to equip all IDF Ground Forces battalions); Australian Army (six systems ordered November 2005); and, as of early 2006, Singapore, Sweden and 'several' other (unidentified) customer countries, possibly including Canada and France. A contract for Skylark I announced in September 2008 was believed to be for Mexico, although the customer was not disclosed by Elbit.

Contractor: Elbit Systems Ltd
Haifa.

Elbit Systems Skylark II

Type: Close-range tactical UAV.

Development: Essentially a scaled-up version of the hand-launched Skylark I, Elbit's Skylark II was first flown in 2005 and unveiled at the Eurosatory defence exhibition in Paris in June 2006.

Description: *Airframe*: Narrow tubular fuselage with underslung, pylon-mounted payload pod; angular tail surfaces; high aspect ratio wings with dihedral on outer panels. Construction mainly of composites.

Mission payloads: Elop Mini-CoMPASS triple-sensor turret (high-definition colour CCD TV camera, thermal imager and laser illuminator), mounted in nose of underfuselage pod.

Guidance and control: Truck-mounted, two-person Elbit UGCS. Fully autonomous flight modes; fully redundant avionics. Skylark II uses same GCS and datalink as Skylark I and is interoperable with its smaller stablemate. It has a data transfer rate of 1.5 Mbytes/s, using a Starlink digital air data terminal.

The Skylark II close-range tactical UAV *(IHS/Patrick Allen)* 1311603

Size comparison of Skylark II (nearest camera) with the smaller Skylark I *(IHS/Patrick Allen)* 1326086

Transportation: Mounted on HMMWV or similar-sized vehicle, system is fully mobile.

System composition: Typical system comprises two air vehicles, payloads and vehicle-mounted GCS with integral launcher.

Launch: Automatic, under own power, from truck-mounted rail launcher.

Recovery: Automatic recovery by parachute and ventral airbags.

Power plant: 4 kW (5.4 hp) electric motor, powered by rechargeable 'plug and play' batteries; two-blade tractor propeller.

Skylark II

Dimensions, External	
Wings, wing span	5.00 m (16 ft 4¾ in)
Weights and Loadings	
Weight, Max launch weight	43 kg (94 lb)
Payload, Max payload	10 kg (22 lb) (est)
Performance	
Altitude	
Operating altitude	150 m to 1,525 m (500 ft to 5,000 ft)
Service ceiling	4,875 m (16,000 ft)
Speed, Max level speed	70 kt (130 km/h; 81 mph)
Radius of operation	32 n miles (59 km; 36 miles) (est)
Endurance	6 hr
Power plant	1 × electric motor

Status: In production and service. Deliveries thought to have started in early 2007; ordered by South Korean Army (contract quoted as "less than USD5 million") in December 2007; other customers not identified.

Contractor: Elbit Systems Ltd
Haifa.

EMIT Dragonfly 2000

Type: Tactical UAV.

Development: History not known.

Description: *Airframe*: Conventional high-wing monoplane configuration; nose-mounted engine. Fixed tailwheel-type landing gear. Composites construction.

Mission payloads: Standard payload is an IAI Tamam POP day/night EO turret. Onboard power for payload is 1.8 kW.

Guidance and control: Automatic and autonomous navigation modes.

Launch: Catapult launch standard; wheeled take-off from runway optional.

Recovery: Wheeled landing standard, with or without arrester hook cable assistance. Parachute for emergency recovery.

Dragonfly 2000 *(EMIT)* 1290333

Power plant: One 18.6 kW (25 hp) Meggitt WAE 342 Hurricane two-cylinder two-stroke engine; two-blade tractor propeller. Fuel capacity 57 litres (15.0 US gallons; 12.5 Imp gallons).

Dragonfly 2000

Dimensions, External	
Overall, length	2.96 m (9 ft 8½ in)
Wings, wing span	5.00 m (16 ft 4¾ in)
Weights and Loadings	
Weight	
Weight empty	84 kg (185 lb)
Max T-O weight	140 kg (308 lb)
Fuel weight, Max fuel weight	40 kg (88 lb)
Payload	
Payload with max fuel	16 kg (35 lb)
Performance	
T-O	
T-O run	260 m (854 ft)
Climb	
Rate of climb, max, at S/L	457 m/min (1,499 ft/min)
Altitude, Service ceiling	4,570 m (15,000 ft)
Speed	
Max level speed, IAS	110 kt (204 km/h; 127 mph) at 1,829 m (6,000 ft)
Loitering speed, IAS	50 kt (93 km/h; 58 mph) at 1,829 m (6,000 ft)
Endurance, at 70 kt (130 km/h; 81mph) at 1,830 m (6,000 ft)	14 hr
Landing	
Landing run	
unarrested	300 m (985 ft)
with arrested hook	50 m (165 ft)

Status: In service with the Israel Defence Force.
Contractor: EMIT Aviation Consult
Petah Tikva.

EMIT Sparrow-N

Type: Tactical mini-UAV.
Development: Described by the manufacturer as a 'sub-tactical' system, the Sparrow-N is the current incarnation of a design that first appeared in the early years of the century, at that time with a nose-mounted engine and more conventional tail unit. On public debut in June 2003, EMIT reported an order for 100 of this version from "a Southeast Asian Navy". The current Sparrow-N appeared in 2005.
Description: *Airframe:* Cylindrical fuselage, with high-mounted constant-chord wings and cruciform tail surfaces; pusher engine. No landing gear.
Mission payloads: Nose-mounted, stabilised Microview day/night EO/IR in surveillance role, with real-time, encrypted, video and data downlinks; 8 kg (17.6 lb) explosive charge in armed version.
Guidance and control: Pre-programmed, with provision to update or amend waypoints during mission. Suitcase-sized GCS, with joystick control via two encrypted RF command uplinks.

Sparrow-N used by UK for loitering munition development work
(IHS/Patrick Allen) 1347078

General appearance of the Sparrow-N 'sub-tactical' UAV
(IHS/Patrick Allen) 1347080

System composition: Four air vehicles, GCS, portable video receiver, launch and recovery system and crew of two. Set-up time 20 minutes.
Launch: Automatic catapult launch.
Recovery: Automatic parachute recovery.
Power plant: One two-cylinder two-stroke piston engine (type and rating not stated); two-blade pusher propeller. See under Weights for fuel details; larger fuel tank optional.
Variants: *Blade:* Name adopted by EMIT and Rafael for armed version bid in 2005 by an Ultra Electronics-led team for the British Army LMCD competition.

Sparrow-N

Dimensions, External	
Overall, length	2.14 m (7 ft 0¼ in)
Wings, wing span	2.44 m (8 ft 0 in)
Weights and Loadings	
Weight, Max launch weight	45 kg (99 lb)
Fuel weight, Max fuel weight	8.5 kg (18.00 lb)
Payload	
Payload with max fuel	12 kg (26 lb)
Performance	
Speed	
Max level speed, IAS	100 kt (185 km/h; 115 mph)
Cruising speed, IAS	70 kt (130 km/h; 81 mph) (est)
Radius of operation	
omni antenna	10.8 n miles (20 km; 12 miles) (est)
directional antenna	65 n miles (120 km; 74 miles) (est)
Endurance	6 hr
Power plant	1 × piston engine

Status: Developed and production-ready. Over time, this UAV has been co-promoted with Cradance Services (Singapore), EDO Defense (US) and Rafael (Israel), among others. One Sparrow-N system was acquired by the British Army in early 2008 for evaluation in the Loitering Munition Concept Demonstration (LMCD) phase of its Indirect Fire Precision Attack (IFPA) programme.
Contractor: EMIT Aviation Consult Ltd
Petah Tikva.

EMIT Sting and Blue Horizon

Type: Long-endurance surveillance UAVs.
Development: EMIT specialises in the design and manufacture of custom-built, long-endurance UAV platforms in metal and composites, and of launch and recovery support equipment.
Description: *Airframe:* Mainly cylindrical (Sting II) or faceted (Blue Horizon) fuselage; low-mounted swept wings, with winglets, at rear; high-mounted swept canard surfaces at front; pusher engine(s); fixed tricycle landing gear. Composites construction.
Mission payloads: To customer's requirements. IAI Tamam POP (which see) specified for Blue Horizon. Up to 2.0 kW of electrical power available for payload and avionics operation.
Guidance and control: Automatic; GCS and ground data terminal.
System composition: Blue Horizon can be fielded in two HMMWVs: one for the GCS/GDT and one for the air vehicles in their transit cases. Ground crew of three.
Launch: Bungee launch or wheeled take-off.
Recovery: Parachute recovery, or wheeled landing with or without optional arrester hook and cable.
Power plant: One (optionally two) two-cylinder piston engine(s) (single 18.6 kW; 25 hp Meggitt WAE 342 two-stroke in Blue Horizon); two-blade pusher propeller(s).
Fuel capacity (Blue Horizon) 60 litres (15.9 US gallons; 13.2 Imp gallons).
Variants: *Sting I:* Large, optionally twin-engined version.
Sting II: Smaller version of Sting I.

Unmanned aerial vehicles > Israel > EMIT Sting and Blue Horizon – IAI Bird Eye 400

EMIT Aviation Sting II canard configuration UAV 0044407

Blue Horizon UAV as purchased by Singapore *(Michael J Gething)* 0137243

Blue Horizon: Modified version of Sting II for Singapore and other customers.

Sting I, Sting II, Blue Horizon

Dimensions, External
 Overall
  length, Blue Horizon..3.20 m (10 ft 6 in)
 Wings
  wing span, Blue Horizon...6.00 m (19 ft 8¼ in)
Weights and Loadings
 Weight
  Weight empty, Blue Horizon.......................................80 kg (176 lb)
  Max T-O weight
   Sting I..1,200 kg (2,645 lb)
   Sting II...130 kg (286 lb)
   with reduced 47 litre/33 kg fuel
    load, Blue Horizon...150 kg (330 lb)
 Fuel weight
  Max fuel weight, Blue Horizon....................................42 kg (92 lb)
 Payload
  Max payload
   Sting I..181 kg (400 lb)
   Sting II...16 kg (35 lb)
   Blue Horizon...17 kg (37 lb)
Performance
 T-O
  T-O run, Blue Horizon..250 m (821 ft)
 Altitude
  Service ceiling, Blue Horizon......................................5,485 m (18,000 ft)
 Speed
  Max level speed, Blue Horizon..........130 kt (241 km/h; 150 mph) at 610 m (2,000 ft)
  Cruising speed
   Sting I..60 kt (111 km/h; 69 mph)
   Sting II...78 kt (144 km/h; 90 mph)
   econ, Blue Horizon...............................70 kt (130 km/h; 81 mph) at 610 m (2,000 ft)
  Loitering speed, Blue Horizon.....................................55 kt (102 km/h; 63 mph)
  Stalling speed
   at 610 m (2,000 ft), Blue Horizon...........................50 kt (93 km/h; 58 mph)
  Radius of operation, datalink.................................81 n miles (150 km; 93 miles)
 Endurance
  Sting I...48 hr
  Sting II..24 hr
  with smaller fuel tank and
   increased payload of 45kg
   (99.2 lb), Sting II..12 hr
  at econ cruising speed and
   1525 m (5,000 ft), Blue Horizon............................24 hr
 Landing
  Landing run
   arrested, Blue Horizon...50 m (165 ft)
   unarrested, Blue Horizon.....................................300 m (985 ft)

Status: EMIT Aviation no longer list Sting as an available system. Development of Blue Horizon complete and the system has been exported. Blue Horizon bid for Singapore armed forces requirement; promoted by Singapore Technologies Dynamics, but there is no evidence that the system was selected by Singapore.

 Customers: Contract valued at approximately US$14 million awarded by Singapore Technologies in November 1998 for unspecified number of Blue Horizon systems. EMIT to assist with development of demonstrator. An unconfirmed report in late 2001 suggested that the Blue Horizon was in use by the Philippine Army. Sri Lanka appears to be another customer, having posted a invitation to tender for "Blue Horizon II" engine spares in 2009.

Contractor: EMIT Aviation Consult Ltd
 Petah Tikva.

IAI Bird Eye 400

Type: Surveillance mini-UAV.

Development: Unveiled at Paris Air Show, June 2005, as newest member of Bird Eye family (now sole member, as projected 100 and 500 variants no longer being promoted). Designed for local-area tactical reconnaissance in support of small, low-echelon operational units.

Description: *Airframe:* Constant-chord, sweptback wings, with flaps and dependent endplate winglets; minimal fuselage; all composite; V tail; pusher engine. No landing gear.

 Mission payloads: Ventrally mounted and gimballed IAI Tamam POP day/night sensor with real-time imagery downlink.

IAI Bird Eye 400 in flight *(IAI)* 1175488

The unorthodox Bird Eye 400, shown inverted *(IHS/Patrick Allen)* 1144681

Bird Eye 400 mini surveillance UAV, underside and payload detail *(IHS/Patrick Allen)* 1144689

Guidance and control: Fully automated flight, with GPS-based waypoint control. Missions planned using digital map referencing and viewed on a computer monitor.
Transportation: Man-portable in two backpacks by two persons.
System composition: Typical system comprises three air vehicles, one backpack control module (BPCM), datalink, one night and two day payloads, plus ground support equipment and spares.
Launch: Hand or bungee launch.
Recovery: Designed for landing on hard surfaces. Aircraft is recovered inverted, to protect sensor, by entering deep stall initiated by flap system, landing impact being absorbed by shock absorbers on its upper surface.
Power plant: Electric motor, driving a hinged two-blade pusher propeller.

Bird Eye 400

Dimensions, External	
Overall, length	0.80 m (2 ft 7½ in)
Wings, wing span	2.00 m (6 ft 6¾ in)
Weights and Loadings	
Weight, Max launch weight	5.6 kg (12.00 lb)
Payload, Max payload	1.2 kg (2.00 lb)
Performance	
Altitude	
Operating altitude, max	610 m (2,000 ft)
Radius of operation	5.4 n miles (10 km; 6 miles)
Endurance	1 hr 20 min (est)
Power plant	1 × electric motor

Status: Development completed in 2005. Manufacturer reports Bird Eye 400 in production and supplied to several customers by July 2007. A newspaper report from Moscow in 2009 announced Russia was buying the system; further reports in 2010 said a repeat order had been placed by Russia, deliveries were continuing in 2011.
Contractor: Israel Aerospace Industries, Malat Division,
Tel-Aviv.

IAI ETOP

Type: Tethered unmanned platform.
Development: After some two years of development, IAI unveiled this new system in June 2010. Its acronym title, signifying Electric Tethered Observation Platform, indicates both its nature and its purpose.
Description: ***Airframe:*** A square-framed body accommodates a quartet of two-blade shrouded rotors; the sensor payload is attached centrally on the underside, and the structure is supported on two looped skids at right angles to one another.
Mission payloads: Multi-sensor capability; a typical payload would be an EO/IR sensor ball turret.
Guidance and control: IAI describes the ETOP as having "single-click operation capability", with no requirement for any dedicated ground crew or maintenance activity. It can be deployed from a static or mobile vehicle or station, on the ground or on board ship, and is said to be stable at its operating height in winds of up to 25 kt (46 km/h; 29 mph). The tether cable serves to provide both the electrical power conduit and the imagery downlink.
Launch: Vertical take-off.
Recovery: Vertical landing
Power plant: Any ground- or ship-based source of electrical energy.

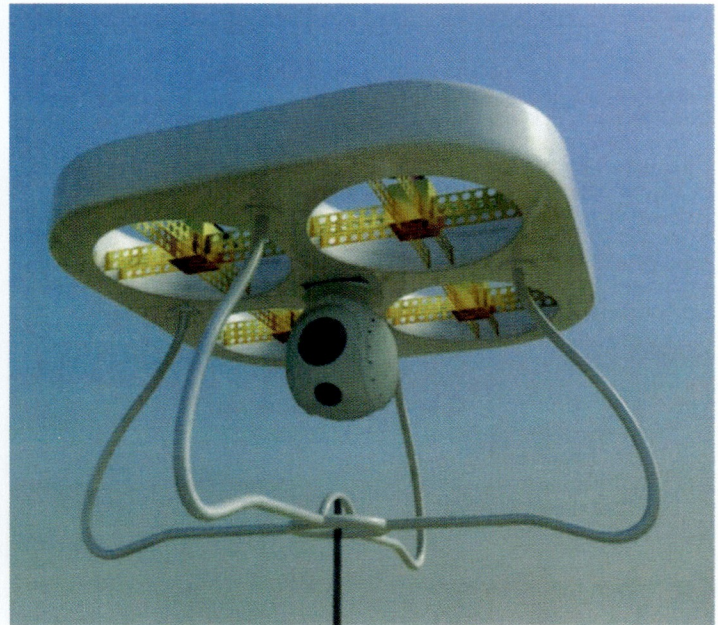

The ETOP tethered unmanned platform *(IAI)* 1395243

ETOP

Dimensions, External	
Overall	
length	1.60 m (5 ft 3 in)
height	0.20 m (7¾ in)
width	1.60 m (5 ft 3 in)
Weights and Loadings	
Weight, Max T-O weight	45 kg (99 lb)
Payload, Max payload	20 kg (44 lb)
Performance	
Climb	
Rate of climb	300 m/min (984 ft/min)
Time to height	30 s to 50 m (30 s to 160 ft)
Altitude, Service ceiling	100 m (320 ft)

Status: Prototypes were under evaluation by 'several clients' in mid-2010; planned to be fielded by or in 2011.
Contractor: Israel Aerospace Industries, Malat Division
Tel-Aviv.

IAI Ghost

Type: Close-range helicopter mini-UAV.
Development: Unveiled at AUVSI North America exhibition in August 2011, the Ghost has a 'mini-Chinook' appearance and has been developed for urban warfare and covert operations. IAI states that it has been optimised for surveillance in built-up areas (including inside buildings), harsh terrain and dense vegetation, and is highly stable even in strong sidewinds and gusts.
Description: ***Airframe:*** Tandem-rotor configuration, with twin three-blade rotors; box-section fuselage. Fixed, four-leg landing gear.
Mission payloads: Flight tests have been undertaken with a NextVision MicroCam D daylight TV camera in a stabilised ventral turret; real-time video downlink. EO system also has a mapping capability that measures indoor space perimeters, permitting safe entry and manoeuvring within a room. IR option confers day/night capability; other sensors under consideration.
Guidance and control: Fully automatic. Power train is synchronised so that both rotors continue to be powered if one engine becomes inoperative. Minimal operator training is said to be needed.
Transportation: Man-portable in two backpacks.
System composition: Two air vehicles, laptop computer GCS, spare batteries and two personnel.
Launch: Automatic vertical take-off.
Recovery: Automatic vertical landing.
Power plant: Twin electric motors (type and rating not revealed).

IAI's 'mini-Chinook' Ghost UAV *(IAI)* 1395349

Ghost tandem-rotor surveillance mini-UAV *(IAI)* 1395350

Ghost

Dimensions, External	
Fuselage, length	1.45 m (4 ft 9 in)
Rotors	
rotor diameter, each	0.75 m (2 ft 5½ in)
Weights and Loadings	
Weight, Max T-O weight	4.0 kg (8.00 lb)
Payload, Max payload	0.5 kg (1.00 lb)
Performance	
Speed, Max level speed	35 kt (65 km/h; 40 mph)
Endurance	30 min
Power plant	2 × electric motor

Status: As of August 2011, the Ghost was said to be nearing the completion of flight testing and would shortly be declared operation-ready. No customer orders had been reported at that time.

Contractor: Israel Aerospace Industries, Malat Division, Tel-Aviv.

IAI Harop

Type: Lethal UAV.

Development: Development of an upgraded and more capable version of the IAI Harpy anti-radar attack drone has been under way for several years, at one time including a joint venture with Raytheon known as CUTLASS (Combat UAV Target Location And Strike System) in an attempt to secure a US order. Categorised as a loitering munition, the Harop (alternatively referred to as **Harpy 2**) differs significantly from its 'fire and forget' predecessor in having man-in-the-loop operator control.

Harop had not been seen publicly outside Israel until February 2009, when it appeared at the Aero India air show. It was further displayed at the Paris Air Show in June 2009, accompanied by announcement of export orders from Germany and a second, unidentified, customer.

Description: *Airframe:* Harop appears to retain the core delta wing and winglets/fins of Harpy, including its retractable side-force panels that stabilise it in the terminal dive. To this have been added tapered outer wing panels, which fold for stowage in the launch container, and a larger, more bulbous nose section having high-mounted foreplanes and an undernose sensor turret. The fairing over the rear-mounted engine is more rounded than on Harpy, suggesting upgrade to a rotary power plant. Composites construction may reasonably be assumed.

Mission payloads: As displayed in June 2009, the undernose EO/IR turret was deemed to be an IAI Tamam POP-200. A satcom datalink is reported to be fitted. For attack, the aircraft carries a high explosive fragmentation warhead.

Harop's Harpy ancestry is very evident in this view
(IHS/Patrick Allen) 1379158

Harop, with outer wings folded, in its launch container
(IHS/Patrick Allen) 1379165

Guidance and control: Mission control shelter (MCS) exercises vehicle control with a 'man in the loop' able to engage or (to avoid collateral damage) abort the attack capability in real time. Attack can be performed from any direction, and at any attack angle from flat to vertical. The operator monitors the attack until the target is hit. In the event of an abort, Harop reverts to loiter mode until commanded to restart an attack.

Transportation: In transportable launch containers.

System composition: Air vehicles, launchers and MCS.

Launch: By solid-propellant booster rocket from storage container. Outer wings unfold on exit. Launch can be from "a variety of platforms", according to IAI; one unconfirmed report has suggested it can be adapted for air launch.

Recovery: Intentionally expendable. However, according to *IHS Jane's* sources, the aircraft can be equipped with optional wheel landing gear for recovery in certain (unspecified) circumstances.

Power plant: One heavy-fuel (Elbit rotary ?) engine, driving a two-blade pusher propeller

Harop

Dimensions, External	
Overall, length	2.50 m (8 ft 2½ in)
Wings, wing span	3.00 m (9 ft 10 in)
Performance	
Endurance	6 hr

Status: IAI announced on 10 June 2009 that the German MoD and armed forces had approved 'an operational requirement' using the Harop system. Adaptation to German requirements, already under way at the time of the announcement, is being undertaken in conjunction with Rheinmetall Defence, and a follow-on contract was planned for later in 2009.

Simultaneously, IAI announced a contract, estimated at over USD100 million, to supply a Harop system to another foreign customer. Although not named, it is thought that this client may be either India or Turkey, both of which have been reported as evaluating the Harop in recent years.

Contractor: Israel Aerospace Industries, MBT Systems Missiles and Space Group
Beer Yacov.

IAI Harpy and Cutlass

Type: Anti-radar attack UAV.

Development: Existence of the Harpy loitering weapon has been known since the late 1980s, although by its nature it remained a largely 'black' programme until the late 1990s. Designed to combat hostile SAMs and radars in all weathers, by day or night, it utilised a basic Dornier air vehicle known as DAR, a joint Dornier/IAI version of which was proposed to the German MoD in an earlier stage of the DAR programme. Negotiations with the US Air Force in 1988 led to an MoU between IAI and General Dynamics to enter the UAV in the DoD's Foreign Weapons Evaluation (FWE) programme in 1989 as an alternative to the cancelled AGM-136 Tacit

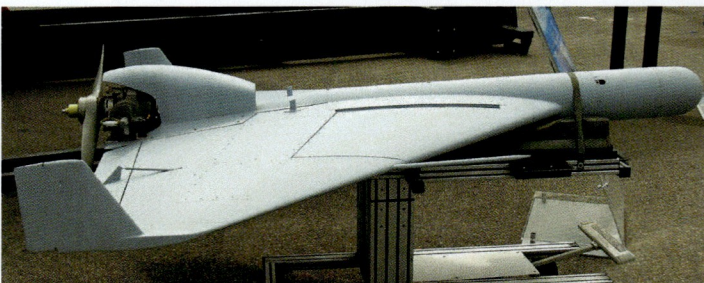

Harpy display at the Paris airshow in 2007 *(IAI)* 1326169

Harop launch and (inset) impacting a target *(IAI)* 1356222

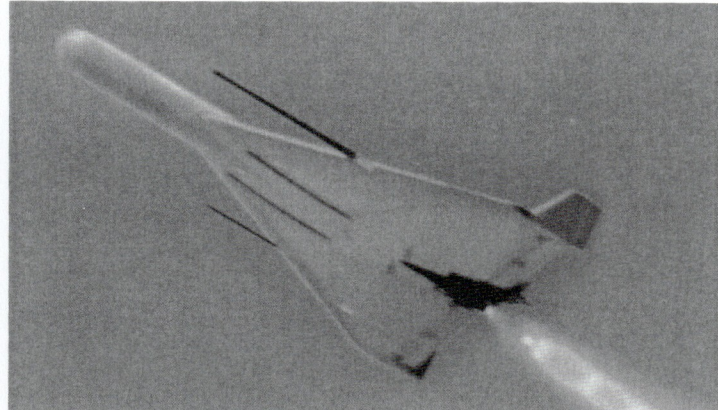

Harpy at launch (left) and in terminal dive on to a radar target

Rainbow defence suppression missile, but this demonstration apparently did not take place. Harpy has been in production for the Israel Defence Force/Air Force and other customers since at least 1988.

Description: *Airframe:* Mid-mounted delta wings with full-span elevons, tip-mounted fins and rudders and retractable side-force panels (two in each wing). Built in two GFRP/CFRP half-shells with integral fittings.

Mission payloads: Israeli-developed passive radar seeker (recently upgraded to cover a wider range of frequencies) and high-explosive warhead. IAI was reported, in late 2000, to be developing an upgraded version equipped with a dual (electromagnetic and EO) sensor and datalink, to allow Harpy to receive updates on potential targets and to be directed against a specific emitter.

Guidance and control: Fully autonomous. Preprogrammed flight profiles, but if target radar is switched off during its approach, Harpy can abort approach before reaching a 'commit' altitude and continue its search flight. Side-force panels are deployed to stabilise air vehicle during terminal dive.

System composition: Typical battery believed to comprise three truck-mounted launcher units, each with 18 containerised air vehicles.

Launch: By booster rocket from ground- or truck-mounted 18-round container.

Recovery: Non-recoverable.

Power plant: One 20.5 kW (27.5 hp) two-cylinder two-stroke engine; two-blade pusher propeller.

Variants: *Harpy:* As described.

Cutlass: (Combat UAV Target Location And Strike System): Developed jointly by IAI and Raytheon Systems of the US and optimised for SEAD role. Combined Harpy air vehicle with a commercial UHF video datalink and two Raytheon subsystems: the IR seeker head from the AIM-9X Sidewinder AAM and Raytheon's automatic target recognition and classification (ATR/C) algorithms. Said to have a 1.83 m (6 ft 0 in) wing span, launch weight of 125 kg (275 lb), 100 kt (185 km/h; 115 mph) cruising speed, 4,575 m (15,000 ft) ceiling and 162 n mile (300 km; 186 mile) range, carrying a 16 kg (35 lb) HE fragmentation warhead. Reportedly already used by IDF in a combat environment; was under consideration by US DoD in 2000, but no US order ensued and IAI/Raytheon joint venture dissolved.

Harpy

Dimensions, External	
Overall length	2.30 m (7 ft 6½ in)
height	0.36 m (1 ft 2¼ in)
Wings, wing span	2.00 m (6 ft 6¾ in)
Weights and Loadings	
Weight, Max launch weight	120 kg (264 lb)
Payload	
Max payload, warhead	32 kg (70 lb)
Performance	
Altitude, Service ceiling	3,000 m (9,840 ft)
Speed, Max level speed	135 kt (250 km/h; 155 mph)
Radius of operation	270 n miles (500 km; 310 miles) (est)
Endurance	
216 n mile (400 km; 248 mile) radius	2 hr

Status: In production as required and in service. Development of a ship-launched version has also been reported.

Customers: Israel; Harpy exports to several countries, including China (reported USD100 million order in late 2002); India; South Korea (contract reportedly worth USD45 million for 100 air vehicles, deliveries beginning in 1999); Spain (ordered in mid-2000); and Turkey (108 initially, with options on more, delivered from mid-2002; 2005 plan to acquire further 48 cancelled in early 2006). Continued in production in 2010.

Contractor: Israel Aerospace Industries, MBT Systems Missiles and Space Group
Yehud.

IAI Heron 1

Type: Medium-altitude, long-endurance UAV.

Development: Existence of the Heron (Hebrew name **Machatz**: 'Crusher') was revealed in October 1994 with news of its first flight (18 October). It then competed, though unsuccessfully, in the US DoD's contemporary MAE (Medium-Altitude Endurance, formerly Tier II) programme, partnered by TRW Inc. During early flight tests, an altitude of 9,750 m (32,000 ft) was achieved on 20 February 1995; and an endurance flight of 51 hours 21 minutes was made on 2 May 1995 at a gross weight of 1,150 kg (2,535 lb) including 600 kg (1,323 lb) of payload. The type's growth potential has spawned a number of projected variants, among them a short-span version which did not come to fruition, but the order book received a significant boost in 2005 with follow-on orders from India and new ones from Israel and Turkey.

A Heron with a 'recently integrated' 134 kW (180 hp) heavy-fuel (Diesel) engine, reportedly modified from a Volkswagen motorcar engine, was displayed at the Paris Air Show in June 2005. Flight trials with this power plant were said to be due to begin near the end of that year, and were

Harpy is launched from behind the battle zone (IAI)

Heron 99+21 is the first for Germany, pictured in early 2010 (Rheinmetall Defence)

Unmanned aerial vehicles > Israel > **IAI Heron 1**

Bristling with antennas, Heron 046 overflies the Asian Aerospace show in Singapore in February 2006 *(IHS/Patrick Allen)* 1177517

A closer look at the front end of 046. Bulged nose indicates presence of satcom datalink *(IHS/Patrick Allen)* 1177513

Heron 1 takes off at Changi *(IHS/Patrick Allen)* 1177503

IAI Heron c/n 007 demonstrating as Machatz 1/Shoval multiple-sensor UAV for the Israel Air and Space Force *(IAI)* 1122524

Another view of Machatz 1/Shoval endurance UAV c/n 007 *(IAI)* 1122539

Heron festooned with comint and other antennas *(IAI)* 1175489

Heron with heavy-fuel engine at Paris Air Show, June 2005 *(IHS/Patrick Allen)* 1144674

presumably seen as a potential retrofit for the original Mk I version.

Heron is the English export name for the UAV. To the Israel Air and Space Force it is known as the **Shoval** (Trail).

Description: *Airframe:* High-wing monoplane with very high-aspect ratio wings fitted with full-span slotted flaps. Twin-boom tail unit with inward-canted fins and rudders. Pusher engine installation. All-composites construction. Fully retractable tricycle landing gear; nose unit retracts rearward into fuselage nacelle, main units forward into front of tailbooms. Hydraulic mainwheel brakes. The wings, booms and tail unit of the Heron were used in the Hunter II derivative of the Northrop Grumman/IAI RQ-5A.

Mission payloads: Large fuselage volume available for a wide variety of single or multiple payloads for day and night operation. Retractable landing gear permits unobstructed coverage by onboard sensors. Heron can perform a wide variety of real-time missions including intelligence collection, surveillance, target acquisition/tracking, elint, comint and communications/data relay.

Standard payload is dual (TV/IR) or triple sensor (TV/IR/laser rangefinder) undernose IAI Tamam MOSP (which see). In addition, can have Elta EL/M-2055 SAR/GMTI or EL/M-2022U maritime surveillance radar in large ventral radome, capable of multi-target track-while-scan of up to 32 targets. Other payloads can include Elta EL/K-7071 comint; EL/L-8385 ESM/elint; Ku-band satcom; data/voice relay packages; or customer-furnished payloads. Single real-time data and video downlink.

Guidance and control: Fully digital avionics, interoperable with Northrop Grumman/IAI Hunter system and adapted for similar compatibility with IAI Searcher. Direct LOS datalink; airborne and/or ground-based data relays for BLOS missions; dual real-time command uplink; satcom datalink integration capability.

Missions can be preplanned before take-off; flight is automatic, with the ability to transmit changes in real time, and route changes can be introduced during flight. Control can be handed over to a second ground or maritime control station. Main subsystems have built-in redundancy and the UAV has a 'return home' capability.

The GCS, housed in a command and control shelter, is a derivative of the GCS-3000 developed for the Hunter UAV, using common workstations equipment for specified functions and receiving uplink and downlink data from the ground data terminal (GDT) operational control centre. The GCS consists basically of two operator consoles (based on workstations) and one command/control console, and provides for all mission planning, control, command and processing functions to operate the UAV and its payloads. Workstations can be configured for various software packages selected by the operator. The datalinks provide RF communication to transmit data, commands and video signals between the GDT and the UAV.

Launch: Conventional and automatic wheeled take-off.
Recovery: Automatic wheeled landing.
Power plant: One 73.5 kW (98.6 hp) turbocharged Rotax 914 F four-cylinder four-stroke engine; two-blade variable-pitch pusher

IAI Heron 1 – IAI Heron TP Eitan < Israel < Unmanned aerial vehicles

Heron displaying its full-span slotted flaps 0087923

Heron prototype with Elta maritime radar at the 1995 Paris Air Show *(IHS/Kenneth Munson)* 0518205

propeller. Wet-wing integral fuel tank, plus fuselage tank; combined capacity 720 litres (190 US gallons; 158 Imp gallons).

Variants: ***Heron 1:*** Current version; *detailed description applies to this model.*

Heron 2: Over time, a number of proposed derivatives have been reported, often conflictingly, under 'Heron 2' (or II) designations. One such was said to be under development in 2001–02 as a launch platform for the Israel Boost Phase Intercept (BPI) anti-ballistic missile programme, with preliminary details that included a 26 m (85.3 ft) wing span, 15 m (49.2 ft) length, 250 kg (551 lb) payload capability and 50 hour endurance. This development has now been identified as the Heron TP. More recently, Northrop Grumman teamed with IAI to bid a Heron-based 'Strike Heron' version, dubbed Hunter II, for the US Army's ER/MP competition (won in 2006 by the GA-ASI Sky Warrior) to replace the RQ-5 Hunter.

Heron TP: Enlarged development of Heron 1. *Described separately.*

Eitan: Israel Air and Space Force variant of Heron TP. *Described separately.*

Heron 1

Dimensions, External
Overall
length..8.50 m (27 ft 10¾ in)
height...2.30 m (7 ft 6½ in)
Fuselage, length..5.20 m (17 ft 0¾ in)
Wings
wing span..16.60 m (54 ft 5½ in)
wing aspect ratio..21.2
Dimensions, Internal
Payload bay, volume..0.80 m³ (28.3 cu ft)
Areas
Wings, Gross wing area..13.00 m² (139.9 sq ft)
Weights and Loadings
Weight, Max T-O weight...1,100 kg (2,425 lb)
Fuel weight, Max fuel weight..430 kg (947 lb)
Payload
Max payload...250 kg (551 lb)
useful, fuel + payload...500 kg (1,102 lb)
Performance
Climb
Rate of climb, max, at S/L...198 m/min (649 ft/min)
Altitude
Operating altitude, max...9,145 m (30,000 ft) (est)
Service ceiling...8,075 m (26,500 ft)

Speed
Cruising speed, max..................................125 kt (232 km/h; 144 mph) at 6,096 m (20,000 ft)
Loitering speed..80 kt (148 km/h; 92 mph) (est)
Stalling speed..40 kt (75 km/h; 46 mph) (est)
Range..540 n miles (1,000 km; 621 miles) (est)
Radius of operation
LOS, mission..108 n miles (200 km; 124 miles)
BLOS, mission...189 n miles (350 km; 217 miles)
Endurance, max, has demonstrated
52 h..40 hr (est)
Time on station
at 81 n miles (150 km; 93 miles)...35 hr
at 270 n miles (500 km; 310 miles)..30 hr

Status: In production and service. No reliable production total known, but the aircraft exhibited at the Asian Aerospace show in Singapore in February 2006 bore constructor's number 046.

Customers: *Australia:* Heron operated by Australia in Afghanistan from January 2010. Leased from Canada.

Canada: Heron system leased under a two year agreement with IAI and MacDonald Dettwiler and Associates announced in August 2008, deliveries from October 2008. Canadian designation is CU-170.

France: Heron airframe selected in early 2001 as basis for the EADS DCS Eagle/Harfang (which see), being developed for French Air Force SIDM (*Système Interimaire de Drone MALE*) programme.

Germany: Three air vehicles and two GCS leased under agreement with IAI and Rheinmetal announced in October 2009. Lease runs for one year, with a two year option, deployment scheduled to start in Afghanistan during March 2010.

India: Ordered by Indian Air Force (four or more) in mid-2000; further 12 (possibly of Heron II standard) ordered in 2005. Now serve also with Indian Navy, whose first squadron, INAS(U) 342 'Flying Sentinels', was activated at Kochi on 6 January 2006 with four Herons and eight Searcher IIs. These aircraft are interfaced with control and imagery interpretation stations aboard three Indian Navy 'Delhi' class destroyers. The Navy is said to plan two further shore-based Heron/Searcher units, one at Porbander and the other at Port Blair in the Andaman Islands. India has declared a long-term larger requirement for possibly as many as 50 Heron systems, and was understood to be discussing further procurement and possibly local manufacture with IAI in early 2006.

Israel: First production order for Machatz 1/Shoval was announced on 11 September 2005, with deliveries to Israel Air and Space Force (IASF) beginning shortly afterwards. Quantity not revealed, but value stated to be in excess of USD50 million, suggesting two or three systems only. Eventual requirement said to be for 'several dozen' systems, including 'a few' to be operated by IASF for the Navy.

Turkey: Contract (USD183 million) signed 18 April 2005 for 10 systems, to be delivered over following 24 to 30 months, with some 30 per cent of manufacture by Tusas Aerospace Industries (TAI) under Israeli UAV Partnership (IUP) between IAI/Elbit and TAI. Further USD50 million supply and support contract later that year. Elbit Systems (Israel) providew GCS; Aselsan (Turkey) supplies EO/IR payloads. Turkish Army and Air Force each to receive four systems and Navy two. First deliveries started February 2010, completion was scheduled for July 2010.

UK: One EADS Eagle, based on Heron airframe and allocated Royal Air Force serial number ZJ989, was deployed to British Army Training Unit at Canadian Forces Base Suffield, Ontario, in October 2003, for trials as part of UK's Joint UAV Experimentation Programme (JUEP). Further details in entry for Eagle.

In 2010-2011 demonstrations of Heron 1 were reported to have been made in Angola and Kenya.

Contractor: Israel Aerospace Industries, Malat Division, Tel-Aviv.

IAI Heron TP Eitan

Type: High-altitude, long-endurance UAV.

Development: Reference to the Eitan (Hebrew for 'steadfast'), by name, first appeared in the Israel Air Force's monthly magazine in April 2004, at which time it was said to be in the 'advanced stages' of development with two prototypes then flying. These were understood to be the aircraft referred to previously as the Machatz ('Crusher') Mk 2 or Heron TP, powered by a turboprop engine and being nearly four times larger than the Heron 1 with a wing span of about 26 m (85.3 ft) and gross weight in the 4,000 kg (8,818 lb) class. In March 2005, Aurora Flight Sciences of the US announced a joint agreement with IAI to market a version of the Eitan under the name **Orion**, with a possible first flight during 2007, but no further news of this project had emerged by mid-2007.

Reports to emerge since 2004 have been both sketchy and contradictory. The accomplishment of a first flight on 15 July 2006 led to speculation by at least one source that the aircraft had undergone a further increase in wing span, to around 35 m (114.8 ft), in an effort to attain a ceiling of 15,240 m (50,000 ft) compared with the 13,720 m (45,000 ft) of the earlier machines. According to the same source, useful load (payload plus fuel) had been increased from the Heron 1's 500 kg (1,102 lb) to around 1,800 kg (3,968 lb) (more than the *gross* weight of the Heron 1) in the Eitan. Other unconfirmed sources have quoted an Eitan

Unmanned aerial vehicles > Israel > IAI Heron TP Eitan – IAI Mosquito

A Heron TP prototype dwarfs the standard Heron on its debut at Paris, June 2007 *(IHS/Kenneth Munson)* 1290192

The Heron TP's bulged boom-ends may allow for fitment of a defensive aids subsystem *(IHS/Kenneth Munson)* 1290193

Pratt & Whitney Canada PT6A-67 in pusher installation on the Heron TP *(IHS/Patrick Allen)* 1326186

Flight tests of Heron TP continued through 2008 *(IAI)* 1293598

Heron TP side number 102, canoe fairing underneath *(IAI)* 1415530

Induction ceremony for Heron TP at Tel Nof on 21 February 2010 *(IAI)* 1415506

gross weight of 5,000 kg (11,023 lb), an endurance of 50 hours, and a cruising speed of 240 kt (444 km/h; 276 mph) at 15,240 m (50,000 ft). During a three hour test flight in 2008 the air vehicle climbed to 12,192 m (40,000ft) in less than one hour and maintained that altitude for a "substantial period", according to IAI.

In June 2007, the Heron TP made its public debut at the Paris Air Show, when it was revealed that 'several' prototypes had been built and tested, and that it was now ready for production. Brief size and weight details were also released at that time. These confirmed the original estimates of size, although IAI officials declined to say whether the Israel Air and Space Force version differed in any way from the displayed prototype. They did, however, confirm that the TP was the largest UAV the company has so far built.

Jane's and other sources have suggested the Eitan is under consideration for ISR; for strike missions, as part of Israel's anti-ballistic missile Boost Phase Intercept (BPI) programme; and as an aerial refuelling tanker.

In May 2008 the European companies Dassault, Thales and Indra signed an agreement to offer Heron TP to meet French and Spanish needs for a MALE UAV in a short timeframe, entering service in 2012. In early 2011 Dassault is again reported to have offered a version of Heron TP to France under a lease agreement. In June 2008 Rheinmetall Defence of Germany announced a coooperation agreement with IAI to market the Heron TP for the Bundeswher SAATEG programme, which requires a MALE surveillance UAV system for initial operational capability from 2010. Five systems are sought.

Description: *Airframe:* High-wing monoplane with very high aspect ratio wings fitted with de-icing system and full-span slotted flaps. Twin-boom tail unit with inward-canted fins and rudders. Pusher engine installation. Fully retractable tricycle landing gear.

Mission payloads: Multiple ISTAR payload capability, including underwing stores/weapons.

Guidance and control: Fully autonomous and triple-redundant. Multiple LOS and satcom (BLOS) datalinks.

Launch: Conventional, automatic, wheeled take-off.

Recovery: Conventional, automatic, wheeled landing.

Power plant: One 895 kw (1,200 shp) Pratt & Whitney Canada PT6A-67 turboprop, driving a four-blade pusher propeller.

Heron TP

Dimensions, External	
Overall, length	14.00 m (45 ft 11¼ in)
Wings, wing span	26.00 m (85 ft 3½ in)
Weights and Loadings	
Weight, Max T-O weight	4,650 kg (10,251 lb)
Payload, Max payload	1,000 kg (2,204 lb)
Performance	
Altitude, Operating altitude	13,715 m (45,000 ft)
Endurance	36 hr
Power plant	1 × turboprop

Status: Under development, reportedly for both the Israel Navy and Air Force. According to IAI, 'several' development Heron TPs had then been built and flown by June 2007. First deliveries in 2008. Initial deployment by Israel Air Force on operations in 2009, official entry into service February 2010 with 210 Squadron at Tel Nof.

Customers: Israel Air and Space Force; Israeli Navy.

Contractor: Israel Aerospace Industries, Malat Division, Tel-Aviv.

IAI Mosquito

Type: Micro UAV.

Development: Development started in 2001, and a Mosquito 1 prototype was flown for the first time on 1 January 2003. Subsequent development has led to the heavier Mosquito 1.5 model.

Description: *Airframe:* Flying wing configuration, with curved leading- and trailing-edges, elevons and twin endplate fins.

Mission payloads: Video camera with real-time downlink.

Guidance and control: Radio remote control for Mosquito 1. Mosquito 1.5 has fully automated flight, with GPS-based in-flight waypoint control.

Mosquito 1.5 prototype *(IAI)* 1122537

The developed Mosquito 1 *(IAI)* 1122538

Mosquito Micro UAV during flight *(IAI)* 1464057

Missions planned using digital map referencing and viewed on a computer monitor. Mission can be updated or changed during flight.
  Launch: Bungee launch.
  Recovery: Parachute landing.
  Power plant: Electric motor, driving a two-blade propeller.

Mosquito 1, Mosquito 1.5

Dimensions, External	
Overall, length	0.30 m (11¾ in)
Wings	
wing span	
Mosquito 1	0.30 m (11¾ in)
Mosquito 1.5	0.34 m (1 ft 1½ in)
Weights and Loadings	
Weight	
Max launch weight	
Mosquito 1	0.25 kg (0.00 lb)
Mosquito 1.5	0.5 kg (1.00 lb)
Payload	
Max payload, Mosquito 1.5	0.020 kg (0.00 lb)
Performance	
Altitude, Operating altitude	150 m (500 ft)
Radius of operation	0.54 n miles (1 km; miles)
Endurance	
Mosquito 1	30 min
Mosquito 1.5	40 min
Power plant	1 × electric motor

Status: Developed and available on request.
Contractor: Israel Aerospace Industries, Malat Division, Tel-Aviv.

IAI NRUAV

Type: Helicopter UAV.
Development: Israel Aerospace Industries' Malat Division has developed a Helicopter Modification Suite (HeMoS) which can be applied to existing, proven, manned helicopters to convert them into unmanned systems. The concept achieved public notice at the Aero India show in early 2007, and in 2008, at the instigation of India's Chief of Defence Staff, Admiral Sureesh Mehta, IAI began discussions with the Indian aerospace and defence industry, with a proposal to convert in-service Chetak or Chetan helicopters to meet an Indian Navy requirement for shipborne unmanned surveillance. By February 2009, Aero India show reports indicated that contractual go-ahead for the project was close to achievement, with a demonstration prototype expected to fly in about September 2009.

Initially, the system is known simply as NRUAV (Naval Rotary UAV), although a name may be selected later. Indian Navy tasks would include real-time ISR, day and night OTH targeting, battle and damage assessment, and communications relay. Hindustan Aeronautics Ltd (HAL) would be the Indian prime partner.

Description: *Airframe:* The existing Chetak (TM 333) and Chetan (Shakti) differ principally in engine type. Apart from that, and fairing-over of fuselage transparencies, principal modifications would involve use of a fly by wire control system, including a 'return home' mode; installation of additional fuel tanks in the cockpit area; automatic take-off and landing capability; and a deck securing system.

Mission payloads: In general terms, NRUAV potential payloads are viewed as including EO/IR radar, sigint and ESM packages. As proposed for the Indian Navy, these are reported to include a belly-mounted Elta EL/M-2022H(V)2 multi-mode radar that includes SAR/GMTI maritime radar, air-to air and navigation and weather avoidance. It can include a nose-mounted IAI Tamam MOSP sensor turret, Indian Defence Avionics Research Establishment RWR, secure two-way datalinks and an HAL Mk 12 Mode S IFF transponder. A future option may be the carriage of one or two torpedoes.

Guidance and control: Indian NRUAVs would utilise a flight control system and GCS configuration derived from those employed on the country's existing fleet of IAI Heron UAVs.

Launch: Conventional, automatic, helicopter take-off.
Recovery: Conventional, automatic, helicopter landing. Rear-mounted harpoon decklock securing system.
Power plant: One 807 kw (1,082 shp) Turbomeca TM 333-2M2 or 1,068 kw (1,432 shp) HAL Shakti (licence Turbomeca Ardiden 1H) turboshaft.

NRUAV

Dimensions, External	
Overall	
length, rotors turning	12.84 m (42 ft 1½ in)
height, to top of rotor head	2.97 m (9 ft 9 in)
width, main rotor blades folded	2.60 m (8 ft 6¼ in)
Fuselage	
length, tail rotor turning	10.17 m (33 ft 4½ in)
Rotors, rotor diameter	11.02 m (36 ft 1¾ in)
Tail rotor, tail rotor diameter	1.91 m (6 ft 3¼ in)

Deleted transparencies characterise the NRUAV conversion *(IAI)* 1412006

NRUAV

Weights and Loadings	
Weight, Max T-O weight	2,200 kg (4,850 lb)
Payload, Max payload	220 kg (485 lb)
Performance	
Altitude, Service ceiling	4,570 m (15,000 ft)
Speed	
Max level speed	100 kt (185 km/h; 115 mph)
Loitering speed	60 kt (111 km/h; 69 mph)
Radius of operation, mission	64 n miles (118 km; 73 miles)
Endurance, max	6 hr
Power plant	1 × turboshaft

Status: IAI has flight tested the HeMoS conversion kit on Alouette III and Bell JetRanger testbeds; it is also suited for application to other helicopter types such as the MD 500 Defender and Eurocopter EC 120.

As of 2009, the NRUAV was under negotiation as a joint venture between IAI and HAL to fulfill a requirement of the Indian Navy and, possibly, Coast Guard. No numerical requirement had then been quoted, but collectively the Indian armed forces have an inventory of more than 250 Chetak/Chetan helicopters on which to draw.

Contractor: Israel Aerospace Industries, Malat Division, Tel-Aviv.

IAI Panther

Type: Tiltrotor tactical UAV.

Development: Unveiled in Israel in early October 2010, making public debut at AUSA exhibition in the US later that month.

Description: *Airframe:* Low-mounted, high aspect ratio wings; blunt-ended cylindrical fuselage pod; twin-boom tail unit. Fixed tricycle landing gear. Composites construction.

Mission payloads: Named payloads are: day/night camera (IAI Tamam Mini-POP in Panther; Micro-POP in Mini Panther), combined with laser rangefinder, pointer or laser designator. Other sensors optional.

Guidance and control: Automatic flight control system controls transitions between vertical take-off and landing phases and forward flight. GCS has a crew of three: air vehicle controller, station controller and mission commander.

Transportation: Panther: Air vehicles, GCS shelter and ground equipment are transported in a medium-sized truck.

Mini Panther: Air vehicles and C² unit can be carried in backpacks by two soldiers.

System composition: Panther: Three air vehicles, GCS, ground datalink, support equipment and spares.

Mini Panther: Two air vehicles plus a command and control unit.

Launch: Fully automatic vertical take-off.

Recovery: Fully automatic vertical landing.

Power plant: Three battery-powered electric motors (ratings not revealed), each driving a four-blade propeller/rotor. Wing-mounted pair tilt for transition between vertical and horizontal flight; fixed rear unit is used only for vertical flight.

Variants: *Panther:* Standard-size version.

Mini Panther: Scaled-down, man-portable version.

Panther, Mini Panther

Dimensions, External	
Wings	
wing span	
Panther	8.00 m (26 ft 3 in)
Mini Panther	2.00 m (6 ft 6¾ in)
Weights and Loadings	
Weight	
Max T-O weight	
Panther	65 kg (143 lb)
Mini Panther	12 kg (26 lb)
Payload	
Max payload	
Panther	8 kg (17.00 lb)
Mini Panther	1 kg (2.00 lb)
Performance	
Altitude	
Service ceiling, Panther	3,050 m (10,000 ft)
Speed	
Cruising speed, max, Panther	70 kt (130 km/h; 81 mph)
Radius of operation, mission, Panther	26 n miles (48 km; 29 miles) (est)
Endurance	
max, Panther	6 hr
max, Mini Panther	2 hr
Power plant	3 × electric motor

Status: Flight testing coupled in 2010, operational in 2011.

Contractor: Israel Aerospace Industries, Malat Division, Tel-Aviv.

IAI Searcher

Type: Long-endurance multirole UAV.

Development: Existence of Searcher (Hebrew name Meyromit) was announced in the third quarter of 1989, and it made its public debut at the Asian Aerospace exhibition in early 1990. One prototype had been completed at that time; a second was completed shortly afterwards. First production Searcher rolled out November 1991. Deliveries of Searcher I to the Israel Defence Force began in mid-1992.

Description: *Airframe:* Shoulder-wing monoplane with pusher engine and twin-boom tail unit, built largely of composites. Generally similar to AAI/IAI Pioneer, but larger, with redesigned tail unit and high-aspect ratio tapered wing with new aerofoil section and Fowler flaps, contributing to much enhanced payload/endurance performance. Non-retractable tricycle landing gear.

Mission payloads: Normal EO payloads are IAI Tamam POP (TV/IR) or MOSP (TV/IR/laser rangefinder) with single real-time data and video downlink; 1.5 kW onboard electrical power supply. Elta EL/M-2055

Panther in forward-flying configuration *(IAI)*

First-released Panther image was of this 6-metre span prototype *(IAI)*

IAI Searcher < Israel < Unmanned aerial vehicles

Searcher II in flight *(IAI)* 1175495

Searcher II displayed at the Paris airshow *(IHS/Patrick Allen)* 1024120

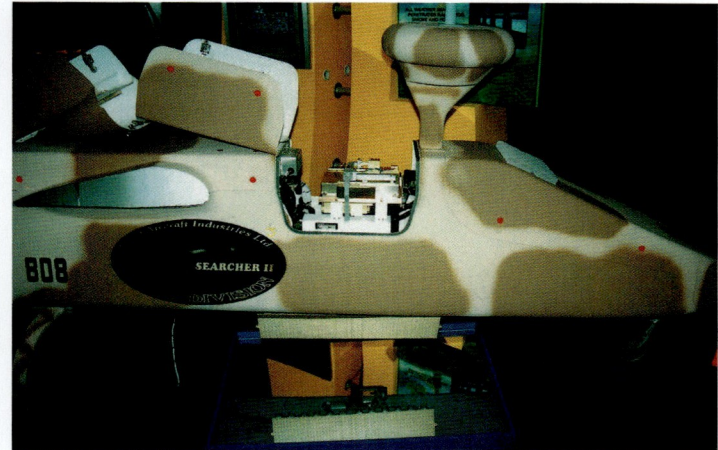
Searcher II payload bay *(Paul Jackson)* 0087644

Searcher II with belly-mounted EL/M-2055 SAR/MTI radar *(IHS/Kenneth Munson)* 0109993

Early production Searcher I *(Paul Jackson)* 0528604

Searcher I *(IAI)* 0579553

synthetic aperture radar, introduced on Searcher II in 2001, enhances night/all-weather capability. Comint and ESM integration capability. Other payloads to customer's choice.

Guidance and control: Searcher can be controlled from a variety of GCSs which command, control, track and communicate with it and/or its payload via direct LOS datalink, dual real-time command uplink, single real-time data and video downlink and airborne or ground-based data relay for BLOS missions. System has GPS-based airborne mission controller mode with real-time manual interrupt capability; autonomous return-home mode if datalink is lost. The IAI GCS-2003 is a major subsystem, centralising all comments sent by system operators to the UAV and all reports received from the UAV (including relay and mission UAVs). Three system operators man the GCS, using computer-driven panels and other units installed in the control station bays. The GCS is housed in a shelter accommodating four such bays and, optionally, a mission commander's desk. Searcher is compatible with other Malat ground stations, avionics and datalinks.

Launch: Wheeled take-off, or by pneumatic catapult or JATO booster rocket. Automatic for Searcher II.

Recovery: Wheeled landing to arrester hook and cable. Automatic for Searcher II.

Power plant: Searcher I: One 26.1 kW (35 hp) Sachs engine initially, with two-blade pusher propeller; replaced by 35.0 kW (47 hp) Limbach L 550 flat-four and three-blade propeller. See under Weights for fuel details.

Searcher II: One 55.9 kW (75 hp) UEL AR 682 rotary engine and three-blade pusher propeller.

Variants: *Searcher I:* Initial production model.

Searcher II: Improved version; public debut at Singapore Air Show, February 1998. Main configuration changes are extended-span wings with modest sweepback, and a rotary engine with a three-blade propeller. Can be configured for tactical surveillance or as a communications relay aircraft. Several payloads can be carried simultaneously. Existing Searcher Is can be upgraded to Mk II standard.

Searcher I, Searcher II

Dimensions, External	
Overall	
length	
Searcher I	5.15 m (16 ft 10¾ in)
Searcher II	5.85 m (19 ft 2¼ in)
height, Searcher I	1.16 m (3 ft 9¾ in)
Wings	
wing span	
Searcher I	7.22 m (23 ft 8¼ in)
Searcher II	8.55 m (28 ft 0½ in)
Engines, propeller diameter	1.40 m (4 ft 7 in)
Weights and Loadings	
Weight	
Max T-O weight	
Searcher I	372 kg (820 lb)
Searcher II	426 kg (939 lb)
Fuel weight	
Max fuel weight	
Searcher I	102 kg (224 lb)
Searcher II	110 kg (242 lb)
Payload	
Max payload	
Searcher I	372 kg (820 lb)
Searcher II	426 kg (939 lb)
Performance	
(A: with L 550 engine)	
Altitude	
Service ceiling	
Searcher I (A)	4,570 m (15,000 ft)
Searcher II	6,095 m (20,000 ft)
Speed	
Cruising speed	
max, Searcher I (A)	105 kt (194 km/h; 121 mph)

Unmanned aerial vehicles > Israel > IAI Searcher – IMI ADM-141 TALD and ITALD

Searcher II of the Sri Lanka Air Force *(SLAF)* 1122540

Searcher I, Searcher II

Loitering speed	
Searcher I (A)	60 kt (111 km/h; 69 mph)
Range	
LOS, Searcher I (A)	65 n miles (120 km; 74 miles)
LOS, Searcher II	108 n miles (200 km; 124 miles)
BLOS, Searcher I (A)	119 n miles (220 km; 136 miles)
BLOS, Searcher II	135 n miles (250 km; 155 miles)
Endurance	
Searcher I (A)	14 hr
Searcher II	15 hr
Time on station at 3,050 m (10,000 ft) at 54 n miles (100 km; 62 miles) from base, Searcher I (A)	12 hr

Status: In production and service.

Customers: Israel (first Searcher I deliveries mid-1992, but operated for IDF by manufacturer until at least 1997). Equip No 200 Squadron at Palmachim; possibly also Nos 146 (Radom) and 155 (Hatserim). Searcher II operational with IDF from May 1999. Some reports from Israel suggest the Searcher force will be run down in favour of Herons.

Foreign customers are not disclosed by IAI but have included India (Searcher II, 10 delivered from 2001; eight more ordered in 2002), Singapore (Searcher II, 42 for No 128 Squadron), Sri Lanka (original quantity lost but now replaced), Taiwan and Thailand (Searcher II, one system; no longer in service). Spain joined the list of operators with an order for four Searcher II air vehicles in 2007 reportedly worth USD23 million, with participation by Indra and EADS. A deployment of these UAVs to Afghanistan, supported by 36 Spanish personnel, was planned in 2008. Indian Navy's first UAV squadron, INAS(U) 342 at Kochi, activated 6 January 2006 with four Herons and eight Searcher IIs.

Searcher sales reportedly exceed 100 air vehicles and 20 GCSs.

Contractor: Israel Aerospace Industries, Malat Division, Tel-Aviv.

Air launch of a TALD from an F/A-18 Hornet 0518020

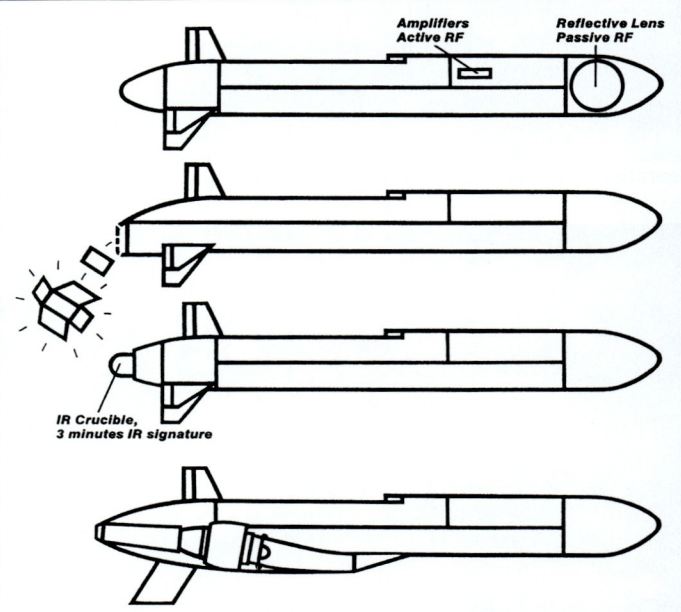

Top to bottom: RF, chaff and IR TALD, and ITALD 0518021

IMI ADM-141 TALD and ITALD

Type: Air-launched expendable decoys.

Development: TALD (Tactical Air-Launched Decoy) was developed in the USA by Brunswick Defense as an improved version of the unpowered IMI Samson, after the import of about 2,250 of the latter for use by the US Navy and Marine Corps. Brunswick later ceased TALD production and withdrew from the ITALD programme. The latter is a jet-powered version with terrain-following capability, development and flight test of which was completed in 1996.

Description: *Airframe:* Fuselage of basically square cross-section with rounded corners. Three tailfins on TALD; ITALD has conventional horizontal tail surfaces plus dorsal and ventral vertical fins. Ventral engine air intake duct on ITALD.

Mission payloads: Active (amplifiers) and passive (Luneberg lens) payloads in RF TALD augment air vehicle RCS, enabling it to be used to saturate air defence systems or as a training target for radar-guided air-to-air and surface-to-air missiles. Chaff TALD can dispense up to 36.3 kg (80 lb) of chaff, either in a single curtain or in 40 incremental ejections. IR TALD has a three-minute, tailcone-mounted IR source for use as a training target for IR-guided missiles.

An upgraded RF payload for ITALD was developed by IMI in 2001 to enable it to simulate the RCS of specific fighter aircraft regarded as posing the greatest threat to ground-based air defence systems.

Guidance and control: Preprogrammed flight profiles permit a variety of flight manoeuvres to simulate manned aircraft or missiles. Each mission can be programmed into the autopilot at the option of the field commander. Programmable variables include launch altitude, attack velocities, flight profiles and manoeuvrability. Can be used in conjunction with other countermeasures to increase the survivability of strike aircraft in heavily defended areas.

In September 2000, IMI received a USD6.35 million US Navy contract for upgrade and retrofit of ITALD with a GPS receiver.

Launch: Air-launched from standard triple or multiple ejector rack, enabling some aircraft to carry as many as 20 TALDs per mission. Air vehicle is designed to be launchable from any stores station that can accommodate a 500 lb Mk 82 bomb, and requires no modification of the parent aircraft. Both types are currently certified on A-4, A-6, A-7, AV-8B, F-4, F-16, F/A-18 and S-3 aircraft. Launch modes include low-level toss and high-speed toss dive, as well as all-altitude airdrop.

Recovery: Non-recoverable.

Power plant: None in TALD; ITALD powered by one 0.79 kN (177 lb st) Teledyne Continental 312 (J700-CA-400) turbojet.

Variants: *TALD (ADM-141A):* Unpowered version, produced in three main mission configurations: RF (principal version), chaff and IR (see drawing and Mission payloads paragraph).

ADM-141C ITALD jet-powered air-launched decoy *(Paul Jackson)* 0087645

ITALD has received more than USD90 million in orders *(IMI)* 0530461

ITALD (ADM-141C): Improved TALD, with turbojet engine and radar altimeter for low-altitude navigation. Increased speed and range; more realistic flight profiles; enhanced IR signature; improved versatility and tactics. Developed under 1991 US$23.7 million contract involving 25 Brunswick-converted TALD air vehicles. First flight 1996, from US Navy F-4 at the Pacific Missile Test Center, Point Mugu, California. Flight trials of launch from F/A-18 Hornet took place in late 1996 at NAS Patuxent River, Maryland.

TALD, ITALD

Dimensions, External
Overall
 length...2.34 m (7 ft 8¼ in)
 height..0.56 m (1 ft 10 in) (est)
Wings, wing span..1.55 m (5 ft 1 in)
Weights and Loadings
Weight
 Max launch weight
  TALD ....181.5 kg (400 lb)
  ITALD ....172.5 kg (380 lb)
Payload
 Max payload, chaff..36.3 kg (80 lb)
Performance
Range
 launched at 250 kt IAS (typical, depending on launch altitude and glide ratio), at 610 m (2,000 ft),
  TALD..14 n miles (25 km; 16 miles)
 launched at 250 kt IAS (typical, depending on launch altitude and glide ratio), at 10,670 m (35,000 ft),
  TALD...68 n miles (125 km; 78 miles)
 at M0.73 at 150 m (500 ft), ITALD...........100 n miles (185 km; 115 miles) (est)
 at M0.80 at 6,100 m (20,000 ft),
  ITALD...160 n miles (296 km; 184 miles) (est)
Endurance, ITALD..35 min
Power plant, ITALD...1 × turbojet

Status: *TALD:* In production and service. Combined Brunswick/IMI production for the US Navy had reached 6,000 by the end of 2001. Some 137 were launched during the first 72 hours of the 1991 Gulf War by US Navy and Marine Corps A-6, A-7, F/A-18 and S-3 aircraft, exposing Iraqi radars to quick destruction by strike aircraft. They were also used to 'soak up' Iraqi surface-to-air missiles; TALDs are believed to account for more than 100 of the 'allied aircraft' losses claimed by the Iraqi forces. Several were also intercepted by Iraqi fighters. In the mid-1990s, TALDs were launched from US Navy A-6 Intruders during NATO air strikes against Bosnian Serb radars and SAM command and control sites; more recently, they were in use during Operation 'Iraqi Freedom' in March/April 2003.

ITALD: In production and service. Initial production contract (USD21.2 million for 98) announced October 1996; second order (USD14 million for 110) awarded 3 March 1998; third (USD21.6 million for 140) in January 2003; fourth (USD12.5 million for an undisclosed quantity) in April 2003.

Customers: *TALD:* Israel (Air Force); US Navy and Marine Corps (all carrier air wings); plus one unidentified customer.
ITALD: US Navy.
Contractor: Israel Military Industries Ltd
Ramat Hasharon.

Four-round Delilah ground launch container 0518012

Delilah mounted underwing on an IDF Air Force F-4E Phantom 0518013

Delilah multirole UAV/strike weapon *(IMI)* 0577375

IMI Delilah

Type: Air- or surface-launched expendable UAV.
Development: Delilah was developed in the 1980s, its existence first becoming known in 1988. Initial deployment was in the role of decoy, and it is possibly still so employed. However, other roles have since been developed, and more recent company literature has described it variously as a stand-off powered UAV, air-to-surface missile or precision strike weapon. A statement on 11 May 2007 by IMI described its first operational use as a weapon and reported how a Delilah launched at stand-off range hit a vehicle in Lebanon.

Description: *Airframe:* Mid-mounted, short-span tapered wings; blunt-nose cylindrical fuselage with ventral engine air intake duct; four sweptback tailfins, indexed in X configuration. Modular construction.

Mission payloads: In the decoy role, Delilah simulates the presence of an attacking aircraft by the use of active and passive means of RCS augmentation. Active repeater elements provide radar feedback in the A (0.1 to 0.25 GHz), C (0.5 to 1 GHz) and L (40 to 60 GHz) frequency bands; passive enhancement is provided by a nose-mounted Luneberg lens reflector which 'captures' enemy radar and transmits a highly augmented image back to the radar detection and fire-control centres.

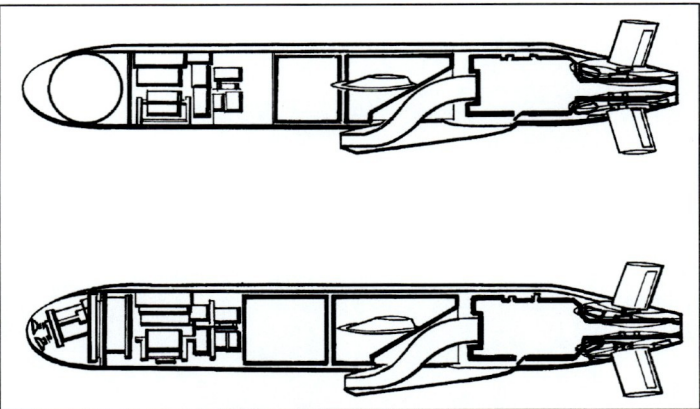

Internal features of the Delilah decoy (top) and Delilah-AR

	Delilah	Delilah-AR
Nose section		
Payload	x	-
GPS antenna	x	x
AR seeker	-	x
Avionics section		
Power supply	x	-
Flight control		
Computer	x	x
Avionics	x	x
GPS receiver	x	x
Central section		
Fuel tank	x	x
Payload	x	-
Warhead and fuze	-	x
Rear section		
Engine	x	x
Alternator	x	x
Control surfaces	x	x
Control servos	x	x

(IHS/John W Wood)

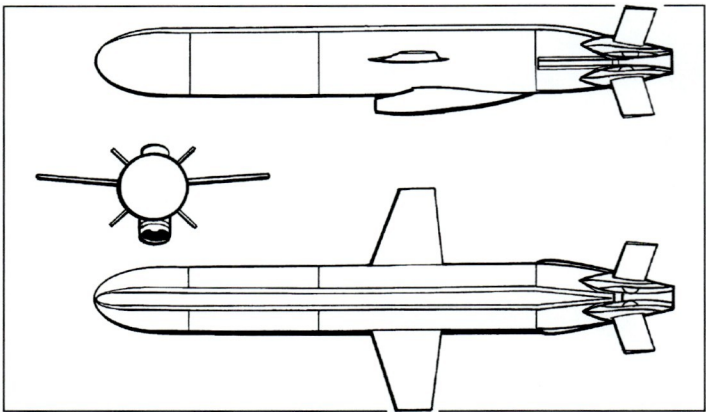

IMI Delilah general arrangement *(IHS/John W Wood)*

In its alternative main decoy function, the Delilah disrupts and neutralises enemy air defence systems by saturating the mission area with chaff before the arrival of an attack force.

Alternative payloads for ground attack (high-explosive warhead), ECM, reconnaissance (EO/IR, with target autotracking) or aerial target roles can be incorporated to suit customer requirements.

Guidance and control: Digital flight control computer; GPS/INS navigation and nose-mounted antenna for accurate transit to, and positioning in, mission area. Delilah flies preprogrammed profiles, and is completely automatic once launched.

Launch: Can be carried by, and launched from, aircraft such as A-4 Skyhawk, F-4 Phantom, F-16 Fighting Falcon and F/A-18 Hornet, by means of a standard MAU 12 underwing pylon with an ejector unit. No aircraft modification is required. Also suitable for air launch from under stub wings on such helicopters as Sikorsky S-70; was test-fired from a US Navy S-70B Seahawk in 2004. Preprogramming, final testing and installation on the carrying aircraft can be accomplished in less than 5 minutes.

Surface launch from ground or at sea, with booster rocket assistance, can be made singly, in ripple mode, or at any desired interval, from a four- or eight-round sealed container. Launch container also serves as storage unit, in which Delilah can be held in operation-ready condition for up to 5 years.

Recovery: Non-recoverable.

Power plant: One 0.73 kN (165 lb st) Noel Penny Turbines NPT 151-4 turbojet initially; current power plant is either a 0.76 kN (170 lb st) Williams J400-WR-401 turbojet or a 0.89 kN (200 lb st) Bet-Shemesh BS 175.

Variants: Delilah: Standard decoy. *Detailed description applies to this version.*

Delilah-AR: Standoff anti-radiation missile version (also known as **Star-1**); broadband (2 to 18 GHz) passive radar seeker capability. Can loiter in target zone waiting for a radar emission before going in for an attack. Unveiled at 1995 Paris Air Show.

Delilah-GL: Ground-launched, long-range, loitering strike missile or reconnaissance version, revealed in mid-2004. Carried by F-4E-2000 Kurnass Phantoms and F-16s of IDF/Air Force. Details in *Jane's Air-Launched Weapons.*

Delilah-SL: Sea-launched version of GL.

Delilah

Dimensions, External	
Overall, length	2.71 m (8 ft 10¾ in)
Fuselage	
height, max	0.33 m (1 ft 1 in)
Wings, wing span	1.15 m (3 ft 9¼ in)
Tailplane, tailplane span	0.82 m (2 ft 8¼ in)
Weights and Loadings	
Weight, Max launch weight	187 kg (412 lb)
Fuel weight	
Max fuel weight with reduced payload	54.4 kg (119 lb)
Payload, Max payload	30.0 kg (66 lb)
Performance	
Climb	
Rate of climb, max, at S/L	1,219 m/min (3,999 ft/min)
Speed, Max operating speed	430 kt from S/L to 6,096 m (796 km/h from S/L to 20,000 ft; 495 mph from S/L to)
Range	135 n miles (250 km; 155 miles) (est)

Status: Decoy version in service with two squadrons of IDF Air Force (Nos 146 and 155) at Ramon and Hatzerim.

Customers: Israel Defence Force/Air Force.

Contractor: Israel Military Industries Ltd
Ramat Hasharon.

Innocon MicroFalcon I

Type: Tactical mini-UAV.

Development: Begun in 2007 as a mini system for 'over the hill' ISTAR missions.

Description: *Airframe:* Unconventional 'boxkite' configuration of sharply sweptback wings and non-swept horizontal tail, connected at their tips by vertical fins and rudders. Fuselage attached at top to apex of swept wings and at rear to tailplane centre. Pusher engine. No landing gear. All-composites construction.

Mission payloads: Controp D-STAMP standard, in ventral location.

Guidance and control: Fully autonomous, including launch and recovery, utilising Innocon's own microNaviator flight control system and Naviator GCS. One-person operation.

Transportation: Can be carried in a backpack by one person.

Launch: Automatic launch by bungee catapult.

Recovery: Parachute automatic recovery.

Power plant: Electric motor (rating not stated), driving a two-blade pusher propeller.

MicroFalcon I

Dimensions, External	
Wings, wing span	1.60 m (5 ft 3 in)
Weights and Loadings	
Weight, Max launch weight	6.0 kg (13.00 lb)
Payload, Max payload	0.75 kg (1.00 lb)

Unorthodox configuration of the MicroFalcon I mini-UAV *(Innocon)*

MicroFalcon I

Performance
Endurance, max..2 hr
Power plant..1 × electric motor

Status: Being marketed in 2008-09.
Contractor: Innocon Ltd
Holon.

Innocon MiniFalcon I and II

Type: Small tactical UAV.
Development: Begun in 2002. Public debut at NAS Patuxent River in August 2003.
Description: *Airframe*: High-mounted wings, tapered on leading-edges; bullet-shaped fuselage pod, with dorsally mounted tailboom carrying inverted V tail surfaces. Fixed tricycle landing gear on MiniFalcon II. Composites construction.
Mission payloads: EO and/or IR surveillance sensors in ventral turret (IAI Tamam POP 200 typical), with real-time imagery downlink.
Guidance and control: Fully automatic, using Innocon's own Miniature Integrated Avionics Suite (MIAS) and Naviator flight control syatem and GCS.
Launch: Automatic launch by catapult.
Recovery: Automatic belly skid or wheel landing, according to version. Can be recovered on land or on board ship.
Power plant: *MiniFalcon I*: One 11.2 kW (15 hp) 3W 157 piston engine; two-blade pusher propeller. Fuel capacity 45 litres (11.9 US gallons; 9.9 Imp gallons).
MiniFalcon II: One 19.4 kW (26 hp) Herbrandson DH 290 flat twin or 26.8 kW (36 hp) UEL AR 741 rotary piston engine; two-blade pusher propeller. Fuel capacity 65 litres (17.2 US gallons; 14.3 Imp gallons).
Variants: *MiniFalcon I*: Baseline version.
MiniFalcon II: Enlarged version, with more powerful engine, increased payload and greater endurance.

MiniFalcon I, MiniFalcon II

Dimensions, External
Overall
  length
    MiniFalcon I... 3.50 m (11 ft 5¾ in)
    MiniFalcon II.. 4.20 m (13 ft 9¼ in)
  height
    MiniFalcon I... 1.00 m (3 ft 3¼ in)
    MiniFalcon II... 1.20 m (3 ft 11¼ in)
Wings
  wing span
    MiniFalcon I... 5.00 m (16 ft 4¾ in)
    MiniFalcon II.. 5.50 m (18 ft 0½ in)
Weights and Loadings
Weight
  Weight empty
    MiniFalcon I.. 30 kg (66 lb)
    MiniFalcon II.. 65 kg (143 lb)
  Max launch weight
    MiniFalcon I.. 90 kg (198 lb)
    MiniFalcon II.. 150 kg (330 lb)
Payload
  Max payload
    MiniFalcon I.. 15 kg (33 lb)
    MiniFalcon II.. 35 kg (77 lb)
Performance
Speed
  Max level speed
    MiniFalcon I.. 90 kt (167 km/h; 104 mph)
    MiniFalcon II.. 120 kt (222 km/h; 138 mph)
  Loitering speed
    MiniFalcon I... 50 kt (93 km/h; 58 mph)
    MiniFalcon II.. 55 kt (102 km/h; 63 mph)
Endurance
  MiniFalcon I.. 12 hr (est)
  MiniFalcon II... 15 hr (est)
Power plant..1 × piston engine

Status: In production and service.
Contractor: Innocon Ltd
Holon.

Steadicopter Black Eagle 50/STD-5 Helivision

Type: Small rotary-wing UAV.
Development: Steadicopter Ltd was created in 1999 (from a company that had previously been established by the Technion Institute of Technology in Haifa) and produced its first helicopter UAV during 2004. This was designated as the STD-5 Helivision and is described as having been a technology demonstrator. A larger and heavier developed version of the STD-5 is called the Black Eagle 50 and is designed to meet both military and civil applications, with the latter including media image gathering, agriculture, high-voltage line inspection and environmental monitoring. The flight control system used aboard these AVs is part of an unmanned kit which can be adapted to any size and type of hovering platform, and which allows manned helicopters to fly autonomously. In terms of programme events, Black Eagle 50 development was completed in mid-2009 and the type has been displayed at the October 2009 Israel Defence International Army and Police Exhibition in Tel Aviv and at the AUVSI International trade show during March 2012.
Description: *Airframe*: STD-5 Helivision: Based on a COTS Bergen design
Black Eagle 50: Tadpole-shaped fuselage carrying a balanced two-bladed main rotor and a tail rotor offset to port. Swept back dorsal and ventral tail surfaces mounted at the rear of the AV's tailboom. Fixed skid undercarriage. Provision for a nose-mounted sensor turret.
Mission payloads: A wide variety of payloads can be fitted, including IR, digital cameras, sensors and mission-oriented devices.
Guidance and control: Pre-programmed, fully autonomous operation, including take-off and landing. Can be programmed in flight via a point and click digital map interface. Four flight modes available:

The Black Eagle 50 in its original configuration *(Steadicopter)*

MiniFalcon II making an automatic landing *(Innocon)*

An in-flight view of the latest (as of October 2012) Black Eagle 50 configuration *(Steadicopter)*

MiniFalcon II small TUAV. MiniFalcon I is similar, but lacks wheel landing gear *(Innocon)*

- Fully autonomous via preset co-ordinates;
- Semi-autonomous (changing flight plan during mission by indicating new co-ordinates);
- Manual navigation by ground operator
- Emergency (fully autonomous return to base and landing.

Provision for in-flight GPS navigation and remote control if appropriate to the particular mission. (PC-based navigation software enables control and navigation via GPS, including a continuous location tracer on a digital map or aerial photograph.) Fully autonomous stabilisation system controls all flight parameters and permits prolonged hovering over a selected area. Real-time video can be displayed on the computer screen or on an external TV screen, depending upon customer requirements. A daylight-readable screen is provided.

Launch: Conventional, automatic, helicopter take-off; manual offset optional.

Recovery: Conventional, automatic, helicopter landing; manual offset optional.

Power plant: STD-5 Helivision: One 46 cc two-cylinder (paired Zenoah G-23s) piston engine. Fuel capacity 3.8 litres (1.0 US gallon; 0.8 Imp gallon).

Black Eagle 50: One 120 cc two-cylinder liquid-cooled piston engine. Fuel capacity 8 litres (2.1 US gallon; 1.76 gallon).

STD-5 Helivision, Black Eagle 50

Dimensions, External
 Overall
  height
   STD-5 Helivision .... 0.56 m (1 ft 10 in)
   Black Eagle 50 .... 0.86 m (2 ft 9¾ in)
 Fuselage
  length
   STD-5 Helivision .... 1.50 m (4 ft 11 in)
   Black Eagle 50 .... 2.00 m (6 ft 6¾ in)
 Rotors
  rotor diameter
   STD-5 Helivision .... 1.80 m (5 ft 10¾ in)
   Black Eagle 50 .... 2.00 m (6 ft 6¾ in)
Weights and Loadings
 Weight
  Weight empty
   STD-5 Helivision .... 8.2 kg (18.00 lb)
   Black Eagle 50 .... 17 kg (37 lb)
  Max T-O weight
   STD-5 Helivision .... 13.6 kg (29 lb) (est)
   Black Eagle 50 .... 35 kg (77 lb) (est)
 Payload, Max payload .... 5.4 kg (11.00 lb)
Performance
 Range
  control, STD-5 Helivision .... 2.7 n miles (5 km; 3 miles)
  control, Black Eagle 50 .... 5.4 n miles (10 km; 6 miles)
 Endurance
  STD-5 Helivision .... 1 hr (est)
  Black Eagle 50 .... 3 hr
 Power plant .... 1 × piston engine

Status: Most recently, Steadicopter has continued to promote the Black Eagle 50 rotary-wing UAV.

Contractor: Steadicopter Ltd
Yokneam Ilit

Top I Vision Casper 200 and 250

Type: Surveillance mini-UAV.

Development: Introduced in 2004.

Description: *Airframe:* Pylon-mounted ('parasol'), high aspect ratio wings, with large centre-section flaps, outer-panel dihedral and raked tips; telescopic fuselage; simple tail unit; pod-mounted engine. Composites construction. No landing gear.

Mission payloads: Options (nose-mounted) include day TV; thermal imager; or radiology payload.

Guidance and control: Portable computer and vehicle-mounted GCSs. Pre-programmed flight and mission profiles; GPS navigation.

Transportation: Can be transported in a backpack and assembled in the field, without special tools, by two-person crew.

System composition: Two or three air vehicles; two GCS.

Launch: Hand or bungee catapult launch.

Recovery: Automatic deployment of wing flaps decelerates aircraft for belly landing.

Power plant: One battery-powered electric motor, driving a two-blade propeller.

Variants: *Casper 200:* Initial production version.

A Black Eagle 50 prototype photographed during flight testing *(Steadicopter)* 1377215

An in-flight view of the STD-5 Helivision rotary-wing UAV *(Steadicopter)* 1042588

The STD-5 Helivision is based on a commercial Bergen design *(Steadicopter)* 1042589

Casper 200 *(IHS/Patrick Allen)* 1144953

Casper 200 *(IHS/Patrick Allen)* 1183454

Casper 200 in flight with wing flaps deployed *(Grzegorz Holdanowicz)* 1039987

Casper 250: Upgraded version. Modified, three-axis stabilised nose payload module.

Sofar: Version adapted by WB Electronics (Poland) in 2004, originally as part (with Rufus aerostat) of **Calineczka** system proposed (unsuccessfully) to Polish Army in 2005 as support for WB's Topaz artillery fire-control system and Trop battlefield management system. C-band digital datalink. Selected by Hungarian Army in December 2006.

Casper 200, Casper 250, Sofar

Dimensions, External	
Overall	
length, Casper 200, Sofar	1.30 m (4 ft 3¼ in)
Wings	
wing span	
Casper 200	2.00 m (6 ft 6¾ in)
Sofar	2.50 m (8 ft 2½ in)
Weights and Loadings	
Weight	
Max launch weight	
Casper 200	2.3 kg (5.00 lb)
Sofar	2.9 kg (6.00 lb)
Performance	
Climb	
Rate of climb	
max, at S/L, Casper 200	420 m/min (1,378 ft/min)
Altitude	
Operating altitude	
min, Casper 200, Sofar	70 m (230 ft)
max, Casper 200, Sofar	250 m (820 ft)
Service ceiling, Casper 250	610 m (2,000 ft)
Speed	
Cruising speed	
max, Casper 200	45 kt (83 km/h; 52 mph)
max, Sofar	38 kt (70 km/h; 44 mph)
Radius of operation	
mission, Casper 250, Sofar	5.4 n miles (10 km; 6 miles)
Endurance	
max, Casper 200	2 hr 20 min
max, Casper 250	1 hr 30 min
max, Sofar	1 hr
Best glide ratio, Casper 200	15
Power plant	1 × electric motor

Status: In production and in service.

Customers: 'A number' (unspecified) of international customers, including several Casper 200 systems delivered to Italian Carabinieri in 2004. Two Sofar systems ordered from WB Electronics by Hungarian Army on 19 December 2006 (contract value EUR800,000; USD1.05 million).

Contractor: Top I Vision
Holon.

Top I Vision Casper 350

Type: Short-range tactical UAV.

Development: Designed for such applications as tactical surveillance and reconnaissance, target acquisition, special operations, law enforcement and perimeter security.

Description: *Airframe:* High-mounted wings with tapered outer panels; pod fuselage; single tailboom, elevated on dorsal pylon and supporting inverted V tail surfaces (two configurations have been shown, see images below); pusher engine. Fixed tricycle landing gear. Composites construction.

The Casper 350 tactical UAV *(Top I Vision)* 1374765

Casper 350 in flight *(Top I Vision)* 1374764

Mission payloads: EO and IR cameras in underfuselage turret. Real-time, continuous video and telemetry downlink.
Guidance and control: No information provided.
Launch: Conventional wheeled take-off or catapult launch.
Recovery: Conventional wheeled landing or parachute recovery.
Power plant: One 50 cm³ (3.05 cu in) piston engine (type and rating not stated); two-blade pusher propeller.

Casper 350

Dimensions, External	
Overall, length	4.20 m (13 ft 9¼ in)
Wings, wing span	2.20 m (7 ft 2½ in)
Weights and Loadings	
Weight, Max T-O weight	32.0 kg (70 lb)
Payload, Max payload	5.5 kg (12.00 lb)
Performance	
Altitude, Service ceiling	3,660 m (12,000 ft)
Speed	
Cruising speed, max	60 kt (111 km/h; 69 mph)
Radius of operation	27 n miles (50 km; 31 miles)
Endurance	4 hr
Power plant	1 × piston engine

Status: Under development in 2008-09. The company showed a vehicle of similar size with extended horizontal tail surfaces, designated Casper 800, in March 2010.

Contractor: Top I Vision
Holon.

Top I Vision Rufus

Type: Helium-filled, tethered, non-rigid surveillance aerostat.

Development: Thought to have been introduced in about 2001, and entered service in 2002. In May 2005, WB Electronics of Poland proposed a UAV system known as **Calineczka**, based on the Rufus aerostat and a modified version of Top I's Casper 200 UAV known as Sofar, for evaluation by the Polish armed forces as an adjunct to its Topaz artillery fire control and Trop battlefield management systems, but this was not adopted.

Description: *Airframe:* Helium-filled envelope with sweptback, inverted Y inflatable tail surfaces.
Mission payloads: 3 CCD digital processing day camera or infra-red thermal sensor with three-axis gyrostabilisation; field of view 360° in azimuth, +8/-105° in elevation; ×20 power zoom and ×5 digital zoom.
Guidance and control: Truck-mounted GCS.
Transportation: Aerostat can be stowed in a 1 × 0.5 × 1 m (3.05 × 0.15 × 3.05 ft) box and transported in a small vehicle when deflated or on trailer when semi-inflated. Set-up and deployment in 30 minutes by ground crew of two or three.
Power plant: None.

Rufus aerostat as used in Polish Calineczka system *(Grzegorz Holdanowicz)* 1039985

Top I Vision Rufus

Dimensions
- Envelope: Length .... 7.62 m (25 ft 0 in)
- Max diameter .... 2.90 m (9 ft 6 in)
- Volume .... 28.3 m3 (1,000 cu ft)

Weights
- Typical payload .... 4.1 kg (9.0 lb)

Performance
- Operating altitude .... 183 m (600 ft)

Status: In service.

Customers: Include Israel and Poland. Selected annually from 2002 by Israeli Police for traffic and anti-crime surveillance; operated also by Israeli TV in March 2006 during election coverage in Tel-Aviv. Polish system reportedly used mainly by unidentified security forces and/or civilian security agencies, including coverage of the Pope's visit to Poland in May 2006; proposed (unsuccessfully) to Polish MoD in 2005. Top I Vision aerostats regularly deployed by Israeli Police for crowd control and monitoring during concerts, state visits and security exercises between 2006-2010.

Contractor: Top I Vision
Holon.

Urban Aeronautics AirMule

Type: Optionally piloted cargo and medevac technology demonstrator.

Development: The AirMule is a derivative of UrbanAero's X-Hawk (see *IHS Jane's All the World's Aircraft: In service*) promoted as an armed troop carrier, and is intended for supply transport and medevac applications. For the latter role, it would be man-rated and is being promoted as being able to transport two casualties while carrying enough fuel for a 2-hour flight. The internal rotor (modified ducted fan) technology underlying the design was previously demonstrated by the company's CityHawk prototype.

A claimed advantage of the system is the ability to fly in cities and restricted areas where conventional rotorcraft are unable to operate. First called Mule and subsequently renamed Urban Aeronautics AirMule, the first engine start was 2 June 2009, and the air vehicle was operated "light on the skids" at an airfield in central Israel during September of that year. Tethered tests against two 500 kg ballast weights started in January 2010, with free hovers scheduled for later that year. By October 2010, 40 test hovers and 10 flying hours had been reported. A systems and structural upgrade lasting four months followed. These comprised an expanded sensors suite, wheeled landing gear to FAR 27 standard and revised lower fuselage contours for improved control response in gusty conditions. Flight testing resumed in 2011.

Description: *Airframe:* Broadly boat-shaped body, fore and aft portions of which each house a ducted lift-fan. Power plant and payload bays amidships. Body sidewalls rise towards rear to support two outrigged, ducted thrust fan units driven by power take-off from central engine. Thruster units bridged by central horizontal stabiliser with twin elevators. Twin-skid landing gear, optionally wheeled.

Mission payloads: Cargo or casualties. MedEvac version also projected to carry active and passive self-protection systems.

A general view of an AirMule AV during tethered tests in January 2010 *(Urban Aeronautics)* 1395408

An AirMule AV equipped with wheeled landing gear and a revised lower fuselage configuration *(Urban Aeronautics)* 1398561

An in-flight view of a 2012 configuration AirMule AV *(Urban Aeronautics)* 1365373

Guidance and control: Preprogrammed or remotely piloted, depending upon mission. Patented aerodynamic technologies include vane control system (VCS) comprising two vertical ducts, each with a set of vanes top and bottom, independent deflection of which, between top and bottom and between ducts, provides manoeuvrability. Rearward deflection of thrust from ducts provides forward motion; thruster fans contribute additional speed and assist manoeuvring at the hover. Normal operation relying on autonomous, GPS guided flight to target area with subsequent optional (ground placed) beacon assisted autoland. STOVL operation became possible with the wheeled landing gear fitted in 2011.

Launch: Vertical take-off.
Recovery: Vertical landing.
Power plant: One r 546 kw (732 shp) Turbomeca Arriel 1D1 turboshaft

AirMule

Dimensions, External
Overall
- length
  - incl thrusters .... 5.50 m (18 ft 0½ in)
  - excl thrusters .... 5.30 m (17 ft 4¾ in)
- height .... 2.00 m (6 ft 6¾ in)
- width, over thrusters .... 3.10 m (10 ft 2 in)
- Fuselage, width .... 2.15 m (7 ft 0¾ in)

Weights and Loadings
Weight
- Weight empty .... 635 kg (1,399 lb)
- Max T-O weight .... 1,089 kg (2,400 lb)

Performance
Altitude
- Operating altitude, max .... 3,660 m (12,000 ft) (est)
- Speed, Max level speed .... 100 kt (185 km/h; 115 mph) (est)
- Power plant .... 1 × turboshaft

Status: Under development. Two prototypes planned.
Contractor: Urban Aeronautics Ltd
Yavne.

UVision Air Wasp

Type: Close-range tactical mini-UAV.

Development: Exhibited at Paris Air Show in June 2011. Designed to achieve local situational awareness for front-line troops. Stated features include excellent manoeuvrability and very low noise signature.

Description: *Airframe:* Blunt-nosed cylindrical body. Four wing and four tail surfaces, indexed in X-X arrangement; rear-mounted engine. No landing gear.

Mission payloads: Stabilised miniature sensor (daylight colour TV or night imaging system) with electronic tracker. Dual datalinks. Real-time continuous transmission of video and telemetry data.

Guidance and control: Fully autonomous flight; command and control by hand-attached micro computer. One-person operation.

Transportation: Man-portable in backpack.

The cruciform-wing Wasp mini-UAV *(IHS/Patrick Allen)* 1465665

Ground control station for the UVision Wasp *(IHS/Patrick Allen)*

Launch: From custom-made miniature launcher.
Recovery: Operator-controlled landing.
Power plant: Electric motor (rating not stated), driving a two-blade pusher propeller.

Air Wasp

Dimensions, External	
Overall, length	0.70 m (2 ft 3½ in)
Fuselage, height	0.08 m (3¼ in)
Wings, wing span	0.67 m (2 ft 2½ in)
Weights and Loadings	
Weight, Max launch weight	2.5 kg (5.00 lb)
Payload, Max payload	1.0 kg (2.00 lb)
Performance	
Speed	
Cruising speed, max	100 kt (185 km/h; 115 mph)
Radius of operation	2.7 n miles (5 km; 3 miles)

Status: Developed and available.
Contractor: UVision Air Ltd
Zur Igal.

Italy

Alenia Molynx

Type: Medium/high altitude, long endurance UAV.
Development: Revealed in October 2006 in the form of a quarter-scale model at an exhibition in Turin, the Molynx is envisaged as entering service for government civil and commercial applications in the second decade of this century. It derives its name from a basic air vehicle (Lynx), associated with that of Carlo Mollino, a noted designer of aerobatic aircraft. Military versions will be named BlackLynx.

Further details were released at the Paris Air Show in June 2007, at unveiling of the company's smaller Sky-Y, which will serve as a technology demonstrator for the larger aircraft. First flight of the Molynx is now targeted for 2011. Sky-Y will work as a technology demonstrator for Molynx.

Description: *Airframe:* High aspect ratio mid-wings; tapered, oval-section fuselage with T tail and underfin; twin, wing-mounted pusher engines (heavy fuel diesel, type unspecified). Retractable tricycle landing gear. Composites construction.

Mission payloads: Planned to include nose-mounted Galileo EO/IR sensors, synthetic aperture radar and hyperspectral sensors and Selex Communications MMW radar. Selex LOS and satcom real-time datalinks.

Guidance and control: GCS will contain newly developed Galileo Avionica mission operator station, a Selex Communications datalink operator station and Quadrics mission and data fusion computers. Air vehicle operation will be essentially autonomous, based on an Alenia flight control system derived from that of the Aermacchi M-346 jet trainer; other avionics include Selex ATC radio, radio altimeter, DGPS navigator and Mode 3 Level 2 IFF.

Launch: Conventional, autonomous, wheeled take-off.
Recovery: Conventional, autonomous, wheeled landing.
Power plant: Two 186 kW (250 hp), 2,400 cc Diesel engines, each driving a three-blade pusher propeller. Fuel (see under Weights) in fuselage and integral wing tanks.

Variants: *Molynx:* Civil version for government and commercial applications. These are foreseen as including law enforcement, border patrol, fire detection and rescue, disaster relief, traffic control, atmospheric and agricultural monitoring, fisheries patrol, communications relay, powerline and pipeline patrol, aerial survey and mapping.

BlackLynx: Military version for ISR, target acquisition, maritime and border surveillance, strike co-ordination and communications relay. Underwing hardpoints for external stores.

Molynx, BlackLynx

Dimensions, External	
Overall	
length	
Molynx	12.27 m (40 ft 3 in)
BlackLynx	13.00 m (42 ft 7¾ in)
Fuselage	
length, Molynx	11.07 m (36 ft 3¾ in)
Wings	
wing span	
Molynx	25.00 m (82 ft 0¼ in)
BlackLynx	28.00 m (91 ft 10¼ in)
Wheels	
wheel track, Molynx	3.60 m (11 ft 9¾ in)
Weights and Loadings	
Weight	
Operating weight, empty	
Molynx	1,600 kg (3,527 lb)
BlackLynx	1,700 kg (3,747 lb)
Max T-O weight	
Molynx	3,000 kg (6,613 lb)
BlackLynx	3,500 kg (7,716 lb)
Fuel weight	
Max fuel weight	
Molynx	800 kg (1,763 lb)
BlackLynx	1,000 kg (2,204 lb)
Payload	
Max payload	
Molynx	600 kg (1,322 lb)
BlackLynx	800 kg (1,763 lb)
Performance	
Altitude, Service ceiling	13,715 m (45,000 ft) (est)
Speed	
Cruising speed, Molynx	220 kt (407 km/h; 253 mph) (est)
Radius of operation	2,000 n miles (3,704 km; 2,301 miles) (est)
Endurance	
Molynx	30 hr (est)
BlackLynx	36 hr (est)

Status: Development under way in 2007, continuing in 2009.
Contractor: Alenia Aeronautica SpA
Rome.

Artist's impression of Molynx in flight *(Alenia)*

Molynx model at the June 2007 Paris Air Show *(IHS/Patrick Allen)*

Alenia Sky-X

Type: Experimental multirole UAV.

Development: Programme initiated January 2003, reportedly against an Italian Air Force requirement for a UCAV able to carry two JDAM or similar weapons. Half-scale 'integration technology' mockup rolled out 13 March 2003. First details released 30 May 2003, with mockup being displayed at Paris Air Show in following month. Definition phase began on 16 October 2003, and design underwent revision during 2003-04, manufacture of full-size Sky-X demonstrator beginning in October 2004. This prototype was rolled out on 11 April 2005, made its maiden flight at Vidsel in northern Sweden on 29 May, and appeared at the Paris Air Show in June 2005. Due to delays in runway completion in Sardinia, plans to continue trials at the Salta di Quirra range were abandoned, and flight testing was instead resumed at Vidsel in October 2005. A total of 17 flights had been made by early November 2006, the last three of these demonstrating Automatic Take-Off and Landing (ATOL), use of auto-throttle, high-load turns and an approach for go-around with low-altitude fly-by on the runway. After this development, flying continued from the Italian Air Force air base at Amendola in Italy, starting in February 2007 and continuing in 2008, including flight refuelling simulations.

Description: *Airframe:* Stealthy, modular design, with shoulder-mounted wings having 35° leading-edge sweepback and W contour trailing-edge. Bullet-shape fuselage, with dorsal intake for jet engine; V tail surfaces. Mixed construction (metal and CFRP). Retractable tricycle landing gear.

Mission payloads: Payload bay designed to accept modular payload packages such as cameras, radar systems, IR or other sensors, or weapons. LOS two-way datalink.

The XD-001 Sky-X prototype *(Alenia)*

Guidance and control: GCS developed by Alenia UCAV Lab. Athena GuideStar GS-311 flight/mission control system.

Launch: Conventional wheeled take-off.

Recovery: Conventional wheeled landing. Butler parachute for emergency recovery.

Power plant: One 4.41 kN (992 lb st Microturbo TRI 60-5/268 turbojet.

Sky-X

Dimensions, External	
Overall	
length	6.94 m (22 ft 9¼ in)
height	2.00 m (6 ft 6¾ in)
Wings	
wing span	5.78 m (18 ft 11½ in)
wing aspect ratio	4.25
Dimensions, Internal	
Payload bay	
length	2.13 m (6 ft 11¾ in)
width	0.86 m (2 ft 9¾ in)
depth	0.48 m (1 ft 7 in)
volume	0.45 m³ (15.9 cu ft)
Areas	
Wings, Gross wing area	7.75 m² (83.4 sq ft)
Weights and Loadings	
Weight	
Operating weight, empty	1,000 kg (2,204 lb)
Max T-O weight	1,450 kg (3,196 lb)
Fuel weight	
Max fuel weight, internal	350 kg (771 lb)
Payload, Max payload	150 kg (330 lb)
Performance	
T-O	
T-O run	300 m (985 ft)
Altitude, Service ceiling	7,620 m (25,000 ft)
Speed	
Max level speed	350 kt (648 km/h; 403 mph)
Cruising speed	260 kt (482 km/h; 299 mph)
g limits	+5
Radius of operation, mission	50 n miles (92 km; 57 miles)
Endurance	2 hr
Landing	
Landing run	700 m (2,297 ft)
Power plant	1 × turbojet

Sky-X in automatic landing configuration *(Alenia)*

Sky-X rolled out before its inaugural flight *(Alenia)*

Sky-X takes off for its maiden flight in Sweden, 29 May 2005 *(Alenia)*

Sky-X making a first automatic take-off *(Alenia)*

Status: Development continuing, and Sky-X has contributed to the Neuron programme. An Alenia Automatic Take-Off and Landing (ATOL) system was validated in late 2006 at the Vidsel range in Sweden. The programme for 2007 included testing of a Mission Management System (MMS), Air Traffic Control (ATC) links, a new datalink and an initial EO sensor. In mid-2008, plans were to remove the aircraft's tail surfaces, following this towards the end of that year by introduction of a thrust-vectoring control system. An integrated weapons bay was scheduled to be flight tested in mid-2009. Join-up manoeuvres with an Alenia C-27J were completed in 2008 to simulate flight refuelling. Some 29 demonstration flights were logged by Sky-X up to April 2009.

Data resulting from the various Sky-X technologies will, like those from the company's newer Sky-Y prototype, be fed into the development programme for the eventual Molynx medium/high altitude endurance UAV.

Contractor: Alenia Aeronautica SpA
Rome.

Alenia Sky-X *(IHS/Patrick Allen)*

Alenia Sky-Y

Type: Technology demonstrator.

Development: Whereas the Sky-X is a technology demonstrator intended to lead to a combat UA system, the Sky-Y fulfils that function on behalf of a MALE system for intelligence, surveillance and reconnaissance, although findings from both demonstrations will feed into Alenia's longer-term project, the Molynx.

Specifically, the technologies that the Sky-Y programme will mature comprise a heavy-fuel (Diesel) engine; autonomous take-off and landing; a mission management system; sensor payloads; plus satcom and other datalinks. Wind tunnel tests took place in October 2006, and details were released at the Paris Air Show on 21 June 2007, the day after the prototype had made its maiden flight at the Vidsel test range in northern Sweden. Test results are expected to identify, and significantly reduce, risks associated with the development of a future MALE system such as Molynx. In November 2008 Selex Galileo reported that the Alenia EOST electro-optical sensor had been tested on Sky-Y at Vidsel.

Simultaneous with the Sky-Y unveiling, it was announced that Alenia Aeronautica, Dassault Aviation and Saab Aerosystems would co-operate on a potential new European MALE UAV development programme to replace the defunct EuroMALE project. This new effort, aimed at fielding a production-ready system by 2011, will be led by Alenia, but other participants, such as those involved in the Dassault-led Neuron programme, are being invited to join. Support for this venture is being sought from the French, Italian and Swedish governments and the European Defence Agency. Sky-Y and Molynx will also support this effort, as well as continuing to develop in their own right. The European Defence Agency's MIDCAS Mid-Air Collision Avoidance System is to be flown on Sky-Y by the end of 2012.

Description: *Airframe:* Low-wing, pod and twin tailboom configuration, with sweptback fins and rudders bridged by double-T horizontal tail surfaces. Nose bulge houses 76 cm (29.9 in) diameter satcom antenna. Pusher engine. Retractable tricycle landing gear. Carbon fibre composites construction.

Mission payloads: Galileo EO/IR sensor and image exploitation system.

Guidance and control: Pre-programmed, via Quadrics flight control system, from new two-person GCS (remote operator and mission controller). Selex Communications wide-band datalinks; satcom; ATC radio integration.

Launch: Autonomous conventional wheeled take-off.

Recovery: Autonomous conventional wheeled landing.

Power plant: One 149 kW (200 hp) Fiat/Diesel Jet TDA CR five-cylinder, liquid-cooled in-line engine initially, with single-stage turbocharger and FADEC; three-blade pusher propeller. Intention is to replace this later with a 186 kW (250 hp) variant of the same engine with a two-stage turbocharger.

Sky-Y

Dimensions, External	
Overall, length	9.725 m (31 ft 10¾ in)
Wings, wing span	9.937 m (32 ft 7¼ in)
Weights and Loadings	
Weight	
Operating weight, empty	850 kg (1,873 lb)
Max T-O weight	1,200 kg (2,645 lb)
Fuel weight, Max fuel weight	200 kg (440 lb)
Payload, Max payload	150 kg (330 lb)
Performance	
Altitude, Service ceiling	7,620 m (25,000 ft) (est)
Speed, Cruising speed	140 kt (259 km/h; 161 mph) at 6,096 m (20,000 ft) (est)
Range	500 n miles (926 km; 575 miles) (est)
Radius of operation, LOS	50 n miles (92 km; 57 miles) (est)
Endurance	14 hr (est)

Status: Under development.

Contractor: Alenia Aeronautica SpA
Rome.

Sky-Y prototype roll-out *(Alenia)* 1290197

Sky-Y maiden flight, 20 June 2007 *(Alenia)* 1290198

Sky-Y mockup at the Paris Air Show, June 2007 *(IHS/Patrick Allen)* 1325688

IAS Archimede

Type: Multirole UAV.

Development: The Archimede platform aircraft is adapted from a fully certified commercial tandem-seat lightplane named Sky Arrow, produced by Iniziative Industriali Italiane. Apart from fairing over the former cockpit area, the other major changes have been to upgrade the usual Rotax 912 engine to a turbocharged Rotax 914, for increased altitude performance, and introduction of a US proprietary flight control system. It is aimed primarily at a wide range of potential civil and parapublic applications.

Description: *Airframe:* Strut-braced high-wing monoplane, with pod and boom fuselage, pusher engine and T tail unit. Fixed tricycle landing gear. Carbon fibre sandwich construction.

Mission payloads: The UAV is suitable for the carriage of a wide range of sensor or other payloads including EO, IR, radar, radiation detectors, atmospheric samplers, photogrammetry cameras and broadcasting equipment.

Guidance and control: L-3 Geneva Aerospace flightTEK autopilot/flight control computer with built-in GPS/INS navigation enables missions to be fully or semi-autonomous or under autopilot-assisted manual control.

Launch: Conventional wheeled take-off.

Recovery: Conventional wheeled landing.

The Archimede unmanned adaptation of a commercial lightplane *(IAS)* 1290339

Archimede alongside its mobile GCS and support vehicle *(IAS)* 1290340

Power plant: One 84.5 kW (113.3 hp) Rotax 914 UL flat-four engine with electronic fuel injection; Hoffmann two-blade constant-speed pusher propeller. Fuel capacity 68 litres (18.0 US gallons; 15.0 Imp gallons).

Archimede

Dimensions, External	
Overall	
length	7.60 m (24 ft 11¼ in)
height	2.60 m (8 ft 6¼ in)
Wings, wing span	9.70 m (31 ft 10 in)
Areas	
Wings, Gross wing area	13.50 m² (145.3 sq ft)
Weights and Loadings	
Weight	
Weight empty	400 kg (881 lb) (est)
Max T-O weight	650 kg (1,433 lb)
Performance	
T-O	
T-O run	150 m (493 ft)
Altitude, Service ceiling	6,095 m (20,000 ft)
Speed	
Cruising speed	89 kt (165 km/h; 102 mph)
Stalling speed, flaps down	40 kt (75 km/h; 46 mph)
Radius of operation, mission	172 n miles (318 km; 197 miles)
Endurance, max	30 hr
Landing	
Landing run	135 m (443 ft)

Status: Introduced in 2008 and being promoted. As of 2010, no production or customer status had been revealed.
Contractor: International Aviation Supply SRL
Brindisi.

IAS Pico

Type: Surveillance mini-UAV.
Development: Introduced in 2008. Offered for both civil and military applications.
Description: *Airframe:* High-wing monoplane with tractor engine and T tail. No landing gear. Composites construction.
Mission payloads: EO/IR pan/tilt/zoom sensor, with image stabilisation, on gyrostabilised gimbal with target tracking.
Guidance and control: Fully automatic, with GPS waypoint navigation; provision to update or amend mission profile during flight.
Transportation: Can be transported in backpack or by small vehicle. Set-up time less than 10 minutes.
Launch: Hand launched.
Recovery: Deep stall or net recovery.
Power plant: Electric motor, driving a two-blade tractor propeller.

Pico

Dimensions, External	
Overall, length	1.00 m (3 ft 3¼ in)
Wings, wing span	2.00 m (6 ft 6¾ in)
Weights and Loadings	
Weight, Max launch weight	5.0 kg (11.00 lb)
Payload, Max payload	1.0 kg (2.00 lb) (est)
Performance	
Altitude	
Operating altitude, typical	300 m (980 ft)
Speed	
Cruising speed, max	27 kt (50 km/h; 31 mph)
Radius of operation	16 n miles (29 km; 18 miles)
Endurance	1 hr 30 min
Power plant	1 × electric motor

Status: Being promoted; production/service status not yet revealed. Under plans announced in 2009, Pico would be manufactured in a new IAS facility in Malta.
Contractor: International Aviation Supply SRL
Brindisi.

Computerised impression of the Pico mini-UAV *(IAS)* 1290335

IAS Pitagora

Type: Experimental UAV.
Development: This highly unorthodox UAV was revealed in October 2007 following the maiden flight of a smaller (2 m span, 20 kg max T-O weight) prototype then known as the Pitagora-1A. The description applies to the developed version being marketed in early 2009.

Pitagora-1A prototype on debut at 2007 Dubai air show
(Robert Hewson) 1290336

Pitagora viewed with propulsor in intermediate position
(Robert Hewson) 1290337

Description: *Airframe:* Triangular main wing, with approx 45° dihedral winglets; cylindrical forebody, with approx 30° anhedral canards. Ducted propulsor in centre of main wing, with pair of four-blade coaxial, counter-rotating rotors.
Mission payloads: Sensor ball turret under forward part of forebody, on stabilised gimbal mount. Payloads can include EO, IR or others of customer's choice, for which 2 kW of onboard 28 V DC power is available.
Guidance and control: Fully or semi-autonomous, or autopilot-assisted manual control. Guided Systems Technologies Inc flight control system selected in mid-2008. Same two-person GCS as Raffaello.
Launch: Propulsor lies flush within wing for vertical take-off, pivoting upwards through 90° for transition to forward flight. Can also make STOL take-offs and landings with propulsor in intermediate position.
Recovery: Vertical or cable-arrested STOL landing, depending upon propulsor position.
Power plant: One 29.1 kW (39 hp) Zanzottera 498i two-cylinder two-stroke engine. Fuel capacity 72 litres (19.0 US gallons; 15.8 Imp gallons).

Pitagora

Dimensions, External	
Overall, length	3.20 m (10 ft 6 in)
Wings, wing span	6.00 m (19 ft 8¼ in)
Weights and Loadings	
Weight	
Weight empty	84 kg (185 lb)
Max launch weight	180 kg (396 lb)
Payload, Max payload	35 kg (77 lb)
Performance	
T-O	
T-O run, STOL	10 m (33 ft)
Altitude	
Operating altitude	1,525 m to 4,570 m (5,000 ft to 15,000 ft)
Service ceiling	5,485 m (18,000 ft)
Speed	
Cruising speed, max	90 kt (167 km/h; 104 mph)
Loitering speed	55 kt (102 km/h; 63 mph)
Radius of operation, datalink	81 n miles (150 km; 93 miles)
Endurance, STOL mode	16 hr
Landing	
Landing run, STOL, arrested	50 m (165 ft)
Power plant	1 × piston engine

Status: First flight of Pitagora-1A October 2007. Developed version being promoted in early 2009.
Contractor: International Aviation Supply SRL
Brindisi.

IAS Raffaello

Type: Short-range tactical UAV.
Development: This UAV was first seen at the 2007 Dubai air show. Promotion was continuing in 2009.

IAS Raffaello – Selex Galileo Falco < Italy < Unmanned aerial vehicles

Raffaello on display at the 2007 Dubai air show *(Robert Hewson)*
1290338

Description: *Airframe:* Low, rear-mounted main wings, with winglets/fins at tips; shoulder-mounted foreplanes; box-section fuselage with pusher engine. Fixed tricycle landing gear. Composites construction.
  Mission payloads: Can include EO, IR and/or synthetic aperture radar.
  Guidance and control: Fully or semi-autonomous, with provision for in-flight manual override.
  Transportation: Mobile GCS, based on HMMWV or similar vehicle.
  System composition: Three air vehicles, payloads, GCS, LOS analogue datalink and GCS crew of two.
  Launch: Conventional wheeled take-off.
  Recovery: Conventional wheeled landing or parachute recovery.
  Power plant: One 29.1 kW (39 hp) Zanzottera 498i two-cylinder two-stroke engine; two-blade pusher propeller. Fuel capacity 72 litres (19.0 US gallons; 15.8 Imp gallons).

Raffaello

Dimensions, External	
Overall, length	3.20 m (10 ft 6 in)
Wings, wing span	6.00 m (19 ft 8¼ in)
Weights and Loadings	
Weight	
Weight empty	84 kg (185 lb)
Max launch weight	180 kg (396 lb)
Payload, Max payload	35 kg (77 lb)
Performance	
Altitude	
Operating altitude	1,525 m to 4,875 m (5,000 ft to 16,000 ft)
Service ceiling	5,485 m (18,000 ft)
Speed	
Max level speed	90 kt (167 km/h; 104 mph)
Loitering speed	55 kt (102 km/h; 63 mph)
Radius of operation	
radio control, mission, omni antenna	54 n miles (100 km; 62 miles)
datalink, mission	81 n miles (150 km; 93 miles)
Endurance, max	16 hr
Landing	
Landing run, with arrester cable	50 m (165 ft)

Status: No production or service status revealed. Promotion continuing in early 2009.
Contractor: International Aviation Supply SRL, Brindisi.

Piaggio Aero P.1HH HammerHead
Type: ISR and security MALE UAV
Development: Piaggio Aero's P.1HH HammerHead ISR and security MALE UAV was launched at the February 2013 IDEX 2013 trade show in Abu Dhabi in the United Arab Emirates. At the time of its public unveiling, the prototype P.1HH had been rolled out and had 'successfully completed' its first engine start and taxi trials on 14 February 2013 at an 'Italian Air Force base'. As such, the P.1HH is derived from Piaggio's P.180 Avanti II business and VIP transport aircraft and is billed as being suitable for a 'wide range of [ISR] and security' applications in both the military and civil domains. Here, specifically identified roles include aerial, land, coastal, maritime and off-shore security together with COMINT and ELINT collection and EW missions. The HammerHead UAV programme involves Selex ES in addition to Piaggio Aero and as of February 2013, the type was scheduled to have made its maiden flight by the end of 2013. P.1HH is STANAG USAR 4671 compliant.
Description: *Airframe:* The P.1HH HammerHead UAV features a streamlined fuselage that is fitted with forward canards, aft- and shoulder-mounted wings and a conventional vertical tail surface that is surmounted by swept back port and starboard horizontal surfaces. Again, the AV incorporates conventional control surfaces and a retractable tricycle undercarriage. The type's engines are mounted in nacelles that protrude fore and aft from the AV's inboard wing sections and drive pusher propellers. A pair of stabilising strakes are mounted beneath the AV's rear fuselage and there is a faired-in SATCOM antenna radome located above its forward fuselage.
  Mission payloads: The P.1HH HammerHead's mission suite is derived from Selex ES's skyISTAR architecture that can include radar, EO, hyperspectral, COMINT and ELINT sensors. Devices specifically identified with the P.1HH comprise Selex ES sourced radar (the Seaspray

The P.1HH HammerHead UAV's airframe is derived from that of Piaggio's P.180 Avanti II business and VIP transport aircraft *(Piaggio Aero)*
1365384

7300E Active Electronically Scanned Array (AESA) equipment) communications and datalink 'solutions'.
  Guidance and control: The P.1HH AV incorporates automatic take-off and landing capabilities together with Selex ES sourced mission management, vehicle management and control, air data terminal and ground control (with LOS and beyond LOS facilities) segments. The type also features a SATCOM capability.
  Transportation: The P.1HH incorporates a 'removable external section wing' to facilitate ground transportation.
  System composition: The P.1HH UAS architecture comprises the AV, a mission equipment/systems package and a GCS.
  Launch: Conventional take-off using a tricycle undercarriage.
  Recovery: Conventional landing using a tricycle undercarriage.
Power plant: Two × 950 shp Pratt & Whitney Canada PT6-66B turboprops driving five-bladed, low-noise Hartzell pusher propellers.

P.1HH Hammerhead

Dimensions, External	
Wings, wing span	15.5 m (50 ft 10¼ in)
Performance	
Altitude, Service ceiling	13,715 m (45,000 ft)
Endurance	16 hr
Power plant	2 × turboprop

Status: Most recently, Piaggio Aero has continued to promote the P.1HH HammerHead ISR and security MALE UAV.
Contractor: Piaggio Aero Industries SpA, Genoa.

Selex Galileo Falco
Type: Medium-altitude endurance tactical UAV.
Development: The Falco (Falcon) was revealed in February 2002, simultaneously with the Nibbio. The turned-down wingtips of the original design were deleted before the first flight, which took place in November 2003 at the Salto di Quirra range in Sardinia. The Falco was designed for target detection, localisation, identification and designation, and is seen as a potential replacement for the Mirach 26 and similar class UAVs. It is also expected to become a platform for stand-off sensors and/or weapons such as the Multipurpose Air-Launched Payload (MALP) being developed by Galileo Avionica.

In June 2005, Galileo announced that two Falcos had been awarded a Permit to Fly by the Italian Civil Aviation Authority (ENAC), acting on behalf of EASA. These appear on the Italian civil register as I-RAIE and I-RAIF.

In June 2006, the company revealed plans to develop a ship-launched version of the Falco by 2008, combining its EOST 45 EO/IR sensor with a Selex Gabbiano X-band multimode maritime radar.

By 2009 Selex Galileo were considering a Falco with increased span and payload, and this emerged as Falco EVO in 2011. First flight is scheduled for the first quarter of 2012. The extended span EVO

Falco EVO at Paris, 2011 *(IHS)*
1465573

Unmanned aerial vehicles > Italy > Selex Galileo Falco

Falco during its first flight in late 2003 *(Galileo)*

Falco at Aberporth in 2009 *(Mark Daly)*

Falco I-PTFC at Aberporth *(Mark Daly)*

Falco ground handling is enhanced by wheel brakes and nosewheel steering *(Selex Galileo)*

On test in Finland, landing after catapult launch *(Selex Galileo)*

The Falco GCS shelter *(IHS/Patrick Allen)*

Falco GCS payload control console *(Galileo)*

Falco about to take off *(IHS/Patrick Allen)*

configuration is available as an upgrade package for earlier air vehicles. Later envelope testing for Falco has been flown at ParcAberporth in West Wales.

Description: *Airframe:* Shoulder-mounted, sweptback gull wings with slotted flaps; central fuselage nacelle; twin tailbooms with sweptback fins and rudders; pusher engine with spinnered propeller. Non-retractable tricycle landing gear.

Mission payloads: Designed to accommodate wide variety of payloads including EO, SAR, ESM, self-protection equipment and droppable external stores. Typical items can include Galileo EOST 45 or other colour TV camera, thermal imager, spotter and laser designator.. General Atomics Lynx radar also said to be under consideration. Up to 30 kg (66.1 lb) of stores could be carried under each wing, including additional fuel. Test flights in 2009 carried the PicoSAR Active Electronically Scanned Array Radar or combined EO/IR payloads.

Guidance and control: Preprogrammed, with en-route updating option. GCS can control two aircraft at a time and allows operator to manage payloads, sensors and collected data in real time. Consoles are functionally interchangeable to permit interactive mission planning, retasking and playback, plus operator training. All control systems, digital buses, control link equipment, automatic area surveillance modes and image processing are fully redundant.

Launch: Automatic, wheeled STOL take-off. Pneumatic catapult option under joint development by Galileo and Robonic in 2006-07, based on latter's MC2555LLR. A catapult launch of a Falco air vehicle was made December 2008 at Cheshnegirovo air base in Bulgaria, following earlier trials in Finland. Catapult trials continued in Finland in 2009.

The Falco medium-altitude tactical UAV *(Galileo)* 1151517

Recovery: Conventional wheeled landing; parachute for emergency recovery.

Power plant: One UEL 682 rotary engine; three-blade variable-pitch pusher propeller. A mid-term choice of a heavy-fuel engine is under consideration. Fuel tanks in wings and fuselage. External fuel can be carried.

Falco, Falco EVO

Dimensions, External	
Overall	
length	5.25 m (17 ft 2¾ in)
height	1.80 m (5 ft 10¾ in)
Fuselage, width	0.60 m (1 ft 11½ in)
Wings	
wing span	
Falco	7.20 m (23 ft 7½ in)
Falco EVO	12.50 m (41 ft 0¼ in)
Weights and Loadings	
Weight	
Max T-O weight	
Falco	320 kg (705 lb) (est)
Falco EVO	650 kg (1,433 lb)
Payload	
Max payload	
Falco	70 kg (154 lb)
Falco EVO	100 kg (220 lb)
Performance	
T-O	
T-O run	60 m (197 ft) (est)
Altitude, Operating altitude	6,500 m (21,320 ft) (est)
Speed, Max level speed	78 kt (144 km/h; 90 mph) (est)
Radius of operation, datalink	81 n miles (150 km; 93 miles) (est)
Endurance	
Falco	8 hr (est)
Falco EVO	18 hr (est)

Status: Four or five systems ordered by Pakistan Air Force; said to have been due for delivery in December 2006. These deliveries were completed in 2008 and became operational in 2009. Another order was placed that year by a second (unidentified) customer.

Contractor: Selex Galileo
Ronchi dei Legionari.

Japan

Gen H-4

Type: Small helicopter UAV.

Development: The H-4 is a commercially available single-seat ultra-light helicopter, first flown in 1999 and subsequently marketed in homebuilder kit form. Small numbers have been sold in Japan and the US. Gen Corporation has also developed a radio-controlled unmanned version (first flight 22 March 2002), which is offered for civil applications such as agricultural use, training or rescue.

Description: *Airframe:* Twin coaxial, contra-rotating two-blade rotors. Skeletal airframe of aluminium tube; rotor blades of carbon fibre and Kevlar with foam core. Landing gear comprises four small 'wheels' each made up of multiple rollers in a circular arrangement.

Mission payloads: Customer choice, depending upon mission.

Guidance and control: Radio-controlled, with a Gen-developed automatic attitude and altitude control system. Two-axis control by tilting of engine/rotor platform; yaw control by differential rotor speeds, actuated electrically.

Launch: Conventional helicopter take-off.

Recovery: Conventional helicopter landing. Ballistic recovery parachute optional.

Power plant: Four 7.5 kW (10 hp) 124 cm³ (7.56 cu in) Gen 125 two-cylinder two-stroke engines, positioned symmetrically on a pivoting platform beneath rotor head. Fuel capacity (petrol/oil mixture) 19 litres (5.0 US gallons; 4.2 Imp gallons).

H-4

Dimensions, External	
Overall, height	2.44 m (8 ft 0 in)
Rotors	
rotor diameter, each	4.00 m (13 ft 1½ in)
Areas	
Rotor disc, each	12.57 m² (135.3 sq ft)
Weights and Loadings	
Weight	
Weight empty	70 kg (154 lb)
Max T-O weight	190 kg (418 lb)
Performance	
Altitude, Service ceiling	3,000 m (9,840 ft)
Speed	
Cruising speed, max	54 kt (100 km/h; 62 mph)
Endurance, max	1 hr

Status: Developed and available.

Contractor: Gen Corporation
Nagano.

Yamaha R-50 and RMAX

Type: Agricultural and environmental unmanned helicopters.

Development: *R-50:* Yamaha's involvement in unmanned helicopters began in 1983, when it was asked by the Japan Agricultural Aviation Association to develop an engine for a research project known as the Remote Control Automatic Spray System (RCASS) to improve ways of spraying pesticides. It ended up developing the complete helicopter, which was powered by a 20.9 kW (28 hp) snowmobile engine and used vertical and directional gyros and a height sensor to control the aircraft's attitude and altitude. An RCASS prototype with co-axial rotors was completed in November 1987, but proved to be unduly complex and expensive, and the programme was terminated in March 1988. Meanwhile, in 1985 Yamaha had begun a parallel design, the R-50, with single main and tail rotors, this making its public debut in its initial L09 version in November 1987. Starting in the following month, a preproduction batch of 20 was built for trials, leading in October 1988 to limited production of the developed L092 version, which operated using

Gen H-4 homebuilt helicopter in single-seat form *(Gen)* 1416718

Chinese 'Tianying-3' RMAX on show in Beijing, 2006 *(People's Daily)* 1151638

Unmanned aerial vehicles > Japan > Yamaha R-50 and RMAX

The Yamaha RMAX in crop-spraying configuration *(Yamaha)* 0527022

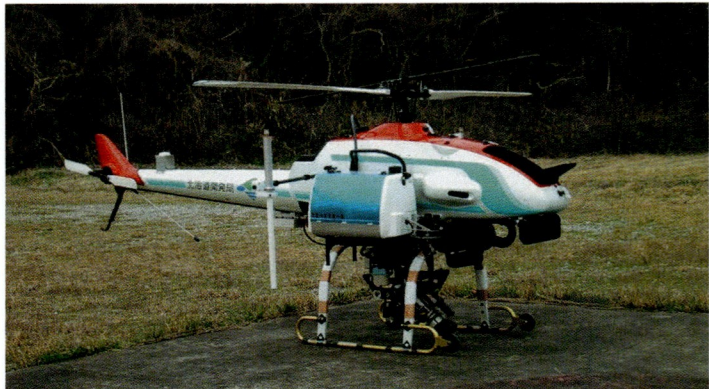

RMAX G-0 for Japanese government *(Yamaha)* 1148708

RMAX G-1, introduced in 2005 *(Shinichi Kiyotani)* 1116804

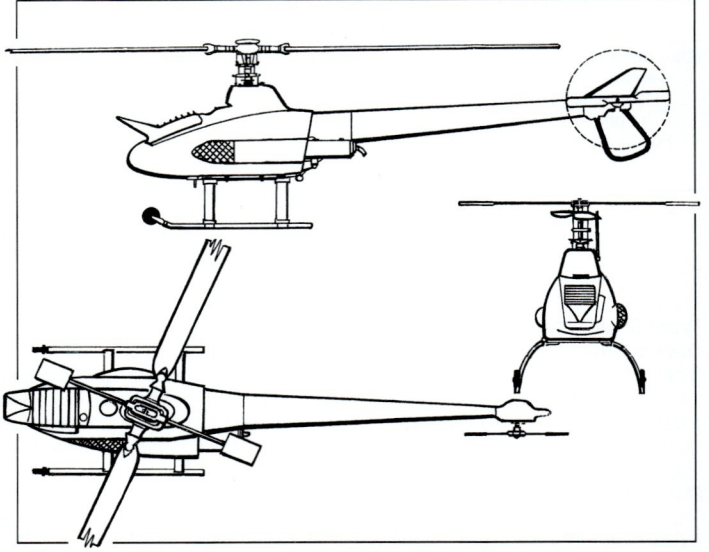

Yamaha R-50 general arrangement *(IHS/John W Wood)* 0518234

a safety tether. Reliability and performance were improved in the L12 version, launched in April 1990, which used a Yamaha Operator Support System (YOSS) that controlled air vehicle height by use of a laser sensor. In FY91, some 106 of these aircraft sprayed about 2,600 ha (6,425 acres) nationwide; by FY93 these figures had risen to about 70,400 ha (173,960 acres) treated by 395 R-50s then in service. The YOSS was in turn replaced from 1995 by the current Yamaha Attitude Control System (YACS) which uses a fibre optic gyro originally developed for car navigation systems. More than 1,100 R-50s have been produced to date.

RMAX: The larger RMAX was introduced (version L15) in late 1997, with YACS as standard, Yamaha's first horizontally opposed engine, and incorporating improvements based on the growing experience of R-50 users. The last-named include enhanced spraying efficiency and tank capacity, fail-safe reliability and greater growth potential. Yamaha has since developed preprogrammed autonomous flight and real-time speed control systems for both the R-50 and RMAX, for potential application to short-range reconnaissance, surveillance, observation and law enforcement missions. Equipment on the RMAX L17 Type IIG version, launched in March 2003, includes an attitude sensor, a DGPS sensor, a communications modem and improved YACS-G control system. As of early 2007, combined R-50/RMAX production had exceeded 1,600 aircraft. By that time, some 10,000 operators had been trained at the Sky Tech Academy training facility run by Yamaha dealers. The R-50 and RMAX systems are not marketed outside Japan, as, in the words of the company, they are designed for "uniquely Japanese farming conditions", and Yamaha Skytech sells the systems through 19 dealers across Japan. There have been exceptions to this (see below) and the RMAX has been used for agricultural duties in South Korea.

Description: *Airframe:* Two-blade main and tail rotors (main rotor removable); twin-skid landing gear with ground handling wheels. All-metal construction. Detachable spraybooms for liquid chemical application. Horizontal stabiliser added to RMAX Mk II.

Mission payloads: R-50 can be fitted with belly tank for liquid (10 litres; 2.6 US gallons; 2.2 Imp gallons) or granular pesticide or fertiliser, plus spreader gear (see Yamaha L09A/F entry for details). With 50 per cent payload and using low-volume liquid spraygear, it can treat 1 ha (2.47 acres) in one flight.

Observation equipment carried by the RMAX includes a digital film camera and a digital still camera, plus three miniature video cameras for navigation purposes. Images from these are downlinked in real time to the GCS, where they are viewed on a four-window split screen monitor. Autonomous RMAX carries a triple-sensor payload turret (daylight colour TV camera, IR camera and laser range-finder).

Guidance and control: The R-50 was controlled originally by radio command, using a hand-held transmitter, with GPS navigation for accurate delivery of chemical, and both types can still be operated in this mode. Now available for both is a choice of three autonomous modes (waypoint navigation, dialogue and stick input), using a Yamaha attitude control system (YACS, which see) in which three fibre optic rate gyros and three accelerometers are fitted to the helicopter body to supply data to an onboard computer unit that regulates all stick operations. An automatic 'return home' mode is also available in case of emergency. GCS optional for aerial RMAX, standard for autonomous RMAX (none for agricultural RMAX).

Launch: Conventional helicopter take-off.
Recovery: Conventional helicopter landing.
Power plant: R-50: One 8.8 kW (11.8 hp) 98 cc Yamaha L 12 water-cooled two-cylinder two-stroke engine.
RMAX: One 15.4 kW (20.7 hp) 246 cc Yamaha L 15 water-cooled two-stroke.
Fuel capacity (petrol/oil mixture) 4 litres (1.06 US gallons; 0.88 Imp gallon) in R-50; 6 litres (1.59 US gallons; 1.32 Imp gallons) in RMAX.

Autonomous RMAX Type III with features that include a horizontal stabiliser and a three-camera sensor ball *(Kenneth Munson)* 0558389

Autonomous RMAX Type II *(Yamaha)* 1043409

Variants: *R-50:* Operational model (configurations L09 and L12) since 1988.

RMAX: Enlarged and improved model, introduced (configuration L15) 1997. More powerful engine with easier starting; modular spraytanks with twice the R-50's capacity; flight stabilisation system; automatic rotor stop; operation monitoring system; greater swath width. Three specialised **Mk II** models now available, starting in March 2003 with latest agricultural version (L17 configuration), which has single 2 Hz GPS and improved speed control. Aerial photography version (L18 configuration) followed in September 2003; has speed and hover control, is operable within 200 m (660 ft) range of controller and, with modified (5 Hz) GPS, at heights up to 150 m (500 ft) above ground level. Third version is fully autonomous RMAX, which has DGPS, speed and position control, can be programmed both before and during flight and flown beyond LOS.

In October 2005, Yamaha revealed a new and upgraded variant, the **G-1**, a militarised version of which (designated **Mk IIG**) was then in use by the Japan Ground Self-Defence Force for surveillance missions in support of its troops deployed in the Samawah area of Iraq. This system comprised four air vehicles and a GCS, the former being equipped with an upgraded navigation system, a CCD video camera and a thermal imager. The JGSDF contract value was quoted as JPY360 million (USD3.1 million).

In 2007 the recommended retail price of a RMAX G-1 system comprising two air vehicles and a base station was JPY136.5 million (USD1.2 milllion)

R-50, RMAX, RMAX Mk IIG

Dimensions, External
Overall
  length
    rotors turning, R-50 .... 3.58 m (11 ft 9 in)
    rotors turning, RMAX .... 3.63 m (11 ft 11 in)
  height .... 1.08 m (3 ft 6½ in)
  width
    spray-boom span, R-50 .... 2.57 m (8 ft 5¼ in)
    RMAX .... 1.64 m (5 ft 4½ in)
Fuselage
  length
    R-50 .... 2.655 m (8 ft 8½ in)
    RMAX .... 2.75 m (9 ft 0¼ in)
Skids
  skid track
    R-50 .... 0.70 m (2 ft 3½ in)
    RMAX .... 0.72 m (2 ft 4¼ in)
Rotors
  rotor diameter
    R-50 .... 3.07 m (10 ft 0¾ in)
    RMAX .... 3.115 m (10 ft 2¾ in)
Tail rotor
  tail rotor diameter
    R-50 .... 0.52 m (1 ft 8½ in)
    RMAX .... 0.545 m (1 ft 9½ in)
Areas
  Rotor disc
    R-50 .... 7.40 m² (79.7 sq ft)
    RMAX .... 7.62 m² (82.0 sq ft)
  Tail rotor disc
    R-50 .... 0.212 m² (2.3 sq ft)
    RMAX .... 0.25 m² (2.7 sq ft)
Weights and Loadings
Weight
  Weight empty
    R-50 .... 47 kg (103 lb)
    agricultural, RMAX .... 62 kg (136 lb)
    aerial photo, RMAX .... 63 kg (138 lb)
    autonomous Type I, RMAX .... 80 kg (176 lb)
    autonomous Type II, RMAX .... 72 kg (158 lb)
    autonomous Type III, RMAX .... 71 kg (156 lb)
    RMAX Mk IIG .... 84 kg (185 lb)
  Max T-O weight
    R-50 .... 67 kg (147 lb)
    RMAX .... 95 kg (209 lb)
Payload
  Payload with max fuel
    R-50 .... 20 kg (44 lb)
    RMAX .... 30 kg (66 lb)
Loading
  Max disc loading
    R-50 .... 9.05 kg/m² (1.85 lb/sq ft)
    RMAX .... 12.47 kg/m² (2.55 lb/sq ft)
Performance
Altitude
  Service ceiling, RMAX .... 2,000 m (6,560 ft)
Speed
  Cruising speed, max, RMAX .... 39 kt (72 km/h; 45 mph)
Radius of operation
  control range, RMAX Mk IIG .... 2.7 n miles (5 km; 3 miles)
Endurance
  R-50 .... 30 min
  with 30 kg (66.1 lb) payload,
    RMAX .... 1 hr 30 min
  with 10 kg (22.0 lb) payload,
    RMAX .... 2 hr 30 min
Power plant .... 1 × piston engine

Status: Some 1,687 R-50s and RMAXs had been registered and 1,281 completed by the end of 2002 and more than 2,000 UAV helicopters were operational in Japan by 2007.

Customers: Japanese crop-spraying and crop-dusting contractors; NASA; JGSDF (one system).

In April 2000, an RMAX equipped with video, still cameras and GPS navigation was chartered by Japan's Public Works Research Institute to monitor lava flow and other peripheral hazards following eruption of the Mount Usu volcano on Hokkaido. As a result, the Hokkaido Development Agency ordered two RMAX in November 2000. Further monitoring of volcanic activity followed in February 2001, following the eruption of Mount Oyama in the previous September. Meanwhile, another RMAX took part in Earth environment remote sensing demonstrations in the US and Canada during June and July 2000.

These UAVs are not generally sold outside Japan, but a number have gone to the United States. One or two were used in 1996 by Carnegie Mellon University of Pittsburgh in the US to develop an autonomous operating system utilising a CCD camera and digital processors. NASA purchased two RMAXs in late 2000. In September and October 2004, an RMAX was used as a surrogate by Northrop Grumman to demonstrate technologies for its entrant in the US Army's (subsequently abandoned) Unmanned Combat Armed Rotorcraft (UCAR) programme. The trials took place at Fort Rucker, Alabama, and Camp Pendleton, California. Northrop Grumman had acquired a number of RMAX vehicles, as In February 2009 the company announced that three RMAX Type II air vehicles were being donated to three universities in the US. In the third quarter of 2005, the US Air Force Research Laboratory was revealed to have been testing one or more RMAXs for potential use in applying pesticide for troop protection in hostile areas.

In January 2006 it became known that nine RMAX UAVs had been supplied to a Chinese company named Beijing BVE Technology since 2001. According to Yamaha, these were understood to be purely for agricultural purposes, but concerns (challenged by Yamaha) had been raised about the legality of the sale on the grounds of the UAVs' potential for the carriage of military payloads, possibly as the result of one — given the name Tianying-3 by China — having appeared at an international defence exhibition in Beijing in April 2006. Yamaha tightened its rules for export controls in the following month.

Also in 2006, an RMAX was in use by French research agency ONERA, in a programme named RESSAC (*Recherche et sauvetage par un système autonome coopérant*), to develop an automatic precision landing system usable by a UAV in a search and rescue role.

Jordan

JARS Jordan Falcon

Type: Tactical UAV.

Development: JARS is a joint venture between Jordan Aerospace Industries (JAI) and King Abdullah II Design and Development Bureau (KADDB). JAI began researching and developing UAVs in 2000, transitioning in February 2004 to a joint programme with KADDB. The Jordan Falcon is intended for use for a wide spectrum of military and civilian tasks involving real-time, round the clock aerial surveillance at distances of up to 27 n miles (50 km; 31 miles).

Description: *Airframe:* Pod fuselage; shoulder-mounted, non-swept constant-chord wings support twin tailbooms bridged by inverted V tail unit; pusher engine; fixed tricycle landing gear. Three-piece wing, connected by removable carbon fibre tubes. Each portion comprises a main and rear spar of epoxy resin fibre cloth reinforced by carbon fibre,

116 Unmanned aerial vehicles > Jordan – Korea, South > JARS Jordan Falcon – KAI K-UCAV

Truck-mounted Jordan Falcon *(IHS/Patrick Allen)* 1137866

JARS Jordan Falcon tactical UAV with original landing gear *(IHS/Patrick Allen)* 1037164

Jordan Falcon, current configuration *(JARS)* 1140495

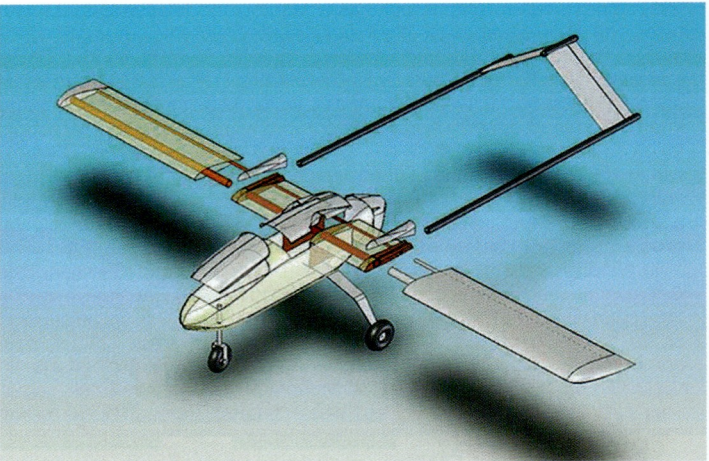
Exploded view of the Jordan Falcon *(JARS)* 1140497

plus moulded ribs and skin. Fuselage is moulded from epoxy resin and glass cloth with longitudinal and lateral reinforcement. Forward fuselage bay accommodates mission equipment, integral fuel tanks and spring shock absorbers for parachute landing.

Mission payloads: Gimbal-mounted, switchable, two-FoV camera standard, with recording and snapshot capability; controlled from GCS by joystick. Alternative payloads can include pan-tilt-zoom daylight colour TV camera or pan-and-tilt IR sensor; communications jammer; communications relay package; or other, according to mission. A 5 W microwave downlink transmits real-time video imagery and telemetry to GCS. Onboard power from 900 W engine-driven generator; 240 W-h emergency storage battery for use in the event of generator failure.

Guidance and control: Table-top or shelter-mounted GCS, incorporating microwave receiver, amplified uplink transmitter and laptop computer, video recorder and crew intercom; can operate on 12 V DC or 220 V AC power. Autopilot or operator control of aircraft heading, altitude, airspeed and GPS waypoint navigation; UAV can be retasked during flight. Flight data HUD, autotracking antenna array and RVT optional.

System composition: Air vehicles as required; launcher; transportation and storage container. Ground crew of three or four. System is integrated on two light all-terrain trucks for higher mobility and fast deployment. Set-up and turnaround time less than 20 minutes, including UAV assembly, subsystems checking and mission planning.

Launch: From mobile launcher in automatic mode; or by conventional wheeled take-off.

Recovery: Automatic parachute recovery or conventional runway landing.

Power plant: One 11.2 kW (15 hp) 200 cc two-cylinder two-stroke engine (type not stated), driving a two-blade fixed-pitch pusher propeller.

Jordan Falcon

Dimensions, External	
Overall	
length	2.95 m (9 ft 8¼ in)
height	0.99 m (3 ft 3 in)
Fuselage, length	2.00 m (6 ft 6¾ in)
Wings, wing span	4.00 m (13 ft 1½ in)
Weights and Loadings	
Weight	
Weight empty	40 kg (88 lb)
Max T-O weight	60 kg (132 lb)
Max launch weight	60 kg (132 lb)
Fuel weight, Max fuel weight	14 kg (30 lb)
Payload, Max payload	6 kg (13.00 lb)
Performance	
Altitude	
Operating altitude, max	3,000 m (9,840 ft)
Speed	
Max level speed	97 kt (180 km/h; 112 mph)
Cruising speed	65 kt (120 km/h; 75 mph)
Loitering speed	54 kt (100 km/h; 62 mph)
Stalling speed	43 kt (80 km/h; 50 mph)
Range	
datalink	30 n miles (55 km; 34 miles)
max	243 n miles (450 km; 279 miles)
Endurance	4 hr

Status: Development continuing in 2005. Trials were scheduled with the Jordan Armed Forces in mid-2005. Marketing continuing.
Contractor: Jordan Advanced Remote Systems
Amman.

Korea, South

KAI K-UCAV

Type: Technology demonstrator.
Development: The K-UCAV was launched as a company-funded project in 2008. It was unveiled at the International Aerospace and Defence Exhibition in Seoul in October 2009, by which time a 20 per cent scale version had already been flight testing for at least a year. It is said to have been conceived for both air-to-air and air-to-ground missions, including SEAD and ISR.

In early 2010, South Korea's Agency for Defence Development (ADD) issued an RfP for a stealthy UCAV demonstrator, bids for which were requested by late March with a down-selection scheduled for the following June. Entitled UCAV Configuration Design Technology Research, this USD15 million programme would be a four-year effort, requiring completion and flight testing by 2013 of two twin-engined prototype/demonstrators. Requirements include airframe use of radar-absorbing materials. KAI's design is one of two national candidates.

Description: *Airframe:* Wings are swept back at about 45 degrees and mid-mounted, with the inboard sections exhibiting much sharper sweep and blending with a fuselage longitudinal chine. Fuselage has a dorsal engine inlet and internal weapons bay. Wide-angle V tail. Retractable tricycle landing gear. Composites construction.

Mission payloads: Not yet specified in detail, but subscale demonstrator tests said to have included dropping of 'stores' from weapons bay. Expected also to have an undernose EO/IR turret for ISR missions.

Guidance and control: Fully autonomous (fly by wire) operation is planned.

Launch: Conventional and automatic wheeled take-off.

Recovery: Conventional and automatic wheeled landing.
Power plant: One (unspecified) turbofan engine.

K-UCAV

Dimensions, External	
Overall, length	8.40 m (27 ft 6¾ in)
Wings, wing span	9.10 m (29 ft 10¼ in)
Weights and Loadings	
Weight	
Weight empty	4,050 kg (8,928 lb)
Performance	
Altitude, Service ceiling	12,000 m (39,380 ft) (est)
Radius of operation	156 n miles (288 km; 179 miles) (est)
Endurance	5 hr (est)
Power plant	1 × turbofan

Status: Development continuing in 2011.
Contractor: Korea Aerospace Industries Ltd
Seoul.

KAI Night Intruder 100N

Type: Small tactical UAV.
Development: Confusingly, although it has an entirely different airframe configuration, KAI has given this new project a similar designation to its already established Night Intruder designs. First news of it was presented at the Singapore Air Show in February 2010.
Description: *Airframe:* Sharply swept wings with endplate fins and rudders; bulbous fuselage; rear-mounted engine. Tricycle landing gear.
Mission payloads: Not yet specified, but expected to include a ventral EO/IR turret.
Guidance and control: Assumed to be similar to that for Night Intruder 100.
Launch: Conventional wheeled take-off; optional catapult launch.
Recovery: Conventional wheeled landing; optional parachute/airbag recovery.
Power plant: One 15.7 kW (21 hp) piston engine, driving a two-blade pusher propeller.

Night Intruder 100N

Dimensions, External	
Overall, length	2.50 m (8 ft 2½ in)
Wings, wing span	3.70 m (12 ft 1¾ in)
Weights and Loadings	
Weight, Max T-O weight	100 kg (220 lb)
Payload, Max payload	16 kg (35 lb)
Performance	
Altitude, Service ceiling	3,000 m (9,840 ft)
Speed, Max level speed	113 kt (209 km/h; 130 mph)
Radius of operation	32 n miles (59 km; 36 miles)
Endurance	6 hr
Power plant	1 × piston engine

Status: Under development in 2011.
Contractor: Korea Aerospace Industries Ltd
Seoul.

KAI Night Intruder 100 and 300

Type: Short-range tactical UAV.
Development: The Night Intruder 300 is the definitive version of what began life in 1991 as the Daewoo (now KAI) Doyosae, developed with support from the Agency for Defence Development (ADD) and first flown in 1993.

At the Seoul Air Show in late 1996 Daewoo briefly showed a more fully developed Doyosae, designated XSR-1. No details were released, but press reports at the time suggested a possible service entry with the South Korean Army in 1998; however, development of the Night Intruder 300 was not completed until August 2000, after more than 100 flight tests. A production contract was awarded in 2001. It made its public debut at Seoul in October 2001, wearing the designation XKRQ-101.

In February 2010, KAI revealed the development of a smaller version, the Night Intruder 100.
Description: *Airframe:* Typical pod and twin boom configuration, with high-mounted wings, pusher engine and mainly composites construction. Fixed tricycle landing gear, with self-sprung mainwheels and oleo nose leg. Wings, booms and tail surfaces detachable for storage and transportation.
Mission payloads: Stabilised, gimbal-mounted, non-retractable dual-sensor ventral turret containing daylight TV and FLIR sensors, both with continuous zoom capability. The TV camera has a 2 to 42° field of view; the FLIR is a third-generation, 3 to 5 micron band thermal imager with 2 to 50° field of view.
Guidance and control: GCS comprises a mission planning bay, a flight control bay and an observation bay. Ground Data Terminal (GDT) relays commands from GCS to air vehicle and relays combined telemetry and

RoK Army Night Intruder 300 *(KAI)* 1142589

Night Intruder 300 in flight *(KAI)* 1142588

The swept-wing Night Intruder 100N small UAV *(KAI)* 1395269

Rear view of the Night Intruder 100N *(KAI)* 1395270

Display model of NI 100 Night Intruder at Singapore Air Show, February 2010 *(IHS/Michael J Gething)* 1356294

Unmanned aerial vehicles > Korea, South > KAI Night Intruder 100 and 300 – KAL KUS-7

NI 300 parafoil recovery

Night Intruder 300 of the Republic of Korea Army in October 2001 *(Peter R Foster)*

Night Intruder 300 rail launch

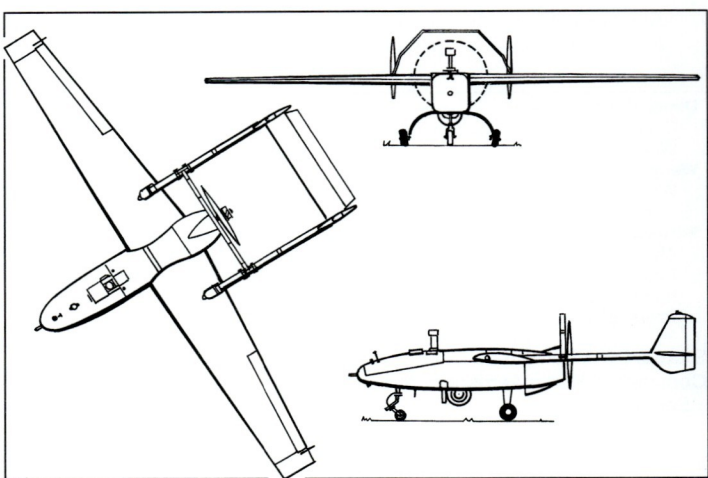
Night Intruder 300 general arrangement *(IHS/John W Wood)*

Recovery: Conventional wheeled landing (both versions); NI 300 has parafoil for emergency recovery. If command link fails, air vehicle can be landed safely at preset location by preprogrammed parafoil flight. Parafoil is ejected by rocket motor and can be repacked into a bag. NI 100 has net system for emergency recovery.

Power plant: *NI 300:* One 38.8 kW (52 hp) UEL AR 801R rotary engine, driving a four-blade fixed-pitch wooden pusher propeller.

NI 100: One 15.7 kW (21 hp) piston engine (type not identified); four-blade pusher propeller.

Variants: *NI 100:* Scaled-down version, unveiled early 2010.

NI 300: In-service version; *description applies to this version except where indicated.*

NI 100, NI 300	
Dimensions, External	
Overall	
length	
NI 100	3.10 m (10 ft 2 in)
NI 300	4.80 m (15 ft 9 in)
height	
NI 100	0.90 m (2 ft 11½ in)
NI 300	1.47 m (4 ft 9¾ in)
Wings	
wing span	
NI 100	4.10 m (13 ft 5½ in)
NI 300	6.40 m (21 ft 0 in)
Weights and Loadings	
Weight	
Weight empty, NI 300	215 kg (473 lb)
Max T-O weight	
NI 100	100 kg (220 lb)
NI 300	300 kg (661 lb)
Payload	
Max payload	
NI 100	16 kg (35 lb)
NI 300	45 kg (99 lb)
Performance	
Altitude	
Service ceiling	
NI 100	3,000 m (9,840 ft)
NI 300	4,500 m (14,760 ft)
Speed	
Max level speed	
NI 100	97 kt (180 km/h; 112 mph)
NI 300	100 kt (185 km/h; 115 mph)
Radius of operation	
mission, NI 100	32 n miles (59 km; 36 miles)
datalink, max, NI 300	65 n miles (120 km; 74 miles)
with relay system, NI 300	194 n miles (359 km; 223 miles)
Endurance, max	6 hr
Power plant, NI 100	1 × piston engine

compressed imagery data to GCS from Aerial Data Terminal (ADT) after decompression of imagery data. Ground Relay Station (GRS), which extends mission radius in mountainous areas, consists of two racks: one datalink receiving commands from GCS/GDT or Launch Control Station (LCS) and uplinking it to air vehicle; the other datalink receives telemetry and imagery data from the air vehicle for transmission to the GCS/GDT or LCS. The LCS is a duplication of the combined GCS/GDT, with similar features except for a shorter operational range. When take-off and landing site is remote from the GCS/GDT, the LCS controls the air vehicle en route towards, and hands over control to, the GCS. It can also be used as a back-up in the event of GCS or GDT failure.

KAI quotes fully autonomous operation from a 'next generation' portable/manpack GCS for the NI 100 version.

System composition: Typically, six air vehicles and one each of GCS, GDT, GRS, LCS and launcher vehicle, plus ground support equipment.

Launch: Conventional wheeled take-off; optionally, from truck-mounted hydraulic/pneumatic catapult.

Status: NI 300 ordered by South Korean Army in 2001 (deliveries completed December 2004) and Navy in 2003; remains in service with both customers. NI 100 being promoted in 2011.

Customers: NI 300: Republic of Korea Army (five systems) and Navy.

Contractor: Korea Aerospace Industries Ltd
Seoul.

KAL KUS-7

Type: Close-range surveillance UAV.

Development: Korean Air Lines was delegated by the Korean government to develop a close-range reconnaissance/surveillance UAS for such parapublic applications as maritime border patrol and wildfire monitoring. This was undertaken by its Aerospace Division (KAL-ASD), which responded with the KUS-7, designed by its R & D centre, Korea Institute of Aerospace Technology (KIAT). A prototype flew for the first time in August 2007.

A prototype of the KUS-7 close-range UAV *(Aceteam)*

Description: *Airframe:* Shoulder-mounted, tapered wings; pod fuselage with pusher engine; twin-boom tail unit. Fixed tricycle landing gear. Composites construction.

Mission payloads: Day/night (EO/IR) sensor in ventral ball turret.

Guidance and control: Automatic (pre-programmed) or manual control, with GPS/INS waypoint navigation; C-band primary uplink/downlink; UHF secondary data uplink. 'Return home' capability in the event of a system failure or loss of the command link.

Launch: Catapult-assisted take-off.

Recovery: Net recovery.

Power plant: One 8.2 kW (11 hp) two-stroke piston engine, driving a two-blade pusher propeller.

KUS-7

Dimensions, External	
Overall	
length	3.10 m (10 ft 2 in)
height	1.10 m (3 ft 7¼ in)
Wings, wing span	3.40 m (11 ft 1¾ in)
Weights and Loadings	
Weight, Max T-O weight	70 kg (154 lb)
Payload, Max payload	7.0 kg (15.00 lb)
Performance	
Altitude, Service ceiling	2,000 m (6,560 ft)
Speed	
Max level speed	81 kt (150 km/h; 93 mph)
Cruising speed	59 kt (109 km/h; 68 mph)
Radius of operation, mission	27 n miles (50 km; 31 miles)
Endurance, max	3 hr
Power plant	1 × piston engine

Status: Development thought to be continuing in 2011.

Contractor: Korean Air Lines (Aerospace Division), Seoul.

KAL KUS-9

Type: Close-range surveillance UAV.

Development: The KUS-9 is a follow-on to KAL-ASD's KUS-7 (which see) in the original five-year (2004-09) programme to develop a close-range surveillance UAV system, and was designed to fulfil applications for both the South Korean Army and in the civil sector. In September 2010, KAL-ASD was selected by Korea's Defence Acquisition Programme Administration (DAPA) as the preferred bidder to develop a domestic MALE UAV, and has proposed the KUS-9 to meet this requirement. According to local reports, the four-year programme is valued at about WON30 billion (USD26 million), and envisages the production of about 30 UAVs for division-level deployment by the South Korean Army from 2014. It is understood that two new customer-specific prototypes will be built for this programme, with the first due to fly during 2013.

Description: *Airframe:* Blended wing/body design with twin-boom tail unit and pusher engine. No landing gear. Composites (CFRP) construction.

Mission payloads: Specific details not released.

A prototype of the KUS-9 on its catapult launch rail *(Aceteam)*

Guidance and control: Believed to be generally as described for KUS-7.

Launch: Catapult-assisted launch.

Recovery: Net recovery.

Power plant: One two-stroke piston engine (rating not quoted), driving a two-blade pusher propeller.

KUS-9

Dimensions, External	
Overall, length	3.40 m (11 ft 1¾ in)
Wings, wing span	4.20 m (13 ft 9¼ in)
Weights and Loadings	
Weight, Max launch weight	150 kg (330 lb)
Payload, Max payload	20 kg (44 lb)
Performance	
Altitude, Service ceiling	4,000 m (13,120 ft)
Speed	
Max level speed	108 kt (200 km/h; 124 mph)
Cruising speed	76 kt (141 km/h; 87 mph)
Radius of operation	32 n miles (59 km; 36 miles)
Endurance	6 hr
Power plant	1 × piston engine

Status: Development continuing in 2011.

Contractor: Korean Air Lines Aerospace Division, Seoul.

KARI Smart

Type: Tilt-rotor UAV technology demonstrator.

Development: The Smart UAV development programme was launched in July 2002 as one of the '21st Century frontier R&D programmes' budgeted for 10 years by the South Korean government. Its objectives are not only to develop a VTOL UAV system with payloads, but to explore and implement innovative technologies to be used in future UAV systems. The 'smart' technologies under study include: automatic take-off and landing; collision avoidance; fault diagnosis and reconfiguration; and active flow separation, vibration and noise controls.

Stage 1 of the programme was completed in March 2005. Results of conceptual and preliminary designs led to the selection of a tilt-rotor aircraft, after comparison with a canard rotor/wing and other configurations. Wind tunnel tests and various performance simulations were made. Key technologies were explored and validated, and control algorithms were verified using 40 per cent scale flying models. During Stage 2, which continued until March 2009, the total system was manufactured, ground tested and prepared for flight testing. The full scale flight test is planned at Stage 3 and the 'smart' technologies will be implemented and validated during this phase, with a completion date scheduled for March 2012. KARI is the nominated prime contractor in charge of design, integration and flight test of the total system; private companies will develop and manufacture subsystems and components.

Some KRW91 billion (USD81 million) funding has reportedly been allocated by the South Korean Ministry of Commerce, Industry and Energy to cover R&D and prototype flight testing, with a further KRW10 billion (USD8.9 million) coming from the private sector.

Mockup of the Smart tilt-rotor UAV displayed in late 2005 *(IHS/Robert Karniol)*

KARI Smart 40 per cent scale model in flight *(KARI)*

Description: *Airframe:* Mid-wing aircraft with flaperons and T tail, of mainly composites construction. Single turboshaft in rear fuselage, with dorsal air intake, drives a three-blade proprotor in a rotating nacelle at each wingtip. Mission payload turret under nose. Front and rear fuselage sections detachable for transportation and maintenance. Retractable main landing gear units under forward and rear fuselage; auxiliary wheel attached at rear of each engine nacelle, pivoting with proprotor system.

Guidance and control: Sheltered GCS with pilot bay, observation bay and navigation bay for mission planning and control and as redundant bay for pilot. Digital flight control systems, dual flight control computers and dual actuators, with redundant GPS-aided INS. Ku-band primary datalink and UHF secondary datalink for command, control and telemetry.

Launch: Vertical take-off (manual or autonomous) with local precision positioning system.

Recovery: Vertical landing (manual or autonomous) with local precision positioning system.

Power plant: One 410 kw (550 shp) Pratt & Whitney Canada PW206C turboshaft, driving twin three-blade proprotors. Fuel capacity 250 litres (66.0 US gallons; 55.0 Imp gallons) in fuselage and 100 litres (26.4 US gallons; 22.0 Imp gallons) in wing tanks.

Smart

Dimensions, External	
Overall	
length	5.00 m (16 ft 4¾ in)
height, nacelles vertical	1.90 m (6 ft 2¾ in)
width, rotors turning	6.90 m (22 ft 7¾ in)
Wings, wing span	4.00 m (13 ft 1½ in)
Engines	
propeller diameter, each	2.90 m (9 ft 6¼ in)
Weights and Loadings	
Weight, Max T-O weight	995 kg (2,193 lb)
Payload, Max payload	90 kg (198 lb)
Performance	
Altitude, Service ceiling	5,000 m (16,400 ft) (est)
Speed	
Max level speed	270 kt (500 km/h; 311 mph) (est)
Cruising speed, max	238 kt (441 km/h; 274 mph) (est)
Loitering speed	135 kt (250 km/h; 155 mph) (est)
Radius of operation, mission	108 n miles (200 km; 124 miles) (est)
Endurance	5 hr (est)

Status: In September 2009, the full scale SUAV system had been assembled and ground tests started. Flight testing was scheduled between 2009-2012. An upgraded vehicle with sense and avoid technology has been designed and tested in 2011.

Contractor: Korea Aerospace Research Institute (KARI)
Daejeon.

Uconsystem RemoEye 002

Type: Close-range mini-UAV.

Development: Uconsystem was formed in June 2001 and has since developed its own ground control systems, flight control computers, small UAVs and aerial targets.

Description: *Airframe:* High-mounted wings with constant chord centre-section and tapered outer trailing-edges; pod and boom fuselage; pusher engine; T tail. No landing gear. Composites construction.

Mission payloads: CCD TV daylight camera or IR night camera with single-axis scanning.

RemoEye 002 hand-launched UAV *(Robert Hewson)*

The portable GCS for the RemoEye series *(IHS/Patrick Allen)*

Guidance and control: Preprogrammed flight mode with updatable waypoint navigation, automatic homing, altitude and speed hold; air vehicle and target position displayed on video monitor. Portable GCS.

System composition: Basic system comprises four air vehicles and containers, portable or mobile GCS, battery and charger, optional video receiving stations.

Launch: Hand-launched.

Recovery: Deep stall to belly landing.

Power plant:
Battery-powered electric motor, driving a two-blade pusher propeller.

RemoEye 002

Dimensions, External	
Overall, length	1.30 m (4 ft 3¼ in)
Wings, wing span	1.50 m (4 ft 11 in)
Weights and Loadings	
Weight, Max launch weight	2.4 kg (5.00 lb)
Performance	
Speed, Max level speed	43 kt (80 km/h; 49 mph)
Radius of operation	5.4 n miles (10 km; 6 miles)
Endurance	1 hr
Power plant	1 × electric motor

Status: Developed and available. Demonstrated to RoK Army in 2006 and in several other countries by 2008.

Contractor: Uconsystem Company Ltd
Daejeon.

Uconsystem RemoEye 006

Type: Short-range mini-UAV.

Development: Uconsystem was formed in June 2001 and has since developed its own ground control systems, flight control computers, small UAVs and aerial targets.

Description: *Airframe:* Pylon-mounted high wings with constant chord centre-section and tapered outer trailing-edges; pod and boom fuselage; pusher engine; T tail. No landing gear. Composites construction.

Mission payloads: CCD TV daylight camera or IR night camera with two-axis scanning.

RemoEye 006 in military camouflage *(IHS/Patrick Allen)*

RemoEye 006 hand-launched UAV *(Robert Hewson)* 1151526

RemoEye series ground control station *(IHS/Patrick Allen)* 1138517

Guidance and control: Preprogrammed flight mode with updatable waypoint navigation, automatic homing, altitude and speed hold; air vehicle and target position displayed on video monitor. Portable or shelter-mounted GCS.

System composition: Basic system comprises four air vehicles and containers, portable or mobile GCS, battery and charger.

Launch: Hand launch or by bungee catapult.

Recovery: Deep stall to belly skid landing.

Power plant: Battery-powered electric motor, driving a two-blade pusher propeller.

RemoEye 006

Dimensions, External	
Overall, length	1.55 m (5 ft 1 in)
Wings, wing span	2.72 m (8 ft 11 in)
Weights and Loadings	
Weight, Max launch weight	6.0 kg (13.00 lb)
Performance	
Speed	
Max level speed	40 kt (74 km/h; 46 mph)
Stalling speed	22 kt (41 km/h; 26 mph)
Radius of operation	8.1 n miles (15 km; 9 miles)
Endurance	1 hr 30 min
Power plant	1 × electric motor

Status: Development complete and available. Demonstrated to RoK Army in 2006 and to several overseas countries by 2008. Delivered to RoK Marine Corps in 2009. In service.

Contractor: Uconsystem Company Ltd
Daejeon.

Uconsystem RemoH-C100

Type: VTOL UAV.

Development: The RemoH-C100 is South Korea's first dedicated unmanned helicopter for crop-dusting. Rotor downwash, and low application height, enable dusting to be precise and economical.

Description: *Airframe:* Two-blade main and tail rotors; composites fuselage with detachable spraybooms. Twin-skid landing gear with ground handling wheels.

Mission payloads: Hopper for agricultural chemical, capacity 30 litres (7.9 US gallons; 6.6 Imp gallons).

Guidance and control: Flights are conducted in automatic or semi-automatic mode. Digital flight control system (computer and navigation sensors): GPS, rate gyros, accelerometers and magnetometer. GCS comprises control board, LCD module, RF module, joystick and battery. Operating functions include flight mode control; airspeed/altitude control; operational range set-up and control, spray system control, switch control and voice check. Monitoring functions include datalink status; airborne equipment operating status; airborne sensor data and operating status; helicopter position and range; flight time and status; and voice messaging for warning situations such as fuel remaining and engine condition. Automatic features include limiting flight to within operating radius; spraying/dusting halted when helicopter speed is at zero or below a pre-set figure; and landing if communications equipment fails and link is lost. Communications interference between air vehicle and GCS are also avoided. Operators receive training package on purchase of system.

Transportation: Complete system, including two operators, transportable in a 1-ton truck.

Launch: Conventional, automatic, vertical take-off.

Recovery: Conventional, automatic, vertical landing.

Power plant: One 23.9 kW (32 hp) two-cylinder two-stroke engine with fuel injection and electric starting. Fuel capacity 12 litres (3.2 US gallons; 2.6 Imp gallons).

RemoH-C100

Dimensions, External	
Overall, height	1.05 m (3 ft 5¼ in)
Fuselage	
length	3.50 m (11 ft 5¾ in)
width	0.66 m (2 ft 2 in)
Rotors, rotor diameter	3.20 m (10 ft 6 in)
Weights and Loadings	
Weight	
Weight empty	60 kg (132 lb)
Max T-O weight	100 kg (220 lb)
Performance	
Speed	
Max level speed	32 kt (59 km/h; 37 mph)
Max operating speed	11 kt (20 km/h; 13 mph) (est)
Radius of operation	0.2 n miles (km; miles)
Endurance	1 hr
Power plant	1 × piston engine

Status: Entered production in 2007.

Contractor: Uconsystem Company Ltd
Daejeon.

RemoH-C100 and operating crew *(Uconsystem)* 1368424

Malaysia

CTRM Aviation Aludra

Type: Short-range tactical UAV.

Development: Although developed primarily for battlefield surveillance and reconnaissance, the Aludra is regarded as also suitable for maritime patrol, law enforcement, border patrol and monitoring of forest fires or other hazardous situations. Its name — also that of a star — is coined from Alliance Unmanned Developmental Research Aircraft.

After its involvement in the 2001 Eagle 150 ARV programme, CTRM began developing an indigenous unmanned aircraft, the EX-01, in 2003. This had a 2.44 m (8 ft) wing span, gross weight of 15 kg (33 lb), and was remotely piloted. In 2004, CTRM teamed with Ikramatic Systems to develop an autopilot for the EX-01, resulting in its redesign as the SR-01, of which eight were built. with increased wing span, more powerful engine and gross weight of 35 kg (77 lb). Elsewhere in 2004 a third company, System Consultancy Services (SCS), had separately developed a similar but smaller own-design UAV named Nyamok, and in December 2006 these three Malaysian companies formed a CTRM-led partnership entitled Unmanned Systems Technology (UST) to develop the SR-01 into the SR-02, which was subsequently renamed Aludra. The MX-01 prototype of this design reportedly made its maiden flight (minus payload) in September 2006, followed by the first fully equipped flight on 15 May 2007. First flight of a prototype equipped with the Agile sensor intended for the Mk 2 took place shortly before the LIMA defence exhibition in late 2007.

One of the 2005 CTRM TUAV prototypes, from which the Aludra has been developed *(MDIC)* 1290331

Aludra Mk 1 prototype at the 2007 LIMA defence show *(UTS/Zulkhas)* 1290332

Description: *Airframe:* High-wing, twin-tailboom configuration, with pusher engine. Fixed tricycle landing gear. Composites construction (glass and carbon fibre, foam and epoxy resin).
  Mission payloads: Four-axis EO/IR sensor in gyrostabilised turret in Mk 1 (and initially in Mk 2); to be upgraded to Thales Agile 2 in Mk 2. Real-time video and telemetry downlink. Ikramatic Imran 400 (Mk 1) or 800 (Mk 2) autopilot.
  Guidance and control: Pre-programmed or remotely operated. Dual (C-band and S-band) datalinks.
  Launch: Conventional wheeled take-off; catapult launch option to be developed for Mk 2.
  Recovery: Conventional wheeled landing.
  Power plant: Auldra Mk 1: One 37.3 kW (50 hp) two-cylinder two-stroke engine, driving a two-blade pusher propeller.
Variants: *Aludra Mk 1:* Developmental version. *Description applies to this version except where indicated.*
  Aludra Mk 2: Intended service version. Increased wing span, upgraded sensor and new Limbach engine.

Aludra Mk 1, Aludra Mk 2

Dimensions, External	
Overall, length	4.27 m (14 ft 0 in)
Wings	
wing span	
Aludra Mk 1	4.88 m (16 ft 0¼ in)
Aludra Mk 2	6.10 m (20 ft 0¼ in)
Weights and Loadings	
Weight	
Max T-O weight	
Aludra Mk 1	200 kg (440 lb)
Aludra Mk 2	250 kg (551 lb)
Payload, Max payload	50 kg (110 lb)
Performance	
T-O	
T-O run	250 m (821 ft)
Altitude, Service ceiling	3,660 m (12,000 ft)
Speed	
Cruising speed, max	120 kt (222 km/h; 138 mph)
Loitering speed	
Aludra Mk 1	50 kt (93 km/h; 58 mph)
Aludra Mk 2	55 kt (102 km/h; 63 mph)
Radius of operation	
LOS, Aludra Mk 1	81 n miles (150 km; 93 miles)
LOS, Aludra Mk 2	27 n miles (50 km; 31 miles)
Endurance	6 hr
Power plant, Aludra Mk 1	1 × piston engine

Status: Officially commissioned by the Malaysian Defence Ministry in 2006, the Aludra was elevated to a "national project with immediate effect" in a 21 October 2008 statement by the country's Prime Minister. At that time, it was understood to be funded jointly by the three UTS members (total of MYR15 million) and the Defence Ministry (MYR5 million). Intention is for Malaysian armed forces to buy contractor services with the system rather than outright purchase. It will reportedly be operated along the coast of East Malaysia from an airfield in Sabah. Deployed on operational trials with Malaysian Joint Forces in 2009 .
Contractor: CTRM Aviation Sdn Bhd
Melaka.

Mexico

Hydra Technologies E1 Gavilán

Type: Surveillance mini-UAV.
Development: The Gavilán made its public debut in mid-2008. Its development history was not disclosed.
Description: *Airframe:* Constant-chord, high-mounted wings; pod and boom fuselage; pusher engine; conventional tail surfaces. No landing gear.
  Mission payloads: Dual-sensor (EO and IR), nose-mounted turret for day and night operations.
  Guidance and control: Radio control.
  Transportation: Man-portable.
  System composition: Air vehicle(s), portable GCS. One-man operation.
  Launch: Hand or bungee catapult launch.
  Recovery: Belly landing.
  Power plant: Battery-powered electric motor (rating not stated), driving a two-blade pusher propeller.

E1

Dimensions, External	
Wings, wing span	1.50 m (4 ft 11 in)
Weights and Loadings	
Weight, Max launch weight	5.0 kg (11.00 lb)
Performance	
Altitude, Service ceiling	2,440 m (8,000 ft) (est)
Radius of operation, mission	5.4 n miles (10 km; 6 miles)
Endurance, max	1 hr 30 min
Power plant	1 × electric motor

Gavilán displayed at the 2008 Farnborough Air Show *(IHS/Patrick Allen)* 1367736

Gavilán mini-UAV in flight *(Hydra Technologies)* 1290314

Status: The Gavilán is understood to be produced for, and used by, the Mexican Department of Homeland Security for monitoring border zones, drug plantations and trafficking, illegal immigration, and for disaster relief. The Jalisco state government is reported to use it for such ecological missions as flood reporting and pollution monitoring. Exports have also been reported, though customer details were not known as of late 2009.
Contractor: Hydra Technologies de Mexico
Zapopan.

Hydra Technologies S4 Ehécatl

Type: Surveillance tactical UAV.

Development: Hydra Technologies was formed in 2001. Named after the Aztec god of the wind, the Ehécatl was unveiled at the Paris Air Show in June 2007, after five years of development, supported by various Mexican agencies and the Mexican National Polytechnics Institute. Prototype flight testing began in 2003, and the S4 version made its maiden flight in 2006. Applications include surveillance (land or maritime), border patrol, anti-smuggling patrol, disaster support, target acquisition and tracking, resource monitoring and force protection.

Description: *Airframe:* Pod and twin tailboom configuration; high wings with tapered leading-edges; double T tail unit. Fixed tricycle landing gear. Construction of composites (carbon fibre, S-glass and Kevlar), aluminium alloy and titanium.
   Mission payloads: Gimballed ventral turret (360° pan, 100° tilt) housing an EO camera with ×26 optical zoom, and a FLIR Systems Photon 320 × 240 IR camera. Options include an 8 Megapixel high-resolution still camera with ×8 optical zoom. Video and telemetry downlinks. Customised payloads to customer's requirements, via partnership with General Dynamics.
   Guidance and control: Mobile GCS, installed in a specially equipped three-man ground vehicle.
   System composition: Two air vehicles, one GCS and crew of three. Set-up time 20 minutes.
   Launch: Conventional wheeled take-off.
   Recovery: Conventional wheeled landing.
   Power plant: One 60 or 80 cm^3 (3.7 or 4.9 cu in) two-stroke piston engine (type and rating not stated), driving a three-blade tractor propeller. Fuel capacity 21 litres (5.5 US gallons; 4.6 Imp gallons).

Ehécatl *(Hydra Technologies)* 1290313

The Ehécatl has standard runway take-off and landing modes *(Hydra Technologies)* 1169651

Ehécatl underside view, displaying sensor turret and wing flaps *(IHS/Patrick Allen)* 1367738

S4 Ehécatl

Dimensions, External	
Wings, wing span	3.70 m (12 ft 1¾ in)
Weights and Loadings	
Weight, Max T-O weight	55.0 kg (121 lb)
Payload, Max payload	9.0 kg (19.00 lb)
Performance	
Altitude, Service ceiling	4,570 m (15,000 ft)
Speed	
Max level speed	90 kt (167 km/h; 104 mph)
Cruising speed	38 kt (70 km/h; 44 mph)
Loitering speed	32 kt (59 km/h; 37 mph)
Radius of operation, LOS	52 n miles (96 km; 59 miles)
Endurance, at cruising speed	8 hr
Power plant	1 × piston engine

Status: Delivery of the first operational system to the Mexican Public Security Secretariat, for use by the Federal Police, was reported in February 2008. At that time, according to *IHS Jane's* sources, a second system was ready for delivery to the Mexican Navy, and potential orders were being discussed with agencies in Colombia, the Dominican Republic and Panama.
Contractor: Hydra Technologies de Mexico
Zapopan.

Netherlands

Dutch Space SPADES

Type: Aerial cargo delivery system.

Development: The Small Parafoil Autonomous Delivery System (SPADES), unveiled in 2003, is a prototype aerial cargo delivery system which uses a ram-air parachute and GPS positioning information to fly a payload to a preprogrammed delivery point. It was developed in collaboration with the NLR (Dutch National Aerospace Laboratory) and funded by the Royal Netherlands Army; the parafoil is supplied by Aérazur of France, based on that company's G9 design as used by the Netherlands Special Forces' Korps Commando Troepen.

Demonstrations have validated the concept, parafoil deployment procedure, in-flight system characteristics and behaviour, remote control (radio link) back-up and autopilot performance. The system has high level of reusability, claimed to be 50 times without overhaul. A first generation system has evolved, in development, into two weight classes enabling payloads ranging from 100-1000 kg (220-2,200 lbs)

Description: *Airframe:* Parafoil, and payload container with shock-absorbing base.
   Mission payloads: Miscellaneous cargo.
   Guidance and control: Autonomous. Control unit processor guides and controls the system via GPS navigation and varying steering lines. Wind information is measured in real time during flight and used by the autopilot for flight path updating.

SPADES 300 Mk2 *(Dutch Space/EADS)* 1410036

SPADES 1000 Mk2 *(Dutch Space/EADS)* 1410035

Launch: By air drop from helicopter or transport aircraft.
Recovery: Soft landing by parafoil.

Specifications

Weights
Payload weight, SPADES 300 Mk2 ....100–225 kg (220–500 lb)
SPADES 1000 Mk2 .... 350-100 kg (750-2,200 lb)
Performance
Drop altitudes .... Up to 10,668 m (35,000 ft)
Standoff ranges .... 13.5–21.5 n miles (25–40 km; 15.5–25 miles)
Landing accuracy .... CEP Within 50 m (165 ft)

Status: Participated in NATO-sponsored Precision Airdrop Capability Demonstration near Bordeaux in July 2006 with 500 kg payloads. The development of what is described as the second generation of SPADES, was completed by 2009, and the system is available in the two weight classes described above. More than 250 test drops completed. SPADES is reported to be undergoing certification and is scheduled to enter service with Special Forces in the Netherlands in 2010.
Contractor: Dutch Space BV (owned by EADS Astrium since July 2006) Leiden.

Geocopter GC-201

Type: Helicopter UAV.
Development: The GC-201 was developed specifically for such civil applications as aerial photogrammetry and for rapid response in rescue and recovery situations. Features include automatic camera triggering, as demonstrated in 2006 in collaboration with the Dutch government; this provides geo-referenced imagery to enhance the effectiveness of first responders to emergency situations. Certification for flight in Dutch civil airspace was granted in May 2009, just ahead of delivery of the first production system to a customer.
Description: *Airframe:* Three-blade main and two-blade tail rotors. Pod and boom fuselage with dependent fin. Twin-skid landing gear. Carbon fibre fuselage and rotor blades.
  Mission payloads: Various 'plug and play' payloads, including: gimbal with stabilised video, thermal or still cameras; digital camera with IMU; lidar scanner; passive microwave radiometer; multispectral camera; or miniature SAR. Real-time video downlink.
  Guidance and control: Pre-programmed. Laptop GCS with flight control system and mission planning software (updatable in flight);

The twin-turbine GC-201 *(IHS/Patrick Allen)* 1380024

GC-201 on display at the Paris Air Show, June 2009 *(IHS/Patrick Allen)* 1380025

GPS-based navigation; real-time video and map displays. Power generator for remote-site operations.
  System composition: One air vehicle and GCS.
  Launch: Conventional, automatic, helicopter take-off.
  Recovery: Conventional, automatic, helicopter landing. Automatic 'return home' mode if command link is lost.
  Power plant: Two 8 kw (10.7 shp) turboshafts. Fuel capacity 9.5 litres (2.5 US gallons; 2.1 Imp gallons).

GC-201

Dimensions, External
Overall, height .... 0.90 m (2 ft 11½ in)
Fuselage, length .... 2.90 m (9 ft 6¼ in)
Rotors, rotor diameter .... 3.30 m (10 ft 10 in)
Tail rotor, tail rotor diameter .... 0.53 m (1 ft 8¾ in)
Dimensions, Internal
Payload bay, volume .... 0.03125 m³ (1.1 cu ft)
Weights and Loadings
Weight
  Weight empty .... 35 kg (77 lb)
  Max T-O weight .... 100 kg (220 lb)
Payload, Max payload .... 30 kg (66 lb)
Performance
Altitude, Service ceiling .... 1,830 m (6,000 ft)
Speed
  Cruising speed, max .... 39 kt (72 km/h; 45 mph)
Endurance .... 4 hr
Power plant .... 2 × turboshaft

Status: In production. The Dutch National Aerospace Laboratory (NLR) ordered two aircraft and two ground stations in June 2008, the first GC-201 (PH-X1A) making its maiden flight in mid-2009. The second system is to be used in the NLR's Facility for Unmanned Rotorcraft Research (FURORE) programme to develop automatic and safety procedures for rotary-wing UAVs.
Contractor: Geocopter BV Culemborg.

Pakistan

ACES Eagle Eye PI/PII and Uqab

Type: Short-range tactical UAVs.
Development: Under development since 2002, as indicated under Variant headings below.
Description: *Airframe:* Strut-braced, high-wing design; pod and twin-tailboom configuration, with pusher engine. Fixed tricycle landing gear. Composites construction.

  Mission payloads: Pan-tilt-zoom camera for real-time telemetry downlink of digital video.
  Guidance and control: Pre-programmed, with real-time GPS-based navigation and tracking and manual override option. Can be re-programmed in flight. Moving map displays in GCS.
  Launch: Conventional wheeled take-off.
  Recovery: Conventional wheeled landing.

General appearance of the Eagle Eye PII *(ACES)* 1290362

Uqab seen at naming ceremony in March 2008 *(APP)* 1290363

Power plant: PI: One 16.4 kW (22 hp) two-cylinder two-stroke engine; two-blade pusher propeller. Fuel capacity 40 litres (10.6 US gallons; 8.8 Imp gallons).
P1T: One 11.9 kW (16 hp) two-cylinder two-stroke engine; two-blade pusher propeller. Fuel capacity 10 litres (2.6 US gallons; 2.2 Imp gallons).
PII: One 28.3 kW (38 hp) rotary engine; two-blade pusher propeller. Fuel capacity 60 litres (15.9 US gallons; 13.2 Imp gallons).

Variants: Eagle Eye PI: Initial version for real-time reconnaissance. First flight 2002.
Eagle Eye P1T: Modified, smaller variant of PI for UAV operator training or aerial target use. Remotely piloted control only, within LOS.
Eagle Eye PII: Larger, more capable version of PI. First flight in 2005. User trials with Pakistan Army began in 2007 and were completed in March 2008. Selected for continuing development as Uqab.
Uqab: Name (Eagle) bestowed on configuration of Eagle Eye PII as at completion of customer trials in March 2008. Development to continue of both tactical and longer-range/endurance strategic versions.

Eagle Eye PI, Eagle Eye PII, Eagle Eye P1T, Uqab

Dimensions, External	
Overall	
length	
Eagle Eye PI	3.81 m (12 ft 6 in)
Eagle Eye P1T	3.12 m (10 ft 2¾ in)
Eagle Eye PII	4.11 m (13 ft 5¾ in)
Wings	
wing span	
Eagle Eye PI	5.00 m (16 ft 4¾ in)
Eagle Eye P1T	3.48 m (11 ft 5 in)
Eagle Eye PII	5.72 m (18 ft 9¼ in)
Weights and Loadings	
Weight	
Max T-O weight	
Eagle Eye PI	130 kg (286 lb)
Eagle Eye P1T	40 kg (88 lb)
Eagle Eye PII	175 kg (385 lb)
Payload	
Max payload	
Eagle Eye PI	30 kg (66 lb)
Eagle Eye PII	40 kg (88 lb)
Performance	
T-O	
T-O run	
Eagle Eye P1T	200 m (657 ft)
Eagle Eye PI	300 m (985 ft) (est)
Eagle Eye PII	400 m (1,313 ft) (est)
Altitude	
Service ceiling	
Eagle Eye PI, Eagle Eye PII	3,050 m (10,000 ft)
Eagle Eye P1T	455 m (1,500 ft)
Speed	
Max level speed	
Eagle Eye PI	64 kt (119 km/h; 74 mph)
Eagle Eye P1T	70 kt (130 km/h; 81 mph)
Eagle Eye PII	81 kt (150 km/h; 93 mph)
Cruising speed	
Eagle Eye PI	43 kt (80 km/h; 49 mph)
Eagle Eye P1T	38 kt (70 km/h; 44 mph)
Eagle Eye PII	65 kt (120 km/h; 75 mph)
Radius of operation	
mission, Eagle Eye PI	43 n miles (79 km; 49 miles)
mission, Eagle Eye PII	54 n miles (100 km; 62 miles)
tactical, Uqab	81 n miles (150 km; 93 miles)
strategic, Uqab	189 n miles (350 km; 217 miles)
Endurance	
max, Eagle Eye PI	3 hr
max, Eagle Eye P1T	1 hr
max, Eagle Eye PII	4 hr

Status: Uqab development for Pakistan Army continuing in 2008-09.
Contractor: Advanced Computing & Engineering Solutions (Pvt) Ltd (ACES)
Rawalpindi.

Albadeey Hud Hud

Type: Short-range tactical UAV.
Development: History not known; possibly related to similar design by Air Weapons Complex.
Description: *Airframe:* High-wing, pod and twin-tailboom configuration with rectangular (Hud Hud I) or sweptback (Hud Hud II and III) fins and rudders; tractor (Hud Hud I) or pusher engine (Hud Hud II and III). Elongated fuselage nacelle on Hud Hud III. Non-retractable tricycle landing gear.
Mission payloads: Reconnaissance and surveillance sensors to customer's requirements. Typically, a high-resolution colour CCD camera with ×26 zoom, in ventral turret, and real-time, frequency modulated L-band video transmitter. Dual-band datalink. Alternatives could include thermal imagers, EW or other EO sensors.
Guidance and control: Autonomous, with standby manual mode. Digital datalink with autopilot and 12-channel GPS navigation. Vehicle-mounted GCS cabin accommodates video data receiver, computerised scan map for GPS data, autotracking antenna and processing and recording equipment. Portable GCS also available.
System composition: Four air vehicles and GCS.
Launch: Wheeled take-off.
Recovery: Wheeled landing.
Power plant:
Hud Hud I: One 100 cm^3 front-mounted piston engine, with two-blade tractor propeller.
Hud Hud II: One rear-mounted 150 cm^3 engine with two-blade pusher propeller.
Hud Hud III: One rear-mounted 275 cm^3 engine with four-blade pusher propeller.
Variants: *Hud Hud I:* Basic, tractor-engined early version.
Hud Hud II: Larger, rear-engined and more capable version.
Hud Hud III:
Further increases in dimensions, weights, engine power and capability.

The more capable Hud Hud III *(Albadeey)* 1290358

Hud Hud II short-range tactical UAV *(Albadeey)* 1290357

Hud Hud I, Hud Hud II, Hud Hud III

Dimensions, External
Overall
length
Hud Hud I .... 2.19 m (7 ft 2¼ in)
Hud Hud II .... 2.84 m (9 ft 3¾ in)
Hud Hud III .... 3.35 m (11 ft 0 in)
Wings
wing span
Hud Hud I .... 3.05 m (10 ft 0 in)
Hud Hud II .... 3.66 m (12 ft 0 in)
Hud Hud III .... 4.88 m (16 ft 0¼ in)

Weights and Loadings
Weight
Max T-O weight
Hud Hud I .... 35 kg (77 lb)
Hud Hud II .... 60 kg (132 lb)
Hud Hud III .... 90 kg (198 lb)
Payload
Max payload
Hud Hud I .... 20 kg (44 lb)
Hud Hud II .... 25 kg (55 lb)
Hud Hud III .... 40 kg (88 lb)

Performance
T-O
T-O run
Hud Hud I .... 50 m (165 ft)
Hud Hud II .... 100 m (329 ft)
Hud Hud III .... 150 m (493 ft)
Altitude
Service ceiling
Hud Hud I .... 3,660 m (12,000 ft)
Hud Hud II .... 4,875 m (16,000 ft)
Hud Hud III .... 6,095 m (20,000 ft)
Speed
Max level speed .... 81 kt (150 km/h; 93 mph)
Loitering speed .... 32 kt (60 km/h; 37 mph)
Radius of operation
Hud Hud I .... 27 n miles (50 km; 31 miles)
Hud Hud II .... 32 n miles (59 km; 36 miles)
Hud Hud III .... 54 n miles (100 km; 62 miles)
Endurance
Hud Hud I .... 2 hr 30 min
Hud Hud II .... 3 hr 30 min
Hud Hud III .... 4 hr

Status: Hud Hud II in service. Hud Hud III competed unsuccessfully against IDS Huma-I/II in 2006.
Customers: Pakistan Army, Navy and Air Force; Yemen Air Force. Promotion of Hud Hud II and III continuing in 2009.
Contractor: Albadeey Technologies
Karachi.

IDS Huma-1

Type: Short-range tactical UAV.
Development: Designed as a low-cost, easy to use system capable of remote sensing or other aerial robotic applications and operable from all kinds of terrain. In late 2007, IDS was understood to be working on a larger tactical version offering heavier payload and greater endurance; at that time, it anticipated a further 1½ years of development for the Huma-1.
Description: *Airframe:* Typical pod and twin tailboom configuration, with unswept, constant-chord high wings and pusher engine. Twin-skid landing gear. Composites construction.
Mission payloads: EO/IR and similar sensors, with real-time imagery and telemetry downlink.
Guidance and control: By GPS-based autopilot system, with 'return home' mode in case of GPS loss.
Launch: By booster rocket from free-standing or vehicle-mounted zero-length launcher.
Recovery: Parachute recovery to skid landing.

Huma-1 at the IDEX exibition, Abu Dhabi, February 2009 *(IHS/Patrick Allen)* 1376561

The Huma-1 tactical UAV, under development in 2008 *(IDS)* 1328316

Huma-1 on its zero-length launcher. Landing is on shock-absorbing underfuselage twin skids *(IDS)* 1328317

Power plant: One 16.4 kW (22 hp) two-cylinder two-stroke engine, driving a two-blade pusher propeller.

Huma-1

Dimensions, External
Overall
length .... 3.76 m (12 ft 4 in)
height, excl shock damper .... 0.79 m (2 ft 7 in)
Wings, wing span .... 4.40 m (14 ft 5¼ in)
Weights and Loadings
Weight
Weight empty .... 122.6 kg (270 lb)
Max launch weight .... 130 kg (286 lb)
Payload, Max payload .... 20.0 kg (44 lb)
Performance
Altitude
Operating altitude
min .... 1,000 m (3,280 ft)
max .... 3,000 m (9,840 ft)
Speed, Max level speed .... 97 kt (180 km/h; 112 mph)
Range, max .... 270 n miles (500 km; 310 miles)
Radius of operation, mission .... 81 n miles (150 km; 93 miles)
Endurance .... 5 hr (est)
Power plant .... 1 ×

Status: Development and marketing continuing, exhibited at IDEX 2009.
Contractor: Integrated Defence Systems
Islamabad.

Integrated Dynamics Border Eagle

Type: Low-altitude, short-range surveillance UAV.
Development: The Border Eagle system is understood to have been introduced in 2003. Current version (2010) is designated **Mk II**.
Description: *Airframe:* Pod and twin-tailboom configuration, with pusher engine, constant-chord high wings, and triangular fins and rudders with bridging tailplane. Fixed tricycle landing gear. All-composites construction, including carbon-reinforced mainwheel legs.
Mission payloads: Typical payloads can include GSP-100 daylight TV camera; onboard video recorder; still camera module; or elint or chemical monitoring packages. Combinations of payload can be carried and activated as required.
Guidance and control: GCS-1200 portable ground station with UHF command uplink; ATPS-1200 antenna tracking and positioning system; ID-TM6 GPS and data telemetry module with software graphics interface display. Options for ID-AP2000 or -AP5000 digital autopilots with navigation and stability modules.

Border Eagle is believed to be Integrated Dynamics' principal export system *(Robert Hewson)* 1290263

Explorer UAV for civilian R & D and patrol applications *(Robert Hewson)* 1290264

Transportation: Complete system transportable in HMMWV or similar vehicle. Booms and tail unit detachable for transportation.

System composition: Typical system comprises four air vehicles, payloads, GCS and ground support equipment. Three-person crew (flight controller, camera operator and technician).

Launch: Conventional wheeled take-off.

Recovery: Conventional wheeled landing standard; parachute recovery optional.

Power plant: One 1.5 kW (2 hp) single-cylinder or 4.5 kW (6 hp) twin-cylinder piston engine; two-blade pusher propeller. Fuel capacity 7 litres (1.8 US gallons; 1.5 Imp gallons).

Border Eagle Mk II

Dimensions, External	
Overall	
length	1.75 m (5 ft 9 in)
height	0.48 m (1 ft 7 in)
Wings, wing span	3.10 m (10 ft 2 in)
Areas	
Wings, Gross wing area	0.98 m² (10.5 sq ft)
Weights and Loadings	
Weight	
Weight empty	9.0 kg (19.00 lb)
Max T-O weight	15.0 kg (33 lb)
Payload, Max payload	4.0 kg (8.00 lb)
Performance	
Altitude	
Operating altitude, typical	305 m (1,000 ft)
Speed	
Cruising speed, max	86 kt (159 km/h; 99 mph)
Loitering speed	16 kt (30 km/h; 18 mph)
Radius of operation, mission	16.2 n miles (30 km; 18 miles)
Endurance	3 hr (est)
Power plant	1 × piston engine

Status: In production and in service. As of December 2006, company reported deliveries of between 15 and 20 aircraft to unidentified customers in Australia, South Korea, Libya, Spain and the US. The US customer is understood to have used the aircraft for Mexican border patrol in the vicinity of Los Angeles. These exports may have been of airframes only, rather than complete systems. No reports of sales within Pakistan had been seen by April 2008, when the system continued to be promoted. The system remained listed as current by the manufacturer in 2012.

Contractor: Integrated Dynamics
Karachi.

Integrated Dynamics Explorer

Type: Patrol, research and payload test platform.

Development: Launched simultaneously with the Rover in September 2007, the Explorer is the larger of Integrated Dynamics' two current offerings aimed at the civilian UAV market. It is a more advanced system, designed for use as a sensor testbed or for scientific research programmes.

Description: *Airframe:* High-mounted wings, fuselage pod, low-set tubular tailboom and V tail surfaces; rear-mounted pusher engine. Fixed tricycle landing gear. Composites construction.

Mission payloads: Typical payloads can include GSP-100 gyrostabilised daylight camera, IR camera, digital still camera, and research payloads. Combinations of payload can be carried and activated as required.

Guidance and control: GCS-1200 ground station; ATPS-1200 antenna tracking and positioning system; ID-AP5000 digital autopilot with navigation and stability modules; ID-TM6 GPS and data telemetry module.

System composition: Two air vehicles with GSP-100 payloads; GCS-1200 ground station with programming and moving map mission display software; digital spread spectrum telecommand link; antennas, cables and operational spares.

Launch: Conventional wheeled take-off; can also be modified for vehicle rooftop launch.

Recovery: Conventional wheeled landing; or parachute recovery to belly landing on a Kevlar-reinforced ventral pan.

Power plant: One 4.5 kW (6 hp) single- or two-cylinder piston engine, driving a two-blade pusher propeller. Fuel capacity 10 litres (2.6 US gallons; 2.2 Imp gallons).

Explorer

Dimensions, External	
Overall	
length	2.75 m (9 ft 0¼ in)
height	0.80 m (2 ft 7½ in)
Wings, wing span	4.51 m (14 ft 9½ in)
Areas	
Wings, Gross wing area	1.94 m² (20.9 sq ft)
Weights and Loadings	
Weight	
Weight empty	9.0 kg (19.00 lb)
Max T-O weight	20.0 kg (44 lb)
Payload, Max payload	8.0 kg (17.00 lb)
Performance	
Altitude, Service ceiling	2,000 m (6,560 ft) (est)
Speed	
Cruising speed, max	86 kt (159 km/h; 99 mph)
Loitering speed	16 kt (30 km/h; 18 mph)
Radius of operation, mission	10.8 n miles (20 km; 12 miles)
Endurance, max	4 hr (est)
Power plant	1 × piston engine

Status: The contractor reported in May 2010 that an Explorer system had been ordered by an unnamed US customer for test and evaluation in law enforcement and security roles.

Contractor: Integrated Dynamics
Karachi.

Integrated Dynamics Hawk

Type: Short-range surveillance tactical UAV.

Development: The current Hawk Mk V evolved from "several years" of field testing of the company's proven Hornet air vehicle.

Description: *Airframe:* High-mounted wings with leading-edge taper; box-section fuselage; sweptback fin and rudder. Twin, fixed, mainwheels and undertail bumper. Composites construction.

Mission payloads: Typical payloads include EO/IR cameras or a GSP-900 surveillance camera with real-time data transmission. Wing hardpoints for external stores.

Hawk Mk V *(Integrated Dynamics)* 1290268

Guidance and control: Pre-programmed, with IFCS-6000 integrated flight control system and GPS navigation. GCS-2000 ground station; ID-TM6 GPS and data telemetry module; ATPS-2000 antenna tracking and positioning system.
Launch: Conventional wheeled take-off.
Recovery: Wheeled landing or parachute recovery.
Power plant: One 16.4 kW (22 hp) two-cylinder two-stroke engine; two-blade tractor propeller. Fuel capacity 25 litres (6.6 US gallons; 5.5 Imp gallons).

Hawk Mk V

Dimensions, External	
Overall	
length	2.95 m (9 ft 8¼ in)
height	0.89 m (2 ft 11 in)
Wings, wing span	4.51 m (14 ft 9½ in)
Areas	
Wings, Gross wing area	3.09 m² (33.3 sq ft)
Weights and Loadings	
Weight	
Weight empty	25.0 kg (55 lb)
Max T-O weight	65.0 kg (143 lb)
Payload, Max payload	15.0 kg (33 lb)
Performance	
Speed	
Cruising speed, max	129 kt (239 km/h; 148 mph)
Loitering speed	71 kt (131 km/h; 82 mph)
Radius of operation	54 n miles (100 km; 62 miles)
Endurance	6 hr
Power plant	1 × piston engine

Status: Continuing to be promoted.
Contractor: Integrated Dynamics
Karachi.

Integrated Dynamics Rover

Type: Close-range mini-UAV.
Development: Unveiled in September 2007. Aimed at providing an affordable civilian system for electronic news gathering and academic research applications. The designation Rover Mk II was later applied.
Description: *Airframe:* Cylindrical fuselage, tapered front and rear; high-mounted wings, forward of which pusher engine is pylon-mounted above fuselage; T tail unit. No landing gear. Composites construction.
Mission payloads: Daylight TV camera standard; IR or digital still camera optional.
Guidance and control: Laptop PC ground station; ID-AP5000 digital autopilot with navigation and stability modules; video and data telemetry modules.
System composition: Four air vehicles, GCS, moving map mission display software, digital spread spectrum telecommand link, antennas and cables.
Launch: Hand launch.
Recovery: Deep stall to belly landing.
Power plant: Electric motor, powered by lithium-ion batteries; two-blade pusher propeller.

Rover

Dimensions, External	
Overall	
length	1.37 m (4 ft 6 in)
height	0.30 m (11¾ in)
Wings, wing span	1.95 m (6 ft 4¾ in)
Areas	
Wings, Gross wing area	0.62 m² (6.7 sq ft)
Weights and Loadings	
Weight	
Weight empty	2.5 kg (5.00 lb) (est)
Max launch weight	5.0 kg (11.00 lb)
Payload, Max payload	1.0 kg (2.00 lb)
Performance	
Altitude	
Operating altitude, typical	610 m (2,000 ft) (est)
Speed	
Cruising speed, max	54 kt (100 km/h; 62 mph)
Loitering speed	16 kt (30 km/h; 18 mph)
Radius of operation, mission	2.7 n miles (5 km; 3 miles) (est)
Endurance	1 hr
Power plant	1 × electric motor

Status: A Rover Mk II system was ordered by a US customer for law enforcement and security tests in 2010.
Contractor: Integrated Dynamics
Karachi.

Integrated Dynamics Shadow

Type: Surveillance tactical UAV.
Development: According to the manufacturer, the Shadow system was developed specifically "to cover a customer requirement". The current version is designated Mk II.
Description: *Airframe:* Typical pod and twin tailboom configuration, with unswept, high-mounted, wire-braced wings and pusher engine. Composites construction, including Kevlar moulded fuselage pans, Kevlar-reinforced equipment bays and side stress panels. Fixed tricycle landing gear has mainwheel legs of aramid-reinforced high-tensile steel.
Mission payloads: Typical payloads include EO/IR, GSP-900 surveillance real-time camera, or experimental mission packages. Available onboard power supply can support a variety of payloads. Within overall capacity, payload combinations can be carried, most of them simultaneously.
Guidance and control: GCS-2000, IFCS-7000 integrated flight control system, ID-TM6 GPS and data telemetry module and ATPS-2000 antenna tracking and positioning system.
System composition: Four air vehicles, GCS, ATPS-2000, programming and moving map mission display software, spares and ground support equipment.
Launch: Conventional wheeled take-off.
Recovery: Conventional wheeled landing or parachute recovery.
Power plant: Typically, one 16.4 kW (22 hp) two-cylinder two-stroke engine with two-blade pusher propeller. Fuel capacity 30 litres (7.9 US gallons; 6.6 Imp gallons).

Shadow Mk II

Dimensions, External	
Overall	
length	2.95 m (9 ft 8¼ in)
height	0.89 m (2 ft 11 in)
Wings, wing span	5.20 m (17 ft 0¾ in)
Areas	
Wings, Gross wing area	3.09 m² (33.3 sq ft)
Weights and Loadings	
Weight	
Weight empty	55.0 kg (121 lb)
Max T-O weight	90.0 kg (198 lb)
Payload, Max payload	25.0 kg (55 lb)
Performance	
Speed	
Cruising speed, max	112 kt (207 km/h; 129 mph)
Loitering speed	40 kt (74 km/h; 46 mph)
Radius of operation	54 n miles (100 km; 62 miles)
Endurance	6 hr
Power plant	1 × piston engine

Status: It is thought that two systems were supplied in 2006 to 'a Spanish government customer' for agricultural research. By 2009 Shadow systems had also been sold to two customers in Australia, one of which is the Australian Research Centre for Aerospace Automation.
Contractor: Integrated Dynamics
Karachi.

The Shadow Mk I short-range surveillance UAV (Integrated Dynamics) 1290269

The Rover hand-launched civilian UAV (Robert Hewson) 1290266

Integrated Dynamics Vision

Type: Short-range surveillance tactical UAV.
Development: Development history not known.
Description: *Airframe:* Pod and single tailboom fuselage; constant-chord, unswept, high wings (with mid-span flaps on Mk II, for low-speed loiter capability); T tail. Fixed tricycle landing gear. Composites construction.

Integrated Dynamics Vision – NDC Vector < Pakistan < Unmanned aerial vehicles

Vision Mk II *(Integrated Dynamics)* 1290272

Vision Mk I *(Integrated Dynamics)* 1290271

Mission payloads: *Mk I:* Surveillance camera module with real-time data transmission via L-band downlink.
Mk II: As Mk I, plus experimental payloads of customer's choice. Payloads can be mixed and most carried simultaneously.
Guidance and control: *Mk I:* UHF command uplink. Stabilisation by ID-AP4 digital gyro autopilot with heading and height lock. Tracking and telemetry via ID-TM6 GPS and data telemetry module with software graphics interface display. Can be remotely piloted or autonomous.
Mk II: Stabilisation by IFCS-7000 integrated flight control system Tracking and telemetry as for Mk I.
Transportation: Tailboom and tail unit detachable for transportation.
System composition: Four air vehicles, portable GCS-1200 ground station, ATPS-1200 antenna tracking and positioning system, programme and moving map mission display software, and GSE-1200 ground support equipment.
Launch: Conventional wheeled take-off (both versions).
Recovery: Conventional wheeled landing or parachute recovery (both versions).
Power plant: *Mk I:* One 4.85 kW (6.5 hp) single-cylinder piston engine and two-blade pusher propeller. Fuel capacity 4.5 litres (1.1 US gallons; 1.0 Imp gallon).
Mk II: One 16.4 kW (22 hp) two-cylinder piston engine and two-blade pusher propeller.
Variants: *Vision Mk I:* Smaller, simpler and shorter-range version.
Vision Mk II: Enlarged from Mk I, with increased fuel and payload and greater range/endurance.

Vision Mk I, Vision Mk II

Dimensions, External	
Overall	
length	
Vision Mk I	2.25 m (7 ft 4½ in)
Vision Mk II	2.90 m (9 ft 6¼ in)
height	
Vision Mk I	0.64 m (2 ft 1¼ in)
Vision Mk II	0.80 m (2 ft 7½ in)
Wings	
wing span	
Vision Mk I	3.22 m (10 ft 6¾ in)
Vision Mk II	4.51 m (14 ft 9½ in)
Areas	
Wings	
Gross wing area	
Vision Mk I	1.44 m² (15.5 sq ft)
Vision Mk II	3.04 m² (32.7 sq ft)
Weights and Loadings	
Weight	
Weight empty	
Vision Mk I	18.0 kg (39 lb)
Vision Mk II	55.0 kg (121 lb)
Max T-O weight	
Vision Mk I	35.0 kg (77 lb)
Vision Mk II	80.0 kg (176 lb)
Payload	
Max payload	
Vision Mk I	10.0 kg (22 lb)
Vision Mk II	15.0 kg (33 lb)
Performance	
T-O	
T-O run, Vision Mk I	20 m (66 ft)
Speed	
Cruising speed, max	112 kt (207 km/h; 129 mph)
Loitering speed	40 kt (74 km/h; 46 mph)
Radius of operation	
mission, optical tracking, Vision Mk I	5.4 n miles (10 km; 6 miles)
mission, GPS/autonomous, Vision Mk I	27 n miles (50 km; 31 miles)
mission, GPS/autonomous, Vision Mk II	54 n miles (100 km; 62 miles)
Endurance	
Vision Mk I	3 hr
Vision Mk II	4 hr
Power plant	1 × piston engine

Status: A Vision Mk I system was supplied to a Spanish government customer for agricultural research in April 2007, and a Vision Mk II system to an Italian research institute in May 2007. The system remained available in 2011.
Contractor: Integrated Dynamics
Karachi.

NDC Vector

Type: Short-range tactical UAV.
Development: Design began in about 1989–90; first flight was made in about 1995 or 1996. Production of the **Mk 1** is believed to have started in 2001. A **Mk 2** was exhibited in Dubai in 2002; performance quoted for this version included a ceiling of 3,660 m (12,000 ft), maximum range of 108 n miles (200 km; 124 miles) and maximum endurance of five hours.
Description: *Airframe:* Typical pod and twin tailboom configuration with shoulder-mounted wings, rectangular tailfins and pusher engine; all-composites construction. Non-retractable tricycle landing gear.
Mission payloads: Daylight or LLTV video camera with real-time imagery transmission. Alternatives could include IR, sigint or chemical detection sensors.
Guidance and control: ID-AP4 digital gyro autopilot with heading and height lock; ID-TM6 GPS and data telemetry module with software graphics interface display.
System composition: Four air vehicles, GCS and 15 personnel.
Launch: Wheeled take-off.
Recovery: Wheeled landing or parachute recovery.
Power plant: One 18.6 kW (25 hp) two-cylinder piston engine; two-blade pusher propeller. Fuel capacity 25 litres (6.6 US gallons; 5.5 Imp gallons).

Vector Mk 1

Dimensions, External	
Overall	
length	3.54 m (11 ft 7¼ in)
height	1.04 m (3 ft 5 in)
Wings, wing span	7.09 m (23 ft 3¼ in)
Areas	
Wings, Gross wing area	4.78 m² (51.5 sq ft)
Weights and Loadings	
Weight	
Weight empty	66 kg (145 lb)
Max T-O weight	105 kg (231 lb)
Performance	
T-O	
T-O run	37 m (120 ft)
Altitude, Service ceiling	4,570 m (15,000 ft)
Speed	
Max level speed	111 kt (206 km/h; 128 mph)
Loitering speed	40 kt (74 km/h; 46 mph)
Range	
optical tracking	5.4 n miles (10 km; 6 miles)
GPS/autonomous	81 n miles (150 km; 93 miles)
Endurance	4 hr 30 min

Status: Domestic deliveries reported to have begun in 2001. Available for export, but no such customers known by 2010.
Customers: Pakistan Army.
Contractor: National Development Complex
Islamabad.

The NDC Vector surveillance UAV 0131504

Satuma Flamingo

Type: Medium-range tactical UAV.

Development: The Flamingo UAV has been developed as a larger and more capable derivative of the company's Jasoos.

Description: *Airframe:* High wings with turned-down tips; box-section fuselage pod; twin tailbooms, fins and rudders; pusher engine. Fixed tricycle landing gear. Composites construction.

   Mission payloads: Promoted with Satuma iHawk Gen 2 sensor (daylight pan-tilt-zoom video camera with on-screen position display).

   Guidance and control: Pre-programmed or remotely operated; updatable in flight; auto navigation and tracking. Dual real-time datalinks. Two-console, truck-mounted GCS includes multiple computers, moving map display, 'point and click' mission planning, joystick control of onboard camera, and video recorder.

   Launch: Conventional wheeled take-off.

   Recovery: Conventional wheeled landing.

   Power plant: One (unidentified) 37.3 to 44.7 kW (50 to 60 hp) four-cylinder two-stroke engine; two-blade pusher propeller.

Flamingo

Dimensions, External	
Overall, length	5.18 m (17 ft 0 in)
Wings, wing span	7.315 m (24 ft 0 in)
Weights and Loadings	
Weight, Max T-O weight	245 kg (540 lb)
Payload, Max payload	35 kg (77 lb)
Performance	
Altitude, Service ceiling	4,265 m (14,000 ft)
Speed, Max level speed	70 kt (130 km/h; 81 mph)
Radius of operation	108 n miles (200 km; 124 miles) (est)
Endurance	8 hr
Power plant	1 × piston engine

Status: Being promoted.

Contractor: Surveillance and Target Unmanned Aircraft (Satuma) Islamabad.

Flamingo is currently the largest of Satuma's assorted family of UAVs *(Satuma)* 1356205

The Flamingo medium-range TUAV *(Satuma)* 1356206

Satuma Jasoos

Type: Medium-range tactical UAV.

Development: The Jasoos system is understood to be a development of the Air Weapons Complex Bravo+ which has been in Pakistan Air Force service since 2004.

Description: *Airframe:* High-mounted wings; box-section fuselage; twin tailbooms, fins and rudders; pusher engine. Fixed tricycle landing gear. Composites construction.

   Mission payloads: Customer-specified. Also promoted with Satuma iHawk Gen 1 or Gen 2 pan-tilt-zoom video camera.

   Guidance and control: Pre-programmed or remotely operated; auto navigation and tracking. Dual real-time datalinks. Truck-mounted GCS with two consoles, multiple computers, 'point and click' mission planning, in-flight update capability, moving map display, joystick control of onboard sensor, real-time video panel with flight status display, and video recorder.

   Launch: Conventional wheeled take-off.

   Recovery: Conventional wheeled landing.

   Power plant: One (unidentified) 28.3 kW (38 hp) two-cylinder two-stroke engine; two-blade pusher propeller.

Variants: *Jasoos:* Operational version; *as described.*

Jasoos air vehicle with its ground control station truck *(Satuma)* 1356208

Jasoos air vehicle — possibly the prototype *(Satuma)* 1356207

FST: Full-scale trainer, for training of GCS crews; flown within visual range only; portable data display station. Jasoos airframe, but with sensor payload omitted; remote control only, plus wing leveller.

Jasoos, Jasoos FST

Dimensions, External	
Overall, length	4.27 m (14 ft 0 in)
Wings, wing span	4.92 m (16 ft 1¾ in)
Weights and Loadings	
Weight	
Max T-O weight	
Jasoos	145 kg (319 lb)
Jasoos FST Vulture	110 kg (242 lb)
Payload, Max payload	25.0 kg (55 lb)
Performance	
Altitude, Service ceiling	3,050 m (10,000 ft)
Speed, Max level speed	70 kt (130 km/h; 81 mph)
Radius of operation, mission	54 n miles (100 km; 62 miles)
Endurance, max	5 hr

Status: Jasoos and FST in production and in service.

   Customers: Pakistan Air Force.

Contractor: Surveillance and Target Unmanned Aircraft (Satuma) Islamabad.

Satuma Mukhbar

Type: Short-range tactical UAV.

Development: Scaled-down, shorter-range version of Jasoos (which see).

Description: *Airframe:* High-wing, pod and twin tailboom design with twin rectangular fins and rudders; pusher engine. Fixed tricycle landing gear. Composites construction.

   Mission payloads: Customer-specified EO.

   Guidance and control: Generally as described for Jasoos.

   Launch: Conventional wheeled take-off.

   Recovery: Conventional wheeled landing.

   Power plant: One (unidentified) 120 cm^3 two-cylinder two-stroke engine; two-blade pusher propeller.

Pakistan Air Force Mukhbar, a smaller-scale version of the Jasoos II *(Pakistan MoD)* 1356209

The Stingray flying-wing mini-UAV *(Satuma)* 1356212

HST Parwaz operator training variant of the Mukhbar *(Satuma)* 1356211

Stingray in flight *(Satuma)* 1356213

Variants: ***Mukhbar:*** Operational version; *as described*.
 HST: Half-scale trainer; simplified version for operator training. Remote operation only. Pakistan Air Force name **Parwaz**.

Mukhbar

Dimensions, External	
Overall, length	2.86 m (9 ft 4½ in)
Wings, wing span	3.56 m (11 ft 8¼ in)
Weights and Loadings	
Weight, Max T-O weight	40 kg (88 lb)
Payload, Max payload	5.0 kg (11.00 lb)
Performance	
Altitude, Service ceiling	2,135 m (7,000 ft)
Speed, Max level speed	64 kt (119 km/h; 74 mph)
Radius of operation, mission	27 n miles (50 km; 31 miles)
Endurance, max	1 hr 30 min

Status: Mukhbar and HST in production and in service.
 Customers: Pakistan Air Force.
Contractor: Surveillance and Target Unmanned Aircraft (Satuma) Islamabad.

Satuma Stingray

Type: Short-range mini-UAV.
Development: First flight of the Stingray was made in July 2008.
Description: ***Airframe:*** Sweptback flying wing with endplate fins; pusher engine. No landing gear. Composites construction.
 Mission payloads: No details known.
 Guidance and control: Pre-programmed, with GPS-based autotracking. Portable GCS, with dual real-time datalinks.
 Launch: By bungee catapult.
 Recovery: Parachute recovery.
 Power plant: One BLDC outrunner electric motor (output not stated); two-blade pusher propeller.

Stingray

Dimensions, External	
Overall	
length	1.50 m (4 ft 11 in)
height	0.40 m (1 ft 3¾ in)
Wings, wing span	3.00 m (9 ft 10 in)
Weights and Loadings	
Weight, Max launch weight	7.5 kg (16.00 lb)
Payload, Max payload	1.5 kg (3.00 lb)
Performance	
Speed, Max level speed	48 kt (89 km/h; 55 mph)
Endurance	1 hr
Power plant	1 × electric motor

Status: Being promoted.
Contractor: Surveillance and Target Unmanned Aircraft (Satuma) Islamabad.

Poland

ITWL Koliber

Type: Close-range surveillance mini-UAV.
Development: The quad-rotor Koliber (Hummingbird) is a kit-built system for urban-area surveillance on behalf of military, police, border guard, fire and rescue service agencies.
Description: ***Airframe:*** Barrel-shaped central module for power plant and payload. Four arms at 90° intervals, each with small two-blade rotor at its tip. Fixed tripod landing gear.
 Mission payloads: Video camera with single-axis FoV. Data and imagery downlinks.
 Guidance and control: Fully automatic (autopilot with mission planning capability) or manual control; GPS navigation.
 Transportation: Man-portable.
 Launch: Vertical take-off.
 Recovery: Vertical landing. Emergency parachute recovery.
 Power plant: Electric motor.

An in-flight view of the Koliber mini-UAV showing the type's four rotor arms and tripod landing gear *(ITWL)* 1509990

Koliber

Dimensions, External	
Overall	
length	0.87 m (2 ft 10¼ in)
height	0.32 m (1 ft 0½ in)
width	0.87 m (2 ft 10¼ in)
Weights and Loadings	
Weight, Max launch weight	3.1 kg (6.00 lb)
Performance	
Speed	
Cruising speed, max	32 kt (59 km/h; 37 mph)
Radius of operation	1.6 n miles (3 km; 1 miles)
Endurance	25 min
Power plant	1 × electric motor

Status: Most recently, the ITWL has continued to promote the Koliber mini-UAV.
Contractor: Instytut Techniczny Wojsk Lotniczych (Air Force Institute of Technology), Warsaw.

ITWL Nietoperz-3L

Type: Close-range mini-UAV.
Development: The Nietoperz-3L (Bat-3L) is characterised as being a 'mobile, light unmanned system' that is suitable for military and civilian applications and which provides 'air-surveillance and reconnaissance of [terrestrial] objects [within a 8.1 n miles (15 km; 9.3 miles) radius]'.
Description: *Airframe:* The Nietoperz-3L features high-mounted swept wings (with endplate fins) and a circular-section, tapering fuselage. Again, the design has no horizontal tail surfaces or landing gear.
   Mission payloads: The Nietoperz-3L incorporates a sensor turret that can house 'controlled and stabilised' daylight video or IR cameras or a digital camera with a real-time image transmission system.
   Guidance and control: AV guidance and control is by means of a laptop GCS via an autopilot. There is provision for pre-programming and manual override if required.
   Transportation: Man-portable.
   Launch: Hand launch with bungee catapult.
   Recovery: Belly landing or parachute recovery.
   Power plant: Electric motor, powered by lithium-polymer batteries; rating of up to 1.5 kW (2.0 hp), depending on battery fit (22.2 V, 10 to 15 Ah). Two-blade pusher propeller.

Nietoperz-3L

Dimensions, External	
Overall, length	1.28 m (4 ft 2½ in)
Wings, wing span	2.41 m (7 ft 11 in)
Weights and Loadings	
Weight	6.5 kg (14.00 lb)
Performance	
Altitude, Operating altitude	100 m to 1,000 m (320 ft to 3,280 ft)
Speed	
Cruising speed	
lower limit	27 kt (50 km/h; 31 mph)
upper limit	54 kt (100 km/h; 62 mph)
-	8.1 n miles (15 km; 9 miles)
Endurance, max	1 hr 30 min
Power plant	1 × electric motor

Status: Most recently, the ITWL has continued to promote the Nietoperz-3L close-range mini-UAV.
Contractor: Instytut Techniczny Wojsk Lotniczych (Air Force Institute of Technology) Warsaw.

An in-flight view of the Nietoperz-3L close-range mini-UAV *(ITWL)*

ITWL UAV surveillance system

Type: Close-range surveillance mini-UAV.
Development: This mini-UAV has been designed for shipboard use by the Polish Navy.

An artist's impression of the ITWL's naval UAV surveillance system AV *(ITWL)*

Description: *Airframe:* Low-mounted wings with central fuselage nacelle; pusher engine; twin tailbooms, with fintips bridged by single tailplane. No landing gear.
   Mission payloads: Stabilised turret containing either a digital EO camera or an IR imager. Real-time imagery downlink.
   Guidance and control: Pre-programmed autopilot flight, with operator take-over capability.
   Launch: Rail launch.
   Recovery: Net barrier recovery. Also capable of safe splashdown landings.
   Power plant: Single piston engine; two-blade pusher propeller.

ITWL UAV surveillance system

Dimensions, External	
Overall, length	2.15 m (7 ft 0¾ in)
Wings, wing span	2.70 m (8 ft 10¼ in)
Weights and Loadings	
Weight, Max launch weight	18 kg (39 lb)
Performance	
Speed	
Max level speed	73 kt (135 km/h; 84 mph)
Cruising speed	49 kt (91 km/h; 56 mph)
Radius of operation	10.8 n miles (20 km; 12 miles)
Endurance	1 hr 30 min
Power plant	1 × piston engine

Status: Most recently, the ITWL has continued to promote the described UAV surveillance system.
Contractor: Instytut Techniczny Wojsk Lotniczych (Air Force Institute of Technology) Warsaw.

Flytronic FlyEye

Type: Surveillance mini-UAV.
Development: Unveiled at the June 2010 Eurosatory defence exhibition (Paris), Flytronic's FlyEye mini-UAV is designed for a range of applications including artillery spotting, convoy overwatch, border surveillance, disaster response, search and rescue and event monitoring. In terms of programme events, the architecture was awarded an Amber Medallion in the June 2012 Balt Military Expo's Grand Prix contest. The FlyEye has a real-time video transmission range of up to 26 n miles (50 km; 31 miles) from its GCS and can operate in temperatures of between –20°C and +50°C (–4°F to +122°F) and a humidity value of up to 95% at a temperature of 50°C (122°F).
Description: *Airframe:* The FlyEye mini-UAV features high-mounted wings (with outer panel dihedral), a pod and boom fuselage and conventional tail surfaces. Again, the AV is not fitted with a landing gear and is constructed from composites.
   Mission payloads: The FlyEye AV is equipped with a twin imager (daylight EO (with optical zoom) and IR according to *IHS Jane's* sources) installation and a real-time video and telemetry downlink. The type's video output can be switch between the two sensors as required.

The FlyEye AV as exhibited at the June 2010 Eurosatory trade show *(IHS/Patrick Allen)*

An impression of the FlyEye AV in-flight *(Flytronic)* 1395247

Guidance and control: Pre-programmed (with GPS navigation, in-flight updating and provision for manual control). Optional 'follow the camera' (to maintain focus on a target) and 'hold over the target' flight modes. Automatic 'return home' if communication link is lost. Light GCS. System can be operated by a crew of two.

Transportation: Man-portable (air vehicle backpack). Set-up time less than 10 minutes.

Launch: Hand launch.

Recovery: Automatic. Belly landing or parachute recovery, depending upon payload carried.

Power plant: Electric motor (powered by rechargeable lithium-polymer batteries). Two-blade propeller.

FlyEye

Dimensions, External	
Overall, length	1.9 m (6 ft 2¾ in)
Wings, wing span	3.6 m (11 ft 9¾ in)
Weights and Loadings	
Weight, Max launch weight	11.0 kg (24 lb)
Payload, Max payload	4.0 kg (8.00 lb)
Performance	
Altitude	
Operating altitude	1,000 m (3,280 ft)
Service ceiling	4,000 m (13,120 ft)
Speed	
Max level speed	65 kt (120 km/h; 75 mph)
Max, without payload	27 kt (50 km/h; 31 mph)
Radius of operation	
Min from GCS and under control	5 n miles (9 km; 5 miles)
Max, from GCS and under radio control	27 n miles (50 km; 31 miles)
Min, beyond LOS radio range	81 n miles (150 km; 93 miles)
Max, beyond LOS radio range	108 n miles (200 km; 124 miles)
Upto, weather dependent	3 hr
Power plant	1 × electric motor

Status: Most recently, Flytronic has continued to promote the FlyEye mini-UAV.

Contractor: Flytronic Sp. zo.o. (a part of the WB Group), Gliwice.

Russian Federation

A-Level Aerosystems ZALA 421-02

Type: Small helicopter UAV.

Development: A-Level Aerosystems began developing its range of small fixed- and rotary-wing tactical UAVs in 2003, all with the objective of low-cost, autonomous civil or military operation; at least eight different designs have been announced. In 2007, the company was also developing an 'intellectual' system designed to need no human interface and not be reliant on GPS.

Description: *Airframe:* Two-blade main rotor and two-blade, starboard-side tail rotor. Pod and boom all-composites fuselage; upper and lower tailfins on 421-05H. Fixed, twin mainwheels and tailwheel landing gear on 421-02; twin-skid landing gear on 421-05H.

Mission payloads: Daytime video camera in standard configuration. Able to carry other payloads, with real-time data downlinks, for such civil and military missions as Earth remote sensing, oil and gas pipeline inspection, aerial survey, ground and maritime surveillance, targeting and communications relay, according to customer's requirements.

Guidance and control: All ZALA 421 series UAVs, both fixed- and rotary-wing, utilise a common GCS. They can fulfil their entire mission automatically under control of the flight computer and A-Level Vostok 228 autopilot. Alternatively, the operator can assume partial or full control at any point during the flight, and can add, delete or modify waypoints via a clearly arranged interface. The air vehicles are stabilised automatically in all three control modes, and all have a 'return home' capability if the command signal is lost and emergency recovery in the event of battery failure. The suitcase-sized GCS is based on two portable computers – one for flight and mission planning and control, the other for payload control and data analysis. The command system is designed to control the UAV via either a standalone network or through real-time integration

ZALA 421-05H mini VTUAV *(A-Level Aerosystems)* 1290237

The ZALA 421-02 helicopter UAV *(A-Level Aerosystems)* 1290234

into an existing tactical network. Software provides access to any data required in real time.

Launch: Conventional, automatic, helicopter take-off.

Recovery: Conventional, automatic, helicopter landing.

Power plant: ZALA 421-02: One 14.9 kW (20 hp) two-cylinder two-stroke engine with dual electric ignition.

Variants: *ZALA 21-02:*
Full-size helicopter version.
ZALA 421-05H:
Miniature version, introduced in 2007.

ZALA 21-02, ZALA 421-05H

Dimensions, External	
Overall height	
ZALA 21-02	0.94 m (3 ft 1 in)
ZALA 421-05H	0.67 m (2 ft 2½ in)
Fuselage length	
ZALA 21-02	2.64 m (8 ft 8 in)
ZALA 421-05H	1.57 m (5 ft 1¾ in)
width	
max, ZALA 21-02	0.67 m (2 ft 2½ in)
max, ZALA 421-05H	0.40 m (1 ft 3¾ in)
Rotors	
rotor diameter	
ZALA 21-02	3.065 m (10 ft 0¾ in)
ZALA 421-05H	1.77 m (5 ft 9¾ in)
Weights and Loadings	
Weight	
Weight empty	
ZALA 21-02	40 kg (88 lb)
ZALA 421-05H	6.0 kg (13.00 lb)
Max T-O weight	
ZALA 21-02	95 kg (209 lb)
ZALA 421-05H	10.0 kg (22 lb) (est)
Payload	
Max payload	
ZALA 21-02	50 kg (110 lb)
ZALA 421-05H	3.5 kg (7.00 lb)

ZALA 21-02, ZALA 421-05H

Performance	
Altitude	
Service ceiling	
ZALA 21-02	4,000 m (13,120 ft)
ZALA 421-05H	2,000 m (6,560 ft)
Speed	
Max level speed	
ZALA 21-02	81 kt (150 km/h; 93 mph)
ZALA 421-05H	75 kt (139 km/h; 86 mph)
Cruising speed, max	43 kt (80 km/h; 49 mph)
Radius of operation	
mission, ZALA 21-02	27 n miles (50 km; 31 miles)
Endurance	
ZALA 21-02	6 hr
ZALA 421-05H	3 hr

Status: Prototypes completed by mid-2007. As at early 2008, validation and flight testing of pre-production 421-02 was in progress. Customers are believed to include Gazprom and other Russian utility agencies. Production discontinued by 2011.

Contractor: A-Level Aerosystems (Zala Aero)
Izhevsk.

A-Level Aerosystems ZALA 421-04

Type: Short-range reconnaissance and surveillance mini-UAV.
Development: Refer to entry for ZALA 421-02.
Description: *Airframe:* All-swept flying wing configuration, with triangular endplate fins and rectangular trailing-edge cutout. No landing gear. Composites construction.
  Mission payloads: No details known.
  Guidance and control: Compact GCS. Autonomous operation, including launch and recovery. For further details, see entry for ZALA 421-02.
  Transportation: Transportable in a 900 × 900 × 300 mm (35.4 × 35.4 × 11.8 in) container.
  Launch: Catapult launch.
  Recovery: Belly landing.
  Power plant: One 1.5 kW electric motor, driving a two-blade propeller. Piston engine optional.

ZALA 421-04

Dimensions, External	
Overall	
length	1.05 m (3 ft 5¼ in)
height	0.15 m (6 in)
Wings, wing span	2.23 m (7 ft 3¾ in)
Weights and Loadings	
Weight	
Weight empty	
electric motor	6.95 kg (15.00 lb)
piston engine	7.9 kg (17.00 lb)
Max launch weight	
electric motor	8.0 kg (17.00 lb)
piston engine	9.0 kg (19.00 lb)
Payload, Max payload	1.0 kg (2.00 lb)
Performance	
Altitude, Service ceiling	3,000 m (9,840 ft)
Speed	
Cruising speed, max	76 kt (141 km/h; 87 mph)
Loitering speed	32 kt (59 km/h; 37 mph)
Radius of operation	
mission, pin antenna	5.4 n miles (10 km; 6 miles) (est)
directional antenna	13.5 n miles (25 km; 15 miles) (est)
servo-control directional antenna	22 n miles (40 km; 25 miles) (est)
Endurance	
max, electric motor	1 hr
max, piston engine	3 hr
Power plant	1 × electric motor
optional	1 × piston engine

Status: The ZALA 421-04 entered service with the Russian Interior Ministry in 2006, when it was used for security overflights of the G8 summit conference in St Petersburg. More may have been ordered in 2007, and promotion was continuing in 2008. In January a version with folding wings designated 421-07 was ordered by the Russian Ministry of Public Defence and Emergency Management for search and rescue, with delivery scheduled for mid-2009. A version designated 421-04M was being listed from 2010, which appears to be a highly revised or new design.

Contractor: A-Level Aerosystems (Zala Aero)
Izhevsk.

A-Level Aerosystems ZALA 421-06

Type: Small helicopter UAV.
Development: Development believed to have started in about 2007. Flight testing and demonstration in 2008.
Description: *Airframe:* Two-blade main rotor and two-blade, starboard-side tail rotor. Pod and boom all-composites fuselage; sweptback upper and lower tailfins. Fixed, twin mainwheels and tailwheel landing gear. A version with skid landing gear has also been shown.
  Mission payloads: Interchangeable payloads include video or IR camera in undernose turret on gyrostabilised gimbal.
  Guidance and control: Autonomous operation, including take-off and landing; GPS/GLONASS navigation. For further details, see entry for ZALA 421-02.
  Launch: Conventional, automatic, helicopter take-off.
  Recovery: Conventional, automatic, helicopter landing.
  Power plant: Option of either electric motor or piston engine; ratings not stated.

ZALA 421-06

Dimensions, External	
Overall, height	0.67 m (2 ft 2½ in)
Fuselage	
length	1.57 m (5 ft 1¾ in)
width	0.40 m (1 ft 3¾ in)
Rotors, rotor diameter	1.77 m (5 ft 9¾ in)
Weights and Loadings	
Weight, Max T-O weight	12.5 kg (27 lb)
Payload, Max payload	2.0 kg (4.00 lb)
Performance	
Altitude, Service ceiling	2,000 m (6,560 ft)
Speed	
Max level speed	37 kt (69 km/h; 43 mph)
Cruising speed	27 kt (50 km/h; 31 mph)
Radius of operation	8.1 n miles (15 km; 9 miles) (est)
Endurance	
electric	40 min
piston engine	1 hr 30 min
Power plant	1 × piston engine

ZALA 421-04 short-range tactical mini-UAV *(A-Level Aerosystems)*

ZALA 421-06 *(A-Level Aerosystems)*

Status: Trials with emergency services successfully completed in 2008; shipboard trials on an icebreaker in Russian Arctic July 2008. Ten reportedly completed by mid-2008; some possibly to Russian police. Marketing continuing in 2011.
Contractor: A-Level Aerosystems (Zala Aero)
Izhevsk.

A-Level Aerosystems ZALA 421-08

Type: Short-range reconnaissance and surveillance MAV.
Development: Refer to entry for ZALA 421-02. Scaled-down ZALA 421-11 version introduced in 2007.
Description: *Airframe:* Flying wing configuration, with leading-edge sweepback and endplate fins; minimal fuselage. No landing gear. Composites construction.
   Mission payloads: Forward- (black and white) and side-looking (colour) gyrostabilised cameras (video or IR). Onboard storage of imagery. Steerable two-axis stabilised camera available.
   Guidance and control: Compact GCS. Autonomous operation. Capability for simultaneous control of more than one air vehicle. For further details, see entry for ZALA 421-02.

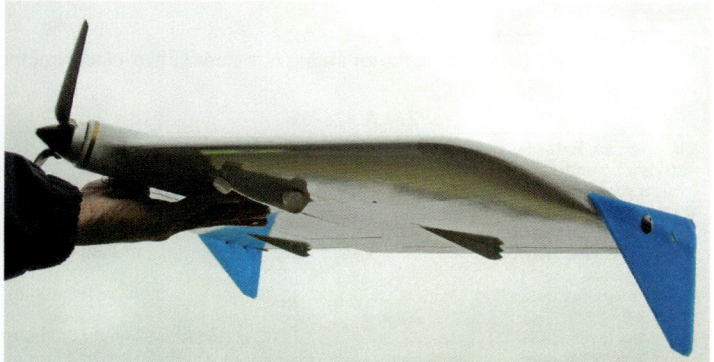

A ZALA 421-08 'bird in the hand' (IHS/Patrick Allen) 1326937

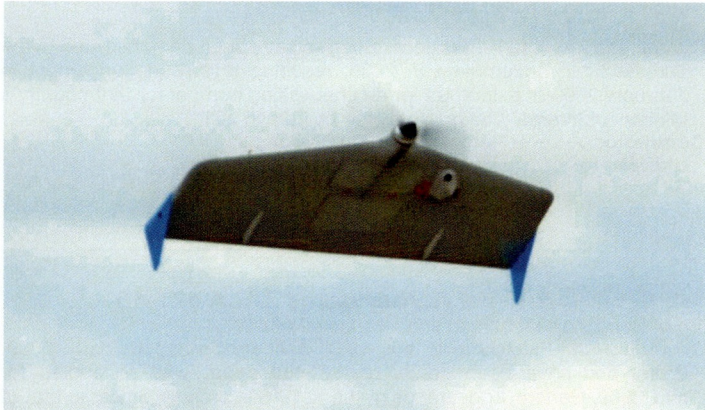

The ZALA 421-08 in flight at Aberporth in the UK
(IHS/Patrick Allen) 1326924

The ZALA 421-08, its operators, the suitcase GCS and a self-powered hand launch (IHS/Patrick Allen) 1326923

421-08 parachute recovery (IHS/Patrick Allen) 1327045

Transportation: Complete 421-08 system transportable in an 820 × 450 × 210 mm (32.3 × 17.7 × 8.3 in) backpack; total weight 9 kg (19.8 lb).
   System composition: Two air vehicles, compact GCS, reserve UAV power supply and backpack container, plus two operators.
   Launch: Hand launch under own power.
   Recovery: Belly landing, or by parachute where recovery space is limited. Aircraft has 'return home' capability if command signal is lost.
   Power plant: One electric motor, driving a two-blade propeller.
Variants: ZALA 421-08: Initial version; *as described.*
   ZALA 421-11: Approximately half-size version; first flight May 2007.

ZALA 421-08, ZALA 421-11

Dimensions, External	
Overall	
length, ZALA 421-08	0.41 m (1 ft 4¼ in)
height, ZALA 421-08	0.07 m (2¾ in)
Wings	
wing span	
ZALA 421-08	0.80 m (2 ft 7½ in)
ZALA 421-11	0.40 m (1 ft 3¾ in)
Weights and Loadings	
Weight	
Weight empty, ZALA 421-08	1.7 kg (3.00 lb)
Max launch weight	
ZALA 421-08	1.9 kg (4.00 lb)
ZALA 421-11	0.79 kg (1.00 lb)
Payload	
Max payload, ZALA 421-08	0.2 kg (0.00 lb)
Performance	
Altitude	
Operating altitude, ZALA 421-08	15 m to 3,600 m (40 ft to 11,820 ft)
Speed	
Max level speed, ZALA 421-11	81 kt (150 km/h; 93 mph)
Cruising speed	
max, ZALA 421-08	81 kt (150 km/h; 93 mph)
max, ZALA 421-11	54 kt (100 km/h; 62 mph)
Loitering speed	
ZALA 421-08, ZALA 421-11	32 kt (59 km/h; 37 mph)
Radius of operation, ZALA 421-08	2.7 n miles (5 km; 3 miles) (est)
Endurance	
ZALA 421-08	1 hr
ZALA 421-11	30 min
Power plant	1 × electric motor

Status: The ZALA 421-08 is understood to have entered service with the Russian Interior Ministry in 2006, and more may have been ordered in 2007. The system was also displayed at the UK's ParcAberporth UAV centre in July 2007, when a pair were flown together. One was recovered conventionally by parachute; the other then demonstrated self-recovery to the launch area after (deliberate) loss of the command signal. The 421-08 continued to be promoted in 2011; the 421-011 may have been withdrawn. The company say to 421-08 has been in large-scale production and is in widespread service with Russian government agencies and abroad.
Contractor: A-Level Aerosystems (Zala Aero)
Izhevsk.

A-Level Aerosystems ZALA 421-09

Type: Endurance tactical UAV.
Development: Introduced in 2009.
Description: *Airframe:* Fuselage pod with front-mounted engine; high-mounted wings, from which twin tailbooms extend to support inverted V tail unit. Fixed landing skids in 'tricycle' arrangement. Composites construction.

ZALA 421-09 in flight *(A-Level Aerosystems)* 1378873

The ZALA 421-09 has an endurance of more than 10 hours
(A-Level Aerosystems) 1356218

   Mission payloads: Standard payloads, which are interchangeable, include still, video or thermal imaging cameras, and radar, on a gyrostabilised gimbal. Options include gas analysers or other specialised modules to customer's requirements.
   Guidance and control: All A-Level Aerosystems UAVs, whether fixed- or rotary-wing, utilise a common GCS, a description of which is given under the entry for the ZALA 421-02. The 421-09 has GPS/GLONASS satellite navigation.
   Launch: Conventional runway take-off; may also be capable of catapult launch.
   Recovery: Skid landing.
   Power plant: One flat-twin piston engine (type and rating not stated); two-blade tractor propeller.

ZALA 421-09

Dimensions, External	
Overall, length	2.50 m (8 ft 2½ in)
Wings, wing span	3.90 m (12 ft 9½ in)
Weights and Loadings	
Weight, Max T-O weight	70 kg (154 lb)
Payload, Max payload	10.0 kg (22 lb)
Performance	
Speed	
Max level speed	70 kt (130 km/h; 81 mph)
Cruising speed	49 kt (91 km/h; 56 mph)
Radius of operation	135 n miles (250 km; 155 miles)
Endurance	10 hr 30 min
Power plant	1 × piston engine

Status: Being promoted from early 2009, discontinued by 2011.
Contractor: A-Level Aerosystems (Zala Aero)
  Izhevsk.

A-Level Aerosystems ZALA 421-12
Type: Short-range reconnaissance and surveillance mini-UAV.
Development: Developed in or by 2008; originally designated 421-04M.
Description: *Airframe:* Sweptback flying wing configuration, with dependent, triangular endplate fins. No landing gear. Composites construction.
   Mission payloads: Built-in, gyrostabilised digital stills camera, interchangeable with video or IR camera modules as required.
   Guidance and control: All A-Level Aerosystems UAVs, whether fixed- or rotary-wing, utilise a common GCS, a description of which is given in the entry for the ZALA 421-02. The 421-12 has GPS/GLONASS satcom navigation.

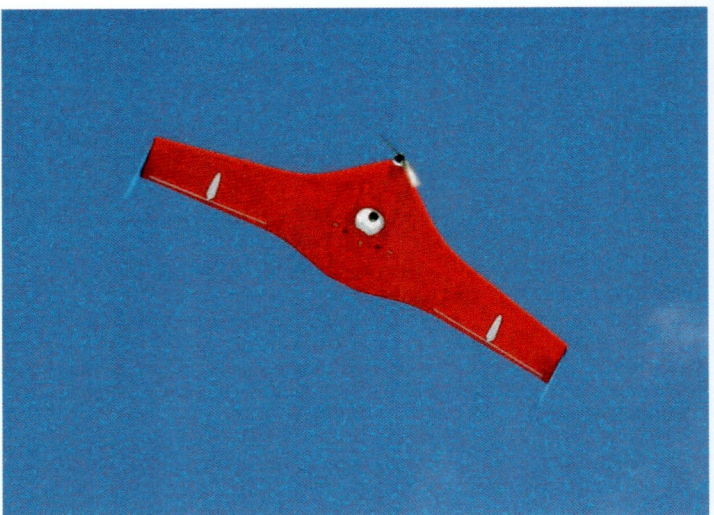

ZALA 421-12 in flight *(A-Level Aerosystems)* 1378872

   Launch: By bungee or pneumatic catapult. Can be launched in wind speeds of up to 10 m/s (22 mph).
   Recovery: Parachute recovery.
   Power plant: One electric motor (rating not stated); two-blade tractor propeller.

ZALA 421-12

Dimensions, External	
Overall, length	0.62 m (2 ft 0½ in)
Wings, wing span	1.60 m (5 ft 3 in)
Weights and Loadings	
Weight, Max launch weight	4.2 kg (9.00 lb)
Payload, Max payload	1.0 kg (2.00 lb)
Performance	
Altitude, Service ceiling	3,595 m (11,800 ft)
Speed	
Max level speed	64 kt (119 km/h; 74 mph)
Cruising speed	35 kt (65 km/h; 40 mph)
Radius of operation, mission	22 n miles (40 km; 25 miles) (est)
Endurance, max	2 hr
Power plant	1 × electric motor

Status: The ZALA 421-12 is understood to have been demonstrated to the Turkmenistan government in 2008, resulting in the announcement in February 2009 of a contract for an unspecified number for that country's Ministry of Internal Affairs. Not listed by Zala in 2011.
Contractor: A-Level Aerosystems (Zala Aero)
  Izhevsk.

A-Level Aerosystems ZALA 421-16
Type: Reconnaissance and surveillance mini-UAV.
Development: Existence of this UAV was announced in early 2009. Prototypes have been flown, and public debut was scheduled for mid-2009. General appearance of the 421-16 is broadly similar to that of the ZALA 421-03.
Description: *Airframe:* Tapered, sweptback flying wing configuration, with much-increased sweep on inboard leading-edge; dependent fin at each wingtip. No landing gear. Composites construction.
   Mission payloads: High-resolution, gyrostabilised video and IR cameras, plus a built-in 10 Megapixel digital camera. Onboard data recording and encrypted downlink. Tests planned with an atmosphere analysing module for applications in the civil sector.
   Guidance and control: As described in entry for ZALA 421-02.
   Launch: Launched from mobile catapult on ground or on board ship.
   Recovery: Parachute recovery standard; net recovery optional.
   Power plant: Brushless electric motor, driving a two-blade tractor propeller.

ZALA 421-16

Dimensions, External	
Overall, length	0.62 m (2 ft 0½ in)
Wings, wing span	1.65 m (5 ft 5 in)
Weights and Loadings	
Weight, Max launch weight	15.0 kg (33 lb)
Payload, Max payload	3.0 kg (6.00 lb)
Performance	
Altitude, Service ceiling	3,600 m (11,810 ft)
Speed	
Cruising speed, max	81 kt (150 km/h; 93 mph)
Loitering speed	38 kt (70 km/h; 44 mph)
Radius of operation	
datalink, mission	37 n miles (68 km; 42 miles)
Endurance, max	5 hr (est)
Power plant	1 × electric motor

A-Level Aerosystems ZALA 421-21

Type: VTOL micro-UAV.
Development: Launch of the six-rotor ZALA 421-21 was announced on 8 June 2010, by which time flight testing had already begun. The UAV can land on structures ('perch and stare') to preserve power, while still monitoring and 'listening in' on a target.
Description: *Airframe:* Central module housing power plant and avionics, with payload suspended beneath. Six arms radiate from the module, each supporting a two-blade rotor at its outer end. Twin-skid landing gear.
Mission payloads: Stills, video or IR camera, with real-time imagery transmission. Combined day video/IR payload also planned.
Guidance and control: Fully autonomous operation with GPS navigation in open areas; or semi-autonomous, without GPS, for use inside buildings.
Transportation: Man-portable.
Launch: Vertical take-off.
Recovery: Vertical landing.
Power plant: Electric motor (power rating not stated).

ZALA 421-21

Weights and Loadings	
Payload, Max payload	0.5 kg (1.00 lb)
Performance	
Altitude, Operating altitude	10 m to 1,000 m (40 ft to 3,280 ft)
Radius of operation, mission	2.7 n miles (5 km; 3 miles)
Endurance, max	25 min
Power plant	1 × electric motor

Status: Development continuing in mid-2010 with the objective of increasing mission radius to 8.1 n miles (15 km; 9.3 miles) and endurance to 40 minutes. Planned to become available before the end of 2010.
Contractor: A-Level Aerosystems (Zala Aero)
Izhevsk.

The multirotor ZALA 421-21 *(A-Level Aerosystems)* 1395226

ENICS Berta

Type: Multirole tactical UAV.
Development: No development history provided.
Description: *Airframe:* Canard layout, with small foreplane; mid-mounted main wing at rear of blunt-nosed cylindrical fuselage. Twin fins mounted at approximately mid-span on main wing, projecting aft of the trailing-edge; sweptback underfin. No landing gear.
Mission payloads: Standard payloads include colour TV, digital or IR cameras, signal repeater, jamming system, IR flares and decoys, Luneberg lens and corner reflector. Other customer-specified payloads

Pulsejet version of the Berta UAV *(ENICS)* 1395217

Berta on its pneumatic launcher *(ENICS)* 1395218

E95Y ground control station truck, with a launch-ready Berta alongside 0 *(ENICS)* 1395219

can also be carried. Video downlink has a range of 32 n miles (60 km; 37 miles).
Guidance and control: Manual or automatic flight modes; GPS and GLONASS navigation. GCS has two operator consoles and is installed in a truck based on a GAZ-3308 four-wheel drive chassis. Two air vehicles can be controlled simultaneously, or up to four by interlinking with a second station.
System composition: Up to four air vehicles; payloads; GCS truck; launcher. As is typical among Russian unmanned systems, different elements have separate designations. In the Berta system, the air vehicle is designated E08 and launcher is E08P; the GCS truck is E95Y, indicating commonality with that used with ENICS' E95 aerial target system.
Launch: Launched by pneumatic catapult.
Recovery: Automatic parachute recovery.
Power plant: One 1.47 kN (331 lb st) ENICS M135 pulsejet. Can also be fitted with (an unspecified) rear-mounted piston engine driving a pusher propeller.

Berta

Dimensions, External	
Overall, length	4.15 m (13 ft 7½ in)
Wings, wing span	5.00 m (16 ft 4¾ in)
Weights and Loadings	
Weight, Max launch weight	150 kg (330 lb)
Payload, Max payload	25 kg (55 lb)
Performance	
Altitude, Operating altitude	100 m to 3,000 m (320 ft to 9,840 ft)
Radius of operation	
datalink, mission	38 n miles (70 km; 43 miles)
Endurance	
max, pulse-jet	30 min
max, piston engine	8 hr
Power plant	1 × piston engine

Status: Being promoted in 2010. Thought to be in production; possibly also in service, but no customer information provided.
Contractor: ENICS JSC
Kazan.

ENICS Eleron

Type: Tactical mini-UAV.
Development: Development of the Eleron (aileron) reportedly began in 2004. The T25 civil version first appeared at the MAKS air show in Moscow in August 2005.
Description: *Airframe:* Tail-less delta with shoulder-mounted wings; pusher engine. No landing gear.
Mission payloads: Roll-stabilised CCD colour TV camera or digital stills camera with 40° FoV.
Guidance and control: T23U ground station. Autonomous or remotely piloted operation; GPS navigation (up to 99 waypoints, 25 m; 82 ft accuracy); 'return home' mode if link is lost.
Transportation: Dismantles for transportation and can be carried by one person in a customised backpack.

The Eleron delta-wing mini-UAV *(ENICS)* 1290203

System composition: Set-up time 5 minutes, turnaround time 10 minutes.
Launch: Automatically, by T23P bungee catapult.
Recovery: Automatic parachute recovery.
Power plant: One 100 W DC brushless electric motor; two-blade pusher propeller.
Variants: **T23:** Military surveillance version.
T25: Civil version of T23. Applications include environmental monitoring and close-range remote surveillance of fires, floods, earthquakes and other natural disasters. Subvariants T25D (day TV camera) and T25N (IR night vision camera).

T23, T25

Dimensions, External	
Overall, length	0.45 m (1 ft 5¾ in)
Wings, wing span	1.47 m (4 ft 9¾ in)
Weights and Loadings	
Weight	
Max launch weight	
T23	2.8 kg (6.00 lb)
T25	3.2 kg (7.00 lb)
Performance	
Altitude, Service ceiling	3,000 m (9,840 ft)
Speed	
Max level speed	56 kt (104 km/h; 64 mph)
Loitering speed	35 kt (65 km/h; 40 mph)
Radius of operation	
TV	5.4 n miles (10 km; 6 miles)
photo	13.5 n miles (25 km; 15 miles)
Endurance	
T23	1 hr 15 min
T25	1 hr 0 min
Power plant	1 × electric motor

Status: T25 said to have been operated by the Russian Border Guard in the Caucasus in late 2005. Eleron deployed and flown from drifting ice floe in Arctic 2008 in tests to monitor ice conditions.
Contractor: ENICS JSC
Kazan.

ENICS Igla

Type: Tactical mini-UAV.
Development: Development history not stated.
Description: *Airframe:* Mid-mounted swept wings and tail surfaces; 'fighter-like' fuselage; composites construction. No landing gear. Air vehicle designation is E26T.
Mission payloads: No details provided.
Guidance and control: Autonomous or under radio command manual control. The truck-mounted GCS is designated E2Y; complete Igla system is E2T.
Launch: Launched by ENICS E95P pneumatic catapult.
Recovery: Parachute recovery.
Power plant: Electric motor (rating not stated); two-blade pusher propeller.

Igla's E2Y GCS truck, flanked by the E95P launcher and E26T air vehicle *(ENICS)* 1395221

E95P launcher, with an E26T 'at the ready' *(ENICS)* 1395222

The E26T air vehicle of the Igla system *(ENICS)* 1395220

Igla

Dimensions, External	
Overall, length	0.40 m (1 ft 3¾ in)
Wings, wing span	1.40 m (4 ft 7 in)
Weights and Loadings	
Weight, Max launch weight	3.20 kg (7.00 lb)
Performance	
Altitude, Service ceiling	2,000 m (6,560 ft)
Speed	
Max level speed	59 kt (109 km/h; 68 mph)
Loitering speed	38 kt (70 km/h; 44 mph)
Endurance, max	35 min
Power plant	1 × electric motor

Status: Being promoted in 2012; production and service status not stated.
Contractor: ENICS JSC
Kazan.

Irkut Irkut-1A

Type: Tethered aerostat surveillance system.
Development: Development history not provided.
Description: *Airframe:* Conventional helium-filled blimp with cruciform tail surfaces.
Mission payloads: Stabilised EOS-VB with colour TV and IR cameras.
Guidance and control: Mission payload control and video downlink channels.
Transportation: By 1½-tonne truck.
System composition: Four aerostats and portable GCS. Set-up time up to 1 hour.
Power plant: None.

Specifications

Dimensions	
Length overall	10.0 m (32.8 ft)
Max diameter	4.0 m (13.1 ft)
Volume	45 to 50 m³ (1,590 to 1,765 cu ft)
Weights	
Aerostat weight	15.0 kg (33.1 lb)
Max payload	6.0 kg (13.2 lb)
Max launching weight	22.0 kg (48.5 lb)

Irkut-1A surveillance aerostat *(Irkut)* 1375446

Specifications

Performance
- Max operating altitude .... 300 m (985 ft)
- Data transmission range .... Up to 2.7 n miles (5 km; 3.1 miles)
- Max time on station .... Unlimited

Status: In production in early 2009. Customers include MChS (Russian Ministry for Emergency Situations).
Contractor: Scientific Production Corporation, Irkut JSC
Irkutsk.

Irkut Irkut-2F and -2T

Type: Remote sensing UAV.
Development: Utilises an air vehicle designed by Moscow-based Novik-XXI.
Description: *Airframe:* High-wing monoplane; pod and boom fuselage; conventional tail surfaces. No landing gear. Composites construction. Can be rapidly assembled and dismantled without special tooling.

Mission payloads: Digital camera in Irkut-2F, imagery from which is stored on board and downloaded to GCS notebook PC after landing for visual and computerised processing and analysis.

TV camera, TV transmitter or IR camera in Irkut-2T, with real-time imagery downlink.

Guidance and control: Irkut-2F: Preprogrammed and automatic, controlled by notebook PC.

Irkut-2T: Truck-mounted GCS has two workstations (pilot and payload operator), and is designed for manual control of system, but UAV can also be operated in preprogrammed, automatic mode. Central control console has two LCD displays and a TV monitor, providing an area map with UAV location, video imagery of surveillance area, flight and navigation data and onboard systems condition.

Transportation: Man-portable.
System composition: Two air vehicles and portable or mobile GCS; plus ground support equipment and (for -2F) charger for UAV main battery. Two personnel for -2F, three for -2T. Set-up time 5 minutes or less.
Launch: Hand-launched.
Recovery: Belly landing, within 100 m (330 ft), on any flat land surface.
Power plant: Battery-powered electric motor in Irkut-2F; not stated for -2T; two-blade propeller in each case.
Variants: Irkut-2F: Electric-powered version for stills photography.
Irkut-2T: More capable version with real-time imagery transmission.

Irkut-2T, Irkut-2F

Dimensions, External
Overall
- length .... 1.00 m (3 ft 3¼ in)
- height .... 0.30 m (11¾ in)
- Wings, wing span .... 2.00 m (6 ft 6¾ in)

Weights and Loadings
- Weight, Max launch weight .... 2.8 kg (6.00 lb)
- Payload, Max payload .... 0.3 kg (0.00 lb)

Performance
Altitude
- Operating altitude .... 100 m to 400 m (320 ft to 1,320 ft)
- Service ceiling .... 2,500 m (8,200 ft)

Irkut-2 remote sensing UAV *(Irkut)* 1151599

The electric-powered Irkut-2 *(Irkut)* 1151600

- Speed, Cruising speed .... 43 kt (80 km/h; 49 mph)
- Radius of operation .... 21.5 n miles (39 km; 24 miles)
- Endurance .... 1 hr

Status: Developed and available. Production and service status not known.
Contractor: Scientific Production Corporation, Irkut JSC
Irkutsk.

Irkut Irkut-2M

Type: Remote sensing UAV.
Development: The Irkut-2M apparently replaces the earlier Irkut-2F and -2T designs, which were of different airframe configuration. The -2M was developed jointly by Irkut and ENICS, and appears to be based on the latter company's Eleron.
Description: *Airframe:* Delta-wing monoplane with tall fin and rudder. No landing gear. Composites construction. Can be assembled and dismantled without special tooling.

Mission payloads: Stabilised EOS-2M turret with two (digital black and white and colour) cameras.

Guidance and control: Autopilot with automatic and manual control modes.
Transportation: Man-portable.
System composition: Two air vehicles and portable GCS. Set-up time 15 minutes or less.
Launch: By bungee rubber catapult.
Recovery: By parachute.
Power plant: Battery-powered electric motor (rating not stated); two-blade pusher propeller.

Irkut-2M

Dimensions, External
Overall
- length .... 1.00 m (3 ft 3¼ in)
- height .... 0.30 m (11¾ in)
- Wings, wing span .... 2.00 m (6 ft 6¾ in)

Weights and Loadings
- Weight, Max launch weight .... 2.8 kg (6.00 lb)
- Payload, Max payload .... 0.3 kg (0.00 lb)

Irkut-2M portable GCS *(Irkut)* 1375440

Irkut-2M in flight *(Irkut)* 1375448

Irkut-2M

Performance	
Altitude	
Operating altitude	100 m to 400 m (320 ft to 1,320 ft)
Service ceiling	3,000 m (9,840 ft)
Speed, Cruising speed	43 kt (80 km/h; 49 mph)
Radius of operation	11 n miles (20 km; 12 miles) (est)
Endurance	1 hr 30 min (est)
Power plant	1 × electric motor

Status: In production in early 2009; customer details not disclosed.
Contractor: Scientific Production Corporation, Irkut JSC Irkutsk.

Irkut Irkut-3

Type: Surveillance mini-UAV.
Development: Introduced in 2011; displayed at LAAD 2011 exhibition in Rio de Janeiro.
Description: *Airframe:* Constant-chord high wings, pod and boom fuselage, pusher engine and conventional tail unit. No landing gear. Composites construction. Rapid assembly or dismantling without need of special tools.
  Mission payloads: TV, IR or stills cameras, or other sensors, to customer's requirements. Real-time imagery downlink.
  Guidance and control: Radio command link.
  System composition: Two air vehicles, suitcase-size GCS and support truck. Set-up time 15 minutes or less.
  Launch: Hand launch.
  Recovery: Parachute recovery within 25 × 25 m (82 × 82 ft) landing area.
  Power plant: One piston engine (type and rating not stated); two-blade pusher propeller.

Irkut-3

Dimensions, External	
Overall	
length	0.90 m (2 ft 11½ in)
height	0.30 m (11¾ in)
Wings, wing span	2.00 m (6 ft 6¾ in)
Weights and Loadings	
Weight, Max launch weight	3.0 kg (6.00 lb)
Payload, Max payload	0.5 kg (1.00 lb)
Performance	
Altitude	
Operating altitude	100 m to 500 m (320 ft to 1,640 ft)
Service ceiling	3,000 m (9,840 ft)
Speed	
Max level speed	48 kt (89 km/h; 55 mph)
Cruising speed, max	32 kt (59 km/h; 37 mph)
Radius of operation, mission	8.1 n miles (15 km; 9 miles)
Endurance, max	1 hr 15 min
Power plant	1 × piston engine

Status: Being promoted in 2011.
Contractor: Scientific Production Corporation, Irkut JSC Irkutsk.

Irkut-3 hand-launched mini-UAV *(Irkut)* 1395325

Irkut Irkut-10

Type: Short-range surveillance UAS.
Development: Development history not provided.
Description: *Airframe:* All-swept flying wing configuration with triangular endplate fins; pusher engine. No landing gear. Composites construction. Can be assembled and dismantled easily without special tooling.
  Mission payloads: Gyrostabilised and interchangeable EO systems (TV, TV/still camera or IR).
  Guidance and control: Fully autonomous (no other details provided).
  System composition: Two air vehicles, GCS and ground support equipment. Set-up time 15 minutes. Operable by a crew of one or two.
  Launch: From lightweight, portable catapult.
  Recovery: By parachute.
  Power plant: Electric motor (rating not stated); two-blade pusher propeller.

Irkut-10

Dimensions, External	
Overall	
length	1.00 m (3 ft 3¼ in)
height	0.60 m (1 ft 11½ in)
Wings, wing span	2.30 m (7 ft 6½ in)
Weights and Loadings	
Weight, Max launch weight	7.0 kg (15.00 lb)
Payload, Max payload	1.0 kg (2.00 lb)
Performance	
Altitude	
Operating altitude	100 m to 500 m (320 ft to 1,640 ft)
Service ceiling	3,000 m (9,840 ft)
Speed	
Max level speed	64 kt (119 km/h; 74 mph)
Cruising speed	43 kt (80 km/h; 49 mph)
Radius of operation, mission	37 n miles (68 km; 42 miles)
Endurance	2 hr 30 min (est)
Power plant	1 × electric motor

Status: Reported to be in series production early 2009; customers not disclosed.
Contractor: Scientific Production Corporation, Irkut JSC Irkutsk.

Irkut-10 on its portable launcher *(Irkut)* 1375443

Irkut-10 in flight *(Irkut)* 1375444

Irkut Irkut-20

Type: Remote sensing UAV.
Development: Utilises the Novik-XXI Grant air vehicle as basis.
Description: *Airframe:* High-wing monoplane with rounded box-section fuselage and conventional tail surfaces. No landing gear. Construction mainly of composites. Can be assembled and dismantled easily without special tooling.
  Mission payloads: TV camera, TV transmitter, IR camera or digital stills camera. Imagery downlink in real time from TV and IR, in near-realtime from digital camera.
  Guidance and control: Generally as described for Irkut-2T.

System composition: Two air vehicles, GCS, transporter/launcher vehicle and ground support equipment. Set-up time 5 minutes or less. Operable by a crew of three.
Launch: Rail-launched from roof of transporter/launcher vehicle.
Recovery: Belly landing within 150 m (500 ft) on any flat land area.
Power plant: One flat-twin piston engine (type and rating not stated); two-blade propeller.

Irkut-20

Dimensions, External
Overall
  length...2.35 m (7 ft 8½ in)
  height...0.60 m (1 ft 11½ in)
Wings, wing span...3.00 m (9 ft 10 in)

Weights and Loadings
Weight, Max launch weight...................................20 kg (44 lb)
Payload, Max payload..3.0 kg (6.00 lb)
Performance
Altitude
  Operating altitude........................100 m to 400 m (320 ft to 1,320 ft)
  Service ceiling...2,500 m (8,200 ft)
Speed
  Max level speed............................97 kt (180 km/h; 112 mph)
  Cruising speed...............................65 kt (120 km/h; 75 mph)
Radius of operation..............................38 n miles (70 km; 43 miles)
Endurance...3 hr
Power plant...1 × piston engine

Status: Developed and available from 2007. Subsequent production status not known.
Contractor: Scientific Production Corporation, Irkut JSC
Irkutsk.

Irkut-20 GCS truck *(Irkut)* 1151603

Irkut-20 on vehicle launcher *(Irkut)* 1151602

Irkut-20 air vehicle *(Irkut)* 1151601

Irkut Irkut-200

Type: Surveillance tactical UAS.
Development: Under way in 2008-09.
Description: *Airframe:* Tapered fuselage with high-mounted, non-swept wings; T-tail, with conventional rudder and elevator; engine mounted in fin leading-edge. Fixed tricycle landing gear. Composites construction.
Mission payloads: Standard payloads include TVIR sensor turret; synthetic aperture radar optional.
Guidance and control: Standard mode is autonomous flight except for take-off and landing, which are operator controlled from GCS. Fully automatic take-off and landing optional.
System composition: Three air vehicles, GCS and ground support equipment.
Launch: Conventional wheeled take-off standard; short take-off optional.
Recovery: Conventional wheeled landing standard; parachute for emergency recovery.
Power plant: One (unidentified two-stroke piston engine (rating not given); tractor propeller.

Irkut-200

Dimensions, External
Overall
  length...4.50 m (14 ft 9¼ in)
  height...1.68 m (5 ft 6¼ in)
Weights and Loadings
Weight, Max T-O weight.......................................200 kg (440 lb)
Payload, Max payload...50 kg (110 lb)
Performance
T-O
  T-O run...250 m (821 ft) (est)
Altitude, Operating altitude.........500 m to 5,000 m (1,640 ft to 16,400 ft) (est)
Speed
  Max level speed................................113 kt (209 km/h; 130 mph) (est)
  Cruising speed..................................76 kt (141 km/h; 87 mph) (est)
Range......................................756 n miles (1,400 km; 870 miles) (est)
Radius of operation, radio control.......108 n miles (200 km; 124 miles) (est)
Endurance...12 hr (est)
Landing
  Landing run...250 m (821 ft) (est)
Power plant...1 × piston engine

Status: Under development. Flight testing due to begin during first half of 2009.
Contractor: Scientific Production Corporation, Irkut JSC
Irkutsk.

Computer-generated image of the Irkut-200 *(Irkut)* 1375447

Irkut Irkut-850

Type: Optionally piloted vehicle.
Development: Basic air vehicle is the German Stemme S10-VT two-seat motor glider; remote control equipment package is installed in right-hand seat. Similar in concept to Sagem/Onera Busard programme in France, which also uses S10-VT.

Unmanned aerial vehicles > Russian Federation > Irkut Irkut-850

Irkut-850 on runway *(Irkut)*

Description: Airframe: Typical motor glider configuration of high aspect ratio, shoulder-mounted wings, streamlined fuselage and T tail; provision for pod-mounted store under each wing. Twin, narrow-track mainwheels and tailwheel landing gear. Composites construction.

Mission payloads: TV and/or IR camera on gyrostabilised turret; automatic high-resolution digital stills camera; 3-D laser (lidar) mapping system; or relay unit. Real-time downlink of acquired data. Two underwing hardpoints for external payloads such as medical kits or food supplies.

Guidance and control: Generally as described for Irkut-2T.

System composition: Two air vehicles, GCS and ground support equipment. Crew of five.

Launch: Conventional runway take-off.
Recovery: Conventional runway landing.
Power plant: One 84.5 kW (113.3 hp) Rotax 914 flat-four engine; two-blade propeller.

Irkut-850

Dimensions, External	
Overall	
length	8.42 m (27 ft 7½ in)
height	1.80 m (5 ft 10¾ in)
Wings, wing span	23.00 m (75 ft 5½ in)
Weights and Loadings	
Weight, Max launch weight	860 kg (1,895 lb)
Payload, Max payload	200 kg (440 lb)
Performance	
T-O	
T-O run	300 m (985 ft)
Altitude	
Service ceiling	
piloted	6,000 m (19,680 ft)
unmanned	9,000 m (29,520 ft)
Speed	
Max level speed	145 kt (269 km/h; 167 mph)
Cruising speed	89 kt (165 km/h; 102 mph)
Range, max	928 n miles (1,718 km; 1,067 miles)
Radius of operation, datalink	108 n miles (200 km; 124 miles)
Endurance, unmanned	12 hr (est)
Landing	
Landing run	300 m (985 ft)

Irkut-850 EO turret *(Irkut)*

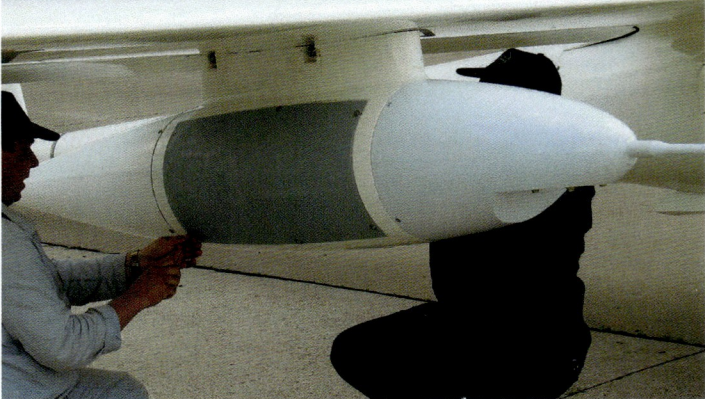

Irkut-850 underwing pod *(Irkut)*

Status: Demonstrated (piloted) in flight at August 2005 Moscow Air Show, when it overflew show area for security surveillance, equipped with a combined TV/IR sensor (reportedly a Controp DSP-1). Trials for MChS (Russian Ministry of Emergency Situations) reported to have begun in October 2005. The contractor described the system as ready for production in early 2009.

Contractor: Scientific Production Corporation, Irkut JSC Irkutsk.

A standard Stemme S10-VT motor glider, on which the Irkut-850 is based *(Paul Jackson)*

Irkut-850 GCS vehicle *(Irkut)*

Irkut-850 GCS workstations *(Irkut)*

IHS Jane's All The World's Aircraft: Unmanned 2013-2014

Lutch Tipchak

Type: Tactical UAV.

Development: The Lutch Design Bureau was established in 1955, originally as an R & D centre for radio and electronic equipment. Current activities now include research into unmanned aircraft systems on behalf of the Russian Ministries of Defence, Internal Affairs and Emergency Situations, the Federal Patrol Service and other agencies, as well as small-batch manufacture. Tipchak (Nomad) was declassified in 2005, and underwent government trials in 2005-06.

Description: *Airframe:* Mid-mounted, slightly swept wings; pod and twin tailboom fuselage; pusher engine. No landing gear.

Mission payloads: Dual-band (visible and infra-red) linescanner standard. Options include sigint, chemical monitoring and communications relay packages. SHF imagery and telemetry downlink.

Guidance and control: Pre-programmed, with manual control option via UHF uplink. Three-person GCS (mission commander and two operators) can control two air vehicles simultaneously.

System composition: Complete system (designation 1K132) comprises six 9M62 air vehicles, GCS truck, launcher/transporter truck, antenna truck and maintenance vehicle. Set-up time 15 minutes.

Launch: Rail launch from truck-mounted pneumatic catapult. Preparation time 15 minutes.

Recovery: Parachute recovery.

Power plant: One 9.7 kW (13 hp) piston engine; two-blade pusher propeller.

Tipchak

Weights and Loadings	
Weight, Max launch weight	50 kg (110 lb)
Payload, Max payload	14.5 kg (31 lb)
Performance	
Altitude, Operating altitude	200 m to 3,000 m (660 ft to 9,840 ft)
Speed	
Max level speed	108 kt (200 km/h; 124 mph)
Loitering speed	49 kt (91 km/h; 56 mph)
Radius of operation	38 n miles (70 km; 43 miles)
Endurance	2 hr
Power plant	1 × piston engine

Status: Series production started in 2008.
Contractor: Lutch Design Bureau JSC
Rybinsk, Yaroslavl Region.

MiG Skat

Type: UCAV technology demonstrator.

Development: Development of the Skat (skate or ray) is said to have begun in about 2005. A full-size mockup was revealed to selected representatives of the Russian media at the Gromov flight test centre, Zhukovsky, during the course of the MAKS air show, Moscow, in August 2007. It has been wind-tunnel tested by CAHI (TsAGI), and an optionally piloted prototype was stated to be the next stage of the development programme, before proceeding to a fully combat-capable demonstrator.

A competing design is believed to be under development by the Sukhoi design bureau, though as of September 2007 no details of this had been released. It may also be noted that, in early 2006, Yakovlev (now associated with Irkut Corporation) unveiled some details of its plans for a Proryv (Breakthrough) family of UAVs, which included a Proryv-U strike version of broadly similar appearance to the Skat.

Description: *Airframe:* Tail-less, blended wing/body stealth design, with planform generally similar to those of the US X-47B and Dassault Neuron; approximately 50° leading-edge sweepback. Central, bifurcated, overwing engine intake. Composites construction. Retractable tricycle landing gear.

Mission payloads: Each internal weapons bay was said to be capable of accommodating an anti-radar or anti-ship missile, or a 250 or 500 kg bomb. On debut, the mockup was displayed with Kh-31 (AS-17 'Krypton') ASMs (actually longer than the stated bay length) and KAB-500R TV-guided bombs. Weapon management system is credited to GosNIIAS. A report in *Jane's Missiles and Rockets* in November 2007 suggested that the new Kh-58UShKE anti-radiation missile may be intended for the Skat.

Guidance and control: No details released, but would be assumed to be autonomous operation. During the initial presentation, the Vega group and Russkaya Avionika were respectively credited with the mission and navigation systems.

Launch: Conventional wheeled take-off, probably planned to be automatic.

Recovery: Conventional wheeled landing, probably also automatic.

Tipchak launcher/transporter, showing catapult and containers for six dismantled UAVs *(Lutch)* 1290217

Tipchak battlefield surveillance UAV *(Lutch)* 1418149

Tipchak GCS vehicle *(Lutch)* 1290216

View of Skat mockup showing the split upper/lower trailing-edge control surfaces; Kh-31 missile in foreground *(RSK MiG)* 1295031

Detail of Skat's divided nose intake *(RSK MiG)* 1290204

Power plant: One 49.4 kN (11,100 lb st) Klimov RD-5000B non-afterburning turbofan (RD-93 derivative). Possibility of thrust vectoring has been speculated.

Skat

Dimensions, External	
Overall	
length	10.25 m (33 ft 7½ in) (est)
height	2.70 m (8 ft 10¼ in) (est)
Wings, wing span	11.50 m (37 ft 8¾ in) (est)
Dimensions, Internal	
Payload bay	
length, two, each	4.40 m (14 ft 5¼ in) (est)
width, two, each	0.65 m (2 ft 1½ in) (est)
depth, two, each	0.75 m (2 ft 5½ in) (est)
Weights and Loadings	
Weight, Max T-O weight	10,000 kg (22,046 lb) (est)
Payload	
Max payload, internal	2,000 kg (4,409 lb) (est)
Performance	
Altitude, Service ceiling	12,000 m (39,380 ft) (est)
Speed, Max level speed	432 kt (800 km/h; 497 mph) at S/L (est)
Radius of operation	1,078 n miles (1,996 km; 1,240 miles) (est)
Power plant	1 × turbofan

Status: Development was reported to be continuing in 2007 but as of early 2011 there had been no confirmation of a first flight by the optionally manned or unmanned versions.

Contractor: Russian Aircraft Corporation (RSK) MiG
Moscow.

Mil Mi-34BP

Type: Helicopter UAV.

Development: This is an unmanned (*bespilotnyi:* without pilot) version of the Mi-34S civil light helicopter, proposed by Mil for mainly civilian or parapublic duties such as aerial survey, crop-spraying, search and rescue, ice patrol, or monitoring of road traffic, oil and gas pipelines, nuclear power stations and environmental issues. It was first mooted in 2005, but progress has apparently been hampered by a lack of funding. However, more recently Mil has indicated its expectation to begin preparatory flight trials, probably in 2010, using an airframe carrying the control and datalink systems for the unmanned version but with a safety pilot on board.

Description: *Airframe:* Typical pod and boom fuselage; four-blade main and two-blade tail rotors; sweptback fin with T tailplane. Twin-skid landing gear. Composites rotor blades; fuselage of metal construction.
Mission payloads: Still camera, TV and/or IR sensors for day and night imagery collection and transmission.
Guidance and control: Radio command uplink and downlink.
Launch: Conventional helicopter vertical take-off.
Recovery: Conventional helicopter vertical landing.
Power plant: Mi-34BP1: One 280 kW (375 hp) VOKBM M-9V piston engine.
Mi-34BP2: One 335 to 373 kW (450 to 500 shp) class Turbomeca Arrius, Ivchenko/Progress AI-450 or Klimov VK-450 turboshaft.

Variants: *Mi-34BP1:* Piston-engined version.
Mi-34BP2: Turbine-powered version.

Mi-34BP1, Mi-34BP2

Dimensions, External	
Overall	
length	8.715 m (28 ft 7 in)
height	2.75 m (9 ft 0¼ in)
Fuselage, width	1.42 m (4 ft 8 in)
Skids, skid track	2.18 m (7 ft 1¾ in)
Rotors, rotor diameter	10.00 m (32 ft 9¾ in)
Tail rotor, tail rotor diameter	1.48 m (4 ft 10¼ in)
Weights and Loadings	
Weight	
Weight empty	
Mi-34BP1	920 kg (2,028 lb)
Mi-34BP2	790 kg (1,741 lb)
Max T-O weight	1,450 kg (3,196 lb)
Payload	
Max payload	
Mi-34BP1	360 kg (793 lb)
Mi-34BP2	520 kg (1,146 lb)
Performance	
Altitude	
Hovering ceiling	
IGE, Mi-34BP1	1,550 m (5,080 ft)
IGE, Mi-34BP2	4,350 m (14,280 ft)
OGE, Mi-34BP1	4,200 m (13,780 ft)
OGE, Mi-34BP2	6,000 m (19,680 ft)
Speed	
Max level speed	
Mi-34BP1	121 kt (224 km/h; 139 mph)
Mi-34BP2	143 kt (265 km/h; 165 mph)
Cruising speed	
Mi-34BP1	105 kt (194 km/h; 121 mph)
Mi-34BP2	119 kt (220 km/h; 137 mph)

Model of the Mi-34BP on display in September 2008
(IHS/Patrick Allen) 1370132

Radius of operation	
Mi-34BP1	334 n miles (618 km; 384 miles)
Mi-34BP2	529 n miles (979 km; 608 miles)
Endurance	
Mi-34BP1	3 hr 30 min
Mi-34BP2	5 hr 30 min
Power plant	
Mi-34BP1	1 × piston engine
Mi-34BP2	1 × turboshaft

Status: Under development.
Contractor: Mil Helicopters
Moscow.

NII Kulon BLA-06 Aist

Type: Multirole UAV.

Development: Development of the Aist (Stork), it is understood to have been authorised by the Russian Ministry of Defence in 2005, with its existence being revealed in model form at a Moscow exhibition in early 2007. At that time, IHS Jane's sources reported its technical design as "in the process of being completed". The BLA designation, introduced by the Vega Corporation UAV organisation to which NII Kulon belongs, indicates *bespilotnyi letatelnyi apparat* (pilotless aerial vehicle). One Russian source has suggested the name 'Yulia' for the system of which the Aist forms the air component. According to a Vega spokesman in January 2009, Aist prototypes were then beginning a two-year test programme in which they would be evaluated as target designators for the Iskander-M ground-launched missile system.

In addition to military roles, Aist is seen as a contender for such civil missions as oil and pipeline monitoring by Gazprom and other Russian utility agencies.

Description: *Airframe:* Low-wing monoplane, with twin engines mounted on overwing pylons; tapered fuselage; V tail unit. Rearward-retracting tricycle landing gear.
Mission payloads: Potential military payloads could include RSTA sensors, EW systems or weapons. For civilian use, the BLA-06 is quoted as carrying a SON-100 dual-sensor turret containing a TV camera and thermal imager, with an X-band SAR or gas analyser as optional alternatives.
Guidance and control: As of early 2009, no details of the Aist system's ground station appeared to have been released.
Launch: Conventional wheeled take-off.
Recovery: Conventional wheeled landing.
Power plant: Two piston engines (type and rating not revealed), each driving a three-blade tractor propeller.

2007 display model of the NII Kulon BLA-06 Aist 1290341

BLA-06

Dimensions, External
- Overall, length..4.70 m (15 ft 5 in)
- Wings, wing span...8.00 m (26 ft 3 in)

Weights and Loadings
- Weight, Max T-O weight..................................500 kg (1,102 lb)

Performance
- T-O
  - T-O run...150 m (493 ft)
  - Altitude, Service ceiling..............................6,000 m (19,680 ft)
- Speed
  - Max level speed...135 kt (250 km/h; 155 mph)
  - Loitering speed..70 kt (130 km/h; 81 mph)
- Radius of operation..125 n miles (231 km; 143 miles)
- Endurance..12 hr
- Landing
  - Landing run..150 m (493 ft)
- Power plant..2 × piston engine

Status: Development and evaluation continuing in 2009.
Contractor: Institut Kulon NII OAO
Moscow.

Transas Dozor

Type: Surveillance UAV.

Development: The original Dozor-2 (now Dozor-50) version of this UAV, which had an inverted-V tail unit, was shown at the MAKS exhibition in Moscow in August 2007. It has since been refined into the slightly larger current version, which has a more conventional empennage and is designated Dozor-4 (now Dozor-85). The name translates as 'patrol' or 'watch'. It is said to be aimed primarily at civil applications on behalf of state utilities and private industry.

The Dozor-5 (now Dozor-100) version has greater wing area, range and endurance and a maximum take-off weight of about 100 kg (220 lb). Transas is also developing other unmanned aerial platforms with gross weights of up to about 500 kg (1,102 lb).

Dozor-3 has since been renamed Dozor-600.

Description: *Airframe:* Typical high-wing, pod and twin tailboom configuration; pusher engine. Fixed tricycle landing gear standard; also planned with ski landing gear.

Mission payloads: Standard payloads of video (forward-looking or oblique), IR or 12 MP digital photo camera. Options include gas analysers, magnetometers and scanners. Data can be stored on board or downlinked in real time.

Guidance and control: Fully autonomous, pre-programmed or manual operation, with GPS/satcom navigation. Original TeKnol control system of Dozor-2 replaced by Transas own-design system in Dozor-4.

Transportation: Mobile GCS mounted on Land-Rover Defender 110 or similar vehicle, plus trailer with two air vehicles in 1.8 × 0.6 × 0.6 m (5.9 × 2.0 × 2.0 ft) containers. Set-up time 30 minutes.

System composition: Two air vehicles, payloads, GCS, transport vehicle, trailer and crew of three (pilot, data specialist and engineer).

Launch: Conventional wheel or ski take-off. Option of catapult launch under consideration.

Recovery: Conventional wheel or ski landing.

Dozor-100 *(Transas)* 1464746

Dozor-50 (ex Dozor-2) *(Transas)* 1464749

Dozor-100 (ex Dozor-5) *(Transas)* 1464752

Dozor-100 (ex Dozor-5) *(Transas)* 1464751

Dozor-2, Dozor-5, Dozor-3

Dimensions, External
- Overall
  - length
    - D-4 RD Dozor-2...2.60 m (8 ft 6¼ in)
    - Dozor-5..3 m (9 ft 10 in)
    - Dozor-3..6.7 m (21 ft 11¾ in)
  - height
    - D-4 RD Dozor-2...0.9 m (2 ft 11½ in)
    - Dozor-5..1.1 m (3 ft 7¼ in)
    - Dozor-3..2.3 m (7 ft 6½ in)
- Wings
  - wing span
    - D-4 RD Dozor-2...4.4 m (14 ft 5¼ in)
    - Dozor-5..6 m (19 ft 8¼ in)
    - Dozor-3..12 m (39 ft 4½ in)

Weights and Loadings
- Weight
  - Max T-O weight
    - D-4 RD Dozor-2...50 kg (110 lb)
    - Dozor-5..130 kg (286 lb)
    - Dozor-3..840 kg (1,851 lb)
- Fuel weight
  - Max fuel weight
    - D-4 RD Dozor-2...10 kg (22 lb)
    - Dozor-5..39 kg (85 lb)
    - Dozor-3..240 kg (529 lb)
- Payload
  - Max payload
    - D-4 RD Dozor-2...5 kg (11.00 lb)
    - Dozor-5..15 kg (33 lb)
    - Dozor-3..120 kg (264 lb)

Performance
- Altitude
  - Service ceiling
    - D-4 RD Dozor-2, Dozor-5............................3,000 m (9,840 ft)
    - Dozor-3..7,000 m (22,960 ft)
- Range
  - D-4 RD Dozor-2...270 n miles (500 km; 310 miles)
  - Dozor-5..648 n miles (1,200 km; 745 miles)
  - Dozor-3..1,998 n miles (3,700 km; 2,299 miles)
- Endurance
  - D-4 RD Dozor-2...4 hr 30 min
  - Dozor-5..10 hr
  - Dozor-3..24 hr

Status: As of 2011, the Dozor-50 (ex Dozor-2), Dozor-85 (ex Dozor-4) and Dozor-600 (ex Dozor-3) have all completed their development with production being halted for the time being. The Dozor-100 (ex Dozor-5) made its first flight in 30 November 2010.

Contractor: Transas Avia
St Petersburg.

Tupolev Tu-143 and Tu-243 Reis

Type: Jet-powered tactical reconnaissance systems.

Development: The Reis (flight) unmanned aerial reconnaissance system was developed by the Tupolev design bureau in the late 1960s to replace the earlier TBR-1 (La-17R) tactical reconnaissance UAV, a modified variant of the Lavochkin aerial target (which see), which had been in service since the early 1960s.

Tupolev Tu-143 and Tu-243 Reis

Tu-143 (VR-3) Reis UAV and its SPU-143 (BAZ 135) transporter and launch container *(Steven J Zaloga)*

The Tu-143 (service designation **VR-3**) first flew in December 1970, entered pre-series production by Kumertau in 1973 and remains a standard Russian tactical reconnaissance/surveillance UAV system. It is used for photographic, television or other surveillance in both military and civil capacities. Capabilities include reconnaissance of troop and facilities deployments, engineering works, or natural or ecological calamities; to reveal areas and extent of forest fires and gas or oil pipeline damage; and to define areas of radiation contamination. The former Soviet Army deployed five VR-3 systems in East Germany, two in the former Czechoslovakia, one in Mongolia and 17 in the USSR. A leaflet-dropping version for psychological operations (psyops) was developed in the late 1970s and early 1980s, but did not go into production. An **M-143** (or VR-3VM) target drone version was tested successfully in the mid-1980s.

A longer-range Tu-243 variant, with day and night capability, was unveiled in 1995 and entered service in 1999. Latest known variant is the Tu-300.

Description: *Airframe:* Tu-143 is of mixed construction (aluminium alloy and GFRP), with low-mounted delta wings (leading-edge sweepback of 58°), small canard surfaces and a dorsal intake for the turbojet engine. Main recovery parachute housed beneath clamshell doors forward of tailfin; brake parachute in large ogival fairing above exhaust nozzle. Retractable 'tricycle' landing skids. Tu-243 generally similar except for longer fuselage.

Mission payloads: Tu-143 payload (normally a PA-1 wet-film camera or a Chibis-B real-time TV camera; optionally a Sigma ground-mapping radar or radiation detection equipment) is mounted in the detachable nose compartment. Camera film is removed and transferred to the data acquisition and processing station after landing; data acquired by TV sensor and radiation detectors can be downlinked in real time.

Tu-243 normally equipped for day and/or night operation with AP-402M panoramic and Aist-M (stork) TV cameras or still camera plus Zima-M (winter) IR sensor. Sigma-R radiometric sensor was an alternative payload. Information collected can be stored on board or downlinked via a radio datalink.

Guidance and control: Preprogrammed mission profiles; onboard navigation and control guidance said to be accurate to approximately 200 m (656 ft) at a range of 38 n miles (70 km; 43.5 miles). Onboard avionics include ABSU-143 automatic flight control system, DISS-7 Doppler speed and drift indicator and A-032 radar altimeter. New NPK-243 flight control and navigation system in Tu-243.

Standard Tu-143 Reis reconnaissance UAV, photographed at Khodinka in 1993 *(Steven J Zaloga)*

Tu-243 Reis-D at Moscow Air Show, August 1995 *(Paul Jackson)*

SPU-243 self-propelled launcher/transporter for the Tu-243 *(Steven J Zaloga)*

Tupolev Tu-143 and Tu-243 Reis – Yakovlev Pchela/Shmel

Clearly a descendant of the Tu-141/143/243 family, Tupolev's Tu-300 prototype was displayed at the 1995 and 1997 Moscow Air Shows *(Paul Jackson)*

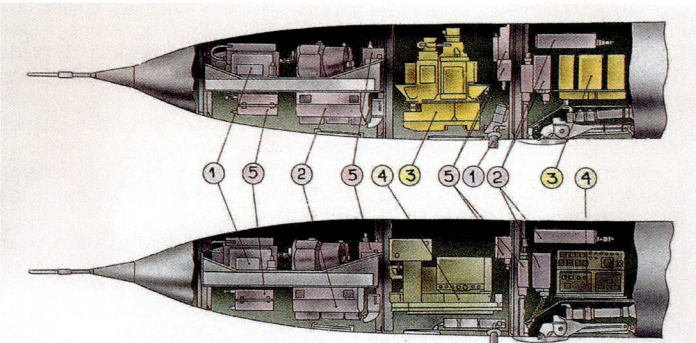

Nose compartment layout of the TV (top) and IR versions of the Tu-243 Reis-D:
1 data downlink
2 AP-402M camera
3 TV equipment
4 IR equipment
5 avionics

System composition: The complete VR-3 system comprises two eight-wheeled ground vehicles: an SPU-143 self-propelled transporter/launcher vehicle and a TZM-143 transporter/refueller (both based on the BAZ 135MB truck), plus a KPK-143 checking/testing system and POD-3 data processing and decoding centre. (Designations with Tu-243 are SPU-243, TZM-243, KPK-243 and POD-3D respectively.) Soviet deployment was in squadrons each operating 12 air vehicles and four SPU-143 launchers. System can self-deploy up to 270 n miles (500 km; 311 miles) from base. Set-up time 35 to 40 minutes after arrival on site.

Launch: Launched from large, truck-mounted cylindrical container, inclined at 15° from the horizontal, by means of an SPRD-251 underfuselage solid-propellant rocket booster (RDTT-243 in Tu-243). This accelerates the air vehicle for the first 550 m (1,805 ft) of flight, at which point it is jettisoned and the turbojet takes over.

Recovery: Recovery sequence begins with engine shutdown and zoom climb, followed by deployment of a brake parachute to slow the UAV to a speed of about 156 kt (290 km/h; 180 mph), when the main parachute deploys. Landing is made on a 'tricycle' gear of three retractable landing skids, after firing of a small braking rocket just before touchdown.

Power plant: One Klimov TR3-117 turbojet (5.79 kN; 1,301 lb st in early Tu-143s, otherwise 6.28 kN; 1,411 lb st); fuel capacity 190 litres (50.2 US gallons; 41.8 Imp gallons).

Variants: *Tu-143:* Initial **Reis** production model. *Description applies to this version except where indicated.* According to *IHS Jane's* sources in May 2006, the Tu-143 is scheduled to be upgraded to Tu-243 standard and redesignated **Reis-M**.

Tu-243: Upgraded version with same airframe, power plant and basic systems but more sophisticated payloads; developed between 1981 and 1987, making first flight in July 1987, although existence not revealed until Paris Air Show, June 1995, followed by first public appearance at Moscow Air Show in August. Designated **Reis-D** (indicating *Dalnyi*: long range) or **VR-3D**. Manufactured since 1996 by Kumertau Aircraft Production Enterprise (KAPP). Latest known order, for 20, placed in third quarter of 1999.

Tu-300 Korshun: (black kite): Further-developed UCAV version, first seen at Moscow Air Shows in 1995 and 1997. Main features include a fatter fuselage with internal weapons bay and a modified nose (see photograph) accommodating new sensors. The central thimble radome is thought to house an MMW radar; below this is a port for a forward-looking TV camera for sighting purposes, while the bullet fairing above the radome may be for a datalink antenna. The enlarged dorsal intake may signify a turbofan power plant, and the Tu-300 designation has been linked with the system name **Reis-F** (*frontovoy*: tactical). No other details appear to have emerged, and present programme status of this version is uncertain.

Tu-143, Tu-243

Dimensions, External	
Overall length	
Tu-143	8.06 m (26 ft 5¼ in)
Tu-243	8.29 m (27 ft 2½ in)
height	
excl rocket booster, Tu-143	1.545 m (5 ft 0¾ in)
excl rocket booster, Tu-243	1.576 m (5 ft 2 in)
Fuselage, height	0.61 m (2 ft 0 in)
Wings, wing span	2.24 m (7 ft 4¼ in)
Areas	
Wings, Gross wing area	2.90 m² (31.2 sq ft)
Weights and Loadings	
Weight	
Max launch weight	
with PA-1 camera, Tu-143	1,390 kg (3,064 lb)
with TV camera, Tu-143	1,400 kg (3,086 lb)
excl booster, Tu-243	1,410 kg (3,108 lb)
incl booster, Tu-243	1,600 kg (3,527 lb)
Max landing weight, Tu-143	1,012 kg (2,231 lb)
Fuel weight	
Max fuel weight, Tu-143	150 kg (330 lb)
Performance	
Altitude	
Operating altitude	
Tu-143	200 m to 1,000 m (660 ft to 3,280 ft)
Tu-243	50 m to 5,000 m (165 ft to 16,400 ft)
Speed	
Max level speed, Tu-143	472 kt (874 km/h; 543 mph) (est)
Radius of operation	
Tu-143	49 n miles (90 km; 56 miles)
Tu-243	97 n miles (179 km; 111 miles)
Endurance, Tu-143	13 min
Power plant	1 × turbojet

Status: In service. Tu-143 production (950 built) took place at Kumertau between 1973 and 1989. According to *Jane's* sources in 2006, 'hundreds' of Tu-143s remain in stock and are to be upgraded to Tu-243 standard and redesignated Reis-M.

Customers: Original VR-3/Tu-143 Reis system: Russian, Ukrainian and other armed forces of the former USSR; Czech Republic (No 345 Reconnaissance Drone Squadron at Pardubice); Romania; Slovak Republic; Syria. Those of Czech and Slovak Republics are no longer in service. Russia is only known customer for Tu-243.

Contractor: Tupolev PSC
Moscow. Production by Kumertau Aviation Production Enterprise.

Yakovlev Pchela/Shmel

Type: Short-range surveillance and tactical UAV.

Development: The Yakovlev OKB's current involvement in UAVs began on 28 June 1982 in response to a Soviet Ministry of Aircraft Production requirement for a small battlefield tactical system for surveillance and ECM roles. From 1982 to 1991, two types of UAV were developed, manufactured and tested. Izd.60C (*Izdeliye:* article) was part of an experimental system; Pchela-1T (Bee), or Shmel-1 (Bumblebee) for export, was the air vehicle element of the Stroy-P (Front Line - Regiment) system (Malakhit for export). Two other subvariants of the Pchela-1 have been developed subsequently.

Description: *Airframe:* High-mounted wings with turned-down tips; cylindrical fuselage, with rear-mounted ducted propeller; four non-retractable leaf spring landing legs. Mainly composites construction.

Pchela-1T tactical UAV, as used in the Stroy-P surveillance system *(Paul Jackson)*

Unmanned aerial vehicles > Russian Federation – Singapore > Yakovlev Pchela/Shmel – ST Aero Skyblade II/III

The complete Stroy-P system: modified tracked vehicle with Pchela on the launch rail *(Mark Lambert)* 0044409

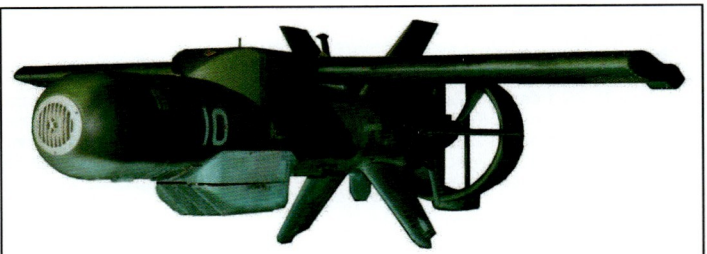

The Yakovlev Izd.60C experimental UAV, predecessor of the Pchela-1 *(Yakovlev)* 0068844

Mission payloads: Onboard sensors are a TV camera with zoom lens (viewing angle 3 to 30°) on Pchela-1T or IRLS on Pchela-1IK. Data transmission in real time.

Guidance and control: The flight control station, inside the launch vehicle, accommodates control posts for two operators, allowing two air vehicles to be controlled simultaneously. Flights can be preprogrammed and/or operator-controlled. The aircraft is autopilot-stabilised, with control surfaces and engine throttle controlled by small electric motors. Communications between the control station and UAV can be maintained at distances of up to 27 n miles (50 km; 31 miles).

System composition: Standard Stroy-P system comprises 10 air vehicles, one GCS/launch vehicle mounted on a BTR-D air-droppable APC, one GAZ-66 loader/transporter truck, one Kamaz 43101 maintenance truck, and a crew of eight. Deployment time 20 minutes.

Launch: The UAV is operated from a BTR-D tracked vehicle chassis, on top of which it is transported, with wings folded, in a drum-shaped container. Alongside this container is a launch rail, also folded during transportation. For deployment, the rail and UAV wings are unfolded, and the Pchela-1T is launched from the rail assisted by a pair of solid-propellant rocket boosters.

Recovery: By parachute descent to a landing on the four spring-loaded landing legs.

Power plant: One 23.9 kW (32 hp) Samara P-032 two-cylinder two-stroke engine; three-blade ducted pusher propeller.

Variants: Izd.60C: Experimental system. First flight 17 June 1983; 14.9 kW (20 hp) Samara P-020 piston engine. Twenty-five launches made, of which 20 were successful. GCS and system integration by NII Kulon; launch equipment by Horizont MOKB. GCS capable of controlling only one air vehicle at a time. Launched from rail by pair of solid-propellant rocket boosters. Recovery by parachute and shock-absorption system (inflatable rubberised nylon bag) located in centre-fuselage. Airframe construction of glass fibre.

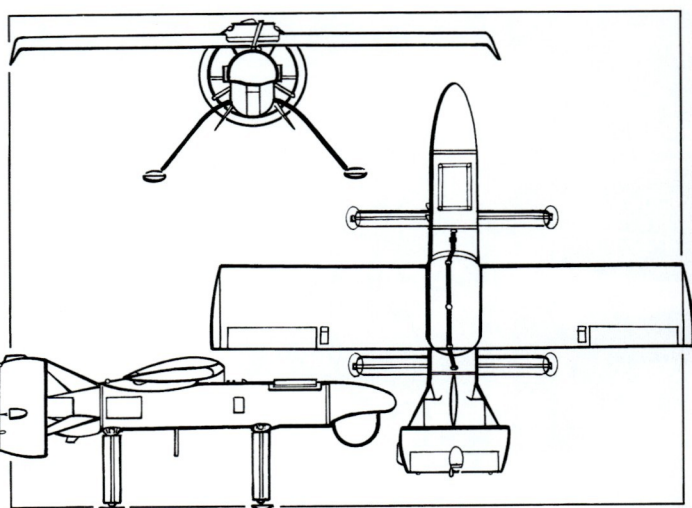

Yakovlev Pchela/Shmel battlefield surveillance UAV *(IHS/John W Wood)* 0518076

Pchela-1T: Larger than Izd. 60C, with more powerful P-032 engine and improved guidance. First flight 26 April 1986; trials programme ended September 1989 after 68 launches, 52 of them successful. Entered production mid-1991, and service in early 1994. *Following description applies to Pchela-1T/Stroy-P except where indicated.*

Pchela-1IK: As Pchela-1T, but with IRLS sensor. State acceptance trials completed successfully in March 2001.

Pchela-1VM: Aerial target version of Pchela-1T (minus TV camera), used for training of PVO (anti-aircraft defence) crews. State acceptance tests completed in 1999.

Pchela-2/Shmel-2: Projected improved version, said to have been under development in late 1999 but not proceeded with.

Pchela-1T

Dimensions, External	
Overall	
length	2.78 m (9 ft 1½ in)
height	1.11 m (3 ft 7¾ in)
Wings, wing span	3.25 m (10 ft 8 in)
Areas	
Wings, Gross wing area	1.83 m² (19.7 sq ft)
Weights and Loadings	
Weight, Max launch weight	138 kg (304 lb)
Performance	
Altitude, Operating altitude	100 m to 2,500 m (320 ft to 8,200 ft)
Speed	
Max level speed	97 kt (180 km/h; 112 mph)
Cruising speed	65 kt (120 km/h; 75 mph)
Radius of operation	27 n miles (50 km; 31 miles)
Endurance	2 hr

Status: Three Pchela-1T systems in service by late 1999. Used operationally against rebel forces in Chechnya in April 1995 and from October 1999. Production is complete, but the system remained in service in 2005 and after, Pchela 1T being listed as an active programme by Yakovlev in 2010, with the suggestion that the system is being upgraded.

Customers: Russian Army, Navy and Marines (Pchela); North Korea (one Malakhit system plus 10 Shmel-1 air vehicles ordered in 1994).

Contractor: A S Yakovlev OKB
Moscow.

Singapore

ST Aero Skyblade II/III

Type: Close-range surveillance mini-UAV.

Development: Skyblade II reportedly first flown in 2003, including trials during Singapore Army exercises in Australia in November of that year. Public debut at Asian Aerospace show, Singapore, in February 2006. A revised and enlarged Skyblade III was shown in 2008. Developed by ST Aerospace in collaboration with DSO National Laboratories, the Defence Science & Technology Agency and the Singapore Army.

Description: Airframe: Pod and boom fuselage (engine pod above wing, sensor pod below in Skyblade II); tapered outer wing panels; conventional tail unit. No landing gear. Skyblade III combines engine and sensors in single pod beneath wing.

Mission payloads: EO and IR sensors in pod.

Guidance and control: Autonomous operation. Notebook PC ground control station with two-person crew. GPS waypoint navigation; real-time situational update via digital RF datalink.

Skyblade III *(ST Aerospace)* 1356296

Skyblade III in tractor-propeller form, as exhibited in 2008 *(IHS/Patrick Allen)* 1343132

Skyblade II *(IHS/Patrick Allen)* 1177892

Launch: Hand launch or bungee catapult.
Recovery: Automatic parachute recovery.
Power plant: Electric motor (details not given); two-blade propeller. Piston engine also originally offered as option.

Skyblade II, Skyblade III

Dimensions, External	
Overall, length	1.40 m (4 ft 7 in)
Wings, wing span	2.60 m (8 ft 6¼ in)
Weights and Loadings	
Weight, Max launch weight	5.0 kg (11.00 lb)
Performance	
Altitude, Operating altitude	90 m to 455 m (300 ft to 1,500 ft)
Speed	
Max level speed	35 kt (65 km/h; 40 mph)
Stalling speed	18 kt (34 km/h; 21 mph)
Radius of operation, mission	4.3 n miles (8 km; 4 miles)
Endurance	1 hr (est)
Power plant	1 × electric motor

Status: First deliveries to Singaporean Army began in January 2006. Four air vehicles were scheduled for delivery during the year, possibly a trials batch of Skyblade II. Deliveries of an undisclosed number of production Skyblade III systems started in 2009, and was scheduled to be completed in 2010, when 44 operators had been trained. Six army units were reported to be equipped, including the 3rd Battalion Singapore Infantry Regiment.
Contractor: ST Aerospace Ltd
Paya Lebar.

ST Aero Skyblade IV

Type: Multirole tactical UAV.
Development: First flight early 2005; public debut at Asian Aerospace show, Singapore, in February 2006, jointly funded with Singapore Ministry of Defence.
Description: Airframe: In 2006 Skyblade IV was configured with unswept, constant chord wings with upturned tips, mounted atop circular-section fuselage of constant diameter; cruciform tail surfaces; pusher engine. No landing gear. By 2008 the design had been revised with a new fuselage of varying section, with shoulder-mounted wings and a butterfly tail. This version may have wings with conventional or upturned tips.
Mission payloads: Video and other sensors appropriate to such missions as reconnaissance, surveillance, BDA, search and rescue, artillery fire support, target tracking and maritime or coastal patrol.
Guidance and control: Fully autonomous, with real-time imagery and telemetry downlinks; based on system for Skyblade II.
Launch: By catapult.
Recovery: Options for parachute/airbag or net recovery, or belly landing.
Power plant: One piston engine (type and rating not stated); two-blade pusher propeller.

Skyblade IV as displayed at Asian Aerospace exhibition in 2006 *(IHS/Patrick Allen)* 1177898

Skyblade IV as displayed in 2010 *(IHS/Patrick Allen)* 1414618

2010 current brochure picture of Skyblade IV *(ST Aerospace)* 1356297

Skyblade IV

Dimensions, External	
Fuselage, length	2.40 m (7 ft 10½ in)
Wings, wing span	3.70 m (12 ft 1¾ in)
Weights and Loadings	
Weight, Max launch weight	70 kg (154 lb)
Payload, Max payload	12 kg (26 lb)

Unmanned aerial vehicles > Singapore – South Africa > ST Aero Skyblade IV – ATE Roadrunner

Skyblade IV

Performance	
Altitude, Service ceiling	4,570 m (15,000 ft)
Speed	
Cruising speed, max	80 kt (148 km/h; 92 mph)
Loitering speed	50 kt (93 km/h; 58 mph)
Radius of operation	27 n miles (50 km; 31 miles)
Endurance	6 hr (est)
Power plant	1 × piston engine

Status: Under development. Prototype had flown some 40 hours by February 2006, including one flight of 6 hours. Revised version flew in 2009. Flight testing by ST Aerospace continuing during 2010; tests by the Singapore Army planned late 2011.
Contractor: Singapore Technologies Aerospace Ltd
Paya Lebar.

Skyblade IV as configured in 2008 *(IHS/Patrick Allen)* 1342678

South Africa

ATE Kiwit

Type: Mini-UAV.
Development: Revealed at Africa Aerospace and Defence exhibition in October 2006. For both military and civil applications.
Description: *Airframe:* Tubular fuselage with ventral sensor pod and V-tail surfaces; composites, high-mounted swept wings.
  Mission payloads: Kiwit can support a range of optical and EO sensors such as colour CCD video cameras, stills cameras and IR cameras. Other optional sensors can include meteorological temperature, humidity, density and pressure detectors and volumetric air data gathering.
  Guidance and control: Man-portable GCS. Fully automatic flight control includes pre-programmed flight plan; activation of loiter pattern at target; in-flight changes to flight plan; automatic return to launch point or other designated landing point; and automatic landing. Video is relayed in real time to GCS, which displays co-ordinates of the centre of the image, as well as those of the UAV with a North indicator. Flight plan is overlaid on a digital map, together with camera footprint indication and UAV position indicator.

The 3 kg (6.6 lb) hand-launched Kiwit *(ATE)* 1165097

ATE Kiwit mini-UAV *(ATE)* 1198453

  Transportation: In either hard cases, padded nylon bags or standard backpacks, depending upon application.
  System composition: One or two air vehicles, plus GCS and other ground equipment.
  Launch: Hand-launched.
  Recovery: Automatic landing on ventral pod, following engine shutdown.
  Power plant: Electric motor, driving a two-blade propeller.

Kiwit

Dimensions, External	
Overall, length	1.20 m (3 ft 11¼ in)
Wings, wing span	2.50 m (8 ft 2½ in)
Weights and Loadings	
Weight, Max launch weight	3.0 kg (6.00 lb)
Performance	
Speed	
Cruising speed, max	27 kt (50 km/h; 31 mph)
Radius of operation, LOS	2.7 n miles (5 km; 3 miles)
Endurance	45 min (est)
Power plant	1 × electric motor

Status: Initial development completed by October 2006, with various configurations flight tested. Low-rate production starting in early 2007. First order, for undisclosed Asian customer, reported in June 2009.
Contractor: Advanced Technologies & Engineering Co (Pty) Ltd (ATE)
Halfway House.

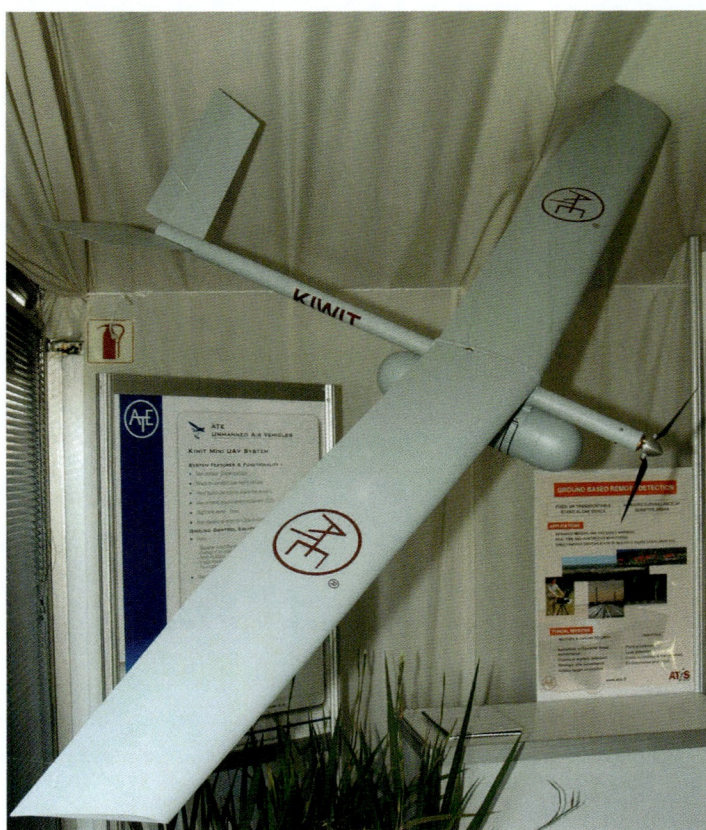

Kiwit on display, October 2006 *(IHS/Patrick Allen)* 1196103

ATE Roadrunner

Type: Close-range mini-UAV.
Development: The Roadrunner mini-UAV was designed in 2008, and a concept demonstrator has completed its initial flight tests to validate aerodynamic performance. Very little in the way of specific data had been released by early 2009, when development was continuing. Potential applications, as its name suggests, could include pursuit of high-speed vehicles or path clearance ahead of convoys.
Description: *Airframe:* Rhomboid wing configuration (low-mounted front pair, high-mounted rear pair), joined at tips by endplate fins; pusher engine. No landing gear.
  Mission payloads: Modular design allows for fitment of a variety of sensors for day and/or night reconnaissance and surveillance.
  Guidance and control: Aircraft has a fully automated RF flight control and tracking system, with provision for manual override. It uses the same GCS as ATE's other mini-UAVs such as Kiwit, Sentinel and Vigil.
  Launch: Hand or foldable catapult launch.
  Recovery: Belly landing.
  Power plant: Electric motor, driving a two-blade pusher propeller.

Roadrunner general appearance *(IHS/Patrick Allen)* 1369156

Roadrunner rear aspect *(IHS/Patrick Allen)* 1369157

Roadrunner

Dimensions, External	
Wings, wing span	1.50 m (4 ft 11 in)
Weights and Loadings	
Weight, Max launch weight	6.0 kg (13.00 lb)
Payload, Max payload	1.0 kg (2.00 lb)
Performance	
Range, LOS, max	2.7 n miles (5 km; 3 miles)
Endurance, max	1 hr
Power plant	1 × electric motor

Status: In development.
Contractor: Advanced Technologies & Engineering Co (Pty) Ltd
  Halfway House.

ATE Sentinel series

Type: Multirole UAVs.
Development: The Sentinel air vehicles were designed in 2008, and a demonstrator prototype has been flown. Development and marketing were continuing in early 2009, although full data had not then been publicly released and Sentinel had not gained the status of a product listing by 2010. The three-sizes family is described as 'learning from' ATE's Vulture TUAV system, but offering an extended range of mission applications combined with greater range and endurance.
Description: *Airframe:* All three variants share a common configuration, namely a high wing, rectangular section fuselage pod, twin tailbooms supporting an inverted V tail unit, and a pusher engine installation. The two larger versions have a fixed tricycle-type landing gear; the smaller 500M has none.
  Mission payloads: A wide variety of sensor or other payloads is possible, due to the air vehicle's modular design.
  Guidance and control: All versions are operated from identical GCS, utilising the same avionics as the ATE Vulture, and are compatible with the ground control segment of that system.
  Launch: Sentinel 500M is launched by the same pneumatic launcher as the Vulture. The 620 and 800 make conventional and automatic wheeled take-offs.
  Recovery: Sentinel 500M is retrieved using the Vulture mobile recovery system. The 620 and 800 make conventional and automatic wheeled landings.
  Variants: *Sentinel 500M:* Smallest member of family, with similar mobility level to Vulture but increased endurance.
  Sentinel 620: Mid-sized, runway-operated version, with more than double the endurance of the 500M.
  Sentinel 800: Larger runway-operated version, with greater endurance than 620.

Sentinel 500M, Sentinel 620, Sentinel 800

Dimensions, External	
Wings	
wing span	
Sentinel 500M	5.00 m (16 ft 4¾ in)
Sentinel 620	6.20 m (20 ft 4 in)
Sentinel 800	8.00 m (26 ft 3 in)
Weights and Loadings	
Weight	
Max T-O weight	
Sentinel 620	200 kg (440 lb) (est)
Sentinel 800	250 kg (551 lb) (est)
Max launch weight, Sentinel 500M	150 kg (330 lb) (est)
Performance	
Endurance	
max, Sentinel 500M	5 hr
max, Sentinel 620	12 hr (est)
max, Sentinel 800	20 hr (est)

Status: Development and marketing.
Contractor: Advanced Technologies & Engineering Co (Pty) Ltd
  Halfway House.

ATE Vigil series

Type:
Development: According to the ATE website in early 2009, the Vigil range is being offered in response to potential commercial or civilian users' requests for a lower ownership cost alternative to the Vulture system. Two versions (Vigil SR and Vigil EE) are proposed, based on the design of ATE's Sentinel. As with that series, data so far published have been minimal.
Description: *Airframe:* Rectangular section fuselage pod, with high-mounted wings and twin tailbooms supporting an inverted V tail unit; pusher engine installation. Fixed tricycle landing gear.
  Mission payloads: Wide variety of sensors can be fitted, according to mission requirements.
  Guidance and control: Generally as for Vulture and Sentinel systems. Both versions have a common GCS, which can optionally be installed in a light utility vehicle.
  Launch: Conventional wheeled take-off.
  Recovery: Conventional wheeled landing.
  Power plant: Details not yet released.

Vigil SR, Vigil EE

Dimensions, External	
Wings, wing span	4.00 m (13 ft 1½ in)
Weights and Loadings	
Weight	
Max T-O weight	
Vigil SR	40 kg (88 lb) (est)
Vigil EE	80 kg (176 lb) (est)
Payload	
Max payload	
Vigil SR	10 kg (22 lb) (est)
Vigil EE	20 kg (44 lb) (est)
Performance	
Endurance	
Vigil SR	4 hr
Vigil EE	8 hr

Status: Being promoted in early 2009. It was not stated at that time whether any examples had yet been completed and flown.
Contractor: Advanced Technologies & Engineering Co (Pty) Ltd
  Halfway House.

ATE Vulture

Type: Tactical UAV.
Development: The Vulture tactical UAV was developed for brigade-level operations, focused on battlefield surveillance, target localisation and artillery fire correction. It first flew on 17 March 1995, and made its first fully autonomous flight on 31 January 1998. Variants are focused on day and night reconnaissance and surveillance.
  Operational experience led to the South African Army Artillery identifying the need for an automated airborne forward observer to perform artillery fire correction of the 127 mm Multiple Rocket Launcher, the G5-155 mm Towed Gun Howitzer and the G6-155 mm Self-propelled Gun Howitzer. The explicit requirement was for a tactical UAV that would be owned by, and operated under the direct command and control of the Artillery. The system would be operated at brigade level, with a similar degree of high mobility and rapid deployment to that of the SA Artillery itself. ATE was contracted on 20 March 2003 for the acquisition and production of Vulture systems, following an open-tender process and successful evaluation of the Vulture system by the South African Army

ATE Vulture

Vulture on its vacuum-atmospheric launcher *(IHS/Patrick Allen)* 1196118

Vulture air vehicle *(ATE)* 1132405

Vulture can be deployed in unprepared terrain within 30 minutes *(ATE)* 1047885

Vulture air vehicle automated launch *(ATE)* 1047884

during a regimental engagement on 21 November 2001. Within a fixed-price, fixed-delivery framework, ATE was contracted to complete development, to establish the logistics support system, and to begin production deliveries in March 2006.

Description: *Airframe:* High-wing monoplane; pod and boom fuselage; T tail unit; pusher engine. Composites structure of glass fibre and carbon fibre, built up from interchangeable modules optimised for rapid assembly and dismantling, ease of transportation, modular parts replacement following recovery damage, and payload versatility.

Mission payloads: Vulture payload is a stabilised M-Tek turret mounting an electro-optic day sight using the near-IR band. Real-time downlink of video imagery and fall-of-shot data. Multisensor payload in Vulture Upgrade Mk I; combination payload (FLIR, colour daylight TV, laser range-finder and autotracker) or miniature SAR under development for Vulture Upgrade Mk II.

Guidance and control: The ATE Vulture GCS (which see) is the mission command and control centre for the main artillery function. Flight is fully automated from launch to recovery, and features preprogrammed waypoints; reprogramming of waypoints; inertial navigation with DGPS; selectable loiter patterns; and automated procedures with return to recovery area in the event of communication loss. Digital communication is by Tellumat CBACS C-band datalink (which see).

Transportation: Entire system transported on three all-terrain, 10-ton, 6 × 6 Samil 100 trucks (GCS vehicle, launch vehicle and recovery system vehicle); GCS includes microwave data and video link mast and antenna assembly. Launch rail is configured to fold for transport; launch truck also carries fold-out ISO standard containers which accommodate two complete air vehicles with payloads, other spares and support equipment. System is also air-transportable in a C-130 Hercules.

System composition: Comprises three vehicle-mounted subsystems: (1) Launcher vehicle with hydraulically extendable/retractable launch rail, container for two air vehicles, spares, operating equipment and two operators; (2) GCS vehicle with command and control centre, communication to gun battery fire control post, crew cabin and four operators (navigator, observer, artillery technical assistant and signaller); and (3) Recovery vehicle with energy absorption device, airbag, laser tracker and two operators.

Launch: Rail-launched automatically by dedicated ATE atmospheric catapult launcher (see entry for ATE Vulture launch and recovery systems).

Recovery: Automated recovery via a laser-based approach to energy-absorbing strap system and touchdown on an inflatable airbag (see entry for ATE Vulture launch and recovery systems).

Power plant: One 28.3 kW (38 hp) two-cylinder, electronic fuel-injection piston engine; two-blade pusher propeller.

Variants: *Vulture:* Initial version, being produced for South African Army Artillery. Deployment in 30 minutes in unprepared terrain. *Detailed description applies to this version.*

Vulture Upgrade Mk I (Night Vulture): Similar size and characteristics to Vulture, but with range increase to 108 n miles (200 km; 124 miles), endurance increase to 4 hours, and multisensor payload; payload weight increased to 35 kg (77.2 lb) and max launching weight to 135 kg (298 lb).

Vulture Upgrade Mk II (Endurance Vulture): Similar size and characteristics to Vulture Upgrade Mk I, but powered by a direct-injection multifuel engine to increase endurance to 6 to 7 hours, and equipped with modular payloads; payload weight as for Upgrade Mk I, but max launching weight increased to 150 kg (330 lb).

Vulture Upgrade Mk III (Civil Vulture): Similar size and characteristics to other Vulture versions, but air vehicle is equipped with a civil kit to support certification by the South African Civil Aviation Authority. Onboard equipment to allow communication with neighbouring air traffic, monitoring by air traffic control, and avoidance of unco-operative air traffic such as balloons, gliders and other non-powered aircraft, comprises the following: VHF radio, Mode C transponder, anti-collision lights, ballistic recovery parachute, distinguishable paint scheme and see-and-avoid function. This allows Civil Vulture to perform such missions as marine or coastal surveillance, anti-poaching, border patrol and weather monitoring.

Vulture

Dimensions, External	
Overall	
length	3.10 m (10 ft 2 in)
height	0.70 m (2 ft 3½ in)
Wings, wing span	5.10 m (16 ft 8¾ in)
Weights and Loadings	
Weight, Max launch weight	125 kg (275 lb)
Payload, Max payload	25 kg (55 lb)
Performance	
Altitude, Service ceiling	5,000 m (16,400 ft)
Speed	
Cruising speed, max	86 kt (159 km/h; 99 mph)
Loitering speed	65 kt (120 km/h; 75 mph)
Stalling speed	48 kt (89 km/h; 56 mph)
Radius of operation	32.4 n miles (60 km; 37 miles)
Endurance, max	3 hr

Status: First production contract, for undisclosed number of systems, awarded 26 February 2003. First system was delivered to South African Army on 23 March 2006.

Customers: South African Army. To be operated with G5 and G6 artillery as part of an Artillery Target Engagement System (ATES), or as a self-sufficient system. At the end of May 2008 an export order was announced to a undisclosed customer.

Contractor: Advanced Technologies & Engineering Co (Pty) Ltd
Halfway House.

Denel Seeker

Type: Reconnaissance, target location and artillery fire correction UAV.

Development: The development of Seeker was funded by the South African National Defence Force (SANDF) to meet a requirement issued in 1982. The first production systems were delivered to the SANDF for evaluation in 1986. Since its service entry in 1991, the system has been progressively improved, based on experience gained in the field. It was first displayed in public at the FIDA defence exhibition in Chile in early 1988.

The original **Seeker I** system utilised an air vehicle designated Seeker 2D, an improved version of which (Seeker 2E) was introduced in 1995. Denel (formerly Kentron) announced the upgraded **Seeker II** system in November 1999, stating that production of 10 systems was planned initially, with deliveries starting in 2000.

Description: *Airframe:* All-composites, low-drag, low-signature airframe of pod-and-twin-tailboom configuration with a pusher engine, fixed tricycle landing gear and a remotely deployable arrester hook. Communication, flight control and navigation equipment and the mission payload are installed in a spacious bay in the fuselage forward of the wings, on top of which is a jamming-resistant, automatically steerable G/H-band directional antenna. Wings, booms, tail and landing gear are detachable for ease of transportation, storage, field assembly and dismantling.

Mission payloads: Provision for multiple payloads. Those currently in use are fully steerable, gyrostabilised high-resolution colour TV with ×36 zoom and an Eloptro thermal imager. Target recognition slant ranges of 3 n miles (5.5 km; 3.4 miles) for TV and 1.5 n miles (2.8 km; 1.7 miles) for FLIR are typical. Denel Goshawk in Seekers for Abu Dhabi, Algeria and (in specially customised version) South Africa. Saab Avitronics ESP (which see) electronic surveillance payload in export Seeker II (reportedly for Algeria).

Guidance and control: Primary ground/air link of the tracking and communication unit (TCU) is a multichannel C-band tracking system, using an omnidirectional antenna for communicating with the air vehicle in the vicinity of the ground station; and a narrow beamwidth, high-gain dish antenna for tracking and long-range communication up to a range of 108 n miles (200 km; 124 miles). The TCU is an unmanned unit which can be located up to 100 m (330 ft) away from the mission control unit (MCU), enabling it to be positioned for optimum line of sight communication with the air vehicle. A back-up UHF command link, as well as an autonomous 'return to base' capability, permit retrieval of the air vehicle even under adverse conditions.

Current production Seeker II, September 2006 *(IHS/Patrick Allen)*

Seeker operating under South African civil register markings (ZU-RPA)

Seeker II (Seeker 3A) air vehicle

Seeker 400 mock up carrying Impi missiles shown September 2010

Seeker II ground control station vehicle *(IHS/Patrick Allen)*

Unmanned aerial vehicles > South Africa > Denel Seeker

Seeker 400 displayed at the African Aerospace and Defence eixhibition in September 2008 *(IHS/Patrick Allen)* 1369107

Seeker 2E: note the modified nosewheel leg *(Steven J Zaloga)* 0054209

The MCU has three operator workstations for the mission commander, air vehicle pilot and dedicated payload operator/observer. The hardware of the workstations is identical, and each is configured to its purpose by means of workstation-specific software and graphically presented display/control functions. Real-time imagery coverage of the reconnaissance area is relayed to the payload station via the communication downlink. Image enhancement techniques aid the observer with target detection and identification. Air vehicle and designated target positions are determined and superimposed on the imagery and plotted simultaneously on maps. Artillery fire correction data are calculated for display in the MCU and at the mobile receiver unit (MRU). The air vehicle can be programmed to perform a sequence of manoeuvres autonomously.

The MRU aids command and control integration of the Seeker system with deployed ground forces. It receives the imagery picture from the air vehicle and has a voice communication channel to the ground station via the air vehicle. The MRU is typically installed in fire-control or command vehicles. An engine-driven 1.5 kW alternator and an emergency battery provide electrical power to the avionics and datalink equipment.

Transportation: The system can be deployed to a state of operational readiness within 3 to 5 hours of arriving on site. Assembly of the air vehicle is achieved in about 10 minutes, followed by a preflight test. Rapid deployment is also enhanced by a comprehensive self-test capability of the MCU and TCU.

System composition: Four to six air vehicles, three ground vehicles, one mobile GCS, two or three generators, one field support shelter and 10 personnel (four to operate, six to transport and set up). The complete Seeker system consists of a mission control unit, a tracking and communication unit, the air vehicles, payloads, a Field Support Subsystem (FSS), and a mobile receiver unit. The FSS is designed to provide complete self-contained support to the system for considerable periods of operational use. It includes back-up spares and consumables, calibration and test equipment, UAV storage containers, a UAV service tent, UAV handling equipment and generators.

Launch: The aircraft takes off from paved, gravel or grass runways under remote control. For rough, short runways, take-off is aided by means of a winch launcher; can also be used with a zero-length launcher. Night take-offs and landings are performed routinely to enable night reconnaissance under any light conditions.

Recovery: Wheeled landing, shortened by means of arrester cable engaged by retractable underfuselage hook.

Power plant: One 37.3 kW (50 hp) Limbach L 550E four-cylinder two-stroke engine; two-blade pusher propeller. Fuel capacity 64 litres (16.9 US gallons; 14.1 Imp gallons) standard; 79 litres (21 US gallons; 17.5 Imp gallons) maximum.

Variants: Seeker 2D: Initial Seeker I system production air vehicle, in service from 1985.

Seeker 2E: Improved Seeker I system version, first flown in September 1995. Modified wings with trailing-edge flaps and integral fuel tanks; digitally controlled, fuel-injection Limbach two-stroke engine; ceiling, payload capacity and endurance all increased. All SANDF Seeker Is have been ugraded to 2E air vehicle standard. *Detailed description applies to this version*.

Seeker II: Enhanced system, introduced in late 1999. Air vehicle for this system, identified as **Seeker 3A**, has a wider fuselage and a straight nosewheel shock-strut. Capability expanded to include electronic surveillance in the 0.5 to 18 GHz frequency range. Payload 50 kg (110.2 lb); ceiling 5,485 m (18,000 ft); communications radius 135 n miles (250 km; 155 miles), extendable to 216 n miles (400 km; 248 miles) with tactical ground station; time over target at 250 km is 5 hours; total flight endurance 12 hours. Digital avionics and new PC-based ground station; GPS/INS navigation.

Seeker 400: First disclosed early 2008, when it was reported to be in development. Shown African Aerospace and Defence Show September 2008. A mock up was displayed again in 2010, this time armed with two Denel Dynamics Impi laser guided missiles. At this time a launch customer was said to exist for Seeker 400, which is some 30 percent larger than the existing Seeker II. An endurance of some 16 hours and a payload of up to 100 kg plus fuel is claimed. First flight expected 2011-2012.

Seeker 2E

Dimensions, External	
Overall	
length	4.438 m (14 ft 6¾ in)
height	1.30 m (4 ft 3¼ in)
Fuselage, length	3.09 m (10 ft 1¾ in)
Wings, wing span	7.00 m (22 ft 11½ in)
Tailplane, tailplane span	1.60 m (5 ft 3 in)
Wheels, wheel track	1.20 m (3 ft 11¼ in)
Dimensions, Internal	
Payload bay, volume	0.12 m³ (4.23 cu ft)
Areas	
Wings, Gross wing area	0.12 m² (1.3 sq ft)
Weights and Loadings	
Weight	
Weight empty	151 kg (332 lb)
Max launch weight	240 kg (529 lb)
Fuel weight, Max fuel weight	61 kg (134 lb)
Payload, Max payload	50 kg (110 lb)
Performance	
T-O	
T-O run	300 m (985 ft)
Climb	
Rate of climb, at S/L	305 m/min (1,000 ft/min)
Altitude, Service ceiling	5,485 m (18,000 ft)
Speed	
Max level speed	120 kt (222 km/h; 138 mph)
Cruising speed	70 kt (130 km/h; 81 mph)
Radius of operation	
without TGS	135 n miles (250 km; 155 miles)
with TGS	216 n miles (400 km; 248 miles)
Endurance	
standard payload	15 hr
max payload	12 hr
Time on station	
at 135 n mile (250 km; 155 mile) radius	5 hr
at 54 n mile (100 km; 62 mile) radius	4 hr
Landing	
Landing run, arrested	70 m (230 ft)

Status: The Seeker began production in 1985 and entered SANDF service in 1991. It is operated on behalf of the SANDF by the South African Air Force and was used operationally by No 10 Squadron during the closing years of the Namibia/Angola border conflict. In the second quarter of 1994, after being specially cleared to operate in civilian-controlled airspace, three Seekers were modified and deployed to monitor polling stations during South Africa's first free elections. Since 1996, Seekers have been operated by the SANDF in co-operation with the South African Police Services and other non-government organisations (NGOs) in surveillance missions, mostly in controlled, non-military airspace, to combat urban crime and illegal immigration and for other law enforcement activities. They have also been deployed on counter-insurgency surveillance patrols in the Kwa-Zulu area of South Africa and elsewhere.

Workstations inside the Seeker II GCS *(IHS/Patrick Allen)* 1196239

Spain

Aerovision Fulmar

Type: MALE mini-UAV.

Development: Aerovision was formed in 2003, and the Fulmar made its first flight in 2004; production-standard development was completed in early 2007.

Description: *Airframe:* Flying wing design, with diamond-shaped central section and sweptback outer panels; pod fuselage, with dorsally mounted pusher engine; twin, inward-canted, independent fins and rudders on short booms from inner/outer wing intersections. No landing gear. Composites construction (Kevlar and carbon fibre). Airframe is watertight to permit recovery at sea.

Mission payloads: Gyrostabilised two-axis gimbal mounting under aircraft's nose for an Aerovision 360° pan/100 tilt daylight video camera with ×18 zoom; a 7.5 to 13.5 micron IR video camera; or a 7 to 14 MP digital camera with optional IR filter. Real-time 2.4 GHz video downlink.

Guidance and control: Pre-programmed and fully automatic, including take-off and landing; GPS waypoint navigation and AP-04 autopilot; 900 MHz command uplink.

Transportation: System is man-portable by two persons, or can be transported by MPV and trailer. Set-up time by two persons is less than 30 minutes.

Launch: By elastic catapult from 6 m (19.7 ft) ramp. Can be launched from ground or on board ship.

Recovery: On land, into a net whose supports inflate automatically when aircraft enters landing mode. At sea, can also be landed on a pneumatically actuated skid.

Power plant: One 2.05 kW (2.75 hp) single-cylinder two-stroke engine, driving a two-blade pusher propeller. Fuel capacity 6.5 litres (1.7 US gallons; 1.4 Imp gallons).

Fulmar

Dimensions, External	
Overall, length	1.23 m (4 ft 0½ in)
Fuselage, height	0.20 m (7¾ in)
Wings, wing span	3.10 m (10 ft 2 in)
Dimensions, Internal	
Payload bay, volume	0.000116 m³ (0.0 cu ft)
Weights and Loadings	
Weight, Max launch weight	20 kg (44 lb)
Payload, Max payload	1.0 kg (2.00 lb)
Performance	
Altitude, Service ceiling	5,000 m (16,400 ft)
Speed	
Launch speed	25.20 m/s (83 ft/s)
Max level speed	81 kt (150 km/h; 93 mph)
Cruising speed	54 kt (100 km/h; 62 mph)
Loitering speed	32 kt (59 km/h; 37 mph)
Radius of operation, mission	216 n miles (400 km; 248 miles)
Endurance, max	8 hr

Status: Developed and available.

Contractor: Aerovision Vehículos Aereos SL, San Sebastian.

The Fulmar medium-range endurance UAV *(IHS/Patrick Allen)* 1325914

Fulmar with a typical undernose sensor *(IHS/Patrick Allen)* 1325916

Alpha Unmanned Systems Atlantic

Type: Short-range tactical UAV.

Development: The Atlantic was designed for tactical ISTAR military or maritime applications, or civil tasks such as search and rescue, environmental monitoring, power or pipeline inspection, cartography and scientific research.

Description: *Airframe:* Semi-elliptical high wing; pod and boom fuselage, with V tail; pusher engine. Fixed tricycle or skid landing gear. Composites construction (carbon fibre and Kevlar).

Mission payloads: EO/IR dual-sensor payload in retractable, gyro-stabilised ventral turret; 2.4 GHz video downlink with optional encryption. Payload options include three-sensor turret and radio or laser altimeter.

Guidance and control: Fully redundant flight control system, based on UAV Navigation AP04 autopilot. Laptop or rack-type GCS computer, with joystick control of air vehicle and cameras. Video receiver, automatically steered high-gain directional antenna and independent power supply. Option for simultaneous control of multiple air vehicles.

Transportation: Air vehicle can be easily dismantled and stowed in a specialised case for transportation.

System composition: Two air vehicles, payloads, GCS and launch catapult. Two-person operation.

Launch: Fully automatic, either by wheeled take-off or by launch from pneumatic catapult.

Recovery: Fully automatic wheel or skid landing. Parachute option for emergency recovery.

Power plant: One 6.7 kW (9 hp) single-cylinder two-stroke piston engine, driving a three-blade pusher propeller. Fuel capacity 12 litres (3.2 US gallons; 2.6 Imp gallons).

The short-range fixed-wing Atlantic complements the company's VTOL types *(Alpha Unmanned Systems)* 1395262

Atlantic

Dimensions, External	
Overall, length	2.80 m (9 ft 2¼ in)
Wings, wing span	3.80 m (12 ft 5½ in)
Dimensions, Internal	
Payload bay, volume	0.009 m³ (0.3 cu ft)
Weights and Loadings	
Weight	
Max T-O weight	31 kg (68 lb)
Max launch weight	31 kg (68 lb)
Payload, Max payload	10 kg (22 lb)
Performance	
Altitude, Service ceiling	4,000 m (13,120 ft)
Speed	
Max level speed	97 kt (180 km/h; 112 mph)
Cruising speed	65 kt (120 km/h; 75 mph)
Radius of operation, mission	54 n miles (100 km; 62 miles)
Power plant	1 × piston engine

Status: Being promoted in 2010.
Contractor: Alpha Unmanned Systems
Alcobendas, Madrid.

Alpha Unmanned Systems Commando

Type: Helicopter mini-UAV.
Development: Designed for both civil and military applications, particularly in urban environments.
Description: *Airframe:* Two-blade main and tail rotors; pod and boom fuselage; twin-skid landing gear. Aluminium alloy and carbon fibre construction.
Mission payloads: Day, low-light, night vision or IR payload in undernose, gyro-stabilised and geo-referenced installation; or as specified by customer. Real-time video downlink (900 MHz, 1.3 GHz or 2.4 GHz available, plus optional encryption).
Guidance and control: Fully autonomous, including take-off and landing. One-person operation; otherwise as described for company's Atlantic UAV.
Transportation: Man-portable in a weatherproof 2.0 × 0.5 × 0.7 m (6.56 × 1.64 × 2.30 ft) carrying case.
System composition: Two air vehicles, payloads and GCS.
Launch: Automatic vertical take-off.
Recovery: Automatic vertical landing.
Power plant: One three-phase brushless electric motor, powered by a lithium-polymer battery.

Commando

Dimensions, External	
Overall, height	0.50 m (1 ft 7¾ in)
Fuselage, length	1.50 m (4 ft 11 in)
Rotors, rotor diameter	1.80 m (5 ft 10¾ in)
Weights and Loadings	
Weight	
Weight empty	6.5 kg (14.00 lb)
Max launch weight	8.0 kg (17.00 lb)
Payload, Max payload	1.5 kg (3.00 lb)
Performance	
Altitude, Service ceiling	3,000 m (9,840 ft)
Speed	
Max level speed	64 kt (119 km/h; 74 mph)
Loitering speed	35 kt (65 km/h; 40 mph)
Radius of operation, mission	25 n miles (46 km; 28 miles)
Endurance, max	1 hr
Power plant	1 × electric motor

Status: Company states that Commando is in operation with 'worldwide' police and security services. It is also used as a trainer for the larger Sniper.
Contractor: Alpha Unmanned Systems
Alcobendas, Madrid.

The autonomous, electric-powered rotary-wing Commando *(IHS/Patrick Allen)* 1419027

Alpha Unmanned Systems Sniper

Type: VTOL mini-UAV.
Development: Developed for wide range of military and civil roles, including ISR, law enforcement and fire detection.
Description: *Airframe:* Two-blade main and tail rotors; aluminium alloy pod-and-boom fuselage. Twin-skid landing gear.
Mission payloads: TV or IR camera (standard or customer-specified) in gyro-stabilised, geo-referenced undernose turret real-time video downlink (900 MHz, 1.3 GHz or 2.4 GHz options, including encryption).
Guidance and control: Fully automated, including take-off and landing. One-person operation; otherwise as described for company's Atlantic UAV.
Transportation: Man-portable in 2.0 × 0.7 × 0.8 m (6.56 × 2.30 × 2.62 ft) weatherproof carrying case.
System composition: Two air vehicles, payloads and GCS.
Launch: Automatic vertical take-off.
Recovery: Automatic vertical landing.
Power plant: One two-cylinder two-stroke engine (rating not stated).

Sniper

Dimensions, External	
Overall, height	0.70 m (2 ft 3½ in)
Fuselage, length	1.60 m (5 ft 3 in)
Rotors, rotor diameter	1.80 m (5 ft 10¾ in)
Weights and Loadings	
Weight, Max T-O weight	14.0 kg (30 lb)
Payload, Max payload	4.0 kg (8.00 lb)
Performance	
Altitude, Service ceiling	3,000 m (9,840 ft)
Speed, Max level speed	81 kt (150 km/h; 93 mph)
Radius of operation, mission	25 n miles (46 km; 28 miles)
Endurance	3 hr (est)

Status: Stated to be in operation with 'security forces worldwide'. Civil customers include Iberdrola, which received two for oil pipeline inspection role.
Contractor: Alpha Unmanned Systems
Alcobendas, Madrid.

Partial view of the Sniper VTOL UAV *(Alpha Unmanned Systems)* 1395264

Indra Albhatros

Type: Tactical mini-UAV.
Development: Unveiled in mid-2009 as a low-cost ISTAR and training asset with a low logistic footprint.
Description: *Airframe:* Shoulder-wing design, with flaps; pod and boom fuselage with pusher engine mounted atop pod; T tail. Fixed tricycle landing gear. Composites construction.
Mission payloads: Options include gyrostabilised or gimballed IAI Tamam MiniPop or MicroPop EO; non-stabilised EO/IR or multispectral cameras; laser illuminator and telemeter; and Blue Force Tracking device. Real-time downlinking of colour video.
Guidance and control: GCS comprises a light, compact, dust- and waterproof PC for mission control and planning, plus an RVT, and is equipped for video digital recording and screen-shot capture. Operation can be fully automatic, assisted or manual; GPS/INS-based navigation. Command and control, video, telemetry and payload data transmitted within LOS by 900 to 2,500 MHz digital datalinks. One-man operation.
Transportation: Complete system is transportable by small ground vehicle or helicopter, in containers with a combined weight of less than 250 kg (551 lb).
System composition: Three air vehicles, payload(s), GCS, datalinks, flight simulator, spares and support equipment.

Indra's 45 kg Albhatros mini-UAV *(Indra)* 1356275

Launch: Conventional wheeled take-off from unprepared airstrips.
Recovery: Conventional wheeled landing. Aircraft incorporates a programmed recovery system if communications link is lost.
Power plant: One 5.2 kW (7 hp) rotary piston engine; two-blade pusher propeller. Gasoline or 'heavy' fuel.

Albhatros

Dimensions, External	
Overall, length	2.47 m (8 ft 1¼ in)
Wings, wing span	3.98 m (13 ft 0¾ in)
Weights and Loadings	
Weight, Max T-O weight	45 kg (99 lb)
Payload, Max payload	10 kg (22 lb)
Performance	
T-O	
T-O run	100 m (329 ft) (est)
Altitude, Service ceiling	3,000 m (9,840 ft)
Speed	
Max level speed	91 kt (169 km/h; 105 mph)
Loitering speed	50 kt (93 km/h; 58 mph)
Stalling speed	34 kt (63 km/h; 40 mph)
Radius of operation, mission	54 n miles (100 km; 62 miles)
Endurance, max	8 hr
Landing	
Landing run	100 m (329 ft) (est)
Power plant	1 × piston engine

Status: Reported as being promoted in 2009.
Contractor: Indra Sistemas SA
  Torrejón de Ardoz.

Indra Mantis

Type: Close-range tactical mini-UAV.
Development: Unveiled in mid-2009.
Description: Airframe: Parasol-wing design; pod and boom fuselage; pusher engine; T tail. No landing gear. Composites construction.
  Mission payloads: Interchangeable noses for daylight (colour video with ×10 zoom) or night (IR) cameras, or laser illuminator. Telemetry and Blue Force tracking device.
  Guidance and control: One-person operation from light, compact, PC-based GCS. GPS/INS navigation (automatic/pre-programmed, assisted or all-manual). Digital datalink (900 to 2,500 MHz) for C² and payload operation; digital video recording and screenshot capture.
  Transportation: Entire system can be stowed in waterproof and dust-resistant backpacks and carried by two people. Set-up/knockdown time 5 to 10 minutes.
  System composition: Standard system comprises three air vehicles, payloads, GCS, datalink, batteries and charger, flight simulator software, plus spares and support equipment.
  Launch: Hand launch.
  Recovery: Automatic, to flap-assisted, steep descent belly landing.
  Power plant: Battery-powered electric motor (rating not stated); two-blade pusher propeller.

Mantis

Dimensions, External	
Overall, length	1.32 m (4 ft 4 in)
Wings, wing span	2.10 m (6 ft 10¾ in)
Weights and Loadings	
Weight, Max launch weight	3.0 kg (6.00 lb)
Performance	
Speed	
Max level speed	54 kt (100 km/h; 62 mph)
Loitering speed	38 kt (70 km/h; 44 mph)
Stalling speed	17 kt (32 km/h; 20 mph)
Radius of operation, LOS, mission	5.4 n miles (10 km; 6 miles)
Endurance, max	1 hr 15 min
Power plant	1 × electric motor

Status: Reported as being promoted in 2009.
Contractor: Indra Sistemas SA
  Torrejón de Ardoz.

Indra Pelicano

Type: Multirole helicopter UAV.
Development: Announced in October 2009, the Pelicano is a collaborative R & D project with CybAero AB of Sweden, financed jointly by Indra and the Spanish Ministry of Industry, Commerce and Tourism. Development is aimed primarily at an operational system to meet Spanish Navy requirements, for introduction into service in 2012. Intended maritime and ground military applications include reconnaissance, anti-piracy patrol, troop protection and humanitarian missions. Additional roles are expected to be homeland security (border and traffic control, emergency management, search and rescue) as well as such civil applications as cartography and pipeline inspection.
Description: Airframe: Pelicano is based on a modified version of the Swedish CybAero APID helicopter UAV (which see).
  Mission payloads: Gyrostabilised EO/IR camera. Growth potential for Ka-band lightweight SAR radar, EW devices, NBC sensors, or light armament and weapons.
  Guidance and control: Autonomous; from common ground station, as described for Indra's Albhatros and Mantis UAVs. Two air vehicles can be controlled simultaneously.
  Transportation: Transportable in 4 × 4 vehicles or by air.
  System composition: One GCS and three or four air vehicles.
  Launch: Automatic vertical take-off from land or ship deck.
  Recovery: Automatic vertical landing on land or ship deck (with deck lock in latter case).
  Power plant: One heavy-fuel engine (type and rating not stated). Fuel capacity 52 litres (13.7 US gallons; 11.4 Imp gallons).

Pelicano

Dimensions, External	
Overall	
length, rotors turning	4.00 m (13 ft 1½ in)
height	1.20 m (3 ft 11¼ in)
Fuselage	
length	3.40 m (11 ft 1¾ in)
width	0.95 m (3 ft 1½ in)
Rotors, rotor diameter	3.30 m (10 ft 10 in)
Weights and Loadings	
Weight, Max T-O weight	200 kg (440 lb)
Payload, Max payload	30 kg (66 lb)
Performance	
Altitude, Service ceiling	3,600 m (11,820 ft) (est)

The extremely lightweight Mantis UAV *(Kenneth Munson)* 1356276

Indra's Pelicano is based on the CybAero APID 60 *(Indra)* 1395223

Pelicano

Speed	
Cruising speed, max	49 kt (91 km/h; 56 mph)
Radius of operation	27 n miles (50 km; 31 miles)
Endurance	4 hr (est)

Status: Development continuing in 2010.
Contractor: Indra Sistemas SA
Torrejón de Ardoz.

INTA HADA

Type: Technology demonstrator.
Development: The HADA (Helicopter Adaptive Aircraft) is part of a larger Spanish national programme known as Platino (*Plataforma Ligera Aérea de Tecnologías Innovadoras*, or Light Aerial Platform for Innovative Technologies), backed by the Spanish Ministries of Defence, Science and Education and involving more than 40 companies and 15 research centres in Spain. The HADA air vehicle takes the form of a helicopter-configured VTOL machine that can 'morph' in flight by deploying fixed wings for more efficient cruising performance. Design of the HADA began in November 2006 and a prototype configured for ISR and security missions was scheduled to make its maiden flight in 2009. INTA is R&D leader of the programme, with Aries Complex leading the industry team. Two sub-scale prototypes, Colibri (Hummingbird) and Alondra (Lark), were expected to proceed to Libélula (Dragonfly), a full-scale vehicle.

Other elements of the larger Platino programme are known by the acronyms COBOR (optical links for diffuse infrared and fibre-optic avionics); MINISARA (miniature SAR for UAVs and light aircraft); SANAS (automatic safe air navigation system); and SATA (VTOL aircraft automatic landing system).

Description: *Airframe:* Two-blade main rotor and shrouded, Fenestron-type tail rotor; V tail surfaces. Two half-span wings attached to fuselage underside, which are retracted for flight in helicopter mode and deployed to full span in 'fixed-wing' aircraft mode. Pusher propeller at rear of fuselage, to which power is transferred from engine when in aircraft mode, with both rotors disengaged. Fixed skid-type landing gear. Composites construction. Advantages claimed for the configuration are:
- wings not exposed to rotor downwash in helicopter mode, avoiding negative lift effects
- wings not exposed to drag effects in low-speed helicopter flight mode
- transition to aircraft mode is performed at a speed corresponding to the maximum lift/drag ratio of the helicopter, enabling wing to be of small dimensions and requiring only ailerons for roll control
- tail rotor can be open or closed for reasons of safety, compactness and noise levels
- mode transition phases are safer than those of tiltrotor designs
- underfuselage portion of wings can be utilised to house additional fuel
- hardpoints on outer wings can be equipped for weapons or auxiliary fuel tanks.

Elements of the flight control system relating to transitional flight phase were in development and stability characteristics were reported to be under analysis in 2007. Particular design challenges include flight dynamics in pitch, and weight associated with the rotor folding mechanism.

Mission payloads: COBOR and MINISARA (see *development*).
Guidance and control: No details provided, but autonomous operation would be assumed.
Launch: Conventional helicopter take-off, climbing to operational altitude and reaching horizontal transition speed before deploying wings, transferring power to pusher propeller and accelerating to aircraft-mode cruising speed.
Recovery: By reversing take-off procedure and converting back from aircraft to helicopter mode for vertical landing.
Power plant: One 130 kw (174 shp) turboshaft or piston engine.

HADA

Dimensions, External	
Wings, wing span	6.00 m (19 ft 8¼ in)
Rotors, rotor diameter	6.00 m (19 ft 8¼ in)
Areas	
Wings, Gross wing area	4.00 m² (43.1 sq ft)
Rotor disc	28.27 m² (304.3 sq ft)
Weights and Loadings	
Weight, Max T-O weight	360 kg (793 lb)
Payload, Max payload	100 kg (220 lb)
Performance	
Speed, Max level speed	230 kt (426 km/h; 265 mph)
Endurance	6 hr (est)
Power plant	1 × turboshaft

Status: Development of the HADA was started in 2007. Near-term market for the unmanned version was seen as navies and civil agencies worldwide. A larger (20-passenger) manned version with 300 n mile (555 km; 345 mile) range was also envisaged. The project was scheduled for completion in 2010. No flight testing of prototypes had been reported by that date.
Contractor: Instituto Nacional de Técnica Aeroespacial
Madrid.

INTA Milano

Type: MALE UAV.
Development: Milano is an R & D programme, which began in 2006 based on experience from the smaller SIVA delivered to the Spanish Army. About three times the size of SIVA, it is aimed at both military and civil missions — in the former case, for RSTA and UAV operator training. Envisaged civil roles include border control, forest fire detection and monitoring and as an aeronautical systems testbed. By mid-2009, contracts worth an approximate total of EUR690,000 had been placed for airframe manufacture by Sofitec Ingenieria (wings) and Avio Composites (fuselage and tail unit). At that time, a maiden flight had been anticipated for the end of 2009.
Description: *Airframe:* Low wing, with sweepback on outer leading-edges; tapered fuselage, with dorsal fairing over satcom antenna; V tail. Retractable tricycle landing gear. Composites construction.
Mission payloads: EO/IR cameras in undernose turret; SAR in underwing pod.
Guidance and control: From truck-mounted GCS and trailer-mounted GDT with satcom two-way datalinks.
Launch: Conventional, automatic wheeled take-off.
Recovery: Conventional, automatic wheeled landing.
Power plant: Single engine (type and rating not revealed), driving a three-blade tractor propeller.

Milano

Dimensions, External	
Overall	
length	8.21 m (26 ft 11¼ in)
height	1.41 m (4 ft 7½ in)
Wings, wing span	12.54 m (41 ft 1¾ in)
Weights and Loadings	
Weight, Max T-O weight	900 kg (1,984 lb)
Performance	
Altitude, Service ceiling	7,000 m (22,960 ft)
Speed	
Cruising speed, max	124 kt (230 km/h; 143 mph)
Radius of operation, LOS	81 n miles (150 km; 93 miles)
Endurance	20 hr

Status: When construction was reported to have started in mid-2009, a prototype maiden flight was anticipated to take place before the end of that year.
Contractor: Instituto Nacional de Técnica Aeroespacial
Madrid.

Mid-2009 model of the INTA Milano *(Kenneth Munson)* 1356234

INTA SIVA

Type: Surveillance UAV.
Development: SIVA (*Sistema Integrado de Vigilancia Aérea*) was developed jointly by Spain's Instituto Nacional de Técnica Aeroespacial, which

HADA in winged aircraft mode *(INTA)* 1290206

INTA SIVA – Tekplus Aerospace Centauro < Spain < Unmanned aerial vehicles

SIVA on Aries pneumatic launcher *(IHS/Kenneth Munson)* 0527019

SIVA with wings fully folded *(IHS/Kenneth Munson)* 0527020

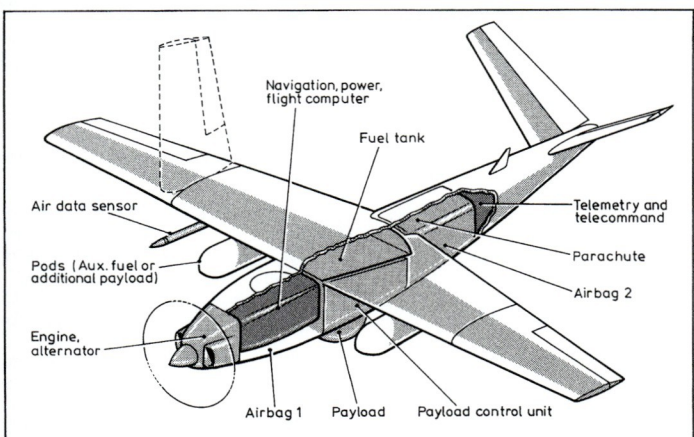
Main features of the SIVA air vehicle *(Mike Keep)* 0518224

SIVA development aircraft with fixed landing gear *(IHS/Kenneth Munson)* 0527018

formed a consortium in the early 1990s with Ceselsa of Spain and Dornier of Germany. It was first shown publicly at the Paris Air Show in June 1993. INTA designed the system, Ceselsa built the GCS and Dornier was responsible for the original flight navigation subsystem. Six air vehicle prototypes were built; flight testing of these began in the last quarter of 1995.

Description: *Airframe:* High-wing monoplane with V tail (included angle 90°); tapered outer portion of each wing folds upward through 90° for parachute deployment, to stabilise air vehicle prior to descent, and fully inward for storage and transportation. Pod under inboard portion of each wing can accommodate additional payload or auxiliary fuel. Composites construction (carbon and glass fibres), with some aluminium or titanium inserts. Optional (1B) fixed tricycle landing gear.

Mission payloads: Centrally located payload bay accommodates a gimballed, steerable and retractable turret, gyrostabilised in azimuth and elevation, upon which are mounted a combined FLIR camera and CCD TV camera with boresighted lines of view for target detection and tracking. The FLIR operates in the 3 to 5 micron IR band and has three fields of view: 1.7, 6 and 23°; the TV camera has continuous zoom between 1.7 and 27°. A total of 2 kW of onboard electrical power is available for these or such other payloads as SAR or elint/EW packages.

Guidance and control: Missions can be autonomous (preprogrammed or with DGPS-updatable waypoints), or ground-controlled (semi-automatic or manual mode). A Modular GCS which is installed in a standard ISO shelter, controls and tracks the air vehicle via a UHF radio-command uplink, receiving real-time imagery data and tracking information via an S-band (2.2 to 2.4 GHz) telemetry downlink. The GCS has three workstations (mission controller, air vehicle controller and payload controller), plus equipment for debriefing, tactical communications and image processing. Navigation and tracking are accurate to 25 m (82 ft) CEP.

Transportation: The entire system is fully transportable by road, sea or air.

System composition: The complete system comprises four air vehicles, four ground vehicles (mobile GCS, GDT trailer, launch vehicle and recovery/maintenance vehicle), and 12 personnel (seven to transport and set up, five to operate).

Launch: By conventional wheeled take-off or by pneumatic catapult.

Recovery: By conventional wheeled landing, or by parachute system which deploys under command (or automatically in the event of an emergency). Twin underfuselage airbags cushion landing impact.

Power plant: One 19.4 kW (26 hp) Sachs SF2-350 two-cylinder two-stroke engine initially; since replaced by a 37.0 kW (49.6 hp) Rotax 503 UL flat-twin; two-blade fixed-pitch propeller. Main fuel tank in centre of fuselage; provision for auxiliary fuel in underwing pods.

SIVA

Dimensions, External	
Overall	
length	4.025 m (13 ft 2½ in)
height	1.15 m (3 ft 9¼ in)
width, wings folded	2.20 m (7 ft 2½ in)
Fuselage	
height, max	0.55 m (1 ft 9¾ in)
width, max	0.42 m (1 ft 4½ in)
Wings, wing span	5.81 m (19 ft 0¾ in)
Engines, propeller diameter	0.71 m (2 ft 4 in)
Dimensions, Internal	
Payload bay, volume	0.015 m³ (0.5 cu ft)
Weights and Loadings	
Weight, Max launch weight	300 kg (661 lb)
Fuel weight, Max fuel weight	60 kg (132 lb)
Payload, Max payload	40 kg (88 lb)
Performance	
Climb	
Rate of climb, max, at S/L	300 m/min (984 ft/min)
Altitude, Service ceiling	6,000 m (19,680 ft)
Speed	
Max level speed	92 kt (170 km/h; 106 mph)
Cruising speed, normal	76 kt (141 km/h; 87 mph)
Radius of operation, mission	81 n miles (150 km; 93 miles)
Endurance	
with max payload	6 hr
with max fuel	10 hr

Status: Contractor development continued through 2002–03; one system was handed over to Spanish Army for operational evaluation July 2003. One production system was completed and was delivered to the Spanish Army in September 2006 for training. The SIVA system was displayed in Madrid during 2008, assigned to 63rd Field Artillery Regiment's Target Identification and Acquisition Group. The system remains in service in 2011.

Customers: Spanish Army.

Contractor: Instituto Nacional de Técnica Aeroespacial Madrid.

Tekplus Aerospace Centauro

Type: Helicopter UAV.

Development: Formed in 2005, Tekplus Aerospace has undertaken aerostructures design work for Airbus and Boeing. The Centauro, its first UAS venture, is targeted at civil applications in Latin America and was

Overhead view of the Centauro *(Tekplus)* 1395323

unveiled in April 2011 at the LAAD exhibition in Rio Janeiro. Development is reported as having been supported by the Spanish MoD and the Centre for Industrial Technological Development (CDTI). Three different-sized variants are projected.

Description: *Airframe:* Two-blade main and tail rotors; pod and boom fuselage; upper and lower vertical fins. Three-leg fixed landing gear. Construction carbon fibre, aluminium alloy and titanium.

Mission payloads: Can include video or IR cameras, SAR, laser designator, EW packages or communications relay. Real-time imagery downlink.

Guidance and control: Pre-programmed with autopilot.

Transportation: Air vehicle(s) and GCS transported on URO Vamtac truck.

System composition: One or more air vehicles, GCS, data terminal and remote video terminal.

Launch: Vertical take-off.

Recovery: Vertical landing.

Power plant: One 19.4 kW (26 hp) two-stroke piston engine.

Variants: *C 18:* Smallest version, optimised for urban missions, fast set-up and easy operation. Payload 7 kg (15.4 lb), endurance 2 hours.

C 30: Short-range, more capable version, with wider range of payload options. Main description applies to this version.

C 50: Largest of current versions; medium range, with payload maximum of more than 75 kg (165 lb) and endurance of up to 8 hours.

In-flight impression of the Centauro *(Tekplus)* 1395324

C 30

Dimensions, External	
Overall	
length	3.60 m (11 ft 9¾ in)
height	1.60 m (5 ft 3 in)
Rotors, rotor diameter	3.50 m (11 ft 5¾ in)
Weights and Loadings	
Weight, Max T-O weight	100 kg (220 lb)
Payload, Max payload	30 kg (66 lb)
Performance	
Altitude, Service ceiling	3,500 m (11,480 ft)
Speed, Max level speed	65 kt (120 km/h; 75 mph)
Radius of operation	108 n miles (200 km; 124 miles) (est)
Endurance	6 hr
Power plant	1 × piston engine

Status: A C30 prototype has been successfully test-flown, and the design is described as fully developed. Tekplus reportedly began production of an initial batch of 20 Centauros by the end of 2011.

Contractor: Tekplus Aerospace
Tres Cantos, Madrid.

Sweden

CybAero APID 55/60

Type: Multirole helicopter UAV.

Development: The APID concept (Autonomous Probe for Industrial Data acquisition) was begun in the early 1990s as a joint development by Scandicraft Systems AB (CybAero's former name) and the FAO (Swedish National Defence Research Establishment). Its early history has been described in previous editions; APID 60 with a revised fuselage is its latest incarnation.

APID is a fully autonomous, multipurpose VTOL UAV, designed to carry a wide variety of sensors and/or other equipment. Civil applications include aerial photography, forest fire assessment, environmental monitoring, powerline inspection, search and rescue, border patrol and day/night traffic surveillance. Military applications include electronic warfare, surveillance, target acquisition and designation, and ordnance survey, especially in high-threat environments. Modular design of both the mechanical and the electronic subsystems affords flexibility and adaptability to specific customer needs. CyAero and Spanish company Indra teamed to develop a rotary wing UAV for naval vessels under the Pelicano project, based on the APID 60. APRID flew A demonstration mission which was flown from a Pakistan Navy frigate in 2008 and the APID air vehicle has also participated in a Norwegian research programme to detect and classify oil spills in the Arctic.

Description: *Airframe:* Two-blade main rotor with Bell-Hiller stabiliser bar and paddles; two-blade tail rotor. Pod and boom fuselage. Construction is of carbon fibre, titanium and aluminium. Landing gear on APID 60 comprises skids; APID 55 has twin mainwheels on cantilever legs plus V-shaped skid under tailfin.

Mission payloads: Can include stabilised cameras, infra-red sensors, microphones, laser scanners, radio scanners and jammers, antennas and other equipment. Large underfuselage mounting area for standard gyrostabilised sensors or other customer-specified payloads, for which 700 W of onboard electrical power is available.

Guidance and control: Aircraft is stabilised and navigated by an advanced, self-contained FCS-52 flight control system. Highly accurate positioning and self-navigation between programmed waypoints, which can be altered during flight via a user-friendly graphic interface. Alternatively, can be operated semi-manually, with operator navigation via a joystick. Optional, customised payload control system. Both systems normally housed in the transport vehicle, but can be operated at any desired location. In all operating modes, the onboard control system provides automatic attitude stabilisation and keeps the aircraft within safe

APID 60 has skid landing gear *(CybAero)* 1415693

APID 55 in 2006 configuration *(CybAero)* 1194903

The multirole APID 55 *(CybAero)*

APID 55 in flight *(CybAero)*

Aircraft Maintenance Company (GAMCO). Seven were ordered and deliveries were scheduled for completion in 2009.
Contractor: CybAero AB
Linköping.

Saab Skeldar

Type: Helicopter UAV.
Development: Developed on the basis of the CybAero APID 55 airframe (which see). Conceptual work was started in late 2004; first flight of Skeldar 5 POC prototype took place in May 2006. Both military and civil applications were foreseen, and a maritime version is also being considered. By 2008 the designation Skeldar V-200 was being applied to a developed version.
Description: *Airframe:* Typical pod and boom fuselage of carbon fibre, titanium and aluminium. Two-blade main and tail rotors. Bell/Hiller type main rotor system with stabiliser bar and paddles. Fixed twin-skid landing gear.
Mission payloads: Customer-specific COTS payloads such as EO/IR sensors, SAR sensors combining radar and target indicator capabilities, and advanced EW suite. Sensor data transmitted via wide-band datalink and other data via narrow band.
In February 2007, Skeldar was flying with a PolyTech Cobolt 275 EO sensor. Other options under consideration at that time included the Selex PicoSAR miniature SAR and Saab Avitronics ESP electronic surveillance equipment. Megaphone, searchlight and light cargo hook also optional.
Guidance and control: Fully autonomous management of air vehicle, payload and mission; GPS navigation. Modular GCS incorporates power supply, environmental control system and vehicle and sensor operator workstations. It can control more than one UAV at a time, has UHF radio link with air traffic control before and during flight, and can be integrated into various platforms such as vehicles, trailers or ground-based containers.
Transportation: Can be dismantled for transportation.
System composition: Two air vehicles, payloads, GDT and GCS.
Launch: Vertical take-off.
Recovery: Vertical landing.
Power plant: One unspecified 41.0 kW (55 hp) water-cooled, fuel-injected, two-cylinder two-stroke engine.
Variants: *Skeldar 5:* Proof of concept prototype.
Skeldar V-150: Proposed initial production model. Unveiled at Eurosatory defence exhibition, June 2006.
Skeldar V-200: Revised and enlarged version with redesigned electrical system, transmission and avionics flying in 2008.

operating limits. Due to its full autonomy, the air vehicle can maintain total radio silence. Two operators (one for air vehicle control and monitoring, one for sensor payload control). System requires a minimum of operator training.
Transportation: Can be transported in a hatched trailer, hatchback or pick-up truck.
System composition: Two air vehicles, GCS, payloads and two HMMWV or similar ground vehicles.
Launch: Autonomous vertical take-off.
Recovery: Autonomous vertical landing.
Power plant: One 41.0 kW (55 hp) water-cooled two-cylinder two-stroke engine with fuel injection and electric starting. Power train comprises drive shaft, drive belt and main gearbox.
Fuel capacity 60 litres (15.9 US gallons; 13.2 Imp gallons).

APID 55, APID 60

Dimensions, External	
Overall	
length, rotors turning	4.00 m (13 ft 1½ in)
height	1.20 m (3 ft 11¼ in)
Fuselage	
length	3.20 m (10 ft 6 in)
width, max	0.95 m (3 ft 1½ in)
Rotors, rotor diameter	3.30 m (10 ft 10 in)
Weights and Loadings	
Weight	
Weight empty	95 kg (209 lb)
Max T-O weight	160 kg (352 lb)
Payload	
Max payload, incl fuel	55 kg (121 lb)
Performance	
Altitude	
Service ceiling	
depending upon payload	3,000 m (9,840 ft)
Speed	
Max level speed	48 kt (89 km/h; 55 mph)
Cruising speed	32 kt (59 km/h; 37 mph)
Radius of operation, max	27 n miles (50 km; 31 miles)

Status: APID system had been deployed in both military and civilian applications.
Customers: Has been used by Swedish armed forces and several civilian customers. Ordered by United Arab Emirates in June 2004, with deliveries of knock-down kits from Sweden for local assembly by Gulf

Skeldar V-200 *(Saab)*

Saab AeroSystems Skeldar in flight

Skeldar V-150 mockup at Eurosatory, June 2006 *(IHS/Patrick Allen)* 1182447

Mockup of proposed V-150 production Skeldar *(Kenneth Munson)* 1151636

Skeldar V-150, Skeldar V-200

Dimensions, External	
Overall	
length, rotors turning	4.00 m (13 ft 1½ in)
height	
Skeldar V-150	1.20 m (3 ft 11¼ in)
Skeldar V-200	1.30 m (4 ft 3¼ in)
Fuselage	
length	
Skeldar V-150	3.20 m (10 ft 6 in)
Skeldar V-200	4.00 m (13 ft 1½ in)
Rotors	
rotor diameter	
Skeldar V-150	3.30 m (10 ft 10 in)
Skeldar V-200	4.7 m (15 ft 5 in)
Weights and Loadings	
Weight	
Weight empty	95 kg (209 lb)
Max T-O weight	
Skeldar V-150	150 kg (330 lb)
Skeldar V-200	200 kg (440 lb)
Payload	
Payload with max fuel	55 kg (121 lb)
Performance	
Altitude	
Service ceiling	
Skeldar V-150	3,500 m (11,480 ft)
Skeldar V-200	4,500 m (14,760 ft)
Speed	
Max level speed	
Skeldar V-150	54 kt (100 km/h; 62 mph)
Skeldar V-200	70 kt (130 km/h; 81 mph)
Radius of operation	
datalink, Skeldar V-150	54 n miles (100 km; 62 miles)
datalink, Skeldar V-200	90 n miles (166 km; 103 miles)
Endurance	4 hr (est)
Power plant	1 × piston engine

Status: Development and testing continuing with Skeldar V-200 in 2008, at which time Saab was reporting discussions with three potential customers. Promotion continuing.

Contractor: Saab Aerosystems AB
Linköping.

Switzerland

RUAG Ranger

Type: Reconnaissance, surveillance and target acquisition UAV.

Development: Ranger was developed to meet specific Swiss Army requirements following the 1985–86 evaluation of four Israeli IAI Scout UAVs. The air vehicle was developed by the Swiss Federal Aircraft Factory (now RUAG Aerospace) in association with Israel Aircraft Industries. Two early prototypes were flight tested (first flight December 1988), followed by the first of six preproduction Rangers in September 1989. Extensive troop trials of these ADS 90 systems were conducted by the Swiss Army in 1990. Swiss user designation ADS indicates *Aufklärungs Drohnen System* (reconnaissance drone system); the operational version is **ADS 95**.

Derived from initial developments in the UAV-SR family, Ranger was designed for real-time day and night observation, reconnaissance and surveillance in demanding topographical and climatic conditions.

Description: *Airframe:* Low-wing monoplane with twin-boom tail unit and sweptback fins with single (port side) rudder. The UAV is equipped as standard with three retractable, hydraulically damped skids, but a wheeled version is also available. Airframe construction is mainly of composites.

Mission payloads: Payloads are in interchangeable modular packages, gimbal mounted and gyrostabilised to permit full hemispherical coverage beneath the air vehicle, and can be retracted into the fuselage to avoid damage on landing. Standard sensor is an IAI Tamam MOSP TV camera with zoom capability and a FLIR sensor; options are a laser

Swiss Air Force ADS 95 Ranger 0024908

range-finder/target designator, a SAR/GMTI sensor, an elint sensor, communications relay or an EW payload.

Guidance and control: Advanced GCS, normally housed in a shelter, consists of two identical consoles for the pilot and observer, with a navigation console in between. Different modes (manual, automatic, and programmed flight control) are available, including emergency modes. In all modes, environmental parameters are updated continuously and fed back to the GCS. The data transmission is performed by a secured microwave primary and UHF back-up uplink, with a microwave band video and telemetry downlink (operational range of up to 97 n miles; 180 km; 112 miles) for real-time data transfer.

System composition: Three to six air vehicles, one GCS, one Remote Communication Terminal (RCT), one launcher, one RAPS autoland position sensor, one or more stand-alone mobile receiving units, plus training, maintenance and logistics equipment. Operational crew of three or four, plus two mechanics and two electronics technicians.

Launch: Launch is by RUAG mobile hydraulic catapult (Robonic pneumatic launcher for Finnish systems), regardless of the type of landing gear fitted.

Recovery: Skid landing system enables Ranger to land on any more or less flat surface (grass, snow, ice, roads or concrete runway). For safety reasons only, it is equipped in peacetime with an emergency recovery parachute; if required, this can be replaced by an additional fuel tank. Ranger Autoland Position Sensor (RAPS) system available optionally.

Power plant: One 31.5 kW (42.2 hp) Göbler-Hirth F 31 two-cylinder two-stroke engine; two-blade fixed-pitch pusher propeller. Fuel capacity 60 litres (15.9 US gallons; 13.2 Imp gallons) standard; a 20 litre (5.3 US

Ranger in flight *(IAI)* 1175494

RUAG Ranger – SwissCopter Tip-Jet DragonFly < Switzerland < Unmanned aerial vehicles

A Finnish Defence Forces Ranger on its Robonic launcher *(Robonic)* 1122519

Preproduction Ranger in flight with payload extended 0505132

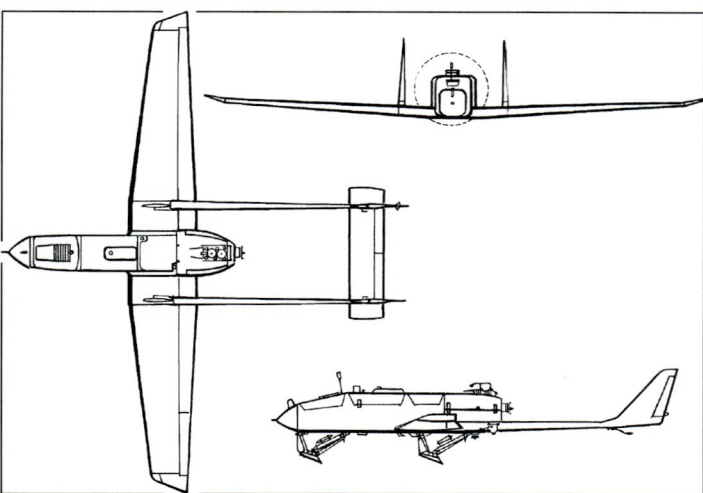

Ranger reconnaissance/surveillance and target acquisition UAV *(IHS/John W Wood)* 0518080

gallon, 4.4 Imp gallon) auxiliary tank can be carried when the emergency parachute is omitted.

ADS 95

Dimensions, External	
Overall	
length	4.611 m (15 ft 1½ in)
height	1.125 m (3 ft 8¼ in)
Fuselage	
height	0.47 m (1 ft 6½ in)
width	0.42 m (1 ft 4½ in)
Wings, wing span	5.708 m (18 ft 8¾ in)
Tailplane, tailplane span	1.553 m (5 ft 1¼ in)
Engines, propeller diameter	0.90 m (2 ft 11½ in)
Areas	
Wings, Gross wing area	3.41 m² (36.7 sq ft)
Weights and Loadings	
Weight, Max launch weight	280 kg (617 lb)
Payload, Max payload	45 kg (99 lb)
Performance	
Altitude, Service ceiling	5,485 m (18,000 ft)
Speed	
Max level speed	130 kt (241 km/h; 150 mph)
Cruising speed	97 kt (180 km/h; 112 mph)
Loitering speed	70 kt (130 km/h; 81 mph)
Stalling speed	49 kt (91 km/h; 57 mph)
Range, command link	97 n miles (179 km; 111 miles)
Endurance	9 hr (est)
Power plant	1 × piston engine

Status: Deliveries to Swiss Air Force (four ADS 95 systems) began mid-1998. Finnish Defence Forces' first system delivered in October 2001; second system ordered in November 2003 and delivered in 2005. An upgrade of the Ranger system was in prospect by 2009. Development of the Super Ranger discontinued.

Customers: Switzerland (operated by Drohnenstaffel 7 of 31 Brigade, Swiss Air Force); Finland (two six-aircraft systems, operated by Artillery Brigade at Niinsalo, possibly 11 operational by 2007).

Contractor: RUAG Aviation
Emmen.

SwissCopter NT 150

Type: Short-range tactical UAV.

Development: This new SwissCopter UAV was unveiled at the IDEX defence exhibition in February 2009. Test flights had been successfully completed at that time.

Description: *Airframe:* Tapered, shoulder-mounted wings; pod and twin tailboom/tailfin configuration; pusher engine. Fixed tricycle landing gear. Composites construction.

Mission payloads: Air vehicle is able to carry multiple payloads to support a variety of ISTAR missions. Standard payload is a lightweight (0.65 kg; 1.4 lb), gyrostabilised, high-resolution colour CCD camera with ×10 optical zoom and an RS-232 datalink. A wide variety of optional FLIR cameras can be installed for both military and civil applications.

Guidance and control: Pre-programmed or remotely operated, including take-off and landing phases, with S-band (2.4 to 2.6 GHz command uplink). Control is exercised using a joystick and a helmet-mounted 'virtual cockpit' that features a headset tracking system which displays live images from the UAV. The pilot's helmet controls the camera in the air vehicle and the pilot is able to switch between autopilot and remote control at any time.

Launch: Conventional, automatic wheeled take-off.

Recovery: Conventional, automatic wheeled landing. Parachute for emergency recovery.

Power plant: One 14.9 kW (20 hp) unidentified flat-twin piston engine; two-blade pusher propeller.

NT 150

Dimensions, External	
Overall	
length	3.10 m (10 ft 2 in)
height	1.00 m (3 ft 3¼ in)
Wings, wing span	4.50 m (14 ft 9¼ in)
Weights and Loadings	
Weight, Max T-O weight	95 kg (209 lb)
Payload, Max payload	40 kg (88 lb)
Performance	
Altitude, Service ceiling	3,500 m (11,480 ft)
Speed	
Max level speed	108 kt (200 km/h; 124 mph)
Cruising speed	65 kt (120 km/h; 75 mph)
Radius of operation, datalink	54 n miles (100 km; 62 miles) (est)
Endurance	5 hr (est)
Power plant	1 × piston engine

Status: Being marketed.

Contractor: SwissCopter AG/UASystems Ltd
Murten.

NT 150 *(SwissCopter)* 1356216

SwissCopter Tip-Jet DragonFly

Type: VTOL UAV.

Development: The Tip-Jet DragonFly has been under development for some years as a dual-mode machine — a single-seat, manned ultralight helicopter and a UAV. Its main feature is the Tip-Jet engine, one of which is located at the end of each rotor blade and runs on an eco-friendly fuel named Perosin (patent pending), based on hydrogen peroxide plus additives. Fuel is mixed with air as it is fed through each rotor blade, reaching a combustion chamber at the tip, where it is ignited; exhaust gases are then ejected through nozzles to drive the rotor. R & D funding for

164 | Unmanned aerial vehicles > Switzerland > **SwissCopter Tip-Jet DragonFly – Swiss UAV Neo S-300**

Geneal appearance of the lightweight Tip-Jet DragonFly *(SwissCopter)*
1290299

The compact Koax X-240 is said to be extremely stable in sustained hover *(Swiss UAV))*
1356287

the unmanned version has been provided by the Swiss Federal Department of Economic Affairs.

Description: *Airframe:* Two-blade main rotor; no tail rotor (propulsion system eliminates torque). Unclad, mainly metal airframe; tailboom supports small tailplane with twin endplate fins. Twin-skid landing gear.
   Mission payloads: Not specified.
   Guidance and control: Remotely piloted via 'monotrail' steering and manoeuvring system.
   Launch: Conventional helicopter take-off.
   Recovery: Conventional helicopter landing.
   Power plant: Two 12.7 daN (28.7 lb st) Tip-Jet engines. Fuel capacity 60 litres (15.9 US gallons; 13.2 Imp gallons) standard; optional second tank, raising total to 120 litres (31.7 US gallons; 26.4 Imp gallons).

Tip-Jet DragonFly

Dimensions, External	
Rotors, rotor diameter	6.00 m (19 ft 8¼ in)
Weights and Loadings	
Weight	
Weight empty	100 kg (220 lb)
Max T-O weight	350 kg (771 lb)
Payload, Max payload	150 kg (330 lb)
Performance	
Climb	
Rate of climb, max, at S/L	650 m/min (2,132 ft/min)
Altitude, Service ceiling	4,000 m (13,120 ft) (est)
Speed	
Max level speed	86 kt (159 km/h; 99 mph)
Cruising speed	30 kt (56 km/h; 35 mph)
Radius of operation	
normal	32 n miles (59 km; 36 miles)
max	108 n miles (200 km; 124 miles)
Endurance	
standard fuel	45 min
with second tank	1 hr 30 min

Status: As of 2009, in production by Peters Engineering, Munich. Launch order received in December 2008 from an unnamed Middle Eastern country for 20, which are a mix of manned and unmanned versions, for "a homeland security project"; reported contract value of USD15 million. No further details have since been released.
Contractor: SwissCopter AG/UASystems Ltd
   Murten.

Swiss UAV Koax X-240 Mk II

Type: Short-range VTOL UAV.
Development: The Koax X-240 is thought to have made its first flight in early 2009. In May of that year, Swiss UAV entered a marketing agreement with Saab of Sweden for joint marketing of a family of rotary-wing UAVs that also includes Saab's Skeldar and the Swiss UAV Neo S-300, using a common Saab-developed ground station. All three systems were demonstrated in Switzerland in August 2009 and displayed at the DSEi defence exhibition in London in the following month.
Description: *Airframe:* Twin co-axial three-blade rotors. Bulbous fuselage with four-leg fixed landing gear. Composites construction.
   Mission payloads: EO/IR or other sensors as appropriate for such missions as law enforcement, border and harbour patrol, base or convoy protection, and traffic monitoring. Has also been promoted as suitable for military roles such as detection of improvised explosive devices (IEDs).

   Guidance and control: Fully autonomous, including take-off and landing. Saab-developed, STANAG 4586 compliant GCS with in-flight updatable GPS waypoint navigation. Autopilot and Ultra Electronics datalinks are dual-redundant and identical to those in Neo S-300.
   Launch: Conventional and automatic helicopter take-off.
   Recovery: Conventional and automatic helicopter landing.
   Power plant: One 8 kw (10.7 shp) heavy-fuel turboshaft. Fuel capacity (Jet A1) up to 22 litres (5.8 US gallons; 4.8 Imp gallons).

Koax X-240 Mk II

Dimensions, External	
Fuselage	
length	1.65 m (5 ft 5 in)
height	0.95 m (3 ft 1½ in)
width	0.50 m (1 ft 7¾ in)
Rotors	
rotor diameter, each	2.65 m (8 ft 8¼ in)
Weights and Loadings	
Weight	
Weight empty	22 kg (48 lb)
Max launch weight	45 kg (99 lb)
Payload, Max payload	8.0 kg (17.00 lb)
Performance	
Altitude, Service ceiling	1,500 m (4,920 ft)
Speed, Max level speed	43 kt (80 km/h; 49 mph)
Radius of operation	16.2 n miles (30 km; 18 miles)
Endurance	1 hr 30 min
Power plant	1 × turboshaft

Status: Development completed in 2009; in production. As of September 2009, the Koax X-240 was understood to already be in service with some unnamed civil customers.
Contractor: Swiss UAV
   Niederdorf.

Swiss UAV Neo S-300

Type: VTOL UAV.
Development: The Neo S-300 was developed from an earlier company design known as the CamClone Mk III. Development began in 2004, maiden flight of the prototype took place in the second quarter of 2006, and it was unveiled at the Dubai Air Show in November 2007. By the end of that year it had accumulated about 1,400 hours of test flying. The Neo S-300 air vehicle is used by EMT of Germany in the Museco system.
Description: *Airframe:* Three-blade main and two-blade tail rotors; streamlined pod and boom fuselage; curved and sweptback horizontal tail surfaces and similar-shaped underfin. Twin-skid landing gear. Construction is of carbon fibre sandwich (fuselage) and aluminium alloy.
   Mission payloads: Typical payloads can include a stabilised camera, IR or UV sensors, radar, gas-laser scanners or other equipment of customer's choice.
   Guidance and control: Ground control station comprises two suitcase-sized elements (GCS-GAIA-AP and GCS-GAIA-VI), which can

Neo S-300 helicopter UAV *(Swiss UAV)*
1290227

The 75 kg MTOW Neo S-300 *(Swiss UAV)* 1290228

Unmanned Systems' CT-450, yet another addition to the ranks of twin-boom, pusher-engined UAVs *(Paul Jackson)* 1395380

The CT-450 is a medium range/endurance tactical system *(Paul Jackson)* 1395379

be used in mobile form or as a fixed, vehicle-mounted installation. Between them, they provide a full overview of the flight location, position and any air vehicle error messages.

The GCS-GAIA is equipped as standard with an RL-Echo remote-control radio link with 900 MHz and 2.4 GHz band default coverage. This, in turn, supports a WLAN satcom telemetry link and, optionally, a GPRS/Edge radio modem for flights over mountainous terrain. If any of these three links is lost, GCS-GAIA switches automatically to one of the other two. If all communication with the UAV is lost, the aircraft's pre-programmed onboard MC-Iason mission controller can autonomously decide whether to continue or abort the mission.

Built-in visualisation software illustrates status of cell, drive and onboard power supply. A communication interface between operators enables bidirectional transmission of all data from the AP-Dedalo 4 autopilot, sensor and telemetry. In a vehicle-mounted installation, the GCS-GAIA can be connected to a high-resolution monitor.

Launch: Conventional helicopter take-off.
Recovery: Conventional helicopter landing.
Power plant: One 14 kW electric motor or, optionally, an equivalent (approx 20 shp) turboshaft engine.

Neo S-300

Dimensions, External	
Overall	
length	2.75 m (9 ft 0¼ in)
height	0.95 m (3 ft 1½ in)
Fuselage	
width, max	0.56 m (1 ft 10 in)
Skids, skid track	0.85 m (2 ft 9½ in)
Rotors, rotor diameter	3.00 m (9 ft 10 in)
Tail rotor, tail rotor diameter	0.65 m (2 ft 1½ in)
Weights and Loadings	
Weight	
Weight empty	25.0 kg (55 lb)
Max T-O weight	75.0 kg (165 lb)
Payload, Max payload	20.0 kg (44 lb)
Performance	
Radius of operation	
mission, control limits	32 n miles (59 km; 36 miles)
Power plant	1 × electric motor

Status: Being promoted from 2008. As of December 2007, described as "already fully usable as a remote-controlled basic model"; development continuing towards fully autonomous operation. In 2010 Swiss UAV was reporting joint integration efforts with EMT of Germany based on the Neo S-300 (Museco) and the Luna GCS.
Contractor: Swiss UAV
Niederdorf.

Unmanned Systems CT-450 Discoverer I

Type: Medium range/endurance tactical UAV.
Development: Public debut at Dubai Airshow in November 2011. Optimised for ISTAR operations in military, law enforcement and civil applications.
Description: Airframe: Shoulder-mounted wings; pod and twin-tailboom fuselage pod; pusher engine. Fixed tricycle landing gear. Composites construction.

Mission payloads: Standard payload of FLIR Systems Ultra Force 275-B EO/IR sensor in underfuselage turret. Options include SAR/GMTI or maritime radar, SIGINT, EW, chemical/biological sensor, communications relay and 3-D mapping system. Piloting camera in nose.

Guidance and control: Suitcase-size, portable (11.5 kg; 25.4 lb) or truck-mounted, STANAG 4586-compliant GCS and SkyView flight management system. Independent air vehicle and camera controllers, throttle controls, plus integrated computer with four reconfigurable I/O switches. Manual or fully autonomous operation.

System composition: Air vehicle and payload(s); truck-mounted GCS and towed datalink terminal and/or man-portable equivalents.
Launch: Conventional wheeled take-off.
Recovery: Conventional wheeled landing.
Power plant: One 13.4 kW (22 hp) Zanzottera rotary engine, driving a two-blade pusher propeller.

CT-450 Discoverer I

Dimensions, External	
Overall	
length	3.10 m (10 ft 2 in)
height	1.00 m (3 ft 3¼ in)
Wings, wing span	4.50 m (14 ft 9¼ in)
Weights and Loadings	
Weight, Max T-O weight	95 kg (209 lb)
Payload, Max payload	40 kg (88 lb)
Performance	
Altitude, Service ceiling	3,500 m (11,480 ft)
Speed	
Max level speed	108 kt (200 km/h; 124 mph)
Cruising speed	102 kt (189 km/h; 117 mph)
Radius of operation	81 n miles (150 km; 93 miles)
Endurance	10 hr

Status: Promotion began in 2011 and continues through 2012.
Contractor: Unmanned Systems AG
Zug.

Unmanned Systems Orca

Type: VTOL tactical UAV.
Development: Public debut at Dubai Airshow in November 2011, when said to be suitable for a range of ISTAR missions, with particular emphasis on maritime operation.
Description: Airframe: Tadpole-shape fuselage with inverted-V tailfins; fixed, twin-skid landing gear. Two-blade rotor; no tail rotor. Engine gas generator is mounted above rotor and rotates with it.

Mission payloads: Standard payload is a FLIR Systems Ultra Force 350 EO/IR multisensor thermal imaging system in an undernose turret. Wide range of optional sensors includes SIGINT or ELINT packages; synthetic aperture, ground penetrating or ground moving target indicating radars (SAR, GPSAR or GMTI); Lidar; hyperspectral cameras; and advanced, stabilised EO/IR systems. Suitable radars quoted include

The Orca tip-jet-powered rotary-wing UAV *(Paul Jackson)* 1395378

Selex SeaSpray 5000e maritime system and Selex PicoSAR. Orca can also carry NBC sensor, sonobuoys, megaphone or loudspeaker, searchlight, transponder, or light cargo hook.

   Guidance and control: Suitcase-size, portable (11.5 kg; 25.4 lb), STANAG-compliant GCS and SkyView flight management system. Independent air vehicle and camera controllers, throttle controls, plus integrated computer with four reconfigurable I/O switches. Manual or fully autonomous operation

   System composition: Two air vehicles with payloads; GCS; one remote data terminal.
   Launch: Vertical take-off.
   Recovery: Vertical landing.
   Power plant: One 160 kW (215 ehp) 'specially designed' turbojet, exhausting through hollow rotor blades to nozzles at blade tips. See under Weights for fuel details; engine can run on several different types of fuel.

Orca

Dimensions, External	
Overall, height	2.10 m (6 ft 10¾ in)
Fuselage	
length	4.30 m (14 ft 1¼ in)
width	1.50 m (4 ft 11 in)
Rotors, rotor diameter	6.20 m (20 ft 4 in)
Weights and Loadings	
Weight, Max T-O weight	350 kg (771 lb)
Fuel weight, Max fuel weight	120 kg (264 lb)
Payload, Max payload	120 kg (264 lb)
Performance	
Altitude, Service ceiling	3,500 m (11,480 ft)
Speed, Max level speed	108 kt (200 km/h; 124 mph)
Radius of operation	54 n miles (100 km; 62 miles)
Endurance	2 hr 30 min
Power plant	1 × turbojet

Status: Promotion began in 2011 and continued through 2012.
Contractor: Unmanned Systems AG
   Zug.

Taiwan

Aeroland AL-4

Type: Short-range tactical UAV.
Development: Introduced in early 2010.
Description: *Airframe:* Parasol wing with rear-mounted engine; pod and boom fuselage; T tail. No landing gear. Composites construction.
   Mission payloads: Nose-mounted colour TV camera with real-time imagery downlink. Other sensors and payloads available
   Guidance and control: Remotely operated from truck-mounted GCS. Autopilot with GPS/IMU.
   Launch: Hand launch.
   Recovery: Belly landing.
   Power plant: One brushless electric motor, powered by three 700 mAh batteries; two-blade pusher propeller.

AL-4

Dimensions, External	
Overall	
length	1.40 m (4 ft 7 in)
height	0.28 m (11 in)
Wings, wing span	2.00 m (6 ft 6¾ in)
Areas	
Wings, Gross wing area	0.5 m² (5.4 sq ft)
Weights and Loadings	
Weight	
Weight empty	2.8 kg (6.00 lb)
Max launch weight	4.2 kg (9.00 lb)
Payload, Max payload	1.0 kg (2.00 lb)
Performance	
Altitude, Service ceiling	3,000 m (9,840 ft)
Speed	
Max level speed	54 kt (100 km/h; 62 mph)
Cruising speed	30 kt (56 km/h; 35 mph)
Radius of operation	13.4 n miles (24 km; 15 miles)
Endurance	1 hr
Power plant	1 × electric motor

Status: Being marketed to civil sector; one unnamed customer in Taiwan by February 2010.
Contractor: Aeroland UAV Inc
   Tainan.

AL-4 at Singapore Air Show, February 2010 *(IHS/Patrick Allen)* 1414714

Aeroland AL-20

Type: MALE surveillance UAV.
Description: *Airframe:* Mid-mounted sweptback wings with winglets or endplate fins; bullet-shaped body, pusher engine. No landing gear. Composites construction.
   Mission payloads: Fuselage-mounted optical sensors with real-time imagery downlink. Other sensors and payloads available.
   Guidance and control: Remotely operated from mobile GCS. Autopilot with GPS/IMU.
   Launch: Catapult launch.
   Recovery: Belly landing.
   Power plant: One 2.6 kW (3.5 hp) two-stroke engine, driving a two-blade pusher propeller. Fuel capacity 10 litres (2.6 US gallons; 2.2 Imp gallons).

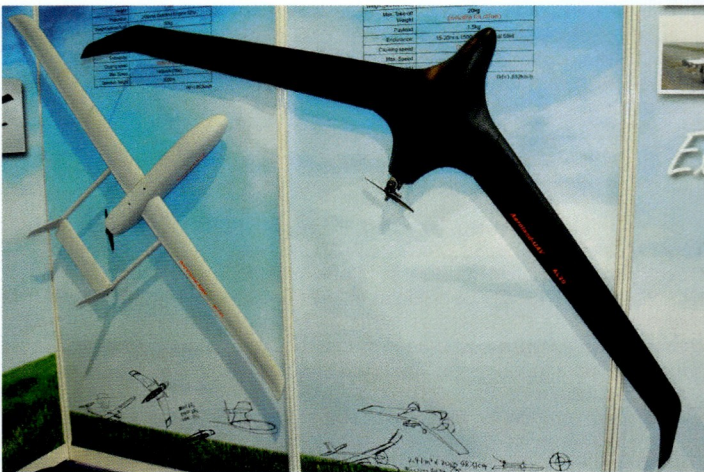

AL-20 on display in early 2010 *(IHS/Patrick Allen)* 1414710

AL-4 rear aspect *(IHS/Patrick Allen)* 1414713

AL-20

Dimensions, External	
Overall	
length	0.70 m (2 ft 3½ in)
height	0.40 m (1 ft 3¾ in)
Wings, wing span	4.00 m (13 ft 1½ in)
Areas	
Wings, Gross wing area	1.0 m² (10.8 sq ft)
Weights and Loadings	
Weight	
Weight empty	10.5 kg (23 lb)
Max launch weight	20.0 kg (44 lb)
Payload, Max payload	1.5 kg (3.00 lb)
Performance	
Altitude, Service ceiling	5,000 m (16,400 ft)
Speed	
Max level speed	70 kt (130 km/h; 81 mph)
Cruising speed	55 kt (102 km/h; 63 mph)
Radius of operation, mission	540 n miles (1,000 km; 621 miles)
Endurance, max	20 hr

Status: Being promoted in 2010. Construction against customer order started November 2010.

Contractor: Aeroland UAV Inc
Tainan.

Aeroland AL-40

Type: Small surveillance UAV.

Development: Aeroland produced around 1,600 target drones for the Taiwan armed forces from 1989 until 2003, when it began development of UAVs, introducing and delivering the first AL-40s in 2006.

Description: *Airframe*: High-wing 'cabin lightplane' configuration, with front-mounted engine and conventional tail surfaces. Fixed tricycle landing gear.

Mission payloads: Fuselage-mounted colour TV camera with real-time imagery downlink.

Guidance and control: Remotely operated from truck-mounted GCS. Autopilot with GPS/IMU.

Launch: Conventional wheeled take-off.

Recovery: Conventional wheeled landing.

Power plant: One 8.2 kW (11 hp) two-stroke engine, driving a two-blade propeller. Fuel capacity 5.5 litres (1.5 US gallons; 1.2 Imp gallons).

AL-40

Dimensions, External	
Overall	
length	2.50 m (8 ft 2½ in)
height	0.50 m (1 ft 7¾ in)
Wings, wing span	3.00 m (9 ft 10 in)
Areas	
Wings, Gross wing area	1.5 m² (16.1 sq ft)
Weights and Loadings	
Weight	
Weight empty	22.5 kg (49 lb)
Max launch weight	40.0 kg (88 lb)
Payload, Max payload	10.0 kg (22 lb)
Performance	
Altitude, Service ceiling	5,000 m (16,400 ft)
Speed	
Max level speed	70 kt (130 km/h; 81 mph)
Cruising speed	55 kt (102 km/h; 63 mph)
Radius of operation, mission	121 n miles (224 km; 139 miles)
Endurance, max	4 hr 30 min

Status: In service. To be replaced with enlarged AL-20 (see separate entry). Production completed in August 2010.

Contractor: Aeroland UAV Inc
Tainan.

Aeroland AL-150

Type: Tactical surveillance UAV.

Development: Largest of current product line. Started under the designation AL-120, uprated specification leading to first flight as the AL-150 in August 2010.

Description: *Airframe*: High-wing, pod and twin-tailboom configuration; rear-mounted engine. Fixed tricycle landing gear, Composites construction.

Mission payloads: Fuselage-mounted optical sensors with real-time imagery downlink; other payloads available.

Guidance and control: Remotely operated from truck-mounted GCS; autopilot with GPS/IMU.

Launch: Conventional wheeled take-off.

Recovery: Conventional wheeled landing.

AL-120 on display in early 2010 *(IHS/Patrick Allen)* 1414718

Power plant: One 23.9 kW (32 hp) two-stroke piston engine, driving a two-blade pusher propeller. Fuel capacity 100 litres (26.3 US gallons; 22.0 Imp gallons) when carrying 10 kg mission payload.

AL-120

Dimensions, External	
Overall	
length	3.50 m (11 ft 5¾ in)
height	1.10 m (3 ft 7¼ in)
Wings, wing span	8.00 m (26 ft 3 in)
Areas	
Wings, Gross wing area	3.1 m² (33.4 sq ft)
Weights and Loadings	
Weight	
Weight empty	50 kg (110 lb)
Max T-O weight	150 kg (330 lb)
Payload, Max payload	40 kg (88 lb)
Performance	
Altitude, Service ceiling	5,000 m (16,400 ft)
Speed	
Max level speed	75 kt (139 km/h; 86 mph)
Cruising speed	59 kt (109 km/h; 68 mph)
Radius of operation	405 n miles (750 km; 466 miles)
Endurance	
with 10 kg payload	16 hr
with 40 kg payload	8 hr
Power plant	1 × piston engine

Status: First flight 2010; production scheduled for 2011.

Contractor: Aeroland UAV Inc
Tainan.

CSIST Blue Magpie

Type: Close-range mini-UAV.

Development: Began in 2006. Unveiled June 2008 at Eurosatory exhibition in Paris.

Description: *Airframe*: Flying wing design, with minimal fuselage and single tailfin. No landing gear. Composites construction.

Mission payloads: Two CCD TV cameras, one forward- and one sideways-looking. Real-time imagery transmission.

Guidance and control: Pre-programmed or remotely piloted; GPS navigation. Utilises same suitcase-sized GCS as CSIST Cardinal, with a hand-held unit for short-range radio control.

The flying-wing Blue Magpie *(IHS/Patrick Allen)*

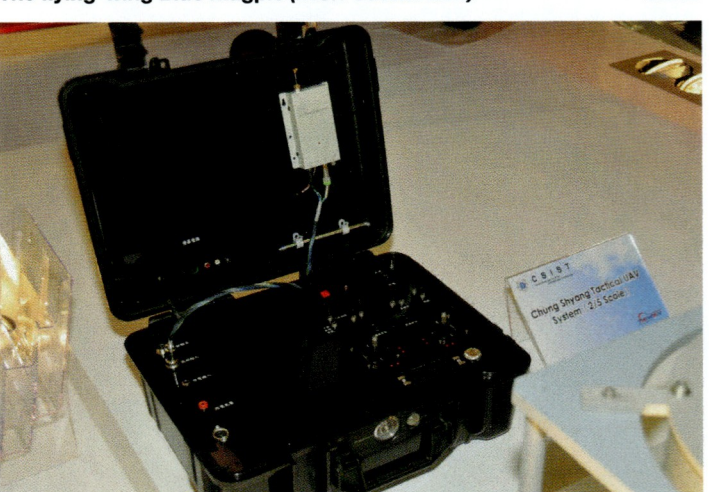

Encased GCS for Blue Magpie and Cardinal mini-UAVs *(IHS/Patrick Allen)*

Transportation: Man-portable.
Launch: Hand launch.
Recovery: Glide landing.

Blue Magpie

Power plant
Electric motor (rating not stated); two-blade tractor propeller.

Status: Developed and available. Displayed in Taipei in 2009 when the programme was still reported to be active, but no orders reported at that time.
Contractor: Chung Shan Institute of Science and Technology Taoyuan.

CSIST Cardinal

Type: Short-range tactical mini-UAV.
Development: Began in 2006, for military RSTA missions. Development testing completed in July 2009; public debut in following month at Taipei Aerospace and Defence Technology Exhibition. Shown again at Taipei in 2011.
Description: *Airframe:* High-wing, pod and boom design with T tail. No landing gear. Composites construction.
Mission payloads: CCD TV daylight camera or thermal imager. Real-time imagery transmission.
Guidance and control: Pre-programmed or remotely piloted.
Transportation: Man-portable in rucksack.

Cardinal mini-UAV on display in 2009 *(IHS/Andrew White)*

System composition: Three air vehicles, portable GCS, RVT and NVGs.
Launch: Hand launch.
Recovery: Belly landing.
Power plant: Electric motor (rating not stated); two-blade pusher propeller.

Cardinal

Dimensions, External	
Overall, length	1.50 m (4 ft 11 in)
Wings, wing span	1.00 m (3 ft 3¼ in)
Weights and Loadings	
Weight, Max launch weight	2.1 kg (4.00 lb)
Performance	
Altitude, Service ceiling	4,500 m (14,760 ft)
Speed	
Cruising speed, max	54 kt (100 km/h; 62 mph)
Radius of operation	10.8 n miles (20 km; 12 miles) (est)
Endurance, max	1 hr 30 min
Power plant	1 × electric motor

Status: Ten pre-production air vehicles completed by August 2009, at which time one system had been delivered to Taiwan Army for evaluation. A production order was reportedly placed by 2011.
Contractor: Chung Shan Institute of Science and Technology Taoyuan.

CSIST Chung Shyang IIC

Type: Multirole tactical UAV.
Development: Original Chung Shyang II was first shown at Asian Aerospace show, Singapore, February 2002. Version seen in 2006 was designated Chung Shyang IIC, with upgrades that include a 'more excellent' (but still unspecified) engine, improved aerodynamics, enhanced payloads and a 'newly optimised' GCS. The Taiwan Coast Guard has been cited as a possible customer. Believed to have STOL runway requirements.
Description: *Airframe:* Low-mounted wings, pod fuselage, and twin tailbooms carrying tailplane and twin fins and rudders. Composites construction. Non-retractable tricycle landing gear.
Mission payloads: Can include daylight TV camera, FLIR and laser rangefinder/designator for day and night surveillance missions, target acquisition and designation, BDA or communications relay; electronic anti-jamming devices. Can also be equipped for border and coastal patrol, geosurvey, traffic control, environmental monitoring or search and rescue. Single real-time encrypted data and video downlink.

Chung Shyang II *(Paul Jackson)*

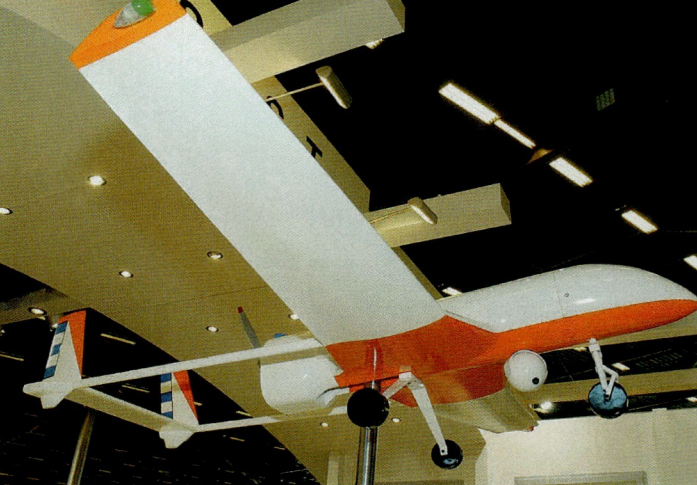

Chung Shyang IIC on display in mid-2006 *(IHS/Patrick Allen)*

Guidance and control: Can be preprogrammed, automatic or manual control; GPS navigation. Dual real-time command uplinks and LOS direct datalink. Autonomous recovery if datalink is lost.
  Launch: Conventional wheeled take-off.
  Recovery: Conventional wheeled landing; arrester hook and cable optional.
  Power plant: One piston engine (type and rating not specified); three-blade pusher propeller.

Chung Shyang IIC

Performance	
Speed, Cruising speed	60 kt (111 km/h; 69 mph)
Range	54 n miles (100 km; 62 miles)
Endurance	8 hr
Power plant	1 × piston engine

Status: Testing complete in 2005. Unofficial report states five prototypes built. CSIST was continuing to promote the IIC version in mid-2006. No activity reported until 2009, when a Chung Shyang was displayed at an exhibition in Taipei. At that time a CSIST official was quoted as saying that the army was considering ordering 20 vehicles in 2010.
Contractor: Chung Shan Institute of Science and Technology
  Taoyuan, Taichung.

Uaver Accipiter

Type: Short-range mini-UAV.
Development: Unveiled August 2009 at Taipei Aerospace and Defence Technology Exhibition; first flight 16 November 2009.
Description: *Airframe:* Mid-wing design, with bullet-shaped fuselage and V tail unit. Fixed tricycle landing gear. All-composites construction.
  Mission payloads: Gimballed CCD TV camera with ×10 zoom.
  Guidance and control: Pre-programmed or remotely operated; flight profiles updatable en route.
  Launch: Bungee launch from stand.
  Recovery: Parachute recovery.
  Power plant: One two-stroke engine (type and rating not stated), driving a two-blade pusher propeller.

Accipiter

Dimensions, External	
Overall, length	1.20 m (3 ft 11¼ in)
Wings, wing span	2.80 m (9 ft 2¼ in)
Weights and Loadings	
Weight, Max launch weight	20 kg (44 lb)
Payload, Max payload	3.0 kg (6.00 lb)
Performance	
Speed	
Cruising speed, max	50 kt (93 km/h; 58 mph)
Radius of operation, mission	27 n miles (50 km; 31 miles)
Endurance, max	6 hr

Status: Under development in 2009; planned for series production, for civil and military applications, in 2010.
Contractor: Carbon-Based Technology Inc, Uaver Division
  Taichung.

Accipiter short-range mini-UAV *(Uaver)* 1356268

Uaver Avian

Type: Close-range tactical mini-UAV.
Development: Began in 2007; unveiled August 2009 at Taipei Aerospace Defence and Technology Exhibition.
Description: *Airframe:* Shoulder-wing configuration; twin wing-mounted engines; single vertical fin. No landing gear. Composites construction. Breaks down into five modules for storage and transportation.
  Mission payloads: Dual CCD TV cameras: one fixed, one with ×10 zoom. IR imager optional. S-band real-time imagery download.
  Guidance and control: Pre-programmed or manual; GPS navigation, updatable in flight. Suitcase-sized GCS with U-band command uplink, imagery and map displays.
  Transportation: Man-portable.
  System composition: Air vehicle, payload(s), GCS and accessories. One-person operation.
  Launch: By hand or bungee catapult.

Avian ground control station *(Uaver)* 1356272

Avian air vehicle *(Uaver)* 1356270

  Recovery: Parachute/airbag recovery.
  Power plant: Twin brushless electric motors (rating not stated), powered by 12 V 30 Ah rechargeable lithium-polymer batteries; two-blade composites tractor propellers.

Avian

Dimensions, External	
Overall, length	0.75 m (2 ft 5½ in)
Wings, wing span	1.60 m (5 ft 3 in)
Weights and Loadings	
Weight, Max launch weight	3.0 kg (6.00 lb)
Payload, Max payload	1.0 kg (2.00 lb)
Performance	
Speed	
Cruising speed, max	30 kt (56 km/h; 35 mph)
Radius of operation, mission	5.4 n miles (10 km; 6 miles)
Endurance, max	1 hr
Power plant	2 × electric motor

Status: Development completed. Five reportedly built by August 2009, with pre-series production then about to begin.
Contractor: Carbon-Based Technology Inc, Uaver Division
  Taichung.

Turkey

TAI Anka

Type: MALE UAV.

Development: The Anka (Phoenix) UAV forms part of Turkey's TİHA (Turkish UAV) programme and is designed to provide the country's armed forces with a MALE ISR UAV with a 24 hour endurance, a 9,144 m (30,000 ft) ceiling, a 108 n mile (200 km; 124 miles) range and the ability to carry EO/IR and SAR radar payloads. Key programme events include:

December 2004 The TİHA programme was launched by the Turkish MoD's Defence Undersecretariat (local acronym SSM) during December 2004.

2007 The Anka's system requirement review was completed during 2007.

2009 Anka prototyping was begun during mid-2009.

July 2010 The prototype Anka UAV was officially rolled-out during July 2010.

December 2010 The prototype Anka UAV made its maiden flight from Sivihisar Air Force Base during December 2010. Thereafter, the Anka programme was re-structured to reflect a block approach, with Anka Block A and B configurations now being intended to provide the type's IOC (with a basic EO/IR payload) and a contractually compliant standard (fitted with a SAR radar) respectively.

Mid-2012 As of mid-2012, the Anka UAV is reported to have completed a series of flight test campaigns during which, the type's performance and those of its sub-systems (including Automatic Take-Off and Landing (ATOL) and EO/IR payload functionality) were demonstrated.

July 2012 IHS Jane's sources were reporting that the SSM had approved the launch of contract negotiations with TAI with regard to the development of a 'long-range, armed' variant of the Anka UAV

27 September 2012 Anka UAV 001 was lost in a crash, an event that is said to have seen a pause in type acceptance testing until the following December.

December 2012 IHS Jane's sources were reporting Turkish contractor TEI as having been awarded an Anka engine development contract. Here, the new unit was to weigh 230 kg (507 lb), offer 123 kW (165 hp) of power and allow the Anka AV to operate at an altitude of 9,144 m (30,000 ft). At the time of this report, development of the new engine was expected to have been completed by 2016.

22 January 2013 Acceptance testing of the Anka Block A UAV was completed. Here the campaign included an 18 plus hour sortie on 20-21 January 2013 and night ATOL functionality during a mission on 22 January 2013. As of the given date, the Anka UAV had accumulated more than 140 flight hours since its maiden flight in December 2010.

Description: *Airframe:* High-wing and V-tail mounted on a mainly graphite/epoxy monocoque fuselage. Pusher engine and retractable tricycle undercarriage. Optional ice protection for the AV's wing and tail leading edges.

Mission payloads: ISR payloads include the Aselsan 300T EO/IR imager (Block A/B), a SAR radar (Block B, with GMTI and inverse SAR operating modes), an onboard recording capability and provision for SIGINT and communications relay equipment.

Guidance and control: Redundant vehicle management system with a segregated mission systems architecture. Communication via dual-band, high-rate datalinks (with an ATC radio relay facility). ATC Mode 3/C transponder aboard the Block A configuration, Mode 5 IFF aboard the Block B AV. Beyond LOS capability with narrow and/or wideband SATCOM options

Transportation: Wing and tail surfaces detachable for storage and transportation.

System composition: A basic Anka system comprises three to four AVs and one or two sets of ground segment elements (GCS, GDT, RVT and ATOL unit) depending on the CONOPS in use.

Launch: Automatic, conventional wheeled take-off.

Recovery: Automatic, conventional wheeled landing.

Power plant: One 114 kW (155 hp) Centurion 2.0S four-cylinder, four-stroke, turbo-diesel engine driving a three-bladed pusher propeller.

An in-flight view of the Anka Block A MALE UAV *(TAI)* 1168645

Anka Block A AV 001 made its maiden flight from Sivihisar AFB during December 2010 *(TAI)* 1423712

A general view of the Anka GCS during high-speed taxi trials with the AV *(TAI)* 1365378

Anka

Dimensions, External	
Overall	
length	8.00 m (26 ft 3 in)
height	3.40 m (11 ft 1¾ in)
Wings, wing span	17.40 m (57 ft 1 in)
Tailplane, tailplane span	4.50 m (14 ft 9¼ in)
Engines, propeller diameter	2.03 m (6 ft 8 in)
Wheels	
wheelbase	3.20 m (10 ft 6 in)
wheel track	2.30 m (7 ft 6½ in)
Weights and Loadings	
Weight	
Weight empty	1,030 kg (2,270 lb)
Max T-O weight	1,680 kg (3,703 lb)
Fuel weight, Max fuel weight	420 kg (925 lb)
Payload, Max payload	230 kg (507 lb)
Performance	
T-O	
T-O run	1,500 m (4,922 ft)
Altitude, Service ceiling	7,925 m (26,000 ft)
Speed	
Max level speed	140 kt (259 km/h; 161 mph)
Loitering speed	88 kt (163 km/h; 101 mph)
Radius of operation	
datalink, LOS datalink	108 n miles (200 km; 124 miles)
Endurance	24 hr
Landing	
Landing run	1,500 m (4,922 ft)
Power plant	1 × piston engine

Status: As of late 2012, TAI was reporting contract negotiations for LRIP of the Anka Block A as being 'underway' and Anka Block B demonstration as being planned for 2013.

Contractor: Turkish Aerospace Industries Inc
Kavaklidere.

TAI R-300

Type: Multi-role helicopter UAV.

Development: As part of the internally funded R-İHA (the Turkish acronym for rotary-wing UAV) programme, TAI has converted a Mosquito XE commercial light helicopter into an autonomous operation proof-of-concept demonstrator. Key events in the R-300 (previously known as the Sivrisinek) effort include:

A mock-up of the R-300 helicopter UAV that was shown at the May 2011 IDEF trade shoe *(IHS/Huw Williams)* 1424115

An early iteration of the R-300 helicopter UAV *(TAI)* 1395345

23 December 2010 The R-300 made its maiden flight.
May 2011 An R-300 AV was on static display at the May 2011 IDEF trade show in Istanbul.
2012 During 2012, the R-300 helicopter UAV is reported to have been undergoing 'developmental testing'.
Description: *Airframe:* The R-300 features an enclosed pod and boom fuselage, two-bladed main and tail rotors and a twin-skid undercarriage.

The R-300 helicopter UAV made its maiden flight during December 2010 *(TAI)* 1395346

Mission payloads: EO/IR turret (mounted beneath the AV's nose).
Guidance and control: Fully autonomous.
Launch: Automatic vertical take-off from land or ship.
Recovery: Automatic vertical landing on land or aboard ship.
Power plant: One 48 kW (64 hp) two-stroke, two cylinder engine.

R-300

Dimensions, External	
Overall, length	5.00 m (16 ft 4¾ in)
Rotors, rotor diameter	5.94 m (19 ft 5¾ in)
Tail rotor	
tail rotor diameter, Mosquito XE	1.02 m (3 ft 4¼ in)
Weights and Loadings	
Weight	
Weight empty	195 kg (429 lb)
Max launch weight	325 kg (716 lb)
Payload, Max payload	130 kg (286 lb)
Performance	
Altitude, Service ceiling	3,050 m (10,000 ft) (est)
Speed	
Cruising speed, max	70 kt (130 km/h; 81 mph) (est)
Radius of operation, mission	81 n miles (150 km; 93 miles) (est)
Endurance	4 hr 30 min (est)
Power plant, piston engine	1 ×

Status: As of late 2012, the R-300 multi-role helicopter UAV was under development. In more detail (and as of the given date), TAI was reporting R-300 developmental testing as being ongoing, with integration of an autonomous take-off and landing system, a 'more capable' payload and improved datalinks being scheduled for 2013. Elsewhere, 2013 was expected to see the R-300 achieve its initial operational capability and over time, TAI is understood to have been in discussions with a range of potential R-300 customers including Turkey's Navy, Gendarmerie and National Police.
Contractor: Turkish Aerospace Industries Inc
Kavaklidere.

Ukraine

SIS A-3 Remez
Type: Close-range multirole mini-RPV.
Development: Development of the Remez was begun in June 1997 by NPS Vzlet (Take-off) design bureau, whose earlier designs included Oko-1 (Eye) and Sinitsa (Titmouse) mini-UAVs. Remez received a design award in Bulgaria, 1998; was exhibited in Ukraine 1999; and made its public debut outside Ukraine at the Eurosatory defence exhibition, Paris, in June 2000. It is promoted for civilian and paramilitary tasks such as environmental patrol, ecological monitoring and disaster reporting in addition to battlefield observation. Alternative **Remez-1 and -2** variants differ only in the degree of payload sophistication. Variants include the **Remez-3T**, which can carry two disposable loads of 1 kg (2.2 lb) each. This, and a so-called 'universal' version, **Remez-3U**, was flight tested in 2005.
Description: *Airframe:* Short, bulbous fuselage with non-swept, nose-mounted canards and rear, low-mounted, slightly swept wings; overwing fins and rudders at approximately 60 per cent span; pusher engine with shrouded propeller. Fixed tricycle landing gear.
Mission payloads: Two TV cameras (one fixed, one rotatable) with real-time video downlink.

Remez-3 on catapult launcher *(SIS)* 1110031

Display model of the NPS Remez-3 multirole RPV *(Michael J Gething)* 0106051

Guidance and control: Remotely piloted from laptop computer with map display, TV tuner and video recorder; GPS navigation. Remote control system developed by Kharkov Proton design bureau.

Transportation: System can be dismantled and transported in three 15 kg (33 lb) boxes.

System composition: Two air vehicles plus ground control and information reception unit with crew of two.

Launch: Conventional wheeled take-off or catapult launch.

Recovery: Parachute recovery.

Power plant: One 1.85 kW (2.5 hp) piston engine; four-blade, shrouded, pusher propeller. See under Weights for fuel details.

Variants:

Remez-3: Standard version; *as described.*

Remez-3T: Version capable of carrying two 1 kg (2.2 lb) releasable payloads. Developed in conjunction with Luch design bureau; undergoing flight testing in second half of 2004. Possible special forces applications.

Remez-3

Dimensions, External	
Overall, length	0.78 m (2 ft 6¾ in)
Wings	
wing span	2.00 m (6 ft 6¾ in)
wing aspect ratio	8.3
Areas	
Wings, Gross wing area	0.48 m² (5.2 sq ft)
Weights and Loadings	
Weight, Max T-O weight	10.0 kg (22 lb)
Fuel weight, Max fuel weight	1.5 kg (3.00 lb)
Payload, Max payload	3.0 kg (6.00 lb)
Performance	
Speed, Max level speed	57 kt (106 km/h; 66 mph)
Range, control	2.7 n miles (5 km; 3 miles)
Endurance	2 hr
Power plant	1 × piston engine

Status: In service.

Customers: Ukrainian Army.

Contractor: Scientifically Industrial Systems Ltd, a division of NPS Vzlet, Kharkov.

SIS A-4 Albatros

Type: Tactical surveillance mini-UAV.

Development: SIS was founded in March 1996 and began UAV development in June 1997. The prototype Albatros-4 made its first flight 5 June 2000 and production was undertaken. The catapult-launched **Albatros-4K** version began flight testing on 12 December 2001 and, as of mid-2005, was undergoing further evaluation with a catapult launching system. **Albatros-4B** is a version for operation in 'hot and high' locations.

Description: *Airframe:* High-wing monoplane with pod and boom fuselage, pusher engine with shrouded propeller and sweptback vertical tail. Quadricycle landing legs with 'feet' on original Albatros-4; the -4K version has fixed, tricycle, wheeled landing gear.

Mission payloads: One fixed, direct TV camera and one controllable TV camera; or one Vikhr thermal imaging system. 1.1 GHz real-time video downlink and 410 MHz digital radio telemetry.

Guidance and control: Microprocessor or CPU-686 remote control system and GPS-35-based satellite navigation.

Albatros-4K parachute recovery *(SIS)*

Albatros-4K in flight *(SIS)*

Albatros-4K production line *(SIS)*

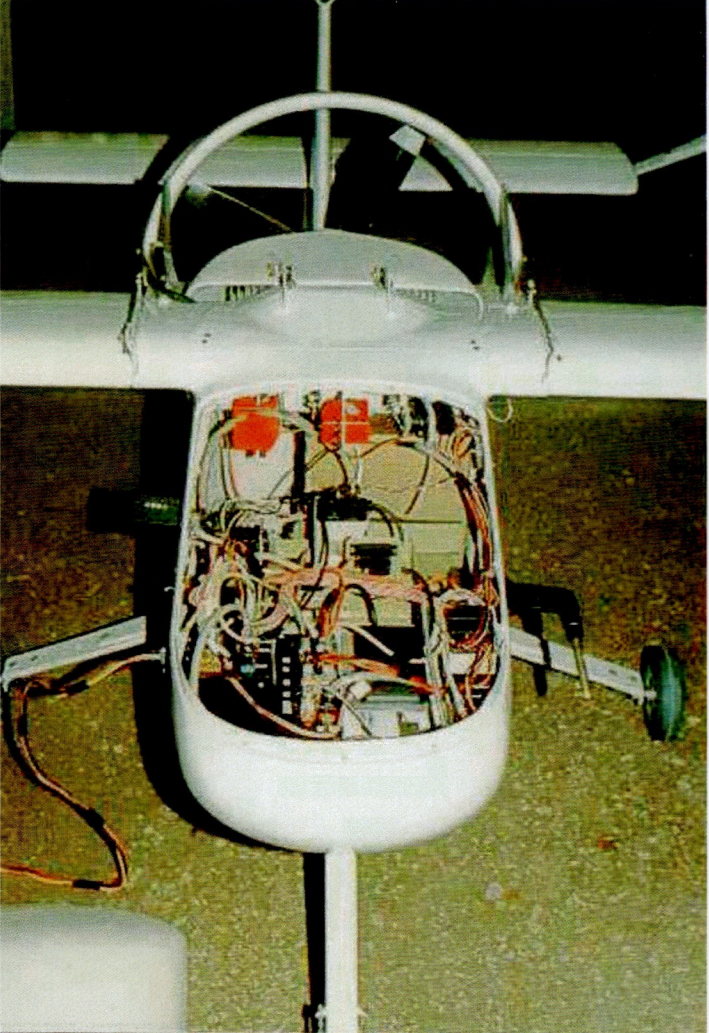

Albatros-4K payload bay *(SIS)*

Transportation: Entire system can be transported in a UAZ-469, GAZ-66 or similar truck.
System composition: Two air vehicles, GCS, antenna system and K-4 catapult.
Launch: Conventional runway take-off or catapult launch.
Recovery: Parachute recovery following engine shutdown.
Power plant: One 2.3 kW (3.1 hp) piston engine; two-blade pusher propeller. See under Weights for fuel details.

Albatros-4

Dimensions, External	
Overall, length	1.425 m (4 ft 8 in)
Wings, wing span	2.475 m (8 ft 1½ in)
Areas	
Wings, Gross wing area	0.68 m² (7.3 sq ft)
Weights and Loadings	
Weight, Max T-O weight	18.3 kg (40 lb)
Fuel weight, Max fuel weight	2.0 kg (4.00 lb)
Payload, Max payload	3.0 kg (6.00 lb)
Performance	
Speed	
Never-exceed speed	67 kt (124 km/h; 77 mph)
Stalling speed	32 kt (60 km/h; 37 mph)
Range, control	10.8 n miles (20 km; 12 miles)
Endurance	2 hr
Power plant	1 × piston engine

Status: In production and service. Company literature implies use for civil tasks such as ecological monitoring, oil and gas pipeline patrol and accident/disaster zone exploration.
Contractor: Scientifically Industrial Systems Ltd, a division of NPS Vzlet, Kharkov.

SIS A-5 Sea Eagle
Type: Small surveillance UAV.
Development: The A-5 Sea Eagle is essentially a scaled-up, twin-engined version of the company's A-4 Albatross, developed for applications involving flights over mountains and large areas of water. Typical roles include area supervision; ecological monitoring; oil and gas pipeline patrol; accident and disaster area investigation.
Description: *Airframe:* Constant-chord, non-swept wings; pod and boom fuselage; conventional tail unit; twin pusher engines mounted on wing trailing-edge. Fixed tricycle landing gear.
Mission payloads: Two TV cameras (one for piloting) or single IR imager; real-time data downlink. Alternative payloads of customer's choice.
Guidance and control: CPU-686E processor-based control; flight management system; digital radio modem; GPS-35 satcom navigation; moving map display; video receiver.
Transportation: Complete system transportable by truck.
System composition: Typical 'Scythian-1' system comprises two air vehicles, GCS, antenna system and catapult.
Launch: Conventional wheeled take-off. Catapult launch optional.
Recovery: Conventional wheeled landing or parachute recovery.
Power plant: Two 1.9 kW (2.5 hp) piston engines, each driving a four-blade pusher propeller.

A-5

Dimensions, External	
Overall, length	1.425 m (4 ft 8 in)
Wings, wing span	3.00 m (9 ft 10 in)
Weights and Loadings	
Weight	
Max T-O weight	28.0 kg (61 lb)
Max launch weight	28.0 kg (61 lb)
Fuel weight, Max fuel weight	4.0 kg (8.00 lb)
Payload, Max payload	7.0 kg (15.00 lb)

Performance	
Speed	
Max level speed	67 kt (124 km/h; 77 mph)
Stalling speed	32 kt (60 km/h; 37 mph)
Radius of operation, mission	21.5 n miles (39 km; 24 miles)
Endurance	5 hr (est)
Power plant	2 × piston engine

Status: Being promoted from 2008. Production/service status not stated.
Contractor: Scientifically Industrial Systems Ltd
Kharkov.

SIS A-10 Phoenix
Type: Multirole endurance UAV.
Development: The A-10 is an unmanned version of the Phoenix-01 two-seat ultralight developed in the late 1990s. Potential applications said to include strategic reconnaissance, elint, ECM, border protection, geodesy and ecological monitoring.
Description: *Airframe:* High-mounted, strut-braced wings and low-set unbraced tailplane, both of constant chord with rounded tips; slightly swept vertical fin with curved leading-edge; pod and boom fuselage; pusher engine. Fixed tricycle landing gear.
Mission payloads: Two TV cameras (one for piloting) or single IR imager, plus real-time data downlink, are standard. Alternative payloads of customer's choice.
Guidance and control: CPU-686E processor-based control; flight management system; digital radio modem; GPS-35 satcom navigation; moving map display; video receiver.
System composition: Typical 'Scythian-1' system comprises two air vehicles, GCS and antenna system.
Launch: Conventional wheeled take-off.
Recovery: Conventional wheeled landing. Recovery parachute optional.
Power plant: One 84.5 kW (113.3 hp) Rotax 914F four-cylinder four-stroke turbocharged engine, driving a pusher propeller.

A-10

Dimensions, External	
Overall	
length	5.80 m (19 ft 0¼ in)
height	2.20 m (7 ft 2½ in)
Wings, wing span	12.83 m (42 ft 1 in)
Weights and Loadings	
Weight, Max T-O weight	750 kg (1,653 lb)
Fuel weight, Max fuel weight	200 kg (440 lb)
Payload, Max payload	150 kg (330 lb)
Performance	
Altitude, Service ceiling	9,000 m (29,520 ft)
Speed	
Max level speed	135 kt (250 km/h; 155 mph)
Cruising speed	105 kt (194 km/h; 121 mph)
Landing speed	43 kt (80 km/h; 49 mph)
Endurance, max	15 hr

Status: Being promoted from 2008. Production/service not stated.
Contractor: Scientifically Industrial Systems Ltd
Kharkov.

SIS A-12 Hurricane
Type: VTOL mini-UAV.
Development: The A-12 was designed for operations in constrained urban, woodland or mountainous conditions. Intended applications include Ministry of Internal Affairs anti-terrorist and border patrol operations.
Description: *Airframe:* Doughnut-shape, with rotating fan blades within a shrouded duct. Central engine and/or payload pod.

A-5 Sea Eagle surveillance mini-UAV *(SIS)*

A-12 Hurricane *(SIS)*

Mission payloads: Two TV cameras (one for piloting) or single IR imager, plus real-time data downlink, are standard. Alternative payloads of customer's choice.
Guidance and control: CPU-686E processor-based control; flight management system; digital radio modem; GPS-35 satcom navigation; moving map display; video receiver.
Transportation: Man-portable.
System composition: Typical 'Scytrhian-1' system comprises two air vehicles, GCS and antenna system.
Launch: Vertical take-off.
Recovery: Vertical landing.

Status: Being promoted in 2008. Production/service status not stated.
Contractor: Scientifically Industrial Systems Ltd
Kharkov.

SIS A-160

Type: Surveillance UAV.
Development: Outwardly, the A-160 is generally similar to the A-5 Sea Eagle, but is larger and has different landing gear. Intended applications are also generally as described for the A-5, also including naval operations.
Description: *Airframe:* Constant-chord, non-swept wings; pod and boom fuselage; conventional tail unit; twin pusher engines mounted on wing trailing-edge. Fixed four-leg landing gear.
Mission payloads: Two TV cameras (one for piloting) or single IR imager; real-time data downlink. Alternative payloads of customer's choice.
Guidance and control: CPU-686E processor-based control; flight management system; digital radio modem; GPS-35 satcom navigation; moving map display; video receiver.
System composition: Typical 'Scythian-1' system comprises two air vehicles, GCS and antenna system.
Launch: Conventional take-off.
Recovery: Conventional landing or parachute recovery.
Power plant: Two 10.0 kW (13.4 hp) piston engines, each driving a four-blade pusher propeller.

A-160	
Dimensions, External	
Overall, length	3.00 m (9 ft 10 in)
Wings, wing span	5.00 m (16 ft 4¾ in)
Weights and Loadings	
Weight, Max T-O weight	160 kg (352 lb)
Fuel weight, Max fuel weight	20.0 kg (44 lb)
Payload, Max payload	50.0 kg (110 lb)
Performance	
Speed	
Max level speed	112 kt (207 km/h; 129 mph)
Stalling speed	35 kt (65 km/h; 41 mph)
Radius of operation	81 n miles (150 km; 93 miles)
Endurance	4 hr (est)
Power plant	2 × piston engine

Status: Being promoted. Production/service status not stated.
Contractor: Scientifically Industrial Systems Ltd
Kharkov.

United Arab Emirates

Adcom Systems Smart Eye 1

Type: Short-range tactical UAV.
Development: This UAV was promoted, though not exhibited, at the Dubai air show in November 2011, at which time it was understood to have been test-flown as a reduced-scale proof of concept demonstrator for the company's much larger Smart Eye 2 (which see). It is quite unrelated to Adcom's much earlier, pod-and-twin-boom Yabhon-Smart Eye (also described separately), despite the similarity of name. Performance is said to include good stability and resistance to turbulence, docile stall, and an excellent rate of climb, even at high angles of attack.
Description: *Airframe:* High aspect ratio, high-mounted front wings and low-mounted rear wings in tandem arrangement; laminar flow aerofoil section, with conventional control surfaces. Serpentine, circular-section fuselage; sweptback single fin and rudder. Engines mounted on pylons above rear pair of wings. Retractable tricycle landing gear
Mission payloads: Not specified, but options may be expected to include EO, IR, thermal imager and laser rangefinder/designator, as with other Adcom UAVs.
Guidance and control: Fully autonomous, via Adcom-3D flight control unit with 'point and click' digital INS/GPS navigation; capability for in-flight programme updates. Adcom's Adnav customised GCS, with two operators, can control single or multiple (up to seven) air vehicles over a radius of up to 150 km (93 miles).
Launch: Conventional, automatic, wheeled take-off.
Recovery: Conventional, automatic, wheeled landing.
Power plant: Two 12.7 kW (17 hp) unidentified piston engines, each driving a two-blade propeller. Fuel capacity 20 litres (5.3 US gallons; 4.4 Imp gallons).

Smart Eye 1	
Dimensions, External	
Overall	
length	3.26 m (10 ft 8¼ in)
height	0.90 m (2 ft 11½ in)
Fuselage, length	3.00 m (9 ft 10 in)
Wings, wing span	4.40 m (14 ft 5¼ in)
Areas	
Wings, Gross wing area	2.40 m² (25.8 sq ft)
Weights and Loadings	
Weight	
Weight empty	50 kg (110 lb)
Max T-O weight	100 kg (220 lb)
Payload, Max payload	40 kg (88 lb)
Performance	
Altitude, Service ceiling	3,000 m (9,840 ft)
Speed	
Cruising speed, max	81 kt (150 km/h; 93 mph)
Stalling speed	38 kt (71 km/h; 44 mph)
Endurance	2 hr
Power plant	2 × piston engine

Status: Development thought to be continuing in 2011-12.
Contractor: Adcom Systems
Industrial City, Abu Dhabi.

Smart Eye 1's tandem-wing layout contributes to very stable flight *(Adcom)* 1395374

Side view of Smart Eye 1 emphasises its snake-like fuselage profile *(Adcom)* 1395375

Adcom Systems United 40

Type: Strategic MALE UAV.
Development: This new large UAV (originally Smart Eye 2, but renamed to acknowledge the 40th anniversary of the UAE) was unveiled at the Dubai air show in mid-November 2011, at which time it was predicted to make its maiden flight within the next few weeks. Development (via subscale test models including the smaller Smart Eye 1, which see) was said to have begun in 2007.

Some inconsistencies have been noted between data released at the November 2011 debut and that appearing later. The specification details given below are those appearing on the company website in January 2012, possibly reflecting the intended production version. Up to that time, no announcement had been made of a maiden flight.

Description: *Airframe:* High aspect ratio, high-mounted front wings and low-mounted, equal-span rear wings in tandem arrangement; laminar flow aerofoil section, with conventional, electrically actuated control surfaces. Front wings have variable incidence. Serpentine, circular-section fuselage; sweptback single fin and rudder. Rear-mounted main engine with pusher propeller, augmented by electric motor for cruise power at altitude. Retractable tricycle landing gear.

The radical new United 40 at Dubai in November 2011 (Paul Jackson) 1395372

United 40 underside detail (Paul Jackson) 1395373

Mission payloads: Gimballed, gyrostabilised platforms for wide range of sensor payloads, including EO, IR, thermal imager, laser rangefinder/designator, synthetic aperture radar and sonar terrain avoidance. Fitted initially with Adcom ADFLIR multisensor turret.

Internal bay and four underwing hardpoints for weapons and/or additional fuel tanks. Internal rotary launcher for up to eight Namrod or similar missiles, plus a further pair on each underwing pylon.

Guidance and control: As described for Smart Eye 1.
Launch: Conventional, automatic, wheeled take-off.
Recovery: Conventional, automatic, wheeled landing. Parachute recovery system planned.
Power plant: Primary power source is an 84.5 kW (113.3 hp) Rotax 914 UL flat-four engine, driving a Woodcomp three-blade pusher propeller. Aircraft is also equipped with a 59.7 kW (80 hp) electric motor for sustained cruise at altitude. Fuel capacity 900 litres (238 US gallons; 198 Imp gallons) in fuselage and wing tanks.

United 40

Dimensions, External	
Overall	
length	11.13 m (36 ft 6¼ in)
height	4.38 m (14 ft 4½ in)
Wings, wing span	20.00 m (65 ft 7½ in)
Areas	
Wings, Gross wing area	24.3 m² (261.6 sq ft)
Weights and Loadings	
Weight	
Weight empty	520 kg (1,146 lb)
Max T-O weight, clean	1,500 kg (3,306 lb)
Payload, Max payload	1,000 kg (2,204 lb)
Performance	
Altitude, Service ceiling	7,000 m (22,960 ft)
Speed	
Cruising speed	
max, Rotax engine	119 kt (220 km/h; 137 mph)
max, electric motor only	81 kt (150 km/h; 93 mph)
Loitering speed	40 kt (74 km/h; 46 mph)
Stalling speed	27 kt (51 km/h; 32 mph)
Range, glide, at 6,000 m (19,680 ft) in event of total power loss	162 n miles (300 km; 186 miles)
Endurance	120 hr
Best glide ratio	43

Status: Under development in 2011-12; three prototypes in test programme.
Contractor: Adcom Systems
Industrial City, Abu Dhabi.

Adcom Systems Yabhon-M

Type: Short-range surveillance UAV.
Development: History not known. Larger derivative of Yabhon-H.
Description: Airframe: As described for Yabhon-H, plus wing flaps. Retractable tricycle landing gear with steerable nosewheel and individually controlled mainwheel brakes. Glass fibre/epoxy construction.

Yabhon-M surveillance UAV (IHS/Patrick Allen) 1137209

The canard configuration Yabhon-M (Robert Hewson) 1151623

Impression of Yabhon-M in flight (Adcom) 1151632

Mission payloads: Nose-mounted gimbal for EO, IR, thermal imaging or laser rangefinder/designator, to customer's requirements. UHF and S-band (2.4 GHz) LOS video and data downlinks.
Guidance and control: Not stated.
Launch: Conventional wheeled take-off.
Recovery: Conventional wheeled landing. Parachute in fuselage for emergency recovery.
Power plant: One four-cylinder two-stroke engine (type and rating not stated); pusher propeller. Fuel capacity 130 litres (34.3 US gallons; 28.6 Imp gallons).

Yabhon-M

Dimensions, External	
Overall	
length	4.30 m (14 ft 1¼ in)
height	1.80 m (5 ft 10¾ in)
Wings, wing span	5.70 m (18 ft 8½ in)
Areas	
Wings, Gross wing area	5.46 m² (58.8 sq ft)
Weights and Loadings	
Weight	
Weight empty	180 kg (396 lb)
Max T-O weight	280 kg (617 lb)
Payload, Max payload	30.0 kg (66 lb)

Yabhon-M

Performance	
T-O	
T-O run	120 m (394 ft)
Climb	
Rate of climb, max, at S/L	420 m/min (1,378 ft/min)
Altitude, Service ceiling	4,500 m (14,760 ft)
Speed	
Max level speed	129 kt (239 km/h; 148 mph)
Cruising speed	113 kt (209 km/h; 130 mph)
Stalling speed	52 kt (97 km/h; 60 mph)
Landing speed, flaps down	46 kt (85 km/h; 53 mph)
Endurance	12 hr
Landing	
Landing run	150 m (493 ft)

Status: Being promoted as of end 2012.
Contractor: Adcom Systems
Industrial City, Abu Dhabi.

Adcom Systems Yabhon-R

Type: Medium-altitude endurance UAV.
Development: Designed for tactical missions including real-time day and night reconnaissance/surveillance, search and rescue, border patrol, remote area monitoring and similar missions.
Description: *Airframe:* Blended wing/body design with mid-mounted, sweptback, laminar flow wings, with flaps; narrow chord unswept canard surfaces; rear-mounted engine; twin, swept fins and rudders at wingtips. Tricycle landing gear with steerable nosewheel; fixed main units on R; on R2, these retract into inboard underwing pods. Glass fibre/epoxy construction.
 Mission payloads: Wide range of sensors can include EO, IR, thermal imager and laser rangefinder/designator. Capability to update third-party synthetic aperture radar.
 Guidance and control: Fully autonomous operation with Adcom 3D flight control unit for automatic navigation. Mission pre-planning capability, online mission profile update and recording.
 Launch: Conventional wheeled take-off.
 Recovery: Conventional wheeled landing. Parachute in fuselage for emergency recovery.
Variants: *Yabhon-R:* Baseline version, as described.
 Yabhon-R2: Projected larger version. Being promoted in late 2011, but none apparently completed at that time.

Power plant: Yabhon-R: One 59.7 to 74.6 kW (80 to 100 hp) four-cylinder two-stroke engine (unidentified); three-blade pusher propeller.
 Yabhon-R2: One 84.5 kW (115 hp) Rotax 914 UL3 flat-four.
 Fuel capacity 240 litres (63.4 US gallons; 52.8 Imp gallons) in R; 270 litres (71.3 US gallons; 59.4 Imp gallons) in R2.

Yabhon-R, Yabhon-R2

Dimensions, External	
Overall	
length	
Yabhon-R	5.00 m (16 ft 4¾ in)
Yabhon-R2	5.50 m (18 ft 0½ in)
height	
Yabhon-R	2.00 m (6 ft 6¾ in)
Yabhon-R2	2.30 m (7 ft 6½ in)
Fuselage	
length, Yabhon-R	4.00 m (13 ft 1½ in)
Wings	
wing span	
Yabhon-R	6.50 m (21 ft 4 in)
Yabhon-R2	8.50 m (27 ft 10¾ in)
Areas	
Wings	
Gross wing area	
incl foreplanes, Yabhon-R	8.00 m² (86.1 sq ft)
incl foreplanes, Yabhon-R2	12.00 m² (129.2 sq ft)
Weights and Loadings	
Weight	
Weight empty	
Yabhon-R	270 kg (595 lb)
Yabhon-R2	385 kg (848 lb)
Max T-O weight	
Yabhon-R	570 kg (1,256 lb)
Yabhon-R2	650 kg (1,433 lb)
Payload	
Max payload	
Yabhon-R	210 kg (462 lb)
Yabhon-R2	270 kg (595 lb)
Performance	
Altitude, Service ceiling	6,700 m (21,980 ft)
Speed	
Cruising speed	
max, Yabhon-R	129 kt (239 km/h; 148 mph)
max, Yabhon-R2	108 kt (200 km/h; 124 mph)
Stalling speed	
Yabhon-R	50 kt (93 km/h; 58 mph)
clean, Yabhon-R2	51 kt (95 km/h; 59 mph)
Endurance	
Yabhon-R	27 hr
Yabhon-R2	30 hr
Power plant	1 × piston engine

Status: Yabhon-R may be in service. Both versions continued to be promoted in 2012.
Contractor: Adcom Systems
Industrial City, Abu Dhabi.

Adcom Systems Yabhon-RX

Type: Medium-altitude endurance UAV.
Description: *Airframe:* Mid-mounted, tapered wings with increased sweep on inboard leading-edges; pod fuselage with pusher engine; twin tailbooms; twin fins and rudders bridged by one-piece tailplane. Conventional aerodynamic control surfaces, including wing flaps. No landing gear. Sandwich construction, mainly of glass fibre/epoxy skins with high-density structural foam filling.
 Mission payloads: Nose-mounted EO sensor with wide FoV; bay in fuselage pod for wide variety of sensors and other payloads, to customer's requirements.
 Guidance and control: Not stated.
 Launch: Fully automatic, from rail launcher.
 Recovery: Fully automatic belly landing. Emergency recovery by parachute in the event of power or command link failure.

The pusher-engined Yabhon-R *(Robert Hewson)* 1151624

Yabhon-R endurance UAV *(Robert Hewson)* 1151625

Yabhon-RX *(Adcom)* 1395370

Yabhon-RX *(Adcom)*

Power plant: One 37.3 kW (50 hp) piston engine (type not specified); two-blade pusher propeller. Fuel capacity 50 litres (13.2 US gallons; 11.0 Imp gallons).

Yabhon-RX

Dimensions, External	
Overall	
length	3.75 m (12 ft 3¾ in)
height	1.00 m (3 ft 3¼ in)
Wings, wing span	5.80 m (19 ft 0¼ in)
Weights and Loadings	
Weight	
Weight empty	70 kg (154 lb)
Max launch weight	160 kg (352 lb)
Payload, Max payload	50 kg (110 lb)
Performance	
Climb	
Rate of climb, max	1,620 m/min (5,315 ft/min)
Altitude, Service ceiling	5,500 m (18,040 ft)
Speed	
Max level speed	129 kt (239 km/h; 148 mph)
Cruising speed	110 kt (204 km/h; 127 mph)
Stalling speed	41 kt (76 km/h; 48 mph)
Endurance	6 hr
Power plant	1 × piston engine

Status: After much revision of design in recent years, this UAV was being promoted in 2011 for day and night reconnaissance and surveillance, search and rescue, border control, environmental monitoring and similar missions. Production and/or operational status not known.

Contractor: Adcom Systems
Industrial City, Abu Dhabi.

Adcom Systems Yabhon-Smart Eye

Type: MALE UAV.

Development: Specific development history of this strategic mission UAV was not revealed, though it is clearly based upon Adcom's years of experience in building what it has described as "hundreds" of its Yabhon series of piston- and jet-engined aerial targets. It was unveiled at the IDEX exhibition in Abu Dhabi in February 2009, reappearing at the Dubai Air Show in November of the same year. According to *Jane's* sources, three prototypes were then planned, the first of which was expected to fly in early 2010. Design features include what is effectively a motor glider

Yabhon-Smart Eye is designed to achieve a five-day endurance *(Robert Hewson)*

configuration, which contributes to its long endurance. As originally exhibited, it featured a secondary jet engine to serve as a back-up if the primary power plant failed, or to boost altitude performance. However, as promoted (though not exhibited) at the November 2011 Dubai show, this feature had evidently been discarded in favour of a higher-powered primary propulsion engine.

The following description applies to the UAV as quoted in company literature at that time. Mission capabilities were said to include reconnaissance, surveillance, border patrol, humanitarian aid, communications relay and special operations.

Confusingly, at the same show, Adcom introduced two other — and entirely different — new UAVs, named Smart Eye 1 and Smart Eye 2. These are the subject of separate descriptions.

Description: *Airframe:* High aspect ratio wings, with tip sections upswept in a continuous curve to form winglets, are low-mounted on a central fuselage nacelle with a rear-mounted engine. Twin-boom tail unit. Retractable tricycle landing gear with steerable nosewheel. Composites construction (glass fibre, carbon fibre and Kevlar).

Mission payloads: Capability can include two separate gimballed cameras, sigint packages or communications relay equipment. Sensors can include EO, IR, thermal imager, laser rangefinder/designator, synthetic aperture radar and sonar terrain avoidance, in up to four underwing pods, each of 100 kg (220 lb) capacity. Typical payload excluding fuel is 70 kg (154 lb) while the maximum payload including fuel is 550 kg (1,213 lb).

Guidance and control: Fully autonomous operation, via advanced flight control unit. Mission pre-planning capability, with online profile updating and recording.

Launch: Conventional wheeled take-off.

Recovery: Conventional wheeled landing.

Power plant: One 86 kW (115 hp) piston engine (type unspecified), driving a three-blade pusher propeller.

Fuel capacity 900 litres (238 US gallons; 198 Imp gallons), in two wing tanks and a fuselage tank.

Yabhon-Smart Eye

Dimensions, External	
Overall	
length	7.00 m (22 ft 11½ in)
height	2.00 m (6 ft 6¾ in)
Wings, wing span	21.00 m (68 ft 10¾ in)
Weights and Loadings	
Weight	
Weight empty	450 kg (992 lb)
Max T-O weight	1,000 kg (2,204 lb)

The long-endurance Yabhon-Smart Eye *(Robert Hewson)*

Yabhon-Smart Eye

Performance	
Altitude, Service ceiling	7,300 m (23,960 ft)
Speed	
Max level speed	119 kt (220 km/h; 137 mph)
Cruising speed, max	70 kt (130 km/h; 81 mph)
Stalling speed	30 kt (56 km/h; 35 mph)
Endurance	120 hr
Best glide ratio	30
Power plant	1 × piston engine

Status: According to a senior company spokesman in November 2009, the Yabhon-Smart Eye was then already 'production ready', although not due to make its maiden flight until February 2010. As of late 2011, its production and/or service status was still uncertain.

Contractor: Adcom Systems
Industrial City, Abu Dhabi.

United Kingdom

Aesir Embler

Type: Technology demonstrator.

Development: Unveiled in July 2009, Embler is one of a quartet of VTOL UAVs developed by Aesir which exploit the Coanda effect to obtain their lift (see entry for Aesir Odin for details). Some 40 per cent smaller than Odin, Embler is being proposed for use in conjunction with a Northrop Grumman Remotec Mk 8 Plus II Wheelbarrow UGV to provide safer detection of improvised explosive devices (IEDs). In this concept, the Embler would be transported to a suspected IED location, then launched to establish exact position of the device without endangering personnel.

Although sufficient to prove the viability of the concept, the 10 minute endurance provided by the electric motor would be inadequate for a fielded system. However, as of the end of July 2009, *IHS Jane's* understood that a small rotary engine being developed by Aeris's engine supplier offered the prospect of increasing this to at least 1 hour with a 1 kg (2.2 lb) payload.

Description: *Launch:* Automatic vertical take-off.
Recovery: Automatic vertical landing.
Power plant: Electric motor initially, driving a ducted fan. Possibly to be replaced later by a small rotary piston engine.

Embler

Dimensions, External	
Fuselage, height	0.6 m (1 ft 11½ in)
Performance	
Endurance, with electric motor	10 min
Power plant	1 × electric motor

Status: Under development in 2009.

Contractor: Aesir Unmanned Autonomous Systems
Peterborough, Cambridgeshire.

Embler in flight *(Aesir)* 1356239

Embler mounted on a Mk 8 Plus Wheelbarrow *(Aesir)* 1293754

Aesir Hoder

Type: Cargo carrying VTOL UAV.

Development: See entry for Aesir Odin. Hoder is primarily intended as a cargo carrier, but adaptable to endurance or other roles by trade-off between payload and fuel weight. The following description, as stated in mid-2009, should be regarded as provisional.

Description: *Airframe:* General description of Odin applies, but configured to incorporate multiple ducted fans.
Mission payloads: Cargo containers, detachable for loading and unloading. Provision for stores release unit for air drops.
Guidance and control: Generally similar in principle to that described for Odin, but from larger and more advanced, mobile and STANAG 4586-compliant GCS with MIL-STD 1553 standard datalinks. Multiple rotors eliminate need for external flaps; instead, flight path control exercised by varying rotor speeds in combination with actuation of aerodynamic control surfaces.
Launch: Automatic vertical take-off.
Recovery: Vertical landing, using enhanced, UCARS-type automatic recovery system.
Power plant: Two rotary piston engines, each driving a ducted fan. JP-8 fuel.

Mockup of a prospective Hoder configuration, 2009 *(Aesir)* 1356238

Hoder

Weights and Loadings	
Weight	
Weight empty	1,500 kg (3,306 lb)
Max T-O weight	2,500 kg (5,511 lb)
Payload, Max payload	1,000 kg (2,204 lb)
Performance	
Endurance	8 hr
Power plant	2 × piston engine

Status: As of mid-2009, Hoder was still in the early stages of conceptual design.
Contractor: Aesir Unmanned Autonomous Systems
Peterborough, Cambridgeshire.

Aesir Odin

Type: VTOL mini-UAV.
Development: Aesir was formed in January 2009, after purchasing the assets of GFS Projects Ltd ('Geoff's Flying Saucers', named after designer Geoffrey Hatton, who developed a VTOL concept obtaining lift by exploiting the Coanda effect, in which a stream of air or fluid tends to 'attach' itself to a nearby curved body instead of following its original path. This, says the company, results in an air vehicle with increased stability, sustained hover capability, and the ability to carry more payload, compared with a conventional ducted fan design). Practicality of the concept was demonstrated in a tethered rig in 2002; the first outdoor free flight was made in 2007.

As of mid-2009, Aesir was marketing two such unmanned VTOL products (Odin and the smaller Vidar, named — as is the company — after a family of Norse gods), and two others (Embler and Hoder) were under development. Prospective applications for Odin include ISTAR, communications relay, electronic warfare, asset protection and IED detection. It can also be used as a loitering munition.

Description: Airframe: Ducted fan, surrounded by a curved outer 'canopy'. Design has no external rotating parts, enabling it to survive low-speed impact with fixed objects.

Mission payloads: 'Plug and play' surveillance or other payloads. Also has option to perform as a stores release unit for air-drop of (for example) medical supplies or explosives.

Guidance and control: Tablet-size ground control unit with touch screen, moving map display and video feed window. Pre-programmed flight profiles, with two-way datalink for operator input of waypoints or action commands. Can also be flown manually.

Utilising Coanda effect, air is expelled radially from fan around canopy to create lift. At canopy edges, four orthogonally mounted flaps extend and retract to vary angle of airflow as it leaves canopy, changing distribution of lift, creating both rotation and ability for pitch and roll control. (Body rotates in opposite direction to fan, neutralising torque.) Yaw control and steering achieved via a series of fixed and movable fish-tailed vanes located either on the canopy or in the duct. (As movement into the lift airflow changes, lift causes the craft to tilt and move in the direction of the tilt.) Aircraft has built-in anti-stall and 'sense and avoid' systems. It also has a natural neutral dynamic buoyancy, making it very stable in flight and in hover.

Transportation: Man-portable.
Launch: Automatic vertical take-off.
Recovery: Automatic vertical landing.
Power plant: One rotary piston engine (type and rating not stated), driving an axial fan via reduction gear. JP-8 fuel.

Odin

Dimensions, External	
Fuselage, width	1.00 m (3 ft 3¼ in)
Weights and Loadings	
Weight	
Weight empty	10.0 kg (22 lb)
Max T-O weight	20.0 kg (44 lb)
Payload, Max payload	10.0 kg (22 lb)
Performance	
Endurance	1 hr
Power plant	1 × piston engine

Status: Developed and available in 2009.
Contractor: Aesir Unmanned Autonomous Systems
Peterborough, Cambridgeshire.

Aesir Vidar

Type: VTOL micro UAV.
Development: Background as for Aesir Odin. Designed to provide surveillance and situational awareness inside buildings and other confined spaces.
Description: Airframe: Similar to that of Odin, but smaller and electric-powered.
Mission payloads: COTS video camera standard; other payloads to customer's requirements.
Guidance and control: As described for Odin.
Transportation: Backpackable.

The tiny Vidar VTOL UAV *(Aesir)* 1356240

Launch: Automatic vertical take-off.
Recovery: Automatic vertical landing.
Power plant: Electric motor (rating not stated), powered by lithium-polymer batteries, with direct drive to a centrifugal fan.

Vidar

Dimensions, External	
Fuselage, height	0.3 m (11¾ in)
Weights and Loadings	
Weight	
Weight empty	0.4 kg (0.00 lb)
Max T-O weight	0.5 kg (1.00 lb)
Payload, Max payload	0.1 kg (0.00 lb)
Performance	
Endurance, max	15 min
Power plant	1 × electric motor

Status: Developed and available in 2009.
Contractor: Aesir Unmanned Autonomous Systems
Peterborough, Cambridgeshire.

Allsopp Skyhook

Type: Tethered aerostats.
Development: For both civil and military applications. Civil roles can include advertising, photography, video surveillance, radio relay, scientific missions and position marking. Also seen as effective, low-cost means of airfield anti-helicopter defence, surveillance, communications

A 7m Allsopp system being tested by the British Army during Urbex 2005 as an aerial camera platform for urban operations *(Allsop)* 1155681

jamming and other roles. Ten standard sizes are available (see Specifications); other sizes can be produced to customers' requirements.

Description: Airframe: Helium-filled oblate spherical balloon, to which is attached a kite-like surface for additional lift. Balloon shape is optimised to minimise effects of rain and temperature changes. Envelopes available in range of colours, including fluorescent. Design protected by UK and US patent.

Mission payloads: Depending upon role, can include solar- or battery-powered day/night video cameras, uncooled thermal imagers, image intensifiers, synthetic aperture radar, data relays, listening devices, radio jamming antennas, mine detection sensors or broadcast aerials. In airfield (anti-aircraft defence) role, Skyhooks could be disposed at a density of about one per square kilometre, lifting loose lines of polyethylene or Kevlar to entangle either the rotor blades of attack helicopters, or the canopy lines of parachutes; or by carrying small bombs to bring down fixed-wing aircraft.

Guidance and control: The Skyhook uses a combination of helium and wind to provide reliable lift. Greater lift and stability can be achieved by 'stacking' several Skyhooks on a common tether. System requires only one person for set-up, and can be left unattended for use in most weathers. Time needed to deploy varies from two to 15 minutes, depending upon type of helium valve used; pack-up time, for storage in backpack, is three minutes. One person can carry three Skyhooks, with their helium and flying line, in a backpack.

Launch: On tether, after inflation from Air Products H10 helium cylinder. Tether can be paid out from a simple wooden kite handle, a Penn-Senator manual reel or a suitable motorised reel.

Recovery: By reeling in tether.

Data applies to standard sizes

Specifications

Dimensions
Volume: Vigilante	0.15 m3 (5.3 cu ft)
Lightweight	0.15 m3 (5.3 cu ft)
Skyhooks: i	1.00 m3 (35.3 cu ft)
ii	1.30 m3 (45.9 cu ft)
iii	2.50 m3 (88.3 cu ft)
iv	5.00 m3 (176.6 cu ft)
v	10.00 m3 (353.1 cu ft)
vi	20.00 m3 (706.3 cu ft)
vii	30.00 m3 (1,059.4 cu ft)
viii	50.00 m3 (1,765.7 cu ft)

Performance
Approx max altitude (unladen)
Vigilant	305 m (1,000 ft)
Lightweight	395 m (1,300 ft)
Skyhooks: i	610 m (2,000 ft)
ii	760 m (2,500 ft)
iii	1,525 m (5,000 ft)
iv	2,135 m (7,000 ft)
v	3,050 m (10,000 ft)
vi	6,100 m (20,000 ft)
vii	7,620 m (25,000 ft)
viii	9,140 m (30,000 ft)

Lift in zero wind: Vigilant ... 0.03 kg (0.07 lb)
Lightweight ... 0.06 kg (0.13 lb)
Skyhooks: i ... 0.3 kg (0.7 lb)
ii ... 0.6 kg (1.3 lb)
iii ... 1.1 kg (2.4 lb)
iv ... 2.5 kg (5.5 lb)
v ... 5.0 kg (11.0 lb)
vi ... 10.0 kg (22.0 lb)
vii ... 15.0 kg (33.1 lb)
viii ... 25.0 kg (55.1 lb)

Lift in wind speed indicated (approx)
Vigilant in 22 kt (40 km/h; 25 mph) ... 0.15 kg (0.33 lb)
Lightweight in 22 kt (40 km/h; 25 mph) ... 0.18 kg (0.40 lb)
Skyhooks: i in 24 kt (44 km/h; 28 mph) ... 1.5 kg (3.3 lb)
ii in 27 kt (50 km/h; 31 mph) ... 2.5 kg (5.5 lb)
iii in 30 kt (56 km/h; 35 mph) ... 5.5 kg (12.1 lb)
iv in 35 kt (64 km/h; 40 mph) ... 12.5 kg (27.6 lb)
v in 39 kt (72 km/h; 45 mph) ... 25.0 kg (55.1 lb)
vi in 43 kt (80 km/h; 50 mph) ... 50.0 kg (110.2 lb)
vii in 52 kt (96 km/h; 60 mph) ... 75.0 kg (165.3 lb)
viii in 56 kt (105 km/h; 65 mph) ... 125.0 kg (275.6 lb)

Dimensions (2.5 m3 Skyhook (typical))
Balloon: Length	2.70 m (8 ft 10.3 in)
Max width	1.70 m (5 ft 6.9 in)
Max height	1.30 m (4 ft 3.2 in)
Helium volume	2.50 m3 (88.3 cu ft)
Helium bottle: Height	0.70 m (2 ft 3.6 in)
Deflated pack size: Length	0.80 m (2 ft 7.5 in)
Width	0.20 m (7.9 in)
Height	0.20 m (7.9 in)

Weights
Weight empty	1.6 kg (3.5 lb)
Helium payload	1.4 kg (3.1 lb)
Wind-assisted payload	Approx 3–4 kg (6.6–8.8 lb)
Helium bottle	13 kg (28.7 lb)
Total system weight	14.5 kg (32 lb)

Performance
Ceiling (single Skyhook)	Approx 2,000 m (6,560 ft)
Time to 305 m (1,000 ft), inflated	1 min
Flying line angle: zero wind	Vertical
any wind up to 30 kt (56 km/h; 35 mph)	45°
On-station unattended endurance without helium top-up	14 days

Status: In production and service. Production of all types runs at approximately 1000 each year.

Customers: Has been used to lift radiosondes from Halley base in Antarctica by British Antarctic Survey team. Other known users have included DERA/QinetiQ (UK); Singapore (Signals Regiment); Sandia National Laboratories, CECOM (US), CUV (US). Last-named (Carolina Unmanned Vehicles) received an SBIR contract from the US Air Force in 2003 for feasibility studies into using a Skyhook to suspend a communications relay payload at low altitude to cover a large area such as the offshore missile range at Eglin AFB, Florida; programme was known as Helikite Elevated Platform — Test Relay (HEP-TR). A second version, with a surveillance payload, was also reportedly under development at that time. In October 2007 a small Low Visibility Skyhook Helikite lifted an ITT Spearnet radio to 200 ft (61 m) over the company's test grounds, acting as a relay to create a mobile, ad-hoc, internet-protocol network. Flying at this the legal altitude limit the company say the system exploited line of sight reception across an area about 100 times greater than the radio would perform alone. US Navy for lifting gyro-stabilised cameras. CENETIX and Riverine group operate Helikites for radio-relay, as do BAE Systems.

Contractor: Allsopp Helikites Ltd
Fordingbridge, Hampshire.

BAE Systems Demon

Type: Technology demonstrator.

Development: Flight testing of the Demon demonstrator will mark the culmination of a five-year, GBP6.5 million research programme launched in June 2005. Known as FLAVIIR (Flapless Air Vehicle Integrated Industrial Research), it has been jointly funded by BAE Systems and the UK Engineering and Physical Sciences Research Council (EPSRC). The FLAVIIR project is led jointly by BAE Systems and Cranfield University, and has involved engineering contributions from nine other universities across England and Wales. Airframe configuration of the Demon is based on that of the 40 kg (88.2 lb), 2.5 m (8 ft 2.4 in) span Eclipse, an earlier BAE Systems experimental UAV; manufacture and assembly was undertaken jointly by Cranfield's Composites Manufacturing Centre and BAE Systems apprentices.

The main focus of FLAVIIR is to develop a range of technologies that could be applied to a low-cost UAV having no conventional control

The Demon demonstrator during final assembly in September 2009 *(BAE Systems)* 1356252

Demon wind tunnel model *(BAE Systems)* 1356267

The diamond-wing Eclipse, on which the Demon design is based *(BAE Systems)* 1367976

surfaces, without suffering any performance penalties by comparison with more conventional flying machines.

Description: *Airframe:* Diamond-shaped low wings; single fin and rudder; dorsal intake for jet engine. Retractable tricycle landing gear. Carbon fibre composites construction.

Mission payloads: Test and telemetry equipment.

Guidance and control: Fully reprogrammable flight control system, in which novel 'control effectors' take the place of conventional three-axis control surfaces. Fluidic thrust vectoring of the main engine exhaust is employed for pitch control, while control in all three axes is exercised by blowing air over purpose-designed circulation control devices on the wing trailing-edges. By the end of the programme, it is hoped to demonstrate completely 'flapless' flight using only the fluidic effectors.

Launch: Conventional wheeled take-off.
Recovery: Conventional wheeled landing.
Power plant: One (unspecified) turbojet.

Demon

Dimensions, External	
Wings, wing span	2.70 m (8 ft 10¼ in)
Weights and Loadings	
Weight, Max T-O weight	80 kg (176 lb)
Performance	
Power plant	1 × turbojet

Status: Under development. In final assembly during third quarter of 2009, with maiden flight anticipated by end of that year.

Contractor: BAE Systems Air Systems
Warton, Lancashire.

BAE Systems GA22

Type: Unmanned airship.

Development: BAE Systems revealed brief details of its GA22 project in July 2008, indicating that it was aimed primarily at civil applications such as event or incident surveillance, communications relay, border security, detection and monitoring of forest fires and floods, marine and fisheries protection, and Earth observation. The basic airship was manufactured by Lindstrand Technologies, which has worked with BAE on its development. Registered G-CFKN, it made its first lift-off, in the Lindstrand hangar at Birkenhead, in mid-October 2008.

Lindstrand had previously supplied a similar unmanned airship, with a 42 kg (93 lb) payload, to the Spanish Ministry of Defence for a classified surveillance mission. This GA22 remained in active service in 2008.

GA22 under test *(Lindstrand Technologies/BAE Systems)* 1418106

GA22 built for BAE Systems *(Lindstrand Technologies/BAE Systems)* 1418108

The GA22 delivered to Spain in 2004 *(Lindstrand Technologies)* 1290305

Description: *Airframe:* Conventional-shape helium non-rigid envelope; cruciform tail surfaces with rudders and elevators; small gondola. Single landing/ground handling wheel under gondola, plus small wheel on lower fin tip to avoid damage on landing. Gondola houses fuel tank, propulsion units, control system and two lead-acid batteries that provide 12 V and 24 V DC onboard electrical power supply. It is attached to two aluminium rails running within 6.5 m (21.3 ft) sleeves within the envelope; the main (lower) portion is rotatable.

Mission payloads: Various, according to assigned mission.

Guidance and control: Initial testing is under radio control; objective is to integrate BAE's existing unmanned autonomous systems technology to make the GA22 more attractive to potential users. Ground handling crew of three during take-off and landing.

Launch: Conventional airship take-off.
Recovery: Conventional airship landing.

Specifications

Power plant	Two piston engines (type and rating not known), each driving a two-blade, vectoring, ducted propeller.
Dimensions	
Envelope: Length overall	22.00 m (72.2 ft)
Max diameter	5.50 m (18 ft)
Volume	300 m³ (10,600 cu ft)
Weights	
Payload: current	60 kg (132 lb)
max	150 kg (331 lb)
Fuel weight	45.4 kg (100 lb)
Performance	
Max level speed	>50 kt (92 km/h; 57 mph)
Ceiling	>1,980 m (6,500 ft)
Mission radius	690 n miles (1,277 km; 794 miles)
Max endurance	24 h

Status: Under development. According to BAE Systems in mid-2008 the GA22 "will be deployed incrementally in line with emerging airspace regulations to be fully autonomous by 2010". The first remotely piloted flight took place on 5 April 2009.

Contractor: BAE Systems Air Systems,
Warton, Lancashire.

BAE Systems HERTI

Type: Developmental MALE UAV.

Development: HERTI (High Endurance Rapid Technology Insertion) was developed as a fully autonomous, lightweight UAV able to carry a BAE Systems ICE surveillance payload. Basis for the HERTI-1D first prototype was a British-owned J5 Marco motor glider by Polish designer Jarosław Janowski, with its piston engine replaced by a small turbojet. The same designer's larger J6 Fregata was selected as basis for the HERTI-1A, which reverted to piston power to provide greater payload capacity, range and endurance. This version is aimed at the civil/commercial marketplace for such roles as coastal watch, border surveillance, pipeline inspection and convoy patrol. The production standard HERTI was launched at Farnborough in July 2008. The is a Slingsby manufactured airframe which is based on the original design. This airframe offers increased MTOW, higher fuel capacity, Triplex FCS and in-built lightening protection. The production standard also includes an enhanced sensor package (new WFOV sensor, multi-function turret by Polytech and the integration of SAR and COROP capabilities). Various trials have taken place which has enabled the HERTI system to progress to production standard, eg. the vehicle is now fully fatigue tested, cleared, qualified and released with full RTS.

Description: *Airframe:* Low/mid-wing monoplane with pod and boom fuselage, pusher engine and V tail. Single mainwheel and tailwheel on HERTI-1D; twin, self-sprung main units and tailwheel on HERTI-1A. Glass fibre composites construction. Wings detachable for trailer storage and transportation.

Mission payloads: HERTI-1D: BAE Systems ICE (Imagery Collection and Exploitation) package.

HERTI-1A: BAE Systems second-generation ICE II, comprising two fixed, wide FoV EO cameras and one narrow FoV EO camera, all with IR and LL capability, mounted in a gimballed turret. Low-bandwidth satcom downlink. Selex lightweight SAR also under consideration. Imagery can be stored onboard or downlinked, on demand, in real time.

Guidance and control: Fully mobile GCS, mounted on specially adapted Land-Rover; based upon four standard PCs and capable of controlling up to four air vehicles simultaneously. Flight profiles fully autonomous, including take-off and landing.

Launch: Fully automatic wheeled take-off (grass or paved runway). Trials using a Robonic launcher are also said to be planned.

Recovery: Fully automatic wheeled landing.

Power plant: One 84.5 kW (113.3 hp) Rotax 914 F turbocharged four-cylinder four-stroke engine, driving a three-blade pusher propeller.

Variants: HERTI-1D: First prototype, modified from J5 Marco motor glider. Wing span 8.04 m (26 ft 4.5 in); max T-O weight 350 kg (771 lb); 38.8 kW (52 hp) Janowski 3PZ-800 (modified Honda BF45) three-cylinder four-stroke marine engine replaced by small turbojet. Programme started June 2004; first flight made in December 2004 at Woomera range in Australia. Subsequently flew several fully autonomous missions at altitudes of up to 1,525 m (5,000 ft), carrying ICE payload.

HERTI-1A: Operational development version, using 'principal components' of J & AS Aero J6 Fregata as airframe basis. Sets for four aircraft purchased in 2005, of which first two each powered by a two-cylinder modified BMW automotive engine and second pair (designated **XPA-1B**) by turbocharged Rotax 914. First flight 18 August

Herti on high-speed taxi trials at Walney Island *(BAE Systems)* 1408329

The Land-Rover-based HERTI ground control station *(BAE Systems)* 1151541

The short-span HERTI-1D prototype *(BAE Systems)* 1132829

The jet-powered HERTI-1D in flight *(BAE Systems)* 1151543

Operator terminals in the HERTI GCS *(BAE Systems)* 1151544

HERTI-1A landing *(BAE Systems)* 1151542

2005 from Campbeltown Airport, Macrihanish, Scotland. Acquisition of components for a further six aircraft was planned by the end of 2006.

HERTI-1A

Dimensions, External
Overall
length..5.11 m (16 ft 9¼ in)
height..1.70 m (5 ft 7 in)
Wings, wing span..12.55 m (41 ft 2 in)
Areas
Wings, Gross wing area......................................9.135 m² (98.3 sq ft)
Weights and Loadings
Weight, Max T-O weight......................................500 kg (1,102 lb)
Payload, Max payload..145 kg (319 lb)
Performance
T-O
T-O run..300 m (985 ft) (est)
Altitude, Service ceiling......................................6,095 m (20,000 ft) (est)
Speed
Max level speed..120 kt (222 km/h; 138 mph)
Cruising speed..90 kt (167 km/h; 104 mph)
Radius of operation..270 n miles (500 km; 310 miles) (est)
Endurance, max..25 hr (est)
Landing
Landing run..300 m (985 ft) (est)

Status: In September 2006, the HERTI/ICE combination was signed into an agreement with the UK Royal Air Force's Air Warfare Centre (AWC) UAV Battlelab to integrate the system into RAF exercises in order to develop techniques, tactics and procedures for the training of personnel to work with UAVs in Iraq and Afghanistan. A further series of trials took place at the Woomera test range in southern Australia in November and December 2006 using the third operational development XPA-1B. These featured fully autonomous flight tests as well as target searches, including some for 'operationally representative' targets designated by the AWC UAV Battlelab. A fourth XPA-1B was due to join the evaluation fleet in April 2007, with three more to follow by the end of the year. Trails from Walney Island airfield in northern England in late 2008 of the production standard Herti, fitted with a Polytech AB electro-optical sensor turret, were followed by flight trials in South Australia in 2009.

Contractor: BAE Systems Air Systems
Warton, Lancashire.

Mantis airborne, over southern Australia. First flight was made in October 2009 *(BAE Systems)* 1293892

Full-size Mantis mockup at Farnborough air show, July 2008 *(IHS/Patrick Allen)* 1303252

BAE Systems Mantis

Type: Technology demonstrator.

Development: Mantis is the subject of a fast-track ACTD programme to develop an autonomous UAS that will help shape technology development to meet the UK's future unmanned aircraft requirements. The name derives from the initials of Multispectral Adaptive Networked Tactical Imaging System. Its objective is to provide an independent capability for persistent ISTAR or long-range precision strike, or a combination of the two.

Development work is understood to have started in early 2007, beginning 'in earnest' in the fourth quarter of that year. The programme was officially unveiled on 14 July 2008 at the Farnborough International air show, with the concurrent signing of a jointly funded (50/50 BAE Systems and the UK MoD) Phase 1 contract and display of a full-size mockup of the aircraft; contract value was not disclosed. At that time, wind tunnel tests had been conducted, design and manufacture of the prototype and its associated ground control infrastructure were under way. Mantis was shipped to south Australia for trials in 2009. The first flight was made in October 2009. Key participants in the programme include Rolls-Royce (integrated propulsive and electrical power system), QinetiQ (communications and flight termination), GE Aviation, Selex Galileo (sensor payloads) and Meggitt.

Phase 2, if approved, would begin in the second half of 2009 and would primarily concern sensor and weaponry integration. No weight or performance specifics were released at the unveiling, but in broad terms the general performance of an operational Mantis would be expected to include a maximum cruising speed of between 200 and 300 kt (370 to 555 km/h; 230 to 345 mph), operating height band of 7,620 to 15,240 m (25,000 to 50,000 ft) and endurance of 24 to 30 hours.

Description: *Airframe:* Low-mounted, unswept, high aspect ratio wings; oval-section fuselage; high-mounted stub-wings on rear fuselage, each supporting pusher engine pod; sweptback T tail. Retractable tricycle landing gear (nose gear from Jetstream 31, main from Piaggio P.180). Composites construction. All-electric systems.

Mission payloads: Sensor payloads to be examined during a Phase 2 are expected to include a wide range of 'plug and play' EO, IR, SAR/GMTI, laser designator, sigint and elint packages. Mantis will also feature onboard processing of collected data.

Six underwing stores hardpoints were shown on BAE's mid-2008 mockup, illustrating a capability to carry up to six GBU-12 Paveway IV laser-guided bombs or 12 Brimstone ASMs on triple launchers. L-3 Wescam MX-20 EOI/IR camera and BAE's imagery collection and exploitation system flown on the development airframe.

Guidance and control: Fully autonomous, including take-off and landing. Triplex flight control system. Over-nose bulge provides space for satcom antenna.

Launch: Conventional, fully automatic wheeled take-off.

Recovery: Conventional, fully automatic wheeled landing.

Power plant: Two Rolls-Royce 250-B17B turboprops, pod-mounted on sides of rear fuselage and each driving a four-blade pusher propeller. Other engines being assessed for any production version.

MANTIS

Dimensions, External
Overall, length..11.6 m (38 ft 0¾ in) (est)
Wings, wing span..21.9 m (71 ft 10¼ in) (est)
Performance
Power plant..2 × turboprop

Status: Development continuing. Prototype configuration does not necessarily represent any eventual production version. Choice of a twin-engine configuration and triple-redundant FCS indicates intention to obtain certification for potential civil applications. Initial flight tests completed 2009-2010. Air vehicle returned to UK June 2010 for further ground tests.

Contractor: BAE Systems Air Systems, Military Air Solutions.
Warton, Lancashire.

Mantis weapon loads could include Paveway LGBs and Brimstone missiles *(IHS/Tracy Johnson)* 1290287

BAE Systems Raven and Corax

Type: Developmental UAVs or UCAVs.

Development: First details of BAE Systems' current UAV programmes were only revealed in February 2006, but they are the result of an R & D programme started about a decade earlier. During that time, BAE has been working on no fewer than six such aircraft, two of which, known as HERTI, are MALE systems and are the subject of a separate description. The other four have all been directed towards the realisation of an agile, low-observable UAV or UCAV, and much of that work still remains classified. The first to actually fly, in about 2001 after a period of research, computer modelling and operational analysis, was a tiny (approximately 1.5 m; 5 ft wing span) model flying-wing glider known as Soarer, openly test-flown from a hilltop not far from BAE's military aircraft facility in Lancashire. This was followed by the twin-jet, blended wing/body Kestrel, which first flew in March 2003, Raven (first flight 17 December 2003) and Corax (first flight January 2005). Differences between these aircraft are summarised under the Variants heading below.

Description: *Airframe:* See under Variants for individual descriptions.

Mission payloads: No details yet known.

Guidance and control: Kestrel flights under control of a remote pilot. Raven and Corax flights fully autonomous, including take-off and landing phases, using BAE-developed duplex digital AFCS.

Launch: Conventional (and automatic for Raven and Corax) wheeled take-off.

Recovery: Conventional (and automatic for Raven and Corax) wheeled landing.

Variants: *Corax:* HALE derivative of Raven, using the same centrebody, power plant, landing gear and flight control system, but allied to high aspect ratio outer wings instead of Raven's shorter, sweptback outer

Corax airborne *(BAE Systems)*

sections, to provide greater endurance/range for projected ISTAR missions. One estimate puts wing span at more than 9.1 m (30 ft). First flight, at Woomera, was made in January 2005. As of mid-2006, there were no plans for further flights by this one-off demonstrator.

Kestrel: Delta planform blended wing/body (BWB) configuration, with winglets (span 5.5 m; 18 ft 0 in); twin outward-canted tailfins; powered by two 0.29 kN (65 lb st) AMT turbojets; fixed tricycle landing gear. Gross weight reportedly 140 kg (308 lb). Composites construction; manufactured by Tasuma (UK) Ltd. Developed in collaboration with Cranfield University. First of several successful and fully autonomous

The BWB-configured, remotely piloted Kestrel prototype A weight-on-wheels switch applies the wheel brakes and airbrakes *(BAE Systems)*

Raven in flight *(BAE Systems)*

One of the two short-span, sweptwing Raven UCAV demonstrators. *(BAE Systems)*

Wing design of Corax is optimised for high altitude and long endurance *(BAE Systems)*

Raven take-off *(BAE Systems)*

Corax taking off *(BAE Systems)*

flights was made in March 2003 from Campbeltown Airport at Macrihanish, Scotland.

Raven: Stealth design, highly unstable aerodynamically, and said by BAE to be the only finless configuration of its type outside of the United States to have been flown. Retractable tricycle landing gear. Carbon fibre composites construction. Size is possibly similar to that of Kestrel. Only nine months after acquisition of data from Kestrel programme, the prototype made its maiden flight on 17 December 2003, this taking place at the Woomera range in Australia. A second Raven began flight trials in November 2004, and Jane's sources report that this programme was still active in mid-2006. Flight trials may have continued into 2007. Data gained contributed to the Taranis programme.

Few precise details of these UAVs have been released. Those known, or estimated, are given within the individual Variant paragraphs.

Status: Flight trials continued in 2006-2007.
Contractor: BAE Systems Air Systems
Warton, Lancashire.

Radar cross-section test of the Replica stealth fighter mockup in 1994 *(BAE Systems)* 0577582

BAE Systems Taranis

Type: UCAV technology demonstrator.
Development: Project Taranis (named after the Celtic god of thunder) was announced on 7 December 2006 with the news of a four-year, GBP124 million contract awarded by the UK MoD's Defence Procurement Agency (DPA) in the preceding month to a BAE Systems-led team for a UCAV technology development programme. Major team partners are Rolls-Royce, QinetiQ and Smiths Aerospace. Funding breakdown is understood to be 75 per cent from government and 25 per cent from industry. According to the MoD announcement, the demonstrator "will be stealthy, fast, and be able to test deploy a range of munitions over a number of targets and be able to defend itself against manned and other unmanned enemy aircraft". It is also intended to provide the evidence needed to inform UK Royal Air Force decisions about a future long-range offensive aircraft and to evaluate how UAVs will contribute to the RAF's future mix of manned/unmanned fixed-wing aircraft. A production decision was once expected in 2011 and eventual fielding in or by 2018. Primary missions are foreseen as deep strike and ISTAR.

Taranis is a key programme in the DPA's Strategic Unmanned Air Vehicles (Experiment), or SUAV(E), created in May 2005 to replace a previously terminated programme known as Future Offensive Air System (FOAS). QinetiQ will contribute software relevant to autonomous capability, plus the communications and flight safety subsystems, with Rolls-Royce responsible for the power plant and Smiths Aerospace for the air vehicle's electrical power and fuel measurement systems. Programme leader BAE Systems will develop the aircraft's low observables (LO) capability, systems integration, control infrastructure, full autonomy elements and (via BAE Systems Australia) flight control computing. A 'significant number' of other UK suppliers are expected to provide supporting technology and components.

It became known in 2003 that BAE has been conducting classified LO research since at least 1994, when it tested a full-size engineering mockup of a manned stealthy combat aircraft known as Replica in its RCS outdoor range at Warton; and in mid-2005 similar testing was reported of an MoD-funded stealthy UCAV configuration named Nightjar. Other recent associated programmes have involved the Corax and Raven subscale UCAV prototypes (which see).

Taranis was rolled out at the BAE Systems Warton facility on 12 July 2010 with flight trials expected to start in 2011.

Nightjar undergoing an RCS test *(BAE Systems)* 1289671

Description: *Airframe:* Blended wing/body design of 'swept delta' configuration, with no vertical tail surfaces. Retractable tricycle landing gear. Composites construction.

Mission payloads: In ISTAR role, advanced EO/IR, radar and tactical datalinks for intelligence gathering, communications and imagery exploitation. For deep strike role, two internal bays for kinetic and non-kinetic laser-guided weapons. No weapon releases are envisaged during initial test flights, but emulated drops of ground attack weapons will be done.

Guidance and control: "Fully integrated autonomous [decision-making] systems".

Launch: Conventional, automatic, wheeled take-off.
Recovery: Conventional, automatic, wheeled landing.
Power plant: One non-afterburning 28.9 kN (6,500 lb st) Rolls-Royce Adour 951 turbofan with low observable intake and jetpipe arrangements.

Taranis	
Weights and Loadings	
Weight, Max T-O weight	8,000 kg (17,636 lb) (est)
Performance	
Power plant	1 × turbofan

Status: Development continuing
Contractor: BAE Systems Air Systems
Warton, Lancashire.

Blue Bear Blackstart

Type: Close-range surveillance mini-UAV.
Description: *Airframe:* Broad-chord, low-mounted wings; slim cylindrical fuselage; sweptback vertical tail. No landing gear. Composites construction. Specification is for standard version, but Blue Bear notes that design is scaleable according to customer requirements.

Mission payloads: Various, including EO and IR cameras, on a two-axis gimbal. Real-time imagery downlink.

Guidance and control: Pre-programmed, using Blue Bear SNAP autopilot; provision for in-flight redirection if required.

Transportation: Man-portable in two cases: one for air vehicles (1.14 × 0,64 × 0.42 m; 3.74 × 2.10 × 1.38 ft) and one for GCS (0.62 × 0.49 × 0.22 m; 2.03 × 1.61 × 0.72 ft)

Taranis unveiled at Warton, 12 July 2010 *(BAE Systems)* 1398139

Blue Bear iSTART

Type: Close-range surveillance micro-UAV.

Description:: Airframe: Short-span, shoulder-mounted, slightly swept wings; close-coupled conventional tail surfaces; narrow, deep fuselage. No landing gear. Composites construction.

    Mission payloads: Various EO, IR or other sensors, to customer's requirements.

    Guidance and control: As described for Blue Bear Blackstart.

    Transportation: Man-portable.

    System composition: Two air vehicles, GCS and one operator.

    Launch: Hand launch.

    Recovery: Automatic belly skid landing.

    Power plant: Brushless electric motor (rating not revealed); two-blade propeller.

iSTART

Dimensions, External	
Overall	
length	0.66 m (2 ft 2 in)
height	0.24 m (9½ in)
Wings, wing span	0.75 m (2 ft 5½ in)
Weights and Loadings	
Weight, Max launch weight	1.5 kg (3.00 lb)
Performance	
Radius of operation	
datalink, mission	1.6 n miles (3 km; 1 miles)
Endurance, max	40 min
Power plant	1 × electric motor

Status: Being promoted.

Contractor: Blue Bear Systems Research, Clapham, Bedfordshire.

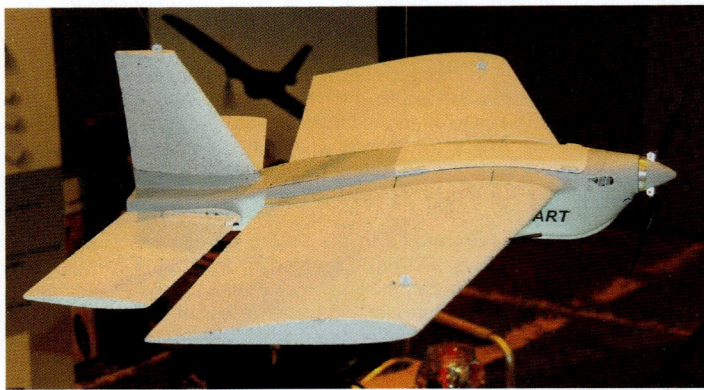

Compact lines of the Blue Bear iSTART micro-UAV *(Paul Jackson)* 1395340

Cyberflight CyberEye I

Type: Tactical mini-UAV.

Development: Unveiled in September 2007 and made its first flight shortly afterwards.

Description: Airframe: Blended wing/body design, with endplate fins; pusher engine mounted on short central fin. No landing gear. Composites construction (expanded polypropylene, or EPP).

    Mission payloads: Gimballed (360° pan, 90° tilt) and gyrostabilised CCD TV camera with ×36 zoom, 30 micron thermal imaging camera, and MPEG 2 video recorder with real-time analogue/digital encrypted downlink. Underslung pods can be fitted, to carry buoyancy aids, telephones, strobes or transponders.

    Guidance and control: With Procerus Technologies Kestrel autopilot, fully autonomous launch, navigation and landing, with in-flight reprogramming (including loiter) via 'click and drag' on map overlay.

CyberEye air vehicle, minus central fin and engine/propeller installation *(Cyberflight)* 1374646

Blackstart GCS *(IHS/Patrick Allen)* 1465272

Mission planning for Blackstart is by a simple 'point and click' interface *(Paul Jackson)* 1395342

    System composition: Two air vehicles, GCS and one operator; optional RVT.

    Launch: Hand launch.

    Recovery: Automatic landing on belly skid

    Power plant: One brushless electric motor (rating not revealed); two-blade propeller.

Blackstart

Dimensions, External	
Overall	
length	0.98 m (3 ft 2½ in)
height	0.35 m (1 ft 1¾ in)
Wings, wing span	1.50 m (4 ft 11 in)
Weights and Loadings	
Weight, Max launch weight	4.0 kg (8.00 lb)
Performance	
Speed, Cruising speed	40 kt (74 km/h; 46 mph) (est)
Radius of operation	
datalink, mission	2.2 n miles (4 km; 2 miles)
Endurance, max	1 hr 30 min
Power plant	1 × electric motor

Status: Introduced and available from 2010.

Contractor: Blue Bear Systems Research, Clapham, Bedfordshire.

Fully rigged CyberEye on bungee launcher *(Cyberflight)* 1374648

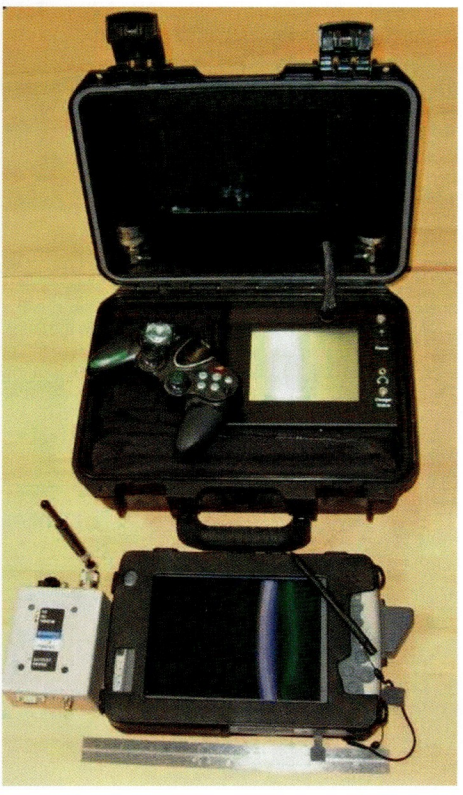

Components of the CyberEye GCS *(Cyberflight)* 1374649

Single-person operation. Collision avoidance software optional. GCS (total weight 2 kg; 4.4 lb) comprises a robust touch-screen laptop computer; digital video receiver (L-band/S-band/256 AES encryption) in a waterproof case; antennas; battery charger (12 to 36 V DC input); and flight/camera controller. Air vehicle has a 'return home' capability if control link is lost for any reason.

   Transportation: System is transportable in backpacks by two persons.
   Power plant: Electric motor (type and rating not stated); two-blade pusher propeller.

CyberEye I

Dimensions, External	
Overall	
length	1.60 m (5 ft 3 in)
height	0.20 m (7¾ in)
Wings, wing span	2.70 m (8 ft 10¼ in)
Weights and Loadings	
Weight, Max launch weight	10.0 kg (22 lb)
Payload, Max payload	2.0 kg (4.00 lb)
Performance	
Altitude, Operating altitude	60 m to 400 m (200 ft to 1,320 ft)
Speed	
Max level speed	59 kt (109 km/h; 68 mph)
Loitering speed	30 kt (56 km/h; 35 mph)
Radius of operation	22 n miles (40 km; 25 miles)
Endurance	3 hr
Power plant	1 × electric motor

Status: Available. UK customers have included Strathclyde Police, who have used CyberEye in marine and mountain rescue operations.
Contractor: Cyberflight Ltd
Moira, Derbyshire.

Cyberflight DM-65

Type: Close-range VTOL mini-UAV.
Development: Developed and produced for such applications as reconnaissance, surveillance, 'perch and stare' crime prevention and search and rescue.
Description: *Airframe:* Quad-rotor configuration, comprising central module for power plant, avionics and payload, and four radiating arms each with rotor at its outer end. Rotor arms fold to permit storage in carrying tube.

   Mission payloads: Low-light (fog penetrating) TV camera or 25 micron thermal imager. Real-time encrypted video downlink. Can also be used to air-drop search and rescue equipment such as inflatable lifejacket or 2-way radio.

   Guidance and control: One-person operation from suitcase-size GCS (total weight 9 kg; 19.8 lb), using touch-screen tablet PC and autopilot system with in-flight programming via 'click and drag' on map overlay. GPS waypoint navigation. Fully autonomous or all-manual operating options. Able to control multiple air vehicles via single datalink. Waterproof case contains digital video receiver (L-, S- or C-band, or encrypted) and communications modem. Battery charging time 12 minutes.

   Transportation: Man-portable in storage tube. Set-up time less than 2 minutes.
   Launch: Fully automatic vertical take-off.
   Recovery: Fully automatic vertical landing.
   Power plant: Electric motor (rating not stated), driving four two-blade rotors.

DM-65

Dimensions, External	
Overall	
length, folded	0.61 m (2 ft 0 in)
height	0.13 m (5 in)
width	0.65 m (2 ft 1½ in)
Weights and Loadings	
Weight, Max T-O weight	2.0 kg (4.00 lb)
Payload, Max payload	1.0 kg (2.00 lb)
Performance	
Altitude	
Operating altitude, normal	100 m (320 ft) (est)
Service ceiling	3,000 m (9,840 ft)

DM-65 storage tube *(Cyberflight)* 1395249

DM-65 air vehicle *(Cyberflight)* 1395248

DM-65

Speed, Max level speed	32 kt (59 km/h; 37 mph)
Radius of operation, mission	1.6 n miles (3 km; 1 miles)
Endurance, max	40 min
Power plant	1 × electric motor

Status: In production and in service. Customer information not disclosed.
Contractor: Cyberflight Ltd
 Moira, Derbyshire.

Cyberflight E-Swift Eye

Type: Short-range surveillance mini-UAV.
Description: *Airframe:* Sweptback flying wing, with winglets and full-span flaperons; bulbous fuselage with nose-mounted sensor. Composites/foam plastics construction.
 Mission payloads: Gimbal-stabilised CCD colour camera with ×10 optical zoom and onboard video recorder; or 30 micron thermal imaging camera; or other, to customer choice. Real-time analogue or digital encrypted video downlink.
 Guidance and control: Autonomous flight, with GPS guidance and tracking; in-flight programming via 'click and drag' on map overlay. GCS housed in waterproof case whose contents include a GPS receiver, an L- or S-band encrypted digital video receiver, a sunlight-viewable touch-screen tablet, antennas and 12 to 36 V battery charger. One-person operation. Multiple air vehicles can be 'swarmed' via single datalink.
 Transportation: Man-portable by two persons in two backpacks. Total weight of GCS, 9 kg (19.8 lb). Set-up time 5 minutes.
 Launch: Hand launch.
 Recovery: Belly skid landing.
 Power plant: One brushless electric motor, powered by lithium-polymer batteries; two-blade pusher propeller.

E-Swift Eye

Dimensions, External	
Overall	
length	0.82 m (2 ft 8¼ in)
height	0.12 m (4¾ in)
Wings, wing span	1.50 m (4 ft 11 in)
Weights and Loadings	
Weight	
Weight empty	1.9 kg (4.00 lb)
Max launch weight	2.2 kg (4.00 lb)
Payload, Max payload	0.3 kg (0.00 lb)
Performance	
Altitude	
Operating altitude	
min	60 m (200 ft)
max	365 m (1,200 ft)
Service ceiling	4,000 m (13,120 ft)
Speed	
Max level speed	51 kt (94 km/h; 59 mph)
Loitering speed	25 kt (46 km/h; 29 mph)
Radius of operation, mission	12 n miles (22 km; 13 miles) (est)
Endurance, max	1 hr
Power plant	1 × electric motor

Status: In production in 2009. UK trials with Strathclyde Police in 2008.
Contractor: Cyberflight Ltd
 Moira, Derbyshire.

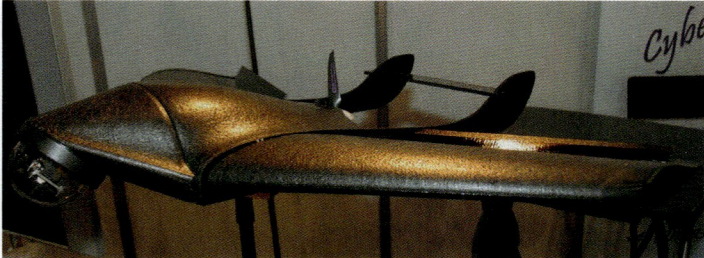
E-Swift Eye surveillance mini-UAV *(IHS/Patrick Allen)* 1328619

Cyberflight Midge

Type: Close-range mini-UAV.
Development: Introduced in 2010. Said by company to be particularly appropriate for convoy protection prosecution roles.
Description: *Airframe:* Slender tubular fuselage, with high-mounted wings (detachable) and V tail. No landing gear.
 Mission payloads: Standard sensor is a gimbal-stabilised TV camera with 360° pan × 90° tilt. Options include 30-micron 320 × 240 thermal imaging camera; NextVision digital PTZ-D; gyrostabilised NextVision MicroCam-D; and MPEG 2 onboard recorder. Real-time video downlink (analogue standard; encrypted and digital available).
 Guidance and control: The Midge is designed for one-person operation. The autopilot provides fully autonomous flight, with

The Midge hand-launched mini-UAV *(Cyberflight)* 1395307

man-in-the-loop provision and in-flight programming via 'click and drag' on map overlay. The GCS, which is stowed in a waterproof case, is based on a sunlight-viewable tablet PC with an aircraft comms modem and L-/S-band or encrypted 256 AES digital video receiver. A GPS receiver allows the ground station to travel with a convoy while UAVs follow. Total GCS weight 2.5 kg (5.5 lb). Multiple air vehicles can be operated via one datalink.
 Transportation: The GCS and air vehicle (with wings detached) are man-portable in a single backpack. Set-up time is approximately 4 minutes and the battery can be charged in one hour.
 Launch: Hand launch.
 Recovery: Belly landing.
 Power plant: One battery-powered 700 W (0.9 hp) electric motor; two-blade propeller.

Midge

Dimensions, External	
Overall	
length	0.95 m (3 ft 1½ in)
height	0.10 m (4 in)
Wings, wing span	1.80 m (5 ft 10¾ in)
Weights and Loadings	
Weight, Max launch weight	2.3 kg (5.00 lb)
Payload, Max payload	0.6 kg (1.00 lb)
Performance	
Altitude	
Operating altitude	60 m to 365 m (200 ft to 1,200 ft)
Service ceiling	3,000 m (9,840 ft)
Speed	
Max level speed	46 kt (85 km/h; 53 mph)
Loitering speed	22 kt (41 km/h; 25 mph)
Radius of operation	5.4 n miles (10 km; 6 miles)
Endurance, extendable	50 min
Power plant	electric motor

Status: Available, and being promoted.
Contractor: Cyberflight Ltd
 Moira, Derbyshire.

Cyberflight Zygo

Type: Short-range surveillance mini-UAV.
Development: Developed for applications in the defence and security markets.
Description: *Airframe:* Constant-chord high wings, with outer panel dihedral; pod and boom fuselage; rectangular tail surfaces. No landing gear. Composites construction.
 Mission payloads: Two-/four-axis gyro-stabilised payloads include Sony FCB-EX series day camera; or FLIR Systems, Thermoteknix or Opgal IR or UV camera. Options include laser rangefinder and gas or NBC sensor.

Zygo hand-launched mini-UAV *(Cyberflight)* 1395253

Zygo in-flight impression (Cyberflight) 1395252

Midge, a downsized variant of its Zygo parent vehicle (Cyberflight) 1395251

Guidance and control: Generally as described for Cyberflight DM-65.

Transportation: Complete system can be carried by a two-man crew in a small vehicle. Set-up time 5 minutes.

Launch: Hand launch.

Recovery: Belly landing.

Power plant: Electric motor (*Zygo* and *Midge*) or piston engine (*Zygo*) no details or ratings provided.

Variants: *Zygo:* Current production version, as of mid-2010. Description applies to this version except where indicated.

Midge: Much scaled-down version; under development in third quarter of 2010.

Zygo, Midge

Dimensions, External	
Overall	
length, Zygo	1.94 m (6 ft 4½ in)
height, Zygo	0.40 m (1 ft 3¾ in)
Wings	
wing span	
Zygo	3.00 m (9 ft 10 in)
Midge	1.00 m (3 ft 3¼ in)
Weights and Loadings	
Weight	
Max launch weight	
Zygo	7.0 kg (15.00 lb)
Midge	0.95 kg (2.00 lb)
Payload, Max payload	2.5 kg (5.00 lb)
Performance	
Altitude	
Operating altitude, Zygo	60 m to 500 m (200 ft to 1,640 ft)
Service ceiling, Zygo	4,500 m (14,760 ft)
Speed	
Max level speed, Zygo	54 kt (100 km/h; 62 mph)
Loitering speed, Zygo	27 kt (50 km/h; 31 mph)
Radius of operation, Zygo	21.5 n miles (39 km; 24 miles)
Endurance	
electric, Zygo	2 hr
piston, Zygo	4 hr
Midge	40 min (est)
Power plant	
optional	1 × electric motor
optional, Zygo	1 × piston engine

Status: In production and in service. Customer information not disclosed.

Contractor: Cyberflight Ltd
Moira, Derbyshire.

FanWing UAVs

Type: Technology demonstrators.

Development: Following the first flight of a small radio-controlled model in Italy in September 1998, FanWing Ltd was formed in the UK on 3 August 1999 to continue developing, patenting and licensing the concept. A number of other small-scale models were wind tunnel tested and flown during the next few years, leading to a 1 m (3.3 ft) span, 2.2 kg (5 lb) version by mid-2002, each gradually improving its predecessor's general efficiency and glide ratio. In August 2002, the company received a UK Department of Trade and Industry grant for a commercial surveillance UAV able to carry a 2 kg (4.4 lb) payload, and this began flight testing in September 2003.

'Open-ended' enlargement has continued since then, this prototype being reflown in November 2003 at 17.5 kg (38.6 lb) gross weight, including 8 kg (17.6 lb) of payload, following the addition of winglets. By June 2004 these weights had risen to 21 and 12 kg (46.3 and 26.5 lb) respectively. Plans included doubling the present weight to allow carriage of heavier payloads, including radar; and working towards the achievement of vertical take-off. (The company estimates that VTO is achievable with about 89.5 kW (120 hp) for a 400 kg (882 lb) FanWing.)

Wind tunnel testing has indicated that the FanWing concept offers 20 kg (44 lb) of lift per engine horsepower, and that lifting efficiency increases with vehicle size. Initial tests on 2009 modifications indicate the possibility of up to 26 kg (57 lb). The company is currently seeking support for the construction of commercial applications, including a first manned version, the FW 8 ultralight, and hopes to progress towards larger manned and unmanned versions after that for such applications as crop-spraying, firefighting, traffic monitoring, cargo lift and rescue operations. Potential UAV applications include reconnaissance, surveillance, mine detection, remote sensing and pipeline or border patrol.

In November 2005, vertical tethered thrust tests were completed on a new FanWing VTOL demonstrator with a subsequent new patent in the national phase of the PCT process in 2009. Flown at ParcAberporth June 2008.

Description: *Airframe:* The FanWing UAV features a conventional fuselage, tail unit and fixed tricycle landing gear, but a conventional wing is replaced by a pair of multiblade, backward-rotating cylinders (rotors) each side, driven by the centrally mounted engine. These cross-flow rotors at the leading-edge pull air in at the front and accelerate it over the trailing-edge, generating lift even when the aircraft is stationary.

Mission payloads: Tests have been conducted using weights to simulate gradually increasing sizes of payload.

Guidance and control: Radio-controlled from ground. Aerodynamically, lift and thrust are controlled by narrow leading-edge

Phase 1/2 FanWing prototype STOL UAV (FanWing) 0583943

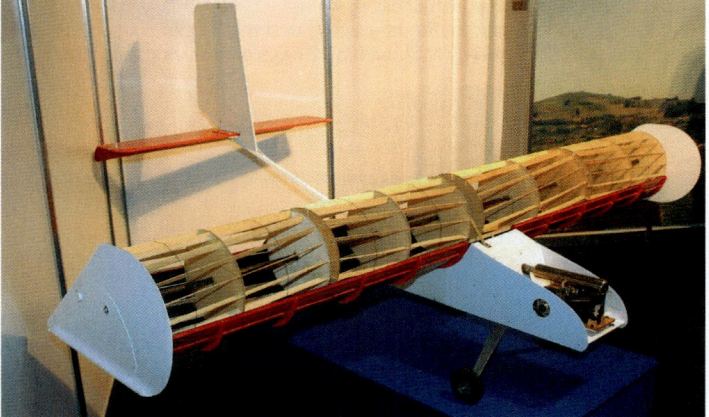

FanWing STOL display model (IHS/Patrick Allen) 1097188

flaps in front of the rotors. Increasing thrust and lift on one wing and decreasing them on the other permits the aircraft to turn without adverse yaw. The wing's small surface area is not greatly affected by the angle of attack, making the aircraft stable in flight, and less affected by turbulence than a conventional wing. It is optimised for low-speed flight (though higher-speed variants are under study), and cannot stall.

Launch: Wheeled take-off.
Recovery: Wheeled landing.
Power plant: *A:* One 11.5 cc glow-plug engine initially; subsequently replaced by a 35 cc four-stroke petrol engine.
B: One 4.5 cc glow-plug engine.
C: One 40 cc two-stroke petrol engine.

FanWing Phase 1/2, FanWing VTOL, FanWing Phase 3

Weights and Loadings	
Weight	
Weight empty	
FanWing Phase 1/2	8.0 kg (17.00 lb)
FanWing Phase 3	15.0 kg (33 lb)
Max T-O weight	
FanWing Phase 1/2	20.0 kg (44 lb)
FanWing VTOL	6.0 kg (13.00 lb)
FanWing Phase 3	40.0 kg (88 lb)
Payload	
Max payload	
FanWing Phase 1/2	12.0 kg (26 lb)
FanWing Phase 3	15.0 kg (33 lb)
Performance	
T-O	
T-O run, at 18 kg (39.7 lb) AUW, FanWing Phase 1/2	20 m (66 ft)
Speed	
Max level speed	
FanWing Phase 1/2	29 kt (54 km/h; 33 mph)
FanWing VTOL	40 kt (74 km/h; 46 mph) (est)
FanWing Phase 3	39 kt (72 km/h; 45 mph)
Loitering speed	
FanWing Phase 1/2	13.5 kt (25 km/h; 16 mph)
FanWing Phase 3	16 kt (30 km/h; 18 mph)
Endurance	
with 4 kg (8.8 lb) fuel, FanWing Phase 1/2	8 hr
with 8 kg (17.6 lb) fuel	10 hr

Status: Development continuing in 2010.
Contractor: FanWing Ltd
43-45 Dorset St, London.

MBDA Fire Shadow

Type: Lethal UAV.

Development: MBDA heads the large Team LM (signifying Loitering Munition) developing the Fire Shadow as a low-cost, all-weather weapon able to carry out precision attacks, day or night, against time-sensitive and/or hard-to-engage surface targets. Other members of the team are Blue Bear Systems Research, Cranfield Aerospace, Cranfield University, Lockheed Martin UK Insys, Marshall Specialist Vehicles (MSV), Meggitt, QinetiQ, Roxel, Selex S&AS, Thales UK, Ultra Electronics and VEGA. The system is jointly funded by MBDA and the UK MoD, and was sole-sourced to the MBDA-led Team CW (indicating Complex Weapons), of which Team LM is a part.

Fire Shadow's function would be to launch over a battle zone, loiter for several hours, but be instantly ready to strike any suitable target that emerged — valuable capability in scenarios where the enemy only reveals its position very briefly. However, it retains 'man in the loop' control for the decision on whether and when to make a strike. Meanwhile, during its loiter period, the missile would fulfil a secondary function of ISTAR data-gathering. A requirement for such a system, based on recent combat experience, was realised during the assessment phase of the UK's Indirect Fire Precision Attack (IFPA) programme which ended in 2005. In British Army (Royal Artillery) service, it would be networked with other ISTAR assets such as the Watchkeeper UAS, and will also be made compatible with 'other battlefield systems'.

Design work began in January 2007, and the first of six prototype air vehicles was delivered by MSV within 12 days of receiving a contract. The

Fire Shadow displayed at DSEi exhibition in London, September 2008 *(IHS/Patrick Allen)* 1328848

Fire Shadow initial launch at Aberporth, April 2008 *(MBDA)* 1331097

first test launch, at the Aberporth test range in Wales on 30 April 2008, was made with the engine started before launch and the wings pre-deployed, drive to the propeller being engaged after jettisoning the booster rocket. The munition climbed to altitude, flew a pre-programmed flight path, then executed a terminal dive (incorporating a high-*g* manoeuvre) to represent a simulated target engagement. Onboard video imagery and positional data, transmitted via datalink, were successfully received and displayed in the Aberporth GCS.

Description: Airframe: Blunt-ended, box-section fuselage; four tailfins, in X configuration; high-mounted, forward-swept, flip-out wings; pusher engine. No landing gear.

Mission payloads: Nose-mounted IR seeker and HE warhead. Imagery and data downlink.

Guidance and control: High integrity datalink (HIDL) adapted from that originally developed for the Thales Watchkeeper UAS.

Launch: Rail launch on ground, with Roxel booster rocket assistance. As of late 2008, no consideration for air launch had been indicated.

Recovery: Non-recoverable.

Power plant: One rotary piston engine (type and rating not revealed) with electronic fuel injection; two-blade fixed-pitch pusher propeller.

Fire Shadow

Dimensions, External	
Overall, length	4.00 m (13 ft 1½ in) (est)
Wings, wing span	4.00 m (13 ft 1½ in) (est)
Weights and Loadings	
Weight, Max launch weight	200 kg (440 lb) (est)
Payload	
Max payload, warhead	10 kg (22 lb) (est)
Performance	
Range	81 n miles (150 km; 93 miles) (est)
Endurance	10 hr (est)
Power plant	1 × piston engine

Status: Under development. As of mid-2008, six development airframes had been completed by MSV. Ground and captive-carry air tests (latter as part of seeker development programme) were due to begin in second half of 2008. Two test flights completed by 2010, with 'main gate' production decision expected in 2011 and IOC in 2012.

Contractor: MBDA
London.

QinetiQ Zephyr

Type: Solar-powered HALE research UAV.

Development: The Zephyr was originally conceived for tethered release from the huge QinetiQ 1 manned balloon which was to have made an attempt on the world altitude record in 2003, with the UAV expected to attain an altitude of 40,230 m (132,000 ft). After the balloon venture failed, partial funding was provided by the UK MoD to evaluate the Zephyr separately for such military applications as reconnaissance and communications relay. At this time, the UAV was quoted as having a 12 m (39.4 ft) wing span and a launching weight of 12 kg (26.5 lb), and was initially flight tested in early 2005 at the Woomera weapons test range in Australia. Further proving flights by two 12 m prototypes (air vehicles 5.1 with battery power only and 5.2 with combined battery/solar power) took place in December 2005 at the White Sands Missile Range in New Mexico. These reached 8,230 m (27,000 ft), and were of 4½ and 6 hours duration respectively, the maximum times permitted under range restrictions. In further trials at White Sands in late July 2006, a third Zephyr prototype (assumed to be one with lithium sulphur batteries) achieved a flight duration of 18 hours, including 7 hours during darkness, and extended the altitude then reached to 10,970 m (36,000 ft).

Meanwhile, in mid-2005, QinetiQ had received a contract from Verhaert, a Belgian space systems company, for a larger (16 m span, 27 kg launch weight) version, to form part of its Mercator environmental

Zephyr take-off at White Sands Missile Range, December 2005 *(QinetiQ)*

monitoring system for the Pegasus HALE programme of the Flemish Institute for Technical Research.

Description: *Airframe:* Carbon fibre framework with Mylar skins on rectangular, detachable, wings and tail surfaces; dihedral on outer wing panels. No landing gear.

Mission payloads: As of mid-2006, these were understood to consist of either a lightweight EO sensor (a 2.5 kg; 5.5 lb equipment is under development by QinetiQ) or a communications relay payload, mounted in a detachable pod.

Guidance and control: Autonomous flight, with GPS-based waypoint navigation.

Launch: Human-assisted launch.

Recovery: Belly landing.

Power plant: Solar panels in majority of upper wing surface provide approx 1.5 kW of electrical power, via rechargeable lithium batteries, to electric motors driving two two-blade propellers mounted at wing leading-edge. Original battery packs featured lithium polymer cells; one Zephyr was flight tested in July 2006 with Sion Power lithium sulphur cells, offering greater endurance.

Zephyr

Dimensions, External	
Wings	
wing span	
first prototypes	12.00 m (39 ft 4½ in)
Mercator prototype	16.00 m (52 ft 6 in)
Areas	
Wings	
Gross wing area, first prototypes	16.00 m² (172.2 sq ft)
Weights and Loadings	
Weight, Max launch weight	30 kg (66 lb) (est)
Payload, Max payload	2.0 kg (4.00 lb)
Performance	
Altitude, Service ceiling	15,240 m (50,000 ft) (est)
Speed	
Cruising speed, econ	12 kt (22 km/h; 14 mph) (est)
Power plant	electric motor

Status: Up to nine aircraft reported to have been built by mid-2006, in various configurations, for some of which details have not been released. At that time, the Mercator civil version had been completed and was awaiting delivery. Objective of Zephyr development is to achieve an endurance of three months.

Contractor: QinetiQ
Farnborough, Hampshire.

Selex Galileo Asio

Type: Close-range VTOL mini-UAV.

Development: Development started in 2005, and first flight was made in early 2006. First details were revealed in July 2007.

Description: *Airframe:* 'Tail-sitter' ducted fan with cylindrical central pod housing sensor (at top), propulsion unit and tailfins (below). Three fixed, curved, landing legs. Composites construction; manufactured by UTRI of Italy (acquired by Selex Galileo in 2011).

Asio at ParcAberporth in 2009 *(Mark Daly)*

Mission payloads: CCD day/night (switchable) video camera with autofocus (×10 optical and ×10 digital zoom, 7 and 62° FoV); or 7.5 to 13.5 micron IR camera with ×2 digital zoom and 36° FoV. Payloads can be controlled manually during flight.

Guidance and control: Can be operated in automatic or semi-automatic mode, with waypoint navigation. Aircraft is tilted to 45° angle for forward flight. Same GCS as for Otus and Strix.

Transportation: The Asio system, including its GCS and related antennas, is man-portable in a lightweight backpack.

System composition: One air vehicle, two camera payloads, GCS (including ground datalink and related antennas) for video and telemetry, user manuals and backpack. Set-up time less than 10 minutes. Total system weight approximately 20 kg (44.1 lb)

Launch: Automatic or semi-automatic vertical take-off.

Recovery: Automatic or semi-automatic vertical landing.

Asio sensor detail *(IHS/Patrick Allen)*

The Asio ducted fan VTOL UAV *(IHS/Patrick Allen)*

Power plant: Electric motor.

Asio

Dimensions, External	
Overall, height	0.75 m (2 ft 5½ in)
Weights and Loadings	
Payload, Max payload	0.5 kg (1.00 lb)
Performance	
Speed	
Cruising speed, max	25 kt (46 km/h; 29 mph)
Radius of operation, datalink	5.4 n miles (10 km; 6 miles)
Endurance	40 min (est)
Power plant	1 × electric motor

Status: Being marketed in 2008. Scheduled for demonstrations in Australia, and other countries, during 2010. First reported sale to Italian Army, 2011.

Contractor: Selex Galileo
Basildon, Essex.

SELEX Galileo Damselfly

Type: VTOL tactical UAV.

Development: The Damselfly, unveiled in model form at ParcAberporth, Wales, on 11 July 2007, was designed for specific VTOL operations where launch and recovery space is restricted, such as from ships or in small urban areas. An electric-powered subscale demonstrator, with a wing span of about 1 m (3.28 ft), began a successful series of hover trials in early April 2007. After conversion to an internal combustion engine, further flight trials, including transition, were planned to begin in November 2007. The ParcAberporth model showed the general configuration of this testbed, which featured wing flaps, all-moving V tail surfaces and wingtip-mounted sensor pods. It was said to have a take-off weight of some 9 kg (19.8 lb) and potential top speed of 150 kt (277 km/h; 172 mph). Further ahead, an 80 kg (176 lb) 'full size' pre-production model was forecast for testing in 2009. However, in its debut announcement, Selex made clear that the design can be built to mini or larger configurations, depending upon specific military or civil requirements.

Description: *Airframe:* Unswept, constant-chord high wings, with flaps and tip-mounted sensor pods; elongated conical fuselage having large fan intake with central bullet at front; all-moving V tail surfaces. Twin-skid landing gear. Composites construction.

Mission payloads: Various, depending upon configuration and mission requirements. The model displayed in July 2007 showed sensors (one EO and one IR) mounted in wingtip pods.

Guidance and control: The Damselfly uses the same common backpackable GCS as Selex's Asio, Otus and Strix family of mini-UAVs. Aerodynamic control for VTOL and forward flight makes use of a vectored-thrust combined propulsion and lift system based on a high-speed fan. The aircraft lifts and hovers on pillars of air expelled through four independently directed nozzles, each incorporating steerable vanes and co-ordinated by extremely advanced avionics. The configuration enables it to combine the hover capability of a helicopter with excellent gust resistance and a small take-off and landing footprint. Once airborne, the thrust nozzles swing back to allow a smooth transition to high-speed forward flight on conventional wings.

Launch: Automatic or semi-automatic vertical take-off.
Recovery: Automatic or semi-automatic vertical landing.

Status: Development continuing in 2007-08. No prototype had emerged by 2010.

Contractor: Selex Galileo Ltd
Basildon, Essex.

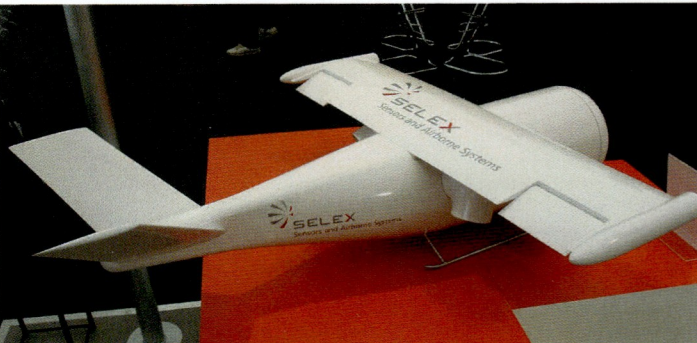

Rear view of Damselfly emphasises conical fuselage and shows wing flaps and V tail *(IHS/Patrick Allen)*

Damselfly model, showing fan intake, thrust control nozzles and wingtip sensor pods *(IHS/Patrick Allen)*

Selex Galileo Otus

Type: Close-range mini-UAV.

Development: Development began in September 2006. First flight was made in December 2006, and first details were released in July 2007.

Description: *Airframe:* Parasol-wing monoplane, with dihedral on outer wing panels; pod and boom fuselage; T tail; pusher engine. Supplied by UTRI of Italy (acquired by Selex Galileo in 2011)..

Mission payloads: CCD day/night (switchable) video camera (autofocus, ×10 optical and ×10 digital zoom, 7 and 62° FoV); or 7.5 to 13.5 micron IR camera (×2 digital zoom and 36° FoV). Payloads can be controlled manually during flight.

Guidance and control: Can be operated in automatic or semi-automatic mode, with waypoint navigation. Same GCS as Asio and Strix.

Transportation: Air vehicle, GCS and antennas are man-portable in a compact, lightweight backpack.

System composition: One air vehicle, three payload cameras, GCS (including ground datalink and related antennas for video and telemetry), user manuals and backpack. Set-up time less than 10 minutes. Total system weight approximately 20 kg (44.1 lb).

Launch: Hand or bungee hand-held catapult launch.

Recovery: Low-speed spiral descent to belly landing.

Power plant: Electric motor, driving a two-blade pusher propeller.

Otus

Dimensions, External	
Overall, length	0.91 m (2 ft 11¾ in)
Fuselage, height	0.305 m (1 ft 0 in)
Wings, wing span	1.52 m (4 ft 11¾ in)
Weights and Loadings	
Payload, Max payload	1.0 kg (2.00 lb)
Performance	
Climb	
Rate of climb, max, at S/L	238 m/min (780 ft/min)
Speed	
Max level speed	45 kt (83 km/h; 52 mph)
Loitering speed	15 kt (28 km/h; 17 mph)
Radius of operation, datalink	5.4 n miles (10 km; 6 miles)
Endurance	1 hr (est)
Power plant	1 × electric motor

Status: Marketed for sale 2011.

Contractor: Selex Galileo
Basildon, Essex.

Tasuma CSV 65

Type: Multipurpose UAV.

Development: New conceptual design; introduced at ParcAberporth exhibition in mid-2006.

Description: *Airframe:* High-mounted, constant-chord, non-swept wings; pod fuselage with pusher engine; pylon-mounted tailboom above wings, supporting inverted V tail surfaces. Fixed, twin-skid (optionally, wheeled) landing gear. Composites construction.

Mission payloads: To customer's choice. Onboard generator optional.

Guidance and control: No details provided.

Transportation: Supplied in modular form with joined wings and separate tailboom, enabling complete aircraft to be transported in a carton less than 2.5 m (8.2 ft) long.

Launch: From Tasuma launcher, or conventional wheeled take-off.

Recovery: Skid or wheeled landing.

Power plant: One 28.3 kW (38 hp) UEL AR 741 rotary engine, driving a two-blade pusher propeller.

CSV 65

Dimensions, External	
Overall, length	3.20 m (10 ft 6 in)
Wings	
wing span	
standard	3.75 m (12 ft 3¾ in)
optional	4.50 m (14 ft 9¼ in)
Weights and Loadings	
Weight	
Weight empty	48 kg (105 lb)
Max T-O weight	85 kg (187 lb) (est)
Max launch weight	85 kg (187 lb) (est)
Payload, Max payload	35 kg (77 lb)
Performance	
Speed	
Cruising speed, max	97 kt (180 km/h; 112 mph)
Loitering speed	40 kt (74 km/h; 46 mph)
Endurance, max	5 hr

Status: One built. Available to customer specification 2009.

Contractor: Tasuma (UK) Ltd
Blandford Heights, Dorset.

Tasuma CSV X0

Type: Adaptable platform UAV.

Development: Introduced at ParcAberporth exhibition in July 2007 as conceptual design "to provide a multicapability platform for system and payload manufacturers to develop their products". Two versions (CSV 40 and CSV 50) offered at that time.

Description: *Airframe:* High-mounted, constant-chord, non-swept wings; pod fuselage with pusher engine; twin tailbooms with rectangular tail surfaces. Fixed Tricycle landing gear.

Mission payloads: As required by customer. Onboard generator optional.

Guidance and control: No details provided.

Transportation: Modular design, with two-piece wings, twin booms and tail assembly. Unassembled airframe can be transported in a carton less than 1.8 m (5.9 ft) long.

Launch: Conventional wheeled take-off, or from Tasuma launcher.

Recovery: Conventional wheeled or belly landing.

CSV X0 *(Tasuma)*

Tasuma CSV 65 conceptual UAV *(IHS/Patrick Allen)*

CSV 40 at ParcAberporth in mid-2007 *(IHS/Patrick Allen)*

The CSV 65, currently Tasuma's largest product *(Tasuma)*

Power plant: One piston engine, driving a two-blade pusher propeller. Choice dependent upon desired configuration; possibilities include 3-W 50 cc, 3-W 150 cc Twin, or 13.4 kW (18 hp) UEL rotary.

Variants: *CSV 40:* Basic version.

CSV 50: Higher-powered engine and operating weights. Dimensionally as for CSV 40, but with option of larger wing if required.

CSV 40, CSV 50

Dimensions, External	
Overall, length	3.20 m (10 ft 6 in)
Wings, wing span	3.10 m (10 ft 2 in)
Weights and Loadings	
Weight	
Weight empty	
CSV 40	18 kg (39 lb)
CSV 50	35 kg (77 lb)
Max launch weight	
CSV 40	28 kg (61 lb) (est)
CSV 50	65 kg (143 lb) (est)
Payload	
Max payload	
CSV 40	5 kg (11.00 lb)
CSV 50	25 kg (55 lb)
Performance	
Speed	
Cruising speed	
max, CSV 40	81 kt (150 km/h; 93 mph)
max, CSV 50	89 kt (165 km/h; 102 mph)
Loitering speed	
CSV 40	35 kt (65 km/h; 40 mph)
CSV 50	38 kt (70 km/h; 44 mph)
Endurance	5 hr
Power plant	1 × piston engine

Status: One built. Available for demonstration 2009.
Contractor: Tasuma (UK) Ltd
Blandford Heights, Dorset.

Tasuma Hawkeye

Type: Close-range mini-UAV.

Development: Begun in 2005; Hawkeye III introduced in 2006.

Description: *Airframe:* Standard Hawkeye has slender tubular fuselage with conventional tail unit; constant-chord, unswept wing, with marked dihedral on outer panels; ventral payload pod. Hawkeye III has more substantial pod and boom fuselage and pusher engine installation. No landing gear.

Mission payloads: Daylight TV or IR sensor, and other avionics, in quickly demountable and interchangeable ventral pod.

Guidance and control: Aircraft is self-stabilised and capable of autonomous flight, following pre-programmed GPS waypoints; an alternative waypoint can be introduced at any time during flight. GCS with receiver/transmitter assembly and video display. Separate hand controls for manual flight operation, plus tripod stand for command and receiver aerials. Aircraft has automatic return to base if command signal is lost.

Transportation: Man-portable in backpack containing single aircraft and GCS for completely self-contained operation. Pack size approximately 1,350 × 500 × 350 mm (53 × 19.7 × 13.8 in), weight about 20 kg (44.1 lb).

System composition: Standard system comprises two air vehicles (three for Hawkeye III), payloads and a GCS, and is operable by two people.

Launch: Hand or catapult launched.

Recovery: Belly landing.

Power plant: 300 W brushless, three-phase DC electric motor, powered by rechargeable lithium-polymer battery pack; two-blade tractor (pusher in Hawkeye III) propeller.

Variants: *Hawkeye:* Original (2005) version.

Hawkeye nose and pod detail *(IHS/Patrick Allen)* 1149577

Hawkeye standard configuration *(Tasuma)* 1290224

Hawkeye III on show in mid-2007 *(IHS/Patrick Allen)* 1327067

Hawkeye III: Introduced mid-2006. More substantial airframe and pusher engine configuration.

Hawkeye, Hawkeye III

Dimensions, External	
Overall	
length	
Hawkeye	1.20 m (3 ft 11¼ in)
Hawkeye III	1.40 m (4 ft 7 in)
Wings	
wing span	
Hawkeye	2.20 m (7 ft 2½ in)
Hawkeye III	1.70 m (5 ft 7 in)
Weights and Loadings	
Weight	
Max launch weight	
Hawkeye	3.8 kg (8.00 lb)
Hawkeye III	3.2 kg (7.00 lb)
Performance	
Altitude	
Service ceiling, Hawkeye	300 m (980 ft)
Speed	
Max level speed	
Hawkeye	45 kt (83 km/h; 52 mph)
Hawkeye III	43 kt (80 km/h; 49 mph)
Loitering speed, Hawkeye	27 kt (50 km/h; 31 mph)
Radius of operation	
mission, standard	2.7 n miles (5 km; 3 miles)
mission, optional	5.4 n miles (10 km; 6 miles)
Endurance, max	1 hr
Power plant	1 × electric motor

Status: Enhanced Hawkeye used by Blue Bear Systems consortium in the UK MoD 2008 Grand Challenge.
Contractor: Tasuma (UK) Ltd
Blandford Heights, Dorset.

Thales Watchkeeper

Type: Close-range/short-range intelligence, surveillance, target acquisition and reconnaissance (ISTAR) UAV.

Development: Studies of UAVs to meet land commanders' tactical reconnaissance requirements after the BAE Systems Phoenix retires were approved by the UK MoD in June 1998, and were allocated the programme names Sender and Spectator. The risk reduction phase was handled for DPA by the Defence Evaluation and Research Agency (DERA, now QinetiQ) at Farnborough.

Four teams bidding for the Sender programme were shortlisted in the second quarter of 2000, and were then asked to study possible

Thales Watchkeeper < United Kingdom < Unmanned aerial vehicles

Watchkeeper made its first flight in April 2008 *(Thales)* 1330645

Watchkeeper completes its maiden flight in the UK, April 2010 *(Thales)* 1398051

Arrested landings are mandatory for Watchkeeper *(Thales)* 1369170

Workstations inside the Watchkeeper GCS *(Thales)* 1289793

Watchkeeper ground control vehicle *(Thales)* 0583315

extrapolation of their proposals to encompass the Spectator requirement as well. The new combined requirement was given the programme name Watchkeeper. Industry spokespersons subsequently indicated that Watchkeeper could result in an acquisition of a range of assets, possibly including high- and medium-altitude air vehicles and some with maritime capability. Sender entered Assessment Stage 1 in May 2000 with the selection of four teams, led by BAE Systems, Lockheed Martin, Northrop Grumman and Thales, to submit their proposals in June 2001 to meet the MoD's ISTAR capability requirement. See earlier editions for details of these team members. Contracts worth GBP3.1 million (USD4.3 million) to each of the four team leaders followed in September 2000, to deliver one-year, risk-reduction assessment phase studies to the MoD in 2001.

Assessment Stage 1 ended in September 2001, and was followed by six-month (from November 2001) study contracts leading to a 12- to 18-month Systems Integration and Assurance Phase (SIAP) stage starting in the second quarter of 2002. Bids by the four teams, whose composition had by then undergone some changes, were submitted by the 14 March 2002 deadline and followed by formal presentations to the DPA on 14 June. Acceleration of the programme was announced by the MoD in July, and down-selection to two semi-finalists was then expected in late August 2002. However, announcement of the choice of these did not occur until 7 February 2003, when it was revealed that the Northrop Grumman and Thales UK teams had been selected to proceed to the next stage. Make-up of these teams then underwent further changes.

A Joint Services Trials Unit (JSTU) was created in 2003, comprising No 32 Regiment Royal Artillery, No 792 Squadron Royal Navy and No 100 Squadron Royal Air Force. The SIAP semi-finalist teams delivered their technical submissions to the DPA on 18 December 2003. Their responses to the Invitation To Tender (ITT) were followed by operational capability assessment (OCA) flight demonstrations at the QinetiQ Aberporth range in March 2004.

Watchkeeper requirements include ability to work within the British Army's Bowman communications network; interoperability with the service's Apache attack helicopters and the RAF's ASTOR surveillance aircraft, and 100 per cent offset manufacturing in the UK of the chosen system. According to the MoD in February 2003, programme cost is in the region of GBP800 million, and the winning system is expected to have an in-service life of some 30 years.

The Thales team was selected as preferred bidder on 20 July 2004. Negotiations began immediately, with the object of achieving IOC from around the end of 2006 and full operational capability in 2007. A GBP6 million interim contract was awarded to the Thales team in January 2005, ahead of the so-called 'main gate' award, which was received six months later. Announcement of the latter, however, indicated that the service entry date for Watchkeeper had been postponed to 2010. In the immediate future, Watchkeeper is now in the demonstration and manufacturing phase, for which Thales and Elbit formed a joint venture company, U-TacS (UAV Tactical Systems Ltd), with a manufacturing facility at Leicester, UK, to achieve the necessary transfer of technology. First flight in the UK of a Hermes 450 development prototype took place at Parc Aberporth in early September 2006, with a 5-hour, first 'full' flight following on 7th of that month. The first flight of Watchkeeper took place at Megido, northern Israel, on 16 April 2008. In July automated take-off and landing flights using Thales's MAGIC ATOLS were performed and validated in the presence of the UK Ministry of Defence at Megido.

Trials in Israel were completed in June 2009, and the Watchkeeper programme moved to Parc Aberporth in West Wales for further flight trials and ground systems tests. Watchkeeper's first UK flight took place on 14 April 2010 from Parc Aberporth. During 2010 the Watchkeeper training centre was scheduled to start operations at Larkhill, Salisbury, where the simulator is based. The UK MoD awarded Thales the initial three-year support contract for the Watchkeeper programme in April 2010. Two air vehicles were taking part in the test programme in 2011 and 400 hours had been accumulated in 270 flights in Israel and the UK , and the first operational deployment to Afghanistan was planned at the end of the year.

Thales (UK) team:
U-TacS (Thales and Elbit Systems)
(WK450 air vehicle and subsystems)
UEL (air vehicle design and engines)
LogicaCMG (command and battlespace
management systems and applications)
Cubic Defense Applications (datalinks)

Vega Group (training)
Marshall Specialist Vehicles (GCS shelters and support)
QinetiQ (airworthiness consultancy and image data management)
Praxis Critical Systems (programme safety consultancy)

Description: *Airframe:* Generally as described for Elbit Hermes 450. APPH nosewheel unit.

Mission payloads: Fully stabilised EO/IR (Elop CoMPASS) and SAR sensors.(Thales Aerospace I-Master SAR/GMTI). Laser target marketing, ranging and designation. VHF/UHF and GSM communications rebroadcast. Options include communications EW/comint and counter IED payloads. Extended endurance fuel tanks can be fitted.

Guidance and control: Intended to be compatible with British Army Bowman system. Cubic Defense Applications received a USD52 million contract in 2006 to develop an advanced datalink that will incorporate both tactical common datalink (TCDL) and high integrity datalink (HIDL) features. Under subcontract to Cubic, Ultra Electronics will assist in this development. In January 2007, Athena Technologies' GuideStar combined INS/GPS and air data sensor suite was chosen for Watchkeeper.

Launch: Conventional and automatic wheeled take-off.
Recovery: Conventional and automatic wheeled landing.
Generally similar to those for Elbit Hermes 450.

Status: Demonstration and manufacturing phase, followed by Test and evaluation towards an initial operational capability in 2011-2012.

Customers: Intended for British Army. Some 54 Watchkeeper air vehicles are on order.

Contractor: U-TacS (UAV Tactical Systems Ltd), Thales UK and Elbit Systems joint company,
Leicester

The UTSL Vigilant mini-UAV *(UTSL)* 1290225

UTSL Vigilant

Type: Short-range surveillance mini-UAV.
Development: Details released during DSEi defence exhibition in London, September 2007. Maiden flight took place in first half of same year. First public flight at ParcAberporth Unmanned Systems 2008.
Description: *Airframe:* Constant-chord, unswept wings with marked dihedral on outer panels; slender tubular fuselage with conventional tail surfaces; payload pod pylon-mounted below fuselage. No landing gear.
Mission payloads: Fixed, forward-looking CCD camera, plus gimbal-mounted daylight camera or thermal imager.
Guidance and control: Fully automatic pre-programmed flight, in both day and night time conditions. GPS/INS navigation and tracking, with full telemetry including flight data recording.
Launch: Automatic, from lightweight bungee catapult.
Recovery: Automatic recovery by steep descent or skid landing.
Power plant: Battery-powered electric motor; two-blade tractor propeller.

Vigilant	
Dimensions, External	
Overall	
length	1.52 m (4 ft 11¾ in)
height	0.61 m (2 ft 0 in)
Wings, wing span	2.18 m (7 ft 1¾ in)
Areas	
Wings, Gross wing area	0.50 m² (5.4 sq ft)
Weights and Loadings	
Weight, Max launch weight	15 kg (33 lb) (est)
Performance	
Altitude, Service ceiling	5,000 m (16,400 ft)
Speed	
Max level speed	57 kt (106 km/h; 66 mph)
Loitering speed	32 kt (59 km/h; 37 mph)
Radius of operation	
electronic	54 n miles (100 km; 62 miles) (est)
surveillance	32.4 n miles (60 km; 37 miles) (est)
Endurance	2 hr (est)
Power plant	1 × electric motor

Status: Reported in service in 2008.
Contractor: Universal Target Systems Ltd
Challock, Kent.

United States

AAI Aerosonde Mark 4.7

Type: Small tactical UAV.
Development: Part of AAI's scalable fleet of Aerosonde UAS, the Mark 4.7 is stated to have increased capacity compared to earlier versions, although very few specifics of this had been released by mid-2009. It took part in flying demonstrations at the US Army's Yuma Proving Ground in Arizona in late June 2008, completing its test flying there in February 2009, and was publicly unveiled at the Australian International Air Show later that month and at the Bahrain show during the following year. Launched and recovered from the M80 Stiletto ship during 2009

Description: *Airframe:* General configuration similar to earlier Aerosonde versions; all-new design fuselage pod and increased wing span.
Mission payloads: EO, IR and laser designator sensors, with real-time imagery downlink.
Guidance and control: Fully automated, with GPS navigation and tracking. Compatible with AAI's Expeditionary GCS; also integrated into AAI's interoperability network of common GCS technologies, including the STANAG 4586-compliant OSGCS and OSRVT.
Launch: Automated rail launch.
Recovery: Automated belly landing or net recovery.
Power plant: Single piston engine (type and rating not provided); two-blade pusher propeller.

Aerosonde Mark 4.7	
Dimensions, External	
Wings, wing span	3.60 m (11 ft 9¾ in)
Performance	
Altitude, Service ceiling	4,570 m (15,000 ft)
Endurance	12 hr
Power plant	1 × piston engine

Status: In development in 2009. Was offered as a contender for the US Navy and Marine Corps' Small Tactical Unmanned Aircraft System (STUAS)/Tier II programme. Trials and demonstrations continuing during 2010.

Contractor: AAI Corporation
Hunt Valley, Maryland.

Aerosonde Mark 4.7 in flight *(AAI)* 1386609

Aerosonde Mark 4.7 launch *(AAI)* 1356228

AAI Aerosonde Mark 5

Type: Tactical UAS.

Development: Development of this UAS has been under way since AAI's acquisition of Australian manufacturer Aerosonde and its North American subsidiary in 2006, and a prototype flew for the first time on 10 December 2008 at the Yuma Proving Ground in Arizona. Announcing this event in early January 2009, AAI observed that the flight had "met or exceeded all test objectives and validated the aerodynamic capabilities". The continuing test programme was planned to explore and validate key performance and payload capabilities, with a view to entering the Mark 5 in forthcoming US Navy STUAS (small tactical UAS) and US Marine Corps Tier II UAS competitions.

Description: *Airframe*: General configuration is essentially similar to that of the AAI RQ-7 Shadow 200 (see separate entry).

Mission payloads: Interchangeable underfuselage turret for ISR, target acquisition and other sensors, according to mission.

Guidance and control: Fully autonomous operation is assumed.

Launch: Conventional, automatic wheeled take-off.

Recovery: Conventional, automatic wheeled landing.

As of mid-January 2009, specification details had not been released, although the January 2009 release described the Mark 5 as having "impressive endurance". General size and performance parameters are expected to be closer to those of the RQ-7 Shadow 200 than to the smaller, earlier members of the Aerosonde family.

Status: Under development in 2009.

Contractor: AAI Corporation
Hunt Valley, Maryland.

Computer-generated image of the Aerosonde Mk 5 *(AAI)*

AAI RQ-7 Shadow 200

Type: Surveillance and target acquisition UAV.

Development: AAI Corporation, in addition to being the US partner on the Pioneer programme, has built a number of own-design UAV test vehicles, which have amassed many hundreds of hours of flight testing. This experience led to development of the company's Shadow family, of which the 200 is the smallest member. It first flew in 1992 and was selected by the US Army in 1999, being ordered into production as the RQ-7A. An earlier variant, the T-tailed Shadow 200T, did not go into production.

Following a number of incidents in the second half of 2003, in-service Shadows received upgrades to their engines, emergency recovery and landing systems. In addition, incremental planned upgrades are under way. Initially, these concerned bringing the prototype and LRIP air vehicles up to full-rate production (FRP) standard and integrating them and the FRP Shadows with the Tactical Common Datalink (TCDL). Beyond this, major improvements have concentrated on increasing payload/range capability and introducing more versatile sensors. This includes an extended wing with additional fuel and hardpoints for additional payloads, including a laser designator.

Description: *Airframe*: Small, shoulder-wing monoplane, with pusher engine, twin tailbooms and inverted V tail unit. Construction is mainly (more than 85 per cent) of composites (graphite and carbon fibre epoxy). Optionally detachable tricycle landing gear.

Mission payloads: IAI Tamam POP-200 standard for RQ-7A; POP-300 for RQ-7B. Latter version due to receive laser rangefinder/designator in 2006. Also under evaluation in late 2006, with a view to possible deployment in 2007, was the BAE Systems Adaptive Unsupervised Real-time Optical Reconnaissance Array (AURORA) hyperspectral imager.

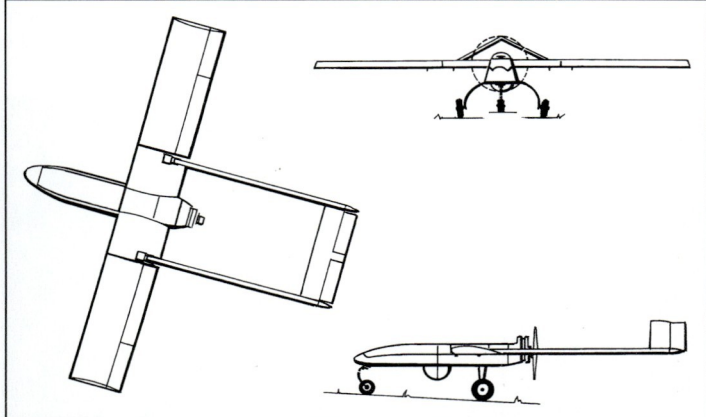

Shadow 200 general arrangement *(IHS/John W Wood)*

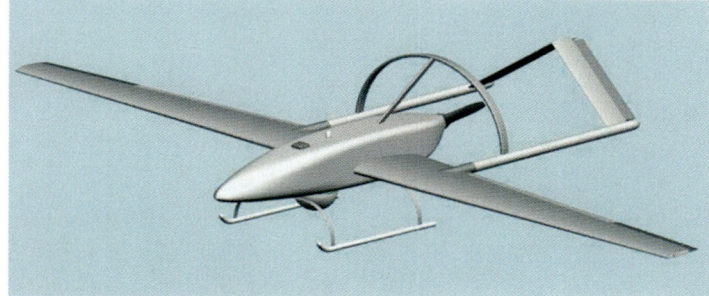

Impression of proposed Shadow 300 *(AAI)*

Guidance and control: AAI 'One System' mobile GCS. Preprogrammed or remotely controlled flight profiles with Athena GuideStar GS-211e GPS-based autopilot navigation. AAI C-band UHF command uplink and video downlink in RQ-7A; TCDL uplink in RQ-7B. Imagery and telemetry data can be transmitted directly to Joint STARS, All Sources Analysis Systems (ASAS) and US Army Field Artillery Targeting and Direction System (AFATDS) in near real time.

Transportation: Air vehicle can be dismantled and stored/transported in a 0.61 × 0.61 × 1.65 m (2 × 2 × 5.4 ft) container; can be carried by two people. Complete system air-transportable in three C-130 aircraft.

System composition: Four air vehicles with payloads; six HMMWVs; two mobile GCSs and two GDTs; four RVTs and antennas; one portable GCS and one portable GDT. One HMMWV transports the UAVs and the hydraulic launch trailer; two others each transport one GCS, two are troop and equipment carrier vehicles, and the sixth carries associated support equipment. US Army ground crew of 22.

Launch: Automatic ground launch by hydraulic catapult, or by conventional wheeled take-off.

Recovery: Automatic wheeled landing.

Powerplant: One 28.3 kW (38 hp) UEL AR 741 rotary engine (RQ-7A and B); two-blade fixed-pitch wooden pusher propeller.

Fuel (40 litres; 10.5 US gallons; 8.7 Imp gallons in RQ-7A) in fire-retardant, explosion-proof wing cells; increased to 57 litres (15.0 US gallons; 12.5 Imp gallons in RQ-7B). Growth option for eventual heavy fuel power plant.

Variants: **RQ-7A**: Initial production version from 2002. *Following description applies to this version except where indicated.*

RQ-7B: Improved version, introduced on to production line August 2004. Changes include longer-span wing with more efficient aerofoil section, to accommodate TCDL; enlargement of tail unit, to balance changes to wing; 'significantly increased' fuel capacity for greater endurance; Athena Technologies GS-211e flight controller for more accurate navigation and attitude sensing; and increased target location accuracy.

Shadow 300: Larger version, with heavy-fuel engine. Selected by Boeing and SAIC in July 2005, with 10-month development contract, as

RQ-7A Shadow 200 of 101st MI Battalion on duty in Iraq *(US Army)*

US Army personnel preparing to launch an RQ-7A *(US DoD)*

Unmanned aerial vehicles > United States > AAI RQ-7 Shadow 200 – AAI Shadow 400 and 600

Inside the One System GCS *(US Army)* 1151561

one of four potential candidates for US Army's Class III (battalion level) FCS requirement.

RQ-7A, RQ-7B

Dimensions, External
Overall
 length..3.40 m (11 ft 1¾ in)
 height..0.91 m (2 ft 11¾ in)
Wings
 wing span
  RQ-7A..3.89 m (12 ft 9¼ in)
  RQ-7B..4.29 m (14 ft 1 in)
 Engines, propeller diameter...........................0.66 m (2 ft 2 in)
Areas
Wings
 Gross wing area, RQ-7A.................................2.14 m² (23.0 sq ft)
Weights and Loadings
Weight
 Weight empty..91.0 kg (200 lb)
 Max launch weight
  RQ-7A...154 kg (339 lb)
  RQ-7B...170 kg (374 lb)
 Fuel weight
  Max fuel weight
   RQ-7A..23.1 kg (50 lb)
   RQ-7B..33.1 kg (72 lb)
 Payload
  Max payload, RQ-7A....................................25.3 kg (55 lb)
Performance
T-O
 T-O run..250 m (821 ft)
Altitude
 Service ceiling
  limited by engine fuel/air
  mixture...4,570 m (15,000 ft)
Speed
 Max level speed
  RQ-7A...123 kt (228 km/h; 142 mph) at S/L
  RQ-7B...118 kt (219 km/h; 136 mph) at S/L
 Cruising speed
  RQ-7A...85 kt (157 km/h; 98 mph)
  RQ-7B...90 kt (167 km/h; 104 mph)
 Loitering speed
  RQ-7A...65 kt (120 km/h; 75 mph)
  RQ-7B...60 kt (111 km/h; 69 mph)
 Stalling speed, at S/L...................................55 kt (102 km/h; 64 mph)
 g limits..+3.6
 Radius of operation.....................................43 n miles (79 km; 49 miles)
  datalink, max..67.5 n miles (125 km; 77 miles)
Endurance
 max, RQ-7A..5 hr 30 min
 max, RQ-7B..7 hr

Status: Selected as winner of US Army Tactical UAV (TUAV) competition in December 1999; initial LRIP contract of USD41.8 million for four Shadow 200 systems for opeval, delivered from November 2000; option for further four LRIP systems exercised with award of USD19.4 million contract on 11 April 2001. Field qualification tests at Fort Huachuca, Arizona, completed in March 2001; IOT&E at Fort Hood, Texas, began at the end of April 2001 and was completed successfully in May 2002, at which time some 1,700 hours in 900 flights had been completed. Milestone C (approval for full-rate production) was achieved on 1 October 2002.

The RQ-7A was fielded in October 2002 to the US Army's 1st and 2nd Stryker Brigade Combat Teams (BCTs) at Fort Lewis, Washington, the 4th Infantry Division at Fort Hood, Texas, and for crew training to Fort Huachuca. Nine Shadow 200 systems were deployed during Operation 'Iraqi Freedom' in 2003; by September 2004, Shadows in Iraq (by then reduced to six systems) had flown some 10,000 hours in about 2,500 sorties, units at that time including the 312th and 313th Military Intelligence Battalions of the 82nd Airborne Division. First recipient of the RQ-7B was the US Army's 172nd Stryker BCT in Alaska.

The 23 systems delivered by early 2005 were distributed between the 1st, 2nd, 3rd, 4th and 25th Infantry Divisions (two, two, three, two and two systems respectively); 172nd Infantry Brigade (one); 1st Cavalry Division (three); 82nd Air Division (two); Maryland and Pennsylvania National Guard (one each); and the TUAV Training Center at Fort Huachuca (four). On

In August 2006, the US Army awarded a USD11.7 million contract for integration and demonstration of a new, advanced tactical common datalink (TCDL) for use with the RQ-7B, to be subcontracted to L-3 Communications — West and Cubic Corporation. In October that year, AAI received a USD13.5 million US Army contract covering upgrade of the EO/IR payload in the existing Shadow 200 fleet to POP-300 standard; additional engines for fleet retrofit; and enhancement of One System ground stations to a single configuration "consistent with the current production baseline".

Customers: US Army (RQ-7A and RQ-7B), US Marines Corps and National Guard; initial requirement for 41 systems (164 aircraft) attained by 2006. New orders up to the start of 2010 bought the total to 116 systems, with 91 systems delivered at that time and deliveries extending to March 2011. In addition 100 extended wing kits were ordered in 2010.

In July 2006, the US DoD recommended (subject to Congressional approval) the FMS sale of two RQ-7B systems to the Polish armed forces, for use in support of European Union brigades and NATO operations. This would comprise 10 (including two spare) air vehicles; their EO/IR payloads; four GCSs; associated GDTs (including two portable); two AN/APX-100 IFFs; two GPS receivers; two launchers; two maintenance vehicles; 12 Thales AN/PRC-148 multiband inter-team radios with waveform SINCGARS and 14 M1152 HMMWVs; and would have a reported contract value of up to USD73 million. The Polish Shadows would be operated by a single ISTAR squadron to be created within the airmobile forces or assigned to an independent reconnaissance regiment.

Italy is buying four Shadow systems under plans announced in 2010.

Contractor: AAI Corporation
Hunt Valley, Maryland.

AAI Shadow 400 and 600

Type: Multirole UAVs.
Development: The Shadow 400 and 600 incorporate state-of-the-art technologies derived from AAI's participation as prime contractor for the US Pioneer UAV programme.

Hardware and software upgrades were made as a result of lessons learned during combat deployment of the Pioneer in Operations Desert Shield and Desert Storm.

Description: *Airframe:* Shoulder-wing monoplane with (on Shadow 600) 15° sweepback on outer panels; pusher engine and twin-boom tail unit; tricycle landing gear with retractable nosewheel. Construction is 95 per cent composites and 5 per cent metal; graphite epoxy fuselage; graphite epoxy/honeycomb sandwich wings with internal ribs; aluminium tube tailbooms; tail surfaces of graphite and Kevlar epoxy. Servo-actuated hydraulic mainwheel brakes. Under-fuselage arrester hook.

Mission payloads: Multiple payloads are offered for both vehicles. EO/IR sensor with laser pointer is the standard payload package: examples include a high-resolution CCD daylight camera with a 10:1 continuous zoom lens; an LLTV CCD camera with 10:1 zoom lens; a Kollmorgen high-resolution FLIR; or a FLIR Systems FLIR. A companion sensor such as a meteorological data or NBC detection package can be carried with any of these payloads. Those for Romania have an FLIR Systems (ex-Inframetrics) 445G dual-sensor payload. To date AAI has integrated more than 32 payloads for customers.

Guidance and control: The Shadow 400 and 600 are operated by AAI's One System GCS which allows operation of multiple air vehicles. The airborne flight computer performs altitude, heading and speed maintenance functions as well as all enhanced aerodynamic damping and navigation. As such, the UAV can be flown in a semi-autonomous mode in which course, speed and altitude changes are updated from the MPCS; or, in conjunction with a navigation system (for example, inertial navigation, Loran or GPS), in a fully autonomous mode. The ground station will automatically transmit commands to the UAV to fly the flight plan; or the flight plan can be transmitted to the UAV, which will then fly the plan autonomously.

The straight-winged Shadow 400 0528766

The console is designed to be utilised either by a single operator controlling both the UAV and sensor payloads or in a multiple-operator/multiple-console configuration. The ability to reconfigure permits each identical console to be assigned to different operator roles such as UAV pilot, payload operator, mission planner, mission commander or combined UAV/payload operator.

The One System portable GCS is also available. This can be used at a remote launch and recovery site, with the GCS located closer to the target area, thus providing increased operational range and flexibility.

The GCS is typically installed in standard military shelters, and normally operates on 110 V 60 Hz AC, but can also operate on 28 V DC in the event of AC power failure.

In late 2001, the Raytheon Tactical Control System (TCS, which see) launched and recovered a Shadow 600 air vehicle, performing flight control, payload control and receipt of data from the UAV.

System composition: Shadow 600 systems for Romania each comprise six air vehicles, five payloads, one GCS, one GDT, six RVTs, three nose cameras, plus ground support equipment, spares and manuals.

Launch: Conventional wheeled take-off standard; automatic take-off or catapult launch optional.

Recovery: Wheeled landing (with remotely activated disc brakes) standard; parachute or autoland recovery system optional. Net recovery option for Shadow 400.

Power plant:
Shadow 400: One 27.6 kW (37 hp) UEL AR 731 rotary engine.
Shadow 600: One 38.8 kW (52 hp) UEL AR 801 rotary engine and four-blade fixed-pitch wooden pusher propeller. Rain propeller available for use in inclement weather and other adverse conditions.

Fuel in fuselage and wing centre-section tanks of Shadow 600, combined capacity 85.5 litres (22.5 US gallons; 18.7 Imp gallons); no wing fuel in Shadow 400.

Variants: *Shadow 400:* Shorter-span wing, without sweepback and containing no fuel; otherwise generally similar to Shadow 600.

Shadow 600: Major version to date. *Description applies to this version except where indicated.*

Shadow 400, Shadow 600

Dimensions, External	
Overall	
length	
Shadow 400	4.34 m (14 ft 2¾ in)
Shadow 600	4.77 m (15 ft 7¾ in)
height, Shadow 600	1.24 m (4 ft 0¾ in)
Wings	
wing span	
Shadow 400	5.05 m (16 ft 6¾ in)
Shadow 600	6.83 m (22 ft 5 in)
Engines	
propeller diameter, Shadow 600	0.74 m (2 ft 5¼ in)
Areas	
Wings	
Gross wing area, Shadow 600	3.754 m² (40.4 sq ft)
Weights and Loadings	
Weight	
Weight empty	
Shadow 400	147 kg (324 lb)
Shadow 600	148.4 kg (327 lb)
Max T-O weight	
Shadow 400	201 kg (443 lb)
Shadow 600	265 kg (584 lb)
Payload	
Max payload	
Shadow 400	30.0 kg (66 lb)
Shadow 600	41.0 kg (90 lb)
Performance	
T-O	
T-O run, Shadow 600	549 m (1,802 ft)
Climb	
Rate of climb	
max, at S/L, Shadow 600	206 m/min (675 ft/min)
Altitude	
Service ceiling	
Shadow 400	3,660 m (12,000 ft)
Shadow 600	4,875 m (16,000 ft)
Speed	
Max level speed	
Shadow 400	100 kt (185 km/h; 115 mph)
Shadow 600	104 kt (193 km/h; 120 mph)
Cruising speed	75 kt (139 km/h; 86 mph)
Loitering speed	65 kt (120 km/h; 75 mph)
Radius of operation	
datalink, max, Shadow 400	100 n miles (185 km; 115 miles)
datalink, max, Shadow 600	108 n miles (200 km; 124 miles)
Endurance	
Shadow 400	5 hr
Shadow 600	12 hr (est)

Status: One 600 system delivered to Turkey for evaluation in 1993; one 600 system ordered by Romania in 1997 and delivered in April 1998, achieving IOC in mid-1999; one Romanian aircraft lost in crash in April 2000. Second 600 system ordered by Romania in November 2000 (USD7.5 million contract). In September 2000, AAI announced a USD22 million contract for Shadow 400 from an Asian customer, believed to be South Korea; this was for a single ship-based system with multiple air vehicles, a GCS and a hydraulic launcher.

Customers: Romania (2nd Romanian Air Force Corps: two Shadow 600 systems); South Korean Navy (one Shadow 400 system).

Contractor: AAI Corporation
Hunt Valley, Maryland.

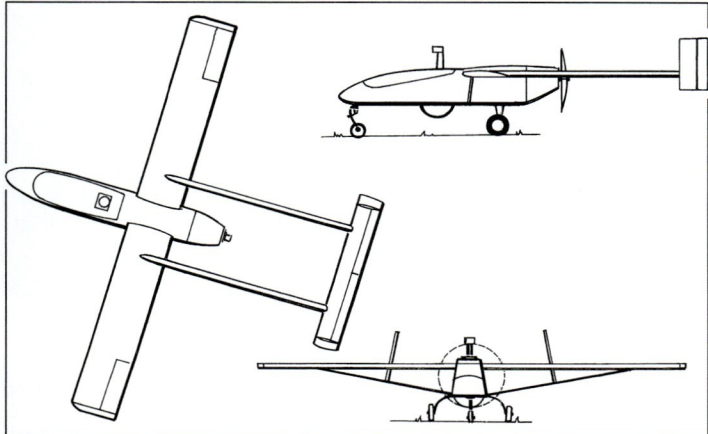
Shadow 400 general arrangement *(John W Wood)*   0528770

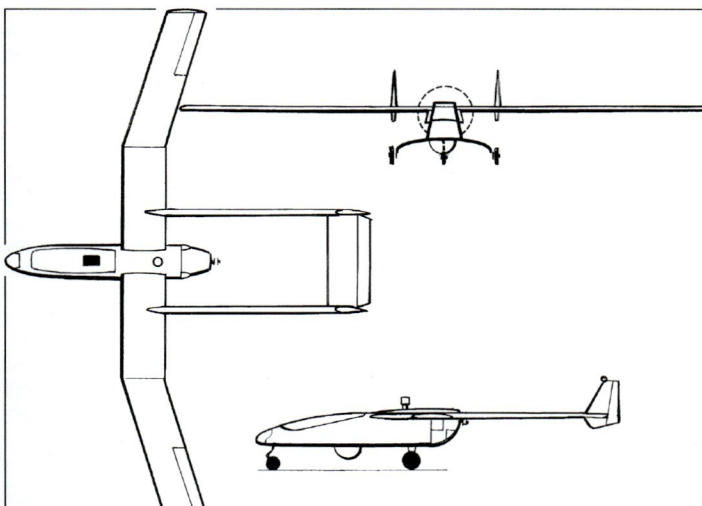
AAI Shadow 600 multirole UAV *(John W Wood)*   0518218

AAI Shadow 600 in flight   0528768

Acuity Technologies AT-3 Owl

Type: Air-launched multirole UAS.

Development: In January 2005, US Naval Air Systems Command (NAVAIR) initiated a development programme for a Wing and Bomb Bay Launched (WBBL) unmanned aircraft system, aimed at extending the reach and enhancing the effectiveness of its maritime patrol aircraft while increasing crew safety and reducing fatigue. The Acuity AT-3 Owl is a candidate in this programme; development began in November 2006, and it was

The Owl has a wide-track gear for landing at recovery sites *(Acuity)*   1356281

View of AT-3 Owl showing its rear-mounted ducted fan *(Acuity)*
1356280

unveiled in mid-2008. It can be launched from wing pylons or the weapons bay of the P-3C Orion and P-8A Poseidon. Further development and testing is continuing under a 2½-year NAVAIR contract; fleet introduction of a selected WBBL system is envisaged for 2012.

Description: *Airframe:* Box-section fuselage with internal payload bay and rear-mounted engine. High-mounted, variable-sweep wings are aligned over rear of fuselage during pylon or bomb bay carriage, pivoting forward after launch and in flight relative to changing CG position as fuel is consumed and sensors deployed. Fixed tricycle landing gear. Composites (glass and carbon fibre) construction.

Mission payloads: Nose-mounted camera head provides EO capability for all missions. Internal payload bay has plug-in racks that allow various payload/fuel combinations. Payloads can include sonobuoys (five A-size or 10 G-size), EO/IR cameras, MAD sensors, miniature SAR, or communications relay packages. Real-time video and data downlinks.

Guidance and control: Pre-programmed, via control and data links from host aircraft or other STANAG 4586-compliant GCS.

Launch: Air launch from NATO-standard 14-inch bomb rack. In addition to P-3C and P-8A, this enables carriage also by such other US Navy types as SH-60B/F and MH-60R helicopters.

Recovery: By wheeled landing at a selected location.

Power plant: One 26.8 kW (36 hp) UEL AR 741 rotary engine (gasoline version in test vehicles; heavy-fuel version proposed for production version); two-blade pusher propeller in duct at rear.

AT-3

Dimensions, External
Overall
  length
    incl duct .... 2.27 m (7 ft 5¼ in)
    stored .... 2.98 m (9 ft 9¼ in)
Wings
  wing span .... 4.18 m (13 ft 8½ in)
  wing chord, constant .... 0.39 m (1 ft 3¼ in)
Dimensions, Internal
Payload bay
  length .... 1.04 m (3 ft 5 in)
  width .... 0.38 m (1 ft 3 in)
Areas
Wings, Gross wing area .... 1.64 m² (17.7 sq ft)
Weights and Loadings
Weight
  Weight empty .... 68 kg (149 lb)
  Max launch weight .... 186 kg (410 lb)
  Fuel weight, Max fuel weight .... 118 kg (260 lb)
  Payload, Max payload .... 91 kg (200 lb)
Performance
Speed, Launch speed .... 141.47 m/s (464 ft/s)
Endurance
  standard fuel .... 8 hr (est)
  with max fuel .... 30 hr (est)

Status: The AT-3 was undergoing high-speed taxi tests in December 2009, with flight testing planned to begin in early 2010.

Contractor: Acuity Technologies Inc
Menlo Park, California.

Acuity Technologies AT-10 Responder

Type: Tilt-wing VTOL UAV.

Development: Being developed with company funding, the AT-10 Responder is an unorthodox design featuring a hybrid propulsion system and tilting inboard wing sections. It has a nose-mounted camera and large internal payload bay. Development began in December 2008, with first flights (both vertical and horizontal) being made in June 2009.

Description: *Airframe:* High-wing design; inboard wing sections, on which engine nacelles are mounted, tilt for vertical take-off and landing. Descending winglets with skids protect propellers in the event of roll during T-O and landing phases. Box-section fuselage; angular tail surfaces. Fixed tricycle landing gear. Composites (glass and carbon fibre) construction.

Mission payloads: Typical payloads could include cameras, miniature SAR or communications relay packages.

Guidance and control: Planned to be fully automatic, but initial test flights being conducted under direct pilot control with gyro-based attitude stabilisation. Inner section of each wing rotates for transition between

Configuration of AT-10 Responder in forward flight *(Acuity)*
1356283

AT-10 Responder in VTOL/hover mode *(Acuity)*
1356282

vertical and forward flight. GPS or DGPS navigation. GCS being developed, adapting COTS software to ensure compliance with STANAG 4586; avionics and radio systems also to be compatible with US Army (AAI) OSRVT and L-3 TCDL.

Transportation: Entire system can be transported in a 0.61 × 0.61 × 2.44 m (2 × 2 × 8 ft) container weighing less than 68 kg (150 lb). Air vehicle wings detach from fuselage in one piece; wingtips can also be removed for more compact storage. Tail unit and landing gear assembly also removable by undoing two bolts.

Launch: Vertical or conventional take-off.

Recovery: Vertical or conventional landing.

Power plant: Propulsion provided by twin 8.95 kW (12 hp) electric motors, each with three-blade propeller, and batteries installed in the wings. For flights of 2 h or less, additional batteries in upper fuselage provide all-electric propulsion. For longer flights, a small generator, driven by a four-stroke internal combustion engine, can be installed in the fuselage rear. This can be turned on and off in flight to provide electric power for cruise propulsion, recharging batteries and high-power payloads.

A power conditioning system has been designed for in-flight or ground charging of the lithium batteries; this also continuously controls generator engine speed and charging, and flight power current, according to battery level and power loads.

AT-10

Dimensions, External
Overall, length .... 1.98 m (6 ft 6 in)
Wings, wing span .... 2.86 m (9 ft 4½ in)
Dimensions, Internal
Payload bay
  length .... 0.91 m (2 ft 11¾ in)
  width .... 0.254 m (10 in)
  depth .... 0.178 m (7 in)
Weights and Loadings
Weight
  Weight empty
    wing batteries only .... 25.0 kg (55 lb)
    wing and fuselage batteries .... 31.8 kg (70 lb)
  Max T-O weight
    VTOL .... 45.4 kg (100 lb)
    CTOL .... 54.4 kg (119 lb)
  Payload, Max payload .... 13.6 kg (29 lb) (est)

AT-10

Performance	
Climb	
Rate of climb, max, at S/L	671 m/min (2,201 ft/min)
Altitude, Service ceiling	4,875 m (16,000 ft)
Speed	
Max level speed	90 kt (167 km/h; 104 mph)
Cruising speed, econ	57 kt (106 km/h; 66 mph)
Loitering speed	38 kt (70 km/h; 44 mph)
Stalling speed	34 kt (63 km/h; 40 mph)
Endurance	2 hr
Power plant	2 × electric motor

Status: Flight testing was continuing in late 2009/early 2010, at which time R & D or EMD funding was being sought to develop the Responder into an operational VTOL system.

Contractor: Acuity Technologies Inc
Menlo Park, California.

ADFS AD-150

Type: VTOL maritime UAV.

Development: The AD-150 is being designed to meet the unique requirements of a high-speed, maritime-capable VTOL UAS. Development was started to achieve the technology required to meet emerging requirements for such capability as stated by the US Marine Corps for a Tier III VUAS and the US Coast Guard for its ICGS programme. It is a third-generation improvement over an earlier basic design by the company, the BattleHog 100x. In 2010 the company was reported to be construction an iron-bird ground rig for full-scale static testing of the propulsion system including T53 turboshaft, transmission, drive train and ducted fans. Wind tunnel testing of the ducted-fan propulsion system was being carried out University of Maryland in 2010. Wind tunnel testing of a 3/10th-scale model of the AD-150 was completed during the previous year.

Description: *Airframe:* Dual, counter-rotating High Torque Aerial Lift (HTAL) lift/propulsion systems at wingtips; mid-mounted, tapered wings; streamlined fuselage; V tail. Non-retractable tricycle landing gear. Composites construction (carbon fibre and Kevlar).

Mission payloads: Standard payloads can include EO/IR/laser designator, SAR or a communications relay package. Modular mission payload design supports additional internal payloads and external wing-mounted stores.

Guidance and control: Fully stabilised electromechanical flight control system, with STANAG 4586-compliant control architecture and redundant GPS/INS and GPS/IMU navigation. Each HTAL unit is gimballed to provide directional stability and control in vertical flight, pivoting from vertical to horizontal position for transition to and from forward flight. Interoperable datalinks (TCDL LOS communications) and GCS interfaces.

Launch: Automatic vertical take-off and transition to forward flight.

Recovery: Automatic transition to vertical flight and vertical landing.

Power plant: One turboshaft engine (Textron Lycoming T53 in prototype, Pratt & Whitney PW200 series for production version), driving dual, wingtip-mounted, eight-blade, counter-rotating HTAL lift/propulsion units.

AD-150, minus HTAL units, on display in August 2007 *(ADFS)*

Models of AD-150 and HTAL shrouded fan *(ADFS)*

Impression of AD-150 in VTOL ship landing mode *(ADFS)*

AD-150

Dimensions, External	
Overall	
length	4.42 m (14 ft 6 in)
height	1.45 m (4 ft 9 in)
Wings, wing span	5.33 m (17 ft 5¾ in)
Weights and Loadings	
Weight, Max launch weight	1,021 kg (2,250 lb)
Payload, Max payload	227 kg (500 lb)
Performance	
Altitude, Service ceiling	6,095 m (20,000 ft) (est)
Speed, Max level speed	300 kt (556 km/h; 345 mph) (est)
Endurance, internal fuel	4 hr (est)
Power plant	1 × turboshaft

Status: Development and prototype construction under way in 2007-08 in preparation for flight test; described by manufacturer in October 2007 as "rapidly maturing".

Contractor: American Dynamics Flight Systems
Jessup, Maryland.

AeroMech Fury

Type: Multirole mini-UAV.

Development: AeroMech Engineering has manufactured several types of UAV for various aerospace companies, delivering the 1,000th example in March 2008. Phase 1 flight testing of the Fury was completed in September 2008, proceeding then to payload integration and customer demonstrations. Attributes include a low-observable design and long endurance in an air vehicle with a small logistics footprint.

Description: *Airframe:* Tail-less, sweptback flying wing configuration, with rear-mounted engine. No landing gear. Composites construction.

Mission payloads: Modular 'plug and play' multi-intelligence and/or EW payloads, according to mission. Was demonstrated in July 2009 with a Chesapeake Technology International (CTI) Thunderstorm electronic attack payload. Up to 400 W of continuous power available for payloads.

Guidance and control: Fully autonomous operation, from AeroMech SharkFin, a STANAG 4586 compatible and portable GCS capable of controlling multiple air vehicles. LOS and BLOS satcom datalinks.

Launch: From pneumatic launcher.

Recovery: Into collapsible/portable recovery net. Can be recovered on land or at sea.

Power plant: One 2.2 kW (3 hp) Cosworth AE-1 single-cylinder heavy-fuel piston engine; two-blade pusher propeller. Can also be flown with gasoline engines. See under Weights for fuel details.

Unmanned aerial vehicles > United States > AeroMech Fury – Aerostar International tethered aerostats and blimps

Fury air vehicle on its pneumatic launcher (AeroMech) 1356292

A ground view of the TIF-25K™ tethered aerostat and its associated MMP (Aerostar International) 1365360

Fury

Fury	
Dimensions, External	
Overall, length	1.37 m (4 ft 6 in)
Wings, wing span	3.66 m (12 ft 0 in)
Weights and Loadings	
Weight	
Weight empty	29.5 kg (65 lb)
Max launch weight	56.7 kg (125 lb)
Fuel weight	
Max fuel weight	
standard	12.2 kg (26 lb)
with auxiliary tanks	16.8 kg (37 lb)
Payload	
Max payload	
with auxiliary fuel	5.9 kg (13.00 lb)
with standard fuel	10.4 kg (22 lb)
Performance	
Altitude, Service ceiling	5,485 m (18,000 ft)
Speed	
Max level speed	125 kt (232 km/h; 144 mph)
Cruising speed, max	80 kt (148 km/h; 92 mph)
Loitering speed	55 kt (102 km/h; 63 mph)
Radius of operation	1,000 n miles (1,852 km; 1,150 miles)
Endurance	18 hr 30 min
Power plant	1 × piston engine

Status: Developed and available; being promoted.
Contractor: AeroMech Engineering Inc
San Luis Obispo, California.

Aerostar International tethered aerostats and blimps

Type: Family of unmanned tethered aerostats and blimps.
Development: Aerostar International Inc (a wholly owned subsidiary of Raven Industries) was established in 1986 and has developed a portfolio of tethered aerostat persistent surveillance platforms, blimps, high-altitude balloons, airships and parachute systems.

Selected aerostat programme events include:

15 October 2007 Aerostar International announced that it had 'successfully' deployed its TIF-25K™ aerostat for the first time on the 12th of the preceding September.

16 August 2010 Aerostar International announced that it had successfully completed flight and adverse weather testing of its High Strength Laminated Aerostat Material (HSLAM) on a TIF-25K™ aerostat. Here, the contractor claimed that use of HSLAM facilitated a 100 per cent increase in payload for a given envelope size and that the product was 'more than' 50 per cent lighter and 71 per cent stronger that the company's traditional aerostat material.

9 March 2012 Aerostar International announced that an example of its TIF-25K™ aerostat had been used in a recent US Department of Homeland Security Science and Technology Directorate sponsored persistent surveillance demonstration at Nogales, Arizona. Designed to facilitate evaluation of a new border surveillance system for use by US Customs and Border Protection, the particular application formed part of a Flexible Area Surveillance Technology (FAST) package that incorporated the cited Aerostar International aerostat, an L-3 Wescam EO/IR imaging sensor and (specifically for the described trial) a Logos® Technologies Kestrel wide-area aerostat persistent surveillance sensor. Developed jointly by Aerostar International and L-3 Communications Integrated Sensor Systems, the FAST application utilised in the Nogales demonstration is said to have generated a 'positive' response from the user community and to have participated in the detection and apprehension of 'numerous' individuals suspected of illegal activity.

10 May 2012 Aerostar International announced that it had 'successfully integrated and deployed' an X-band (8 to 12.5 GHz) band Vista Research Smart Sensing Radar System (SSRS) aboard its TIF-25K™ aerostat.

Description: *Airframe:* Aerostar International's aerostat portfolio makes use of conventional aerostat envelopes of varying sizes matched to inverted Y-shaped tail surfaces (cruciform stabilising fins in the cases of the TIF-250™ and -460™ AVs). Use of Aerostar's HSLAM in the construction of these AVs is said to increase their payload lifting capabilities, reduce their weight and increase the overall strength of their envelope strength (see *Development*).

Mission payloads: A variety of payload types (including EO/IR imagers, surveillance radars, communications relay equipment, atmospheric research sensors and COMINT collection systems) can be integrated with Aerostar International's aerostat family (see *Development*).

Guidance and control: Aerostar International aerostats make use of a guidance and control package that includes mobile (TIF-17K™ and TIF-25K™ aerostats) and heavy-duty (TIF-56K™ through TIF-75KH™ aerostats) mooring platforms; a command and control module; a proprietary Integrated Telemetry System™ (ITS™ - incorporating instantaneous AV monitoring facilities, GPS location, temperature and, wind speed/direction measurement); launch and recovery winches; an Integrated Flying Platform (IFP - TIF-4500™/-5500™/-6500™ aerostats) and EO and/or fibre-optic tethers. The TIF-17K™ and TIF-25K™ aerostats require a team of four to five individuals to set them up, together with two to three people to launch and recover them. For its part, the IFP incorporates an integral winch and tether.

Launch: By winch.
Recovery: By winch.

Variants: *TIF-17K™:* Capable of accommodating interchangeable payloads (including EO/IR sensors, radars, communications repeaters, atmospheric research sensors and Automatic Identification System (AIS) repeaters), a complete TIF-17K™ tethered aerostat system comprises the AV, a US Department of Transportation compliant Mobile Mooring Platform (MMP), a launch and recovery winch, an electro-optic tether and a C2 module that can be customised to meet specific mission requirements. Use of HSLAM material in the construction of the TIF-17K™ facilitates an increase in its payload lift capability. TIF-17K™ has a volume of 481.4 m^3 (17,000 cu ft), overall length of 20.1 m (66 ft), diameter of 6.4 m (21 ft), useable payload of 104.3 kg (230 lb) and a maximum altitude of 609.6 m (2,000 ft).

TIF-25K™: The TIF-25K™ tethered aerostat is designed for rapid deployment in the persistent surveillance and communications relay roles. System features include use of an MMP, an EO tether and the ITS™. The type's envelope can be constructed from HSLAM for greater payload capacity and strength. TIF-25K™ has a volume of 707.9 m^3 (25,000 cu ft), overall length of 23.2 m (76ft), diameter of 7.6 m (25 ft), useable payload of

220.5 kg (486 lb) at 609.6 m (2,000 ft,) and maximum altitude 914.4 m (3000 ft).

TIF-56K™: The TIF-56K™ tethered aerostat architecture includes power and communications through a fibre-optic tether and an AV endurance of 14 days. TIF-56K™ has a volume of 1,585.7 m³ (56,000 cu ft), overall length of 30.8 m (101 ft), diameter of 10.1 m (33 ft), useable payload of 446.8 kg (985 lb) at 914.4 m (3,000 ft) and a maximum altitude of 1,524 m (5,000 ft).

TIF-75KH™: The TIF-75KH™ tethered aerostat receives power from and communicates with the ground via its tether line and has a volume of 2,123.8 m³ (75,000 cu ft), an overall length of 36.3 m (119 ft), diameter of 11.9 m (39 ft), useable payload of 693.9 kg (1,530 lb) at an altitude of 914.4 m (3,000 ft), and a maximum altitude of 1,524 m (5,000 ft).

TIF-250™/-460™: Aerostar International characterises its TIF-250™/TIF-450™ series of blimps as being suitable for accurate location and wind direction marking. TIF-250™ has an overall length of 4.9 m (16.2 ft), diameter of 1.9 m (6.2 ft), useable payload of 2.3 kg (5 lb) and maximum altitude of 142.4 m (500 ft). The TIF-460™ has an overall length of 6.1 m (19.8 ft), diameter 2.3 m (7.5 ft), useable payload of 4.9 kg (11 lb) and maximum altitude of 152.4 m (500 ft).

TIF-900™/-1600™: Aerostar International characterises its TIF-900™/TIF-1600™ series of 'easy-to-deploy' tethered blimps as having a compact footprint and as being supplied as standard with load lines and a 39.6 m (130 ft) tether. For flight times of more than eight to 12 hours or to achieve altitudes of more than 152.4 m (500 ft), both these AVs require a dilation panel. The TIF-900™ has an overall length of 8.4 m (27.4 ft), diameter of 2.7 m (9 ft), useable payload of 7.7 kg (17 lb) and maximum altitude of 152.4 m (500.0 ft - standard configuration). The TIF-1600™ has an overall length of 8.9 m (29.5 ft), diameter of 3.3 m (10.8 ft), useable payload of 13.6 kg (30 lb) and maximum altitude of 152.4 m (500 ft - standard configuration).

TIF-2675™/-3750™: Aerostar International characterises its TIF-2675™/TIF-3750™ series of blimps as being suitable for surveillance, communications relay and targeting applications in addition to location marking and other civilian tasks. Both AVs are supplied with load and payload lines and 39.6 m (130 ft) of nylon tether line as standard.The TIF-2675™ has an overall length of 10.7 m (35 ft), diameter of 3.9 m (12.7 ft), useable payload of 28.1 kg (62 lb) and a maximum altitude of 152.4 m (500 ft). The TIF-3750™ has an overall length of 11.9 m (38.9 ft), diameter of 4.3 m (14.2 ft), useable payload of 45.4 kg (100 lb) and a maximum altitude of 152.4 m (500 ft).

TIF-4500™/-5500™/-6500™: Aerostar International describes its TIF-4500™/TIF-5500™/TIF-6500™ series of tethered aerostats as being suitable for emergency response, communications relay and ISR applications with the design incorporating load and payload lines and an IFP. The use of a dilation panel is required for missions of more than 12 hours duration and/or for mission to be flown at more than 152.4 m (500 ft). The TIF-4500™ has a volume of 127.4 m³ (4,500 cu ft), overall length of 13.1 m (43 ft), diameter of 4.3 m (14 ft), and a useable payload of 58.9 kg (130 lb) at an altitude of 152.4 m (500 ft). TIF-5500™ has a volume of 155.7 m³ (5,500 cu ft), overall length of 14 m (46 ft), diameter of 4.6 m (15 ft), and useable payload of 81.7 kg (180 lb) at an altitude of 152.4 m (500 ft). TIF-6500™ has a volume of 184.1 m³ (6,500 cu ft), overall length of 14.9 m (49 ft), diameter of 4.9 m (16 ft), and a useable payload of 102 kg (225 lb) at an altitude of 152.4 m (500 ft).

Status: Most recently, Aerostar International has continued to promote all the tethered aerostats and blimps described in the foregoing.

Contractor: Aerostar International Inc,
Sioux Falls, South Dakota.

Aerovel Flexrotor

Type: Technology demonstrator.

Development: Aerovel is a new company formed by Tad McGeer, previously associated with the successful Aerosonde (now an AAI product) and Insitu (now Boeing) ScanEagle designs. The Flexrotor, unveiled in March 2010, is an attempt to overcome the relatively short range and endurance capabilities usually associated with vertical take-off aircraft. Both civil and military roles are envisaged. Prospective applications include weather and environmental monitoring, geological survey, and land- or ship-based imaging reconnaissance.

Description: *Airframe:* Mid-mounted wings; mainly cylindrical fuselage, with ogival nosecone, short tapering tailboom and high aspect ratio V tail. No landing gear. Composites construction.

Mission payloads: Not yet specified, but include surveillance or geophysical sensors, housed in compartment in non-rotating nosecone.

Guidance and control: According to the company website in early 2010, "Automation encompasses the full ground-handling cycle, from retrieval through to parking, fuelling and launch, which will be possible from an unattended and portable base station". Aerodynamic control of the Flexrotor in forward flight is by conventional wing and tail surfaces; thrust-borne flight has normal cyclic and collective pitch control of the prop/rotor, whose torque is counteracted by a small electric-powered roll thruster at each wingtip. Aircraft transitions to wing-borne forward flight after launch, reversing the procedure for vertical landing.

Launch: Automatic vertical take-off; variety of launch platforms possible, on land or at sea.

The Flexrotor prototype as seen at its unveiling *(Aerovel)* 1395203

Recovery: Automatic vertical landing.

Power plant: One single-cylinder, 28 cm³ two-stroke engine, with reduction-gear drive to a two-blade propeller/rotor.

Flexrotor

Dimensions, External	
Overall, length	1.60 m (5 ft 3 in)
Fuselage, width	0.18 m (7 in)
Wings, wing span	3.00 m (9 ft 10 in)
Engines, propeller diameter	1.85 m (6 ft 0¾ in)
Areas	
Wings, Gross wing area	0.70 m² (7.5 sq ft)
Weights and Loadings	
Weight	
Max T-O weight, vertical T-O	19.2 kg (42 lb)
Payload, Max payload	0.9 kg (1.00 lb)
Performance	
Climb	
Rate of climb	
max, forward flight, at S/L	229 m/min (751 ft/min)
max, vertical flight, at S/L	61 m/min (200 ft/min)
Speed	
Max level speed	78 kt (144 km/h; 90 mph)
Cruising speed, normal	43 kt (80 km/h; 49 mph)
Range, max	1,620 n miles (3,000 km; 1,864 miles) (est)
Endurance, max	40 hr (est)

Status: Under development; first flight of a test article targeted for mid-2010 and first user trials in late 2011.

Contractor: Aerovel Corporation
White Salmon, Washington.

Arcturus T-15 and T-16

Type: Multipurpose UAV.

Description: *Airframe:* High-wing design with T tail. No landing gear. Monocoque composites construction.

Mission payloads: Can be customised with military or civil payloads for surveillance/reconnaissance, targeting, communications relay, environmental monitoring, area mapping, fire detection and a range of similar applications.

Guidance and control: Autonomous or manual operation. Cloud Cap Piccolo II autopilot.

Launch: From portable pneumatic launcher.

Recovery: Belly landing.

Power plant: Standard power plant for both types is a 57 cm³ Honda two-cylinder four-stroke engine, driving a two-blade tractor propeller.

The baseline T-15 *(US Navy)* 1356229

T-16, identified by its revised wing contours *(Arcturus)*   1356230

Options include a Cosworth heavy-fuel engine or an electric motor.
Variants: *T-15:* Baseline version. Smaller, faster, has been used as a high altitude return vehicle.

T-16: Larger and more capable version. Extended-span wings have higher aspect ratio with multiple tapers. Current version (2010) is designated T-16XL.

T-15, T-16

Dimensions, External	
Overall length	
T-15	1.83 m (6 ft 0 in)
T-16	2.08 m (6 ft 10 in)
Wings	
wing span	
T-15	3.05 m (10 ft 0 in)
T-16	3.94 m (12 ft 11 in)
Weights and Loadings	
Weight	
Weight empty	
T-15	12.7 kg (27 lb)
T-16	18.1 kg (39 lb)
Max launch weight	
T-15	20.4 kg (44 lb)
T-16	38.6 kg (85 lb)
Payload	
Max payload	
T-15	4.5 kg (9.00 lb)
T-16	13.6 kg (29 lb)
Performance	
Speed	
Max level speed	
T-15	90 kt (167 km/h; 104 mph)
T-16	75 kt (139 km/h; 86 mph)
Cruising speed	50 kt (93 km/h; 58 mph)
Endurance	
with 9.1 kg (20 lb) payload, T-16	16 hr
Power plant	1 × piston engine

Status: In production and in service. Both types have been used by various agencies for reconnaissance and surveillance, or as payload test vehicles and UAV trainers. One T-15 mission involved air launch from a balloon over the McGregor Range in New Mexico to demonstrate deployment and autonomous recovery of a payload from an altitude of 19,800 m (65,000 ft). The T-16 has received an FAA Experimental Airworthiness Certificate (EAC) for operation in unrestricted airspace.
Contractor: Arcturus UAV
Rohnert Park, California.

Arcturus T-20

Type: Surveillance UAV.
Development: The T-20 was developed in late 2008. The prototype made its first flight on 28 January 2009 at Edwards AFB, and flew its first mission at the US Army Dugway Proving Ground in Utah on 13 May 2009.
Description: *Airframe:* High-wing monoplane with T tail. No landing gear. Monocoque composites construction.

The T-20 Tier II class surveillance UAV *(Arcturus)*   1356231

Arcturus's portable launcher dismantles into sections no more than 1.83 m (6 ft) long *(Arcturus)*   1356232

Mission payloads: PTZ EO/IR cameras, targeting lasers and multiple sensors, mounted on a modular quick-change pallet in the main payload bay. Hardpoints allow the use of wing pods which can house WMD sensors, and guided parafoil delivery munitions.

Guidance and control: Can be operated autonomously or manually; Cloud Cap Piccolo GPS autopilot with waypoint navigation can accept multiple flight plans from GCS. Aircraft can return autonomously to a specified location. Is also satcom and DGPS compatible.

Launch: From portable pneumatic launcher. Set-up time less than 10 minutes.
Recovery: Belly landing.
Power plant: One 190 cm^3 Honda four-stroke single-cylinder engine; two-blade tractor propeller.

T-20

Dimensions, External	
Overall, length	2.87 m (9 ft 5 in)
Wings, wing span	5.26 m (17 ft 3 in)
Dimensions, Internal	
Payload bay	
length	0.86 m (2 ft 9¾ in)
width	0.27 m (10¾ in)
depth	0.27 m (10¾ in)
volume	0.0064 m^3 (0.2 cu ft)
Weights and Loadings	
Weight	
Weight empty	36.3 kg (80 lb)
Max launch weight	74.8 kg (164 lb)
Payload, Max payload	29.5 kg (65 lb)
Performance	
Altitude, Service ceiling	4,570 m (15,000 ft) (est)
Speed	
Max level speed	90 kt (167 km/h; 104 mph)
Cruising speed, max	80 kt (148 km/h; 92 mph)
Loitering speed	50 kt (93 km/h; 58 mph)
Endurance	
with 15.9 kg (35 lb) payload	16 hr

Status: In production since 2009.
Contractor: Arcturus UAV
Rohnert Park, California.

Aurora Flight Sciences Excalibur

Type: Tactical strike demonstrator UAV.
Development: Being developed under contract from US Army Aviation Directorate of Applied Technology. Prototype made its first vertical takeoff and landing test in June 2009. Developed versions are intended to be compatible with munitions such as Hellfire, APKWS and Viper Strike.
Description: *Airframe:* Bath-shaped fuselage with central, vertically orientated, forward-tilting turbofan and inward-canted twin tailfins.

Wind tunnel model of Excalibur *(Aurora)*   1197352

High-mounted constant-chord wings, with slight forward sweep, at rear. Three electrically powered lift-fans: one in nose and one at outer end of each wing.

Mission payloads: Various (not specified by manufacturer).

Guidance and control: Details not provided.

Launch: VTOL, STOVL and CTOL. Two-thirds of lift derived from turbofan, one-third from lift-fans. After vertical or short take-off, aircraft transitions to forward flight by forward tilting of turbofan and retraction of wing lift-fans.

Recovery: VTOL, STOVL and CTOL, two first-named by reverse of take-off procedure.

Power plant: Three electrically powered lift-fans and one Williams turbofan engine.

Excalibur

Dimensions, External	
Overall, length	7.01 m (23 ft 0 in)
Fuselage	
width, max	2.13 m (6 ft 11¾ in)
Wings, wing span	6.40 m (21 ft 0 in)
Weights and Loadings	
Weight, Max T-O weight	1,179 kg (2,599 lb)
Payload	
Max payload, short take-off	544 kg (1,199 lb)
Performance	
Altitude, Service ceiling	12,190 m (40,000 ft)
Speed	
Cruising speed, max	400 kt (741 km/h; 460 mph)
Endurance	3 hr
Power plant	1 × turbofan

Status: Under development.

Customers: Boeing; Mississippi State University; US Army Space and Missile Defense Command.

Contractor: Aurora Flight Sciences Corporation
Manassas, Virginia.

Aurora Flight Sciences GoldenEye

Type: VTOL UAV.

Development: Began under DARPA Clandestine UAV programme. Development now concentrated on Goldeneye 80.

Description: *Airframe:* Barrel-shaped core body plus five component modules: wings (two, pivoting), tail surfaces (four), propulsion unit, avionics, and thrust-vectoring module. All modules mate to core body with standard interfaces and can be removed or replaced using standard tools. Construction is of graphite and glass fibre composites with an epoxy matrix. Landing gear leg attached to tip of each tail surface. Low acoustic (GoldenEye 50 and GoldenEye 80 are less than 59 dBa at 150 m; 500 ft) and IR signatures.

Mission payloads: Specifically designed for carriage of surveillance or RSTA sensors and/or chemical agent detectors, in internal payload bay or dedicated sensor turret. GoldenEye 80 to carry payload developed by US Army Night Vision Lab and FLIR Systems which includes high resolution video and high resolution infra-red cameras, laser rangefinder, laser tracker and laser designator.

Guidance and control: Athena Technologies GuideStar GS-111 autonomous flight control system; integrated INS/GPS navigation, magnetic compass and air data sensor suite. For OAV variant, General Dynamics Robotic Systems (GDRS) is responsible for design and integration of technologies for collision avoidance system, EO/IR sensors and datalinks; and for scalable Thor Warfighter Machine interface with aircraft's GCS.

GoldenEye 80 first flight *(Aurora Flight Sciences)* 1343573

Aerodynamic control (pitch, yaw, partial roll and transition to/from hover) by variable-geometry, vectored thrust exhaust flow nozzle. Wing angle of attack controlled by outboard elevons; pitch angle is independent of fuselage pitch during transition. Wing can be trimmed to zero lift in hover, to minimise gust response; duct and wing can both be trimmed for high lift and/or low drag for efficient cruise. One-person operability.

Launch: Unassisted vertical take-off in 'tail-sitter' attitude, followed by transition to horizontal flight.

Recovery: Unassisted vertical landing.

Variants: *GoldenEye 50:* Smaller-scale version. Twelve prototypes built, first of which made maiden flight in July 2004; first autonomous transitions April 2005; 100th flight in early March 2006. Experimental Airworthiness Certificates awarded in May and August 2007 allowing operation in US National Airspace System. Described as a technology development platform for the Goldeneye 80.

GoldenEye 80 (formerly OAV): Mid-size version. Built for DARPA Organic Air Vehicle (OAV II) competition under USD2.4 million Phase 1 contract of 29 November 2004; sized for FCS Class II UAV requirements. Team GoldenEye (Aurora, General Dynamics Robotic Systems and Northrop Grumman) selected, with Honeywell, as semi-finalist in Class II competition with award of further USD5.7 million contract increment of 8 June 2006. Team GoldenEye selected for OAV II Phase III under a USD23.6 million contract in June 2006. First flight 6 November 2006.

GoldenEye 100: First in GoldenEye series. Maiden flight 8 September 2003. Proof of concept air vehicle flown under the Clandestine UAV project.

Power plant: GoldenEye 50: One Desert Aircraft DA-50 piston engine, driving a seven-blade ducted fan propeller.

GoldenEye 80: One heavy-fuel rotary engine.

GoldenEye 80 night operations *(Aurora Flight Sciences)* 1343572

GoldenEye 50 in flight *(Aurora)* 1197353

GoldenEye 50, GoldenEye 80, GoldenEye 100

Dimensions, External
  Overall
    height
      GoldenEye 50 .... 0.85 m (2 ft 9½ in)
      GoldenEye 80 .... 1.65 m (5 ft 5 in)
  Fuselage
    height
      GoldenEye 50 .... 0.46 m (1 ft 6 in)
      GoldenEye 80 .... 0.91 m (2 ft 11¾ in)
  Wings
    wing span
      GoldenEye 50 .... 1.37 m (4 ft 6 in)
      GoldenEye 80 .... 2.92 m (9 ft 7 in)

Weights and Loadings
  Weight
    Max T-O weight
      GoldenEye 50 .... 7.7 kg (16.00 lb)
      GoldenEye 100 .... 57.6 kg (126 lb)
  Payload
    Max payload
      GoldenEye 50 .... 0.9 kg (1.00 lb)
      GoldenEye 80 .... 7.2 kg (15.00 lb)

Performance
  Altitude
    Service ceiling
      GoldenEye 50 .... 1,525 m (5,000 ft)
      GoldenEye 80 .... 3,050 m (10,000 ft)
  Speed
    Max level speed
      GoldenEye 50 .... 100 kt (185 km/h; 115 mph)
      GoldenEye 80 .... 139 kt (257 km/h; 160 mph)
    Cruising speed
      GoldenEye 50 .... 55 kt (102 km/h; 63 mph)
      GoldenEye 80 .... 60 kt (111 km/h; 69 mph)
  Radius of operation
    mission, at cruising speed,
      GoldenEye 50 .... 27 n miles (50 km; 31 miles)
    mission, at cruising speed,
      GoldenEye 80 .... 180 n miles (333 km; 207 miles)
  Endurance
    at cruising speed, GoldenEye 50 .... 1 hr
    at cruising speed, GoldenEye 80 .... 8 hr
  Power plant, GoldenEye 50 .... 1 × piston engine

Status: Development continuing in 2009, when GoldenEye 80 was displayed at Virginia AeroSpace Legislative Reception.
  Customers: Athena Technologies reported as launch customer for two GoldenEye 50s. One GoldenEye-OAV and two GoldenEye 50s built for DARPA.
Contractor: Aurora Flight Sciences Corporation
Manassas, Virginia.

Aurora Flight Sciences Orion HALL
Type: HALE UAV.
Development: The Orion HALL (high altitude long loiter) UAV, being developed as an ACTD in association with Boeing's Phantom Works, is intended for use on ISR and atmospheric science missions. A prototype is under construction at Columbus Mississippi with first flight planned for 2010. A limit load test on the wing was completed in 2009.
Description: *Airframe:* High-mounted, high aspect ratio wings; pod and boom fuselage; conventional tail surfaces. Construction of graphite composites with epoxy matrix. Retractable tricycle landing gear.
  Mission payloads: Various (not specified by manufacturer) up to 182 kg (400 lb).
  Guidance and control: No details provided.
  Launch: Conventional wheeled take-off.
  Recovery: Conventional wheeled landing.
  Power plant: Liquid hydrogen fuelled, modified and supercharged Ford automobile engine; four-blade propeller. Conventional powerplant also under study.

Orion HALL

Dimensions, External
  Overall, length... 11.91 m (39 ft 1 in)
  Fuselage, width... 3.05 m (10 ft 0 in)
  Wings, wing span... 33.76 m (110 ft 9¼ in)
  Engines, propeller diameter.. 4.9 m (16 ft 1 in) (est)
Weights and Loadings
  Weight, Max T-O weight.. 2,358 kg (5,198 lb)
Performance
  Altitude, Service ceiling... 19,810 m (65,000 ft) (est)
  Speed
    Cruising speed, max... 250 kt (463 km/h; 288 mph)
  Endurance, at 19,810 m (65,000 ft)... 100 hr

Status: Development began in 2006. Contract signed with US Army in July 2007 for continued development.
  Customers: Will be operated by Mississippi State University (Raspet Flight Research Laboratory); funded by US Army Space and Missile Defense Command.
Contractor: Aurora Flight Sciences Corporation
Manassas, Virginia.

Aurora Flight Sciences Skate
Type: VTOL micro-UAV.
Development: Aurora unveiled the initial version of the Skate SUAS in mid-2010, a prototype making the design's air show debut at Farnborough International in July, where it was flown in the indoor flight demonstration area. Earlier, in April 2010, Aurora had received a six-month, Phase 1 SBIR contract from DARPA to undertake a design study for a day/night version that would harness solar and thermal energy sources to extend endurance by recharging the onboard batteries. If the design study results appear feasible, a potential Phase 2 contract, to build and fly a new prototype, could follow. The system is designed to have the agility for use in urban warfare situations, including flight inside buildings.
Description: *Airframe:* Prototypes exhibited at Farnborough were of rectangular flying-wing configuration, with twin angular fins, and were built of composites. The airframe is hinged laterally and longitudinally, enabling it to be folded for storage and transportation after detaching the fins.
  In the version proposed under the DARPA contract, the thin-film lithium batteries will themselves be shaped to form the aerofoil-section aircraft body. Additionally, the wing's upper surface will be covered with solar cells to recharge the batteries by day, while the undersurface will have infra-red photovoltaic cells to do likewise during the hours of darkness. Aurora calculates that 95 per cent of the required 40 W of power would be generated by solar energy, and the remaining 5 per cent from thermal energy.
  Mission payloads: EO or IR camera in an interchangeable pod.
  Guidance and control: Fully autonomous flight. Independently articulating engine pods allow rapid transition between vertical and horizontal flight, and high manoeuvrability by means of thrust vectoring.
  Transportation: Man-portable. The folded-up air vehicle and its hand-held control unit fit into a carry case that can be transported in a single backpack.
  Launch: Automatic vertical take-off.
  Recovery: Automatic vertical landing.
  Power plant: Two 20 W electric motors, each driving a three-blade propeller.

Full-size mockup of the hydrogen-fuelled Orion HALL *(Aurora)*

Skate on show at Farnborough, July 2010 *(IHS/Patrick Allen)*

Skate

Dimensions, External	
Overall	
length	
stowed	0.33 m (1 ft 1 in)
deployed	0.66 m (2 ft 2 in) (est)
height	
stowed	0.127 m (5 in)
deployed	0.051 m (2 in) (est)
width, stowed	0.305 m (1 ft 0 in)
Wings	
wing span, deployed	0.61 m (2 ft 0 in) (est)
Weights and Loadings	
Weight, Max T-O weight	0.9 kg (1.00 lb)
Payload, Max payload	0.2 kg (0.00 lb)
Performance	
Speed, Max level speed	50 kt (93 km/h; 58 mph)
Endurance, max	1 hr
Power plant	2 × electric motor

Status: Development continuing in 2010.
Contractor: Aurora Flight Sciences Corporation
  Manassas, Virginia.

AV Global Observer

Type: Experimental HALE UAV.
Development: Designed as an all-latitude, very high altitude, long-endurance platform for communications relay and remote sensing missions. Initial flight tests of a subscale prototype, each of more than an hours duration, were made on 26 May and 2 June 2005: the first known flights by a UAV powered by a liquid hydrogen propulsion system. The company has projected that a developed 4,536 kg (10,000 lb) gross weight full-size version could carry up to 454 kg (1,000 lb) of payload, have a wing span of between 45.7 and 76.2 m (150 and 250 ft) and be able to fly at 19,810 m (65,000 ft) for at least a week. AV is building three Global Observer air vehicles, the first of which, GO1, was shipped to Edwards Air Force Base in December 2009, making its maiden flight in August 2010; final assembly of the second was being completed in mid-2010 and a third is on order. GO1 has four engines, a payload of 181 kg (400 lb) and endurance of seven days.
Description: *Airframe:* Sailplane configuration, with high-mounted braced wings and pod and boom fuselage for subscale prototype. GO1 has wider, full length fuselage, unbraced cantilever wings. Composites construction. Fixed tricycle landing gear.
  Mission payloads: Would be mission-specific for such roles as communications relay, weather monitoring and tracking, fire detection and support, homeland security, environmental and agricultural monitoring or aerial imaging and mapping. Payload weight is projected to be between 180 to 454 kg (400 to 1,000 lb) in the three full size vehicles.

The subscale prototype on a test flight *(AV)*

  Guidance and control: Tests flights made under both manual control and autonomous GPS waypoint navigation.
  Launch: Conventional runway take-off.
  Recovery: Conventional runway landing.
  Power plant: Subscale prototype: Eight electric motors spaced along the wing leading-edges, driving two-blade propellers and powered by a propulsion system based on liquid hydrogen fuel cells and housed in the fuselage.
  GO1: Four electric motors pylon-mounted under wings inboard, driving two-blade propellers.

Specifications

Dimensions	
Wing span (subscale prototype)	15.24 m (50 ft 0.0 in)
Global Observer	Between 45.7 and 76.2 m (150 and 250 ft)

Status: Under development from 2005. Ground vibration, structural and taxi testing of GO1 was completed by May 2010. First flight was made on 5 August 2010; this air vehicle crashed on 1 April 2011 during its ninth flight at Edwards. At this time the second airframe was almost complete.
Contractor: AV Inc (formerly AeroVironment Inc)
  Simi Valley, California.

Global Observer first flight 5 August 2010 *(AV Inc)*

First Global Observer prepares for first flight in 2010 *(AV Inc)*

AV Mercury

Type: Nano UAV technology demonstrator.
Development: Mercury is being developed under Phase 2 of the DARPA Defense Sciences Office's Nano Air Vehicle (NAV) programme to design and demonstrate an extremely small, ultra-lightweight UAS with the potential to perform indoor and outdoor military missions in challenging environments. Four contractors (AeroVironment, Charles Stark Draper Laboratory, Lockheed Martin Technology Laboratories and MicroPropulsion Corporation) contested Phase 1, which ended in March

Mercury Phase 2 NAV *(AV)*

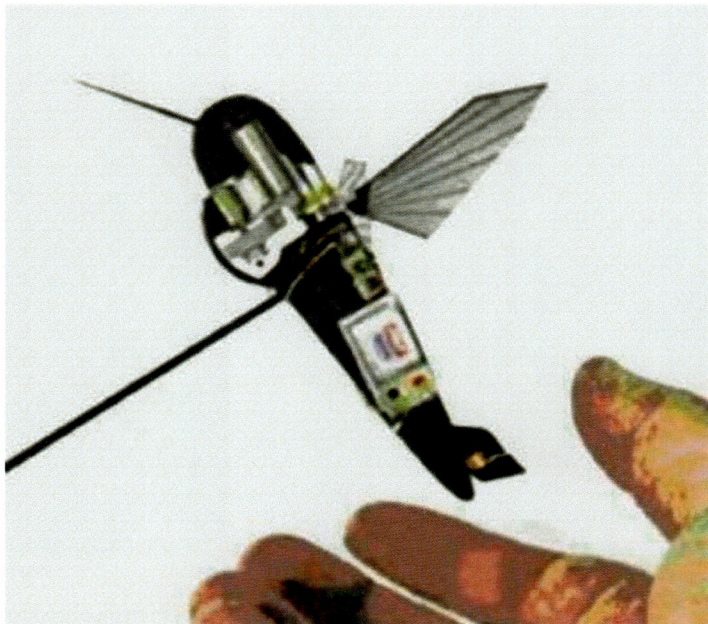
An early Phase 1 Mercury test vehicle *(AV)* 1356265

Hand-launching a Puma AE *(AV)* 1356223

Puma AE can be operated on land or at sea *(AV)* 1356224

2008. As well as completing preliminary design of their vehicle concepts, each developed technologies for lift generation and vehicle control via rotating and flapping wings; advanced stability and control using video images; and analytical tools for developing low Reynolds number aircraft. Based on the Phase 1 results, DARPA selected AeroVironment's NAV design for continued development, awarding an 18-month, USD2.1 million Phase 2 contract in April 2009. In this, according to the NAV programme manager, DARPA is looking for "a system that has 20 minutes of flight time, can withstand 2.5 m (8.2 ft)/s wind gusts, can operate inside buildings and have up to a kilometre command and control range". It would also be able to hover while carrying a payload.

Phase 1 activity by Mercury and predecessor trials machines (originally called Nano SCOUT: Sensor Covert Observer in Urban Terrain) spanned from September 2007 to January 2009, culminating in a successful 20-second hovering flight in December 2008. It encompassed stability control in all three axes; altitude control; ability to hover and climb; forward, backward and sideways flight; and take-off from ground.

Description: *Airframe:* Mercury is a tail-less, flapping-wing design, modelled on the hummingbird in nature.
  Guidance and control: Remotely piloted. Aircraft flight is controlled via manipulation of the flapping wings.
  Launch: Vertical take-off.
  Recovery: Vertical landing.
  Power plant: Flapping wings have a battery-powered electric power train for both lift and propulsion.
(Phase 2 objective)

Mercury

Dimensions, External	
Wings, wing span	0.076 m (3 in) (est)
Weights and Loadings	
Weight, Max launch weight	0.01 kg (0.00 lb) (est)
Payload, Max payload	0.002 kg (0.00 lb)
Performance	
Speed	
Cruising speed, max	19 kt (35 km/h; 22 mph)
Radius of operation	0.5 n miles (km; miles)
Endurance	20 min

Status: Development continuing in 2009-10. Phase 2 targets include reductions in size and weight, lowering noise levels and increasing flight endurance.

Contractor: AV Inc (formerly AeroVironment)
Simi Valley, California.

Composites construction. Airframe is fully waterproofed for use in land-based or maritime operations.
  Mission payloads: Mechanically and digitally stabilised, modular turret with combined EO, IR camera and IR illuminator, gimballed for 360° continuous pan and +10/–90° tilt.
  Guidance and control: Puma AE is operated from the same common Ground Control System (GCS) as the company's Raven and Wasp systems.
  Transportation: System is man-portable.
  System composition: Air vehicles (number unspecified), GCS, spares and ground support equipment.
  Launch: Hand launch.
  Recovery: Automatic or manually controlled deep-stall landing.
  Power plant: One piston engine (type and rating not identified), driving a two-blade tractor propeller.

Puma AE

Dimensions, External	
Overall, length	1.40 m (4 ft 7 in)
Wings, wing span	2.81 m (9 ft 2¾ in)
Weights and Loadings	
Weight, Max launch weight	5.9 kg (13.00 lb)
Performance	
Altitude, Operating altitude	150 m (500 ft)
Speed	
Cruising speed, max	45 kt (83 km/h; 52 mph)
Loitering speed	20 kt (37 km/h; 23 mph)
Range, max	8.1 n miles (15 km; 9 miles)
Radius of operation	8.1 n miles (15 km; 9 miles)
Endurance, max	2 hr
Power plant	1 × piston engine

Status: As of 2011, in production.
The Puma AE was the subject of a 1 July 2008 one-year contract from US Special Operations Command (SOCOM) for an initial 10 systems, valued at about USD6 million.
In April 2011, an indefinite delivery indefinite quantity (IDIQ) contract fixed-price delivery order valued at USD11,500,000 was awarded for Puma AECV from SOCOM.

Contractor: AV Inc (formerly AeroVironment).
Simi Valley, California.

AV Puma AE

Type: Hand-launched UAV.
Development: Introduced in 2008, the Puma AE was designed for use in both terrestrial and maritime applications (AE thus indicating 'all environments'). Compared with AV's other Puma variants, it has an all-new airframe proofed against landings in salt water, but is interoperable using the same ground control segment. In addition, unlike most other hand-launched UAS, the Puma AE carries both and EO and IR gimballed payload cameras which can focus continuously on, and track, a target. The Puma AE is modular in design and quiet to avoid detection. No auxiliary equipment is required for launch or recovery operations.
Description: ***Airframe:*** High-wing monoplane, with dihedral on outer panels; tapered fuselage; conventional tail surfaces. No landing gear.

AV RQ-11 Raven

Type: Close-range RSTA mini-UAV.
Development: Developed to supplement or replace company's Pointer UAV for use by ground troops at company level or below, to enable them to oversee battle space beyond LOS and for military operations in urban terrain (MOUT). POC prototypes, named Flashlite, were funded in April 2001 and first flown in October 2001; advanced testing and military evaluation led to further development into Raven, funded in March 2002 under a US Army ACTD programme known as Pathfinder. First LRIP Block 1 deliveries were made in May 2003, followed four months later by the Block 2 version, which was evaluated operationally in Afghanistan in October 2003. The designation RQ-11A was allocated in late 2004. From 2009 deliveries started of the digital Raven with an improved data link.

Wasp III hand launch *(IHS/Patrick Allen)* 1326897

AV Wasp III Micro Unmanned Air Vehicle (UAV) *(AeroVironment)* 1295124

Status: In December 2006, AV Inc was awarded a USAF Indefinite Delivery, Indefinite Quantity (IDIQ) contract for the Wasp III with a potential value of USD45 million over a five-year period. On 6 August 2007, AV announced the 'recent' delivery of a first tranche of 30 systems, believed to be for deployment to Iraq with the US Marine Corps (USMC). Plans were understood to include acquisition of 75 more systems in FY08 and the remaining 116 in FY09. Continuing improvement is planned over the lifetime of the system.

Selected and first flown in December 2006, the Wasp III underwent operational evaluation by, and began deliveries to, the US Air Force (USAF) in mid-2007. Deliveries exceeded 1,200 air vehicles by 2010.

As of 2011, in full rate production.

Contractor: AV Inc (formerly AeroVironment Inc)
Simi Valley, California.

BAE Systems Coyote

Type: Surveillance mini-UAV.

Development: The Coyote UAS was developed by Advanced Ceramics Research (ACR - acquired by BAE Systems during 2009) in response to a 2004 specification issued by the US Office of Naval Research (ONR) that called for an expendable, low-cost sonochute-launched UAV (SL-UAV) that could fit into, and be launched from, a standard air-launch sonobuoy dispenser. Manufactured under a contract from US Naval Air Systems Command (NAVAIR), it was envisaged for carriage by the US Navy's P-3C Orion maritime patrol aircraft or SH-60 Seahawk anti-submarine helicopter to extend the reach of the host aircraft's surveillance capability. During the period 2007-2009, Coyote was successfully test launched from a C-12 utility transport and a P-3C maritime patrol aircraft.

Description: *Airframe:* Fuselage of basically box section, rounded at front. Unswept, high-mounted wings folding back, tailplanes low-mounted and folding forward); tapered twin fins, folding forward and downward; pusher engine with folding propeller blades. No landing gear.

Mission payloads: Colour (×10 optical zoom) and dual polarity (×2 digital zoom) imagers.

Guidance and control: Military or Industrial, Scientific and Medical (ISM) band radio modem. Narrow and wideband operating modes available and Coyote can use its host aircraft's tactical station hardware plus proprietary software. Mission profiles are programmable (from either

The Coyote tube-launched AV *(BAE Systems)* 1472962

Description: *Airframe:* Wasp III uses the company's original Wasp (which see) as its core, to which have been added wing extensions, an all-moving vertical tail and a ventral pod to house the payload cameras.

Mission payloads: Ventral pod houses integrated forward- and side-looking EO colour cameras, plus a modular (swappable) forward- or side-looking high-resolution EO camera with electronic pan-tilt-zoom or an IR imager.

Guidance and control: Wasp III uses the same common hand-held ground control unit as the company's Raven, Puma and Swift UAVs, providing autonomous or remotely operated flight and GPS navigation.

Transportation: The Wasp III system is backpackable, and can be operated by one person.

System composition: Two air vehicles, GCS, interchangeable payloads, batteries and field-repair kit.

Launch: Hand-launched.

Recovery: Belly landing on land.

Power plant: One electric motor, driving a two-blade tractor propeller.

Wasp III

Dimensions, External	
Overall, length	0.38 m (1 ft 3 in)
Wings, wing span	0.72 m (2 ft 4¼ in)
Weights and Loadings	
Weight, Max launch weight	0.43 kg (0.00 lb)
Performance	
Altitude, Operating altitude	15 m to 305 m (50 ft to 1,000 ft)
Speed	
Cruising speed, max	35 kt (65 km/h; 40 mph)
Loitering speed	22 kt (41 km/h; 25 mph)
Radius of operation, LOS	2.7 n miles (5 km; 3 miles)
Endurance	45 min
Power plant	1 × electric motor

The Coyote AV alongside its launch canister (rear right), with a standard AN/SSQ-53E DIFAR sonobuoy for size comparison *(BAE Systems)* 1472961

the tactical officer's or the pilot's station or while the AV is still in the sonobuoy chute.

Launch: Coyote is launched from standard A-size sonobuoy tube (length 92 cm; 36 in, outside diameter 12.7 cm; 5 in) and is stowed within the dispenser in its own canister. On ejection, the AV deploys a parachute and discards the canister, whereupon its wings, fins and propeller blades unfold and its engine is started. The parachute is jettisoned after a further 10 seconds and the AV begins its climb-out.

Recovery: The Coyote is designed to be expendable.

Power plant: One 200 W, 12 V DC brushless electric motor, powered by lithium-polymer batteries and driving a two-blade, folding, pusher propeller.

Coyote

Dimensions, External	
Overall, height	0.30 m (11¾ in)
Fuselage	
length	0.79 m (2 ft 7 in)
height, max	0.30 m (11¾ in)
Wings, wing span	1.47 m (4 ft 9¾ in)
Engines, propeller diameter	0.33 m (1 ft 1 in)
Weights and Loadings	
Weight	
Weight empty	5.9 kg (13.00 lb)
Max launch weight	6.4 kg (14.00 lb)
Payload, Max payload	1.36 kg (2.00 lb)
Performance	
Altitude	
Operating altitude	150 m to 365 m (500 ft to 1,200 ft)
Service ceiling	6,095 m (20,000 ft)
Speed	
Max level speed	85 kt (157 km/h; 98 mph)
Cruising speed	60 kt (111 km/h; 69 mph)
Stalling speed	50 kt (93 km/h; 58 mph)
Radius of operation, LOS, mission	20 n miles (37 km; 23 miles)
Endurance	1 hr 30 min
Power plant	1 × electric motor

Status: Most recently, BAE Systems has continued to promote the Coyote UAS.

Contractor: BAE Systems Unmanned Aircraft Programs, Tucson, Arizona.

BAE Systems Manta

Type: Small tactical UAV.

Development: Originally developed by Advanced Ceramics Research (ARC - acquired by BAE Systems during 2009), the Manta UAS is known to have been procured by the USN and by the US National Oceanic and Atmospheric Administration (NOAA). Here, NOAA is reported to have employed the Manta 'continuously' since February 2005 during which month, the Administration trialled the AV in the small vessel tracking/identification/observation and whale spotting roles off Hawaii. Thereafter, the type was used in a NOAA campaign in Greenland during 2007-2008. Elsewhere, a trio of such AVs are reported to have been used by the Scripps Institute of Oceanography (San Diego, California) to study cloud pollution over the Indian Ocean's Maldive Islands.

Description: Airframe: The Manta UAS incorporates high-mounted tapered wings, a pod fuselage (with a large payload bay), twin tailbooms (supporting inverted-V tail surfaces), a pusher engine and a fixed tricycle landing gear.

Mission payloads: The Manta UAS's mission payloads are configurable to meet specific customer requirements.

Guidance and control: Military or Industrial, Scientific and Medical (ISM) radio modem with narrow and wideband operating modes are available.

Launch: The Manta UAS's standard means of take-off is via its fixed undercarriage. As an option, the AV can also be launched from a pneumatic launcher.

The Manta small tactical UAV (BAE Systems) 1472964

A general view of the Manta AV mounted on its optional pneumatic launcher (BAE Systems) 1472963

Recovery: Wheeled landing.

Power plant: Two-cylinder two-stroke engine (type and rating not stated); three-blade pusher propeller. Nominal mission fuel capacity 7.9 litres (2.1 US gallons; 1.8 Imp gallons)

Manta

Dimensions, External	
Overall	
length	1.90 m (6 ft 2¾ in)
height	0.62 m (2 ft 0½ in)
Wings, wing span	2.67 m (8 ft 9 in)
Dimensions, Internal	
Weapons bay, volume	0.013 m³ (0.5 cu ft)
Weights and Loadings	
Weight, Max T-O weight	27.7 kg (61 lb)
Payload, Max payload	7.2 kg (15.00 lb)
Performance	
Altitude, Service ceiling	4,875 m (16,000 ft)
Speed	
Max level speed	65 kt (120 km/h; 75 mph)
Loitering speed	45 kt (83 km/h; 52 mph)
Stalling speed	38 kt (71 km/h; 44 mph)
-	20 n miles (37 km; 23 miles)
-	6 hr
Power plant	1 × piston engine

Status: Most recently, BAE Systems was describing the Manta UAS as being 'developed and available'.

Contractor: BAE Systems Unmanned Aircraft Programs, Tucson, Arizona.

BAE Systems Silver Fox

Type: Tactical mini-UAV.

Development: Originally known by the acronym SWARM (Smart Warfighter Array of Reconfigurable Modules - a designation that recognised the AV's ability to be flown in 'swarms' from a single GCS), the BAE Systems (formerly Advanced Ceramics Research (ARC) - see following) Silver Fox UAS was designed and sponsored by the US Office of Naval Research (ONR). Over time, Silver Fox AVs are known to have been procured by 'military, government and private sector' customers, with identified clients including Canada (as the CU-167), the US National Oceanic and Atmospheric Administration (NOAA) and the USMC. Again, BAE Systems describes the architecture as having the logistic footprint of a Group 1 Small UAS (SUAS) platform with capabilities approaching those of a Group 3 Tactical UAS (TUAS) and as having a low probability-of-detection in both the visual and aural spectra. Overall, the contractor further claims that the Silver Fox AV's small logistical footprint and affordability facilitates 'targeted deployments that are responsive to ever-changing mission needs'. Significant programme events have included:

A Silver Fox AV shown disassembled in its shipping container (BAE Systems) 1472968

A forward view of the Silver Fox AV on its launcher *(BAE Systems)*
1472967

A rear view of the Silver Fox AV on its gas spring catapult launcher *(BAE Systems)*
1472965

The Silver Fox AV *(BAE Systems)*
1472966

Description: *Airframe:* The Silver Fox AV features a simple tubular fuselage, low-mounted constant chord wings, conventional aerodynamic control surfaces, GFRP construction and a belly skid.

Mission payloads: As of 2012, the baseline Silver Fox AV was equipped with COTS monochrome or colour daytime TV or IR sensors with real-time imagery downlink. Up to 25 W of onboard electrical power available.

Guidance and control: Military or Industrial, Scientific and Medical (ISM) radio modem (with narrow and wideband operating modes available).

Launch: By gas spring catapult.

Recovery: Autonomous belly skid or net recovery.

Power plant: One model aircraft engine (type and rating not stated); two-blade propeller. Fuel capacity 3.40 litres (0.90 US gallon; 0.75 Imp gallon).

System composition: The composition of a Silver Fox UAS system is configurable to meet specific customer needs and can include a specified number of AVs, a GCS, an antenna unit and a launcher.

Silver Fox

Dimensions, External	
Overall, length	1.47 m (4 ft 9¾ in)
Wings, wing span	3.05 m (10 ft 0 in)
Weights and Loadings	
Weight, Max launch weight	16.30 kg (35 lb)
Fuel weight, Max fuel weight	2.54 kg (5.00 lb)
Payload, Max payload	2.20 kg (4.00 lb)
Performance	
Altitude, Service ceiling	6,095 m (20,000 ft)
Speed, Cruising speed	45 kt (83 km/h; 52 mph)
Endurance	8 hr
Power plant	1 × piston engine

Status: Most recently, BAE Systems has continued to promote the Silver Fox UAS.

Contractor: BAE Systems Unmanned Aircraft Programs, Tucson, Arizona.

2002 The Silver Fox UAS was first publicly revealed during 2002.

Early 2003 6 × Silver Fox AVs were deployed in support of Operation 'Iraqi Freedom' and were used to undertake Reconnaissance, Intelligence, Surveillance and Target Acquisition (RISTA) sorties. None of these AVs were lost during their time in Iraq.

10 December 2004 A Silver Fox AV demonstrated its ability to detect Improvised Explosive Device (IED) detonation wires as part of an ONR sponsored counter IED (C-IED) programme.

March 2005 A Silver Fox AV equipped with an 2.4 m (8 ft) Ø transmitting antenna and a fibre-optic link to tow a gradiometer array 21.3 m (70 ft) behind it, successfully detected a range of IED detonation wire gauges from 'low altitude'.

February 2006 The Silver Fox UAS was evaluated by NOAA in a six day trial off Hawaii, during which the AV is understood to have undertaken small vessel observation and whale spotting.

August 2006 A Silver Fox AV equipped with an onboard gradiometer demonstrated a tunnel detection capability.

June 2009 BAE Systems acquired Advanced Ceramics Research (ARC - the Silver Fox's original manufacturer) during June 2009.

2012 As of 2012, BAE Systems was reporting Silver Fox as being deployed in the Middle East in support of US forces in the region.

1 May 2012 On 1 May 2012, the USAF reported a 'Speckles' Silver Fox AV as being the last UAV to take-off and land in Iraq in support of the withdrawal of US forces from Iraq at the end of Operation 'New Dawn'. The Silver Fox 'Speckles' configuration was funded by the US Joint Improvised Explosive Device Defeat Organization (JIEDDO), was built by BAE Systems and featured EO and short- or long-wave IR sensors for use in the Route Clearance Patrol (RCP) role. Again, the AV was equipped with a direct video downlink and had an endurance of eight hours, with the 'Speckles' programme (which is understood to have begun during 2009) as a whole being managed by the USAF's Air Force Research Laboratory.

Boeing A160/A160T Hummingbird

Type: Long-endurance helicopter UAV.

Development: The latest A160T Hummingbird long-endurance VTOL UAV has its origins in the Frontier Systems (acquired by Boeing in May 2004) A160 platform that was powered by a 224 kW (300 hp) four-cylinder modified Subaru car (automobile) engine driving a three-bladed (later four-bladed) main rotor. For its part, the turboshaft-powered A160T is designed for ISR, direct attack, communications relay and precision re-supply applications together with remote delivery of unmanned ground vehicles and sensors. Significant programme events include:

March 1998 DARPA awarded Frontier Systems an Advanced Technology Demonstration (ATD) contract with regard to the development of a low-observability, rotary-winged surveillance UAV that would have a mission endurance of up to 48 hours. Here, the first year of the programme is understood to have been devoted to ground tests and trials of the proposed vehicle's flight control system. In all, Frontier is believed to have constructed three flying A160 prototypes together with a Robinson R22 helicopter that it modified to Maverick avionic and flight control testbed configuration.

7 December 2001 A160 AV number A001 underwent its first hover test.

29 January 2002 A160 AV number A001 made its maiden flight at Victorville, California. Here the AV ascended vertically, hovered (at an altitude of 15 m [50 ft]), retracted its main undercarriage members, achieved an altitude and speed of 366 m (1,200 ft) and 45 kt (83 km/h; 52 mph) respectively and landed vertically. Subsequently, A001 was involved in two crashes and was mothballed, while airframes A002 and A003 were re-engineered with a new four-bladed main rotor.

27 November 2002 A160 AV number A002 made the type's first hover flight with the four-bladed main rotor.

10 February 2003 An A160 AV made the type's maiden flight with a four-bladed main rotor.

October 2003 A160 AV number A003 is reported to have been involved in a crash.

May 2004 Frontier Systems was acquired by Boeing, with final development of the A160 now being undertaken by the latter's 'Phantom Works' prior to the design's transfer to Boeing's then Integrated Defense Systems business unit for production.

17 September 2004 A160 AV number A002 made its maiden flight under Boeing ownership. As of the given date, Boeing was also understood to be producing five × 'Phase 1' A160 AVs. Here, four of these were to be powered by an 'improved' 290 kW (389 hp) piston engine, with the fifth acting as the prototype for turboshaft-powered A160T.

August 2005 An A160 AV was flown round a 1,043 n mile (1,931 km; 1,200 mile) course (at a speed and altitude of 60 kt (111 km/h; 69 mph) and 1,219 m (4,000 ft) respectively before crashing due to mechanical failure.

June 2007 The prototype A160T made its maiden flight.

Boeing A160/A160T Hummingbird

An A160T AV with its FORESTER radar antenna deployed *(Boeing)*
1416301

An A160T mock-up fitted with stub wings for the carriage of air-to-surface missiles *(IHS/Patrick Allen)*
1410121

An in-flight view that emphasises the A160T Hummingbird's streamlined design *(Boeing)*
1416300

The A160T was preceded by the piston-engined A160 configuration shown here *(DARPA)*
1122548

27 September 2007 The A160T prototype completed an eight hour sortie during which, it carried a 454 kg (1,000 lb) payload.

10 December 2007 An A160T AV crashed due to a flight computer failure that caused it to depart from controlled flight.

9 May 2008 An A160T AV demonstrated the type's ability to hover out of ground effect at altitudes of 4,572 m (15,000 ft) and 6,096 m (20,000 ft).

14 May 2008 An A160T AV clocked up an 18 hour, 41 minutes and 28 seconds flight which was adjudged by the Fédération Aéronautique Internationale (FAI) to be a world record for an autonomously controlled 500 kg (1,102 lb) to 2,500 kg (5,512 lb) class UAV

13 November 2008 Boeing announced the first use of an EO/IR sensor (believed to be an L-3 Wescam MX-15D equipment) aboard an A160T AV.

20 November 2008 The A160T completed its 100th flight hour since June 2007.

25 November 2008 An A160T AV utilised its two-speed transmission to 'change gear' in flight for the first time.

16 May 2009 Boeing announced that Subaru-powered A160s had flown 31 sorties during the period 2001-2005, that the 'improved' piston engined A160s had flown five times during the period 2005 to 2007 and that the A160T had completed 18 sorties between 2007 and May 2009.

31 August - 8 October 2009 The A160T completed a series of 20 flight characteristics validation sorties with the antenna for the FOLiage penetrating REconnaissance, Surveillance, Tracking and Engagement Radar (FORESTER) installed.

16 October 2009 *IHS Jane's* sources reported US Special Operations Command (SOCOM) as being in the process of acquiring up to 10 × A160Ts under the designation YMQ-18A.

March 2010 Boeing launched company-funded production of 21 × 'white tail' A160Ts at its Mesa, Arizona helicopter manufacturing facility.

28 July 2010 A160T AV number A007 crashed at Victorville, California during the course of optimum-speed rotor trials. As of the given date (and apart from A007), one A160T AV was owned by DARPA, with a further eight being assigned to SOCOM.

1 August 2010 SOCOM began a 45 day long trial of the A160T(YMQ-18A)/FORESTER combination in Belize.

4 September 2010 On 4 September 2010, the US Embassy in Belize announced that one of the two A160T/YMQ-18A AVs deployed to that country for the FORESTER test programme had been lost in a crash at the Central farm Airfield at Cayo (17° 2' 21.05" N; 88° 57' 15.21" W). At this time, 28 sorties had been flown and 90 per cent of the planned test events completed.

2 December 2010 The USN's Naval Air Systems Command awarded Boeing subsidiary Frontier Systems a then year USD30,000,000 firm, fixed-price, Government-Owned/Contractor Operated (GOCO) contract with regard to the procurement of two × A160T AVs (suitably 'role modified'), three × GCSs, operator training and support services and pre-deployment readiness activities in the continental US relating to a six month duration USMC Cargo Unmanned Aircraft System (CUAS) deployment to Afghanistan. While not confirmed, it is believed that one of these AVs suffered an electrical fire and required a 'major rebuild' and in the event, the A160T is understood not to have participated in the CUAS effort.

7 December 2010 DARPA awarded Frontier Systems a then year USD12,800,000 contract modification with regard to planning, payload integration and AV 'improvements' relating to an operational demonstration of the A160T's military utility when equipped with a modified Autonomous Real-time Ground Ubiquitous Surveillance Imaging System (ARGUS-IS) EO imaging system pod that housed both the ARGUS-IS sensor and a SIGINT payload. Here, the intent was to deploy the A160T/ARGUS-IS to Afghanistan during June 2012 (see following).

17 April 2012 An ARGUS-IS equipped A160T crashed while under test at Victorville, California, with the accident resulting in what is said to have been 'excessive damage to the ARGUS payload and [its host] aircraft'.

8 June 2012 Following the effective loss of an A160T/ARGUS-IS combination on the 17th of the preceding April, the US Army issued a stop-work notice on the effort, with the citation being reported by usually reliable sources as noting that "given the challenges experienced on the A160 program and the probability of continued technical and schedule delays, the projected cost and risk involved in completing the ARGUS A160 effort with Boeing has increased significantly [to the point where] program continuation is no longer in the best interests of the [US] Government. This combined with indications that the desired result of deployment with [the] ARGUS payload can no longer be achieved as planned has resulted in the [US] Army issuing a stop-work order on June 8".

February 2013 As of February 2013, *IHS Jane's* sources were reporting that the US military was 'currently evaluating'' its options with regard to the integration of the ARGUS-IS sensor aboard 'other aircraft and airships'. Here (and aside from the A160T), the sensor was noted as having been tested aboard a YEH-60A helicopter, the Proteus high-altitude testbed aircraft and a DHC-8 platform.

The given specification data is Boeing sourced and refers to the A160T.

Description: *Airframe:* Four-blade (originally three-blade on early 160 AVs) main and two-blade tail (to port) rotors, with the former being of a hingeless, 'rigid' type with low tip speeds and low disc loading. Again, blade stiffness and thickness/chord ratio reduces from root to tip and the rotor rpm can be slowed down to 40% of maximum. Elsewhere, the A160/160T features a smoothly contoured, low drag/low observability, teardrop-shaped fuselage. A rectangular ventral fin (incorporating a fixed tailwheel) is mounted at the end of a slender tail boom and the AV's construction makes use of an ultra-lightweight carbon fibre material. A retractable main landing gear provides 84 cm (33 in) ground clearance for underfuselage payloads and the A160 incorporated interchangeable nose sections with a payload volume of at least 1.1 m³ (38.8 cu ft). For its part (and from nose to tail), the A160T features a main payload space in its removable nose section (suitable for EO/IR and radar sensors), detachable wing mounts (up to eight × air-to-surface missiles), a forward belly bay (nose camera and/or an EO/IR sensor), ventral hardpoints for external stores (such as the FORESTER radar's 6.6 m (21.5 ft) long antenna housing), an aft belly bay (payload items and/or avionics) and a tailboom space (lightweight GMTI/SAR radar, a Laser Imaging Detection And Ranging (LIDAR) sensor, EO/IR or SIGINT equipment or avionics). In addition, the A160T can be equipped with a stationary, above-the-rotor housing that can accommodate SATCOM equipment or a FLIR.

Boeing A160/A160T Hummingbird – Boeing Phantom Eye

An A160T display model showing the type's logistics pod (IHS/Patrick Allen)

The Phantom Eye HALE UAV demonstrator photographed during its first medium-speed taxi test at Edwards AFB, California on 10 March 2012 (NASA)

Mission payloads: Mission payloads that have been associated with the A160T include DARPA's Affordable Adaptive Conformal Electronic scanning array Radar (AACER), BAE Systems' ARGUS-IS wide-area EO sensor, SRC's FORESTER foliage penetration radar, L-3 Wescam's MX-15D EO/IR imager and Northrop Grumman's Vehicle And Dismount Exploitation Radar (VADER). Of these, ARGUS-IS, FORESTER and MX-15D are known to have been test flown on the type. In terms of re-supply applications, the A160T can be equipped with a ventral logistics pod and can be configured for slung loads. Up to eight × Hellfire-type air-to-surface missiles can be accommodated on detachable port and starboard stub wings.

Guidance and control: Frontier Systems flight control system, with GPS waypoint navigation. Autonomous operation, with manual override.

Launch: Conventional and autonomous helicopter take-off.

Recovery: Conventional and autonomous helicopter landing.

Power plant: **A160** One 224 kW (300 hp) four-cylinder modified Subaru car (automobile) engine; heavy-fuel engine was to have been demonstrated. **Phase 1 A160** 290 kW (389 hp) piston engine. **A160T** One 485 kW (650 shp) Pratt & Whitney Canada PW207D turboshaft

A160T

Dimensions, External	
Overall, length	10.7 m (35 ft 1¼ in)
Rotors, rotor diameter	11.0 m (36 ft 1 in)
Weights and Loadings	
Weight	
Weight empty	1,134 kg (2,500 lb)
Max T-O weight	2,948 kg (6,499 lb)
Fuel weight, Max fuel weight	1,179 kg (2,599 lb)
Payload	
Max payload, maximum	454 kg (1,000 lb)
Performance	
Altitude	
Service ceiling	
Cruise mode	6,095 m (20,000 ft)
Maximum	9,145 m (30,000 ft)
Hovering ceiling	4,570 m (15,000 ft)
Speed	
Cruising speed, max	121 kt (225 km/h; 140 mph)
Range	2,250 n miles (4,167 km; 2,589 miles)
Endurance	>20 hr
Power plant	1 × turboshaft

Status: Most recently, Boeing has continued to promote the A160T long-endurance helicopter UAV.

Contractor: Boeing Defense, Space and Security
St Louis, Missouri.

Boeing Phantom Eye

Type: High-altitude, long-endurance UAV.

Development: Boeing's internally funded Phantom Eye HALE UAV demonstrator is designed to validate a hydrogen-powered AV as a persistent ISR and communications node and is described by the company as being a 'natural evolution' from its earlier piston-engined Condor HALE UAV that set a number of altitude and endurance records during the late 1980's. Alongside Boeing, Aurora Flight Sciences (wings), Ball Aerospace (aluminium and foam fuel tanks), the Ford Motor Company (baseline engines), MAHLE Powertrain (propulsion controls) and Turbo Solutions Engineering (turbochargers) have been associated with the programme as have DARPA (stake holder) and NASA (provision of flight test facilities). Significant programme events include:

June 2010 The Phantom Eye AV's fuselage, wings and empennage were reported as being in the 'final stages of assembly' at Boeing's 'Phantom Works' during June 2010.

12 July 2010 Boeing unveiled the completed Phantom Eye AV on 12 July 2010 at which time, the capability was being pitched to fulfil a US Missile Defense Agency (MDA) requirement for an airborne IR ballistic missile detection system.

25 March 2011 The disassembled Phantom Eye AV was delivered to NASA's Dryden Flight Research Center (Edwards AFB, California - 34° 54' 1.47" N; 117° 52' 9.94" W) for re-assembly prior to initial ground tests.

Mid-June 2011 As of mid-June 2011, the Phantom Eye AV is reported to have completed 12 days of ground vibration/structural mode interaction testing and to have had its fuel tanks filled with liquid hydrogen in order to evaluate re-fuelling procedures and validate the AV's fully fuelled configuration.

10 March 2012 The Phantom Eye AV underwent its first medium-speed taxi test at Edwards AFB during which, it reached a speed of 30 kt (56 km/h; 35 mph).

1 June 2012 The Phantom Eye AV made its maiden flight from Edwards AFB during which it remained airborne for 28 minutes and reached an altitude and cruising speed of 1,244 m (4,080 ft) and 62 kt (115 km/h; 72 mph) respectively. During the ensuing landing run, the AV dug into the landing surface and snapped off its nosewheel.

January 2013 Boeing was suggesting that an operational development of the Phantom Eye demonstrator would be suitable for a range of customers including the US DoD, the US Department of Homeland Security and 'various' telecommunications operators.

6 February 2013 Following repairs and modifications to its landing gear and implementation of upgrades to its autonomous flight system, the Phantom Eye AV made a second series of taxi test at Edwards AFB on 6 February 2013. Here, the AV reached a maximum speed of 40 kt (74 km/h; 46 mph).

The quoted specification data is sourced from Boeing.

Description: *Airframe:* The hydrogen-powered Boeing Phantom Eye HALE UAV demonstrator features a bulbous, near circular section forward fuselage that tapers into a tail boom that carries three tail surfaces installed in an inverted Y configuration. The platform's long span, high aspect ratio wings are shoulder mounted and are braced by wing-to-fuselage struts to both port and starboard. Each wing incorporates a high set engine nacelle that incorporate dorsal and ventral air scoops and four-bladed tractor propellers. Conventional control surfaces are let into the platform's tail and wing (not confirmed) surfaces. Take-off is by means of a detachable four-wheel trolley, with landing being achieved on a centrally-mounted main landing skid and a nosewheel.

Mission payloads: The Phantom Eye HALE demonstrator is configured to carry a 204 kg (450 lb) representative payload.

Guidance and control: The Phantom Eye AV is equipped with an autonomous flight system that utilises Boeing's Common Open-mission Management Command and Control (COMC2) software package

Launch: Conventional take-off using a detachable four wheel trolley.

Recovery: Conventional landing using an integral main skid and a nosewheel.

Power plant: Two × 112 kW (150 hp) four-cylinder, 2.3 litre Ford Motor Company automotive engines that have been modified to run on hydrogen, feature three-stage supercharging and intercooling and drive (via wide-ratio single-stage reduction gears) Boeing-developed four-bladed tractor propellers. Two × 2.4 m (7.9 ft) Ø spherical fuel tanks holding a total of 839 kg (1,850 lb) of liquid hydrogen.

Phantom Eye

Dimensions, External	
Wings, wing span	46 m (150 ft 11 in)
Weights and Loadings	
Weight	
Weight empty	3,402 kg (7,500 lb)
Max T-O weight	4,445 kg (9,799 lb)
Fuel weight, Max fuel weight	839 kg (1,849 lb)
Payload, Max payload	204 kg (449 lb)
Performance	
Altitude, Service ceiling	19,810 m (65,000 ft)
Speed	
Max level speed	200 kt (370 km/h; 230 mph)
Cruising speed	150 kt (278 km/h; 173 mph)
Endurance	96 hr

Status: As of 2013, the Phantom Eye HALE UAV demonstrator's test programme was continuing.

Contractor: The Boeing Company, Boeing Defense, Space and Security
St Louis, Missouri.

Unmanned aerial vehicles > United States > Boeing Little Bird – Boeing Phantom Ray

Boeing Little Bird

Type: Dual-mode manned/unmanned helicopter.

Development: Known in its UAV form as Unmanned Little Bird, or **ULB**, this programme was launched by Boeing in October 2003. It was revealed at the AUSA exposition in October 2004, having made its first flight on 8 September and first autonomous flight on 16 October that year. These flights carried an onboard safety pilot, and also demonstrated the carriage of 227 kg (500 lb) external cargo loads. In early 2005, Boeing began a new USD1.6 million joint programme with the US Army's Aviation Applied Technology Directorate (AATD) aimed at further refining the requirements for safe and accurate UAV weapons deployment. Some 180 hours of flight testing had been completed by mid-2005, at which time it was being proposed to the US Army both as a near-term solution for the FCS Class IV requirement, pending introduction of the RQ-8B Fire Scout, and for that service's Armed Reconnaissance Helicopter (ARH) programme. On 30 June 2006, the ULB prototype made its first true unmanned flight without the onboard human presence, having accumulated more than 450 hours' flying by that time in autonomous demonstrations that included target identification, precision resupply, communications relay and weapon firing.

A second prototype, designated **A/MH-6X**, was flown (piloted) for the first time on 20 September 2006. This combines the proven performance of the A/MH-6M Mission Enhanced Little Bird (MELB) manned helicopter, in service with the US Army's 160th Special Operations Aviation Regiment, with the UAV technologies of the ULB demonstrator. This second aircraft is being used to expand both the manned and unmanned envelopes. Aircraft performance is similar to that of the ULB demonstrator, which has a payload capacity of more than 1,089 kg (2,400 lb), but the A/MH-6X can carry an additional 454 kg (1,000 lb) that can be used for increased range, endurance or mission hardware.

Description: *Airframe:* ULB prototype modified from a standard MD Helicopters MD 530F helicopter (N7032C).

Mission payloads: Initial ULB flight tests were conducted using an L-3 Wescam MX-15 EO/IR sensor and L-3 Communications TCDL datalink. Successful launch of 2.75 inch unguided rockets was demonstrated by the ULB in August 2005, and plans were in hand in 2007 for firings of the Northrop Grumman Viper Strike PGM and integration of the 0.50 inch GAU-19 Gatling gun.

Guidance and control: Boeing ground control station and navigation/guidance system. The A/MH-6X prototype is described as having "a 'glass' cockpit that provides system redundancy and additional technologies in digital maps and data fusion. It also has many network-centric features like Ku-band communication, digital radios, Internet Protocol-addressable aircraft systems and onboard, high-bandwidth data processing and storage". The unmanned hardware and capability developed for this programme is said to be adaptable to other helicopter types.

Launch: Conventional, automatic helicopter take-off.

Recovery: Conventional, automatic helicopter landing.

Power plant: One 485 kw (650 shp) Rolls-Royce 250-C30 turboshaft, derated to 317 kw (425 shp). Standard usable fuel capacity 242 litres (64.0 US gallons; 53.3 Imp gallons); optionally increasable to 322 litres (85.0 US gallons; 70.8 Imp gallons).

Unmanned Little Bird, A/MH-6X

Dimensions, External	
Overall	
length, rotors turning	9.94 m (32 ft 7¼ in)
height	2.67 m (8 ft 9 in)
Fuselage, length	7.49 m (24 ft 7 in)
Skids, skid track	1.91 m (6 ft 3¼ in)
Rotors, rotor diameter	8.33 m (27 ft 4 in)
Tail rotor, tail rotor diameter	1.42 m (4 ft 8 in)
Weights and Loadings	
Payload	
Max payload	
Unmanned Little Bird	1,089 kg (2,400 lb) (est)
A/MH-6X	1,542 kg (3,399 lb) (est)
Performance	
Altitude, Service ceiling	5,485 m (18,000 ft)
Speed	
Cruising speed, max	126 kt (233 km/h; 145 mph)
Radius of operation, mission	300 n miles (555 km; 345 miles)
Endurance, max	7 hr
Power plant	1 × turboshaft

Status: Development was continuing in 2007 for offer to domestic and international markets, but more recently the system has found a role as a test vehicle, mostly flown with a safety pilot. While the US Army continues to show interest in the programme, Thales Aerospace announced in February 2007 its intention to use the Little Bird system to help define French Army and Navy VTOL UAV requirements under a two-year contract from the DGA. In 2009 DGA awarded DCNS and Thales the second phase of the study to design and demonstrate an automatic take-off, landing, and deck landing system for rotary-wing UAVs, using Little Bird based in Mesa Arizona. Tests on fixed and moving platforms are planned before deck trials on a French warship, planned for 2011.

Contractor: The Boeing Company, Boeing Defense, Space and Security. St Louis, Missouri.

Boeing Phantom Ray

Type: Technology demonstrator.

Development: The company-funded Phantom Ray project was launched by Boeing's 'Phantom Works' during October 2008 and is described as being an 'evolution' of the company's X-45C entry for the DARPA/USAF/USN Joint Unmanned Combat Air Systems (J-UCAS) programme. In October 2004 DARPA awarded Boeing a then year USD767 million contract to construct and test three X-45C AVs, the first of which was publicly rolled-out on 2 March 2006. After a period of testing and demonstrations the J-UCAS effort was terminated and the X-45C prototypes placed in storage. Subsequently, Boeing is understood to have utilised one of these X-45C airframes as the basis for its privately funded Phantom Ray AV which was designed to be used as a flying testbed for "advanced technologies". Significant events in the Phantom Ray programme included:

2010 *IHS Jane's* sources report construction of the Phantom Ray as having been completed during 2010.

February 2010 As of February 2010, Boeing was outlining a Phantom Ray test programme that would begin during the following December and would involve 10 sorties over a period of six months. Here, the round as a whole was billed as possibly including investigations into the type's suitability for ISR, SEAD, electronic attack, hunter/killer and autonomous aerial refuelling applications.

December 2010 December 2010 saw the Phantom Ray demonstrator being flown from the Boeing facility in St Louis, Missouri to Edwards AFB, California (34° 54' 1.47" N; 117° 52' 9.94" W) on top of NASA's Boeing 747 Shuttle Carrier Aircraft.

27 April 2011 The Phantom Ray AV made its maiden flight from NASA's Dryden Flight Research Center at Edwards AFB. Unconfirmed sources report the AV as having been airborne for 17 minutes and as having reached an altitude and speed of 2,286 m (7,500 ft) and 178 kt (330 km/h; 205 mph) respectively.

X-45C underside detail *(IHS/Patrick Allen)*

A general view of the X-45C mockup that was displayed at the 2004 Farnborough International Air Show *(IHS/Patrick Allen)*

September 2011 As of September 2011, unconfirmed sources were reporting Boeing as having placed the Phantom Ray demonstrator in flyable storage.

The given specification data is Boeing sourced.

Description: *Airframe:* Stealthy, 'arrow-head' flying wing design with about 55° leading-edge sweepback and W-planform trailing-edge; blended fuselage, with internal bay(s); buried turbofan engine with serpentine dorsal intake; no vertical tail surfaces. Retractable tricycle landing gear. All-composites construction.

Mission payloads: Specific to individual test flights.

Guidance and control: Fully autonomous, including taxi, take-off and landing phases.

Launch: Conventional and autonomous wheeled take-off.

Recovery: Conventional and autonomous wheeled landing.

Power plant: One approx 51.2 kN (11,500 lb st) General Electric F404-GE-102D non-afterburning turbofan.

Phantom Ray

Dimensions, External	
Overall, length	10.9 m (35 ft 9¼ in)
Wings, wing span	15.2 m (49 ft 10½ in)
Weights and Loadings	
Weight, Max T-O weight	16,556 kg (36,499 lb)
Performance	
Altitude, Operating altitude	12,190 m (40,000 ft)
Speed, Cruising speed	534 kt (988 km/h; 614 mph)
Power plant	1 × turbofan

Status: As of 2013, the status of the Phantom Ray demonstrator was uncertain.

Contractor: Boeing Defense, Space and Security, St Louis, Missouri.

Boeing/Insitu ScanEagle

Type: Endurance mini-UAV.

Development: ScanEagle is a variant of the Insitu SeaScan, built under a 15-month original agreement announced on 11 February 2002 and being developed separately by Boeing's Phantom Works in collaboration with Insitu. Boeing equipped the prototype air vehicle using its own systems integration, communications and payload technologies. It made its first, remotely controlled, flight in April that year and its first autonomous flight on 19 June 2002, following a pneumatic catapult launch. Made at Boeing's Boardman, Oregon, facility, the latter flight lasted 45 minutes, flying a preprogrammed course at a maximum altitude of 457 m (1,500 ft). A number of test waypoints were completed using DGPS, as well as the ability to make real-time updates to the flight plan from the GCS. The UAV was retrieved using the patented Insitu Skyhook technique.

In late January 2003, the ScanEagle made five flights, totalling more than 20 hours, as a communications relay platform in a US Navy exercise ('Giant Shadow') in the Bahamas. Small-scale production has ensued, as described under the 'Status' heading below. Boeing has quoted potential applications as including persistent ISR, US Navy SEAL operations escort, low-cost sea lane protection, sentinel and sentry guard duty, battlefield communications relay network node, and convoy support.

In June 2003, Boeing and Insitu signed a longer-term contract to continue their collaboration and begin production of the ScanEagle, three A models of which are understood to have been completed at that time. On 15 September 2003, Boeing announced that in recent flights two of these had been launched simultaneously, the first completing a 15.2-hour endurance flight while, in its latter stages, being monitored by the other, which relayed real-time video to the GCS. At that time the three A models had completed a total of 70 sorties. Persistent ISR was demonstrated between December 2003 and June 2004 during US Joint Forces Command's exercise 'Forward Look III', providing imagery and targeting to UAVs and other airborne assets, ground stations, command centres and ships at sea. In August 2004, a Boeing-owned aircraft completed a 16

ScanEagle handling *(USMC)* 1151558

ScanEagle launch in Iraq *(USMC/Sgt G M Deanda)* 1151559

hour 45 minute flight off the US west coast in Puget Sound, Washington, after launch from a fishing boat, thought to be the longest flight up to that time by a UAV launched and retrieved at sea; and in December 2004 another ScanEagle successfully demonstrated high-speed wireless communications relay (voice and video).

ScanEagles have participated in a number of military demonstrations, experiments and exercises, some of which are described under the *Status* and *Customers* headings below.

Description: *Airframe:* Mainly cylindrical fuselage; mid-mounted, sweptback wings with tall endplate fins and rudders; pusher engine. Modular internal avionics bay. No landing gear. Folding-wing version for air launch under study in 2006.

Mission payloads: The standard payload until 2006 comprised either an EO or an IR camera in a gimballed and inertially stabilised turret; the EO camera has ×25 zoom, the IR camera had an 18° FoV and ×2.5 fixed zoom. In August 2006, as part of the Block D reconfiguration, Insitu completed an upgrade to the sensor turret enabling it to house larger cameras. The IR sensor was replaced by a new DRS Technologies E6000 unit offering ×7.5 digital zoom and enhanced resolution, and a Mode C transponder was added.

Other payloads that have been studied include biochemical sensors, laser illuminators and a magnetometer (see following).

Guidance and control: Athena GuideStar GS-111m DGPS waypoint navigation and autonomous object tracking initially, but new software demonstrated in March 2005, enabling ScanEagle to map its route autonomously while in flight and to complete 'a series of manoeuvres' (unspecified). UHF (900 MHz) datalink and S-band (2.4 GHz) video downlink. GCS has two consoles, each able to control up to four air vehicles.

Transportation: Dismantles to pack into storage box 171 × 45 × 45 cm (67.3 × 17.7 × 17.7 in).

Launch: On land, from a Mk 2 'wedge' pneumatic catapult; at sea, from a low-pressure (3 bar; 43.5 lb/sq in), 12 *g* pneumatic catapult. Launch velocity approximately 26 m (85 ft)/s.

Recovery: On land or at sea, by Insitu-developed patented Skyhook retrieval system. Aircraft flies into a single line suspended over the water from a 15.2 m (50 ft) boom. A hook on the wingtip then engages the line to arrest the aircraft. On land, ScanEagle can also be recovered and landed conventionally within an area of 30.5 × 183 m (100 × 600 ft).

Power plant: One 1.9 kW (2.5 hp) 3W-28 gasoline piston engine; two-blade pusher propeller; also Sonnex Research HFE conversion.

ScanEagle

Dimensions, External	
Overall, length	1.22 m (4 ft 0 in)
Fuselage, width	0.178 m (7 in)
Wings, wing span	3.05 m (10 ft 0 in)
Weights and Loadings	
Weight	
Weight empty	9.1 kg (20.00 lb)
Max launch weight	18.1 kg (39 lb)
Payload	
Payload with max fuel	5.6 kg (12.00 lb)
Performance	
Altitude, Service ceiling	4,875 m (16,000 ft) (est)
Speed	
Max level speed	70 kt (130 km/h; 81 mph)
Cruising speed	48 kt (89 km/h; 55 mph)
Loitering speed	41 kt (76 km/h; 47 mph)
Endurance, no reserves	15 hr (est)
Power plant	1 × piston engine

Status: By March 2007, US Navy and Marine Corps ScanEagles had exceeded 30,000 combat flight hours. Development continued in 2006 with improvements to the sensor turret (see Mission payloads above), addition of a new video transmitter interoperable with the L-3 Communications ROVER III (Receive-Only Video Enhanced Receiver) downlink, and incorporation of an in-flight fuel measurement system. The

Payload nose of a ScanEagle A-15 *(US Navy)*

resultant version is known as the Block D configuration. A pre-production Block D aircraft flew a 22-hour mission at Boeing's Boardman test centre in November 2006. This was exceeded in January 2007 when a ScanEagle made a 28 hour 44 minute flight after its gasoline engine had been converted by Sonex Research to run on JP-5 aviation fuel. By August 2009 more than 2,500 combat hours and more than 300 shipboard sorties had been flown with this heavy fuel engine (HFE) since March 2008.

Other developments under way in the second half of 2006 included Boeing studies into an air-launched version that could be carried by such host aircraft as the C-130 Hercules or V-22 Osprey; and studies by Insitu into the use of the 0.45 kg (1 lb) ImSAR NanoSAR miniature radar, with the capability of 35 cm (13.8 in) resolution at a range of 1 km (0.6 miles). In January 2007, the US Air Force awarded Boeing a contract to integrate the ScanEagle with a ShotSpotter sniper detection and location system, for trials by the 820th Security Forces Group at Moody AFB, Georgia. In February 2007, NATO declared the ScanEagle system compliant with its STANAG 4586 interoperability standard. ScanEagle had surpassed 200,000 operational flight hours by August 2009, when more than 1,000 air vehicles had been produced.

Moving forward, 5 April 2012 saw Insitu announce that an expeditionary Group 2 ScanEagle AV had successfully completed a two and half hour long test flight using a United Technologies' 1,500 W hydrogen-powered fuel cell and a USN Naval Research Laboratory "hydrogen fuelling solution". Elsewhere in the world, 16 May 2012 saw Insitu announce that the ScanEagle AV had exceeded 600,000 combat flight hours and had achieved a 99 per cent mission-readiness rate during those hours. Three months later (7 August 2012), Insitu reported that it had begun field evaluation of a new mid-wave IR/EO turret and the Hood Technology Corporation's SuperEO imager as potential sensors for the ScanEagle. On the same day, Insitu announced that a limited military aircraft-type classification certificate had been issued by the Dutch Military Aviation Authority that allowed ScanEagle to be flown by the Netherlands military. Identified US ScanEagle contracting activity during 2012 is as follows:

15 May 2012: Insitu (Bingen, Washington) was awarded a then year USD35,507,379 modification (to the previously awarded firm, fixed-price contract N00019-11-C-0061) with regard to the provision of additional operational and maintenance services in support of the ScanEagle UAS. Here, the manifest included the provision of real-time EO/IR and mid-wave IR imagery and data in support of Operation 'Enduring Freedom'. At the time of its announcement, work on the effort was to be undertaken at Bingen and it was scheduled for completion by the end of December 2012. The programme's contracting activity was the USN's Naval Air Systems Command (Patuxent River, Maryland).

21 August 2012: Insitu (Bingen, Washington) was awarded a then year USD23,401,476 modification (to the previously awarded firm, fixed-price contract N00019-11-C-0061) with regard to the provision of additional operational and maintenance services in support of the ScanEagle UAS. The cited services were to provide real-time EO/IR and mid-wave IR imagery and data in support of Operation 'Enduring Freedom'. At the time of its announcement, work on the effort was to be undertaken at Bingen and it was scheduled for completion by the end of August 2013. The programme's contracting activity was the USN's Naval Air Systems Command (Patuxent River, Maryland).

Customers: *Australia* Under an AUD11.5 million (USD9 million) contract, ScanEagle services were leased from Boeing by the Australian Army for use by its Overwatch Battle Group (West) 2 in Operation 'Catalyst' in southern Iraq during the period November 2006 to June 2007. On 20 June 2012, Insitu Pacific announced that it was under contract to fly ScanEagle on behalf of the Queensland Government's Department of Agriculture, Fisheries and Forestry during a Siam weed detection trial in the far north of the State. Elsewhere, 12 July 2012 saw the same contractor announce that the Australian Army was extending and expanding an existing contract with it to include provision of ScanEagle systems for trials with the Royal Australian Navy (RAN). Here, ScanEagle was to be installed aboard a number of RAN vessels, with flight trials from a frigate platform to take place during September 2012. Thereafter (and in keeping with the service's NA2020 plan), the RAN was intending to have stood-up a dedicated UAS unit by 2020.

Japan On 11 July 2012, Insitu Pacific signed a contract with Japan's Mitsubishi Heavy Industries with regard to the supply of ScanEagle systems for use in a "comprehensive operational evaluation" by the Japanese Ground Self-Defence Force.

Malaysia On 15 April 2012, Insitu Pacific and Composites Technology Research Malaysia (CTRM) announced that they had signed a contract under which the ScanEagle AV was to be supplied to CTRM for use by its Unmanned Systems Technology (UST) subsidiary.

Netherlands On 19 March 2012, Insitu announced that it had signed a contract with the Dutch Ministry of Defence with regard to the supply of an HFE version of its ScanEagle UAV for use both "domestically and abroad". At the time of the announcement, the Dutch ScanEagle capability was scheduled to enter service during the second half of 2012 and was a replacement for an earlier programme that had ended in the middle of 2011 (see also *Status*).

Singapore On 9 July 2012, Insitu Pacific announced that the Republic of Singapore Navy had awarded it a contract with regard to the supply of ScanEagle AVs for use aboard its missile corvettes.

US DoD: On 5 June 2006, Boeing announced award of a USD8.2 million contract from the Defense Threat Reduction Agency (DTRA) to modify two ScanEagles for a programme known as Biological Combat Assessment System (BCAS). In this, one aircraft would be adapted for the carriage of a payload that can detect, track and collect samples of airborne biological warfare agents for recovery and analysis, while the other would collect and record meteorological data and plume tracks with its standard EO/IR equipment. Upon completion of initial testing, the contract provides an option to order four more, two-UAV, similarly configured systems.

US Marine Corps: Two ScanEagle systems were ordered by the USMC in July 2004 for the First Marine Expeditionary Force (1 MEF) in Iraq; these were delivered in August 2004 and the contract was renewed in July 2005. USMC ScanEagles had flown more than 20,000 hours in-theatre by mid-2006.

US Navy: On 25 April 2005, Boeing announced receipt of a USD14.5 million contract from the US Navy for an unspecified number of ScanEagles, to be used for ship-based Expeditionary Strike Group (ESG) missions providing persistent ISR and to increase oil platform security in the Persian Gulf. Successful sea trials were completed aboard the USS *Cleveland* (LPD-7) in the Gulf area some two months later, and a USD13 million contract modification was awarded in September 2005 to provide ScanEagle system support for USN high-speed vessels and an at-sea forward staging base. The system entered USN service in July 2005, and had amassed 1,600 flight hours by the end of February 2006. Deployment on other USN ships was due to follow. As of July 2006, these had included USS *Trenton* (LPD-14), USS *Oak Hill* (LSD-51) and the High-Speed Vessel (HSV) *Swift*.

ScanEagle Skyhook recovery *(US Navy)*

ScanEagle in flight *(Thales)*

US Air Force: In March 2007, the first military graduates of Boeing's UAV Training Center at Clovis, New Mexico, were assigned to the USAF's 820th Special Forces Grooup at Moody AFB, Georgia.

UK MoD: In June 2004, Thales UK's Team Joint UAV Experimentation Programme (JUEP), which included Boeing and QinetiQ, received a contract to provide two ScanEagle systems. This was followed in November 2004 by a UK MoD contract to conduct trials in the first quarter of 2005 as part of the Phase 2 maritime evaluation element of JUEP. These took place in March 2005 at the Benbecula range off the Outer Hebrides islands in north-west Scotland, the UAVs being launched from a land-based catapult and handed over to a ship-based control on board the Type 23 frigate HMS *Sutherland*. Demonstrations included co-operative tasking with a Royal Navy Sea King ASaC Mk 7 helicopter of No. 849 Squadron from RNAS Culdrose, Cornwall. From *Sutherland* in March 2006, in the final Phase 3 of the JUEP programme, ScanEagle made a number of autonomous take-offs, flights and recoveries during Trial 'Vigilant Viper'. During these, the Sea King used a high-gain, wide beamwidth antenna developed by QinetiQ that enabled it to receive imagery at 'operationally useful ranges' from the UAV.

Contractors: The Boeing Company, Boeing Defense, Space and Security St Louis, Missouri.Insitu Group Inc (a wholly owned subsidiary of the Boeing Company),
Bingen, Washington.

Boeing X-48B/X-48C

Type: Blended Wing Body (BWB) technology demonstrator.

Development: Investigation of the BWB design concept dates back at least to 1997, when a 5.18 m (17 ft) span McDonnell Douglas model was test-flown, and over the past decade NASA and its partners have tested several variously-sized BWB models in the Langley Research Center wind tunnels. In early 2000, in partnership with NASA, Boeing began building a 14 per cent scale BWB-LSV (Low Speed Vehicle) that received an X-48A designation in late 2001, but cuts in the NASA budget caused this to be abandoned. The X-48B, which first flew in July 2007, has been developed by Boeing's Phantom Works at Huntington Beach, California, in partnership with NASA and the US Air Force Research Laboratory (AFRL), to explore the structural, aerodynamic and operational advantages of the BWB concept. Tests were designed to gather detailed information about the design's stability and flight control characteristics, especially during take-off and landing. Later tests, after modification, were planned to cover such other aspects as low noise footprint and handling at transonic speeds.

Advantages claimed for the BWB configuration are the potential to save 20 to 30 per cent on fuel consumption and costs compared with conventional 'tube and wing' designs, an obvious attraction for commercial operation. USAF is interested in the technology for potential long-range multirole military applications such as tanker/transport or C2ISR.

Description: *Airframe:* Hybrid configuration resembles a flying wing, but differs in that the wing blends smoothly into a wide, flat, tail-less, aerofoil-shaped fuselage, allowing entire aircraft to generate lift while offering less drag than a conventional circular fuselage. A total of 20 elevon surfaces on the trailing-edge; rudders on wingtip fins (replaced by inboard vertical surfaces on X-48C). Leading-edge slats were replaced by slatless leading-edge after sixth flight. Engines are pylon-mounted high above rear end, to minimise the acoustic signature. Fixed tricycle landing gear. Composites construction.

Guidance and control: Remotely piloted, aided by forward-looking camera in aircraft's nose. GCS operator uses conventional aircraft controls and instrumentation.

Launch: Conventional wheeled take-off.

Recovery: Conventional wheeled landing.

Power plant: X-48B: Three 0.24 kN (55 lb st) JetCat P200 model aircraft turbojets
X-48C: Two 0.35 kN (80 lb st) geared turbofan engines (X-48C)

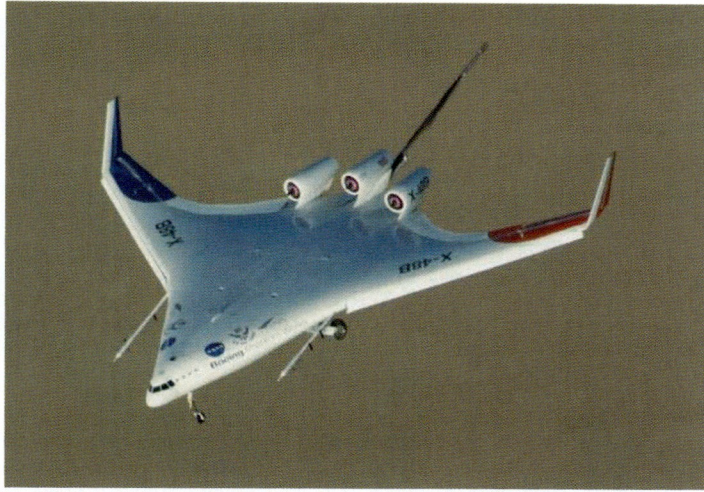
X-48B No. 2 above Rogers Dry Lake during its fifth flight, 14 August 2007 *(NASA/Carla Thomas)*

View from below as the X-48B overflies Edwards AFB in California *(NASA/Carla Thomas)*

X-48B prototype No. 2, showing its spin parachute tail 'sting' *(NASA/Tony Landis)*

The X-48C is characterised by twin engines and inboard vertical tail surfaces *(NASA)*

X-48B

Dimensions, External
  Wings, wing span..6.22 m (20 ft 5 in)
Weights and Loadings
  Weight, Max T-O weight..237 kg (522 lb)
Performance
  Altitude, Service ceiling...............................3,050 m (10,000 ft) (est)
  Speed, Max level speed.....................118 kt (219 km/h; 136 mph) (est)
  Endurance...40 min (est)
  Power plant..3 × turbojet

Status: Cranfield Aerospace of the UK was contracted by Boeing to build two 8.5 per cent unmanned scale models of the X-48B, the first of which began wind tunnel testing at NASA Langley on 7 April 2006. In mid-May 2006, after completing 250 hours of tunnel tests, it was shipped to Dryden Flight Research Center at Edwards AFB to serve as a flight test back-up machine.

Ship No. 2, the ground, taxiing and flight test article (call sign "Skyray 48"), made a 31-minute maiden flight at Edwards AFB, California, on 20 July 2007, reaching an altitude of 2,280 m (7,500 ft). Five additional flights had been made by 28 August 2007. The next (seventh) flight, on 18 January 2008, was to check out flight control software changes connected with ability to fly after removal of the fixed leading-edge slats. The latter mode was expected to begin with the 10th flight. At least three flights were completed with slats removed in 2008, and the air vehicle completed a total of 20 flights for envelope expansion. Between July 2008 until December 2009 some 52 flights were made to examine aircraft performance, followed by eight further flight tests described as "limiter assaults" up to March 2010. After 80 flights, the air vehicle received a new computer and up to 12 further flights were planned between September-November 2010. A switch to twin geared turbofan engines will then be made. Initially a single turbofan will be flown between the two outer turbojets. The winglets will be replaced by vertical surfaces outboard of the engines for noise shielding under trials in the Environmentally Responsible Aviation (ERA) programme. In this configuration the designation is X-48C.

Contractor: The Boeing Company, Boeing Defense, Space and Security. St Louis, Missouri.

Brock Technologies AV8-R

Type: Close-range mini-UAV.

Development: The AV8-R is intended for training, scientific research, ISR and military applications. It was designed at the request of the US Army's Picatinny Arsenal, as a low-cost platform for payload test flights.

Description: *Airframe:* High-wing monoplane with conventional tail surfaces and pusher power plant. Forward section of fuselage, containing payload bay, pulls out for ease of access to sensor compartment. No landing gear. Composites construction and tool-less assembly: wings and payload section secured by thumbscrews.

Mission payloads: As required by customer. Basic version is equipped with a 2.4 GHz wireless colour camera, imagery downlink and telemetry transmitter.

Guidance and control: Autonomous, based on Cloud Cap Piccolo LT or Procerus Kestrel autopilot.

Transportation: Man-portable.

Launch: Hand or catapult launch.

Recovery: Automatic, to belly skid landing.

Power plant: One 275 W (0.37 hp) electric motor, powered by lithium-polymer batteries; two-blade pusher propeller.

The fully assembled Brock AV8-R *(Brock Technologies)*

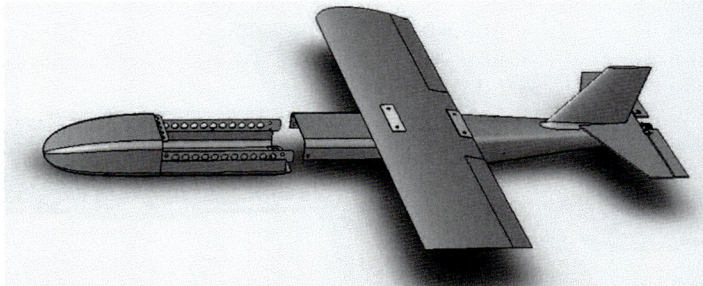

AV8-R, showing the front section pulled out for access to the payload bay *(Brock Technologies)*

AV8-R

Dimensions, External
  Overall
    length..0.909 m (2 ft 11¾ in)
    height..0.226 m (9 in)
  Wings
    wing span...0.914 m (3 ft 0 in)
    wing chord, constant.....................................0.19 m (7½ in)
Dimensions, Internal
  Payload bay
    length..0.48 m (1 ft 7 in)
    width..0.10 m (4 in)
    depth..0.08 m (3¼ in)
    volume..0.0037 m³ (0.1 cu ft)
Areas
  Wings, Gross wing area......................................0.0017 m² (0.0 sq ft)
Weights and Loadings
  Weight, Max launch weight................................2.4 kg (5.00 lb)
  Payload, Max payload..0.45 kg (0.00 lb)
Performance
  Altitude, Service ceiling......................................3,050 m (10,000 ft)
  Speed
    Max level speed..60 kt (111 km/h; 69 mph)
    Stalling speed...32 kt (60 km/h; 37 mph)
  Endurance, max..45 min (est)
  Power plant..1 × electric motor

Status: In service. Customers include unnamed research, commercial and US military agencies.

Contractor: Brock Technologies Inc, Vail, Arizona.

Brock Technologies BT20 Eel

Type: Close-range surveillance mini-UAV.

Development: Development history not known.

Description: *Airframe:* High-mounted constant-chord wings; cylindrical fuselage; conventional tail surfaces. No landing gear. Composites construction; tool-less assembly in less than 1 minute.

Mission payloads: Retractable pan and tilt colour TV camera standard; others according to mission and customer requirements.

Unassembled components of the BT20 Eel *(Brock Technologies)*

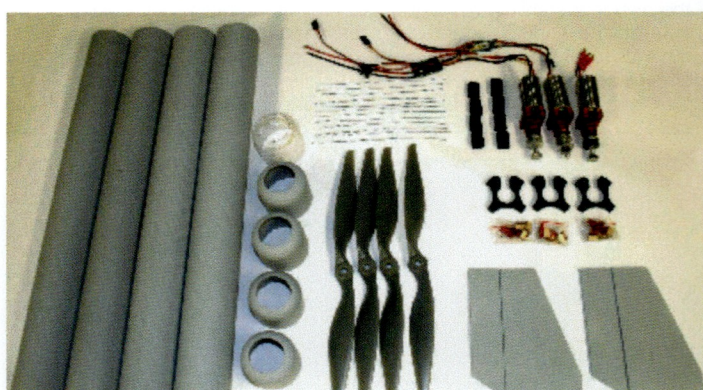

Fully assembled Eel air vehicle *(Brock Technologies)*

Guidance and control: Fully autonomous, via Procerus Technologies Kestrel autopilot; GPS navigation. Compatible with Falcon View flight planning software.
Transportation: Man-portable in backpack.
System composition: Four air vehicles, payloads and GCS.
Launch: Automatic take-off.
Recovery: Automatic landing.
Power plant: One 300 W (0.4 hp) electric motor; two-blade propeller.

BT20

Dimensions, External	
Overall	
length	0.66 m (2 ft 2 in)
height	0.17 m (6¾ in)
Wings, wing span	0.99 m (3 ft 3 in)
Weights and Loadings	
Weight, Max T-O weight	0.7 kg (1.00 lb)
Performance	
Speed	
Max level speed	40 kt (74 km/h; 46 mph)
Stalling speed	15 kt (28 km/h; 18 mph)
Radius of operation	13 n miles (24 km; 15 miles)
Endurance	1 hr 30 min
Power plant	1 × electric motor

Status: In service. Customers include unnamed research, commercial and US military agencies.
Contractor: Brock Technologies Inc
Vail, Arizona.

Brock Technologies Havoc

Type: Endurance UAV.
Development: Havoc was designed under a February 2009, USD100,000 SBIR contract from the US Air Force, to "support varying mission applications". A USD750,000 Phase 2 contract, to complete and fly a prototype, followed in January 2010.
Description: *Airframe:* Shoulder-mounted wings; pod and twin tailboom configuration; pusher engine. Optional fixed or retractable tricycle landing gear. Dismantles into shipping case that doubles as an assembly table. As with other Brock UAVs, no tools are required for assembly.
Mission payloads: Not specified. Havoc has two modular payload bays in fuselage pod and a 1 kW starter/generator to power onboard systems.
Guidance and control: No details provided. However, company states that Havoc carries all equipment required by the FAA for operation in US national airspace (NAS).
Launch: Conventional wheeled take-off or from trailer-mounted pneumatic catapult.
Recovery: Wheeled or belly skid landing.
Power plant: One 13.4 kW (18 hp) electric motor; two-blade pusher propeller.

Havoc

Dimensions, External	
Overall, length	3.27 m (10 ft 8¾ in)
Wings, wing span	4.88 m (16 ft 0¼ in)
Dimensions, Internal	
Payload bay	
volume, front	0.0329 m³ (1.2 cu ft)
Payload bay	
volume, rear	0.0201 m³ (0.7 cu ft)
Weights and Loadings	
Weight	
Weight empty	38 kg (83 lb)
Max launch weight	79 kg (174 lb)
Payload, Max payload	13.5 kg (29 lb)
Performance	
Speed	
Max level speed	103 kt (191 km/h; 119 mph)
Stalling speed	32 kt (60 km/h; 37 mph)
Endurance	39 hr
Power plant	1 × electric motor

Status: First flight July 2010. Developed and available.
Contractor: Brock Technologies Inc
Vail, Arizona.

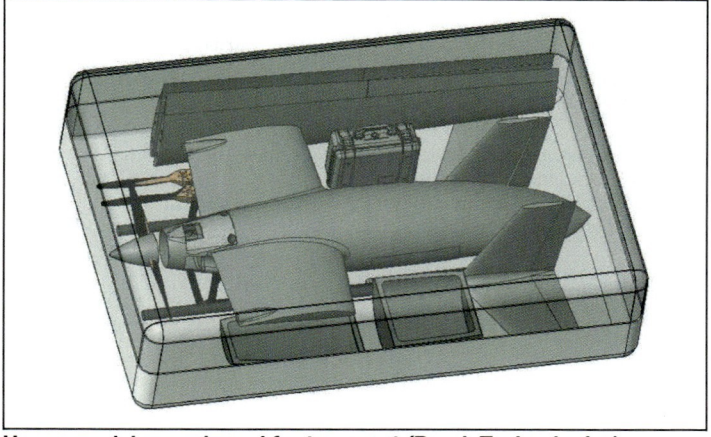
Havoc modules packaged for transport *(Brock Technologies)*
1418591

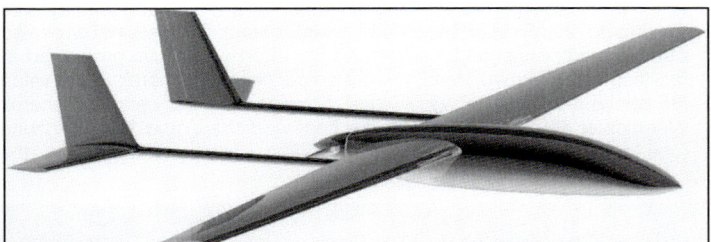
Artist impression of the twin-boom Havoc *(Brock Technologies)*
1418590

Havoc on a trolley-mounted catapult *(Brock Technologies)*
1418592

Brock Technologies Spear

Type: General-purpose mini-UAV.
Development: Designed as a lightweight, portable and robust airframe for UAS training, scientific research and ISR or other military applications. Flight demonstrations are understood to have been made during January to March 2010.
Description: *Airframe:* Wing and fuselage pod can be mounted, respectively (and independently of each other), above and below a tubular spine at a variety of positions depending upon disposition of the payload and energy source, eliminating the need for ballast. Conventional tail unit. No landing gear. Composites construction.
Mission payloads: Two modular, removable payload bays. Pan-tilt-zoom camera standard; for ISR missions, front of fuselage can be equipped with a Brock BASE gimbal mounting. Other payloads as required by customer. 500 W of onboard power available in piston-engined version.
Guidance and control: Automatic, based on Cloud Cap Piccolo LT or Procerus Kestrel autopilot.
Transportation: Dismantles for carriage in a 143 × 41.5 × 29.5 cm (56.3 × 16.3 × 11.6 in) Ameripack 13527 shipping case. Set-up time 5 minutes.
Launch: Can be launched by hand or from a vehicle or catapult.
Recovery: Belly landing.

Spear general appearance demonstrates its simplicity *(Brock Technologies)*
1418595

Spear prototype during a test flight *(Brock Technologies)*
1418593

Specifications

Power plant
Can be powered by either a 1.5 kW (2 hp) electric motor or a piston engine (rating not stated); three-blade folding propeller.

Status: In service with various unnamed research, commercial or US military agencies.
Contractor: Brock Technologies Inc
Vail, Arizona.

BTC Scythe

Type: Multirole mini-UAV.
Development: The Scythe (named for its wing shape) was preceded by a smaller (1.12 m; 3 ft 8 in wing span) design, known as the Swarm I, built for the US Office of Naval Research.
Description: *Airframe:* Mid-mounted crescent wings (modified Clark 2411 thin-section aerofoil); mainly cylindrical fuselage; angular tail surfaces, with slight sweepback on vertical fin. No landing gear. Thermoplastics construction.
Mission payloads: Offered with BTC-40 CCD TV camera as standard payload; small black and white camera for piloting, housed in blister on starboard wing leading-edge.
Guidance and control: Pre-programmed; no further details known.
Transportation: System is transportable in three packing cases.
System composition: Three air vehicles, payloads, GCS and launcher. Turnaround time less than 5 minutes.
Launch: By 7.3 m (24 ft) bungee catapult.
Recovery: By ram-air parafoil, deployed under autopilot or manual control, to short-field landing.
Power plant: One two-cylinder two-stroke engine (rating not stated), with two-blade tractor propeller; automotive (Mogas) fuel.

Scythe

Dimensions, External	
Wings, wing span	1.80 m (5 ft 10¾ in)
Areas	
Wings, Gross wing area	0.30 m² (3.2 sq ft) (est)
Weights and Loadings	
Weight	
Weight empty	1.8 kg (3.00 lb)
Max launch weight	7.7 kg (16.00 lb)
Payload, Max payload	2.7 kg (5.00 lb)
Performance	
Speed	
Max level speed	125 kt (232 km/h; 144 mph)
Loitering speed	60 kt (111 km/h; 69 mph)
Stalling speed	35 kt (65 km/h; 41 mph) (est)
Endurance	6 hr (est)
Power plant	1 × piston engine

Status: Being marketed in 2008, still listed by maker in 2010.
Contractor: Brandebury Tool Company Inc (BTC)
Gaithersburg, Maryland.

Wing-mounted Scythe piloting camera *(BTC)* 1290283

Scythe *(BTC)* 1290282

L-3 CUS/NASC TigerShark

Type: Testbed and tactical UAV.
Development: Devised as a testbed for a long-range, low-cost airframe with a potential useful load of more than 45.4 kg (100 lb), TigerShark was designed and prototyped in less than 90 days. Key programme events include:

2002-2005 IHS Jane's sources report TigerShark AV deliveries to a 'prime customer' (thought to have been the US Special Operations Command).

August 2004 The TigerShark UAV is reported to have been first 'publicly revealed' during August 2004

2005 The USN's Naval Air Systems Command launched the US Joint Improvised Explosive Device Defeat Organization (JIEDDO) funded TigerShark-based 'Copperhead' Counter-IED (C-IED) programme as part of its Special Surveillance Program (SSP) effort. Elsewhere in the world, 2005 is also reported as having seen the TigerShark AV deployed to Afghanistan in support of US Special Forces operating in the region.

October 2008 The 'Copperhead' capability underwent a 'tactically realistic' trial at the US Army's Yuma Proving Ground in Arizona where it is reported to have generated 'successful' results.

April 2009 The 'Copperhead' capability was undergoing its operational assessment.

May 2009 JIEDDO reported 'Copperhead' as having 'tested very well against [its] specified IED threat' and as having been fielded 'to the warfighter'.

8 June 2011 The USN's Naval Air Systems Command announced that the 'Copperhead' AV had gone 'from the drawing board to being operational within seven months' and that during the period 2006 to June 2011, the deployed capability had logged 'more than 20,000 flight hours on more than 4,000 missions' over southwest Asia. The Command further noted that in-theatre, the 'Copperhead' capability was 'overseen' by USN Reservists and that as of June 2011, the 'Copperhead' AV was equipped with the 16.8 GHz centre frequency JIEDDO/Sandia National Laboratories MiniSAR SAR radar.

28 September 2011 The USN's Naval Air Warfare Center's Aircraft Division (Lakehurst, New Jersey) awarded the Navmar Applied Sciences Corporation (NASC - Warminster, Pennsylvania) a then year USD12,336,876 order (against the previously issued basic ordering agreement N68335-10-G-0026) with regard to the procurement of 10 × 'appropriately configured' TigerShark UAVs (for use in 'precision weaponised system testing') together with the assessment, procurement and deployment of 'innovative' ISR and communications systems, micro UAV technology and related hardware in support of Phase 3 of the Persistent Ground Surveillance System (PGSS) programme. Work on the effort was to be undertaken at facilities in Warminster, Pennsylvania (48 per cent workshare), Patuxent River, Maryland (20 per cent workshare), Avon Park, Florida (19 per cent workshare), Washington DC (seven per cent workshare) and Tampa, Florida (six per cent workshare) and at the time of its announcement, the programme was scheduled for completion by the end of September 2013.

26 September 2012 The USN's Naval Air Warfare Center's Aircraft Division (Lakehurst, New Jersey) awarded the NASC a further then year USD41,463,854 cost plus fixed fee, firm, fixed-price, cost reimbursable contract with regard to a Phase III Small Business Innovation Research (SBIR) project under Topics N92-170 (*Laser Detection and Ranging Identification Demonstration*), N94-178 (*Air Deployable Expendable Multi-parameter Environmental Probe*) and AF083-006 (*Low Cost ISR UAV*). In more detail, this effort included the procurement of 21 × TigerShark Block 3 (TSB 3) XTS persistent surveillance UAVs (together with options for a further 30 × TSB 3 AVs), TSB 3 XTS product upgrades, associated GCSs, research and engineering support and deployment training in support of persistent threat detection systems deployed outside the continental US. Work on the effort was to be performed at facilities in Johnstown, Pennsylvania (55 per cent workshare); Afghanistan (35 per cent workshare); Hollywood, Maryland (six per cent workshare); Warminster, Pennsylvania (three per cent workshare) and Yuma, Arizona (one per cent workshare) and at the time of its announcement, the

The FoxCar is a slightly enlarged version of the TigerShark AV *(BTC)* 1290281

The TigerShark AV displays a configuration that has been adopted by an increasing number of tactical UAVs *(BTC)* 1290280

programme was scheduled for completion by the end of September 2015.

October 2012 As of October 2012, usually reliable sources were reporting the TigerShark UAV as having been deployed to Afghanistan (the JIEDDO and US Special Forces) and to have been used in support of the US DoD's Counter-Narcoterrorism Technology Program Office.

1 November 2012 On 1 November 2012, General Dynamics Ordnance and Tactical Systems announced that a TigerShark AV had been used to air launch three 81 mm mortar rounds equipped with General Dynamics Roll Control Fixed Canard (RCFC) control system and a US Army Armament Research and Development Engineering Center (ARDEC) developed fuzing solution as part of the US Army's Air Drop Mortar (ADM) programme.

December 2012 IHS Jane's sources were reporting at least one 'Copperhead' TigerShark UAV as being in service as a C-IED surveillance tool in Afghanistan's Regional Command East area-of-operations.

Readers should further note that the TigerShark AV has been identified as the launch platform for the Blacktip lethal micro UAV and that over time, the type has been associated with both the Brandebury Tool Company (BTC - subsequently Brandebury Aerostructures, BAI Aerosystems and then L-3 Communications Unmanned Systems - L-3 CUS) and the NAVMAR Applied Sciences Corporation (NASC). The given TigerShark specification data is after L-3 Communications Unmanned Systems.

Description: *Airframe:* Conventional pod and twin tailboom configuration with high-mounted wings; pusher engine. Non-retractable tricycle landing gear (with disc brakes on FoxCar). Laminated plywood fuselage; wings and tail surfaces of thermoplastics composites construction.

Mission payloads: Over time, TigerShark UAVs are known to have been equipped with the JIEDDO/Sandia National Laboratories MiniSAR SAR radar and to have test launched guided 81 mm mortar rounds.

Guidance and control: Programmable, with GPS waypoint navigation. UHF bi-directional command link; L-band imagery downlink. Portable Generation IV GCS with moving map display and automatic tracking.

Launch: Conventional wheeled take-off.

Recovery: Conventional wheeled landing.

Power plant: One 18.6 kW (25 hp) two-cylinder two-stroke engine; two-blade composites pusher propeller. Fuel capacity (Mogas) 64 litres (17.0 US gallons; 14.2 Imp gallons). At one time, an 8 to 14 kW brushless electric motor was under consideration for use aboard the FoxCar AV.

Variants: *TigerShark:* Baseline version; *as described.*

FoxCar: FoxCar combines fuselage of TigerShark with enlarged tailbooms plus increased wing and tail surfaces with optimised aerofoil sections.

TigerShark, FoxCar

Dimensions, External	
Overall	
length, TigerShark	4.1 m (13 ft 5½ in)
height	1.2 m (3 ft 11¼ in)
Wings	
wing span	
TigerShark	5.3 m (17 ft 4¾ in)
FoxCar	6.4 m (21 ft 0 in)
Weights and Loadings	
Weight	
Weight empty	95 kg (209 lb)
Max T-O weight	
with fuel & max payload	144 kg (317 lb)
Performance	
Speed	
-	70 kt (130 km/h; 81 mph)
Cruising speed	56 kt (104 km/h; 64 mph)
Endurance	
max, TigerShark	10 hr
max, FoxCar	9 hr 30 min
Power plant	1 × piston engine

Status: Over time, design and development of the TigerShark AV has been variously associated with the BTC, BAI Aeosystems, L-3 CUS and the NASC. In this latter context, L-3 CUS and the NASC are the latest identified TigerShark manufactures (with the latter receiving TigerShark related contracts during September 2011 and September 2012 - see *Development*). Again, IHS Jane's sources report BTC as having produced a 'significant quantity' of TigerShark AVs for the NASC in the past. As of 2013, the 'Copperhead' TigerShark C-IED configuration was in service with the US military in Afghanistan (at the US Army's Forward Operating Base Shank in Afghanistan's Logar province), while the status of the FoxCar platform was uncertain.

Contractor: L-3 Communications Unmanned Systems, Carrollton, Texas and Easton, Maryland. Navmar Applied Sciences Corporation, Warminster, Pennsylvania.

CIRPAS Pelican

Type: Optionally piloted vehicle.

Development: CIRPAS was established by the US Office of Naval Research (ONR) in the second quarter of 1996 to provide UAV flight services to the RDT&E communities. Its prime contractor is the California Institute of Technology. In addition to the Pelican, the CIRPAS fleet includes two Predators and an Altus I.

The Pelican OPV is a modified Cessna 337 Skymaster first developed by the ONR for low-altitude, long-endurance atmospheric and oceanographic sampling; this effort is supported by NASA's Environmental Research Aircraft and Sensor Technology (ERAST) programme. Aircraft conversion was undertaken by General Atomics under a March 1996 contract from CIRPAS. First flight as a UAV was made on 10 December 1998, at which time a second Pelican was also undergoing conversion. Two aircraft were converted, Cessna 337H (N84NX, US Navy 167782) and ex-USAF O-2A (68-111550, US Navy 167783). CIRPAS capabilities were later expanded to include military exercises using Pelican.

Pelican is intended to function as a manned or unmanned sensor platform embodying the flight control package developed for the Predator UAV. It also functions as a surrogate vehicle for nose-mounted Predator payloads, in which form it took part in March 1999 in the US Navy urban warfare exercise Fleet Battle Experiment Echo. Video images were transmitted to GCSs via a satcom datalink and to the command ship USS *Coronado* via a COTS converter. This imagery was then transmitted to a comsat relay van and to other surface vessels. Since 2001 the Pelican aircraft has been used as a surrogate UAV in six major DoD training

Pelican is highly modified Cessna 337 Skymaster for flying nose mounted payloads *(CIRPAS)* 1369166

A Predator EO/IR turret carried on the Pelican's nose station *(CIRPAS)* 1369168

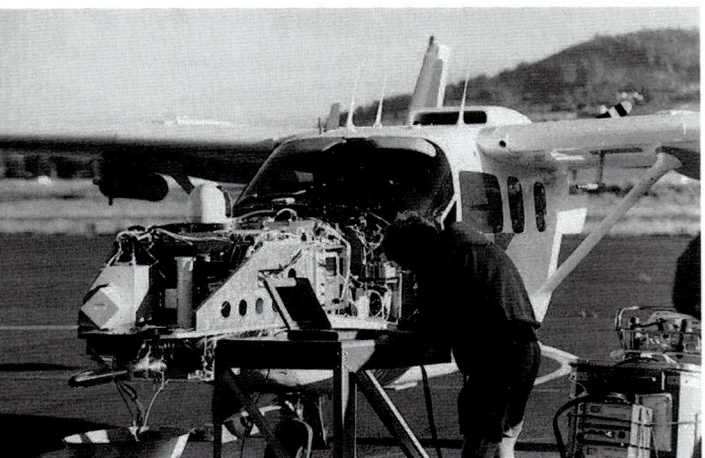

Installing a payload in Pelican's modified nose *(NPS/CIRPAS)*

exercises that require a UAV capability for ground forces, but where a true UAV cannot fly because of FAA restrictions.

Description: *Airframe:* Cessna 337 modified by deletion of nose-mounted engine and replacement of rear-mounted one by a more powerful unit.

Mission payloads: Main payload bay in nose. Cabin space and four underwing hardpoints available for additional payloads. Onboard power of 2 kW available for payload operation. Typical payload is L3 Wescam 14TS.

Guidance and control: Aircraft control functions are provided by a General Atomics GCS.

Launch: Conventional wheeled take-off.
Recovery: Conventional wheeled landing.

Pelican

Dimensions, External	
Wings, wing span	12.80 m (42 ft 0 in)
Weights and Loadings	
Weight, Max T-O weight	2,086 kg (4,598 lb)
Payload	
Max payload, nose	159 kg (350 lb)
Performance	
Altitude	
Service ceiling	
max	4,570 m (15,000 ft)
preferred	3,050 m (10,000 ft)
Speed	
Cruising speed, normal, IAS	90 kt (167 km/h; 104 mph)
Endurance	
max	8 hr
typical	6 hr

Status: Conducting research missions and exercise duties as described above. The aircraft remained in regular use in 2009.

Contractor: The Center for Interdisciplinary Remotely Piloted Aircraft Studies
Marina, California.

Cyber Defense Systems CyberBug

Type: Close-range mini-UAV.

Development: The CyberBug was developed for military, law enforcement and commercial applications, including search and rescue, traffic monitoring, environmental missions, research, border patrol and drug interdiction. In June 2007, a CyberBug became only the 12th UAV, and the first of less than 45.4 kg (100 lb) weight, to be awarded an FAA experimental airworthiness certificate (EAC) allowing it to fly in the US national airspace.

Description: *Airframe:* Two-piece box-section fuselage, tapered at front and rear; angular conventional tail unit with movable control surfaces. Delta-shaped Rogallo-type wing mounted above fuselage on pylon. Spring landing skid beneath front fuselage.

Mission payloads: All versions can be equipped with a gyrostabilised pan and tilt colour daylight camera; the Medium and Large sizes have option to replace this with a pan and tilt IR camera for night surveillance. All versions have video and data downlink transmission.

Guidance and control: Pre-programmed or manual operation from a portable (laptop) ground station. Autonomous modes include waypoint navigation, convoy following, GPS camera tracking and automatic landing. Operator can control both aircraft and cameras from remote locations.

Transportation: Two MicroBugs can be carried in a standard ALICE rucksack. The two larger versions can be carried singly in a waterproof soft case.

Launch: Hand launch.
Recovery: Glide approach to automatic landing on underfuselage skid.
Power plant: Battery-powered electric motor; two-blade propeller.

Variants: *MicroBug:* Smallest version, foldable for maximum portability. Said to have been developed for use by 'air force battlefield airmen'.
MediumBug: Promoted as 'an effective mix of size and performance'.
LargeBug: For carriage of larger or heavier payloads. Often used as an R&D platform.

MediumBug, MicroBug, LargeBug

Dimensions, External	
Overall	
length, folded, MicroBug	0.38 m (1 ft 3 in)
Wings	
wing span	
MicroBug	0.76 m (2 ft 6 in)
MediumBug	1.22 m (4 ft 0 in)
LargeBug	1.52 m (4 ft 11¾ in)
Weights and Loadings	
Weight	
Max launch weight	
MicroBug	1.2 kg (2.00 lb)
MediumBug	3.9 kg (8.00 lb)
LargeBug	6.6 kg (14.00 lb)
Performance	
Speed	
Cruising speed	
max, MicroBug	19 kt (35 km/h; 22 mph)
max, MediumBug, LargeBug	22 kt (41 km/h; 25 mph)
Radius of operation, mission	2.7 n miles (5 km; 3 miles)
Endurance	
MicroBug	40 min
MediumBug, LargeBug	1 hr 5 min
Power plant	1 × electric motor

Status: CyberBugs have been widely demonstrated to domestic law enforcement and military forces in several countries.

Contractor: Cyber Aerospace
Tulsa, Oklahoma.

Cyber Defense Systems CyberScout

Type: Experimental VTOL UAV.

Development: The prototype CyberScout made its first hovering flight in September 2004. According to company literature, it is designed "to operate innovative clandestine reconnaissance, IED [improvised explosive device] search and destroy surveillance, and target acquisition missions". Other potential applications include convoy protection, BDA and operations in urban terrain. In fully developed form, the design is planned to be turbojet-powered.

Description: *Airframe:* Pod and boom fuselage. Foreplanes, with flaps, mid-mounted on pod; main wing at rear of tailboom, with central cutout having upper and lower vertical fins at inboard ends. Fixed, vertically

The flexwing CyberBug mini-UAV *(IHS/Patrick Allen)*

Prototype CyberScout three-engined VTOL air vehicle *(IHS/Patrick Allen)*

CyberScout prototype in hovering flight *(Cyber Defense)* 1290195

mounted ducted fan engine in forward part of pod; twin engines, in swivelling nacelles, between fins in wing trailing-edge cutout. 'Tricycle' arrangement, non-retractable landing legs.

Mission payloads: None in prototype. Developed version to have a continuous forward-looking camera, and be capable of carrying a wide variety of cameras, sensors, weapons and instruments. Typical fit could be gyrostabilised pan-tilt-zoom camera, IR camera or customer-specified payloads.

Guidance and control: Prototype is remotely piloted. Developed version will be semi-autonomous, with manual override, controlled via a laptop GCS.

Transportation: Designed for quick and easy deployment from HMMWVs and other easily mobile launch platforms.

Launch: Vertical take-off on all three engines, with rear pair tilted downwards, swivelling to horizontal for transition to forward flight.

Recovery: Automatic vertical or conventional skid landing.

Power plant: Prototype: Three ducted-fan piston engines (type and rating not stated), one mounted vertically in forward fuselage and two in swivelling nacelles at rear.

Developed version: Three turbojets.

CyberScout prototype, CyberScout developed version

Dimensions, External	
Overall length	
CyberScout prototype	1.52 m (4 ft 11¾ in)
CyberScout developed version	2.29 m (7 ft 6¼ in)
Wings	
wing span, CyberScout prototype	1.52 m (4 ft 11¾ in)
	2.29 m (7 ft 6¼ in)
Weights and Loadings	
Weight	
Max T-O weight	
CyberScout prototype	5.4 kg (11.00 lb)
CyberScout developed version	31.8 kg (70 lb)
Performance	
Speed	
Max level speed	
CyberScout developed version	350 kt (648 km/h; 403 mph)
Cruising speed	
CyberScout prototype	100 kt (185 km/h; 115 mph)
CyberScout developed version	250 kt (463 km/h; 288 mph) (est)
Radius of operation	
CyberScout developed version	100 n miles (185 km; 115 miles)
Endurance	
CyberScout prototype	10 min
CyberScout developed version	1 hr
Power plant	
CyberScout prototype	piston engine
CyberScout developed version	turbojet

Status: Development complete.
Contractor: Cyber Aerospace
Tulsa, Oklahoma.

DARPA Vulture programme

Type: Ultra-long endurance UAV.

Development: Objective of the Vulture programme is to develop a 'pseudo-satellite' aircraft capable of remaining on-station uninterrupted for five years or more, with a 454 kg (1,000 lb) payload, a 5 kW power supply and a 99 per cent probability of maintaining its on-station position against the prevailing winds, to perform ISR and communication missions over an area of interest. The name VULTURE was coined from Very high altitude, Ultra endurance, Loitering Theatre Unmanned Reconnaissance Element.

Energy management and technology reliability are among the programme's greatest challenges. Potential civil applications could include climate change research, weather monitoring and regional telecommunications. The 12-month Phase 1, launched with contracts to three competitors in April 2008, covers conceptual system definition; formal reliability and mission success analysis and conceptual designs for sub- and full-scale demonstrators. It will conclude with a system requirements review (SRR). The risk reduction and development (RRD) Phase 2 that would follow this from 2009 to mid-2012 would involve building and testing a subscale vehicle able to — and proving by demonstration — fly for an uninterrupted period of three months. If the programme continues beyond this, Phase 3 would require a full-scale demonstrator to prove itself with a 12-month uninterrupted flight.

In September 2010 The Boeing Company signed an agreement with DARPA to develop and fly the SolarEagle unmanned aircraft for the Vulture II demonstration program. Under the terms of the $89 million contract, SolarEagle will make its first demonstration flight in 2014.

Description: *Airframe:* Various; refer to individual Variants below.
Launch: Runway take-off.
Recovery: Runway landing.

Variants: *Aurora Odysseus:* Phase 1 contract, value of which was not stated, announced by Aurora Flight Sciences 14 April 2008. Team partners included BAE Systems (payloads, sensors and conops); C S Draper Laboratories (electronics and control systems); and Sierra Nevada Corporation (SNC). Designed to take off as three separate, approximately 48.8 m (160 ft) span aircraft, formating in the stratosphere to join up at the wingtips to form a single vehicle spanning 152 m (500 ft). Advantages claimed for this concept are easier transition to high altitudes, and the facility for individual units to be recalled for repair or replacement. SNC's expertise in autonomous refuelling systems was intended to assist in developing the docking manoeuvre. Odysseus was designed to use solar energy and fuel cells to power the aircraft during daylight, and stored solar energy to power it at night. To maximise its capture of solar energy, it would fly in a Z shape during daylight, morphing to a lower-drag, straight-wing configuration at night.

Boeing design: Phase 1 contract (USD3.8 million) announced 21 April 2008. Boeing divisions involved are Integrated Defense Systems Advanced Systems (leader), Phantom Works, Spectrolab and Space and Intelligence Systems. Team partners include QinetiQ, C S Draper

DARPA image of a possible Vulture configuration *(DARPA)* 1290284

Laboratories and Versa Power Systems. Boeing's straight-wing SolarEagle design is influenced by QinetiQ's Zephyr, which set an unofficial world endurance record of 54 hours in September 2007, but will store solar energy in regenerative fuel cells instead of using the latter's battery power.

Lockheed Martin design: The Skunk Works' design was a straight-winged single vehicle powered by electric ring motors with direct drive to 10 propellers. Spanning more than 91.5 m (300 ft), it features three fuselages interconnected by a bracing strut for increased rigidity. Each fuselage supports a rotating tail unit, these and the wing upper surface being covered in solar panels to gather maximum available solar energy.

Status: First flight of the Boeing SolarEagle demonstrator is scheduled for 2014.

Contractor: Defense Advanced Research Projects Agency
Arlington, Virginia.

DP-5X with nose-mounted GE40 grenade launcher *(Dragonfly)*
1151614

Dragonfly Pictures DP series

Type: Helicopter UAVs.

Development: Began 1992, initially with a 'hobby' helicopter (the DP-1), which was used to verify the capability to carry and stabilise an imaging payload. A slightly larger DP-2 was developed in 1992–93, but proved less successful, giving way instead to the more effective DP-3 in 1993–94. Design effort in 1994–96 focused on more advanced modular concepts, resulting in the DP-4, which made its first flight in the second quarter of 1995. Since then, Dragonfly Pictures has further developed the modularity concept for unmanned VTOL aircraft with its patented Advanced Multi-Mission Platform or AMMP. First example in this form is the DP-5X.

Metal Storm Ltd of Australia announced in February 2004 that, under a USD325,000 contract from the UAV manufacturer, it was to undertake live firing trials of its GE40 40 mm grenade launcher mounted on a Dragonfly DP-4X. Ground firings of two and four inert rounds, from a machine fitted with twin GE40s, were completed in May 2004. Plans for in-flight trials, including the firing of live rounds, were then scheduled for 2005, but did not occur until the following year. These eventually took place on 27 and 28 September 2006, this time using a DP-5X prototype as the trials vehicle. Conducted at the Warren Grove Air National Guard Bombing Range in New Jersey as part of a DARPA-contracted activity, initial tests were made from the DP-5X in strap-down mode; untethered hover sighting shots were also made from various altitudes. The demonstrations concluded with forward-flight 'strafing runs' that targeted a vehicle sited on the firing range. The live firings utilised Metal Storm's FC440 remotely operated fire-control unit and MK16-KE 40 mm kinetic energy projectiles, as well as the, GE40 grenade launcher.

Description: *Airframe:* All-metal fuselage with skid landing gear; two-blade main and tail rotors with composites blades, driven by two V belts via three-stage gearbox. Main rotor blades have offset flapping hinges. Engine, gearbox and skids all have four-bolt attachment for ease of assembly (in DP-4, takes about 5 minutes).

Mission payloads: Compatible with various real-time aerial imaging (film, video or IR), communications and environmental sensors. Stability augmentation system (computer and vertical gyro) standard.

Guidance and control: Programmable and fully automatic flight control system, with radio command uplink (DP-3) or two-way digital communications link (DP-4); GPS navigation. Telemetry downlink.

Launch: Vertical take-off.

Recovery: Vertical landing. Full autorotation in event of power or command link interruption.

Variants: *DP-3:* Small, lightweight unmanned VTOL vehicle.

DP-4X: Larger vehicle, with higher endurance and more than 15.9 kg (35 lb) payload capacity. Baseline version and prototype for future development. Two updated versions were launched in 2007, the DP-4X and longer-range DP-4XT.

DP-5X: Advanced modular (AMMP) design with payload capacity increased to more than 45.4 kg (100 lb) and extended range option. Three other versions were projected in 2007: DP-5, DP-5T and DP-5XT. As of November 2011 the DP-5X and the DP-5XT were on offer according to the companies website.

DP-3, DP-4X, DP-4XT, DP-5, DP-5T, DP-5X, DP-5XT

Dimensions, External
  Overall
    length, rotors turning, DP-3 .... 2.56 m (8 ft 4¾ in)
    height, DP-4X .... 1.07 m (3 ft 6¼ in) (est)
  Fuselage
    length
      DP-3 .... 1.75 m (5 ft 9 in)
      DP-4X .... 2.13 m (6 ft 11¾ in) (est)
  Rotors
    rotor diameter
      DP-3 .... 2.24 m (7 ft 4¼ in)
      DP-4X .... 3.05 m (10 ft 0 in) (est)

Weights and Loadings
  Weight
    Weight empty, DP-3 .... 22.2 kg (48 lb)
  Max T-O weight
    DP-3 .... 38.1 kg (83 lb)
    DP-4X .... 63.5 kg (139 lb) (est)
  Fuel weight
    Max fuel weight
      with 7.7 kg (10 lb) payload, DP-4X .... 31.8 kg (70 lb)
      with 7.7 kg (10 lb) payload, DP-4XT .... 77.1 kg (169 lb)
      with 7.7 kg (10 lb) payload, DP-5 .... 24.9 kg (54 lb)
      with 7.7 kg (10 lb) payload, DP-5T,
        DP-5X .... 74.8 kg (164 lb)
      with 7.7 kg (10 lb) payload, DP-5XT .... 176.9 kg (389 lb)
      with 15.9 kg (35.0 lb) payload, DP-4X .... 20.4 kg (44 lb)
      with 15.9 kg (35.0 lb) payload, DP-4XT .... 65.8 kg (145 lb)
      with 15.9 kg (35.0 lb) payload, DP-5 .... 13.6 kg (29 lb)
      with 15.9 kg (35.0 lb) payload, DP-5T,
        DP-5X .... 63.5 kg (139 lb)
      with 15.9 kg (35.0 lb) payload, DP-5XT .... 166 kg (365 lb)
      with 34.0 kg (75.0 lb) payload, DP-4XT .... 47.6 kg (104 lb)
      with 34.0 kg (75.0 lb) payload, DP-5T,
        DP-5X .... 45.4 kg (100 lb)
      with 34.0 kg (75.0 lb) payload, DP-5XT .... 147 kg (324 lb)
      with 68.0 kg (150 lb) payload, DP-4XT .... 13.6 kg (29 lb)
      with 68.0 kg (150 lb) payload, DP-5T,
        DP-5X .... 11.3 kg (24 lb)
      with 68.0 kg (150 lb) payload, DP-5XT .... 113 kg (249 lb)
  Payload
    Max payload
      DP-3 .... 13.6 kg (29 lb)
      DP-4X, DP-4XT .... 15.9 kg (35 lb) (est)
      DP-5T, DP-5X, DP-5XT .... 68 kg (149 lb)

Performance
  Climb
    Rate of climb
      typical, at S/L, DP-3 .... 91 m/min (300 ft/min) (est)
  Altitude
    Service ceiling, DP-3 .... 3,050 m (10,000 ft) (est)
  Speed
    Max level speed, DP-3 .... 52 kt (96 km/h; 60 mph) (est)
  Range
    with 4.5 kg (10 lb) payload, DP-4X .... 680 n miles (1,259 km; 782 miles)
    with 4.5 kg (10 lb) payload, DP-4XT .... 880 n miles (1,629 km; 1,012 miles)
    with 4.5 kg (10 lb) payload, DP-5 .... 370 n miles (685 km; 425 miles)
    with 4.5 kg (10 lb) payload, DP-5T .... 580 n miles (1,074 km; 667 miles)
    with 4.5 kg (10 lb) payload, DP-5X .... 685 n miles (1,268 km; 788 miles)
    with 4.5 kg (10 lb) payload, DP-5XT .... 990 n miles (1,833 km; 1,139 miles)
    with 15.9 kg (35 lb) payload, DP-4X .... 435 n miles (805 km; 500 miles)
    with 15.9 kg (35 lb) payload, DP-4XT .... 740 n miles (1,370 km; 851 miles)
    with 15.9 kg (35 lb) payload, DP-5 .... 170 n miles (314 km; 195 miles)
    with 15.9 kg (35 lb) payload, DP-5T .... 480 n miles (889 km; 552 miles)
    with 15.9 kg (35 lb) payload, DP-5X .... 575 n miles (1,064 km; 661 miles)
    with 15.9 kg (35 lb) payload, DP-5XT .... 920 n miles (1,703 km; 1,058 miles)
    with 34.0 kg (75 lb) payload, DP-4XT .... 530 n miles (981 km; 609 miles)
    with 34.0 kg (75 lb) payload, DP-5T .... 340 n miles (629 km; 391 miles)
    with 34.0 kg (75 lb) payload, DP-5X .... 410 n miles (759 km; 471 miles)
    with 34.0 kg (75 lb) payload, DP-5XT .... 810 n miles (1,500 km; 932 miles)

Dragonfly's baseline model, the DP-4
0528823

DP-3, DP-4X, DP-4XT, DP-5, DP-5T, DP-5X, DP-5XT

with 68.0 kg (150 lb) payload, DP-4XT	150 n miles (277 km; 172 miles)
with 68.0 kg (150 lb) payload, DP-5T	80 n miles (148 km; 92 miles)
with 68.0 kg (150 lb) payload, DP-5X	100 n miles (185 km; 115 miles)
with 68.0 kg (150 lb) payload, DP-5XT	630 n miles (1,166 km; 725 miles)
Endurance	
DP-3	2 hr
with 7.7 kg (10 lb) payload, DP-4X	12 hr 54 min
with 7.7 kg (10 lb) payload, DP-4XT	11 hr 12 min
with 7.7 kg (10 lb) payload, DP-5	6 hr 0 min
with 7.7 kg (10 lb) payload, DP-5T	6 hr 42 min
with 7.7 kg (10 lb) payload, DP-5X	9 hr 12 min
with 7.7 kg (10 lb) payload, DP-5XT	9 hr 54 min
with 15.9 kg (35.0 lb) payload, DP-4X	8 hr 6 min
with 15.9 kg (35.0 lb) payload, DP-4XT	9 hr 24 min
with 15.9 kg (35.0 lb) payload, DP-5	2 hr 30 min
with 15.9 kg (35.0 lb) payload, DP-5T	5 hr 36 min
with 15.9 kg (35.0 lb) payload, DP-5X	7 hr 42 min
with 15.9 kg (35.0 lb) payload, DP-5XT	9 hr 18 min
with 34.0 kg (75 lb) payload, DP-4XT	6 hr 48 min
with 34.0 kg (75 lb) payload, DP-5T	3 hr 36 min
with 34.0 kg (75 lb) payload, DP-5X	5 hr 30 min
with 34.0 kg (75 lb) payload, DP-5XT	8 hr 18 min
with 68.0 kg (150 lb) payload, DP-4XT	1 hr 54 min
with 68.0 kg (150 lb) payload, DP-5T	1 hr 0 min
with 68.0 kg (150 lb) payload, DP-5X	1 hr 24 min
with 68.0 kg (150 lb) payload, DP-5XT	6 hr 18 min

Status: DP-3, DP-4 and DP-5X developed and available; Services offered.
Contractor: Dragonfly Pictures Inc
Essington, Pennsylvania.

Dragonfly Pictures DP-6 Whisper

Type: VTOL tactical UAV.
Development: Developed to provide covert and positive identification of high-value targets and their vehicles, the Whisper is nearly silent and maintenance-free due to its electric propulsion system. It can provide mobile ad hoc Wi-Fi networking, eavesdropping, facial recognition or close-range imagery for superior situational awareness and understanding during mobile operations in complex terrain. The system was introduced in August 2006, and accumulated some 50 hours of test flights during its first year.
Description: *Airframe:* Tandem-rotor configuration, with three-blade rotors; fixed, four-legged (optionally four-wheeled) landing gear Composites construction.

Mission payloads: Directional microphone and speaker; Secnet 11 secure communication relay; EO/IR day/night sensors with three-axis stabilisation; facial recognition payload. All are field-changeable without need for tools.

Falcon 11 screen and DP-6 Whisper GCS monitor displays *(IHS/Patrick Allen)* 1328626

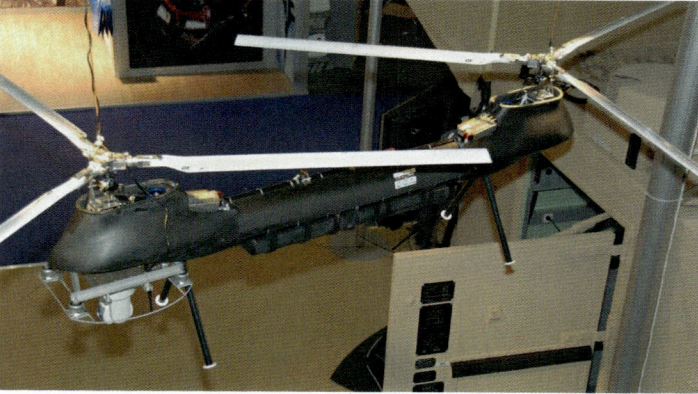

Whisper on show in September 2007, equipped with Harris Falcon 11 networking system payload *(IHS/Patrick Allen)* 1328633

Guidance and control: MicroPilot autonomous control system with a STANAG 4589 compliant GCS. For continuous endurance, operator can tether and hover the DP-6 above a ground power source. Aircraft is targeted for ease of adoption due to its minimal noise, maintenance, training, logistics and piloting skills.
Transportation: Transportable, assembled, in a trunk. Set-up time 5 minutes, turnaround time 2 minutes.
System composition: Two aircraft, two sensors, automation system, battery management system and GCS. Deployable by one person.
Launch: Autonomous vertical take-off. Can be launched and recovered in crosswinds up to 30 kt (56 km/h; 35 mph).
Recovery: Autonomous vertical landing. Full autorotation in event of power loss under development. In the event of loss of the command link, flight control system climbs aircraft to recover link; if link recovery not achieved, aircraft flies a pre-specified path to a recovery point.
Power plant: Two battery-powered 4.85 kW (6.5 hp) electric motors standard; batteries can be replaced in 3 minutes or recharged in 15 minutes. A turbine-powered version is also available.

DP-6

Dimensions, External	
Overall, height	0.76 m (2 ft 6 in)
Fuselage	
length	1.80 m (5 ft 10¾ in)
width, max	0.49 m (1 ft 7¼ in)
Engines	
propeller diameter, each	2.26 m (7 ft 5 in)
Weights and Loadings	
Weight	
Weight empty	10.9 kg (24 lb)
Max T-O weight	36.3 kg (80 lb)
Payload	
Max payload	
mission, for 1 h loiter	1.1 kg (2.00 lb)
mission, for 24 h on tether	11.8 kg (26 lb)
useful	14.1 kg (31 lb)
Performance	
Climb	
Rate of climb, max, at S/L	259 m/min (849 ft/min)
Altitude, Service ceiling	4,570 m (15,000 ft)
Speed	
Max level speed	66 kt (122 km/h; 76 mph)
Cruising speed	
max	45 kt (83 km/h; 52 mph)
econ	30 kt (56 km/h; 35 mph)
Radius of operation, mission	45 n miles (83 km; 51 miles)
Endurance, on station at 5 n miles (9.3 km; 5.75 miles) from GCS	1 hr
Power plant	2 × electric motor

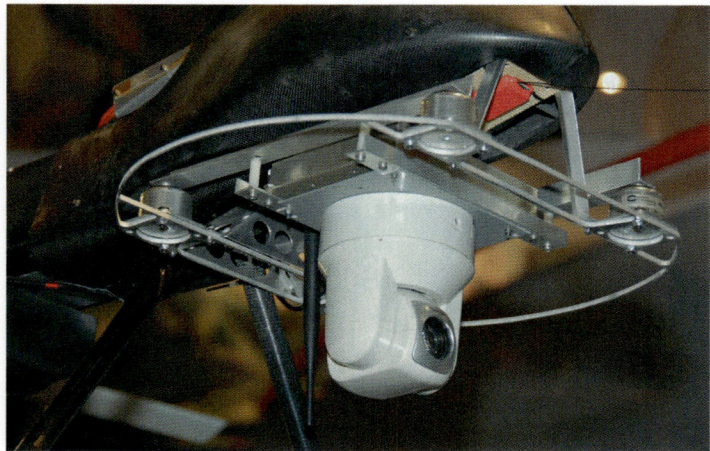

The Falcon 11's Falcon Watch undernose EO/IR sensor turret *(IHS/Patrick Allen)* 1328631

Status: Displayed at DSEi defence exhibition in London, UK, in September 2007, associated with Harris Corporation Falcon 11 networking system. The first aircraft was sold to L-3 for payload integration and another was used in joint testing with Northrop Grumman for DARPA in 2009. DP-6 aircraft have served as a flight controls testbed with the US Army and other customers, and in support of contracts with military, security, university researchers and defence system integrators. The DP-6 was reported to be participating in a US Army Medical Research and Material Command research programme for combat medics UAS (CM-UAS) in 2010-2011.

Contractor: Dragonfly Pictures Inc
Essington, Pennsylvania.

Dragonfly Pictures DP-11 Bayonet

Type: VTOL tactical UAV.

Development: The DP-11 Bayonet has been designed to combine the endurance, speed and simplicity of a fixed-wing aircraft with VTOL capability, able to provide persistent ISR support for tactical-level manoeuvre decisions and unit-level force defence and force protection. Main users at which it is aimed are naval vessels and marines' land forces, for maritime patrol, security, interdiction and anti-terrorist operations.

Description: *Airframe:* Simple aluminium airframe, with upper and lower skins of durable ABS plastics panels attached directly to the frame. Captive quick-disconnect fasteners allow removal of the skins for maintenance access and shipboard maintenance.

Mission payloads: Universal payload interface to enable field-changeable 'plug and play' payloads, including: mini-SAR; three-axis, stabilised, day/night EO/IR; laser designator; secure communication relay; Mode 3C IFF; One System, ROVER and OSRVT.

Guidance and control: The DP-11 is equipped for autonomous waypoint navigation, precision autonomous take-off and landing. Upgrades to provide higher levels of autonomy are planned.

Transportation: All components are two-man portable. System can be transported in a single HMMWV. Set-up time 5 minutes; turnaround time 5 minutes.

System composition: Three air vehicles, payloads, one GCS and support equipment.

Launch: Autonomous vertical take-off. System operable in crosswinds of up to 30 kt (56 km/h; 35 mph).

Recovery: Autonomous vertical landing. In the event of losing command link, flight control system climbs the aircraft to recover link; if unable to recover link, aircraft flies a pre-specified path to a recovery point.

Power plant: Two 10.1 kW (13.6 hp) unidentified piston engines, each driving a three-blade rotor. JP-5, JP-8 or Diesel fuel.

DP-11

Dimensions, External	
Overall	
height	2.23 m (7 ft 3¾ in)
stored	1.98 m (6 ft 6 in)
width	0.40 m (1 ft 3¾ in)
Fuselage, length	1.68 m (5 ft 6¼ in)
Rotors	
rotor diameter, each	1.19 m (3 ft 10¾ in)
Weights and Loadings	
Weight	
Weight empty	35.8 kg (78 lb)
Max T-O weight	60.3 kg (132 lb)
Performance	
Altitude	
Operating altitude, mission	3,050 m (10,000 ft) (est)
Service ceiling	6,095 m (20,000 ft) (est)
Speed	
Max level speed	112 kt (207 km/h; 129 mph) (est)
Cruising speed	
max	68 kt (126 km/h; 78 mph) (est)
econ	52 kt (96 km/h; 60 mph) (est)
Range	479 n miles (887 km; 551 miles) (est)
Endurance	
on station at 50 n miles (93 km; 57 miles) from base	7 hr 24 min (est)
max	9 hr (est)
Power plant	2 × piston engine

Status: The DP-11 system has completed conceptual design, and was planned for development during 2008. First flight had not been reported by 2011.

Contractor: Dragonfly Pictures Inc
Essington, Pennsylvania.

DRS RQ-15A Neptune

Type: Maritime tactical UAV.

Development: This UAV from DRS was revealed publicly in July 2002. According to the company, it was developed specifically for tactical military operations over land or water without need of a runway. It can be deployed from surface vessels and recovered for day or night operations. The designation RQ-15A was allocated in the second quarter of 2007.

Description: *Airframe:* Blended wing/body design; pusher engine mounted above rear of main fuselage, flanked by two short tailbooms culminating in large fins bridged by a central tailplane. Conventional ailerons, rudders and elevator. Composites construction. Can be dismantled into three subassemblies. Fuselage has built-in 'receiver tubes' that allow it to slide on to its launch rails to stow inside its transport container.

Mission payloads: Primary payload is an electro-optical camera and an uncooled infra-red camera mounted on the DRS GS-207 two axis stabilised gimbal. Payload can be changed in the field. Avionics based on a digital autopilot, with interfaces to payload control, environmental sensor and communications modules. Electronic and sensor systems are protected from water intrusion above the waterline.

Guidance and control: GCS is a ruggedised laptop computer based system, interfaced with the communications module and air vehicle hand controller. It can undertake mission planning, flight plan updating and handle more than 100 waypoints. Flight operations are primarily autonomous, but with provision for operator intervention to manage the payload when over areas of interest. The digital datalink, which has a range of more than 40 n miles (74 km; 46 miles), is optimised for secure operation over open water. A remote terminal receiver is available as an option.

The Neptune is optimised for at-sea launch and recovery *(DRS)*

Neptune on its launcher at the 2006 Farnborough International Air Show *(IHS/Patrick Allen)*

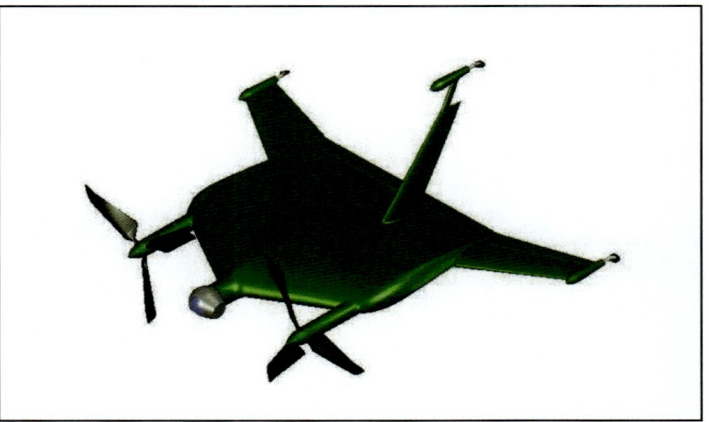

Computerised impression of the Dragonfly DP-11 Bayonet *(DPI)*

Neptune engine detail *(IHS/Patrick Allen)* 1183900

System composition: Three air vehicles, their payloads, one GCS, one RVT (optional), a launcher/transporter and field support kits.
Launch: From zero-length pneumatic launcher, on land or on board ship.
Recovery: Autonomous parachute recovery; manual skid landing. Recoverable on land or water.
Power plant: One 11.2 kW (15 hp) two-cylinder two-stroke engine; two-blade pusher propeller.

RQ-15A

Dimensions, External	
Overall	
length	1.83 m (6 ft 0 in)
height	0.51 m (1 ft 8 in)
Wings, wing span	2.13 m (6 ft 11¾ in)
Weights and Loadings	
Weight, Max launch weight	59.0 kg (130 lb)
Payload	
Max payload, incl parachute	9.1 kg (20.00 lb)
Performance	
Altitude	
Operating altitude, typical	915 m (3,000 ft)
Service ceiling	2,440 m (8,000 ft)
Speed	
Max level speed	85 kt (157 km/h; 98 mph)
Cruising speed	70 kt (130 km/h; 81 mph)
Loitering speed	60 kt (111 km/h; 69 mph)
Radius of operation, datalink	40 n miles (74 km; 46 miles) (est)
Endurance	4 hr

Status: Contracts with a combined value of USD5 million were announced on 26 March 2003, to provide 'several' Neptune systems to the US Navy. Delivery of these began in 2003 and was completed April 2004. Operational evaluation followed by planned product improvement programme leading to planned low rate initial production in 2008.
Customers: US Navy.
Contractor: DRS Defense Solutions
Bethesda, Maryland.

Emmen Aerospace Condor

Type: Helicopter SUAV.
Development: The Condor was developed for both military and civil applications, both as a UAV and as a target for live-fire or simulated

Condor in flight *(USAF/Staff Sgt Stephen J Otero)* 1151657

Condor with fully enclosed fuselage *(Emmen)* 1290165

Condor internal details *(USAF/Staff Sgt Stephen J Otero)* 1151656

training. Specification data applies to the standard version, but other sizes are available on request.
Description: *Airframe:* Two-blade main and tail rotors; pod and boom fuselage; twin-skid landing gear.
Mission payloads: Standard payload is a 7 to 47° FoV colour zoom camera. Options include 35° FoV infra-red camera; combined LLTV and IR cameras with mini-spotlight or laser pointer; or combined colour zoom and IR cameras with mini-spotlight or laser pointer. All payloads are fully gimballed and stabilised.
Guidance and control: Fully automatic one-person operation, with different control modes input via a laptop computer. GPS waypoint navigation. Compatible with the US Army's L-3 Remotely Operated Video Enhancement Receiver (ROVER).
Launch: Conventional and automatic vertical take-off.
Recovery: Conventional and automatic vertical landing; touchdown accuracy of 2.1 to 4.6 m (7 to 15 ft).
Power plant: Electric motor, powered by rechargeable batteries.

Condor

Dimensions, External	
Overall, length	1.19 m (3 ft 10¾ in)
Fuselage, width	0.20 m (7¾ in)
Rotors, rotor diameter	1.19 m (3 ft 10¾ in)
Weights and Loadings	
Weight, Max T-O weight	6.35 kg (13.00 lb)
Payload, Max payload	1.8 kg (3.00 lb)
Performance	
Speed, Max level speed	52 kt (96 km/h; 60 mph)
Radius of operation	26 n miles (48 km; 29 miles)
Endurance	
standard	1 hr (est)
with extended flight batteries	1 hr 40 min (est)
Power plant	1 × electric motor

Status: In April 2007, the Condor was one of a number of SUAVs flown during the exercise 'Atlantic Strike 5' at Avon Park, Florida, where it was operated by joint terminal attack controllers (JTACs) from the US Army, Air Force, Marine Corps and Canadian Air Force. Continues in production 2011.
Contractor: Emmen Aerospace Inc
Mount Pleasant, South Carolina.

Emmen Aerospace Super Swiper

Type: Surveillance mini-UAV.

Development: The Super Swiper is designed to be both affordable and of good performance. Applications include damage assessment, disaster relief and situational awareness.

Description: *Airframe:* Flying wing design, with fuselage and fin; tractor engine. No landing gear.

Mission payloads: Stabilised, retractable, pan and tilt daylight zoom camera, or IR; automatic video tracking and automatic target following available as options.

Guidance and control: Super Swiper can be flown manually, semi-manually or fully automatically, via a 1.3 GHz command uplink (L-, S- or C-band transmitter), by a single operator. It is compatible with the L-3 Communications Remotely Operated Video Enhancement Receiver (ROVER).

Transportation: Man-portable.
Launch: Hand launch.
Recovery: Belly landing.
Power plant: Electric motor (rating not stated), with rechargeable batteries; three-blade propeller.

Super Swiper

Dimensions, External	
Overall, length	1.88 m (6 ft 2 in)
Wings, wing span	1.93 m (6 ft 4 in)
Weights and Loadings	
Weight, Max launch weight	4.08 kg (8.00 lb)
Payload	
Max payload	
standard	1.05 kg (2.00 lb)
droppable	0.54 kg (1.00 lb)
Performance	
Speed	
Max level speed	61 kt (113 km/h; 70 mph)
Loitering speed	26 kt (48 km/h; 30 mph)
Radius of operation, mission	8.7 n miles (16 km; 10 miles) (est)
Endurance	
standard	1 hr 30 min (est)
with droppable payload	45 min
Power plant	1 × electric motor

Status: In production and available. Emmen report 'several' systems in use with US armed forces; deployed in Iraq in 2009.

Contractor: Emmen Aerospace Inc
Mount Pleasant, South Carolina.

Emmen Aerospace Swiper

Type: Lightweight surveillance UAV.

Development: The Swiper is designed to be both affordable and expendable. Applications include damage assessment, disaster relief and situational awareness.

Description: *Airframe:* Parasol wing; pod and boom fuselage; pusher engine; conventional tail surfaces. No landing gear.

Mission payloads: Stabilised pan-and-tilt daylight or low-light colour camera standard; fixed IR, auto-video tracking and auto-target following optional.

Guidance and control: Depending upon version and price, Swiper can be flown manually, semi-manually or automatically via an L-band command uplink by a single operator. It is compatible with the L-3 Remotely Operated Video Enhancement Receiver (ROVER).

Transportation: Man-portable.
Launch: Hand launch.
Recovery: Belly landing.
Power plant: Electric motor with rechargeable batteries driving a two-blade propeller.

Swiper

Dimensions, External	
Overall, length	0.86 m (2 ft 9¾ in)
Wings, wing span	1.07 m (3 ft 6¼ in)
Weights and Loadings	
Weight, Max launch weight	1.5 kg (3.00 lb)
Performance	
Speed	
Max level speed	39 kt (72 km/h; 45 mph)
Loitering speed	13 kt (24 km/h; 15 mph)
Radius of operation, mission	1.7 n miles (3 km; 2 miles)
Endurance	30 min (est)
Power plant	1 × electric motor

Status: In production and service. The manufacturer reports that several dozen are in service with US Armed Forces and that the Swiper has been deployed in Iraq and Afghanistan. In April 2007, the Swiper was flown during exercise 'Atlantic Strike 5' at Avon Park, Florida, where it was operated by joint terminal attack controllers (JTACs) from the US Army, Air Force, Marine Corps and Canadian Air Force. Swiper remained available and in production in 2011, alongside Super Swiper and Condor (see separate entries). More than 150 systems of the three types have been produced.

Contractor: Emmen Aerospace Inc
Mount Pleasant, South Carolina.

Super Swiper air vehicle *(Emmen)*

Hand-launching a Swiper *(Emmen)*

Super Swiper hand launch *(Emmen)*

The Swiper lightweight SUAV *(Emmen)*

GA-ASI Altair

Type: Scientific research platform.

Development: Altair is the civil version of the company's Predator B, having been completed as the third prototype (B-003) of that aircraft. It was co-funded from January 2000 by NASA's ERAST (now Earth Science Enterprise) programme to explore the requirements for a conventionally fuelled aircraft for high-altitude atmospheric and scientific missions. Design parameters included a mission endurance of 24 to 48 hours at a primary altitude range of 12,200 to 19,800 m (40,000 to 65,000 ft) with a payload of at least 300 kg (661 lb), and to achieve the required altitude Altair differs from the standard Predator B in having a 3.05 m (10 ft) extension to each wing. The aircraft, alternatively known as Predator B-ER (for extended range), was also required to demonstrate OTH command and control beyond LOS radio capability via a satellite link; 'sense and avoid' operation in unrestricted airspace; and the ability to communicate with FAA air traffic controllers.

Altair flew for the first time on 9 June 2003 and has since been used for a series of environmental science missions and military technology demonstrations. On 25 August 2005 it was awarded the FAA's first-ever Experimental category airworthiness certificate issued for a UAS, having received civil registration N8172V six days earlier.

Description: *Airframe:* Fuselage and power plant as for Predator B but with enlarged wings and tail surfaces.

Mission payloads: Sensors tested successfully in the April/May 2004 Mariner Demonstrator configuration included Raytheon's AN/AAS-52 Multi-Spectral Targeting System (MTS) and SeaVue maritime surveillance radar. For Canadian Forces trials later that year, these were replaced by an L-3 Wescam MX-20 EO/IR turret and Telephonics APS-143 radar respectively. An ocean colour sensor, gas chromatograph and ozone photometer, and a Cirrus digital mapping camera were installed for the NOAA trials in 2005. A fire mapping infrared pod was developed with NASA in 2006. The Altair has six underwing stations and one on the fuselage centreline available for the carriage of external payloads.

Altair wearing its new N8172V civil registration *(NASA/Carla Thomas)*

Under-view of the NOAA colour sensor and vertical sounder ports *(NASA/Tom Tschida)*

Altair N8172V in operation for NOAA in 2005 *(NASA/Carla Thomas)*

Guidance and control: Generally as described for Predator B.
Launch: Conventional and automatic wheeled take-off.
Recovery: Conventional and automatic wheeled landing.
Power plant: One 708 kw (950 shp) Honeywell TPE331-10T turboprop, flat rated at 522 kw (700 shp) for max continuous operation; McCauley three-blade variable-pitch aluminium pusher propeller. See under Weights for fuel details.

Altair

Dimensions, External	
Overall	
length	10.82 m (35 ft 6 in)
height	3.96 m (13 ft 0 in)
Fuselage	
width, max	1.13 m (3 ft 8½ in)
Wings	
wing span	26.21 m (86 ft 0 in)
wing chord	
at root	1.65 m (5 ft 5 in)
at tip	0.73 m (2 ft 4¾ in)
wing aspect ratio	24.4
Tailplane, tailplane span	6.83 m (22 ft 5 in)
Engines, propeller diameter	2.90 m (9 ft 6¼ in)
Wheels	
wheelbase	3.35 m (11 ft 0 in)
wheel track	3.66 m (12 ft 0 in)
Dimensions, Internal	
Payload bay	
volume, total	1.30 m^3 (46 cu ft)
Areas	
Wings, Gross wing area	28.15 m^2 (303.0 sq ft)
Weights and Loadings	
Weight, Max T-O weight	3,265 kg (7,198 lb)
Fuel weight, Max fuel weight	1,361 kg (3,000 lb)
Payload	
Max payload	
internal	299 kg (659 lb)
external	1,363 kg (3,004 lb)
Performance	
Climb	
Rate of climb, max, at S/L	762 m/min (2,500 ft/min)
Altitude, Operating altitude	15,850 m (52,000 ft)
Speed	
Max level speed	225 kt (417 km/h; 259 mph)
Cruising speed	170 kt (315 km/h; 196 mph)
Range, max	5,000 n miles (9,260 km; 5,753 miles)
Endurance	30 hr (est)
Power plant	1 × turboprop

NOAA ocean colour sensor installation and satcom antenna *(NASA/Tom Tschida)*

Status: In service. In addition to its roles for NASA, Altair has been viewed by GA-ASI as a likely platform for maritime patrol missions, and during the first half of 2004 was reconfigured as a demonstrator for the company's Mariner variant of Predator B, proposed to the US Navy as the UAV constituent for that service's Broad Area Maritime Surveillance (BAMS) programme. In this configuration it carried a Raytheon SeaVue multimode maritime radar, housed in a company-designed ventral pod; a Raytheon MTS-B in an undernose turret; and C-band LOS and Ku-band satcom datalinks. Its maiden flight in this configuration was made on 22 April 2004.

In July 2004 it flew forest fire patrol missions over the Alaskan interior, where its infra-red sensor successfully located a previously undetected outbreak. Later that month it flew contract missions for the US Coast Guard over the Bering Sea from King Salmon air base, Alaska, searching for foreign fishing boats operating illegally in US waters. It was then loaned to the Canadian Forces in August 2004 for participation in exercise ALIX (Atlantic Littoral ISR Experiment), during which it received the Canadian military designation CU-163 and was equipped with a Telephonics maritime surveillance radar and an L-3 Wescam turret.

Between April and November 2005 Altair was deployed on behalf of the US National Oceanic and Atmospheric Administration (NOAA) off the coast of southern California with sensors designed to measure ocean colour, atmospheric composition and temperature, and surface imaging. During 2006, NASA's Ames Research Center evaluated Altair further in the forest fire detection role in the western United States, in a collaborative venture with the US Forest Service known as WRAP (Wildfire Research and Applications Partnership under a lease arrangement). The centreline-mounted fire pod was used to map the Esperanza, California, fire in December 2006. NASA bought their own Predator B in November 2006, which has been used for science and fire missions instead of Altair. Subsequently, Altair has been used for flight training at GA-ASI's Gray Butte Flight Operations Facility.

Contractor: General Atomics Aeronautical Systems Inc
San Diego, California.

GA-ASI Gnat

Type: Medium-altitude, multimission long-endurance UAV.

Development: Except for a low-mounted wing configuration, the Gnat 750 has a basically similar airframe to its Amber 1 predecessor. It was designed in 1988 and first flight-tested in 1989. Initial production started in October that year. In 1992 the Gnat 750 became the first UAV to be flown remotely via satellite link, and in July of that year it successfully made a continuous flight of more than 40 hours from the company's El Mirage flight test facility in Adelanto, California. Successful handover of a Gnat 750 from one GCS to another, without loss of control, was demonstrated in California in March 1997. In a week-long exercise in November 1997, a TCS configured for and installed aboard USS *Tarawa* controlled a DoD Gnat 750 from a surface ship.

Upgrades to this version and to the I-Gnat (or improved Gnat) have included increasing wing hardpoint capacities from 68 kg (150 lb) to 113 kg (250 lb); incorporating the Lynx high-resolution SAR; a turbocharged engine and glycol-based de-icing.

A US Army I-Gnat ER air vehicle, AI-001, had flown over 8,000 hours by the end of August 2007, logging over 620 flights during three and a half years.

Description: *Airframe*: Low-wing monoplane with a slender fuselage, inverted V tail, rear-mounted engine, and retractable tricycle undercarriage with steerable nosewheel, differential mainwheel braking and anti-skid capability. The airframe is made of carbon/epoxy composites, and is stressed for 6 *g* manoeuvres. Payloads are accommodated in the nose, the contours of which vary according to the particular sensor carried. Five hardpoints available for external stores (I-Gnat): one on centreline and two under each wing; 35 A 28 V DC payload power and RS-422 serial bus for payload control at each station.

Mission payloads: Mission payloads can be carried for surveillance (radar, stabilised FLIR, TV and LLTV), reconnaissance (IR linescanner),

I-Gnat medium-altitude endurance UAV

I-Gnat flying with undernose Lynx radar *(GA-ASI)*

First flight of the US Army I-Gnat ER *(GA-ASI)*

ESM, direction-finding, radio and datalink relay, NBC detection, air-delivered payloads, or custom designs. Onboard power supply 3.5 kW.

Guidance and control: The DGCS 87 digital ground control station, installed in an S-280 shelter, is fully programmable and configurable to a variety of UAVs, trackers and datalinks, and can control multiple UAVs and payloads. It has four displays (one head-up, two head-down and a map display) with touchscreens, and interfaces with a DFCS 50 digital flight control system on board the UAV. A portable GCS, and a modular GCS based on the portable GCS, are optional variants. Navigation is automatic, with GPS/INS options; the datalink operates in C-band.

System composition: Up to eight air vehicles, one GCS, plus ground support equipment.

Launch: Wheeled take-off from suitable terrain. Automatic take-off introduced in 2003.

Recovery: Wheeled landing where terrain is suitable.

Variants: *Gnat 750*: Initial version, in service. *Detailed description applies mainly to Gnat 750 except where indicated.*

I-Gnat: Improved US military and international version, with increased dimensions, provision for external hardpoints, more powerful turbocharged engine, radar altimeter, capability for air-to-air datalink relay payload, and a more versatile mission planning potential. Otherwise generally similar to Gnat 750. Flew to 9,295 m (30,500 ft) on 19 October 1998; later that week remained airborne for 38 hours and landed with approximately 10 hours' fuel remaining.

Gnat-XP: Specially equipped version of I-Gnat for 'a three-symbol customer' (alias the CIA). Further details classified.

I-Gnat ER/Sky Warrior Alpha: Multimission ISR version for US Army. Stretched fuselage and increased wing span; power plant as for I-Gnat; payload and fuel capacities increased; LOS-only datalink. It should be noted that US Army I-Gnats adopt the Predator military designation **RQ-1L** and are Hellfire (AGM-114K) capable. Later dubbed Sky Warrior Alpha by US Army. See note below.

Gnat 750, I-Gnat ER, I-Gnat

(**A**: I-Gnat, Rotax 912 **B**: I-Gnat, Rotax 914)

Dimensions, External
  Overall
    length
      Gnat 750 .... 5.33 m (17 ft 5¾ in)
      I-Gnat .... 5.76 m (18 ft 10¾ in)
      I-Gnat ER .... 8.13 m (26 ft 8 in)
  Wings
    wing span
      Gnat 750 .... 10.76 m (35 ft 3½ in)
      I-Gnat .... 12.80 m (42 ft 0 in)
      I-Gnat ER .... 16.76 m (54 ft 11¾ in)
  Engines
    propeller diameter
      Gnat 750 .... 1.52 m (4 ft 11¾ in)
      Gnat 750 .... 1.52 m (4 ft 11¾ in)
      (A) .... 1.83 m (6 ft 0 in)
      (B) .... 1.73 m (5 ft 8 in)
Weights and Loadings
  Weight
    Weight empty
      Gnat 750 .... 254 kg (559 lb)
      I-Gnat .... 385 kg (848 lb)
      I-Gnat ER .... 513 kg (1,130 lb)
    Max T-O weight
      Gnat 750 .... 511 kg (1,126 lb)
      Gnat 750 .... 511 kg (1,126 lb)
      (A) .... 703 kg (1,549 lb)
      (B) .... 748 kg (1,649 lb)
      I-Gnat ER .... 1,043 kg (2,299 lb)

Gnat 750, I-Gnat ER, I-Gnat

Fuel weight	
Max fuel weight	
Gnat 750	193 kg (425 lb)
I-Gnat	227 kg (500 lb)
I-Gnat ER	283 kg (623 lb)
Payload	
Max payload	
internal, Gnat 750	63.5 kg (139 lb)
internal, I-Gnat	91 kg (200 lb)
internal, I-Gnat ER	204 kg (449 lb)
underslung, each wing, I-Gnat, I-Gnat ER	45.5 kg (100 lb)
underslung, fuselage, I-Gnat, I-Gnat ER	68 kg (149 lb)
Performance	
T-O	
T-O run	
(A)	366 m (1,201 ft)
(B)	320 m (1,050 ft)
Climb	
Rate of climb	
max, at S/L, Gnat 750	335 m/min (1,099 ft/min)
max, at S/L (A)	244 m/min (800 ft/min)
max, at S/L (B)	396 m/min (1,299 ft/min)
Altitude	
Service ceiling	
Gnat 750, I-Gnat ER	7,620 m (25,000 ft)
Gnat 750, I-Gnat ER	7,620 m (25,000 ft)
(B)	9,145 m (30,000 ft)
Speed	
Max level speed	
Gnat 750	140 kt (259 km/h; 161 mph)
I-Gnat	125 kt (232 km/h; 144 mph) (est)
I-Gnat ER	120 kt (222 km/h; 138 mph)
Cruising speed	
long range, Gnat 750	46 kt (85 km/h; 53 mph)
long range, I-Gnat ER	73 kt (135 km/h; 84 mph)
Stalling speed, Gnat 750	36 kt (67 km/h; 42 mph)
g limits	+6
Range, LOS, Gnat 750, I-Gnat ER	108 n miles (200 km; 124 miles)
Radius of operation	
max, Gnat 750, I-Gnat	1,500 n miles (2,778 km; 1,726 miles)
Endurance, at 1,525 m (5,000 ft)	40 hr (est)
Time on station, at 1,080 n miles (2,000 km; 1,243 miles) radius, Gnat 750	12 hr
Landing	
Landing run	
(A)	305 m (1,001 ft)
(B)	320 m (1,050 ft)
I-Gnat ER	518 m (1,700 ft)

Status: In production (2008) and service. Has been used in drug interdiction operations. Ordered by Turkish Army (one Gnat 750 system) in 1992. In 1993 the US Army awarded a USD1.4 million contract to a Questech/General Atomics/Marconi team for a joint precision strike demonstration using a telecom satellite to relay piloting and imagery transmission commands to and from a Gnat 750. The one hour demonstration took place on 2 December 1993, encouraging a further development of the design which eventually became the Predator.

The Gnat 750 is deployed with the US government under the Interim Medium Altitude Endurance (I-MAE, alias Tier I) programme. Under this approximately USD6 million programme, contracted by the CIA, two modified Gnat 750s equipped with off-the-shelf EO and IR sensors were acquired. Reconnaissance overflights of Bosnia from a base in Albania were conducted for a brief period in February 1994 before being halted due to bad weather and datalink problems. They were resumed in late 1994, from bases in Croatia, after the UAVs had been refitted with a new Sigint sensor and a high-resolution Mitsubishi thermal imager.

Flight testing of General Atomics Aeronautical Systems' AN/APY-8 Lynx synthetic aperture radar in an I-Gnat began in March 1999, following earlier trials in a Twin Otter testbed.

Customers: *Gnat 750*: Turkey (Army, six AVs and two GCSs); US (Army, CIA, Department of Environment and other agencies).

I-Gnat: More than 12 delivered to Turkey and two unnamed customers in 1998 and 1999.

I-Gnat ER: US Army ordered one three-AV system in May 2003 (contract value USD7.99 million), to be delivered in May 2004 and used by Aviation and Missile Command at Redstone Arsenal, Alabama, for Objective Force transformation CONOPS. From March 2004, these aircraft were on deployment in Iraq, supporting the war on terrorism, with operations and maintenance provided by GA-ASI personnel. US Army placed a further contract on 1 February 2005, valued at USD4.22 million, for two additional AVs, three 'legacy hardware' modification kits and the development and integration of TALS. The Army subsequently ordered two additional I-Gnat ERs and then renamed future I-Gnat ERs Sky Warrior Alfa and ordered an additional 12 AVs to complete delivery in 2007.

Contractor: General Atomics Aeronautical Systems Inc
San Diego, California.

GA-ASI MQ-1B and RQ-1A Predator

Type: MALE armed surveillance UAV.

Development: *Note: RQ-1A and RQ-1B (now MQ-1B) designate the complete system. At first, individual subsystems were consistently referred to using other suffix letters, including RQ-1K and RQ-1L/MQ-1L for the air vehicles and RQ-1P and RQ-1Q/MQ-1Q for the ground stations. However, this practice was discontinued, probably in late 2006, when designations were standardised to RQ-1A for armed and MQ-1B for unarmed airborne subsystems, with MD designations being allocated to the ground stations.*

Development of the Predator began with on 7 January 1994 with the award of a USD31.7 million ACTD contract from DARPA for a MALE UAV, which developed as a highly advanced version of Gnat-750, making its maiden flight on 3 July of the same year. UHF satcom was introduced in 1994 for BLOS communications between the aircraft and GCS; this was upgraded to Ku-band satcom in June 1995, enabling significantly higher data rates and lower latencies. SAR was successfully integrated in March 1996, providing photographic quality imagery capable of 'seeing' and photographing targets, day or night, through inclement weather. The Thirty-month ACTD was concluded 30 June 1996.

Predator designated RQ-1A by the USAF in 1997; the first LRIP contract was awarded on 19 August 1997. In May 1998, a more powerful turbocharged engine and wing de-icing system was installed as part of the USAF Block 10 upgrade contract, resulting in redesignation as RQ-1B. Laser designators were added in 1999, to provide targeting data to manned strike aircraft during Operation 'Allied Force' in Kosovo. In February 2001, Predator became the first UAV to be weaponised, destroying a ground target with a live Hellfire missile during flight tests; its designation transitioned to MQ-1B in 2002 to reflect armed capability.

The first FRP contract was awarded in August 2003 and in February 2005, two years ahead of schedule, Predator became first ACTD unmanned aircraft programme to graduate into an operational system.

Several Predators have been used for demonstrations of other techniques and technologies. In December 1995, a GA-ASI Predator took part in the US Navy's Composite Training Unit Exercise (COMPTUEX), successfully demonstrating its ability to provide reconnaissance support to USN Carrier Battle Groups. In June 1996, GA-ASI participated in a US Navy demonstration in which a Predator was flown and controlled from a partially submerged nuclear-powered submarine, the USS *Chicago*. During this exercise, the submarine's onboard control station received real-time surveillance footage from the UAV. In January 2001, Predator successfully performed its first communications relay in support of the joint US/UK Extendor project. During this exercise, the aircraft passed real-time operational data from a ground-based forward air controller to an RAF Jaguar attack aircraft. In August 2002, a Predator successfully air-launched a US Navy FINDER mini-UAV — the first-ever air launch of one UAV from another while in flight. In November 2003, two US Navy CIRPAS Predators completed a six-flight, four-day deployment with the US Coast Guard to the King Salmon area of Alaska, where they carried out surveillance and communications relay missions in collaboration with the cutter *Hickory*. In May 2006, the US Federal Aviation Administration (FAA)

Hellfire-armed MQ-1B Predator *(USAF)*

The last MQ-1 for USAF delivered in March 2011 *(GA-ASI)*

Unmanned aerial vehicles > United States > GA-ASI MQ-1B and RQ-1A Predator

Preflight checks before an 'Enduring Freedom' mission *(USAF/Tech Sgt Scott Reed)*

issued a Certificate of Authorization (COA) to the Predator, allowing it to be used within US civilian airspace to aid in disaster recovery operations.

Description: *Airframe:* Low-wing monoplane with slender fuselage, high aspect ratio wing and inverted-V tail. Advanced low-speed aerodynamic configuration and computer-designed low Reynolds number aerofoils provide high aerodynamic efficiency. Inverted-V tail is a unique feature, providing propeller protection and keeping tail control surfaces clear of wing turbulence. Advanced graphite/epoxy construction results in very sturdy airframe (more than 5 g ultimate capability at normal T-O weight) and very light structure. Single stores hardpoint under each wing, capable of carrying 68 kg (150 lb) stores load. Fuselage contains majority of aircraft systems. Tricycle landing gear retracts into fuselage, reducing drag and clearing FoV for optical sensors.

Mission payloads: Predator is designed to carry 204 kg (450 lb) of internal payload with maximum fuel. Sensor imagery is downlinked to a GA-ASI GCS. Numerous mission payloads have been integrated, including EO/IR, SAR, classified sigint systems, communications relay, specialised/scientific sensors and weapons. Principal operational systems are summarised below; further details can be found in Issue 32 and earlier editions.

Sensors: Initial fit of L-3 Wescam 14TS Skyball in undernose ball turret. Northrop Grumman AN/ZPQ-1 Tactical Endurance Synthetic Aperture Radar (TESAR) was added from early 1996. Both subsequently replaced by, respectively, Raytheon MTS-A multisensor turret (from 2002) and/or GA-ASI Lynx synthetic aperture radar, both of which are fully described in the Payloads section.

Weapons: Primary armament of MQ-1B Predator is the AGM-114K Hellfire C laser-guided anti-tank missile (one under each wing). USAF has also evaluated the Viper Strike PGM version of the Northrop Grumman Bat and the Air-To-Air Stinger (ATAS) air-launched version of Raytheon's FIM-92 missile.

Guidance and control: Predator missions are conducted from the same GA-ASI mobile/portable GCS that controls the company's MQ-9 Reaper/Predator B family of UAS; a full description of this is given within the latter entry.

System composition: A basic Predator system comprises several air vehicles (one to four, depending on concept of operations), their payloads, one GCS, communications and support equipment, spare parts and support.

Launch: Conventional wheeled take-off.

Recovery: Conventional wheeled landing.

Powerplant: One 78.3 kW (105 hp) Rotax 914 four-cylinder four-stroke turbocharged engine; two-blade variable-pitch pusher propeller.

Variants: *RQ-1A Predator:* Initial system for US Air Force.
RQ-1B Predator: USAF Block 10 upgrade version.

Inside a Predator ground control station at Creech AFB *(USAF)*

A Predator of the Italian Air Force *(Italian Air Force)*

MQ-1B Predator: Designation applied to USAF Predators equipped to carry weapons.

MQ-1B

Dimensions, External	
Overall	
length	8.23 m (27 ft 0 in)
height	2.06 m (6 ft 9 in)
Fuselage	
height	
front	1.07 m (3 ft 6¼ in)
rear	0.66 m (2 ft 2 in)
width	
front	1.09 m (3 ft 7 in)
rear	0.74 m (2 ft 5¼ in)
Wings, wing span	16.76 m (54 ft 11¾ in)
Tailplane, tailplane span	4.37 m (14 ft 4 in)
Engines, propeller diameter	1.83 m (6 ft 0 in)
Wheels	
wheelbase	2.77 m (9 ft 1 in)
wheel track	3.10 m (10 ft 2 in)
Weights and Loadings	
Weight	
Weight empty	567 kg (1,250 lb)
Max T-O weight	1,020 kg (2,248 lb)
Fuel weight, Max fuel weight	249.5 kg (550 lb)
Payload	
Max payload	
internal	204 kg (449 lb)
external	136 kg (299 lb)
Performance	
T-O	
T-O run	1,829 m (6,000 ft)
Altitude, Service ceiling	7,620 m (25,000 ft)
Speed	
Max level speed	120 kt (222 km/h; 138 mph)
Loitering speed	73 kt (135 km/h; 84 mph)
Range	3,500 n miles (6,482 km; 4,027 miles) (est)
Endurance	
typical	24 hr
max	40 hr

Status: Production for USAF is complete. Predators passed the 100,000 flight hours milestone on 27 September 2004, of which nearly 70,000 hours were in combat theatre operations. As of June 2007 the Predator series of aircraft, including the larger B versions, surpassed 300,000 flight hours. By 12 August 2007 the Predator A fleet alone had achieved more than 300,000 hours, with this milestone being passed by air vehicle P-137 on an armed reconnaissance mission over Iraq. The Predator UAS family was accumulating approximately 10,000 flying hours each month (most on A type vehicles) during 2007. As of February 2011, Predator family aircraft had accumulated 1.3 million hours and monthly flying hours for Predators were running between 33,000-35,000.

Balkans: The first operational deployment of Predator overseas took place from July to October 1995. Based in Gjader, Albania, it supported NATO air strikes against Serbian forces. This marked the first time the aircraft provided real-time targeting data and BDA. From March 1996 to September 1997, Predator deployed to Taszar, Hungary, where it flew reconnaissance missions in support of another NATO operation. In March 1999, Predator became the first UAV to play a major role in a full-scale NATO conflict when it supported Operation 'Allied Force' to remove Serbian forces from Kosovo. During this four-month air campaign, Predator flew over 600 sorties over Kosovo, performed real-time surveillance of Serbian forces, and provided critical time-sensitive targeting data and BDA.

Middle East: In December 2002, while performing reconnaissance over the Iraqi no-fly zone in support of the United Nations' Operation 'Southern Watch', a Predator engaged an Iraqi MiG-25 with a Stinger air-to-air missile, marking the first combat engagement between a UAV and a manned aircraft. On 3 November 2002, another Predator, reportedly flying from Djibouti, used its Hellfire missiles to destroy a civilian vehicle containing suspected al-Qaeda terrorists some 161 km (100 miles) east of the Yemeni capital of Sana'a. US Air Force Predators deployed to Iraq in March 2003 in support of Operation 'Iraqi Freedom', serving with the

An 'Iraqi Freedom' MQ-1B of the 46th Expeditionary Reconnaissance Squadron lands at Tallil AB *(USAF/Staff Sgt Suzanne Jenkins)*
0572234

Predator with Lynx SAR and 14TS EO/IR turret *(GA-ASI)* 1127309

46th Expeditionary Reconnaissance Squadron at Balad AB. Italian Air Force Predators deployed to southern Iraq in January 2005.

Asia: In September 2001, Predator deployed to Afghanistan to provide long-endurance ISR for Operation 'Enduring Freedom' following the 11 September 2001 terrorist attacks in the US. The following month, it became the first UAV ever to fire offensive weapons in action, targeting and destroying enemy forces in the theatre. According to one unconfirmed source, some 40 operational Hellfire launches had been made by the end of that year. From December 2001, Predators were also tasked with relaying real-time video and targeting information to the crews of AC-130 Spectre gunships operating against Taliban and al-Qaeda forces.

Customers: *US Air Force:* Allocated from 3 September 1996 to newly formed 11th Reconnaissance Squadron (RS) (57th Reconnaissance Wing) of USAF Air Combat Command, with headquarters at Nellis AFB, Nevada; operational home base is Creech AFB at Indian Springs, Nevada. Support unit is 645 Materiel Squadron, AFMC. USAF inherited five 'residual' systems from the ACTD programme. First two aircraft delivered to 11th RS in November 1996. The 11th RS now has the primary task of training Predator crews for ISR missions.

Second Predator squadron (15th RS) activated 1 August 1997 and received first two aircraft in May 1998. Third unit (17th RS) activated 8 March 2002. As of late 2006, according to Jane's sources, sub-units of the 15th RS were engaged in forward deployments to Iraq (46th Expeditionary Reconnaissance Squadron at Balad), Afghanistan (62nd ERS at Kandahar) and elsewhere; the 17th RS was providing "special capabilities for classified missions"; and a new unit at Creech was the 3rd Special Operations Squadron, originally activated 28 October 2005 by SOCOM at Hurlburt Field, Florida. In August 2006, the AFSOC's 11th Intelligence Squadron was reactivated at Hurlburt to exploit and disseminate the intelligence gathered by the 3rd SOS.

Also announced are the 30th Reconnaissance Squadron, a test unit activated in August 2005 and based at the Tonopah Test Range in Nevada; and the Creech-based 19th Attack Squadron and 79th RS. According to *IHS Jane's* sources, other USAF units at which Predators serve, or have served, include the 26th Weapons School at Nellis, 556th Test and Evaluation Squadron at Creech, and a detachment of the 452nd Flight Test Squadron at Gray Butte.

In November 2006, California Air National Guard's 196th RS (163rd Reconnaissance Wing) became the first US National Guard squadron to stand up 24/7 Predator operations. In January 2007, the North Dakota ANG activated another Predator squadron, the Fargo-based 178th RS (119th RW). In August 2007, the Arizona ANG formally activated the 214th Reconnaissance Squadron, based at Davis-Monthan Air Base in Tucson. Fourth ANG unit is the 111th RS, based at Ellington Field, Texas. The 163rd, 119th, and 214th RWs are all charged with remotely operating Predators deployed in Southwest Asia.

As of 30 June 2009, 220 RQ/MQ-1 Predators, including three for the US Navy, had been delivered to the USAF. The handover of the final MQ-1 to USAF, tail number 268, was on 4 March 2011.

US Navy: TCS Program Office funded acquisition of two RQ-1A Predators (96-3030 and 96-3035) from USAF in the second quarter of 1999, for use by the Center for Interdisciplinary Remotely Piloted Aircraft Studies (CIRPAS) at the Naval Postgraduate School at Marina, California. A third Predator A was later acquired from USAF.

Italian Air Force: An Italian Air Force USD55 million contract for five Predators was awarded in August 2001. Operator is the 28° Gruppo Velivoli Teleguidati (remotely piloted aircraft squadron) of the Italian Air Force's 32° Stormo at Foggia-Amendola, which was activated on 1 March 2002. The first Predator for Italy (serial 32-01) made its maiden flight (in the US) on 31 January 2004; deliveries followed later that year, IOC was declared in December 2004, and the unit deployed to Tallil AB in southern Iraq in January 2005, where it formed part of the 6th Autonomous Unit of the Italian Combined Joint Task Force in that country. It had completed 1,000 in-theatre flight hours by early 2006. Italian Predators were upgraded with a satcom datalink, SAR payload and an improved EO/IR sensor during 2006. As of early August 2009, Italian Air Force Predators had accumulated 5,000 flying hours, including about 1,600 operational hours in Iraq and 3,200 in Afghanistan.

UK Royal Air Force: In early 2004, the RAF formed No. 1115 Flight to train on operating the Predator, this unit being 'embedded' within USAF's 15th RS at Indian Springs. In a statement to the UK Parliament on 20 April 2006, Armed Forces Minister Adam Ingram confirmed that RAF personnel activities had included commanding six Predator missile strike sorties, five in Iraq (the first of which is believed to have been in late 2004) and one in Afghanistan.

Contractor: General Atomics Aeronautical Systems Inc
Poway, California.

GA-ASI MQ-1C Sky Warrior

Type: Armed reconnaissance UAV.

Development: This variant of the Predator was developed to bid for the US Army's Extended Range MultiPurpose (ERMP) UAV programme, launched in 2002 to find a successor to the RQ/MQ-5 Hunter. Performance requirements included a 300 km (186 mile) range, 24-hour endurance, and the ability to loiter for up to 30 hours at 7,620 m (25,000 ft). Prototyping began in July 2004, and in the following month, with AAI Corporation and Sparta, GA-ASI formed Team Warrior to develop its ERMP candidate. This, together with Northrop Grumman's Hunter II, was down-selected as a semi-finalist in January 2005. A System Capabilities Demonstration at Fort Huachuca, Arizona, was completed in March 2005, with one of the two Warrior prototypes being fitted with a 101 kW (135 hp) Thielert Centurion 1.7 heavy-fuel engine (first flight October 2004) and the other with a gasoline engine. Trials included operation by the US Army's AAI 'One System' GCS (OSGCS), as used for the RQ-7 Shadow and other UAVs; transition from LOS datalink to OTH satcom link control; automatic as well as pilot-controlled landings; and flight with four Hellfire missiles.

The Team Warrior candidate aircraft system won the competition and a US Army four-year, USD214.3 million System Development and Demonstration (SDD) contract was awarded on 8 August 2005; designation YMQ-1C and Army name Sky Warrior were allocated in late 2006. The first Sky Warrior Block 1 for the US Army's ERMP programme flew on 31 March 2008. The Sky Warrior Block 1 started tests with the Lynx Block 30 synthetic aperture radar in March 2009.

Description: *Airframe:*

Mission payloads: Sensors: Raytheon AN/DAS-2 EO/IR ELRF/LD ; GA-ASI AN/DPY-1 Lynx II SAR/GMTI radar.

Avionics: AN/ARC-201 and AN/ARC-231 SINCGARS radio comms; AN/APX-118 Mk 12A Mode S IFF.

Weapons: Four AGM-114 Hellfire air-to-surface missiles.

Guidance and control: Preprogrammed, with option for mission change during flight. 'One System' GCS, interoperable with that for RQ-7 Shadow and able to operate multiple air vehicles simultaneously. Operable with or without satcom datalinks.

Main differences between Sky Warrior and Predator are a larger wing, more bulbous nose dome and Diesel propulsion *(GA-ASI)*
1184588

236 Unmanned aerial vehicles > United States > **GA-ASI MQ-1C Sky Warrior – GA-ASI MQ-9 Reaper, Predator B and Mariner**

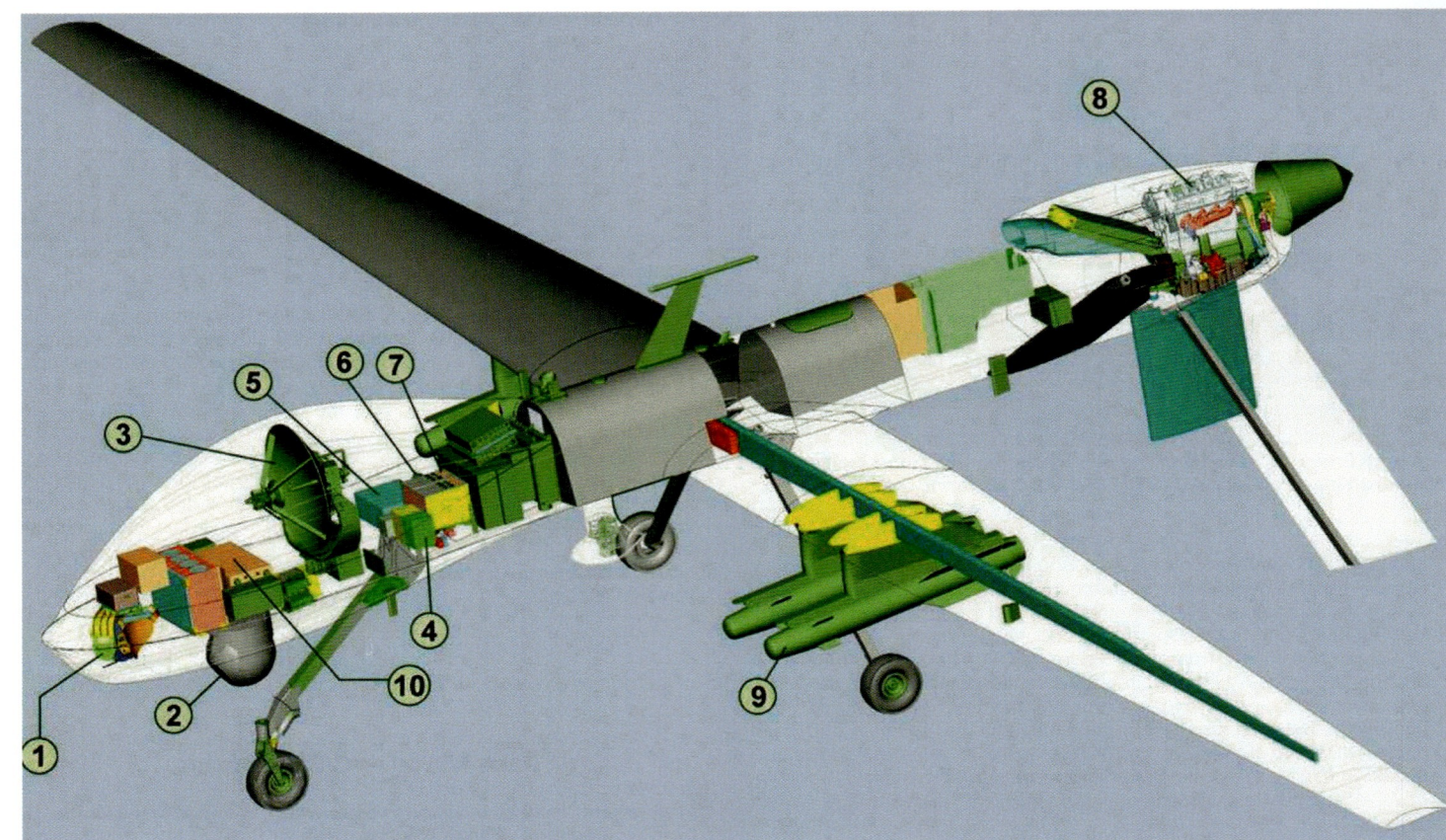

Principal features of the MQ-1C Sky Warrior: (1) DPY-1 Lynx II radar, (2) AAS-52(V) EO sensor, (3) satcom antenna, (4) SINCGARS comms, (5) EPLRS, (6) ARC-210 radio, (7) APX-119 IFF, (8) Centurion 1.7 engine, (9) Hellfire missiles and (10) Hellfire electronics *(US Army)* 1184589

System composition: Twelve air vehicles and their payloads; five GCSs; five GDTs; two portable GCSs; two portable GDTs; one satcom GDT; four TALS auto-land systems and other associated ground support equipment.
Launch: Conventional, automatic wheeled take-off.
Recovery: Conventional, automatic wheeled landing.
Power plant: One 101 kW (135 hp) Thielert Centurion 1.7 Diesel engine; three-blade pusher propeller.
Variants: *Sky Warrior Alpha:* Interim ER/MP version (upgraded I-Gnat ER); EO/IR, satcom link, relay package, synthetic aperture radar and weapons capable. Gross weight 1,066 kg (2,350 lb), ceiling 7,620 m (25,000 ft), mission radius 67.5 n miles (125 km; 78 miles). Deployed to Iraq in November 2006.
Sky Warrior Block 0: Second interim ER/MP version, in SDD configuration, with enlarged wing, four Hellfire stations and Thielert engine. Gross weight 1,451 kg (3,000 lb), ceiling 8,840 m (29,000 ft), mission radius 189 n miles (350 km; 217 miles). Four ordered in USD11.7 million contract modification of 14 February 2007; scheduled for fielding by March 2008.
YMQ-1C Sky Warrior: Designation of SDD aircraft, 17 of which were due to begin delivery in December 2007. Representative of initial production version, with higher gross weight, larger sensor capacity, and de-icing systems.

Sky Warrior Block 0, Sky Warrior Alpha, YMQ-1C, Sky Warrior Block 1

Dimensions, External
  Overall
    length... 8.53 m (27 ft 11¾ in) (est)
    height... 2.21 m (7 ft 3 in) (est)
    Wings, wing span.................................... 17.07 m (56 ft 0 in) (est)

Sky Warrior can carry twice the weapon load of the MQ-1B Predator *(GA-ASI)* 1116445

Weights and Loadings
  Weight
    Max T-O weight
      Sky Warrior Block 0, Sky
        Warrior Block 1.. 1,451 kg (3,198 lb) (est)
      post-2010 (proposed), with
        120 kW (160 hp) engine and up
        to 8 Hellfires.. 1,633 kg (3,600 lb) (est)
  Payload
    Max payload
      internal... 136 kg (299 lb) (est)
      external.. 227 kg (500 lb) (est)
Performance
  Climb
    Rate of climb
      above 6,705 m (22,000 ft)........................ 250 m/min (820 ft/min) (est)
  Altitude
    Service ceiling
      Sky Warrior Alpha... 7,620 m (25,000 ft) (est)
      Sky Warrior Block 0, YMQ-1C .... 8,840 m (29,000 ft) (est)
  Speed, Max level speed.................... 150 kt (278 km/h; 173 mph) (est)
  Radius of operation
    mission, Sky Warrior Alpha...................... 67.5 n miles (125 km; 77 miles) (est)
    mission, Sky Warrior Block 0,
      YMQ-1C.. 189 n miles (350 km; 217 miles) (est)
  Endurance
    on-station, at 7,620 m (25,000 ft) at
      300 km (miles) from base... 24 hr (est)
    max.. 30 hr (est)

Status: The US Army has stated a requirement for 11 Sky Warrior systems (each of 12 air vehicles and five ground stations), planning to deploy them with one company assigned to each of 10 divisions and one to a training unit. Primary missions will be reconnaissance, surveillance, target acquisition, communications relay and attack. The system is planned to start being fielded in 2009. In advance of this, two Sky Warrior Block 0 air vehicles started operations in Iraq during April 2008, the first of seven Block 0 on order in a quick reaction programme.
Customers: US Army.
Contractor: General Atomics Aeronautical Systems Inc
San Diego, California.

GA-ASI MQ-9 Reaper, Predator B and Mariner

Type: Long-endurance, medium/high-altitude UAS.
Development: Development of the Predator B began as a GA-ASI-funded internal R&D effort in 1998 to develop the next generation of the Predator design. Goal for the programme was to take the proven Predator A technology into the jet age by developing an aircraft able to fly higher and faster and carry significantly more payload. Both military and commercial applications were envisaged, and GA-ASI designed two versions — a turboprop aircraft that could climb to 14,630 m (48,000 ft) and have 32

UK Royal Air Force Reaper ZZ201 at Kandahar air base
(IHS/Patrick Allen) 1298683

hours of endurance; and a turbojet that could reach 18,300 m (60,000 ft) and have an 18-hour endurance.

The first turboprop-powered Predator B (PB-001) made its maiden flight on 21 February 2001 and reached a milestone altitude of 15,850 m (52,000 ft) on 16 August. In October 2001 it was acquired by the USAF, by agreement with whom development of the turbojet version (whose PB-002 prototype was flight-ready at that time) was abandoned. This, too, received a turboprop engine and the two prototypes were delivered to the USAF, receiving the designation YMQ-9A in January 2002.

Description: *Airframe:* Of generally similar appearance and construction to RQ/MQ-1 Predator except for tail configuration, enlarged dimensions, general strengthening and power plant. Low-wing monoplane with slender fuselage, V-tail and ventral fin, built of advanced composites for robustness and low structural weight. Each tailplane incorporates two elevator control surfaces; ventral fin incorporates a full-depth rudder. Each wing has dual outboard ailerons and inboard flaps. Additionally, each wing incorporates three underwing hardpoints for external stores: inboard pair each stressed for a 680 kg (1,500 lb) load, two centreboard (340 kg; 750 lb each) and two outboard (68 kg; 150 lb each). Tricycle landing gear retracts into fuselage to reduce drag and clear the viewing field for the optical sensors.

Mission payloads: Predator B can carry multiple mission payloads, including EO/IR, SAR, maritime multimode radar, ESM, sigint, laser capabilities and weapons. Sensor imagery is downlinked to a GA-ASI GCS. Available onboard power was increased to 49 kW in 2007.

Sensors: Specific sensors linked to the MQ-9 Reaper have included GA-ASI AN/APY-8 Lynx SAR/GMTI radar; L-3 Wescam 14TS EO/IR or Raytheon AN/AAS-52 Multispectral Targeting System (MTS-B) (see Payloads section for details); plus Mode IV IFF and ESM. In March 2007, the USAF notified its intention to award GA-ASI a contract to integrate the Northrop Grumman Airborne Signal Intelligence Payload (ASIP) on both the MQ-9 Reaper and MQ-1 Predator.

Weapons: MQ-9 Reaper can carry up to 16 AGM-114P Hellfire missiles; has also been tested and cleared for carriage of two GBU-12 Paveway II laser-guided bombs and the GBU-38 500 lb variant of the Joint Direct Attack Munition (JDAM), and for mixed loads of the aforementioned weapons.

Guidance and control: High mobility/portability GA-ASI GCS allows direct real-time control of Predator/Reaper series aircraft, and can be located on any land base, in an aircraft or on board a ship. A typical GCS consists of two identical pilot/payload operator (PPO) workstations that incorporate control consoles and operator displays, allowing operators to control and monitor the aircraft, its payloads and aircraft subsystems. PPO workstations are redundant and interoperable with all GA-ASI aircraft; multifunction workstations (MFW) support data exploitation and other payloads.

Missions can be flown autonomously under the control of an onboard suite of redundant computers and sensors. These missions are preprogrammed by operators in the GCS and are initiated by an operator once the aircraft is airborne. For both real-time and autonomous missions, the pilot operator is responsible for landing the aircraft following mission completion.

Communication datalinks enable GCS operators to uplink control commands and downlink payload video and telemetry data to and from

NASA's Ikhana Predator B with an Autonomous Modular Sensor under the left wing configured for mapping of wildfires *(NASA)* 1328528

MQ-9 Reaper flying with a mix of Hellfire missiles and GBU-12 bombs *(GA-ASI)* 1184590

the aircraft. The datalink can be C-band LOS at ranges up to 150 n miles 278 km), or autonomously to the range limits of the aircraft. Alternatively, a Ku-band BLOS satcom datalink enables Predator series aircraft to be controlled from anywhere in the world: for example, USAF Predators and Reapers are routinely operated worldwide from GCSs located at a USAF base in Nevada, by beaming commands to the aircraft via satellite link. The aircraft then transmits images and information back to the GCS sensor operator for dissemination. GA-ASI also manufactures an RVT that provides real-time imagery directly from the aircraft to warfighters in the field, on ships or in the air.

In 2008-09, GA-ASI was developing an 'Advanced Cockpit' GCS that will be equipped with numerous new features designed to improve operator efficiency and increase situational awareness. It includes 3-D maps, intuitive touch-screen technology, ergonomic design and wraparound synthetic vision.

System composition: A basic Predator B system consists of several air vehicles (one to four depending on concept of operations), one GCS, communications and support equipment, spare parts and support. USAF states that "mission-specific equipment is employed in a 'plug and play' mission kit concept allowing specific aircraft and control station configurations to be tailored to fit mission needs".

Launch: Conventional wheeled take-off.

Recovery: Conventional wheeled landing.

Powerplant: YMQ-9A prototype One 671 kw (900 shp) Honeywell TPE331-10GD turboprop, derated to 599 kw (750 shp); McCauley three-blade, variable-pitch aluminium pusher propeller.

production Predator B One 701 kw (940 shp) Honeywell TPE331-10GD turboprop; propeller as for A.

See under Weights for fuel details.

Arrival of the first 42nd ATKS MQ-9 Reaper at Creech AFB, 13 March 2007 *(USAF/Senior Airman Larry E Reid)* 1151647

Predator Bs in use by the US Customs and Border Protection service *(GA-ASI)* 1409025

Unmanned aerial vehicles > United States > GA-ASI MQ-9 Reaper, Predator B and Mariner

'Falcon Prowl' Predator B 02-4003 with underwing DB110 pod *(Crown copyright/JUET)* 1151552

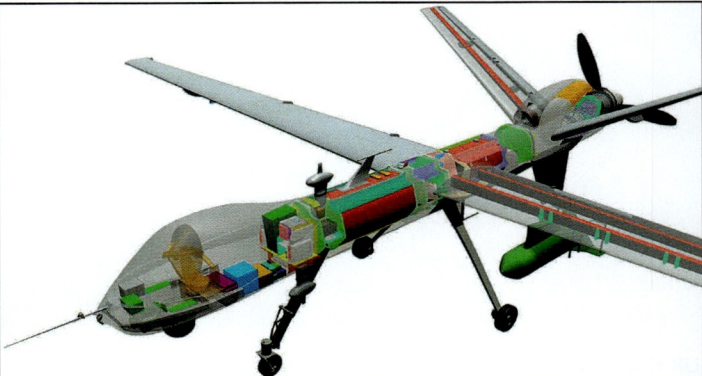

Predator B internal details *(GA-ASI)* 1127311

Underwing pods and wingtip antennas on this YQM-9A indicate carriage of an experimental sigint or ESM system, possibly the Raytheon Batfish *(US Army)* 1184506

Variants: **YMQ-9A**: Two Predator B prototypes (PB-001 and -002) acquired by US Air Force, later receiving USAF serial numbers 02-4001 and 02-4002. These feature a redundant avionics system that provides for increased reliability. Williams FJ44-2A turbofan in second prototype replaced by a TPE331-10 turboprop following USAF acquisition. Two pre-production YQM-9As (PB-003/02-4003 and PB-004/02-4004) ordered in October 2001, first of these making its maiden flight 17 October 2003. Equipment trialled in YQM-9As included AN/APY-8 Lynx radar (completed January 2005), GBU-12 Paveway II laser guided bomb (August 2004, at NAWC China Lake) and AGM-114P Hellfire (May 2005). One also known to have tested an experimental sigint system (thought to be Raytheon Batfish) on behalf of US Navy. The YQM-9A continued in USAF service in 2009.

MQ-9 Reaper: USAF MQ-9 designation was allocated in January 2002, at which time it was known as MQ-9 Predator B Hunter-Killer; new name Reaper was announced in September 2006.

First four production MQ-9s (02-4005 to 02-4008) ordered in December 2002; first of these flown October 2003, was delivered in following month, and deployed to Southwest Asia in January 2004. On 29 March 2004, GA-ASI received a USD17.01 million USAF contract to develop specifications for production and weaponisation of the MQ-9, and on 30 March 2005 received a further USD68.20 million contract, involving retrofit of the first four aircraft for systems development and demonstration of this version. A further five MQ-9s (4009-4013) were ordered for USAF under a USD41.4 million contract of January 2006; and in March 2007 USD43.66 million was allocated for the purchase of two additional MQ-9 systems.

In May 2006, the US Federal Aviation Administration (FAA) issued a Certificate of Authorization (COA) to the MQ-9, allowing it to be used within US civilian airspace to search for disaster survivors.

Mariner Demonstrator: Original Mariner Demonstrator was the GA-ASI Altair aircraft, reconfigured for maritime surveillance. GA-ASI integrated two different maritime radars into this aircraft. For specification and other details, see entry for Altair.

GBU-12 Paveway II laser-guided bomb on a YQM-9A prototype *(GA-ASI)* 0577767

Mariner Demonstrator II: Successor to original Mariner Demonstrator, using a production Predator B on loan from USAF as the platform aircraft. In 2006, this was integrated with a third maritime radar for operations in the US Navy's 'Experiment Trident Warrior' (June 2006) and Australia's 'Northwest Shelf' UAV trial (September 2006). Support was also given in 2008 with further maritime demonstrations to the US Department of Homeland Security (DHS) and the US Coast Guard.

Predator B: Production Predator B is of similar appearance to prototypes, but with increased internal payload, fuel load, MTOW and range (see Specifications below). It is the baseline air vehicle for the MQ-9 Reaper and non-military applications.

Ikhana: Ikhana is a production Predator B delivered to NASA in November 2006; its name derives from a Native American Choctaw word meaning 'intelligent, conscious or aware'. Can carry more than 1,746 kg (3,850 lb) of payload, including 1,361 kg (3,000 lb) on the wings, providing ability for a wide variety of payloads for numerous missions. Is used for HALE firefighting, Earth science missions and aeronautical research. In April 2007, following integration of a NASA Ames Autonomous Modular Sensor — Wildfire (AMS), a thermal/IR imaging system able to generate clear images through heavy smoke, Ikhana was enabled to assist firefighters with the management and containment of forest fires. The AMS pod allows Ikhana to perform high-altitude wildfire surveillance, track wildfire movement, detect fire hot-spots and measure fire temperature fluctuations. Data are then transmitted in real time to firefighting command authorities via Ku-band satcom datalinks, enabling them to better understand and contain a blaze.

YMQ-9A, Predator B

Dimensions, External
Overall
  length...10.97 m (36 ft 0 in)
height
  YMQ-9A..3.60 m (11 ft 9¾ in)
  Predator B...3.81 m (12 ft 6 in)
Fuselage
  width, max..1.13 m (3 ft 8½ in)
Wings
  wing span..20.12 m (66 ft 0¼ in)
  wing chord
    at root, YMQ-9A..1.65 m (5 ft 5 in)
    at root, Predator B..1.74 m (5 ft 8½ in)
    at tip...0.86 m (2 ft 9¾ in)
  wing aspect ratio
    YMQ-9A..17.0
    Predator B...16.8
Tailplane, tailplane span...6.22 m (20 ft 5 in)
Engines, propeller diameter...2.90 m (9 ft 6¼ in)
Wheels
  wheelbase
    YMQ-9A..3.08 m (10 ft 1¼ in)
    Predator B...5.36 m (17 ft 7 in)
  wheel track...3.66 m (12 ft 0 in)
Dimensions, Internal
Payload bay
  volume
    with satcom dish, Predator B ....0.28 m³ (9.9 cu ft)
    without satcom dish, Predator
    B..1.10 m³ (39 cu ft)
Areas
Wings
  Gross wing area
    YMQ-9A..23.78 m² (256.0 sq ft)
    Predator B...24.15 m² (259.9 sq ft)
Weights and Loadings
Weight
  Weight empty, Predator B...1,996 kg (4,400 lb)
  Max T-O weight
    YMQ-9A..3,401 kg (7,497 lb)
    Predator B...4,762 kg (10,498 lb)
Fuel weight
  Max fuel weight
    YMQ-9A..1,361 kg (3,000 lb)
    Predator B...1,814 kg (3,999 lb)

YMQ-9A, Predator B

Payload	
Payload with max fuel	
internal, YMQ-9A	340 kg (749 lb)
internal, Predator B	386 kg (850 lb)
external	1,360 kg (2,998 lb)
Performance	
Climb	
Rate of climb	
max, at S/L, YMQ-9A	1,158 m/min (3,799 ft/min)
max, at S/L, Predator B	838 m/min (2,749 ft/min)
Altitude	
Operating altitude, max	15,240 m (50,000 ft)
Speed	
Max level speed	
YMQ-9A	225 kt (417 km/h; 259 mph)
Predator B	240 kt (444 km/h; 276 mph)
Cruising speed	
YMQ-9A	170 kt (315 km/h; 196 mph)
Predator B	180 kt (333 km/h; 207 mph)
Range	
max, YMQ-9A	4,000 n miles (7,408 km; 4,603 miles)
max, Predator B	4,600 n miles (8,519 km; 5,293 miles)
Endurance	
clean, YMQ-9A	29 hr
clean, Predator B	32 hr
Power plant	1 × turboprop

Status: YMQ-9A and Ikhana in service; MQ-9 Reaper and Predator B in production and in service. In addition to customers listed, demonstrations or offers have been made to Australia (Mariner II), Canada (Predator B) and the US Immigration and Naturalization Service (Predator B).

Customers: *United Kingdom:* The UK took an early interest in Predator B development, notably in connection with the 'persistent surveillance' strand of its three-year Joint UAV Experimentation Programme (JUEP). To this end, some 44 UK Royal Air Force personnel, designated as No. 1115 Flight of the RAF, were embedded with the USAF Predator teams at Nellis and Creech AFBs in the US from early 2004 to gain operating experience, including at least one firing of a Hellfire missile. Between November 2004 and February 2005, during the UAV 'Trial Falcon Prowl', GA-ASI demonstrated to the UK MoD a YMQ-9A prototype (02-4003) fitted with a modified Goodrich DB-110 dual-based sensor in a targeting pod under the port wing for the Hellfire on the opposite side. Flown over southern California, it captured high-resolution imagery of multiple targets over very long ranges.

Negotiations began in the second quarter of 2006 for the purchase of two Reapers, resulting in a foreign military sales contract in March 2007. These were delivered later that year to the re-formed No. 39 Squadron of the Royal Air Force, based at Waddington, Lincolnshire, replacing the elderly Canberra PR Mk 9 aircraft that had been the unit's reconnaissance and surveillance asset until their retirement in July 2006. Plans then were for their deployment in the third quarter of 2007 to Afghanistan and, possibly, Iraq. A third Reaper system was ordered in January 2008.

US Air Force: USAF currently plans to purchase 60 MQ-9s, and to deploy three Reaper squadrons by 2009. As a prelude to this, two YMQ-9A prototypes, armed with 500 lb GBU-12 laser-guided bombs, were revealed to have been deployed for operational use (theatre unspecified) by late 2005. The next four are production-standard MQ-9As (higher gross TOW, more internal fuel, AN/APY-8 Lynx radar and MTS-B sensor). The 42nd Attack Squadron (42nd ATKS), responsible for training pilots and sensor operators assigned to the MQ-9, was reactivated in November 2006 at Creech AFB, Nevada, where the first operational MQ-9 arrived on 13 March 2007, equipped with an interim armament of GBU-12s and Hellfire missiles. The 19th ATKS, specifically charged with operating the MQ-9, was also activated in November 2006. In November 2007, the MQ-9 conducted its first precision strike sortie, targeting Afghan insurgents with a Hellfire missile. IOT&E of the full-standard Hunter-Killer version was achieved in early 2008. As of June 2009, Reaper combat utilisation rate was in the region of 87 per cent. A new Reaper unit, the 33rd Special Operations Squadron, was activated on 31 July 2009 at Cannon AFB, New Mexico, as part of the 27th Special Operations Wing. Miscellaneous other USAF units to which Reapers have been deployed include the 556th Test and Evaluation Squadron and 26th Weapons School, both at Nellis AFB, Nevada; the 30th Reconnaisance Squadron at Tonopah Test Range, Nevada; and the 452nd Flight Test Squadron at Gray Butte, California.

US Navy: The USN has acquired three MQ-9/Predator Bs from GA-ASI (first order December 2005) to support payload integration and testing. Further details of these have not been disclosed. In September 2009 it was reported that the US Navy was to base one or more of its aircraft in the Seychelles in the coming weeks to fly anti-piracy surveillance patrols along the East African coast.

US Customs and Border Protection (CBP): One Predator B system (single air vehicle) ordered and delivered in September 2005 (USD14.1 million contract), to provide long-endurance and surveillance over the southwestern US and protect its borders from terrorism, drug smuggling and illegal immigration. Began operating along the US/Mexican border from Fort Huachuca, Arizona, in October 2005, but lost in crash on 25 April 2006 after some 800 hours of operation. Contract had option for a second system, which was delivered in September 2006; a third aircraft was delivered in July 2007 and a further four aircraft have been delivered since then. A second operational base has been established in Grand Forks, North Dakota, to support operations on the US northern border.

NASA: One Predator B acquired under the name Ikhana for a programme of scientific missions and aeronautical research, including autonomous payload operation, flight control systems and operation in national airspace. It replaced an Altair leased from GA-ASI.

Italy: Two Reapers were ordered by the Italian Air Force on 9 February 2009.

Contractor: General Atomics Aeronautical Systems Inc
Poway, California.

GA-ASI Predator C Avenger

Type: HALE surveillance and attack UAV.

Development: As originally proposed, Predator C was to have been a direct variant of the Predator B, with the latter's turboprop engine replaced by a turbojet or turbofan. However, with the emphasis in recent years on stepping up production of the MQ-1 Predator and MQ-9 Reaper versions, the stretched-out development phase has been used to undertake a more extensive redesign, although the Avenger's shape still shows evidence of its ancestry. Although not completely stealthy, it has been 'stealthed up' to some degree, to reduce radar signature. The body is bulkier, to accommodate an internal weapons bay; the wings have been not only

Avenger features in this view include the thick-section inner wings, wide-angle V tail and tail hook *(GA-ASI)*

Unmanned aerial vehicles > United States > **GA-ASI Predator C Avenger – Griffon MQM-171A BroadSword**

Avenger shows off the compound sweep of its wing trailing edge in this view from the rear *(GA-ASI)* 1293694

swept, but thickened and strengthened inboard and designed to fold if required. The last-named feature, plus provision of a tail-end arrester hook, indicates provision in the design for a carrier-based version if such should be required — as does choice of the name Avenger, previously borne by two earlier US Navy aircraft programmes. However, USAF is more likely to be the launch customer, with the UK Royal Air Force also expected to show interest.

After years of conjecture, images of the Predator C/Avenger were finally released for publication in mid-April 2009, following the prototype's first three flights on 4, 13 and 14 April. The initial flight test programme was expected to last some two to three months, after which it was understood that production deliveries could be made available within a year of customer funding being received , The second prototype, which rolled out in late 2010 has its overall length extended to 13.4 m (44 ft). It has also been suggested that the 'over 400 knots' provisional top speed figure quoted may be a generous under-estimate. A third prototype is also scheduled.

In February 2011, the company reported conclusion of a wind tunnel test on a model of its Sea Avenger which is aimed at the Unmanned Carrier-Launched Airborne Surveillance and Strike (UCLASS) carrier-based air system for the US Navy. The wind tunnel test validated the low-speed characteristics of a new wing, resulting in higher endurance and lower approach speeds and also designed to increase aircraft dash speeds. The goal was to validate the characteristics of an updated wing in the approach, launch, and cruise configurations.

Description: *Airframe:* Wings are mid-mounted and swept back at about 17°; outer sections have constant chord, inboard sections a forward-swept trailing edge. Design includes provision, if required, for outer wings to fold. Fuselage upper and lower lobes roughly semi-circular, meeting in straight-line chine running whole length of body. Arrester hook under rear of fuselage. Semi-circular dorsal intake for engine; boat-tail exhaust. All-moving, constant-chord twin tail surfaces (aspect ratio approx 5), canted outward at about 50°. Retractable tricycle landing gear (main units inward, nosewheel rearward), adapted from that of Northrop F-5 and fitted with anti-skid braking system. Composites construction.

Mission payloads: At the time of launch, sensors were stated to include a GA-ASI Lynx SAR and 'various' EO/IR camera systems. Among the latter then being evaluated were a FLIR system based on that selected for the Lockheed Martin F-35, and a GA-ASI in-house full-motion video sensor. A pure reconnaissance version will be capable of carrying a wide-area surveillance system internally for special mission applications.

Predator C/Avenger design has produced a smaller radar cross-section than its predecessors *(GA-ASI)* 1356202

Onboard avionics include Ku-band satcom and a laser altimeter.

Internal bay, with removable doors, for weapons. Avenger can carry the same mix of weapons as Predator B; other potential stores could include GBU-38 JDAMs and laser-guided bombs, a semi-buried sensor pod or supplementary fuel tanks.

Guidance and control: Fully autonomous, from GA-ASI proprietary GCS. Aerodynamic control from wing flaps and ailerons, plus dual-servo all-moving tail surfaces.

Launch: Conventional and automatic wheeled take-off.
Recovery: Conventional and automatic wheeled landing.
Power plant: One 21.35 kN (4,800 lb st) Pratt & Whitney Canada PW545B turbofan. Fuel tanks in fuselage and inboard wing sections.

Predator C

Dimensions, External	
Overall	
length	
first prototype	12.5 m (41 ft 0¼ in) (est)
second prototype	13.4 m (43 ft 11½ in) (est)
Wings, wing span	20.1 m (65 ft 11¼ in) (est)
Weights and Loadings	
Payload, Max payload	1,360 kg (2,998 lb) (est)
Performance	
Altitude, Service ceiling	18,290 m (60,000 ft) (est)
Speed, Max level speed	400 kt (741 km/h; 460 mph) (est)
Endurance	20 hr (est)
Power plant	1 × turbofan

Status: Undergoing initial flight test programme during second quarter of 2009.
Contractor: General Atomics Aeronautical Systems Inc
Poway, California.

Avenger C second air vehicle with extended fuselage and increased payload *(GA-ASI)* 1464774

Griffon MQM-171A BroadSword

Type: Tactical UAV and aerial target.
Development: A scaled-up version of the company's MQM-170A Outlaw target (which see), the BroadSword was developed as an Unmanned Aircraft System - Target (UAS - T) for the US Army's Targets Management Office at Redstone Arsenal, Alabama. A successful 2-hour flight at that location on 5 April 2007 was said at that time to be the first step towards accomplishing autonomous flight "in the near future" and a critical milestone toward a production decision. The programme involved development of a new aircraft, avionics, launcher and ground support equipment, and the system was scheduled to enter service with the US

BroadSword in flight *(Griffon Aerospace)* 1290251

Preparing a BroadSword for launch *(Griffon Aerospace)* 1290252

Army in FY08. Griffon Aerospace also promote a further enlarged version, the BroadSword XL, as an observation platform and/or a sensor or power plant test and development vehicle.

Description: *Airframe:* Generally as described for MQM-170A Outlaw; XL version able to carry pylon-mounted external payloads, and can utilise 'wet' fuel bays in wings for greater endurance.

Mission payloads: Generally similar to those listed for MQM-170A.

Guidance and control: Standard GCS is as described for MQM-170A, but aircraft is compatible with a variety of airborne and ground control systems. In US Army use, is controlled by the Micro Systems Inc Target Tracking and Control System (TTCS) available on most DoD test ranges.

Launch: Target version flown from pneumatic launcher; wheeled take-off from unimproved runway is standard for XL.

Recovery: Belly skid landing or parachute recovery for target version; wheeled landing for XL.

Power plant: One 44.13 kW (59.2 hp) 3W-Modellmotoren 684i B4 TS flat-four engine; two-blade pusher propeller.

Variants: *MQM-171A BroadSword:* UAS - T version for US Army.
BroadSword XL: Stretched version of MQM-171A.

MQM-171A, BroadSword XL

Dimensions, External	
Overall	
length	
MQM-171A	4.19 m (13 ft 9 in)
BroadSword XL	4.50 m (14 ft 9¼ in)
height	
excl landing gear, MQM-171A	1.24 m (4 ft 0¾ in)
Fuselage	
height, max, MQM-171A	0.51 m (1 ft 8 in)
Wings	
wing span	
MQM-171A	5.18 m (17 ft 0 in)
BroadSword XL	6.86 m (22 ft 6 in)
Weights and Loadings	
Weight	
Max T-O weight, BroadSword XL	249 kg (548 lb)
Max launch weight, BroadSword XL	249 kg (548 lb)
Payload	
Max payload	
MQM-171A	18 kg (39 lb)
BroadSword XL	54 kg (119 lb)
Performance	
Altitude	
Operating altitude, min, MQM-171A	305 m (1,000 ft) (est)
Service ceiling	
MQM-171A	3,660 m (12,000 ft) (est)
BroadSword XL	4,265 m (14,000 ft)
Speed	
Cruising speed	
max, MQM-171A	115 kt (213 km/h; 132 mph)
max, BroadSword XL	110 kt (204 km/h; 127 mph)
econ, MQM-171A	60 kt (111 km/h; 69 mph)
Radius of operation	
mission, MQM-171A	54 n miles (100 km; 62 miles) (est)
Endurance, max, MQM-171A	1 hr (est)

Status: MQM-171A was scheduled to enter US Army service at White Sands Missile Range in March 2008.

Contractor: Griffon Aerospace
Madison, Alabama.

Honeywell RQ-16 T-Hawk™ MAV

Type: VTOL micro UAV.

Development: The Honeywell MAV, now named T-Hawk™ (a contraction of the term Tarantula Hawk, a wasp species), was selected in 2006 as winner of the US Army's Future Combat System (FCS) Class I competition for a platoon-level micro UAV. Here, implementation was to be via the service's Brigade Combat Team Modernization (BCTM) Increment 1 (also known as the Early Infantry Brigade Combat Team (E-IBCT) Increment 1) programme (see following). After several years of preliminary design and testing, a 737 mm (29 in) diameter test vehicle was flown in 2004, followed in January 2005 by announcement of the maiden (tethered) flight of the 330 mm (13 in) diameter prototype. First free flight was announced in early June 2005. Partners in the programme were AAI Corporation (airframe), Avid LLC (modelling and simulation) and Techsburg Inc (testing and acoustics). Key programme events include:

- **October 2005** *IHS Jane's* sources report the US Army's 25th Infantry Division (Light) as having received five × MAV AVs for trials at Schofield Barracks, Hawaii.
- **February 2006** *IHS Jane's* sources were reporting Honeywell as having awarded the AAI Corporation a then year USD1.7 million order for 55 × MAV airframes for use in a DARPA ACTD programme. As of the cited date, delivery was reported as having been set for November 2006.
- **May 2006** *IHS Jane's* sources report Honeywell as having been awarded a then year USD61 million FCS Class I standard MAV System Development and Demonstration (SDD) contract, with the work manifest calling for the production of six × MAV systems (two AVs per system) and their portable ground stations. Preliminary and critical system design reviews were to take place during December 2006 and December 2010 respectively, with the maiden flight of a prototype being set for December 2008.
- **2007** *IHS Jane's* sources were reporting the USN as having acquired 20 × MAVs (under the designation YRQ-16A) for evaluation by the US Multi-Service Explosive Ordnance Disposal Group in Iraq.

MAV is also known by the T-Hawk™ designator *(Honeywell)* 1376958

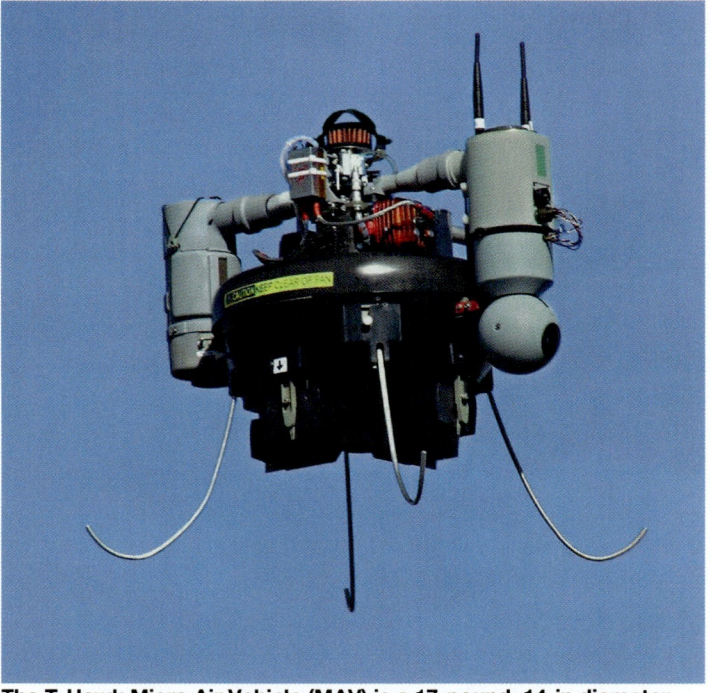
The T-Hawk Micro Air Vehicle (MAV) is a 17-pound, 14-in diameter autonomous vertical take-off and landing system *(Honeywell)* 1294859

Unmanned aerial vehicles > United States > Honeywell RQ-16 T-Hawk™ MAV

Early versions of T-Hawk™ were trialled in Iraq *(Honeywell)*

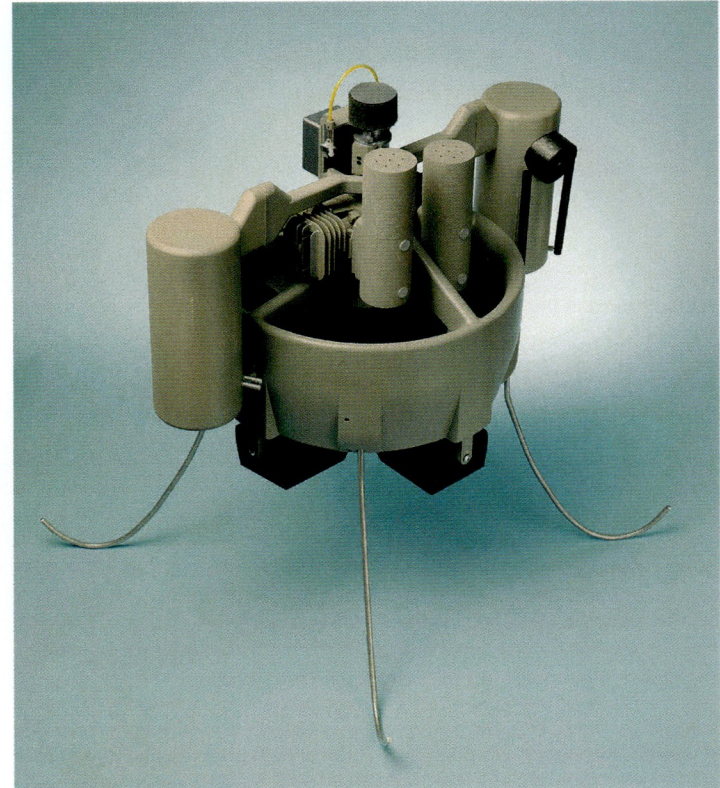
The piston-engined T-MAV development vehicle *(Honeywell)*

- **December 2007** IHS Jane's sources were reporting UK contractor RCV Engines as having been selected to produce a demonstrator heavy-fuel engine for MAV.
- **March 2008** The T-Hawk™ (under the RQ-16A designation) was adopted by the US Geological Survey's (USGS) Unmanned Aircraft Systems Project Office. Here, the USGS signed an initial contract with Honeywell that was to run for 12 months plus 120 days and which contained 12 month options to carry the programme through to June 2010. In addition to Honeywell, support for the USGS's T-Hawk™ programme was provided by the Pennsylvania Army National Guard.
- **3 November 2008** The USN's Naval Air Systems Command (Patuxent River, Maryland) awarded Honeywell a not to be exceeded then year USD65,500,817 undifinitised contract with regard to the supply of 90 × Block II MAV systems, with each system comprising 2 × AVs, 1 × ground control unit, a spares holding, an operator/maintainer training package and field support services. At the time of this effort's announcement, Honeywell was noting that it was scheduled to have completed the necessary hardware deliveries by the end of December 2009, while the USN was specifying programme completion by the end of December 2011.
- **14 June 2010** Honeywell announced that the T-Hawk™ AV had completed its 10,000 flight (an explosive ordinance disposal evaluation sortie over Iraq) since its introduction.
- **24 August 2010** Honeywell announced that it was to provide FCS Class I UAV systems, training and logistical support as part of a then year USD11 million LRIP contract relating to the initial brigade set for the US Army's BCTM/E-IBCT Increment 1 programme. Here, Honeywell was supporting the BCTM/E-IBCT team being led by Boeing, with the first equipment being slated (it is thought) for supply to the US Army's 1st Armoured Division.
- **10 December 2010** The USN's Naval Air Systems Command awarded Honeywell a then year USD6,567,624 contract modification (to the previously awarded firm, fixed-price contract N00019-09-C-0004) with regard to the supply of nine × T-Hawk™ systems (and an associated spares package) to the United Kingdom under the US Foreign Military Sales construct. At the time of its announcement, this effort was scheduled for completion by the end of December 2011.
- **January 2011** Florida's Miami-Dade Police Department is reported to have acquired at least one T-Hawk™ AV for airborne surveillance duties.
- **6 January 2011** The US Army issued Boeing with a 'stop work' notice relating to the BCTM/E-IBCT Increment 1 T-Hawk™ configuration and announced on the 4th of the following February that it had decided to "no longer pursue" either the BCTM/E-IBCT Increment 1 unattended ground sensor or Class I unmanned air system.
- **19 April 2011** Honeywell announced that four T-Hawk™ AV equipped with radiation sensors were being used to monitor the Fukushima Daichi nuclear power plant that had been badly damaged in the earthquake and tsunami that had all but devastated the east coast of Japan's north island during the preceding March. One of these AVs is noted as having had to make a force landing on Fukushima's Number 2 reactor during the course of a monitoring sortie that was flown on 24 June 2011.
- **19 September 2012** The USN's Naval Sea Systems Command's Indian Head Division (Indian Head, Maryland) awarded Honeywell a then year USD8,286,161 contract with regard to the provision of post-production support services for the block II RQ-16B T-Hawk™ MAV. At the time of the announcement, the RQ-16B was reported to be "supporting" Operation 'Enduring Freedom' in Afghanistan and the described support programme was scheduled for completion by the end of September 2013.

Description: *The following specification data should be regard as being generic to the T-Hawk™ architecture as a whole*

Airframe: The T-Hawk™ AV is built around a barrel-shaped ducted fan that provides lift and is surmounted by a T-shaped structure that incorporates the device's engine, ignition unit, air filter, carrying handle and mounts for avionic and payload pods. Directional control is by means of control vanes positioned in the fan's slipstream (see *Guidance and control*) and the AV makes use of four curved wire 'legs' for take-off and landing. Early iterations of the T-Hawk™ featured video link and GPS antennas on top of their payload pods together with a datalink aerial attached to the pod's lower end. Later configurations have had the payload pod antennas re-arranged to accommodate the sensor's gimballed optics.

Mission payloads: T-MAV™ carries either an EO or IR camera, with the latest known examples featuring gimballed optics. As specified by the USGS, early, non-gimballed payload iterations provided 768 × 494 and 324 × 256 resolution at EO and IR respectively, featured ×2 digital (IR) and ×10 optical (EO) zooms and had 36° and 5° to 46° FoVs at IR and EO respectively. Imagery can be downlinked in real time, or up to 10 minutes of imagery can be stored on board. Other payloads can include radio

MAV test vehicle during a November 2006 an opeval search for IEDs *(US Navy/MCS3 Kenneth G Takada)*

The MAV and its ground support vehicle during 2006 opeval *(US Navy/MCS3 Kenneth G Takada)* 1151642

relays and datalinks, or other modular items (such as radiation sensors - see previously) that are compliant with the AV's mechanical and electrical interfaces.

  Guidance and control: Pre-programmed control with dynamic re-tasking and manual intervention, using Honeywell proprietary micro-electrical mechanical systems (MEMS) technology. INS/GPS navigation, with up to 100 waypoints in a flight plan and up to 10 pre-planned flight plans stored in the system's ground control unit. Mission computer activates control vanes to tilt vehicle in transition to forward flight, this being achieved by combination of fan thrust and 'wing' lift from fan duct. GCS can store up to 240 minutes of sensor imagery.

  Transportation: Man-portable and backpackable. Set-up time less than 5 minutes.

  System composition: Two air vehicles, two operators and one ground control unit. Complete system weighs less than 18.1 kg (40 lb), with the individual AVs having a dry weight of 7.9 kg (17.5 lb) with a non-gimballed IR camera installed.

  Launch: Vertical take-off.

  Recovery: Vertical landing.

  Power plant: One 3.0 kW (4 hp) 3W Modellmotoren flat twin piston engine, driving a fixed-pitch ducted fan. Production versions have electronic fuel injection rather than carburettor. 75 dBA acoustic signature at 100 m (328 ft).

Variants: Over time, the T-Hawk™ (Honewell tradename) AV has been referred to as the Gasoline-powered Micro AV (G-MAV - US Army Term), the MAV (Honeywell term) and as the YRQ-16A, RQ-16A and RQ-16B (USN terminology). Further confusion arises from putative Block II and III configurations, with unconfirmed sources linking the Block II nomenclature with the introduction of stabilised gimbaled sensors and the Block III with an encrypted digital datalink for C2 and imagery transmission. Again, the Block II terminology has been associated with the USN's RQ-16B designation (see *Development*) while the Block III nomenclature has been associated with the now cancelled BCTM/E-IBCT Increment 1 configuration.

RQ-16A

Dimensions, External	
Overall, height	0.559 m (1 ft 10 in) (est)
Engines, propeller diameter	0.33 m (1 ft 1 in)
Weights and Loadings	
Weight	
Weight empty	2.3 kg (5.00 lb)
Max T-O weight	7.8 kg (17.00 lb)
Performance	
Climb	
Rate of climb, max, at S/L	457 m/min (1,499 ft/min)
Altitude	
Operating altitude, normal	0 m to 150 m (0 ft to 500 ft)
Service ceiling	3,200 m (10,500 ft)
Speed, Max level speed	50 kt (93 km/h; 58 mph)
Radius of operation, mission	5.4 n miles (10 km; 6 miles)
Endurance, at 1,675 m (5,500 ft)	50 min (est)
Power plant, T-MAV	1 × piston engine

Status: As of 2012, T-Hawk™ configurations were in service and were being promoted.

Customers: Over time, T-Hawk™ configurations are understood to have been procured/leased by DARPA, the British Army, Poland (possibly), the US Army, the US Geological Survey and the USN.

Contractor: Honeywell Defense and Space Electronics Systems Albuquerque, New Mexico.

Insitu RQ-21A Integrator

Type: Multirole surveillance UAV.

Development: Unveiled in August 2007, the Integrator is a next-generation development of Insitu's Insight system, with a somewhat larger air vehicle. Development began in 2004, and a prototype was flown for the first time in 2006. It uses ground system components common to Insight, but special design consideration has been given to easier payload integration in the air vehicle for a wide variety of civil and military missions.

Description: *Airframe:* Cylindrical fuselage; high-mounted, sweptback wings with winglet tips; twin, unconnected tubular tailbooms, each with sweptback vertical fin and half-tailplane; pusher engine. No landing gear. Modular design includes propulsion module, nose and internal CG bays, wings and tail units. Composites construction.

  Mission payloads: Baseline system is equipped with inertially stabilised EO, long-wave or mid-wave infrared (LWIR or MWIR) cameras, IR marker and optional laser rangefinder in a nose-mounted turret. Internal (centre-fuselage) payload options can include a wide range of intelligence, communication and expendable packages. Integrator also has hardpoints to carry external stores. All existing ScanEagle payloads, EO and IR turrets and communications can be carried, and all are

Integrator's internal fuselage payload bay *(Insitu Group)* 1290262

Integrator on Mk4 launcher *(Insitu Group)* 1410040

The Integrator capitalises on Insight/ScanEagle experience *(Insitu Group)* 1290261

Integrator's separate tailbooms aid one-person assembly and disassembly *(Insitu Group)* 1290260

decoupled from the airframe to ease integration. Some 500 W of power is available for up to 11.3 kg (25 lb) of payload. A SINCGARS single-channel ground and airborne UHF/VHF radio relay system payload has been demonstrated on Integrator.

Future potential payloads may include a miniature SAR and Spike missiles.

Guidance and control: Autonomous; generally as described for Insight; customised Athena GuideStar autopilot and differential GPS navigation.

Transportation: Aircraft is completely modular and easily transportable. Centre-section core is designed so that all parts can be adapted and structurally sized for growth. It can be assembled and dismantled by one person, and fits easily into the back of an HMMWV or pick-up truck.

Launch: Autonomous, from new, lighter-weight version of Insitu SuperWedge pneumatic catapult.

Recovery: Autonomous, using new, lighter-weight version of Insitu patented SkyHook recovery system.

Power plant: One 6.0 kW (8 hp) piston engine (type not specified); mogas or Diesel fuel; two-blade pusher propeller.

Integrator

Dimensions, External	
Overall, length	2.23 m (7 ft 3¾ in)
Wings, wing span	4.87 m (15 ft 11¾ in)
Dimensions, Internal	
Payload bay	
length	0.89 m (2 ft 11 in)
width	0.25 m (9¾ in)
Weights and Loadings	
Weight, Max launch weight	61.2 kg (134 lb)
Fuel weight, Max fuel weight	20.4 kg (44 lb)
Payload, Max payload	22.7 kg (50 lb)
Performance	
Altitude, Service ceiling	6,095 m (20,000 ft) (est)
Speed	
Max level speed	90 kt (167 km/h; 104 mph) (est)
Cruising speed	55 kt (102 km/h; 63 mph) (est)
Radius of operation	
LOS	55 n miles (101 km; 63 miles) (est)
BLOS	550 n miles (1,018 km; 632 miles) (est)
Endurance	24 hr (est)
Power plant	piston engine

Status: Selected for STUAS/Tier II in July 2010 by US Navy. Two systems delivered in January 2012 for Early Operational Capability with USMC squadron VMU-3, scheduled to be followed by full-rate production of up to 56 systems for the US Navy and USMC.

Contractor: Insitu Group Inc
Bingen, Washington.

ISL ACE

Type: Unmanned airship.

Development: Existence announced November 2004. Developed to meet short-term requirement for long-endurance communications relay.

ISL Airborne Communications Extender (ACE) *(ISL Bosch)* 1135864

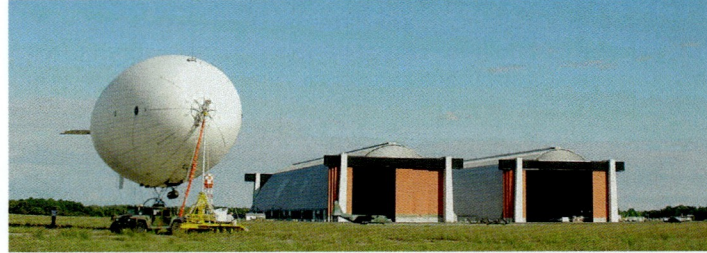

ACE, with ground equipment, alongside its hangar *(ISL Bosch)* 1135866

Description: Airframe: Conventional airship shape, utilising 'latest fabric technologies'; automatic ballast control with patented water recovery system.

Mission payloads: To customer's requirements. Payloads can be distributed anywhere along the envelope to maintain correct balance. Nearly 7 kW of power is available for payloads, provided by three engine-driven generators.

Guidance and control: Operated by typical flight crew of two (pilot and flight engineer) in air conditioned GCS with car-type reclining seats. Airship systems, telemetry and video are monitored and recorded at one station, while pilot has a dedicated autopilot computer. Flights are observed on a real-time moving map display, and flight plan changes can be uploaded to the onboard autopilot. Loiter flight patterns can be moved to new map co-ordinates at the click of a mouse, enabling mission objectives to be altered quickly. All flight plans are stored by the airship's computer as well as in the GCS, enabling it to fly home autonomously if the GCS command link is lost. All systems are redundant, and all power circuits controllable from the ground.

Launch: Conventional airship take-off.

Recovery: Conventional airship landing.

Power plant: Details of engine(s) not provided. Standard fuel capacity 341 litres (90 US gallons; 75 Imp gallons); option for range-extending auxiliary tanks.

ACE

Dimensions	
Length overall	38.1 m (125 ft)
Weights	
Max payload	408 kg (900 lb)
Performance	
Endurance with 181 kg (400 lb)	
payload, standard fuel	30 h

Status: Delivered to US Army.

Contractor: ISL Aeronautical Systems
Brownsboro, Alabama.

ISL BA-131

Type: Unmanned airship.

Development: Developed for border and maritime surveillance for undisclosed foreign military customer.

Description: Airframe: Conventional airship shape, utilising 'latest fabric technologies'; proprietary helium and air valves; four tailfins, indexed in X configuration. Automatic ballast control, with patented water recovery system and proprietary helium and air valve. Underslung pod, with fixed tricycle landing gear, for payload and power plant.

Mission payloads: Maritime surveillance radar and EO/IR sensors.

Guidance and control: Operated by typical crew of two (pilot and flight engineer) in air conditioned GCS with car-type reclining seats. Airship systems, telemetry and video are monitored and recorded at one workstation, while pilot has a dedicated autopilot computer. Payload display computers are also in the GCS. Flights are observed on a real-time moving map display, and flight plan changes can be uploaded to the onboard autopilot. Loiter flight patterns can be moved to new map co-ordinates at the click of a mouse, enabling mission objectives to be altered quickly. All flight plans are stored by the airship's computer as well

ISL BA-131 – Kaman K-MAX < United States < Unmanned aerial vehicles

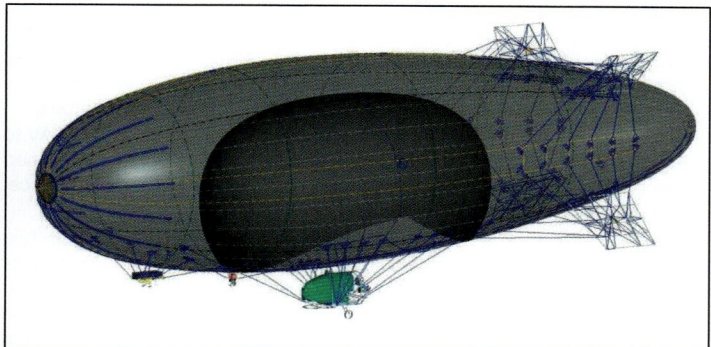

Design configuration of the BA-131 *(ISL)* 1290317

as in the GCS, enabling it to fly home autonomously if the GCS command link is lost. All systems are redundant, and all power circuits controllable from the ground.
  Launch: Conventional airship take-off, using proprietary hand-held controller.
  Recovery: Conventional airship landing, using proprietary hand-held controller.
  Power plant: Two 100 kW (135 hp) Thielert Centurion 2.0 diesel engines; pusher propellers.

BA-131

Dimensions	
Length overall	46.33 m (152 ft 0.0 in)
Envelope volume	3,709.5 m³ (131,000 cu ft)
Weights	
Max payload	Not disclosed
Performance	
Max endurance, standard fuel	30 h

Status: Under development for delivery scheduled in 2009.
Contractor: ISL Aeronautical Systems
  Brownsboro, Alabama.

ISL SASS LITE

Type: Lighter than air surveillance UAV.
Development: Small Airship Surveillance System, Low Intensity Target Exploitation (SASS LITE) is a stable platform that can be equipped with a variety of sensors. The first example flew in June 1989. As of 2009, six of these airships (001 to 006), in five different sizes, are known to have been built.
Description: *Airframe:* Conventional shape non-rigid, helium-filled airship with cruciform tail surfaces fitted with elevators/rudders. Internal ballonets for pressure control. Twin-engined versions have vectored thrust.
  Mission payloads: MTI radar, daylight TV camera and FLIR. Can be equipped with a variety of sensors, radio relays, scientific instruments or EW devices. Multiple daylight and IR cameras can be carried on independent gimbals on a single mission. SASS LITE has also carried a combination multispectral payload that included radar, IR and visible light sensors interfaced by a software bridge. Payload power is provided by two onboard alternators delivering a total of 3.25 kW at 24 V DC.
  Guidance: Triple-redundant radio control; primary control link and video downlink are in the microwave frequency L band. An autotrack dish antenna provides direct control at up to 54 n miles (100 km; 62 miles), and the L-band links are backed up by a function-matched P-band datalink equipped with Yagi directional antennas. These are interfaced to a digital autopilot and GPS receiver to facilitate autonomous navigation beyond radio LOS. An independent, battery-powered multifunction flight termination system (FTS) on the airship's nose-ring allows for engine shutdown, ballast and helium vent control, enabling 'free balloon' movement to a selected landing site. The FTS also incorporates a hot-wire burn system which can be activated to destroy the airship in flight if militarily necessary. Electronics, located in impact-resistant compartments on the envelope or power gondola, are jamming-resistant; the primary system is militarily hardened.
  An advanced digital switching and telemetry system allows the ground operators to control all airship functions and the payload. Three 9,600 baud digital links are embedded in the datalink system to facilitate control of all airship functions. Automated control devices such as pressure control, vents and autonavigation can be instantly overridden by the pilot via the digital link structure.
  Launch: Conventional airship take-off.
  Recovery: Conventional airship landing.
  Power plant: Single 31.3 kW (42 hp) engine (type not known) in 001 to 004, in pod slung beneath belly of airship; twin engines, with vectored thrust, in 005 and 006.
  Fuel capacity 170 litres (45 US gallons; 37.5 Imp gallons).

SASS LITE

Dimensions (envelope)	
Length: 001, 002	18.29 m (60 ft)
003	21.95 m (72 ft)
004	25.00 m (82 ft)
005	28.04 m (92 ft)
006	30.48 m (100 ft)
Max diameter: 001, 002	6.10 m (20 ft)
003	6.40 m (21 ft)
004	6.71 m (22 ft)
005, 006	7.62 m (25 ft)
Volume: 001, 002	339.8 m3 (12,000 cu ft)
003	481.4 m3 (17,000 cu ft)
004	623.0 m3 (22,000 cu ft)
005	849.5 m3 (30,000 cu ft)
006	976.9 m3 (34,500 cu ft)
Weights (typical)	
Weight empty	408 kg (900 lb)
Max payload	227 kg (500 lb)
Max T-O weight	771 kg (1,700 lb)
Performance	
Max level speed: 005, 006	39 kt (72 km/h; 45 mph)
Typical mission speeds: all	22–30 kt (40–56 km/h; 25–35 mph)
Typical mission altitudes	760–1,070 m (2,500–3,500 ft)
Ceiling	1,525–1,830 m (5,000–6,000 ft)
Operational radius	54 n miles (100 km; 62 miles)
Endurance	12–24 h

Status: In service. Has been used for border and waterway surveillance, search and rescue, radio relay and instrumentation for missile data collection.
Customers: US Army.
Contractor: ISL Aeronautical Systems.
  Brownsboro, Alabama.

Single-engined Bosch SASS LITE in flight 0044397

Kaman K-MAX

Type: Optionally piloted helicopter.
Development: The K-MAX UAS is an unmanned adaptation of the Kaman K-1200 helicopter, with the current (2012) K-MAX UAS being the product of a Lockheed Martin-Kaman Aerospace Corporation strategic teaming relationship. The K-MAX UAS has been developed for over a decade using a combination of industry investment and US Government programme funding. Key programme events include:
  June 1999 A K-1200 helicopter equipped with a pan-tilt-zoom video camera and a laser range-finder was used to locate simulated targets for a Dragon Fire semi-autonomous mortar system (which the helicopter had previously airlifted into position), with the trial taking place at Marine Corps Air Station (MCAS) Yuma, Arizona. During the effort, targeting data generated by the K-MAX was downlinked to a US Marine Corps ground station for use as required.
  July 1999 Kaman received funding from the USMC's Warfighting Laboratory to design and install a prototype remote piloting package in one of its K-MAX helicopters, as part of the service's Broad-area Unmanned Responsive Resupply Operations (BURRO) programme.
  Fourth quarter 2000 A BURRO demonstration (at the USMC's Combat Center, Twenty-Nine Palms, California) confirmed the feasibility of using an unmanned VTOL platform to deliver supplies to troop units widely dispersed on a battlefield.
  June-September 2003 The K-MAX BURRO AV demonstrates its potential as both an unmanned logistics UAV that was capable of lifting an up to 2,722 kg (6,000 lb) payload and as one capable of carrying a number of different external loads to differing delivery points during a single sortie. Of the two, the June 2003 single load effort is understood to have taken place at Fort Rucker, Alabama with the multiple load demonstration being

Unmanned aerial vehicles > United States > Kaman K-MAX

A USMC K-MAX unmanned helicopter prepares to lift-off from a main operating base in Helmand Province, Afghanistan *(USMC)* 1356662

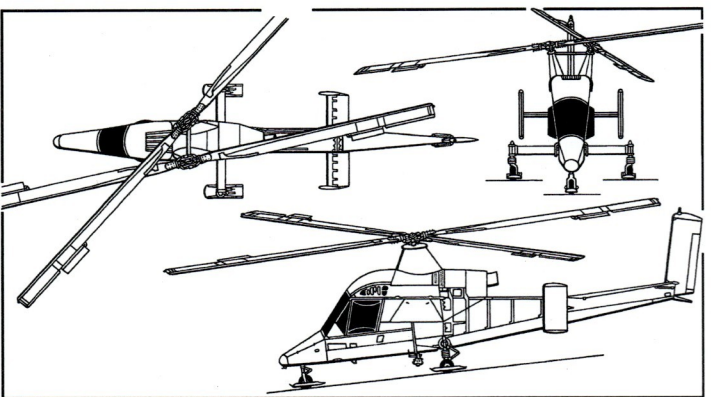

Kaman K-MAX three-view drawing *(Mike Keep)* 0507267

undertaken at Alabama's Redstone Arsenal during the following September.

July 2003 *IHS Jane's* sources describe the K-MAX's intermeshing rotor system as having formed a 'key ingredient' in Northrop Grumman's unsuccessful bid to become one of the two competing finalists in the DARPA/US Army Unmanned Combat Armed Rotorcraft (UCAR) programme. The UCAR effort was cancelled during 2004.

June 2004 Kaman was awarded a US Army contract to improve the K-MAX BURRO configuration by adding an external fuel tankage and a GPS tracking system.

November 2005 Enhanced BURRO capabilities were demonstrated at Huntsville, Alabama when the K-MAX AV autonomously delivered and recovered a small remotely operated UGV (a 'robot transporting a robot') as part of a US Army's Family of Unmanned Systems Experiments (FUSE).

March 2007 Kaman announced that it was entering into a strategic teaming agreement with the then Lockheed Martin Systems Integration to market the K-MAX in both manned and unmanned government applications on a worldwide basis. Again, the partnership was intended to develop an advanced technology unmanned variant for use by the military and in other government applications, with Lockheed Martin's contribution to the enterprise centring on the development of an autonomous military flight control system together with mission-specific avionics, sensors and weapons solutions.

August 2009 Kaman received a contract from the USMC to fly the K-MAX in the service's Immediate Cargo Unmanned Aerial System Demonstration (ICUASD). Here, the USMC was studying the use of UAVs in place of trucks or manned aircraft for resupply missions and required an ability to lift and move 2,721.6 kg (6,000 lbs) of stores in a 24 hour period. For the ICUASD effort, Kaman was teamed with Lockheed (Team K-MAX), with Lockheed integrating LOS and BLOS datalinks, an onboard mission management system and a GCS into the architecture.

February 2010 K-MAX ICUASD flight trials at the Dugway Proving Grounds, Utah included hovering at 3,657 m (12,000 ft) with a 680 kg (1,500 lb) sling load and delivering more than 1,361 kg (3,000 lbs) of cargo within a six hour period and two 278 km (150 n mile, 172 miles) round-trips. As an optional demonstration, Team K-MAX also used a four-hook carousel to facilitate multi-load deliveries in a single flight. Here (and lifting a total cargo weight of 3,450 lbs [1,565 kg]), the AV flew to three pre-programmed delivery co-ordinates and autonomously released a sling load at each location. Another load delivery was performed under manual control by the ground operator.

December 2010 Lockheed Martin was awarded a then year USD48 million competitive, firm, fixed-price contract with regard to the preparation of a pair of unmanned K-MAX helicopters for use in support of USMC operations in Afghanistan. The contract's scope included obtaining Category III US Naval Air Systems Command (NAVAIR) flight clearance, procurement of the AVs by NAVAIR, the conduct of a Quick Reaction Assessment (QRA) prior to their Deployment Readiness Review, contractor and USMC operator training and deployment staging.

26 August 2011 The K-MAX AV completed the previously noted QRA, with the AV being claimed to have exceed its requirements and to have delivered 15,241 kg (33,600 lbs) of cargo over operationally representative distances of between 56 and 171 km (30 to 92 n miles; 35 to 106 miles) in a series of 12 (four by day and eight by night) sorties.

September-December 2011 The decision to deploy the K-MAX AV to Afghanistan was taken on 28 September 2011, with the first unmanned helicopter delivery operation 'in history' starting in Helmand Province, Afghanistan on the 17th of the following December.

17 December 2011-12 May 2012 During the cited period, two unmanned K-MAX helicopters collectively transported in excess of 619,153 kg (1,365,000 lbs) of cargo between main and forward operating bases throughout Helmand Province. The highest daily total was 13,608 kg (30,000 lbs) delivered on 30 March 2012, with the total comprising 9,253 kg (20,400 lbs) of cargo to forward operating bases in six sorties and 4,354 kg (9,600 lbs) retrograde (with the 'retrograde' technique (transport from forward to main bases) forming part of an 'evolving' concept-of-operations). Lockheed Martin further reported NAVAIR and USMC confidence in the system growing to the point where a revised flight clearance was granted that allowed USMC helicopter support teams to manually hook-up retrograde sling loads with the AV in the hover.

27 September 2012 The USN's Naval Air Systems Command (Patuxent River, Maryland) awarded Lockheed Martin a then year USD8,907,747 modification (to the previously awarded firm, fixed-price contract N00019-11-C-0013) with regard to the provision of operational services relating to the two K-MAX cargo UASs that were deployed in support of the USMC. Work on the effort was to be performed at locations outside the continental US (90 per cent workshare) and at Patuxent River (10 per cent workshare) and at the time of its announcement, the programme was scheduled for completion by the end of March 2013.

Description: Airframe: Two two-blade intermeshing, contrarotating main rotors; no tail rotor; slimline fuselage; tailplane amidships with endplate fins; tall rear fin with inset rudder. Fixed tricycle landing gear. Light alloy airframe; GFRP and CFRP rotor blades and flaps.

Mission payloads: Typical unmanned cargo loads include water, food, fuel, ammunition and other supplies for which, a special multihook carousel has been designed having four separate, individually controlled load attachment points suspended from a single long cable. As a surrogate (piloted) UAV, K-MAX took part in a 1998 Office of Naval Research (ONR) exercise carrying a Magic Lantern Adaptation (MLA) minehunting sensor payload operated from a remote ground station. BURRO limited technical assessments were trialled off Hawaii in August 1998 and at MCAS Yuma, Arizona, in June 1999. In addition to autonomous delivery of a number of different external loads for the US Army, BURRO has demonstrated the capability to carry and deliver non-LOS launching systems ('NetFires Box') - thus becoming a robot delivering another robot to the future battlefield. While at Redstone Arsenal, BURRO carried a Multiple Launch Rocket System (MLRS) pod, delivering it precisely to a simulated rearming point.

Guidance and control: Blade angle of attack controlled by trailing-edge flaps and light control linkage, avoiding need for hydraulic power; electric actuator in each flap control run to track blades on ground or in flight. Intermeshing rotors cause pronounced pitch attitude change in response to collective pitch change; tailplane is connected to collective to alleviate this, to reduce blade stresses and to produce touchdown and lift-off in level attitude. Turns at or near the hover are effected by applying differential torque by means of differential collective pitch commanded (in piloted version) from foot pedals; rudder is connected to foot pedals to help balance turns. A small fore-and-aft cyclic pitch change also occurs which, at low powers near autorotation, would produce directional control reversal; to offset this, a non-linear cam between the collective lever and the rotor blade controls phases out differential collective and replaces it progressively with differential cyclic. Guidance and navigation are provided by redundant Embedded GPS Inertial (EGI) and ADC units, with redundant Flight Control Computers (FCC) driving redundant pairs of pitch, roll, direction and collective control actuators. An onboard Mission Management Computer (MMC) interfaces the LOS and BLOS datalinks to the two FCCs, while the GCS consists of a ruggedised laptop computer that is interfaced with the LOS/BLOS datalink ground segment.

Launch: Conventional and autonomous helicopter take-off.

Recovery: Conventional and autonomous helicopter landing, plus the capability to 'hand off' control to a separate ground station.

Power plant: One Honeywell T5317A-1 turboshaft, flat rated at 1,007 kw (1,350 shp). Fuel capacity 865 litres (228.5 US gallons; 190 Imp gallons).

K-MAX

Dimensions, External	
Overall	
length	15.85 m (52 ft 0 in)
height, to centre of rotor hubs	4.14 m (13 ft 7 in)
width, rotors turning	15.67 m (51 ft 5 in)
Fuselage, length	12.73 m (41 ft 9¼ in)
Wheels	
wheelbase	4.11 m (13 ft 5¾ in)
wheel track	3.68 m (12 ft 1 in)
Rotors, rotor diameter	14.73 m (48 ft 4 in)

K-MAX

Areas	
Rotor disc, total	340.9 m² (3,669.4 sq ft)
Weights and Loadings	
Weight	
Operating weight, empty	2,334 kg (5,145 lb)
Max T-O weight	
without jettisonable load	2,948 kg (6,499 lb)
with jettisonable load	5,443 kg (11,999 lb)
Fuel weight, Max fuel weight	705 kg (1,554 lb)
Payload	
Max payload, underslung	2,721 kg (5,998 lb)
Performance	
Climb	
Rate of climb, at S/L	762 m/min (2,500 ft/min)
Altitude	
Service ceiling	4,570 m (15,000 ft)
Hovering ceiling	
IGE, AUW, at 2,721 kg (6,000 lb)	8,015 m (26,300 ft)
OGE, at 2,721 kg (6,000 lb)	8,875 m (29,120 ft)
Speed	
Never-exceed speed	
clean	100 kt (185 km/h; 115 mph)
with external load	80 kt (148 km/h; 92 mph)
Range, max fuel	300 n miles (555 km; 345 miles)
Power plant	turboshaft

Status: As of 2012, two K-MAX UASs were deployed in Helmand Province, Afghanistan in support of USMC logistic requirements. As of the given date, the K-MAX UAS BURRO system was home based at Kaman's Bloomfield, Connecticut facility and was supporting the US Army's Autonomous Technologies for Unmanned Aircraft Systems (ATUAS) and the US Navy's Advanced Autonomous Cargo Unmanned System (AACUS) programmes. Lockheed Martin further notes that additional systems were available for deployment or production.
Customers: US Army and USMC.
Contractor: Kaman Aerospace Corporation, Bloomfield, Connecticut. Lockheed Martin Mission Systems and Sensors, Owego, New York.

L-3 BAI Evolution

Type: Reconnaissance and chemical-biological agent detection mini-UAV.
Development: An offshoot of BAI's involvement in early stages of the NRL Dragon Eye programme, the Evolution first appeared in 2003, and was the smallest UAV being manufactured by the company as of early 2005. Development has continued during that year to incorporate P(Y) code GPS and other user-requested features such as a miniature communications relay payload.
Description: *Airframe:* Back-packable, five-piece quick-connect assembly comprising nose, two wings, fuselage and tail unit.
Mission payloads: Can include switchable colour daylight E-O sensor (dual forward and side-looking); monochrome low-light E-O sensor (dual forward and side-looking); IR camera (various types of microbolometer); communications relay; and chemical detection/biological collection sensor.
Guidance and control: Autopilot-controlled throughout flight, but switchable from autonomous to manual operation via four-channel UHF uplink when using the U-Drive control, allowing the user to steer the AV to left and right and control its altitude.
Transportation: Man-portable in lightweight and waterproof case.

Elements of the portable Evolution GCS *(L-3 BAI)* 1135854

L-3 BAI Evolution mini-UAV *(US Navy)* 1127269

L-3 BAI's Evolution, derived from the Dragon Eye *(L-3 BAI)* 1135853

System composition: Two to three UAVs, one GCS, laptop computer, RVT and accessories. Operable by a two-person team.
Launch: Can be launched by hand, from bungee catapult, from vehicle or by rifle-style pneumatic launcher.
Recovery: Skid landing.
Power plant: Two electric motors (rating not stated), powered by lithium battery and each driving a two-blade folding propeller.
Variants: *Evolution:* Standard version. *Description applies to this version.*
Evolution XT: Endurance increased to more than 1 hour 40 minutes.
Wide-body Evolution: Endurance increased to 2 hours 30 minutes.

Evolution

Dimensions, External	
Overall	
length	0.89 m (2 ft 11 in)
height	0.27 m (10¾ in)
Wings, wing span	1.14 m (3 ft 9 in)
Weights and Loadings	
Weight	
Weight empty	2.50 kg (5.00 lb)
Max launch weight	2.95 kg (6.00 lb)
Payload, Max payload	0.45 kg (0.00 lb)
Performance	
Altitude	
Operating altitude, typical	90 m (300 ft)
Speed, Max level speed	43 kt (80 km/h; 49 mph)
Range	5.4 n miles (10 km; 6 miles)
Endurance	45 min
Power plant	2 × electric motor

Status: In production and service in 2005-06; operationally deployed in several (unspecified) parts of the world. Believed to be out of use 2007-2008, described as obsolete by manufacturer 2009.
Customers: US Department of Defense.
Contractor: L-3 Unmanned Systems Inc (BAI Aerosystems division) Easton, Maryland.

L-3 Cutlass

Type: Tube-launched expendable UAV.
Development: On 5 May 2009, company (then known as L-3 Geneva Aeerospace) announced award of a USD49.7 million contract from the US Air Force Research Laboratory as the first-year instalment of a potentially five-year contract under the USAF's SUAS Research and Evaluation (SURE) programme. Ground launch was successfully demonstrated at Fort Hood, Texas, on 4 June that year. Spokesman at a technical conference in Washington in July 2010 intimated that air-launch tests could take place "within 18 months".
Description: *Airframe:* Oval-section fuselage; flip-out swept wings and non-swept tail surfaces. Composites (GFRP) construction.

Mission payloads: Low-light TV camera, steerable +80/–120° in azimuth, in undernose ball turret; interchangeable with infra-red camera. Video downlink (analogue or digital) 2.4 GHz. Other options include targeting devices, including moving vehicle tracker, and explosive warhead.

Guidance and control: Laptop GCS with L-3 flightTEK flight control/mission management computer and missionTEK control software. Primary command link 902-928 MHz.

System composition: Three air vehicles, laptop GCS and remote video terminal.

Launch: Tube-launched from aerial platform or surface-based (ground or ship) zero-length ramp.

Recovery: Non-recoverable.

Power plant: One Hyperion AXI 2826 electric motor (rating not stated), powered by lithium-polymer batteries; two-blade folding pusher propeller.

Cutlass

Dimensions, External	
Overall	
length	0.83 m (2 ft 8¾ in)
height	0.114 m (4½ in)
Wings, wing span	1.40 m (4 ft 7 in)
Weights and Loadings	
Weight, Max launch weight	6.8 kg (14.00 lb) (est)
Payload, Max payload	1.4 kg (3.00 lb)
Performance	
Speed	
Max level speed, EAS	85 kt (157 km/h; 98 mph)
Cruising speed, max, EAS	65 kt (120 km/h; 75 mph)
Radius of operation, mission	30 n miles (55 km; 34 miles)
Endurance, max	1 hr
Best glide ratio	10
Power plant	1 × electric motor

Status: Follow-on contract options apparently not taken up, but Cutlass was continuing to be promoted in mid-2011.

Contractor: L-3 Unmanned Systems Inc
Carollton, Texas.

L-3 Mobius

Type: MALE UAV.

Development: In 2007, Mobius represented the latest in a decade or more of various attempts to develop unmanned or optionally piloted aircraft based on the well-known canard sportplane designs of Burt Rutan. Mobius is the outcome of trials conducted with an airframe called Berkut, a modified Rutan Long-EZ acquired by Geneva Aerospace (now a division of L-3 Communications) from Berkut Engineering & Design Inc (BEDI). A militarised version of the Berkut, named Bright Eagle, was investigated by the US Air Force Battlelab in 2003 as a potential long-endurance elint platform carrying a targeting pod for the HARM anti-radiation missile, but this was not pursued further.

In January 2006, Geneva announced successful completion of an unmanned test flight at the Yuma Proving Ground by a Berkut outfitted with a full suite of the company's command, control and communication equipments, in a programme supported by US Naval Air Systems Command and the US Air Force UAV Battlelab. One outcome of this developing technology was expected to help land large UAVs autonomously on naval ships. Next step in the evolution of Mobius involved installation and testing of an automatic landing system in one of Geneva's Dakota UAVs as a prelude to incorporation in the Berkut testbed. Successful completion of six consecutive autolandings by the Dakota, also at Yuma, was announced in March 2006.

In April 2007, the US Air Force UAV Battlelab announced that the Mobius, which had been modified by BEDI from the manned Berkut, had successfully performed two fully independent flights at Yuma, including autonomous take-offs and landings, on 1 March 2007. These tests were part of a larger initiative to demonstrate the ability to control two independent payloads via secure remote datalinks.

L-3 subsequently commissioned two more Mobius air vehicles for optionally piloted operation, one of which began flying in 2009.

Specification data for the Berkut are given as an general guide.

Description: *Airframe:* Rear-mounted sweptback wings, with swept winglets; non-swept canard surfaces; pod-shaped fuselage, with cockpit area faired over; rear-mounted pusher engine. All-composites construction. Retractable tricycle landing gear.

Mission payloads: Data acquisition and telemetry systems for early Berkut/Mobius trials supplied by Pi Research. Other) payloads later trialled, including the Wescam MX-15 sensor.

Guidance and control: The January 2006 Berkut test aircraft was outfitted with L-3 Geneva's trademarked flightTEK, linkTEK and missionTEK equipments. The first of these is a flight control computer, featuring the company's Variable Autonomy Control System (VACS) software, that serves as a total mission management system. The linkTEK is a network-centric solution blending both LOS and BLOS communications into a single data router that can handle data and video telemetry needs, while missionTEK is a GCS software product with an intuitive interface giving an operator the power to command and control multiple assorted UAVs with minimal training.

Launch: Fully automatic wheeled take-off.

Recovery: Fully automatic wheeled landing.

Power plant: A: One 186 kW (250 hp) Lycoming IO-540 flat-six engine, driving a two-blade pusher propeller. Standard fuel capacity 220 litres (58.0 US gallons; 48.3 Imp gallons).

Berkut, Mobius

Dimensions, External	
Overall	
length, Berkut	5.64 m (18 ft 6 in)
height, Berkut	2.29 m (7 ft 6¼ in)
Wings	
wing span, Berkut	8.13 m (26 ft 8 in)
Areas	
Wings	
Gross wing area, Berkut	10.22 m² (110.0 sq ft)
Weights and Loadings	
Weight	
Weight empty, Berkut	553 kg (1,219 lb)
Max T-O weight, Berkut	998 kg (2,200 lb)
Payload	
Max payload	
internal and external, Mobius	454 kg (1,000 lb)
Performance	
Altitude	
Service ceiling, Berkut	7,620 m (25,000 ft)
Speed	
Max level speed	
Berkut	239 kt (443 km/h; 275 mph)
Mobius	217 kt (402 km/h; 250 mph) (est)
Stalling speed, Berkut	57 kt (106 km/h; 66 mph)
Endurance, max, Mobius	20 hr (est)

Status: Development continuing. Further trials of the original Mobius were planned for later in 2007 following closure of the UAV Battlelab. A later Mobius air vehicle flew publicly with a safety pilot during the AUVSI air demonstration at Webster Field, Maryland in October 2009, with L-3 Communications being the sponsor.

Contractor: L-3 Unmanned Systems Inc
Carollton, Texas.

L-3 Viking 400

Type: Short-range tactical UAV.

Development: The Viking series replaced some of BAI Aerosystems' (now part of L-3 Unmanned Systems) earlier UAV designs in 2005. Viking 300 replaced the TERN-P; Viking 400 supersedes the Isis. Development of the 400 was initiated in February 2005 in response to a customer requirement for a UAV able to carry a payload of between 22.7 and 34.0 kg (50 and 75 lb) and have an endurance of approximately 12 hours.

Description: *Airframe:* Typical pod and twin tailboom configuration; high-mounted wings; pusher engine; fixed tricycle landing gear. Moulded composites monocoque construction.

Mission payloads: Designed for customer-supplied (unspecified) payload; 250 W of onboard power available.

Guidance and control: Cloud Cap Piccolo guidance system.

System composition: Typically, two air vehicles, one GCS and five operators.

The Berkut Mobius technology demonstrator *(USAF)* 1290169

L-3 Viking 400 *(L-3 BAI)* 1175725

The Viking family of UAVs *(L-3 BAI)*

Launch: Conventional wheeled take-off on semi-prepared surfaces.
Recovery: Conventional wheeled landing, with hydraulic brakes, on semi-prepared surfaces.
Power plant: One 25.4 kW (34 hp) two-cylinder two-stroke with 500 W alternator; two-blade fixed-pitch propeller; 28.3 kW (38 hp) rotary engine optional.
Fuel capacity 106 litres (28.0 US gallons; 23.3 Imp gallons).

Viking 400

Dimensions, External	
Overall	
length	4.44 m (14 ft 6¾ in)
height	1.52 m (4 ft 11¾ in)
Wings, wing span	6.10 m (20 ft 0¼ in)
Weights and Loadings	
Weight	
Weight empty	101 kg (222 lb)
Max T-O weight	204 kg (449 lb)
Payload, Max payload	27.2 kg (59 lb)
Performance	
Altitude	
Operating altitude	915 m to 1,525 m (3,000 ft to 5,000 ft)
Service ceiling	3,050 m (10,000 ft)
Speed	
Never-exceed speed	130 kt (240 km/h; 149 mph)
Max level speed	93 kt (172 km/h; 107 mph)
Cruising speed	60 kt (111 km/h; 69 mph)
Stalling speed	35 kt (65 km/h; 41 mph)
Range	
control	43 n miles (79 km; 49 miles)
max	625 n miles (1,157 km; 719 miles)
Endurance	12 hr
Power plant	1 × piston engine

Status: As of January 2006, the Viking 400 was undergoing developmental testing at the Yuma Proving Grounds, Arizona. SOCOM signed a five-year USD250 million deal for Viking 400 for the Expeditionary Unmanned Aircraft System programme in late 2009.
Contractor: L-3 Unmanned Systems Inc
Carollton, Texas.

Lockheed Martin 420K

Type: AEW and surveillance tethered aerostat.
Development: IHS Jane's sources report that Lockheed Martin Mission Systems and Sensors' (LM MS2) portfolio of aerostats (of which, the 420K is one) were originally developed by Goodyear (subsequently Loral Defense Systems - Akron and then LM MS2), with Lockheed Martin estimating that it (and those lighter-than-air manufacturers that it has absorbed) have fielded at least 8,000 aerostats and 300 airships between them since 1911. Most recently (and apart from the 420K), LM MS2's aerostat portfolio has included the 56K, 64K, 74K, 275K and 595K types, with the alpha-numerical designators referring to the particular aerostat's envelope capacity measured in thousands of cubic feet. Key 420K and related programme events include:

1992 The then Lockheed Martin Training and Technical Services business unit was appointed support contractor for the USAF's Tethered Aerostat Radar System (TARS) that stretches along America's border with Mexico and along the country's Gulf Coast.
29 August 1996 The USAF awarded the then Lockheed Martin Services business unit a then year USD22,694,989 fixed price award fee TARS Operation, Maintenance and Support (OM&S) contract that was to run until 2001 and included the provision of OM&S services (including warehousing) at 11 TARS sites; OM&S support for TV Martí (an aerostat-based US television propaganda system that was directed at Cuba) and support and services for those organisations that made use of TARS derived data, TARS interfaces and "TARS associated" equipment and sensors.
31 August 1998 Lockheed Martin was awarded a then year USD22,882,449 contract modification (to its existing TARS OM&S award) with regard to the operation, maintenance and support of the TARS system and TV Martí during US FY1999 (up to and including 30 September 1999).

A TARS 420K at its mooring mast *(Lockheed Martin)*

First Quarter 2000 Lockheed Martin was awarded a then year USD13.5 million contract with regard to the upgrading of six TARS sites (Deming in New Mexico; Fort Huachuca in Arizona; Horseshoe Beach in Florida (subsequently closed), Matagorda in Texas (subsequently closed); Morgan City in Louisiana (subsequently closed) and Yuma in Arizona) during the period 2000 to 2005. As a part of this, Lockheed Martin was to replace the existing TCOM model 71M® aerostat/E-LASS radar combinations at these sites with a Lockheed Martin 420K aerostat/L-88 radar package.
13 June 2001 The then Lockheed Martin Technology Service Group (Cherry Hill, New Jersey) was awarded a then year USD 210,942,078 firm, fixed-price, award fee contract with regard to the provision of US FY2002 OM&S services for the TARS network.
14 February 2002 Lockheed Martin was awarded a then year USD79,104,464 (maximum) indefinite delivery/quantity contract with regard to the supply of up to 11 × L-88(V)3 radars for use in the TARS system. At the time of its announcement, work on the effort was scheduled for completion by the end of February 2008.
March 2002 As of March 2002, TARS sites that had completed transition to the 420K aerostat/L-88 radar combination are understood to have included those at Deming, Fort Huachuca and Yuma.
30 March 2002 A 420K aerostat located at Rio Grande, Texas broke away from its moorings and drifted more than 261 n miles (483 km; 300 miles) before coming to rest on private land near Burnet in Texas. During its flight, the aerostat's trailing tether damaged a number of power lines, interrupting power supplies in 'several' Texan counties.
16 July 2003 Lockheed Martin was awarded a then year USD4 million order with regard to the supply of an L-88(V) radar for installation aboard the company's 275K aerostat located at the TARS site at Cudjoe Key, Florida. Due to the nature of the site, Cudjoe Key was (and is) the only TARS station to use the 275K vehicle rather than the 420K.
3 October 2007 The then Lockheed Martin Information and Technology Services business unit (Cherry Hill, New Jersey) was awarded a then year USD20,434,996 contract modification with regard to the operation, maintenance and support of nine TARS sites within the continental US during the period 1 October 2007 to 30 September 2008.
9 July 2008 The ITT Systems Division (subsequently ITT Exelis Mission Systems, both headquartered at Colorado Springs, Colorado) was awarded a then year USD33,697,369 fixed-price, cost plus award fee (with reimbursable line items) contract with regard to the operation, maintenance and support of eight operational TARS sites '24 hours a day, seven days per week' together with the provision of "cradle-to-grave" life-cycle management for the entire network.
31 August 2011 ITT (Colorado Springs, Colorado) was awarded a then year USD32,196,975 firm, fixed-price contract modification with regard to

An aerial view of the TARS site at Cudjoe Key, Florida *(Lockheed Martin)*

the provision of "resources" for the TARS programme in order to ensure 'effective and efficient support for [US] counterdrug/counter-narco terrorism and [USAF] North American Aerospace Defense Command's air sovereignty missions'.

1 October 2012 ITT Exelis (Newport News, Virginia) was awarded a then year USD35,794,575 contract modification with regard to the provision of an 'air and surface surveillance capability' (the TARS system) in support of the North American aerospace Defense (NORAD) Command's air sovereignty mission and the US DoD's Counterdrug Program. Work on the effort was to be undertaken at facilities in Cudjoe Key, Florida; Deming, New Mexico; Eagle Pass, Texas; Fort Huachuca, Arizona; Lajas, Puerto Rico; Marfa, Texas; Newport News, Virginia; Rio Grande City, Texas and Yuma, Arizona and at the time of its announcement, the programme was scheduled for completion by 30 September 2013.

Description: *Airframe:* Conventional Class IV envelope shape, with inverted Y tail surfaces and large ventral cover to accommodate the radar antenna.

Mission payloads: Lockheed Martin L-88(V)3 L-band radar standard (12.19 m (40.0 ft) diameter radome); provision for additional payloads, such as IR, LLTV, communications relay or elint equipment, as required. Envelope can accommodate radar, sigint or EO payloads measuring up to 8.84 m (29 ft) wide by 5.18 m (17 ft) in height. Available power: 5.8 kW standard, 7.5 kW optional.

Guidance and control: The aerostats detect, identify and track targets and relay data through a GCS for assessment at a C^3I centre. They can also relay intercept commands to appropriate command assets and provide continuous mission updates and assessment. Ground crew of five.

Launch: Can be launched in winds up to 35 kt (65 km/h; 40 mph).

Recovery: Can be recovered in winds up to 40 kt (74 km/h; 46 mph).

420K

Dimensions, External	
Envelope	
length	63.55 m (208 ft 6 in)
diameter	21.18 m (69 ft 5¾ in)
Dimensions, Internal	
Envelope, volume	11,893 m³ (419,997 cu ft)
Performance	
Altitude, Operating altitude	4,575 m (15,000 ft)
Range, radar detection range	200 n miles (370 km; 230 miles)
Endurance, weather dependent	120-168 hr

Status: As of 2012, the Lockheed Martin 420K aerostat/L-88 radar combination was in service at the TARS sites at Deming, New Mexico; Eagle Pass, Texas; Fort Huachuca, Arizona; Lajas, Puerto Rico; Marfa, Texas; Rio Grande City, Texas and Yuma, Arizona. The TARS site at Cudjoe Key, Florida was equipped with a Lockheed Martin 275K aerostat/L-88 radar package for geographical reasons.

Customers: US Air Force Air Combat Command.

Contractor: Lockheed Martin Maritime Systems and Sensors Akron, Ohio.

Lockheed Martin Desert Hawk

Type: Close-range surveillance mini-UAV.

Development: This small UAV was developed by Lockheed Martin's Skunk Works in 2001 in response to a USAF Electronic Systems Center programme, known as Force Protection Airborne Surveillance System (FPASS), to address a Central Command (CENTCOM) requirement for enhanced security at US overseas bases. Revealed in mid-2002, originally under the manufacturer's name SentryOwl, it was designed for a small footprint (typically 0.9 m²; 10 sq ft). Other applications were foreseen as including indirect fire targeting, BDA, urban site security and convoy escort.

Under the Central Command Air Force (CENTAF) name Desert Hawk, it was quickly adopted (late February 2002) by the US Air Force. Two prototypes were flown within two weeks of contract award, followed by

Desert Hawk III and travel case *(IHS/Patrick Allen)*

The hand-launched Desert Hawk III in service with the British Army *(Crown copyright)*

two others with minor design changes, before the design was frozen in May 2002. The first two systems were delivered in late June or early July 2002 to the Electronic Systems Center at Hanscom AFB, Massachusetts. Six more systems (48 air vehicles), ordered under a March 2002 Lockheed Martin subcontract to AeroMech Engineering Inc, were delivered to USAF in October 2002, and by mid-November were in use in Afghanistan as base protection assets in Operation 'Enduring Freedom'. Lockheed Martin was responsible for ground station development, payload configuration, antenna design, customer support and overall programme management; GCS software and aircraft avionics, including the miniature autopilot, were supplied by Jennings Engineering. The original Desert Hawk I was later upgraded to I+ standard. The completely redesigned and improved Desert Hawk III was introduced in 2006 and entered service in mid-2007.

Description: *Airframe: Desert Hawk I/FPASS:* High-mounted, semi-elliptical wings with elevons; bottle-shaped fuselage; Y tail unit (V tailfins with rudders, plus underfin); conventional aerodynamic control surfaces; pusher engine. No landing gear. Composites construction (mould-injected expanded polypropylene foam); tail unit attached by Velcro straps; Kevlar skid panels under nose and tail. Manufactured by AeroMech Engineering Inc of San Luis Obispo, California.

Desert Hawk III: Redesigned and more robust, with less bulky (more tapered) fuselage, front-mounted engine, turreted EO sensors and conventional tail unit.

Mission payloads: Desert Hawk I: Unstabilised combined EO/IR (colour TV, LLTV and thermal imager), or trio of colour TV cameras (interchangeable), with real time imagery downlink and digitally recorded video for subsequent analysis. Other potential payloads can include sensors for detecting IEDs or buried explosives, or for other specialised missions.

Desert Hawk III: Five different 'plug and play' payloads available: including: a two-axis, inertially stabilised EO turret; a fixed, side-looking combined EO/IR module comprising a colour TV, black and white LLTV and a long-wave thermal imager; and a low-light illuminator combining an EO imager responsive in the near-IR band with a near-IR laser illuminator. In September 2009 tests of Desert Hawk III with a turret offering 360-degree IR and colour EO coverage were completed. The payload weight in this application was reported to be less than 1 kg.

Adjusting the camera in a Desert Hawk I at Tallil AB, Iraq *(US Air Force)*

Lockheed Martin Desert Hawk < United States < Unmanned aerial vehicles

Quad bike-assisted hand launch of a Desert Hawk III *(Crown copyright)* 1293582

Bungee cord launch of a Desert Hawk I in Afghanistan *(US Air Force/Tech Sgt Christopher Gish)* 1151621

Guidance and control: Autonomous or remotely piloted operation via portable, laptop-based, two-person GCS and a remote video terminal; can be reprogrammed in flight. Antennas can be free-standing or concealed in trees or buildings. Air vehicle has terrain avoidance capability, and can be controlled to a position accuracy of less than 25 m (82 ft), a velocity accuracy of less than 2 m (6.6 ft)/s and a landing radius accuracy of less than 50 m (165 ft); it also has a fail-safe 'return home' mode. Downloaded imagery can be digitally recorded for analysis. Power requirements are 250/120 V AC and 50/60 Hz.

Desert Hawk III GCS retains the graphical user interface and advanced software of the I+ system, but has two configurations, each with an optional RVT. First has independent communications and antennas; second has integrated systems. Other features include a touch-screen laptop and digital terrain elevation data. Latter is uploaded to the air vehicle at the time of launch, supporting lost link or non-LOS communications disruptions.

In an August 2008 exercise at NAS Oceana, Virginia, Lockheed Martin successfully demonstrated a technology suite named ICARUS - Intelligent Control and Autonomous Replanning of Unmanned Systems. In this trial, control of the sensors on board the Desert Hawk III and an unmanned surface vehicle was handed off between ICARUS consoles within the tactical operations centre, the mobile C^2 unit and soldiers on the ground. Throughout the exercise, ICARUS planned and replanned vehicle operations to meet task requests, enabling a single operator to act as mission manager.

Transportation: Desert Hawk I/FPASS is shipped in sturdy, waterproof containers. A complete system takes up a space of 2.44 × 2.44 × 1.22 m (8 × 8 × 4 ft) and weighs approximately 236 kg (520 lb). Desert Hawk III is backpack or suitcase portable.

System composition: Standard system comprises six air vehicles with EO/IR payloads; portable ground station; RVT; field repair kit; and GPS satellite tracking and location; GCS displays, communications and software; transportation cases; launch equipment; and two-person crew. Set-up time 10 minutes. British Army Desert Hawk batteries deploy them in three-aircraft detachments.

Launch: Desert Hawk I: Bungee cord catapult. Engine starts when aircraft velocity reaches 15 m (50 ft)/s.

Desert Hawk III: Designed for hand launch. British Army practice is to do this from a moving quad bike rather than a standing start.

Recovery: Automatic approach, engine shutdown and glide to belly landing. Desert Hawk III airframe is designed to reduce impact stresses by intentionally breaking up into nine parts on touchdown.

Power plant: A: 1 kW battery-powered, rear-mounted electric motor; two-blade pusher propeller. Rechargeable lithium-ion batteries.

B: Battery-powered, front-mounted electric motor and two-blade tractor propeller.

Variants: Desert Hawk I: Initial US Central Command Air Force (CENTAF) version, delivered from July 2002. US Central Command (CENTCOM) name FPASS. Developed to enhance security at overseas bases by conducting area surveillance, patrolling base perimeters and runway approach/departure paths, and performing convoy overwatch.

Desert Hawk I+: Upgraded version (FPASS II) for USAF and UK, featuring new communications, imagery manipulation and Tactical Automated Security System (TASS) integration.

Desert Hawk III: Redesigned and more sophisticated version, with entirely redesigned airframe, to new build standard that includes enhanced 'plug and play' IR and colour or black and white EO imager payloads.

Desert Hawk I, Desert Hawk I+, Desert Hawk III

Dimensions, External	
Overall length	
Desert Hawk I, Desert Hawk I+	0.86 m (2 ft 9¾ in)
Desert Hawk III	0.91 m (2 ft 11¾ in)
Wings wing span	
Desert Hawk I, Desert Hawk I+	1.32 m (4 ft 4 in)
Desert Hawk III	1.37 m (4 ft 6 in)
Dimensions, Internal	
Payload bay volume, Desert Hawk III	0.00472 m³ (0.2 cu ft)
Areas	
Wings	
Gross wing area, Desert Hawk III	0.325 m² (3.5 sq ft)
Weights and Loadings	
Weight	
Weight empty, Desert Hawk III	2.95 kg (6.00 lb)
Max launch weight	
Desert Hawk I, Desert Hawk I+	3.2 kg (7.00 lb)
Payload	
Max payload	
Desert Hawk I, Desert Hawk I+	0.5 kg (1.00 lb) (est)
Desert Hawk III	1.0 kg (2.00 lb)
Performance	
Altitude	
Operating altitude, normal	150 m (500 ft)
Service ceiling	
launch density altitude	2,285 m (7,500 ft)
Speed, Max level speed	50 kt (93 km/h; 58 mph)
Radius of operation	
Desert Hawk I, Desert Hawk I+	6 n miles (11 km; 6 miles)
Desert Hawk III	8.0 n miles (14 km; 9 miles) (est)
Endurance	
Desert Hawk I, Desert Hawk I+	1 hr (est)
Desert Hawk III	1 hr 30 min (est)
Power plant	1 × electric motor

Status: In production and service. In addition to USAF and British Army, system has been supplied to other US agencies, including the Department of Homeland Security.

Customers: Desert Hawk I: US Air Force and other US agencies. USAF systems were deployed in Operation 'Enduring Freedom' (Afghanistan) for airbase protection in 2002, eight systems having been delivered by the end of that year, with 12 more scheduled to follow in 2003-04. Also deployed by 332nd Expeditionary Security Forces Squadron (ESFS) from Tallil AB during Operation 'Iraqi Freedom' in 2003, and by other ESFS later. According to a February 2006 release, the USAF inventory then stood at 21 Desert Hawk systems (126 air vehicles). As of late 2006, USAF had 48 systems on order and the type continued in use in both theatres, including 20 systems in Afghanistan. Operator training was given by the 99th Ground Combat Training Squadron Desert Warfare Training Center at Creech AFB, Nevada. Replacement by RQ-11B Ravens began in October 2007, and at that year-end the revised DoD Roadmap stated the

Desert Hawk III *(IHS/Patrick Allen)* 1346421

planned Desert Hawk inventory to be 18 systems, which would represent 108 aircraft.

The UK MoD acquired a small number (reportedly four) of Desert Hawk I systems for use in Iraq in Ocober 2003; these were later upgraded to I+ standard. All Desert Hawk I out of British service by 2008

Desert Hawk I+ and III: British Army (four batteries of 32 Regiment, Royal Artillery: Nos. 18, 22, 42 and 57) and No. 1 Battery, 47 Regiment. In February 2006, the UK MoD awarded Lockheed Martin a USD2.65 million contract to supply four new Desert Hawk III systems, and upgrade the existing fleet to the new build standard, for delivery the following month. Initial deployment was to Camp Bastion, Afghanistan, in July 2006. As of IOC in June 2007, three batteries of 32 Regiment, totalling 25 three-aircraft flying detachments, had converted from Phoenix to Desert Hawk I+ and one battery (eight detachments) had converted to Desert Hawk III. Additional (unspecified) quantities of Desert Hawk III were ordered in April (USD6.3 million) and November 2007 USD4.8 million) contracts. Nos. 18, 42 and 57 Batteries deployed in Afghanistan, No, 22 in Iraq. Total of 27 lost in Afghanistan by early 2008. Use by 47 Regiment's No. 1 Battery in southern Afghanistan was revealed in August 2008. UK MoD reports 187 Desert Hawk III ordered or in service in 2008.

Contractor: Lockheed Martin Aeronautics Company , Palmdale Operations Palmdale, California.

Lockheed Martin HALE-D/HAA

Type: High-altitude unmanned airship demonstrator (HALE-D)/proposed high-altitude unmanned airship sensor platform (HAA).

Development: Lockheed Martin Mission Systems and Sensors has developed a High-Altitude Long-Endurance Demonstrator (HALE-D) unmanned airship to verify the technology needed to create a proposed unmanned High-Altitude Airship (HAA) sensor platform. As understood by *IHS Jane's*, the HALE-D vehicle has a length and diameter of 73.2 m (240.2 ft) and 21.3 m (69.9 ft) respectively, a hull volume of about 14,158 m^3 (499,986 ft^3), a cruising speed of 20 kts (37 km/h; 23 mph) at a station-keeping altitude of 18,288 m (60,000 ft), an endurance of more than 15 days and the ability to carry an up to 36.3 kg (80 lb) 'user-defined' payload. A payload reported as comprising a 30 to 512 MHz band Thales Communications Multi-channel Multi-band Airborne Radio, an ITT Exelis high-resolution EO system and an L-3 Communications Mini Common DataLink [MCDL]). Again, the HALE-D design makes use of a pair of 2 kW electric propulsion motors and an array of thin-film solar cells (rated at 15 kW) and re-chargeable lithium polymer batteries as its power source. Here, the vehicle's battery provision is billed as providing 40 kW/h of power storage. The HALE-D payload is further noted as requiring 150 W of power and the vehicle as whole is both recoverable and reusable.

Key programme events include:

29 September 2003 The US Missile Defense Agency (MDA) selected Lockheed Martin to develop an unmanned High-Altitude Airship (HAA) sensor platform that would provide high-altitude, persistent, anti-ballistic missile surveillance and which would be able to lift a 1,814 kg (3,999 lb) payload into a quasi-geostationary orbit at an altitude of 19,800 m (65,000 ft). According to *IHS Jane's* sources, Lockheed Martin was in competition with the Worldwide Aeros Corporation and Boeing for the HAA programme, with the same sources describing the effort as envisaging a prototype AV with an endurance of approximately one month (the operational HAA was to be able to remain airborne for up to a year)

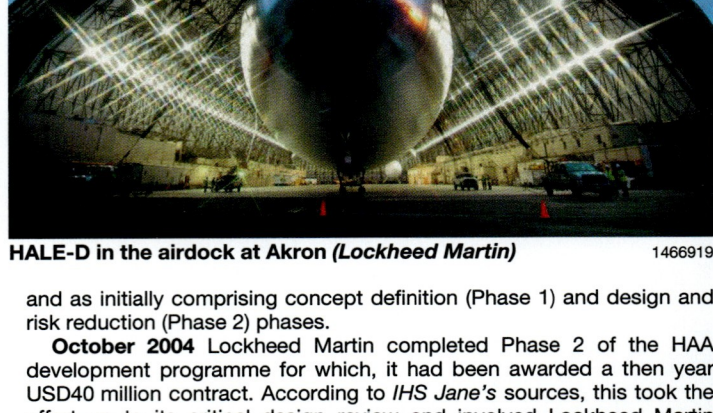

HALE-D in the airdock at Akron *(Lockheed Martin)* 1466919

and as initially comprising concept definition (Phase 1) and design and risk reduction (Phase 2) phases.

October 2004 Lockheed Martin completed Phase 2 of the HAA development programme for which, it had been awarded a then year USD40 million contract. According to *IHS Jane's* sources, this took the effort up to its critical design review and involved Lockheed Martin Advanced Development Programs (Palmdale, California), Lockheed Martin Space Systems (Denver, Colorado) and StratCom International (Keedysville, Maryland).

26 July 2005 The MDA announced its intention of awarding Lockheed Martin a Phase 3 HAA contract which would include completion and testing of a prototype AV (the HALE-D). According to *IHS Jane's* sources, the HALE-D was then specified to demonstrate a one-month loiter time (at 18,300 m (60,000 ft) while carrying an up to 36.3 kg (80 lb) payload and providing 150 W of continuous payload power) as well as effective launch and recovery, station-keeping and autonomous flight control.

8 December 2005 The MDA awarded Lockheed Martin a then year USD145.2 million Phase 3 HAA contract.

February 2007 *IHS Jane's* sources report the MDA as having exercised an HAA technology improvement project contract with regard to the development of 'key technologies' for the operational vehicle.

April 2008 *IHS Jane's* sources report oversight of the HAA programme as having been transferred from the MDA to the US Army's Space and Missile Defense Command.

27 July 2011 The HALE-D technology demonstrator made its maiden flight from the Akron Airdock in Ohio. Take-off was at 05:47 local time and the AV ascended to approximately 9,754 m (32,000 ft) before a helium level anomaly resulted in a decision to terminate the flight. Thereafter, the airship was brought down at a 'pre-determined location' in southwestern Pennsylvania shortly before 08:30 local time. In an associated release, Lockheed Martin's then Vice President, Ship and Aviation Systems Dan Schulz characterised the flight as having not reaching its target altitude but as having demonstrated 'a variety of advanced technologies including launch and control of the airship, [its] communications links, [its] unique propulsion system, solar array electricity generation, [a] remote piloting communications and control capability, in-flight operations and controlled vehicle recovery to a remote unpopulated area'. As of late July 2011, Lockheed Martin was understood to have been recovering the AV and to have been undertaking a 'full evaluation' of the incident.

October 2012 During October 2012, the US Government Accountability Office (GAO) published a report entitled *Defense Acquisitions - Future Aerostat and Airship Investment Decisions Drive Oversight and Coordination Needs* in which it outlined (amongst other things) the status of the HALE-D programme. Here, the report noted that following its 27 July 2011 crash, the HALE-D demonstrator's envelope and solar cells were destroyed and its payload was damaged by fire

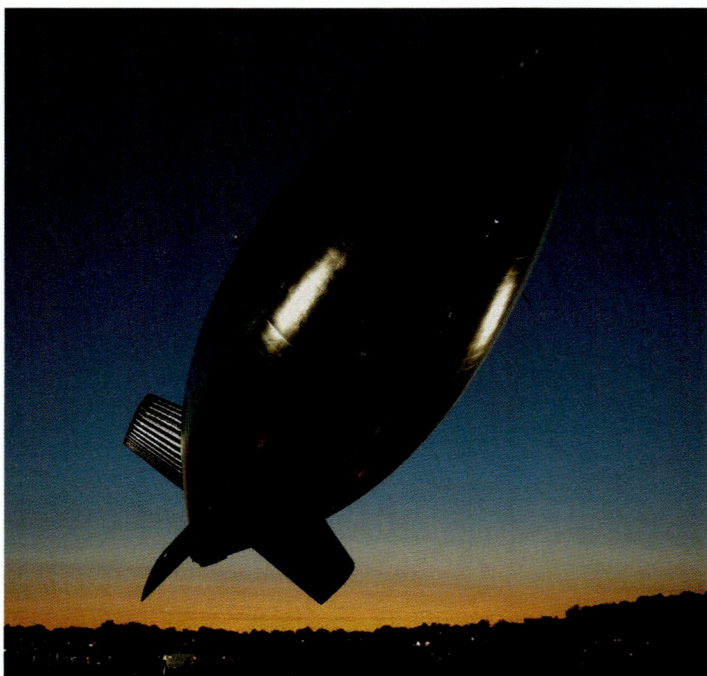

HALE-D pictured on its first flight, 27 July 2011 *(Lockheed Martin)* 1398665

HALE-D at Akron *(Lockheed Martin)* 1466920

HAA duties may include short- and long-range missile warning, surveillance and target acquisition, communications relay, and weather/environmental monitoring *(Lockheed Martin)* 1130303

during recovery operations. At the time of the report, the GAO was noting that the HALE-D programme office was stating that it did not have funding for a 'continued [HALE-D] demonstration effort'. Up to this point, the HALE-D/HAA programme is understood to have cost then year USD36.3 million during the period US FYs 2008 to 2011, with the total breaking down into then year USD16.7 million, then year USD11.5 million, then year USD5.3 million and then year USD2.8 million in RDT&E expenditure during US FYs 2008, 2009, 2010 and 2011 respectively.

The following *Description*, *Airframe*, *Mission payloads*, *Guidance and control*, *Launch* and *Recovery* sub-entries relate to the HALE-D demonstrator unless stated otherwise.

Description: *Airframe:* Typical semi-rigid airship shape, with four tail surfaces in X configuration. Envelope skin of Vectran high strength/weight ratio material for durability.

Mission payloads: One × 2.9 kg (6.6 lb) Thales Communications 30-512 MHz band Multi-channel Multi-band Airborne Radio (MMAR - prototype configuration), one × 17 kg (37.5 lb) ITT Exelis high-resolution EO system and one × <0.9 kg (2 lb) L-3 Communications Mini Common DataLink (MCDL) unit. Potential HAA payloads have been billed as including air and missile surveillance sensors, communications relay packages, weather monitoring equipment and a directed-energy weapon system.

Guidance and control: Single fixed-site GCS with BLOS communications. A mobile GCS is understood to have been contemplated for use with the HAA if it were to be deployed outside the continental US. In the proposed HAA, propulsion units would be used to maintain the AV's geostationary position above the jetstream, propel it aloft and guide its take-off and landing during ascent and descent.

Launch: Conventional airship take-off.

Recovery: Conventional airship recovery.

Status: As of October 2012, the HALE-D programme office was being reported as being unable to fund a 'continued [HALE-D] demonstration effort' following the prototype's 27 July 2011 crash and effective loss (see previously). Accordingly (and as of the given date), the future of the HAA programme was in doubt.

Contractor: Lockheed Martin Mission Systems and Sensors
Akron, Ohio.

Lockheed Martin ISIS

Type: Persistent radar surveillance unmanned airship technology demonstrator.

Development: As its designation suggests, the Lockheed Martin/DARPA/USAF Integrated Sensor Is the Structure (ISIS) unmanned airship demonstrator is intended to validate the concept of an operational, stratospheric, autonomous, unmanned sensor system that would be capable of a time on station that is measured in years (at least one and possibly up to 10) and which would be able to detect and track air and ground targets at ranges of between 162 n miles (300 km; 186 miles - dismounts) and 324 n miles (600 km; 373 miles - 'advanced' cruise missiles) from an altitude of between 19,812 and 21,336 m (65,000 and 70,000 ft).

Key programme events include:

US FY2004 Phase 1 (a feasibility study including an objective system design) of the ISIS programme was launched during late US FY2004 and was completed by the end of US FY2005.

8 August 2006 Raytheon Space and Airborne Systems announced that DARPA had awarded it a then year USD8 million contract with regard to the development of a dual-band ISIS Active Electronically Scanned

A computer-generated impression of an on-station ISIS persistent surveillance airship *(Lockheed Martin)* 1365359

Array (AESA) radar and the means by which the AESA modules would be bonded to the host airship's hull.

26 September 2006 Lockheed Martin's Akron, Ohio business unit announced that it had been awarded a then year, approximately USD10 million, two year duration contract with regard to continuing development of 'advanced material technology and next-generation hull material for [use in the ISIS] stratospheric [airship]'.

US FY2008 A technology readiness review of Phase 2 of the ISIS programme is reported to have taken place in mid-US FY2008.

27 April 2009 Lockheed Martin's Advanced Development Programs business unit in Palmdale, California was awarded a then year USD100 million increment (from an existing then year USD400 million cost plus fixed fee award) with regard to Phase 3 of the ISIS programme. This increment included system design demonstration, large scale integration activities, high fidelity flight test simulation and the construction and flight testing of a 137 (L) × 150 (Ø) m (450 × 150 ft) one third scale demonstrator with an integral 600 m^2 (6,458 ft^2) AESA array.

At the time of the announcement, work on the project was to be undertaken at the following facilities:

- El Segundo, California (37 per cent workshare)
- Palmdale, California (33 per cent workshare)
- Denver, Colorado (14 per cent workshare)
- Akron, Ohio (three per cent workshare)
- Litchfield Park, Arizona (three per cent workshare)
- Mesa, Arizona (three per cent workshare)
- Frederica, Delaware (two per cent workshare)
- Huntsville, Alabama (two per cent workshare)
- Monrovia, California (two per cent workshare)
- Sunnyvale, California (one per cent workshare)

The programme was scheduled for completion by the end of March 2013. The effort's contracting activity was DARPA (Arlington, Virginia).

May 2009 During May 2009, DARPA reported back on some of the properties that the ISIS hull material had achieved against set targets. The material had the following properties:

- An area density of 90.6 g/m^2, against a required value of ≤100 g/m^2.
- A matrix glass transition tempreature of –101°C (–150°F), against a required figure of ≤–90°C (–130°F).
- A fibre strength-to-weight ration of 1,274 kN.m/kg, against a required value of ≥1,000 kN.m/kg.
- Retention of 85 % of fibre strength after a 'life' of 22 years, against a required value of >85% strength retention after five years.

For its part, the evolving AESA had demonstrated a 1.8 kg/m^2 area density (against a required value of ≤2 kg/m^2), a power consumption on receive of 4.7 W/m^2 (against a specified figure of ≤5.0 W/m^2) and viable array-to-hull bonding. Again, the architecture's transceiver modules were reported to have achieved a figure of merit of 1.1 × 10^4 W^{-2} (against a required value of ≥1 × 10^4 W^{-2}) and a mean time to failure value of >1.98 × 10^6 hours (against a specified figure of >10^6 hours). This latter value represented 'demonstrated' Technology Readiness Level 5. Last but not least, the ISIS regenerative power sub-system was rated at 779 W-h/kg against a required demonstrated value of 400 W-h/kg.

25 January 2011 Allied Techsystems (ATK) announced that Lockheed Martin had selected it to provide a Thermal Control Sub-system (TCS) for

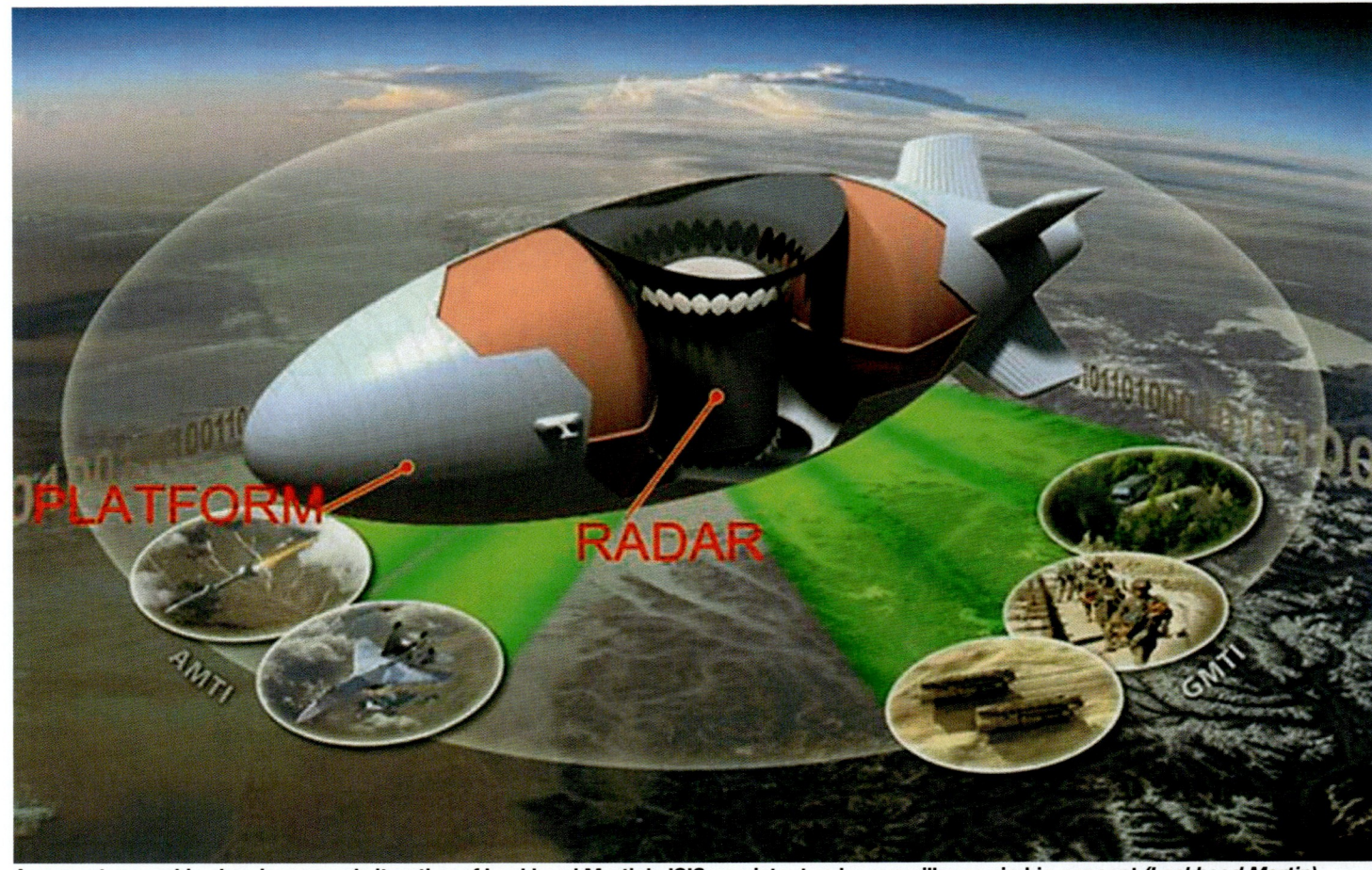

A computer graphic showing an early iteration of Lockheed Martin's ISIS persistent radar surveillance airship concept *(Lockheed Martin)*

the ISIS programme. In an associated release, ATK went on to note that the ISIS TCS contract involved it in the design, development, construction and testing of hardware that would be capable of performing heat acquisition, transport and rejection for two 'complete thermal systems'. again, the baseline design for both systems was described as comprising a pumped, two-phase, fluid loop that interfaced with 'large, lightweight' radiator panels attached to the vehicle's power bay.

October 2012 In its October 2012 report entitled *Future Aerostat and Airship Investment Decisions Drive Oversight and Coordination Needs*, the US Government Accountability Office described ISIS as having an estimated cost of then year USD506.1 million for the period US FY2007 to US FY2014 (broken down into annual payments, of then year dollars, are USD25.1 million in 2007, USD30.5 million in 2008, USD81.2 million in 2009, USD250.3 million in 2010, 0USD24.1 million in 2011, USD60.2 million in 2012, and USD34.7 million in 2013 in Research, Development, Test and Evaluation (RDT&E) funding).

Elsewhere, the report characterised the demonstrator as featuring a 'radar sensor of unprecedented proportions that is fully integrated into a stratospheric airship to support the need for persistent wide-area surveillance, tracking and engagement of time-critical air and ground targets in urban and rural environments.' As such, the ISIS demonstrator was further described as being 'expected to fly for 365 days at an altitude of 65,000 ft [19,812].' In programmatic terms, the report went on to note that ISIS had 'experienced technical challenges stemming from [sub-system] development and radar [antenna] panel manufacturing' issues as a consequence of which, DARPA had 'temporarily' delayed airframe development activities during the early part of 2012 in order to focus on radar risk reduction activities. During this hiatus, the ISIS team was scheduled to formulate an airship risk reduction plan and to conduct 'limited airship activities'. Thereafter, DARPA was expected to 'reassess the future plan for [the programme] with the [USAF].'

Description: *Airframe:* Traditional-shape hull, with four tailfins indexed in X configuration. Specially developed glass fibre composites skin. Lockheed Martin characterises the ISIS demonstrator as being 137 m (450 ft) long, as having a diameter of 46 m (150 ft) and as being equipped with an integral 600 m² (6,458 ft²) AESA array. Again, an operational ISIS AV is intended to be capable of achieving sustained and sprint airspeeds of 60 kts (111 km/h; 69 mph) and 100 kts (185 km/h; 115 mph) respectively.

Mission payloads: A lightweight, dual-band (Ultra High Frequency [300 MHz to 3 GHz] and X-band [8 to 12.5 GHz]) AESA radar that, once full developed, would be able to track 'advanced' cruise missiles and dismounted enemy combatants at ranges of up to 324 n miles (600 km; 373 miles) and 162 n miles (300 km; 186 miles) respectively. To achieve this, the operational architecture would make use of a 6,000 m² AESA (64,583 ft² - 600 m² (6,458 ft²) in the demonstrator) that would be integrated into the host airship's structure and would rely on aperture size rather than brute power to achieve the specified performance requirements. Again, the Raytheon Space and Airborne Systems ISIS radar is intended to be capable of simultaneous Air MTI (AMTI) and GMTI from an altitude of between 19,812 and 21,336 m (65,000 and 70,000 ft). Physically, the architecture makes use of up to four million, low power, transceiver modules that are based on cell phone technology and are bonded to the host AV's hull to provide the necessary aperture.

Raytheon has described the radar's transceiver modules as being approximately one centimetre thick and as being fixed in place using a bonding agent capable of withstanding an ambient temperature of −80°C (−112°F). Such an environment (combined with the sensor's minimised power requirement) eliminates the need for the sort of cooling system usually associated with high performance MTI radars, while the using aperture size rather than output power approach to achieve required performance facilitates the use of solar-regenerative power (with fuel cells instead of batteries) for both the radar and the airship's other sub-systems, most important of these is the station keeping system. Overall, the envisaged ISIS radar represents more than 30 per cent of the overall airship's mass (as against no more than three per cent in a conventional aerostat-sensor combination). For its part, the ISIS radar's GMTI mode is designed to detect 'dismounts' across its entire line-of-sight; to have target detection/location and tracking ranges of 324 n miles (600 km; 373 miles) and 162 n miles (300 km; 186 miles - both at a 3° grazing angle) respectively and to provide 'LSRS [Littoral Surveillance Radar System] - like resolution'. Again, ISIS GMTI functionality will provide wide-area foliage penetration with 'Joint STARS [Joint Surveillance Target Attack Radar System] precision across an extremely large operational area'. In AMTI mode, the ISIS sensor is designed for 'the theoretical limit at the radar horizon' (effectively, 324 n miles (600 km; 373 miles) from the cited operational altitude) and combines the search, track and fire-control functions in a 'single platform'. In terms of deployment, an operational ISIS platform would be launched from a site (or sites) within the continental US (which would also be home to a 'permanent' ISIS ground station/s); would be capable of 'global deployment' within 10 days and would have an up to 10 year service life. As such, the capability would have the potential to replace a range of existing surveillance platforms including the E-2C airborne early warning and control aircraft, the E-3 Airborne Warning And Control System (AWACS) and the E-8 Joint STARS.

Guidance and control: Autonomous operation.
Launch: Conventional, automatic, airship take-off.
Recovery: Conventional, automatic, airship landing.
Power plant: Solar-regenerative main propulsion system, powering fuel cells instead of batteries. Small engine on starboard side of nose to aid station-keeping.

Status: As of October 2012, the US Government Accountability Office (GAO) was characterising the ISIS programme as having encountered technical challenges relating to 'sub-system' development and radar panel manufacture as a result of which, DARPA had 'temporarily' delayed ISIS airframe development activities during the early part of 2012 in order to concentrate on radar risk reduction work. During this lull, the ISIS development team was expected to develop an airship risk reduction plan and to undertake 'limited airship activities'. Thereafter, DARPA and the USAF were expected to 'reassess' their future plans for the programme. The GAO further noted that the ISIS effort was expected to have cost then year USD506.1 million during the period US Fiscal Years 2007 to 2014 (see *Development*).

Contractor: Lockheed Martin Advanced Development Programs (ISIS prime),
Palmdale, California.Lockheed Martin Mission Systems and Sensors (airship construction),
Akron, Ohio.Raytheon Space and Airborne Systems (ISIS radar),
El Segundo, California.

Lockheed Martin MQ-X

Type: Hunter/killer MALE UAV.

Development: The unofficial designation MQ-X is associated with proposals for a next-generation UAS to follow on from the MQ-1 Predator and MQ-9 Reaper. According to Lockheed Martin, the US Air Force is "looking to acquire" some 200 to 250 such assets, although as of mid-November 2009 no formal programme launch had been announced. Nevertheless, in anticipation of such an event, in early October 2009 the company's Skunk Works released the accompanying image of a conceptual design for such a requirement, together with minimal descriptive detail. Performance parameters were said to include an endurance of at least 24 hours and a dash speed of Mach 0.8.

Description: *Airframe:* Stealthy, modular, high-wing design, with V tail surfaces plus a central tailfin; composites construction may be assumed. Design is understood to provide options for both short- and long-span wings, for operations at around 7,620 m (25,000 ft) and upwards of 12,200 m (40,000 ft) respectively. The single turbo-Diesel engine, pod-mounted atop the central fin with its propeller arc shielded from radar by the V tail, would power the aircraft during patrol flights and loiter. High transit and dash speeds would be achieved using the twin jets, while all three engines in unison would be employed for climb to the higher operating altitudes required.

No dimensional or weight data accompanied the Lockheed Martin release, but the fuselage size was said to be comparable to those of the F-22 Raptor and F-35 Lightning II.

Mission payloads: Both ISR and weapon payloads in internal bays.
Guidance and control: Fully autonomous.
Launch: Wheeled take-off.
Recovery: Wheeled landing.
Power plant: One turbo-Diesel engine (approx 224 kW; 300 hp), driving a pusher propeller, plus two small turbojets.

MQ-X

Performance	
Altitude	
Operating altitude, long wing	12,190 m (40,000 ft) (est)
Endurance, long wing	40 hr (est)

Status: At conceptual design stage in 2009.
Contractor: Lockheed Martin Aeronautics Company
Palmdale, California.

LM Skunk Works' conceptual MQ-X design *(Lockheed Martin)*
1293830

Lockheed Martin RQ-170 Sentinel

Type: Developmental HALE UAS.

Development: The identity and out-of-sequence service designation of this hitherto 'black' UAV was officially acknowledged by the US Air Force in a brief statement on 4 December 2009, responding to a media enquiry following the appearance on the internet of indistinct unofficial photographs taken in Kandahar in 2007. Apart from also identifying the aircraft's operating unit, the statement revealed only that it was "a stealthy unmanned aircraft system ..... to provide reconnaissance and surveillance support to forward deployed combat forces".

Given the unknown zoom factor of the camera(s) taking the first unofficial pictures, the Sentinel's size has so far (end of 2009) proved difficult to establish. Original 'guesstimates' of a 20 m (66 ft) wing span were later revised upward to about 27.4 m (90 ft), influenced in part by assessment of the aircraft's robust-looking landing gear, suggesting a somewhat larger airframe. If correct, this would make it dimensionally similar to Lockheed Martin's earlier P-175 Polecat prototype, which was estimated to have a maximum take-off weight of some 4,082 kg (9,000 lb). However, the Sentinel appears rather bulkier, with a thicker wing section. Most likely, this would accommodate a greater fuel capacity than that carried by the Polecat, whose endurance was quoted at a relatively modest 4 hours. An operational HALE aircraft might be expected to require at least double that capability. Not only is there an undeniable family likeness between the two, but in December 2005, a few months after the Polecat's first flight, Lockheed Martin is said to have flown another stealth design, to which was linked the project name 'Desert Prowler'. It seems at least possible that the RQ-170 is, or has evolved from, that aircraft. However, until more official detail or better photographs emerge, all such assessment remains necessarily speculative.

Description: *Airframe:* Highly sweptback, tail-less, flying wing design, with compound sweep on leading- and trailing-edges. Unlike earlier and fully stealthy designs, the RQ-170's wings have blunt leading-edges. Single intake above nose, flanked by 'teardrop' overwing fairing each side, assumed to accommodate sensors and/or control avionics and a satcom antenna. Forward-retracting mainwheels; sideways-retracting nosewheel, offset to port to avoid compromising FoV of underfuselage sensors.

The high intake and low position of the semi-circular exhaust suggest that the aircraft may have an S-shaped intake tunnel to shield the engine face from radar, as (for example) in Lockheed Martin's F-22 Raptor.

Mission payloads: Classified. Aircraft is apparently employed only in an ISR-gathering role, but appears to be large enough to permit an internal bay that could, notwithstanding the RQ designation, accommodate weapons if required to do so.

Guidance and control: No details released. However, given its squadron companions, and the fact that it has been seen to share the Predator/Reaper hangar at Kandahar, the RQ-170 may be interoperable with the GCS used at Creech AFB to control those two systems.

Launch: Wheeled take-off.
Recovery: Wheeled landing.
Power plant: Single unidentified turbojet or turbofan.

RQ-170

Dimensions, External	
Overall, length	12.2 m (40 ft 0¼ in) (est)
Wings, wing span	27.4 m (89 ft 10¾ in) (est)
Performance	
Altitude, Service ceiling	15,240 m (50,000 ft) (est)
Power plant, optional	1 × turbojet
	1 × turbofan

Status: The US Air Force has identified the RQ-170 as being operated by Air Combat Command's 432nd Wing at Creech AFB, Nevada, and flown by the 30th Reconnaissance Squadron, a Predator/Reaper trials unit, at the nearby Tonopah Test Range (also known as Area 52). It has been forward-deployed to Kandahar, Afghanistan, since at least 2007.
Contractor: Lockheed Martin Aeronautics Company, Palmdale Operations
Palmdale, California.

Lockheed Martin Stalker

Type: Surveillance mini-UAV.

Development: Stalker development began as a fast-track programme in early 2006; following a maiden flight in mid-2006, it was officially unveiled in August 2007. Some detail information remains undisclosed or approximate.

Shot Stalker displayed at a US Army exhibition in April 2010
(Jay Miller/Aerofax)
1395205

Description: *Airframe:* High-wing, pod and boom monoplane with T tail. No landing gear. Composites construction.

Mission payloads: Stalker: Retractable ventral turret for daylight TV, LLTV or IR cameras and laser designator 'plug and play' modular payloads, mounted on a common two-axis gimbal.

Shot Stalker: As Stalker, plus ShotSpotter acoustic sensors. These are tiny microphones which, due to the extreme quietness of the aircraft's electric engine, are able to detect the location of hostile weapon fire and cue the aircraft and its EO/IR sensors towards them.

Lockheed Martin also quotes a capability for droppable payloads, but says no weapons have been designed for this purpose.

Guidance and control: Autonomous operation, via Lockheed Martin proprietary ground station.

Transportation: Dismantles into man-portable components.

Launch: Hand launched.

Recovery: Autonomous, to belly landing.

Power plant: Lockheed Martin low-noise ('hush drive') electric motor, driving a two-blade propeller.

Variants: *Stalker:* Initial production version.

Shot Stalker: Joint development begun in 2009 between Lockheed Martin and ShotSpotter Inc, adding acoustic sensors to Stalker's existing detection capability to geolocate hostile gunshots and mortar fire.

Stalker, Shot Stalker

Dimensions, External	
Overall, length	2.29 m (7 ft 6¼ in) (est)
Wings, wing span	2.90 m (9 ft 6¼ in)
Weights and Loadings	
Weight	
Max launch weight	
Stalker	6.4 kg (14.00 lb)
Shot Stalker	7.3 kg (16.00 lb)
Performance	
Altitude, Service ceiling	4,570 m (15,000 ft)
Endurance	2 hr (est)
Power plant	1 × electric motor

Status: Stalker in production and in service. Customer not officially disclosed, but reported as having replaced RQ-11 Raven with US Special Operations Command (SOCOM). Shot Stalker production, and US Army evaluation, anticipated in mid-2010.

Contractor: Lockheed Martin Aeronautics Company, Palmdale Operations Palmdale, California.

MAV6 M1400-I Blue Devil II

Type: Optionally manned, long-endurance, multi-intelligence platform.

Development: Blue Devil II is the second wide-area surveillance programme to bear the name and follows on from the manned Blue Devil I system that was installed aboard a fixed-wing Beechcraft King Air 90. Launched in October 2010 using US Army Rapid Equipping Force funds, Blue Devil II then migrated to US Air Force (USAF) management, where it fell within the Big Safari programme office's overall sphere of activity. Blue Devil II has been supported by the US's Joint Improvised Explosive Device Defeat Organization (JIEDDO), with prime contractor MAV6 (formerly Ares Systems Group) receiving a USD86.2 million USAF contract for the development of Blue Devil II during March 2011; the air vehicle platform is the M1400 airship that has been developed by MAV6 using a 112.78 m (370 ft) hull supplied by TCOM LLP.

Essentially, Blue Devil II is a fast-track, single-vehicle technology and concept demonstrator that was aimed at expeditious fielding in response to urgent Central Operations Command (COCOM) needs – most particularly, the threat from IEDs. Where it differs from Blue Devil I is in the multiplicity of onboard sensors (upwards of 10) that were planned for

Stalker overhead view *(Lockheed Martin)* 1395213

Stalker operational deployment *(Lockheed Martin)* 1395212

A front view of the M1400-I airship used in the Blue Devil II programme *(MAV6)* 1365345

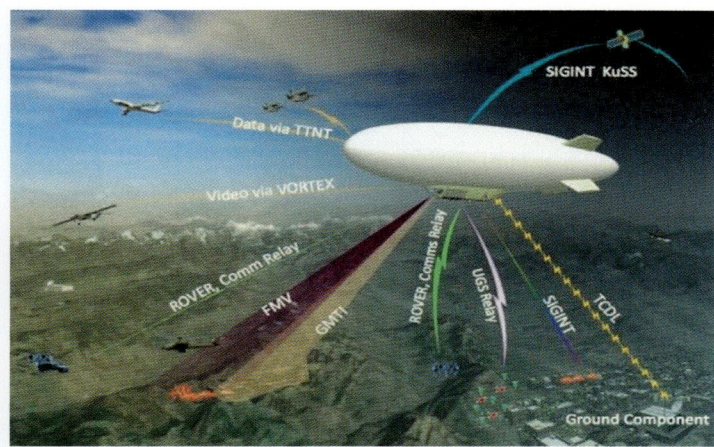

A schematic shown the Blue Devil II's interrelated sensor payloads (USAF)

simultaneous carriage together with a capacity for near-real-time tactical dissemination of fused ISR information; onboard sensor data processing; full motion video sensor control hand-off and long-term data storage architecture for forensic analysis. Again, a 'central node' concept envisaged Blue Devil II pulling in, via Line Of Sight (LOS) links, imagery and other intelligence data from other ISR assets such as the Global Hawk and Reaper UAVs, the LEMV unmanned airship and the E-8C Joint STARS manned platform. Here, received data was to be processed by Blue Devil II's powerful onboard servers, which would separate relevant from irrelevant data before disseminating the relevant content to end-users.

Operation in manned mode was considered for self-deployment to a designated surveillance area and the US Navy's MZ-3A (ABC Lightship A-170) airship has been employed for pilot training and payload testing relating to the Blue Devil II programme.

Description: *Airframe:* The MAV6 M1400 AV is a conventional-shape, helium-filled, non-rigid airship, with four tailfins indexed in X configuration. On the underside are two gondolas, the forward one containing a crew cabin and payload compartment. The smaller rear gondola accommodates the main drive engines and fuel tanks. The non-retractable tricycle landing gear comprises a single wheel below the rear of the front gondola, with the rear pair amidships below the rear gondola.

Mission payloads: Ten or more sensors can be carried simultaneously, employing a payload integration infrastructure (PII) based on a NATO 463L pallet system that allows for integration of new payloads in less than 4 hours without modification to the airframe. L-3 Communications Integrated Systems Group received a USD12.05 million USAF contract in August 2011 to provide integration and test support. Onboard power for systems operation is provided by a 120 kVA generator.

Payload options fall within the following groups.

Communications: AOptix Technologies high-bandwidth laser downlink (up to 40 GB per second transmission rate); Rockwell Collins radio links. Other links said to include Ku-band (12.5 to 18 GHz) satcom; a USAF Tactical Targeting Network Technology (TTNT) application; an unmanned ground system relay; Cubic Defense Tactical Common DataLink (TCDL); and provision to communicate with all US armed forces via L-3 Communications — West's Remotely Operated Video Enhanced Receiver (ROVER).

GMTI radar: Details awaited.

Synthetic aperture radar: Details awaited.

Signals intelligence: Details awaited.

Wide area EO/IR: Said to include Sierra Nevada Corporation Gorgon Stare and a WAAS equipment; *IHS Jane's* sources suggest that the latter may well be the Autonomous Real-time Ground Ubiquitous Surveillance (ARGUS EO) system under development by BAE Systems Electronic Solutions, additional funding for which was being sought in the FY2012 budget request.

High-definition full-motion video: A General Dynamics 38.1 cm (15 in) gimbal axis.

Processing equipment: Details awaited.

Armament: Although essentially an intelligence-gathering asset, MAV6 documentation describes the M1400-I as a "C4ISR aerial fusion node (and weapon system platform)".

Guidance and control: Blue Devil II is designed to be operable in either autonomous, remotely piloted or manned mode. On 18 August 2011, Rockwell Collins announced its selection to provide the Blue Devil II GCS, together with the system's flight control and vehicle control systems, radios, and a "real-time *ad hoc* communications capability".

Launch: Conventional airship take-off.

Recovery: Conventional airship landing.

Power plant: Twin Diesel turboprops (identity and rating not disclosed), with propeller thrust vectoring, in rear gondola; small manoeuvring engine at rear of envelope.

Blue Devil II

Dimensions, External	
Overall, length	112.78 m (370 ft 0¼ in)
Dimensions, Internal	
Hold	
length	
Gondola payload compartment	7.01 m (23 ft 0 in)
width	
Gondola payload compartment	3.05 m (10 ft 0 in)
height	
Gondola payload compartment	2.13 m (6 ft 11¾ in)
volume	
Gondola payload compartment	45.6 m³ (1,610 cu ft)
Payload bay, volume	538 m³ (18,999 cu ft)
Envelope, volume	37,000 m³ (1,306,643 cu ft)
Weights and Loadings	
Payload	
Max payload	
5 day mission	1,134 kg (2,500 lb)
3 day mission	3,402 kg (7,500 lb)
Performance	
Altitude, Operating altitude	6,100 m (20,020 ft)
Speed	
Cruising speed, max	80 kt (148 km/h; 92 mph)
Endurance, max, dependent on environmental conditions	72–216 hr

Status: The USAF issued a 'stop work' notice on the Blue Devil II programme during May 2012. According to unconfirmed media reports this decision was based on cost overruns, programme delays and technical difficulties with the M1400's stabilising fins, flight control system and wiring looms. As of August 2012, a MAV6 spokesman confirmed the 'stop work' notice and noted that MAV6 had placed the Blue Devil II M1400 AV in storage ("awaiting further government direction"); had spun-off an airship specific business that would continue to promote the M1400 and was developing a conventional UAV variant of the Blue Devil II payload architecture.

Contractor: MAV6 LLC
Vicksburg, Mississippi.

MBDA TiGER

Type: Lethal or close-range tactical UAV.

Development: Designed for one-man operation by front-line troops against such threats as snipers or IEDs, the TiGER was first shown publicly at the 2010 Farnborough Air Show. Derived from 'Tactical Grenade Extended Range', its name reflects its primary application as a precision attack weapon, though with the warhead removed it can also perform ISR missions. In November 2010 it was demonstrated to the US Air Force's Special Operations Command (SOCOM) as a candidate in that service's Lethal Miniature Aerial Munition System (LMAMS) programme, from which a winner was expected to be selected in 2012.

Description: *Airframe:* Pod fuselage; tubular tailboom with rectangular vertical upper and lower tailfins. Low aspect ratio high wings, inflated from compressed air bottle forming part of the system kit. No landing gear. Composites construction.

Mission payloads: TiGER has two onboard video cameras: a side-looking pan-tilt-zoom one for en-route navigation, and a front-facing

TiGER and its GCS on display at the 2011 Paris Air Show (IHS/Patrick Allen)

one in the nose for target lock-on. Warhead is a tandem pair of 40 mm grenades.

Guidance and control: Laptop or PDA based GCS, providing GPS/INS guidance with autotracker and man-in-the-loop targeting.

Transportation: Man-portable, the UAV is carried, disassembled, in a launch tube that can be stowed in a backpack or rucksack. It can be assembled and launched in less than 2 minutes.

Launch: Hand launch.

Recovery: Non-recoverable in strike mode. For ISR missions, aircraft can be flown into a recovery net to drop on to a landing mat.

Power plant: Electric motor (rating not stated); two-blade propeller.

TiGER

Dimensions, External	
Wings, wing span	0.61 m (2 ft 0 in)
Weights and Loadings	
Weight, Max launch weight	1.4 kg (3.00 lb)
Payload, Max payload	0.45 kg (0.00 lb)
Performance	
Range	1.7 n miles (3 km; 2 miles)
Endurance	15 min
Power plant	1 × electric motor

Status: Promotion continuing as of end 2012.
Contractor: MBDA Missile Systems Inc
  Arlington, Virginia.

Mission Technologies Buster

Type: Close-range surveillance SUAV system.

Development: Buster is in production and various models have been customised for customer missions and requirements.

Description: *Airframe:* Twin-wing, high-lift configuration (tandem pair of wings joined by endplates at their tips), forming a self-bracing unit that offers a wider flight envelope, smaller footprint, higher payload capacity, longer endurance and a 'stable and sturdy modular platform', said to be lighter than a single wing and some 40 per cent smaller for the same amount of lift. Non-swept fin and rudder with low-set tailplane. Modular structure, built of composites. System options are para, wheeled or skid landing.

Mission payloads: LLTV or EO/IR; 16-channel C-band (4.4 to 4.8 GHz), RS-170 standard real-time video downlink and 225 to 400 MHz communications datalink.

Other payloads have included high-definition TV and Scientific Applications and Research Associates (SARA) Low-cost Scout UAV Acoustic System (LOSAS) sensor. Demonstrated to the US Army's Communications Electronics Command (CECOM) in mid-2003. Alternative configurations available to customer's requirements. Up to 25 W of interface power available at 5 or 15 V DC.

Buster's distinctive endplate tandem wing configuration is claimed to provide structural integrity *(Mission Technologies)* 1329031

Guidance and control: Preprogrammed, with GPS navigation; can be retasked in flight. Laptop-size GCS, weight 3.6 kg (8 lb), provides touch screen operation, 'point-and-click' mission planning, real-time rerouting, moving map display, video display recording, target marking and alarm displays. Up to 65 W of electrical (battery) power available at 5 and 15 V DC, including 25 W for UAV payloads.

Transportation: Man-portable or can be transported in and operated from a HMMWV or a moving vehicle. Set-up time is 10 minutes with a two man crew.

System composition: Two or more air vehicles, GCS and 4.5 kg (10 lb) foldable launcher. One-man operation; total weight (complete system) 15.9 kg (35 lb).

Launch: Catapult launch standard; wheeled take-off optional.

Recovery: Automatic recovery (engine shut-down and parachute deployment) standard (from as low as 23 m; 75 ft); wheeled or skid landing optional.

Power plant: One 1.2 kW (1.6 hp) single-cylinder piston engine; two-blade composites propeller. Heavy-fuel engine optional. Fuel capacity 1.07 litres (0.28 US gallons; 0.24 Imp gallons).

Buster

Dimensions, External	
Overall	
length	1.02 m (3 ft 4¼ in)
height	0.40 m (1 ft 3¾ in)
Wings, wing span	1.25 m (4 ft 1¼ in)
Dimensions, Internal	
Payload bay, volume	0.001 m³ (0.0 cu ft)
Weights and Loadings	
Weight	
Operating weight, empty	4.45 kg (9.00 lb)
Max launch weight	6.35 kg (13.00 lb)
Fuel weight, Max fuel weight	0.82 kg (1.00 lb)
Payload, Max payload	1.76 kg (3.00 lb)
Performance	
Climb	
Rate of climb, max, at S/L	134 m/min (439 ft/min)
Altitude, Operating altitude	60 m to 3,050 m (200 ft to 10,000 ft)
Speed	
Max level speed	65 kt (120 km/h; 75 mph)
Cruising speed	35 kt (65 km/h; 40 mph)
Stalling speed	27 kt (51 km/h; 32 mph)
Range	
control range	21.6 n miles (40 km; 24 miles)
video, C-band	5.4 n miles (10 km; 6 miles)
video, L-band	27 n miles (50 km; 31 miles)
Endurance	4 hr (est)
Power plant	1 × piston engine

Status: Since 2002, has been 'constantly involved' in numerous US Army C4ISR exercises at Fort Dix and airborne assault exercises at Fort Benning. US Army's Night Vision Labs at Fort Belvoir, Virginia is managing the programme. Further exercises at Dahlgren, Virginia (under US Navy auspices) have focused Buster as an 'enabler' in coalition warfare. Systems (believed to be two in number) have been delivered to UK MoD.

Customers: US Army; SOCOM; US Navy; UK MoD.

Contractor: Mission Technologies Inc
  San Antonio, Texas.

The man-portable Buster SUAS 0576760

Buster mounted on its portable launcher 0549871

MLB Bat

Type: Close-range surveillance mini-UAV.

Development: History not known. By 2009 current version was designated Bat 3.

Description: *Airframe:* Constant-chord, shoulder-mounted wings with movable control surfaces; pod fuselage with rear-mounted pusher engine; twin tailbooms, pylon-mounted above rear fuselage and bridged by anhedral tail surfaces connected by short, horizontal centre-section. Construction is of Kevlar, carbon fibre and aluminium, with Dacron wing skins. Disassembled airframe can be stowed in a 1.22 m (4 ft) long, 0.38 m (15 in) diameter tube. Non-retractable tricycle landing gear.

Mission payloads: Colour CCD video camera with 45° FoV, 900 MHz two-way modem and S-band (2.4 GHz) video downlink are standard; option for three-axis, stabilised gimbal mount with 22° and 53° FoV colour cameras; or for IR thermal imager.

Guidance and control: Fully autonomous, preprogrammed mission profiles with GPS waypoint navigation. Can also be controlled manually using eight-channel PCM radio uplink on 72 MHz. Receiving and recording station for downlinked video. Laptop computer with moving map and flight data displays is used to monitor flight and store data. Cloud Cap Piccolo 2 autopilot optional.

Transportation: Entire system is man-portable.

System composition: One air vehicle with three-axis sensor turret, two colour TV cameras, GCS, catapult launcher and support equipment.

Launch: By car top mounted bungee catapult.

Recovery: Automatic 'return home' with parachute recovery; or by conventional and autonomous wheeled landing.

Specifications

Power plant
One 26 cc two-cylinder two-stroke engine (type and rating not stated); two-blade pusher propeller.

Dimensions
Wing span .... 1.83 m (6 ft 0.0 in)
Length overall .... 1.40 m (4 ft 7.0 in)
Height overall .... 0.61 m (2 ft 0.0 in)

Weights
Max payload .... 1.8 kg (4.0 lb)
Max launching weight .... 8.6 kg (19.0 lb)

Performance
Max level speed .... 60 kt (111 km/h; 69 mph)
Cruising speed .... 35 kt (65 km/h; 40 mph)
Max operating altitude .... 762 m (2,500 ft)
Ceiling .... 2,745 m (9,000 ft)
Mission radius (datalink limit) .... 5.2 n miles (9.7 km; 6.0 miles)
Range with max fuel .... 174 n miles (322 km; 200 miles)
Endurance: nominal .... 4 h
  max .... 8 h

Status: In production and in service.

Customers: Demonstrations in collaboration with civilian agencies have included traffic monitoring for the US Department of Transportation, and cold-weather scientific data-gathering as part of NASA's programme for Mars exploration. A small number of Bats was deployed to the Middle East in April/May 2003 as part of Operation 'Iraqi Freedom'.

Contractor: MLB Company
Mountain View, California.

Car-top launch of the Bat UAV *(MLB)* 1174695

The Bat's nose-mounted twin-camera ball turret *(MLB)* 1174694

The Bat aerial vehicle *(MLB)* 0580223

MLB Super Bat

Type: Mini-UAV for surveillance and mapping.

Development: The Super Bat UAV is the latest in a family of mini-UAVs (Bat family) that have been in production for 10 years.

Description: The MLB Super Bat is a small, unmanned aerial vehicle (UAV) that has mission capabilities typically found only in larger UAVs.

The Super Bat is a complete UAV system that can operate autonomously, deliver high-quality video imagery and can be transported in a small SUV. The aircraft operates autonomously, has a 10-hour duration and telemetry range of up to 10 miles. Due to its small size and being light in weight, the Super Bat is launched using a car-top, bungee-powered catapult and can land autonomously on wheels.

A small clearing is adequate to operate the Super Bat. A 2.0 inch^3 gasoline engine powers the aircraft and its muffler reduces the noise level to nearly inaudible at a distance of 305 m (1,000 ft). Typical altitude for operation is 200 to 2,500 ft above the local terrain, with a cruising speed of 40 kt and a maximum speed of 65 kt.

The Bat can easily operate in limited visibility conditions without danger to persons on the ground or possibility of detection. An optional lighting system allows nighttime operation of the Super Bat.

A pusher engine configuration is used to permit unobstructed installation of sensors in the nose and the Super Bat uses a gimbal turret camera system that is automatically aimed at locations specified by the operator. The TASE family of gimbal systems (manufactured by Cloud Cap Technologies) is used in the Super Bat and is available with E/O, IR, and SWIR camera configurations.

Super Bat UAV *(MLB)* 1390324

MLB Super Bat

Super Bat UAV *(MLB)* 1390323

Super Bat UAV *(MLB)* 1390325

The aircraft and its systems are modular in design for simple maintenance and replacement of damaged components. The airframe is constructed of Kevlar, carbon fibre, and aluminium.

The Super Bat has a 2.6 m (103 inch) wingspan and a ready-to-fly weight of 34 lb. Cloud Cap Technologies Piccolo autopilot and ground station are used for the flight control of the Super Bat. A laptop PC with a moving map display shows the aircraft's location, speed, and height in real time and allows the operator to fly the UAV using simple mission commands. System monitoring windows are used to keep track of the Super Bat's critical systems and a live video capture window with optional video stabilization and object tracking capabilities. The flight path is specified as a series of mission legs, each with its own altitude, speed and waypoint. The operator can change the mission plan by 'dragging and dropping' waypoints over the map display and then uploading the new flight plan to the aircraft.

Airframe: The configuration is a twin boom pusher with skid landing gear and a retractable shield to protect the sensor. The airframe is made from Kevlar and carbon fibre composite materials and has a gross weight of 34 lb. The wingspan is 103 inches and the shipping container is 50× 20 × 18 inches in size.

Mission payloads: The Super Bat can carry up to 5 lb sensor payload including stabilized gimbal camera systems (EO/IR) and digital still cameras for mapping. Cloud Cap Technologies TASE or TASE Duo is the standard sensor payloads.

Guidance and control: Fully autonomous operation (launch, landing, waypoint navigation and so on). using the Cloud Cap Technologies Piccolo II flight control system. Optional manual control modes are supported including flight through the autopilot or full manual remote control.

Transportation: Two aircraft and all equipment needed to operate can fit inside a mid-size SUV. The equipment can be shipped in 4 boxes with a 250 lb total shipping weight.

System composition: One air vehicle ready to fly with TASE stabilized gimbal sensor and 12 Mpixel digital mapping camera. Catapult launcher, Piccolo ground station, all ground station equipment and support equipment.

Launch: Bungee-catapult launched system, car-top mounted with transportation box. Catapult folds for transport and mounts on any car roof racks.

Recovery: Autonomous landing on shock absorbing skids.

Power plant: One 26 cc two-stroke engine, gasoline, 40:1 oil mixed in.

Super Bat

Dimensions, External	
Wings, wing span	2.6 m (8 ft 6¼ in)
Weights and Loadings	
Payload	
Max payload, inc fuel and Cloud Cap TASE gimbal sensor	15.4 kg (33 lb)
Performance	
Altitude, Service ceiling	3,050 m (10,000 ft)
Speed	
Max level speed	65 kt (120 km/h; 75 mph)
Cruising speed	40 kt (74 km/h; 46 mph)
Range, max fuel	217 n miles (401 km; 249 miles)
Endurance	10 hr

Status: As of 2011, in production and in service.
  Customers: US Air Force, NASA, US Army, private research companies and universities.
Contractor: MLB Company
  Mountain View, California.

Northrop Grumman Bat

Type: Multirole tactical UAV.
Development: The Bat is a specific line of development stemming from the KillerBee (see separate entry) designed in collaboration with Northrop Grumman by Swift Engineering Inc; the programme was launched with the name Bat in April 2009. In the following month, and for a similar purpose, Raytheon purchased the name KillerBee KB-4 and rights to some technology from Northrop Grumman. Northrop Grumman has retained Swift Engineering to continue to work on design refinement, product line development, flight test support and manufacturing of the Bat.

A family of similar vehicles was planned, with wing spans ranging up to 10.1 m (33.2 ft) and maximum take-off weight of 1,043 kg (2,300 lb). Bat 2, Bat 3 and Bat 4 were earlier versions of shorter wing span with two-blade propellers described in previous editions of JUAV; Northrop Grumman subsequently developed Bat 10 and Bat 12. It is now focusing on Bat 12 (a designation which reflects the wingspan in feet) which flew for the first time in 2010, for which the following data has been released.

Description: *Airframe:* Scaleable design. Three-part blended wing/body flying wing configuration, comprising delta-shaped centre-section and sweptback outer panels with turned-down tips; rear-mounted engine with pusher propeller. Landing gear optional. Composites construction. Winglet antennas on Bat 12.

Mission payloads: EO/IR surveillance, reconnaissance and targeting, communications relay, psychological warfare, SIGINT or counter-IED payloads. The Bat 12 has a payload power of 750 W.

Guidance and control: Fully autonomous; GPS/IMU navigation. Vehicle-mounted GCS and/or portable laptop control unit. Integrating with US Army and Marine Corps One system GCS. Multiple simultaneous air vehicle operation possible.

Transportation: Air-transportable in fixed-wing C-130 or CH-47, CH-53 and H-60 series helicopters. Ground-deployable by HMMWV.

Bat 12 is now the focus of the programme *(Northrop Grumman)* 1421697

The later Bat UAVs have a five-blade propeller *(Northrop Grumman)* 1421696

Nothrop Grumman Bat *(Northrop Grumman)* 1421695

Bat 4 accelerating from launcher *(Northrop Grumman)* 1356244

Bat in flight *(Northrop Grumman)* 1356245

System composition: Three air vehicles and payloads; HMMWV-mounted GCS, tracking antenna, launch/recovery trailer, net recovery and spares. Set-up time 1 hour, tear down time 45 minutes.
 Launch: Automatic, from pneumatic/hydraulic rail launcher.
 Recovery: Net recovery for current version; wheeled landing gear optional.
 Power plant: Bat 12: One 11.8 kW (15.8 hp) Hirth two-stroke piston engine, driving a five-blade composite pusher propeller; heavy fuel 9.6kW (13.0 hp) engine optional

Bat 12

Dimensions, External	
Overall, length	1.90 m (6 ft 2¾ in)
Wings, wing span	3.66 m (12 ft 0 in)
Dimensions, Internal	
Payload bay, volume	0.964 m³ (34.0 cu ft)
Weights and Loadings	
Weight, Max launch weight	74.4 kg (164 lb)
Payload, Max payload	22.7 kg (50 lb)
Performance	
Altitude, Service ceiling	4,570 m (15,000 ft)
Speed, Max level speed	125 kt (232 km/h; 144 mph)
Endurance, with 22.7 kg payload	12 hr
Power plant	1 × piston engine

Status: In June 2009 at the US Naval Air Facility El Centro, California, a Bat 4 made five flight demonstrations with a communications payload for an unnamed 'government customer'. As of August 2009, Northrop Grumman announced that Bat development was continuing with the integration of "several new payloads, a common ground control architecture and air vehicle upgrades that include a new engine, a new launcher capability and several air vehicle capability enhancements". Bat 12 flew in January 2010.
 Contractor: Northrop Grumman Unmanned Systems
 San Diego, California.

Northrop Grumman Firebird

Type: Optionally piloted surveillance aircraft.
Development: Northrop Grumman's Firebird optionally piloted, MALE surveillance aircraft has been produced in Demonstrator Firebird (DF) and Production-Ready Firebird (PRF) configurations. Key programme events are as follows:
 Early 2009 IHS Jane's sources report Northrop Grumman as having commissioned its subsidiary Scaled Composites to develop (as a private venture) the DF configured Firebird optionally piloted aircraft during the early part of 2009. Here, the AV was also known as the Scaled Composites Model 355.
 February 2010 The prototype DF configuration Firebird (US civil registration N355SX) made its maiden flight from the Scaled Composites facility at Mojave Airport, California (35° 3' 32.93" N; 118° 8' 54.74" W) during February 2010.
 October 2010 Firebird AV N355SX was demonstrated to US DoD officials at a facility in Sacramento, California during October 2010. Here, the platform is reported to have been equipped with Full Motion Video (FMV), EO/IR imaging and SAR radar sensors and to have used them simultaneously during the demonstration.
 December 2010 IHS Jane's sources report Firebird AV N355SX as having been test flown with two separate FMV sensor installed.
 May 2011 The DF configured Firebird AV was publicly revealed during May 2011.
 23 May - 3 June 2011 Firebird AV N355SX participated in the US Joint Forces Command's 'Empire Challenge 2011' exercise at Fort Huachuca, Arizona (31° 32' 58.25" N; 110° 21' 59.43" W) where it demonstrated the use of four separate payloads (SIGINT/DF, EO/IR imaging, radar and communications relay), rapid payload switch-out between missions (within 60 minutes), sensor control by operators in different ground locations and simultaneous use of an Electronic Support (ES) and three EO/IR sensors.
 October 2011 Development of the PRF configuration Firebird optionally piloted aircraft is understood to have begun *circa* October 2011
 27 July 2012 Firebird AV N355SX was used to demonstrate Northrop Grumman's SmartNode airborne communications node pod as part of a ground and air trial of the system linked to the US Army's High Antenna for Radio Communications (HARC) system. Here, the demonstration flight launched from Mojave Airport in California.

The Demonstrator Firebird AV in UAV configuration, pictured with a sensor fit that is believed to comprise an AN/DAS-1 MTS-B EO system and a SIGINT or communications relay package *(Northrop Grumman)* 1398556

The Demonstrator Firebird AV in manned configuration *(Northrop Grumman)* 1395308

Unmanned aerial vehicles > United States > Northrop Grumman Firebird

An in-flight view of the Demonstrator Firebird AV carrying Northrop Grumman's SmartNode airborne communications node pod on its centreline stores station *(Northrop Grumman)* 1365364

A ground view of the prototype Production-Ready Firebird AV that emphasises the type's two-seat cockpit *(Northrop Grumman)* 1365363

The prototype Production-Ready Firebird AV photographed during its maiden flight on 11 November 2012 *(Northrop Grumman)* 1365362

The Demonstrator Firebird AV photographed with Northrop Grumman's VADER GMTI/SAR radar on its centreline *(Northrop Grumman)* 1395314

November 2012 As of November 2012, *IHS Jane's* sources were reporting a total of 15 different payload types as having been installed aboard Firebird AV N355SX.

11 November 2012 The prototype PRF configured Firebird AV (US civil registration N241PR) made its maiden flight on 11 November 2012.

Description: *Airframe:* **DF configuration** High aspect ratio shoulder wings, forward-swept, with anhedral, inboard of tailbooms; dihedral outboard, with anhedral tips. Central fuselage nacelle; rear-mounted engine; twin tailbooms, with tips connected by chevron-shaped horizontal surfaces. Large internal avionics bay, with square access panel on port side. One centreline and two wing-mounted hardpoints for external stores. Forward-retracting tricycle landing gear. Mainly composites construction. For unmanned operation, entire forward-upper fuselage section replaced by windowless duplicate and pilot's seat by control equipment.

PRF configuration Similar to the DFC but with a longer fuselage nacelle, a two-person cockpit and revised horizontal tail surface and wing geometries. Internal payload bay ahead of the AV's engine and two wing hardpoints for pod-mounted sensors or precision guided munitions.

Mission payloads: **DF configuration** Payloads identified as having been flown aboard the DF configured Firebird AV N355SX comprise:
- the Raytheon AN/DAS-1 Multi-spectral Targeting System B (MTS-B) EO target detection, ranging and tracking system
- the Ku-band (12.5 to 18 GHz) Northrop Grumman AN/ZPY-1 GMTI/SAR radar
- LOS and Beyond LOS (BLOS) datalinks
- a Northrop Grumman-sourced ES/SIGINT package (possibly the contractor's Common SIGINT System 1500 (CSS 1500) equipment)
- the Northrop Grumman SmartNode airborne communications node pod
- the FLIR Systems Star SAFIRE® 380-HD EO sensor
- the Northrop Grumman Vehicle And Dismount Exploitation Radar (VADER) GMTI/SAR sensor.

As of late 2012, Northrop Grumman was reporting the DF configured Firebird AV's baseline sensor fit as comprising two × FLIR Systems FMV EO sensors, an ES/SIGINT package and the AN/ZPY-1 GMTI/SAR radar.

PRF configuration A typical PRF mission payload might comprise up to three × high-definition EO/IR imagers (including FMV), digital GMTI/SAR radars, ES/SIGINT sensors and communications relay equipment.

Guidance and control: **DF configuration** Described only as "standards-based interfaces and command protocol". Multiple wide-band LOS and BLOS datalinks. In manned operation, the cockpit appears to contain four or more display screens.

PRF configuration In manned configuration, the PRF Firebird AV is equipped with a Garmin G3000™ Electronic Flight Information System (EFIS). LOS and BLOS links (including an L-3 Communications SATCOM package) for data hand-off and unmanned C2.

Launch: **DF and PRF configurations** Conventional wheeled take-off.

Recovery: **DF and PRF configurations** Conventional wheeled landing.

Power plant: **DF and PRF configurations** One 261 kW (350 hp) Lycoming TEO-540E turbocharged flat-six engine, operable on Avgas or Mogas and driving a three-blade constant-speed pusher propeller. Option of a heavy-fuel alternative engine is being investigated.

Firebird

Dimensions, External	
Overall	
length	
DF configuration	10.36 m (33 ft 11¾ in)
PRF configuration	10.82 m (35 ft 6 in)
height	
DF configuration	2.96 m (9 ft 8½ in)
PRF configuration	2.99 m (9 ft 9¾ in)
Wings	
wing span	
PRF configuration, wings folded	16.51 m (54 ft 2 in)
DF configuration	19.81 m (65 ft 0 in)
PRF configuration, wings deployed	22.01 m (72 ft 2½ in)
Weights and Loadings	
Weight	
Max T-O weight	
DF configuration	2,268 kg (5,000 lb)
PRF configuration	3,016 kg (6,649 lb)
Payload	
Max payload	
total, (DF/PRF configurations)	563 kg (1,241 lb)
Performance	
Altitude	
Service ceiling	
DF configuration	9,145 m (30,000 ft)
PRF configuration	9,755 m (32,000 ft)
Speed	
Max level speed	
dash, (DF/PRF configurations)	200 kt (371 km/h; 231 mph)
Endurance	24-40 hr

Status: As of late 2012, Northrop Grumman was continuing to fly the DF configuration Firebird AV N355SX and to have been under contract to produce 10 × PRF configuration Firebird AV for an unidentified customer over a five year period.

Contractor: Northrop Grumman Aerospace Systems
El Segundo, California.

Northrop Grumman LEMV

Type: MALE surveillance UAS.

Development: The LEMV (Long Endurance Multi-intelligence Vehicle) is an airship-based hybrid aircraft for "unblinking eye" long-term persistent surveillance. As such, the LEMV programme was launched during 2009 by the US Army's Space and Missile Defense Command/Army Forces Strategic Command (USASMDC/ARSTRAT) with specific application to US activities in Afghanistan. Following the issue of a formal Request for Proposals (RfP) in early 2010, award of a USD517 million, five-year development contract to Northrop Grumman was announced on 14 June 2010. Under this, the company was to design, develop and test one LEMV within 18 months and then transport it to southwest Asia for military assessment. The contract also included an option for two more LEMV airships.

As of mid-2010, envelope inflation was targeted for April 2011, with a maiden flight at Tillamook, Oregon, to follow three months later. After IOT & E at the Yuma Proving Ground in Arizona, deployment to southwest Asia for flights over Afghanistan was planned for the end of 2011. This ambitious schemata was not achieved, with LEMV number one not making its maiden flight until August 2012 (see Status).

Northrop Grumman is partnered on the LEMV programme by Hybrid Air Vehicles Ltd of the UK (formerly SkyCat), whose HAV-3 sub-scale SkyCat prototype has served to test control laws for operation of the LEMV. Other technology partners across 18 US states include AAI Corporation (ground station), ILC Dover (envelope design and assembly), SAIC (data processing and dissemination) and Warwick Mills (envelope fabric supplier). Systems integration has been undertaken by Northrop Grumman.

Description: *Airframe:* Non-rigid, hybrid buoyant aircraft, deriving 60 per cent of its lift aerostatically from the helium-filled envelope and 40 per cent from the envelope's aerodynamic shape; thrust-vectoring engines provide additional lift and control for take-off and landing. Three-lobe envelope, with rigid structure on the underside, recessed amidships and accommodating the flight deck, payloads, fuel tanks and avionics. Redundant rapid-deflation system. Four tailfins, indexed in X configuration. Landing skids under envelope outer lobes.

Mission payloads: Baseline ISR payloads for the LEMV AV include four EO sensors (2 × L-3 Wescam MX-15HDi and 2 × L-3 Wescam MX-20D), a GMTI radar (possibly Northrop Grumman's GMTI/SAR Vehicle And Dismount Exploitation Radar - VADER), a SIGINT package and an onboard communications capability. In this latter context, IHS Jane's sources associate L-3 Communications (datalinks) and Rockwell Collins equipment with the type's communications fit. Overall, the LEMV's sensor provision takes the form of a plug-and-play architecture (facilitating in-the-field sensor swap out) and there is a minimum of 16 kW of onboard electrical power available, with growth potential to 73 kW.

Guidance and control: The LEMV requirement is for three alternative forms of control: autonomous; remotely from a GCS; or (mainly for ferry flights to a deployment area) by an onboard pilot. Station-keeping on location is required to be accurate to within a 3.5 km (2.2 mile) radius. The C^2 system is also required to be global and interoperable with AAI's One System common ground station.

Launch: Conventional, automatic take-off, with a required take off run of approximately 305 m (1,000 ft).

Recovery: Conventional, automatic, airship-type landing.

Power plant: Four 99 kW (133 hp) Thielert Centurion 2.0 four-cylinder, inline, ducted-propeller turbo-Diesel engines with vectored thrust.

Variants: *HAV304:* Northrop Grumman designation for full-size LEMV.

HAV-3: UK-built hybrid airship; to be used as part of LEMV development programme. First flown on 29 August 2008, and had made about 70 flights by time of LEMV contract award. Length approx 15.2 m (50 ft), max take-off weight 247 kg (544 lb).

LEMV

Dimensions, External	
Overall, length	91.4 m (299 ft 10½ in)
Weights and Loadings	
Fuel weight, Fuel with max payload	6,804 kg (15,000 lb)
Payload, Max payload	3,175 kg (6,999 lb)
Performance	
Altitude, Operating altitude	6,100 m (20,020 ft)
Speed	
Cruising speed	
loiter speed	30 kt (56 km/h; 35 mph)
max	80 kt (148 km/h; 92 mph)
Range, mission radius	1,086 n miles (2,011 km; 1,249 miles)
Endurance	504 hr

Status: LEMV number one made its maiden flight from Joint Base McGuire-Dix-Lakehurst, New Jersey on 7 August 2012. Lasting 90 minutes, this initial sortie was flown with a safety pilot aboard and was designed to demonstrate the vehicle's airworthiness (including safe launch and recovery) together with verification of its flight control system and system level performance. On all counts, the US Army noted that the stated objectives had been met.

Contractor: Northrop Grumman Aerospace Systems
Redondo Beach, California.

LEMV AV number one photographed during its maiden flight on 7 August 2012 *(US Army)* 1365344

A computer generated image of the LEMV AV showing the positioning of its four EO sensor turrets and radar antenna *(Northrop Grumman)* 1395239

Northrop Grumman MQ-8C Fire Scout

Type: VTOL Tactical UAV (VTUAV).

Development: The MQ-8C Fire Scout is a VTUAV that is based on the airframe of the Bell Helicopter 407 utility helicopter and is a development of the private venture Fire-X VTUAV demonstrator that was produced jointly by Northrop Grumman and Bell Helicopter. Key programme events include:

30 April 2010 The USN issued an RFP for a Persistent Ship-Based Unmanned Aircraft System (PSB UAS) that would:

- be capable of supporting irregular warfare operations from the sea
- be compatible with a range of surface combatants including cruisers, destroyers and the service's Littoral Combat Ship (LCS)
- have an 2016 to 2020 IOC
- have a 300 to 1,000 n miles (556 to 1,852 km; 346 to 1,151 miles) mission radius
- have an 8 to 72 hour on-station endurance at mission radius (the 72 hour figure was to be achieved using multiple AVs)
- incorporate a mission suite that could (over time) include an EO/IR sensor, still/full motion video, laser designation/range-finding, a COMINT/ELINT capabilities, wide area surveillance/SAR radar and measurement and signature intelligence capabilities

A computer generated impression of the MQ-8C *(Northrop Grumman)* 1394861

The prototype Fire-X VTUAV demonstrator made its maiden flight at the Yuma Proving Grounds, Arizona on 10 December 2010 *(Northrop Grumman)* 1356627

- be able to carry a 272 to 454 kg (600 to 1,001 lb) payload
- incorporate LOS and satellite communications links
- have an operational ceiling of between 4,572 m and 7,620 m (15,000 and 25,000 ft).

While not confirmed, this RFP probably formed a key driver for the Fire-X VTUAV demonstrator effort.

3 May 2010 The Northrop Grumman/Bell Helicopter consortium first revealed that it was developing the Fire-X VTUAV demonstrator.

17 September 2010 External power was connected to the Fire-X prototype's (US civil registration N91796, constructor's number 53343) main computers and 'other' sub-systems for the first time.

10 December 2010 The prototype Fire-X AV N91796 made its first unmanned flight at the Yuma Proving Grounds in Arizona (33° 0' 56.65" N; 114° 14' 36.19" W).

February 2011 As of February 2011, the USN was planning to procure 12 × MQ-8C AVs that would be based on the Fire-X VTUAV demonstrator.

23 April 2012 On 23 April 2012, the USN's Naval Air Systems Command (Patuxent River, Maryland) awarded Northrop Grumman a then year, not to be exceeded USD262,336,248 undefinitised contract action with regard to the development, manufacture and test of two × VTUAVs, the production of a further six × AVs and the procurement of an associated spares holding, with the whole forming part of the Service's VTUAV Endurance Upgrade Rapid Deployment Capability (an alternative designation for the MQ-8C) programme. Work on the effort was to be undertaken at facilities in Moss Point, Mississippi (47% workshare), San Diego, California (46% workshare) and Yuma, Arizona (seven per cent workshare) and at the time of its announcement, the programme was scheduled for completion by the end of May 2014. In an associated media release, Northrop Grumman noted Bell Helicopter and Rolls Royce as being 'major suppliers' on the programme; that final assembly of the new MQ-8Cs would take place at the company's Moss Point Unmanned Systems Center and that as of the given date, the USN had a requirement for 28 × MQ-8Cs.

31 January 2013 Summit Aviation delivered its first Faraday Cage (an enclosure designed to protect electronic components from electromagnetic interference) into the MQ-8C programme.

Description: Airframe: The MQ-8C utilises a depopulated Bell 407 helicopter pod and boom airframe with a modified nose configuration (to house sensors) and all original glazing plated over. Twin-skid landing gear. Vertical and horizontal tail surfaces. Four-bladed main and two-bladed tail (to port) rotors. Folding main rotor (contained within the AV's height × width × length envelope).

Mission payloads: *IHS Jane's* sources suggest that as initially deployed, the MQ-8C AV would be equipped with the FLIR Systems BRITE Star® II EO imager and the X-band (8 to 12.5 GHz) Telephonics AN/ZPY-4 (RDR-1700B+) multi-mode maritime surveillance radar.

Guidance and control: Generally similar to the architecture installed aboard the MQ-8B Fire Scout configuration (see separate entry). MQ-8C is designed to operate with 'nearly any type of future or current military standards-based control segment, communicating as easily with shipboard controllers using the [USN's Raytheon] TCS as field commanders using the US Army's universal [One System GCS].

Launch: Automatic vertical take-off.

Recovery: Automatic vertical landing.

Power plant: One 606 kw (813 shp) Rolls-Royce 250-C47B turboshaft engine (with FADEC, Fire-X and MQ-8C).

MQ-8C

Dimensions, External	
Overall	
length	
Fire-X and MQ 8C	12.6 m (41 ft 4 in)
height	
Fire-X	3.1 m (10 ft 2 in)
MQ-8C	3.3 m (10 ft 10 in)
width	
MQ-8C	2.4 m (7 ft 10½ in)
Fire X	2.5 m (8 ft 2½ in)
Rotors	
rotor diameter	
Fire-X and MQ-C	10.7 m (35 ft 1¼ in)

Weights and Loadings	
Weight	
Fire-X and MQ-8C	2,722 kg (6,000 lb)
Payload	
Max payload	
MQ-8C, internal	454 kg (1,000 lb)
Fire-X and MQ-8C, slung load	1,200 kg (2,645 lb)
Fire-X	1,361 kg (3,000 lb)
Performance	
Altitude	
Service ceiling	
MQ-8C	5,180 m (17,000 ft)
Fire-X (more than)	6,095 m (20,000 ft)
Speed	
Max level speed	
Fire-X, maximum	133 kt (246 km/h; 153 mph)
MQ-8C	140 kt (259 km/h; 161 mph)
Endurance, Fire-X and MQ-8C	14 hr
Power plant, turboshaft	1 ×

Status: As of 2013, Northrop Grumman was under contract to supply the USN with eight × MQ-8C VTUAVs. Northrop Grumman-sourced specification data.

Contractor: Northrop Grumman Unmanned Systems, San Diego, California.

Northrop Grumman RQ-8A and MQ-8B Fire Scout

Type: VTOL tactical UAV.

Development: According to *IHS Jane's* sources, the MQ-8 Fire Scout VTOL tactical UAV began life as Northrop Grumman's Ryan Aeronautical Center's Model 379 (itself a version of the Schweizer 330SP/333 manned turboshaft helicopter), with the company self-funding a manned (AV P-2, bearing the US civil registration N2119X) and an unmanned (AV P-1) prototype. Key programme events include:

October 1999 *IHS Jane's* sources were reporting a Fire Scout prototype as having completed an initial 39 hour flight test programme.

January 2000 *IHS Jane's* sources were reporting a Fire Scout prototype as having made its first autonomous, GPS-guided flight at the USN's Naval Air Warfare Center at China Lake in California, with the type's first fully autonomous sortie being flown on 12 January 2000.

9 February 2000 The USN's Naval Air Systems Command (Patuxent River, Maryland) awarded Northrop Grumman a then year

An EMD RQ-8A on approach for autonomous landing on USS *Nashville*, January 2006 *(Northrop Grumman)* 1151556

MQ-8B loaded on USAF C-17 for deployment to Afghanistan in April 2011 *(US Navy)* 1398535

Company-owned MQ-8 Fire Scout P-6, registration N393NG, which will carry an EO/IR payload and handle Increment 1 software check flights *(Northrop Grumman)* 1166972

MQ-8B Fire Scout VTUAV at Webster Field, Patuxent River, Md. mounting a flight test air data boom *(Northrop Grumman)* 1343232

USD93,721,957 cost plus incentive fee/award fee contract with regard to the Engineering and Manufacturing Development (EMD) phase of its Vertical Take-off and landing UAV (VTUAV) programme. Here, the work manifest comprised one × VTUAV system (two × AVs, USN Bureau of Aeronautics numbers (BuNo) 166414 (E-1) and 166415 [E-2]), technical manuals, an operations security programme, operator and maintainer training and pre-operational training. Work on the effort was to be undertaken at facilities in San Diego, California (78% workshare); Owego, New Jersey (six per cent workshare); Salt Lake City, Utah (five per cent workshare); Big Flats, New York (four per cent workshare); Baltimore, Maryland (four per cent workshare) and Sparks, Nevada (three per cent workshare). At the time of its announcement, the programme was scheduled for completion by the end of December 2003.

4 November 2000 Fire Scout AV P-1 was lost in a landing accident at the USN's Naval Air Warfare Center at China Lake, California. The loss was determined to have been the result of a faulty radar altimeter reading.

1 May 2001 The USN's Naval Air Systems Command (Patuxent River, Maryland) awarded Northrop Grumman's Ryan Aeronautical Center (San Diego, California) a then year USD14,167,267 modification (to the previously awarded cost plus fixed fee contract N00019-00-C-0277) with regard to the exercising of an option covering the procurement of one Fire Scout AV, its associated support equipment and an initial training package. At the time of its announcement, work on this effort was scheduled for completion by the end of April 2002.

7 May 2001 Northrop Grumman announced that the USN had awarded it a then year USD14.2 million Fire Scout Low-Rate Initial Production (LRIP) contract, with the first LRIP 'system' (three × AVs, 2 × GCSs, a datalink suite, remote data terminals and modular mission payloads) being intended for the USMC. As of the given date, Northrop Grumman's 'VTUAV' team included Israel Aircraft (subsequently Aerospace) Industries' Tamam Division (payloads), the Schweizer Aircraft Corporation (airframe), the then Lockheed Martin Systems Integration, L-3 Communications and the Sierra Nevada Corporation.

3 July 2001 Northrop Grumman announced that it had taken delivery of a second unmanned Fire Scout prototype from Schweizer (AV P-3).

14 September 2001 Northrop Grumman announced that it had taken delivery of the first EMD RQ-8A Fire Scout airframe from Schweizer (AV E-1).

19 May 2002 The third prototype RQ-8A (AV P-3, USN BuNo 166401) completed its first two flights from the USN's Naval Air Weapons Center at China Lake, California.

18 June 2002 Northrop Grumman announced that the Fire Scout GCS had successfully completed its USN acceptance test programme. As so tested, the architecture included an S-280 shelter, Rockwell Collins' AN/ARC-210(V) communications system radio, L-3 Communications' TCDL, the Sierra Nevada Corporation's UAV Common Automatic Recovery System (UCARS), Raytheon Tactical Control System (TCS) software and Northrop Grumman's datalink control processor software.

16 August 2002 *IHS Jane's* sources were reporting a Fire Scout AV as having demonstrated the type's sensor payload in-flight for the first time.

1 October 2002 Northrop Grumman announced that it had rolled-out the first full production standard RQ-8A AV on the 30th of the preceding September.

11 October 2002 *IHS Jane's* sources were reporting a Fire Scout AV as having downlinked real-time, in-flight imagery for the first time.

22 November 2002 *IHS Jane's* sources were reporting Fire Scout AV E-1 as having begun flight testing.

March-June 2003 *IHS Jane's* sources were reporting the period March-June 2003 as having seen the integration and testing of the General Atomics Aeronautical Systems Lynx GMTI/SAR radar aboard a Fire Scout AV, with the first flight test of the Lynx/Fire Scout combination being said to have taken place on 23 June 2003.

1 April 2003 Northrop Grumman announced that it had (in collaboration with Schweizer) successfully tested a four-bladed main rotor upgrade for the Fire Scout RQ-8A AV that was intended to increase the vehicle's payload capacity, speed, range, achievable altitude and endurance. According to *IHS Jane's* sources, initial testing of the new rotor configuration was carried out on the Schweizer 330 series helicopter that bore the US civil registration N330ST. Elsewhere, the same sources were noting that Fire Scout AV P-2 had made its maiden flight with the new four bladed main rotor on the 18th of the month.

16 April 2003 Northrop Grumman announced that it completed its first RQ-8A flight under the control of a USMC vehicle-mounted S-788 GCS during the preceding March. Intended to demonstrate AV control from both the USN's S-280 architecture and the S-788, the described effort was conducted at the NAS Patuxent River's Webster Field annex.

July-October 2003 *IHS Jane's* sources were reporting the Fire Scout AV as having made 13 tests flights with a combined payload of the previously noted Lynx radar, an EO sensor and a communications relay package.

August 2003 *IHS Jane's* sources were reporting the first phase of Fire Scout shipboard compatibility testing aboard the USS *Denver*, while Northrop Grumman used the month to announce the MQ-8B's selection as the US Army's Future Combat System's (FCS) Class IV UAS architecture.

17 December 2003 The Fire Scout UAV successfully completed its 100th consecutive flight.

2 March 2004 The USN's Naval Air Systems Command (Patuxent River, Maryland) awarded Northrop Grumman a then year USD49,000,000 ceiling priced, undifinitised modification (to the previously awarded cost plus fixed fee contract N00019-00-C-0277) with regard to the continuation of RQ-8B development and testing. Here, the manifest included the procurement of 2 × EMD RQ-8B AVs and work on the effort was to be undertaken at facilities in San Diego, California (85% workshare) and Elmira, New York (15% workshare). At the time of its announcement, the programme was scheduled for completion by the end of October 2005.

26 March 2004 The USN's Naval Air Systems Command (Patuxent River) awarded Raytheon's Falls Church, Virginia business unit a not to be exceeded then year USD36,800,000 cost plus award/incentive fee modification (to the previously awarded N00019-00-C-0190 contract) with regard to the supply of TCS software for use in integrating the Fire Scout AV into the USN's Littoral Combat Ship weapon system. Here, the manifest included engineering and test support for the Fire Scout's initial operating capability and work on the effort was to be undertaken at facilities in Falls Church, Virginia (56% workshare); Dahlgren, Virginia (30% workshare); San Pedro, California (10% workshare) and State College, Pennsylvania (four per cent workshare). At the time of its announcement, the programme was scheduled for completion by the end of March 2008.

5 April 2005 The USN's Naval Air Systems Command (Patuxent River, Maryland) awarded Northrop Grumman a then year USD11,748,783 modification (to the previously awarded cost plus fixed fee contract N00019-00-C-0277) with regard to the procurement of Fire Scout hardware (8 × airframes, IFF transponders, radar altimeters, 16 × GPS/INS navigation systems, antenna units, pressure transducers and 'precision differents') in support of the US Army's FCS Class IV UAS programme. Work on the effort was to be undertaken at facilities in Elmira, New York (85% workshare) and San Diego, California (15% workshare) and at the time of its announcement, the programme was scheduled for completion by the end of November 2006.

30 June 2005 The USN's Naval Air Systems Command (Patuxent River, Maryland) awarded Northrop Grumman a then year USD15,241,828 modification (to the existing cost plus fixed fee contract N00019-00-C-0277) with regard to the procurement of two × RQ-8B AVs,

MQ-8B is claimed to be all weather capable over land or sea *(Northrop Grumman)* 1343304

Maiden flight of the US Navy's first production MQ-8B *(Northrop Grumman)* 1165462

their payloads and associated non-recurring engineering work. Work on the effort was to be undertaken at facilities in Elmira, New York (48% workshare); San Diego, California (44% workshare) and Moss Point, Mississippi (eight per cent workshare) and at the time of its announcement, the programme was scheduled for completion by the end of August 2008.

22 July 2005 As part of a support effort for US Army and USN interest in 'weaponizing' the Fire Scout UAV, two 70 mm (2.75 in) Mk 66 unguided rockets were fired from an in-flight RQ-8 AV over the US Army's Yuma Proving Grounds in Arizona.

7 October 2005 The USN's Naval Air Systems Command (Patuxent River, Maryland) awarded Northrop Grumman a then year USD5,789,220 modification (to the previously awarded cost plus fixed fee contract N00019-00-C-0277) with regard to the design, manufacture and test of a shipboard compatible RQ-8 control station that would facilitate the AV being operated from the service's Littoral Combat Ship. Work on the effort was to be undertaken at facilities in Owego, New York (65% workshare) and San Diego, California (35% workshare) and at the time of its announcement, the programme was scheduled for completion by the end of June 2006.

15 December 2005 The USN's Naval Air Systems Command (Patuxent River, Maryland) awarded Northrop Grumman a then year USD8,345,263 modification (to the existing cost plus fixed fee contract N00019-00-C-0277) with regard to shipboard testing of the RQ-8. Here, the manifest included flight testing from and the necessary shipboard installation work aboard the service's High Speed Vessel - 2 (the USS *Swift*) and at the time of its announcement, the effort was scheduled for completion by the end of June 2006.

2006 The MQ-8B nomenclature is said to have been promulgated during 2006. Readers should note that the MQ-8B re-designation of the RQ-8B is claimed by some sources to have taken place in June of the previous year.

3 January 2006 Northrop Grumman's Moss Point, Mississippi UAS Center received the first airframe for the MQ-8B programme from the Schweizer Aircraft Corporation. In an associated media release, Northrop Grumman noted that this AV was the first of a batch of 12 which were to be supplied to both the US Army (eight) and the USN (four).

16-17 January 2006 IHS Jane's sources were reporting that a pair of Fire Scout EMD AVs had made a total of nine autonomous landings aboard the USS *Nashville* as part of the type's ongoing shipboard compatibility testing.

20 March 2006 The USN's Naval Air Systems Command (Patuxent River, Maryland) awarded Northrop Grumman a then year USD29,286,714 modification (to the existing cost plus fixed fee contract N00019-00-C-0277) with regard to continuation of the 'RQ-8' Fire Scout's development and test programme. Work on the effort was to be performed at facilities in San Diego, California (85% workshare) and Elmira, New York (15% workshare) and at the time of its announcement, the programme was scheduled for completion by the end of June 2006.

May-June 2006 IHS Jane's sources were reporting Fire Scout 'risk reduction operations' aboard the USS *Swift*.

28 July 2006 The USN's Naval Air Systems Command (Patuxent River, Maryland) awarded Northrop Grumman a then year USD135,821,763 modification (to a previously awarded cost plus fixed fee award contract) with regard to continuation of 'RQ-8B' development and testing. Work on the effort was to be undertaken at facilities in San Diego, California (81% workshare); Moss Point, Mississippi (seven per cent workshare); Horseheads, New York (six per cent workshare); Wilsonville, Oregon (four per cent workshare) and Wayne, New jersey (two per cent workshare) and at the time of its announcement, the programme was scheduled for completion by the end of August 2008. In an associated media release, Northrop Grumman noted that this award 'definitized' the remaining work needed to complete the RQ-8's SDD phase 'through 2008'.

14 December 2006 The USN's Naval Air Systems Command (Patuxent River, Maryland) awarded Northrop Grumman a then year USD167,242,493 modification (to the previously awarded cost plus fixed fee contract N00019-00-C-0277) with regard to the supply of two × 'RQ-8B' AVs. At the time of its announcement, the effort was scheduled for completion by the end of October 2008.

22 May 2007 Northrop Grumman announced that it had successfully performed an engine run on the US Army's first MQ-8B AV.

31 May 2007 Northrop Grumman announced that the MQ-8B had achieved its Milestone C decision thereby opening the way for LRIP of the configuration.

June 2007 Fire Scout Lot 1 LRIP was authorised by the USN.

August 2007 A pair of US Army and USN MQ-8B UAVs were successfully demonstrated as a single load for a KC-130T tanker/transport aircraft.

19 March 2008 Northrop Grumman announced its intention of flying Telephonics' X-band (9.375 GHz centre frequency) RDR-1700B surveillance radar aboard a company-owned Fire Scout demonstrator as part of a USN initiative to equip the AV with radar.

20 August 2008 Northrop Grumman announced that the Fire Scout UAS had completed its first flight with a BRITE Starl® II EO/IR imager and the TCDL installed.

2 September 2008 Northrop Grumman announced the maiden flight of its company-owned MQ-8B demonstrator (AV P-6). During the same month, the USN authorised Fire Scout Lot 2 LRIP.

23 February 2009 Northrop Grumman announced that the USN's Naval air Systems Command (Patuxent River, Maryland) had awarded it a then year USD40 million contract modification with regard to Lot 3 LRIP of three × MQ-8B AVs (and their associated EO payloads), three × GCSs, three × light harpoon grids, three × UAV automatic recovery systems and six × portable electronic display devices. At the time of the announcement, work on this effort was scheduled for completion by the end of March 2011.

May 2009 Northrop Grumman announced that an MQ-8B had successfully completed a series of at-sea, fully autonomous flight operations from the USS *McInerney* during the period 4-8 May 2009. This round represented the MQ-8B's first autonomous operation from a US surface combatant. Elsewhere in the programme, 5 May 2012 saw Northrop Grumman reporting that the USN had awarded it a then year USD5 million MQ-8B contractor logistics support contract with three one year extension options which, if exercised, would bring the value of the award up to then year USD19 million.

30 June 2009 Northrop Grumman announced that the MQ-8B's operational land-based configuration (in the form of AV P-7) had completed its first flight operations (including both Reconnaissance, Surveillance and Target Acquisition (RSTA)/ISR functionality) at the Yuma Proving Ground in Arizona.

6 October 2009 Northrop Grumman announced that the MQ-8B AV P-7 had completed trials with a new, company-developed, STANAG 4586 compatible GCS during the preceding September. Here, the new GCS was described as incorporating redundancy, multiple voice/ secondary C2 radios, the TCDL, a digital intercommunications system and personal computer-based, COTS workstation components.

24 November 2009 Northrop Grumman announced that it had completed delivery of the first three LRIP production standard MQ-8B AVs to the USN. An accompanying media release went on to note that two of these AVs had been used for the type's Military Utility Assessment aboard the USS *McInerney* and that Fire Scout AVs had been aboard *McInerney* four times since December 2008 during which time, they had completed 110 × shipboard take-offs and landings, accumulated more than 47 flight hours and had made 45 landings using the harpoon grid.

16 December 2009 Northrop Grumman announced that its company-owned MQ-8B AV had completed a maritime sensor demonstration (using the RDR-1700B radar and a FLIR Systems EO sensor) at NAS Patuxent River during the preceding October.

2010 IHS Jane's sources were reporting the cancellation of the US Army's FCS Class IV UAS.

15 February 2010 Northrop Grumman announced that an MQ-8B AV had demonstrated a proof-of-concept autonomous resupply capability during the US Army's then current Army Expeditionary Warrior Experiment (AEWE) at Fort Benning, Georgia. During this trial, the AV was equipped with ruggedised cargo containers that were attached to external pylons and could be released via remote control.

Northrop Grumman RQ-8A and MQ-8B Fire Scout < United States < Unmanned aerial vehicles

RQ-8A rocket-firing trials at Yuma, July 2005 *(Northrop Grumman)*

RQ-8A Fire Scout undergoing Lynx SAR trials for the US Army MQ-8B FCS version *(Northrop Grumman)*

14 July 2010 Northrop Grumman announced that it had completed a series of MQ-8B flight demonstrations in the United Arab Emirates in order to assess the type's performance in extreme environmental conditions.

2 August 2010 Control of an MQ-8B AV was lost for approximately 27 minutes during a test flight from NAS Patuxent River. Investigations pin pointed a software anomaly as having caused the AV to deviate from its pre-programmed flight procedures and the USN grounded its then fleet of six MQ-8Bs untilt he fault was rectified.

January-June 2011 Two MQ-8B AVs were embarked aboard the USS *Halyburton* for a six month deployment that included operations over Libya (Operation 'Unified Protector' - see following), three anti-piracy actions and support (in collaboration with the ship's SH-60B helicopter) for USS *Halyburton's* transit of the Strait of Hormuz. During the cruise, the AVs flew for more than 435 hours and achieved an 80% plus sortie completion rate.

13 April 2011 Three MQ-8B AVs arrived in Afghanistan to meet an 'urgent needs [ISR] requirement' filed by US Central Command.

21 June 2011 One of the MQ-8B AVs deployed aboard the USS *Halyburton* was lost (it is believed) to ground fire during an Operation 'Unified Protector' ISR sortie over central and western Libya.

29 August 2011 The USN's Naval Air Systems Command (Patuxent River, Maryland) awarded Northrop Grumman a then year USD10,474,708 cost plus fixed fee contract with regard to the provision of logistic support services for the MQ-8B. Here, the manifest included logistics management (and management information), technical data updates, maintenance and supply services, AV transportation, training services, flight operations and deployment support. Work on the effort was to be performed at facilities in St Inigoes, Maryland (40% workshare); San Diego, California (20% workshare) and at various other locations outside the continental US (40% workshare).

22 September 2011 The USN's Naval Air Systems Command (Patuxent River) awarded Northrop Grumman a then year USD17,098,027 cost plus fixed fee contract with regard to the implementation of the MQ-8B Rapid Deployment Capability Weaponization Program (RDCWP). Here, the effort included the installation, engineering, manufacture and data development of suitable weapon systems, with the whole to include 12 × stores management systems. Work on the programme was to be undertaken at facilities in San Diego, California (75% workshare) and Grand Rapids, Michigan (25% workshare) and at the time of its announcement, the effort was scheduled for completion by the end of March 2013. In this context, *IHS Jane's* sources report BAE Systems as having received a USN contract to integrate the Advanced Precision Kill Weapon System (APKW) semi-active laser guided rocket into the Fire Scout architecture during September 2012.

23 September 2011 A Patuxent River-based Fire Scout AV made the type's first flight using a 50-50 blend of bio and traditional aviation fuel.

28 September 2011 The USN's Naval Air Systems Command (Patuxent River, Maryland) awarded Northrop Grumman a then year USD18,650,000 cost plus fixed fee contract with regard to the provision of ISR services in support of the deployment of the MQ-8B AV. Work on the effort was to be undertaken in Afghanistan (90% workshare) and at Patuxent River, Maryland (10% workshare) and at the time of its announcement, the programme was scheduled for completion by the end of October 2012.

29 September 2011 The USN's Naval Air Systems Command (Patuxent River, Maryland) awarded Northrop Grumman a then year USD7,550,000 cost plus fixed fee delivery order (against the previously issued basis ordering agreement N00019-10-G-0003) with regard to the provision of MQ-8B software sustainment services. Here, the manifest included analysis of Engineering Change Proposals (ECP), action plan and milestone development, laboratory studies and analysis, software technology insertion, configuration management, quality assurance and the development/review/update of relevant specifications and technical documentation. At the time of its announcement, the programme was scheduled for completion by the end of June 2012.

25 October 2011 An MQ-8B AV demonstrated its ability to transmit acquired imagery to the cockpit of an MH-60 helicopter and a US Coast Guard patrol vessel in real-time.

30 March 2012 An MQ-8B AV deployed aboard the USS *Simpson* was ditched (and recovered) during operations off the west coast of Africa. The accident was caused by the AV's failure to establish a link with its parent vessel's UCARS system.

April 2012 As of April 2012, the USN's MQ-8B fleet had flown more than 5,000 operational hours (3,000 of which had been completed during the preceding 12 months) and was slated to continue supporting US Africa Command until US Fiscal Year 2014.

6 April 2012 One of the three MQ-8B AVs deployed to southwest Asia was lost in a crash in northern Afghanistan. As of April 2012, the USN's trio of Afghan-based Fire Scouts were being reported as having clocked-up 2,800 flight hours and delivered 300 hours of full motion video per month since their arrival in-theatre during April 2011.

8 May 2012 The USN's Naval Air Systems Command (Patuxent River) awarded Northrop Grumman a then year USD25,709,758 modification (to the existing firm, fixed-price contract N00019-07-C-0041) with regard to the procurement of 3 × Lot 5 LRIP Fire Scout AVs and one × GCS. Work on the effort was to be undertaken at facilities in Moss Point, Mississippi (55% workshare) and San Diego, California (45% workshare) and at the time of its announcement, the programme was scheduled for completion by the end of December 2013.

June 2012 *IHS Jane's* sources reported the USN's Office of Naval Research as developing a Fire Scout Multi-Mode Sensor Seeker (MMSS) to facilitate Automatic Target Recognition (ATR) and tracking of pirate targets. The prototype MMSS mated a 17.8 cm (7 in) LADAR telescope and new target recognition software algorithms with the BRITE Star® II EO turret. Elsewhere in the programme, four MQ-8Bs (operated by Helicopter Anti-Submarine Squadron Light 42's (HSL-42) Detachment 7) deployed aboard the USS *Klakring* for a six month cruise in the US Africa Command's area-of-operations on 29 June 2012. Unconfirmed sources report MQ-8B AVs USN BuNos 167784, 167785 and 167787 as having been assigned to the described detachment.

10 July 2012 Northrop Grumman announced the opening of a new Fire Scout Training Center at NAS Jacksonville in Florida. In an accompanying media release the contractor went on to note that MQ-8B AVs had accrued more that 2,800 flight hours over Afghanistan since May 2011.

August 2012 *IHS Jane's* sources were reporting US contractor Cubic as developing a compact ISR datalink system for the MQ-8B AV built around its C- (4 to 8 GHz) and Ku-band (12.5 to 18 GHz) Multi-band Miniature Transceiver (MMT) and RF technology that used standard waveforms to transfer data and stream video.

27 September 2012 The USN's Naval Air Systems Command (Patuxent River, Maryland) awarded Northrop Grumman a then year USD28,126,043 cost plus fixed fee contract with regard to MQ-8B software sustainment, non-recurring engineering support and obsolescence management (including future growth software development). Work on the effort was to be undertaken at facilities in San Diego, California (90% workshare) and Patuxent River, Maryland (10% workshare) and at the time of its announcement, the programme was scheduled for completion by the end of September 2013.

5 October 2012 The USN's Naval Air Systems Command (Patuxent River) awarded Northrop Grumman a then year USD24,476,987 firm, fixed-price delivery order (against the previously issued basic ordering agreement N00019-10-G-0003) with regard to the supply of MQ-8B spares and 'supplies'. Work on the effort was to be undertaken at facilities in San Diego, California (36% workshare); Horseheads, New York (30% workshare); Salt Lake City, Utah (11% workshare); Sparks, Nevada (11% workshare) and in various other locations within the continental US (12% workshare). At the time of its announcement, the programme was scheduled for completion by the end of April 2014.

28 November 2012 The USN's Naval Air Systems Command (Patuxent River, Maryland) awarded Northrop Grumman a then year USD15,010,161 cost plus fixed fee contract with regard to the operation and maintenance of MQ-8B UAVs in Afghanistan (90% workshare) and at Patuxent River (10% workshare). At the time of its announcement, work on the effort was scheduled for completion by the end of November 2013.

December 2012 As of December 2012, *IHS Jane's* sources were reporting the USN as intending to conduct a live-fire assessment of the

Northrop Grumman RQ-8A and MQ-8B Fire Scout

Fire Scout displayed in October 2005 with external cargo pods *(IHS/Michael Sirak)* 1116701

MQ-8B armed with the Advanced Precision Kill Weapon System (APKWS) during March 2013. The same report also noted that the service was 'in the midst' of integrating the MQ-8B with its 'Perry-class' frigates.

1 December 2012 The USN's fourth at-sea MQ-8B deployment (involving HSL-42's Detachment 2) returned to the continental US after five months aboard the USS *Klakring* providing anti-piracy and real-time ISR support in the US Africa Command's area of responsibility. Here, four × MQ-8B AVs were embarked, with their operating Detachment accumulating more than 500 flight hours and 'regularly maintaining' 12 hour days on-station. Again (and during one instance in September 2012), the Detachment provided 'just over' 24 hours of continuous ISR coverage, with the effort requiring 10 separate flights, eight refuellings and the host vessel 'setting flight quarters' for launch or recovery 20 times. Elsewhere, the Detachment is understood to have pioneered dual MQ-8B AV operations.

13 December 2012 An MQ-8B AV deployed aboard the USS *Robert G Bradley* ditched in the Mediterranean after encountering icing conditions during a sortie that had been launched on the given date. At the time, *Robert G Bradley* was operating with three × MQ-8Bs embarked.

20 December 2012 The USN's Naval Air Systems Command (Patuxent River, Maryland) awarded Northrop Grumman a then year USD33,270,000 cost plus incentive fee contract with regard to the development, production, integration and testing of nine X-band (8 to 12.5 GHz) Telephonics AN/ZPY-4 (also designated as the RDR-1700B+) radars aboard MQ-8B AVs. Work on the effort was to be undertaken at facilities in San Diego, California (70% workshare) and Patuxent River, Maryland (30% workshare) and at the time of its announcement, the programme was scheduled for completion by the end of June 2014. On the same day, the same contracting activity awarded Northrop Grumman a second then year USD19,166,627 cost plus fixed fee contract with regard to the provision of logistical services for the MQ-8B system. Work on this second effort was to be undertaken at facilities in San Diego, California (90% workshare) and at Patuxent River, Maryland (10% workshare) and at the time of its announcement, the programme was scheduled for completion by the end of November 2013.

Description: *Airframe:* Tadpole-shaped fuselage carrying a three- (RQ-8A) or four-bladed (RQ/MQ-8B) main rotor (with folding blades) and a two-bladed tail rotor (carried to port). Dorsal and ventral swept vertical tail surfaces at the rear of the vehicle's fuselage and a skid undercarriage. Ventral harpoon landing system. Under nose sensor turret/sensor mounting (with dual payload nose configuration option) and external antennas for the AV's UCARS system, TCDL (two aerials), V/UHF radios (three), radar altimeter (two) and GPS system (two). Drive train and powerplant installation derived from those of the Schweizer 330SP/333 maned helicopters. Forward shielded avionics compartment. Main fuel tank located in a dorsal fairing that also houses the main rotor drive shaft and flight control actuators.

Mission payloads: Sensors: Northrop Grumman ESS/IAI Tamam U-MOSP multi-mission sensor (IR and zoom TV, each with three fields of view) and laser designator/range-finder in RQ-8A. FLIR Systems BRITE Star® II EO sensor for US Navy MQ-8B; possibly Cobra mine detection system later. ASTAMIDS for mine detection and JTRS communications relay planned for US Army FSC MQ-8B. General Atomics AN/APY-8 Lynx GMTI/SAR and Telephonics RDR-1700B radars trialled. X-band (8 to 12.5 GHz) Telephonics AN/ZPY-4 (RDR-1700B+) multi-mode maritime surveillance radars procured for installation on (initially) nine × MQ-8B AVs (see *Development*). Avionics 'based on those used in Global Hawk' UAV. A sigint payload is under consideration.

Weapons and other external payloads: Trials in mid-2003 equipped with two four-round pods of air-to-air Stinger (ATAS) missiles. Further trials in July 2005 with live firing of 2.75 in Mark 66 unguided rockets at Yuma Proving Grounds, Arizona. As of 2012, the BAE Systems APKW) semi-active laser guided rocket was being integrated into the Fire Scout architecture. In October 2005, Northrop Grumman revealed that it was marketing to the US Army an external cargo pod modification. With 113 kg (250 lb) in each pod and a further 136 kg (300 lb) in the nose, such a version would be able to ferry up to 363 kg (800 lb) of supplies to troops in the field. Possible future weapons for the US Navy in spiral development include the APKWS, the Compact Very Light Weight Torpedo, the Low Cost Guided Image Rocket and Viper Strike.

Guidance and control: From US Navy S-280 shelter-mounted mission planning and tactical control station (USMC S-788 HMMWV-mounted shelter also completed); INS/GPS navigation; automatic approach and landing, using Sierra Nevada AN/UPN-51(V) UCARS. L-3 Communications UHF Tactical Common Data Link (TCDL) and AN/ARC-210 command and control link. Fire Scout is also launch demonstration system for Raytheon's TCS. GCS able to control three air vehicles simultaneously; any two can provide up to 12 hours' continuous coverage. Shipboard integration by Lockheed Martin Federal Systems.

Transportation: Transportable by HMMWV. Optional ground handling wheels.

System composition: Two or three AVs, two or three mission payloads, a datalink suite, remote data terminals and two GCSs (incorporating TCDL and TCS) per system; one NATO-standard talon grid for shipboard operations.

Launch: Conventional, autonomous helicopter take-off.

Recovery: Conventional, autonomous helicopter landing, using Sierra Nevada UCARS-V2 system.

Power plant: One 313 kw (420 shp) Rolls-Royce 250-C20W (US military designation T63-A-720) turboshaft, derated to 239 kw (320 shp). Fuel capacity 719 litres (190 US gallons; 158 Imp gallons).

Variants: *RQ-8A:* Initial USN and USMC version with three-blade main rotor.

MQ-8B: USN and US Army FCS configuration with a four-bladed main rotor (with improved blade aerofoil section), uprated engine gearbox, increased fuel and payload capacity, different sensors and avionics, side-mounted sponsons for stores attachment (supply pods or weapon-carrying ability) and other improvements. Critical design review passed in December 2005. US Army requirement lapsed in 2010.

MQ-8B

Dimensions, External	
Overall	
length, rotor folded	9.2 m (30 ft 2¼ in)
height, top of tail antenna	2.87 m (9 ft 5 in)
Fuselage	
length, with dual payload nose	7.3 m (23 ft 11½ in)
width	1.9 m (6 ft 2¾ in)
Skids, skid track	1.75 m (5 ft 9 in)
Rotors, rotor diameter	8.398 m (27 ft 6¾ in)
Tail rotor, tail rotor diameter	1.30 m (4 ft 3¼ in)
Dimensions, Internal	
Payload bay, volume	0.227 m³ (8.0 cu ft)
Areas	
Rotor disc	55.18 m² (594.0 sq ft)
Tail rotor disc	1.32 m² (14.2 sq ft)
Weights and Loadings	
Weight	
Weight empty	661 kg (1,457 lb)
Max T-O weight	1,429 kg (3,150 lb)
Payload, Max payload	272 kg (599 lb)
Performance	
Altitude	
Service ceiling	6,095 m (20,000 ft)
Hovering ceiling, OGE	1,525 m (5,000 ft)
Speed	
Max level speed, IAS	125 kt (232 km/h; 144 mph) at S/L (est)
Radius of operation, mission	110 n miles (203 km; 126 miles)
Endurance	
at 110 n miles (204 km; 127 miles) from base with max payload	3 hr
at 250 n miles (463 km; 287 miles) from base with 45 kg (100 lb) payload	2 hr 12 min
max, with baseline payload	8 hr (est)
max, with 272 kg (600 lb) payload	5 hr

Status: Overtime, sources have identified those AVs bearing the USN BuNos 167784, 167785, 167786, 167787, 167788, 167789, 167790, 167792, 167986, 167987 and 167988 as being MQ-8Bs. As of the summer of 2012, 16 × MQ-8Bs were in service with the USN, with one (thought to be BuNo 167988) being assigned to Air Test and Evaluation Squadron One (VX-1 - NAS Patuxent River, Maryland) and the remaining 15 to the service's Naval Test Wing Atlantic's UAS Test Directorate (also based at NAS Patuxent River). Of the cited AVs, BuNos 167784, 167785 and 167787 are understood to have formed part of the four AV detachment assigned to the USS *Klakring* for its June 2012 cruise. Here, the AVs were assigned to HSL-42's Detachment 7 for the duration of the deployment. Of the USN's RQ-8A AVs, BuNos 166400, 166401, 166414 and 166415 were struck off-charge on 31 December 2008, 31 December 2008, 30 September 2011 and 31 December 2008 respectively. RQ-8A AV BuNo 166416 is reported to have been damaged beyond economical repair and to have been placed in storage. As of 2013, Northrop Grumman was continuing to promote the MQ-8B Fire Scout VTUAV.

Contractor: Northrop Grumman Unmanned Systems
San Diego, California.

Northrop Grumman RQ-4 Global Hawk

Type: High-altitude long-endurance surveillance UAV.

Development: *ACTD phase:* Request for proposals (RFP) issued on 1 June 1994; five contractor teams, headed by Loral, Northrop Grumman, Orbital Sciences, Raytheon and Teledyne Ryan, bid designs to ARPA for the Tier II+ requirement in March 1995. Original intention was to select two teams to compete for Phase 2, but insufficient funds were available and selection of the Teledyne Ryan Aeronautical (TRA, which became part of Northrop Grumman in July 1999) team as sole winner was announced 23 May 1995 with award of a USD164 million, 31-month ACTD contract for the completion of five prototypes.

Air Vehicle 1 (AV-1, USAF serial number 95–2001) was rolled out on 20 February 1997 and engine test runs began 9 May 1997. First flight (56 minutes), from Edwards AFB, California, was made on 28 February 1998, reaching 9,750 m (32,000 ft). Second test flight, on 10 May 1998, lasted 2 hours 24 minutes, reaching 12,500 m (41,000 ft), and demonstrating two-way control handover via satcom command link. Third flight (30 May 1998) lasted 5 hours 22 minutes, reached 15,600 m (51,200 ft), and covered 1,918 n miles (3,540 km; 2,200 miles).

Second prototype (AV-2, 95–2002) flew for the first time on 20 November 1998, with a second flight on 4 December. On 22 January 1999, it employed all three sensors during a single 6 hour 24 minute sortie, capturing 21 EO, three SAR (mapping mode) and two IR scenes. It was lost in a crash on 29 March 1999 after inadvertently receiving a flight termination signal from another UAV some distance away. AV-3, delivered to Edwards AFB on 29 April 1999, suffered minor damage (collapsed nose gear) on 6 December that year when a software problem affecting its taxying speed caused it to veer off the runway; flying was resumed on 11 March 2000.

AV-1, on its 22nd sortie (19/20 October 1999), made Global Hawk's first overwater flight during a 24.8 hour trip to Alaska and back. A UAV endurance record was set on 14/15 April 2000 with a flight of 31.5 hours. Flights over Canadian territory (Alberta and British Columbia), in a joint evaluation programme, were made in early 2000, and on 8/9 May 2000 AV-4 made a 28-hour flight from Eglin AFB, Florida, across the Atlantic to Portugal to take part in a NATO exercise; the flight included 4 hours on station in Portugal. On 21 to 23 April 2001 Global Hawk AV-5, named *Southern Cross II* for the occasion, made a 23 hour 22 minute non-stop trans-Pacific flight from Edwards AFB to RAAF Edinburgh AB, South Australia, to take part in the six-week Exercise 'Tandem Thrust' in May; it returned to Edwards on 7 June 2001.

Global Hawk flying hours by 24 May 2001 had included more than 330 hours in FAA-controlled civil airspace, and on 15 August 2003 Northrop Grumman announced that the aircraft had been granted a national Certificate of Authorization (CoA) by the FAA to fly routinely in US national airspace — the first such certificate to be awarded to a UAV.

EMD phase: In September 2000, Global Hawk was recommended for production by the US Joint Forces Command. Milestone 2 (Defense Acquisition Board review) was completed on 16 February 2001. On 21 March 2001, the system entered the engineering and manufacturing development (EMD) phase with the announcement of a USD45 million contract for two further Global Hawks, AV-6 (00-2006) and AV-7 (00-2007), designed to introduce various improvements compared with the ACTD quintet including, initially, upgraded com radios, a Mode S transponder, TCAS and an ELT. The first EMD aircraft flew on 23 April 2002, and the second was delivered to USAF at Edwards AFB on 14 February 2003.

Since then, notwithstanding the subsequent manufacture and

RQ-4N, marinised Global Hawk, selected for BAMS in April 2008 *(Northrop Grumman)*

deployment of production Global Hawks, EMD has been a continuing process that is at present scheduled to go on until 2011. 'Period 5' of that development, was due for completion by 31 December 2007. To accelerate the programme and achieve full operational capability by 2007 instead of 2009, traditional USAF 'block number' improvement standards for production aircraft were at first replaced by a proposal for so-called 'spiral' enhancements, each ingredient of which (mainly concerning sensor and other equipment upgrades) would be cut into the production line as its development was completed. However, the block numbering system, broadly corresponding to the 'spiral' proposals, was later reinstated, the initial production RQ-4A now being designated the Block 10 version and the RQ-4B Block 20 and beyond, as detailed under the Variants heading below.

In February and March 2006, Northrop Grumman demonstrated a further application for the Global Hawk with three 28-hour flights over known drug-trafficking routes across the southern US border, the Gulf of Mexico and the Caribbean. These confirmed the UAV's ability to detect and track low-flying aircraft and fast-moving small boats from an altitude of 18,280 m (60,000 ft), passing the targets' locations to a US Navy P-3 Orion maritime patrol aircraft. In July 2006, the company also revealed that it is studying the concept of using a Global Hawk, fitted with a nose-mounted, upward-looking IR sensor, as a means of detecting and tracking ballistic missiles.

Global Hawk flew its first civil emergency support sorties in 2007, completing three missions to survey fires in Southern California.

The RQ-4 Global Hawk celebrated the 10th anniversary of its first flight on 28 February 2008. Some 25,000 combat hours had been logged by the fleet on 9 July 2009 (of 32,500 total hours). A considerably more detailed development history can be found in *IHS Jane's C4I: Air*.

Description: *Airframe:* Very high-aspect ratio, low-mounted CFRP wings, built by Vought and ATK Composites; quarter-chord sweepback 5° 54'; fitted with inboard and outboard ailerons plus inboard and outboard upper-surface spoilers. Fuselage is basically of aluminium alloy box section, with rounded corners; dorsally-mounted composites engine pod at rear; large bulged dielectric fairing above nose houses 1.22 m (4 ft) diameter wideband satcom antenna; synthetic aperture radar in underfuselage bulge forward of wingroot; smaller blister under rear fuselage houses common datalink (CDL/LOS) antenna. All three radomes by Marion Composites. Honeywell pressurised payload and avionics compartments and environmental control system. Composites V tail surfaces (dihedral angle 50°), with inboard and outboard ruddervators, plus rear fuselage, nacelle and fairings, are manufactured by Aurora Flight Sciences. Retractable tricycle landing gear (modified Learjet 45 twin-wheel main units and modified CF-5F single-wheel nose unit),

RQ-4 Block 40 arrives for the firs time at Grand Forks, the second main operating base, on 26 May 2011 *(Northrop Grumman/USAF)*

First Euro Hawk completed assembly in July 2009 and was rolled out in October 2009 *(Northrop Grumman)*

04-2015, the first RQ-4B Block 20 Global Hawk *(Northrop Grumman)*

Maiden flight of the first production Block 20 RQ-4B, 1 March 2007 *(Northrop Grumman)*

supplied by Héroux Inc of Québec. Upper front fuselage mounting points for elint antennas standard on production aircraft.

Airframe changes in the RQ-4B include an increased-span wing with slightly reduced sweepback (5° at quarter-chord); 2° dihedral; underwing hardpoints (single 454 kg; 1,000 lb point each side for fuel or sensor pod); slightly longer, deeper and more bulbous fuselage; increased tail surface height and area; and a taller and wider-track main landing gear with single mainwheels that retract rearwards into conformal pods on the underside of the wing trailing-edge. The main and single nose unit wheels and tyres are standard Lockheed Martin F-16 units and are fitted with Goodrich electrically actuated differential brakes.

Mission payloads: Primary sensor is a Raytheon Space and Airborne Systems 'basic' ISS reconnaissance suite in ACTD/EMD aircraft and early production RQ-4As, linked to GHGS via satellite. It comprises a combined electro-optical/infra-red sensor (digital Kodak CCD stills camera and 3 to 5 micron IR sensor based on AN/AAQ-16B helicopter system) mounted under the nose, with an X-band synthetic aperture radar just to the rear. Radar incorporates off-the-shelf hardware from the HISAR commercial reconnaissance system. The integrated nature of the sensor package provides the capability to select radar, IR and visible wavelength imagery as required, and to use the radar simultaneously with either of the other two sensors. Sensor area coverage in 24-hour loiter: 40,000 sq n miles (137,320 km^2; 53,019 sq miles). The ISS also incorporated a Kearfott KN-4072 digital GPS/INS navigation system originally in ACTD aircraft, replaced in later aircraft by dual Northrop Grumman (ex-Litton) LN-100G or LN-211G equipment — one for the EO/IR sensor and one for the radar. A 'see and detect' video camera in the nose acts as the 'eye' for monitoring take-offs and landings. An Enhanced ISS (EISS) suite (for which acceptance tests were completed in October 2005) is introduced on Blocks 20 and 30 RQ-4Bs from Lot 3, and available onboard power is increased by the addition of a 25 kVA AC generator, the 10 kVA unit of earlier aircraft also being retained as a back-up.

Over time, five major upgrades have been noted for the high-time ACTD aircraft AV-3 during its activities in Iraq and Afghanistan, namely: BAE Systems (E&IS) Hyper Wide comint, secure communications, air traffic control (ATC) voice communications, an automatic target recognition 'blue force' tracker, and an advanced information architecture (AIA) for more detailed imagery. Hyper Wide is introduced as a 'clip-in' installation as from Lot 3 RQ-4Bs.

Other planned payload improvements include:

Northrop Grumman Advanced Signals Intelligence Payload (ASIP), an elint package scheduled for introduction on Block 30/Lot 5 aircraft for

In white and blue NASA livery, one of two pre-production Global Hawks made available to NASA when the USAF had finished with them *(NASA Dryden)*

EMD Global Hawk 00-2006 in flight, showing the ventral Hyper Wide comint radome *(Northrop Grumman)*

This detail shows clearly the bifurcated ventral radome for the Hyper Wide comint installation, as fitted on EMD aircraft 00-2007 *(Northrop Grumman)*

Northrop Grumman RQ-4 Global Hawk < United States < Unmanned aerial vehicles

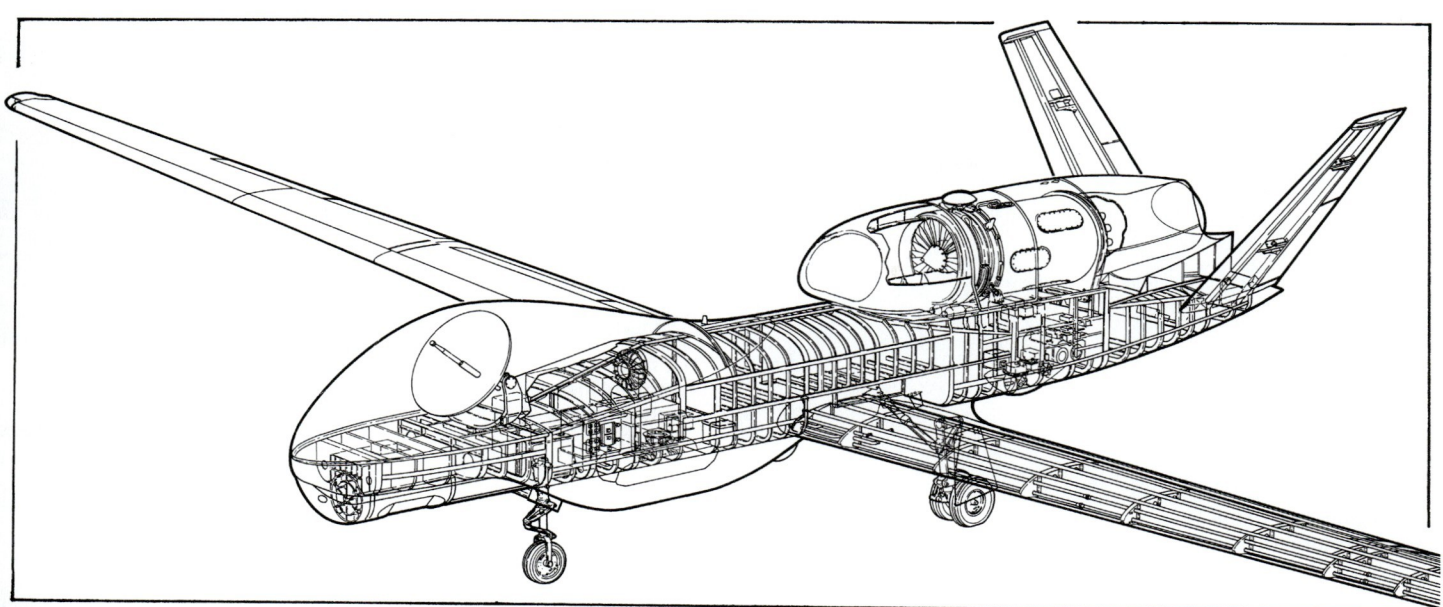

RQ-4A Global Hawk cutaway *(TRA)*

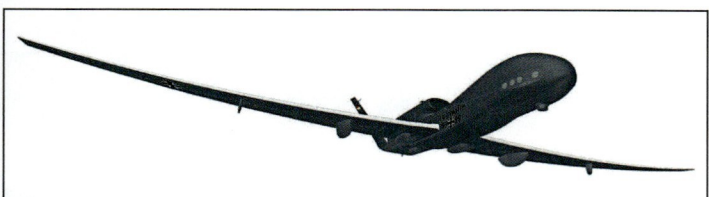

Early 2007 computer graphic of the latest Euro Hawk configuration *(Northrop Grumman)*

fielding in 2008. Acceptance tests of this were completed in October 2005, and in January 2006 Northrop Grumman announced the first flight of its own HBS PCU (high-band system production configuration unit) component of ASIP, installed on EMD aircraft AV-7.

MP-RTIP (Multi-Platform Radar Technology Insertion Program). This upgraded SAR/GMTI radar, under joint development by Northrop Grumman and Raytheon, is slated for introduction on the Block 40 RQ-4B, replacing the EISS. Two demonstration equipments, plus three for installation in Global Hawks, were ordered in April 2004. A design readiness review is scheduled for October 2008 and contract completion is due by 31 May 2010. Meanwhile, a dummy pod to house the radar was flight-tested on 27 April 2006, using the Northrop Grumman-owned Proteus as a surrogate carrier. Flight tests with an actual radar, again aboard the Proteus, began during the last week of September 2006.

A defensive aids system comprising a Raytheon AN/ALR-89(V) radar warning receiver, Raytheon AN/ALE-50 towed radar decoy and a Northrop Grumman LR-100 ESM system.

Consideration was given in 2005 to external carriage of mini- or micro-UAVs such as SilentEyes or KillerBee as supplementary augmentation devices. So far as is known, no commitment to such devices had been made by early 2008.

Northrop Grumman's AEW and EW Systems sectors are developing 'advanced' payloads to meet the particular requirements of the US Navy BAMS programme, the main component of which is a 360 degree Multi-Function Active Sensor (MFAS) active electronically scanned array radar. The electro optical system is Northrop Grumman's Nighthunter II.

See also Euro Hawk description under Variants heading for details of HALE Elint Payload (HEP) for this version.

Guidance and control: Global Hawk is controlled from a Raytheon Intelligence and Information Systems AN/MSQ-131 Global Hawk Ground Segment (GHGS, which see), consisting of a four-person RD-2A Mission Control Element (MCE) and a two-person RD-2B Launch and Recovery Element (LRE), each housed in a transportable tactical shelter complete with power and environmental units. Each GHGS can control up to three air vehicles simultaneously. Flight control and navigation are performed by two dual-redundant Curtiss-Wright (formerly Vista Controls) Integrated Mission Management Computers (IMMC) with BAE (formerly Marconi Integrated) Systems mission planning software and OmniStar GPS/INS. In 2003, Curtiss-Wright began developing a sensor management unit (SMU) to connect the onboard ISS suite with the satellite and LOS datalinks and interface with the IMMCs, enabling payloads to be interchanged without affecting other aspects of the UAV's performance. Other uplink/downlink equipments include an L-3 Communications — West X-band LOS common datalink (CDL), Inmarsat Ku-band satcom link, a UHF satcom back-up, and three Rockwell Collins AN/ARC-210(V) ATC communications transceivers.

System composition: Two air vehicles and two GHGS ground control stations.

Launch: Conventional, autonomous take-off (runway requirement 1,525 m; 5,000 ft).

Recovery: Conventional, autonomous landing (runway requirement 1,525 m; 5,000 ft).

Power plant: One 33.8 kN (7,600 lb st) FADEC-equipped Rolls-Royce F137-AD-100 (military AE 3007H) turbofan. See under Weights for fuel details.

Variants: **YRQ-4A:** Designation of five ACTD and two EMD aircraft (see Development above). Block 0 standard (prototype EO/IR and radar in ISS, limited sigint capability (imint only).

RQ-4A: First nine (Block 10) LRIP examples — seven for US Air Force plus two for US Navy (see table). Basic EO/IR and radar (though more modes) in ISS, limited QRC comint, simultaneous image recording and Inmarsat satcom. The last Block 10 aircraft was delivered to USAF on 23 June 2006.

RQ-4B: Enhanced, larger version, in three-stage (Blocks 20, 30 and 40) upgrade programme. Block 20 introduced on production line from July 2004, under USD299.8 million contract of 31 March 2002. Enhanced (EISS) sensor suite; 'Group A' level sigint; increased maximum T-O weight; increased payload of 1,361 kg (3,000 lb); redesigned and larger wing with stores hardpoints (first wing delivered by Vought in July 2005); modified main landing gear (single-wheel units, farther out on the wing and with electric differential braking); fuel measurement system; greater

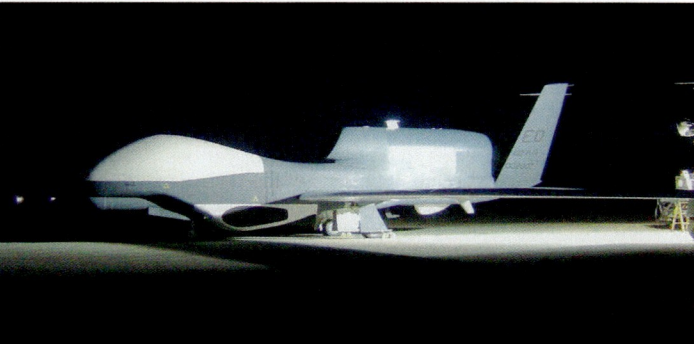

The black 'teardrop' feature on the ventral radome in this view of EMD YRQ-4A 00-2007 is believed to indicate installation of the HBS PCU elint payload tested in late 2005 or early 2006 *(Northrop Grumman)*

YRQ-4A Global Hawk AV-3/98-2003 on duty in Iraq *(USAF)*

Unmanned aerial vehicles > United States > Northrop Grumman RQ-4 Global Hawk

N-1/166509, first of the US Navy GHMD demonstrators *(Northrop Grumman)* 1151564

02-2010, the third production-standard RQ-4A, at Beale AFB in the insignia of USAF's 12th Reconnaissance Squadron, to whom it was delivered on 28 October 2004 *(USAF)* 1047879

The final Block 10 RQ-4A (04-2014), delivered in 3Q06 *(Northrop Grumman)* 1151554

The first RQ-4B Global Hawk (04-2015) in final assembly. The 'teardrop' installation in the upper nose section is the starboard antenna for the ASIP intelligence payload *(Northrop Grumman)* 1122569

RQ-4A Global Hawk 02-2010 from LRIP production Lot 2 *(USAF)* 1122568

endurance; 25 kVA main generator; additional pressurised payload bay in fuselage. ESM capability scheduled from AV-17 (first Lot 3 Block 20 RQ-4B). 'Full' (Northrop Grumman High Band System Production Configuration Unit, or HBS PCU, first air-tested early 2006) sigint capability in Block 30; active electronically scanned array (AESA) radar (Multi-Platform Radar Technology Insertion Program, or MP-RTIP) in Block 40.

Total of 44 planned (six Block 20, 26 Block 30 and 12 Block 40). Lot 3 LRIP includes three Block 20 RQ-4Bs, first of which (AV-17, USAF serial 04-2015) was rolled out 25 August 2006. It began taxi tests 26 September 2006 and made its maiden flight on 1 March 2007. Lot 4 comprises three Block 20 and one Block 30. The Block 30 flew for the first time in November 2007. Lot 5 comprises five Block 30 and Lot 6 five Block 30. Lot 4 deliveries are scheduled for completion by July 2008 and Lot 5 by February 2009. Five production air vehicles were delivered to the USAF during 2007. Air vehicle 26 was in final assembly in mid-2009.

Euro Hawk: The German MoD plans to acquire this version of Global Hawk as a 2010 replacement for the wide area search (WAS) sigint roles currently performed by its ageing 'Peace Peek' Dassault-Breguet Atlantic 1s serving with Marinefliegergeschwader 3 'Graf Zeppelin'. The Euro Hawk project began with an August 2000 agreement between Northrop Grumman and EADS, and received official launch approval by the US Air Force and the BWB (German MoD) on 19 October 2001. Under an MoU with EADS, an EADS Defence Electronics elint (SAR/MTI) module, to locate and classify radar emitters, was trialled in a YRQ-4A in 2002. Known as the HALE Elint Payload (HEP), this radar sensor was designed to identify sources of electromagnetic radiation and intended to verify the feasibility of combining with the Global Hawk's comint to provide a full sigint capability. Following integration of the payload by Northrop Grumman between October 2001 and May 2002, the trials aircraft (Global Hawk AV-3, 98–2003) made two successful test flights with the sensor at Edwards AFB on 17 and 22 November 2002, demonstrating the payload's ability to disseminate information via datalink to the GCS. Following a transatlantic flight from Edwards AFB on 14/15 October 2003, a six-flight demonstration in Germany over the North Sea from the German naval air base at Nordholz was undertaken by AV-1 to evaluate the HEP and the EADS-developed elint ground support station (EGSS) before the aircraft returned to Edwards on 6 November. The elint module is serving as the foundation for a more comprehensive sigint (imint/comint) multiband system to intercept voice and radar communications within a specified region. A possible third stage would envisage use for other ISR missions. To fulfil the Atlantic's wide area surveillance and reconnaissance missions, Euro Hawk would be required to demonstrate a range of at least 1,620 n miles (3,000 km; 1,864 miles) and endurance of 36 hours. It has been estimated that a procurement programme for five Euro Hawks would cost around EUR350 million. The production elint/comint suite would utilise technologies developed for EADS Defence Electronics' Fast Emitter Location System (FELS), already trialled on the country's Tornados.

In March 2005, EADS and Northrop Grumman submitted their combined responses to a September 2004 RFP from the BWB, and details were released in May 2005 of a new 50-50 joint company to manage the programme as national prime contractor. This company, EuroHawk GmbH, formally came into existence on 4 November 2005, co-located with EADS Defence and Security at Friedrichshafen. Plans, confirmed in a US and German defence ministries MoU signed on 16 May 2006, are to replace the Atlantics with one prototype and four production EuroHawks, based on the RQ-4B airframe, with contracts for series production expected in 2009, preceding delivery of the demonstration air vehicle in 2010. Five Euro Hawks were the subject of a EUR430 million (USD555 million) contract awarded on 31 January 2007. The first (demonstration) aircraft is due for delivery in 2010, with four production aircraft following in 2011 to 2014. The wing and fuselage of the first Eurohawk were mated in 2008 at Palmdale, and structural assembly was completed by July 2009. Rollout followed in October 2009.

Maritime version RQ-4N: In April 2008 the US Navy selected the RQ-4N version of the Block 40 Global Hawk for its its Broad Area Maritime Surveillance (BAMS) programme. The contract for System Development and Demonstration of BAMS was reported to be worth USD1.16 billion. The number of air vehicles to be built has not yet been declared. Some reports put the fleet at 48 to 68; during the BAMS competition Northrop Grumman said its RQ-4N proposal was the most efficient system being offered, requiring fewer air vehicles than rival submissions.. BAMS is intended to provide persistent ISR on a global scale where no other naval forces are present. Air vehicles will be based at five locations: Hawaii; Diego Garcia; Florida; Japan and Italy. At each base air vehicles will provide continuous 24 hours a day coverage at ranges out to 2,000 n miles. They will handle a significant proportion of the ISR missions currently flown by P-3 aircraft. The US Navy previously awarded Northrop Grumman a USD500,000 contract in 2000 to consider a maritime surveillance version of Global Hawk, and acquired two RQ-4As in 2005 for a maritime demonstration programme (see under Customers below). At

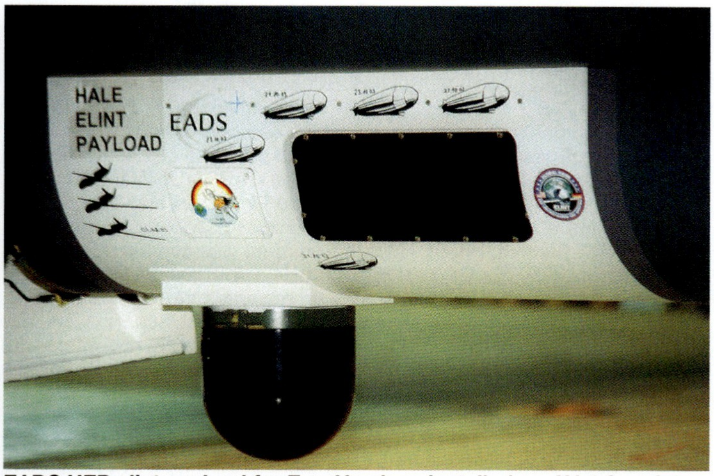

EADS HEP elint payload for EuroHawk as installed on AV-1 for trials in Germany in October 2003 *(IHS/JJL)* 0564043

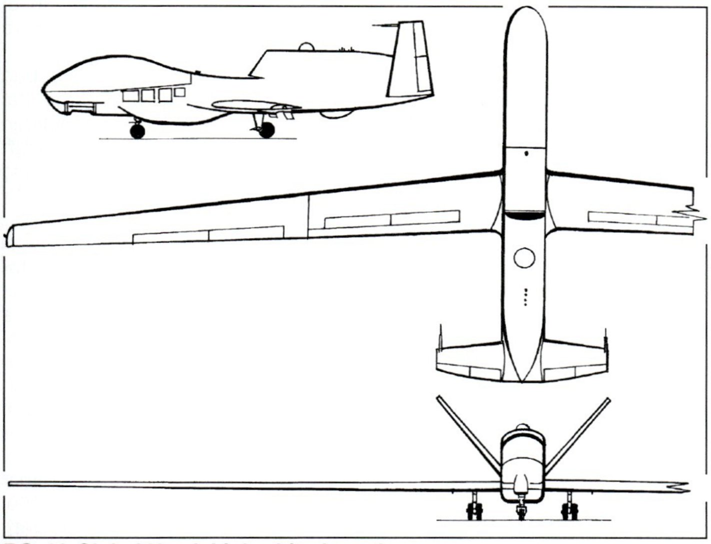

RQ-4A Global Hawk high-altitude endurance UAV *(IHS/John W Wood)* 0518115

the Paris airshow in 2007 the company said it was testing elements of BAMS in parallel using cyber networks and testbeds to bring development forward under a risk reduction programme called Head Start. This was intended to allow the RQ-4N to achieve initial operational capability ahead of the US Navy's target date of the fourth quarter of Fiscal Year 2013.

In early 2008 Northrop Grumman reported that each Block 20/30/40 air vehicle cost approximately USD27.6 million, without the sensor package.

RQ-4A, RQ-4B

Dimensions, External	
Overall	
length	
RQ-4A	13.51 m (44 ft 4 in)
RQ-4B	14.50 m (47 ft 6¾ in)
height	
RQ-4A	4.44 m (14 ft 6¾ in)
RQ-4B	4.70 m (15 ft 5 in)
Fuselage, width	1.45 m (4 ft 9 in)
Wings	
wing span	
RQ-4A	35.41 m (116 ft 2 in)
RQ-4B	39.90 m (130 ft 10¾ in)
wing aspect ratio, RQ-4A	25.0
Tailplane	
tailplane span	
RQ-4A	3.51 m (11 ft 6¼ in)
RQ-4B	3.81 m (12 ft 6 in)
Wheels	
wheelbase, RQ-4A	4.50 m (14 ft 9¼ in) (est)
wheel track	
c/l of shock-struts, RQ-4A	3.28 m (10 ft 9¼ in) (est)
Areas	
Wings	
Gross wing area, RQ-4A	50.17 m² (540.0 sq ft)
Fins	
Tail fin, RQ-4B	4.50 m² (48.44 sq ft)
Weights and Loadings	
Weight	
Weight empty, RQ-4A	4,173 kg (9,199 lb)
Max T-O weight	
RQ-4A	12,111 kg (26,700 lb)
RQ-4B	14,628 kg (32,249 lb)
Fuel weight	
Max fuel weight, RQ-4A	6,577 kg (14,499 lb)
Payload	
Max payload	
RQ-4A	907 kg (1,999 lb)
RQ-4B	1,361 kg (3,000 lb)
Performance	
Altitude	
Operating altitude, loiter	15,240 m to 19,810 m (50,000 ft to 65,000 ft)
Service ceiling	
RQ-4A	19,810 m (65,000 ft) (est)
RQ-4B	18,290 m (60,000 ft) (est)
Speed	
Loitering speed	
RQ-4A	343 kt (635 km/h; 395 mph)
RQ-4B	310 kt (574 km/h; 357 mph)
Range	
ferry, RQ-4A	12,000 n miles (22,224 km; 13,809 miles)
ferry, RQ-4B	12,300 n miles (22,779 km; 14,154 miles)
Endurance	
RQ-4A	35 hr
RQ-4B	36 hr
Time on station, at 1,200 n miles (2,222 km; 1,381 miles)	24 hr
Power plant	1 × turbofan

Status: Two ACTD Global Hawks (98–2004 and 98–2005) were deployed, as the 12th Expeditionary Reconnaissance Squadron in Operation 'Enduring Freedom', for surveillance over Afghanistan, following the terrorist attacks on New York and Washington of 11 September 2001, one being based at Al Dhafra in Abu Dhabi and the other at RAAF Edinburgh AB, Australia. First imagery was relayed on 27 November 2001, but 98-2005 was lost due to a control surface malfunction on 30 December 2001 while returning to Al Dhafra. On 10 July 2002, 98-2004 was also lost, due to an internal fuel nozzle failure, crashing close to the Afghan border while attempting to reach Shamsi AB in Pakistan. Despite these losses, by early 2003 'Enduring Freedom' Global Hawks had provided US Central Command with more than 15,000 images totalling over 1,000 hours. In late February 2003, air vehicles AV-3 and AV-6 were prepared for deployment to Abu Dhabi to support operations over Iraq, though in the event only AV-3 was employed during Operation 'Iraqi Freedom'. Based in the UAE from 8 March 2003, it acquired some 3,655 images with its various sensors which aided location and identification of more than 300 tanks, 13 SAM batteries, more than 70 individual SAM launchers and a number of SAM-associated other assets before it returned to Edwards AFB on 5 May 2003. Its contribution amounted to more than 55 per cent of that operation's surveillance intelligence despite flying only some 5 per cent of the total surveillance missions of the campaign. In early 2008 Global Hawks were operational from Beale AFB (main operating base, home of the 9th reconnaissance wing), Edwards AFB, NAS Patuxent River and other, un-named locations.

Long-lead procurement of parts for the first two (Lot 1) LRIP Block 10 Global Hawks (P1/02-2008 and P2/02-2009) was approved in a USD20.5 million contract, with the full LRIP contract, worth USD101.3 million, following in February 2002. This pair were scheduled to carry X-band SAR in addition to EO/IR. The first LRIP aircraft was rolled out on 1 August 2003; deliveries were made in September and December 2003, preceded in June by one ground station MCE.

LRIP Lot 2 (four Block 10 AVs, three ISS payloads, two EO/IR sensors and one LRE for USAF, plus two AVs, two ISS, two LREs and one MCE for the US Navy) was the subject of a USD302.9 million FY03 contract award announced on 11 February 2003. The first Lot 2 aircraft made its maiden flight on 1 July 2004 and was delivered to USAF's 12th Reconnaissance Squadron at Beale AFB, California, on 28 October 2004. Deliveries were completed in 2005. The US Navy's first Global Hawk (N-1) made its initial flight on 6 October 2004 and both were delivered in 2005.

Four LRIP (Lot 3) aircraft were ordered under a USD207.7 million contract in October 2004. These comprise one Block 10 RQ-4A with an ISS suite and three Block 20 RQ-4Bs, one with ISS, two with EISS and all three with an interim sigint (comint) package. In all, according to stated USAF plans at mid-2003, its first 19 production Global Hawks would be LRIP aircraft, this total being achieved with 10 RQ-4As, four RQ-4Bs procured in 2005 (Lot 4, three Block 20 and one Block 30) and five (four Block 30 and one Block 40) in Lot 5, for which a USD287 million contract was awarded in March 2007. Long-lead items for Lot 6 were ordered in a USD60.62 million contract awarded on 8 May 2006, and the Lot 6 production contract (five Block 30s, three MCEs and three LREs) valued at USD371 million followed in May 2007. Acquisition will be held at five per year until Block 20/30 IOT&E (beginning late 2008) has been completed. Long-lead items for Lot 7 (two Block 30 and three Block 40) were contracted in February 2007 and the USD302.9 million production contract for these air vehicles followed in November 2009. In January 2008, 17 new air vehicles where reported to be in various stages of production and flight test: six Block 20s; 10 Block 30s and one Block 40, together with six ground stations. The first Block 40-configured fuselage was delivered to the Palmdale production site in September 2007 and the first Block 40 Global Hawk, AF-18, made its maiden flight at Palmdale on 16 November 2009.

Although the full scope of the US Navy BAMS award had not been laid out it detail the contract may virtually double Northrop Grumman's existing orderbook for Global Hawk. The first BAMS production air vehicles are expected to emerge in the 2011-2012 timescale.

Customers: US Air Force: Series production of 51 (originally 63) Global Hawks for USAF is envisaged, comprising seven Block 10 RQ-4As and 44 RQ-4Bs (six Block 20, 26 Block 30 and 12 Block 40). Aircraft are delivered initially to the 452nd Flight Test Squadron at Edwards AFB for acceptance and operational check flights, before allocation to the 12th Reconnaissance Squadron, 9th Reconnaissance Wing, at Beale AFB, California. Aircraft deployed to SW Asia are designated as being operated by the 352nd Expeditionary Reconnaissance Squadron. In June 2005, Grand Forks AFB, North Dakota, was named as being the second Global Hawk home base and the first Block 40 Global Hawk arrived at Grand Forks in May 2011. Andersen AFB on Guam has been nominated as one of three forward operating bases.

Preceding the February 2006 return of AV-3 to the US, two production Block 10 Global Hawks (AF-4/02-2011 and AF-5/02-2012) arrived at Al Dhafra in early January, AF-5 flying its first, near-24-hour sortie within 36 hours of arriving in-theatre. By June 2006, this pair had already flown about 50 missions and more than 1,000 combat hours.

US Navy: Two RQ-4As (N-1/166509 and N-2/166510) acquired for evaluation in Global Hawk Maritime Demonstration (GHMD) programme, managed by USN's UAS programme office, PMA-263. N-1 made its maiden flight on 6 October 2004, and both were delivered to the USN in 2005. Early flights were made from Edwards AFB, California, "to characterise the performance of Navy-specific sensor modifications" (understood to be introduction of maritime detection modes for the SAR sensor); N-1 then took part in USN's Experiment 'Trident Warrior 2005'; and supported 'a wide range of tests and training' for USAF, to release latter's assets for activities overseas. One of these tests, in February/March 2006, involved three flights using the ISS radar in air-to-air mode to illustrate the aircraft's potential for an AEW role. After more than 200 hours' flying from Edwards, N-1 was flown to NAS Patuxent River, Maryland, on 27/28 March 2006 for operation by USN test squadron VX-20, first for participation in April 2006 in the USN's Joint Expeditionary Force Experiment 2006 and later to develop operational techniques and tactics for persistent maritime ISR on behalf of US Northern Command and 'other federal agencies'. N-2 flew from Edwards to Patuxent River on 6 December 2006.

In late July 2006, one GHMD aircraft took part in a multinational Rim of the Pacific (RIMPAC) exercise by flying four successful maritime missions from Edwards AFB to Hawaii. Equipped with sensors carrying new maritime software modes, it flew more than 2,172 n miles (4,023 km; 2,500 miles) each way for a total of more than 100 flight hours and provided more than 8 hours on-station time during each mission. Captured imagery included a ship-sinking exercise, expanded maritime interdiction operations, and wide-area search and surveillance to locate target vessels at sea. On return over land, operators were able to switch the sensors from maritime to over-land modes without difficulty.

In April 2008 a maritime Block 30 RQ-4N Global Hawk derivative was selected for the US Navy's Broad Area Maritime Surveillance (BAMS) programme. See above under Maritime Version.

Germany: The German Air Force is a customer for the Euro Hawk, as described under the Variants heading above.

Australia: The Global Hawk is a prospective candidate, together with the GA-ASI Mariner and (possibly) the AeroVironment Global Observer, for the Royal Australian Air Force (RAAF) Project Air 7000 multimission maritime HALE UAV requirement. In 2006 Australia, along with Japan and Singapore, was also considering an alternative proposal by USAF Pacific Command for a regional 'pool' of Global Hawks to be based in Guam.

Under a AUD30 million agreement announced in March 1999, the Australian Defence Science and Technology Organisation (DSTO) collaborated with USAF and Northrop Grumman in 2001 in developing and testing a maritime surveillance mode for the Global Hawk's synthetic aperture radar in order to evaluate the UAV's suitability to meet the Air 7000 requirement. In the ensuing five years, 'at least 10' Global Hawk transits between SW Asia and the US had reportedly been made via Australia, including (13 to 18 February 2006) that by AV-3 returning home at the end of its deployment to the Persian Gulf area. If Global Hawk should be selected for the unmanned element of Air 7000, an Australian-developed ground control station for the RAAF version would be created, according to a March 2005 announcement, by a joint team comprising Northrop Grumman and the Australian arms of Tenix Defence, Saab Systems and L-3 Communications Integrated Systems. Meanwhile, the DSTO awarded Northrop Grumman an AUD4.7 million (USD3.5 million) contract in July 2006 for a virtual trial in October 2006 in the latter's Cyber Warfare Integration Network (CWIN) in San Diego, California, to examine further how the UAV would perform in the Australian environment. Subject to US export approval, the RAAF has said it would seek funding for at least six Global Hawks by 2012. Australia subsequently abandoned this plan, citing the need to cut spending.

NASA: NASA's Dryden Flight Research Center has acquired three of the surviving YRQ-4As (AV-1, AV-6 and AV-7) for use in support of science and sensor technology missions.

NATO: In mid-2006, NATO signified its intention to acquire four Global Hawks equipped with the MP-RTIP radar as part of its Alliance Ground Surveillance (AGS) fleet. The MoU by fifteen NATO nations was signed in September 2009.

Contractor: Northrop Grumman Unmanned Systems
San Diego, California.

Northrop Grumman X-47

Type: Experimental Unmanned Combat Air Vehicle (UCAV).

Development: A Northrop Grumman-led team received a USD635.8 million contract on 8 August 2007, to produce two X-47B demonstrator UCAVs for the US Navy's UCAS-D (Unmanned Combat Air System-Demonstrator) programme, which had been held in competition with Boeing. The contract marked the culmination of a competition dating back to 30 June 2000 when USD2.3 million DARPA/US Navy feasibility study contracts were awarded to Northrop Grumman and Boeing for a future US Naval Unmanned Combat Air Vehicle (UCAV-N), to be operable from aircraft carriers. Phase 1A of this contract was completed in March 2001, followed by the USD25 million Phase 1B; Phase 2A contracts for further studies were awarded to the two manufacturers in March 2002, initially for USD10 million each but increased to USD13 million four months later. In April 2003, DARPA decided that Phase 2B, to continue the development period ahead to mid-2005, should require both designs to address the UCAV joint requirements of the US Air Force and the US Navy. The joint programme was retitled Joint Unmanned Combat Aircraft System (J-UCAS) in July 2003, by which time Lockheed Martin had joined the Northrop Grumman team. The eventual selected type was planned to enter the EMD phase within the 2008 to 2010 timeframe. The original joint-service objectives of the programme included a combat radius of 1,300 n miles (2,407 km; 1,496 miles) with a payload of 2,041 kg (4,500 lb) and the ability to loiter for 2 hours over a target up to 1,000 n miles (1,852 km; 1,151 miles) away.

A full-size mockup of the Northrop Grumman design, later dubbed X-47A Pegasus, was rolled out on 26 February 2001; roll-out of the first flying prototype took place on 30 July the same year, although its maiden flight was not until 23 February 2003. The primary purpose of this aircraft was to establish the suitability of the basic design for carrier use, including arrester wire landings on a simulated carrier deck at the US Naval Air Warfare Center, China Lake, California.

The successor to the X-47A proof of concept is the X-47B; an operational UCAV demonstrator. Two X-47B prototypes were ordered for the US Navy under a Phase 2B (modified Phase 2A) DARPA contract announced on 1 May 2003. The contract was later extended to include a third aircraft as a USAF demonstrator intended to assess the feasibility of meeting both US Air Force and US Navy UCAV requirements. Cutbacks in FY06 budget led to a revised contract on 17 October 2005, under which Northrop Grumman was to deliver only two X-47B prototypes, the first of which was a non-stealthy example, minus mission systems, primarily for carrier compatibility trials. In June 2005 it was announced that GKN Aerospace was commencing construction of the aircraft. Radar cross-section testing was started by Lockheed Martin from October 2005. Mission systems were to be installed in the second prototype aircraft. There was an option to complete a third aircraft after changeover of programme management from DARPA to USAF in mid-2006 which was confirmed in a USD56 million agreement of October 2005; this also allowed for carrier landing trials and an in-flight refuelling demonstration, with the option to proceed further, to a much larger X-47C. However, the number of X-47B prototypes was again reduced to two after the

X-47B first flight from Edwards AFB, 4 February 2011 *(US Navy)*

X-47B AV-2 under dynamic proof testing at Palmdale, January 2011 *(Northrop Grumman)*

Northrop Grumman X-47

X-47B pictured during low-speed taxi trials *(Northrop Grumman)*

the UK MoD. A simulated exercise demonstrating simultaneous control of four X-47Bs was conducted at the Naval Air Warfare Center's Weapons

Launch: Conventional wheeled take-off.
Recovery: Conventional wheeled landing with arrester hook, using US

Power plant:
A: One 14.19 kN (3,190 lb st) Pratt & Whitney Canada JT15D-5C turbofan.
See under Weights for fuel details.

Dimensions, External
Overall
length
X-47A .... 8.50 m (27 ft 10 in)
X-47B .... 11.64 m (38 ft 2 in)
height
X-47A .... 1.86 m (6 ft 1¼ in)
X-47B .... 3.17 m (10 ft 4¾ in)
folded, X-47B .... 5.27 m (17 ft 3½ in)
wing span
folded, X-47B .... 9.42 m (30 ft 10½ in)
Wheels
X-47A .... 3.08 m (10 ft 1¼ in)
outer mainwheels, X-47A .... 3.02 m (9 ft 11 in)
Areas
Wings
Gross wing area, X-47A
Weights and Loadings
Weight empty, X-47A .... 1,740 kg (3,836 lb)
Fuel weight
Max fuel weight
X-47A .... 717 kg (1,580 lb)
mission, X-47A .... 472 kg (1,040 lb)
Payload
Max payload, X-47B .... 2,041 kg (4,499 lb)
Performance
Altitude
Service ceiling, X-47B .... 12,190 m (40,000 ft)
Range, ferry, X-47B .... 3,500 n miles (6,482 km; 4,027 miles)
Power plant .... 1 × turbofan

Status: Initial X-47A engine runs occurred in December 2001, January and March 2002 and were followed by the first autonomous engine start and shutdown in April 2002. The first two low speed taxi tests at China Lake were made on 19 July and 6 September 2002. The first flight of the X-47A was achieved on 23 February 2003.

The development schedule of the X-47B aimed for a first flight in November 2009, but this was delayed until February 2011, when AV1 flew from Edwards AFB. It made two further test flights on 1 and 4 March 2011 when it reached a height of 1,524 m (5,000 ft) and a top speed of 200 kt (370 km/h; 230 mph). Carrier landing trials scheduled for 2013

Contractor: Northrop Grumman Integrated Systems,
San Diego, California.

NRI AutoCopter

Development: NRI was formed in July 2001, and first flight of what is now the AutoCopter was made by designer and company founder Michael Fouche in 2002. It was revealed in July 2003, made its first fully autonomous flight on 11 June 2005, and has since developed into a family of small unmanned helicopters for a variety of civil and military

Explorer, Express E (E15), Express G (G15), Expedition

Endurance
  max, Express E (E15)..12 min
  max, Express G (G15)..2 hr
  max, Explorer, Expedition..50 min
Power plant
  Express E (E15)...1 × electric motor
  Express G (G15)...1 × piston engine
  Explorer, Expedition...1 × turboshaft

Status: In production and service; distributors in Canada, Poland and UK. AutoCopter variants have been sold to civil or military customers in Canada, Colombia, Indonesia, Italy, South Korea, Spain, Sweden, Turkey, the UK, the US and elsewhere. About half of all sales are for export.

US Army customers have included the Defense Threat Reduction Agency at Fort Belvoir, Virginia; Research Laboratory at Aberdeen proving Grounds, Maryland; Army Laboratory at Huntsville, Alabama; and Communications-Electronics Research, Development and Engineering Center (CERDEC). The Colombian Air Force was a customer for AutoCopter A.

Civilian users in the US have included the Agricultural Research Service, Idaho National Laboratories, Google Inc, Sony Corporation and a number of state universities.

Contractor: Neural Robotics Inc (NRI)
Goodlettsville, Tennessee.

NRL FINDER

Type: Air-launched deployable mini-UAV.
Development: FINDER was begun as an ACTD programme on behalf of the US Defense Threat Reduction Agency's Counter-Proliferation II (CP2) activity for the detection of chemical warfare agents. Its role is reflected in its name, which is an acronym for Flight-Inserted Detection Expendable for Reconnaissance. NRL was awarded a three-year development contract in May 1998, and a first flight was made on 22 March 2000.
Description: *Airframe:* Parasol-wing monoplane; mainly circular-section fuselage; conventional tail surfaces; pusher engine. No landing gear. Wings, tail surfaces and propeller blades fold for stowage in launch pod.

Mission payloads: Smiths Spectrometric Point Ionizing Detector Expendable/Recoverable (SPIDER) chemical warfare agent detection (atmospheric sampling) sensor, with data transmission back to GCS via the host aircraft. Possibly, also, sample collection instrumentation. Available onboard power 180 W.

For 2007 US Army Battlelab Spectre demonstration (see Status heading below), two FINDERs each to be equipped with a Sensors Unlimited short-wave infra-red (SWIR) camera, with a 640 × 512 focal plane array detector and two-FoV lens. This will be integrated with a laser pointer in a gimballed mounting.

Guidance and control: From Predator ground control station. FINDER is released remotely from Predator wing pylon to descend to a low level and loiter in the vicinity of a suspected chemical agent cloud for up to 2 hours. Data from the SPIDER sensor can be transmitted back to a Predator Infra-Red Airborne Narrow-band Hyperspectral combat Assessor (PIRANHA) system aboard the Predator host UAV for transmission to warfighters; or the FINDER can collect air samples for recovery by ground forces for analysis. Directed by the GCS, the two

Two FINDERS can be carried by RQ-1 Predator *(GA-ASI)* 1409903

FINDER chemical warfare agent detection UAV *(NRL)* 0109982

Dark rectangle on this FINDER UAV covers the SPIDER sensor package *(IHS/IDR)* 0521319

UAVs then co-operate to track the plume and gather meteorological and chemical data that can be used to predict the plume's path and resulting threatened areas.

Launch: Ejected from storage pod, then deploying folded wings, tail surfaces and propeller blades.

Recovery: On completion of mission, FINDER follows host aircraft to a designated recovery site for an automatic landing.

Power plant: 1.26 kW electric propulsion (no details known); two-blade pusher propeller with folding blades. Fuel capacity 10.6 litres (2.8 US gallons; 2.3 Imp gallons).

FINDER

Dimensions, External
  Overall
    length
      folded...1.65 m (5 ft 5 in)
      deployed..1.60 m (5 ft 3 in)
  Fuselage, width...0.22 m (8¾ in)
  Wings
    wing span, deployed..2.62 m (8 ft 7¼ in)
Areas
  Wings, Gross wing area...0.50 m² (5.4 sq ft)
Weights and Loadings
  Weight, Max launch weight...26.8 kg (59 lb)
  Payload
    Max payload...6.1 kg (13.00 lb)
    SPIDER..3.2 kg (7.00 lb)
Performance
  Altitude, Service ceiling...4,570 m (15,000 ft)
  Speed
    Max level speed..87 kt (161 km/h; 100 mph)
    Cruising speed...70 kt (130 km/h; 81 mph)
    Loitering speed..61 kt (113 km/h; 70 mph)
  Range..521 n miles (964 km; 599 miles)
  Radius of operation, with 2 h loiter...........................43 n miles (79 km; 49 miles)
  Endurance
    typical...6 hr 30 min
    max...10 hr
  Power plant...1 × electric motor

Status: Atmospheric sampling trial flights in July and October 2001 were followed in August 2002 by the air launch at Edwards AFB, California, of a pair of FINDERs, equipped with the SPIDER sensor payload and carried by a Predator UAV in underwing pods. ACTD programme is believed to have ended in 2004, at which time eight FINDER systems (16 air vehicles) were planned to remain as residual assets.

In December 2006, it was announced that a Lockheed Martin AC-130 Spectre special operations gunship was to be adapted to air-launch two FINDERs during 2007, as part of a US Army Battlelab demonstration to provide an off-board sensing capability and targeting support for the GA-ASI MQ-1 Predator.

Customers: US Defense Threat Reduction Agency (DTRA); US Army Battlelab.
Contractor: Naval Research Laboratory
Washington, DC.

NRL Ion Tiger

Type: Technology demonstrator.

Development: The Ion Tiger is believed to have made its maiden flight in early March 2009, its electric motor being powered at that time by batteries. Since then, it has become part of a programme, sponsored by the US Office of Naval Research, to demonstrate quieter and more efficient sources of energy while simultaneously offering significantly greater endurance for surveillance missions, at lower cost and with less risk of detection. In this, the NRL is partnered by Arcturus UAV (manufacturer of the air vehicle), Protonex Technology Corporation, Hawaii University and HyperComp Engineering Inc.

The fuel cell system utilises high-pressure, lightweight hydrogen storage tanks to reduce weight. It produces little noise, emits less infra-red energy than traditional engines or batteries, is pollution-free, and is expected to be twice as efficient in power output as an internal combustion engine. After installation, the Ion Tiger made a 23 hour 17 minute flight at Aberdeen Proving Ground on 9-10 October 2009, setting an unofficial endurance record for a hydrogen-powered fuel cell UAV. Five weeks later, on 16-17 November, it exceeded this with a flight of 26 hours 1 minute duration.

Description: *Airframe:* High-wing monoplane with cruciform tail unit. No landing gear. Composites construction.

Mission payloads: In this programme, the fuel cell power system itself is the mission payload.

Guidance and control: Not disclosed.

Launch: From portable pneumatic launcher.

Recovery: Belly landing.

Power plant: Electric motor, powered by a 550 W (0.74 hp) polymer fuel cell and driving a two-blade propeller.

Ion Tiger

Weights and Loadings	
Weight, Max launch weight	17 kg (37 lb) (est)
Payload, Max payload	2.3 kg (5.00 lb)
Performance	
Endurance	26 hr (est)
Power plant	1 × electric motor

Status: Development continuing, focusing on increasing the power of the fuel cell to 1.5 or 2 hp.

Contractor: Naval Research Laboratory
Washington, DC.

The Ion Tiger as unveiled in 2009 *(NRL)* 1395227

Ion Tiger during a fuel cell powered test flight *(NRL)* 1395228

NRL RQ-14A Dragon Eye and Sea ALL

Type: Multirole, man-portable mini-UAVs.

Development: Following demonstrations of its earlier MITE mini-UAV, Dragon Eye was developed by NRL Electronic Warfare Division and US Marine Corps Warfighting Laboratory (MCWL) as their candidate for the USMC's March 2000 Interim-Small Unit Remote Scouting System (I-SURSS) programme to demonstrate small unit reconnaissance and threat detection. The Dragon Eye first prototype, of similar twin-tailed

Nose-mounted sensors on a Dragon Eye *(E R Hooton)* 0580226

Sea ALL hand launch *(NRL)* 0521424

Readying a Sea ALL for flight *(US Navy)* 0558429

configuration to MITE, made its first flight on 5 May 2000, but was quickly followed by a simpler, single-tailed development which made its maiden flight on 28 July the same year. Industry contracts were awarded in July 2001, and deliveries began in June 2002. The designation RQ-14A was applied to Dragon Eye in 2007. The US Navy variant is known as Sea ALL (Sea Airborne Lead Line); six prototypes were ordered by programme co-sponsor, US Office of Naval Research (ONR). AeroVironment and L-3 BAI Aerosystems each manufactured an initial batch of 40 Dragon Eyes for further trials; L-3 BAI now separately markets its own variant of the UAV under the air vehicle name **Evolution**.

Description: *Airframe:* Constant-chord, high-mounted wing; square-section fuselage; large triangular fin and rudder. No landing gear. Dismantles into six components, storable in package 38 × 38 × 18 cm (15 × 15 × 7 in).

Mission payloads: Interchangeable noses containing two colour or monochrome daylight or LLTV cameras or IR imager; options include communications relay.

Guidance and control: Dragon Eye: Autonomous flight and landing under one-person operation; can fly with MicroPilot autopilot and GPS to preprogrammed waypoints or receive en-route updates from ground control operator. UHF command uplink. GCS is based on a Panasonic CF-34 ToughBook laptop computer (weight 4.2 kg; 9.3 lb excluding battery), plus uplink transmitter and antenna, video antenna and receiver, video overlay card and video goggles.

Sea ALL: Currently remotely piloted; being developed to allow autonomous operation, with control stick steering for man-in-the-loop search.

Dragon Eye received the designation RQ-14A in 2007 *(IHS/Patrick Allen)* 1025574

System composition: Three air vehicles; one lightweight laptop computer ground station; two EO and one IR payload for each air vehicle; two remote receiving stations; batteries for 2 hours of flight.

Launch: Hand- or bungee-launched.

Recovery: Dragon Eye, by autopilot-commanded, deep stall terminal descent to belly landing, or automatic parachute recovery. Sea ALL, steered into net for shipboard recovery.

Power plant: Twin 214 W electric motors (lithium disulphide batteries) in preproduction batches, each driving a two-blade tractor propeller with folding blades. Serial production Dragon Eyes have lithium-ion batteries.

Dragon Eye, Sea ALL

Dimensions, External	
Overall, length	0.91 m (2 ft 11¾ in)
Wings, wing span	1.14 m (3 ft 9 in)
Weights and Loadings	
Weight	
Max launch weight	
Dragon Eye	2.49 kg (5.00 lb)
Sea ALL	2.04 kg (4.00 lb)
Payload, Max payload	0.225 kg (0.00 lb)
Performance	
Altitude, Operating altitude	60 m to 365 m (200 ft to 1,200 ft)
Speed, Cruising speed	35 kt (65 km/h; 40 mph)
Power plant	2 × electric motor

Status: Trials of Dragon Eye during Exercise 'Capable Warrior' at Camp Pendleton, California, in June 2001; Sea ALL trialled with USN's Fifth Fleet a few months later. Manufacturing bids solicited in 2001, to build initial 40 systems (later increased to 80) for delivery to USMC; these manufactured equally by AeroVironment and L-3 BAI Aerosystems. Original plan to achieve IOC in July 2003 pre-empted by deployment of 20 Dragon Eyes and 10 ground stations with 1st Battalion, 7th Marine Regiment, during Operation 'Iraqi Freedom' in March/April 2003.

USMC had eventual requirement for 342 systems (1,026 air vehicles) by FY06, for use at company, platoon and squad level. Initial tranche towards this objective ordered under contract to AeroVironment on 13 November 2003.

Customers: US Navy and Marine Corps.

Contractor: Naval Research Laboratory
Washington, DC.

Piasecki Turais

Type: Air-launched surveillance UAV.

Development: The concept to which the Turais is designed is that of a wing and bomb bay launched unmanned air vehicle (WBBL-UAV), to be deployable in flight from the Lockheed Martin P-3C and other manned aircraft, thus extending their coverage area, effective time on station, and stand-off range from threat environments. It was designed by Piasecki under a January 2007 Phase 2 SBIR contract from US Naval Air Systems Command (NAVAIR). In May 2008, the option phase of this contract was taken up, covering design finalisation, prototype manufacture, and staging of a free-flight demonstration in 2009. Outside of the contract terms, Piasecki is also studying compatibility of the design with stores stations of the P-8A Poseidon and F-18E/F/G Super Hornet. According to the company, the Turais would also be adaptable, with minimal modification, for launch from USN MH-60R/S helicopters and USMC C-130J transports.

Description: *Airframe:* Cylindrical body with flattened topside. High-mounted, rectangular wing and horizontal tail surfaces, plus dependent vertical tail surfaces of similar shape. For carriage, wing pivots to align fore and aft; tail surfaces also fold.

Mission payloads: Could include EO, IR, SAR, sonobuoys or other sensors. Interchangeable payloads for such missions as ISR, target acquisition, command and control, communications relay, sigint, anti-submarine and anti-surface vessel warfare. Real-time imagery downlink.

Guidance and control: From host aircraft; multiple launch and control envisaged.

Launch: Air-launched. Turais has a standardised bomb rack interface that is compatible with the BRU-12/A and BRU-14/A (bomb bay) and BRU-15/A (wing) racks of the P-3C and certain other US Navy or Marine Corps aircraft.

Recovery: Parachute recovery, at sea or on land.

Power plant: One (unspecified) heavy-fuel turbojet.

Turais would extend the surveillance reach of maritime patrollers such as the P-3C or P-8A *(Piasecki)* 1356278

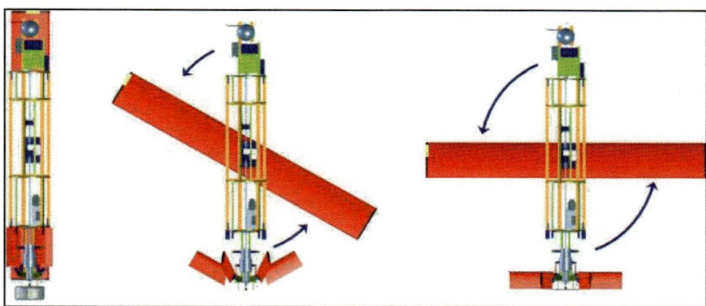
Turais has a pivoting wing and folding empennage for compact stowage before launch *(Piasecki)* 1356279

Turais

Weights and Loadings	
Weight, Max launch weight	340 kg (749 lb)
Payload, Max payload	90 kg (198 lb)
Performance	
Speed	
Cruising speed, max	200 kt (370 km/h; 230 mph)
Radius of operation	150 n miles (277 km; 172 miles)
Time on station, at mission radius	6 hr 30 min
Power plant	1 × turbojet

Status: Under development. First air-drop test of an aerodynamic prototype from a P-3C was due to have taken place in 2009, although by year-end it had not been announced as having done so.

Contractor: Piasecki Aircraft Corporation
Essington, Pennsylvania.

Prioria Robotics Maveric

Type: Close-range tactical mini-UAV.

Development: From initial R & D in the previous year, Prioria began developing and prototyping the Maveric in 2006, continuing into 2007 with field testing in readiness for series production. Novel features include the ability to be removed, fully assembled, from a small-diameter tube and launched in less than 2 minutes. Additional capabilities are being investigated.

Description: *Airframe:* Bird-like configuration, with bendable wings that can be folded around the fuselage for stowage in a 152 mm (6 inch) diameter tube. Carbon fibre fuselage and tail unit; wings of carbon fibre and cloth webbing. Rudder and elevator control surfaces. Pusher engine with folding propeller blades. No landing gear. Airframe is water resistant in light rain.

Mission payloads: Nose- and side-mounted cameras in M100; nose-mounted and retractable ventral cameras in M150. Standard M150 payload comprises a nose-mounted 5 Mp digital camera and a 640 × 480, NTSC output, analogue EO camera on a retractable gimbal. Optional for both versions are: a fixed, side-looking EO camera; or a 320 × 240 resolution, fixed and side-looking IR camera. Onboard video processing

Computer impression of the Turais in operation *(Piasecki)* 1356277

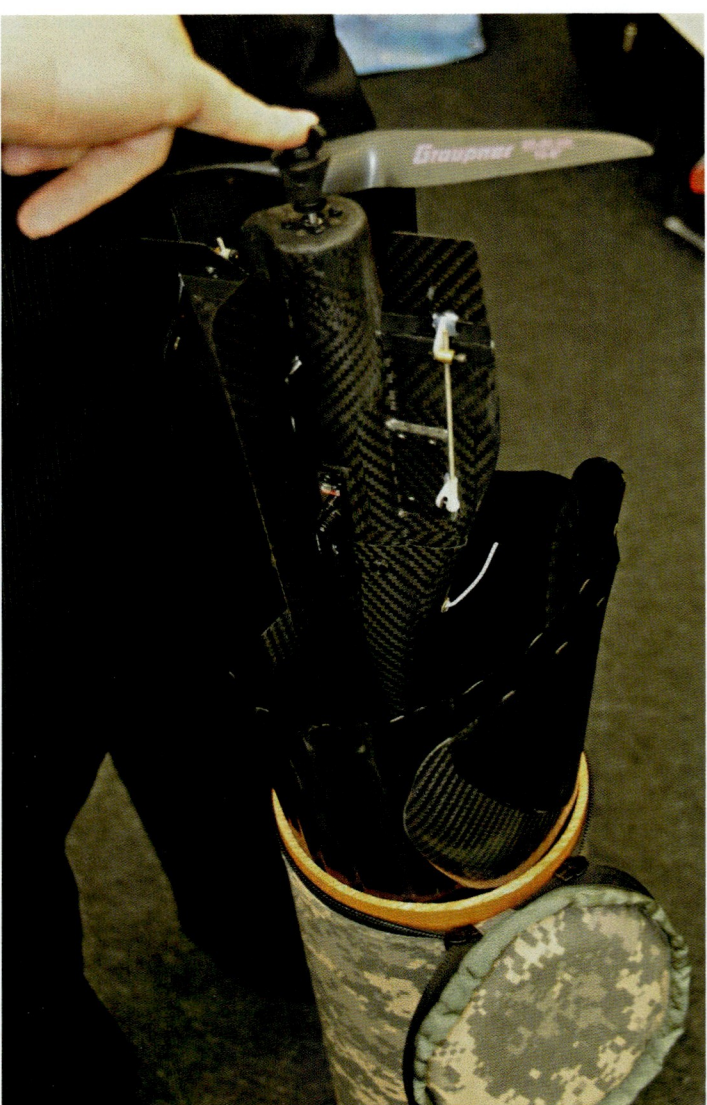

Maveric air vehicle being inserted into its storage tube *(IHS/Patrick Allen)* 1347056

Maveric requires only simple hand launch *(IHS/Patrick Allen)* 1347053

Maveric shows off its bird-like appearance *(IHS/Patrick Allen)* 1347047

M100, M150	
Dimensions, External	
Overall, length	0.67 m (2 ft 2½ in)
Wings, wing span	0.75 m (2 ft 5½ in)
Dimensions, Internal	
Payload bay	
length	0.102 m (4 in)
width	0.063 m (2½ in)
depth	0.038 m (1½ in)
volume	2.46 m³ (87 cu ft)
Weights and Loadings	
Weight, Max launch weight	1.13 kg (2.00 lb)
Payload	
Max payload	
M100	0.100 kg (0.00 lb)
M150	0.300 kg (0.00 lb)
Performance	
Altitude	
Service ceiling	
theoretical	7,620 m (25,000 ft)
tested	4,875 m (16,000 ft)
Speed	
Max level speed	55 kt (102 km/h; 63 mph)
Cruising speed	26 kt (48 km/h; 30 mph)
Stalling speed	18 kt (34 km/h; 21 mph)
Radius of operation	
datalink, mission	2.7 n miles (5 km; 3 miles)
Endurance	50 min (est)
Power plant	1 × electric motor

in M150, with COTS 2.4 GHz imagery downlink. Alternative payloads can be installed to customer's requirements.

Guidance and control: Fully autonomous (M150), including launch, GPS-based navigation and recovery, using Prioria's own Merlin onboard, collision-avoiding processor and Procerus Kestrel autopilot. GPS tracking system developed in partnership with Cyberflight of UK. GCS comprises a laptop computer and a communications box, combined weight 2.5 kg (5.5 lb). Command and control via 900 MHz datalink. Aircraft can also be flown manually, using joystick and autopilot assistance.

Transportation: Man-portable and operable by one person. Deployable in less than 2 minutes.

Launch: Hand launch. Can be launched and operated in wind speeds of up to 20 kt (37 km/h; 23 mph) sustained or 30 kt (55 km/h; 34 mph) gusting.

Recovery: Skid landing, or deep stall to vertical descent.

Power plant: Electric motor, powered by 11.1 V rechargeable lithium-polymer battery; two-blade pusher propeller.

Variants: *M100:* Baseline version; configured with a fixed EO and/or IR payload.

M150: Configured with retractable gimballed EO and forward-looking camera.

Status: Entered production in February 2008. Demonstrated to US Special Operations Command (SOCOM) in November 2008. In 2010 five Maveric systems ordered under a USD2.8 million contract by Canadian DND for operations in Afghanistan, with logistic support by ING Engineering of Ottawa.

Contractor: Prioria Robotics Inc
Gainesville, Florida.

Proxy Aviation SkyRaider

Type: Optionally piloted vehicle.

Development: Unveiled in August 2006, the SkyRaider is essentially a scaled-up version of the company's SkyWatcher OPV (which see), with a larger useful load and internal volume and a more powerful engine. The civil-registered prototype (N95PX) is based on a Firefly airframe manufactured by KARI in South Korea.

Description: *Airframe:* Generally as described for SkyWatcher, but of enlarged dimensions, with more powerful engine and retractable landing gear.

Mission payloads: As described for SkyWatcher. FLIR Systems StarSAFIRE III on the 'live' UAV in the July 2007 CRBRUS tests.

Guidance and control: As described for SkyWatcher.

Launch: Conventional, automatic, wheeled take-off.

The SkyRaider optionally piloted UAS *(Proxy)* 1290164

The SkyWatcher OPV in flight *(Proxy)* 1290201

Recovery: Conventional, automatic, wheeled landing.
Power plant: One 194 kW (260 hp) Textron Lycoming IO-540-C4B5 flat-four engine; three-blade pusher propeller. Heavy-fuel engine offered as alternative option.

SkyRaider

Dimensions, External	
Overall	
length	6.10 m (20 ft 0¼ in)
height	2.36 m (7 ft 9 in)
Foreplanes, foreplane span	4.77 m (15 ft 7¾ in)
Wings, wing span	9.75 m (31 ft 11¾ in)
Dimensions, Internal	
Payload bay	
length	1.22 m (4 ft 0 in)
width	1.22 m (4 ft 0 in)
depth	0.76 m (2 ft 6 in)
Weights and Loadings	
Weight	
Weight empty	816 kg (1,798 lb)
Max T-O weight	1,814 kg (3,999 lb)
Payload	
Max payload	
internal, with 20 hrs fuel	454 kg (1,000 lb)
internal, with 30 hrs fuel	150 kg (330 lb)
external	145 kg (319 lb)
Performance	
Altitude, Service ceiling	7,315 m (24,000 ft)
Speed	
Cruising speed, max	175 kt (324 km/h; 201 mph)
Loitering speed	120 kt (222 km/h; 138 mph)
Endurance	30 hr
Power plant	1 × piston engine

Status: One 'live' SkyRaider and one SkyWatcher, each with its own simulated second example, undertook a series of Co-operative Rules Based Reconnaissance Unmanned System (CRBRUS) demonstrations at Creech AFB, Nevada, between 1 and 11 July 2007. The CRBRUS flights, using Proxy's SkyForce distributed management system under operational control of the US Air Force Battle Lab, were said to mark the first successful demonstration of multiple UAVs performing fully autonomous co-operative flight. One operator controlled all four air vehicles as they performed diverse ISR, target search, 'hunter-killer' and other group tactical scenarios. The trials were the first operational demonstration of the SkyRaider.
Contractor: Proxy Aviation Systems Inc
Germantown, Maryland.

Proxy Aviation SkyWatcher

Type: Optionally piloted vehicle.
Development: Proxy Aviation Systems was formed in 2003. Its SkyWatcher OPV is based on the commercially available Velocity lightplane, and was unveiled in June 2006 following the maiden flight of a piloted prototype earlier that year. It can serve as a low/medium altitude UAV for such applications as persistent ISR, BDA, communications relay and other military tactical or civil missions, with the capacity for both internal and external payloads. The first two prototypes (N20PX and N30PX) were registered in February 2005. A similar but larger version, the **SkyRaider**, is described separately.

Description: *Airframe:* Mid-mounted swept wings and non-swept foreplane; endplate fins and rudders on main wing; pusher engine. Non-retractable tricycle landing gear. Composites construction.
Mission payloads: Can include EO/IR imagers, SAR, sigint, comint, communications relay or other packages. Main payload bay in fuselage, plus external stores point under each wing. FLIR Systems StarSAFIRE III was installed for the July 2007 CRBRUS trials.
Guidance and control: Aircraft is controlled by a Proxy SkyForce distributed management system (DMS), which can control multiple air vehicles simultaneously while providing datalinks to ground-based or other assets. It thus has the ability to make fully independent decisions during flight, without any human intervention. SkyForce can be adapted to work with other manned or unmanned vehicles besides Proxy's own products.

Main ingredients of the DMS are Proxy mission-oriented software, a three-workstation primary mission management GCS and mobile ground control user terminals. Features include single-point management of an autonomous fleet of aircraft with various types of payload; network-centric co-operative flight of up to 12 air vehicles in 'constellation' formation; and end-user tasking. Each UAV is pre-loaded with a flight plan, which can include tactical objectives, to self-manage its mission over an extended period of time. The system automatically generates a queue of the available aircraft and their corresponding payloads, from which a user can select the appropriate aircraft and type of payload for the mission at hand. The user controls the payload, and the virtual pilot manages the aircraft flight based on operator inputs to the payload.
System composition: A typical system would comprise three or four air vehicles, a control station and a recovery unit.
Launch: Conventional, automatic, wheeled take-off.
Recovery: Conventional, automatic, wheeled landing.
Power plant: Prototypes powered by a 149 kW (200 hp) Delta Hawk DW-180VA piston engine, driving a three-blade pusher propeller. Proxy offers a heavy-fuel alternative for improved fuel consumption and endurance.

SkyWatcher at Creech AFB in May 2007 *(Proxy)* 1290163

SkyWatcher payload bay *(Proxy)* 1290202

SkyWatcher

Dimensions, External
- Overall
  - length..6.10 m (20 ft 0¼ in)
  - height..2.36 m (7 ft 9 in)
- Foreplanes, foreplane span..............................4.77 m (15 ft 7¾ in)
- Wings, wing span...9.45 m (31 ft 0 in)

Dimensions, Internal
- Payload bay
  - length..1.22 m (4 ft 0 in)
  - width...0.76 m (2 ft 6 in)
  - depth..0.61 m (2 ft 0 in)

Weights and Loadings
- Weight
  - Weight empty...862 kg (1,900 lb)
  - Max T-O weight..1,315 kg (2,899 lb)
- Payload
  - Max payload
    - internal, with 8 hrs fuel.............................295 kg (650 lb)
    - internal, with 15 hrs fuel...........................150 kg (330 lb)
    - external..145 kg (319 lb)

Performance
- Altitude, Service ceiling.......................................6,095 m (20,000 ft)
- Speed
  - Cruising speed..150 kt (278 km/h; 173 mph)
  - Loitering speed..105 kt (194 km/h; 121 mph)
- Endurance..15 hr
- Power plant..1 × piston engine

Status: A USAF-sponsored demonstration was conducted at the Yuma Proving Ground in Arizona on 13 December 2006 in which the SkyWatcher successfully performed several automatic take-offs and landings and met specific test objectives such as runway centreline tracking, traffic pattern ground tracking, precise touchdown point and proper braking action. This led to a USAF contract for further tests. These took place from 8 to 10 February 2007 at Eglin AFB, Florida, when a single SkyWatcher demonstrated co-operative flight with three simulated SkyWatchers, each of which performed a different role and operated a unique sensor package.

One 'live' SkyWatcher and one SkyRaider, each with its own simulated second example, undertook a series of Co-operative Rules Based Reconnaissance Unmanned System (CRBRUS) demonstrations at Creech AFB, Nevada, between 1 and 11 July 2007. The CRBRUS flights, using Proxy's SkyForce distributive management system under operational control of the US Air Force Battle Lab, were said to mark the first successful demonstration of multiple UAVs performing fully autonomous co-operative flight. One operator controlled all four air vehicles as they performed diverse ISR, target search, 'hunter-killer' and other group tactical scenarios. The system continues in development with the Canadian Patent Office granting a patent to Proxy Aviation in March 2010 for aspects of its Universal Distributed Management System (UDMS) and Virtual Pilot (VP) technology. Proxy report that The VP allows one operator to control up to 12 unmanned aircraft and 20 additional sensors or ground control stations.

Contractor: Proxy Aviation Systems Inc
Germantown, Maryland.

Raytheon Cobra

Type: Company-owned systems testbed UAV.

Development: Prototype first flown in early months of 2006, followed by reservation in April of civil registrations for eight production aircraft, six of which taken up in May/June/July 2006. Received FAA experimental airworthiness certification 29 September 2006, at which time one production aircraft (N601RN) completed and five more under construction. Ten air vehicles had been built by 2010. Developed to serve as a test platform for small weapons and unmanned technology.

Description: *Airframe:* Simple pod and boom fuselage, with mid-mounted wings and conventional tail surfaces. Fixed tricycle landing gear.

Mission payloads: Planned to include various sensors, C³ systems and payloads for other applications.

Guidance and control: Cloud Cap Piccolo autopilot.

Launch: Conventional wheeled take-off.

Recovery: Conventional wheeled landing.

Power plant: One 12.3 kW (16.5 hp) Desert Aviation DA-150 piston engine, driving a two-blade propeller.

Cobra

Dimensions, External
- Overall, length..2.74 m (8 ft 11¾ in)
- Wings, wing span...3.05 m (10 ft 0 in)

Weights and Loadings
- Weight, Max T-O weight......................................45.4 kg (100 lb)

Performance
- Endurance, typical..4 hr
- Power plant..1 × piston engine

Status: Production by Quartus Engineering Inc; six registered in 2006; four more built by 2010, when the systems were being operated in support of other programmes on a regular basis. One Cobra air vehicle damaged during demonstration at USAF Academy, Colorado Springs, July 2008.

Contractor: Raytheon Missile Systems
Tucson, Arizona.

Raytheon KillerBee

Type: Tactical battlefield reconnaissance UAV.

Development: Developed, originally by Swift Engineering Inc in collaboration with Northrop Grumman, with an eye to potential markets such as the US armed forces, Department of Homeland Security, private security companies and other military or civil agencies. According to airframe manufacturer Swift, KillerBee was planned to demonstrate, by the end of 2006, an advanced engine that would provide 30 hours' endurance with a 3.2 kg (7 lb) payload or 8 hours with a 9.1 kg (20 lb) payload. Swift planned to bid KillerBee for the Tier II US Navy/Marine Corps short-range TUAS requirement, in competition with such other designs as the GoldenEye 80, ScanEagle and Aerosonde. Northrop Grumman purchased all rights to KillerBee in 2009 and continues with a development of the design under the name Bat (which see). Raytheon purchased the name KillerBee and rights to some technology from Northrop Grumman in May 2009. Raytheon now has rights to produce, improve and sell KillerBee KB-4 and will offer this 10 ft (3 m) span KillerBee for the US Navy and Marine Corps' TUAS Tier II competition.

Description: *Airframe:* Three-part blended wing/body flying wing configuration, comprising delta-shaped centre-section and sweptback outer panels with turned-down tips. Composites construction. All internal systems accessible from a single panel. No landing gear.

Mission payloads: Standard: high-resolution EO with pan, tilt and zoom; high-resolution IR with IR pointer. Optional payloads can include laser rangefinder/designator, hyperspectral sensor, miniature SAR, chemical/biological detection equipment, flare dispensers, or a searchlight and megaphone. Weapons and sensors can be carried simultaneously. Range can be extended by means of a communications relay package.

Guidance and control: Single laptop- or PDA-based GCS with multiple large-screen imagery workstations, can operate up to four air vehicles simultaneously. Flight profiles are fully autonomous, and sensor data from any AV can be routed to RVTs in real time.

Transportation: KB-2 can be transported on or behind a HMMWV, SUV or commercial van, and launched from a box containing all equipment required for its deployment and recovery. Configuration enables several air vehicles to be stacked in a comparatively small space. KB-4 packs into two major assemblies (a trailer/launcher and a GCS), both of which can be transported internally by C-130, helicopter or V-22 Osprey tiltrotor, or in an ISO-20 standard shipping container.

Launch: Air launch from manned or unmanned platforms; or surface launch by pneumatic catapult.

The first production Cobra testbed UAV *(Raytheon)*

Net recovery is one option for KillerBee - skid landing is also provided for *(Raytheon)*

KillerBee's three part construction, flying wing configuration *(Raytheon)*

Underwing MALDs on a USAF F-16 fighter *(Raytheon)*

Recovery: Multiple options include net retrieval and skid landing. Free-standing net can be erected by two persons in 30 minutes and can withstand winds of up to 39 kt (72 km/h; 45 mph).

Power plant: Single piston engine (types and ratings not stated); two-blade pusher propeller. A version with 17 hp heavy fuel engine driving a five-blade geared propeller and 800 A alternator under development in 2009.

Variants:

KB-2, KB-3

Smaller versions of KillerBee, shorter span, lighter payload. See previous editions of *IHS Jane's All the World's Aircraft: Unmanned*.

KB-4

Version adopted by Raytheon with nearly double the payload of predecessors. First flight Yuma 12 April 2007.

(KB-4)

KB-4

Dimensions, External	
Wings, wing span	3.00 m (9 ft 10 in)
Dimensions, Internal	
Payload bay, volume	0.095 m³ (3.4 cu ft)
Weights and Loadings	
Weight, Max launch weight	61.0 kg (134 lb)
Payload, Max payload	30.0 kg (66 lb)
Performance	
Altitude, Service ceiling	5,850 m (19,200 ft)
Speed	
Cruising speed, max	58 kt (107 km/h; 67 mph)
Power plant	1 × piston engine

Status: KillerBee KB-4 was entered by Raytheon in the US Navy/USMC Small Tactical UAS (Tier II) competition, but was not selected. Designed by Swift Engineering Inc of San Clemente, California. KillerBee KB-4 became a Raytheon product in May 2009.

Contractor: Raytheon Missile Systems Advanced Programs
Tucson, Arizona.

Raytheon MALD

Type: Miniature air-launched decoy.

Development: On 20 May 2003 the USAF's Materiel Command's Air Armament Center's Precision Strike Systems Program Office (Eglin AFB, Florida) awarded Raytheon a then year USD88 million, five-year duration, System Development and Demonstration (SDD) contract. The contract covered the design and construction of a MALD AV with which to replace Northrop Grumman's cancelled ADM-160A device. As originally specified, the Raytheon MALD was to cost between then year USD75,000 and then year USD125,000 per round and was to be available from June 2008. Subsequent programme events have included:

24 October 2006 Raytheon announced that MALD had successfully completed nine free-flight launches from an F-16 over the Eglin AFB test range.

26 February 2008 Raytheon announced that MALD had completed government/contractor verification team flight testing during the preceeding January.

31 March 2008 Raytheon was awarded a then year USD80,295,119 cost plus fixed fee contract with regard to MALD-J Phase II (subsequently Block II and then Increment II - see *Variants*) risk reduction.

5 January 2009 The USAF awarded Raytheon a then year USD12,247,290 contract modification with regard to a 14 month long concept refinement study that centred on the introduction of a datalink and increased effective radiated power into the MALD-J Block II/Increment II configuration.

17 March 2009 Raytheon announced that it had delivered the first LRIP MALD AV to the USAF.

14 June 2009 Raytheon announced that MALD-J had completed its preliminary design review.

13 January 2010 Raytheon announced that MALD-J had completed its initial free-flight test during the preceding December.

18 February 2010 Raytheon announced that MALD-J had completed its critical design review.

31 March 2010 Raytheon announced that it had delivered an "operationally significant" quantity of MALD AVs to the USAF.

30 April 2010 The USAF awarded Raytheon a then year USD53,100,000 contract with regard to the MALD-J programme's EMD phase.

5 May 2010 The USAF awarded Raytheon a then year USD96,744,354 contract with regard to Lot 3 LRIP MALD production. Sources report Lot 3 as having comprised 300 MALD AVs, with Lots 1 and 2 amounting to more than 100 examples.

27 May 2010 The USAF awarded Raytheon a then year USD82,972,665 firm, fixed-price contract modification (to the existing award FA8682-10-C-0007) with regard to Lot 4 LRIP MALD production.

25 May 2011 Raytheon announced that it had launched two MALD instrumented shapes from a MALD® Cargo Air Launch System (MCALS) launcher (see *Variants*)

27 May 2011 The USAF awarded Raytheon a then year USD82,972,665 firm, fixed-price contract modification (to the existing award FA8682-10-C-0007) with regard to Lot 4 LRIP MALD production.

29 June 2011 Raytheon announced that it had undertaken the first powered launch of a MALD-J from a B-52 "earlier" in the month.

6 September 2011 Raytheon announced that MALD-J had completed a test round that had involved multiple AVs operating in both free-flight and captive carry modes.

29 November 2011 Raytheon announced that the USAF had cleared MALD-J for LRIP and had exercised a then year USD5 million option with regard to re-directing some Lot 4 LRIP production from the MALD to the MALD-J configuration. While not confirmed, this modification is understood to have involved 240 AVs.

March 2012 The USAF is reported to have cancelled the MALD-J Block II/Increment II upgrade programme.

6 July 2012 Raytheon announced that it was working with the USN on integrating MALD-J into the F/A-18E/F weapon system, with activities including proofing the AV for carrier take-offs and landings and utilising its modularity to facilitate rapid changes of jamming payloads to meet new threats.

August 2012 August 2012 saw four MALD-Js being launched as part of the type's IOT & E process.

22 August 2012 The USAF awarded Raytheon a then year USD81,839,791 firm, fixed price contract with regard to the MALD-J programme.

MALD triple-mounted on a B-52H *(Raytheon)*

Unmanned aerial vehicles > United States > Raytheon MALD – Scaled Composites Proteus

The first production example was handed over on 16 March 2009 *(Raytheon)* 1386673

A MALD on display at Farnborough in July 2006 *(IHS/Patrick Allen)* 1184084

24 September 2012 Raytheon announced that a total of 13 MALD and MALD-J AVs had been test fired during the period January to September 2012, with all the launches being considered to have been "successful". In an associated media release, the contractor also noted that it had delivered the first MALD-J AV to the USAF on 6 September 2012 and that within the cited test round total, four out of four MALD-J operational test launches had been "successfully" completed.

November 2012 General Atomics Aeronautical Systems Inc (GA-ASI) and Raytheon completed ground verification testing of the latter's MALD decoy aboard the former's Predator® - B/MQ-9 Reaper UAV during November 2012. As of early 2013, full integration of MALD aboard Predator® - B/MQ-9 Reaper was expected to have been completed by the end of 2013.

December 2012 The US Director, Operational Test and Evaluation's US FY2012 annual report characterised MALD as being 'operationally effective for combat but not operationally suitable due to poor matériel reliability in the intended operational environment'. With regard to the MALD-J, the report noted that assessment of the AV could not be completed due to a lack of reliability data derived from MALD testing. Again, the report cited FY2012 as having seen the launch of 14 × MALD/MALD-J without 'additional failures' post completion of MALD-J's EMD phase.

Description: *Airframe:* Sharp-nosed, box-section fuselage with Y configuration tailfins. Flip-out variable-geometry wings, designed to sweep back electromechanically to 35° at high altitude. Composites construction by Composites Engineering Inc (CEI), with Moog control surface actuators. Attaches to host aircraft pylon by standard 35.6 cm (14 inch) suspension lugs. MALD-J is reported to make use of zero sweep at altitude in order to maximise endurance. As illustrated in a 2006 graphic, early iterations of MALD comprised (from nose to tail) a payload bay (located in a 0.69 m (2 ft 3.1 in) nose cone); an integrated, dual-redundant flight termination/telemetry system; an electro-mechanical wing deployment assembly; two dorsal suspension lugs; a MIL-STD-1760 or "unpowered" interface; an electrically or lanyard actuated thermal battery; a GPS-aided inertial navigation system; a turbojet engine and a rear mounted control actuation system.

Mission payloads: Active RF decoy payload (MALD), active jamming/RF decoy payload (MALD-J) or user selected 19.1 kg payload (MALD-V).

Guidance and control: Pre-programmed mission profiles can enable a small group of MALDs to replicate an air strike package by presenting various aircraft signatures that an opposing air defence system might expect to see. If the MALD strike package is not totally engaged on ingress, causing the opposing defences to use manned interceptors, the remaining MALDs will be able to disperse and loiter in the area of interest for an extended period. Onboard avionics, based on those employed in Raytheon PGMs, employ GPS-aided inertial navigation systems (GAINS).

Launch: Air launch.

Recovery: Expendable.

Power plant: One 0.53 kN (120 lb st) Hamilton Sundstrand TJ-120 turbojet.

Variants: AMD-160B MALD Raytheon's ADM-160B AV is described as being designed to present the radar signature and operational flight profile characteristics of a fighter, bomber or strike aircraft with sufficient fidelity to stimulate, deceive and decoy threat integrated air defence systems. As such (and in a 2006 iteration), it had an overall length of 2.85 m (9.35 ft), a wingspan of between 1.37 m (4.49 ft - 35° sweep angle) and 1.71 m (5.61 ft - 0° sweep angle), an overall body width (wings folded and including the device's tail fins) of 500 mm (19.68 in), an overall body height (including tail fins) of 391 mm (15.39 in), a weight of approximately 113.4 kg (250 lb) and was intended to be carried and launched from a range of fixed-wing aircraft that included the A/O-10, B-1B, B-2, B-52H, F-15, F-16, F-22 and F-35 types. As of July 2012, Raytheon was reporting the mature ADM-160B as having a length, weight and range of 3.66 m (12 ft), less than 136 kg (229.8 lb) and approximately 499.7 n miles (925.4 km; 575 miles) respectively. According to *IHS Jane's* sources, the device's flight avionics are derived from Raytheon-developed weapons such as the EGBU-16/-24 GPS/laser-guided bombs, with an active decoy payload (broadcasting a "Radio Frequency (RF) signal that mimics the radar cross-section of a larger aircraft") being provided by Raytheon's Goleta, California-based electronic warfare business unit. Sources suggest that the type's 7.7 kg TJ-120 power plant is intended to facilitate ADM-160B's achievement of speeds of up to Mach 0.93 at an altitude of 12,192 m (40,000 ft). In terms of duration, the specification that ADM-160B is designed to fulfil requires it to be able to fly a sortie of at least 45 minutes at an altitude of 10,668 m (35,000 ft) and one of at least 20 minutes at an altitude of 914 m (3,000 ft), with (in both cases) the AV being able to survive manoeuvres in excess of 2 *g*.

Elsewhere within the AMD-160B programme, Raytheon has funded development of the MALD® Cargo Air Launch System (MCALS) which takes the form of an eight round, in-flight re-loadable, steel framework launcher that is pallet-mounted and is suitable for installation (as its designation suggests) aboard transport aircraft such as the C-130. Functionally, MCALS ejects MALD rounds at a pre-determined altitude, with the AVs initiating a standard wing deployment and engine ignition sequence once clear of their host launcher. An MCALS media release further notes that the system opens the way for the "non-traditional use of a high capacity aircraft to deliver hundreds of MALDs during a single combat sortie".

MALD-J Derived from the AMD-160B (and possibly designated as the AMD-160C), Raytheon describes the MALD Jammer (MALD-J) AV as offering "all [the] capabilities of MALD and adds jamming capabilities". *IHS Jane's* sources report the device as being equipped with a "classified active jamming payload" and as being capable of acting as an "expendable close-in jammer" that is also able to "fulfil the MALD decoy mission". The same sources describe MALD-J's external configuration as being "identical" to that of the AMD-160B. As of March 2012, usually reliable sources were reporting that the USAF had cancelled a proposed MALD-J Block/Increment II upgrade that would have added a datalink capability and increased effective radiated power to the baseline MALD-J Block/Increment I configuration.

MALD-V As of July 2012, Raytheon was briefing the MALD-V concept as being an "empty vehicle" with the ability to carry a user selected 19.1 kg payload. According to Raytheon's then Vice President, Air Warfare Systems (Harry Schulte) customers could install "anything they want in the cargo area", with the cited *IHS Jane's* sources listing possibilities as including "jamming equipment, surveillance equipment or a warhead along with additional datalinks and a seeker". Again, it has been suggested that the putative MALD-V could be used as an Affordable Miniature Airborne Target (AMAT) for air-to-air combat training with beyond visual range weapons. In terms of launch options, Schulte also noted that MALD-V could be launched from a "standard missile rail".

MALD-ADM-160B

Dimensions, External
Overall, length...3.66 m (12 ft)
Weights and Loadings
Weight, Max launch weight.................................... 136 kg (299 lb) (est)
Performance
Altitude
Operating altitude, max............................. 12,190 m (40,000 ft) (est)
g limits...+2 (est)
Range...................................500 n miles (926 km; 575 miles) (est)
Power plant... 1 × turbojet

Status: Most recently, both MALD and MALD-J were in production, with MALD-J deliveries having begun during the autumn of 2012.

Customers: USAF (planned procurement of 596 × MALD and 2,404 × MALD-J).

Contractor: Raytheon Missile Systems
Tucson, Arizona.

Scaled Composites Proteus

Type: Optionally piloted, High-Altitude, Long-Endurance (HALE) vehicle.
Development: Scaled Composites' UAV experience has included design and/or constructional activity for such programmes as the Raptor

Scaled Composites Proteus < United States < Unmanned aerial vehicles

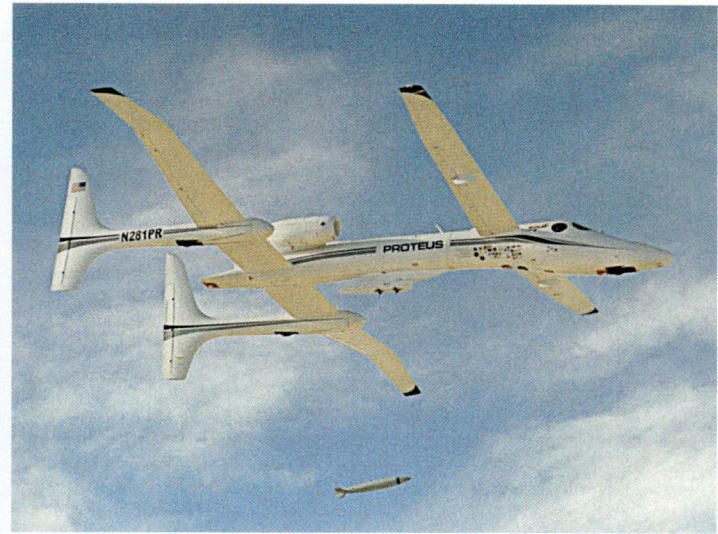
Proteus Mk 82 bomb drop in February 2005 as part of the Model 395 development programme *(Northrop Grumman)* 1042505

Proteus formates closely on NASA Global Hawk during flight refuelling simulation in 2011 *(NASA Dryden Flight Test Center)* 1421691

Undernose Amphitech radar on Proteus in 2003 *(NASA/Tom Tschida)* 0589874

Proteus in flight with dummy pod for MP-RTIP radar *(USAF)* 1151547

Proteus in flight without ventral antenna pod *(Scaled Composites)* 0104604

Computer-enhanced image of Proteus modified as weapon-carrying Model 395 *(Northrop Grumman)* 0590817

Proteus at Mojave Airport *(NASA/Patrick Wright)* 0114423

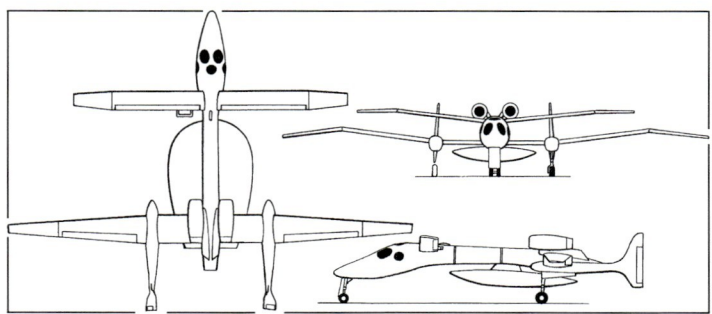
Proteus high-altitude, long-operation multirole aircraft *(James Goulding)* 0084579

Demonstrator, Bell Eagle Eye and Freewing Scorpion. The highly unorthodox Proteus was revealed in October 1997, the prototype (still the only example built) making its first flight on 26 July 1998. In piloted form, it can be configured for telecommunications, atmospheric research, satellite launch, or reconnaissance, surveillance and commercial imaging missions. In the last-named role it can also be operated in UAV form. It made its first flight with a reconnaissance payload on 2 February 1999, carrying NASA's ARTIS, using a high-resolution digital camera, as part of the latter's ERAST programme. In October 2000, with onboard pilots but minus a payload pod, Proteus set three new world altitude records in Class C1e, including a peak of 19,137 m (62,786 ft).

The aircraft is now owned by Northrop Grumman and operated by Scaled Composites as a systems testbed.

Description: *Airframe:* A long, circular-section fuselage with a drooped nose, large dihedral/anhedral foreplanes at the front and rear-mounted main wings. Twin, unbraced tailbooms at the end of the flat wing centre-sections, each ending in a vertical fin and rudder. Outboard of the tailbooms, the wing has dihedral inner and anhedral outer panels. The fuselage centre-section and underslung antenna fairing are interchangeable, depending upon the mission. There is also provision for extending wing and foreplane tips. All-composites construction. Tricycle landing gear retracts into nose and tailbooms.

Mission payloads: Airborne Real-Time Imaging System (ARTIS) initially. Various internal and pod-mounted payloads, including telecommunications, atmospheric sampling and imaging. Telecommunications relay antenna in 4.57 m (15 ft) wide suspended ventral pod, stabilised mechanically in pitch and roll, and liquid cooled, provides 65 n mile (120 km; 75 mile) diameter communications footprint. See under Status heading below for subsequent payloads tested. Up to 18 kW of electrical power available for payload operation.

Guidance and control: No details provided.
Launch: Conventional runway take-off.
Recovery: Conventional runway landing.
Power plant: Two 10.20 kN (2,293 lb st) Williams FJ44-2A turbofans. See under Weights for fuel details.

Proteus

Dimensions, External	
Overall	
length	17.17 m (56 ft 4 in)
height	5.38 m (17 ft 7¾ in)
Foreplanes	
foreplane span	
normal	16.66 m (54 ft 8 in)
extended	19.71 m (64 ft 8 in)
Wings	
wing span	
normal	23.65 m (77 ft 7 in)
extended	27.99 m (91 ft 10 in)
Areas	
Wings	
Gross wing area, normal	27.92 m² (300.5 sq ft)
Weights and Loadings	
(A: OPV version)	
Weight	
Weight empty (A)	2,658 kg (5,859 lb)
Max T-O weight	7,166 kg (15,798 lb)
Fuel weight	
Max fuel weight	
excl fuselage tank (A)	2,676 kg (5,899 lb)
incl fuselage tank (A)	4,082 kg (8,999 lb)
Payload	
Max payload (A)	2,268 kg (5,000 lb)
Performance	
T-O	
T-O run	
AUW of 5,670 kg (12,500 lb)	433 m (1,421 ft)
AUW of 3,629 kg (8,000 lb)	223 m (732 ft)
Climb	
Rate of climb	
at S/L, AUW of 5,670 kg (12,500 lb)	1,036 m/min (3,399 ft/min)
Altitude	
Service ceiling	
AUW of 5,670 kg (12,500 lb)	17,680 m (58,000 ft)
AUW of 3,629 kg (8,000 lb)	19,505 m (64,000 ft)
Speed	
Max level speed	272 kt (504 km/h; 313 mph)
Cruising speed, econ	190 kt (352 km/h; 219 mph) at 6,096 m (20,000 ft)
	280 kt (519 km/h; 322 mph) at 12,192 m (40,000 ft)
Endurance	
at 500 n miles (926 km; 575 miles) radius	22 hr
at 1,000 n miles (1,852 km; 1,150 miles) radius	4 hr
Power plant	2 × turbofan

Status: OTH command and control capability from a remote GCS more than 200 n miles (370 km; 230 miles) away was demonstrated at Mojave on 18 May 2000. At Las Cruces, New Mexico, on 13 March 2002, Proteus was flown under remote control, fitted with a 'detect, see and avoid' suite that enabled its ground controller to manoeuvre it successfully away from a potential collision course with other nearby aircraft, including a T-34C Mentor trainer, Beechcraft Duchess business twin and F/A-18A Hornet combat aircraft. Thought to be the first time a UAV has been enabled to avoid a collision course based on onboard sensors detecting such a possible incident, the Proteus flew about 18 scenarios involving single aircraft or converging groups of two approaching it at a variety of angles, altitudes and speeds. The test was viewed as a significant step towards the eventual ability of UAVs to operate within controlled national airspace.

Throughout 2002, Proteus also continued to fly various ERAST missions for NASA; by mid-year it had accumulated more than 850 flight hours, including about 150 missions above 15,240 m (50,000 ft).

In February 2005, Northrop Grumman projected a company-funded Proteus precision strike derivative dubbed Model 395, able to carry in the region of 2,950 kg (6,500 lb) of weaponry beneath the fuselage in addition to some 450 kg (992 lb) of internal payload. As a stage in this programme, Proteus (in piloted form) demonstrated release of a 500 lb inert general purpose bomb at Nellis AFB, Nevada, on 24 February 2005.

A further series of collision avoidance test flights began on 21 September 2005, Proteus this time being fitted with TCAS II and ADS-B (Automatic Dependent Surveillance Broadcast) equipment.

In 2006, the aircraft was used as a surrogate UAS for test flights of the Northrop Grumman/Raytheon Multi-Platform Radar Technology Insertion Program (MP-RTIP) AESA radar destined for installation in the Block 40 version of the RQ-4B Global Hawk. This is housed in an 8.23 m (27 ft) ventral pod designed and built by Scaled Composites to house the MP-RTIP antenna. Flight trials with a dummy pod began in April 2006, and the first of three developmental radars made a 2-hour first flight aboard the Proteus in late September 2006, testing both SAR and MTI modes at speeds up to 100 kt (185 km/h; 115 mph) at an altitude of 6,700 m (22,000 ft). Proteus continued to fly on MP-RTIP tests during 2009. In early 2011 Proteus was deployed as a manned UAV surrogate on flight refuelling trials part of the DARPA KQ-X programme. In the risk reduction flight test Proteus formated closely with a NASA Global Hawk. Wake turbulence, engine performance and flight control responsivemess were investigated at 45,000 feet.

Contractor: Scaled Composites LLC
Mojave, California.

Swift KillerBee

Description: A tactical battlefield reconnaissance UAV developed by Swift Engineering, originally in collaboration with Northrop Grumman, with an eye to potential markets such as the US armed forces, Department of Homeland Security, private security companies and other military or civil agencies. Northrop Grumman purchased all rights to KillerBee in 2009. Northrop Grumman continues with a development of the design under the name Bat. Swift Engineering has been retained by Northrop Grumman to continue to work on design refinement, product line development, flight test support and manufacturing of Bat. Raytheon purchased the name KillerBee and rights to some technology from Northrop Grumman in May 2009. Raytheon now has rights to produce, improve and sell KillerBee KB-4 (which see) and was to offer the 10 ft (3.00 m) span UAV for the US Navy and Marine Corps' TUAS Tier II competition.

Variants: KB-2: Smallest member of current family.

KB-3: Scaled-up KB-2, with double the payload. Developmental model, demonstrated to US Air Force at Creech AFB, Nevada, in March 2006, and to US Marine Corps at Camp Pendleton, California, on 17 February 2007. Latter occasion was an ONR-sponsored demonstration.

Contractor: Swift Engineering Inc
San Clemente, California.

TAG M2600

Type: Tactical multirole VTOL UAV.
Development: The M2600 is understood to have been developed under an early 2004 contract from 'a US Special Forces organisation', and was unveiled at the Eurosatory defence exhibition in Paris in June of that year. Unlike TAG's range of smaller helicopter UAVs, it is based on the full-size airframe of the Robinson R22 commercial helicopter, differing chiefly in having an extended tailboom. After delivery to its original customer, it began to be offered commercially in about May 2006.

Description: Airframe: Conventional pod and boom configuration with carbon fibre and aluminium fuselage; two-blade main and tail rotors. Fixed twin-skid landing gear.

Mission payloads: Mission-specific to customer requirements. Typical applications include observation and surveillance, border patrol, pipeline inspection and search and rescue. Has also been test-flown with 2.75 in air-to-ground rocket pods and, possibly, other weapons.

Guidance and control: Fully autonomous or remotely piloted flight via telemetry and datalink; GPS navigation.
Launch: Conventional helicopter take-off.
Recovery: Conventional helicopter landing.
Power plant: One 2,600 cm³ four-cylinder piston engine. Fuel capacity 75 litres (19.8 US gallons; 16.5 Imp gallons).

The R22-based M2600 VTOL UAV with underslung rocket pods (TAG) 1290279

M2600

Dimensions, External
- Overall
  - length...6.35 m (20 ft 10 in)
  - height..2.41 m (7 ft 11 in)
- Skids, skid track...................................1.68 m (5 ft 6¼ in)
- Rotors, rotor diameter..........................7.77 m (25 ft 6 in)
- Tail rotor, tail rotor diameter................1.27 m (4 ft 2 in)

Areas
- Rotor disc..47.45 m² (510.7 sq ft)
- Tail rotor disc..0.40 m² (4.3 sq ft)

Weights and Loadings
- Weight, Max T-O weight.......................653 kg (1,439 lb)
- Payload, Max payload..........................227 kg (500 lb)

Performance
- Climb
  - Rate of climb, max, at S/L................350 m/min (1,148 ft/min)
- Altitude
  - Service ceiling..................................3,200 m (10,500 ft)
  - Hovering ceiling, IGE........................2,135 m (7,000 ft)
- Speed
  - Max level speed................................108 kt (200 km/h; 124 mph)
  - Cruising speed, normal.....................89 kt (165 km/h; 102 mph)
- Radius of operation
  - mission, with 5 kg (lb) payload.........95 n miles (175 km; 109 miles)
- Endurance
  - max, with 5 kg (lb) payload..............2 hr
- Power plant...1 × piston engine

Status: Thought to be in production and service. Promotion continuing in 2008, no subsequent reports.

Contractor: Tactical Aerospace Group (TAG)
Beverly Hills, California.

TCOM tethered aerostats

Type: Helium-filled, tethered, non-rigid aerostats.

Development: Since the 1980s, key events in TCOM's aerostat programme have included:

1980s: During the 1980s, TCOM supplied the US Army with a Small Aerostat Surveillance System (SASS) for use in South Korea (one system), the Caribbean and the Pacific Ocean (collectively, three systems designated as Ship-based Aerostat System - SBAS). SASS and SBAS were progressively based on the company's model 31M, 32M® and 38M® AVs.

1980-1983: During the period 1980-1983, TCOM supplied Israel with a mixed fleet of model 71M® and System 365H surveillance aerostats.

1984-1985: During the period 1984-1985, TCOM is understood to have supplied the then US Customs Service (USCS) with the Caribbean Ballon (CARIBALL) 1 surveillance aerostat. Stationed at High Rock, Grand Bahama Island, CARIBALL 1 remained in service until US Fiscal Year (FY) 1992.

1986: During 1986, TCOM supplied the then USCS with four model 71M® surveillance aerostats for use in the service's SOuthWest Region BALloon (SOWRBALL) programme.

July 1990: During July 1990, TCOM installed a model 71M® aerostat-based Low-Altitude Surveillance System (LASS) in Kuwait.

1993: During 1993, Kuwait awarded TCOM a USD39.7 million contract with regard to the supply of an aerostat LASS with which to replace the 1990 system that had been lost during the Iraqi invasion of Kuwait.

April 1997: During April 1997, the Italian Navy introduced its model 32M® aerostat-based South Adriatic Aerostat Coastal Surveillance (SAACS) system into service to monitor illegal activity in the Straits of Otranto.

October 1999: As of October 1999, TCOM was noting the existence of a model 15M aerostat-based USMC Stationary LTA Platform (MCSLaP) and Italy's SAACS together with sales to customers in Europe (seven systems), Israel ('multiple' systems), Kuwait, Saudi Arabia (one LASS system), the United Arab Emirates (one LASS system) and the US ('multiple' systems for the the US Army, the USAF and the then USCS).

2002: During 2002, India procured two model 71M® aerostats for use in an extended range air and surface surveillance programme.

21 November 2003: On 21 November 2003, TCOM was awarded a USD84.8 million contract with regard to the supply of a model 71M® aerostat-based LASS for Kuwait.

October 2004: During October 2004, the US Army was understood to have fielded a TCOM aerostat-based surveillance system as part of its southwest Asian Rapid Aerostat Initial Deployment (RAID) programme.

Mid-2005: As of mid-2005, the USMC is understood to have fielded its model 32M® aerostat-based Marine Airborne Re-Transmission System (MARTS) in Iraq. Elsewhere (and as of the same date), IHS Jane's sources were reporting TCOM as having delivered 34 × model 15M and 15 × model 17M® aerostats to customers around the world, with a further 16 × model 17M®s on order.

12 September 2007: On 12 September 2007, TCOM announced that the US Army had awarded it an aerostat-based Cellular Aerostat Platform System (CAPS) development and demonstration contract.

A TCOM 38M® aerostat in-flight *(TCOM)*

2008: As of 2008, TCOM was reporting that it had supplied 24 × model 17M® aerostats into the US Army's RAIDS programme.

25 August 2009: TCOM model 74M™ aerostat number one made its maiden flight on 25 August 2009. The model 74M™ is the AV used in the the US Army's Joint Land attack cruise missile defense Elevated Netted Sensor (JLENS) programme.

January-June 2010: During the period January-June 2010, TCOM was awarded 'initial US Government quick reaction contracts' with regard to the US Army's Persistent Ground Surveillance System (PGSS) programme.

Mid-2010: As of mid-2010, had delivered 28 × model 17M® aerostats into the RAID programme and was under contract to produce 13 × model 22M™ AVs for the PGSS effort.

19 May 2011: On 19 May 2011, the USN (acting on behalf of the US Army) awarded TCOM a USD14.9 million fixed price contract with regard to the supply of five × model 22M+™ aerostats, five × tether-up kits and an associated spares holding for use in the PGSS programme.

25 May 2011: On 25 May 2011, the USN awarded TCOM a USD41.2 million cost plus fixed fee contract with regard to the supply of a 'prototype' model 22M™ aerostat, technical support personnel and training/test support/payload integration work, with the whole relating to the PGSS effort.

7 July 2011: On 7 July 2011, the USN awarded TCOM a USD10.2 million contract modification with regard to the supply of five × model 22M+™ aerostats, six × spare tether-up kits and two × line replaceable units/site spares, with the whole relating to Phase III of the PGSS programme.

8 August 2011: On 8 August 2011, the USN awarded TCOM a USD12.3 million contract modification with regard to the previously noted PGSS Phase III effort.

2 September 2011: On 2 September 2011, the USN awarded TCOM a USD38.6 million firm, fixed-price contract with regard to the procurement of 15 × model 22M+™ aerostats, spare tether-up kits, line replace-

A model 71M® aerostat equipped with an Elta APR radar *(TCOM)*

A MARTS model 32M® aerostat on its mobile mooring system *(TCOM)* 1145274

A model 17M® aerostat on a truck-mounted winch for off-road, on-the-move operations *(TCOM)* 1145273

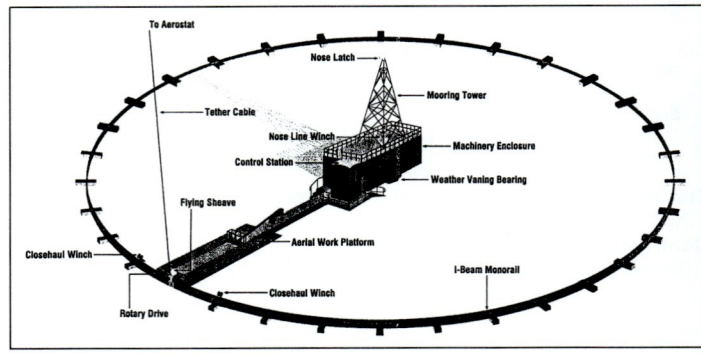

The TCOM aerostat ground handling and mooring system 0518229

able units, component kits and field representatives for use in the PGSS effort.

7 December 2011: On 7 December 2011, the USMC awarded TCOM a USD10.4 million firm, fixed-price contract with regard to the provision of Contractor Engineering and Technical Support (CETS) for Kuwait's LASS architecture.

17 January 2012: On 17 January 2012, the USN awarded TCOM a USD12.2 million contract with regard to the supply of three × model 28M aerostats and three × site spares packages for use in the PGSS programme.

21 March 2012: On 21 March 2012, the USN awarded TCOM a USD13.4 million contract modification with regard to the exercising of an option relating to the procurement of seven × model 22M™ aerostats, site/tether-up spares and line replaceable units in support of the PGSS effort.

4 April 2012: On 4 April 2012, the USN awarded TCOM a USD7.7 million contract modification with regard to the procurement of 15 × AC to DC field modification kits for use with PGSS model 22M™ aerostats.

17 September 2012: On 17 September 2012, the USN awarded TCOM a USD11.7 million firm, fixed-price contract with regard to the supply of model 22M™ and 28M aerostat parts/spares kits and field service representatives in support of the PGSS programme.

October 2012: As of October 2012, a total of 59 × model 22M™/28M aerostats had been procured for use in the PGSS effort.

14 January 2013: On 14 January 2013, the USN awarded TCOM a USD15.6 firm, fixed-price delivery order with regard to the procurement of model 22M™ and 28M aerostat parts and spares in support of the PGSS programme.

Description: *Airframe:* Streamlined, elongated teardrop shape with inverted Y tail surfaces and a ventral air-inflated payload housing. The envelopes are made of multi-layered synthetic fabrics (PLC or polyester, Mylar and Tedlar) laminated together with a specially formulated polyurethane adhesive. The seams are thermally bonded. An internal air ballonet is provided to maintain the internal pressure constant as altitude and temperature change. The tailfins are normally filled with air or, in special cases, with helium for additional lift and altitude performance. The high-strength tether cable is made of Kevlar or Vectran (strength member), with both electrical and fibre optic conductors in the centre core, and is sheathed with copper braid and a polymer protective jacket. The copper braid is an integral part of an overall lightning protection system, designed to conduct to ground the high currents injected by lightning strikes with no permanent damage to the system or tether.

Mission payloads: Over time, the following sensor systems have been associated with TCOM's family of aerostats:
- the TCOM Airborne Imaging System (AIS)
- an aerostat variant of the 6 to 10 GHz band Northrop Grumman AN/APG-66(V) multi-mode airborne radar (designated as AN/APG-66SR)
- an aerostat variant of the D-band (1 to 2 GHz) Northrop Grumman AN/TPS-63 surveillance radar
- the X-band (8 to 10 GHz sub-band) Telephonics APS-143(V)2 surveillance radar (used in TCOM's Maritime Aerostat Tracking and Surveillance System (MATSS))
- the 2 to 18 GHz band ITT Exelis Electronic Systems (formerly EDO Reconnaissance and Surveillance Systems) AR-900 electronic support and direction-finding system
- the D-band Northrop Grumman Advanced Technology Low Altitude Surveillance System (ATLASS) radar
- the Raytheon BBN Technologies Boomerang acoustic small arms fire detection system
- the D-band Northrop Grumman Enhanced Low Altitude Surveillance System (E-LASS) radar
- the Elta Systems EL/I-3330 Multi-Payload Aerostat System (MPAS)
- the Elta Systems EL/M-2083 Aerostat Programmable Radar (APR)
- the Raytheon Integrated Defense Systems JLENS Precision Track Illumination Radar (PTIR)
- the Raytheon Integrated Defense Systems JLENS surveillance radar
- the D-band Lockheed Martin Maritime Systems and Sensors L-88 surveillance radar
- the L-3 Sonoma Model 14 Skyball™ multi-sensor surveillance EO payload
- the L-3 Wescam MX-15 multi-sensor EO payload
- variants of the FLIR Systems Star SAFIRE™ multi-sensor surveillance and targeting EO payload
- the US Army Research Laboratory's Unattended Transient Acoustic Measurement And Signature Intelligence (MASINT) System (UTAMS).

Alongside airborne early warning and surveillance applications, TCOM aerostats have been used in a number of communications applications. By way of illustration, TCOM is reported to have developed a Very Low Frequency (VLF - 3 to 30 kHz) communications aerial (designated as the Tethered Aerostat Antenna Programme - TAAP) that used the tether of the aerostat as the aerial. As such, TAAP was 'successfully' tested by the USAF, the USMC (see following) and the US Defense Communications Agency. Staying with the US military, May 1998 and February 1999 saw the US Army undertaking demonstrations of a Co-operative Engagement Capability (CEC) relay capability using a model 15M aerostat. The same vehicle is further noted as having been used for three other communications system demonstrations (using the service's Single Channel Ground-to-Air Radio System (SINCGARS), an Enhanced Position Location Reporting System (EPLRS) and a Near-Term Digital Radio (NTDR)) during 1998-1999. Here, the SINCGARS effort is said to have increased range by seven times when compared with a terrestrial installation, while the model 15M/EPLRS combination offered a range of 70 km, a figure that extended to 140 km in a potential point-to-point hop mode. Elsewhere in the world, TCOM noted that, as of March 1999, an 'international customer' had 'recently' procured five model 32M® mobile aerostat systems for use in a VLF communications system. In this application, the aerostat's tether was noted as acting as the VLF antenna and the overall system's mobility was noted as allowing the user to set up a communications facility at a selected site within four hours.

Moving forward to 2005, the USMC introduced the model 32M®-based MARTS system for use in Iraq. Here, the MARTS aerostat was flown at an altitude of 1,000 m (3,281 ft) and was used to simultaneously relay EPLRS (AN/ASQ-177, AN/GRC-229 and AN/VSQ-1 and -2), 'Have Quick II' (AN/ARC-164, AN/GRC-240 and AN/PRC-117) and SINCGARS (AN/ARC-210(V), AN/MRC-145A and AN/PRC-119) radio communications signals. Within the architecture, the necessary transceiver antennas were mounted inside the aerostat's envelope, with associated communications security and encryption equipment being ground-based and connected to the aerial array via the vehicle's fibre-optic tether. A year later, 23 October 2006 saw TCOM and telecommunication contractor TECORE Inc announce their Airborne Rapid Deployment Cellular System (ARDCS) which combined TCOM's model 17M® mobile aerostat with TECORE's rapid response cellular telephone technology. Taking these in reverse order, TECORE's rapid response systems are designed to address emergency response, disaster recovery and tactical military communications requirements; facilitate "immediate" deployment of full wireless network functionality; make use of full feature commercial-off-the-shelf technology and incorporate the company's RapidMSC™ mobile switching centre platform to provide

The swept-tailed Sky-Vu version of TCOM's model 15M aerostat *(IHS/Kenneth Munson)* 0558394

(amongst other services) messaging, high-speed data and call priority facilities. Other system features are thought to have included:
- the ability to connect with existing Global System for Mobile communications (GSM)/Code Division Multiple Access (CDMA), Public Switched Telephone Network (PSTN) or satellite/Very Small Aperture Terminal (VSAT) infrastructure
- support for high-speed wireless data (GSM General Packet Radio Service (GPRS)/Enhanced Data for GSM Evolution (EDGE) and Code Division Multiple Access (CDMA) 1× Radio Transmission Technology (RTT)/1 × Evolution-Data Only (EV-DO)
- connectivity back to PSTN facilities via local T1/E1 or satellite
- scalability
- configurability with a variety of radio interfaces for CDMA/GSM support and pico-/macro-cell deployments
- local or remote management/administration.

For its part, the ARDCS model 17M® aerostat application was noted as being able to lift an appropriate Radio Frequency (RF) payload and antenna system to altitudes of up to 305 m (1,000 ft) from which height, the architecture was noted as providing unobstructed cellular coverage over an area greater than 1,554 km². ARDCS was further noted as being "self-contained" in two 6.1 m International Standards Organisation (ISO) - compliant containers and as being deployable in "two hours or less". Subsequent to the unveiling of the ARDCS system, TCOM went on to develop the generally similar CAPS system. Like ARDCS, CAPS made use of the TCOM model 17M® aerostat, was fielded in two 6.1 m ISO containers, was equipped with a 91 kg payload, could be flown up to an altitude of 305 m (1,000 ft) and could provide connectivity for up to 2,000 subscribers operating within a 1,554 km² (600 sq mile) footprint.

Guidance and control: Communication between the aerostat and the surface is effected by optical technology with both electrical and optical fibres embedded within the core of the tether.

Launch and Recovery: The 'weather-vaning' TCOM ground handling and mooring systems provide a high degree of safety for the aerostat and ground personnel during launch and recovery operations, as well as for the ground-moored aerostat. The key feature is the system's ability to rotate through 360° as it follows the aerostat's response to changing wind direction. During recovery, the system can be rotated hydraulically to align the aerostat and mooring system axes. The main tether winch system, with a maximum tension capability of up to 18,144 kg (40,000 lb), has a maximum line speed at reduced tensions of up to 137 m (450 ft) per minute.

Launch or recovery of the large systems requires a five-person crew plus a flight director who controls the crew activities via an intercom linking all personnel. The entire mooring system is centrally controlled by the winch operator from a station at the juncture of the boom and machinery enclosure. Three line handlers ensure that the handling lines are attached to or detached from their respective port, starboard and nose-line close-haul winches. The fifth crew member is the telemetry and command operator, who is remotely located in the site control room.

Variants: Over time, TCOM is understood to have produced/proposed or had previously marketed the following aerostat configurations:

model 15M: As of 2013, TCOM was no longer promoting the model 15M aerostat.

model 17M®: TCOM characterises its model 17M® aerostat as being a 'tactical' system that is 'highly mobile' and is 'operable from a small trailer [that is] designed for towing by any of several different field vehicles or [are] truck-mounted for off-road, on-the-move surveillance'. Again, the 'standard issue' trailer used with such aerostats is modified to contain a mooring device as well as the system's power source. Over time, the model 17M® aerostat has been identified as having been used in the RAID and Off-Road Tactical Aerostat System (ORTAS) architectures.

model 22M™: Like the model 17M®, TCOM characterises the model 22M aerostat as being one of its 'tactical' systems. As such, the model 22M™ AV is used in the US Army's PGSS system.

model 25M: As of 2013, TCOM was no longer promoting the 25 m (82.0 ft) long model 25M aerostat.

model 28M: TCOM characterises the model 28M aerostat as being an 'operational' system that is portable, offers better altitude performance than its 'tactical' products and is applicable to land, shipborne and coastal applications. The model 28M AV is used in the US Army's PGSS programme.

TCOM model 74M™ aerostat number one photographed during its maiden flight on 25 August 2009 *(TCOM)* 1411198

model 31M: As of 2013, TCOM was no longer promoting the 31 m (101.7 ft) long model 31M aerostat.

model 32M®: IHS Jane's sources report TCOM's model 32M® aerostat as having been used in the USMC's MARTS system, the Italian SAACS application and the United Arab Emirates' (UAE) Emirates Coastal Defence System (ECDS). As of 2013, TCOM was no longer promoting the model 32M® AV.

model 38M®: As of 2013, TCOM was no longer promoting the model 38M® aerostat.

model 53M/System 250: As of 2013, TCOM was no longer promoting the 7,100 m³ (250.734 cu ft) volume model 53M/System 250 aerostat.

model 67M/System 365: As of 2013, TCOM was no longer promoting the 11,000 m³ (388,462 cu ft) volume model 67M/System 365 aerostat.

model 71M®: TCOM characterises the model 71M® aerostat as being a 'strategic' system that is designed for aerial surveillance applications (with a duration of up to 30 days) and offers an enhanced payload capacity and altitude performance. Model 71M® applications include an Indian aerostat surveillance system, Israel's Extended Air Defence Aerostat (EADA) programme and the Kuwaiti and UAE LASS architectures.

model 74M™: Like its model 71M®, TCOM characterises the model 74M™ aerostat as being a 'strategic' system. As such, it is used in the US Army's JLENS programme.

model 76M: As of 2013, TCOM was no longer promoting the 20,000 m³ (706,294 cu ft) volume model 76M aerostat.

TCOM tethered aerostats

Dimensions
Length overall
15M	15.0 m (49.2 ft)
17M®	17.0 m (55.8 ft)
22M™	22.0 m (72.2 ft)
28M	28.0 m (91.9 ft)
32M®	32.0 m (104.9 ft)
38M®	38.0 m (124.7 ft)
71M®	71.0 m (232.9 ft)
74M™	74.0 m (242.8 ft)

Maximum diameter
15M	6.2 m (20.3 ft)
17M®	6.2 m (20.3 ft)
32M®	10.0 m (32.8 ft)
38M®	10.2 m (33.5 ft)
71M®	22.0 m (72.2 ft)

Volume
15M	320 m3 (11,301 cu ft)
17M®	340 m3 (12,007 cu ft)
32M®	1,700 m3 (60,035 cu ft)
38M®	2,492 m3 (88.004 cu ft)
71M®	16,700 m3 (589,756 cu ft)

Weights
Maximum payload
15M	17 kg (38 lb)
17M®	90 kg (198 lb)
22M™	181 kg (399 lb)
28M	453 kg (999 lb)
32M®	275 kg (606 lb)
38M®	225 kg (496 lb)
71M®	1,600 kg (3,527 lb)
74™	3,182 kg (7,015 lb)

TCOM tethered aerostats

Hull structural weight	
15M...	144 kg
17M®...	143 kg
32M®...	658 kg
71M®...	4,091 kg
Tether length	
15M...	460 m (1,509 ft)
17M®...	457 m (1,499 ft)
22M™..	1,000 m (3,281 ft)
28M...	1,006-1,585 m (3,301-5,200 ft)
32M®/38M®..	1,067-1,524 m (3,501-5,000 ft)
71M®...	6,100 m (20,013 ft)
74M™..	4,877 m (16,001 ft, nominal)
Performance	
Maximum operating altitude (reduced payload)	
15M...	300 m (984 ft)
17M®...	305 m (1,000 ft)
28M...	1,524 m (5,000 ft)
32M®...	1,400 m (4,593 ft)
71M®...	6,700 m (21,982 ft)
Operating altitude (maximum payload)	
15M...	300 m (984 ft)
17M®...	305 m (1,000 ft)
22M™/28M...	914 m (3,000 ft)
32M®...	900 m (2,953 ft)
38M®...	1,500 m (4,921 ft)
71M®...	4,600 m (15,092 ft)
74M™..	3,048 m (10,000 ft)
Tether break strength	
15M...	1,820 kg (4,012 lb)
17M®/22M™...	3,175 kg (7,000 lb)
28M...	6,350 kg (13,999 lb)
32M®/38M®..	6,400-10,886 kg (14,110-24,000 lb)
71M®...	34,000 kg (74,957 lb)
74M™..	36,300 kg (80,028 lb)
Payload power (via tether)	
15M/17M®..	1 kVA
32M®/38M®..	10 kVA
71M®...	30 kVA
74M™..	80 kVA
Flight duration	
15M...	5 days
17M®...	7 days
22M™..	10 days
32M®/38M®..	14 days
71M®/74M™..	30 days
Operational windspeed	
15M/17M®..	40 kt (74 km/h; 46 mph)
32M®/38M®..	51 kt (94 km/h; 58 mph)
71M®...	70 kt (130 km/h; 81 mph)
Survivable windspeed	
15M/17M®..	55 kt (102 km/h; 63 mph)
32M®/38M®..	70 kt (130 km/h; 81 mph)
71M®...	90 kt (167 km/h; 104 mph)
Source TCOM	

Status: As of 2013, TCOM was promoting the model 17M®, 22M™, 28M, 71M® and 74M™ aerostats. Most recently, *IHS Janes* has identified the following TCOM aerostat-based surveillance systems as being in service (or as having recently been in service):

- **ECDS (UAE military)** model 32M® aerostat, a "maritime" radar, a FLIR Systems Star SAFIRE™ III EO sensor and a V/UHF radio repeater package
- **Indian aerostat surveillance system** model 71M® aerostat and an Elta Systems EL/M-2083 APR
- **EADA (Israeli military)** model 71M® aerostat and an Elta Systems EL/M-2083 APR. Also known as the Airstar system
- **JLENS (US Army)** model 74M™ aerostat, the Raytheon Integrated Defense Systems JLENS PTIR and JLENS surveillance radars and a FLIR Systems Star SAFIRE™ EO sensor application
- **Kuwait LASS** model 71M® aerostat, an ITT Exelis Electronic Systems AR-900 electronic support and direction-finding system and an aerostat configured variant of Northrop Grumman AN/TPS-63 surveillance radar
- **ORTAS (UAE military)** model 17M® aerostat and a FLIR Systems Star SAFIRE™ III EO sensor
- **PGSS (US Army)** model 22M™ and 28M aerostats, a Raytheon BBN Technologies Boomerang acoustic small arms fire detection system, an L-3 Wescam MX-15/MX-15HDi multi-sensor EO payload and/or a FLIR Systems Star SAFIRE™ 380 HD EO sensor
- **RAID (US Army)** model 17M® aerostat, a FLIR Systems Star SAFIRE™ III EO sensor and a communications relay package
- **SAACS system (Italian Navy)** model 32M® aerostat, a Northrop Grumman AN/APG-66SR radar, an L-3 Sonoma Model 14 Skyball™ multi-sensor surveillance EO payload and a General Dynamics URC-200 communications transceiver
- **UAE LASS** a model 71M® aerostat and a Northrop Grumman Advanced Technology Low Altitude Surveillance System (ATLASS) or aerostat configured Northrop Grumman AN/TPS-63 radar

Customers: Over time, *IHS Jane's* sources report Australia, Canada, Egypt, France, Germany, India, Iran, Israel, Italy, Jordan, South Korea, Kuwait, Nigeria, Saudi Arabia, the United Arab Emirates, the United Kingdom and the United States (DoD, Army, Navy, Air Force, Marine Corps, Coast Guard and the then Customs Service) as having procured/operated TCOM aerostats.

Contractor: TCOM LP
Columbia, Maryland.

UAV Factory Penguin B

Type: Multi-purpose, long endurance mini UAV.

Development: *IHS Jane's* sources first identified the Penguin B UAV during the Paris-based, June 2010 Eurosatory defence exhibition. Programme events include:

May 2012 Kansas State University Salina deployed its Penguin B UAV to Fort Riley as part of an ongoing programme (funded by the United States Office of Scientific Research) that was looking at how UAVs interact with manned and other unmanned aircraft within America's national airspace.

July 2012 During the period 5 to 7 July 2012, a Penguin B UAV stayed aloft for a period of 54 hours and 27 minutes, an endurance claimed by UAV Factory as being the then 'longest recorded flight for a mini-class unmanned aircraft'. For the sortie, the AV was loaded with approximately 13 kg of pre-mixed petrol and oil and had a take-off weight of 22.3 kg. Air temperatures and wind speeds during the flight reached more than 30°C and up to 20 m/s respectively and at the end of the sortie, the AV executed a successful belly landing. Again, the AV was fitted with a Piccolo® autopilot and an electronic fuel-injection engine.

Description: *Airframe:* The Penguin B airframe comprises a streamlined fuselage; a shoulder-mounted wing centre section (with high-lift flaps); left and right outer wing panels (with ailerons); twin tailbooms; a two part, inverted V-shaped tail unit; a rear mounted engine (driving a two-bladed pusher propeller) and a demountable tricycle undercarriage.

In more detail, the AV's forward fuselage incorporates a dorsal access hatch and a ventral universal payload mount that features removable ballast slugs and pre-determined payload mounting holes. Here, integrators can design custom mounts for their payloads using associated Computer-Aided Design (CAD) software and the complete unit can be replaced by items such as retractable gimbals if required. Quick release fasteners are used extensively throughout the vehicle's airframe, with examples being the ¼ turn Dzus fasteners that are used to secure its 'oversize' access covers and the 'high-end', industrial grade, push-pull connectors that are used in its tailboom structure. *IHS Jane's* sources describe the Penguin B airframe as being made out of composites (hollow moulded sandwich construction) with carbon fibre

A ground view of the Penguin B multi-purpose mini UAV *(IHS/Patrick Allen)* 1476571

The UAV Factory's second generation portable GCS that is used with company's Penguin B mini UAV *(IHS/Patrick Allen)* 1476570

UAV Factory Penguin B < United States < Unmanned aerial vehicles

Parachute recovery is an available option for the Penguin B mini UAV *(UAV Factory)* 1509457

A close-up of the engine installation used on the Penguin B mini UAV. Note also the dorsal GPS compartment located on the AV's centre wing section *(IHS/Patrick Allen)* 1476566

tailbooms. When fitted, Penguin B's standard tricycle undercarriage can be replaced by a heavy duty option that is designed to facilitate landing on unprepared surfaces and incorporates a higher shock absorbing capacity than the standard arrangement and larger inflatable wheel tyres.

Mission payloads: User specified (see also Airframe).

Guidance and control: GPS navigation (installed in a centre wing compartment) and either a Cloud Cap Technology Piccolo® (recommended) or a Procerus Technologies Kestrel autopilot (both validated for autonomous flight, catapult take-off and runway landings and supported by configuration files). Aerodynamic control by means of inboard high-lift flaps, outboard ailerons and rudders. The AV's tail surface configuration has an option for four movable surface servos to increase system reliability in case of progressive servo failures. Penguin B can also be fitted with a heated or unheated pitot-static tube if required.

Launch: Pneumatic catapult, car-top launcher or conventional wheeled take-off (with option to remove the undercarriage if required).

Recovery: Conventional wheeled or belly (when the wheeled undercarriage is removed) landing. Parachute recovery as an available option.

Power plant: One 1.9 kW (2.5 hp) two-stroke piston engine, driving a two-blade pusher propeller. Maximum fuel capacity 7.5 litres (1.9 US gallons; 1.7 Imp gallons). Engine options comprise a 2.6 kW (3.55 hp) 3W 28 CS unit that offers a shorter take-off run (including from unprepared surfaces) and improved rate of climb or an electronic fuel injected 2.5 kW (3.35 hp) 3W 28i configuration. Other options include an 80 W generator upgrade and a 7,500 cc (457.7 cu in) fuel tank (with either a fuel level and/or a header tank fuel sensor).

Penguin B

Dimensions, External	
Overall	
length	2.286 m (7 ft 6 in)
height	0.901 m (2 ft 11½ in)
Fuselage, ground clearance	0.288 m (11¼ in)
Wings, wing span	3.30 m (10 ft 10 in)
Dimensions, Internal	
Payload bay	
length	0.41 m (1 ft 4¼ in)
width	0.209 m (8¼ in)
depth	0.138 m (5½ in)
volume	0.02 m³ (0.7 cu ft)
Areas	
Wings, Gross wing area	0.79 m² (8.5 sq ft)
Weights and Loadings	
Weight	
Weight empty	
fuel-injected engine & 7500 cc fuel tank - fuel and payload excluded	10.0 kg (22 lb)
Max T-O weight	21.5 kg (47 lb)
Payload, Max payload	10.0 kg (22 lb)
Performance	
T-O	
T-O run, sea level, 15kg Av Weight, 15°C, concrete runway	30 m (99 ft)
Speed	
Max level speed	70 kt (130 km/h; 81 mph)
Cruising speed	43 kt (79 km/h; 49 mph)
Stalling speed	26 kt (47 km/h; 30 mph)
Endurance	
fuel-injected engine & 7500 fuel tank	20 hr
Power plant	1 × piston engine

Status: UAV Factory had delivered more than 70 Penguin B mini UAVs to clients in 17 countries around the world by the end of 2012.

Contractor: UAV Factory USA LLC,
Irvington, New York. UAV Factory Ltd Europe
Jelgava, Latvia.

AERIAL TARGETS

Australia

AAA Phoenix Jet

Type: Jet-powered recoverable aerial target.

Development: Unveiled at Avalon air show in March 2007. Company also manufactures tow targets and towing systems under licence, and provides services with these equipments.

Description: *Airframe:* Mid-mounted delta wing; circular-section fuselage, with sweptback tailfin. No landing gear. Construction of vacuum-formed glass fibre/epoxy sandwich.

Mission payloads: Include smoke, flares, Luneberg lens, stabilised camera and MDI. Smoke plume can be emitted continuously for 4 minutes or intermittently for 10 minutes from a 4.4 litre (1.2 US gallon; 1.0 Imp gallon) tank. Jet exhaust provides IR signature.

Guidance and control: Controlled via UAV Navigation AP04 dual-redundant autopilot; GPS navigation.

Launch: Automatic catapult launch.

Recovery: Parachute recovery.

Power plant: One turbojet engine (type and rating not stated). Fuel capacity (30.4 litres (8.0 US gallons; 6.7 Imp gallons).

Phoenix Jet

Dimensions, External	
Fuselage	
length	2.20 m (7 ft 2½ in)
width	0.255 m (10 in)
Wings, wing span	2.00 m (6 ft 6¾ in)
Weights and Loadings	
Weight	
Weight empty	22 kg (48 lb)
Max launch weight	55 kg (121 lb)
Performance	
Altitude, Service ceiling	6,095 m (20,000 ft)
Speed, Max level speed	300 kt (556 km/h; 345 mph) (est)
Endurance	1 hr 30 min
Power plant	1 × turbojet

Status: In production and in service.

Contractor: Air Affairs Australia
Nowra, New South Wales.

Cyber Technology CyBird

Type: Recoverable aerial target.

Development: Designed as a reusable platform for a variety of payloads. Typical applications include R & D of payloads for target tracking, fire control systems, and radar or EW range trials.

Description: *Airframe:* Low-mounted wings with upturned tips; cylindrical fuselage with dorsal intake; conventional tail surfaces. Fixed tricycle landing gear. All-composites construction.

Mission payloads: As specified by customers.

Guidance and control: Fully autonomous, via UAV Navigation GCS03 ground station and AP04 autopilot.

Launch: Conventional, automatic wheeled take-off.

Recovery: Conventional, automatic wheeled landing.

Power plant: One 0.2 kN (44 lb st) unidentified turbojet. Fuel capacity 36 litres (9.5 US gallons; 7.9 Imp gallons).

CyBird

Dimensions, External	
Overall, length	3.00 m (9 ft 10 in)
Wings, wing span	3.00 m (9 ft 10 in)
Weights and Loadings	
Payload, Max payload	10 kg (22 lb)
Performance	
Speed	
Max level speed, IAS	200 kt (370 km/h; 230 mph)
Stalling speed, IAS	65 kt (121 km/h; 75 mph)
Power plant	1 × turbojet

Status: Understood to be in production and in service; customer(s) not identified.

Contractor: Cyber Technology (WA) Pty Ltd
Bibra Lake, Western Australia.

Phoenix Jet target ready for catapult launch *(AAA)* 1290346

The jet-powered multirole CyBird target *(Cyber Technology)* 1395257

Belgium

Belgian Defence Ultima

Type: Aerial target.

Development: Designed by Belgian Defence personnel; underwent initial flight trials in 1994. Remotely piloted aerial target family for land, sea or rotary-wing aircraft anti-aircraft weaponry (guns or missiles, depending upon version). Produced by Belgian Defence Department solely for use of its own units.

Description: *Airframe:* Conventional high-wing monoplane configuration. No landing gear; fuselage underside reinforced to absorb landing impact.

Mission payloads: Ultima 1: Two 30-second smoke flares and two 60-second IR flares at wingtips; external reflecting stripes for proximity fuze ranging; radar reflectors.

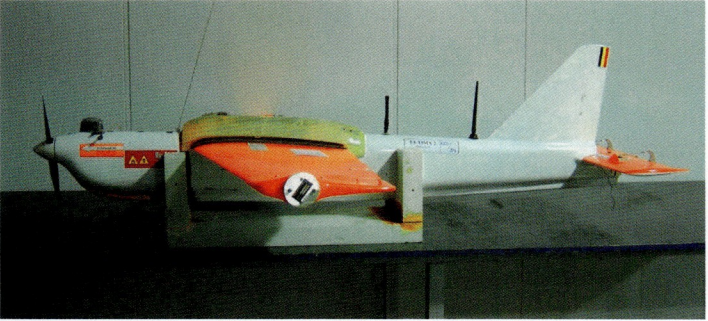

A standard Ultima 2 recoverable target *(Belgian Defence)* 1135096

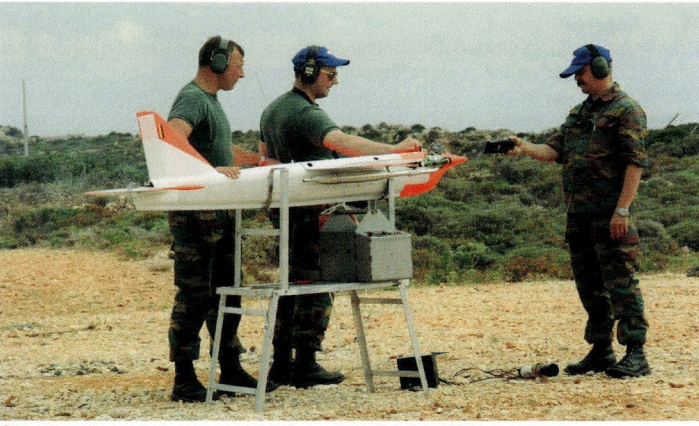

Starting the engine on an Ultima 1 at the NAMFI range on Crete *(Ch-H Spiltoir/DetAA)* 1194679

Aerial targets > Belgium – Brazil > **Belgian Defence Ultima – Aeromot AM 03089**

Ultima hand launch *(Belgian Defence)* 1194678

Special markings denote the 500th Ultima 1 to be produced *(Belgian Defence)* 1135094

Ultima 2: MDI suite comprising sensor, telemetry and radio downlink modem; GPS antennas and receiver and radio downlink modem; single smoke flare for emergency localisation.

Ultima 3: As Ultima 1, but with wiring for GPS suite.

In addition, all versions are equipped with a complete safety suite, mandatory for flying in segregated Belgian airspace.

Guidance and control: Two-stick, 40 MHz band GCS with 12 km (7.5 mile) control range; onboard decoupled receivers; SPCM datalink. MDI results from Ultima 2 are downloaded to GCS in near-realtime for analysis.

Launch: Ultima 1 and 3 hand-launched; Tasuma TML-3 launcher for Ultima 2.

Recovery: Belly skid landing; Ultima 2 has parachute for emergency recovery in case of main system failure or loss of communication.

Power plant: One 3.05 kW (4.1 hp), 34.97 cc single-cylinder engine; two-blade propeller. Fuel capacity (90% 133 octane and 10% nitromethane) Ultima 1;1 litre (0.26 US gallon; 0.22 Imp gallon), Ultima 2 and Ultima 3; 1.5 litres (0.40 US gallon; 0.33 Imp gallon).

Variants: *Ultima 1:* Non-recoverable, single-use target for anti-aircraft IR missiles. First flight October 1994.

Ultima 2: Recoverable, multiple-use target for anti-aircraft gunnery training. First flight September 1997.

Ultima 3: Recoverable, multiple-use target training aircraft. First flight November 2004.

Ultima 1, Ultima 2, Ultima 3

Dimensions, External	
Overall	
length	2.00 m (6 ft 6¾ in)
height	0.47 m (1 ft 6½ in)
Fuselage	
height, max	0.21 m (8¼ in)
width	0.12 m (4¾ in)
Wings, wing span	1.90 m (6 ft 2¾ in)
Engines, propeller diameter	0.42 m (1 ft 4½ in)
Areas	
Wings, Gross wing area	0.80 m² (8.6 sq ft)
Tailplane	
Tailplanes	0.16 m² (1.72 sq ft)
Weights and Loadings	
Weight	
Weight empty	
excl fuel, Ultima 1, Ultima 3	8.5 kg (18.00 lb)
excl fuel, Ultima 2	9.3 kg (20.00 lb)
Max launch weight	15.0 kg (33 lb)
Performance	
Altitude, Operating altitude	1,000 m (3,280 ft)
Speed	
Max level speed	
Ultima 1, Ultima 3	111 kt (206 km/h; 128 mph)
Ultima 2	94 kt (174 km/h; 108 mph)
Cruising speed	65 kt (120 km/h; 75 mph)
Stalling speed	44 kt (82 km/h; 51 mph)
Range	3.8 n miles (7 km; 4 miles)
Endurance	
Ultima 1	15 min
Ultima 2, Ultima 3	20 min
Power plant	1 × piston engine

Status: In service. Initial trials of Ultima 1 conducted in 1994; used as target for Mistral SAMs during live firing exercise in June 1995 (when it achieved IOC) at NATO Missile Firing Installation (NAMFI), Crete (Belgian Army 14th and 35th Air Defence Artillery Battalions and Artillery School). Ultima 2 achieved IOC in September 2004. Total of 591 Ultima 1, 25 Ultima 2 and 15 Ultima 3 produced by mid-2008, when production was continuing. Deployed at Lombardsijde range in Belgium and during live firing at NAMFI. Remained in service 2010-2011.

Customers: Belgian Defence Department units only, except for live firings at NAMFI with French Army 57th Artillery Regiment; not produced for other countries or customers.

Contractor: Belgian Defence Nieuwpoort.

Brazil

Aeromot AM 03089

Type: Aerial tow target.

Development: Developed for gunnery training, using both visual and radar methods of detection.

Description: *Airframe:* The tow body is made of high impact resistant thermoplastic material with aluminium bulkheads. Ogival transparent nosecone housing visual or radar detection device. Six delta-shaped tailfins.

Mission payloads: The -401 has a 600 W lamp in the nose transparency, electrical power for which is provided by a 28 V 35 A generator driven by a rear-mounted turbine and a regulator circuit; the electrical system is actuated automatically at the moment of launch. The -402 and -403 omit the electrical system and turbine, and the lamp is replaced by a trihedral radar reflector in the nose.

Launch: The target is equipped with an internal tow reel located in the mid-section of the tow body. It is locked to an underwing launcher on the towing aircraft until released on command of the pilot. A friction braking system of low alloy steel automatically controls cable reel-out. Tow cable is of high-resistance steel, with stepped-diameter tow lines.

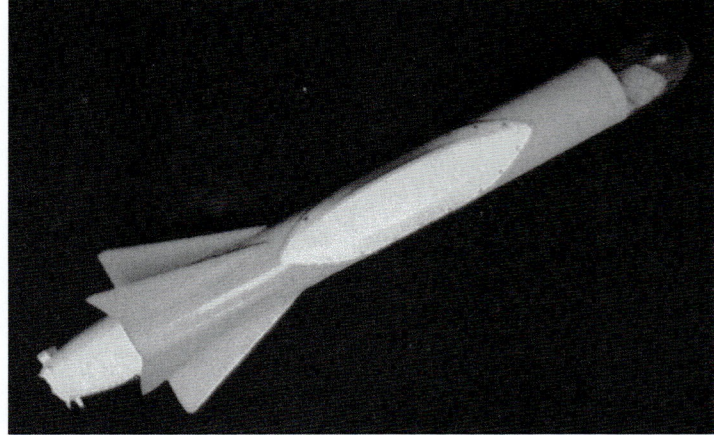

Aeromot AM 03089-401 0517905

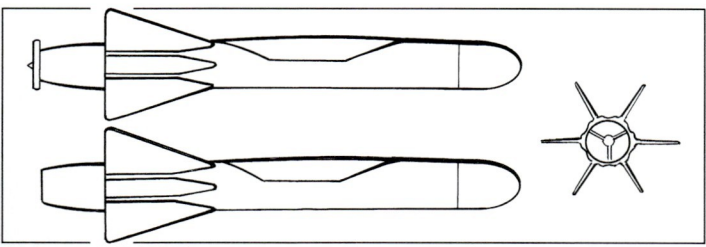

Aeromot AM 03089 towed target. Upper side view shows the -401 version with rear-mounted turbine *(IHS/John W Wood)* 0517906

Recovery: By reeling in tow cable.

Variants: AM 03089-401: Visual detection version, with powerful internal lamp in nose transparency and associated electrical system.

AM 03089-402: Radar detection version, with radar reflector instead of lamp. No electrical system.

AM 03089-403: Lighter weight version of -402, with shorter tow cable. No electrical system.

AM 03089-401, AM 03089-402, AM 03089-403

Dimensions, External	
Overall, length	2.252 m (7 ft 4¾ in)
Fuselage, height	0.229 m (9 in)
Tailplane, tailplane span	0.603 m (1 ft 11¾ in)
Performance	
Speed	
Never-exceed speed target on aircraft	350 kt (648 km/h; 402 mph)
Max operating speed target in tow	350 kt (648 km/h; 403 mph)

Status: In production and service.

Customers: Brazilian armed forces.

Contractor: Aeronaves e Motores SA (Aeromot) Porto Alegre.

Bulgaria

Aviotechnica Yastreb-2

Type: Recoverable aerial target.

Development: The Yastreb (Sparrowhawk) is a ground-launched aerial target system for anti-aircraft artillery, air-to-air missile and surface-to-air missile weapons training. After developing various earlier aerial drones (RUM-2MB, M-200 Siniger, UTRUM-2 and P-200) for the armies of former Warsaw Pact countries, Aviotechnica began development of the Yastreb in 1972. Series production began with the Yastreb-1 in 1981 and continued with the Yastreb-2 family from 1983. Until 1990, the majority of the 1,500 or so produced to date were for the former USSR.

Description: Airframe: Tapered mid-wing monoplane with circular-section fuselage, V tail and two rectangular ventral fins. NACA 64A series wing aerofoil section. Steel tube fuselage frame with duralumin and composites skin; wings and tail surfaces have metal spars. Ailerons located at approximately mid-span of wings; fixed tab on each outer wing trailing-edge; tail surfaces have inset ruddervators. No landing gear, but underfuselage reinforced to absorb landing impact.

Mission payloads: A variety of internal devices can be installed in the fuselage, including miss-distance indicators, flares and radar reflectors. There are external attachments for towed target bodies or banner targets. A single infra-red source can be mounted under the port wing.

Guidance and control: Three-channel autopilot controls engine throttle, ailerons and ruddervators stabilise air vehicle airspeed, height and course. Flight control can be by direct remote piloting from the ground, or automatic via the autopilot with provision for overriding by the ground controller. Telemetry downlink data are displayed in analogue and

Yastreb-2 launch *(Aviotechnica)* 1170760

digital form on a ground station TV monitor. Preprogrammed control provides third option in Yastreb-2MA.

System composition: System comprises air vehicle(s) (designation AL.06. Ya2); launcher (SCT.06); GCS (SP.1B); air vehicle support equipment (SP.1K); and air vehicle functional systems (FTs.1). Set-up time, by four ground crew, is one hour (30 minutes for system assembly, 25 minutes preflight preparation and five minutes prelaunch preparation).

Launch: By small booster rocket motor from truck-mounted zero-length launcher.

Recovery: Parachute recovery to landing on underfuselage airbags. Deployment of parachute is automatic in the event of a control link failure.

Power plant: One 16.4 kW (22 hp) Limbach L 275 E two-cylinder two-stroke engine; two-blade composites propeller. Single fuel tank in fuselage, capacity 8 litres (2.1 US gallons; 1.8 Imp gallons).

Variants: Yastreb-2: Initial production version; 7.2 kW (9.6 hp) DB-150SHE single-cylinder two-stroke engine.

Yastreb-2MA: Improved version, developed from 1990. Principal differences are greater range, third (preprogrammed) method of control and ability to carry larger payloads.

Yastreb-2MB: Similar to -2MA but with 12.3 kW (16.5 hp) DB-250 engine, increasing maximum speed to more than

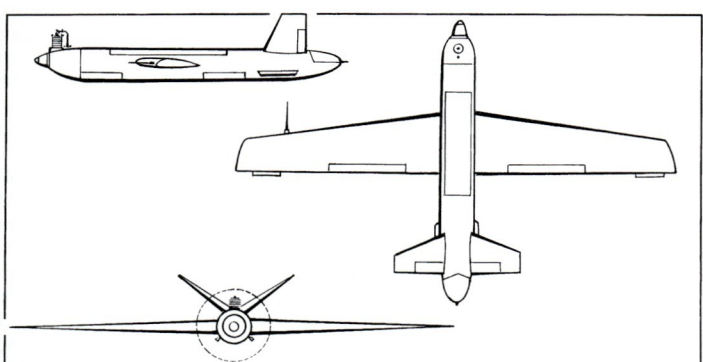

Yastreb-2 general arrangement; –2MA and –2MB are externally similar *(IHS/John W Wood)* 0517909

Yastreb-2 production line *(Aviotechnica)* 1170758

Yastreb-2 gunnery and missile target *(Aviotechnica)* 1170762

129 kt (240 km/h; 149 mph) and permitting wider range of possible missions. Modified navigation and control systems.

Yastreb-2S: Electronic warfare version; described separately.

Yastreb-2MV: Similar to -2MB but with 16.4 kW (22 hp) Limbach L 275 E engine, increasing maximum speed to more than 129 kt (240 km/h; 149 mph) and permitting wider range of possible missions. Modified navigation and control systems. *Description applies mainly to this version except where indicated.*

Yastreb-2, Yastreb-2MV

Dimensions, External
Overall
  length, Yastreb-2MV .... 2.68 m (8 ft 9½ in)
  height, Yastreb-2MV .... 0.533 m (1 ft 9 in)
Fuselage
  width, Yastreb-2MV .... 0.28 m (11 in)
Wings
  wing span, Yastreb-2MV .... 3.52 m (11 ft 6½ in)
Engines
  propeller diameter, Yastreb-2MV .... 0.61 m (2 ft 0 in)
Weights and Loadings
Weight
  Max launch weight, Yastreb-2MV .... 62.5 kg (137 lb)
Payload
  Max payload, Yastreb-2MV .... 15 kg (33 lb)
Performance
Altitude
  Operating altitude
    Yastreb-2 .... 500 m to 2,200 m (1,640 ft to 7,220 ft)
    Yastreb-2MV .... 500 m to 2,400 m (1,640 ft to 7,875 ft)
Speed
  Max level speed
    Yastreb-2 .... 102 kt (189 km/h; 117 mph)
    Yastreb-2MV .... 129 kt (239 km/h; 148 mph) (est)
g limits .... +6/-3
Radius of operation
  Yastreb-2 .... 8 n miles (14 km; 9 miles)
  Yastreb-2MV .... 24 n miles (44 km; 27 miles) (est)
  on a one-way mission,
    Yastreb-2MV .... 86 n miles (159 km; 99 miles) (est)
Endurance .... 1 hr
Power plant .... 1 × piston engine

Status: Still in service in Bulgaria.

Customers: Earlier targets (1969–78) produced for Bulgaria; former East Germany; India; Iraq; and former USSR. Yastreb series produced for Bulgaria; and (until 1990, when suspended due to lack of payment) former USSR. Total Yastreb family production exceeds 1,500.

Contractor: Aviotechnica S.p. Ltd
Plovdiv.

Canada

Bristol Aerospace Black Brant

Type: Ballistic target.

Development: Black Brant was originally developed as a sounding rocket, making its first flight, from Churchill Research Range, in June 1965. Since then, over 900 have been launched, for numerous operators, from ranges around the world and with 98.5 per cent reliability. It was acquired by Bristol from Boeing Canada in 1990.

Black Brant's current primary role is that of a supersonic target for theatre missile defence systems. The vehicle family has grown from the single-stage BB5 to the high-performance, four-stage BB12; principal current models are the BB5, BB9 Mod 1 and BB10 Mod 1.

Description: *Airframe:* Tubular body with ogival nosecone and cruciform tailfins. Vehicles are based on a building block approach, with the BB5 forming the basis of all other configurations (see drawing). Performance is increased by adding various military booster motors and/or Bristol's Nihka upper-stage motor. All core hardware components are common to all vehicles to simplify logistics and operational flexibility.

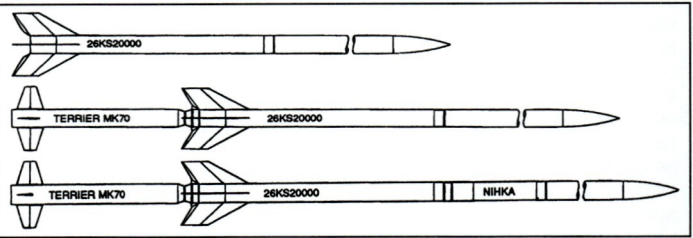

Top to bottom: Black Brant BB5, BB9 Mod 1 and BB10 Mod 1

Mission payloads: Can be developed by Bristol (more than 125 already designed and built) or supplied by user. Black Brants have accommodated payloads weighing from 68 to 680 kg (150 to 1,500 lb) and with principal diameters of up to 58.4 cm (23 in). The vehicle can fulfil a wide range of military target, tracking and training requirements; can re-enter as a single unit; or can deploy multiple targets before re-entry. Specific threats can be simulated by signature augmentation. Requirements imposed on the payload are primarily those of an aerodynamic or structural nature. Payload recovery systems can be supplied.

Guidance and control: Vehicle guidance and attitude control systems are available.

Launch: Ballistic trajectory can be tailored to specific mission requirements by appropriate selection of vehicle, payload weight and launch angle; with multistage vehicles, additional flexibility can be achieved by varying ignition time and attitude of the upper-stage motor. Black Brants are compatible with many of the fixed and mobile launchers currently in use, and have flown from 20 different sites worldwide including fully developed facilities such as White Sands Missile Range and the Wallops Island Flight Facility. Other sites have included Hawaii, Kwajalein, and the San Marco platform off the coast of Kenya. They can be launched on land or at sea; a fully qualified turnkey launch service is available.

Recovery: Target is expendable; payload recovery can be provided as a customer option. A fully qualified flight termination system is also available.

Black Brant launch

BB5, BB9 Mod 1, BB10 Mod 1

Dimensions, External
Overall
  length
    BB5 .... 5.64 m (18 ft 6 in)
    BB9 Mod 1 .... 9.56 m (31 ft 4½ in)
    BB10 Mod 1 .... 11.88 m (38 ft 11¾ in)
Fuselage, width .... 0.44 m (1 ft 5¼ in)
Performance
Range
  BB5 .... 302 n miles (559 km; 347 miles)
  BB9 Mod 1 .... 454 n miles (840 km; 522 miles)
  BB10 Mod 1 .... 1,134 n miles (2,100 km; 1,305 miles)

Status: In production and service.

Customers: No specific information.

Contractor: Bristol Aerospace Ltd
Winnipeg, Manitoba.

Bristol Aerospace Excalibur 1b

Type: Expendable ballistic target.

Development: Designed to simulate the threat of supersonic high- and low-diving missiles, and developed initially as a high-altitude supersonic target for Canadian Navy. Used for training with acquisition radars and high-speed missiles such as SM-2 and Sea Sparrow. Employed in 1999 and 2000, during exercises in New Mexico, to simulate theatre ballistic missile (TBM) threat. Can also be used as a tactical decoy.

Description: *Airframe:* Slim, spin-stabilised tubular steel body with dart-shaped front portion. Four tapered, low-aspect ratio molybdenum fins in X configuration at rear of dart; four cropped-delta fins on rocket body.

Mission payloads: Boeing RF SAS (which see) active electronic radar augmentation and variable amplitude microwave transmitter in dart body; RCS variable within C and X bands (4 to 12 GHz), to reflect variety of hostile ballistic missiles; minimum RCS 0.01 m^2 (0.11 sq ft), maximum RCS 1.0 m^2 (10.76 sq ft). Two 4.8 V Ni/Cd batteries for payload power supply.

Guidance and control: On completion of rocket motor burn, at range of approximately 2,135 m (7,000 ft), dart body separates from rocket and continues on ballistic trajectory. RCS augmentation antennas are integral with two of the four dart tailfins; one tailfin is a transmit antenna and one a receive antenna. They provide relatively constant RCS at aspect angles ranging from 18 to 80°.

The ballistic target's maximum altitude and impact distance are determined by the launch angle; final impact point depends upon meteorological conditions. The trajectory simulation software is used to predict the dart's nominal impact point, based on meteorological data, which comprise windspeed and direction, air temperature and atmospheric pressure, all versus altitude, up to the apogee altitude of the given target trajectory.

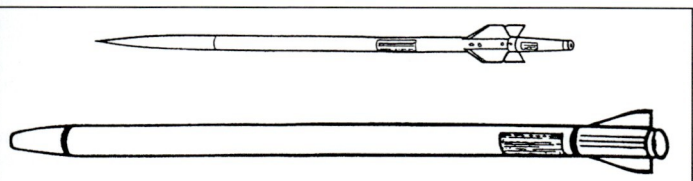

Excalibur dart body (top) and rocket body

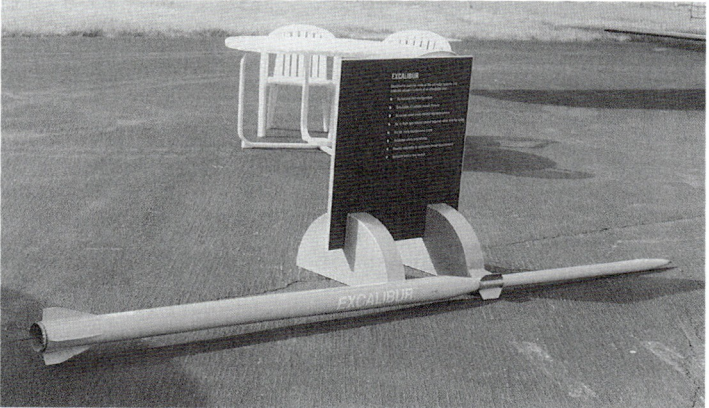

Dart body and rocket section of the Excalibur *(IHS/Kenneth Munson)*

Launch: Rocket launch from variable-angle (20 to 85°) ramp; elevation determines maximum altitude and impact distance of dart body. Launch computer monitors launch angle and azimuth, sending firing pulses when both are correct to separate dart from rocket motor. Three-person operation.

The launcher utilises a spiral rail that spin-stabilises the target during launch, which helps to reduce the size of the dart impact dispersion zone. For shipboard operation, launch rail elevation and heading angles are monitored by the control station computer, which is linked to the ship's gyroscope compass and to a gyroscope mounted on the launch rail. The computer monitors the launcher attitude and the ship's heading, and sends an ignition pulse to the rocket motor igniter when the launch elevation angle and magnetic heading are both within a predetermined launch range. The land launch station comprises two small consoles that control the payload and the rocket motor igniter. The payload console allows payload current and voltage monitoring, payload switching and battery charging.

Recovery: Non-recoverable.

Powerplant: One Space Data Corporation Viper IIIA rocket motor, with 26 kg (57.3 lb) of solid propellant and separable 1 W/1 A double-squib igniter; jettisoned after 2.5 s rocket motor burn.

Specifications

Dimensions
- Dart body: Length .... 1,440 mm (4 ft 8.7 in)
- Diameter .... 54 mm (2.125 in)
- Span over tailfins .... 157 mm (6.2 in)
- Rocket body: Length .... 2,438 mm (8 ft 0.0 in)
- Diameter .... 114 mm (4.5 in)
- Span over tailfins .... 241 mm (9.5 in)
- Launcher: Length (incl rail) .... 3.68 m (12 ft 1.0 in)
- Width .... 1.14 m (3 ft 9.0 in)
- Height (launch rail flat) .... 1.13 m (3 ft 8.6 in)

Weights
- Dart body .... 8.9 kg (19.6 lb)
- Rocket body .... 38 kg (83.8 lb)

Performance
- Max speed M5
- Burnout speed .... 3,187 kt (5,903 km/h; 3,668 mph)
- Apogee: 15° launch .... 3,350 m (11,000 ft)
- 35° launch .... 12,800 m (42,000 ft)
- 58° launch .... 47,000 m (154,200 ft)
- 80° launch .... 112,775 m (370,000 ft)
- Separation altitude: 35° launch .... 1,220 m (4,000 ft)
- 58° launch .... 1,800 m (5,900 ft)
- Range: 15° launch .... 18 n miles (33 km; 20.7 miles)
- 35° launch .... 30 n miles (55 km; 34 miles)
- 58° launch .... 63 n miles (116 km; 72 miles)
- Flight time: 15° launch .... 54 s
- 35° launch .... 1 min 45 s
- 58° launch .... 3 min 20 s

Status: In production and service.

Customers: Canadian Forces Maritime Command, as part of Canadian Patrol Frigate (CPF) programme.

Selected by US Army Aviation and Missile Command in March 2001 as live-fire training target for Patriot air defence missile system; between 1 and 32 Excaliburs per year to be supplied under five-year agreement; FMS sales to Germany and Netherlands.

Contractor: Bristol Aerospace Ltd
Winnipeg, Manitoba.

Excalibur launch

GD OTS TRAP

Type: Ballistic target.

Development: TRAP is an acronym for Target Radar Augmented Projectile. Development, at that time by SNC Technologies Inc (now GD OTS), began in 1983 to meet an air defence training need of the Canadian Navy for a high-speed, gun-launched target able to simulate the threat of missiles such as Harpoon and Exocet. Based on low-cost naval artillery rounds, initial development was of a 5 inch calibre projectile, followed by a 3 inch variant. In 1990, development began of a 76 mm version, type classification of which was achieved in November 1991 with support from the Canadian DND and Royal Danish Navy. Two 57 mm variants were classified in 1995 for use with the Bofors 57 mm Mk 2 naval gun, one based on the gun's Pre-Fragmented High Explosive (PFHE) round and one based on the High Capability Extended Range (HCER) projectile. Other calibres, including 155 mm, have also been developed.

TRAP has the same in-flight characteristics as a standard same-calibre projectile, and has been successfully tested to NATO safety and suitability requirements.

Description: *Airframe:* Typical artillery projectile shape.

Mission payloads: Typical payload is an RF reflecting lens allowing a radar 'energy incident' to reflect back to the emitter at angles of up to 45° from the projectile axis, providing an RCS equivalent to that of a physically larger target. Can also be provided with an IR signature capability.

Guidance and control: System incorporates trajectory and performance generator software for range safety.

Launch: Can be fired in single or multiple presentation modes from standard, unmodified naval guns and howitzers.

Recovery: Target is expendable.

Status: The TRAP target system is understood to still be in service.

Contractor: General Dynamics Ordnance and Tactical Systems — Canada Inc
Le Gardeur, Québec.

Meggitt Pop-up Helicopter

Type: Ground-based aerial target.

Development: The Pop-up Helicopter target system was designed during the 1980s by Boeing Canada to help air defence gunners learn and maintain the skills needed to defeat such pop-up threats as the contemporary Soviet 'Hind-D', 'Havoc' and 'Hokum' helicopters. Since then, the system has been redesigned to meet more current requirements, and provides a highly realistic simulation of attack helicopters for training and evaluation of low-level air defence weapons in day or night operations. Specifications apply to the current production version.

Description: *Airframe:* Target aircraft is a half-scale representation of the Russian Mil Mi-24 'Hind-D' helicopter, constructed from moulded glass fibre and operated by an electrically actuated scissor lift assembly mounted on a trailer.

Mission payloads: Target aircraft can be fitted with radar and other augmentation devices, and other payloads, to represent various attack helicopter systems.

Guidance and control: Up to six Pop-up Helicopter targets can be controlled by radio links from the GCS. By extending to some 12.2 m (40 ft) above ground level in less than 15 seconds, it offers a unique way of replicating the actual speeds of real threats. Although each target is at a fixed location, they can be relocated quickly; the rotating 3-D airframe also enabling profile changes to be presented. At the same time, the ability to pop up, disappear and seemingly reappear at a different location provides a challenging scenario to preclude gunners' complacency. Preprogrammed control is especially valuable in simulating 'swarming' helicopter threats.

Launch: Target aircraft and its three-stage, asymmetric scissor lift assembly are mounted on a four-wheel trailer. Rotation mount on the lift platform houses power supply lines and motor for rotation of the main rotor. A double-acting actuator allows the target to be rotated up to 90° to left or right.

Pop-up with scissor lift fully extended *(Meggitt)* 1290254

Power plant: Scissor lift is equipped with a 35.8 kW (48 hp), 2.4 litre Diesel engine with a 100 litre (26.4 US gallon; 22.0 Imp gallon) fuel supply, generating 6 kW of 208 V, three-phase AC electrical power.

Pop-up Helicopter

Dimensions, External	
Fuselage	
length	8.32 m (27 ft 3½ in)
height	1.45 m (4 ft 9 in)
width, max	1.12 m (3 ft 8 in)
Rotors, rotor diameter	6.60 m (21 ft 7¾ in)
Weights and Loadings	
Weight, Max launch weight	170 kg (374 lb)
Payload, Max payload	55 kg (121 lb)
Performance	
Climb	
Rate of climb, nominal	50.9 m/min (167 ft/min)

Status: The Pop-up Helicopter target system was in production and in use by the Canadian Forces' 4th Air Defence Regiment. Elsewhere in the world, 1 April 2008 saw UK contractor QinetiQ announce that the Pop-up Helicopter was one of two systems being made available (with immediate effect) as part of its Combined Aerial Target Service (CATS) contract with the UK's Ministry of Defence. Initiated during December 2006, CATS provides ground-based air defence training and aerial target services for

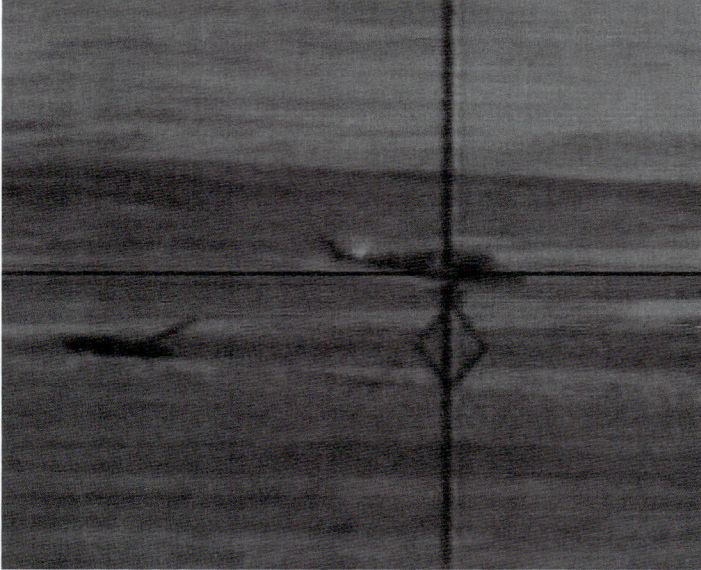
Pop-up viewed by approaching ADATS missile from 5 km range *(Meggitt)* 1290256

Scissor lift fully retracted. Aircraft can be rotated 90° to left or right *(Meggitt)* 1290255

the UK Royal Navy and air-to-air weapons training for the UK Royal Air Force and as of 2012, the Pop-Up Helicopter was understood to remain a part of this effort.

Contractor: Meggitt Training Systems Canada
Medicine Hat, Alberta.

Meggitt Vindicator II

Type: Recoverable aerial target.

Development: The Vindicator II was designed by Bristol Aerospace as a basic training target for the evaluation and testing of anti-aircraft 20 and 40 mm gunnery and low-speed missile systems, and for training personnel in the use of such weapon systems. It has been used as a target for Blowpipe, Javelin, ADATS, Vertically Launched Sea Sparrow, Vulcan (with PIVADS) and Avenger surface-to-air missiles, and for Bofors 40 mm, Oto Melara 76 mm, Skyguard 35 mm and FAADS 20 mm gun systems. It can be used for shipboard and land-based operations, and precise flight profiles can be repeated or varied to meet the exact requirements of individual weapons, sensors and/or tracking systems. In July 1999 the system was acquired from Bristol by Schreiner Target Services, which in turn became part of Meggitt in late 2004.

During AEGIS ship trials with the US Navy at Point Mugu, California, in June 2000, Vindicator was used to replicate a UAV on one mission and a helicopter on a later sortie, in the latter case utilising a Boeing Helicopter Radar Signature System (HRSS). Developed for the Canadian Navy, this is believed to have been a 'first' for aerial target operations. The HRSS provides a radar return for a full-size helicopter and can be configured to represent various rotary-wing threats.

Description: *Airframe:* Bullet-shaped fuselage; low-mounted, low-aspect ratio tapered wings with endplate fins; ventral landing skid. All-composites (glass fibre and epoxy) construction. All major components can be easily removed from the fuselage. Initial batch of 20 built under subcontract by Tasuma (UK) Ltd; subsequent manufacture in Canada.

Mission payloads: Modular construction combined with extensive use of quick-release fittings enables Vindicator to carry a wide range of payloads in nose, mid- and rear-fuselage bays. These can include active radar augmentation (three 63.5 mm; 2.5 in Luneberg lenses and one 190.5 mm; 7.5 in lens in nose bay, plus two 7.5 in lenses in fuselage); radar transponder; Boeing HRSS (see above); radar altimeter; flare pack (maximum 12 flares); visual augmentation (smoke generator and/or continuous or strobe light); Racal Doppler radar or Air Target Sweden acoustic miss-distance indicator; corner reflectors; autostabiliser with height lock; height- and range-tracking systems. The Vindicator incorporates programmable target signatures in E-Prom, and can store up to four different fixed- or rotary-wing signatures, various frequency bands, CW signal-type pulses, time domain data, and skin scintillation. The Vindicator is also capable of being equipped with an IR hot nose in order to increase the target's heat signature.

Guidance and control: Standard control station is the Meggitt Universal Target Control Station (UTCS), used when the Vindicator is equipped with autopilot avionics. The UTCS receives and records vehicle telemetry, can be used to fly waypoint navigation using GPS, controls payloads, and allows night operation. Vehicle position is shown on a moving map display. Radio-controlled air vehicles do not require the UTCS and instead are flown visually by the pilot, using a hand-held encoder and transmitter.

Fully programmable digital autopilot, which continuously monitors command datalink quality, has 16 analogue input channels with 14-bit resolution, eight servo outputs, and eight discrete bits that can be configured as inputs or outputs. Asynchronous serial communications are used for the command and telemetry links (autopilot has a universal asynchronous receiver/transmitter). Standard airframe incorporates GPS tracking; optional radar tracking allows for flights beyond visual range. Target aircraft's heading and altitude can be controlled manually or by an automated waypoint navigation system. Telemetry information is downlinked in real time.

Typical flight profiles include low-altitude attack and crossing patterns, low-altitude attack with a pop-up manoeuvre, continuously varying altitude, jinking, or any combination of these. With active radar augmentation fitted, Vindicator can simulate the signature of an attack

Vindicator II in flight *(Meggitt)*

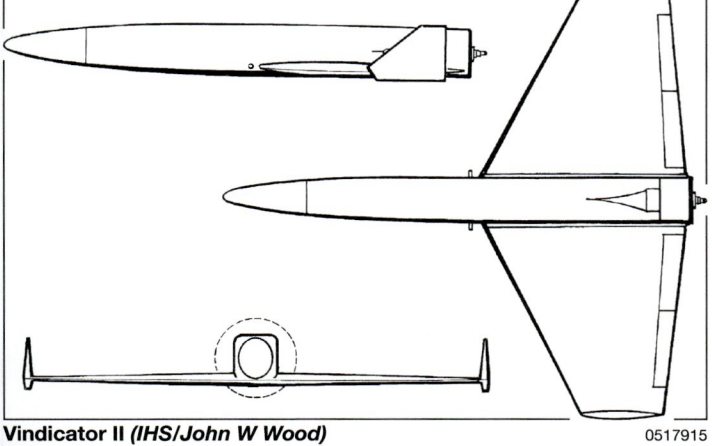
Vindicator II *(IHS/John W Wood)*

Vindicator II ready for launch *(Meggitt)*

Close up of Vindicator II in flight *(Meggitt)*

Vindicator II being launched *(Meggitt)*

helicopter, including blade/hub Doppler and blade flash.

The Vindicator is also capable of replicating a high diver attack threat.

Launch: From simple, low-maintenance pneumatic mobile launcher (4.57 m; 15 ft launch stroke).

Recovery: Conventional skid landing or parachute recovery can be commanded either manually or automatically. Fail-safe system automatically shuts down engine and deploys parachute in the event of autopilot or sensor failure, and initiates automatic recovery if the command link is lost. After recovery, Vindicator can be readied for a new flight within 30 minutes.

Power plant: One 26.1 kW (35 hp) UEL AR 731 rotary engine; two-blade pusher propeller. Fuel capacity 23 litres (6.1 US gallons; 5.0 Imp gallons).

Vindicator II

Dimensions, External	
Overall	
length	2.67 m (8 ft 9 in)
height	0.51 m (1 ft 8 in)
Wings, wing span	2.59 m (8 ft 6 in)
Areas	
Wings, Gross wing area	1.73 m² (18.6 sq ft)
Weights and Loadings	
Weight	
Weight empty	68 kg (149 lb)
Max launch weight, typical	77.1 kg (169 lb)
Payload	
Max payload, incl fuel	24.9 kg (54 lb)
Performance	
Altitude, Operating altitude	5 m to 4,570 m (16.5 ft to 15,000 ft)
Speed	
Launch speed	29.323 m/s (96 ft/s)
Max level speed	174 kt (322 km/h; 200 mph) at 1,000 m (3,280 ft)
g limits	+7
Endurance, at max speed	2 hr

Status: In production and service.

Customers: Canadian Forces, US Navy and other international customers.

Contractor: Meggitt Training Systems Canada
Medicine Hat, Alberta.

China

AVIC TL-8

Type: Recoverable jet-powered aerial target.

Development: This high-subsonic target was publicised by AVIC in late 2010, though no manufacturer was named, no example was exhibited and no photographs of an example had been seen by early 2012. It was referred to as a 'multifunctional' target to simulate third-generation fighters and cruise missiles, prompting suggestions that it could, itself, be adaptable to the latter role.

A published AVIC three-view drawing of the TL-8 reveals an airframe bearing a strong resemblance in size and weight to the Northrop Grumman BQM-74 target (which see). Maximum level speed is M0.85.

Description: ***Airframe:*** Short-span, tapered mid-wings; blunt-nosed, mainly cylindrical fuselage with underslung engine nacelle; inverted Y tail surfaces. No landing gear.

Mission payloads: As appropriate for simulation of aircraft and cruise missiles; no specific details revealed.

Guidance and control: Not revealed.

Launch: Ramp launch.

Recovery: Probable parachute recovery.

Power plant: One turbojet (rating not revealed) in ventral nacelle.

TL-8

Dimensions, External	
Overall, length	3.77 m (12 ft 4½ in)
Fuselage	
height	0.35 m (1 ft 1¾ in)
width	0.35 m (1 ft 1¾ in)
Wings, wing span	1.76 m (5 ft 9¼ in)
Weights and Loadings	
Weight, Max launch weight	250 kg (551 lb)
Performance	
g limits	+6
Endurance	40 min
Power plant	1 × turbojet

Status: Promotion began in 2010 and continued through 2011; current status uncertain.

Contractor: AVIC Aviation Techniques Co Ltd
Beijing.

CAF Chang Kong 1

Type: Recoverable jet-powered aerial target.

Development: Design studies leading to the Chang Kong 1 (Wide Blue Sky) began at Nanjing Aeronautical Institute (now NUAA) in the late 1960s, configuration of the CK1 prototype being finalised in late 1976 and clearly owing much to the contemporary Soviet Lavochkin La-17M (which see). The CK1A followed in 1977 and the CK1B in 1982, but the first definitive model was the CK1C, tested successfully in the third quarter of 1984. This was intended for use as a target for various types of missile, and is able to make high-manoeuvre flights at bank angles of up to 77°. Engines for the Chang Kong series are taken from retired J-6 (MiG-19) fighters. Current production is by Changzhou Aircraft Factory.

Description: ***Airframe:*** Constant-chord mid-wing monoplane with 2° anhedral. Fuselage built in three sections, those at front (housing radio control, telemetry and electrical equipment) and rear (autopilot and flares) being of aluminium alloy. Central portion, made from sheet steel, forms integral fuel tank. Rectangular tail surfaces, with low-set tailplane. Conventional ailerons, elevators and rudder. No landing gear; engine nacelle is reinforced to absorb landing impact, resulting in only minor damage which can be repaired easily before reuse.

Mission payloads: CK1C has miss-distance indicator (antenna at rear of fin-tip); infra-red augmentation pod at each wingtip; five corner reflectors for radar signature augmentation; and three smoke generators (on undersurface of each wing and on rear edge of engine nacelle fairing) to provide visual augmentation to aid tracking by ground-based optical aids. Smoke generators replaced by three flares in CK1E.

Guidance and control: *Flight control:* Four-channel autopilot (pitch, roll, yaw and altitude) in rear of fuselage stabilises aircraft and controls its flight in response to radio commands from ground; it incorporates gyroscope, directional gyro, three-axis rate gyro, programmer, electrical actuator, amplifier and converter. First 85 seconds of flight are programme controlled; mission then comes under preplanned radio command from ground controller.

Uplink: Onboard radar transponder for identification and tracking from ground. Airborne radio equipment comprises receiver/decoder which enables up to 24 command signals to be conveyed to autopilot and other onboard systems.

Downlink: A 52-channel telemetry system provides ground controller with continuous indication of altitude, speed, bank angle, engine rpm and temperature, and other functions.

Main electrical power for avionics and equipment provided by engine-driven generator, with alternator for AC power; emergency battery

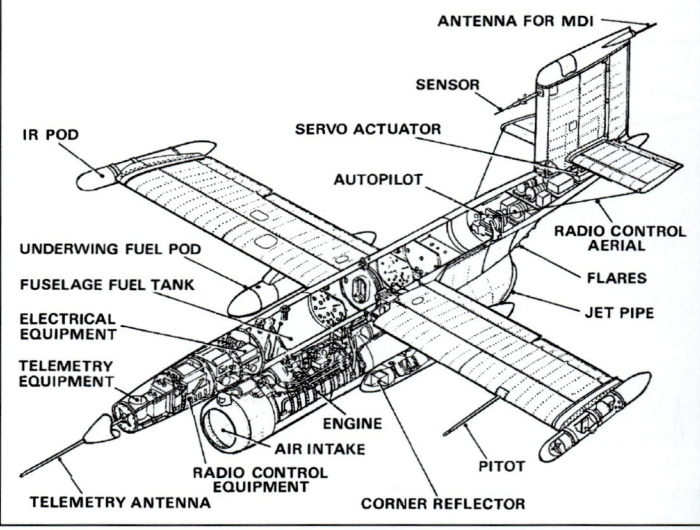

Cutaway drawing showing the main features of the aircraft

Latest known version of this Chinese target is the CK1G *(IHS/Robert Hewson)*

Chang Kong 1 launch 1151539

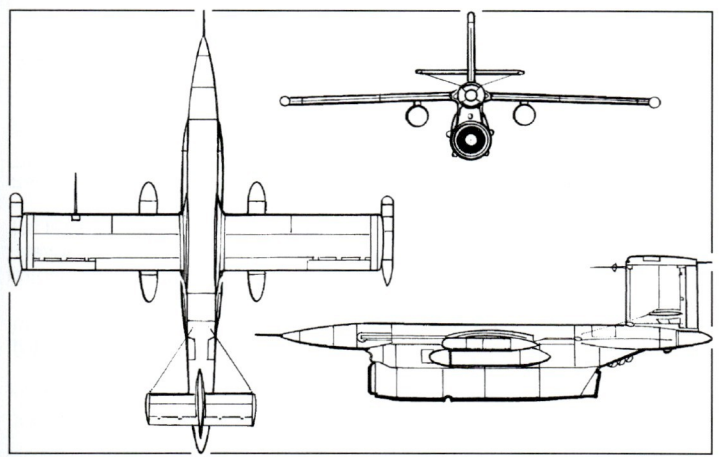

Changzhou Chang Kong 1C (IHS/John W Wood) 0517919

supplies DC power for continued safe flight in the event of main system or engine failure.

Launch: Launched from reusable trolley, upon which drone is mounted on three short guide-rails and attached by a single pin at base of engine nacelle. Complete ensemble accelerates along runway under engine power, connecting pin being ejected pneumatically when speed reaches 151 to 154 kt (280 to 285 km/h; 174 to 177 mph). Drone then lifts off trolley and enters climb-out phase, trolley decelerating and being brought to a halt under radio command by brake-chute and wheel brakes.

Recovery: Target is directed to a preselected landing site, where engine is shut down at a pre-determined speed and altitude and the drone glides to an unpowered landing.

Power plant: One modified Shenyang/Liming WP6 turbojet (afterburner deleted and its variable-area nozzle replaced by fixed-area nozzle with 8° downward deflection) in underslung nacelle beneath centre of fuselage. Engine thrust 25.5 kN (5,732 lb) in CK1C; see under Variants for ratings in other models. Thrust is varied by changing engine rpm, effected by throttle actuator under radio command from ground or via onboard autopilot command. Main fuel (see under Weights for details) in steel integral tank forming central portion of fuselage; auxiliary fuel in underwing pods.

Variants: *CK1:* Prototype. Engine thrust 21.08 kN (4,740 lb); maximum T-O weight 2,060 kg (4,541 lb).

CK1A: Early trials version, with special equipment in underwing pods. Engine thrust 22.06 kN (4,960 lb); maximum T-O weight 2,160 kg (4,762 lb).

CK1B: Low-altitude version. Underwing equipment pods replaced by non-jettisonable auxiliary fuel tanks. Engine thrust 23.04 kN (5,181 lb); maximum T-O weight 2,360 kg (5,203 lb).

CK1C: High-manoeuvrability production version of CK1B. *Detailed description applies to this model except where indicated.*

CK1E: Extra-low-level variant, with reduced wing span. Engine thrust 23.04 kN (5,181 lb); maximum T-O weight as for CK1C.

CK1G: Latest reported variant (2002), but no details had emerged by mid-2006.

CK1C, CK1E

Dimensions, External	
Overall	
length, CK1C	8.439 m (27 ft 8¼ in)
height, CK1C	2.955 m (9 ft 8¼ in)
Fuselage	
width, CK1C	0.55 m (1 ft 9¾ in)
Wings	
wing span	
CK1C	7.50 m (24 ft 7¼ in)
CK1E	6.88 m (22 ft 6¾ in)
Areas	
Wings	
Gross wing area, CK1C	8.55 m² (92.0 sq ft)
Weights and Loadings	
Weight	
Weight empty, CK1C	1,537 kg (3,388 lb)
Max T-O weight, CK1C	2,450 kg (5,401 lb)
Fuel weight	
Max fuel weight	
internal, CK1C	600 kg (1,322 lb)
external, CK1C	280 kg (617 lb)
Performance	
Altitude	
Operating altitude, CK1C	500 m to 16,500 m (1,640 ft to 54,140 ft)
Speed	
T-O speed, CK1C	154 kt (285 km/h; 177 mph) (est)
Max operating speed, CK1C	491 kt (909 km/h; 565 mph) (est)
Range, CK1C	485 n miles (898 km; 558 miles) (est)
Endurance	
at low and medium altitude, CK1C	60 min (est)
Power plant, CK1C	1 × turbojet

Status: In service.

Customers: Chinese People's Liberation Army.

Contractor: Changzhou Aircraft Factory
Changzhou, Jiangsu.

Xian ASN-7

Type: Recoverable aerial target.

Development: The ASN-7 is the highest powered of the Xian ASN Technical Group's current range of target aircraft. It was previously designated B-7.

Description: *Airframe:* Shoulder-wing monoplane with cylindrical fuselage and V tail. Construction is a mixture of glass fibre honeycomb, metal and wood. It is designed to survive up to 50 take-offs and landings; a ventral skid and a semicircular fairing under each wingtip minimise main airframe damage on landing.

Mission payloads: The ASN-7 can be equipped with two tracers or 12 decoys, or can trail a hard or soft tow body, for air defence weapons training.

Guidance and control: The air vehicle can be flown under remote radio control (30 switching and four proportional channels), or can be programmed via the dual-channel autopilot and a flight control and management computer. Direction-finding and distance measuring equipment keep operator informed of bearing and distance from the GCS. Aircraft has 12 analogue and 16 switching channels for telemetering bank and pitch angles, heading, airspeed, altitude, engine rpm and other data.

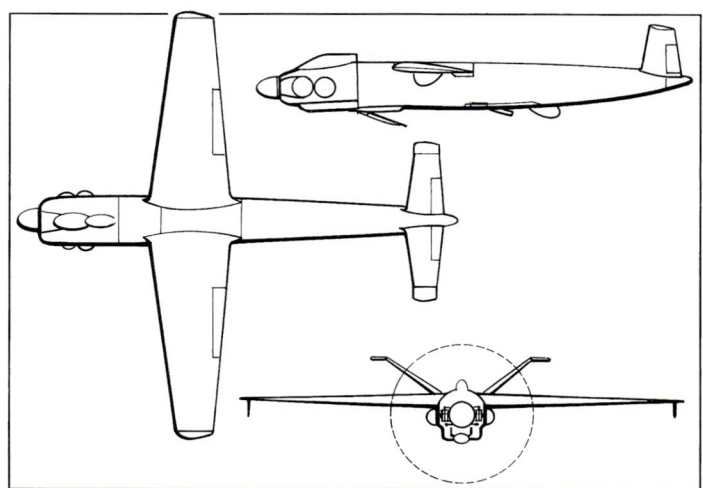

The V-tailed ASN-7 target drone (IHS/John W Wood) 0517922

ASN-7 on its zero-length launcher 0518180

Aerial targets > China > Xian ASN-7 – Xian ASN-9

ASN-7 on its zero-length launcher

Airborne management and control system has several fail-safe functions, including autonomous 'return home' mode if control signal is lost for more than a preset time.

System composition: Air vehicle(s); command vehicle and shelter-mounted GCS; portable remote-control console; launch trailer; DF system; four/five-person ground crew.

Launch: By booster rocket from trailer-mounted zero-length launcher.
Recovery: By parachute descent to skid landing.
Power plant: One 35 kW (47 hp) Xian HS-700 two-stroke flat-four; two-blade fixed-pitch wooden propeller.

ASN-7

Dimensions, External
Overall
length .... 2.65 m (8 ft 8¼ in)
height .... 0.55 m (1 ft 9¾ in)
Wings, wing span .... 2.68 m (8 ft 9½ in)
Areas
Wings, Gross wing area .... 0.831 m² (8.9 sq ft)
Weights and Loadings
Weight, Max launch weight .... 92 kg (202 lb) (est)
Payload, Max payload .... 10 kg (22 lb)
Performance
Climb
Rate of climb, max, at S/L .... 660 m/min (2,165 ft/min)
Altitude
Operating altitude, max .... 5,000 m (16,400 ft)
Speed, Max level speed .... 194 kt (359 km/h; 223 mph)
Radius of operation, radio control .... 21.6 n miles (40 km; 24 miles)
Endurance, max .... 1 hr (est)

Status: In production and service.
Customers: Chinese People's Liberation Army.
Contractor: Xian ASN Technical Group
Xian, Shaanxi.

Xian ASN-9

Type: Recoverable aerial target.
Development: The ASN-9 (originally B-9) was developed primarily for use at sea by the Chinese PLA Navy. It can be launched from the decks of several different types of warship and used as a target for ship-to-air missiles or to tow targets for ships' guns. The barometric altimeter introduced in 1995 enables the ASN-9 to perform sea-skimming level flights down to 7.5 m (25 ft) above the water.

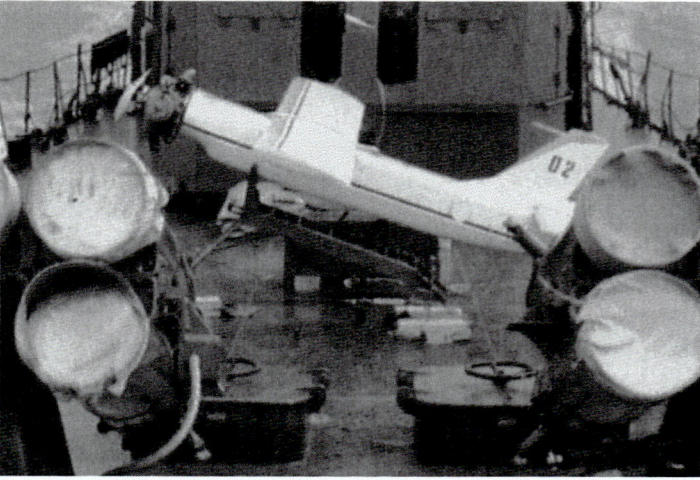

Shipboard zero-length launch of an ASN-9

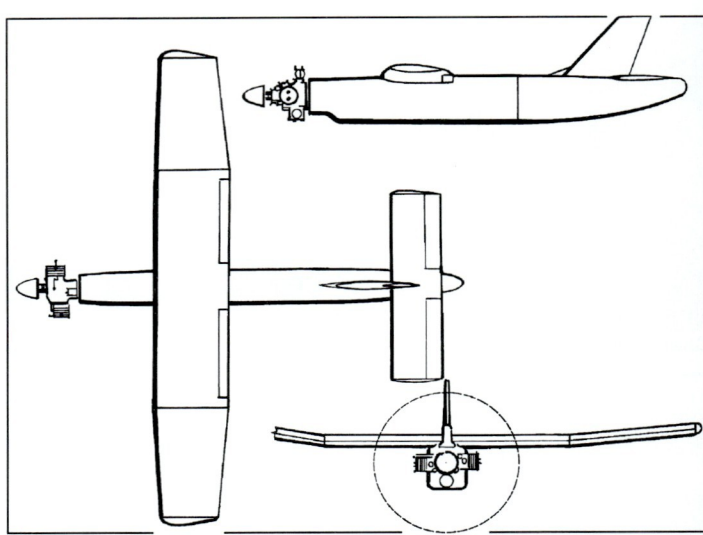

The Xian ASN-9 naval target drone (IHS/John W Wood)

Description: Airframe: High-wing monoplane with sweptback vertical tail. All compartments are watertight and protected against saltwater corrosion. No landing gear; can be fitted with shock-absorbing ventral skid for landings ashore.

Mission payloads: Can include one or two towed targets (each 2 m; 6.56 ft long and 500 mm; 19.7 inches in diameter, deployed sequentially); one or two infra-red sources; or a laser receiver.

Guidance and control: Radio-command uplink. Onboard equipment includes autopilot, vertical gyro, barometric altimeter, radio receiver and antenna.

System composition: Air vehicle(s); GCS; launch system; four/five-person crew.

Launch: By booster rocket from zero-length launcher. Launcher can be secured to ship's deck by steel ropes, adjusted to required launch angle by screws and dismantled easily after use.

Recovery: By parachute descent on to sea surface. Can be recovered on land by parachute descent or skid landing.

Power plant: One 16.4 kW (22 hp at 5,800–6,000 rpm) Xian HS-350 two-stroke flat-twin; two-blade fixed-pitch wooden propeller. See under Weights for fuel details.

ASN-9

Dimensions, External
Overall
length .... 2.52 m (8 ft 3¼ in)
height .... 0.72 m (2 ft 4¼ in)
Fuselage
height .... 0.28 m (11 in)
width .... 0.24 m (9½ in)
Wings, wing span .... 2.82 m (9 ft 3 in)
Tailplane, tailplane span .... 1.24 m (4 ft 0¾ in)
Weights and Loadings
Weight, Max launch weight .... 52 kg (114 lb) (est)
Fuel weight, Max fuel weight .... 8 kg (17.00 lb)
Performance
Altitude
Operating altitude, max .... 4,000 m (13,120 ft)
Speed, Max level speed .... 140 kt (259 km/h; 161 mph)
Radius of operation, radio control .... 8.1 n miles (15 km; 9 miles)
Endurance .... 45 min (est)

Status: In production and service.
Customers: Chinese PLA Navy.
Contractor: Xian ASN Technical Group
Xian, Shaanxi.

Xian ASN-12

Type: Recoverable aerial target.

Development: The ASN-12 can itself serve as an aerial target, fly with towed target bodies, or carry an aerial camera for the training of surveillance UAV crews.

Description: *Airframe:* Mid-wing monoplane with slight dihedral; three-axis flight control surfaces; circular-section fuselage. Twin landing skids under forward fuselage.

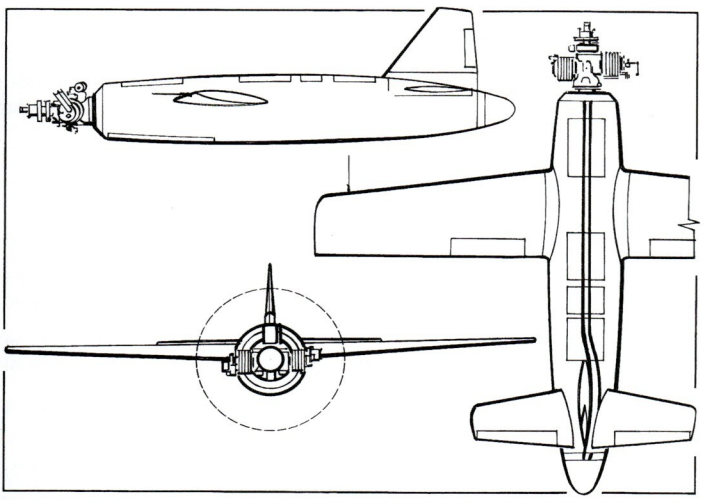

ASN-12 aerial target drone *(IHS/John W Wood)*

Xian ASN-12 on its zero-length launcher

Rocket-boosted ASN-12 launch

The ASN-12 viewed from the rear

Mission payloads: Two towed targets; two infra-red sources; or single aerial camera (frame size 70 × 70 mm; 2.75 × 2.75 in).

Guidance and control: Autopilot, radio receiver and telemetry transmitter.

System composition: Air vehicle(s); GCS; launcher; four/five-person crew.

Launch: By booster rocket from zero-length launcher on ground or mounted on a trailer.

Recovery: Normally by parachute landing; twin underfuselage skids for emergency landings.

Power plant: One 14.9 kW (20 hp) Xian HS-350 two-stroke flat-twin; two-blade fixed-pitch wooden propeller.

ASN-12

Dimensions, External	
Overall	
length	2.83 m (9 ft 3½ in)
height	0.90 m (2 ft 11½ in)
Wings, wing span	2.80 m (9 ft 2¼ in)
Weights and Loadings	
Weight, Max launch weight	75 kg (165 lb)
Payload, Max payload	8 kg (17.00 lb)
Performance	
Altitude	
Operating altitude, max	3,000 m (9,840 ft)
Speed, Max level speed	135 kt (250 km/h; 155 mph) at S/L
Radius of operation, radio control	21.6 n miles (40 km; 24 miles)
Endurance, max	1 hr 10 min

Status: In production and service.

Customers: Chinese People's Liberation Army.

Contractor: Xian ASN Technical Group
Xian, Shaanxi.

Xian ASN-106

Type: Recoverable jet-powered aerial target.

Development: When first shown publicly at Airshow China in November 2010, this high-subsonic target was presented in disparate fashion. On the ASN Group stand was a full-size machine, labelled as an ASN-106, with low-mounted wings, a dorsal engine intake and twin upper tailfins. However, accompanying literature — which reappeared at a Middle East air show a year later, though the machine itself did not — depicted the ASN-106 as being a high-wing design with an underslung engine nacelle and different tail surface arrangement. In the absence of clarification, the following description and specification refer to the latter configuration.

Description: *Airframe:* High-mounted swept wings; torpedo-shape fuselage, with ventral engine nacelle; inverted Y tail surfaces. No landing gear.

Mission payloads: As appropriate for the simulating the EO, IR and radar characteristics of advanced weapon systems.

Guidance and control: Not revealed.

Launch: Ramp launch.

Configuration of the ASN-106 target as shown in company literature in 2010 and 2011

Perhaps of an earlier developmental configuration, this model was also branded as an ASN-106 at the 2010 Zhuhai air show *(Robert Hewson)*

Recovery: Probable parachute recovery.
Power plant: One turbojet (rating not stated) in ventral nacelle.

ASN-106

Dimensions, External	
Overall, length	3.72 m (12 ft 2½ in)
Wings, wing span	2.16 m (7 ft 1 in)
Weights and Loadings	
Weight, Max launch weight	170 kg (374 lb)
Payload, Max payload	20 kg (44 lb)
Performance	
Altitude, Operating altitude	10 m to 10,000 m (40 ft to 32,800 ft)
Speed	
Cruising speed, max	324 kt (600 km/h; 373 mph)
g limits	
max	+7
sustained	+6
Radius of operation, radio control	81 n miles (150 km; 93 miles)
Endurance	1 hr (est)

Status:
Contractor: Xian ASN Technical Group
Xian, Shaanxi.

Yuhe BJ9906

Type: Recoverable aerial target/UAV.
Development: Prototype construction and first flight of the BJ9906 took place in June 1999. Its primary use is as a training target for AAA weapons. It can also be used for missile interception training, reconnaissance, electronic warfare and other missions.

A Yuhe BJ9906 being launched 0068849

Description: *Airframe:* Mid-wing monoplane with 6.4 per cent thickness/chord ratio and 6° dihedral. Circular-section fuselage; conventional tail surfaces. Mixed (metal and composites) construction. Coloured light on each wingtip. No landing gear.
Mission payloads: The BJ9906 can carry camera and aerial photography equipment, EW equipment, IR devices, radar trackers or other payloads. It can also trail one 'hard' or two 'soft' tow target bodies.
Guidance and control: Onboard autopilot and GPS; 30-channel radio command uplink and telemetry downlink (three-channel remote control at 233.3 MHz and 12-channel telemetry transmission at 798.7 MHz).
Launch: By rocket-assisted take-off from zero-length launcher.
Recovery: Normally by skid landing; parachute recovery optional.
Power plant: One 14.9 kW (20 hp) Yuhe YH-350 (Xian HS-350) two-cylinder two-stroke engine; two-blade wooden propeller. See under Weights for fuel details.

BJ9906

Dimensions, External	
Overall	
length	2.544 m (8 ft 4¼ in)
height	
excl landing skid	0.60 m (1 ft 11½ in)
incl landing skid	0.74 m (2 ft 5¼ in)
Fuselage	
height, max	0.28 m (11 in)
Wings, wing span	2.695 m (8 ft 10 in)
Areas	
Wings, Gross wing area	1.14 m² (12.3 sq ft)
Weights and Loadings	
Weight	
Weight empty	46 kg (101 lb)
Max launch weight	
standard	53 kg (116 lb)
incl auxiliary fuel tank	60 kg (132 lb)
Fuel weight, Max fuel weight	7 kg (15.00 lb)
Payload, Max payload	7 kg (15.00 lb)
Performance	
Climb	
Rate of climb, max, at S/L	600 m/min (1,968 ft/min)
Altitude, Operating altitude	10 m to 3,000 m (33 ft to 9,840 ft)
Speed, Max level speed	135 kt (250 km/h; 155 mph)
Radius of operation	
at 1,000 m (3,280 ft)	27 n miles (50 km; 31 miles)
Endurance	
standard fuel	50 min
with auxiliary tank	1 hr 30 min

Status: In service; production status uncertain. Has possibly replaced the earlier Yuhe BJ7104.
Customers: Chinese armed forces.
Contractor: Yuhe Group Company Ltd
Nanjing, Jiangsu.

France

Aviation Design Carine

Type: Jet-powered recoverable target.
Development: Aviation Design specialises in the design and manufacture of jet-powered unmanned vehicles. The Carine (*Cible Autonome Rapide Infrarouge pour tirs Nocturnes avec Evasives*: autonomous high-speed infra-red drone for evasive nocturnal firing) was developed in collaboration with the French Army for training with surface-to-air missiles in a nocturnal environment. Development began in November 1998; it was unveiled at the Paris Air Show in June 1999.

The Carine high-performance jet-powered target *(Kenneth Munson)* 0079579

The company also produces small turbojet engines in the 5 to 100 daN (112 to 2,248 lb st) power range, and undertakes prototyping on behalf of other customers such as Dassault Aviation and Sagem.
Description: *Airframe:* CAD design having low-mounted, sweptback wings with centrally mounted ailerons and pod at each tip; cigar-shaped fuselage. V tail surfaces (included angle approximately 120°) with arrowhead finlet at each tip; underslung jet engine. Fully moulded carbon fibre, glass fibre and epoxy construction.
Mission payloads: Infra-red signature provided throughout flight by a novel system (patent applied for) which uses hot air from the jet engine to heat the target, rendering the carriage of IR flares unnecessary.
Guidance and control: Carine is fitted with an autopilot (electronic computer and gyroscope). Flight track is followed by ground station by means of GPS. Speed, altitude and other flight parameters are relayed to GCS via telemetry downlink.
Launch: By catapult, or from prepared runway.

Carine on launch trolley *(Aviation Design)* 1137103

Recovery: Parachute recovery system, deployed by either autopilot or GCS; or runway landing.
Power plant: One 0.29 kN (66 lb st) JPX miniature turbojet.

Carine

Dimensions, External	
Overall, length	2.60 m (8 ft 6¼ in)
Wings, wing span	2.70 m (8 ft 10¼ in)
Weights and Loadings	
Weight	
Weight empty	30 kg (66 lb)
Performance	
T-O	
T-O run	80 m (263 ft)
Speed, Max level speed	216 kt (400 km/h; 249 mph)
g limits	+8
Radius of operation	
control, tested	19 n miles (35 km; 21 miles)
control, objective	27 n miles (50 km; 31 miles)
Endurance	45 min
Power plant	1 × turbojet

Status: In production and service. Company literature states that "hundreds" of its jet-powered unmanned vehicles are produced per year, with exports to more than 30 countries worldwide.

Customers: Include French Army and Navy; other customers not identified.

Contractor: Aviation Design
Milly La Foret.

EADS DC Eclipse

Type: Non-recoverable ballistic target.

Development: Designed by CAC Systèmes. (became part of EADS, and from 2010 EADS Cassidian division) to simulate supersonic ballistic aerial threats with high or low attack trajectories; as training tools for anti-missile missiles such as Aster and Patriot; for weapon systems qualification; and for radar calibration. Suitable also for civil meteorological and scientific applications.

Description: *Airframe:* Cylindrical body with ogival nosecone and cruciform tailfins.

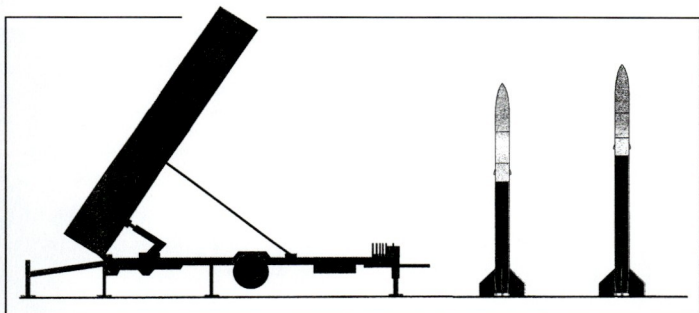

The mobile hydraulic launcher with Eclipse T1 (left) and T2 targets

Mission payloads: Military payloads can include infra-red and smoke flares; a 17.8 cm (seven inch) Luneberg lens; active radar transponder; Doppler or nuclear acoustic MDI; telemetry system; and a radio-controlled destruction system (RCDS).

Civil payloads can include various scientific and meteorological sensors; microgravity measurement equipment; and a multichannel numeric telemetry system.

Guidance and control: Each system includes ASTEC special trajectory software which accurately calculates ballistic trajectories and points of impact of flights with a range of more than 54 n miles (100 km; 62 miles). RCDS optional in military systems for range safety. Operated by a two-person crew.

Launch: From mobile hydraulic launcher on land or at sea. In the latter case, a vertical reference unit controls the launch sequence to maintain trajectory precision.

Recovery: Normally non-recoverable. Options include recovery and remotely controlled destruction.

Power plant: One SNPE solid-propellant booster rocket (rating not known).

Variants: *Eclipse T1:* Basic version.
Eclipse T2: Higher-performance version.

DC Eclipse

Dimensions	
Length overall: T1	3.55 m (11 ft 7.8 in)
T2	3.80 m (12 ft 5.6 in)
Body diameter (max): T1, T2	0.235 m (9.25 in)
Span over tailfins: T1	0.735 m (2 ft 4.9 in)
T2	0.835 m (2 ft 8.9 in)
Weights	
Max payload: T1	20 kg (44.1 lb)
T2	25 kg (55.1 lb)
Max launching weight: T1	153 kg (337 lb)
T2	227 kg (500 lb)
Performance	
Max speed in climb: T1	M2.25
T2	M4.30
Max speed in descent: T1	M1.20
T2	M2.86
Max altitude (87° launch): T1	15,000 m (49,200 ft)
T2	72,000 m (236,220 ft)
Max range: T1 (50° launch)	11 n miles (20 km; 12.5 miles)
T2 (65° launch)	>54 n miles (100 km; 62 miles)
Max impact accuracy: T1	400 m (1,312 ft)
T2	4,000 m (13,125 ft)

Status: In production and service.

Customers: France.

Contractor: Cassidian (an EADS company).
Vélizy-Villacoublay, France.

EADS DC Fox TS1

Type: Recoverable aerial target.

Development: Known until 1991 as Aspic, this aircraft appeared in 1988, development having been started by CAC Systèmes in 1986 in co-operation with ten other French aerospace companies to meet the requirements of the French Army and potential export customers. More than 900 have been built for civil and military applications. Used by French armed forces in presentations for 40 to 127 mm anti-aircraft gun training and Mistral, Sadral, AATCP, Simbad, Crotale, Stinger and Roland missiles.

Description: *Airframe:* High-wing monoplane with pod and boom fuselage, pusher engine and T tail. Constructed of duralumin, glass fibre, carbon fibre and styrofoam; wing and tail surfaces attached by single bolts.

Mission payloads: Various combinations of infra-red or smoke flares (up to 14), hot nose, Luneberg lenses, chaff dispensers, radio altimeter (for 5 m; 15 ft low-altitude or sea-skimming flights), acoustic or Doppler miss-distance indicator, radar or IFF transponder, corner reflector, towed targets, heading sensor or other equipment. Provision for some stores to be mounted under wings or fuselage.

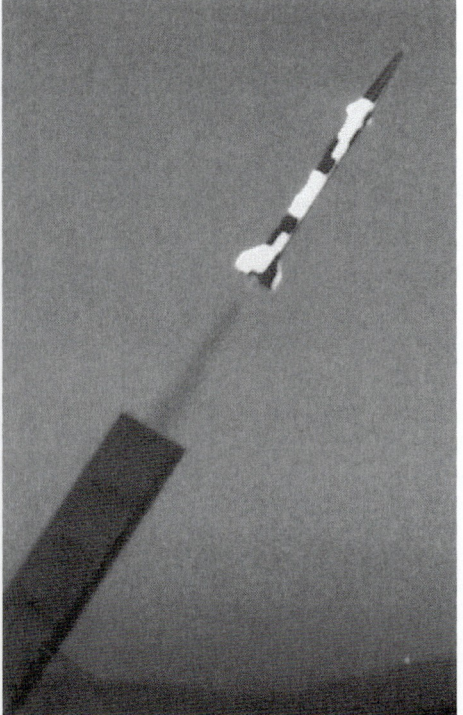

Eclipse being launched

Fox TS1 aerial target *(Paul Jackson)*

306 Aerial targets > France > **EADS DC Fox TS1 – EADS DC Fox TS3**

In-flight view of the Fox TS1 0518249

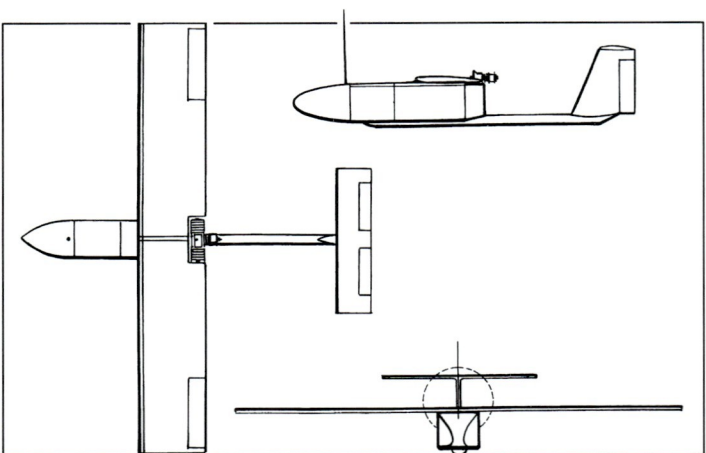
Fox TS1 target *(IHS/John W Wood)* 0517935

Status: Fox TS1 in production and service; at least 900 ordered, of which about 20 per cent for export. Mini Fox also in production as aerial target.
  Customers: French Army; four known export customers, including Romania and two in Middle East (one of which is reportedly Qatar). Romanian TS1s in service since 1996 at Capu Midia AFB.
Contractor: Cassidian (an EADS company).
  Vélizy-Villacoublay, France.

EADS DC Fox TS3

Type: Recoverable aerial target.
Development: Original development by CAC Systèmes (now part of EADS). In use for gunnery training and qualification for such missiles as Mistral, Roland, Crotale, Stinger or other SAMs, and for 20 to 127 mm anti-aircraft artillery.
Description: *Airframe:* Short-span high-wing monoplane with T tail; tractor engine configuration. Construction as for Fox TS1.
  Mission payloads: As described for Fox TS1.
  Guidance and control: As for Fox TS1, plus flight preprogramming management system.
  System composition: Normally four air vehicles per system; otherwise generally as for Fox TS1.

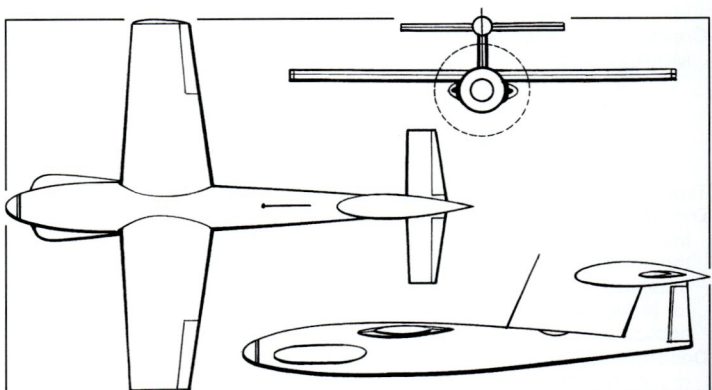
Fox TS3 target *(IHS/John W Wood)* 0518251

The Fox TS3 for the French Army *(Paul Jackson)* 0533542

  Guidance and control: Fixed or mobile GCS. Digital flight computer controls pitch, roll, height, safety and recovery; eight- to ten-channel GPS receiver and/or VHF telemetry system optional. Ni/Cd batteries for onboard power supply. Two or three Fox targets can be piloted simultaneously.
  System composition: Standard system comprises up to eight air vehicles plus mobile launching catapult, transportation container (optionally trailer-mounted), omnidirectional datalinks, piloting control station and setting mini-station. Options include pneumatic launcher, mapping and location system, telemetry storage and display system, mini radar station and MDI receiving station.
  Launch: Automatic day or night launch by mobile bungee catapult is standard; hydraulic launcher optional.
  Recovery: By commanded or automatically deployed Aerazur parachute system, or belly landing in any flat area. Recovered air vehicle can be readied for another flight within 30 minutes.
  Power plant: One 16.4 kW (22 hp) Limbach L 275E flat-twin engine; two-blade pusher propeller.
Variants: *Fox AT:* Close-range reconnaissance/surveillance version; described separately.
  Fox TS1: Original aerial target version. *Description and specifications apply to this model.*
  Fox TS3: Higher-speed target version (more than 250 kt; 463 km/h; 287 mph); described separately.
  Fox TX: Electronic warfare (EW) version of Fox AT.
  Mini Fox: Very low-cost target for small anti-aircraft arms. Has also been sold as small civil reconnaissance UAV.

Fox TS1

Dimensions, External	
Overall	
length	2.75 m (9 ft 0¼ in)
height	0.80 m (2 ft 7½ in)
Wings, wing span	3.60 m (11 ft 9¾ in)
Weights and Loadings	
Weight	
Weight empty	65 kg (143 lb)
Max launch weight	85 kg (187 lb)
Payload, Max payload	15 kg (33 lb)
Performance	
Altitude, Operating altitude	5 m to 3,000 m (15 ft to 9,840 ft)
Speed	
Max level speed	175 kt (324 km/h; 201 mph)
Cruising speed	126 kt (233 km/h; 145 mph)
Range, datalink, extendable	27 n miles (50 km; 31 miles)
Endurance, extendable	1 hr

Fox TS3 leaving its mobile pneumatic launcher 0518252

Launch: Hydraulic or pneumatic launcher.
Recovery: Parachute recovery.
Power plant: One 37.3 kW (50 hp) Limbach L 550E flat-four engine, driving a two-blade propeller.

Fox TS3

Dimensions, External	
Overall, length	3.15 m (10 ft 4 in)
Wings, wing span	2.60 m (8 ft 6¼ in)
Weights and Loadings	
Weight	
Weight empty	74 kg (163 lb)
Max launch weight	120 kg (264 lb)
Payload, Max payload	15 kg (33 lb) (est)
Performance	
Altitude, Operating altitude	5 m to 4,000 m (20 ft to 13,120 ft)
Speed	
Max level speed	252 kt (467 km/h; 290 mph)
Cruising speed	194 kt (359 km/h; 223 mph)
Range, datalink	27 n miles (50 km; 31 miles)
Endurance, extendable	50 min
Power plant	1 × piston engine

Status: In production and service.
Customers: French Army (20 delivered); Netherlands; two other (unidentified) export customers.
Contractor: Cassidian (an EADS company).
Vélizy-Villacoublay, France.

Secapem A3GT

Type: Recoverable aerial target.
Development: Development of the A3GT is thought to have started in 2008. The designation signifies 'Anti-Aircraft Artillery Gunnery Target'.
Description: *Airframe:* Low, unswept wings with slight dihedral; cylindrical fuselage with pointed nosecone; rectangular V tail. No landing gear. Composites construction.
Mission payloads: A Secapem Acoustic Shot Position Indicator MDI (ASPI) is integrated into the aerial target. This automatically detects and measures the passage of supersonic projectiles, transmitting the data to the scoring station in real time via radio downlink.
Guidance and control: Manual LOS radio control with GPS navigation. Aerodynamic control by ailerons and ruddervators.
Two types of scoring station are available: a full-size one or a smaller, hand-held one, each with a touch screen. Latter can be employed by a training officer to display score and target range data, or by flight controllers to display the air vehicle's flightpath and flight parameters. Scoring station provides a real-time display showing the position of each shot around the target. Maximum detection distance depends on the calibre of the projectile, but is generally from 7 to 9 m (23 to 30 ft). GPS data, including a precise target range indicator derived from differential calculations, are also displayed.
System composition: Air vehicle, launcher, GCS, MDI and scoring station.
Launch: By pneumatic catapult.
Recovery: Belly landing.
Power plant: One two-cylinder two-stroke engine (type and power rating not stated); two-blade pusher propeller.

A3GT

Dimensions, External	
Overall, length	2.57 m (8 ft 5¼ in)
Wings, wing span	3.13 m (10 ft 3¼ in)
Weights and Loadings	
Weight, Max launch weight	43 kg (94 lb)
Performance	
Altitude, Operating altitude	5 m to 4,875 m (10 ft to 16,000 ft)
Speed, Max level speed	108 kt (200 km/h; 124 mph)
Endurance, max	1 hr

Status: Developed and available. Being promoted in 2010.
Contractor: Secapem SA
Ozoir-la-Ferrière.

An A3GT aerial target ready for launch *(Secapem)* 1395245

Secapem MDI 154

Type: Aerial towed miss-distance indicator.
Development: The MDI 154 system is designed for surface-to-air and air-to-air gunnery training, and supersedes the company's earlier Type 200.
Description: *Airframe:* Consists of an MDI containing an optional sleeve or IR flares to aid visual acquisition, and tow cable. Sleeve is of elongated cone shape, made from coloured polyamide fabric and attached to rear of MDI.
Mission payloads: The MDI measures firing accuracy by detecting the proximity of supersonic projectiles to their target within spherical volumes centred on the target. For winch towing, it is attached to a towbar equipped with a swivel at the front and another at the back. Alternatively, it can be ejected and towed through a Secapem 1220 container installed under the host aircraft. When fitted with an optional IR flares dispenser, the MDI switches on each of the four flares by remote control.
Guidance and control: The MDI incorporates a GPS system allowing it to follow the position and speed of the target at the receiving station. Real-time feedback of firing accuracy is transmitted to a Multi Receiver System (MRS) on the ground or a Secapem display in the cockpit of the towing aircraft. The MRS can receive simultaneous data from four transmitters using four different radio frequencies.
Primary function of the MRS is to collect data from the Secapem transmitter. This information is immediately processed in relation to the MDI configuration ('global' or 'sector' display, ammunition properties, target/drone characteristics and so on), recorded on the hard disk, displayed on-screen and, if required, transferred to a network. Various configurations such as date and time of mission, MDI, calibre and target configuration can be prepared before firing and then run automatically, all received data being transferred directly to the initiating computer.
Launch: Sleeve by reel-out from towing aircraft.
Recovery: Sleeve by being jettisoned for recovery on ground.

Specifications

Dimensions	
MDI 154: Length	0.50 m (1 ft 7.7 in)
Diameter	0.08 m (3.15 in)
Sleeve: Length	3.50 to 5.00 m (11 ft 5.8 in to 16 ft 4.9 in)
Diameter	0.50 m (1 ft 7.7 in)
MRS: Length	290 mm (11.4 in)
Width	258 mm (10.2 in)
Depth	87 mm (3.4 in)
Weights	
MDI 154	3.0 kg (6.6 lb)
Sleeve	2.2 kg (4.9 lb)
MRS	5.5 kg (12.1 lb)
Performance	
MDI 154: Transmitter	MDI 189 family
Transmitting power	100 mW to 1 W (on request)
Autonomy	2 h
Operating temperature	–30/+60° C (-22/+140°F)
MRS: Hard disk	20 GB
RAM	512 Mbytes
Operating system	Windows XP/SP2

Sleeve-towing MDI 154 carried by a Reims F 406 *(Secapem)* 1279565

Optional sleeve attachment for MDI 154 *(Secapem)* 1279576

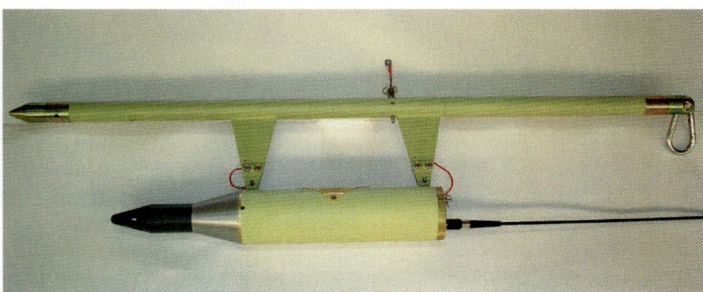

MDI 154 with towbar *(Secapem)* 1279567

Specifications

Interfaces	VGA, serial and USB ports
Operation on battery	3 h
Time to recharge	3 h
External power supply	20 to 28 V (3 A)
External adapter	110/240 V AC, 50/60 Hz

Status: In production and service.
 Customers: French armed forces.
Contractor: Secapem SA
 Ozoir-la-Ferrière.

Secapem PR53 and PR55

Type: Banner-type aerial tow target.
Development: Developed for air-to-air or ground-to-air gunnery training.
Description: *Airframe:* These banner targets are made up of a rectangular nylon square mesh attached to a glass fibre spreader bar ballasted with a stabilising weight and incorporating a radar reflector. A triangular bridle is attached to the spreader bar to provide the link with the tow cable through a swivel.
 Mission payloads: Radar reflector.
 Guidance and control: Generally similar to that of other Secapem targets.
 Launch: For carriage to the firing area, the PR53 is folded and stowed in the lower compartment of a Secapem Type 520 container (see entry for Taxan for details), with the tow cable in the upper compartment. Targets are qualified for use on Mirage, Alpha Jet, Falcon 20, F-5, F-16 and Hawk Series 60/100/200.
 The PR55 target is operated by a 'snatch' method, without a container, directly from the runway.
 Recovery: By cable release at end of mission.
Variants: *PR53:* Manufactured to French standards, with red-coloured mesh.
 PR55: Manufactured to UK standards, with white mesh, black paint edges and black paint patch.

Specifications

Dimensions	
Length: overall	12.40 m (40 ft 8.2 in)
rectangular portion	10.00 m (32 ft 9.7 in)
triangular portion	2.30 m (7 ft 6.6 in)
Width: rectangular portion	1.83 m (6 ft 0.0 in)
Spreader bar length	2.13 m (6 ft 11.9 in)
Weights	
Banner target	26 kg (57.3 lb)
Type 520 container and PR53, complete	210 kg (463 lb)
Performance	
Max towing speed	220 kt (407 km/h; 253 mph)
Altitude	<9,145 m (30,000 ft)

Status: As of end-2006, these targets were in production and service.
 Customers: French armed forces and those of other (unspecified) countries.

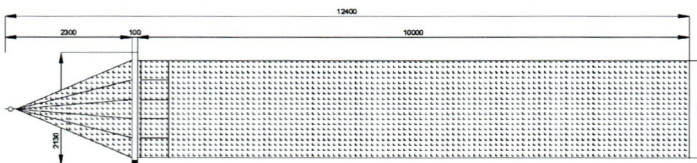
PR53/55 diagram *(Secapem)* 1146794

PR53 banner target on a French Air Force Falcon 20 *(Secapem)* 1279562

Contractor: Secapem SA
 Ozoir-la-Ferrière.

Secapem Taxan

Type: Aerial tow target.
Development: Developed under French Air Force procurement in 1980s as a high-speed, manoeuvrable and reusable towed target for air-to-air and ground-to-air gunnery training.
Description: *Airframe:* Target comprises an aerodynamic body or module containing a Secapem MAE 15 transmitter; connected by a swivel link to a foldable cruciform wing ('Tetraplan') to aid visual acquisition of the target in flight. The body is a steel cylindrical module which houses the MDI; the dart-shaped Tetraplan is constructed from four triangular, ladder-proof nylon mesh panels with four rectangular mesh panels to the rear. The mesh assembly is mounted on four bracing arms and a central mast, manufactured from glass fibre epoxy tubes. Taxan is designed to be carried in the Secapem Type 520 container, which can be installed on a standard NATO pylon or rack.
 Mission payloads: The MDI system consists of a sensor, a transmitter and an antenna installed in the target, and a set of receiving equipment on the ground or on board the host aircraft. An acoustic sensor in the target forebody is 'excited' by a shockwave proportional to the proximity of passing supersonic projectiles; these pressure signals are converted into electrical signals of proportional amplitude and transmitted by radio in real time to the MAE 15 receiver. Signals are then demodulated and

Taxan on a BAE Hawk *(Secapem)* 1146796

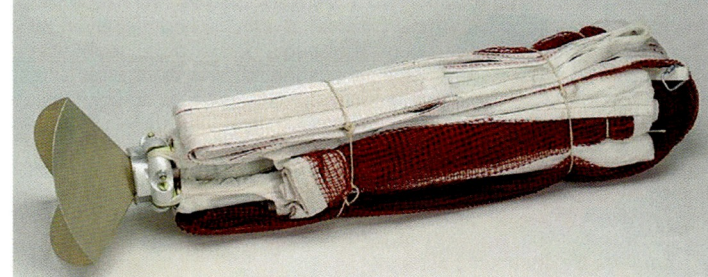
Taxan Tetraplan (stowed) *(Secapem)* 1146813

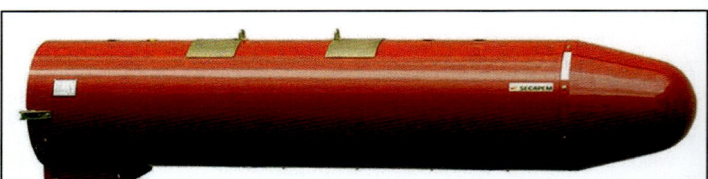
Type 520 container *(Secapem)* 1146807

Taxan tow target on a French Air Force Mirage 2000 *(Secapem)* 1279581

Taxan aerial tow target *(Secapem)* 1146812

Secapem Taxan tow target *(IHS/Kenneth Munson)* 0054180

processed by comparing their amplitude to reference thresholds determined during flight test. Two scores are displayed simultaneously, counting bullets having passed through two spheres of proximity centred on the sensor. The spheres' radii can be adjusted by the operator according to calibre. A trihedral shaped radar reflector is mounted to the rear of the aluminium block to assist radar acquisition. An optional scoring illustration with burst histogram is available. The MDI works in the 138.8 to 150.5 MHz waveband, has a transmitting power of more than 7 W over distances of up to 54 n miles (100 km; 62 miles), and can operate in temperatures from –40 to +50°C (–40 to +122°F). Power supply for the target and container is by 28 V DC battery.

Guidance and control: Can be flown in a variety of gunnery patterns including circular, butterfly, downward spiral, or combat dash (up to 500 kt; 926 km/h; 575 mph CAS and 5 g). The Secapem MAE 15 system can provide real-time scoring at a receiving station, which can be announced to the pilot immediately after each firing pass.

Launch: Stowed with its towing cable in a Type 520 light alloy pod container; can be towed by most fighter aircraft with external stores-carrying and ordnance release capability. The container, designed for installation under the stores pylon of a towing aircraft, is attached to the pylon by twin suspension lugs at standard NATO 356 mm (14 in) positions. The equipment is the same for all types of towing aircraft, and Taxan is qualified for use with F-4, F-5, F-16, Alpha Jet, Falcon 20, Jaguar, Mirage III/5/F1/2000, A-4 Skyhawk, Super Etendard and Hawk Series 60/100/200.

Inside the container is a sliding platform to accommodate the tow cable, a canvas set for cable coiling, a tripping hook, an ejector system and an electrical circuit. Circuit differences are met by a harness dedicated to the towing aircraft, and no aircraft modification is required.

Tow cable lengths are 500 m (1,640 ft) for air-to-air firing and 1,200 m (3,940 ft) for air defence ground gunnery firing. For air-to-air practice, Taxan can be deployed at altitudes between 305 and 10,670 m (1,000 and 35,000 ft); for air defence gunnery practice firing, the tow aircraft flies at about 365 m (1,200 ft).

Recovery: A tripping hook is used to hold the cable during the towing phase of the flight; at the end of a practice mission, the cable/target assembly is released by opening the hook, and can be recovered for refurbishment and reuse.

Specifications

Dimensions	
Target length: dart module	1.113 m (3 ft 7.8 in)
sleeve	5.870 m (19 ft 3.1 in)
overall	6.983 m (22 ft 10.9 in)
Dart module body diameter	130 mm (5.12 in)
Target max width	1.50 m (4 ft 11.1 in)
Type 520 container: Length	2.425 m (7 ft 11.5 in)
Diameter: excl hook	0.482 m (1 ft 7.0 in)
incl hook	0.600 m (1 ft 11.6 in)
Weights	
Taxan	45 kg (99.2 lb)
Taxan + 500 m cable	88 kg (194 lb)
Type 520 container and Taxan, complete	230 kg (507 lb)
Performance	
Max towing speed	500 kt, M0.95
Towing altitude	Up to 10,670 m (35,000 ft)
Max towing speed	
Alpha Jet	300 kt (556 km/h; 345 mph) IAS
Hawk	400 kt (741 km/h; 460 mph) IAS
F-5	450 kt (833 km/h; 518 mph) IAS
Mirage, F-4 and F-16	500 kt (926 km/h; 575 mph) IAS
g limits: Alpha Jet	+1.6
Hawk, F-5	+3
F-4, F-16, Mirage	+5

Status: In production and service.

Customers: French Air Force, Army and Navy; exported to some 20 countries.

Contractor: Secapem SA
Ozoir-la-Ferrière.

Germany

EADS DS Do-DT25, Do-DT35, Do-DT45 and Do-DT55

Type: Jet-powered recoverable aerial targets.

Development: This family of targets is the outcome of feasibility and design studies undertaken by the Military Air Systems business unit of EADS Defence & Security, now EADS Cassidian, and formerly EADS Dornier. The studies were made in collaboration with the University of Stuttgart. They have been produced as a low-cost solution for the need for high-speed threat simulation against existing and future air defence systems, with special emphasis on the characteristics required for target detection, acquisition, evaluation, tracking and interception.

Description: *Airframe*: DT25 and DT35 are mid-wing monoplanes with bulbous, cigar-shaped fuselage and T-tail; DT55 has slim, cylindrical fuselage with mid-mounted canards and rear-mounted delta wings bearing arrowhead endplate fins. No landing gear.

Mission payloads: In DT25, can include smoke, IR or radar augmentors, ECM, IRCM, IFF and SETA-3 MDI; in DT35, can include smoke, IR augmentor or radar repeater and SETA-3 MDI.

Guidance and control: All use the same fixed or mobile GCS for mission planning, command and control of air vehicle and its payloads, and use UHF datalinks and telemetry similar to those employed for the EADS Do-SK6 tow target. They can be operated either manually or autonomously via autopilot and GPS waypoint navigation. GCS can control up to six targets simultaneously.

Launch: Under own power by pneumatic catapult.

Recovery: Parachute recovery.

Power plant: DT25 R and IR: Two 0.16 kN (36 lb st) (optionally 0.24 kN; 54 lb st) turbojets, mounted internally. Fuel capacity 40 litres (10.6 US gallons; 8.8 Imp gallons).

Do-DT25s on the launch ramp with ventrally mounted Do-DT55s *(Robonic)* 1151534

Largest of current family, the twin-engined Do-DT25 *(IHS/Patrick Allen)* 1066240

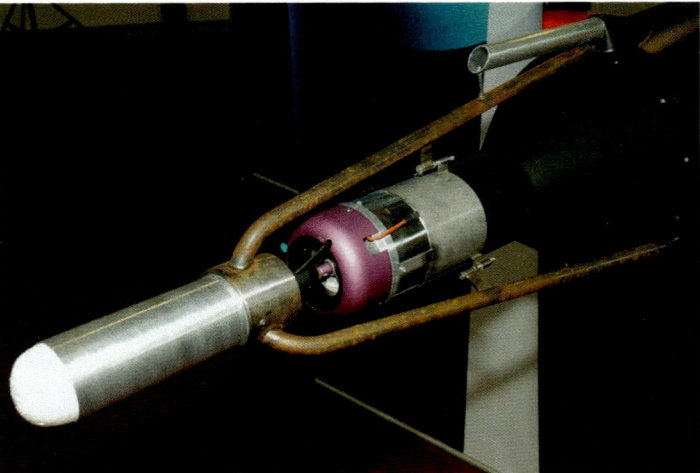

Do-DT25 nose-mounted IR augmentor *(IHS/Patrick Allen)* 1149501

The Do-DT35 *(IHS/Patrick Allen)* 1066241

DT35: Two 0.16 kN (36 lb st) turbojets, pod-mounted externally on rear fuselage.

DT35-200: Two external pod-mounted 0.22 kN (49.5 lb st) turbojets. Fuel capacity 21 litres (5.5 US gallons; 4.6 Imp gallons).

DT55: One 0.24 kN (54 lb st) turbojet, mounted internally.

Variants: **Do-DT25**: Twin-engined, low subsonic target for basic surface-to-air and air-to-air missile training. Subvariants are **Do-T25 R** (radar) and **Do-T25 IR** (infra-red).

Do-DT35/DT-45: Smaller, mid-subsonic target for advanced air defence training. Subvariant **Do-DT35-200** is a sea-skimming (down to 23 m; 75 ft) and faster version. Later this version was re-designated **Do-DT45**.

Do-DT55: Very small, single-jet high-subsonic target; development based on German Air Force requirement for an anti-radiation missile simulator (ARMS). Target is mounted under the fuselage of the Do-DT25 and air-launched at a stand-off distance by radio control on approach to the missile installation. Debut demonstration 8 October 2005 at NAMFI range.

Status: A small production batch of Do-DT25s (15 or more) was ordered in early 2003 for German Air Force air defence training. These were used in live firing trials with MANPADS shoulder-fired weapons in Exercise 'Baltic Fire' at the Polish range near Uska on the Baltic coast in November 2003. Qualification trials of the Do-DT25 and Do-DT35 were completed in 2004, with both types being cleared for service use in July of that year. Demonstrations of the Do-DT55 to the German Air Force and BWB at the NAMFI range on Crete took place in October 2004, resulting in qualification of this target also as an anti-radar missile simulator. Further target simulation demonstrations at NAMFI took place in 2006.

Via EADS North America Defense, the Do-DT35 was demonstrated to the US Army at the McGregor Range in New Mexico in June 2006. Announcing this in mid-August, the company reported that the DT35 was then "in full production for NATO and allied military forces, with an output capacity of up to 200 units per year". The 500th target was produced by June 2008. A Do-DT-35/45 system comprising an undisclosed number of targets, a Robonics launcher, two GCS, spares and support, was delivered to the US Army in October 2008.

Contractor: Cassidian (an EADS company).
Friedrichshafen, Germany.

Do-DT25, Do-DT35, Do-DT55, Do-DT25 R, Do-DT25 IR, Do-DT45

Dimensions, External
- Overall
  - length
    - Do-DT25 .... 3.00 m (9 ft 10 in)
    - Do-DT35 .... 1.80 m (5 ft 10¾ in)
    - Do-DT45 .... 2.15 m (7 ft 0¾ in)
    - Do-DT55 .... 1.70 m (5 ft 7 in)
- Wings
  - wing span
    - Do-DT25 .... 2.60 m (8 ft 6¼ in)
    - Do-DT35 .... 1.50 m (4 ft 11 in)
    - Do-DT45 .... 1.80 m (5 ft 10¾ in)
    - Do-DT55 .... 0.60 m (1 ft 11½ in)

Weights and Loadings
- Weight
  - Max launch weight
    - Do-DT25 .... 125 kg (275 lb)
    - Do-DT35 .... 55 kg (121 lb)
    - Do-DT45 .... 70.0 kg (154 lb)
    - Do-DT55 .... 20.0 kg (44 lb)
- Payload
  - Max payload
    - Do-DT25 .... 16.0 kg (35 lb)
    - Do-DT35 .... 6.0 kg (13.00 lb)
    - Do-DT45 .... 8.0 kg (17.00 lb)

Performance
- Altitude, Service ceiling .... 7,620 m (25,000 ft)
- Speed
  - Max level speed
    - normal engines, Do-DT25 .... 243 kt (450 km/h; 280 mph)
    - optional engines, Do-DT25 .... 291 kt (539 km/h; 335 mph)
    - Do-DT35 .... 360 kt (667 km/h; 414 mph)
    - Do-DT45 .... 427 kt (791 km/h; 491 mph)
    - Do-DT55 .... 486 kt (900 km/h; 559 mph)
- Range
  - datalink .... 54 n miles (100 km; 62 miles)
  - at max speed, Do-DT55 .... 30 n miles (55 km; 34 miles)
- Endurance
  - 50% of max speed, Do-DT25 IR .... 1 hr 0 min
  - .... 1 hr 20 min
  - 50% of max speed, Do-DT35 .... 30 min
  - 30% of max speedv, Do-DT45 .... 40 min
  - at econ cruising speed, Do-DT25 IR .... 1 hr 40 min
  - at econ cruising speed, Do-DT25 R .... 2 hr 0 min
  - at econ cruising speed, Do-DT35, Do-DT45 .... 1 hr 0 min
- Power plant
  - Do-DT25 R, Do-DT25 IR, Do-DT35 .... 2 × turbojet
  - Do-DT55 .... 1 × turbojet

EADS DS Do-SK6

Type: Towed target.

Development: The Do-SK6 family of towed targets is designed for use in surface-to-air and air-to-air gunnery and short-range missile firing.

Description: *Airframe*: Rigid, modularly designed cylindrical body of load-bearing GFRP, subdivided into front, centre and rear sections; hemispherical nosecap; four tailfins in X configuration. Construction based on German armed forces specification calling for a speed

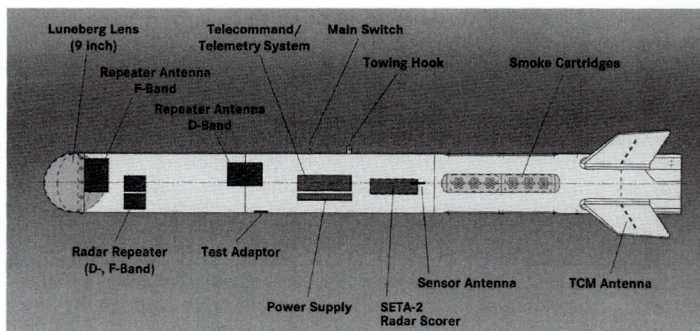

Do-SK6 R configured for Roland missile firings 0518185

Do-SK6 targets and winch pods under the wings of a Learjet 0518186

requirement of 350 kt (648 km/h; 403 mph) or above with a tow cable length of up to 6,000 m (19,685 ft).

Mission payloads: Vary according to customer requirements. Include corner or laser reflectors; Luneberg lens; radar echo repeaters; banner augmentation; up to 36 smoke cartridges; IR tracking or decoy flares; radar or acoustic scorer; parachute recovery system; altitude measuring unit; RF decoy payload; microprocessor-controlled telemetry/telecommand unit. SETA-2 or SETA-3 radar MDI.

Launch: Do-SK6 series is certified for the Learjet 35/36 with the Marquardt MTR 101 winch and tow reel system, and for the Falcon 20 with the Meggitt RM 30A system.

Recovery: See Launch.

Variants: ***Do-SK6:*** Basic augmentation target for surface-to-air and air-to-air applications.

Do-SK6 A: Acoustic MDI target for surface-to-air applications.

Do-SK6 B: Augmentation target for standard air-to-air gunnery; introduced 1995.

Do-SK6 IR: Flare/plume target; surface-to-air applications.

Do-SK6 IRCM: IR signature and decoy target for IR-guided surface-to-air and air-to-air short-range missile firing. In service since 1994.

Do-SK6 LL: Low-level target to simulate sea-skimming anti-ship missiles. Equipped with barometric altimeter which transmits readings to control station onboard the tow aircraft; can be flown down to minimum altitude between 15 and 30 m (50 and 100 ft). Tested successfully in 1995.

Do-SK6 R: For use with RF-guided short-range surface-to-air missile firings. Introduced 1991; more than 400 Roland missiles have been fired against it.

Do-SK6 BSH: Radar scorer target; surface-to-air and air-to-air use with SETA-3 pulse Doppler radar MDI. Rear-section IR flares replaced by 6 × 0.5 m (19.68 × 1.64 ft) sisal string banner towed 10 m (33 ft) behind hard target body. Standard target in German armed forces (more than 600 ordered) for ground-to-air and ship-to-air gunnery.

Do-SK6, Do-SK6 B, Do-SK6 IRCM, Do-SK6 LL, Do-SK6 R, Do-SK6 BSH, Do-SK6 A, Do-SK6 IR

Dimensions, External	
Overall length	
Do-SK6	2.63 m (8 ft 7½ in)
Do-SK6 B	2.81 m (9 ft 2¾ in)
Do-SK6 IRCM	2.82 m (9 ft 3 in)
Do-SK6 LL, Do-SK6 R, Do-SK6 BSH	2.66 m (8 ft 8¾ in)
Fuselage, width	0.24 m (9½ in)
Tailplane tailplane span	
A3GT Do-SK6 A, Do-SK6 B, Do-SK6 BSH, Do-SK6 IR, Do-SK6 IRCM, Do-SK6 R,	
Do-SK6 LL	0.63 m (2 ft 0¾ in)
Do-SK6	0.62 m (2 ft 0½ in)
Weights and Loadings	
Weight	
Weight empty	
Do-SK6 B	37 kg (81 lb)
Do-SK6 IRCM	42 kg (92 lb)
Do-SK6 LL	27 kg (59 lb)
Do-SK6 R	40 kg (88 lb)
Do-SK6 BSH	29 kg (63 lb)
Payload, Max payload	50 kg (110 lb) (est)

Status: In production and service.

Customers: Used in France (CEL and CEM); Germany; Singapore; Sweden.

Contractor: Cassidian (an EADS company).
Friedrichshafen, Germany.

Decoy flares in the forward section of a Do-SK6 *(IHS/Patrick Allen)*

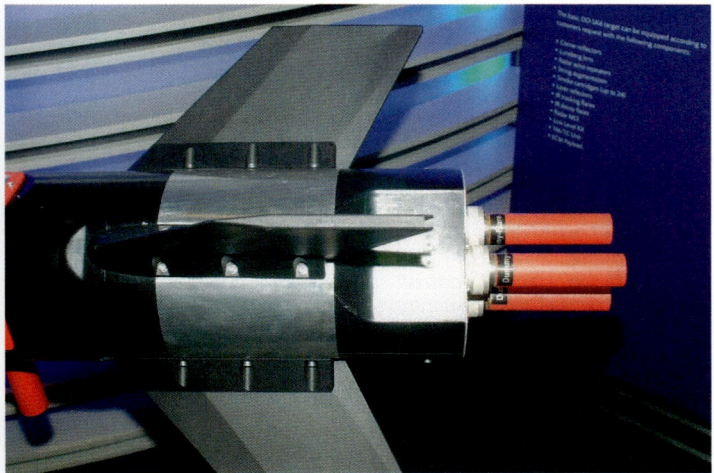
Do-SK6 tail section, showing IR tracking flares *(IHS/Patrick Allen)*

A typical member of the Do-SK6 family *(IHS/Kenneth Munson)*

EADS DS Do-SK10

Type: Tow target.

Development: The Do-SK10 is a sleeve type, dart-shaped air-to-air gunnery target for high-speed, high-*g* manoeuvres in the training and qualification of fighter aircraft crews.

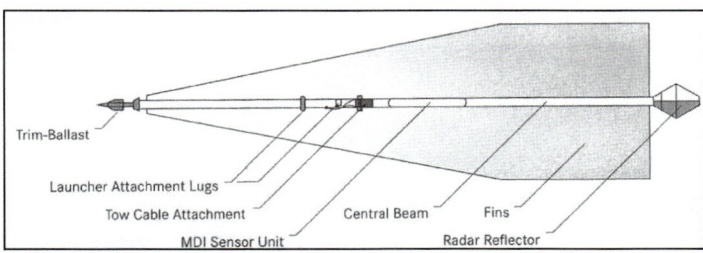
Main components of the Do-SK10

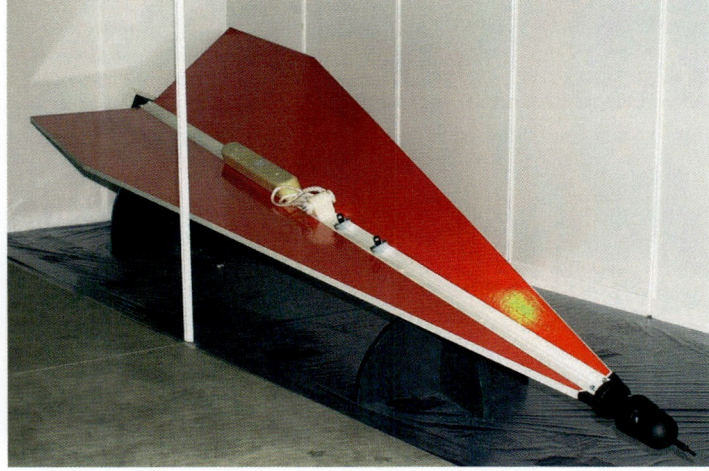
Do-SK10 air-to-air gunnery target *(IHS/Patrick Allen)*

Aerial targets > Germany – Greece > EADS DS Do-SK10 – EADS 3 Sigma Alkyon

Do-SK10 under the wing of a German Air Force F-4 Phantom   0518188

Description: *Airframe:* Central beam, rear-mounted radar reflector, nose-mounted trim ballast, towing hook, and three fins of sandwich construction with a core of paper honeycomb.

Mission payloads: Can be used with SETA-1 pulse Doppler radar scorer, which provides reliable and accurate scoring results in real time to the fighter aircraft, even at rapidly changing speeds and altitudes. The basic SETA-1 has one scoring sphere with a radius of up to 8 m (26.25 ft); further configurations provide scoring sphere radii of up to 16 m (52.5 ft), with two or more hit detection zones or direct distance measuring features. Maximum resolution is 10,000 rds/min; hits are indicated at the receiving station in real time.

Launch: The Do-SK10 can be fitted to the A/A 37-U15 winch system, and is certified for towing by the Tornado, A-4, F-4, F-86, F-100 and F-104.

Do-SK10

Dimensions, External	
Overall	
length	5.75 m (18 ft 10½ in)
width	1.45 m (4 ft 9 in)
Performance	
Altitude, Service ceiling	9,145 m (30,000 ft)

Status: In service since 1987; more than 3,000 produced.
  Customers: Germany and other unspecified countries.
Contractor: Cassidian (an EADS company).
  Friedrichshafen, Germany.

EADS DS SETA-3 S1

Type: Towed target.
Development: The SETA-3 S1 is used for low-level ground-to-air and ship-to-air gunnery training.
Description: *Airframe:* Open-ended conical sleeve.
Mission payloads: SETA-3 real-time pulse Doppler MDI with built-in telemetry/telecommand capability.
Launch: Adapted to MTR 101 launch pod on Learjet 35 and 36, and to Meggitt RM-24 winch system on Pilatus PC-9, and certified on these aircraft. Launch pod can also be mounted on any aircraft with a suitable stores pylon containing a NATO-standard release unit.
Recovery: Target can be dropped via TLM-controlled mechanism after use.

Specifications

Dimensions	
Sleeve: Length	5.79 m (19 ft 0.0 in)
Diameter: min	0.36 m (1 ft 2.2 in)
max	0.50 m (1 ft 7.7 in)
Typical towline length	1,500 – 3,000 m (4,920 – 9,840 ft)
Weights	
Weight incl MDI	6 kg (13.2 lb)
Performance	
Max towing speed	350 kt (648 km/h; 403 mph) IAS
Reel-out velocity	175 kt (324 km/h; 201 mph) IAS

Status: In production and service.
  Customers: German armed forces.
Contractor: Cassidian (an EADS company).
  Friedrichshafen, Germany.

Greece

EADS 3 Sigma Alkyon

Type: Recoverable aerial target.
Development: The Alkyon aerial target was designed and developed to meet training demands imposed mainly by AAA and MANPADS. Originally flying at only visual ranges, its manoeuvrability, endurance and low-speed capability soon led to its adoption to simulate a great variety of air defence threats in an integrated environment. It can be used in several configurations, for AAA firings and for IR-, laser- or RF-guided air defence missiles.
Description: *Airframe:* Low-mounted, double-delta wings; high-mounted canard surfaces; bullet-shaped fuselage; pusher engine; twin, sweptback fins and rudders. No wheeled landing gear, but underfuselage skid pods can be attached for belly landings.
Mission payloads: AAA configuration: Smoke cartridges or a smoke generator.

Alkyon target on launch catapult *(EADS 3 Sigma)*   1136380

Alkyon targets at the NAMFI range on Crete *(EADS 3 Sigma)*   1136381

IR configuration: IR flares or IR enhanced pod (IREP).
Laser configuration: Laser reflector surfaces attached to fuselage sides, and laser lenses at front and rear.
Radar configuration: Aluminium stripes or RF amplifiers.
Combinations of the above configurations can be offered. Additional payloads can be accommodated, depending upon customer requirements.
Guidance and control: GCS/GPS configuration standard; radio-command remote-control system optional.
Launch: By bungee type catapult from land or ship.
Recovery: Parachute recovery or belly skid landing. Parachute recovery system is energised by the onboard flight termination system or by preprogrammed or operator command.
Power plant: One 8.6 kW (11.5 hp) 3W-120iB2 two-cylinder two-stroke engine, driving a two-blade pusher propeller. Fuel capacity 10 litres (2.6 US gallons; 2.2 Imp gallons).

Alkyon

Dimensions, External	
Overall, length	2.15 m (7 ft 0¾ in)
Fuselage, width	0.21 m (8¼ in)
Wings, wing span	2.06 m (6 ft 9 in)
Areas	
Wings, Gross wing area	0.83 m² (8.9 sq ft)
Weights and Loadings	
Weight	
Weight empty	22 kg (48 lb)

Alkyon is a close-range gunnery and missile target *(IHS/Patrick Allen)*   0523962

View emphasising the Alkyon's double-delta wing shape *(IHS/Patrick Allen)* 0523961

Alkyon

Payload, Max payload	14 kg (30 lb)
Performance	
Altitude, Operating altitude	50 m to 4,000 m (165 ft to 13,120 ft)
Speed, Max level speed	135 kt (250 km/h; 155 mph)
Endurance	1 hr 30 min

Status: In production and service. Has been used at NAMFI range on Crete; Meppen test range (Germany); Todendorf firing range (Germany); Andravida firing range (Greece); CEL and CEM (France); Rafael's Shdema proving ground (Israel); and over the Atlantic Ocean, Aegean Sea and Ionian Sea.

Customers: NATO armed forces.

Contractor: EADS 3 Sigma SA
Chania, Crete.

EADS 3 Sigma Iris Jet

Type: Recoverable aerial target.

Development: The Iris Jet aerial target was designed to meet increased performance requirements in terms of speed, manoeuvrability and endurance. Its 'almost unlimited' payload integration capability makes it suitable for single or multiple synchronised threat representation, in either single-type or multiple-type target combinations, exceeding tactical training or testing scenario requirements for air-to-air and surface-to-air

Iris Jet targets on the deck of a warship *(EADS 3 Sigma)* 1136384

Iris Jet being launched from a Robonic MC2555LR at the Vidsel range in Sweden *(Robonic)* 1311295

Iris Jet taking off *(EADS 3 Sigma)* 1136385

firings. The target is certified and in use by leading defence industries for weapon systems development and acceptance tests, and is serving worldwide armed forces' needs for a variety of long- and medium-range air defence weapon systems such as Patriot, NSSM, ESSM, SM1, SM2, Hawk and TOR-M1.

Description: *Airframe:* Low-wing monoplane with flaps; mainly cylindrical fuselage; inverted Y tail surfaces, similar to Iris Prop; dorsal boom incorporates engine air intake. Composites construction. No landing gear.

Mission payloads: Smoke cartridges or a smoke generator for visual detection. In laser configuration, laser reflectors are attached to the sides and laser lenses on the required edges. In radar configuration, a great variety of passive and/or active means of amplification (aluminium stripes, Luneberg lenses, corner reflectors or RF amplifiers) can be integrated to enlarge the target's very low RCS. A wide range of mission payloads such as IFF transponders, radar altimeter, chaff, rejected flares, MDI, EW payloads, cameras and several GFE equipments can be installed, either as stand-alones or in various combinations.

Guidance and control: Digital microprocessor-based autopilot control is programmable, with or without remote control. For accurate navigation and air vehicle position monitoring and control, the Iris Jet is equipped with a GPS terminal. In aerial target configuration, the aircraft is remotely guided from the mobile GCS to follow a preprogrammed four-dimensional flight path. It can also be controlled in semi-automatic mode, or manually via a hand-held control box, to allow instantaneous pattern changes during a mission.

Launch: Can be launched from ground or ship by mobile bungee-type catapult, without need for booster rocket assistance.

Recovery: Parachute recovery system, energised by the onboard flight termination system or by preprogrammed or operator command.

Power plant: One 1.07 kN (240 lb st) variable-speed turbojet (type not stated), with dorsal air intake. Max fuel capacity 79 litres (20.9 US gallons; 17.4 Imp gallons).

Iris Jet

Dimensions, External	
Overall, length	4.055 m (13 ft 3¾ in)
Wings, wing span	2.80 m (9 ft 2¼ in)
Areas	
Wings, Gross wing area	1.105 m² (11.9 sq ft)
Weights and Loadings	
Weight	
Weight empty	100 kg (220 lb)
Max launch weight	200 kg (440 lb)
Payload	
Max payload, incl fuel	100 kg (220 lb)
Performance	
Altitude, Operating altitude	100 m to 12,190 m (330 ft to 40,000 ft)
Speed, Max level speed	464 kt (859 km/h; 534 mph)
Endurance	1 hr

Status: In production and service since 1999. Certified by NAMFI (Crete), Meppen and Todendorf ranges (Germany) and Shdema firing range (Israel), and for open-ocean operations aboard NATO warships.

Customers: Various NATO member countries, air defence manufacturers and Israel Defence Force.

Contractor: EADS 3 Sigma SA
Chania, Crete.

EADS 3 Sigma Iris Prop

Type: Recoverable aerial target.

Development: The Iris Prop aerial target was designed to cover training requirements for target drones with long endurance, high payload capacity and sufficient speed for air defence weapon system firings, short/medium-range surface-to-air guided weapons, air-to-air missiles, ground-to-air gunnery and close-in weapon systems, personnel evaluation, or close-range reconnaissance and BDA.

Description: *Airframe:* Bullet-shaped glass fibre fuselage; pusher engine; low-mounted wings with slight leading-edge sweepback; inverted Y tail surfaces, carried above and behind main body on inclined dorsal fairing and slender boom. No landing gear.

Mission payloads: Smoke cartridges or a smoke generator for visual detection. In IR configuration, carries IR flares or IR enhanced pod (IREP). In laser configuration, laser reflector surfaces are attached to sides and laser lenses on required edges. In radar configuration, a wide variety of passive and/or active means of amplification such as aluminium stripes, Luneberg lenses, corner reflectors or RF amplifiers are integrated to enlarge the target's very low RCS. A large number of mission payloads, such as IFF transponders, radar altimeter, chaff, rejected flares, MDI, cameras and several types of GFE, can be installed, either as stand-alones or in various combinations.

Guidance and control: Digital microprocessor-based autopilot control is programmable, with or without remote control. For accurate navigation and air vehicle position monitoring and control, Iris Prop is equipped with a GPS terminal. In aerial target configuration, the aircraft is remotely guided from the mobile GCS to follow a preprogrammed four-dimensional flight path. It can also be controlled in semi-automatic mode, or manually via a hand-held control box, to allow instantaneous pattern changes during a mission.

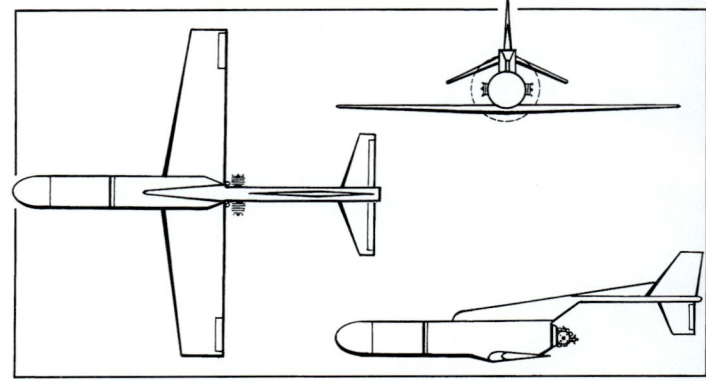

Iris Prop recoverable target *(IHS/John W Wood)* 0517936

Launch: By mobile bungee-type catapult from ground or ship.

Recovery: Parachute recovery or skid landing. The parachute recovery system is energised by the onboard flight termination system or by a preprogrammed or operator command.

Power plant: One 28.3 kW (38 hp) rotary engine; two-blade wooden or composites pusher propeller. Maximum fuel capacity 25 litres (6.6 US gallons; 5.5 Imp gallons).

Iris Prop

Dimensions, External	
Overall, length	3.30 m (10 ft 10 in)
Wings, wing span	2.95 m (9 ft 8¼ in)
Areas	
Wings, Gross wing area	1.22 m² (13.1 sq ft)
Weights and Loadings	
Weight	
Weight empty	55 kg (121 lb)
Max launch weight	100 kg (220 lb)
Payload	
Payload with max fuel	45 kg (99 lb)
Performance	
Altitude	
Operating altitude	100 m to 4,800 m (320 ft to 15,740 ft)
using radar altimeter	5 m to 4,800 m (20 ft to 15,740 ft)
Speed, Max level speed	189 kt (350 km/h; 217 mph)
Endurance	2 hr 30 min

Status: In production and service since 1994. Has been used with a variety of long- and medium-range air defence weapon systems such as OSA-AKM, TOR-M1, SM-1 and NSSM firings. EADS 3 Sigma also operates as services provider for target facilities.

Customers: Various NATO member countries, air defence equipment manufacturers and Israel Defence Force.

Contractor: EADS 3 Sigma SA
Chania, Crete.

Iris Prop on launch catapult *(EADS 3 Sigma)* 1136387

Iris Prop at sea *(EADS 3 Sigma)* 1136386

EADS 3 Sigma Perseas family

Type: Recoverable aerial targets.

Development: The Perseas family of target systems utilises a common platform and avionics with interchangeable propulsion units.

Description: *Airframe:* Bullet-shaped fuselage with single (jet or prop) or twin (jet) interchangeable engine pods; low-mounted, slightly swept wings; conventional sweptback tail surfaces with small, square endplate finlets. No landing gear.

Mission payloads: Smoke cartridges or a smoke generator for visual detection. In laser configuration, laser reflector surfaces are attached to sides and laser lenses on required edges. In radar configuration, a great variety of passive and/or active means of amplification such as aluminium stripes, Luneberg lenses, corner reflectors or RF amplifiers are integrated to enlarge its very low RCS. A large number of mission payloads such as IFF transponders, radar altimeter, chaff, rejected flares, MDI, cameras or

Ditched Iris Prop awaiting recovery from the sea 0001574

The single-jet Perseas SJ *(EADS 3 Sigma)*

At-sea photograph of the twin-jet Perseas TJ *(EADS 3 Sigma)*

various GFE can be installed, either as stand-alones or in various combinations.

Guidance and control: Full microprocessor-based autopilot control, programmable with or without remote control. GPS terminal for navigation and air vehicle position monitoring. Perseas family targets are remotely guided from the mobile GCS to follow a preprogrammed four-dimensional flight path. They can also be controlled in semi-automatic mode, or manually via a hand-held control box, to allow instantaneous pattern changes during a mission.

Launch: By bungee-type catapult from ground or ship; no booster rocket assistance required.

Recovery: Parachute recovery or skid landing. The parachute recovery system is energised by the onboard flight termination system or by a preprogrammed or operator command.

Power plant: One (Perseas SJ) or two (Perseas TJ) 0.20 kN (50 lb st) turbojets (type not stated); or (Perseas Prop) one 16.4 kW (22 hp) unspecified piston engine. Max fuel capacity (all) 30 litres (7.9 US gallons; 6.6 Imp gallons).

Variants: Perseas Prop: The propeller-driven member of the family, with basic performance characteristics. Suitable for prolonged complex or stand-alone tactical scenarios. Capable of carrying almost any payload, it is used to simulate air-breathing threats against air defence gunnery and missiles requiring prolonged flight endurance and ample payload capability.

Perseas SJ: Single-jet member of family, with medium performance. Engine located beneath the wing to provide physical IR/UV emission; combined with improved flying characteristics, this provides excellent representation for weapon systems like Stinger, Mistral, LeFlaSys, Oselot, Avenger and ASRAD that require such a target profile.

Perseas TJ: Twin-jet, medium/high-performance version, with engines mounted each side of rear fuselage. Designed to cover increased speed and manoeuvrability requirements. Multiple flight scenarios make it suitable for weapon systems such as Stinger, NSSM, SM-1, Crotale NG, OSA-AKM and TOR-M1, defending clustered assets.

Perseas TJ, Perseas SJ, Perseas Prop

Dimensions, External	
Overall, length	2.715 m (8 ft 11 in)
Wings, wing span	2.22 m (7 ft 3½ in)
Areas	
Wings, Gross wing area	0.697 m² (7.5 sq ft)
Weights and Loadings	
Weight	
Weight empty	
Perseas Prop	50.0 kg (110 lb)
Perseas SJ	49.5 kg (109 lb)
Perseas TJ	53.0 kg (116 lb)
Max launch weight	
Perseas Prop	70.0 kg (154 lb)
Perseas SJ	85.0 kg (187 lb)
Perseas TJ	95.0 kg (209 lb)
Payload	
Max payload	
incl fuel, Perseas Prop	20.0 kg (44 lb)
incl fuel, Perseas SJ	35.5 kg (78 lb)
incl fuel, Perseas TJ	42.0 kg (92 lb)
Performance	
Altitude	
Operating altitude	
Perseas Prop	100 m to 4,000 m (320 ft to 13,120 ft)
using radar altimeter, Perseas Prop	5 m to 4,000 m (20 ft to 13,120 ft)
Perseas SJ, Perseas TJ	100 m to 6,000 m (320 ft to 19,680 ft)
using radar altimeter, Perseas SJ, Perseas TJ	5 m to 6,000 m (20 ft to 19,680 ft)
Speed	
Max level speed	
Perseas Prop	151 kt (280 km/h; 174 mph)
Perseas SJ	297 kt (550 km/h; 342 mph)
Perseas TJ	226 kt (419 km/h; 260 mph)
Endurance	
Perseas Prop	2 hr
Perseas SJ	50 min
Perseas TJ	1 hr
Power plant	1 × turbojet

Status: In production and service since 1998.

Customers: Various NATO member countries, air defence equipment manufacturers and Israel Defence Force.

Contractor: EADS 3 Sigma SA
Chania, Crete.

India

ADE Lakshya

Type: Recoverable aerial target.

Development: According to *IHS Jane's* sources, development of the Lakshya (True Aim - also known as the Lakshya Pilotless Target Aircraft) AV began during the early 1980s. Other key programme events include:

1983 *IHS Jane's* sources report the first of 18 prototype Lakshya AVs being test launched during 1983. Flight trials with the Hindustan Aeronautics Ltd (HAL) PTAE 7 turbojet are said to have begun with the 16th prototype.

2000 The Lakshya system entered service with the Indian Air Force (IAF) during 2000.

2001 The Lakshya system entered service with the Indian Navy during 2001.

March 2002 *IHS Jane's* sources report flight trials of a Lakshya AV powered by a Defence Research and Development Organisation (DRDO) developed turbojet as having begun at the Chandipur test range during March 2002.

November 2002 *IHS Jane's* report Israeli evaluation of the Lakshya as having taken place during November 2002, with negotiation for the lease or procurement of 20 × Lakshya systems (possibly for an EW or reconnaissance role) continuing into 2004 at which time, Israel is said to have withdrawn from the potential deal.

2003 The Lakshya system entered service with the Indian Army during 2003.

27 April 2003 *IHS Jane's* sources report flight testing of an 'upgraded' Lakshya variant (with an 'improved engine and sub-systems') as having begun on 27 April 2003 (date not confirmed - see also following).

Mid-2004 *IHS Jane's* sources were reporting five Lakshya systems as being in service with the Indian Air Force (IAF - including the service's

Indian Air Force Lakshya jet-powered target *(IHS/Patrick Allen)*

Aerial targets > India > ADE Lakshya – Kadet Defence Systems Javelin X

Lakshya at Aero India show, February 2005 (IHS/Patrick Allen)
1136637

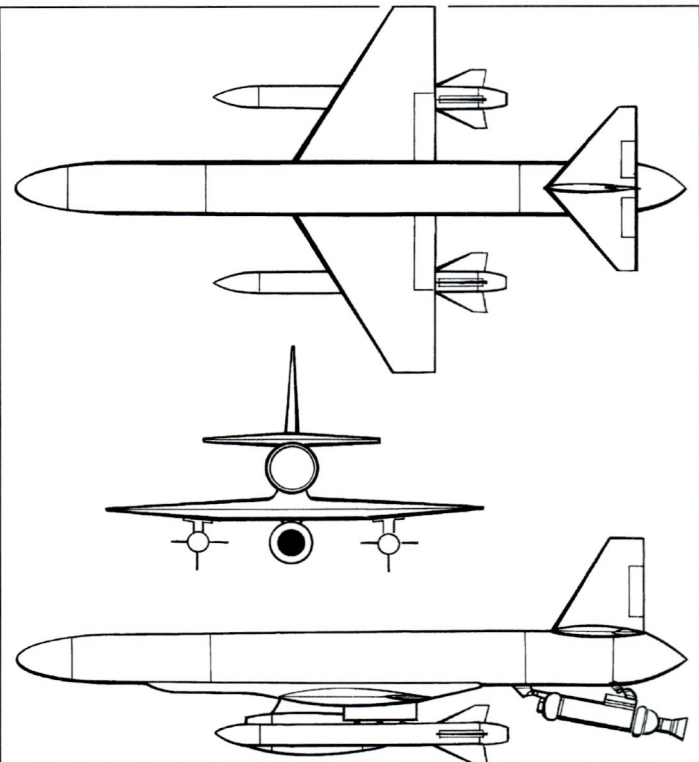

ADE Lakshya variable-speed target (IHS/John W Wood)
0517939

Kalaikunda-based Target Tug Flight) with a further 30 such systems having been supplied to the Indian Army's Artillery Corps.

February 2005 *IHS Jane's* sources were reporting HAL as being due to deliver 10 × Lakshya AVs to the IAF by the end of the following March, with a further 17 being scheduled to have been delivered to both the IAF and the Indian Navy by mid-2006.

June 2006 *IHS Jane's* sources were associating a reported contract for 200 derivatives of the 44.1 kN (9,921 lb st) NPO Saturn TRDD-50MT turbofan with the Lakshya programme. Such a power plant was said to have had the potential to extend the Lakshya AV's endurance to one 'several hours' long.

April 2011 As of April 2011, *IHS Jane's* sources were reporting the existence of a digital variant of the baseline Lakshya AV (identified by the cited sources as the Lakshya-2) which had undergone development trials and was being offered to customers.

14 May 2012 In a written parliamentary reply, India's then Minister of State for Defence Dr M M PallamRajuin reported HAL as having manufactured 30 × Lakshya-1 'UAVs' during the period 2006/7 to 2012. Of the cited number, 15 had been supplied to the IAF, five to the Indian Navy and five to the DRDO for (it is assumed) test purposes. In this context, multiple Indian media sources have reported the Lakshya-1 configuration as incorporating digital engine controls, with an example being noted as having been test launched at the Chandipur Integrated Test Range on 23 August 2012.

Readers should further note that aside from the unidentified DRDO developed turbojet, HAL's PTAE-7 turbojet and the NPO Saturn TRDD-50MT turbofan, the 4.22 kN (948 lb st) Microturbo TRI 60-5 turbojet has been associated with the Lakshya architecture.

Description: *Airframe:* Low-wing monoplane with underslung turbojet, symmetrical-section cropped-delta wing and horizontal tail surfaces; conventional ailerons, elevators and rudder. Modular construction is mainly of light alloy, with GFRP used for the tail unit and ogival nosecone and tailcone. Pylon under each wing for carriage of tow target body. No landing gear. Crushable nosecone, energy dissipating skids under the engine pod, and expendable wingtips, facilitate recovery over land; tailcone is ejected to deploy recovery parachute. Front equipment bay is watertight for recovery at sea. Lakshya has a flotation potential of more than 4 hours.

Mission payloads: Expendable towed subtargets (one under each wing, each with 1,500 m; 4,920 ft of cable), equipped normally with RCS augmentation plus one of the following: analogue MDI and visual cartridges; digital MDI; or IR flares. RCS augmentation devices are oriented to suit either air-to-air or surface-to-air engagements.

Guidance and control: Flight control: Onboard avionics consist of a three-axis autopilot, transponder, encoder/decoder, flight control electronics, sensors, servo amplifiers, actuators, and a telemetry transmitter. Flight control system caters for height and heading hold, programmable launch phase, and emergency recovery routines, and provides for flights in autonomous or manual control modes. Onboard electrical power (28 V DC) is provided by a 70 A engine-driven alternator and power conditioning unit, with two silver-zinc batteries for back-up. A locator RF beacon and flashing light are carried to facilitate recovery after a mission. Flight control system initiates recovery automatically under certain specific system malfunction conditions.

Ground control: GCS comprises telecommand, telemetry and tracking systems, within a transportable shelter for mobility and ease of deployment. Telecommand range 54 n miles (100 km; 62 miles). Ground station includes control and display console, plus a simulator mode of operation for the training of controllers. Air vehicle position is indicated in real time on a monitor, with flight parameters shown on another monitor. Mission profiles are tailored to specific gun and missile crew training requirements; from two to 10 presentations can be flown, depending upon these requirements.

System composition: Includes six air vehicles and one GCS.

Launch: From land or ship by 36.8 kN (8,267 lb st) solid propellant booster rocket from zero-length launcher.

Recovery: Two-stage parachute recovery system, deployed on command or through onboard logic. Descent rate is about 6 m/s (1,181 ft/min). Lakshya can be recovered over land or sea, and can be towed by boats and retrieved by ship, or airlifted by helicopter.

Power plant: One 4.22 kN (948 lb st) HAL PTAE 7 turbojet.

Lakshya

Dimensions, External	
Overall	
length	5.929 m (19 ft 5½ in)
height	1.64 m (5 ft 4½ in)
Fuselage, width	0.425 m (1 ft 4¾ in)
Wings, wing span	3.00 m (9 ft 10 in)
Tailplane, tailplane span	1.658 m (5 ft 5¼ in)
Weights and Loadings	
Weight	
Weight empty	330 kg (727 lb)
Max launch weight	
incl tow bodies and booster	705 kg (1,554 lb)
Fuel weight, Max fuel weight	190 kg (418 lb)
Performance	
(A: clean B: with one tow body stowed and one in tow)	
Climb	
Rate of climb, max, at S/L	1,500 m/min (4,921 ft/min) (est)
Altitude	
Operating altitude	
max (A)	9,000 m (29,520 ft)
max (B)	5,000 m (16,400 ft)
min	300 m (980 ft)
Speed	
Max operating Mach number	
clean configuration	0.65
with one tow body stowed and one in tow	0.60
g limits	
manoeuvring	+3
structural limit	+6
Radius of operation, LOS	54 n miles (100 km; 62 miles)
Endurance	
at M0.56 at 1,500 m (4,920 ft)	
(A)	50 min
Power plant	1 × turbojet

Status: As of 2012, the Lakshya recoverable aerial target was reported as being in 'series production' at HAL's Bangalore facility and as having been procured by the IAF, the Indian Navy, the Indian Army and the DRDO.

Contractor: Aeronautical Development Establishment (system designer), Bangalore.

Kadet Defence Systems Javelin X

Type: Selectively expendable multirole aerial target.

Development: The Javelin X is designed to simulate attack aircraft during land, sea and air combat training exercises, weapon development and weapon evaluation. In its basic architecture, it can operate from fixed ground facilities (ranges) and mobile facilities (ships). It is available in variants for visual, extended visual (neutral weather) and beyond visual range (VR, EVR and BVR) flights of 0.8, 2.7 and 13.5 n miles (1.5, 5 and 25 km; 0.9, 3.1 and 15.5 miles) respectively.

Javelin X multirole target *(Kadet Defence Systems)* 1290242

A Javelin X on its launch cradle *(Kadet Defence Systems)* 1290243

Design and development began on 21 March 2006, with first and second prototypes making maiden flights on 15 August and 12 December 2006. Flight testing of the final production version took place from 17 February to 15 May 2007.

Description: *Airframe:* Low-drag, modular, shoulder-wing monoplane with V tail. No landing gear. All-composite construction.

Mission payloads: Include RCS enhancements and optional IR source. For IR-guided weaponry, the aircraft can also tow a radar-reflective sleeve or a hard tow body with a heat source.

Guidance and control: VR and EVR versions are operated manually in an optically tracked radius under UHF radio control. The BVR version flies autonomously with a three-axis GPS-based autopilot and long-range data downlink, and is tracked by the GCS, with location displayed on map with telemetry, speed, heading and altitude data.

Transportation: Easily transportable by air, land and sea. Deployment time less than 4 hours.

System composition: A typical fielded system consists of three air vehicles, launcher, hand-held ground control unit, binocular-fitted optical tracking chair, four personnel and ground support equipment. In addition, the BVR version has a portable GCS with a computer-based piloting and tracking station, joystick and long-range radio uplink.

Launch: Operator-controlled launch from pneumatic (nitrogen) launcher.

Recovery: Operator-controlled belly landing, on land or at sea.

Power plant: One 0.75 kW (10 hp) Torch 90 single-cylinder two-stroke engine; two-blade tractor propeller. Fuel capacity 4 litres (1.06 US gallons; 0.88 Imp gallons).

Javelin X

Dimensions, External	
Overall, length	2.46 m (8 ft 0¾ in)
Dimensions, Internal	
Payload bay, volume	0.004 m³ (0.1 cu ft) (est)
Weights and Loadings	
Weight	
Weight empty	18 kg (39 lb)
Max launch weight	25 kg (55 lb)
Fuel weight, Max fuel weight	3.5 kg (7.00 lb)
Payload, Max payload	5.5 kg (12.00 lb)
Performance	
Altitude	
Operating altitude	
VR	10 m to 800 m (40 ft to 2,620 ft)
BVR	100 m to 1,500 m (320 ft to 4,920 ft)
Speed	
Max level speed	120 kt (222 km/h; 138 mph)
Cruising speed, max	100 kt (185 km/h; 115 mph)
Loitering speed	40 kt (74 km/h; 46 mph)
Radius of operation	
datalink, mission	13.5 n miles (25 km; 15 miles)
Endurance, max	1 hr 30 min

Status: First sale, to the Military Radar Division of Bharat Electronics, was made in January 2008, following customer evaluation in September 2007. At that time it was also being evaluated by the Air Defence division of the Indian Army.

Contractor: Kadet Defence Systems
Kolkata.

Kadet Defence Systems JX-2

Type: Recoverable aerial target.

Development: Development of the JX-2 began in June 2008, maiden flight taking place on 1 December that year. It is intended for such applications as gunnery, air defence and missile training, and weapon development and evaluation.

Description: *Airframe:* Pod and boom fuselage, with mid-mounted trapezoidal wings and T tail. No landing gear.

Mission payloads: Can include passive reflectors, IR flares, smoke dispensers or tow bodies.

Guidance and control: Radio-controlled.

Launch: From pneumatic launcher.

Recovery: Belly landing.

Power plant: One two-stroke piston engine (type and rating not stated); two-blade tractor propeller.

JX-2

Dimensions, External	
Overall, length	2.60 m (8 ft 6¼ in)
Wings, wing span	3.20 m (10 ft 6 in)
Weights and Loadings	
Weight, Max launch weight	20.0 kg (44 lb)
Performance	
Speed, Max level speed	116 kt (215 km/h; 133 mph)
Radius of operation	
LOS, mission, VR	0.8 n miles (1 km; miles)
LOS, mission, BVR	11 n miles (20 km; 12 miles)
Endurance, max	1 hr (est)
Power plant	1 × piston engine

Status: Following selection of the JX-2 as winner of an Indian MoD competition, the company reported the receipt of a "long-term contract" for the Indian Army on 7 May 2010.

Contractor: Kadet Defence Systems
Kolkata.

JX-2 target, in production for the Indian Army
(Kadet Defence Systems) 1395224

MKU Bullseye

Type: Recoverable aerial target.

Development: In addition to the TERP and Erasmus UAVs, MKU has added the Bullseye aerial target to its range of products.

Description: *Airframe:* Typical low-wing light aircraft configuration. Fixed, tailwheel type landing gear. Balsa and composites construction.

Mission payloads: Radar-reflective banner subtarget, towed behind aircraft on cable.

Guidance and control: Remotely controlled by hand-held, six-channel radio with eight digital servos.

Transportation: Man-portable in backpack (target) and case (GCS).

Launch: Conventional wheeled take-off.

Recovery: Conventional wheeled landing.

Bullseye at IDEX 2009 *(IHS/Patrick Allen)* 1376640

Power plant: One 65 cm³ two-cylinder two-stroke engine, driving a two-blade tractor propeller.

Bullseye

Dimensions, External
Overall, length... 2.06 m (6 ft 9 in)
Wings, wing span... 1.80 m (5 ft 10¾ in)
Weights and Loadings
Weight, Max T-O weight... 7.5 kg (16.00 lb)
Performance
Altitude, Service ceiling.. 1,000 m (3,280 ft) (est)
Speed, Max level speed... 29 kt (54 km/h; 33 mph) (est)
Radius of operation, command link.......................... 0.65 n miles (1 km; miles)
Endurance... 20 min (est)
Power plant... 1 × piston engine

Status: No orders reported by April 2009. Marketing was continuing at that time.
Contractor: MKU Pvt Ltd
New Delhi.

Iran

Qods Saeghe

Type: Recoverable aerial target.
Development: Saeghe (lightning) was developed as an air defence gunnery training target.
Description: Airframe: Low-mounted delta wings, with elevons; bullet-shaped fuselage; sweptback fin and rudder; no horizontal tail surfaces or landing gear. All-composites construction.
Mission payloads: Gyrostabilised radar corner reflector and three IR flares in Saeghe 2.
Guidance and control: Saeghe 1 has simple radio command of flight path and manoeuvres, plus wings-level stabilisation. Portable (or mobile, in small van) GCS for Saeghe 2 has full, microprocessor-based autopilot control, programmable with or without operator control. In-sight flight unit for LOS control without telemetry data; or GPS-based programmable navigation allowing target to be controlled, tracked and viewed on monitor during both LOS and BLOS missions.
Launch: By booster rocket from zero-length launcher, which can be mounted on a pick-up truck or ship's deck.
Recovery: Parachute recovery over land or water standard. Can also be recovered, on land, by belly skid landing.
Power plant: One 18.6 kW (25 hp) flat-twin piston engine (type not stated); two-blade pusher propeller.
Variants: Saeghe 1: Basic model, with simple radio command of flight path and manoeuvres; wings-level stabilisation. Also known as the N-Q-A 100.
Saeghe 2: More advanced version. Telecommand and telemetry links between air vehicle and GCS; GPS navigation. Same airframe as Saeghe 1. *Specification data apply to this version.*

Saeghe 2

Dimensions, External
Overall
 length... 2.81 m (9 ft 2¾ in)
 height... 0.70 m (2 ft 3½ in)
Fuselage, length.. 2.66 m (8 ft 8¾ in)
Wings, wing span... 2.60 m (8 ft 6¼ in)
Weights and Loadings
Weight, Max launch weight.. 60 kg (132 lb)
Performance
Altitude, Service ceiling... 3,355 m (11,000 ft)
Speed, Max level speed... 135 kt (250 km/h; 155 mph)
Radius of operation, max, mission........................... 27 n miles (50 km; 31 miles)
Endurance.. 45 min
Power plant... 1 × piston engine

Status: In production and service.
As of May 2011, the Fars News Agency announced that the Iranian Army plans to increase the range of its home-made UAS and also equip its reconnaissance unmanned aircraft with missile-launching systems, according to an Iranian commander.
Speaking to the Fars News Agency, Lieutenant Commander of the Iranian Army's Self-Sufficiency Jihad, Rear Admiral Farhad Amiri referred to Iran's capability in designing and building different types of unmanned aircraft, and explained that UAS are usually used for collecting intelligence and information, wiretapping, photography and shooting footage, but they can also be used for other purposes on the basis of their power and capabilities. "For example, in naval missions we need long-range UAS or we need the unmanned aircraft which have the capability to launch missiles, each of which needs its own special technology. Of course, all these (projects) are on our working agenda in a move to satisfy our needs during the next two years," he added.
SOURCE IS UASVISION.COM from FARS NEWS AGENCY
Customers: Iranian armed forces. Possibly also exported.
Contractor: Qods Aviation Industries
Tehran.

Saeghe 1 target drone for air defence training   0087737

The Qods Saeghe 2 in flight   0087738

Saeghe's zero-length launcher can be mounted on a small pick-up truck *(Robert Hewson)*   0576821

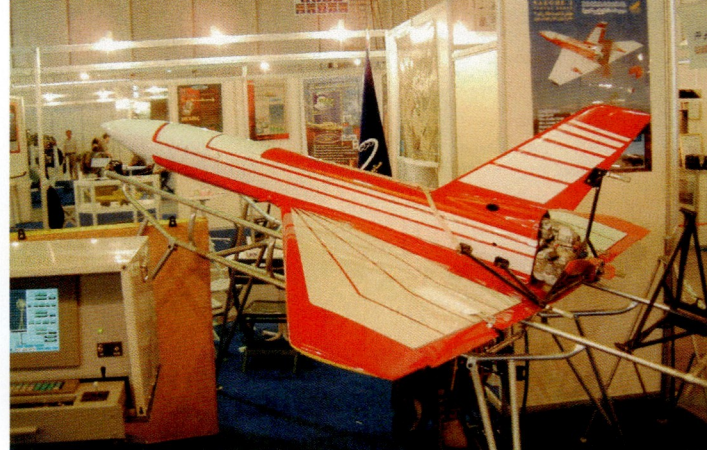

A Saeghe 2 on display in late 2002 *(Robert Hewson)*   0576822

Israel

Rafael Black Sparrow

Type: Air-launched ballistic target.

Development: The Black Sparrow programme was funded jointly by the US Ballistic Missile Defense Organization (BMDO) and the Israeli Missile Defence Organisation (IMDO) in 1996 in order to develop a high-fidelity, threat-representative ballistic target for Israel's Theatre Missile Defence (TMD) requirements. It simulates the Scud-B ballistic missile and has full telemetry and test range safety capabilities.

Description: *Mission payloads:* Modular warhead simulator, which can be adapted to emulate various warhead threats such as high explosive and bulk chemical for lethality analysis.

Guidance and control: The target simulates the trajectory (velocity and altitude profiles) and the thermal and RCS signatures of the Scud-B ballistic missile, being able to simulate such re-entry manoeuvres as ballistic and barrel roll. Its telemetry supports full data for interceptor test analysis.

Black Sparrow on an Israeli Air Force F-15 *(Rafael)* 1128968

Launch: Air-launched.

Recovery: Expendable. Design incorporates a fully redundant flight termination system to meet the range safety requirements governing missile interceptor tests.

Black Sparrow

Dimensions, External	
Overall	
length	4.85 m (15 ft 11 in)
width	0.526 m (1 ft 8¾ in)
Weights and Loadings	
Weight, Max launch weight	1,275 kg (2,810 lb)

Status: Black Sparrow has been proven in missile defence engagement scenarios with the Israeli Arrow weapon system. By 2005 it was fully developed and market-ready, having successfully completed a series of fly-out and interception tests. The Black Sparrow was selected by the French DGA for testing the SAMP/T system.

Contractor: Rafael Armament Development Authority Ltd, Haifa.

The Black Sparrow air-launched target *(Rafael)* 1122521

Italy

Selex Galileo Locusta

Type: Expendable target or subtarget.

Development: Described as "a mini-UAV secondary aerial target", development of the Locusta began in 2004 as a 'live' target for tactical missiles and air defence artillery. It was designed to be deployed either as a subtarget payload for the Mirach 100/5 target or, alternatively, as a stand-alone target launched from other types of host aircraft. In the former application, one Locusta is carried under each wing of the Mirach parent target. Locusta can simulate incoming threats from both surface-to-air and air-to-air missiles. It can also be configured for such operational roles as decoy and jamming.

Description: *Airframe:* Cylindrical body with ogival nosecone and ventral engine air intake. Four front-mounted fins in cruciform configuration and four larger tailfins in X configuration. Mainly carbon fibre construction.

Mission payloads: Modular payloads include radar cross-section augmentors (S-band and Ku-band) and IR augmentors.

Guidance and control: Flight profiles are mainly autonomous, but ground controller can monitor the mission and, if the target survives a firing or exceeds mission safety parameters, can command termination of the mission and recovery by parachute. The Locusta GCS can be supplied as a stand-alone product or integrated into the Mirach 100/5 GCS, and can control up to 12 targets at a time, these being identified on a time-share basis before launch by means of a hardware code selector.

Launch: Air-launched from Mirach 100/5 aerial target or other host aircraft.

Recovery: Locusta is designed to be expendable, but is equipped with a recovery parachute for missions where it exceeds mission or range safety parameters or survives the engagement.

Power plant: One miniature turbojet (type and rating not known).

Locusta

Dimensions, External	
Overall	
length	1.80 m (5 ft 10¾ in)
width, over tailfins	1.00 m (3 ft 3¼ in) (est)
Weights and Loadings	
Weight, Max launch weight	20 kg (44 lb)
Performance	
Altitude, Service ceiling	6,000 m (19,680 ft)
Speed, Max level speed	410 kt (759 km/h; 472 mph)
Radius of operation	
mission, typical pre-programmed	32 n miles (59 km; 36 miles)
Endurance	5 min
Power plant	1 × turbojet

Status: In May 2006, after launch trials from a Mirach 100/5 at the NAMFI firing range in Crete, Locusta received a qualification certificate from the BWB (German Defence Ministry) for use as an anti-radiation missile simulator (ARMS) for tactical firings of the Patriot surface-to-air missile by the German Air Force.

Contractor: Selex Galileo Avionica SpA, Ronchi dei Legionari.

Model of Locusta wing-mounted on a Mirach 100/5 *(IHS/Patrick Allen)* 1327047

Selex Galileo Mirach 100/5

Type: Recoverable aerial target.

Development: Developed by Meteor under 1995 Italian MoD contract for an advanced threat-representative target system. First flown December 1996. Extensively redesigned compared with earlier Mirach 100 versions. Higher operating speeds and ceiling; able to simulate multiple threats, including sea-skimming missile attacks against moving ships. Low radar cross-section and reduced IR signature.

Meteor became a Galileo Avionica company in 2002; Galileo is now a subsidiary of Selex Sensors and Airborne Systems SpA, a Finmeccanica company.

Description: *Airframe:* High-mounted wings with greater span and area than earlier models; twin lateral intakes for higher-powered turbojet; modified tail assembly comprising high-mounted tailplane with 5° dihedral, plus sweptback, outward-canted (32°) twin ventral fins; canard rudder.

Mission payloads: Thales IRCM pod; APL tracking seeker simulator; NAWC AN/DPT-1 radar simulator; Southern California plume generator

320 Aerial targets > Italy – Japan > **Selex Galileo Mirach 100/5 – Fuji J/AQM-1**

Mirach 100/5 launch *(Selex Galileo)* 1151520

Twin-rocket launch of a Mirach 100/5 *(Selex Galileo)* 1151592

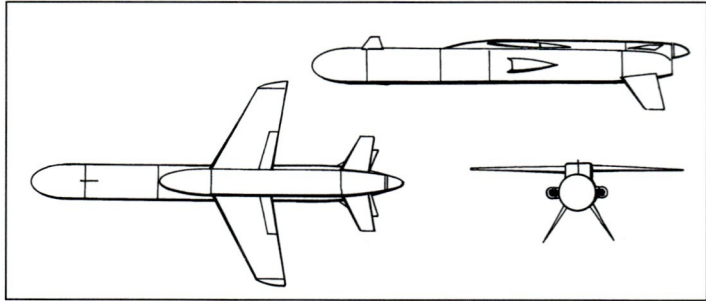

Mirach 100/5 general arrangement *(IHS/John W Wood)* 0089087

pod; Matelec Luneberg lenses; Herley radar transponders; Selex Galileo visual augmentation devices; secondary towed targets; MDI.

Guidance and control: Raytheon Systems digital signal processor; BAE Systems INS/GPS navigation; Alenia radar altimeter. GCS can control up to eight air vehicles simultaneously, flying at high subsonic speeds both solo and in formation. Otherwise similar to that for other Mirach 100 versions.

Launch: Generally as described for earlier Mirach 100s.
Recovery: Generally as described for earlier Mirach 100s.
Power plant: One 1.57 kN (353 lb st) Microturbo TRS 18-1-201-1 turbojet.

Mirach 100/5

Dimensions, External	
Overall	
length	4.065 m (13 ft 4 in)
height	0.89 m (2 ft 11 in)
Fuselage, width	0.49 m (1 ft 7¼ in)
Wings, wing span	2.30 m (7 ft 6½ in)
Tailplane, tailplane span	1.06 m (3 ft 5¾ in)
Dimensions, Internal	
Payload bay, volume	0.08 m³ (2.8 cu ft)
Areas	
Wings, Gross wing area	1.10 m² (11.8 sq ft)
Weights and Loadings	
Weight	
Max launch weight, from ramp	380 kg (837 lb)
Payload, Max payload	60 kg (132 lb)
Performance	
Altitude, Operating altitude	5 m to 12,495 m (10 ft to 41,000 ft)
Speed, Max level speed	540 kt (1,000 km/h; 621 mph) at S/L
g limits	+6
Endurance	1 hr 30 min
Power plant	1 × turbojet

Status: According to the CEO of Selex S & AS in October 2005, the Mirach 100/5 "has become the standard threat simulator used by European armed forces to train and qualify major weapons, replacing all previous-generation systems in use at firing ranges".

Selected by UK Royal Navy to fulfil its (Chukar II) Replacement Aerial Subsonic Target (RAST) requirement. BAE Systems adapted control systems for shipboard use. Contract with QinetiQ in 2003 as Replacement Aerial Target System (RATS) for use on UK ranges as successor to Jindivik. On 14 December 2006, Galileo Avionica received a GBP70 million contract from QinetiQ to provide 50 systems (10 to 15 air vehicles per year for 10 years) as an initial order under the QinetiQ-led Combined Aerial Target Service (CATS) programme to provide land-based air defence, ship-based (Royal Navy) and (for Royal Air Force) air-to-air targeting services for weapons training of the UK armed forces. Mirachs in UK service are operated chiefly from the ranges at Aberporth, South Wales, and in the Hebrides, or from on board Royal Navy ships.

The Mirach 100/5 airframe also forms the basis of EADS Military Aircraft's Carapas demonstrator for the Surveyor-600 UAV programme.

Customers: Known customers include France (37 in service from April 2005 with the Centre d'Essais de Lancement de Missiles at Biscarosse and Levant Island); Italy (Navy: 32 ordered for 2002–03 delivery; for use at Salto di Quirra range in Sardinia); Spain, for use at CEDEA range; the UK (39 targets and three control stations ordered in June 1998 for Royal Navy Fleet Target Group at RNAS Culdrose, achieving IOC in July 2000; 10 systems ordered by QinetiQ in June 2003 with an option for 10 systems a year for 10 years, amended in 2005 to cover the supply of 40 additional systems); and the NATO Missile Firing Installation (NAMFI) in Crete. The Mirach 100/5 has also been employed by the IRIS-T weapon consortium.

Contractor: Selex Galileo Avionica SpA.
Ronchi dei Legionari.

Japan

Fuji J/AQM-1

Type: Expendable air-launched target.
Development: Developed as Fuji Model 820 under Japan Defence Agency contract of July 1983. Initial flight, by the first of 11 prototypes, was made in March 1986, and the flight test programme was completed in December that year. Series production began in FY87, and the first four production examples (two each to the JASDF and the JDA) were delivered in October 1988.
Description: *Airframe:* Mid-wing monoplane; clipped delta wings are interchangeable port/starboard; cylindrical fuselage with ogival nosecone and tailcone; four identical tailfins, indexed in X configuration at 45° to vertical and horizontal datum lines. Engine mounted on centreline pylon under rear fuselage. Construction almost entirely of metal (aluminium alloy and steel) except for glass fibre nosecone and tailcone. No landing gear.

Mission payloads: Mission equipment includes a Ku-band reflector in each of the fore and aft body compartments. Smoke or infra-red generators can be attached to the wingtips. A miss-distance indicator can be installed in the forward section of the fuselage.

Guidance and control: The J/AQM-1 has a preprogrammed guidance system, which can be overridden by radio command from either the launch aircraft or a surface station. The digital flight control system includes an engine throttle control.

Launch: Air-launched from an underwing pylon (one under each wing) of an F-4EJ Phantom or F-15J/DJ Eagle of the JASDF.
Recovery: Non-recoverable.
Power plant: One 1.96 kN (441 lb st) Mitsubishi TJM-3 turbojet in underslung nacelle. Fuel in centre-fuselage tank, capacity 47 litres (12.4 US gallons; 10.3 Imp gallons).

J/AQM-1

Dimensions, External
- Overall
  - length...3.65 m (11 ft 11¾ in)
  - height...0.92 m (3 ft 0¼ in)
- Fuselage, width..0.35 m (1 ft 1¾ in)
- Wings
  - wing span, over smoke pods............................2.07 m (6 ft 9½ in)
- Tailplane, tailplane span................................0.98 m (3 ft 2½ in)

Areas
- Wings, Gross wing area.....................1.20 m² (12.9 sq ft)
- Fins, Tail fin.................................0.30 m² (3.23 sq ft)

Weights and Loadings
- Weight, Max launch weight..........................235.5 kg (519 lb)
- Fuel weight, Max fuel weight............................35 kg (77 lb)
- Payload, Max payload..................................42.4 kg (93 lb)

Performance
- Altitude, Operating altitude....................610 m to 9,145 m (2,000 ft to 30,000 ft)
- Speed, Max level speed...............560 kt (1,037 km/h; 644 mph) at 9,144 m (30,000 ft)
- g limits...+5/–2
- Endurance
  - at M0.91 at 9,150 m (30,000 ft)...............................16 min
- Power plant..1 × turbojet

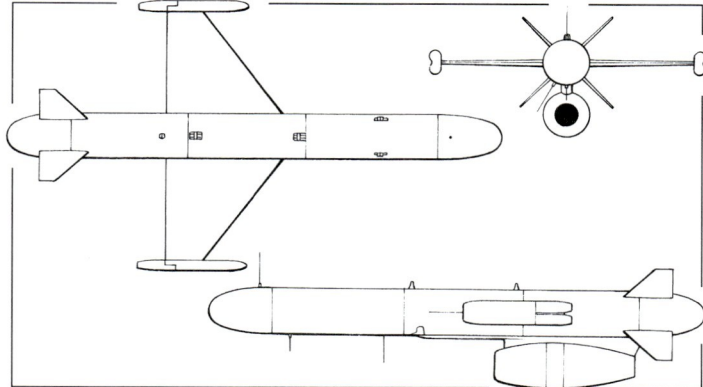

Fuji J/AQM-1 expendable air-launched target 0517951

Status: In production and service. First 20 to the TRDI (Technical Research and Development Institute) of the JDA; some 360 on order; more than 300 delivered to JASDF by March 2005. Deliveries were: 10 in FY00, 12 in FY01, 21 in FY02, 41 in FY03, 42 in FY04, 21 in FY05, 21 in FY06, 13 in FY07 and 26 in FY08.

Customers: Japan (Air Self-Defence Force and TRDI).
Contractor: Fuji Heavy Industries Ltd, Aerospace Company Utsunomiya.

Jordan

JARS Jordan Arrow

Type: Recoverable aerial target.
Development: Revealed at SOFEX exhibition in 2004. Designed for training and simulation with close-in and short-range air defence weapon systems, especially man-portable SAMs and AA artillery.
Description: *Airframe:* Bullet-shaped fuselage with pusher engine; mid-mounted delta wings, from which twin booms extend rearwards to support inverted V tail unit. No landing gear. Modular construction, of GFRP and aluminium alloys.

Mission payloads: Modular, and can include the following:
- IR augmentation by emitter
- hot nose, black-body IR source
- visual augmentation for optical tracking (continuous or strobe light emitter, smoke generator or smoke cartridges)
- up to four smoke tracking flares
- up to four IR flares
- up to four pyrotechnic tracking flares
- active and passive radar augmentation by transponder, up to 2 × 7.5 in Luneberg lenses or corner reflectors
- IR countermeasures (strobe jammer or flare dispenser)
- active and passive radar countermeasures (jammer or chaff dispenser)
- scoring equipment (acoustic or Doppler MDI or hit indicator)
- towed subtargets, deployed from underwing hardpoints; and
- other equipment such as IFF transponder, radar seeker simulator, radar or ultrasonic altimeter, or laser receivers and reflectors.

Payloads can be mixed and most carried simultaneously.

Guidance and control: Autonomous (pre-programmed), via digital automatic control system with GPS-based navigation; air vehicle tracking uses map overlay. Mission programme can be updated during flight. GCS can control up to four air vehicles simultaneously.

System composition: Four to eight air vehicles, GCS, catapult launcher and ground support equipment. System can also include two trucks for mobile operations, requiring a crew of five persons.

Launch: From ground or ship by pneumatic catapult; or from zero-length ramp by solid propellant booster rocket.

Recovery: Parachute recovery, controlled manually from GCS or automatically on termination of mission and in the event of failure of, or damage to, onboard systems. Flotation system and emergency radio beacon for flights over water.

Jordan Arrow preparing for launch *(Jordan Advanced Remote Systems)* 1464791

Jordan Arrow being launched *(Jordan Advanced Remote Systems)* 1464793

Jordan Arrow *(Jordan Advanced Remote Systems)* 1464792

A different interaction of the Jordan Arrow aerial target *(IHS/Patrick Allen)* 1137867

Power plant: One 18.6 kW (25 hp) unidentified piston engine standard, driving a two-blade pusher propeller; 28.3 kW (38 hp) UEL AR 731 or 741 rotary engine optional.

Jordan Arrow

Dimensions, External	
Overall	
length	3.00 m (9 ft 10 in)
height	0.60 m (1 ft 11½ in)
Wings, wing span	2.20 m (7 ft 2½ in)
Weights and Loadings	
Weight, Max launch weight	60 kg (132 lb)
Payload, Max payload	15 kg (33 lb)
Performance	
Altitude, Operating altitude	50 m to 5,000 m (160 ft to 16,400 ft)
Speed	
Max level speed	
with piston engine	189 kt (350 km/h; 217 mph)
with rotary engine	243 kt (450 km/h; 280 mph)
g limits	+8/–3
Radius of operation	
radio control	10.8 n miles (20 km; 12 miles)
autonomous mode	54 n miles (100 km; 62 miles)
Endurance	2 hr
Power plant	1 × piston engine

Status: As of 2011, trials with the Jordanian Air Defense have been completed and initial production was due to commence.
Contractor: Jordan Advanced Remote Systems
Amman.

Pakistan

Albadeey ABJT

Type: Recoverable aerial target.
Development: High-speed target for live-fire use with AAA and surface-launch or air-to-air missiles. Available in standard, autopilot and BVR versions.
Description: *Airframe:* Low-mounted swept wings; bullet-shaped fuselage; close-coupled twin tailbooms with trapezoidal fins/rudders and high-mounted enclosed tailplane/elevator. Landing gear optional. Composites construction.
  Mission payloads: Not specified.
  Guidance and control: Remotely piloted.
  Launch: Catapult launch with rocket assistance; or conventional wheeled take-off.
  Recovery: Belly or conventional wheeled landing.
  Power plant: One 19.6 daN (44.1 lb st) turbojet.

ABJT

Dimensions, External	
Overall, length	2.39 m (7 ft 10 in)
Wings, wing span	2.03 m (6 ft 8 in)
Weights and Loadings	
Weight	
Max T-O weight	20.0 kg (44 lb)
Max launch weight	20.0 kg (44 lb)
Performance	
Speed, Max level speed	243 kt (450 km/h; 280 mph)
Radius of operation	
mission, normal	1.1 n miles (2 km; 1 miles)
mission, autopilot	2.7 n miles (5 km; 3 miles)
mission, BVR	10.8 n miles (20 km; 12 miles)
Endurance	20 min
Power plant	1 × turbojet

Status: In production and service.
  Customers:
  Pakistan Army, Navy and Air Force; Yemen Air Force; manufacturers Bofors (Sweden), Contraves (Italy), Diehl Raytheon Missile Systems (Germany), Oerlikon (Switzerland) and Selex Galileo (Italy).
Contractor: Albadeey Technologies
  Karachi.

The ABJT small jet-powered target *(Albadeey)* 1290355

Albadeey Shahbaz

Type: Recoverable aerial target.
Development: The Shahbaz (Aurora) was developed as a low-cost, expendable aerial target for tracking and live firing with medium-calibre AA artillery and radar-guided missiles. It is available in autopilot and beyond visual range (BVR) versions.

Shahbaz gunnery and missile target on its launcher *(Albadeey)* 1290360

Shahbaz target being readied for use *(Albadeey)* 1290361

Description: *Airframe:* Constant-chord, high-mounted wings and low-mounted tailplane; box-section fuselage. No landing gear. Bears a close resemblance to the PAC Baaz.
  Mission payloads: No details known.
  Guidance and control: Remote or autopilot control.
  Launch: By bungee catapult.
  Recovery: By parachute or belly landing.
  Power plant: One 8.2 kW (11 hp) 100 cc two-stroke piston engine, driving a two-blade propeller.

Shahbaz

Dimensions, External	
Overall	
length	2.18 m (7 ft 1¾ in)
height	0.49 m (1 ft 7¼ in)
Wings, wing span	2.44 m (8 ft 0 in)
Engines, propeller diameter	0.56 m (1 ft 10 in)
Areas	
Wings, Gross wing area	1.00 m² (10.8 sq ft)
Weights and Loadings	
Weight	
Weight empty	17.0 kg (37 lb)
Max launch weight	22.0 kg (48 lb)
Performance	
Speed	
Cruising speed, max	108 kt (200 km/h; 124 mph)

Shahbaz

Radius of operation	
normal	1.6 n miles (3 km; 1 miles)
with autopilot	5.4 n miles (10 km; 6 miles)
Endurance	45 min
Power plant	1 × piston engine

Status: In service.
Customers: Pakistan Army.
Contractor: Albadeey Technologies
Karachi.

Albadeey Shahzore

Type: Recoverable aerial target.
Development: Designed for use with AAA and surface- or air-launched missiles. Available in standard, autopilot and BVR versions.
Description: *Airframe:* Typical 'model aircraft' configuration, with high-mounted wings (swept on leading-edge), box-section fuselage and conventional tail surfaces with swept fin. Tail wheel-type fixed landing gear. Composites construction.
   Mission payloads: Not specified.
   Guidance and control: Remotely piloted; also available in autopilot or BVR versions.
   Launch: From catapult with rocket assistance, or conventional wheeled take-off.
   Recovery: Belly or wheeled landing.
   Power plant: One 150 cm³ two-cylinder two-stroke engine, driving a two-blade tractor propeller.

Shahzore

Dimensions, External	
Overall, length	2.54 m (8 ft 4 in)
Wings, wing span	2.54 m (8 ft 4 in)
Weights and Loadings	
Weight	
Max T-O weight	28.0 kg (61 lb)
Max launch weight	28.0 kg (61 lb)
Performance	
Speed, Max level speed	148 kt (274 km/h; 170 mph)
Radius of operation	
mission, normal	1.6 n miles (3 km; 1 miles)
mission, autopilot	2.7 n miles (5 km; 3 miles)
mission, BVR	10.8 n miles (20 km; 12 miles)
Endurance, max	1 hr 30 min

Status: In production and service.

Shahzore target *(Albadeey)* 1290356

Shahzore missile and gunnery target *(Albadeey)* 1290359

Customers: Pakistan Army, Navy and Air Force; Yemen Air Force; various manufacturers, including Bofors (Sweden), Contraves (Italy), Diehl Raytheon Missile Systems (Germany), Oerlikon (Switzerland) and Selex Galileo (Italy).
Contractor: Albadeey Technologies
Karachi.

Integrated Dynamics Nishan

Type: Aerial target and decoy.
Development: Described as "representing the next generation of air defence training and simulation systems". Developed from 1997.
Description: *Airframe:* Mid-wing double-delta planform with midships-mounted canard surfaces; box-section fuselage with rounded corners; twin fins and rudders at approximately mid-span; pusher engine. No landing gear. Composites construction.
   Mission payloads: Nishan Mk II: Up to 24 smoke or 16 infra-red flares, which can be carried in combinations and activated as required. 'Hot nosecone' IR source; chaff dispenser pods; sea-skimming module; surveillance real-time camera module; acoustic and Doppler radar MDI; passive and active radar enhancements. System payloads can be mixed, and most can be carried simultaneously.
   Nishan TJ-1000: Up to eight smoke or infra-red flares, which can be carried in combinations and activated as required. Radar altimeter module; surveillance module; acoustic and Doppler radar MDI. System payloads can be mixed, and most can be carried simultaneously.
   Guidance and control: ID-TM6 GPS and data telemetry module and ATPS-1200 antenna tracking and positioning system. Nishan Mk II has AP-5000 digital autopilot with heading and height lock; TJ-1000 has IFCS-6000 integrated flight control system.
   System composition: Ten Mk II or six TJ-1000 air vehicles, plus: portable GCS-1200 ground station; ATPS-1200 antenna tracking and positioning system; programming and moving map display software; spares; and GSE-1200 ground support equipment subsystem.
   Launch: Zero-length launcher or pneumatic catapult.
   Recovery: Skid landing or parachute.
   Power plant: Mk II: One 18.6 kW (25 hp), 240 cm³ (optionally 22.4 kW; 30 hp) two-cylinder two-stroke engine, driving a two-blade pusher propeller. Fuel capacity 18 litres (4.75 US gallons; 4.0 Imp gallons).
   TJ-1000: One 26.7 daN (60 lb st) miniature turbojet. Fuel capacity 15 litres (4.0 US gallons; 3.3 Imp gallons).
Variants: *Nishan Mk II:* Piston-engined version.
   Nishan TJ-1000: Jet-powered version, with modified wings.

Nishan Mk II, Nishan TJ-1000

Dimensions, External	
Overall	
length	3.06 m (10 ft 0½ in)
height	0.89 m (2 ft 11 in)
Wings	
wing span	
Nishan Mk II	2.87 m (9 ft 5 in)
Nishan TJ-1000	2.20 m (7 ft 2½ in)
Areas	
Wings, Gross wing area	2.42 m² (26.0 sq ft)
Weights and Loadings	
Weight	
Weight empty	
Nishan Mk II	30.0 kg (66 lb)
Nishan TJ-1000	16.0 kg (35 lb)
Max launch weight	
Nishan Mk II	60.0 kg (132 lb)
Nishan TJ-1000	35.0 kg (77 lb)
Payload, Max payload	12.0 kg (26 lb)
Performance	
Altitude, Service ceiling	2,000 m (6,560 ft)
Speed	
Cruising speed, max	162 kt (300 km/h; 186 mph)
Loitering speed	135 kt (250 km/h; 155 mph)
Radius of operation	
Nishan Mk II	19 n miles (35 km; 21 miles)
Nishan TJ-1000	16 n miles (29 km; 18 miles) (est)
Endurance	
Nishan Mk II	1 hr
Nishan TJ-1000	30 min
Power plant	
Nishan Mk II	1 × piston engine
Nishan TJ-1000	1 × turbojet

Status: Mk II described as 'proven over many years of field use', implying service in Pakistan, but no details quoted of production numbers or customers.
Contractor: Integrated Dynamics
Karachi.

Integrated Dynamics Tornado

Type: Expendable target and decoy.
Development: Designed as a lightweight, high-speed target and decoy; development history not known.

ID's Tornado lookalike target and decoy *(Integrated Dynamics)*

Ababeel hand-launched target *(Paul Jackson)*

Description: *Airframe:* Scale representation of Panavia Tornado multirole combat aircraft. No landing gear. Composites construction, including detachable CFRP vertical tail.

Mission payloads: Corner reflectors and radar-reflecting mesh built into composites airframe. Combinations of flares and/or radar enhancement devices can be carried and activated as desired. IR flares can be carried on each wingtip for signature enhancement.

Guidance and control: Pre-programmed mission profiles, with autonomous navigation and tracking and altitude control. IFCS-6000 integrated flight control system. Multiple Tornados can be controlled simultaneously.

System composition: Eight air vehicles, pneumatic catapult launcher and portable GCS.

Launch: Designed for zero-length or pneumatic catapult launch. Launch velocity 38 m (125 ft)/s.

Recovery: Non-recoverable ('fire and forget').

Power plant: Two 8.9 daN (20 lb st) miniature turbojets. Fuel capacity 10 litres (2.6 US gallons; 2.2 Imp gallons).

Description: *Airframe:* Small, simple high-wing monoplane with GFRP fuselage and a polystyrene foam, epoxy laminated wing. Full-span flaperons and elevator; no rudder. Current models have increased fin area and less tapered wings. No landing gear.

Mission payloads: None carried.

Guidance and control: Radio command (four-channel VHF transceiver). Transmitter power 9.6 V at 1.2 Ah; RF output 1.5 W. Receiver power 4.8 V at 1.2 Ah. Rechargeable battery. Optional ground support equipment consists of a box containing a 2.5 litre (0.66 US gallon; 0.55 Imp gallon) fuel container, electric engine starter, ammeter, 12 V battery to power equipment, and a selection of hand tools.

Launch: Hand-launched or wheeled take-off.

Recovery: Belly landing on any reasonably level terrain, or wheeled landing.

Power plant: One 1.3 kW (1.8 hp) 10 cc single-cylinder two-stroke engine with two-blade propeller. Fuel capacity 0.38 litre (0.10 US gallon; 0.08 Imp gallon).

Tornado

Dimensions, External	
Overall	
length	2.15 m (7 ft 0¾ in)
height	0.89 m (2 ft 11 in)
Wings, wing span	2.36 m (7 ft 9 in)
Areas	
Wings, Gross wing area	0.63 m² (6.8 sq ft)
Weights and Loadings	
Weight	
Weight empty	18.0 kg (39 lb)
Max launch weight	25.0 kg (55 lb)
Payload, Max payload	5.0 kg (11.00 lb)
Performance	
Speed	
Max level speed	250 kt (463 km/h; 288 mph)
Loitering speed	70 kt (130 km/h; 81 mph)
Radius of operation, mission	54 n miles (100 km; 62 miles)
Endurance	20 min
Power plant	2 × turbojet

Status: Being promoted since 2008. No reports found of production/service status.

Contractor: Integrated Dynamics
Karachi.

Ababeel

Dimensions, External	
Overall	
length	1.41 m (4 ft 7½ in)
height	0.20 m (7¾ in)
Wings, wing span	1.60 m (5 ft 3 in)
Tailplane, tailplane span	0.60 m (1 ft 11½ in)
Engines, propeller diameter	0.28 m (11 in)
Weights and Loadings	
Weight	
Max T-O weight	2.8 kg (6.00 lb)
Max launch weight	2.8 kg (6.00 lb)
Performance	
Speed, Max level speed	75 kt (139 km/h; 86 mph)
Range, radio control	0.8 n miles (1 km; miles)
Endurance, max	20 min

Status: In production and service; several hundred produced.

Customers: Pakistan Army; Royal Saudi Air Force (120).

Contractor: Pakistan Aeronautical Complex, Aircraft Manufacturing Factory
Kamra.

PAC Ababeel

Type: Small arms target.

Development: The Ababeel (swallow) was designed as an air defence target for use with small arms and machine guns, entering service with the Pakistan Army in 1988. Also used as basic trainer for operators of the larger and faster Baaz.

PAC also manufactures the Meggitt Banshee target.

PAC Baaz

Type: Recoverable aerial target.

Development: Baaz (eagle) was designed as an air defence gunnery target for small arms, machine guns and short-range AAM systems with optical tracking facilities. It can also be used as a basic trainer for drone operators.

PAC also manufactures the Meggitt Banshee target.

Description: *Airframe:* Simple high-wing monoplane, with parallel chord wings and horizontal tail; GFRP fuselage, epoxy laminated polystyrene wing. Landing gear optional.

Mission payloads: Two smoke generators or IR flares.

PAC Ababeel small arms target *(Paul Jackson)*

PAC Baaz target

Baaz Mk II *(PAC)*

Guidance and control: Radio command, by up to seven-channel VHF/FM transceiver. Transmitter power 9.6 V at 1.2 Ah; RF output 1 W; rechargeable battery. Receiver power 6 V at 1.2 Ah; rechargeable battery. Transceiver has two built-in fail-safe devices. If transmitter fails, engine shuts down after 1.5 seconds; if power fails or receiver malfunctions for any reason, including projectile strikes, an independent auxiliary power supply automatically operates the same fail-safe sequence.
Launch: Normally by bungee-powered catapult; wheeled take-off optional. Set-up time 20 minutes; turnaround time 5 to 10 minutes.
Recovery: Direct landing on reinforced belly. Recovery of Baaz Mk II is by parachute (depending upon available space) or, optionally, on wheeled landing gear.
Power plant: One 6.0 to 7.5 kW (8 to 10 hp), 100 cc Quadra-Aerrow Q100B single-cylinder two-stroke engine in Mk I; two-blade propeller. Alternative 85 cc engine also available for Mk I; Mk II has a 125 cc engine. Fuel is 20:1 petrol/oil mixture.

Baaz Mk I, Baaz Mk II

Dimensions, External	
Overall, length	2.10 m (6 ft 10¾ in)
Wings	
wing span	
Baaz Mk I	2.44 m (8 ft 0 in)
Baaz Mk II	3.36 m (11 ft 0¼ in)
Weights and Loadings	
Weight	
Max launch weight	
Baaz Mk I	22 kg (48 lb)
Baaz Mk II	28 kg (61 lb)
Performance	
Speed, Max level speed	97 kt (180 km/h; 112 mph)
Range	
visual	0.8 n miles (1 km; miles)
optical	1.6 n miles (3 km; 1 miles)
Endurance	
max, Baaz Mk I	30 min
max, Baaz Mk II	40 min

Status: In production and service.
Customers: Pakistan armed forces; Royal Saudi Air Force (70).
Contractor: Pakistan Aeronautical Complex, Aircraft Manufacturing Factory Kamra.

Satuma Shooting Star

Type: Jet-powered recoverable target.
Development: Apparently still in project stage as of mid-2009.
Description: *Airframe:* Low-wing design with sweptback flying surfaces; circular-section fuselage with twin rear-mounted podded engines. No landing gear. Composites construction.
  Mission payloads: Up to eight flares; Luneberg lens; MDI.

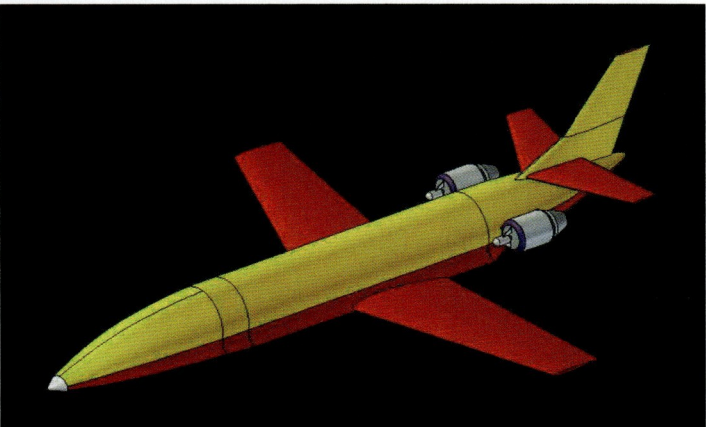
Computer-generated sketch of the proposed Shooting Star *(Satuma)*

Guidance and control: Intended for autonomous operation. Man-portable GCS with single console.
Launch: Assisted by booster rocket.
Recovery: Parachute recovery.
Power plant: Two (unidentified) 20.0 daN (45 lb st) turbojets.

Shooting Star

Dimensions, External	
Overall, length	2.74 m (8 ft 11¾ in)
Wings, wing span	1.07 m (3 ft 6¼ in)
Weights and Loadings	
Weight, Max launch weight	40 kg (88 lb)
Performance	
Altitude, Service ceiling	5,485 m (18,000 ft) (est)
Speed	
Cruising speed, max	297 kt (550 km/h; 342 mph) (est)
Radius of operation	108 n miles (200 km; 124 miles) (est)
Endurance, max	1 hr
Power plant	2 × turbojet

Status: Being promoted from 2009.
Contractor: Surveillance and Target Unmanned Aircraft (Satuma) Islamabad.

Satuma Tunder

Type: Recoverable aerial target.
Development: History not known.
Description: *Airframe:* Low-wing tail-less delta with sweptback fin; pusher engine. No landing gear. Composites construction.
  Mission payloads: Up to 16 smoke and/or IR augmentors.
  Guidance and control: SR by remote control with optical tracking; LR pre-programmed with GPS tracking. Dual, real-time datalinks.
  Launch: Catapult launch.
  Recovery: Parachute recovery or belly landing.
  Power plant: One (unidentified) 28.3 kW (38 hp) two-cylinder two-stroke engine; two-blade pusher propeller.
Variants: *Tunder SR:* Close-range (within LOS) version.
  Tunder LR: Longer-range version.

Tunder SR, Tunder LR

Dimensions, External	
Overall, length	2.83 m (9 ft 3½ in)
Wings, wing span	3.05 m (10 ft 0 in)
Weights and Loadings	
Weight, Max launch weight	68 kg (149 lb)

A Tunder SR alongside its launch trailer *(Satuma)*

Tunder at-sea recovery *(Satuma)*

Tunder SR, Tunder LR

Performance
Speed, Max level speed .... 162 kt (300 km/h; 186 mph)
Radius of operation
    mission, Tunder SR .... 5.4 n miles (10 km; 6 miles)
    mission, Tunder LR .... 21.5 n miles (39 km; 24 miles)
Endurance, max .... 1 hr 30 min

Status: In production and in service.
    Customers: Pakistan Air Force.
Contractor: Surveillance and Target Unmanned Aircraft (Satuma) Islamabad.

Poland

ITWL OCPJ-07x

Type: Recoverable aerial target.
Development: The OCPJ-07x recoverable aerial target is designed for training and as target for anti-aircraft defence forces (specifically for vehicle-launched surface-to-air missiles). The AV may also be known as Turbo Komar and its development is believed to have been completed by 2007. As of 2011, *IHS Jane's* sources were reporting the AV as possibly being 'in production and service'.
Description: *Airframe:* The OCPJ-07x AV features mid-set wings (with sweptback leading-edges), a tapered cylindrical fuselage and a T-shaped tailplane with endplate fins. Again, the AV is not fitted with landing gear.
    Mission payloads: Not stated.
    Guidance and control: Automatic, following a pre-programmed flight profile.
    Launch: Rail launch by catapult.
    Recovery: Parachute recovery.
    Power plant: One 16 daN (36 lb st) miniature turbojet.

OCPJ-07x

Dimensions, External
Overall, length .... 2.25 m (7 ft 4½ in)
Wings, wing span .... 2.04 m (6 ft 8¼ in)
Weights and Loadings
Weight, Max launch weight .... 21 kg (46 lb)
Performance
Climb
    Rate of climb, max, at S/L .... 1,200 m/min (3,937 ft/min)
Altitude, Operating altitude .... 100 m to 3,000 m (320 ft to 9,840 ft)
Speed
    Cruising speed, max .... 194 kt (359 km/h; 223 mph)
Radius of operation, mission .... 10.8 n miles (20 km; 12 miles)
Endurance, max .... 20 min
Power plant .... 1 × turbojet

Status: Most recently, the ITWL has continued to promote the OCPJ-07x recoverable aerial target.
Contractor: Instytut Techniczny Wojsk Lotniczych (Air Force Institute of Technology) Warsaw.

The OCPJ-07x recoverable aerial target shown mounted on its rail launcher *(ITWL)*   1509993

ITWL SMCP-JU Komar

Type: Recoverable aerial target.
Development: The SMCP-JU Komar (Mosquito) was developed as a training target for use with the Strzała-2M and Grom anti-aircraft missile systems and AA artillery. As such, the AV is reported to have entered service with the Polish Army during 2006.
Description: *Airframe:* The SMCP-JU Komar recoverable aerial target features mid-set, 'swept delta' wings, a bullet-shaped fuselage and a broad-chord tailfin. The AV has no landing gear and is constructed from composites.
    Mission payloads: IR emitter as standard (wingtip-mounted flare pods). Radar reflector and optical imagery source as options.
    Guidance and control: Telemetry control system and autopilot.

The SMCP-JU Komar recoverable aerial target *(ITWL)*   1509991

    System composition: A Polish Army SMCP-JU Komar system comprises six × Komar AVs, two × Szerszen targets (see separate entry) and a mobile transport facility (incorporating a launch rail, payload and guidance compartments and visual flight controls).
    Launch: Catapult launch.
    Recovery: Parachute recovery optional.
    Power plant: Single-cylinder two-stroke engine; two-blade tractor propeller.

Komar

Dimensions, External
Wings, wing span .... 2.20 m (7 ft 2½ in)
Weights and Loadings
Weight, Max launch weight .... 24 kg (52 lb)
Performance
Altitude, Service ceiling .... 1,000 m (3,280 ft)
Speed
    Max level speed .... 97 kt (180 km/h; 112 mph)
    Cruising speed, typical .... 43 kt (80 km/h; 49 mph)
Radius of operation .... 1.6 n miles (3 km; 1 miles)
Endurance .... 1 hr
Power plant .... 1 × piston engine

Status: Most recently, the ITWL has continued to promote the SMCP-JU Komar recoverable aerial target.
Contractor: Instytut Techniczny Wojsk Lotniczych (Air Force Institute of Technology) Warsaw.

ITWL SMCP-WU Szerszen

Type: Recoverable aerial target.

The SMCP-WU Szerszen recoverable aerial target shown mounted on its catapult launcher *(ITWL)*   1509994

Development: The SMCP-WU Szerszen (Hornet) was developed as a training target for use with the Strzała-2M and Grom portable anti-aircraft missile systems and AA artillery. As such, it is reported to have entered service with the Polish Army during 2006.

Description: *Airframe:* The SMCP-WU Szerszen recoverable aerial target features mid-set swept wings (with endplate fins) and a bullet-shaped body. Again, the AV is not fitted with landing gear and is constructed of composites.

Mission payloads: The SMCP-WU Szerszen's standard payload is a sleeve target (towed on a 100 m (330 ft) cable) and an MDI system. A radar reflector (3 m² RCS) and an optical imagery source are options.

Guidance and control: Telemetry control system and autopilot.

System composition: A Polish Army SMCP-WU Szerszen system comprises two × Szerszen AVs, six × Komar recoverable aerial targets (see separate entry) and a mobile unit (incorporating a launcher, payload and guidance compartments and visual flight controls).

Launch: Catapult launch.
Recovery: Parachute recovery.
Power plant: Two-cylinder two-stroke engine; two-blade tractor propeller.

Szerszen

Dimensions, External	
Wings, wing span	3.20 m (10 ft 6 in)
Weights and Loadings	
Weight, Max launch weight	35 kg (77 lb)
Performance	
Altitude, Service ceiling	1,000 m (3,280 ft)
Speed	
Max level speed	97 kt (180 km/h; 112 mph)
Cruising speed, typical	38 kt (70 km/h; 44 mph)
Radius of operation	1.6 n miles (3 km; 1 miles)
Endurance	2 hr
Power plant	1 × piston engine

Status: Most recently, the ITWL has continued to promote the SMCP-SU Szerszen recoverable aerial target.

Contractor: Instytut Techniczny Wojsk Lotniczych (Air Force Institute of Technology)
Warsaw.

Romania

Elmec ATT-01

Type: Recoverable aerial target.

Development: Electromecanica Ploiesti (Elmec) was founded in 1981. It is now integrated, together with other Romanian manufacturers, in the national company ROMARM. The ATT-01 target was developed for training and live firing with AA artillery and IR-homing or radar-guided missiles.

Description: *Airframe:* High-wing monoplane with slender fuselage, slightly tapered wings and angular tail surfaces; conventional three-axis flying control surfaces. Fixed 'taildragger' landing gear.

Mission payloads: Visual and IR augmentation.
Guidance and control: By radio remote control within line of sight.
Launch: Conventional wheeled take-off.
Recovery: Conventional wheeled landing.
Power plant: One 3.2 kW (4.3 hp) single-cylinder piston engine; two-blade propeller.

ATT-01

Dimensions, External	
Overall, length	2.20 m (7 ft 2½ in)
Wings, wing span	3.20 m (10 ft 6 in)
Weights and Loadings	
Weight	
Weight empty	18 kg (39 lb)
Max T-O weight	9.07 kg (20 lb)
Performance	
T-O	
T-O run	
prepared terrain	30 m (99 ft)
unprepared terrain	50 m (165 ft)
Altitude, Service ceiling	1,000 m (3,280 ft)
Speed	
Max level speed	76 kt (141 km/h; 87 mph)
Cruising speed	54 kt (100 km/h; 62 mph)
g limits	+2.8
Radius of operation, radio remote control within line of sight	1.4 n miles (2 km; 1 miles)
Endurance	30 min
Landing	
Landing run	50 m (165 ft)
Power plant	1 × piston engine

Status: In service. Available for export.
Customers: Romanian Army.
Contractor: Electromecanica Ploiesti (Elmec)
Prahova.

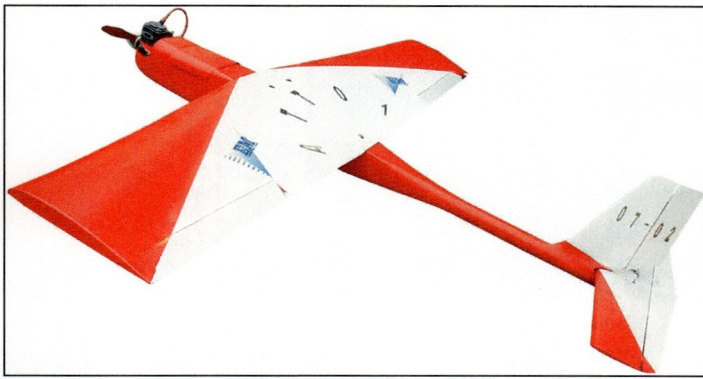

ATT-01 remotely piloted target *(Elmec)*   1139158

ATT-01 remotely piloted target   0533481

Elmec RT-3

Type: Ballistic target.

Development: The RT-3 was developed for live firing use at small-sized ranges against infra-red homing air defence missiles.

Description: *Airframe:* Cylindrical body with conical nosecone and four tailfins in X configuration.

Mission payloads: Not stated.
Guidance and control: By trajectory adjustment.
Launch: Rocket launch.
Recovery: Non-recoverable.
Power plant: Rocket motor(s).

Specifications

Dimensions	
Length overall	2.31 m (7 ft 7.0 in)
Body diameter (max)	141 mm (5.6 in)
Wing span	398 mm (1 ft 3.7 in)
Weights	
Max launching weight	43 kg (94.8 lb)
Performance	
Max flight speed	972 kt (1,800 km/h; 1,118 mph)
Max flight altitude (50° elevation)	4,500 m (22,965 ft)
Max flight distance (50° elevation)	5.4 n miles (10 km; 6.2 miles)
Minimum launch elevation	20°
Max flight time (50° elevation)	50 s
Environment temperature for operation	±50°C (−58 to +122°F)

Status: Remained in service 2011.
Customers: Romanian Army.

ATT-01 Romanian Army gunnery and missile target   0533482

Elmec RT-3 target rocket *(Elmec)*

Contractor: Electromecanica Ploiesti (Elmec) Prahova.

Elmec RT-11D

Type: Ballistic target.
Development: Developed for training and live firing at small-sized ranges against radar-guided ground-to-air missiles.
Description: Airframe: Cylindrical body with conical nosecone; two sets of four clipped-delta fins, one amidships and one at rear; four small vanes/fins at junction of nosecone and main target body.
    Mission payloads: Not stated.
    Guidance and control: By trajectory adjustment.
    Launch: Rocket launch.
    Recovery: Non-recoverable.
    Power plant: Rocket motor(s).

Specifications
Dimensions
    Length overall..11.464 m (37 ft 7.3 in)
    Body diameter (max)...0.654 m (2 ft 1.7 in)
Weights
    Max launching weight...2,200 kg (4,850 lb)

Performance
    Max flight speed...972 kt (1,800 km/h; 1,118 mph)
    Max flight altitude (50° elevation)....................................5,500 m (18,040 ft)
    Max flight distance (50° elevation).....................10.8 n miles (16 km; 12.4 miles)
    Max flight time (50° elevation)..70 s

Status: In service.
    Customers: Romanian Army.
Contractor: Electromecanica Ploiesti (Elmec) Prahova.

The RT-11D expendable rocket-powered target *(Elmec)*

Elmec TPDM-01

Type: Parachute target.
Development: The TPDM-01 was developed to simulate the IR and radar signatures of a full-size attack aircraft. It is used for training and practice firing with IR- or radar-guided air-to-air and surface-to-air missiles, and as a training aid for surveillance radar operators.
Description: Airframe: Target consists of a cylindrical main body, a parachute and an uncoupling system.
    Mission payloads: Polyhedral radar reflector; torch (minimum 2 million Cd intensity at ground level) and priming system.
    Launch: Air-dropped from host aircraft.
    Recovery: Normal parachute descent.

The TPDM-01 parachute target *(Elmec)*

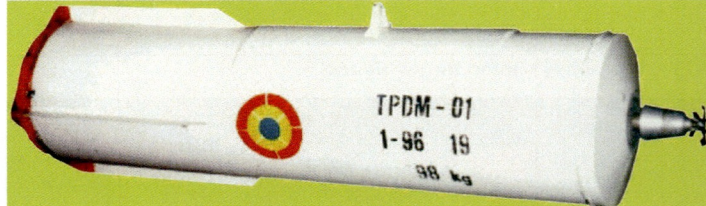

TPDM-01 target in stowed configuration *(Elmec)*

TPDM-01 interception *(Elmec)*

Specifications

Dimensions
Length...1.165 m (3 ft 9.9 in)
Max width..342 mm (1 ft 1.5 in)
Main body diameter..280 mm (11.0 in)

Weights
Total weight..98 kg (216 lb)

Performance
Max air-drop altitude...17,000 m (55,775 ft)
Torch burning time: at S/L...3 min
  at 3,000–6,000 m (9,840–19,680 ft)..4 min
  at 9,000–17,000 m (29,530–55,775 ft)...5 min
Rate of descent with burning torch and deployed reflector
  at 16,000–12,000 m (52,490–39,370 ft)......................12 m (329.4 ft)/s
  at 12,000–8,000 m (39,370–26,245 ft)..........................7 m (23.0 ft)/s
  at 8,000–4,000 m (26,245–13,120 ft)............................5 m (16.4 ft)/s
  below 4,000 m (13,120 ft)..4 m (13.1 ft)/s
Target discovery distance (optical means, clear daylight)..Up to 19 n miles (35 km; 22 miles)
Environment temperature for operation..................±50°C (–58 to +122°F)

Status: In service.
   Customers: Romanian armed forces.
Contractor: Electromecanica Ploiesti (Elmec) Prahova.

Russian Federation

Augur Au-23 and Au-26

Type: Hot-air balloon targets.

Development: These unmanned hot-air balloons are designed to imitate air targets for training of air defence missile and artillery combat units. They do not require a specially equipped launch site or highly qualified ground crews, and can perform round-the-clock flights in any weather, both independently and by command from the GCS.

Description: *Airframe:* Both types consist of the following main parts: balloon envelope, suspension system and rigging. Envelope of the Au-23 is 'natural' spherical shape; that of the Au-26 is of aerodynamic ('barrage balloon' configuration. Envelopes are made of high-strength, heat-resistant fabric. The suspension system supports the fuel system, flight control system components and burners; the rigging attaches the envelope to the suspension system and, in the case of the tethered Au-26, secures the balloon at its working altitude. Radar cross-section varies from 1.8 to 40 m² (19.4 to 430.5 sq ft), depending upon the equipment carried and almost regardless of flight aspect angle.

Mission payloads: Both types can be equipped with: active and passive jamming system; Luneberg lens passive radar reflector; radio transponder for tracking by ground radar stations; checkout equipment for prelaunch tests of target balloons; and automated target scoring and evaluation system.

Guidance and control: The Au-23 is available in three forms. Variant A operates automatically, with flight altitude maintained within the range 50 to 1,000 m (165 to 3,280 ft) by means of a barometric altimeter. Variant B is controlled remotely by the ground operator within a vertical flight profile from ground level up to 3,000 m (9,840 ft), and Variant C is a tethered balloon with an automatic system that maintains it at an altitude of up to 200 m (655 ft).

The Au-26 is available only as a tethered aerostat.

Launch: Conventional hot-air balloon inflation and take-off.

Recovery: Conventional hot-air balloon landing. Safe operation is ensured by an onboard soft-landing system in the event of fuel depletion or exit from the remote control zone. To facilitate search for undestroyed target balloons, each is fitted with a portable radio beacon.

Power plant: None.

Specifications

Dimensions
Envelope max diameter: Au-23.................................6.80 m (22 ft 3.7 in)
  Au-26...6.70 m (21 ft 11.8 in)
Length overall: Au-26..14.60 m (47 ft 10.8 in)
Height overall, incl suspension system: Au-23..........................8.90 m (29 ft 2.4 in)
Envelope volume: Au-23..120.0 m3 (4,240 cu ft)
  Au-26...284.0 m3 (10,030 cu ft)

Weights
Lifting force: Au-23...5.5 kg (12.1 lb)
  Au-26...12.5 kg (27.6 lb)

Status: Understood to be in production and service.
Contractor: Augur Aeronautical Centre Inc Moscow.

Augur RD series

Type: Remotely piloted airship.

Development: Details of the RD-1.5 were revealed by the Augur Aeronautical Centre (Vozdukhoplavatelnyi Tsentr Avgur) at the Moscow Air Show in August 1997, when it was described as 'a target for helicopters.' Augur has since increased the range to include the larger RD-2 and RD-2.5; all are produced and marketed by Augur's RosAeroSystems Division.

Description: *Airframe:* Semi-rigid, helium-filled envelope with small payload gondola; horizontal tail surfaces and ventral fin.

Mission payloads: Banners for indoor advertising (all versions); RD-2 and RD-2.5 can optionally carry an audio broadcasting system.

Guidance and control: Radio controlled. Ground crew of two.

Launch: Conventional airship take-off.

Recovery: Conventional airship landing.

Augur RD-1.5 remotely piloted airship target   0007432

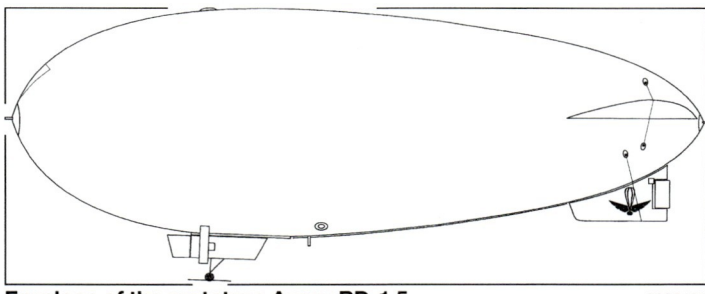

Envelope of the prototype Augur RD-1.5   0007431

Power plant: Two (probably vectored thrust) 59.7 kW (80 hp) Robby Power 600 piston engines, mounted one each side of gondola, plus single rear-mounted 67.1 kW (90 hp) Graupner Speed 600 engine.

Specifications

Dimensions
Envelope: Length: RD-1.5..6.05 m (19 ft 10.2 in)
  RD-2...8.08 m (26 ft 6.1 in)
  RD-2.5...7.44 m (24 ft 4.9 in)
Max diameter: RD-1.5..1.79 m (5 ft 10.5 in)
  RD-2..2.02 m (6 ft 7.5 in)
  RD-2.5..2.48 m (8 ft 1.6 in)
Fineness ratio: RD-1.5..3.38
  RD-2..4.00
  RD-2.5...3.00
Volume: RD-1.5..9.00 m3 (317.8 cu ft)
  RD-2..17.00 m3 (600.4 cu ft)
  RD-2.5...24.00 m3 (847.6 cu ft)
Height overall: RD-1.5...1.95 m (6 ft 4.8 in)
  RD-2..2.30 m (7 ft 6.6 in)
  RD-2.5..2.70 m (8 ft 10.3 in)
Tail surface area (total): RD-1.5...................................0.90 m2 (9.69 sq ft)
  RD-2..1.40 m2 (15.07 sq ft)
  RD-2.5...2.40 m2 (25.83 sq ft)

Weights
Weight empty: RD-1.5...6.2 kg (13.7 lb)
Max payload: RD-1.5..2.8 kg (6.2 lb)
  RD-2..5.1 kg (11.2 lb)
  RD-2.5...8.4 kg (18.5 lb)

Performance
Max level speed: RD-1.5...................................23 kt (43 km/h; 27 mph)
  RD-2, RD-2.5..19 kt (36 km/h; 32 mph)
Cruising speed: RD-1.5....................................19 kt (36 km/h; 32 mph)
  RD-2, RD-2.5..16 kt (29 km/h; 18 mph)
Speed with rear engine only
  RD-1.5...10 kt (18 km/h; 11 mph)
Endurance: at max level speed
  RD-1.5, RD-2..30 min
  RD-2.5..1 h
  at cruising speed above: RD-1.5...1 h 30 min

Status: RD-1.5 prototype flight tested in mid-1997. All three versions have been produced and are in service.
Contractor: Augur Aeronautical Centre Inc
Moscow.

ENICS E95

Type: Jet-powered recoverable aerial target.
Development: Revealed at Moscow Air Show August 1995; reappeared at 2001 event with modified tail unit. Designed for test and evaluation of AA weapons and for aerial gunnery training; simulates small-sized manoeuvring targets such as cruise missile, guided bomb or UAV. Said to be capable of performing 60 to 70 per cent of air defence training requirements. Within the system, the air vehicle is designated E95M, the pneumatic catapult E95P and the ground station E95U; the complete complex has an E95T designation.

ENICS E95 on display alongside its pulsejet engine in 2001 *(Paul Jackson)*

E95M development vehicle on its E95P pneumatic launcher

E95M in flight *(ENICS)*

Landed E95M with its recovery parachute *(ENICS)*

E95P launcher with target uploaded for launch *(ENICS)*

E95M target on E95P launcher *(ENICS)*

E95U and ground control truck *(ENICS)*

Description: *Airframe:* Cylindrical body with rounded nosecone and conical tail end; low-mounted, constant chord wings with upturned tips; original V tail (included angle 100°) superseded by conventional tailplane with twin endplate fins; dorsal pulsejet engine. No landing gear.

Mission payloads: 190 mm (7.5 in) Luneberg lens, corner reflector, or IR and smoke flares. Radar signature can be varied from 0.8 to 1.2 m^2 (8.6 to 12.9 sq ft) with corner reflector and to 7.0 to 8.0 m^2 (75.3 to 86.1 sq ft) with Luneberg lens. Pulsejet engine also provides realistic IR performance from both exhaust and hot metal.

Guidance and control: Radio control or pre-programmed autonomous, with computer-based tracking and GPS positioning. Flight profiles can include steep dives at angles up to 30° and snaking manoeuvres at bank angles up to 45°.

Transportation: Entire system ground-transportable in two 3-ton standard trucks plus GCS and launcher trailers; or by air in transport aircraft with cabin dimensions of 15 × 4 × 4 m (49.2 × 13.1 × 13.1 ft); or by rail or ship.

System composition: Air vehicle(s); shelter-mounted, climate-controlled, ground tracking and control station, carried on a GAZ-3308 all-terrain truck chassis; trailer-mounted launching ramp and pneumatic catapult; maintenance facility in pneumatic tent hangar. Ground crew of five. Turnround time 15 minutes.

Launch: By pneumatic catapult from trailer-mounted ramp.
Recovery: Parachute recovery.
Power plant: One 0.15 kN (33.7 lb st) ENICS M135 pulsejet, with glow-plug ignition.

E95

Dimensions, External	
Overall	
length	2.35 m (7 ft 8½ in)
height	0.64 m (2 ft 1¼ in)
Fuselage, width	0.25 m (9¾ in)
Wings, wing span	2.90 m (9 ft 6¼ in)
Weights and Loadings	
Weight, Max launch weight	70 kg (154 lb)
Performance	
Climb	
Rate of climb, max, at S/L	900 m/min (2,952 ft/min)
Altitude	
Operating altitude	
min	100 m (330 ft)
max	3,000 m (9,840 ft)
g limits	+6/−1
Radius of operation, mission, max	27 n miles (50 km; 31 miles)
Endurance, max	30 min

Status: Continuing in production and service in 2005-06.
Customers: Russian armed forces and manufacturers of air defence gun and missile systems.
Contractor: ENICS JSC
Kazan.

Sokol Dan

Type: Recoverable aerial target.
Development: The Dan (tribute) was first displayed publicly at the 1993 Moscow air show, having made its first flight in January of that year. It entered production to replace the obsolete Lavochkin La-17M, providing target simulation for anti-aircraft artillery and surface-to-air missile crews. A modernised version is now in production and service.
In 2005, Sokol was promoting a derivative known as Danem as an ecological monitoring UAV.
Description: *Airframe:* Mainly cylindrical body with ogival nosecone and dorsal intake for turbojet engine. Non-swept short-span wings, with narrow pod at each tip. Small-area conventional tail surfaces. No landing gear. Average life 10 flights.
Mission payloads: Dan can be fitted with wingtip pods housing smoke or infra-red augmentation devices; radar simulator; radar jammer; MDI; can also be equipped to carry towed subtargets. Nose-mounted EO sensor in Danem.
Guidance and control: Missions can be preprogrammed and/or remotely radio-controlled with radar tracking.
System composition: Aerial target(s); launcher; loader/transporter; ground automatic monitoring system; engine starter; surveillance radar; radio control station; tracking radar; fuel bowser; oil dispenser; airfield mobile ground power unit. Launch set-up time 30 minutes.
Launch: By solid propellant booster rocket from ground-based zero-length launcher. Pneumatic engine starter for turbojet (Danem).
Recovery: Parachute recovery system, landing on pneumatic shock-absorbers.
Power plant: Dan and Dan-M: One Granit MD-120 turbojet (1.18 kN; 265 lb st).
Danem: One UELrotary piston with three-blade, shrouded, pusher propeller; or one turbojet.
Variants: *Dan:* Standard target version, as described.
Dan-M: Upgraded version, revealed in mid-2004 and forecast at that time to enter production in 2005. Improved range and endurance resulting from reduced launch weight and more fuel-efficient version of MD-120 engine. Otherwise outwardly similar to original Dan.
Danem: Revealed 2005, the 'em' added to name signifying ecological monitoring; option of rotary piston engine or turbojet; then targeted for 2007 first flight. Retains general appearance of target, but larger.

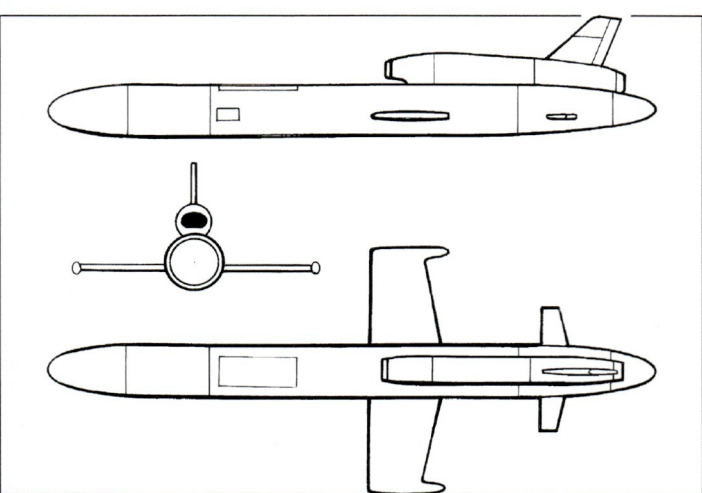
Sokol Dan subsonic jet-powered target (IHS/John W Wood) 0517953

Model of Danem at MAKS 2005 *(Robert Hewson)* 1151532

Dan aerial target on mobile launcher at the 2001 Moscow air show *(Paul Jackson)* 0536715

Latest Dan-M version of the Sokol target *(IHS/Patrick Allen)* 1137015

Dan, Dan-M, Danem

(**A:** rotary **B:** turbojet)
Dimensions, External
  Overall
    length
      Dan, Dan-M .... 4.60 m (15 ft 1 in)
      Danem .... 3.91 m (12 ft 10 in)
    height
      Dan, Dan-M .... 0.815 m (2 ft 8 in)
      Danem .... 0.91 m (2 ft 11¾ in)
  Fuselage
    width, Dan, Dan-M .... 0.45 m (1 ft 5¾ in)
  Wings
    wing span
      Dan, Dan-M .... 2.70 m (8 ft 10¼ in)
      Danem .... 3.76 m (12 ft 4 in)
Weights and Loadings
  Weight
    Max launch weight
      Dan .... 395 kg (870 lb)
      Dan-M .... 375 kg (826 lb)
      Danem (A) .... 180 kg (396 lb)
      Danem (B) .... 280 kg (617 lb)
  Payload
    Max payload, Danem .... 90 kg (198 lb)
Performance
  Climb
    Rate of climb
      max, S/L, Dan, Dan-M .... 1,380 m/min (4,527 ft/min)
  Altitude
    Operating altitude
      Dan, Dan-M .... 50 m to 9,000 m (160 ft to 29,520 ft)
      Danem (A) .... 500 m to 4,000 m (1,640 ft to 13,120 ft)
      Danem (B) .... 500 m to 6,000 m (1,640 ft to 19,680 ft)
  Speed
    Max level speed
      Dan, Dan-M .... 405 kt (750 km/h; 466 mph)
      Danem (A) .... 243 kt (450 km/h; 280 mph)
      Danem (B) .... 351 kt (650 km/h; 404 mph)
    Cruising speed
      Dan, Dan-M .... 162 kt (300 km/h; 186 mph)
      Danem (A) .... 108 kt (200 km/h; 124 mph)
      Danem (B) .... 162 kt (300 km/h; 186 mph)
  g limits
    Dan, Dan-M .... +9/–3
    Danem .... +5/–2
  Range
    Dan .... 216 n miles (400 km; 248 miles)
    Dan-M .... 367 n miles (679 km; 422 miles)
  Endurance
    Dan .... 40 min

Dan, Dan-M, Danem

Dan-M	1 hr 10 min
Danem (A)	3 hr
Danem (B)	1 hr 30 min
Power plant	
Dan, Dan-M	turbojet
Danem	piston engine

Status: In production from late 1993 by Strela factory in Orenburg.

Customers: Dan in service with Russian ground-based air defence forces.

Contractor: Sokol Experimental Design Bureau Kazan.

Strela (Lavochkin) La-17

Type: Aerial target and reconnaissance UAV.

Development: The former USSR's first purpose-designed powered aerial target to enter series production. Development was initiated in June 1950 together with that of a subsonic ramjet engine to power it. The first version entered service in late 1954; series production of final (turbojet-powered) version ended as recently as 1993. No overall production total is known, but in addition to the factories mentioned below the Tbilisi Aviation State Association (TASA) in Georgia is believed to have produced 886 La-17s during approximately 1956 to1962.

See under Variants below for further details. See also entry for essentially similar Chinese NUAA Chang Kong.

Description: *Airframe:* All-metal mid-wing monoplane; wings, and vertical and horizontal tail surfaces, all of constant chord. Circular-section fuselage made up of three welded subassemblies, of which central portion forms an integral fuel tank. No landing gear; instead, underslung engine nacelle cowling reinforced to withstand belly landing impact of 4,000 kg (8,818 lb).

Mission payloads: See under Variants.

Guidance and control: See under Variants.

Launch: La-17 originally air-launched; all other variants ground-launched by twin solid-propellant booster rockets.

Recovery: Belly landing on reinforced engine nacelle.

Power plant: One 5.89 kN (1,323 lb max thrust) RD-900 ramjet in La-17 and La-17A; one 19.12 kN (4,299 lb st) RD-9BK turbojet in La-17M (RD-9BKR in La-17MA and La-17R); one 24.03 kN (5,401 lb st) R-11K-300 in La-17MM.

Variants: *La-17 (Article 201):* Initial version. Airframe designed by I A Merkulov of Lavochkin OKB 301, RD-900 ramjet engine by M M Bondaryuk of OKB 670. Prototypes built 1951; first flight 13 May 1953 (air-launched from modified Tu-4 bomber). Payloads included a radar reflector on each wing and one in the tailcone, and smoke generators or light tracer for night firing capability. MRV-2 guidance system, with tracking by P-30 or SON-4RR radar. Unsuccessful parachute recovery system abandoned in favour of belly landing on reinforced engine nacelle. In series production (reported total of 847) by Orenburg GAZ (now PO Strela) until 1956; entered PVO service shortly after completion of state acceptance trials in September 1954. Ground-launched version (by two solid-fuel booster rockets) designated **La-17A** (Article 201A).

La-17M (Article 203): Redesigned for turbojet power in the light of dissatisfaction with suitability of ramjet as power plant type; RD-9BK engine modified from RD-9B (minus afterburner) as used in MiG-19 fighter; modified fuel system. Ground-launched by two PRD-23M-203 solid-propellant booster rockets from wheeled trolley. Improved radio guidance system (MVR-2M) and AP-73 autopilot. Radar reflectors increased to two per wing. Prototypes built 1954; series production at Orenburg 1958 to1964; entered service 1960. Modified version with RD-9BKR engine introduced 1961 as **La-17MA** (Article 202).

An La-17R, with booster rockets, on its launch trolley *(RART collection, via Nigel Eastaway)* 0075708

View of an La-17R showing underfuselage camera fairing forward of engine intake *(Nigel Eastaway)* 0075709

La-17R (Article 204): Reconnaissance version, developed in 1958–59. Airframe and engine as La-17M, but with two AFBA-40 (high-altitude) or AFBA-21 (low-altitude) photographic cameras in underside of lengthened nosecone. Pre-programmed with new AP-63 autopilot. Air vehicle transported by ZIL 134K truck, combined system being designated SATR-1; launch and recovery as for La-17M. Production undertaken by GAZ in Smolensk; entered service 1962; withdrawn in about 1974.

LA-17MM (Article 202M): Development at Orenburg started 1962, principally to enhance operating height range of La-17M, to simulate low-altitude threats, and to increase endurance. New, replaceable landing skids with improved impact absorption; new automatic landing system developed. Production began in 1964. Power plant originally as for La-17M, but supply of MiG-19 engines virtually exhausted by the mid-1970s and R-11K-300 (adapted from R-11F2S-300 of MiG-21, minus the afterburner) selected instead. This version (still designated La-17MM) continued in production at Orenburg/Strela from 1978 to 1993.

La-17K: Target drone version, with RD-9BKR turbojet replaced by Sokol R-11K-300; some systems and ground equipment also modernised. Entered production at Orenburg 1978, continuing until mid-1993; still in service (including use as surrogate UAV) at turn of century.

La-17, La-17A, La-17R, La-17M, La-17MA, LA-17MM, LA-17K

Dimensions, External	
Overall	
length	
La-17, La-17A, La-17M, La-17MA, LA-17MM, LA-17K	8.435 m (27 ft 8 in)
LA-17R	8.98 m (29 ft 5½ in)
height	
LA-17R	2.98 m (9 ft 9¼ in)
LA-17MM	3.03 m (9 ft 11¼ in)
Fuselage, width	0.55 m (1 ft 9¾ in)
Wings, wing span	7.50 m (24 ft 7¼ in)
Tailplane, tailplane span	2.18 m (7 ft 1¾ in)
Weights and Loadings	
Weight	
Max launch weight	
La-17	1,810 kg (3,990 lb)
La-17A	2,410 kg (5,313 lb)
La-17MA	2,775 kg (6,117 lb)
LA-17R	2,840 kg (6,261 lb)
La-17M	2,760 kg (6,084 lb)
LA-17MM	3,065 kg (6,757 lb)
Performance	
Altitude	
Operating altitude	
La-17M	3,000 m to 17,000 m (9,840 ft to 55,780 ft)
LA-17MM	100 m to 17,500 m (320 ft to 57,420 ft)
La-17, La-17A	10,000 m (32,800 ft)
LA-17MA	17,000 m (55,780 ft)
LA-17R	7,000 m (22,960 ft)
Speed	
Max level speed	
La-17	491 kt (909 km/h; 565 mph)
La-17M	486 kt (900 km/h; 559 mph)
LA-17MM	518 kt (959 km/h; 596 mph)
Radius of operation, LA-17R	216 n miles (400 km; 248 miles)
Endurance	
La-17, La-17A	40 min
LA-17R	45 min
La-17M, La-17MA, LA-17MM	1 hr

Status: Production complete. Later variants were still in service post-2000, may now be out of use.

Customers: Russian/former USSR armed forces; exports and production licence to China.

Contractor: Final production by Strela Orenburg Production Association (PO Strela) Orenburg.

Slovenia

Aviotech RVM01

Type: Recoverable aerial target.
Development: History not known.
Description: *Airframe:* High-wing monoplane with sweptback wing and tail surfaces.
   Mission payloads: Four infra-red flares.
   Guidance and control: Radio-controlled, with visual tracking unit.
   Launch: No information.
   Recovery: No information.
   Power plant: One 4.9 kW (6.6 hp) single-cylinder piston engine; two-blade propeller.

RVM01	
Dimensions, External	
Fuselage, length	1.90 m (6 ft 2¾ in)
Wings, wing span	2.30 m (7 ft 6½ in)
Weights and Loadings	
Weight, Max launch weight	13.5 kg (29 lb)
Performance	
Speed, Max level speed	108 kt (200 km/h; 124 mph)
Radius of operation	
LOS	0.8 n miles (1 km; miles)
with visual tracking unit	2.2 n miles (4 km; 2 miles)
Power plant	1 × piston engine

Status: In production and service from 2005, when new version(s) were reportedly under development, no subsequent information has been received.
   Customers: Slovenian Army.
Contractor: Aviotech Ltd
Ptuj.

Aviotech RVM01 gunnery target   0527013

South Africa

BAE Systems LOCATS

Type: Recoverable aerial target.
Development: LOCATS (LOw Cost Aerial Target System) is designed for the training of anti-aircraft artillery and surface-to-air missile crews.

LOCATS on cradle *(IST)*   0569548

Description: *Airframe:* Low-mounted tapered wings; conventional tail surfaces. Glass fibre modular construction. Typical lifetime of more than 30 missions with minimum repair.
   Mission payloads: Can include telemetry and sensors; towed banner or sleeve target; IR enhancement; radar transponder; smoke flares; and acoustic or radar MDI and scoring systems.
   Guidance and control: Remote pilot control via a 60 MHz nine-channel radio link. Autopilot with altitude and attitude lock; GPS navigation. LOCATS can be equipped as a simple target for flight within LOS only, or as a fully instrumented one capable of OLOS operation on preprogrammed flight paths to simulate a modern attack aircraft.
   System composition: Three air vehicles; trailer-mounted launcher; 4 × 4 configuration support and towing vehicle. Containers are incorporated into the support vehicle and trailer to carry all ground support equipment, a communications system, spares and the airframes. System is crewed by a remote pilot and two technicians; the air vehicle can be readied for launch within an hour of arriving on site.
   Launch: By pneumatic catapult from zero-length launcher.
   Recovery: Integral parachute recovery system or skid landing.
   Power plant: One 17.9 kW (24 hp) two-cylinder two-stroke engine; two-blade propeller.

LOCATS target drone at the South African Army Battle School at Lohathla *(IHS/Michael J Gething)*   0024445

LOCATS

Dimensions, External
Overall
  length..2.90 m (9 ft 6¼ in)
  height..0.90 m (2 ft 11½ in)
Wings, wing span...3.20 m (10 ft 6 in)
Engines, propeller diameter..................................0.60 m (1 ft 11½ in)

Weights and Loadings
Weight
  Weight empty...48 kg (105 lb)
  Max launch weight...70 kg (154 lb)
Payload
  Max payload, incl fuel..22 kg (48 lb)

Performance
Altitude, Service ceiling.......................................2,200 m (7,220 ft)
Speed
  Launch speed..30.35 m/s (100 ft/s)
  Max level speed..167 kt (309 km/h; 192 mph)
  Stalling speed...49 kt (91 km/h; 57 mph)
Radius of operation
  simple target..1.3 n miles (2 km; 1 miles)
  with GPS autopilot...21.6 n miles (40 km; 24 miles) (est)
Endurance, max..1 hr 30 min

LOCATS on launcher *(IST)*

Status: In production.
  Customers: South African Army.
Contractor: BAE Systems, Land Systems Dynamics Waterkloof.

LOCATS catapult launch *(IHS/Michael J Gething)*

Denel Skua

Type: High-speed recoverable target.
Development: Begun by Kentron (now Denel) as a technology demonstration programme, part-funded by the South African Air Force, for a dual-role (reconnaissance and target) high-speed drone. Initial launch and recovery trials took place in 1987 and first of 18 complete all-up tests in January 1990. Prototypes were followed by preproduction batch of six, used at Overburg Test Range during qualification of V3C A-Darter air-to-air missile. The Skua entered SAAF service in 1993. Its primary function is in the development and testing of air-to-air missiles, but it can also be used to simulate high-speed attack aircraft over land or water. In 2009 Skua was being deployed during continuing full-scale development trials of A-Darter missile, which are expected to continue until 2011-2012.

Description: *Airframe:* Long cylindrical body with ogival nose cone, high-mounted swept wings and twin fins and rudders; underslung turbojet engine; wingtip equipment pods. All-composites construction.
  Mission payloads: Can be fitted with radar augmentation, infra-red or visual detection systems and miss-distance scoring equipment. Can also tow a separate target body on a 2,400 m (7,875 ft) cable; passive or active miss-distance measuring equipment and associated telemetry can be installed either in the tow target or in the Skua's fuselage. Other payloads can be installed to customers' requirements.

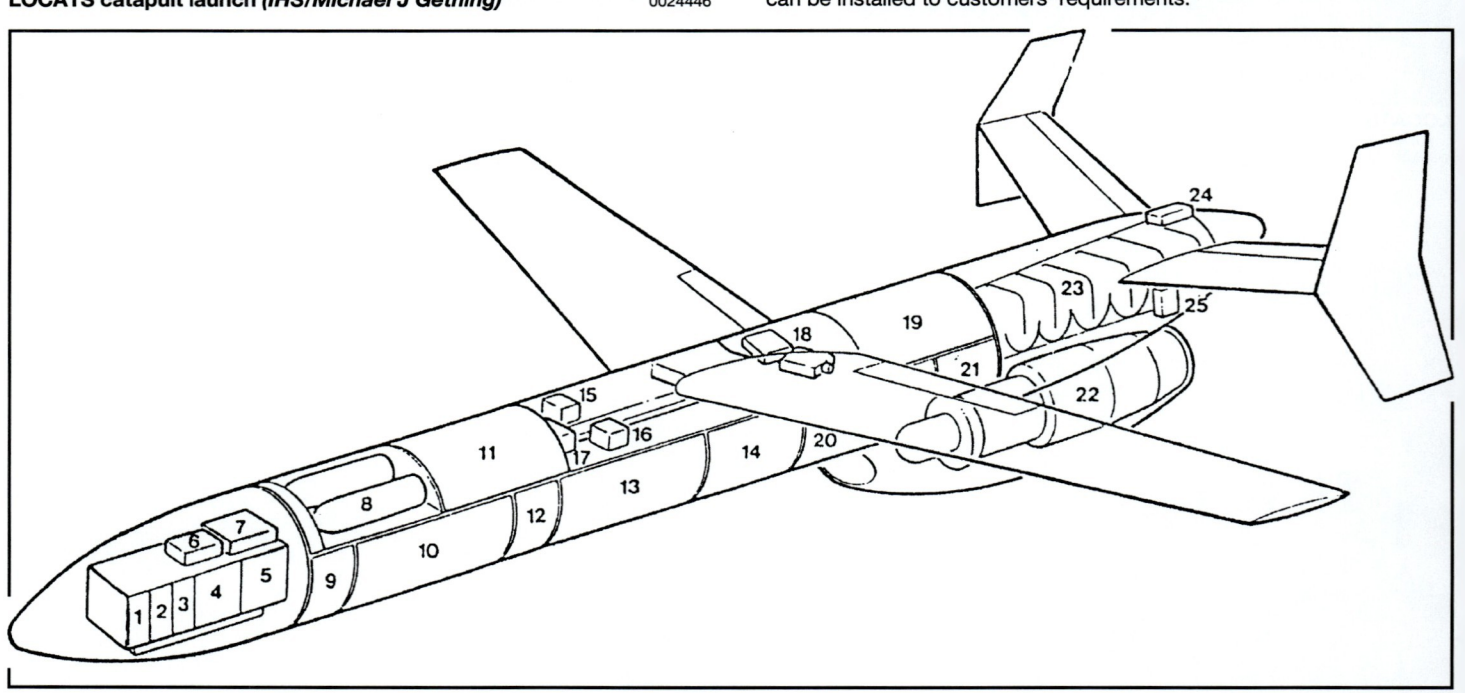

1	Flight control computer	9	Service panel	19	Landing airbag
2	Command communication	10	Payload bay	20	Fuel tank
3	Power control	11	Landing airbag	21	Spare compartment
4	Gyro	12-14	Fuel tanks	22	Turbojet
5	Spare position	15	Magnetometer	23	Main parachute
6	Air data transducer	16	Accelerometer	24	Parachute release
7	GPS receiver	17	Recovery controller	25	Elevator servo
8	High-pressure air cylinder	18	Aileron servo		

Internal features of the Denel Skua

Zero-length launch of a Denel Skua *(Alf Yssel/MediaMakers)*
0079575

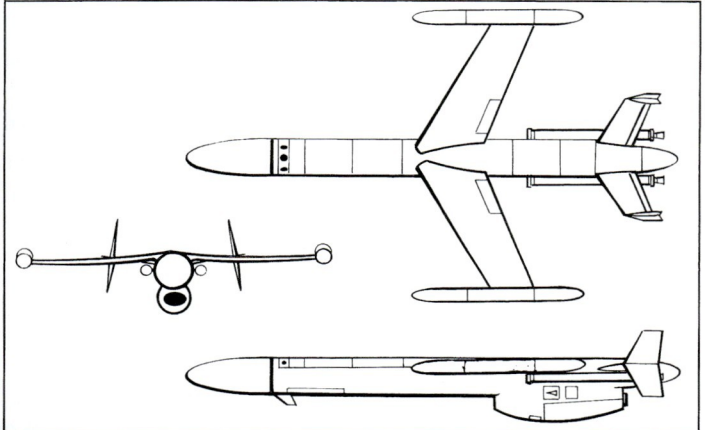
Skua high-speed jet-powered target *(IHS/John W Wood)* 0518008

Skua model on display in September 2006 *(IHS/Patrick Allen)*
1196242

Guidance and control: Missions can be preprogrammed or remotely piloted. An onboard GPS receiver determines the position of the air vehicle and, if the command signal is lost, Skua will fly autonomously by GPS inputs to the recovery area. The GCS-3 ground control station uses UHF telecommand and telemetry links and can control two Skuas simultaneously; a datalink is optional. Tracking is via position feedback from air vehicle's navigation system.

System composition: Four to eight air vehicles; GCS-3 mobile ground control station; LRN-3 launcher unit; ground support equipment.

Launch: By pair of solid-propellant booster rockets from LRN-3 self-contained zero-length launching ramp.

Recovery: By two-stage parachute system to airbag inverted landing. Can be recovered over water.

Power plant: One 4.50 kN (1,012 lb st) turbojet.

Skua

Dimensions, External	
Overall, length	6.00 m (19 ft 8¼ in)
Fuselage, width	0.50 m (1 ft 7¾ in)
Wings, wing span	3.57 m (11 ft 8½ in)
Weights and Loadings	
Weight, Max launch weight	700 kg (1,543 lb)
Payload	
Max payload	
internal	70 kg (154 lb)
external	160 kg (352 lb)
Performance	
Altitude, Operating altitude	10 m to 10,700 m (40 ft to 35,100 ft)
Speed, Max level speed	497 kt (920 km/h; 572 mph) at S/L
	500 kt (926 km/h; 575 mph) at 10,000 m (32,808 ft)
Stalling speed, IAS	150 kt (278 km/h; 173 mph) (est)
g limits	+6
Radius of operation, LOS	108 n miles (200 km; 124 miles)
Endurance	
at M0.75 at 10,000 m (32,800 ft)	1 hr 25 min
Power plant	1 × turbojet

Status: In production and service.

Customers: South African Air Force; United Arab Emirates Navy.

Contractor: Denel Aerospace Systems
Centurion.

Spain

INTA ALBA

Type: Lightweight, recoverable, gunnery and missile target.

Development: Until 1994, Spanish Army AAA training was conducted with relatively high-specification drones towing a secondary target, but in the following year it launched a programme to reduce costs and improve efficiency by developing simpler, low-speed targets using experience gained from the SIVA surveillance UAV programme. This resulted in the ALO close-range reconnaissance UAV (which see) and the ALBA (*Avión Ligero Blanco Aéreo:* lightweight aerial target), designed by INTA and manufactured by Sistemas de Control Remoto (SCR). Successful live firing trials with Mistral missiles were first conducted in October 1996 and have continued routinely since then.

Description: *Airframe:* Simple high-wing monoplane with V tail (included angle 110°); no landing gear. Modular construction (fuselage, tail unit and wing) of GFRP with some carbon fibre reinforcement.

Mission payloads: Two smoke canisters under each wing and two IR flares at each wingtip; MDI; small internal bay for additional payload or equipment. Autopilot has altitude and heading mode for flight stabilisation during OLOS operation. Optional azimuth control system for commanded OLOS navigation.

(INTA) ALBA air target being prepared, launched and flown during a demonstration flight at CEDEAR Test Range fitted with smoke and IR flares *(IHS/Patrick Allen)*
1021242

Aerial targets > Spain > INTA ALBA – INTA Diana

Close-up of engine on (INTA) Alba Air Target *(IHS/Patrick Allen)* 1021313

ALBA in flight *(IHS/Patrick Allen)* 0522128

ALBA and Mirach 100/4 targets in store at Spain's CEDEA test range *(IHS/Patrick Allen)* 0523963

Guidance and control: Futaba 16 channel UHF (35 to 40 MHz) PCM radio control transmitter, with booster capable of dual transmission for fail-safe operation. GPS navigation with real-time datalink. A UHF uplink based on a Becker MCS-30 12-channel transmitter, with a double 'follow-on' switcher for activation of the IR and smoke flares, is available optionally.

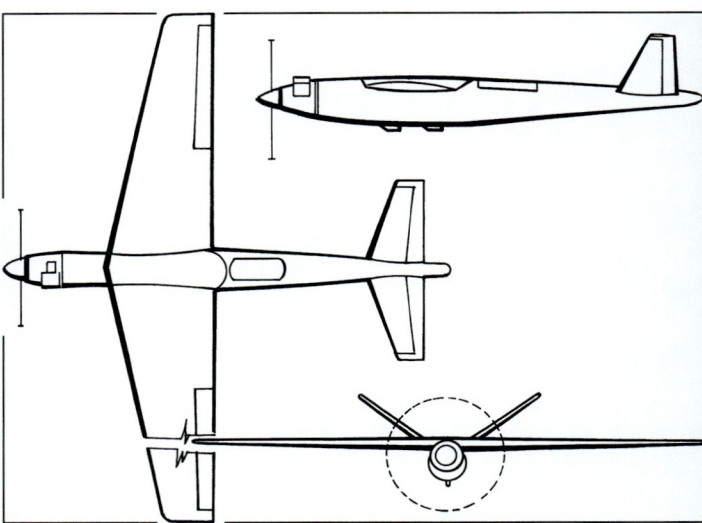

The V-tailed ALBA gunnery target *(IHS/John W Wood)* 0054182

Transportation: Aircraft dismantles into fuselage, wing and tail sections and is man-portable by one or two persons or transportable in a light vehicle.
Launch: From foldable elastomeric launching ramp (6 m; 19.7 ft long when deployed), transportable by lightweight truck. Turnaround time between flights approximately 15 minutes.
Recovery: By cruciform parachute.
Power plant: One Quadra Aerrow 75 cc (4.6 cu in) single-cylinder two-stroke engine (6.0 kW; 8 hp at 7,500 rpm); two-blade propeller.

ALBA

Dimensions, External	
Overall, length	1.80 m (5 ft 10¾ in)
Wings, wing span	2.23 m (7 ft 3¾ in)
Weights and Loadings	
Weight	
Weight empty, equipped	12 kg (26 lb)
Payload	
Max payload, internal	3 kg (6.00 lb)
Performance	
Speed	
Cruising speed, max	124 kt (230 km/h; 143 mph) (est)
Radius of operation, LOS	2.7 n miles (5 km; 3 miles)
Endurance	60 min (est)
Power plant	1 × piston engine

Status: As of 2011, in production and service. Standard Spanish Army gunnery training target since 1999.
Customers: Spanish Army.
Contractor: Instituto Nacional de Técnica Aeroespacial
Madrid.

INTA Diana

Type: Jet-powered, recoverable aerial target.
Development: Introduced at Eurosatory defence exhibition in Paris, June 2006. Described as highly manoeuvrable, with capability for evasive or sea-skimming flights; also suitable as a technology R & D platform in addition to target use.
Description: Airframe: Mid-mounted, short-span wings; cylindrical fuselage, with ogival nose- and tailcones; four tailfins in X configuration (upper pair short and stubby, lower pair larger, with movable control surfaces); engine in underslung pod amidships. No landing gear. Carbon fibre and epoxy construction.

Mission payloads: Various. Internal bays in forward and rear fuselage (capacities 10 and 15 kg; 22.0 and 33.1 lb respectively). Payloads can also be fitted in underwing or wingtip pods; tow-body subtargets can also be carried.

Diana high-speed aerial target *(IHS/Patrick Allen)* 1182164

Guidance and control: No details provided.
Launch: Ramp-launched under own power.
Recovery: By parachute stowed in fuselage aft of rear payload bay.
Power plant: One turbojet (type and rating not stated) in underfuselage pod. Fuel tank in centre-fuselage.

Diana

Dimensions, External	
Overall	
length	3.47 m (11 ft 4½ in)
height	0.675 m (2 ft 2½ in)
Fuselage	
height, max	0.355 m (1 ft 2 in)
Wings	
wing span, excl pods	1.845 m (6 ft 0¾ in)
Tailplane, tailplane span	0.97 m (3 ft 2¼ in)
Weights and Loadings	
Weight	
Weight empty	90 kg (198 lb)
Max launch weight	160 kg (352 lb)
Payload, Max payload	25 kg (55 lb)
Performance	
Altitude, Service ceiling	6,000 m (19,680 ft)
Speed, Max level speed	388 kt (719 km/h; 447 mph) (est)
Radius of operation, mission	48 n miles (88 km; 55 miles)
Endurance, max	1 hr (est)
Power plant	1 × turbojet

Status: As of early 2008, the status of this system was uncertain.
Contractor: Instituto Nacional de Técnica Aeroespacial
Madrid.

SCR Scrab I

Type: Jet-powered recoverable aerial target.
Development: Designed as low-cost target for use with missile firings for weapon evaluation and acceptance testing, and for anti-aircraft air defence training. Development began in or about 2000, and the target was first demonstrated to the Spanish MoD at the CEDEA test range in southern Spain on 30 May 2001.
Description: Airframe: Mid-wing configuration, with cylindrical fuselage; outer wings swept back; rectangular wing centre-section extends to rear of aircraft and supports hexagonal endplate fins. Dorsally mounted engine. No landing gear.
 Mission payloads: Can include IR or smoke flares, Luneberg lenses, corner reflectors, laser-reflective strips and MDI system.

Guidance and control: Choice of two operating modes: manual, by telecommand within LOS; or BLOS autonomous, using autopilot and GPS navigation with preprogrammable waypoints.
 System composition: Five air vehicles; portable control station; portable launching ramp; replacement parts and documentation.
 Launch: By elastomeric catapult.
 Recovery: Parachute recovery.
 Power plant: One 0.16 kN (35.3 lb st) turbojet.

Scrab I

Dimensions, External	
Overall, length	1.95 m (6 ft 4¾ in)
Wings, wing span	1.62 m (5 ft 3¾ in)
Weights and Loadings	
Weight	
Weight empty	12 kg (26 lb)
Payload, Max payload	6 kg (13.00 lb)
Performance	
Speed, Cruising speed	194 kt (359 km/h; 223 mph)
Radius of operation	
LOS	0.8 n miles (1 km; miles)
BLOS	32.4 n miles (60 km; 37 miles)
Endurance	40 min
Power plant, 0.16 kN (35.3 lb st)	1 × turbojet

Status: In production for, and in service with, Spanish Army and Navy.
 Customers: Initial contracts from Spanish Army and Navy in 2003, with deliveries beginning in same year. Further contracts placed by both services in 2004, 2005, 2006, 2007 and 2008.
Contractor: Sistemas de Control Remoto
Madrid.

SCR Scrab II

Type: Jet-powered recoverable target.
Development: Scrab II, developed at the request of the Spanish Army's artillery headquarters, is essentially an enlarged and twin-engined version of the Scrab I, designed for increased payload, speed and autonomy. A prototype made its maiden flight in June 2004.
Description: Airframe: Generally as for Scrab I except for twin dorsally mounted jet engines.
 Mission payloads: Can include smoke generator; IR flares; Luneberg lenses; corner reflectors; MDI system.
 Guidance and control: Autonomous, as for Scrab I, with autopilot and GPS navigation.
 System composition: As for Scrab I, except includes five air vehicles.
 Launch: By elastomeric ramp.
 Recovery: Parachute recovery.
 Power plant: Two 0.22 kN (48.5 lb st) AMT turbojets.

The long-nosed, single-jet Scrab I *(SCR)*

Scrab II with launcher on Spanish warship *(SCR)*

Scrab I jet-powered target on launching ramp *(IHS/Patrick Allen)*

Scrab II (at rear), with its smaller sibling *(IHS/Patrick Allen)*

Scrab II

Dimensions, External	
Overall, length	2.94 m (9 ft 7¾ in)
Wings, wing span	2.52 m (8 ft 3¼ in)
Weights and Loadings	
Weight	
Weight empty	31 kg (68 lb)
Payload, Max payload	10 kg (22 lb)
Performance	
Speed, Cruising speed	233 kt (432 km/h; 268 mph)
Range	43 n miles (79 km; 49 miles)
Endurance	50 min
Power plant	2 × turbojet

Status: Demonstrated to Spanish Army in 2004; production started in 2005. In service with Spanish Army and Navy.

Customers: First system (three aircraft, launcher and GCS) delivered to INTA of Spanish Ministry of Defence in 2005; second system delivered in March 2007. Spanish Navy ordered a four-aircraft system, delivered in 2007.

Contractor: Sistemas de Control Remoto
Madrid.

Sweden

Bülow PM-8

Type: Towed target.
Development: The PM-8 is a small, dart-shaped tow-body target similar in configuration to the range produced by Meggitt in the USA.
Description: *Airframe:* Main body is of phenolic material, offering a good radar response from the sides. Nose- and tailcones are of ABS plastics; tailfins are aluminium. Towline is attached at the tow-body's CG.
Mission payloads: Target is equipped with two internal spherical radar reflectors and one large forward-pointing reflector. Options include a built-in MDI, laser prisms or tapes, and altitude reporting equipment for sea-skimming operation.
Launch: Compatible with Bülow BTW 832 winch (which see).
Recovery: See above.
Power plant: None.

PM-8

Dimensions	
Length overall	2.08 m (6 ft 9.9 in)
Body diameter	0.225 m (8.9 in)
Span over tailfins	0.595 m (1 ft 11.4 in)
Weights	
Total weight	15–30 kg (33.1–66.1 lb)
Performance	
Max towing speed	400 kt (741 km/h; 460 mph)

Status: Not known.
Contractor: Bülow Air Target Systems AB
Vätö.

The Bülow PM-8 radar reflecting tow target 0024447

Bülow sleeve targets

Type: Towed targets.
Development: Bülow has been producing air-towed targets since 1947; its production of sleeve targets alone exceeds 20,000. Recent innovations include targets with constant pressure devices, IR targets, altitude reporting and helicopter-towed sea-skimming sleeve targets.
Description: *Airframe:* See individual descriptions of typical variants. All sleeve targets can be manufactured to individual customer requirements. Figures in designations indicate front diameter, rear diameter and sleeve length in millimetres. Towline is attached to rear of sleeve.
Mission payloads: All sleeve targets can be fitted with MDI equipment, including Bülow's own BTS 900 scoring system.

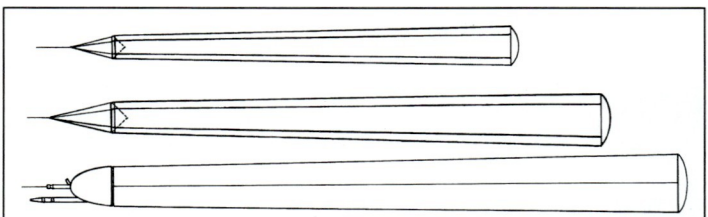

Top to bottom: the Bülow KRN 27-45-450, KRZ 34-60-550 and NCP 46-66-600 0024448

The HRL 110 sea-skimming helicopter-towed target 0024450

System composition: All Bülow targets and MDIs can be automatically launched and docked in the Bülow BTW 832 tow target winch.
Recovery: See above.
Power plant: None.
Variants: *HRL 110:* Large sleeve target, designed to simulate fast attacking motor torpedo-boats, but suitable also for gunnery training of coastal and AA artillery. Options include built-in MDI, IR emission and altimeter.
IR-65-85-600: Designed to imitate the IR emission generated by the heat from an approaching missile. Soft sleeve incorporates radar reflecting mesh; solid nose portion equipped with laser prisms and bracket for MDI transmitter.
KRN 27-45-450: Radar sleeve target. Mantles and bottom in mesh-net; amplified radar reflection by means of copper net cone in front of sleeve.
KRN 27-60-1000: As described for KRN 27-45-450.
KRZ 27-45-450: Radar sleeve target for actuating of proximity fuzed ammunition by radar reflecting mesh in mantle sections.
KRZ 34-60-550: As described for KRZ-27-45-450.
KRZ 51-95-800: As described for KRZ 27-45-450; also suitable as target for missiles.
L 29-45-450: Laser and radar reflecting sleeve target; GFRP nosecone with laser and radar reflecting prisms (radial reflection 0 to 360°, axial reflection 0 to 270°). Heated area 0.1 m² (1.07 sq ft), waveband 8 to 12 microns. Detection range with damped laser is approximately 8.6 n miles (16 km; 9.9 miles).
NCP 46-66-600: High-speed target, suitable also for missiles; actuates proximity fuzed ammunition by radar reflecting mesh in mantle sections. GFRP nosecone with constant pressure regulator maintains shape and

Nose end of the laser reflecting L 29-45-450 0024449

Performance

	Max towing speed	RCS by waveband in m² (sq ft)					
		Incoming			Passing		
	kt (km/h; mph)	C	X	Ku	C	X	Ku
HRL 110	100(185;115)			?			
IR 65-85-600	180(333;207)			?			
KRN 27-45-450	350(648;403)	18.0 (193.8)	1.6 (17.2)	8.0 (86.1)	0.9 (9.69)	0.9 (9.69)	1.0 (10.8)
KRN 27-60-1000	200(370;230)	18.0 (193.8)	1.6 (17.2)	8.0 (86.1)	0.9 (9.69)	0.9 (9.69)	1.0 (10.8)
KRZ 27-45-450	350(648;403)	0.3 (3.23)	0.2 (2.15)	0.2 (2.15)	3.0 (32.3)	3.0 (32.3)	3.0 (32.3)
KRZ 34-60-550	200(370;230)	0.3 (3.23)	0.2 (2.15)	0.2 (2.15)	3.0 (32.3)	3.0 (32.3)	3.0 (32.3)
KRZ 51-95-800	200(370;230)	0.3 (3.23)	0.2 (2.15)	0.2 (21.5)	3.0 (32.3)	3.0 (32.3)	3.0 (32.3)
L 29-35-450	350(648;403)	0.3 (32.3)	0.7 (7.53)	0.6 (6.46)	2.0 (21.5)	0.8 (8.61)	0.8 (8.61)
NCP 46-66-600	450(833;518)	0.3 (3.23)	0.2 (2.15)	0.2 (2.15)	3.0 (32.3)	3.0 (32.3)	3.0 (32.3)
NCP 42-95-800	400(741;460)	0.3 (3.23)	0.2 (2.15)	0.2 (2.15)	3.0 (32.3)	3.0 (32.3)	3.0 (32.3)
NCP 46-72-800	450(883;518)	0.3 (3.23)	0.2 (2.15)	0.2 (2.15)	3.0 (32.3)	3.0 (32.3)	3.0 (32.3)
NCP 34-74-1000	400(741;460)	0.3 (3.23)	0.2 (2.15)	0.2 (2.15)	3.0 (32.3)	3.0 (32.3)	3.0 (32.3)
NCP 46-800-1000	450(833;518)	0.3 (3.23)	0.2 (2.15)	0.2 (2.15)	3.0 (32.3)	3.0 (32.3)	3.0 (32.3)
NKRZ 34-52-450	350(648;403)	1.0 (10.8)	3.0 (3.23)	16.0 (172.2)	3.0 (32.3)	2.0 (21.5)	2.0 (21.5)

stability of sleeve even after several hits. Forward-pointing corner reflector and laser reflectors optional.

NCP 42-95-800: As described for NCP 46-66-600.
NCP 46-72-800: As described for NCP 46-66-600.
NCP 34-74-1000: As described for NCP 46-66-600.
NCP 46-80-1000: As described for NCP 46-66-600.
NKRZ 34-52-450: Radar sleeve target. Radar-reflecting mesh in sections to actuate proximity-fuzed ammunition; amplified reflection by means of corner reflector in nosecone.

Specifications – see table above

Status: In production and service.
   Customers: Argentina; Austria; Belgium; Chile; Denmark; Finland; France; Germany; Italy; Netherlands; Norway; Singapore; Sweden; Switzerland; UK; USA; former USSR; Yugoslavia.
Contractor: Bülow Air Target Systems AB
Vätö.

FMV SM3B

Type: Infra-red tow target.
Development: Towed by a target drone; designed to avoid direct missile hit on the drone. Compatible with MQM-107B (which see), which can carry up to four individual SM3Bs at wingtips or on underwing pylons.
Description: *Airframe:* Bullet-shaped pod in two parts; smaller rear portion has four sweptback wings/fins.
   Mission payloads: Two to four infra-red flares in rear portion.
   Guidance and control: Nose-towed by target drone; towline cut at launcher by telecommand at end of mission.
   Launch: Telecommand signal releases tow body by a pyrotechnic device which simultaneously ignites the flares and activates the one-way inertial reeler.
   Recovery: Winged rear portion normally survives the mission. A two-way reeling machine is under development, to enable this portion to be reeled back into the pod after a mission.

SM3B
Dimensions
Length overall..0.36 m (1 ft 2.2 in)
Body diameter (max)...0.15 m (5.9 in)
Max towline length...50 m (164 ft)
Weights
Basic weight...5 kg (11.0 lb)

Status: In production and service.
   Customers: Swedish armed forces.
Contractor: Enator Miltest AB (now supported by FMV)
Vidsel.

FMV SM6

Type: Infra-red tow target.
Development: Designed in late 1980s as a low-cost target for training with IR-guided missiles, and has since developed into a seven-version family.
Description: *Airframe:* Bullet-shaped body with four sweptback tailfins indexed in X configuration.
   Mission payloads: As described under Variants above, for IR, optical or radar augmentation. Space for additional equipment such as MDI, transponder, telemetry unit or spotlight.
   Guidance and control: CG-towed by manned aircraft or target drone; command signals can be transmitted either from the tow aircraft or, preferably, from a GCS. Currently equipped with six-channel, 35 MHz TKM 1 command receiver; 400 MHz TKM 3 recommended for future production.
   Launch and Recovery: SM6 is compatible with the Meggitt MRL-25A reeling machine, which in turn is compatible with the MQM-107B target drone.
Variants: *SM-6B:* Basic version: four flares in fin-tip position; corner reflector or 13.3 cm (5.25 in) Luneberg lens.
   SM6B/Rb70: Version of SM6B specially designed for use with laser-guided ground-to-air missiles.
   SM6F: As SM6B, plus two countermeasures flares.
   SM6R: Four flares in fin-tip position; two 17.8 cm (7 in) Luneberg lenses (in nose and aft).
   SM6F/R: As SM6R, plus two countermeasures flares.
   SM6IR: Specially designed with a plume burner to simulate cruise missile IR radiation.
   SM6L: As SM6B, plus special nose flare installation.

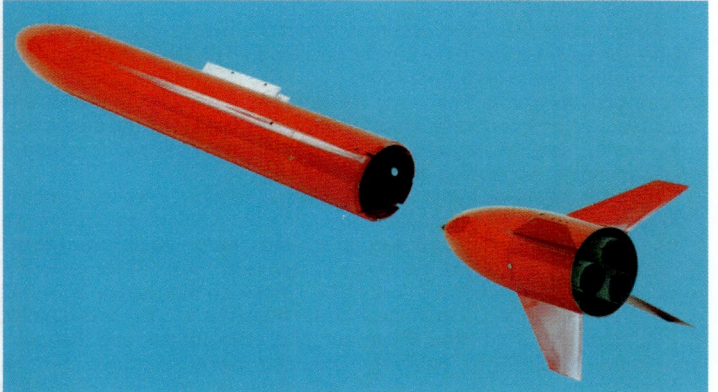

Front and rear elements of the SM3B pod

Baseline member of the SM6 family is the SM6B

SM6

Dimensions	
Length overall	1.80 m (5 ft 10.9 in)
Body diameter (max)	0.175 m (6.9 in)
Radar cross-section	Approx 1.0 m2 (10.76 sq ft)
Space for additional payload	8.0 dm3 (0.28 cu ft)
Weights	
Basic weight	27 kg (59.5 lb)

Status: In production and service.
  Customers: Swedish armed forces.
Contractor: Enator Miltest AB (now supported by FMV)
  Vidsel.

FMV SM9A

Type: Tow target.
Development: Designed as a multiple-choice tow target for development and training involving either infra-red or radar-guided missiles. Modular design permits easy change from one configuration to the other.
Description: *Airframe:* Cylindrical body with hemispherical nosecone and tailcone; four forward-swept tailfins in X configuration.
  Mission payloads: Four to eight fixed flares at fin-tip position. Other options include two externally mounted countermeasures flares; three Luneberg lenses covering a sector of 280°; two flares mounted under nosecone; a laser beam reflector; or an ejectable banner aft of the tailfins. Space provision for such additional equipment as MDI, transponder, telemetry unit or spotlight.
  Guidance and control: By 24-channel, 400 MHz TKM 3 digital command receiver; command signal transmitted either from towing aircraft or, preferably, from GCS.

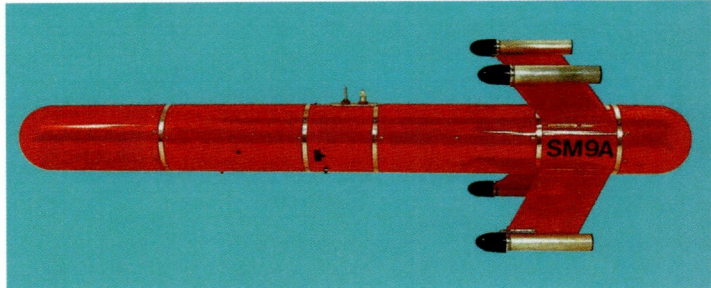

The modular, multirole SM9A 0079572

  Launch and Recovery: Compatible with Meggitt MRL-25A reeling machine and MQM-107B target drone.

SM9A

Dimensions	
Length overall	1.97 m (6 ft 5.6 in)
Body diameter (max)	0.205 m (8.1 in)
Radar cross-section: J-band	Approx 1.0 m2 (10.76 sq ft)
G-band	Approx 0.4 m2 (4.31 sq ft)
Space for additional payload	12.0 dm3 (0.42 cu ft)
Weights	
Basic weight	25 kg (55.1 lb)
Weight of additional payload	10 kg (22.0 lb)

Status: In production and service.
  Customers: Swedish armed forces.
Contractor: Enator Miltest AB (now supported by FMV)
  Vidsel.

Turkey

TAI Simsek

Type: Recoverable jet-powered high-speed aerial target.
Development: The Simsek (Lightning) high-speed aerial target has been developed to meet 'emerging Turkish Armed Forces advanced training requirements'. Key programme events include:
  2009 Development of the Simsek aerial target was launched during 2009 with developmental funding being provided by TAI and the Turkish Scientific and Technological Research Council (local acronym TÜBİTAK).
  2011 IHS Jane's sources report Simsek first phase development (including system design and integration) as having been completed during 2011.
  Mid-2012 IHS Jane's sources report Simsek system and sub-system testing as having been completed by mid-2012.
  4 August 2012 The Simsek aerial target made its maiden flight.
Description: *Airframe:* The Simsek aerial target features high-mounted wings and conventional horizontal and vertical tail surfaces mounted on a cylindrical fuselage with an elliptical nose and a dorsal engine air intake. No landing gear.
  Mission payloads: Passive RCS augmentation (Luneberg Lens), IR flares, an acoustic miss distance indicator, a countermeasures dispensing system and a smoke generator.
  Guidance and control: Fully autonomous, including take-off and landing. Manual or autonomous guidance using waypoints that can be updated during flight. Fail-safe and 'return home' modes for automatic recovery to a predetermined point.
  Launch: Land-based or shipboard pneumatic launcher.
  Recovery: Parachute recovery, on land or at sea.
  Power plant: One 393 N TEI TJ-2X turbojet.

Simsek

Dimensions, External	
Overall, length	2.70 m (8 ft 10¼ in)
Wings, wing span	1.50 m (4 ft 11 in)

Performance	
Altitude, Operating altitude	10 m to 4,570 m (33 ft to 15,000 ft)
Speed, Max level speed	400 kt (741 km/h; 460 mph)
Radius of operation	
datalink, mission	54 n miles (100 km; 62 miles)
Endurance, max	1 hr
Power plant	1 × turbojet

Status: As of late 2012, TAI was characterising the Simsek high-speed aerial target as continuing to be 'under development'.
Contractor: Turkish Aerospace Industries Inc
  Kavaklidere.

TAI Turna

Type: Recoverable aerial target.
Development: The Turna (Crane [bird]) recoverable aerial target is designed to meet Turkish Air Force and Land Forces requirements for gunnery and target tracking training. Key programme events include:
  August 1995 The Turkish MoD's R&D Department (local acronym MSB-ARGE) awarded TAI a contract with regard to the development of the Turna aerial target system.
  10 September 1996 IHS Jane's sources report the prototype Turna aerial target as having made its maiden flight on 10 September 1996.
  1997 Turna prototype testing was completed during 1997.
  1998 The Turkish Army and Navy completed operational testing of the Turna aerial target during 1998.
  2001 IHS Jane's sources report the first deliveries of the Turna aerial target to the Turkish military as having taken place during 2001.
  2003 The introduction of a TAI-developed digital autopilot provided the Turna AV with automatic flight modes (including roll, pitch, altitude, speed, waypoints and heading hold).
  4 April 2008 IHS Jane's sources report the maiden flight of a Turna aerial target powered by a locally developed TEI-TP-1X turboprop as having taken place on 4 April 2008.
Description: *Airframe:* The Turna aerial target is a low-wing monoplane and features cropped-delta wings, a cylindrical fuselage (with an elliptical nosecone), detachable sweptback V-configuration tail surfaces and a pusher engine. No landing gear.
  Mission payloads: Passive RCS augmentation (Luneberg Lenses), visual and IR signature augmentation (up to eight × smoke generators in a belly 'sledge' and up to four × wingtip-mounted flares) and a towed target (152 m (500 ft) cable and the ability to house an MDI sensor or four × optional flares). All onboard flares and smoke cartridges are activated from an associated GCS to which, the MDI transmits scoring data directly.
  Guidance and control: Basic autopilot hold modes and waypoint navigation, with flight planning/monitoring being executed from a shelter-mounted or portable GCS. Automatic fail-safe procedures are available (including a 'return home' mode and parachute recovery).

The Simsek high-speed aerial target pictured on its pneumatic launcher *(TAI)* 1365379

The Turna recoverable aerial target *(IHS/Patrick Allen)* 1097538

Transportation: The Turna's wings, fuselage, nosecone and belly sledge are detachable for air or land transportation in carrying box.

System composition: A typical Turna system comprises six × AVs, a GCS, a foldable launcher trailer and ground support equipment. Turnaround time is approximately 30 minutes.

Launch: By bungee cord catapult (mounted on the ground support trailer).

Recovery: Cruciform parachute recovery on land or at sea.

Power plant: One 28.3 kW (38 hp) Wankel AR-731 engine driving a two-bladed pusher propeller.

Turna

Dimensions, External	
Overall	
length	2.66 m (8 ft 8¾ in)
height	0.55 m (1 ft 9¾ in)
Fuselage	
width, max	0.30 m (11¾ in)
Wings, wing span	2.24 m (7 ft 4¼ in)
Engines, propeller diameter	0.66 m (2 ft 2 in)
Weights and Loadings	
Weight	
Weight empty	45 kg (99 lb)
Max launch weight	75 kg (165 lb)
Payload	
Max payload	
internal	15.0 kg (33 lb)
external, towed	10.0 kg (22 lb)
Performance	
Altitude, Service ceiling	3,660 m (12,000 ft)
Speed	
Cruising speed, max	180 kt (333 km/h; 207 mph)
Radius of operation	27 n miles (50 km; 31 miles)
Endurance	1 hr 30 min

Status: As of late 2012, TAI was reporting the Turna aerial target as having been in production since 2000 and as being in service with the Turkish Army and Air Force. Here, the AV is used as a target/target tracking trainer for air-to-air, Evolved Sea Sparrow Missile (ESSM), Hawk, Rapier, Sea Sparrow, SM-1 and Stinger missiles together with 7.62 mm to 76 mm calibre barrelled gun systems.

Customers: The Turkish Army and Air Force.

Contractor: Turkish Aerospace Industries Inc, Kavaklidere.

Ukraine

SIS A-11 Swift

Type: Recoverable aerial target.

Development: Developed as gunnery and missile target for use with land-based air defence weapons.

The A-11 Swift jet-powered air defence target *(SIS)* 1368735

Description: ***Airframe:*** Blunt-nosed cylindrical fuselage; rear-mounted, unswept low wings and sweptback canard surfaces; dorsally mounted pulsejet engine. Four fixed landing legs under centre-fuselage.

Mission payloads: Active or passive radar reflectors. Thermal radiation capacity 11 kW.

Guidance and control: No details provided.

Transportation: Transportable by standard military off-road vehicle or small truck.

Launch: Rail-launched by CGM-6C catapult.

Recovery: Conventional landing.

Power plant: One approx 7.4 daN (16.5 lb st) pulse-jet.

A-11

Weights and Loadings	
Weight, Max launch weight	30.0 kg (66 lb)
Fuel weight, Max fuel weight	7.5 kg (16.00 lb)
Payload, Max payload	5.0 kg (11.00 lb)
Performance	
Altitude, Operating altitude	100 m to 4,000 m (320 ft to 13,120 ft)
Speed	
Max level speed	194 kt (359 km/h; 223 mph)
Stalling speed	60 kt (112 km/h; 69 mph)
Radius of operation	32 n miles (59 km; 36 miles)
Endurance	20 min
Power plant	1 × pulse jet

Status: Being promoted. Production/service status not stated.

Contractor: Scientifically Industrial Systems Ltd, Kharkov.

United Arab Emirates

Adcom Systems Yabhon-GRN 1 and 2

Type: These new high-speed targets were displayed at the Dubai air show in November 2011.

Development: Designed as enhanced performance targets for air-to-air or ground-to-air missile systems.

Description: ***Airframe:*** Blended wing/body design with forward-swept outer wing panels; slight dihedral on wings, more marked dihedral on tailplane; sweptback single fin and rudder; rear-mounted engine(s). No landing gear. Composites construction.

Mission payloads: Large internal volume for wide range of sensors including active or passive radar augmenters, Luneberg lens, MDIs, IR and smoke cartridges, low-altitude flight devices and transponders.

Guidance and control: Preprogrammed, with programmable digital autopilot, automatic navigation, stability control and heading hold.

Transportation: Yabhon-GRN 2 is transportable in a 3.85 × 1.20 × 0.47 m (12.6 × 3.9 × 1.5 ft) container.

Yabhon-GRN 1 *(Paul Jackson)* 1395366

Adcom Systems Yabhon-GRN 1 and 2

Yabhon-GRN 2 *(Paul Jackson)* 1395367

Launch: Ramp-launched by catapult.
Recovery: Parachute recovery.
Power plant: Single (GRN 1) or twin (GRN 2) turbojets (type and ratings not stated). Fuel capacity 125 litres (33.0 US gallons; 27.5 Imp gallons) in GRN 1; 400 litres (106 US gallons; 88 Imp gallons) in GRN 2.
Variants: *Yabhon-GRN 1:* Single-engine version.
Yabhon-GRN 2: Larger, twin-engine version.

Yabhon-GRN 1, Yabhon-GRN 2

Dimensions, External
Overall
 length
  excl. pitot, Yabhon-GRN 1 .... 3.72 m (12 ft 2½ in)
  excl. pitot, Yabhon-GRN 2 .... 5.00 m (16 ft 4¾ in)
 height
  Yabhon-GRN 1 .... 0.71 m (2 ft 4 in)
  Yabhon-GRN 2 .... 1.00 m (3 ft 3¼ in)
Wings
 wing span
  Yabhon-GRN 1 .... 2.31 m (7 ft 7 in)
  Yabhon-GRN 2 .... 3.30 m (10 ft 10 in)

Weights and Loadings
Weight
 Weight empty
  Yabhon-GRN 1 .... 80 kg (176 lb)
  Yabhon-GRN 2 .... 180 kg (396 lb)
 Max launch weight
  Yabhon-GRN 1 .... 220 kg (485 lb)
  Yabhon-GRN 2 .... 680 kg (1,499 lb)
Payload
 Max payload
  Yabhon-GRN 1 .... 40 kg (88 lb)
  Yabhon-GRN 2 .... 200 kg (440 lb)

Performance
Climb
 Rate of climb
  Yabhon-GRN 1 .... 2,520 m/min (8,267 ft/min)
  Yabhon-GRN 2 .... 1,440 m/min (4,724 ft/min)
Altitude
 Service ceiling
  practical, Yabhon-GRN 1 .... 9,000 m (29,520 ft)
  theoretical, Yabhon-GRN 1 .... 11,000 m (36,080 ft)
  practical, Yabhon-GRN 2 .... 8,000 m (26,240 ft)
  theoretical, Yabhon-GRN 2 .... 9,000 m (29,520 ft)
Speed
 Max level speed
  Yabhon-GRN 1 .... 540 kt (1,000 km/h; 621 mph)
  Yabhon-GRN 2 .... 502 kt (930 km/h; 578 mph)
 Cruising speed
  Yabhon-GRN 1 .... 428 kt (793 km/h; 493 mph)
  Yabhon-GRN 2 .... 464 kt (859 km/h; 534 mph)
 Stalling speed
  Yabhon-GRN 1 .... 88 kt (163 km/h; 102 mph)
  Yabhon-GRN 2 .... 113 kt (210 km/h; 130 mph)
Endurance
  Yabhon-GRN 1 .... 1 hr 30 min
  Yabhon-GRN 2 .... 2 hr

Status: Being promoted in 2011-12.
Contractor: Adcom Systems,
 Industrial City, Abu Dhabi.

Adcom Systems Yabhon-HMD

Type: Recoverable aerial target.
Development: Derivative of company's earlier Yabhon-HM target, with enhanced performance.
Description: *Airframe:* Rear-mounted low wings with sweptback endplate fins and rudders; rear-mounted turbojet engine. No landing gear. Construction of CFRP, GFRP and epoxy resin. Wings detachable for storage and transportation.
 Mission payloads: Active and passive radar augmentors, Luneberg lens, IR flares or smoke generators; MDIs; low-altitude flight devices; transponders.
 Guidance and control: Preprogrammed, with programmable digital autopilot, automatic navigation stability control and heading hold.
 Transportation: Transportable in 3.26 × 1.08 × 0.46 m (10.7 × 3.5 × 1.5 ft) container.
 Launch: Catapult launch from land or ship, under own power.

Yabhon-HMD jet-powered weapons target *(Paul Jackson)* 1395368

Recovery: Parachute recovery.
Power plant: Rear-mounted turbojet (type and rating not stated). Fuel capacity 110 litres (29.1 US gallons; 24.2 Imp gallons).

Yabhon-HMD

Dimensions, External
Overall
 length .... 4.32 m (14 ft 2 in)
 height .... 0.66 m (2 ft 2 in)
Wings, wing span .... 3.38 m (11 ft 1 in)

Weights and Loadings
Weight
 Weight empty .... 105 kg (231 lb)
 Max launch weight .... 220 kg (485 lb)
Payload, Max payload .... 25 kg (55 lb)

Performance
Climb
 Rate of climb, at S/L .... 1,800 m/min (5,905 ft/min)
Altitude
 Service ceiling
  practical .... 8,000 m (26,240 ft)
  theoretical .... 9,500 m (31,160 ft)
Speed
 Max level speed .... 350 kt (648 km/h; 403 mph)
 Cruising speed .... 270 kt (500 km/h; 311 mph)
 Stalling speed .... 60 kt (112 km/h; 69 mph)
Endurance .... 1 hr
Power plant .... 1 × turbojet

Status: Being promoted in 2011 for use with surface-to-air and air-to-air missiles, and AA artillery.
Contractor: Adcom Systems
 Industrial City, Abu Dhabi.

Adcom Systems Yabhon-N

Type: Recoverable aerial target.
Development: Yabhon-N appears to be a development of, and successor to, this company's SAT-400 target.
Description: *Airframe:* Rear-mounted lmid-wings with elevons and sweptback endplate fins and rudders; triangular planform fuselage; pusher engine installation. No landing gear. Composites construction.
 Mission payloads: Include active and passive radar augmenters, Luneberg lens, IR flares or smoke generators.
 Guidance and control: Preprogrammed, with programmable digital autopilot, automatic navigation, stability control and heading hold.
 Launch: Catapult launch.
 Recovery: Parachute recovery.

Yabhon-N's body shape contributes to the total lifting area *(Paul Jackson)* 1395369

Power plant: One 37.3 kW (50 hp) piston engine; two-blade pusher propeller. Fuel capacity 20 litres (5.3 US gallons; 4.4 Imp gallons).

Yabhon-N

Dimensions, External
Overall
length...3.00 m (9 ft 10 in)
height..0.53 m (1 ft 8¾ in)
Wings, wing span..2.75 m (9 ft 0¼ in)
Areas
Wings, Gross wing area.. 2.92 m² (31.4 sq ft)
Weights and Loadings
Weight
Weight empty.. 55 kg (121 lb)
Max launch weight... 100 kg (220 lb)
Payload, Max payload... 40 kg (88 lb)

Performance
Climb, Rate of climb...720 m/min (2,362 ft/min)
Altitude, Service ceiling...6,000 m (19,680 ft)
Speed
Launch speed..26.94 m/s (88 ft/s)
Max level speed...194 kt (359 km/h; 223 mph)
g limits...+6
Endurance..3 hr

Status: Thought to be in service in UAE for weapon training and exercise applications.
Contractor: Adcom Systems Industrial City, Abu Dhabi.

United Kingdom

Flight Refuelling Falconet II

Type: Recoverable subsonic target drone.

Development: Falconet was designed to provide effective training by simulating the speed, manoeuvres and signature of modern low-level ground attack aircraft, and was developed as an advanced subsonic aerial target (ASAT) to a UK MoD requirement. Prototype flight testing began on 14 February 1982, and deliveries to the British Army followed from late 1983, full service entry being achieved in June 1986. More than 450 Falconets have been produced. The upgraded Falconet II was introduced in 2001.

Description: *Airframe:* Low/mid-wing monoplane, mainly of stressed skin aluminium alloy construction with some composite components. Wings, each attached to fuselage by four bolts, have plain ailerons, and are interchangeable port/starboard. Fuselage centre-section contains fuel and smoke trail oil tanks. Crushable polystyrene nosecone is expendable and replaced after each flight. Cylindrical canister in rear fuselage houses recovery parachute. Engine pod attached under fuselage by two bolts. Robust modular construction, designed to be assembled from kits by local technicians; typical assembly time 2 hours.

Mission payloads: Payloads can be carried either as underwing stores or within the fuselage. Internal payloads include smoke trail equipment, radar transponders, miss-distance indication equipment and passive radar signature enhancement. Wing-mounted payloads include thermal or visual wavelength flares, subtargets, chaff and infra-red decoy dispensers and special-to-mission pods. Subtargets can be configured to suit a wide variety of weapon systems including HAWK, Roland, Crotale, Shahine II, Stinger, Redeye, Skyguard, close-in weapon systems and anti-aircraft artillery.

Guidance and control: Manoeuvres are effected by aileron, elevator and engine thrust control; there is no rudder. External commands are detected by an antenna/receiver system and actioned by the onboard flight control computer. A telemetry transmitter relays data on air vehicle status back to the GCS. An onboard transponder is fitted to assist radar tracking. GPS navigation in Falconet II.

Falconet currently uses a fixed-base GCS at the Hebrides Range, operated by QinetiQ Aerial Target Services. However, on 4 March 2002, an upgraded Falconet II was successfully flown using a portable, GPS-based GCS known as Montage, developed by US company Micro

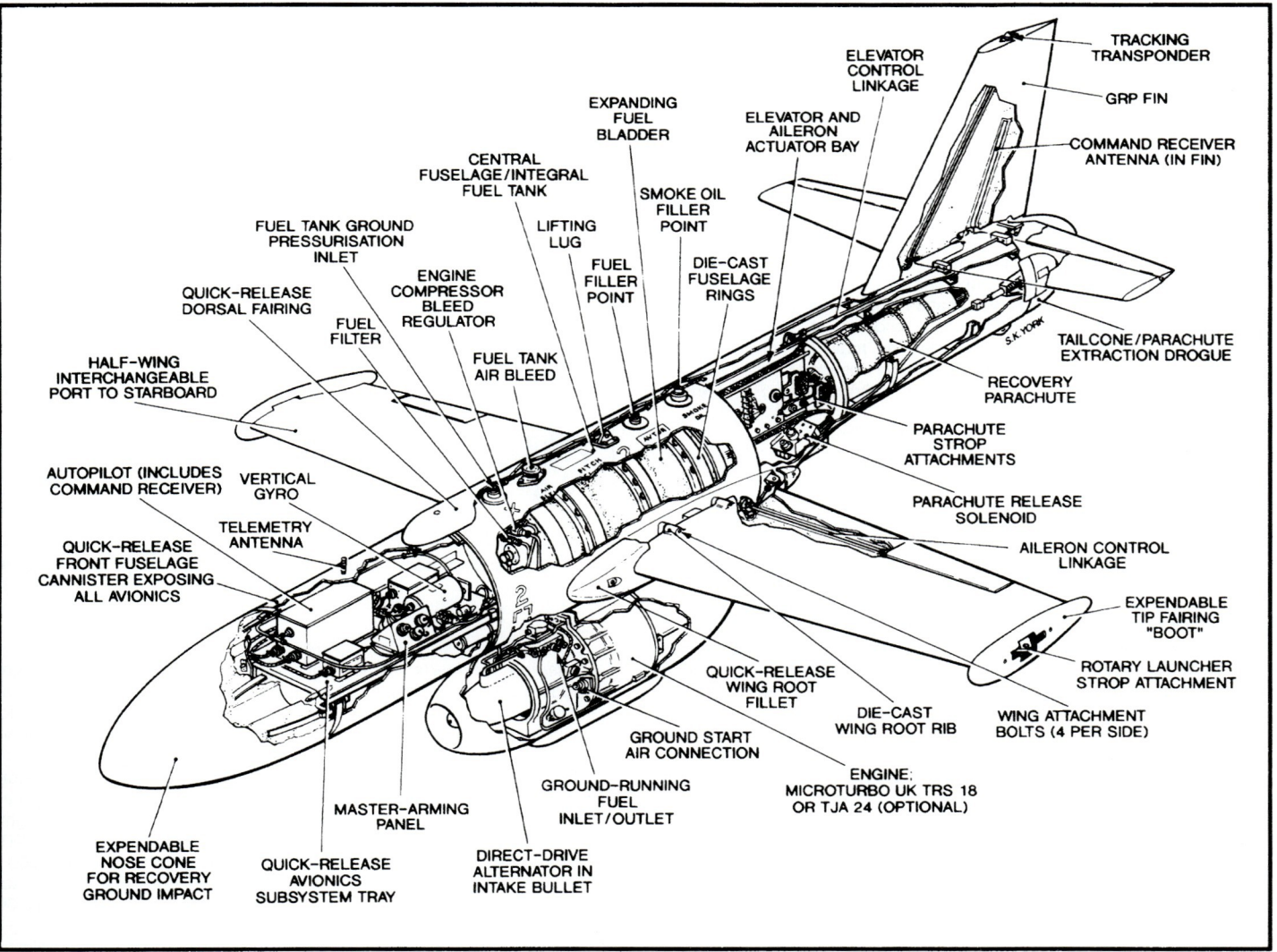

Main internal and external features of Falconet

Falconet on the launch trolley for a 'carousel' take-off 0517957

Falconet zero-length launch 0517959

Mk 2 sea-skimming version of the Modular Rushton Towed Target 0517963

System Inc and already in use by the US Army and Navy as its primary target control system.

Launch: (1) For fixed-base operation, Falconet launches under its own power without rocket boost, from a launch trolley tethered to the centre of a circular runway. This runway permits take-off in any wind direction at the cost of only 3 litres (0.8 US gallon; 0.7 Imp gallon) of fuel. Engine is started using compressed air from the ground start trolley; the trolley also provides fuel until immediately before launch, enabling the Falconet to take off with a full fuel tank.

(2) For mobile or shipboard operation, Falconet can be fitted with twin booster rockets for zero-length launching.

(3) By Robonic MC-2045-H pneumatic rail launcher (Falconet II).

Recovery: Parachute recovery on land or at sea, air vehicle descending vertically nose-down under the main parachute; surface impact is absorbed by crushable, replaceable nosecone.

Status: Continues in service for Rapier firings at British Army Royal Artillery Hebrides Range.

Customers: UK (British Army and Royal Air Force).

Contractor: Flight Refuelling Ltd
Wimborne, Dorset.

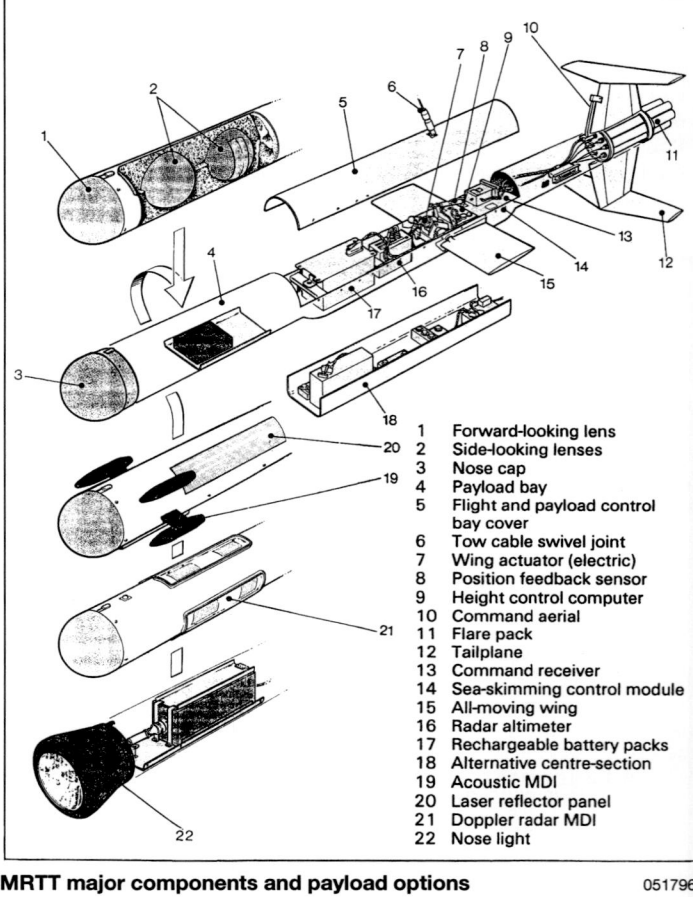

1 Forward-looking lens
2 Side-looking lenses
3 Nose cap
4 Payload bay
5 Flight and payload control bay cover
6 Tow cable swivel joint
7 Wing actuator (electric)
8 Position feedback sensor
9 Height control computer
10 Command aerial
11 Flare pack
12 Tailplane
13 Command receiver
14 Sea-skimming control module
15 All-moving wing
16 Radar altimeter
17 Rechargeable battery packs
18 Alternative centre-section
19 Acoustic MDI
20 Laser reflector panel
21 Doppler radar MDI
22 Nose light

MRTT major components and payload options 0517962

Flight Refuelling MRTT

Type: Towed target.

Development: MRTT (Modular Rushton Towed Target) is the current version of the original Rushton towed target produced for many years by Flight Refuelling Ltd. The first sea-skimming version developed from the Rushton system was known as LLHK, for Low-Level Height Keeper. The MRTT generation was launched, after an 18-month development period, with a UK MoD production contract of early 1989. It was used in the development programmes for Goalkeeper and Sea Wolf, and is in regular use by the Royal Navy with the Phalanx close-in weapon system and anti-aircraft gunnery of all calibres. Other weapon systems with which MRTT has been used include Satan, Seadart, Seacat and Otobreda 76/62. Towing aircraft have included A-4 Skyhawk, F-4 Phantom, Il-28/H-5, Canberra TT Mk 18, Falcon 20, Learjet and Westwind. May also be known as Black Kite.

Description: *Airframe:* Slim cylindrical body with all-moving stub wings and double T tailplane. Modular construction; composites nose cap. Up to 8 km (5 miles) of towline.

Mission payloads: Variety of payloads available (see diagram), including up to six smoke, IR or visual flares; Luneberg lenses; laser reflecting panel; acoustic or radar miss-distance indicator. Forward-facing 600 W visual acquisition lamp with independent power supply.

Guidance and control: Target behaviour is ordered by UHF two-tone commands from the tow aircraft for height control command, equipment and MDI command; height-keeping parameters are preset before flight.

Launch: From pylon-mounted winch pack on the tow aircraft. MRTT can be towed behind a variety of aircraft at tow lengths of up to 8,000 m (26,250 ft), depending on aircraft type and winch pack fitted. Usual mounting (except Westwind) is underwing. Compatible with DX6G, Marquardt, RFB, RM30A, RMK19 and Rushton winch packs.

Recovery: By cable reel-in and securing target against crutch points of winch pack.

Flight Refuelling MRTT – Meggitt BTT-1 Imp < United Kingdom < Aerial targets

Much-shot-at MRTT at the 1998 Farnborough Air Show *(IHS/Kenneth Munson)* 0054186

Imp being hand launched 0517968

Variants: MRTT Mk 1: With wingless lower section; used for non-sea-skimming roles.

MRTT Mk 2: Equipped with radar altimeter and height control computer; designed specifically for sea-skimming and ditching prevention. A range of heights can be selected, from 152 m (500 ft) down to 5 m (15 ft) above sea level, with a height-keeping accuracy of ±2.1 m (±7 ft).

MRTT

Dimensions
- Wing span..0.61 m (2 ft 0.0 in)
- Length: max...3.05 m (10 ft 0.0 in)
- min..2.74 m (9 ft 0.0 in)
- Height overall..0.69 m (2 ft 3.0 in)
- Payload bay: Length..0.61 m (2 ft 0.0 in)
- Diameter...0.19 m (7.5 in)

Weights
- Max payload..7.5 kg (16.5 lb)
- Typical weight..45.4 kg (100 lb)
- Max weight..59.9 kg (132 lb)

Performance
- Design operating speed..........................300 kt (555 km/h; 345 mph)
- Towing speed: max..................................400 kt (741 km/h; 460 mph)
- min...230 kt (426 km/h; 264 mph)
- Min heights to minimise ditching probability
  - Sea State 0–4..5 m (15 ft)
  - Sea State 4–7..15 m (50 ft)
  - above Sea State 7..30 m (100 ft)
- Max carriage altitude..12,200 m (40,000 ft)

Status: Remains in service with Royal Navy, navies worldwide, and target towing contractors.

Customers: Have included Australia; China; France; Germany; India; Italy; Netherlands; UK (Royal Navy).

Contractor: Flight Refuelling Ltd
Wimborne, Dorset.

Meggitt BTT-1 Imp

Type: Training and all-arms air defence target.

Development: More than 500 Imps have been produced since 1983 for a variety of customers. They are used as targets for a variety of light air defence weapons, including 0.50 in general purpose machine guns and a range of visually controlled gun systems of up to 25 mm/1 in bore. Imps are also used to train operators of Meggitt's BTT-3 Banshee target (which see). Imps can be used on land or at sea.

Description: *Airframe:* Mid-wing monoplane of cropped-delta planform with sweptback fin; no horizontal tail surfaces. Fuselage is of glass fibre-reinforced resin with additional strengthening; wings are of plywood over a polystyrene foam core. All flying surfaces are further covered with tough vinyl. Primary components are interchangeable to simplify replacement.

Guidance and control: Radio-command system (hand-held Meggitt transmitter), incorporating automatic fail-safe system, giving independent proportional control of elevons, engine and (optionally) recovery parachute. Variety of VHF and UHF frequencies available. Additional high-power transmitter can be used for flights out to extended ranges.

Launch: Hand launched (optionally by lightweight bungee catapult). Can be readied for flight 10 minutes after arrival on site.

Imp on lightweight catapult 0517969

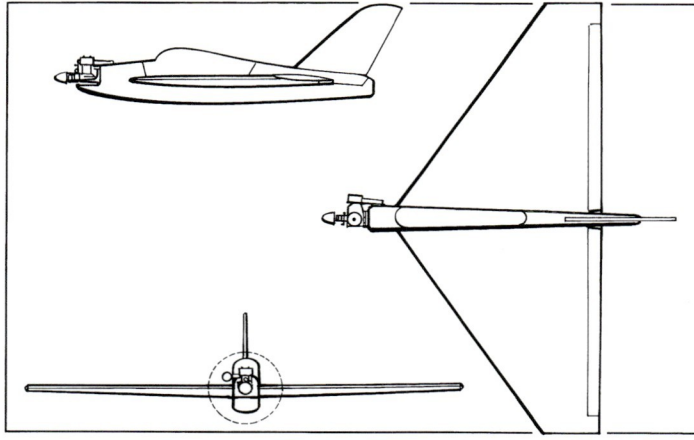

Meggitt BTT-1 Imp *(IHS/John W Wood)* 0518004

Recovery: By conventional ventral skid landing or, optionally, by parachute. Latter can be commanded manually or, in the event of target damage or command link failure, automatically.

Power plant: One 1.57 kW (2.1 hp) 15 cc glow-plug engine; two-blade fixed-pitch propeller.

BTT-1

Dimensions, External
- Overall
  - length...1.09 m (3 ft 7 in)
  - height..0.43 m (1 ft 5 in)
- Wings, wing span..1.83 m (6 ft 0 in)
- Engines, propeller diameter.................................0.33 m (1 ft 1 in)

Weights and Loadings
- Weight
  - Weight empty..4.5 kg (9.00 lb)
  - Max launch weight..5.9 kg (13.00 lb)
- Fuel weight, Max fuel weight.................................0.45 kg (0.00 lb)

Performance
- Speed, Max level speed...82 kt (152 km/h; 94 mph)

346 Aerial targets > United Kingdom > **Meggitt BTT-1 Imp – Meggitt BTT-3 Banshee**

Imp/Banshee operator training   0517970

BTT-1

Range	
visual	1.1 n miles (2 km; 1 miles)
optical tracking	3.2 n miles (5 km; 3 miles)
Endurance	30 min

Status: Production complete. Remained in service with some operators in 2006-2007. May be out of service.

Customers: More than 20 countries, including Denmark (Air Force and Navy); Egypt; Indonesia; Norway (Navy); Oman (Army, Air Force, Navy and Royal Guard); Qatar; Saudi Arabia; Turkey (Air Force); UK (Army, Royal Air Force and Royal Navy).

Contractor: Meggitt Defence Systems Ltd UK
Ashford, Kent.

Meggitt BTT-3 Banshee

Type: Recoverable aerial target.

Development: First flown in January 1983, Banshee was designed as a low-cost target system to simulate the threat of missiles and aircraft for gun and missile air defence systems. It has since become one of the most widely used such systems in the world, with over 5,000 produced since 1984 for at least 42 user countries. Weapon systems with which Banshees have been used include Blowpipe, Chaparral, Crotale, Javelin, Phalanx, Rapier and Sea Sparrow; and gun systems from 0.50 in to 76 mm.

The initial production version of Banshee first flew in June 1984 and entered British Army service in July 1986; the Banshee 300, introduced from late 1988, can be equipped with a radar altimeter enabling it to fly down to 5 m (15 ft) above the water to simulate a sea-skimming missile. The 3,000th Banshee, a 400, was handed over to the UK MoD on 7 December 2000.

Description: *Airframe:* Low-wing monoplane with cropped-delta wings and sweptback fin; no horizontal tail surfaces. Wings have 50° sweepback and approximately 3° dihedral. Entire airframe is constructed from GFRP. Payload bays in detachable nosecone and in front, centre and rear of fuselage. All electronic subsystems are watertight for recovery from sea.

The current-standard Banshee 400 (IHS/Patrick Allen)   1182265

Preparing a Banshee 400 for launch at the Spanish Army's CEDEA test range (IHS/Patrick Allen)   0523969

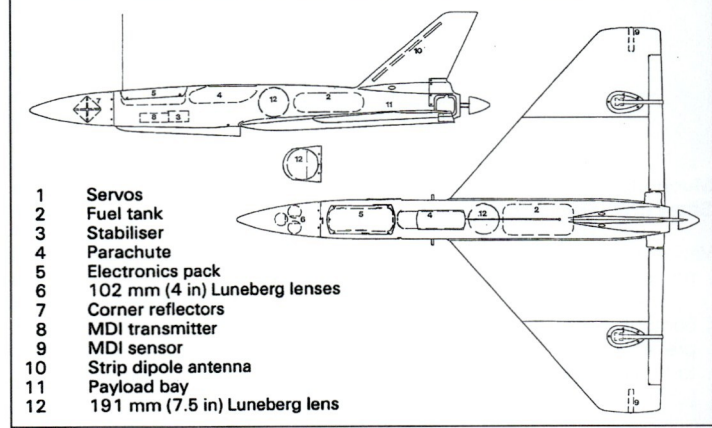

1	Servos
2	Fuel tank
3	Stabiliser
4	Parachute
5	Electronics pack
6	102 mm (4 in) Luneberg lenses
7	Corner reflectors
8	MDI transmitter
9	MDI sensor
10	Strip dipole antenna
11	Payload bay
12	191 mm (7.5 in) Luneberg lens

Location of Banshee internal features   0517972

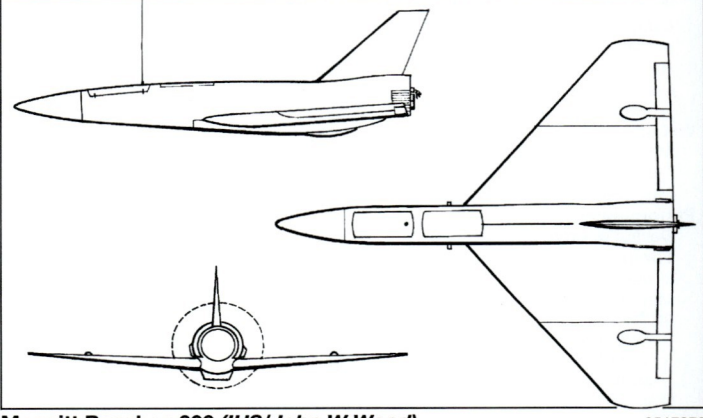

Meggitt Banshee 300 (IHS/John W Wood)   0517973

Vertical fin is detachable for storage/transportation and for easier access to rear payload bay.

Mission payloads: Alternative mission loads include a flare pod able to contain up to 16 smoke flares, eight IR flares or a chaff dispenser. Corner reflector sets can be carried, consisting of an optimised triple array giving a spherical cardioid polar diagram with phase-interacted global fine structure and echo peaks tuned to customer requirements. Up to three Luneberg lenses can be carried, providing radar enhancement in a range of bandwidths and attitudes. Acoustic and Doppler radar miss-distance indicators can be carried simultaneously. Banshee can also tow sleeve and banner targets or a tow body that can be fitted with the MDI system and pyrotechnics.

New 'hot nose' reusable IR thermal source introduced on Banshee 400 in 1996; enhances target's IR signature without affecting performance, or other payloads or enhancements. IR signature can also be augmented by a 'hot' wing leading-edge, an option taken by three European user countries. Has operated successfully in live firings in UK and overseas. Reconnaissance capability introduced 2001 by availability of interchangeable, 'plug-in' nosecone with built-in camera; Pakistan first customer for this version.

Guidance and control: Command control system manufactured by Meggitt is available on a selection of VHF and UHF frequencies. For short-range operation (up to 5.4 n miles; 10 km; 6.2 miles), the target is tracked using optical equipment, (gyrostabilised for optical tracking on board ship). Where flight to extended ranges is required, a Meggitt digital autopilot and telemetry system are fitted to the target, and one of two tracking options is used: radar or satellite GPS.

IHS Jane's All The World's Aircraft: Unmanned 2013-2014   © 2013 IHS

Meggitt BTT-3 Banshee < United Kingdom < Aerial targets

Banshee 400 parachute recovery (IHS/Patrick Allen)

Nosecone of Banshee Recce, as ordered by Pakistan (IHS/Kenneth Munson)

Banshee 400 is the current production version (IHS/Patrick Allen)

The Meggitt radar tracking system enables the target to be operated at ranges in excess of 16.2 n miles (30 km; 18.6 miles), day or night. It uses a coded radar transponder enabling multiple targets to be tracked simultaneously, and can be configured for use with other radars such as a ship's surveillance radar. Real-time telemetry information such as target height, heading and onboard battery voltage is displayed in the Banshee telemetry ground station. Specifically designed for low-level use, the radar altimeter is installed for sea-skimming missions at altitudes down to 5 m (15 ft), and is also available for use in other applications.

For operating Banshee at ranges in excess of 21.6 n miles (40 km; 24.8 miles), the Meggitt DGPS tracking system is used. This system, which displays the target tracking and telemetry information on a CRT monitor, also enables multiple target flights to be conducted simultaneously.

Launch: From elastic power-band catapult launcher which is normally mounted on a ground support trailer but which can be removed and used separately. Can also be launched by booster rocket from zero-length launcher.

Recovery: By belly skid landing on land or water, or by deploying Irvin-GQ cruciform parachute. Underfuselage flare housing acts also as landing skid. Parachute mode can be commanded manually or automatically by a fail-safe system in the event of in-flight damage or failure of the radio-command link. Maximum turnround time on land or at sea is 30 minutes after return to the launch site.

Variants: *ASR-4 Spectre:* Surveillance UAV version.

Banshee: Initial version, produced mainly for UK, some NATO countries and Middle Eastern customers.

Banshee 300: Improved aerodynamics include streamlining formerly separate underfuselage smoke pod into fuselage contours and deepening fuselage slightly to offer alternative internal location for Luneberg lens. Fin detachable, to facilitate storage/transportation and access to rear payload bay. Improved command and control systems include radar altimeter height-lock module to permit simulation of sea-skimming missiles. Option of more powerful engine for higher performance. *Description applies to this version except where indicated.*

Banshee 400: Enhanced development of Banshee 300, introduced in mid-1996. Main innovation is introduction of a 'hot nose' thermal nosecone IR augmentor, fuelled by liquid propane gas, as an alternative to short-burn pyrotechnic flares.

Banshee 500: Higher-performance version, with new 38.8 kW (52 hp) Meggitt Tempest engine; introduced in 2001.

Banshee Jet: Introduced as a customer option after completion of desert performance trials in US in April 2003. Powered by two 0.2 kN (45 lb st) small turbojets and able to use 'heavy' fuel, either kerosene Diesel or Jet A1; dash speed quoted as 250 kt (463 km/h; 287 mph); endurance approx 45 minutes on standard Banshee fuel tank; alternative larger tank available to increase this to more than 1 hour. Most components interchangeable with standard Banshee, and all normal Banshee payloads can be carried. Launched using standard MDS Mk 4 catapult and controlled by a standard Meggitt Wizard GCS. As of early 2007, no known customers.

348 Aerial targets > United Kingdom > Meggitt BTT-3 Banshee – Meggitt Petrel

Banshees of the Royal Brunei armed forces during an exercise in the US *(US Navy)* 1151523

Banshee Recce: Interchangeable, camera-carrying nosecone available from 2001; initial order from Pakistan.

Crecerelle (Kestrel): French (Sagem) surveillance UAV variant of Spectre. *Described separately.*

Hawkeye: Surveillance version of Spectre for civil customer. Brief details in early editions.

Status: Continuing in production and service in 2006-07. As of 2011, still listed on manufacturer website.

Customers: With production currently exceeding 5,000, Banshee has been purchased by, or operated by TTL and Meggitt under contract in, at least 42 countries including Abu Dhabi; Botswana; Brunei; Denmark; Egypt; Finland; France; Indonesia; Italy; Jordan; Kuwait; Malaysia; Norway (Navy); Oman; Pakistan (Navy); Qatar; Romania; Saudi Arabia; Spain (Army); Thailand; Turkey; UK (700 or more since 1987, including 565 Series 300s); and the US.

Banshee is one of the three principal targets adopted for the UK's 20-year Combined Aerial Target Service (CATS) programme, for which Meggitt received a contract on 14 December 2006 from QinetiQ, the overall provider of the CAT service for UK armed forces weapon training. The UK's Banshees are operated chiefly from the ranges at Manorbier, Aberporth and in the Hebrides.

Banshee has also been supplied to a number of aerospace companies worldwide, including BAE Systems, Boeing, MBDA, Oerlikon Contraves and Thales, for weapon systems evaluation.

Contractor: Meggitt Defence Systems Ltd UK
Ashford, Kent.

Meggitt Petrel

Type: Non-recoverable ballistic target.

Development: Petrel was developed in the UK by BAJ (Bristol Aerojet) Ltd. It has been used by the Royal Navy for prototype weapons trials and in-service practice for Seawolf and Sea Dart defence missile systems.

Description: *Mission payloads:* Payload bay allows various radar enhancement systems to be fitted, to suit the particular weapon system being tested.

Guidance and control: See under Launch below.

System composition: The Petrel system, including all equipment needed for rocket preparation, can be supplied in a 6.1 m (20 ft) long ISO shipping container which is used as a magazine and preparation room. The system requires only a small crew and can be made operational in less than a day.

Launch: Petrel is fired from a free-standing launch tube by three small, high-thrust booster motors which separate after about a quarter of a second. The target accelerates for 30 seconds during main motor burn and then continues ballistically to impact. Since its inception, the system has been developed to provide a multiple launch capability allowing up to six targets in the air simultaneously.

The launch tube can be used on land or on the deck of a ship, in which case the instant of fire is controlled by a gyrostabilised reference unit. Trajectories (see diagram for typical examples) are achieved by selecting mass and drag before launch, and terminal dives can closely resemble known anti-ship and tactical ballistic missile threats. In the event of a launch failure the rocket will fall short and within the safety trace; in well over 600 launches to date, no Petrel has yet fallen outside the predicted safety trace.

Recovery: Non-recoverable.
Power plant: Solid-propellant rocket motor.

Petrel

Dimensions, External	
Overall	
length	3.00 m (9 ft 10 in) (est)
height	0.180 m (7 in)
width	0.180 m (7 in)
Performance	
Power plant	1 × rocket

Status: In production from 2006. Remains in service.
Customers: UK (Royal Navy); French armed forces.
Contractor: Meggitt Defence Systems Ltd UK
Ashford, Kent.

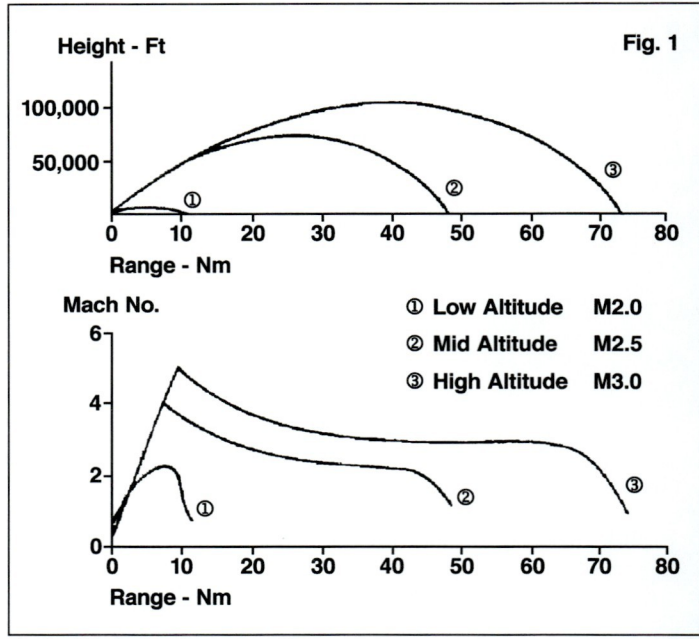

Typical Petrel trajectory patterns 0517975

① Low Altitude M2.0
② Mid Altitude M2.5
③ High Altitude M3.0

Multiple launch of Petrel targets 0517976

Meggitt Petrel ballistic target 0002105

IHS Jane's All The World's Aircraft: Unmanned 2013-2014

Meggitt Snipe

Type: Recoverable aerial target.

Development: The Mks 4, 5 and 15 are the current versions of Snipe; the earlier Mks I, II and III developed originally by AEL Ltd are no longer in service. The Mk 15, which entered service in April 1989, is a low-cost system used both as a target for small arms gunnery and Blowpipe missiles, and as a training aircraft for operators of more expensive targets. Mk 5 can be configured to suit such weapon systems as Javelin, Stinger, RBS 70, SA-7 'Grail', Crotale, Roland, and air defence guns with calibres between 20 and 76.2 mm.

Description: *Airframe:* Typical 'model aircraft' mid-wing monoplane; high-lift wing section on Mk 4, high-speed section on Mk 5. Kevlar-reinforced GFRP fuselage; composites wings on Mks 4/5, foam core with marine plywood veneer on Mk 15. Ventral landing skid. All variants fully modular for ease of maintenance.

Mission payloads: Can include 45-second smoke flares (up to 16 on Mks 4/5, four on Mk 15) and/or 60-second infra-red flares (eight on Mks 4/5, four on Mk 15); chaff dispensers (Mks 4/5); up to two Luneberg lenses (Mks 4/5); acoustic or Doppler radar miss-distance indicator. Various towed target bodies and sleeves, and advanced tracking and telemetry system, available for Mks 4/5.

Guidance and control: Flight: Triple rate stabilised device controls air vehicle in pitch and roll, coupled with pressure-controlled altitude. Meggitt digital autopilot with heading and height lock.

Snipe Aerial Target System in flight *(Meggitt)* 1295681

Uplink: Digital proportional radio command with independent control of all onboard equipment. Standard power output 15 W; UHF and VHF frequencies available.

Transportation: Mks 4/5 by trailer/launcher which can accommodate three air vehicles (dismantled), all control equipment, spares and tools for extended operation. Trailer is 5.85 m long including drawbar, 1.79 m wide and 1.90 m high (19.2 × 5.9 × 6.2 ft) and weighs approximately 1,750 kg (3,858 lb). Launcher assembly is semi-fixed to trailer but can be easily removed if required for independent mounting. Launch rail length is 4.87 m (16 ft) folded, 9.12 m (29.9 ft) unfolded.

Launch: Mks 4/5 from rail by bungee catapult with electric winch; Mk 15 catapult or hand launched.

Recovery: Parachute or belly skid landing.

Power plant: One 18.6 kW (25 hp) Meggitt MDS 342 flat-twin engine in Mks 4 and 5; Mk 15 has 62 cc single-cylinder two-stroke engine. Two-blade propeller. Fuel capacity (Mk 15) 1 litre (0.25 US gallon; 0.22 Imp gallon).

Snipe Mk 4, Snipe Mk 5, Snipe Mk 15

Dimensions, External	
Overall length	
Snipe Mk 4, Snipe Mk 5	2.67 m (8 ft 9 in)
Snipe Mk 15	1.60 m (5 ft 3 in)
Wings wing span	
Snipe Mk 4	3.215 m (10 ft 6½ in)
Snipe Mk 5	3.06 m (10 ft 0½ in)
Snipe Mk 15	2.20 m (7 ft 2½ in)
Weights and Loadings	
Weight empty	
Snipe Mk 4, Snipe Mk 5	60 kg (132 lb)
Snipe Mk 15	5 kg (11.00 lb)
Performance	
Speed Max level speed	
Snipe Mk 4	135 kt (250 km/h; 155 mph)
Snipe Mk 5	156 kt (289 km/h; 180 mph)
Snipe Mk 15	108 kt (200 km/h; 124 mph)
Stalling speed	
Snipe Mk 5	46 kt (86 km/h; 53 mph)
Snipe Mk 15	32 kt (60 km/h; 37 mph)

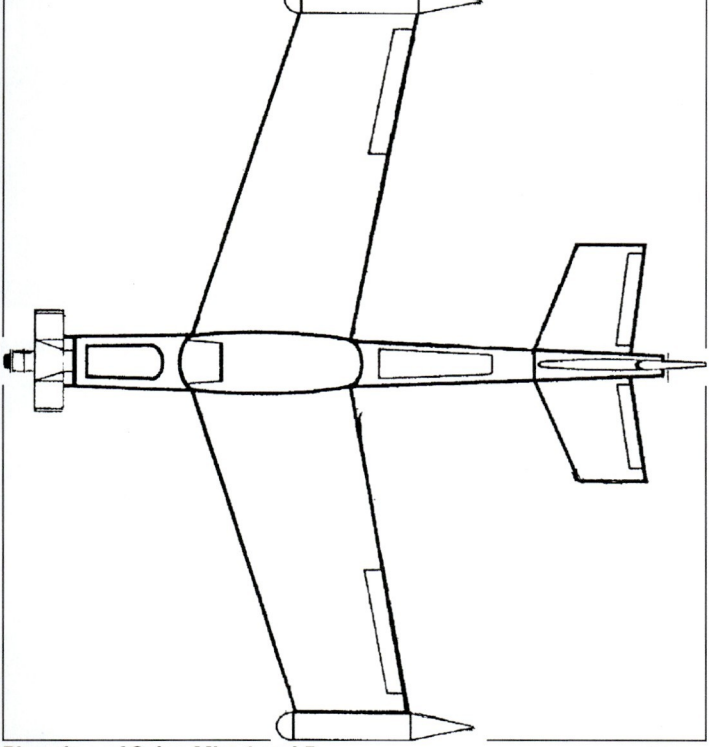

Plan view of Snipe Mks 4 and 5 0518193

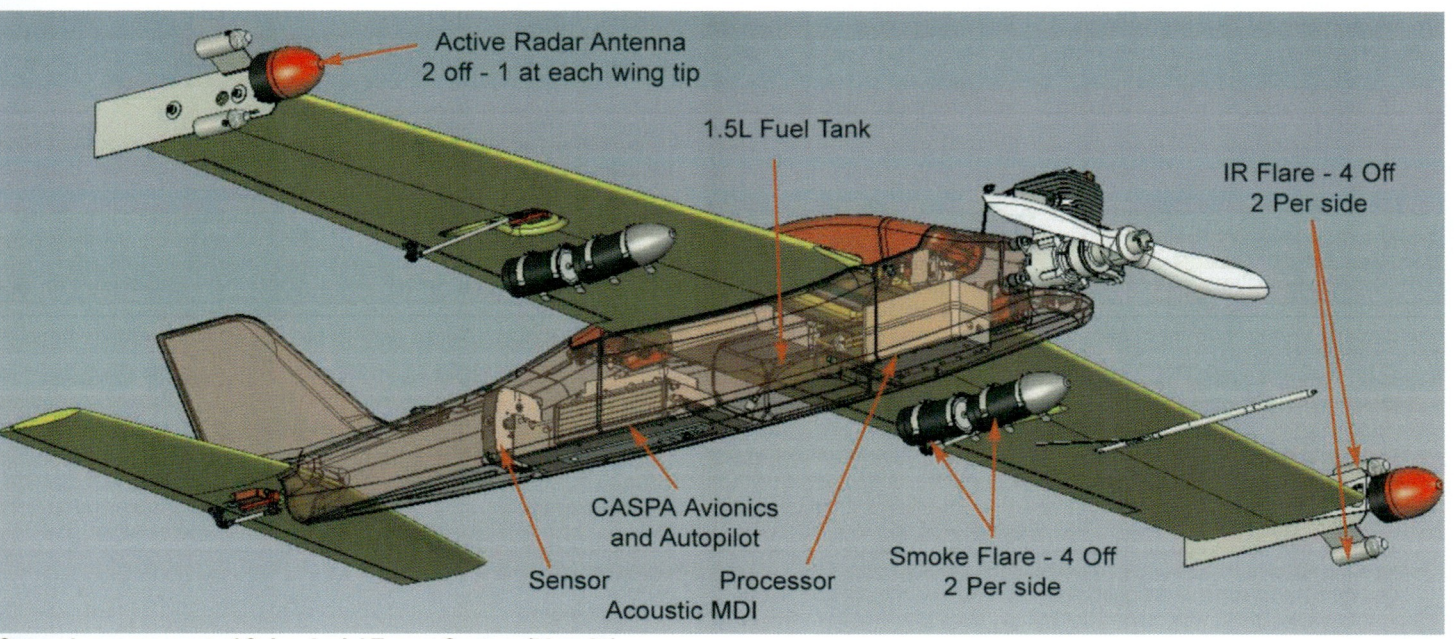

General arrangement of Snipe Aerial Target System *(Meggitt)* 1295682

Aerial targets > United Kingdom > Meggitt Snipe – Meggitt Voodoo

Snipe Aerial Target System in ready to launch position *(Meggitt)*
1295680

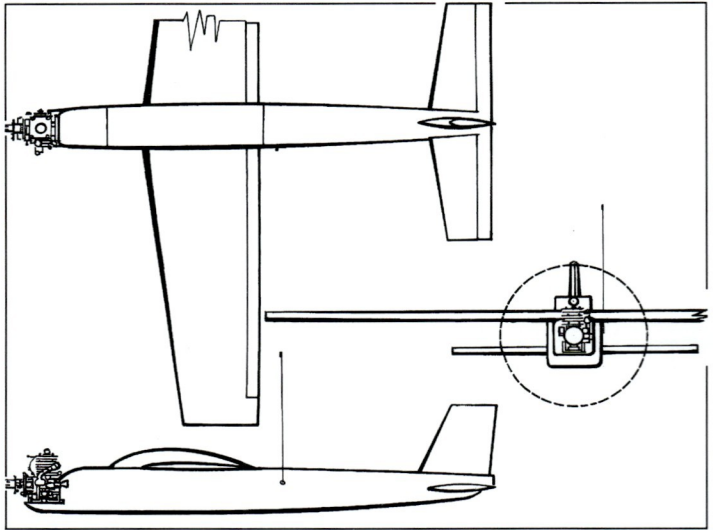
Snipe Mk 15 low-cost small arms target *(IHS/John W Wood)*
0518192

Snipe Mk 4, Snipe Mk 5, Snipe Mk 15
Range
depending upon tracking system,
Snipe Mk 4, Snipe Mk 5 .... 43.5 n miles (80 km; 50 miles)
visual tracking, Snipe Mk 15 .... 4.3 n miles (8 km; 4 miles) (est)
Endurance
typical, Snipe Mk 4 .... 1 hr 20 min
typical, Snipe Mk 5 .... 1 hr 40 min
Snipe Mk 15 .... 30 min (est)

Status: In service in, or operated under contract in, more than 15 countries. As of 2011, available.

Customers: Several NATO countries; Finland; Middle East; Asia.

Contractor: Meggitt Defence Systems Ltd UK
Ashford, Kent.

Voodoo and GT-400 targets on display at 2006 Farnborough Air Show *(IHS/Patrick Allen)*
1182259

UK-serialled Voodoo on a Robonic launcher. Underwing pods can contain a forward-looking Luneberg lens, decoy dispensers, IR flares or a variety of other augmentation devices *(Meggitt)*
1395207

Meggitt Voodoo

Type: Recoverable aerial target.

Development: Designed to utilise proven aerospace/motor sport manufacturing techniques to provide high performance at low cost. Public debut at Paris Air Show, June 2001, following maiden flight early that year and further testing in desert conditions.

Description: *Airframe:* Shoulder-wing monoplane with tapered wings, tapered circular-section fuselage and V tail; largely of carbon fibre construction. No landing gear. Manoeuvres can include dives at angles of up to 55°.

Mission payloads: Can include RCS, thermal, laser, laser reflectivity and visual (smoke plume) enhancement devices, NATO chaff/flare dispensers, radar altimeter, sea-skimming module and electronic threat simulation, plus electronic, acoustic and Doppler scoring systems. Underwing hardpoints enable Voodoo also to tow a subtarget. Payloads can be mixed, and most can be carried simultaneously.

Guidance and control: Meggitt digital autopilot with track and height lock. Integrated DGPS with automatic flight and digital telemetry systems. Utilises same GCS as Banshee (which see).

Launch: Automatic launch by pneumatic catapult at velocity of 55 m (164 ft)/s. British Voodoos use a Finnish Robonic launcher; those for Spain use one produced by OKT Norge of Norway.

Recovery: Automatic parachute recovery standard; airbag optional.

Power plant: One 108 kW (145 hp) 955 cc Ilmore Fury three-cylinder, water-cooled, four-stroke in-line engine; two-blade propeller. Fuel capacity 45 litres (11.9 US gallons; 9.9 Imp gallons) standard, in one mid-fuselage and two in-wing tanks. Auxiliary internal tanks can be fitted, containing a further 10 litres (2.6 US gallons; 2.2 Imp gallons) of fuel or smoke fluid.

Voodoo
Dimensions, External
Overall
length .... 3.65 m (11 ft 11¾ in)
height .... 1.03 m (3 ft 4½ in)
Wings, wing span .... 3.90 m (12 ft 9½ in)
Areas
Wings, Gross wing area .... 3.48 m² (37.5 sq ft)
Weights and Loadings
Weight
Weight empty .... 155 kg (341 lb)
Max launch weight .... 210 kg (462 lb)

Meggitt Voodoo – UTSL GSAT-200 NG < United Kingdom < Aerial targets

One of the first batch of Voodoos produced for the UK's CATS programme *(Meggitt)*

Voodoo

Performance
Climb
  Rate of climb, max, at S/L .... 1,524 m/min (5,000 ft/min)
  Altitude, Operating altitude .... 5 m to 1,525 m (16.5 ft to 5,000 ft) (est)
Speed
  Max level speed .... 330 kt (611 km/h; 380 mph)
  Stalling speed .... 97 kt (180 km/h; 112 mph)
g limits, all axes .... +5
Radius of operation .... 65 n miles (120 km; 74 miles) (est)
Endurance, typical .... 1 hr 30 min (est)
Power plant .... 1 × piston engine

Status: As of 2011, in production and in service.

Customers: Six systems (12 aircraft) ordered by launch customer AIDC of Taiwan in mid-2002, delivery of which is believed to have been made in 2003. Also sold to Spain (four in 2005 and two more in 2009), where it continues in service at the Cedea range. Others equip Meggitt's own Field Services team, which has provided Voodoo target services to the Danish, Norwegian and Portuguese navies, as well as in Germany and with the US Air Force at Tyndall AFB in Florida.

First UK order received 14 December 2006, when this target and company's Banshee formed part of contract received from QinetiQ as ingredients in 20-year Combined Aerial Target Service (CATS) programme for UK armed forces weapon training. Initial block of eight UK serials (ZZ421 to ZZ428, c/n 026 to 033) allocated to Voodoo targets, of which first four were declared fit for service in early 2010.

Contractor: Meggitt Defence Systems Ltd UK
Ashford, Kent.

UTSL GSAT-200 NG

Type: Close- to mid-range recoverable aerial target.

Development: The Gunnery Subsonic Aerial Target (GSAT) NG is the latest in Universal Target Systems' line of low-cost, lightweight aerial target aircraft. It is designed for LOS daylight use and BVR day and night operations over land and sea against close-range, 'fire and forget', heat-seeking SAMs or close-range gunnery.

Description: *Airframe:* High-wing monoplane incorporating a GFRP fuselage with Kevlar reinforcement and epoxy and glass cloth wings. Large access panels allow all components to be accessible quickly and easily.

Mission payloads: The GSAT-200 NG can be fitted with industry standard enhancements to facilitate correct weapon function in a 'whole range' of firing scenarios. Here, enhancements include four smoke tracking flares; four IR flares; acoustic or radar MDI and laser reflector augmentation, with the particular application being designed to meet specific weapon functions. Radar-reflective mesh is fitted as standard and customer specified variants can also be supplied.

Guidance and control: The target is controlled autonomously by a target operator using a ruggedised laptop variant of UTSL's TCTS (see separate entry).

Launch: By simple, low-cost, low-maintenance, pneumatic-powered launcher from any site on land or on board ship.

Recovery: Re-usable parachute retrieval.

Power plant: One 5.2 kW (7 hp) 62 cc single-cylinder two-stroke petrol engine (type not stated), driving a two-blade propeller.

GSAT-200 NG

Dimensions, External
Overall
  length .... 1.90 m (6 ft 2¾ in)
  height .... 0.44 m (1 ft 5¼ in)
Wings, wing span .... 2.20 m (7 ft 2½ in)
Areas
Wings, Gross wing area .... 0.99 m² (10.7 sq ft)
Weights and Loadings
Weight
  Weight empty .... 10 kg (22 lb)
  Max launch weight .... 19 kg (41 lb) (est)
Payload
  Max payload, incl fuel .... 9 kg (19.00 lb) (est)
Performance
Altitude, Service ceiling .... 2,000 m (6,560 ft) (est)
Speed
  Max level speed .... 101 kt (187 km/h; 116 mph)
  Stalling speed .... 13 kt (25 km/h; 15 mph)
Range
  optical tracking .... 2.7 n miles (5 km; 3 miles) (est)
  BLOS .... 44 n miles (81 km; 50 miles) (est)
Endurance
  WOT .... 70 min
  maximum .... 90 min

Universal Target Systems' GSAT-200 NG aerial target *(UTSL)*

Universal Target Systems' GSAT-200 NG launcher *(UTSL)*

Aerial targets > United Kingdom > UTSL GSAT-200 NG – UTSL MSAT-500 NG

Status: In production and service in 2006. Available as outright sale, as operator flight training programme, or as supplier-supported hire service.
  Customers: Supplied to unidentified customers in Europe, the Middle East and Asia.
Contractor: Universal Target Systems Ltd
  Challock, Kent.

UTSL MSAT-500 NG

Type: Long-range, high-performance, medium-speed, recoverable, autonomous aerial target.
Development: The Missile Subsonic Aerial Target (MSAT) - 500 NG is UTSL's latest low-cost, high-performance aerial target aircraft.
Description: *Airframe:* High-wing delta monoplane with endplate fins; GFRP fuselage with Kevlar reinforcement and epoxy and glass cloth wings. Large panels permit quick and easy access to all components.
  Mission payloads: Can include 16 smoke tracking or infra-red flares; acoustic or radar miss-distance scoring systems; laser reflector augmentation; chaff dispensers; 127 or 191 mm (5 or 7.5 in) Luneberg lenses; corner arrays; radar transponder; sleeve or other tow targets; or UTSL-TE (which see) thermal hard body enhancement device. Variants for EW or surveillance requirements can also be provided.
  Guidance and control: The MSAT-500 NG is automatically controlled via UTSL's Command, Tracking, telemetry and Control Station (CTCS). A radar altimeter is used for low-level operations. Using the latest (2012) autonomous flight control system, the target is configured in automatic mode for automatic launch and recovery, with OLOS operation utilising a UTLS proprietary digital autopilot and GPS- or radar-based tracking and telemetry. MSAT-500 NG is usable by both day and night.
  Launch: By simple, reliable and robust, low-cost, low-maintenance bungee catapult launcher from any site on land or on board ship.
  Recovery: Re-useable parachute retrieval on land or at sea. At sea, the MSAT-500 NG floats to allow ease of recovery.

Power plant: One UTSL 27.9 kW (38 hp) rotary engine; two-blade pusher propeller, or UTSL 36 kW (50 hp) rotary engine. Fuel capacity 23 litres (6.07 US gallons; 5.00 Imp gallons).

MSAT-500 NG

Dimensions, External	
Overall	
length	2.7 m (8 ft 10¼ in)
height	0.50 m (1 ft 7¾ in)
Wings, wing span	2.75 m (9 ft 0¼ in)
Areas	
Wings, Gross wing area	2.5 m² (26.9 sq ft)
Weights and Loadings	
Weight	
Weight empty	65 kg (143 lb)
Max launch weight, in excess of	89 kg (196 lb)
Payload	
Max payload, incl fuel	24 kg (52 lb)
Performance	
Altitude, Operating altitude	5 m to 5,000 m (16 ft to 16,400 ft)
Speed	
Max level speed	200 kt (370 km/h; 230 mph)
Stalling speed	46 kt (86 km/h; 53 mph)
Range	
optical tracking	5.4 n miles (10 km; 6 miles)
electronic tracking	46 n miles (85 km; 52 miles)
Endurance, in excess of	1 hr 45 min

MSAT-500 NG during flight *(UTSL)* 1464069

MSAT-500 NG being launched *(UTSL)* 1464067

MSAT-500 NG being launched from a ship *(UTSL)* 1464068

MSAT-500 NG *(UTSL)* 1464071

MSAT-500 on catapult launcher *(UTSL)* 1148767

MSAT-500 on display in September 2005 *(UTSL)* 1179147

IHS Jane's All The World's Aircraft: Unmanned 2013-2014

Status: In production and service since 2006. Upgraded to NG (Next Generation) in 2010.

Customers: Supplied to unidentified customers in Europe, the Middle East and Asia.

Contractor: Universal Target Systems Ltd
Challock, Kent.

United States

Alliant Techsystems GQM-173 MSST

Type: Jet-powered supersonic aerial target.

Development: The MSST (Multi-Stage Supersonic Target) is being developed to simulate the threat of next-generation surface to surface anti-ship cruise missiles – such as the Novator 34M-54E Klub, for example (SS-N-27, NATO reporting name 'Sizzler') – that cruise at subsonic speeds, initiate a separation, then make a supersonic dash to their intended target. A final RFP was issued in November 2007 and, competing with a bid from a Northrop Grumman/Orbital Sciences team, ATK was awarded a four-year, USD97 million development contract by US Naval Air Systems Command (Navair) in early September 2008. Two evaluation and seven development ZGQM-173A prototypes are to be built under this contract, with a first flight expected in the second quarter of 2011. Development completion is due by October 2012, after which a production decision will be taken with a view to service entry in 2014. ATK is partnered in the development by Composite Engineering Inc (CEi), whose front end of the Orbital Sciences GQM-163 Coyote target (which see) will form the sprint stage of the MSST. ATK will develop the target's cruise stage, mission planning and C² systems.

Description: Airframe: Needle-nosed sprint stage at front, joined to rear subsonic stage by a 'strongback' fairing on which are mounted small, low aspect ratio wings; rear element has ventral intake for jet engine and conventional tail surfaces.

Mission payloads: Details not yet released.

Guidance and control: Requirements include programmable navigation and the ability to execute terminal weaves.

System composition: Air vehicle, launcher and associated support equipment.

Launch: Surface launch.

Recovery: Non-recoverable.

Power plant: Turbojet (to be selected) for subsonic stage; new rocket motor for sprint stage.

GQM-173 MSST

Performance (requirement)	
Min sprint speed at 5 m (16 ft) above S/L	M2.2
Range: Overall	>100 n miles (185 km; 115 miles)
Subsonic cruise portion	>90 n miles (166 km; 103 miles)

Status: Design and development.
Contractor: Alliant Techsystems Inc (ATK)
Minneapolis, Minnesota.

Impression of the MSST *(ATK)* 1290306

BAE Systems QF-4

Type: Full-scale aerial target.

Development: Continuing programme since early 1970s to utilise ex-operational Phantoms as high-speed FSATs for US Air Force and Navy. In mid-1998, two QF-4s, one equipped with TCAS and one without (for comparison) were used in US Air Force Battlelab trials as part of the US effort towards the ability to certify UAVs as safe to operate in civil airspace. US Navy use of QF-4 (see earlier editions for details) ended in July 2004.

According to a US Defense Science Board report of September 2005, the inventory of QF-4s could be exhausted by 2011 at the current attrition rate of around 25 a year. Coupled with the rising cost of conversion, it advocated urgent attention to the need for a replacement FSAT. Meanwhile, QF-4 production continues, at least until mid-2009.

Description: Airframe: Twin-jet low-wing monoplane; all-metal semi-monocoque construction. Internal gun and fire-control radar removed.

Mission payloads: Provisions for large internal or externally mounted payloads including target augmentation devices and auxiliary systems. Meggitt scoring systems. Primary mission is to provide full-scale representative threat presentations for weapon systems test and evaluation. QF-4 can also be used for manned operations including subscale target launch, tow target presentations, electronic warfare and weapons system fire-control tracking exercises.

A pair of QF-4Es of the USAF's 82nd Aerial Targets Squadron *(Jamie Hunter)* 1172911

Guidance and control: All-attitude manoeuvring capability. USAF QF-4s mostly preprogrammed.

Launch: Conventional runway take-off.

Recovery: Conventional runway landing.

Power plant: Two 48.5 kN (10,900 lb st) General Electric J79-GE-8 turbojets (75.6 kN; 17,000 lb st with afterburning).

Variants: QRF-4C and QF-4E/G: Programme launched 1991 to convert surplus USAF F-4Es, F-4Gs and RF-4Cs for US Air Force use. Three-year development contract awarded to Tracor Flight Systems (now BAE Systems Flight Systems) February 1992 for 10 preproduction aircraft (AF101 to 110: three QRF-4Cs, five QF-4Es and two QF-4Gs; first flight September 1993). First to be delivered was AF106, a QF-4E, on 31 October 1995. Subsequent contracts have been in multiyear procurement (MYP) programmes, as follows:

MYP 1: Lots 1 to 3 ordered in 1995, 1996 and 1997 for 36 (AF111 to 146), 36 (AF147 to 182) and 24 (AF183 to 206) respectively. Made up of two QRF-4Cs, 17 QF-4Es and 77 QF-4Gs.

MYP 2: Lots 4 to 9, comprising five batches of 12 and one of 24 ordered in 1998 (AF207 to 218), 1999 (AF219 to 230), 2000 (AF231 to 242), 2001 (AF243 to 254), 2002 (AF255 to 266) and 2003 (AF267 to 290). Made up of 66 QF-4Es and 18 QF-4Gs.

MYP 3: Lots 10 to 15, of which orders as of April 2007 comprise 13 in Lot 10 (AF291 to 303) ordered in April 2004; 22 in Lot 11 ordered in January 2005; 20 in Lot 12 (AF321 to 340); and 20 in Lot 13 (AF346 to AF365). Contract values are USD17 million, 21 million, 25.1 million and 26.5 million respectively. The two remaining contract options can be taken at any time up to 2013.

Deliveries from Lots 10, 11, 12 and 13 were due for completion by July 2006, August 2007, July 2008 and July 2009 respectively. Lot 13 sees the return of the RF-4C as a candidate airframe, due to dwindling stocks of the F-4E and G and the transition to the QRF-4C was made in July 2008, with deliveries starting in October. Ahead of this, conversion of the first two QRF-4Cs (AF-324 and -325) within Lot 12 was ordered in a separate USD5.65 million contract awarded on 21 December 2005. Lot 14 was awarded in January 2008, with Lot 15 (14 targets) on option for deliveries between 2010-2011. Lots 16 and 17 planned before stocks are depleted and the QF-16 starts to take over inFY15.

QF-4N

Dimensions, External	
Overall length	19.20 m (62 ft 11¾ in)
height	5.02 m (16 ft 5½ in)
Wings, wing span	11.77 m (38 ft 7½ in)
Areas	
Wings, Gross wing area	49.2 m² (530.0 sq ft)
Weights and Loadings	
Weight empty	13,757 kg (30,328 lb)
Max T-O weight	28,030 kg (61,795 lb)
Performance	
Climb Rate of climb, at S/L	8,534 m/min (28,000 ft/min)
Altitude, Service ceiling	16,575 m (54,380 ft)
Speed, Max level speed	1,290 kt (2,389 km/h; 1,485 mph) at 14,630 m (48,000 ft)
Radius of operation, typical	347 n miles (642 km; 399 miles)

Status: QRF-4C and QF-4E/G: US Air Force versions. In service with 82nd Aerial Targets Squadron at Tyndall AFB, Florida, with a detachment at Holloman AFB, New Mexico. Initial batch of 36 delivered May 1996 to May 1997. Some 217 reported as completed by April 2007. At that time the cumulative MYP order total had reached approximately 250. Wastage is at a rate of about 25 per year.

Customers: US Air Force (QRF-4C/QF-4E/QF-4G in production and service in 2006); US Navy (QF-4B/QF-4N/QF-4S; no longer in service).

Contractor: BAE Systems Flight Systems
Mojave, California.

Boeing QF-16

Type: Full-scale aerial target.

Development: Although it continues in full rate production, US Air Force stocks of the long-serving QF-4 Phantom aerial target are expected to be exhausted by about 2015. The early-model F-16 Fighting Falcon has long been regarded as its likely successor and, following a risk reduction assessment, a Request for Proposal (RfP) was issued on 25 June 2009 calling for a "fourth generation threat" replacement for the QF-4. This resulted in an 8 March 2010, USD69.7 million Phase 1 development contract being awarded to Boeing's Defense, Space & Security division for initial Engineering, Manufacturing and Development (EMD) of the QF-16 FSAT to replace the diminishing QF-4 fleet. The contract calls for completion of two prototype QF-16s, and contains provision for the eventual conversion of up to 126 Falcons to QF configuration. Boeing is partnered in the QF-16 programme by the Platform Solutions unit of BAE Systems Electronics, Intelligence & Support at Johnson City, New York.

Airframes are expected to be drawn from stocks of stored Block 15, 25 and 30 F-16As and Cs with 50 to 300 hours of available life remaining. Subject to passing the customary preliminary and critical design reviews in FY10 and FY11 respectively, the QF-16 programme is planned to enter the Low-Rate Initial Production (LRIP) phase (Milestone C) in the third or fourth quarter of FY13, probably with an initial batch of 20 aircraft. Production deliveries should begin by the end of FY14, with Initial Operational Capability (IOC) due to be achieved about a year later. After design and development at St Louis, and ground and flight testing, production will be at Boeing's Cecil Field, Florida, facility. User unit will be the 691st Armament Systems Squadron (ARSS) at Eglin AFB, Florida.

Description: *Airframe:* Single-jet monoplane. Detailed descriptions in *Jane's All the World's Aircraft: In Service* and *Jane's All the World's Aircraft: In Production.*

Mission payloads: Details not yet specified.

Guidance and control: Full details not yet released. Requirement includes capability for both unmanned (NULLO) and manned operation.

Launch: Conventional, automatic, wheeled take-off.

Recovery: Conventional, automatic, wheeled landing.

Power plant: One Pratt & Whitney turbofan: 100.5 kN (22,600 lb reheat thrust) F100-PW-200D in F-16A, 105.7 kN (23,770 lb thrust) F100-PW-220F in F-16C.

QF-16

Dimensions, External	
Overall	
length	15.09 m (49 ft 6 in)
height	5.09 m (16 ft 8½ in)
Wings, wing span	9.45 m (31 ft 0 in)
Weights and Loadings	
Weight	
Weight empty	7,070 kg (15,586 lb)
Max T-O weight, clean	10,800 kg (23,809 lb)
Performance	
Altitude, Service ceiling	15,240 m (50,000 ft) (est)
Radius of operation	500 n miles (926 km; 575 miles) (est)
Power plant	1 × turbofan

2001 picture of F-16As of the North Dakota Air National Guard *(USAF/Staff Sgt Greg L Davis)* 1165199

Status: In initial EMD development phase for US Air Force. By May 2010, work had begun on an initial six prototype conversions.

Contractor: Boeing Defense, Space & Security
St Louis, Missouri.

CEi BQM-167 Skeeter

Type: Recoverable aerial target.

Development: The BQM-167 aerial target was first flown during 2001 and was Composite Engineering Inc (CEi) submission to meet the USAF's Air Force Subscale Aerial Target (AFSAT) requirement. Key programme events include:

July 2002 The USAF selected BMQ-167 as its AFSAT, with the type replacing legacy BQM-34A and MQM-107D/E targets in the air defence weapon and AAM test and evaluation role.

2003 The USAF awarded CEi a contract to build and test a series of BQM-167 prototypes.

August 2004 CEi was awarded a then year USD8.7 million BQM-167 Lot 1 Low Rate Initial Production (LRIP) contract. According to *IHS Jane's* sources, this effort involved the procurement of 10 × BQM-167 targets for the Luftwaffe (German Air Force).

January 2011 Aside from the noted Lot 1 LRIP tranche, *IHS Jane's* sources report the USAF as having procured 226 six additional BQM-167 aerial targets, with the total being broken down into individual batches of 40 (Lot 2 production), 24 (Lot 3), 42 (Lot 4), 50 (Lot 5), 30 (Lot 6) and 40 (Lot 7).

6 September 2012 The USAF's Air Force Life Cycle Management Center (Eglin AFB, Florida) awarded CEi a then year USD7,317,122 indefinite delivery/quantity, firm, fixed-price contract with regard to the procurement of 54 type specific reparable spares for use in the service's BQM-167 targets. At the time of its announcement, work on this effort was scheduled for completion by 9 November 2015.

Description: *Airframe:* Mainly cylindrical fuselage, tapered at each end, with ventral turbojet; sweptback wing and tail surfaces. Construction mainly of carbon fibre composites.

Mission payloads: BQM-167A and -167i targets can be equipped with towed IR or RF sources, fixed IR or RF sources, a CounterMeasures Dispensing System (CMDS), smoke generation equipment (19 litres capacity) and pod or internally mounted active Electronic Attack (EA) payloads. All USAF BQM-167 targets are fitted with the AN/ALE-47 CMDS, EA system controls, wing tip IR source pods and nose-mounted Luneburg lens radar cross section augmentation. BQM-167 targets can also be fitted with a proprietary tow target launch system, with *IHS Jane's* sources identifying the Meggitt TRX-4A as a candidate payload for such an application. Here, the noted launch system is capable of deploying targets at variable distances of up to 2,438 m (8,000 ft). Within both the BQM-167A and -167i weapon systems, the various payloads are accommodated on underwing hardpoints (68 to 90 kg (150 to 198 lb) payload/station), in wing tip pods (45.4 kg (100 lb) payload/pod) and internally (model dependent 113 to 293 kg (249 to 645 lb) payload).

Guidance and control: Rockwell Collins (formerly Athena Technologies) Athena 111m integrated flight control system (airspeed, altitude, climb rate, angle-of-attack, sideslip, 3-D automatic navigation (GPS/INS) and automatic take-off and landing). Overall BQM-167 C2 provision is described as being "configurable (up to [eight] simultaneous targets)".

Launch: Ground launch using RATO (zero length launch rail, one or two RATO bottles) or pneumatic techniques.

Recovery: Drogue and main parachute recovery.

Power plant: One 4.45 kN (1,000 lb st) Microturbo TR 60-5+ turbojet in ventral nacelle. BMQ-167A standard fuel capacity 435 litres (115.0 US gallons; 95.8 Imp gallons).

A wind tunnel model of the BQM-167 *(CEi)* 0576350

Variants: BQM-167A: Standard USAF production version.
BQM-167i: Export version of the BQM-167A.
BQM-167X: Re-designed, higher-speed version, flight tested in 2008. Subsequently formed the basis for company's SuperSonic Aerial Target for USN (see entry for SSAT).

BQM-167

(A: BQM-167A B: BQM-167i)
Dimensions, External
  Overall
    length .... 6.15 m (20 ft 2¼ in)
    height .... 1.32 m (4 ft 4 in)
  Fuselage, width .... 0.61 m (2 ft 0 in)
  Wings, wing span .... 3.18 m (10 ft 5¼ in) (est)
Dimensions, Internal
  Payload bay, volume .... 0.1984 m³ (7.0 cu ft) (est)
Weights and Loadings
  Weight
    Max launch weight
      (A) .... 930 kg (2,050 lb)
      (B) .... 649 kg (1,430 lb)
Performance
  Altitude
    Operating altitude
      (A) .... 15 m to 15,545 m (50 ft to 51,000 ft)
      (B) .... 15 m to 15,240 m (40 ft to 50,000 ft)
  Speed, Cruising Mach number 0.93
  g limits .... +9/–2
  Range .... 400 n miles (740 km; 460 miles) (est)
  Endurance .... 3 hr (est)
  Power plant .... 1 × turbojet

Status: As of 2012, the BMQ-167 aerial target was being reported as being in service.
  Customers: The Luftwaffe and the USAF.
Contractor: Composite Engineering Inc
Sacramento, California.

CEI Firejet

Type: Multirole aerial target.
Development: Unveiled in mid-2007.
Description: Airframe: Mid-mounted delta wings with curved tips; circular-section fuselage with pointed nosecone; sweptback fin and rudder. No landing gear. Carbon fibre composites construction.
  Mission payloads: Luneberg lens or smoke oil canisters.
  Guidance and control: Autonomous and manual (up to three aircraft simultaneously). Waypoint navigation (up to 100 points), re-programmable in flight.
  Launch: By pneumatic catapult.
  Recovery: Parachute recovery (deployment altitude 305 m; 1,000 ft). Can be recovered from land or sea.
  Power plant: Twin 0.20 kN (45 lb st) turbojets. Fuel capacity 57.35 litres (15.15 US gallons; 12.61 Imp gallons).

Firejet

Dimensions, External
  Overall
    length .... 2.47 m (8 ft 1¼ in)
    height .... 0.57 m (1 ft 10½ in)
  Wings, wing span .... 2.00 m (6 ft 6¾ in)
Weights and Loadings
  Weight
    Weight empty .... 38.6 kg (85 lb)
    Max launch weight .... 84.0 kg (185 lb)
  Fuel weight, Max fuel weight .... 45.4 kg (100 lb)
Performance
  Altitude, Service ceiling .... 7,620 m (25,000 ft)

The Firejet delta-winged target *(CEI)* 1290194

Speed
  Max level speed .... 400 kt (741 km/h; 460 mph) (est)
  Cruising speed, normal .... 300 kt (556 km/h; 345 mph)
g limits
  tested .... +4
  planned .... +6
Endurance .... 2 hr (est)
  300 kt at 4,570 m (15,000 ft) .... 30 min
Power plant .... 2 × turbojet

Status: Development continuing in mid-2007. Expected to be market-ready by end of that year.
Contractor: Composite Engineering Inc
Sacramento, California.

CEI MQM-107 Streaker

Type: Recoverable variable speed aerial target.
Development: As the Beechcraft Model 1089, the Streaker was the 1975 winner of the US Army's Variable Speed Training Target (VSTT) competition, and has since been produced in large numbers for the US services and for export; total orders now exceed 1,500. In addition to design, development and delivery of the air vehicles, ground support equipment, spares and ancillaries, the company's contracts include operation and maintenance of the system in US service. The MQM-107 programme was acquired by CEI from Raytheon Missile Systems in October 2003.

The MQM-107 is the US Army's primary subsonic missile training target, providing a variety of threat simulations for development, testing and training on the MIM-104 Patriot, MIM-23 Improved Hawk, FIM-92 Stinger and Patriot PAC-3 Interceptor. It has also been used extensively by the US Air Force for air-to-air combat training with the AIM-9 Sidewinder, AIM-7 Sparrow and AIM-120 AMRAAM.

Second-source producer Marconi North America (now BAE Systems Flight Systems) teamed with AeroSpace Technologies of Australia (ASTA) in October 1995 to bid the MQM-107E for the Australian Defence Forces AATS (ADF Aerial Target System) requirement for a Jindivik replacement for the Royal Australian Air Force and Navy; it was selected as the preferred system in late 1996.

Description: Airframe: All-aluminium low-wing monoplane with sweptback wings and tail unit; engine suspended on pylon beneath centre of mainly cylindrical fuselage. Modular design throughout, with flat-section wing and tail surfaces of bonded honeycomb (fixed surfaces) or foam-filled aluminium (moving surfaces). Ogival nosecone and tailcone. Improved waterproofing on 107B and later models. Air vehicle is designed to operate at altitudes from sea level to 12,200 m (40,000 ft) and at speeds of up to M0.8 (107E, M0.85). Longer fuselage of 107B and subsequent models provides a larger payload section, easier access to payload and electronics, and waterproofing provisions for sea-water recoveries. Target has established an in-service record of more than 25 missions per operational loss.

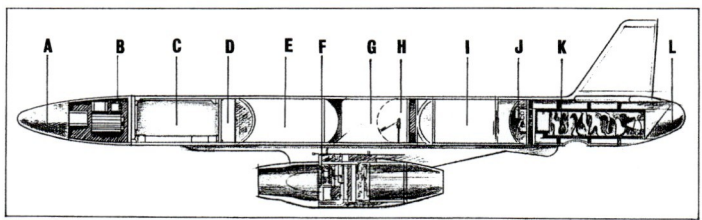

(A) nosecone; (B) guidance and control equipment; (C) optional payload; (D) smoke/oil tank; (E) forward fuel tank; (F) power plant; (G) centre fuel tank; (H) fuel quantity sensor; (I) aft fuel tank; (J) fuel management controls; (K) recovery parachutes; (L) tailcone

MQM-107D Streaker internal details 0517991

First MQM-107E target produced by Tracor Flight Systems (now BAE Systems) 0518256

Aerial targets > United States > CEI MQM-107 Streaker

MQM-107D with external payloads
1 pyrotechnic optical plume simulator
2 flare dispenser pod
3 scorer pod
4 foam cone radar reflector
5 infra-red augmentor boom
6 flare/chaff dispenser
7 bullet scorer/tow banner
8 bi-static radar reflective pod
9 radar tow target
10 infra-red tow target
11 infra-red wingtip pod

Royal Australian Navy Kalkara with underwing TPT7 tow bodies *(Paul Jackson)* 0533543

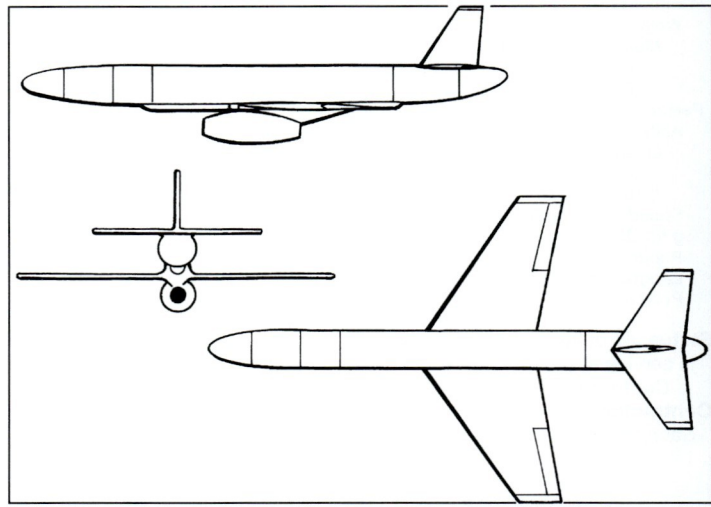

MQM-107 variable speed training target *(IHS/John W Wood)* 0517992

Mission payloads: Principal function is to tow a variety of subtargets for missile training and evaluation. Two radar, infra-red or visually augmented tow targets, or a 2.23 m² (24 sq ft) banner with a corner reflector, can be carried on each mission and towed separately up to 2,440 m (8,000 ft) behind the MQM-107. Wingtip and pod-mounted augmentation and scoring devices can be carried. A 26.5 litre (7 US gallon; 5.8 Imp gallon) smoke/oil tank (15 litres; 4 US gallons; 3.3 Imp gallons in 107A) is carried for visual augmentation.

Typical external payloads can include pyrotechnic optical plume simulators, TA-8 radar or IR augmentation tow subtargets, flare/chaff dispensers, passive and active radar augmentation, bistatic radar reflector pods, gunnery tow banners, and miss-distance indicator scoring pods.

Guidance and control: Guidance and control systems provide for ground control within the trajectory pre-programmed for the first 20 seconds after rocket-assisted take-off (RATO) ignition. Flight controller is then in control and is provided with all pertinent flight information by radio link from air vehicle's onboard sensors, enabling command of vehicle manoeuvres and recovery. In flight, guidance and control system stabilises automatically around the roll, yaw and pitch attitudes, and provides an altitude hold mode. Command and control is GPS-based in the MQM-107E.

Flight control developments include a terrain-following guidance capability which has demonstrated extremely low-altitude flight profiles. The MQM-107 has provisions for a high-*g* autopilot, to extend the manoeuvring and high-*g* envelope of the vehicle. The flight controller can select either constant airspeed or constant altitude high-*g* manoeuvres, and 6 *g* manoeuvres can be maintained during use with air-to-air or surface-to-air weapon systems.

Launch: Surface launched, using RATO booster, from lightweight zero-length launcher. Booster is jettisoned about 2 seconds after ignition, after accelerating target to approximately 220 kt (407 km/h; 253 mph).

Recovery: Command recovery system using 1.65 m (5.4 ft) diameter drogue and 15.24 m (50 ft) main parachute stowed in rear of fuselage. Recovery cycle can be initiated by remote command, by a 6 second loss of command link, or by electrical power loss.

Power plant: Single turbojet, as listed under Variants heading, in underfuselage pod. Standard fuel capacity 246 litres (65 US gallons; 54 Imp gallons), with provision in longer fuselage versions to increase this to 284 litres (75 US gallons; 62.5 Imp gallons). Optional wing insert fuel tanks can add a further 113 litres (30 US gallons; 25 Imp gallons).

Variants: *MQM-107A:* Initial production version with 2.85 kN (640 lb st) Teledyne Continental J402-CA-700 turbojet and short fuselage. Produced for US Army (385 delivered from April 1976 to early 1979); supplied also to Abu Dhabi (with 3.73 kN; 838 lb st Microturbo TRI 60-2 turbojet), Jordan, South Korea, Sweden (designation **Rb06 Girun**) and Taiwan.

MQM-107B: Introduced 1982, with TRI 60-2 Model 074 engine as standard, permitting increase in maximum operating speed and manoeuvring capability. Longer fuselage, subsequently adopted for all later models, allowed greater payload volume; systems improvements based on experience with 107A. Delivered to US Army (139) and US Air Force (70) between 1984 and 1986; exported to Egypt, Sweden (Rb06B) and Taiwan. International models retain TRI 60-2 engine for commonality, but incorporate same improvements as later US versions.

MQM-107C: Non-standard model, combining surplus J402-CA-700 engines with long fuselage of 107B; 69 to US Army 1985.

MQM-107D: As 107B, but initially with 4.27 kN (960 lb st) Teledyne Continental 373-8 engine. First US Army contract for 90 awarded October 1985 (delivered from January 1987); follow-on contracts have increased total US procurement to 691 (453 for Army, 221 for Air Force, 17 for Navy), of which those ordered from 1989 have 4.23 kN (950 lb st) TRI 60-5 turbojets.

Super MQM: Improved version of MQM-107D, designed to meet or exceed requirements of US Subsonic Subscale Aerial Target (SSAT) programme, including increased high-*g* manoeuvres for USAF/USN air-to-air combat training.

Dreem: Modified version of MQM-107D (Drone RF Electronic Enhancement Mechanism), developed by Boeing's Phantom Works under a two-year USAF contract as a possible cheaper alternative to such full-scale aerial targets as the QF-4. First test flights were made in November and December 1998. Gross weight 642 kg (1,415 lb), cruising speed 300 kt (556 km/h; 345 mph), endurance approximately 30 minutes. One underwing pod (see accompanying photograph) carries Boeing radar augmentation, the other contains US Navy ECM.

MQM-107E: Improved performance US Army/Air Force version, with TRI 60-5 or Teledyne Continental 373-8B turbojet; reshaped wing leading-edges, rudder and elevators; decreased tailplane incidence; new digital autopilot. Qualification flight tests conducted 1992 at White Sands Missile Range, New Mexico. Reported requirement for up to 200. Initial US Army contract, for seven preseries and 80 production articles, awarded July 1994 to Tracor (now BAE Systems Flight Systems). First example completed July 1996 and made first flight 10 December 1996; entered USAF service in 1998.

Kalkara (storm bird): Version for Royal Australian Air Force and Navy, based on MQM-107E (TRI 60-5 engine) but with different GPS-based command and control system. Initial nine-year, USD35 million contract awarded in February 1997 for 20 air vehicles plus tow targets, two GPS-equipped ground stations and support services. All now delivered and operational; scheduled to remain in service until 2008.

MQM-107D, MQM-107B, MQM-107A, MQM-107E

Dimensions, External	
Overall length	
MQM-107B, MQM-107D, MQM-107A	5.51 m (18 ft 1 in)
MQM-107E	5.13 m (16 ft 10 in)
height	
MQM-107B, MQM-107D, MQM-107A	1.47 m (4 ft 9¾ in)
MQM-107E	1.54 m (5 ft 0¾ in)
Fuselage, width	0.38 m (1 ft 3 in)
Wings wing span	
MQM-107B, MQM-107D, MQM-107A	3.01 m (9 ft 10½ in)
MQM-107E	3.02 m (9 ft 11 in)
Tailplane, tailplane span	1.58 m (5 ft 2¼ in)

MQM-107D, MQM-107B, MQM-107A, MQM-107E

Dimensions, Internal	
Payload bay volume	
MQM-107B, MQM-107D, MQM-107E	0.1357 m³ (4.8 cu ft)
MQM-107A	0.0924 m³ (3.3 cu ft)
Weights and Loadings	
Weight empty	
MQM-107A	218 kg (480 lb)
MQM-107B, MQM-107D	261 kg (575 lb)
Max launch weight	
excl booster, MQM-107B	664 kg (1,463 lb)
excl booster, MQM-107D, MQM-107E	662 kg (1,459 lb)
Payload	
Max payload	
internal, MQM-107A	43 kg (94 lb)
internal, MQM-107B, MQM-107D, MQM-107E	45 kg (99 lb)
external, MQM-107A	113 kg (249 lb)
external, MQM-107B	160 kg (352 lb)
external, MQM-107D, MQM-107E	91 kg (200 lb)
Performance	
Altitude, Operating altitude	15 m to 12,190 m (50 ft to 40,000 ft)
Endurance	
MQM-107A	3 hr (est)
MQM-107B	2 hr 18 min
MQM-107D, MQM-107E	2 hr 15 min
Power plant	1 × turbojet

Status: In production and service. Serves as aerial target for air defence systems such as Chaparral, Crotale, HAWK, Improved HAWK, Mistral, Patriot, Rapier, Skyguard/Sparrow, Skyguard/35 mm and Stinger, and the Vulcan air defence gun system. Can also be used with air-to-air missiles such as AIM-7 Sparrow and AIM-120 AMRAAM. USAF user is 82nd Tactical Air Targets Squadron, 475th Weapons Evaluation Group, at Tyndall AFB, Florida. US Air Force was reported in 2004 as phasing out all of its remaining MQM-107s in favour of the new BQM-167 Skeeter.

Kalkara is operated from Jervis Bay Range Facility and from HMAS *Stirling* on Garden Island, near Perth, plus deployments from Darwin, Woomera and elsewhere.

Customers: Australia; Egypt; Jordan; South Korea; Sweden; Taiwan; UAE; USA (Army, Navy and Air Force).

Contractor: Composite Engineering Inc
Sacramento, California.(MQM-107E: BAE Systems Flight Systems Mojave, California.)

CEI SSAT

Type: Recoverable subsonic aerial target.

Development: Winner of a competition launched by the US Navy in early 2009, the SSAT (SubSonic Aerial Target) programme intended to result in an all-altitude, high-subsonic target intended as a one-type replacement for the service's current AQM-37C/D, BQM-34S and BQM-74E. On 28 January 2011, CEI was selected, in preference over bids from Northrop Grumman (BQM-74X) and DRS Technologies/Selex Galileo (Mirach 100/X), to receive a USD31.5 million EMD contract to demonstrate its own SSAT candidate.

Basis for this is the BQM-167X, a modified version of the company's existing BQM-167A Skeeter that was flight tested at the USN's Point Mugu Test Range between September 2007 and February 2008. It differs from the standard Skeeter chiefly in having a redesigned wing and its jet engine mounted within an area-ruled fuselage.

Requirements for the eventual SSAT include the ability to simulate threats from modern fast-jets and anti-ship cruise missiles, and to perform fixed or programmable weaving manoeuvres. (For other parameters, see Specifications below.) During the five-flight 2007-08 demonstration, the BQM-167X achieved Mach 0.95 at low altitude; sustained supersonic speed (without use of full engine thrust); altitude of more than 12,200 m (40,000 ft); endurance of more than 1 hour; and sustained turns at more than 6 *g*.

Artist's impression of SSAT *(NAVAIR)* 1364878

BQM-167X being launched at US Navy's Point Mugu Test Range in 2008 *(CEI)* 1395347

Description: *Airframe:* Area-ruled, cylindrical fuselage; internal turbojet; sweptback wing and tail surfaces. Construction mainly of carbon composites. Between 60 and 70 per cent of components are common with the BQM-167A.

Mission payloads: Required augmentation and other systems include AN/DPT-2B and -2C radar transmitters and AN/ULQ-21 ECM.

Guidance and control: By UHF command link from Micro Systems Inc SNTC (System for Naval Target Cpntrol).

Launch: Surface launch, assisted by booster rocket. Unlike AQM-37, which it will replace, air launch is not required.

Recovery: Parachute recovery.

Power plant: One 4.20 kN (944 lb st) Microturbo TR 60-5+ turbojet.

SSAT

Performance	
Altitude, Operating altitude	5 m to 10,670 m (10 ft to 35,000 ft)
Radius of operation	
minimum	150 n miles (277 km; 172 miles)
control link, normal	200 n miles (370 km; 230 miles)
with relay	330 n miles (611 km; 379 miles)
Power plant	1 × turbojet

Status: BQM-167X has reportedly been exported to Taiwan. SSAT is on a course of three-year (2011-13) engineering and manufacturing developments, which includes funding for first two production lots. Requirement is for USN service entry in 2015, with full fielding by 2017.

Contractor: Composite Engineering Inc
Sacramento, California.

Griffon MQM-170A Outlaw

Type: Recoverable aerial target.

Development: History not known. Understood to have entered series production about mid-2005.

Description: *Airframe:* Slightly tapered, low-mounted wings; tubular fuselage with pointed nosecone; V tail surfaces. Composites construction. Fixed tricycle landing gear optional.

Mission payloads: Variety of EO/IR sensors and other specialised payloads can be carried in multiple bays. Mid-fuselage bays can be configured interchangeably for payloads or fuel.

In target use for US Army, standard payloads include MILES (Multiple Integrated Laser Engagement System), IR enhancer, precision gunnery

An MQM-170A Outlaw with drop payload *(Griffon Aerospace)* 1290253

Outlaw landing *(Griffon Aerospace)* 1151522

system (PGS) reflectors, smoke generator, night-time lighting kit and Doppler-based scoring system.

Guidance and control: Manual (within LOS or BLOS, console being able to control several aircraft simultaneously); or preprogrammed BVR, with autopilot and GPS waypoint navigation. If control link is lost, automatic flight termination system allows controller to end mission by means of glide landing, stall landing or deployment of a recovery parachute. Standard GCS comprises one or more laptop computers to display air vehicle health and status and mission execution. Telemetry suitcase contains aircraft and sensor transceivers and laptop interfaces. Datalinks to drive sensor displays are available via the telemetry suitcase.

Launch: Pneumatic catapult launch standard; wheeled take-off optional. Launcher can be used with either wheeled or non-wheeled versions.

Recovery: Belly skid or conventional wheeled landing.

Power plant: One 12.7 kW (17 hp) 3W-Modellmotoren 150i two-cylinder two-stroke engine; two-blade pusher propeller.

Variants: **MQM-170A Outlaw:** For standard LOS and 'over the hill' missions.

MQM-171A BroadSword: Enlarged dual-use (tactical UAV and target) version; *described separately.*

MQM-170A

Dimensions, External	
Overall	
length	2.70 m (8 ft 10¼ in)
height, excl landing gear	0.70 m (2 ft 3½ in)
Wings, wing span	4.15 m (13 ft 7½ in)
Dimensions, Internal	
Payload bay, volume	0.425 m³ (15.0 cu ft)
Weights and Loadings	
Weight	
Weight empty	33.6 kg (74 lb) (est)
Max T-O weight	59 kg (130 lb) (est)
Max launch weight	59 kg (130 lb) (est)
Payload	
Max payload, incl fuel	18.1 kg (39 lb) (est)
Performance	
Altitude, Operating altitude	5 m to 4,875 m (10 ft to 16,000 ft)
Speed	
Max level speed	104 kt (193 km/h; 120 mph)
Cruising speed	70 kt (130 km/h; 81 mph)
g limits	±8
Radius of operation, BLOS, mission	60 n miles (111 km; 69 miles)
Endurance	
standard	1 hr
max	4 hr (est)

Status: As of February 2008, the MQM-170A was in production and service as a standard US Army target, mainly for use with JDAM and Stinger missiles. Other users include US Air Force, Navy and Marine Corps, and Foreign Military Sales customers.

Contractor: Griffon Aerospace
Madison, Alabama.

Meggitt GT-400

Type: Releasable gliding aerial tow target.

Development: Described as the world's first launchable gliding tow body, the GT-400 is designed for use with air-to-air and surface-to-air missiles, to provide a multirole glide target/platform that meets the needs for a manoeuvring threat simulator or payload-carrying vehicle for weapons testing. It is carried and released as a standard towed target during target acquisition, but then released from the towline when ready for deployment into preprogrammed flight. It can be used with standard reeling machines such as the Meggitt RM-30A. A successful second phase flight demonstration for the US Air Force was carried out at Tustin on 10 December 2005, when the GT-400 was launched from a Phoenix Air Learjet 35 and engaged by F-15 and F-16 fighters with AIM-120 missiles. GT-400 has been certificated by the EASA for carriage on the Falcon 20.

Description: *Airframe:* Mainly cylindrical body; high-mounted wings, with ailerons; cruciform tailfins with elevators on horizontal pair.

GT-400 mounted on a Learjet *(Meggitt)* 1151577

Mission payloads: Radar augmentation with 163 mm (6.4 in) diameter monostatic reflector lenses; visual augmentation with flares optional.

Guidance and control: Controlled via UHF telemetry link from a command and control station to an onboard three-axis autopilot that uses inputs from position, rate and pressure sensors to keep target on a preplanned course at desired speed. Roll control by ailerons, pitch control by elevators. GPS navigation.

Launch: Target is reeled out a short distance before launch/release from towline. Once deployed, a command is sent to target by command and control station; it then releases from towline and glides, following its preset flight plan. It is compatible with current business jet tow aircraft such as the Learjet 35, Falcon 20 and Gulfstream G100.

Recovery: Flight termination system and parachute recovery in the event of unsuccessful engagement.

Variants: **GT-400(-1):** Rated for flight speeds of 210 to 400 kt (389 to 741 km/h; 242 to 460 mph) IAS; no augmentation. Communication by 900 MHz hopping frequency.

GT-400(-2): Rated for flight speeds of 210 to 400 kt (389 to 741 km/h; 242 to 460 mph) IAS; augmented with 163 mm (6.4 in) diameter monostatic radar reflector lenses. Communication by 450 MHz hopping frequency.

GT-400(-3): As described for GT-400(-1), but with 163 mm (6.4 in) diameter monostatic radar reflector lens augmentation.

GT-400

Dimensions	
Wing span	0.72 m (2 ft 4.5 in)
Body diameter (max)	0.19 m (7.5 in)
Length overall	2.64 m (8 ft 7.8 in)
Weights	
Max launching weight	81.6 kg (180 lb)
Performance	
Towing speed	190-400 kt (352-741 km/h; 219-460 mph) IAS
Flight speed	220-400 kt (407-741 km/h; 253-460 mph) IAS
Max towing altitude	12,200 m (40,000 ft)
Glide range	>40 n miles (74 km; 46 miles)

Status: In production for US and international customers. Export launch customer reported to be Australian Army (contract value approximately AUD280,000); delivered in 2006. More than 50 were delivered by 2008. Other customers in the Middle East and Europe.

Contractor: Meggitt Defense Systems Inc
Tustin, California.

Meggitt SGT-20

Type: Air-launched, recoverable tow target.

Development: Developed as a sleeve-type self-contained target system for use with the MQM-107 Streaker, which can carry one under each wing.

Description: *Airframe:* Lightweight (aluminium) cylindrical forebody with conical nose and four sweptback tailfins in cruciform configuration, plus a non-metallic string sleeve visual augmentor. Forebody has an impact resistant nose and an internal, inertia-braked towline reel accommodating up to 366 m (1,200 ft) of towline. Internal reel eliminates the need for a reeling machine on the tow aircraft.

Mission payloads: Forebody has provision to accommodate a variety of available scoring systems. Minimal aerodynamic drag ensures maximum number of target presentations per mission.

Launch: Reel-out and sleeve deployment are fully automatic after release of the SGT-20 from the launcher.

Recovery: Following ground recovery and retrieval, refurbishment consists of replacing the expended towline, sleeve deployment bag, and launcher interface ball-lock pin.

SGT-20 self-contained sleeve target for use with the MQM-107 0518266

SGT-20

Dimensions, External
Overall
length
forebody .... 1.37 m (4 ft 6 in)
sleeve, standard .... 4.57 m (15 ft 0 in)
height
forebody .... 2.13 m (7 ft 0 in)
sleeve, standard .... 0.61 m (2 ft 0 in)
Weights and Loadings
Weight
Weight empty
excl scoring equipment .... 18 kg (39.0 lb)

Status: In production and service.
Customers: Canada (Air Force); Finland (Air Force); US (Air Force, Army and Navy); and others.
Contractor: Meggitt Defense Systems Inc

Meggitt TDK-39

Type: Air-launched, recoverable tow target.
Development: Developed to replace the TDU-10B Dart target and the interim Secapem 90B (A/A37U-33), the TDK-39 (USAF designation **A/A37U-36**) is the target portion of the AGTS-36 aerial target system and is operated from a Meggitt RMK-35 reeling machine/launcher. It is the standard gunnery target for the US Air Force and Air National Guard, compatible with F-4, F-15 and F-16 aircraft and readily certifiable with other types.
Description: *Airframe:* Target consists of an aluminium forebody, with four low-aspect ratio tailfins in X configuration, and a prepacked, non-rigid string sleeve visual augmentor.

Mission payloads: Meggitt Defense Systems RADOPS Doppler radar scoring system installed in forebody, projecting a conical antenna pattern around the target sleeve for real-time scoring realism. Sensor, operating at 3.245 GHz, features self-contained telemetry transmitter and can detect 20 mm and larger projectiles passing through the scoring zone. Replacement of visual augmentor permits quick system turnround.

Guidance and control: Programmed control is automatic, reeling out and recovering target in response to simple discrete commands. Airborne receiving station installed in reeling machine contains telemetry receiver and microprocessor-based signal processor. Signals are processed and displayed as a count on host aircraft cockpit display panel.

Launch: Automatic. Deployment initiated by simple command from tow aircraft, reeling out 610 m (2,000 ft) of standard aerial steel towline. Forebody contains visual augmentor automatic deploy/release mechanism. Target can be deployed at altitudes from 305 to 7,620 m (1,000 to 25,000 ft) and an airspeed of 250 kt (463 km/h; 288 mph) CAS; for captive carriage, and after deployment, the envelope extends to 12,200 m (40,000 ft) and M0.9. Target flies at negative angle of attack to balance towline forces for aerodynamic stability throughout flight envelope; scoring accuracy is not affected by airspeed or altitude. Manoeuvres can include race track, figure-of-eight, butterfly and high-angle combat dart patterns.

Recovery: By reel-in at 305 to 7,620 m (1,000 to 25,000 ft) altitude and 230 to 250 kt (426 to 463 km/h; 265 to 288 mph) CAS airspeed. RMK-35 safety system with dual pyrotechnic firing circuits can be used to jettison target and towline in case of emergency. Visual augmentor is released when forebody is recovered back to launcher.

TDK-39

Dimensions, External
Overall
length
forebody, incl augmentor bag .... 1.91 m (6 ft 3 in)
visual augmentor .... 9.14 m (30 ft 0 in)
system, packed .... 2.39 m (7 ft 10 in)
system, deployed .... 7.01 m (23 ft 0 in)
width
forebody .... 0.27 m (10½ in)
visual augmentor .... 0.79 m (2 ft 7 in)

Status: In production and service. Has been manufactured in Greece for the Hellenic Air Force by Miltech Hellas under a contract signed in 2000.
Customers: US (Air Force and Air National Guard) and six international users, including Greece.
Contractor: Meggitt Defense Systems Inc

Meggitt tow targets

Type: Aerial tow targets.
Development: Meggitt towed aerial targets include radar, infra-red, naval and gunnery targets, together with compatible launchers and support equipment. All can be utilised on any UAV, commercial or military aircraft capable of carrying external stores. The targets are suitable for surface-to-air, air-to-air and naval weapons.
Description: *Airframe:* The targets consist of a protruded glass fibre airframe, thermoplastic tailfins, and various nosecones and tailcones depending upon mission requirements. All have the same body diameter and fin span. Reeling machine targets have a mounting swivel for attachment to a tow reel cable. One-way (non-recoverable) targets for use with Meggitt's LTC launchers have an internally mounted reel and cable located in the approximate centre of the target body.

Mission payloads: See under Variants for individual payloads. All basic types can also be fitted with Doppler radar or acoustic miss-distance indicating (MDI) systems, and with a Meggitt flare/chaff dispenser.

Meggitt also produces wingtip IR augmentation pods for target drones which provide subscale drones with a realistic IR signature. When used against non-warhead (training) missiles, the pods increase the target drones' survivability by up to 90 per cent.

Guidance and control: Meggitt tow targets are designed to be flown in either a one-way configuration from the company's LTC series of launchers, drone aircraft launchers, or in a two-way (recoverable) configuration from all makes of reeling machine. In the one-way configuration, the subtarget is fitted with an internally mounted tow reel. This is controlled mechanically by the reel brake assembly, which is designed to provide towline pay-out when the target is launched. The towline is released at the launcher when the mission is completed or the target is destroyed. The 'deploy' and 'release' commands for the launcher are hard-wired from a cockpit control panel. Cable pay-out, acceleration and deceleration of one-way targets used with the LTC launchers are controlled automatically by the reel mechanism.

Launch: The Meggitt LTC series of one-way launchers can be carried by any aircraft with standard underwing store or bomb shackles. The launcher, which stores the target during ferry and provides the tow cable attachment point, contains the mechanism for deployment of the target and release of the cable after the mission. Tow cable lengths of up to 6,705 m (22,000 ft) are available with the LTC launchers.

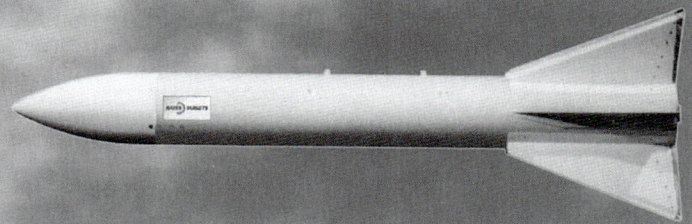

TGT aerial gunnery tow target

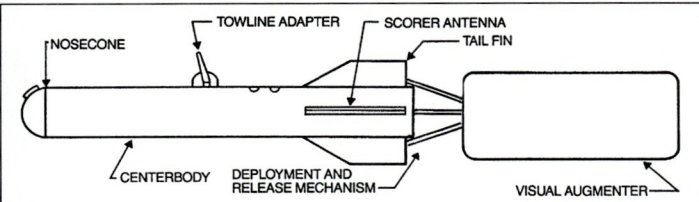

Main features of the TDK-39 target

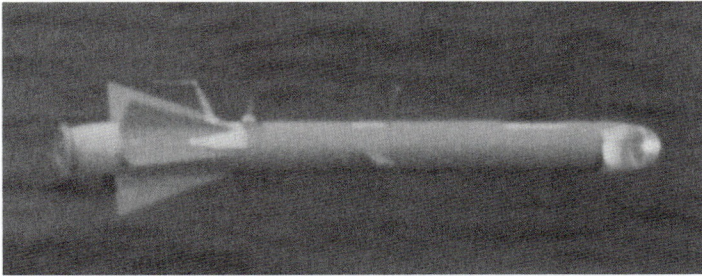

TLX-1 sea-skimming tow target *(Meggitt)*

The TDK-39 on its RMK-35 reeler/launcher

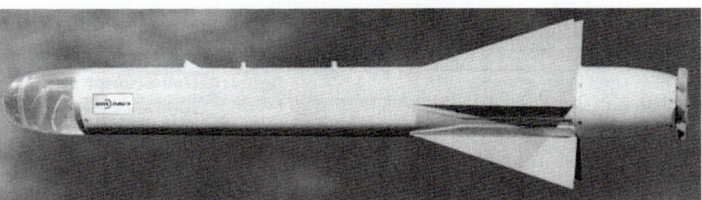

The TGX light-augmented radar tow target *(Meggitt)*

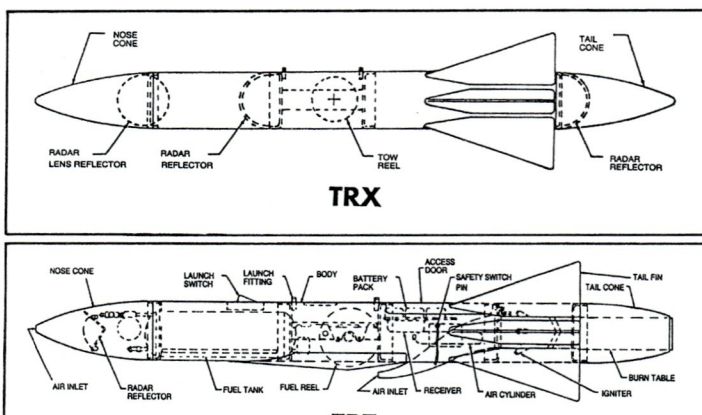

Meggitt target family internal details

Recovery: The targets can be provided in a one-way (non-recoverable) configuration for use with a Meggitt LTC launcher, or in a two-way (recoverable) configuration for use with a reeling machine. The one-way system is appropriate for users needing only a small number of targets, or to provide target operations at the lowest possible initial cost. The one-way system is also used with a reeling machine to provide a back-up capability. The targets can be used with reeling machines which can recover the targets if they are not destroyed during the firing mission. Tow cable lengths in excess of 6,705 m (22,000 ft) are available with reeling machines.

Variants: *TGT:* The TGT target is used primarily for air-to-air and surface-to-air gunnery. It is visually augmented with a rope or solid banner which deploys after target launch. The banner is used for visual acquisition and identification; radar reflectors are used for weapons system acquisition. Banner length can vary from 3 to 12 m (9.8 to 39.4 ft).

TGX: Radar-augmented target. Primary feature is a ram-air powered, coaxially mounted lamp in a clear Plexiglas nosecone providing 60,000 cd of light for visibility at up to 5.4 n miles (10 km; 6.2 miles) in daylight. Proprietary Ku-band radar reflector mounted coaxially in front of lamp to provide forward aspect radar augmentation to 5 m² (53.8 sq ft) while allowing light to shine through. A 28 V alternator, driven continuously via a ram-air turbine in the tailcone, provides power for the lamp and any special payloads such as scoring systems, active radar augmentation, missile seeker simulation or EW packages. Variants include **TGX-2**, for use from manned aircraft with RM-30A or RM-30B tow reel; this has no internal towline or reel, and is equipped with a mesh corner reflector in the nose for passive augmentation. The **TGX-[towline length]** is similar, but has an internal reel for one-way operation under an LTC type launcher. Towline lengths of up to 3,660 m (12,000 ft) can be achieved from an LTC-2 launcher; more than 5,485 m (18,000 ft) is possible from an LTC-6 containing a supplementary tow reel.

TIX: IR-augmented target, for use with weapon systems or trackers that use heat-seeking guidance technology. A propane burner operating at approximately 1,135°C provides IR output for 30 to 45 minutes; the burner is operable at up to 6,100 m (20,000 ft) and M0.9. The IR output is 400 W per steradian in the 1.8 to 3 micron band and 250 W in the 3 to 5 band; Ku-band radar cross-section is 5 m² (53.8 sq ft) in forward aspect, 2 m² (21.5 sq ft) all-aspect.

TLX-1: Low-level, height-keeping (sea-skimming) target. Altitude is governed via an autopilot, radar altimeter and two ailerons, and can be commanded in flight over an RF command and control link. Altitude settings are fully adjustable during flight. Telemetry provides the host aircraft with altitude, wing angle and acceleration information. Payloads include MDI systems and active radar augmentation. Passive radar enhancement provided by Ku-band screen reflector mounted coaxially in the nose and providing forward aspect RCS augmentation to 10 m² (107.6 sq ft). Reflector is used in conjunction with a 600,000 cd lamp in clear Plexiglas nosecone for visual light enhancement. Power for lamp, payloads and flight control system provided by a 1 kW 28 V DC generator driven by a ram-air turbine in the tailcone. TLX can also be fitted with an IR augmentation plume device which replicates an air-breathing jet engine. The plume is compatible with 76 mm IR fuzed gun ammunition. Another option is an all-aspect, laser retroreflective augmentation package. Scoring/MDI is available for all types of missile and gun, and can be fitted in conjunction with the various augmentation packages. Maximum tow cable length is 8,500 m (28,000 ft).

TRX radar-augmented tow target *(Meggitt)*

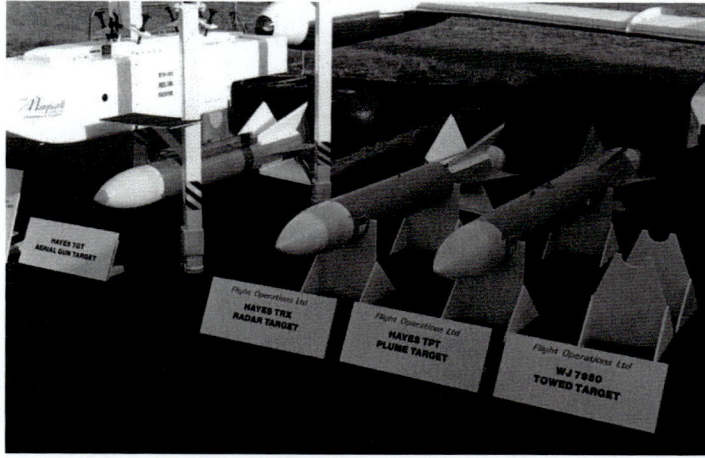

Left to right: TGT, TRX and TPT *(Paul Jackson)*

Meggitt TPT infra-red plume target *(IHS/Patrick Allen)*

TPT: Infra-red plume (jet engine exhaust simulation) target. Utilises SMU-114/A IR augmenter originally developed for US Navy. Uses readily available jet fuel (Jet A, JP-4, JP-5 and JP-8) which, when burned, produces an IR signature very similar to that of a jet aircraft engine exhaust. Pilot light ignition is by pyrotechnic igniter, independently activated on command from a laptop control station via a telemetry link to the target's flight computer. Once lit, the plume can be commanded 'on' for full IR output only when needed. Fuel capacity allows for a pilot-on time of 35 minutes with approximately 8 minutes of plume, or any combination thereof. Target has a second igniter for redundancy, or if pilot light is turned off. Variants are the **TPT-4**, for use with manned aircraft using RM-30A or RM-30B tow reel; and **TPT-[towline length]** for use with drone aircraft equipped with an LTC type launcher.

TRX: Radar-augmented target, designed for use with any gun or missile system using radar for acquisition, tracking, ranging, guidance and/or fuzing. Augmentation is normally passive, with the radar signature varied to meet customer requirements by the arrangement of lenses or reflectors. The **TRX-3** variant is for use with manned aircraft equipped with the Meggitt LTC-2 launcher; standard configuration is with 3,660 m (12,000 ft) of steel towline for one-way operations, with corner reflectors at front and rear of target for radar augmentation. The **TRX-4A**, with a one-way reel and 2,440 m (8,000 ft) of towline, is for use with drone aircraft using LTC-3, LTC-4 or other launchers designed for targets with ball-lock pins. For manned aircraft using the RM-30A or RM-30B tow reel, the **TRX-9** has corner reflectors at front and rear, and no internal reel or line; this also comes in a 'large RCS' version with Luneberg lenses added at the target's front and mid-section. TRX users include the Royal Australian Navy on its Kalkara (MQM-107) targets.

Specifications

Dimensions	
Length: TGT	2.06 m (6 ft 9.0 in)
TGX	2.235 m (7 ft 4.0 in)
TIX	2.32 m (7 ft 7.5 in)
TPT	2.59 m (8 ft 6.0 in)
TRX	2.51 m (8 ft 3.0 in)
Body diameter: all	0.228 m (9.0 in)
Tailfin span: all	0.603 m (1 ft 11.7 in)
Weights	
Nominal weight excl scoring and towline	
TGT	28 kg (61.7 lb)
TGX	30 kg (66.1 lb)
TIX	29 kg (63.9 lb)
TPT, fuelled	40 kg (88.2 lb)
TRX	19 kg (41.9 lb)

Status: In production and in service in 2006.

Customers: Over time, these subtargets have been supplied to the United States (Army, Navy, Air Force) and 22 other countries including Australia, Canada and Netherlands.

Contractor: Meggitt Defense Systems Inc
Tustin, California.

Northrop Grumman BQM-34 Firebee I

Type: Air- or surface-launched recoverable aerial target.

Development: The Firebee I was developed as a joint US Army, Navy and Air Force project, making its first flight in mid-1951. Early models, lacking the now familiar nosecone, were produced for the USAF (645 Q-2A/B), USN (570 KDA-1/4), US Army (35 XM-21) and Royal Canadian Air Force (30 KDA-4). Development of the current BQM-34A improved Firebee (then designated Q-2C) for the two major users began on 25 February 1958; this flew for the first time on 19 December that year. The first production BQM-34A flew on 25 January 1960, and the Firebee I has been in virtually continuous production by Teledyne Ryan (now part of Northrop Grumman) and by Fuji in Japan ever since, bringing the total order book to more than 7,400, including the early Q-2/KDA models. The much modified

BQM-34-53 Firebee mounted on DC-130A drone control aircraft *(Northrop Grumman)* 1122559

supersonic Firebee II (283 BQM-34E/F/T, produced from 1969 to 1976) is no longer in service.

As well as Canada, Firebee I has been exported to Israel and Italy, and built under licence in Japan; others have been provided in support of NATO missile test and evaluation programmes, and almost 1,000 were converted to AQM-34 series RPVs during the Vietnam war. Target presentations have been made to virtually every surface-to-air and air-to-air weapon in the US arsenal.

A competition for a successor target (Target 21), to replace the Northrop Grumman BQM-74 (which see) as well as the BQM-34, was initiated in March 1999 with the award of two USD1.8 million, eight month conceptual design contracts to Northrop Grumman and Lockheed Martin. This resulted in selection of the Composite Engineering (CEI) BQM-167A as the single next-generation Air Force Subscale Aerial Target (AFSAT) in July 2002.

Current US Navy Firebees are receiving an upgrade under a USD2.5 million contract announced by Northrop Grumman on 8 December 2003. This introduces the same avionics suite, autopilot and command/control architecture currently used in the Navy's BQM-74E targets, enabling autonomous waypoint navigation, preprogrammed manoeuvres, satellite command and control, and 'plug and play' common digital architecture for payloads. First flight of this updated version was announced by Northrop Grumman on 17 August 2005, with production deliveries starting in 2009.

Description: *Airframe:* Mid-wing monoplane; two-spar constant chord wing with 45° sweepback and drooped leading-edges; dihedral and incidence 0°; sweptback tail unit (fin 48°, tailplane 45°); ventral fin under tailcone; oval-section tapered fuselage, with keel under central portion to absorb landing impact. Ailerons actuated by electromechanical servos. Construction mainly of aluminium alloy; glass fibre for fin-tip housing guidance and control antenna, nosecone, tailcone and tailplane tips. Wingtips are detachable to permit use of optional extended-span panels.

Mission payloads: Specialised mission equipment can include a wide range of 'building block' units, among which are visual or radar-reflecting banner targets, radar or IR Towbee towed targets, wingtip-mounted IR or 45.4 kg (100 lb) ECM pods, internal or underwing chaff dispensers, wingtip tow launchers, camera pods, scoring equipment, flares, or other forms of IR augmentation, IFF or locator beacons or reflector pods for radar augmentation. The BQM-34A/S can be equipped with adjustable TWT or solid-state amplifiers for use as radar echo enhancers in a wide range of frequencies.

To reduce the vulnerability of the Firebee itself and decrease the number 'killed', extensive use is made of IR or radar enhanced subtargets, either towed behind on cables or mounted on the wingtips.

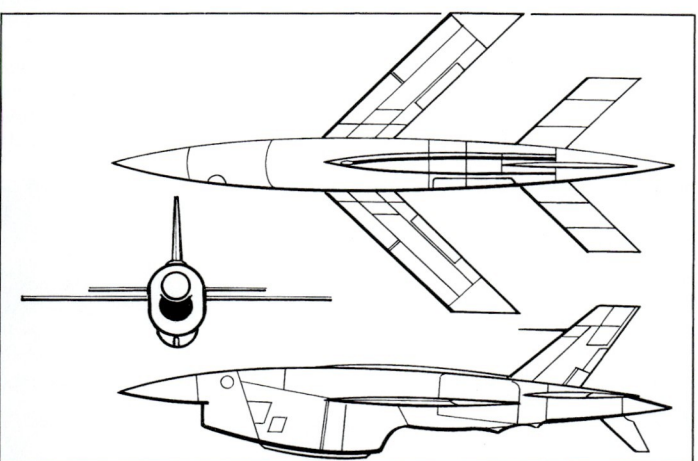

Northrop Grumman BQM-34A Firebee I *(IHS/John W Wood)* 0517997

US Navy BQM-34A with extended flare holders at wingtips 0517996

Northrop Grumman BQM-34 Firebee I

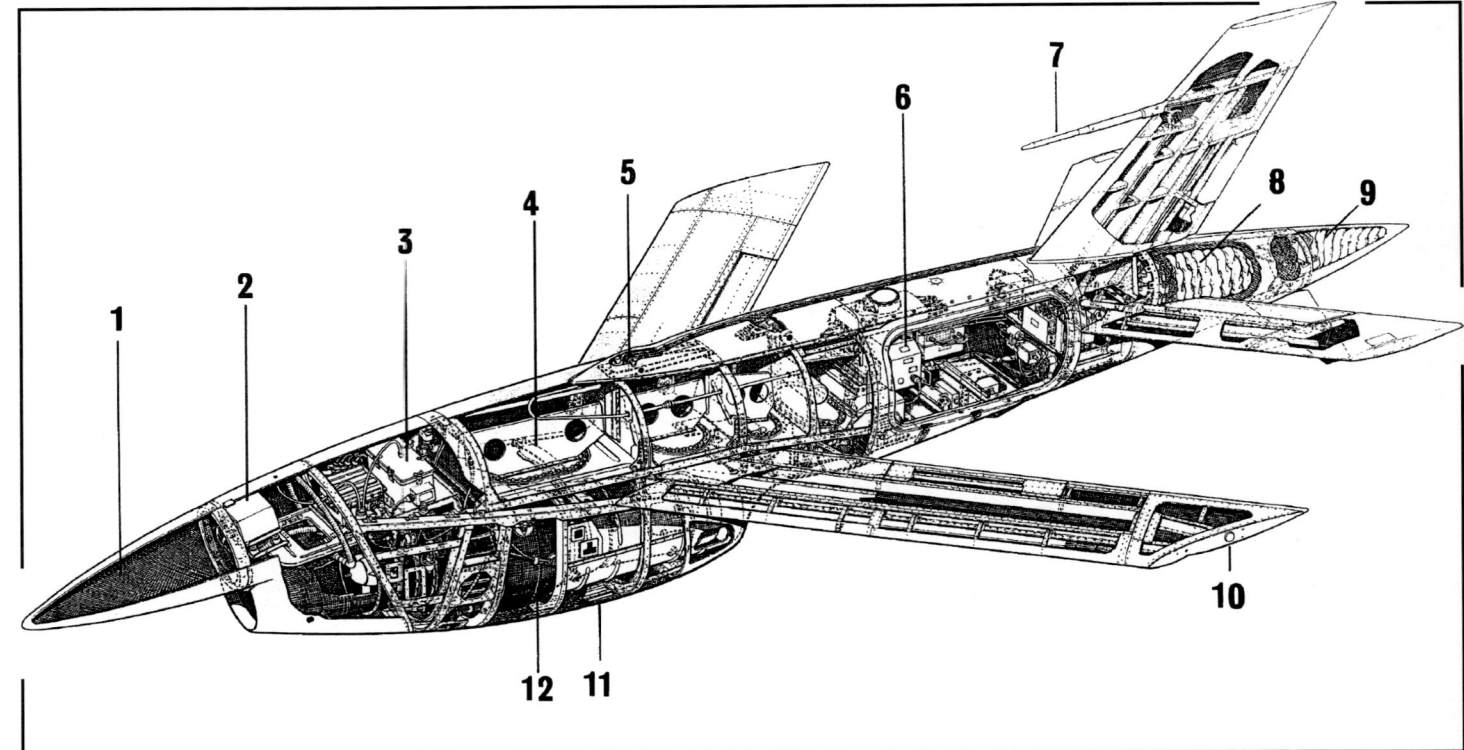

1	Nose radome	5	Parachute riser attachment	9	Drogue parachute
2	Forward equipment bay	6	Rear equipment bay	10	Wingtip augmentation stores
3	Battery/power compartment	7	Pitot static tube	11	Impact keel
4	Fuel compartment	8	Main parachute	12	Power plant

Firebee I internal details

Guidance and control: Flight control: The BQM-34 is equipped with a three-axis flight control system for tactical air combat simulation which gives the target an ability to perform 4, 5 or 6 *g* manoeuvres. A Radar Altimeter Low-Altitude Control System (RALACS), when added to the Firebee I's control system, permits precision low-altitude flight down to 3 m (10 ft) over water and 30 m (100 ft) over land. Electrical power is derived from a 28 V 200 A DC engine-driven generator, with power for the control system vertical gyro furnished by a 400 Hz 115 V 250 W inverter. A 28 V lead-acid battery provides power for the electrical components of the recovery system and for control during the pre-landing glide phase.

Uplink: Remote-control methods include a choice of radar or UHF radio. Normal method for BQM-34A is by Herley-Vega Drone Tracking and Control System (DTCS) or (USAF from 1989) Ryan/Northrop Grumman Microprocessor Flight Control System (MFCS); or by Motorola Integrated Target Control System (ITCS) for US Navy BQM-34S. The target can be controlled from either a manned aircraft or a surface station. Remote command includes the activation of special scoring and augmentation equipment on board the Firebee. Basic commands consist primarily of on/off functions uplinked to an onboard receiver unit and relayed to the appropriate subsystem.

Other types of remote command and tracking systems can include a microwave command and guidance system to control the Firebee beyond the line of sight from a ground station via an airborne relay station. Command and control of US Navy BQM-34S version being upgraded in 2003–04, as described under Development heading above.

Launch: Can be launched from a suitably modified aircraft, or surface-launched from a zero-length launcher using a 50.3 kN (11,300 lb) nominal thrust RATO bottle with 2.2 second burn time. US Navy has launched BQM-34S from ships under way at 15 kt (27.5 km/h; 17 mph).

Recovery: A two-stage parachute recovery system deploys automatically in the event of engine failure or loss of radio-command link, or can be initiated by operator command. The system incorporates a disconnect which releases the parachute from the target on contact with the ground or water.

Variants: (Target versions only)

BQM-34A: Standard US Air Force version, also supplied initially to US Navy; air-launched.

BQM-34J and BQM-34AJ(Kai): Designations of Japanese-built BQM-34A. Deliveries of initial BQM-34J ended in 1996; improved BQM-34AJ(Kai) delivered from 1993 (44 by July 2005); redesigned by Fuji with preprogrammed flight path control function to simulate terminal flight phase of anti-ship missiles.

BQM-34S: Standard US Navy version, first ordered in 1976; ship or ground launched. Some also to US Army.

BQM-34-53: On 20 December 2002, Northrop Grumman conducted a 36-minute first flight demonstration of a much-modified and fully autonomous BQM-34S to deliver a variety of unspecified payload packages to multiple, preprogrammed locations. The fuselage was refitted with bomb rack-equipped wings from a 1970s-era AQM-34L and newly designed composites payload pods manufactured by Grove Aircraft. Other changes included modified software and incorporation of a

Recovering a Firebee I from the sea

US Air Force BQM-34A in flight

portable range safety flight termination system. In February 2003, the designation BQM-34-53 was applied to five similarly modified aircraft, completed and delivered in 17 days for use in Operation 'Iraqi Freedom', as described under the 'Status' heading below.

MQM-34D: Principal US Army version; ground-launched; extended-span wings and longer-burning rocket booster, permitting higher maximum launching weight.

Model 232: International model, supplied to Israel and Italy.

MQM-34D, BQM-34A, BQM-34S

Dimensions, External	
Overall	
length	6.98 m (22 ft 10¾ in)
height	2.04 m (6 ft 8¼ in)
Fuselage, height	0.94 m (3 ft 1 in)
Wings, wing span	3.93 m (12 ft 10¾ in)
Areas	
Wings, Gross wing area	3.34 m² (36.0 sq ft)
Weights and Loadings	
Weight	
Weight empty	680 kg (1,499 lb)
Max launch weight	
BQM-34A, BQM-34S	1,134 kg (2,500 lb)
MQM-34D	1,542 kg (3,399 lb)
Fuel weight, Max fuel weight	295 kg (650 lb)
Performance	
Climb	
Rate of climb	
max, at S/L, AUW of 1,000 kg (2,200 lb), J69-T-41A engine	5,486 m/min (18,000 ft/min)
max, at S/L, AUW of 1,000 kg (2,200 lb), J85-GE-7 engine	6,706 m/min (22,000 ft/min)
Altitude, Operating altitude	5 m to 18,290 m (10 ft to 60,000 ft)
Speed	
Never-exceed speed	635 kt (1,176 km/h; 730 mph)
Max level speed	600 kt (1,111 km/h; 690 mph) at 1,981 m (6,500 ft)
Cruising speed	
max, AUW of 816 kg (1,800 lb)	547 kt (1,013 km/h; 629 mph) at 15,240 m (50,000 ft)
Stalling speed	
power on, AUW of 816 kg (1,800 lb)	101 kt (188 km/h; 117 mph)
g limits	+7
Range, standard fuel	692 n miles (1,281 km; 796 miles)
Endurance	1 hr 55 min (est)
Power plant	1 × turbojet

Status: In service. Further 23 BQM-34As ordered for US Air Force in April 1999 and 32 in January 2000; production of these began in July 2000 and was completed in April 2002.

In February 2003, Northrop Grumman delivered five BQM-34-53s for use in Operation 'Iraqi Freedom'. Modified with the larger wings of the Vietnam-era AQM-34L for extended range, these were used for chaff dispensing over Baghdad during the early stages of that campaign. Three were air-launched from US Navy DC-130A Hercules drone control aircraft, the other two being ground-launched.

The US Navy Firebee inventory was reported to be about 200 in December 2003; these are being upgraded with BQM-74-standard avionics and command/control equipment, as described above. Upgraded BQM-34 started to be re-delivered from July 2009, with 15 vehicles being retrofitted initially.

Customers: Israel (27 Model 232); Italy (13 Model 232); Japan (JASDF 20; JMSDF more than 60 BQM-34AJ and 44 BQM-34AJ(Kai)); USA (Army 601 MQM-34D/BQM-34S, Navy over 3,500 BQM-34A/S, Air Force 1,895 BQM-34A).

Contractor: Northrop Grumman Unmanned Systems
San Diego, California.

Northrop Grumman BQM-74F

Type: Turbojet-powered recoverable aerial target.

Development: Northrop Grumman first announced proposals for an upgraded version of the BQM-74E, then referred to as 'Target 2000', in 1997. Suggested improvements included new avionics, a 300 lb thrust turbojet, swept wings, GPS-based flight control system and adjustable radar signatures. The proposals became fact when the company was awarded a three-year, USD24.9 million system development and demonstration (SDD) contract for the BQM-74F from the US Navy on 28 February 2002. The target passed its critical design review in July 2003. The first of six prototypes was rolled out on 22 August 2005 and made its maiden flight seven days later. Objective of the programme was to provide more realistic threat simulation that more effectively emulates the latest anti-ship cruise missiles and fighter aircraft. Improvements include higher speed, greater manoeuvrability, and longer range and endurance than the current in-service BQM-74E. A six-month flight test programme was forecast, with a production decision then expected to be taken in FY06. However, this was still in abeyance as of mid-2007.

Description: *Airframe:* Sweptback, tapered, shoulder wings; circular-section fuselage with underslung air intake duct; inverted Y tail unit.

The new, sweptwing BQM-74F *(Northrop Grumman)* 1122557

Mission payloads: Payload bay in forward fuselage can accommodate active or passive radar augmentation; AN/DPT-2B seeker simulator; IR augmentation (flares or plumers); AN/DPQ-90 radar tracker; T-1438D locator beacon; RT-1378 radar altimeter; tow system; AN/DSQ-50A scalar scoring system; AN/DPN-88 IFF; ULQ-21 ECM; and battery.

Guidance and control: AN/DKW-3B(V) command and control system. Six linkable missions, each with up to 70 waypoints, can be pre-programmed and selected either pre- or post-launch. Mission profiles can be changed and/or reloaded via the command and control datalink. Waypoint navigation is standard operating mode, using the integrated IMU/GPS avionics. Mission planning capability, integrated into the PC-based support equipment, provides detailed mission plans verified with embedded six-degrees-of-freedom simulation capability for pre-flight verification.

Weave capability includes pre-programmed, fixed, circular and flat-weave manoeuvres and user-programmable weaves. PC-based field test equipment provides real-time simulation of programmable weaves before downloading into the air vehicle avionics, and any pre-loaded weave manoeuvres can be selected after launch.

Launch: Air launch.
Recovery: Guardian recovery parachute, stowed in rear fuselage.
Power plant: One 1.33 kN (300 lb st) Williams WR24-8 (J400-WR-404) turbojet. Fuel tank in centre-fuselage.

BQM-74F

Dimensions, External	
Overall, length	4.57 m (15 ft 0 in)
Fuselage, width	0.355 m (1 ft 2 in)
Wings, wing span	2.13 m (6 ft 11¾ in)
Dimensions, Internal	
Payload bay, volume	0.00142 m³ (0.1 cu ft)
Areas	
Wings, Gross wing area	0.74 m² (8.0 sq ft)
Weights and Loadings	
Weight	
Max launch weight	
air-launch, excl boosters	267 kg (588 lb)
Fuel weight, Max fuel weight	84.4 kg (186 lb)
Payload, Max payload	45.4 kg (100 lb)
Performance	
Altitude, Operating altitude	0 m to 12,190 m (7 ft to 40,000 ft)
Speed, Max level speed	600 kt (1,111 km/h; 690 mph) at S/L (est)
g limits	
sustained	+5
instantaneous	+8
Range	
max, at M0.9 at 15 m (50 ft)	225 n miles (416 km; 258 miles)
max, at 300 kt at 6,100 m (20,000 ft)	500 n miles (926 km; 575 miles)
Endurance	1 hr 55 min (est)
Power plant	1 × turbojet

Status: Six targets produced for flight test programme. Production decision, originally planned for FY06, had not been announced by mid-2007.

Contractor: Northrop Grumman Unmanned Systems
San Diego, California.

Northrop Grumman BQM-74/MQM-74/Chukar

Type: Turbojet-powered recoverable aerial target.

Development: Descended from the Northrop NV-105A design of 1965, this long-serving target has been a US Navy mainstay for more than 30 years and continues in use into the 21st century. It has undergone progressive improvements in payloads, power plant and overall capability, even figuring as an EW decoy in the early stages of the 1991 Gulf War. Others have figured as test vehicles in past USN and USAF programmes for surveillance, decoy and strike UAVs. Its primary function, however, is as an aerial target for anti-aircraft gunnery, surface-to-air and air-to-air

Aerial targets > United States > Northrop Grumman BQM-74/MQM-74/Chukar

BQM-74E ready for air launch from a DC-130 Hercules 0518011

Nearer target is the extended range Chukar IIIER *(Northrop Grumman)* 1122558

BQM-74C simulating a sea-skimming anti-ship missile 0517984

Shipboard launch of a Chukar III 0525106

missile training (particularly as a cruise missile simulator) and weapon system evaluation. More than 7,000 BQM-74s have been produced. International versions, of which more than 1,150 have been built, are named Chukar (pronounced 'chucker'), after a rock partridge of the western USA. Details of earlier BQM/MQM-74As and their Chukar I and II counterparts can be found in previous editions of *Jane's Unmanned Aerial Vehicles and Targets*.

A competition for a single type of target to replace both the BQM-74 and the Northrop Grumman BQM-34 Firebee in US service was launched in March 1999 with the award of two USD1.8 million, eight month conceptual design contracts to Northrop Grumman and Lockheed Martin. However, the US Air Force later opted for the Composite Engineering BQM-167A as its next subsonic aerial target, while the US Navy continues to use the BQM-74E for more than 80 per cent of its airborne target training missions. Production of the BQM-74E continues into 2008. As of mid-2007, no production decision had been announced regarding its proposed successor, the BQM-74F.

Description: *Airframe:* BQM-74E and earlier variants have detachable, constant-chord, shoulder-mounted tapered wings, without dihedral; Northrop G-9224-080 wing section with 8 per cent thickness/chord ratio. Tapered, circular-section fuselage with underslung air intake duct; inverted Y tail unit with 30° tailplane anhedral; nose and tail skins removable for access to equipment and power plant. Ailerons and elevators actuated electrically; aluminium alloy and GFRP construction.

Mission payloads: Main payload compartment is in front fuselage, between control equipment bay and fuel tank. Onboard acquisition and tracking aids can include fore and aft Luneberg lenses, passive radar augmentation, four wingtip-mounted Mk 28 Mod 3 IR flares, pyrotechnic IR plume augmentors, active L-band augmentation, and a smoke system to improve visual detection. A low-cost IR tow target can be attached to each wingtip and towed approximately 30 m (100 ft) behind the target. A reusable active RF augmentation tow target system can be used for various RF guided missile systems. Japan's Chukar IIs can carry a Lockheed Martin laser tracker pod at each wingtip.

BQM-74C and E are equipped with Luneberg lenses, passive radar augmentation and smoke system as in earlier models; other equipment includes locator beacon, radar altimeter, seeker simulator (to duplicate cruise missile emissions), radar transponder for IFF, and scoring equipment. Other provisions include a flight profile programmer with UHF command override; active L- and Ku-band radar augmentation; and Tacan receiver. Can also have payload kit with flotation gear for mobile sea range operations, can be launched from beyond horizon of target ships, and have been flown at height of 3 m (10 ft) above water while simulating cruise missile flight profiles. Electrical power provided by an engine-driven alternator through a rectifier/regulator; 28 V Ni/Cd battery as secondary power source.

Guidance and control: Guidance and control of BQM versions can be either manual or automatic. In manual mode, an operator guides the target from a ground-based command and control station via an onboard receiver/transmitter; operator controls target pitch and bank angles, altitude and engine rpm, and can initiate altitude hold and parachute

Northrop Grumman BQM-74/MQM-74/Chukar – Orbital Sciences GQM-163A Coyote < United States < Aerial targets

Recovering a BQM-74E from the sea *(US Navy)* 1122556

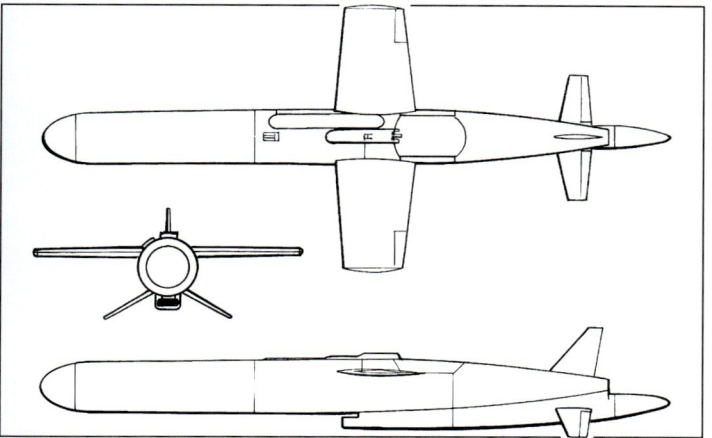

Northrop Grumman BQM-74C target *(IHS/John W Wood)* 0517983

1. Power control unit
2. Vertical gyro
3. Digital avionics processor
4. Batteries
5. Tracking transmitter
6. Tracking antenna
7. Command antennas
8. Telemetry transmitter
9. Receiver
10. Altitude transducer
11. Forward air-launch lug
12. Fuel tank
13. Aileron actuator
14. Rear air-launch fitting and lug
15. Recovery parachute
16. Fuel control
17. Power plant
18. Smoke/oil tank
19. Elevator actuator
20. Engine oil mist generator
21. Converter voltage regulator
22. Airspeed transducer

Internal details of the BQM-74C/Chukar III 0517985

recovery. In automatic mode, target is preprogrammed for an autonomous mission.

Onboard equipment includes a Northrop Grumman digital avionics processor, vertical and yaw rate gyros, an AN/DKW-3 or AN/DKW-4 target control system, aileron and elevator servos, and an altitude hold pressure transducer. An improved manoeuvrability package has a closed loop control device installed in the flight control system, enabling the air vehicle to perform constant g manoeuvres at any of five selected levels up to and including 6 g.

US Navy BQM-74Es began to be upgraded in 1998 with a new integrated avionics unit (IAU), replacing outdated vertical and yaw rate gyros, altitude and airspeed sensors with new state-of-the-art avionics including a GPS navigation receiver. The first upgraded BQM-74E was redelivered in 1999. A similar upgrade package was applied to the USN's BQM-34 Firebees in 2004.

Launch: Surface launch, from ground or ship's deck, is accomplished by means of two jettisonable RATO booster rockets from a ZL-5 or similar zero-length launcher. BQM variants can also be air-launched.

Recovery: Normal recovery is by automatic pull-up followed by main parachute deployment, initiated automatically in emergencies such as interruption of continuous radio signal or loss of parachute command channel. An alternative method is by direct main parachute deployment, initiated automatically on loss of electrical power. Main parachute is housed in fuselage immediately aft of wing, and disconnects automatically on impact. Target can be recovered over land or from water.

Power plant: One 1.07 kN (240 lb st) Williams WR24-8 (J400-WR-404) turbojet.

Variants: *BQM-74E/Chukar III:* Current US Navy production version, delivered from November 1992; principal changes are uprated engine and software improvements. First five contracts totalled 786 (150, 50, 202, 291 and 93), all of which had been delivered by mid-1997. April 1996 contract for 120 (USD26.5 million) was followed by further 119 (contract value USD28.8 million) ordered in December 1997; 90 more (USD23.6 million) ordered in January 1999, 71 in February 2000 (USD20.1 million); 78 in April 2001 (USD23.6 million); 109 in February 2002 (USD29.7 million); 40, including 13 extended-range, in January 2004 (USD10.2 million); 60, including 48 extended-range, in March 2005 (USD24.33 million); and 60 in February 2006, bringing overall BQM-74E total to 1,533. Deliveries of the three last-named batches were due for completion by July 2006, December 2006 and January 2008 respectively. In February 2009 a contract for 160 more BQM-74E targets was placed by the US Navy, with production expected to be completed in 2011.

Chukar III licence-built by Fuji in Japan since 1994, totalling more than 200 by July 2007. In FY 07 production was reported to be 12 and 18 were scheduled for FY 08.

BQM-74F: Modernised, sweptwing and swept-tailed version; described separately.

Specifications

Dimensions
Wing span	1.75 m (5 ft 9.0 in)
Wing area	0.74 m2 (8.0 sq ft)
Length overall	3.95 m (12 ft 11.5 in)
Body diameter (max)	0.36 m (1 ft 2.0 in)
Height overall	0.71 m (2 ft 4.0 in)
Tail unit span	0.80 m (2 ft 7.5 in)

Weights
Weight empty	133 kg (294 lb)
Fuel weight	50.3 kg (111 lb)
Max launching weight, incl boosters	270 kg (595 lb)
Max air-launch flying weight, excl boosters	206 kg (455 lb)

Performance
Max level speed at 6,100 m (20,000 ft)	
Max level speed at S/L	>515 kt (953 km/h; 592 mph)
Operating height band min	2 m (7 ft)
max	12,200 m (40,000 ft)
Max range: standard	520 n miles (963 km; 598 miles)
extended	640 n miles (1,185 km; 736 miles)
Endurance: standard	1 h 18 min
extended	1 h 36 min
g limit	+6

Status: BQM-74C and MQM-74C production completed, but both still in service; BQM-74E in production and service; BQM-74F under development.

Customers: US Navy is main customer for BQM-74; some MQM-74s may still be used by US Army and Army National Guard.

Chukar IIs and/or IIIs are currently (2007) operated by Japan, Singapore, Spain and Taiwan. Earlier customers included Argentina; Belgium; Brazil; Brunei; Canada (Navy); Chile; Denmark; Finland; France; Germany; Greece; IIndia; Iran; Israel; Italy; Jordan; South Korea; Malaysia; Netherlands; Nigeria; Norway; Saudi Arabia; Sweden; and UK. Chukars are also in service at the NATO Missile Firing Installation (NAMFI) in Crete, where they are used to train crews of radar and non-radar directed air defence guns, active and semi-active radar-guided, visual and IR-guided surface-to-air and air-to-air missiles. Galileo Avionica (Italy) provides Chukar services at the Salto di Quirra range in Sardinia.

Contractor: Northrop Grumman Unmanned Systems
San Diego, California.

Orbital Sciences GQM-163A Coyote

Type: Supersonic Sea-Skimming Target (SSST).

Development: The Orbital Sciences Corporation's (Chandler, Arizona) GQM-163A Coyote is an analogue for supersonic, sea-skimming anti-ship cruise missiles such as the Raduga 3M80 (NATO Reporting Name 'Sunburn') and is designed to support ship defence system RDT&E and fleet training exercises. The Coyote High Diver (HD) sub-variant is a modification of the baseline device that incorporates additional thermal protection and guidance software that facilitates operation of the AV at altitudes of up to 18,288 m (60,000 ft). The GQM-163A's avionics are based on the Orbital Launch Systems Group's multi-programme Modular Avionics Control Hardware (MACH) package that incorporates a Power PC flight computer that hosts a real-time operating system and flight proven software that is based on the contractor's common object oriented C++ application framework. Over time Coyote has also benefited from Orbital's new generation Front End Sub-system (FES - see following) which includes a new navigation capability and a new autopilot/flight software package. *IHS Jane's* sources report the new generation FES as having been fitted to Phase 3 FRP and up GQM-163A AVs, with the introduction being said to represent a block upgrade of the AV's existing avionic fit and as including an Orbital designed flight computer, a radar altimeter (in place of the legacy laser unit), a new design of Lithium-Ion battery and a telemetry transmitter that has three times the bandwidth of its predecessor. Concurrent with the development of the new generation FES, Orbital has fielded a Target Support Test Set (TSTS) that is designed to support either legacy GQM-163A AVs or those equipped with new generation FES. Orbital is known to have delivered at least four TSTSs to the USN (for use in field processing) and makes use of the equipment in to support vehicle production. Again, the TSTS represents an upgrade in

Aerial targets > United States > Orbital Sciences GQM-163A Coyote

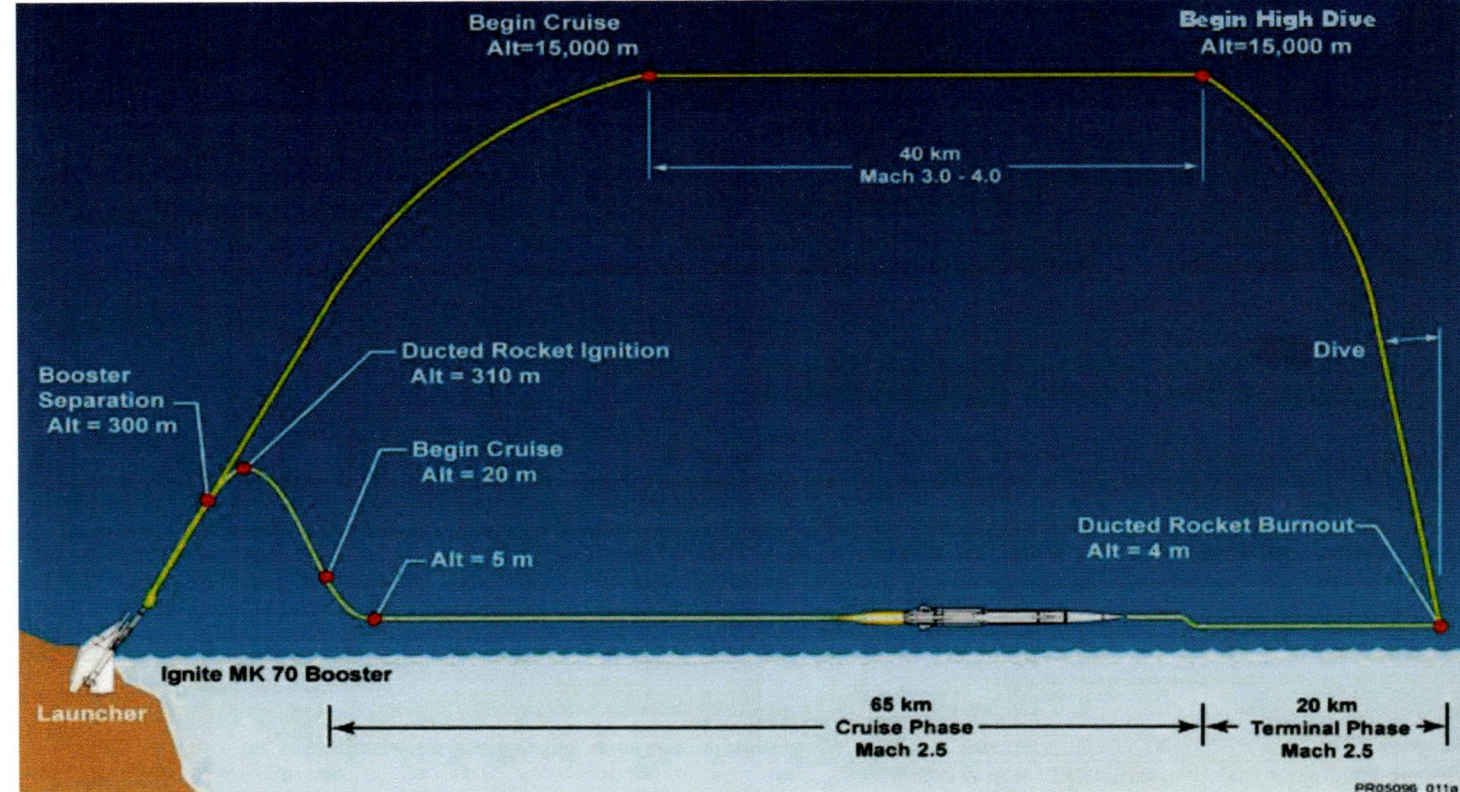

The Coyote high-diver flight profile combines a Mach 3-4 cruise phase followed by a Mach 2.5 near-vertical dive. Its normal profile is a low-altitude cruise and terminal attack, both flown at Mach 2.5 *(NAVAIR)*

reliability (as well as a reduction in testing time) over the earlier Aerial Target Test Set (ATTS). Key programme events include:

29 June 2000 The USN awarded the Orbital Sciences Corporation (Chandler, Arizona) a then year USD34 million GQM-163A EMD contract that included LRIP, FRP and launch operation support options. Here, the LRIP option was taken up and resulted in the production of 20 AVs. Elsewhere, Phases 1 and 2 of the FRP option are said to have seen the production of a further 32 GQM-163As, with the delivery of which was originally scheduled for completion by the end of 2011.

October 2005 The GMQ-163A made its first operational flight during the month.

March 2007 During March 2007, it was announced that France was to acquire a single Coyote target in a then year USD9.2 million US Foreign Military Sales deal with the USN's Naval Air Systems Command. While not confirmed, *IHS Jane's* sources suggest that the AV's procurement was in connection with the Principle Anti-Air Missile System (PAAMS) programme. PAAMS is installed aboard British (Daring-class), French (Horizon-class) and Italian (Orrizante-class) air warfare vessels.

2008 During 2008, Orbital Sciences is reported to have begun development and qualification of its new generation GQM-163A FES and to have emphasised the new design's cost advantage and enhanced reliability when compared with the legacy assembly.

8 July 2010 Orbital Sciences is reported to have successfully test launched a Coyote HD target from the San Nicolas Island, California test site. During this event, the AV climbed to an altitude of 10,668 m (35,000 ft), accelerated to approximately Mach 3.3, flew an approximately 96 n miles (177 km; 110 miles) long test route and executed a pre-planned 40° terminal dive. According to Orbital, this launch validated the Coyote HD configuration's suitability for high-altitude naval threat simulation and anti-missile response testing.

8 December 2010 Orbital Sciences is reported to have successfully launched a Coyote test round equipped with the new generation FES.

15 January 2011 Orbital Sciences is reported to have launched a GQM-163A AV from the US Pacific Missile Range Facility (PRMF - Barking Sands, Hawaii) in order to validate the location as alternative site for Coyote operations.

March 2011 As of March 2011, *IHS Jane's* sources were reporting the USN as having ordered 89 × GQM-163A AVs (with an estimated value of then year USD340 million) and FRP-III as seeing the launch of production Coyotes fitted with the new generation FES. As of the given date, Orbital Sciences was noted as having launched GQM-163A rounds from both San Nicolas Island and the PMRF with "100 per cent success".

29 August 2012 The USN Naval Air Systems Command (Patuxent River, Maryland) awarded Orbital Sciences a then year USD26,401,290 firm, fixed-price contract with regard to FRP-VI production of the GQM-163A. Work on the effort was to be undertaken at facilities in Chandler, Arizona (71 per cent workshare); Camden, Arkansas (24 per cent workshare); Vergennes, Vermont (three per cent workshare) and Hollister, California (two per cent workshare) and at the time of its announcement, the programme was scheduled for completion by the end of June 2015.

Description: *Airframe:* Two-stage (booster and sustainer) cylindrical body with Von Karmen nosecone, making use of existing residual and current missile and target assets to minimize cost. Rear section is a MK 70 Mod 1 Standard Missile booster, with an NSROC 1 fin set. The ducted rocket/ramjet sustainer is based on technology developed by ARC for the USAF's variable-flow ducted rocket (VFDR) program; nosecone and front end avionics for phase one and two of the FRP contact are supplied by Composite Engineering Inc (CEi) (ex-Raytheon) AQM-37D target. While future targets involving phase three and beyond of the FRP contact will be equipped with an Orbital designed OFES structure, avionics and software which are backwards compatible with the ducted rocket and Fin Actuator Control System (FACS).

Orbital Sciences' GQM-163A Coyote supersonic sea-skimming target overflies a target barge at the conclusion of the type's fifth and final development test flight *(US Navy)*

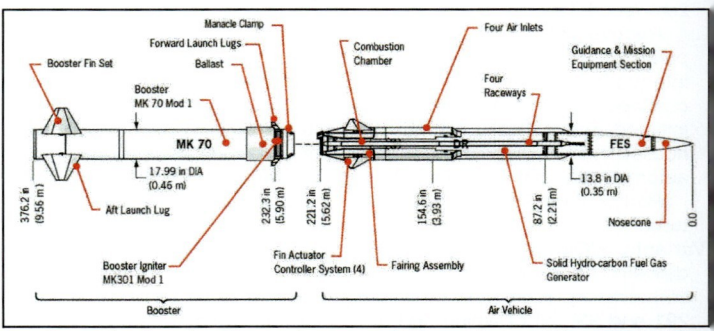

A schematic showing the make-up and dimensions of the GQM-163A Coyote supersonic sea-skimming target and its launch booster *(Orbital Sciences)*

Mission payloads: Herley AN/DPN-90 X-band (8 to 12.5 GHz) radar transponder; Meggitt AN/DSQ-50A miss distance indicator; NAWC AN/DPT-2B radar jamming transmitter (8 to 10 GHz or 10 to 20 GHz) and NAWC C/D/E/F/G-band (500 MHz to 6 GHz) AN/ULQ-21 radar jamming system.

Guidance and control: Inertial Measurement Unit (IMU) and GPS blended navigation solution. Altitude control accomplished by laser or radar altimeter with default to GPS in event of altimeter failure. Vehicle attitude control accomplished via fin actuator control system (FACS). Manoeuvres are skid to turn. Vehicle speed control accomplished by throttleable DR.

Launch: Surface-launched from US Navy standard MK 7 launcher.

Recovery: Non-recoverable.

GQM-163A

Dimensions, External
Overall, length	9.56 m (31 ft 4½ in)
Fuselage length	
sustainer	5.62 m (18 ft 5¼ in)
booster	3.94 m (12 ft 11 in)

Weights and Loadings
Weight
Max launch weight	
sustainer	771 kg (1,699 lb)
with booster	1,814 kg (3,999 lb)

Performance
g limits
terminal phase	+12
elevation	+7
combined plane	+11.2

Range
during cruising phase at M2.50	36 n miles (66 km; 41 miles)
terminal phase	11 n miles (20 km; 12 miles)

Status: As of August 2012, GQM-163A Coyote FRP-VI production had been launched and the AV was in service with the USN.

Customers: USN.

Contractor: Orbital Sciences Corporation, Launch Systems Group Chandler, Arizona. ***Sub-contractor:*** Major sub-contractors include Aerojet (previously ARC: ducted rocket/ramjet) and Goodrich (Fin Actuator Control System).

CONTROL AND COMMUNICATIONS

Canada

MicroPilot MP1x28/MP2x28

Type: Family of miniature UAV autopilots and complementary products.

Description: Over time, identified members of the MicroPilot miniature UAV autopilot family comprise the MP1028^g, MP2028^g, MP2028^{XP}, MP2128^g, MP2128^{HELI}, MP2128^{LRC}, MP2128$^{HELI-LRC}$ and the MP2128^{3x}. Known details of these various equipments are as follows:

MP1028^g: Described as being the lowest cost member of the MicroPilot UAV autopilot family, MP1028^g incorporates the 'most important' features (and reliability level) of the MP2028^g equipment, has the same weight/size footprint as the MP2028^g and is intended for 'entry level' UAV applications where cost is the 'overriding' consideration.

MP2028^g: Described as the 'smallest UAV autopilot in the world', MP2028^g is a single board equipment whose capabilities include airspeed/altitude hold, turn co-ordination, GPS navigation and autonomous launch and recovery. The architecture also supports 'extensive' data logging and manual overrides. Other system features include:

- user programmable feedback loop gain and flight parameters
- control of up to 24 servos (including flaps, flaperons, elevons, V-tail, X-tail, split rudders, split ailerons and flap/aileron mixes)
- the ability to hold up to 1,000 GPS waypoints
- full integration of 3-axis gyros/accelerometers, GPS, pressure altimeter and pressure airspeed sensors on a single board
- the ability to repeat or skip groups of waypoints, with the function based on sensor or other external inputs
- support for manually directed and autonomous flight modes
- integrated radio control override
- programmable error handling for loss of GPS/control signal, engine failure, datalink loss and low battery voltage
- the incorporation of MicroPilot's HORIZON^{mp} ground control software (see following).

MP2028^{XP}: MP2028^{XP} is an expendable equipment that is designed specifically for disposable UAV applications, a use that is reflected in its price point. MP2028^{XP} is flight enabled via MicroPilot supplied software keys that are provided via the internet.

MP2128^g: MP2128^g features a 'more powerful' processor than other members of the family and is designed for 'more advanced' MAV and UAV applications. When compared with other family members, MP2128^g provides faster update rates for feedback loops, improved altitude estimation and a platform for future improvements/new features.

MP2128^{HELI}: Capable of controlling helicopters and fixed-wing/VTOL UAVS, MP2128^{HELI} provides 150 MIPS of processing power, weighs 28 g (1 oz), measures 40 × 100 mm (1.6 × 3.9 in). Price includes a compass and an ultrasonic altitude sensor to facilitate autonomous take-off and landing.

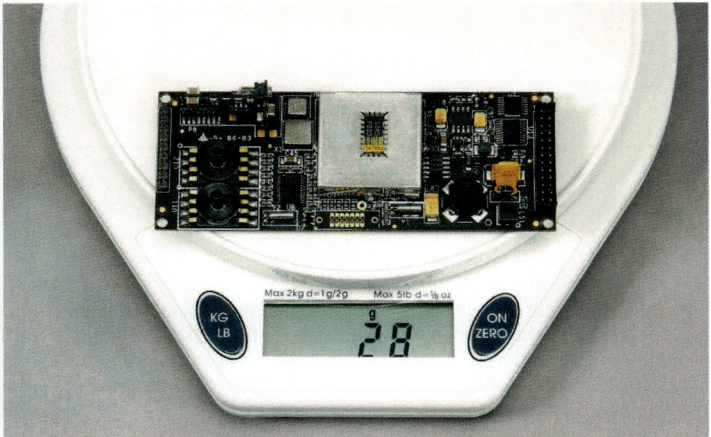

The MP2028^g miniature UAV autopilot displayed on a weighing scale *(MicroPilot)* 0589012

MP2128^{LRC}: The MP2128^{LRC} is described as being an integrated autopilot that offers redundant communication and as incorporating an autopilot and a modem that are contained in a low-weight, ruggedised aluminium enclosure that is designed to protect sensitive electronics and is deemed 'convenient' to install in a 'variety' of airframes.

MP2128^{3x}: MicroPilot describes MP2128^{3x} as being the 'first triple redundancy autopilot' that has been designed 'specifically' for small UAV applications. As such, it is based on the MP2128^{HELI} and combines three such autopilots in a single system. Here, each of the three has a full set of sensors that include three axis gyros, three axis accelerometers, airspeed pressure transducers, altitude pressure transducers and GPS receivers. Each of the three autopilots is 'fully capable' of flying a UAV on its own should one or both of its compatriots fail. This capability is facilitated by a separate, redundant, voting mechanism that compares the outputs from the three autopilots and if one or more are found to have failed, it/they are disabled and the system continues to fly under the control of the remaining functioning element/s. MP2128^{3x} is also capable of managing redundant datalinks.

MicroPilot also notes that its miniature UAV autopilot family includes the HORIZON^{mp} ground control software that provides users with a 'friendly' point and click interface for mission planning, parameter adjustment, flight monitoring and mission simulation. The company further reports that it offers the 'only' UAV autopilots to feature open architectures.

trueHWIL^{mp} Matlab-based simulator: MicroPilot describes the trueHWIL^{mp} simulator as offering a 'dramatic improvement in simulator fidelity' by means of the electrical simulation of 'all' sensor outputs using analogue-to-digital conversion, signal conditioning and PulseWidth Modulation (PWM) interface boards. Again, trueHWIL^{mp} allows users to replicate the conditions their UAVs experience in flight and offers 'superior' on-the-ground validation of autopilot set-up and integration.

Specifications

Dimensions	
Board: Length	100 mm (3.9 in)
Width	40 mm (1.6 in)
Height	15 mm (0.6 in)
Weights	
Board (incl GPS, receiver, gyros and all sensors)	28 g (1 oz)
Performance	
Supply voltage	4.2-26 V
GPS update rate	4 Hz
User defined telemetry:	
Transmission	100 fields/s
Update rate	5-30 Hz
Onboard datalog fields	52 (3 MB)
Servo: Update rate	50-200 Hz
Resolution	11-bit
PID loop update rate	30 Hz
Sensors: Max airspeed	270 kt (500 km/h; 311 mph)
Altitude above launch point	up to 12,000 m (39,370 ft)
Max angular rate	150°/s

Status: MicroPilot reports the use of its miniature MAV/UAV autopilots by more than 600 clients in more than 60 countries around the world and has (most recently) continued to promote the MP1028^g, MP2028^g, MP2028^{XP}, MP2128^g, MP2128^{HELI}, MP2128^{LRC} and MP2128^{3x} autopilots. *Jane's* sources suggest that the company's autopilots are/have been used aboard the ALBA, Bird Eye 100, Blueye, Boomerang, Casper 250, MASS, Micro B, Mosquito, Scrab and SkyLite B vehicles. MicroPilot also produces a range of complementary products including the MP Compass, the MP AGL ultrasonic altimeter, the MP ADC, the MP-DAYVIEW^{PTZ} and MP-NIGHTVIEW^{PTZ} camera systems, the MP-Vision^e electric and MP-Trainer^g UAVs, and the MP CAT lightweight pneumatic catapult.

Contractor: MicroPilot
Stony Mountain, Manitoba.

China

Xian ASN-104 and ASN-105B GCS

Type: Ground control station for ASN-104 and ASN-105B UAVs.

Description: The ground control station for these Chinese UAVs consists of a truck-mounted main command station and a smaller, vehicle-mounted or man-portable mini-station. It is manned by two operators, one for command and navigation, one for flight control. The system includes equipment for radio remote control, radio telemetry, radio tracking and video image transmission. Command station monitors display flight parameters, flight path, a map of the flying area, and reconnaissance video. Remote control and telemetry antennas are erected separately at the front and rear of the command truck; the tracking unit is tripod-mounted away from the truck and its own antenna.

The remote control system has 30 commands for switch functions and four for proportional functions (32 and eight commands, respectively, for the ASN-105B). The telemetry system has 12 continuous and 16 discrete channels (ASN-105B, 22 and 36 respectively). Principal measured parameters include bank, pitch and heading angles, altitude, and engine rpm. Tracking, by polar co-ordinates, can be performed automatically, according to programmed waypoints, or manually. An X-Y plotter records the air vehicle flight path on a map display. Video imagery can be downlinked by a UHF datalink (6 to 10 MHz, selectable manually to suit the needs of different TV cameras).

Status: Over time, *IHS Jane's* sources report the ASN-104 and ASN-105B UAVs as having been procured by the Chinese military.

Contractor: Xian ASN Technical Group
Xian, Shaanxi.

Van-based mini-mobile GCS 0062360

Truck-mounted main command station and tripod-mounted tracking unit 0062361

Interior of ASN-105 main control station 0062359

Denmark

Sky-Watch Huginn X1 ground station

Type: Ground station for the Huginn (mythological Raven) X1 VTOL mini UAV.

Description: The Huginn X1 ground station is 10.5 cm wide by 24.0 cm long by 6.5 cm high and weighs 1,260 g when fitted with its main lithium polymer battery. Here, this externally mounted power source requires an approximately 60 minute recharge period following a typical Huginn X1 AV flight and is supported by a back-up battery for the station's GPS capability (see following) within its casing. Other system features include:
- one data and two video antennas
- a water and dust resistant casing
- a graphical user interface
- a software package that supports Windows 7 (or newer) operating system
- a way point navigation capability that supports Google maps
- Universal Serial Bus (USB) connectivity
- a video output
- a sensor sub-system that incorporates three × solid state Micro Electro-Mechanical System (MEMS) gyros, three × solid state MEMS accelerometers, three × magnetometers, barometric pressure and temperature (internal and external) sensors and a GPS receiver
- a hand controller for manual operation of the Huginn X1 AV (the hand controller is connected to the ground station by cable and features one-handed functionality, a 6 Hz update rate and GPS aided flight and position lock).

Status: As of February 2013, Sky-Watch has continued to promote the Huginn X1 ground station.

Contractor: Sky-Watch,
Støvring.

France

Secapem MDI189

Type: Miss distance indicator for gunnery and missile proficiency training.

Description: The Secapem MDI189 is an automatic scoring system that acquires and processes practice shots fired at targets. As such, it is integrated with a target (towed or drone) and acoustically detects and measures the passage of supersonic projectiles and transmits the acquired data by radio to an associated receiver system. MDI189 can be configured for both global and sector scoring, with the former using a single sensor and displaying scores achieved in different radii and the distance of passage of each projectile fired. The sector scoring configuration makes use of four sensors and shows the same data as the global option together with angular information.

The MDI189 sensor takes the form of an electromagnetic microphone that is positioned in the rear of a resonating cavity. Here, the sensor detects air pressure disturbances and/or shock waves, with the amplitude of the disturbance depending on a combination of the atmospheric pressure at the operating altitude, the distance at which the projectile passes the target and the velocity and calibre of the projectile. The device's transmitter is connected to the system sensor/s and an antenna and comprises acquisition and radio cards. Here, the acquisition card facilitates the treatment and reading of data from the acoustic sensors and its integration with that from GPS, altimeter and clock capabilities. For its part, the radio card transmits the information to the architecture's receiver system which 'immediately' processes acquired data, records it to hard disk, visually displays it and (if required) transfers it to a network for multiple examination. Receiver system components include a processing mother card and up to four receiver cards (multi-receiver system configuration), thereby facilitating data restitution from four transmitters simultaneously on four different frequencies.

Specifications

Dimensions	
Transmitter	
Length	130 mm (5.2 in)
Width	60 mm (2.4 in)
Depth	37 mm (1.5 in)
Sensor	
Length	120 mm (4.7 in)
Diameter	430 mm (16.9 in)
Multi-receiver system (without screen)	
Length	290 mm (11.4 in)
Width	245-258 mm (10.0-10.2 in)
Depth	87 mm (3.4 in)
Weights	
Transmitter	0.210 kg (0.463 lb)
Sensor	0.154 kg (0.340 lb)
Multi-receiver system	5.500 kg (12.125 lb)
Performance	
Transmitter	
Frequency	138.8-150.5 MHz
Transmitting power	100 mW-1 W (on request)
Power supply	12 V DC
Power consumption	<500 mA
Operating temperature	–30 to +60°C (–22 to +140°F)
Sensor: Cable length	300 mm (11.8 in)
Multi-receiver system	
External power supply	20 - 28 V (3 A)
External adapter	110/240 V AC (50/60 Hz)
Operation on battery	3 h
Time to recharge	3 h
Detection radius: Air-to-air mode	6-15 m (19.7-49.2 ft)
Surface-to-surface/air-to-ground modes	4-12 m (13.1-39.4 ft)
Surface-to-air mode	4-36 m (13.1-118.1 ft)

Status: Over time, the MDI189 miss distance indicator is reported to have been procured by the French Army and offshore customers (including Finland) and has been associated with the Avartek, MSAT-500 and SQ20 targets. Most recently, Secapem has continued to promote the MDI189 equipment.

Contractor: Secapem SA
Ozoir-la-Ferrière.

Thales France MAGIC ATOLS

Type: Automatic take-off and landing system.
Development: During August 2008, Thales announced that MAGIC ATOLS had been 'validated' as part of the 'Watchkeeper' UAV architecture.
Description: The Thales France Microwave And GPS Integrated Co-operative Automatic Take-Off and Landing System (MAGIC ATOLS) is described as making use of "revolutionary" radar design and as being a compact, automatic, stand-alone system that ensures "maximum" safety (in all weather conditions) during the take-off and landing phases of UAV operations. Suitable for use with both fixed- and rotary-winged AVs, MAGIC ATOLS comprises airborne and ground elements, with system features including:

- accurate location of the AV both on the runway and in the glide path
- the provision of steering command information during both AV take-off and landing
- the ability to handle 'more than' 5 AVs simultaneously
- 'instantaneous' acquisition of AVs within a 'wide' FoV and with 'high' probability-of-intercept
- 'no' risk of lock-off during AV tracking
- use of an AV safety recognition protocol based on a handshake and the AV's tail number
- 'high' location accuracy based on 'sub-decimetric' technology
- dual datalinks and beacons to ensure data security through redundancy
- a ground element that can be set-up by a single operator without any 'specific' training and on 'any' kind of landing strip
- an airborne segment that has no external antennas and has a 'very low' power consumption
- 'full' self-test and automatic calibration of the system's ground element
- Ethernet wired, optical fibre and World Geodetic System 84 (GPS-like, for direct hybridisation with an inertial navigation system) interfaces between the system's ground element and the AV GCS
- a human-safe low radiated power value
- suitability for deck landing and trajectography applications
- an options package that includes a dual altimeter and a radar embedded secure datalink that is redundant to the system's other links.

Specifications

Dimensions	
Airborne segment	120 × 120 × 50 mm (4.8 × 4.8 × 1.9 in)
Ground segment	130 × 40 × 20 cm (51.2 × 15.8 × 7.9 in)
Weights	
Airborne segment	500 g (1.1 lb)
Ground segment	<20 kg (44.1 lb)
Performance	
Power: Airborne segment	1 W
Ground segment	150 W

Status: Most recently, Thales France has continued to promote the MAGIC ATOLS system and it has been selected for use in the UK's 'Watchkeeper' UAV programme.
Contractor: Thales France
Pessac.

International

CDA/Ultra Electronics HIDL

Type: Digital High Integrity DataLink (HIDL).
Development: HIDL is understood to have been originally developed for use in NATO's Maritime Unmanned Vehicle programme.
Description: Developed jointly by Cubic Defense Applications (CDA) and Ultra Electronics, HIDL is described as being a 'robust digital datalink' that is designed to assist operators in making safe AV take-offs and landings and to facilitate the transfer of imagery to surface terminals for analysis and dissemination. Again, HIDL allows a single surface control station to simultaneously manage multiple AVs and CDA claims that the link's waveform provides 'spectral versatility' in order to avoid interference with radio traffic in dense radio frequency environments. Other system features include:

- use of any available radio frequency channel/s even if non-contiguous
- a networked configuration to facilitate multiple simultaneous use
- use of time and frequency diversity for resistance to jamming
- low latency for safe control during AV take-off and landing
- a voice channel for communication with ATC authorities
- an over-the-horizon relay function
- variable data rates (3 Kbps to 20 Mbps)
- the ability to act as a back-up for CDL applications.

Specifications

Dimensions
No data
Weights
No data
Performance
See Description

Status: Over time, HIDL has continued to be promoted and has been selected for installation aboard the U-TacS WK450 UAV.
Contractor: Cubic Defense Applications
Orlando, Florida. Ultra Electronics
Greenford, United Kingdom.

Israel

Aeronautics control systems

Type: Range of UAV control systems.

Description: Over time, Israeli contractor Aeronautics Defense Systems has produced a range of UAV control systems, the known details of which are as follows:

GCS: Aeronautics' Ground Control Station (GCS) is a generic equipment that enables the user to perform various UAV/USV/UGV missions in real-time. As such, the system consists of 'advanced' software and 'user-friendly' displays that are designed for 'comprehensive' mission management, simulation and mission briefing/debriefing. Again, the generic GCS is noted as having been 'operationally proven' while operating sensors installed aboard a variety of platforms including fixed-wing aircraft, helicopters, UAVs and USVs. Aeronautics' GCS can be supplied in a number of configurations and can be installed in shelters, onboard vehicles and inside headquarters as required.

LRS: Aeronautics' Launch and Recovery Station (LRS) is described as being a compact, easily-transportable, UAV GCS that is used to operate a UAV during the opening and closing stages of a mission. LRS is usually deployed close to a runway to ensure LOS radio and visual contact with the air vehicle and its components are enclosed in two cases based on standard 48 cm (19 in) racks. Here, subsystems include a system computer (running the same operational software as Aeronautic's GCS - see previously), a fold-away 48 cm (19 in) display screen (for digital map and video display), RF and control electronics, a control panel (with keyboard), an intercom (for co-ordination the UAV team's activities) and a digital video recorder. The LRS can also be used for air vehicle pre-flight testing and pre-flight/in-flight mission planning.

RPCS: Aeronautics' Remote Payload Control Station (RPCS) is described as being a three-level Remote Video Terminal (RVT) and payload controller that is designed to receive real-time video imagery directly from unmanned platforms. As such, it functions independently of

A general view of one of the Aeronautics GCS's control console configurations *(Aeronautics Defense Systems)*

Aeronautics' RVT video terminal *(Aeronautics Defense Systems)*

the previously described MCS and is usually operated in the vicinity of the mission area. RPCS allows the user full payload control and has been configured for use by special forces, police units and the like.

RVT: Aeronautics' RVT is designed to receive real-time video imagery directly from both manned and unmanned platforms. As such, it is designed to function independently of the crew controlling the platform/payload that is generating the imagery and is usually located within the vicinity of the mission area. As with the company's RPCS, RVT is configured for use by special forces, police units and the like.

UMAS™: Aeronautics' Unmanned Multi-Application System™ (UMAS) is a software-based package that is designed to provide 'advanced' control of a 'variety' of manned and unmanned applications. As such, it is described as incorporating proprietary artificial intelligence and 'unique' interfaces and as offering 'unrivalled' levels of system reliability and performance. System functions include artificial intelligence-based electronic/mechanical failure prediction; real-time decision making support; the facilitating of data transfer between systems; and hierarchical handling of fixed/variable order data strings as a function of tool safety (using an integral time domain camouflage interface). UMAS™'s open architecture is further noted as turning an unmanned vehicle from being a closed system into one that can participate in a network as part of a network centric warfare application.

Status: Over time, Aeronautics Defense Systems has promoted its UAV GCS technology. Of the cited equipments, the GCS is reported to have been deployed with 'numerous customers' worldwide, while the RPCS is noted as having been used operationally by the US Navy (PMA 263) during 2001 and to have been in service at Naval Air Station Fallon, Nevada, during the period 2004-2005. Most recently, Aeronautics Defense Systems has continued to promote its GCS, LRS, RPCS, RVT and UMAS™ products.

Contractor: Aeronautics Defense Systems Ltd
Yavne.

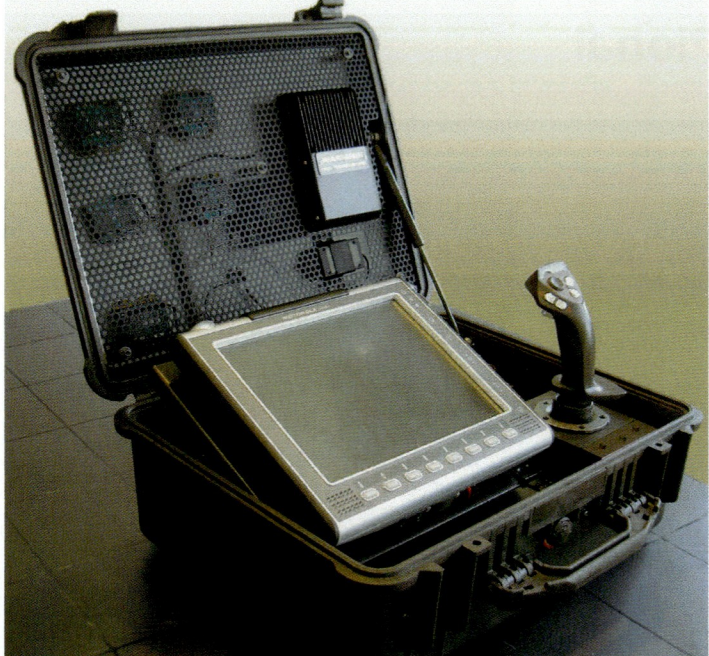

Aeronautics' RPCS display/payload control system *(Aeronautics Defense Systems)*

Elbit Systems UGCS

Type: UAV ground control station.

Description: Elbit's Universal Ground Control Station (UGCS) provides mission planning, control, exploitation/dissemination, debriefing and simulation capabilities within an overall UAV architecture. Design concepts used include:

- PC-based, fully redundant, client-server, open architecture hardware
- object-oriented software that makes use of commercial tools and a Windows-type man-machine interface
- a STANAG ready interoperability capability
- a range of safety features that are designed to DEFence STANdard (DEFSTAN) 56, are Standard 882C compliant and provide flight and mission critical failure redundancy
- use of an open architecture that facilitates the 'easy' integration of new payloads, datalinks and air vehicles

Looking in more detail at the architecture as a whole, the UGCS configuration makes use of military standard shelters or modified commercial containers, is suitable for both ground-based and shipboard applications, is transportable by air (transport aircraft and helicopters), sea and land and has an associated Remote Video Terminal (RVT) that is man-portable and comprises a control/display unit and a tripod-mounted reception array. Functionally, the UGCS is noted as featuring a high level of automation that facilitates same operator control of both the air vehicle and its EO/IR/laser payload. When linked to two ground data terminals, the UGCS can be used to control two UAVs that are performing two independent missions concurrently.

Status: Elbit notes that its UGCS has been supplied to customers, 'mainly as part of the Hermes® UAV system'.

Contractor: Elbit Systems Ltd
Haifa.

Elisra In-Tact

Type: Tactical remote receiver system.

Description: Incorporating both COTS and proprietary hardware and software, In-Tact is a robust, transportable intelligence workstation for the reception and display of real-time video and telemetry data from UAVs and other sensors and payloads. Raw data can be converted into a graphic intelligence overlay on a digital map by various manipulation methods, after which the processed data can be distributed via tactical radio networks, either in graphic map overlay form or as compressed video stills. It can monitor the camera footprint on to the video display and the digital map, and can interface to a wide variety of RF data links.

The system is suitable for installation in ground vehicles, airborne or naval platforms.

Specifications

Dimensions	
Height	625 mm (2 ft 0.6 in)
Width	535 mm (1 ft 9.1 in)
Depth	680 mm (2 ft 2.8 in)
Weights	
Total weight	Approx 50 kg (110 lb)
Performance	
Reception range	Up to 54 n miles (100 km; 62 miles)
System processor	Pentium III 800 MHz with 20 Gbyte hard disk
Interfaces: video input	CCIR and RS-170
video output	CCIR (up to 600 m (1968.5 ft) cable connection) and PG-170
serial	10 base T and RS-232/-422A
Downlink channels	Bi-phase modulated or selectable rate (L/S/C-band, X-band as option)
Power supply	18–32 V DC or 110/220 V 50/60 Hz AC
Power consumption	150 W

Status: Over time, In-Tact has continued to be promoted.

The compact Tadiran In-Tact portable workstation *(Elisra)* 0101614

Contractor: Elisra Electronic Systems Ltd (Intelligence, COMINT and C4I Division)
Holon.

Elisra MRS-2000 and MRS-2000M

Type: Man-pack receiving systems.

Description: The MRS-2000 transportable intelligence computer can receive and display real-time video and telemetry from manned and unmanned platforms or other sensors and payloads. Various picture and data manipulation tools convert raw data into a graphic intelligence overlay (real-time video and platform telemetry, plus UAV/camera footprint) correlated with a digital map. Electronic map capability, an integral GPS receiver and compass provide the operator with the means for system operation, mission management, orientation and a three-dimensional view of the battlefield terrain. Target information and other standard tactical symbols help to complete the situation awareness picture.

The MRS-2000 is powered by four COTS rechargeable lithium-ion battery packs, but can also be connected to a 12 to 32 V DC external power source. The MRS 2000 is an all-digital unit which can be used in autonomous or integrated mode, the latter linked to other operators through a local area network. It consists of a back-pack with receiver and antennas, computers based upon Pentium III 500 MHz with 20 Gbyte hard drive, a GPS antenna and digital compass. There is a TFT hand-held display with 10.4 in (264.2 mm) 1024 x 768 pixel screen. The system operates in the C and L/S reception bands with an option of Ku-band. For long-distance tracking the hardware is placed upon a tripod. A remote antenna terminal is available as an option.

The MRS-2000M configuration is described as being an all-digital system that features:
- portable, man-pack, vehicular and remote data terminal formats
- the ability to extract telemetry data to any electronic map for situation picture build-up
- image processing and video enhancement capabilities
- graphic overlays and a frozen image storage facility
- automatic target report generation
- an artillery correction capability
- integral graphic overlay building tools
- optional antenna auto-track and wireless data dissemination capabilities.

A general view showing the receiver/antenna/computer/digital compass backpack and the hand-held display that are associated with the MRS-2000 man-pack receiving system *(Elisra)* 1120704

Specifications

Weights
- System and harness... 12 kg (26.5 lb)
- Tripod... 3 kg (6.6 lb)

Performance
- Reception band... 1–4 and 4–8 GHz (12.5–18 GHz option)
- LOS range with omni antenna
  - 1–4 GHz band... 6.5 n miles (12 km; 7.5 miles)
  - 4–8 GHz band... 4.3 n miles (8 km; 5.0 miles)
- LOS range with directional antenna
  - 1–4 GHz band... 27 n miles (50 km; 31 miles)
  - 4–8 GHz band... 18.9 n miles (35 km; 21.7 miles)
- Operating time on batteries... 4 h
- Video archive capacity... 10–20 s
- Temperature: operating... 0°C to 50°C (32°F to 122°F)
- storage... –20°C to +60°C (–4°F to +140°F)
- Shock and vibration... MIL-STD-810D/E ground-mobile compliant

MRS-200M Weights
- System and harness... 12 kg (26.5 lb)
- Tripod... 3 kg (6.6 lb)

Performance
- Reception band... 2–4 or 4–8 GHz (1–2 GHz option)
- LOS range with omni antenna
  - 2–4 GHz band... 5.4 n miles (10 km; 6.2 miles)
  - 4–8 GHz band... 4.3 n miles (8 km; 5.0 miles)
- LOS range with directional antenna
  - 2–4 GHz band... 24.3 n miles (45 km; 28.0 miles)
  - 4–8 GHz band... 18.9 n miles (35 km; 21.7 miles)
- Operating time on batteries... At least 3.5 h (continuous use)
- Video archive capacity... Up to 30 minute session (hard disc capacity dependent)
- Temperature: Operating... –5°C to +50°C (23°F to 122°F)
- Storage... –20°C to +60°C (–4°F to +140°F)
- Shock and vibration... MIL-STD-810D/E ground-mobile compliant

Status: Over time, the MRS-2000 man-pack receiving system has been used to provide telemetry support for Hermes, Pioneer, Ranger, Searcher and 'other' UAVs. Users are known to have included the US Marine Corps (four MRS-2000 systems delivered by June 2003 for evaluation). Most recently, MRS-2000 series equipment have continued to be promoted.

Contractor: Elisra Electronic Systems Ltd (Intelligence, COMINT and C4I Division)
Holon.

Elisra SL-RAMBO

Type: Personal video reception system.

Description: Elisra's SL-RAMBO is a wrist-wearable personal digital video receiving system that is capable of downloading live video imagery from a variety of aerial, surface and ground sources in real-time. As such, it provides "vital" field intelligence, supports tactical operations and has been optimised to communicate with and display video (and other types of information) from small/mini-UAVs in real-time and close to the area-of-interest. System features include:
- use of a rechargeable battery
- 'low' weight and volume
- the ability to display video and telemetry on 'any' Personal Digital Assistant (PDA), goggles or command, control, communications, computers and intelligence system
- direct reception from a mini-UAV at ranges in excess of 2.7 n miles (5.0 km; 3.1 miles)
- use of a common visual language to ease data dissemination across a theatre-of-operations
- Advanced Encryption Standard (AES) encryption as an option
- the ability to operate under 'severe jamming and interference' conditions
- omni-directional (flexible/foldable - the foldable type is packaged in a compact unit that can be carried in a pouch on the user's combat vest) and directional antennas
- integral built-in test and GPS facilities
- bandwidth economy.

Specifications

Dimensions
- Depth... 65 mm (2.6 in)
- Width... 170 mm (6.7 in)
- Height... 85 mm (3.4 in)

Weights
- System... ~1,400 gr (3.1 lb)

Performance
- Frequency band... 2–4 GHz
- Data rate... 1.4 Mbps (video and telemetry)
- Operating modes... Single frequency or frequency hopping (selectable)
- Space diversity... 2 or 3 antennas
- Receiver sensitivity... –95 dBm
- Monitor... 'Smart' monitor and keypad
- Keypad controlled functions... Brightness, contrast, frequency selection, GPS display, left/right hand flip-over, on/off, text on/off and zoom
- Continuous operation... ~4 h

Status: Over time, SL-RAMBO has been promoted.

Contractor: Elisra Electronic Systems Ltd (Communication Division)
Holon.

Elisra SRST

Type: Shipboard receiving station.

Description: The Elisra Shipboard Receiving Station (SRST) is designed for the gathering and processing of imint in real-time and can receive telemetry and video imagery from sensor systems installed aboard UAVs, helicopters and fixed-wing aircraft. Imagery is displayed on a black and white monitor and frozen images can be processed off-line and displayed on a high-resolution colour graphics monitor. Here, the monitor displays digital maps that are correlated to the real-time imagery. Downlinked data are received by a pair of stabilised Antenna Subsystems (AS) that auto-track incoming signals and convert them into baseband video and telemetry data. Functionally, the AS architecture comprises a rotary RF receiver and dish and omni-directional antennas (protected by a radome) and can operate in up to Sea State 3. Available map scales comprise 1:25,000, 1:50,000 and 1:100,000 and cover an area of 54 × 54 n miles (100 × 100 km; 62 × 62 miles). New maps can be field loaded via a local area network interface with an external digital map source. System operation is by means of MMI direct access functions, switches, displayed buttons and menus. Data input is by means of a keyboard and trackball. SRST interfaces with its host vessel's gyrocompass and GPS system and individual installations can comprise up to two equipments per ship. Each installed SRST can receive data from a single source, with both systems being able to function simultaneously without mutual interference. System options include a multi-platform telemetry support protocol, data dissemination by tactical radio link (including e-mail, situational overlays and frozen imagery), CCIR/PAL video input, an up to 600 m coax composite video line driver (CCIR/NTSC/PAL/RS-170A) and a swappable map/video display format (upper/lower screen).

Specifications

Performance
- Reception frequency... 4.4–5.1 GHz (L/S/Ku-band options)
- Range... Up to 81 n miles (150 km; 93 miles)
- Video input standards... NTSC and RS-170A
- System processor... Pentium III (800 MHz; 20 Gbyte hard disk)

Sunlight-readable displays
- MMI/map... 38 mm (1.5 in) TFT 1280 × 1024 colour
- Video... 229 mm (9 in) TFT black and white
- Data terminal coverage... DTED, 1 or 2 map
- Temperature: operating... 0/+50°C (32/122°F)
- storage... –20/+60°C (–4/+140°F)
- Sea State... Up to 3

Status: Over time, SRST is reported as having been procured by 'several' navies around the world. Again (and over time), SRST has continued to be promoted.

Contractor: Elisra Electronic Systems Ltd (Intelligence, COMINT and C4I Division)
Holon.

The SL-RAMBO wrist monitor *(Elisra)* 1367967

Elisra StarLink

Type: Digital datalink for small and mini-UAVs.

Description: Based on COTS components, Elisra's StarLink datalink is claimed to be 'rugged', 'reliable', 'very affordable', 'easy to operate' and 'extremely compact and lightweight'. System features include:
- the ability to display on a 'wide variety' of ruggedised, handheld computers and Personal Digital Assistants
- variable rate video transmission (using 'advanced' compression/expansion techniques')
- use of on-reception space diversity for non-LOS communications
- 'high' reception sensitivity
- multipath and channel fading protection
- 'high' spectral efficiency (differential quadrature phase shift keying modulation)
- low probability-of-intercept/detection (via use of 'SpSp' techniques)
- 'immunity' from jamming and interference
- channel separation sufficient to facilitate 'several' mini-UAVs operating in the same area
- an automatic tracking capability
- Internet protocol, RS-232 and RS-422 interfaces.

Specifications

Weights
Air data terminal..<180 g (0.4 lb)
Performance
Range: max..8.1 n miles (15 km; 9.3 miles)
Power consumption: battery...<7 W

Status: Over time, the Elisra StarLink datalink system has continued to be promoted.

Contractor: Elisra Electronic Systems Ltd (Communication Division) Holon.

The StarLink air data terminal *(Elisra)* 1367965

Elisra TDDL

Type: Digital datalink system

Description: Elisra's Tactical Digital Data Link (TDDL) is, as its designation suggests, a digital datalink system that is described as being 'compact' and 'lightweight' and has been designed specifically for UAV (and 'other' unmanned platform) applications where there is a requirement for secure communications, extended range and a high degree of reliability. As such, TDDL provides point-to-point, full duplex, jamming resistant, digital, microwave communications between sensor platforms and control terminals. Equally, TDDL is described as being an evolutionary product and its use of software-defined radio technology as facilitating coding modulation and data rate flexibility/programmability together with interoperability with Tactical Common Data Link and STANAG 7085 systems. Other equipment features include:
- 'high' data rates
- provision for airborne relay
- forward error correction, variable rate interleaving, direct sequence spread spectrum and frequency hopping (as an option)
- anti-jamming provision (spread spectrum and optional frequency hopping)
- provision for encryption
- digital video compression, MPEG 2/4 compatibility and interfaces for SAR and FLIR sensors
- convection cooling
- an airborne relay configuration for OTH functionality
- built-in test and calibration.

Specifications

Dimensions
Not stated
Weights
System..<6.00 kg (<13.2 lb)

Performance
Frequency band.................................14.40–15.35 GHz (2–4 GHz, 4–8 GHz and 8–12.5 GHz available as options)
Data rates: uplink..9.6–200 kb/s
downlink......................1.6–10.71 Mb/s (can be upgraded to 45 Mb/s, symmetric or asymmetric)
Modulation: uplink...BPSK
Error correction codes....................Convolution, concatenated or interleaving
Control interfaces..............MIL-STD-1553B, RS-422, 10/100 Base-T and remote operation via F/O interface (option)
Transmitter power output...2 W or 10 W
Power consumption...<200 W
Antenna options
airborne...Omni, sector or 1-axis steerable
ground (with integrated pedestal
with monopulse or GPS tracker)................Omni, 1-axis or 2-axis steerable

Status: Over time, TDDL has been promoted and Elisra is known to have won a tender to upgrade US RQ-2 Pioneer UAVs with a Ku-band (12.5–18 GHz) digital datalink. As of 2008, Elisra datalinks had accumulated in excess of 500,000 operational flight hours aboard more than 350 aerial vehicles of 14 different types operated by 14 countries around the world.

Contractor: Elisra Electronic Systems Ltd (Communication Division) Holon.

Elisra V-TVR

Type: Vehicle-mounted Tactical Video Receiver (V-TVR).

Description: As its designation suggests, V-TVR is a vehicular configuration of Elisra's TVL II system and is packaged in the shell of a 'standard' military vehicle radio set. System features include:
- the ability to receive video from airborne, surface and ground platforms (stationary or mobile)
- suitability for installation aboard a range of vehicle types including jeeps, armoured personnel carriers and main battle tanks
- on-panel 16-channel selection
- frequency uploading using an external, personal computer-based loader
- omni, directional and hemispheric antenna options
- two video ports
- provision for the reception of telemetry
- provision for scrambling/encryption
- an optional magnetic antenna base for ultra-fast installation.

Specifications

Performance
Frequency..2-4 and 4-8 GHz

Status: Over time, V-TVR has been promoted. The system is understood to have been fielded by the Israel Defence Forces and to have been evaluated by the armed forces of countries in the Americas, Europe and Asia.

Contractor: Elisra Electronic Systems Ltd (Communication Division) Holon.

Elta EL/K-1861 GDT

Type: Ground data terminal.

Development: The EL/K-1861 Ground Data Terminal (GDT) is a C- (4 to 8 GHz) or X-band (8 to 12.5 GHz), high-performance, mobile, microwave communications terminal that is designed to provide a datalink interface between ground or shipboard command and control stations and airborne reconnaissance platforms such as unmanned aerial vehicles. It has a modular architecture (using a set of standard interfaced building blocks) and can be customised for a wide range of ground-to-air, ship-to-air or ground-to-ground applications. The system is line-of-sight (LOS), handles one-way or full duplex communications and one of (or a mixture of) command or telemetry data (up to 200 kbits/s), RS-170 or CCIR video (black and white or colour) and 64 kbits/s to 280 Mbits/s data and voice. Other system features include a self-sustained AC/DC power supply, a back-up navigation capability (in range and azimuth), 'full' remote control (via an RS-422 interface), online and off-line built-in test and customised operational frequencies 'on request'. Physically, Jane's sources report the baseline EL/K-1861 GDT as including an antenna assembly (made up of directional monopulse and an omni-directional antennas, with the latter's beam pattern optimised for close range use), a radio frequency box (mounted on the rear face of the main antenna and incorporating all necessary transceiver/control functions), an antenna positioner (taking the form of a heavy-duty azimuth/elevation pedestal with manual and automatic positioning/tracking modes), an electrical system (incorporating a 3 kVA generator; an AC acceptance panel (generator or 110/230 V AC mains), a DC power subsystem (including a 100 VA back-up battery for uninterrupted power supply) and a 2-ton, High-Mobility, Multi-purpose Wheeled Vehicle (HMMWV) compatible trailer mounting. System options include fibre-optic remote control (at ranges of up to 5 km; 3.1 miles) and an electronic counter-countermeasures/low probability of intercept operating mode.

The EL/K-1861 GDT *(Elta)*

EL/K-1861 is air-transportable as a helicopter load (MIL-STD-209) or aboard a C-130/C-141 class fixed-wing transport aircraft.

Specifications

Dimensions
- Antenna diameter: C-band..1.83 m (6 ft 0 in)
- X-band..2.13 m (7 ft 0 in)

Weights
- Weights: towing..Approx 1,910 kg (4,211 lb)

Performance
- LOS range...>162 n miles (300 km; 186 miles)
- Frequency: range..............................4-8 GHz, 8-12.5 GHz and 12.5-18 GHz
- span...Up to 1,500 MHz (600 MHz baseline)
- tuning step.......................................1 MHz (separate up- and downlinks)
- Transmit: power...........................2 W (baseline; 10, 20 or 50 W options)
- command data mode data rate..........Up to 200 kbits/s (7.3 kbits/s baseline)
- command data mode modulation...FSK
- medium rate data mode data rate.........Up to 50 Mbits/s (standards = 1.544, 2.048, 8.448 or 10.71 Mbits/s)
- medium rate data modulation....................................BPSK or QPSK
- Receive: video with telemetry telemetry...............Up to 200 kbits/s (7.3 kbits/s baseline)
- video with telemetry video.........................CCIR or RS-170 (1 Vptp)
- video with telemetry modulation....................FM with FSK sub-carrier
- telemetry only telemetry................Up to 200 kbits/s (7.3 kbits/s baseline)
- telemetry only modulation..FSK
- medium rate data mode data rate.........Up to 50 Mbits/s (standards = 1.544, 2.048, 8.448 or 10.71 Mbits/s)
- medium rate data mode modulation.....................................BPSK or QPSK
- medium rate data mode forward error correction.......................Viterbi 1/2, 3/4 or 7/8
- Signal quality: video...............S/N ≥45 dB (typical, ≥18 dB at receiver threshold)
- data................................BER ≤10–8 dB (typical, ≤10–5 dB at receiver threshold)
- Navigation: range accuracy.....................................200 m (656.12 ft)
- azimuth accuracy...2 mrad
- Power consumption...................<1 kW (system, typical, 2.5 kW max)

Status: Understood to have been procured. Customer(s) not stated. Over time, the EL/K-1861 GDT has continued to be promoted for 'airborne reconnaissance' (presumed to include UAVs) applications.

Contractor: Elta Systems Ltd
Ashdod.

Elta EL/K-1862 CDT

Type: Compact data terminal.

Description: The EL/K-1862 Compact Data Terminal (CDT) is a low-weight, transportable, communications terminal that is designed to provide a datalink interface between ground or shipboard command and control stations and airborne reconnaissance platforms such as unmanned aerial vehicles. According to *IHS Jane's* sources, the baseline EL/K-1862 configuration includes an antenna assembly (incorporating directional monopulse reflector and omni-directional antennas), a Radio Frequency (RF) unit (accommodates all required transceiver and control functions), an antenna positioner (located to minimise RF signal losses to the antenna assembly) and a quadripod support. User-selected options (including

The EL/K-1862 CDT *(Elta)*

fibre-optic remote control and an electronic counter-countermeasures/low probability of intercept mode) can be used to customise the system's configuration to meet specific requirements. The baseline equipment can be set up in less than 15 minutes by a two-man team.

Specifications

Weights
- Baseline weight...≤100 kg (220 lb)

Performance
- LOS range...>100 n miles (185 km; 115 miles)
- Frequency: range..........................300 MHz-3 GHz, 1-2 GHz, 2-4 GHz, 4-8 GHz and 8-12.5 GHz
- span...Up to 1,500 MHz (600 MHz baseline)
- tuning step.......................................1 MHz (sepatate up- and downlinks)
- Transmit power.................................2 W (10, 20 or 50 W options)
- Transmit data rate: command data mode..................Up to 200 kbits/s (7.3 kbits/s baseline)
- medium rate data mode...............Up to 50 Mbps (standards = 1,544, 2.048, 8.448 or 10.710 Mbits/s)
- Transmit data modulation: command data mode..FSK
- medium rate data mode.....................................BPSK or QPSK
- Receive mode: telemetry........................Up to 200 kbits/s (7.3 kbits/s baseline)
- telemetry video...............................CCIR or RS-170 (1 Vptp)
- telemetry modulation.........................FM with FSK sub-carrier
- telemetry only..........................Up to 200 kbits/s (7.3 kbits/s baseline)
- telemetry only modulation..FSK
- Medium rate data mode data rate.........Up to 50 Mbits/s (standards = 1,544, 2.048, 8.448 or 10.710 Mbits/s)
- medium rate data mode modulation.....................................BPSK or QPSK
- medium rate data mode forward error correction.......................Viterbi 1/2, 3/4 or 7/8
- Signal quality: video...............S/N ≥45 dB (typical, ≥18 dB at receiver threshold)
- data..................................BER ≤10–8 (typical, ≤10–8 dB at receiver threshold)
- Navigation: range accuracy.....................................200 m (656.17 ft)
- azimuth accuracy...3 mrad
- Power consumption..250 V AC (baseline)

Status: Understood to have been procured. Customer(s) not stated. Over time, the Elta EL/K-1862 CDT has continued to be promoted for 'airborne reconnaissance' (presumed to include UAVs) applications.

Contractor: Elta Systems Ltd
Ashdod.

Elta EL/K-1865 ADT

Type: Air Data Terminal (ADT).

Description: Designed for installation (in single- or dual-configurations) aboard UAVs and manned special mission aircraft, the baseline EL/K-1865 ADR includes an omni-directional antenna, a transceiver (with a synthesised local oscillator), a solid-state power amplifier, a front-end filter/pre-amplifier and a DC power supply module. Other system features include:

- on- and off-line built-in test
- one-way or full-duplex communications functionality
- provision of a back-up navigation capability (range and azimuth)
- 'full' remote controllability (via an RS-422 line)
- an integral directional antenna pedestal drive
- the ability to carry command/telemetry data (up to 200 kbits/s), black and white/colour video (CCIR or RS-170), data (64 kbits/s to 280 Mbits/s) and voice singly or in combination
- an options package that includes a spread spectrum modem, a video compression unit, a digital data modem, a data encryption module and a planar directional antenna (controlled in azimuth and elevation).

Specifications

Dimensions
- Transceiver: Height..160 mm (6.29 in)
- Width...210 mm (8.27 in)
- Depth..300 mm (11.81 in)
- Front end: Height..160 mm (6.29 in)
- Width...382 mm (15.03 in)
- Depth..260 mm (10.24 in)

Weights
- System...2.5-10.0 kg (5.5-22.1 lb)

Performance
- Frequency: range....................................300 MHz-3 GHz, 1-2 GHz, 2-4 GHz, 4-8 GHz, 8-12.5 GHz and 12.5-18 GHz
- span......................................Up to 1,500 MHz (600 MHz baseline)
- tuning step...................................1 MHz (separate up- and downlinks)
- Antenna: transmit power... 2, 10 or 25 W
- Receive mode: command data mode data rate....................................Up to 200 kbits/s (7.3 kbits/s baseline)
- command data mode modulation..FSK
- Transmit mode: video with telemetry telemetry.................................Up to 200 kbits/s (7.3 kbits/s baseline)
- video with telemetry video...............................CCIR or RS-170
- video with telemetry modulation................FM with FSK sub-carrier
- video with telemetry IF bandwidth.......................20 MHz (max)
- telemetry only telemetry..................up to 200 kbits/s (7.3 kbits/s baseline)
- telemetry only modulation...FSK
- telemetry only IF bandwidth.................................60 kHz (at TM = 7.3 kbits/s)
- medium data rate mode data rate.........Up to 50 Mbits/s (standards = 1.544, 2.048 or 10,710 Mbits/s)
- medium rate data mode modulation...BPSK or QPSK
- medium rate data mode forward error correction..Viterbi 1/2, 3/4 or 7/8
- high capacity data mode data rate.............................70, 140 or 280 Mbits/s
- high capacity data mode modulation... 8-PSK
- high capacity data mode forward error correction...Viterbi 5/6
- Signal quality: data.......................BER >10–8 dB (typical, 10–5 dB at receiver threshold)
- Power: source................................28 V DC (MIL-STD-704 compliant)
- consumption......................................100-250 W (system, amplifier dependent)

Status: Over time, the EL/K-1865 ADT has continued to be promoted.
Contractor: Elta Systems Ltd
Ashdod.

The DGDT's tripod-mounted antennas *(Elta)*
0079342

Elta EL/K-1871 DGDT

Type: Down-sized ground data terminal.
Description: The EL/K-1871 Down-sized Ground Data Terminal (DGDT) is a lightweight, transportable, communications terminal that is designed to provide a datalink interface between ground or shipboard command and control stations and airborne reconnaissance platforms such as unmanned aerial vehicles. As such, *IHS Jane's* sources report the baseline equipment as comprising an antenna assembly (incorporating C-band (4.4 to 5.85 GHz sub-band) directional monopulse planar and omni-directional antennas), a transceiver unit (housing all the required radio frequency transmit, receive and control modules), an antenna positioner (a lightweight, elevation-over-azimuth pedestal with manual/automatic positioning and antenna tracking modes), an interface and control unit (incorporating an AC/DC power supply interface and a built-in test status display/operator panel) and a tripod mount. System options include installation aboard a Silver Eagle (or equivalent) trailer, a 4.5 to 20 m (14.8 to 65.6 ft) telescopic antenna mast, a stabilised antenna/radome assembly for shipboard applications, fibre-optic remote control (at ranges of up to 500 m; 1,640 ft), up to 35 Mbits/s data (1.544, 2.048, 8.448 or 10.71 Mbits/s for standard configuration), BPSQ or QPSK modulation, Viterbi ½, ¾ or ⅞ FEC and a 57.6 kbits/s RS-422A remote control interface.

Specifications

Weights
- Baseline weight..59 kg (130 lb)

Performance
- LOS range...>162 n miles (300 km; 186 miles)
- Frequency: range......................................300 MHz-3 GHz, 1-2 GHz, 2-4 GHz, 4-8 GHz, 8-12.5 GHz and 12.5-18 GHz
- span......................................Up to 1.5 GHz (600 MHz baseline)
- tuning step...................................1 MHz (separate up- and downlinks)

Telescopic mast of the EL/K-1871 *(Elta)*
0079341

- Transmit: power..1 W (baseline; 2, 10 or 25 W options)
- command data mode data rate..........Up to 200 kbits/s (7.3 kbits/s baseline)
- Receive: video with telemetry telemetry..................................Up to 200 kbits/s (7.3 kbits/s baseline)
- video with telemetry video...............................CCIR or RS-170
- video with telemetry modulation................FM with FSK sub-carrier
- video with telemetry IF bandwidth.......................18 MHz
- telemetry only telemetry..................Up to 200 kbits/s (7.3 kbits/s baseline)
- telemetry only modulation...FSK
- telemetry only IF bandwidth.................................60 kHz (TM = 7.3 Kbits/s)
- digital data mode data rate....................Up to 35 Mbits/s (standards = 1.544, 2.048, 8.448 or 10.710 Mbits/s)
- digital data mode modulation...BPSK or QPSK
- digital data mode forward error correction..Viterbi 1/2, 3/4 or 7/8

Specifications

Remote control	RS-422A (57.6 kbits/s)
Signal quality: video	S/N ≥45 dB (typical, ≥18 dB at receiver threshold)
data	BER ≤10–9 dB (typical, ≤10–5 dB at receiver threshold)
Navigation: range accuracy	200 m (656.12 ft)
azimuth accuracy	3 mrad
Power: supply	19-32 V DC; 90-270 V AC (48-62 Hz)
consumption	250 W (baseline max)

Status: Understood to have been procured. Customer(s) not stated. Over time, the Elta EL/K-1871 DGDT has continued to be promoted for 'airborne reconnaissance' (presumed to include UAVs) applications.

Contractor: Elta Systems Ltd
Ashdod.

Elta EL/K-1891 SCS

Type: X- (8 to 12.5 GHz) or Ku-band (12.5 to 18 GHz) Satellite Communications System (SCS).

Description: The Elta Systems EL/K-1891 is a full duplex satellite communications system that operates in either the X- or the Ku-band and is described as being 'ideally suited' for wideband, uninterrupted, over-the-horizon functionality. The subscriber ports within an EL/K-1891 network can be either static or mobile (vehicular, airborne or shipboard) with the only proviso being that they must be within the satellite's footprint with the system antenna tracking the relevant satellite. Communications between any two subscribers is bi-directional, is routed via standard geo-stationary commercial satellite links and can include voice, data and compressed Joint Photographic Experts Group (JPEG) video. Physically, EL/K-1891 comprises platform and ground station elements, with the former comprising an antenna assembly, transceiver and a high-power travelling wave tube amplifier and the latter, a 9.3 m diameter (or larger) commercial dish antenna. Other system features include:

- mobile application, dual-axis, stabilised antenna pedestals that can accommodate either a 25 × 10 cm planar array or an up to 80 cm diameter dish antenna (antenna formats can be tailored/adapted to meet specific customer requirements)
- orthogonal polarisation between the system's transmit and receive elements
- autonomous antenna pointing (based on inputs from external navigation aids)
- the use of commercial-off-the-shelf components 'wherever possible'
- coherent reception
- Time Division Multiple Access (TDMA) over Code Division Multiple Access (CDMA) modulation (with the option of tailoring to a specific user's requirements)
- CDMA over TDMA point-to-point linkage (complying with international satellite communications standards)
- turbo cooling.

Specifications

Dimensions	
Mobile planar array antenna: Length	250 mm (9.84 in)
Depth	100 mm (3.94 in)
Mobile dish antenna: Diameter	Up to 800 mm (31.50 in)
Weights	
Mobile system	24.0 kg (52.9 lb)
Performance	
Frequency	8-12.5 or 12.5-18 GHz
Amplifier output	100 W
Data rate	Up to 128 Kbit/s
Bit error rate	Better than 10–8

Status: IHS Jane's sources report the EL/K-1891 as being installed aboard the Heron UAV where it makes use of the system's planar array antenna assembly.

Contractor: Elta Systems Ltd
Ashdod.

IAI AGCS

Type: Family of UAV ground control stations.

Description: Malat's Advanced Ground Control Stations (AGCS) are designed with sub-system modularity and commonality, an approach that enables users to extend mission range, capability and endurance by adding a common series of modules. The basic AGCS can control a full UAV mission with a single operator. The more commonly used configuration uses two operators either in a split control role (one controlling the UAV and one the mission payload) or in a mission commander and a mission operator mode. Each GCS can control one airborne platform (or two in relay operation) equipped with a variety of payloads.

Computerised workstations and menu-driven software are used to facilitate the man/machine interface. Built-in test equipment reports line replaceable unit failures that can be rectified without interrupting a mission. In addition the AGCS incorporates an embedded simulator for

A general view of an AGCS series console (IAI) 1375076

operator proficiency training, mission planning and as a debriefing tool.

The inherent modularity of the AGCS enables optional workstations to be added in order to create a mission commander bay, an intelligence interpreters bay and the like. For controlling UAVs with multiple payloads, additional workstations can be integrated to serve the needs of operators of mission specific payloads such as synthetic aperture and maritime patrol radars, electronic support and communications intelligence systems and the like.

Multiple AGCS configurations have been fielded, with various additional functions such as embedded C^4I connectivity (the reception and distribution of relevant mission data to and from external C^4I entities) and the capability to control two UAVs, using two separate datalinks. Here, the UAV range includes the Ranger, Searcher Mk I, Searcher Mk II, Heron, Heron TP and I-View types.

Depending on the customer's operational concept, the whole mission cycle (including Automatic Take-Off and Landing - ATOL) can be performed from the AGCS. If required, a dedicated Advanced Launch and Recovery System (ALRS) can also be provided. ALRS performs the ATOL functions and hands-off UAV control to the AGCS which is deployed at a forward site. ATOL is a standard AGCS function that minimises human error during take-off and landing and reduces the cost of operator training. A variety of ATOL sensors including DGPS, RAPS (a laser tracker sensor) or radar can be used with the architecture.

Multiple operational ATOL modes are available to suite the particular UAV system or operational scenario being used. Here, the options include wheeled runway take-off, launcher take-off, rocket assisted take-off, wheeled runway landing, skid landing or precision controlled parafoil landing in an unprepared zone. Another optional control station within the architecture is the AMRS (Active Mobile Receiving Station) that facilitates the allocation of EO payload control to forward forces.

Status: Most recently, the AGCS has continued to be promoted.

Contractor: Israel Aerospace Industries Ltd, Malat UAV Division
Tel-Aviv.

Rafael DLV-52

Type: Airborne reconnaissance and surveillance communication system.

Description: DLV-52 is a two-way, digital, communication system that supports reconnaissance and surveillance missions and comprises Airborne Data Link (ADL), Ground Data Link (GDL) and Ground Control Shelter (GCS) segments. In more detail, the ADL downlinks real-time day/night sensor imagery and radar, sigint and annotated telemetry data and includes a transceiver, a controller, a power supply and two (forward and aft) selectable, gimbaled antennas. The GDL automatically acquires and tracks ADL signals, serves as a command and control uplink and includes a transceiver, an acquisition and tracking subsystem, an antenna and a trailer. The GCS is optional and functions as a command, control and maintenance centre, is connected to the GDL via fibre-optic cabling, transmits acquired sensor data to a remote imagery exploitation control centre and includes an air-conditioned operator/equipment shelter (incorporating operator consoles, fibre-optic cabling, standard long-haul communications equipment, a power distribution unit, a work bench and a storage cabinet), a power generator and trailer or truck mounting.

Specifications

Dimensions	
ADL package: Length	760 mm (29.9 in)
Width	540 mm (21.3 in)
Height	240 mm (9.5 in)
DLS (incl trailer): Length	4.7 m (15.4 ft)
Width	2.5 m (8.2 ft)
Height	4.0 m (13.1 ft)
GCS: Length	3.8 m (12.3 ft)
Width	2.0 m (6.8 ft)
Height	2.2 m (7.2 ft)
GDL antenna: Diameter	2.1 m (7.0 ft)

Specifications

Weights
- ADL package.. 40 kg (88 lb)
- GDL (incl trailer)... 5,500 kg (12,125 lb)
- GCS.. 650 kg (1,433 lb)

Performance
- Frequency... 12.5–18 GHz (options available)
- Operational range........................... >200 n miles (370 km; 230 miles)
- Data rate: Downlink imagery.. Up to 120 mbps
- Downlink annotation/telemetry.. 19.2 kbps
- Uplink command and control... 180 kbps
- Bit error rate: Imagery... Better than 10^{-9}
- Annotation/telemetry.. Better than 10^{-5}
- Command and control... Better than 10^{-5}
- GDL antenna: Azimuth... 360°
- Elevation... –5 to +85°
- Power supply: ADL... 115 V AC (3-phase, 400 Hz)
- GDL/GCS... 220 V AC (3-phase, 50 Hz)

Status: Over time, the DLV-52 airborne reconnaissance and surveillance communication system has continued to be promoted for use with airborne reconnaissance pods and UAV/aerostat surveillance systems.

Contractor: Rafael Advanced Defense Systems Ltd
Haifa.

Italy

Selex Galileo Alamak

Type: Ground control station.

Development: Developed as a common GCS for the Galileo Mirach 20 UAV and Mirach 100 aerial target. Designed specifically to integrate, in a single system, all command, control, tracking and telemetry functions previously performed by a series of separate equipments.

Description: Installed in a truck-mounted, environmentally controlled shelter, the Alamak GCS incorporates the main command, control, tracking and telemetry system; auxiliary telemetry data equipment and associated recording system; and a large plotting station. Provision is made for installing other, optional, equipment. The GCS and the air vehicle are connected by a UHF-band radio command uplink and a D-band tracking and telemetry downlink which, between them, provide a continuous flow of signals, telemetry data and sensor information and allow real-time display of air vehicle co-ordinates.

Specifications

Performance
- Acquisition range... Up to 67.5 n miles (125 km; 78 miles)
- Acquisition angle.. ±5.5°
- Coverage: azimuth.. 360° continuous
- elevation.. –10/+85°
- Acceleration: azimuth.. 60°/s2
- Targeting accuracy... 150 m (492 ft) CEP
- Measurement accuracy: azimuth... 1 mrad RMS
- slant range.. ±30 m (98.5 ft) RMS

Plotter accuracy
- 0–5.4 n miles (0–10 km; 0–6.2 miles)................................... 20 m (66 ft) RMS
- 5.4–54 n miles(10–100 km 6.2–62 miles)........................... 150 m (492 ft) RMS
- 54–67.5 n miles(100–125 km; 62–78 miles)...................... 200 m (656 ft) RMS
- 67.5–135 n miles (125–250 km; 78–155 miles)............... 400 m (1,312 ft) RMS

Status: Over time, the Alamak GCS is reported as having been procured.

Contractor: Selex Galileo
Ronchi dei Legionari.

A general view of the truck-mounted Alamak ground control shelter *(Selex)* 0062356

The interior of the Alamak GCS *(Selex)* 0062355

Selex Galileo Falco GCS

Type: GCS for use with the Falco medium-altitude, tactical endurance UAV.

Development: The Falco GCS was first identified by *Jane's* during 2002/3.

Description: The Falco GCS provides control of AV payloads and sensors; handles collected data in real-time; can be used to pre-programme AV taskings and can 'enhance' the AV's 'autonomous operational features'. Again, the architecture is designed to facilitate mission planning and re-tasking, mission simulation (for operator training) and mission rehearsal and play back.

Using the station, Falco AVs can be flown manually or in a fully automatic mode that includes automatic take-off and landing. Post-mission, the GCS is capable of off-line target data evaluation and processing for report dissemination through a Command, Control, Communications, Computers and Intelligence (C4I) net. An associated ground data terminal provides a more than 108 n miles (200 km; 123 miles) redundant, jam-resistant, real-time link between the GCS and the Falco AV.

Elsewhere, the Falco GCS is supported by the Falco UAV Battle Laboratory (UAVBL) that incorporates an AV/sensor simulator that is linked to a real-world or virtual GCS. UAVBL functions include operational analysis of missions, system requirement definition, technology demonstration, operator briefing/debriefing and the simulation and validation of route plans, terrain/obstacle avoidance schematas, sensor coverage verification, radio link verification, LOS verification and platform sensor optimisation.

A complete baseline Falco UAV system comprises four AVs, a customer specific sensor suite, the described GCS, ground support equipment and the noted ground data terminal.

Status: Over time, the Falco GCS has continued to be promoted.

Contractor: Selex Galileo
Ronchi dei Legionari.

Japan

Yamaha YACS

Type: Attitude control system.

Description: The Yamaha Attitude Control System (YACS) was developed to enhance ease of control of the company's R-50 and RMAX remotely controlled helicopters (which see). Three fibre optic gyroscopes and three accelerometers are fitted to the helicopter body to supply data to an onboard computer unit that regulates all three-axis and engine throttle stick operations. It enables the helicopter to maintain a stable flight pattern without unduly reacting to irregularities in the terrain, thus eliminating the need for the operator to make constant remote-control adjustments.

The basic remote-control operation remains unchanged, but because YACS allows the helicopter to assume a horizontal flight attitude automatically, it takes off and lands smoothly with almost no need for complicated remote-control stick operations; instead, the operator need concentrate only on throttle control.

Furthermore, the adoption of modern fibre optic gyroscope technology means that the amount of stabilisation time for the gyroscopes (the time needed to produce readings on fuselage angle in flight) has been greatly reduced, from the 2 to 3 minutes required by conventional units to about 10 seconds. Thus, as soon as the electronic control unit is switched on, the engine can be started and operations begun without delay.

The YACS also benefits handling of the RPH. By making the stabilising function a supportive one, with system priority given to handling enhancement, both forward-rear and left-right movement can be more easily achieved with simple 'aileron' and 'elevator' operations. The handling enhancement control function is also adjustable to three levels to allow for differences in operating conditions and operator skills.

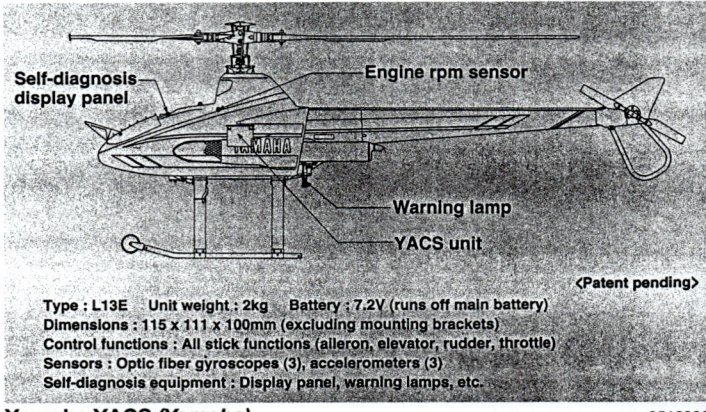

Yamaha YACS *(Yamaha)*

Specifications

Dimensions (excl mounting brackets)	
Length	115 mm (4.5 in)
Width	111 mm (4.4 in)
Height	100 mm (3.9 in)
Weights	
Unit weight	2.0 kg (4.4 lb)

Status: Over time, the YACS attitude control system is reported to have been procured by 'Japanese operators'.

Contractor: Yamaha Motor Co Ltd
Shizuoka.

South Africa

ATE Vulture GCS

Type: Dedicated GCS.

Development: This ground control station was developed by ATE specifically for the company's Vulture tactical UAV (which see).

Description: The Vulture GCS is staffed by a crew of four: navigator, observer, signaller and artillery technical assistant. It serves as the mission command and control centre for the Vulture system, including communication with the gun battery's fire control post, enabling the system to perform the following main artillery functions:

(a) Detection of targets within an area of 2 × 2 km (1.24 × 1.24 miles), to an accuracy of 30 m (100 ft) at an elevation of 2,000 m (6,560 ft) above ground level

(b) Identification of a target within an area of 100 × 100 m (330 × 330 ft)

(c) Fixation and relaying of target co-ordinates within 30 seconds, with an accuracy of 30 m (100 ft)

(d) Detection and fixation of fall of shot in a 2 × 2 km (1.24 × 1.24 mile) area in less than 10 seconds

(e) Relaying of fire correction co-ordinates within 30 seconds after impact.

Communication between the GCS and the air vehicle is via a Tellumat CBACS (which see) digital video and datalink.

Status: The ATE Vulture GCS has been procured by the South African Army as part of the Vulture tactical UAV system.

Contractor: Advanced Technologies & Engineering Co (Pty) Ltd
Halfway House.

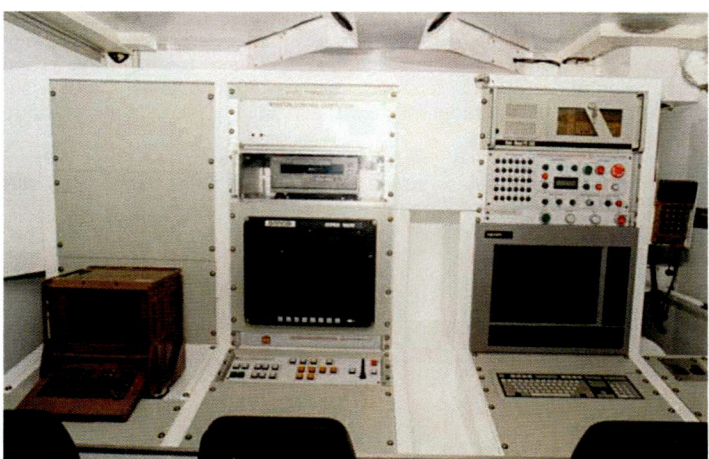

Vulture GCS interior, showing (left to right) the artillery computer, observer panel and navigator panel *(ATE)*

The Vulture ground control station vehicle *(ATE)*

Tellumat CBACS

Type: Telemetry and video communications link.

Description: This C-band (G/H frequency band) communication system was designed for military UAV, helicopter and terrestrial applications, with main features of long range, low latency and excellent security. It combines a fully duplex, multiple-channel digital telemetry uplink (ground node transceiver, or GNTC) with in-flight channel selection and a secure video downlink (airborne node transceiver, or ANTC). Narrowband or direct sequence spread spectrum modulation (DSSS) are available on the GNTC uplink, with Reed-Solomon forward error correction on transferred data. The ANTC downlink has single-carrier, high (FM) modulation; the interface is asynchronous RS-422. Video data are available in high-resolution, colour or monochrome, low-latency, encrypted or scrambled format. Its compactness offers low airborne and ground remote mast-mounted mass, steered narrow-beam antennas and resistance to jamming, and ranging accuracy is better than 15 m (50 ft).

Specifications

Dimensions (ANTC)	
Height	300 mm (11.8 in)
Width	210 mm (8.3 in)
Depth	160 mm (6.3 in)
Weights	
Weight complete: ANTC	8.0 kg (17.6 lb)
GNTC (mast)	11.0 kg (24.25 lb)
Performance	
Frequency band: ANTC	H
GNTC	G
Gross data rate: ANTC	47 kbits/s
GNTC: with video	47 kbits/s
without video	Up to 4 Mbits/s
Bit error rate at specified sensitivity	
ANTC, GNTC	Better than 10^{-6}
Range	108 n miles (200 km; 124 miles)
Power requirements (nominal, 27.5 V DC)	
ANTC	<240 W
GNTC	<90 W
Temperature requirements: ANTC	–20/+55°C
GNTC	–15/+50°C

Status: Most recently, Tellumat has continued to promote the CBACS telemetry and video link. *IHS Jane's* sources report the system as having been installed aboard the ATE Vulture tactical UAV.

Contractor: Tellumat (Pty) Ltd, Defence Division
Tokai.

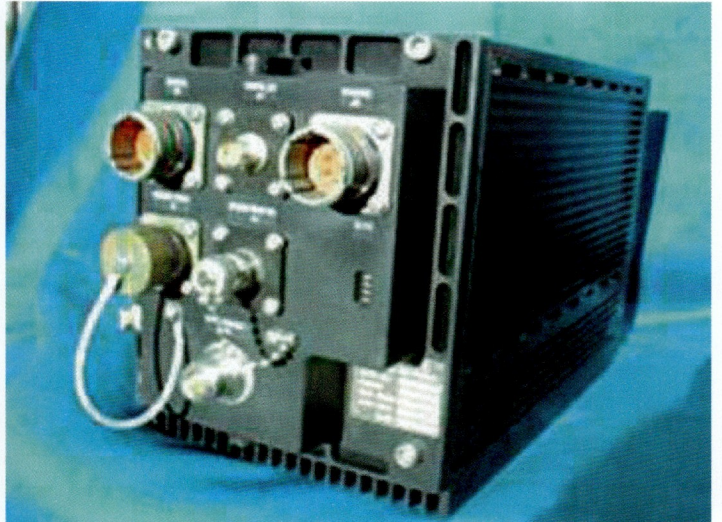

CBACS airborne node transceiver *(Tellumat)*

CBACS outdoor ground node transceiver *(Tellumat)*

Spain

UAV Navigation AP04

Type: Family of UAV autopilots.

Description: The AP04 series is described as being a "fully integrated" autopilot that features manual override, a radio link and payload control capabilities. Known details of identified members of the family are as follows:

AP04: AP04 is designed for use aboard UAVs that range in size from micro-UAVs with a wingspan of less than 1 m (3.3 ft) to medium sized AVs with a wingspan of 4 m (13.1 ft) and a payload capacity of 36.3 kg (80.0 lb). Again, AP04 is noted as being suitable for both fixed- and rotary-winged AVs; it is capable of fully automatic take-off (including catapult launches that are in excess of 15 *g*), hover control, flight plan following and landing (parachute and standard runway), and configures the AV 'at each stage of flight'. The device's integrated radio link has a range of more than 52.1 n miles (96.5 km; 60.0 miles) and a manual override mechanism allows operators to exercise 'full' AV control via a joystick arrangement. AP04 features dual redundant microprocessors to enhance safety and the device's payload control functions include flare, parachute, camera and antenna activation. An RS-232 interface is used for communications with ground stations and the unit can be integrated with an engine management/control system. Overall, AP04 can control up to 16 servos or peripherals.

AP04R: The AP04R configuration is designed for use aboard UAVs that range in size from micro-UAVs with a wingspan of less than 1 m (3.3 ft) to 70.0 kg (154.3 lb) target drones capable of speeds of up to 350 kt (648 km/h; 403 mph). As with AP04, AP04R is reported as being suitable for both fixed- and rotary-winged AVs. AP04R can control up to seven servos or peripherals.

AP04OEM2: AP04OEM2 is described as being a fully integrated, miniaturised air data-INS system and is billed as being suitable for a variety of applications including: electric micro UAV control; payload gyro-stabilisation and geo-referencing; and acting as an air data and attitude heading reference system for manned aircraft. AP04OEM2 'inherits' the design architecture and reliability features of the AP04, while its single processor architecture is claimed to reduce cost, size and power consumption. Software redundancy facilitates survival of individual sensor failures while maintaining accurate estimates of attitude and position.

AP04SDK: The AP04SDK configuration is an AP04 application that is designed to facilitate the use of customer developed/specified flight and payload control software. AP04SDK is supplied with a skeleton application that is written in C and which is intended to be modified by the customer 'from the outset'.

Specifications

(AP04/AP04R/AP04SDK) Dimensions	
AP04/AP04SDK: Length	74.0 mm (2.9 in)
Width	68.0 mm (2.7 in)
Height	59.7 mm (2.3 in)
AP04R: Length	74.0 mm (2.9 in)
Width	68.0 mm (2.7 in)
Height	46.7 mm (1.8 in)
Weights	
AP04/AP04SDK	300 gr (10.6 oz)
AP04R	200 gr (7.1 oz)
Performance	
Servo or I/O lines: AP04/AP04SDK	16
AP04R	7
Communications: Baud rate	115.2 Kbit/s
AP04 range	More than 52.1 n miles (96.5 km; 60.0 miles)
Frequency	902-928 MHz (1.3 GHz option)

Specifications

I/O: PWM rate	50 or 200 Hz
PWM signal	1-2 ms (1 µs steps)
RS-232 Baud rate	4.8, 9.6, 19.2 or 38.4 K
Airspeed: LSP	25-150 kt (46-278 km/h; 29-173 mph)
normal	35-250 kt (65-463 km/h; 40-288 mph)
HSP	45-450 kt (83-833 km/h; 52-518 mph)
Altitude	0-6,096 m (0-20,000 ft)
Maximum acceleration	10 g (vertical)
Maximum angular rate	300°/s
GPS: Channels	12 (differential)
Temperature	–40°C to +85°C (–40°F to +185°F)
Power: Supply	7-36 V (unregulated)
AP04/AP04SDK consumption	2.5 W
AP04R consumption	300 mA
(AP04OEM2) Dimensions	
Length	51.0 mm (2.0 in)
Width	36.0 mm (1.4 in)
Height	15.4 mm (0.6 in)
Weights	
Unit	20 g (0.7 oz)
Performance	
Angular rate: Range	± 300°/s (±150°/s for Z axis)
Non-linearity	0.1% of FS
Noise	0.1°/s/√Hz (0.05°/s/√Hz for Z axis)
3 dB bandwidth	40 Hz
Accelerometers: Range	±2 g (±10 g for Z axis)
Non-linearity	0.2% of FS
Noise	500 µg/√Hz (200 µg/√Hz for Z axis)
Magnetometer: Range	±2 gauss
Non-linearity	1% of FS
Air data: Static range	0-110 kPa
Non-linearity	1% of FS
Pitot range	0-10 kPa
Pitot non-linearity	1% of FS
Accuracy: Pitch and roll	<1° (with/without GPS)
Heading	<1° (with/without GPS)
Horizontal position	2 m CEP position drifts (WAAS/EGNOS, 2.5 m CEP without) at 14.8 m/minute (with air data system - with/without GPS)
Velocity	0.1 m/s (with GPS); 2.5 m/s (without GPS)
Power requirement: Input voltage	7.5-36 V DC
Current	100 mA
Temperature	–40°C to +85°C (–40°F to +185°F)
Humidity	up to 90% (relative, non-condensing)
Shock	500 g, 8 m/s, ½ sine
GPS antenna: Type	active or passive
Power supply	3 V
Sensitivity	–144 dB
Data output rates	20, 50 or 100 Hz

Status: Most recently, the AP04, AP04OEM2, AP04R and AP04SDK autopilots have continued to be promoted.
Contractor: UAV Navigation
Madrid.

UAV Navigation GCS03
Type: GCS interface unit.
Description: UAV Navigation's GCS03 GCS interface unit is designed to communicate with the contractor's AP04 autopilot family (which see). As such, it takes the form of a 'compact and robust' communication system that facilitates communication between a UAV and its GCS, real-time manual control of the UAV and automatic antenna steering. Again, GCS03 acts as a bi-directional communication relay between the UAV and the GCS (via an integrated radio link), performs low-level message integrity verification (guaranteeing that no unnecessary communications clog the link) and is capable of data communication at ranges of up to 53.9 n miles (100 km; 62.1 miles). On-the-move UAV tracking and antenna pointing (in pan and tilt and using a standard PWM control signal) make use of the architecture's GPS facility.

Elsewhere, GCS03 includes an Ethernet port (to facilitate telemetry distribution over a computer network) and incorporates a stand-by battery that allows for up to two hours of communication with a UAV in the event of external power loss. UAV Navigation further notes that its VISIONAIR software package makes a 'perfect' partner for both the GCS03 and its AP04 autopilots. Here, the tool is claimed to be the 'most intuitive product [in] its category' and is billed as allowing 'anyone to be in command of a UAV in a matter of minutes [without the need for] computer training and/or programming skills'. Data handled includes video imagery, vehicle location, AV attitude, AV engine parameters and vehicle dynamics. VISIONAIR runs on a personal computer that incorporates a Pentium 4 processor (operating at 1.5 GHz), a 256 Mb video card, 1 Gb of RAM and a minimum of 10 Gb of free hard drive space.

Specifications

Dimensions	
Unit: Length	123.0 mm (4.8 in)
Width	103.0 mm (4.1 in)
Height	30.0 mm (1.2 in)
Weights	
Unit	300 gr
Performance	
Control: Rate	50 Hz
Resolution	1 mrad
Channels	4
Fixed-wing functions	Aileron, elevator, rudder and throttle
Helicopter functions	Altitude, forward/lateral motion and heading
Communications: Type	Frequency hopping spread spectrum
Baud rate	115.2 Kb/s (full duplex)
Range	53.9 n miles (100 km; 62.1 miles)
Frequency	902-928 MHz (1.3 GHz option)
Antenna steering: PWM rate	50 Hz
PWM signal	1-2 ms (1 µs steps)
GPS: Type	Differential
Channels	12
Power: Supply	9-36 V (unregulated)
Consumption	2 W

Status: Most recently, both the GCS03 and VISIONAIR products have continued to be promoted.
Contractor: UAV Navigation
Madrid.

Sweden

Air Target Sweden Marque and Mini Marque
Type: Scoring stations.
Description: The Marque and Mini Marque are scoring stations for the collection, calculation and presentation of firing results from up to six targets simultaneously. The data are presented in real time. They use a Windows XP operating system and SLQ database. The Marque system is the basic one and the Mini Marque is a more mobile unit and the two

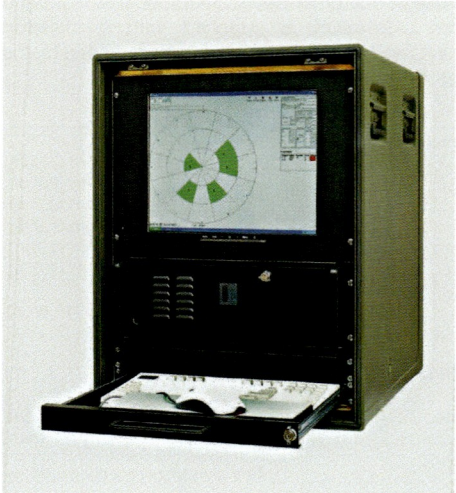

The Air TargetMarque scoring station (Air Target Sweden)
1323731

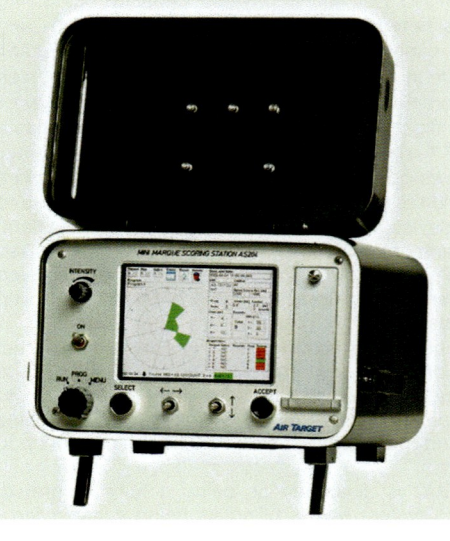

The Air Target Mini Marque scoring station (Air Target Swea
1323732

configurations are linked to Air Target Sweden's universal 12 sector Miss Distance Indicators (which see) by radio. The stations display a picture of each hit on the target. The system can also be used for enemy fire indicators when used with a UAV.

Status: Over time, the Marque and Mini Marque equipments are noted as having been procured.

Contractor: Air Target Sweden AB
Kista.

Air Target Sweden scoring systems

Type: Family of target scoring systems.

Description: Air Target Sweden has been a specialist in scoring systems since 1956 and has sold its products into more than 30 countries around the world. The company's universal 12 sector Miss Distance Indicators (MDI) assess fire being directed at a target coming from any direction, with the target-towing or UAV operator being able to change the target's course during the mission. The systems are compatible with a full range of weaponry from 5.56 mm small arms to 127 mm (5 in) plus shells and missiles. Air Target Sweden's MDIs are battery operated and make use of pressure sensors (that are installed on or close to the target) to detect the shock wave generated by a passing supersonic projectile or missile. The amplitude and time differences between the different sensors are used to calculate the shortest distance between the projectile and the target and the angular relationship between the two. Scoring data from the MDI is radioed to a ground-based scoring station (using a telemetry signal within the 400 to 470 MHz frequency band), with each station being able to receive data from six target-mounted MDIs simultaneously. Air Target Sweden notes that its MDIs and scoring stations can be customised to meet specific customer requirements, with its over time product range including the following equipments:

AS-113/12U: The AS-113/12U MDI is described as being a 'universal' equipment that is designed for UAV applications and incorporates two 'small' microphone clusters that are installed on each of the air vehicle's wingtips.

AS-133/12U: Like the AS-113/12U, AS-133/12U is described as being a 'universal' MDI that is suitable for hard/stiff target and UAV applications.

AS-135/12U: The AS-135/12U is another 'universal' MDI that is suitable for hard/stiff target and UAV applications and features a constellation of six microphones that are flush mounted on the target/UAV's body.

Marque Scoring Station: The ground-based, 432 mm (17 in) Marque Scoring Station is capable of calculating and displaying real-time firing results from up to six target/UAV-mounted MDIs simultaneously. The equipment makes use of Windows XP-based software and a structured query language database.

The AS-135/12U *(Air Target Swedem)*

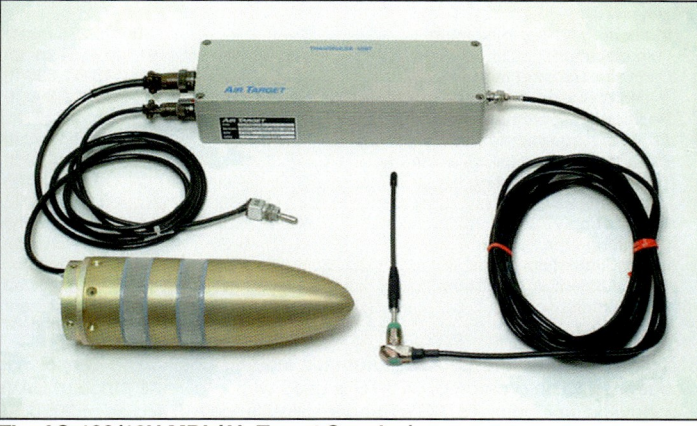

The AS-133/12U MDI *(Air Target Sweden)*

The ground-based Marque Scoring Station *(Air Target Sweden)*

The components that make up the AS-113/12U MDI *(Air Target Sweden)*

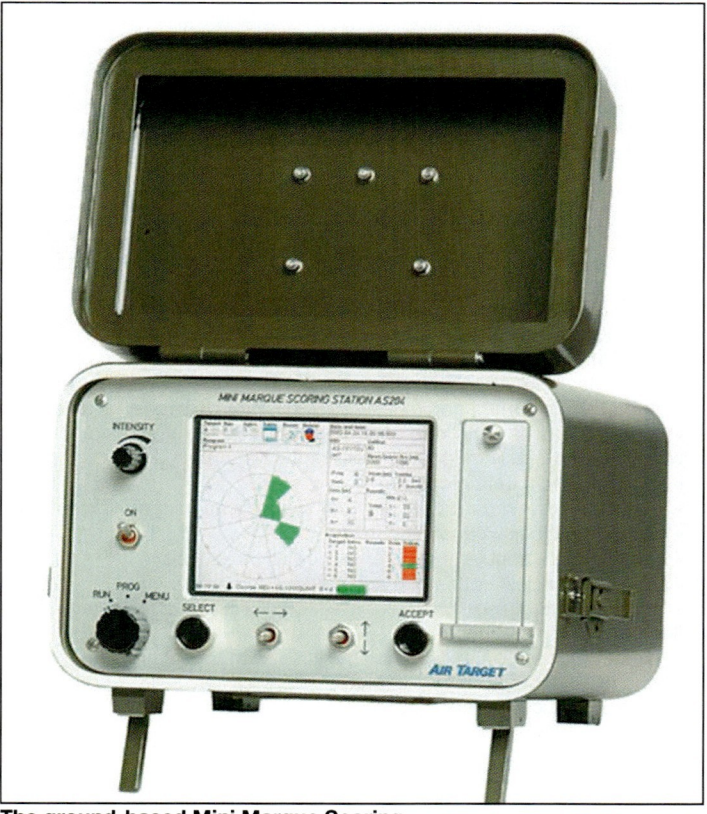

The ground-based Mini Marque Scoring Station *(Air Target Sweden)*

Mini Marque Scoring Station: The ground-based Mini Marque Scoring Station is a portable equipment that is capable of calculating and displaying real-time firing results from up to six target/UAV-mounted MDIs simultaneously. As such, it makes use of Windows XP-based software and a structured query language database.

Status: Over time, Air Target Sweden has continued to promote its family of target scoring systems.
Contractor: Air Target Sweden AB
Kista.

Switzerland

RUAG OPATS
Type: Automatic landing positioning sensor.
Development: The Object Position and Tracking System (OPATS - previously known as RAPS) was developed as a major element of a UAV landing system. As such, it was designed to alleviate the high cost of training ground-based pilots and to improve the safety and reliability of UAV landing procedures. OPATS was originally developed for use with the Ranger AV and can facilitate the automatic and safe landing of 'any' type of UAV, irrespective of its size.
Description: OPATS consists of a motorised sensor platform mounted on a tripod, an electronic unit and a battery pack platform. It is deployed in close vicinity to the landing area, remote from the GCS. It can be fully operated from the GCS, connected by fibre optic video and computer communications links. The air vehicle is equipped with two parallel, lightweight, 60 mm (2.4 in) diameter passive retro reflectors mounted on the air vehicle's nose or wings. A laser radar and TV camera on the OPATS platform point towards the approaching UAV, illuminating it by IR light pulses, which are echoed back to the OPATS by the retroreflectors. From these echoes, OPATS determines UAV distance and azimuth or elevation angles, while the TV camera gives the operator visual information. During the landing approach, OPATS continuously measures UAV positions (at a 'high rate' of 40 m/s) and transmits them to the GCS, where they are used as feedback in the automatic landing servo control loop; this enables the UAV to be precisely guided on to a predetermined landing point. OPATS operating modes comprise 'manual point' (manual pointing for search or tracking purposes), 'auto point' (GCS directed pointing), 'scan' (an automatic function over a 200 × 120 mrad window), 'auto track' (post acquisition automatic tracking) and 'standby' (system powered up, no laser emission).

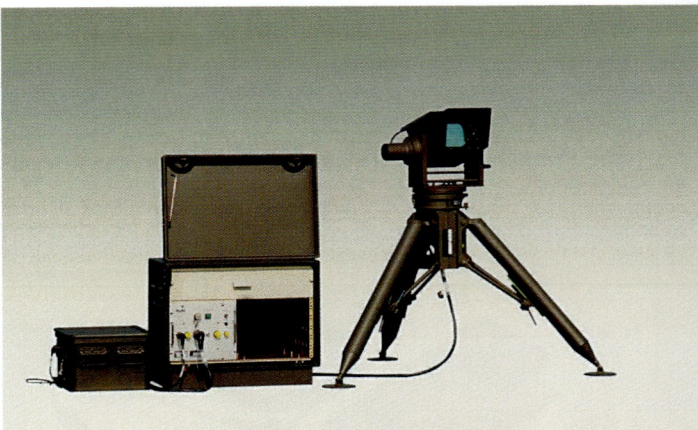

A general view of the OPATS electronic unit (left) and the system's tripod-mounted sensor *(RUAG)* 1329021

A close-up of the OPATS system's retro-reflectors mounted in the nose of a UAV *(RUAG)* 1329019

Specifications
Weights	
Weight, incl batteries	142 kg (313 lb)
Performance	
Measuring range: in azimuth	300° (5,236 mrad)
in elevation	−35 to +55° (−611 to +960 mrad)
Distance: tracking mode	35–3,500 m (115–11,500 ft)
search mode	150–3,500 m (492–11,500 ft)
Position measurement interval	40 ms
Angle measurements accuracy	±1 mrad (0.5 mrad typical)
Distance measurements accuracy	±1.5 m (4.9 ft)
Temperature: operating	−35 to +60°C (−31 to +140°F)
storage	−35 to +71°C (−31 to +160°F)
Power supply	24 V DC (battery)
Operating time	6–20 h (single battery charge)
Lateral object speed	20°/s (max)
Lateral object acceleration	5°/s2
Object radial speed	distance accuracy requirement dependent

Status: Over time, the OPATS automatic landing positioning sensor has continued to be promoted. OPATS is used with a 'diverse range of UAVs worldwide' (including RUAG's Ranger surveillance AV) and as of 2010, 90 OPATS systems were in service with more than 20 different end-users.
Contractor: RUAG Aviation
Emmen.

United Arab Emirates

Adcom ADCOM-3D
Type: UAV Flight Control Unit (FCU).
Description: Adcom describes the ADCOM-3D FCU as making use of the 'latest electronic technology' and 'advanced digital processing techniques' for control and measurement. System features include:
- altitude and heading reference correction (using a strap-down IMU with GPS and 3-axis magnetometer feedback)
- onboard flight data recording
- stabilisation and control of the host vehicle's angular movements
- an automatic navigation capability
- independent angular measurement on all three axes
- measurement of both static and dynamic pressure
- the ability to change pre-programmed flight plans at 'any stage in the flight'
- automatic take-off and landing facilities (with the latter making use of GPS and an optional laser altimeter)
- an automatic recovery function
- fail-safe functionality in case of uplink interruption
- fault identification
- automatic guidance to a pre-set landing point in case of emergency
- sensor thermal compensation
- use of an eCAN bus configuration to facilitate interfacing with servos, a transceiver unit, a laser altimeter and 'other peripherals'
- a 3-axis magnetometer interface
- the availability of two RS-232 serial ports
- control of up to 16 payloads.

Specifications
Dimensions	
Length	200 mm (7.87 in)
Width	110 mm (4.33 in)
Depth	50 mm (1.97 in)
Weights	
Unit	0.9 kg (1.98 lb)
Performance	
Angular speed	±150°/s (maximum)
Acceleration	±2 g (maximum)
Power supply	9-18 V (300 mA with over voltage and polarity protection)
Operating temperature	−20°C to +70°C

Status: Over time, ADCOM-3D has continued to be promoted.
Contractor: Adcom Systems
Abu Dhabi.

Adcom ADNAV GCS

Type: UAV GCS.

Development: IHS Jane's All the Worlds's Aircraft: Unmanned first became aware of the Adcom ADNAV GCS during July 2009.

Description: Manned (in baseline form) by two operators, the Adcom ADNAV GCS tracks, commands, controls and communicates with AVs and their payloads. As such, a single ADNAV GCS is able to simultaneously control up to seven AVs and an integral 'enhanced mission planner' provides 'flexible' tactical mission planning together with four modes of automatic AV navigation. In all, ADNAV incorporates three control bays that provide accommodation for a pilot, an AV navigator and (if required) an observer. The pilot's bay (as its designation suggests) provides AV flight control, while the navigator station is equipped with a digital map display that shows AV flight paths and allows the progress of missions to be monitored. The optional observer's position is used to monitor 'flight mission data'. The available operator displays include windows for data processing, AV parameter monitoring and GPS and INS status checking, among other things. The GCS system also features built-in redundancy, datalink redundancy, a dual power interruption safety mode, an artillery fire adjustment capability, a flight data recording facility (with playback capability), the provision of 'full' situational awareness, point and click AV navigation, automatic terrain avoidance warning and automatic fault monitoring.

Specifications

Performance
Control range.. Up to 80.9 n miles (150 km; 93.2 miles)

Status: As of mid-2009, Adcom (formerly Advanced Target Systems) is known to have been actively marketing the ADNAV UAV GCS.

Contractor: Adcom Systems
Abu Dhabi.

United Kingdom

Meggitt CASPA

Type: Integrated UAV avionics and autopilot solution.

Development: CASPA was originally developed for installation aboard Meggitt's Banshee and Voodoo AVs.

Description: Meggitt's Combined Autopilot and Surveillance Payload Avionics (CASPA) unit incorporates: a command receiver; a telemetry transmitter; a 3-axis, solid-state attitude sensor; airspeed and altitude pressure transducers and a GPS receiver. Autopilot functions/features include:
- auto-launch/auto-recovery
- automatic height/turn/throttle, height/turn/airspeed and height/heading/airspeed
- a sea-skimming radar altimeter mode
- waypoint, height hold and heading hold auto-navigation features
- backup via portable manual hand control unit.

Other system features include:
- programmable flight control settings (elevator, aileron and rudder)
- automatic software (flight termination, engine rpm counter, ignition kill relay, marine (option) and battery under-voltage detection)
- a direct data interface for local test/control and in-the-field software updating
- analogue and digital payload monitoring interfaces.

CASPA is available in standard and mini configurations, the mini-CASPA being described as smaller, lighter, cheaper and providing 'reduced complexity' when compared to the standard model. Accordingly, Meggitt promotes mini-CASPA as being an 'ideal' solution for semi-expendable target drone applications.

Specifications

Dimensions
No data available
Weights
No data available
Performance
Communications: Range........................ 54 n miles (100 km; 62 miles)
Typical carrier frequency... 400-465 MHz
Carrier spacing... 25 kHz
Modulation... FSK
Baud rate: Command receiver................................... 2,400
Telemetry transmitter... 2,400 or 9,600
Command receiver sensitivity.................................... <–110 dBm
GPS: Channels... 12 (simultaneous)
Update.. 1 s
Gyro: 3-axis rate sensor... ±300°/s
3-axis accelerometer sensor...................................... ±5 g/axis
Air data: Altitude.. 0-9,144 m (0-30,000 ft)
Airspeed.. 0-350 kt (0-648 km/h; 0-403 mph)
Power: Avionics supply... 12 V (<600 mA)
Temperature: Operating.......................–20°C to +55°C (–4°F to +131°F)

Status: Most recently, Meggitt has continued to promote CASPA for use aboard UAVs and targets.

Contractor: Meggitt Defence Systems Ltd UK
Ashford, Kent.

Meggitt Wizard

Type: Aerial target C2 ground station.

Description: The Wizard aerial target C2 ground station is capable of controlling up to four targets while simultaneously tracking and displaying up to four more at ranges of over 54 n miles (100 km; 62 miles). As such, the equipment is mounted in a 483 mm (19 in) rack and features a telemetry reception station, an LCD monitor, a data recorder, a command transmitter, a ruggedised PC, an uninterruptible power supply, a GPS receiver, an embedded trainer/simulator and post-flight telemetry analysis software. Displays include real-time telemetry/status and map and/or imagery and mission waypoints can be dragged and dropped using the architecture's mouse.

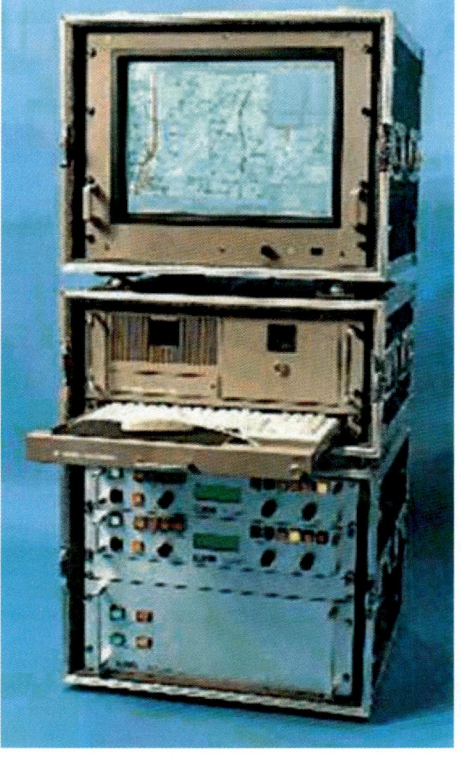

The Wizard command and control ground station (Meggitt)
0573766

Functionally, entire missions (including launch and recovery) can be performed automatically, with a combination of the air vehicle's autopilot, DGPS and pre-determined waypoints facilitating the use of pre-programmed 'accurate and reliable' flight profiles. Here, GPS waypoints and height and speed commands can be altered in-flight using the system's keyboard. Positional and telemetry data are displayed to the operator via the equipment's monitor, and laptop computer-generated mission scenarios can be run on the architecture's embedded trainer/simulator for mission planning and training purposes.

Status: Over time, the Wizard architecture is understood to have been supplied to the Turkish Air Force and Bahrain Defence Force for use with the Banshee target system. Most recently, Meggitt Defence Systems has continued to promote the Wizard C2 ground station.

Contractor: Meggitt Defence Systems Ltd UK
Ashford, Kent.

UTSL TCTS

Type: Tracking, Command and Telemetry System (TCTS).

Description: UTSL's low cost TCTS command, tracking and telemetry system is designed to facilitate target/UAS surveillance operations at ranges of more than 43.2 n miles (80 km; 49.7 miles) and utilises GPS/INS for positional data and platform sensors for air vehicle operational status. Data is transmitted to and from the ground station via a high-speed, spread spectrum, duplex, UHF (300 MHz to 3 GHz band) RF link, with received and processed information being displayed on an operator's high-definition colour monitor. An uninterruptable power supply can be incorporated if required and flight data is recorded on a disk for post-mission evaluation. The architecture provides audible warnings of critical system failures; facilitates sensor payload 'lock-on' to targets of interest; allows for air vehicles to be held in a holding pattern while

The UTSL TCTS stand-alone configuration *(UTSL)* 1179148

imagery is being recorded; and features a tracking/telemetry control console that provides the operator with mission programming, pre-flight check-out, multiple AV in-flight control and mission data recording/playback facilities. Mission programming options include flight pattern, target presentation legs and alternative courses, with all the relevant data being presented to the operator via menu-driven displays. In-flight control capabilities include launch, mission execution and vehicle recovery initiation. Associated data displayed to the operator include flight pattern, air vehicle position, air vehicle telemetry, air vehicle status, RF link condition and GPS satellite status. According to *IHS Jane's* sources, the described architecture is 483 mm (19 in) equipment racking compatible and is typically housed in a fixed-site or mobile protected environment.

Specifications

Performance	
Operating range (LOS/conditions dependent)	>43.2 n miles (80 km; 49.7 miles)
GPS: Channels	12
Band	L1
Update rate	Once/s
Frequency: Transmit carrier	Typically 300 MHz–3 GHz band (customer specified)
Receive carrier	Typically 300 MHz–3 GHz band (customer specified)
Link type	Full duplex
Modulation	Narrow bandwidth FM
Temperature: Operational	–5° to +35°C (23° to +95°F)
Storage	–10° to +55°C (14° to +131°F)
AV tracking data	Landing point, scaled map, take-off point, target/UAV position, target/UAV speed, target/UAV track heading, target/UAV tracking history and waypoints
AV status data	Altitude, battery levels, engine rpm, flight time, GPS status, satellite status, spare channels, target/UAV course bearing, target/UAV receive status, target/UAV speed and target/UAV telemetry transmit status

Status: Most recently, UTSL reports TCTS (as part of UTSL MSAT and GSAT aerial target systems) as being in service with both naval and land force customers.
Contractor: Universal Target Systems Ltd
Challock, Kent.

United States

AAI OSGCS

Type: UAV ground control station.
Description: Originally designed for use with AAI's Shadow® Tactical UAS, the One System® (the term 'One System' is a US Army registered trademark) Ground Control Station (OSGCS) has evolved into a NATO STANdardisation AGreement (STANAG) 4685 compliant architecture that is capable of handling a range of air vehicle types (see Status). As such, the OSGCS makes use of COTS components, is shelter-mounted and provides UAV control, data reception and data dissemination facilities. Over time, two OSGCS configurations have been identified, with known details including:

S-280 configuration: The S-280 configuration OSGCS is installed in a US Army S-280 climate-controlled shelter that is carried on a five ton, six-wheel, Family of Medium Tactical Vehicles (FMTV) chassis. As such, it was the One System® baseline during 2007. It provides air vehicle command and control, payload control, weapons launch control and imagery reception/dissemination facilities. Other system features include:
• net centric compatibility
• US Army Extended Range/Multi-Purpose (ER/MP) UAV system (MQ-1C Grey Eagle) compliance
• a modular, open design
• interoperability compliant (including US Army and Joint Information Exchange (JIE) requirements)
• use of a common, US Army Unmanned Aircraft System (UAS) compliant controller that is multiple UAS capable
• multiple datalink connections (including TCDL)
• simultaneous multiple mission control
• the ability to support air data relay, attack and reconnaissance/armed reconnaissance missions
• flight critical redundancy
• integrated automatic air vehicle launch/landing (by day or at night)
• an embedded training facility
• US Army software Block 2 compliant
• support for EO/IR, WAS, SAR and GMTI payloads
• real-time digital video imagery dissemination (with associated archive and retrieval facilities and including low bit rate functionality)
• automatic fault and air vehicle health monitoring
• transportability by land, air and sea
• support for 'full' US Army Unit Level Logistics System - Aviation (ULLS-A) and Integrated Electronic Technical Manual (IETM) implementation.

A general view of AAI's UGCS *(AAI Corporation)* 1414453

A close-up of the S-788 OSGCS configuration *(AAI Corporation)* 1414448

A general view of the S-788 OSGCS configuration (AAI Corporation) 1120721

S-788 configuration: Understood to be the system configuration used with the Shadow® UAV, the S-788 OSGCS configuration makes use of a US Army climate-controlled S-788 shelter that is mounted on a HMMWV chassis. As such, it is modular in design, incorporates redundant hardware and air vehicle operator and mission payload operator terminals and makes use of a UNIX-based operating system running 'mature' software. Other system features include:
- transportability by land, sea and air
- use of modular software
- Joint Technical Architecture - Army (JTA-A) compliance
- use of voice-over-Internet intercommunications
- an embedded training facility
- electronic pre-flight checklist and vehicle health monitoring
- automatic target search ('point at co-ordinates')
- rate position and automatic tracking
- automated marking and search areas
- an artillery correction capability (via in-video 'point and click')
- automatic terrain avoidance warning
- automatic fault monitoring
- a 'full' mission and payload planning capability
- flight data recording and mission playback
- 'point and click' navigation
- integrated automatic air vehicle launch/landing (by day or at night)
- a digital video image archive and retrieval/dissemination capability.

Status: AAI's One System® architecture entered service with the US Army during 2001, has been used operationally in Operations 'Iraqi Freedom' and 'Enduring Freedom' and has accumulated hundreds of thousands of hours of service use. UAVs supported by the system are reported as including the AAI Corporation Aerosonde® and Shadow 200, the AAI/IAI Pioneer®, the General Atomics ER/MP system, the Northrop Grumman Fire Scout and the Northrop Grumman/IAI Hunter. Notable programme events include expansion of the capability to accommodate the Hunter UAV during 2004, a late 2005 then year USD30 million award covering the supply of OSGCS technology into the General Atomics ER/MP (Grey Eagle) programme and a 2007 hard/software refresh to meet additional US Army unmanned aircraft interoperability requirements.

Elsewhere, AAI introduced a Universal GCS (UGCS) during 2009 that: is NATO STANAG 4586 compliant; meets 'emerging' joint service interoperability requirements; is capable of controlling multiple unmanned aircraft, land vehicles or surface vessels; is Conex/S-280/S-788 shelter configurable; and can be packaged for shipboard use.

Contractor: AAI Corporation (an operating unit of Textron Systems, a Textron Inc company)
Hunt Valley, Maryland.

AAI OSPGCS

Type: Portable UAV ground control station.
Description: AAI's One System® Portable Ground Control Station (OSPGCS) is described as performing all the functions of the 'full' One System Ground Control Station (OSGCS) but in a smaller, more portable package. As such, the architecture is designed to facilitate remote launch and recovery operations, thereby freeing the OSGCS for operations at tactical operations centres and/or forward operating bases. System features include:
- common OSGCS/OSPGCS displays
- an integral interface for a portable ground data terminal
- automated mission planning
- an air vehicle pre-flight capability
- use of 'mature' OSGCS software
- air vehicle status and combined map and video displays
- 'quick' set-up/tear-down times.

Status: AAI's One System architecture entered service with the US Army in 2001. It has been used operationally in southwest Asia and has accumulated hundreds of thousands of hours in service use. UAVs supported by the system are reported as including the AAI Corporation Aerosonde® and Shadow® 200, AAI/IAI Pioneer®, General Atomics ER/MP (Grey Eagle), Northrop Grumman Fire Scout and the Northrop Grumman/IAI Hunter.

Contractor: AAI Corporation (an operating unit of Textron Systems, a Textron Inc company)
Hunt Valley, Maryland.

AAI OSRVT

Type: Modular video and data system.
Description: AAI's One System® Remote Video Terminal (OSRVT) is described as being a modular video and data system that is capable of receiving surveillance imagery and geospatial data from joint operations TUAVs and the manned aircraft Litening electro-optic sensor system. In its manpack configuration, OSRVT incorporates a tablet display (using the Windows operating system), UHF/L/S/C and Ku band (300 MHz to 3 GHz, 1 to 2 GHz, 2 to 4 GHz, 4 to 8 GHz and 12.5 to 18 GHz) antennas and a multi-band receiver unit. Other system features include:
- modular hardware that can be configured for manpack, vehicle-mounted, stationary, airborne and maritime applications
- a joint annotated map/video display (with real-time video and map windows, providing 'intuitive presentations')
- portable software that will run on any Windows XP platform
- designed for use with touchscreen, touchpad, mouse or trackball interfaces ('no keyboard required')
- use of FalconView maps (with seamless export of target data to FalconView software on other platforms)
- a reception range of up to 43.2 n miles (80.0 km, 49.7 miles) when used with an associated extended-range antenna
- the ability to create Joint Photographic Experts Group (JPEG) snapshots for display in FalconView or any standard Windows image viewer.

Status: AAI's One System architecture entered service with the US Army in 2001. It has been used operationally in Operations 'Iraqi Freedom' and 'Enduring Freedom' and has accumulated hundreds of thousands of hours in service use. UAVs supported by the system are reported as including the AAI Corporation Aerosonde® and Shadow® 200, the AAI/IAI Pioneer®, the General Atomics ER/MP (Sky Warrior), the Northrop Grumman Fire Scout and the Northrop Grumman/IAI Hunter.

Design, development and demonstration of the OSRVT was covered by a USD7.5 million US Army contract. Elsewhere, 25 October 2006 saw the service award the company a full-rate production contract for 51 OSRVTs and 38 extended-range antennas. This award also included the provision for spares, training services and contractor logistic support. In 2007, the contract's period of delivery was expanded through July 2008. As of February 2008, the value of the contract stood at then year USD39.6 million. On 14 September 2007, AAI was awarded a then year USD13,197,998 modification to an existing firm, fixed-price contract with regard to the supply of OSRVTs and MultiDirectional Antenna Systems (MDAS). At the time of its announcement, work on the effort was scheduled for completion by 31 December 2009 and its contracting activity was the US Army's Aviation and Missile Command (Redstone Arsenal, Alabama). During late 2007, AAI received an additional then year USD20.2 million US Army award covering the production of 613 additional OSRVTs and 208 MDASs, with options for a further 1,517 OSRVT production units and 411 MDAS systems. As of February 2008 (and including the described order), the US Army has procured a total of 946 OSRVT production units and 361 MDAS units.

In April 2008, AAI received an additional then year USD14.5 million US Army award for additional spares, training services and contractor logistics support covering the period 1 May 2008 to 30 April 2009. In September 2008, AAI received another then year USD4.5 million US Army award relating to the production of 91 OSRVT and 52 MDAS units, bringing the total quantity procured to 1,037 production OSRVTs and 413 MDASs. In October 2008, the US Army began procurement of a further 1,358 OSRVTs, 207 MDASs and miscellaneous spares. The final negotiated award value was USD18.9 million, with then total procurement standing at 2,395 OSRVTs and 620 MDASs. In April 2009, AAI received a then year USD29.2 million award relating to contractor logistic support during the period 1 May 2009 to 30 April 2010. Moving forward, 8 March 2011 saw the US Army's Contracting Command (Redstone Arsenal, Alabama) awarded AAI a then year USD7,420,466 firm, fixed-price contract with regard to the acquisition of new receiver components for integration into the OSRVT baseline. At the time of its announcement, this effort was scheduled for completion by 31 August 2011. On the same day, the US Army Contracting command awarded AAI a second then year USD6,773,664 cost plus fixed fee contract with regard to the supply of 1,184 × OSRVTs. At the time of its announcement, this second programme was scheduled for completion by 31 October 2012. In terms of identified 2012 contracting activity, 29 August 2012 saw the US Army's Contracting Command award AAI a then year USD7,288,966 contract modification (to an existing cost plus fixed fee award) with regard to the provision of technical services relating to the OSRVT. At the time of its announcement, work on the effort was scheduled for completion by 20 August 2013.

Contractor: AAI Corporation (an operating unit of Textron Systems, a Textron Inc company)
Hunt Valley, Maryland.

BAE Systems NSU

Type: Navigation sensor unit.

Description: The BAE Systems NSU is part of the MIAG family of products and has been 'widely successful' when navigation, guidance and control are required and sensor size and weight are limited. The baseline configuration includes AHRS, embedded GPS receiver, embedded air data sensors, processing, and a variety of analogue/digital system interfaces.

Specifications

Dimensions (core)	
Height	127 mm (5.0 in)
Width	142 mm (5.6 in)
Depth	156 mm (6.2 in)
Weights	
Core weight	<2.5 kg (5.5 lb)
Performance	
Bus interface	1553B, RS-422, RS-232
Software	"C"
Temperature range	–54/+70°C
Acceleration	±17.6 g

Status: Over time, BAE Systems' NSUs have been procured for use aboard BQM-74, CL-327, Mirach 100/5E and other UAVs/targets.

Contractor: BAE Systems Electronics, Intelligence and Support
Santa Monica, California.

BAE Systems NSU *(BAE Systems)* 0079266

CDA TCDL

Type: X- (8 to 12.5 GHz) and Ku-band (12.5 to 18 GHz) Tactical Common DataLink (TCDL).

Development: The CDA TCDL is believed to have been originally developed for use in naval UAV applications.

Description: The Cubic Defense Applications (CDA) TCDL is described as being a digital, signal processing-driven datalink that is designed to provide a "cost-effective and highly accurate" means of transferring radar imagery and data from intelligence, surveillance and reconnaissance AVs to surface terminals. System features include:

- modular packaging to optimise installation aboard AVs that have volume and weight constraints ("multiple packaging form factors")
- programmable data rates (including the CDL 200 Kbps and 10.71 Mbps rates; up to 50 Mbps; in-flight adjustable)
- "flexible" interfaces (including Ethernet, fibre-optic, RS-232 and RS-422)
- "superior" receiver sensitivity to facilitate a 10-8 bit error rate
- a "clean" output spectrum for "improved" battlefield frequency management
- availability as a full-duplex transceiver or as a receive-only remote data terminal
- use of waveforms, message formats and data rates that are compatible with other CDL compliant terminals
- integral CDL Continuously Variable Slope Delta (CVSD) audio
- interchangeable RF modules to facilitate specifically required RF output powers
- availability of Modular Interoperable DataLink (MIDL) and Interoperable DataLink (IDL) control formats
- spread spectrum functionality for jamming resistance and low probability-of-intercept
- automatic acquisition and tracking of airborne platforms
- the ability to locate the architecture's surface antenna up to 500 m (1,640 ft) from the operator's terminal (other remote configuration options available)
- direct/alternating current or battery (optional) ground terminal functionality
- an options package that includes communications security, flexible input/output and video processing.

Specifications

Dimensions	
Unit volume	<6 litres (0.21 cu ft - excluding antenna/diplexer assembly)
Weights	
Unit	<4.5 kg (9.9 lb - excluding antenna/diplexer assembly)
Performance	
Frequency	8-12.5 GHz and 12.5-18 GHz or user selectable
Bit error rate	10-8
Power: Input	<125 W at 28 V DC
Output	2 W RF (higher RF outputs available)

Status: Over time, the CDA TCDL has continued to be promoted. Known applications include the Northrop Grumman MQ-8B Fire Scout and U-TacS WK450 (in collaboration with the United Kingdom contractor Ultra Electronics) UAVs. In the Fire Scout context, 16 June 2008 saw CDA announce that it had delivered three ground and nine airborne initial production TCDL terminals into the Fire Scout programme.

Contractor: Cubic Defense Applications
Orlando, Florida.

Curtiss-Wright IMMC

Type: Integrated mission management computer.

Development: Developed (originally by Vista Controls) for Global Hawk high-altitude endurance UAV (which see).

Description: The IMMC, a derivative of Vista's flight control navigation unit (FCNU), performs the aircraft's flight control and navigation function. There are two IMMCs per airframe.

The FCNU was designed to meet the emerging need for low-cost, open architecture, integrated GPS/INS hardware for navigation, guidance and flight control. It incorporates a VMEbus precise position service (PPS) or standard position service (SPS) GPS receiver and a full complement of VMEbus processor and input/output components. The unit is integrated with a solid-state IMU to provide all the basic inertial navigation system data: position, velocity, time, attitude, heading, angular rate and acceleration. By integrating the GPS receiver and the inertial navigation system, a higher performance is achieved with minimum size, weight and cost. The FCNU can also share inertial sensor data with other vehicle systems, which eliminates costly redundant sensors.

The GPS receiver function is implemented using a multichannel, L1, L2 P(Y) code and a C/A code-capable receiver with RTCM-104 differential capability. The GPS and inertial navigation solutions are then coupled using a centralised Kalman filter which aids the code tracking loop of the GPS receiver. The resulting GPS/INS navigation information is made available on the bus for other vehicle management functions.

Coupling the GPS and INS subsystems provides improved accuracy and stability, increased reliability, jamming immunity, and higher dynamic operation over the complete military operational environment. Data from other avionics sensors can also be used to aid the navigation solution. Vehicle attitude data are also available via an optional 1553 bus interface for use in ballistic computations and similar applications. The IMMC hosts the UAV flight control and vehicle management software and communications with other vehicle subsystems via a MIL-STD-1553B datalink, Ethernet, RS-422 and RS-232 serial datalinks, ARINC-429 interface and a variety of analogue and discrete input/output to perform flight control and management of the vehicle subsystems. The IMMC is designed to meet the rigorous environmental conditions demanded by the mission profile and is a prime application of commercial-off-the-shelf (COTS) technology, tailored for the military environment.

Status: The Curtiss-Wright IMMC has been installed aboard RQ-4 Global Hawk UAVs.

Contractor: Curtiss-Wright Controls Embedded Computing
Santa Clarita, California.

GA-ASI Advanced Cockpit GCS

Type: UAV GCS.

Description: The General Atomics Aeronautical Systems Inc (GA-ASI) Advanced Cockpit GCS is optimised for use with UAVs such as the company's MQ-1 Predator and MQ-9 Reaper and is claimed to offer 'significantly' improved situational awareness and reduced pilot workload. Innovations include an ergonomically optimised workstation with integrated and intuitive interfaces that are designed to make potentially hazardous situations easier to identify and to generally improve the decision making process. Other system features include:

The GA-ASI Block 50 Advanced Cockpit GCS provides an 'intuitive and ergonomically optimised work environment' that incorporates multiple touch screens of the types shown here *(GA-ASI)* 1166023

- the ability to be retrofitted into existing USAF, USN, US Department of Homeland Security and NASA fixed-site facilities and already fielded GCS shelters
- being based on a STANAG 4586 architecture to facilitate interoperability across various types of UAV
- an optimised crew station that features intuitive controls and information displays
- touch-screen technology
- the inclusion of anthropometric features such as adjustable displays, controls and ergonomic seating
- fused situational awareness data to facilitate a common operating picture on a single display
- compliance with MIL-STD-1472 and other human factor standards
- a synthetic 'out of the window' view that features graphical overlays, a 120° FoV with terrain avoidance and threats/special use air space indications embedded in 3-D graphics
- support for high-definition video feeds
- parallel, STANAG 4586-based core UAV control system/vehicle specific module development that will support quick-reaction system upgrade requirements.

GA-ASI is also noted as having established a 'strong, collaborative relationship' with national laboratories and academic institutions to validate the design of its Advanced Cockpit anthropometric and human-system interface.

Status: GA-ASI initiated internal Advanced Cockpit research and development *circa* March 2006 and is reported to have demonstrated a prototype during tests undertaken in July 2007. Such tests took place at GA-ASI's Gray Butte Flight Operations Facility (Palmdale, California), involved a mission-configured MQ-1 Predator® and demonstrated a Level 4 control capability. The manual launch and recovery mode was also tested, with the pilot 'hand flying' the AV using the Advanced Cockpit's stick and throttle commands. Hold modes were also tested. Elsewhere, June 2007 saw the USAF issue GA-ASI with a sole-source award with regard to an 'advanced cockpit' development effort for its MQ-1 and MQ-9 systems. GA-ASI further notes that the USAF has designated the Advanced Cockpit as the 'Block 50 Advanced Cockpit' and has funded the remaining system development and integration, with the work to be completed during the course of 2012.

Contractor: General Atomics Aeronautical Systems Inc
San Diego, California.

GA-ASI RVT

Type: Video receiver.
Description: The General Atomics Aeronautical Systems Inc (GA-ASI) Remote Video Terminal (RVT) is designed to provide the ground-based, shipboard or airborne warfighter with real-time imagery from a range of UAVs including GA-ASI's I-GNAT ER/Sky Warrior Alpha, Predator, Predator B and Sky Warrior. As such, the architecture comprises a receiver, antennas (including a 6 dBi, C-band (4 to 8 GHz) omni and an 18 dBi, 150 × 150 mm (5.91 × 5.91 in) directional unit), a PC display, a carrying case and a number of power options (10 to 28 V operation; BA-5590 type, AC adaptors or a DC/AC adaptor). It is backpack portable and incorporates an options range that includes an analogue/digital receiver, a test transmitter, a tracking antenna, additional antenna cabling, customisation kits and a moving map display format. Other system features include a GUI display, NTSC/RS-170 video display, automatic frequency scan and aircraft/target positioning decoding and display.

Specifications

Dimensions	
Receiver: Length (with battery)	274 mm (10.8 in)
Width (with battery)	145 mm (5.7 in)
Height (with battery)	84 mm (3.3 in)
Weights	
Receiver: with battery	3.2 kg (7.05 lb)
without peripherals	1.8 kg (3.97 lb)
Performance	
Frequency range	5.25-5.85 GHz (1.0 MHz steps)
Range	108 n miles (200 km; 124 miles)
Immersion	91 cm (3 ft) of water
Operating altitude	<4,575 m (<15,000 ft)
Operating temperature	–20°C to +50°C (–4°F to +122°F)

Status: Most recently, GA-ASI has continued to promote RVT-type technology.
Contractor: General Atomics Aeronautical Systems Inc
San Diego, California.

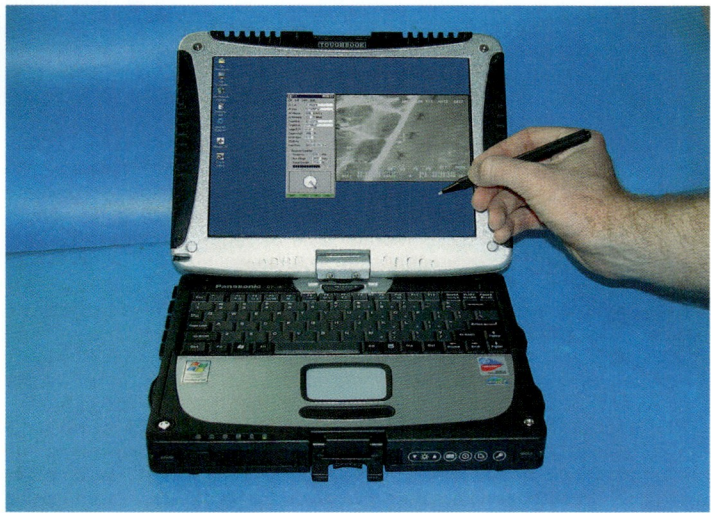

A general view of GA-ASI's RVT video terminal *(GA-ASI)* 1166022

Herley C² systems

Type: Family of UAV C² systems.
Description: Over time, Herley Industries (formerly Vega Precision Laboratories) has produced an ongoing series of UAV orientated C² systems in a wide variety of configurations that have ranged from visual control stations to computerised facilities. Here, ground stations have been created for fixed sites, transportable shelter, mobile and portable

Typical appearance of a Herley G-band GCS shelter 0518140

applications, with associated airborne equipment being electrically and mechanically configured for specific vehicle types. System operating range (up to 200 n miles (370 km; 230 miles) or up to 400 n miles (741 km; 460 miles) with a relay) and sophistication (relating to enhanced protection in electronic threat environments) has been dictated by user requirements. Herley Industries is further noted as having produced UAV applicable radar transponders, telemetry systems, flight termination systems, PCM encoding/decoding systems and radar altimeters.

Status: Over time, Herley C² systems have been procured to support a wide range of UAVs and targets including the AQM-37, BQM-34 Firebee, BQM-45, BQM-74/Chukar III, MQM-107 Streaker, MQM-107E and QF-4 types.

Contractor: Herley Industries Inc
Lancaster, Pennsylvania.

L-3 Communications CMDL

Type: 1.71 GHz to 15.35 GHz band airborne datalink modem.

Development: l-3 Communications' Compact Multi-band DataLink (CMDL) has been developed to provide duplex or simplex links for airborne (including UAV) and ground applications.

Description: The L-3 Communications CMDL modem is described as operating within the L- (1 to 2 GHz), S- (2 to 4 GHz), C- (4 to 8 GHz) or Ku- (12.5 to 18 GHz) bands and as providing a duplex or simplex link for ground or airborne platforms. As such, it is noted as being "small and rugged", as offering both analogue and digital waveforms and as being interoperable with the TCDL system and Remotely Operated Video Enhanced Receiver (ROVER) terminals (specifically, the ROVER III and "future ROVER terminals"). In terms of specific usage, L-3 Communications notes CMDL as being "ideal" for use in unmanned aerial vehicle, targeting pod and other airborne terminal (that co-ordinate with ground forces using ROVER and TCDL equipment) applications. Again, band and configuration are determined through Radio Front-End (RFE) and antenna selection and system configurations include duplex, transmit-only and receive-only. Functionally, CMDL provides the ability to transmit and receive sensor data, imagery, streaming video and discrete and command and control data. It is further claimed to "enhance" air-to-ground co-ordination; "reduce" time-on-target for close air support and to be able to provide situational awareness, targeting, surveillance and battle damage assessment. As such, functional applications include convoy protection and other time critical operations where "eyes on target" are required. Other system features include a "compact" and "low" size, weight and power profile; a data rate of up to 45 Mbit/s (data rate can be customer specified); "multiple" video formats; forward error correction; a live streaming video capability and an encryption facility. RFE options comprise L- (transmit only, 1 W power amplifier (external amplifier can be added if required for higher output and duplex operation), internal to modem configuration), S- (transmit only, 2 W power amplifier (external amplifier can be added if required for higher output and duplex operation), internal to modem configuration), C- (transmit only, 25 W power amplifier, external configuration, 100 W power consumption or duplex, 25 W power amplifier, external configuration, 120 W power consumption) and Ku-band (duplex, 15 W power amplifier, external configuration, 120 W power consumption) assemblies.

Specifications

Dimensions	
Modem: Length	135 mm (5.3 in)
Width	109 mm (4.3 in)
Depth	43 mm (1.7 in)
Weights	
Modem	1.27 kg (2.8 lb)
Performance	
Frequency bands	1.71-1.85 GHz (0.5 MHz steps), 2.20-2.50 GHz (0.5 MHz steps), 4.40-5.00 GHz (1.0 MHz steps), 5.25-5.85 GHz (1.0 MHz steps) and 14.40-15.35 GHz (1.0 MHz steps)
Modulation formats	BPSK, FM analogue video, FSK, O-QPSK and QPSK
Transmit/receive data rates	200 kbit/s, 400 kbit/s, 455 kbit/s (H.261, FSK only), 466 kbit/s (MPEG-4, FSK), 1.6 Mbit/s, 2.0 Mbit/s, 3.2 Mbit/s, 10.71 Mbit/s, 21.42 Mbit/s, 44.73 Mbit/s and up to 45 Mbit/s (custom maximum)
Forward error correction	Rate ½ concatenated (with Reed-Solomon and byte interleaver), rate ½ convolutional (with bit interleaver) and turbo codes
Video	H.261 (decode only), H.264 (encode/decode), MPEG-2 and -4 (encode/decode), NTSC/PAL (decode only) and RS-170 (encode/decode)
Networking	CDI Annex B (Packet Mux compliant) and IPv4/IPv6 (future release)
Encryption	AES (future release) and triple DES
Interfaces	10/100 base-T Ethernet, limited discreet I/O (platform, RFE and antenna status), RS-170 (video), RS-232 (navigation/GPS data) and RS-422 (antenna control, future release)
Humidity	0-100% (condensing)
Altitude	Up to 15,240 m (50,000 ft)
Vibration	MIL-STD-810F compliant
Temperature	–40°C to +65°C (–40°F to +149°F)
EMI	MIL-STD-461 compliant
Power	30 W (modem)
Input power	28 V DC (modem, MIL-STD-704 compliant)

Status: Over time, L-3 Communications has continued to promote the CMDL.

Contractor: L-3 Communication Systems-West
Salt Lake City, Utah.

L-3 Communications Global Hawk ICS

Type: Integrated communications system for the RQ-4 Global Hawk UAV.

Description: The Global Hawk ICS (Integrated Communications System) comprises a Common Airborne Modem Assembly (CAMA), a satcom Radio Frequency Assembly (RFA), a High Voltage Power Supply (HVPS), a High Power Amplifier (HPA), a satcom antenna, a LOS RFA, a LOS dual-band antenna, two UHF transceivers, two UHF Power Amplifiers (PA), two LNA/diplexers and two UHF antennas. When paired with the appropriate surface and satellite terminals, the Global Hawk ICS provides the RQ-4A with a 'complete' communications package that includes a CDL compatible, full duplex, wideband, air-to-ground datalink; a full duplex satcom datalink; and a redundant, full duplex, UHF satcom/LOS command and control link.

Specifications

Dimensions	
CAMA: Length	437 mm (17.2 in)
Width	427 mm (16.8 in)
Height	488 mm (19.2 in)
Satcom RFA: Length	412 mm (16.2 in)
Width	264 mm (10.4 in)
Height	302 mm (11.9 in)
HVPS: Length	475 mm (18.7 in)
Width	259 mm (10.2 in)
Height	196 mm (7.7 in)
HPA: Length	572 mm (22.5 in)
Width	264 mm (10.4 in)
Height	236 mm (9.3 in)
Satcom antenna: Length	1,387 mm (54.6 in)
Width	1,240 mm (48.8 in)
Height	1,240 mm (48.8 in)
LOS RFA: Length	539 mm (21.2 in)
Width	356 mm (14.0 in)
Height	132 mm (5.2 in)
LOS antenna: Length	371 mm (14.6 in)
Width	259 mm (10.2 in)
Height	259 mm (10.2 in)
UHF transceiver: Length (each)	249 mm (9.8 in)
Width (each)	127 mm (5.0 in)
Height (each)	142 mm (5.6 in)
UHF PA: Length (each)	254 mm (10.0 in)
Width (each)	127 mm (5.0 in)
Height (each)	178 mm (7.0 in)
UHF LNA/diplexer: Length (each)	152 mm (6.0 in)
Width (each)	137 mm (5.4 in)
Height (each)	31 mm (1.2 in)
UHF satcom antenna: Length	404 mm (15.9 in)
Width	330 mm (13.0 in)
Height	213 mm (8.4 in)
Weights	
CAMA	38.6 kg (85.0 lb)
Satcom RFA	11.8 kg (26.0 lb)
HVPS	18.1 kg (40.0 lb)
HPA	25.4 kg (56.0 lb)
Satcom antenna	20.9 kg (46.0 lb)
LOS RFA	15.9 kg (35.0 lb)
LOS antenna	4.1 kg (9.0 lb)
UHF transceiver (each)	5.9 kg (13.0 lb)
UHF PA (each)	6.4 kg (14.0 lb)
UHF LNA/diplexer (each)	1.4 kg (3.0 lb)
UHF satcom antenna	3.6 kg (8.0 lb)
Performance	
Frequency: CDL LOS link	8–12.5 GHz
Ku-band satcom link	12.5–18 GHz
UHF satcom/LOS link	300 MHz–3 GHz
Forward link: CDL LOS link	200 Kb/s (composite rate)
Ku-band satcom link	200 Kb/s (composite rate)
Forward/return link	
UHF satcom link	1.2 Kb/s (composite rate)
UHF LOS link	9.6 Kb/s (half duplex)
Return link	
CDL LOS link	10.71, 137 or 274 Mb/s (data rates)
Ku-band satcom link	1.544–50 Mb/s (data rate)
Power: CAMA	310 W
Satcom RFA	78 W
HVPS	1,800 W
HPA	33 W
Satcom antenna	35 W
LOS RFA	495 W
LOS antenna	7 W
UHF transceiver (each)	150 W
UHF PA (each)	700 W
UHF LNA/diplexer (each)	7 W

Status: L-3 Communications' Global Hawk ICS is known to have been installed aboard the RQ-4 Block 0/Block 10 (formerly RQ-4A) Global Hawk UAV and (over time) has continued to be promoted.

Contractor: L-3 Communication Systems-West
Salt Lake City, Utah.

L-3 Communications KuSDL

Type: Ku-band (12.5 to 18 GHz) satellite communications data link system.

Description: Designed specifically for use with the MQ/RQ-1 Predator, the L-3 Communications Predator KuSDL (Ku-band Satellite communications Data Link) utilises commercial, geostationary satellites to effect full-duplex satellite communications linking the MQ/RQ-1 to a remote control/exploitation complex. The command link provides real-time control and data, while the return link transfers real-time electro-optic, infra-red or synthetic aperture radar motion video (VQ compressed) to the exploitation facility. As such, the complete architecture comprises ground, satellite and UAV terminals, with the former providing the necessary mission control and exploitation; the satellite terminal, 'bent pipe' relay facilities and the UAV equipment, low-level Earth resources collection.

Specifications

Dimensions	
Aperture: ground terminal	5.5–6.2 m (18–20 ft)
satellite aperture	Existing satellite dependent
UAV	762 mm (30 in)
Performance	
Channel format	
all terminals forward	SS-QPSK; ½ FEC; 1024 sym; intlv depth (8,000 Mchip/s, spread)
all terminals return	O-QPSK; concatenated FEC; RS-231/-247; intlv 8 × RS words; conv (½ 7)
Bandwidth requirement	
all terminals forward	Spread spectrum command within 9 MHz allocation
all terminals return	Coded telemetry/imagery within 5 MHz allocation
Information rate	
all terminals forward	200 kb/s (includes TBD reserve capacity)
all terminals return	2 × T1 (3,200 Kb/s) composite TLM/imagery or T1 (1,600 Kb/s)
ERIP: ground terminal forward	< 67 dBW
satellite terminal forward	≈ 31 dBW
satellite terminal return	≈ 23 dBW
UAV terminal return	< 55.5 dBW
GT: ground terminal return	30–31 dB/K
satellite terminal forward/return	4.5 dBW
UAV terminal forward	12 dB/K
Carrier frequency	
ground terminal forward Tx	14–14.5 GHz
ground terminal return Rx	10.95–12.75 GHz
UAV terminal forward Rx	10.95–12.75 GHz
UAV terminal return Tx	14–14.5 GHz
Transmit power: ground terminal	125 W
satellite terminal	35–50 W
UAV terminal	50 W
Signal format	
ground terminal	RS-422 I/F, custom (command/telemetry); RS-422 I/F, 5 Mb/s par (SAR); NTSC, analogue (EO/IR video)
UAV terminal	RS-422 I/F, custom (command/telemetry); RS-422 I/F, 5 Mb/s par (SAR); NTSC, analogue (EO/IR video)
Microwave sensing	
ground terminal	Display, exploitation console and database
UAV terminal	WEC/SAR at 16.4 GHz

Status: L-3's KuSDL architecture is known to have been installed aboard the MQ-1 Predator UAV and (over time) has continued to be promoted.

Contractor: L-3 Communication Systems-West
Salt Lake City, Utah.

L-3 Communications MUDL

Type: 1,755 to 1,850 MHz band datalink.

Development: The L-3 Communications Mini UAV DataLink (MUDL) was first identified by *Jane's* during 2008.

Description: The MUDL is a small, lightweight, affordable, modular and scalable datalink that is designed for mini UAV applications with wing spans down to 610 to 914 mm (24 to 36 in). As such, it can be tailored in terms of mechanical packaging and functionality to meet specific requirements and is billed as providing 'guaranteed UAV control, high quality Moving Picture Experts Group (MPEG) - 2/4 or other format video transmission, data security, flexible data types and adaptive data rates with flexible bandwidths for extended range'. Other system features include:

- provision of in-flight communications to and from the host UAV (with the information footprint including waypoint updates, AV collaborative messaging, transmitted sensor data, host AV position and status date and other required custom data)
- LOS and BLOS functionality
- 'adaptable and expandable' data rates across the 1 kbit/s to 10.71 Mbit/s range
- CDL compatibility
- use of a software programmable architecture that facilitates flexibility in waveform choice and the ability to introduce new waveforms without hardware changes
- secure data transfer
- an options/growth path package that includes a Peripheral component interconnect Mezzanine Card (PMC) daughter card for video and sensor imagery compression; custom form factors; a Joint Tactical Radio System (JTRS) waveform capability; a BLOS satellite communications capability; TCDL compatibility; ROVER remote video terminal compatibility and 30 MHz to 3 GHz and 12.5 to 18 GHz frequency options.

Specifications

Dimensions	
Unit: Length	125.7 mm (4.95 in)
Width	77.5 mm (3.05 in)
Depth	17.8 mm (0.70 in)
Weights	
Unit	<0.2 kg (0.5 lb)
Performance	
Baseline frequency	1,755–1,850 MHz
Baseline data rate	30 kbit/s (full duplex, expandable)
Interfaces	Ethernet, GPS (1 pulse/s), PMC connector, RS-232, RS-422 (× 2), USB and video (growth option)

Status: Over time, the L-3 Communications MUDL has continued to be promoted.

Contractor: L-3 Communication Systems-West
Salt Lake City, Utah.

L-3 Communications ROVER series

Type: Family of portable sensor data terminals.

Description: Known details of the L-3 Communications-West Remotely Operated Video Enhanced Receiver (ROVER) family are as follows:

ROVER 1: ROVER 1 is described as being an experimental iteration that was based on a C-band (4 to 8 GHz) receiver and which facilitated the relay (via an AC-130 gunship aircraft) of Predator UAV derived imagery to a command post 'in the rear'.

ROVER 3: The receive-only ROVER 3 terminal is described as providing real-time, full-motion video for situational awareness, targeting, Battle Damage Assessment (BDA), surveillance and convoy support applications where an 'eyes on target' capability is required. System features include:

- L- (1 to 2 GHz), C- and Ku- (12.5 to 18 GHz) functionality
- Ku-band and L/C-band omni-directional and Global Positioning System (GPS) antennas
- rucksack portability
- a laptop computer-based man-machine interface
- a software package that facilitates pre-mission configuration, automatic frequency acquisition, multiple platform access, user position identification (via an integral GPS capability) and Moving Picture Experts Group (MPEG) - 2 and H.261 video display
- a wireless access point (to facilitate un-tethered operation)
- an options package that includes a directional antenna (for increased range), integrated communications security, integrated MPEG-2 and a ruggedised integrated display.

ROVER 4: The receive-only ROVER 4 terminal provides real-time, full motion video for situational awareness, targeting, BDA, surveillance, convoy operations and other situations where 'eyes on target' are required. Again, ROVER 4 has 'proven' interoperability with datalinks in the L-, S- (2 to 4 GHz), C- and Ku-bands; UAVs such as the Dragon Eye, Predator and Shadow; the LITENING EO targeting pod (when used as a surveillance tool) and 'other joint and coalition assets'. ROVER 4 forms a complete, ready-to-use system that is housed in a rugged transit case and includes the necessary antennas and cabling, a video display, a recording capability and a wireless access point. Other system features include:

An view of the ROVER 3 receive-only sensor data terminal that shows the architecture's laptop computer display (centre) and receiver unit (lower left) *(L-3 Communications)*

- digital (C- and Ku-band) and analogue (L-, S- and C-band) reception
- an antenna package that comprises Ku-band and L-, S- and C-band omni-directional elements (both with an integral low noise amplifiers and Direct Current (DC) power via a radio frequency cable)
- pre-mission configuration software and automatic frequency acquisition
- software-configurable waveforms, bands/frequencies and video protocols
- analogue, H.261, MPEG-2 and MPEG-4 video display software (with an integral H.261/MPEG-2/MPEG-4 decoder)
- a digital video recording capability (standard .wmv file format)
- Key Length and Value (KLV) Metadata mapped to Falcon View
- a BA-5590 form factor battery or DC power
- a directional C-band antenna as an option.

ROVER 5: The handheld ROVER 5 transceiver device is described as being a 'small, lightweight and rugged' software defined radio that provides a digital capability for full motion video, situational awareness, targeting, BDA, surveillance, convoy overwatch and other situations where 'eyes on target' are required. The device is further noted as having both forward (via radio and video codec software uploads) and backward (with both the ROVER 3 and 4 terminals; the Dragon Eye, Predator and Shadow UAVs such terminals support; EO targeting pods such as the LITENING equipment and 'other joint and coalition assets') compatibility, as being capable of transmitting time sensitive targeting data and as being able to display sensor data from 'multiple airborne platforms'. Other system features include:
- an encryption optimised design
- an integrated 142 mm (5.6 in), sunlight readable video display
- a menu-driven user interface (with touch screen and button control)
- forward error correction
- digital and analogue waveforms
- support for KLV Metadata
- an 'intuitive control' graphical interface that facilitates pre-mission configuration and waveform, band and frequency control.

ROVER 5i: The handheld ROVER 5i device is described as being a 'revolutionary handheld transceiver' that 'transforms the traditional ISR receive-only device into a two-way, wideband, [Internet Protocol (IP) enabled] tool'. As such, ROVER 5i is further noted as providing 'enhanced' air and ground co-ordination, shortened talk-on target times, IP networking, secure interoperability, backwards compatibility with ROVER III/4 and digital full motion video.

ROVER 6: The tri-service ROVER 6 transceiver is described as providing communications relay and real-time full motion video (and other data) for situational awareness, targeting, BDA, surveillance, convoy overwatch and 'eyes on target'. Other system features include:
- Ultra High Frequency (UHF - 400 to 470 MHz sub-band), L-, S-, C- and Ku-band functionality
- a transmit capability with external transmitter control and transmitter amplifier blanking/enabling signal capabilities
- two simultaneous reception channels (same or different bands, diversity reception with two receive antennas, single data source and two external receiver interfaces)
- AES, Triple DES and Type 1 (pending) secure digital communications
- BA-5590 battery or a battery eliminator for AC or DC (10-32 V DC, 40 W, approximate) input
- Web-browser graphical user interface control.

Specifications

Dimensions
 Receiver: ROVER 3/4............97 × 140 × 394 mm (3.8 × 5.5 × 15.5 in - including battery)
 Terminal: ROVER 5/5i............241 × 142 × 57 mm (9.5 × 5.6 × 2.25 in - including antenna)
Weights
 Receiver: ROVER 3............<3.6 kg (<8 lb - excluding peripherals); 5.4 kg (12 lb - with battery); 22 kg (48 lb - system, approximate)
 ROVER 4............3.6 kg (8 lb - excluding peripherals); 4.7 kg (10.25 lb - with battery); 22 kg (48 lb - system, approximate)
 Terminal: ROVER 5............1.6 kg (3.5 lb)
 ROVER 5i............1.8 kg (3.9 lb)
 ROVER 6............4.5 kg (9.9 lb, without battery)
Performance
 Frequency range: ROVER 3....1,710-1,840 MHz (0.5 MHz steps), 4.40-5.85 GHz (1 MHz steps) and 14.40-15.35 GHz (5 MHz steps)
 ROVER 4............1.71-1.85 GHz (0.5 MHz steps), 2.30-2.50 GHz (1.0 MHz steps), 4.40-5.85 GHz (1.0 MHz steps), 5.25-5.58 GHz (1.0 MHz steps) and 14.40-15.35 GHz (5.0 MHz steps)
 ROVER 5............400-470 MHz, 1.71-1.85 GHz (0.5 MHz steps), 2.20-2.50 GHz (0.5 MHz steps), 4.40-4.950 (1.0 MHz steps), 5.25-5.85 GHz (1.0 MHz steps) and 14.40-15.35 GHz (1.0 MHz steps)
 ROVER 5i............400-470 MHz, 1.71-1.85 GHz (0.5 MHz steps), 2.20-2.50 GHz (0.5 MHz steps), 4.40-4.940 (1.0 MHz steps), 5.25-5.85 GHz (1.0 MHz steps) and 14.40-15.35 GHz (1.0 MHz steps)
 ROVER 6............400-470 MHz (1 kHz steps), 1,625-1,850 MHz (0.5 MHz steps), 2.20-2.50 GHz (0.5 MHz steps), 4.40-4.95 GHz (1.0 MHz steps), 5.25-5.85 GHz (1.0 MHz steps), 14.40-14.83 GHz (1.0 MHz steps) and 15.15-15.35 GHz (1.0 MHz steps)
 Antenna gain: ROVER 3............0 dBi (L-band) and >3 dBi (C- and Ku-band)
 ROVER 4............>0 dBi (S-band, 0°-60° above the horizon), >1 dBi (L-band, 0°-60° above the horizon and Ku-band, 20°-60° above the horizon) and ≥1 dBi (C-band, 0°-60° above the horizon)
Interfaces: ROVER 3/4............BA-5590 battery connector, coaxial TNC antenna RF connector, COMSEC fill connector, Ethernet (10/100 Base-T), front panel (four control buttons, eight digit display), NEMA-0183 (GPS interface), RS-170 (NTSC video) and RS-485 (directional antenna control)
 ROVER 5............audio input/output, BNC external video input/output, Ethernet (10/100 Base-T), external power connector, RS-232 (× 2), RS-422 (× 2, controllable direction) and USB 2.0 (file storage)
 ROVER 5i............audio input/output, BNC external video input/output, Ethernet (10/100 Base-T), power connector, RS-232 (× 2, configurable, 1 × RS-422 as alternative), RS-422 (× 2) and USB 2.0 (file storage)
 ROVER 6............DC bias RF receive (dual, for external LAN), Ethernet (10/100 Base-T, IPv4 networking), external directional antenna control (dual), external power amplifier, local GPS input, remote GPS output, RS-232 (1 × user channel) and RS-422 (1 × user channel), triaxial video in/out
Power: ROVER 3............1 × BA-5590 battery (10-12 h operation) or AC/DC adapter
 ROVER 4............1 × BA-5590 battery (10-12 h operation) or battery eliminator (95-270 V AC (47-440 Hz) or 11-36 V DC)
 ROVER 5............1 × lithium-polymer battery (2.5-3.0 h operation, estimated) or battery eliminator (95-270 V AC (47-440 Hz) or 9-36 V DC)
 ROVER 6............1 × BA-5590 battery or battery eliminator (AC or DC (10-32 V DC, 40 W, approximate) input)
Altitude: ROVER 3/4............<4,572 m (<15,000 ft, operating)
 ROVER 6............9,144 m (30,000 ft, operating)
Temperature: ROVER 3/4............–20°C to +70°C (–4°F to +158°F, operating)
 ROVER 5............0°C to +45°C (+32°F to +113°F)
 ROVER 6............–20°C to +60°C (–4°F to +140°F, ambient, operating), –20°C to +70°C (–4°F to +158°F, cold plate or forced air, operating) or –20°C to +85°C (–4°F to +185°F, non-operating)
Shock: ROVER 3/4............9 g (11 ms, half-sine, operating)
 ROVER 6............20 g (11 ms, terminal sawtooth peak, operating)
Immersion: ROVER 3/4............up to 0.91 m (3.00 ft) of water for 30 minutes
 ROVER 6............1 m (3.28 ft) of water for 30 minutes

Status: As of June 2006, in excess of 700 ROVER 3 systems are reported to have been delivered to the United States Air Force, Army, Marine Corps and Special Forces, with the equipment being able to receive imagery from sources such as the Predator UAV, the C-130 mounted 'Scathe View' EO sensor system and the SNIPER EO targeting pod. In terms of specifically identified ROVER contracting activity, 31 August 2007 saw L-3 Communication Systems - West being awarded a then year USD16,328,300 firm, fixed-price contract with regard to the supply of ROVER 3 terminals. At the time of its announcement, work on the effort was scheduled for completion by 30 April 2008 and its contracting activity was the US Army's Aviation and Missile Command (Redstone Arsenal, Alabama). Moving forward, 7 October 2008 saw L-3 being awarded a then year USD26,649,683 firm, fixed-price contract with regard to the supply of an unspecified number of ROVER equipments. At the time of its announcement, work on the effort was scheduled for completion by 30 December 2009 and as before, its contracting activity was the US Army's Aviation and Missile Command. Returning to the ROVER 3 configuration, September 2007 saw L-3 reporting that approximately 2,000 such equipments had been delivered to customers worldwide (with the offshore procurement base including Australia, Canada, Norway (40 units) and the United Kingdom (40)). As of the given date, an additional 1,000 ROVER 3 terminals were noted as being 'under contract'.

In terms of ROVER 6 activity, 18 November 2010 saw the US Army's Aviation and Missile Command award L-3 Communication Systems - West a then year USD38,532,487 firm, fixed-price contract with regard to the supply of 1,181 ROVER 6 'upgrade kits'. At the time of its announcement, work on this effort was scheduled to have been completed by 31 January 2012. Five months later (8 April 2011), the US Army's Contracting Command (Redstone Arsenal, Alabama) awarded L-3 Communication Systems - West a then year USD34,299,296 firm, fixed-price contract with regard to the procurement of 1,184 ROVER 6 terminals. At the time of its announcement, the effort was scheduled for completion by 30 April 2012.

Contractor: L-3 Communication Systems-West
Salt Lake City, Utah.

MSI MONTAGE

Type: Ground control station.

Description: The Micro Systems Inc (MSI) MOdular Networked TArGet control Equipment (MONTAGE) provides control for target drones based in South Korea, Spain, Taiwan, the UK and the USA. It has a networked approach which permits operators to fly different targets with the same system, even simultaneously. Customers can start with a single target control system and add additional modules later. It can control up to eight targets.

MONTAGE is a multiple target GPS-band ground control station with UHF datalink, transponders, RF relays, installation kits and test equipment. It is available in fixed, portable, shelter-mounted, transportable and airborne fixed mount configurations. It features PC-control consoles using Windows operating system as well as moving map displays. Differential GPS provides a typical accuracy of 1 m (3 ft) and >10 g operation without fixed site range tracking.

The UHF datalink is a low-power (5 W) unit operating with a PRF of 10 Hz and programmable channel assignments. The frequencies are selectable, and non-UHF frequencies can be incorporated. With an airborne relay the range of the system can be extended to 400 n miles (740 km; 461 miles) but the target transponder can also act as a relay as well as providing drone control.

Micro Systems provides a complete range of support and logistics including installation and training.

Status: Over time, the MONTAGE GCS has been procured by the USN (as System for Naval Target Control: SNTC), the US Army (the service's 'primary' air target control system) and customers in South Korea, Spain, Taiwan and the United Kingdom (Flight Refuelling Falconet at QinetiQ's Hebrides range). The system operates with a variety of targets including BQM-74E, Chukar III, BQM-34S, MQM-107, QUH-1 and Falconet. Again (and over time), MONTAGE has continued to be promoted.

Contractor: Micro Systems Inc (a Herley Industries company)
Fort Walton Beach, Florida.

Northrop Grumman LN-100G

Type: INS/GPS navigation sensor.

Description: Northrop Grumman's LN-100G INS/GPS navigation sensor combines the company's Zero-Lock™ laser gyro with INS and GPS electronics and is capable of providing hybrid GPS/INS, free inertial and GPS-only navigation solutions simultaneously. As such, the LN-100G architecture can accommodate a number of GPS receiver solutions including Standard Positioning System (SPS), Precise Positioning Service (PPS) and GPS receiver application module/selective availability anti-spoofing module options. Other system features include:
- two space card slots for the addition of analogue input/output modules, ARINC interfaces and 'other' expansion modules
- ARINC 429, dual Military Standard 1553B (× 2) and RS-422 digital interfaces (2-wire synchro/3-wire analogue and range/bearing options)
- validated, Ada-based software
- free convection cooling
- use of a non-dithered ring laser gyro (no acoustic noise, no SAR jitter).

Specifications

Dimensions
Inertial navigation unit: Length	279 mm (11.0 in)
Width	178 mm (7.0 in)
Depth	178 mm (7.0 in)
Optional mount: Length	348 mm (13.7 in)
Width	183 mm (7.2 in)
Height	19 mm (0.75 in)

Weights
Inertial navigation unit	9.8 kg (21.61 lb)
Optional mount	1.1 kg (2.43 lb)

Performance
Position
4-minute gyrocompass align; inertial only	0.8/0.6 n miles/h (1.5/1.1 km/h; 0.9/0.7 miles/h)
4-minute gyrocompass align; GPS/inertial	10 m (32.8 ft) CEP
4 + 4-minute EIA align; inertial only	0.5 n miles/h (0.9 km/h; 0.6 miles/h)
4 + 4-minute EIA align; GPS/inertial	10 m (32.8 ft) CEP
After loss of GPS; GPS/inertial	120 m (393.7 ft)/20 minutes
Velocity/axis; inertial only	0.76 m/s (2.5 ft/s) RMS
Velocity/axis; GPS/inertial	0.015 m/s (0.049 ft/s) RMS
Attitude (pitch/roll/azimuth); inertial only	0.05° RMS
Attitude (pitch/roll/azimuth); GPS/inertial	0.02° RMS
All axes acceleration	16 g
All axes attitude	Unlimited
Roll/pitch/azimuth rate	>400°/s
Roll/pitch/azimuth acceleration	>1,500°/s2
Power	37.5 W (28 V DC)
Altitude	640-21,336 m (2,100-70,000 ft)
Temperature	–54/+71°C (–65/+160°F)
Vibration: Random	8.1 g RMS
Sine	±5 g (5-2,000 Hz)
Shock	21 g (40 ms)
Acoustic noise	140 dB

PPS/SPS GPS receiver: Operating
frequency	L1/L2
Antispoof/enhanced	P(Y) code/receiver
Anti-jam	Aiding
Channels	12 or all-in-view
Service life	20+ years/8,100 h
MTBF	14,400 h

Status: Over time, LN-100G has continued to be promoted for UAV applications, with known installations including those aboard the Mariner, MQ-9/Predator B, Predator, and RQ-4 Global Hawk air vehicles.

Contractor: Northrop Grumman Corporation, Navigation Systems Division
Woodland Hills, California.

Northrop Grumman LN-200

Type: Fibre-optic Inertial Measurement Unit (IMU).

Description: The LN-200 IMU incorporates inertial fiber-optic gyro and micro-machined accelerometer technology and is billed as being suitable for space vehicle stabilisation, missile guidance, radar/EO/FLIR stabilisation, motion compensation, unmanned underwater vehicle/UAV guidance and control, camera/mapping and IMU-for-higher order system applications. Other system features include:
- hermetic sealing
- no moving parts or gaseous cavities
- an RS-485 serial databus input/output
- use of three by solid-state fibre-optic gyros and three by solid-state silicon accelerometers.

Specifications

Dimensions
Unit: Depth	89 mm (3.50 in)
Height	85 mm (3.35 in)

Weights
Unit	<750 g (1.65 lb)

Performance
Accelerometer: Bias repeatability	300 μg-3.0 mili-g
Scale factor accuracy	300-5,000 ppm
Gyro: Bias repeatability	1°/h-10°/h
Scale factor accuracy	100-500 ppm
Random walk	0.07-0.15°/sq root h
Angular rate	up to ±11,459°/s
Angular acceleration	±100,000°/s2
Acceleration	>70 g
Angular attitude	Unlimited
MTBF	>20,000 h
Nominal steady-state power	12 W
Temperature	–54 to +71°C (–65 to +160°F)

Status: Over time, LN-200 has continued to be promoted for UAV applications, with known installations including those aboard the BQM-74E target and the Predator and RQ-4 Global Hawk (LN-211G configuration) UAVs.

Contractor: Northrop Grumman Corporation, Navigation Systems Division
Woodland Hills, California.

Northrop Grumman LN-251

Type: Integrated GPS/INS navigation system.

Description: LN-251 is described as being a 'complete' integrated GPS/INS navigation system that incorporates an embedded, 12-channel, 'all-in-view', selective availability anti-spoofing module, P(Y) code GPS element. As such it offers blended GPS/INS, INS-only and GPS-only navigation solutions and other system features include:
- passive cooling (MIL-E-5400, Class 1A compatible)
- bolt-down or rack mounting
- 'full' built-in test
- standard GPS key loaders
- gyrocompass, stored heading, manual-at-sea, GPS aided, automatic EIA_1 and IFA_2 operating modes
- MIL-STD-1553B, RS-422 and RS-485 interfaces.

Specifications

Dimensions
Volume	5,358.6 cm3 (327 cu in)

Weights
Unit (with GPS)	5.8 kg (12.7 lb)

Performance
Position: GPS aided	<16 m (52.5 ft) SEP
Inertial	0.8 m (2.6 ft) CEP
Velocity: GPS aided	<0.03 m (0.09 ft)/s RMS
Inertial	<0.80 m (2.62 ft)/s RMS
Heading: GPS aided	<0.02° RMS
Inertial	<0.10° RMS
Pitch and roll: GPS aided	<0.02° RMS
Inertial	<0.05° RMS
Acquisition time: GPS aided	<2 min TTFF3
Power: Typical	25 W
Maximum	30 W
Temperature	–54/+71°C (–65/+160°F)
Vibration	8.9 g RMS
Velocity	12,000 m (39,370 ft)/s
Angular rate	1,000°/s
Angular acceleration	1,500°/s2
MTBF	>20,000 h

Status: Over time, LN-251 has continued to be promoted for use in UAV applications.

Contractor: Northrop Grumman Corporation, Navigation Systems Division
Woodland Hills, California.

Raytheon AN/MSQ-131 GHGS

Type: Mission control station.

Development: The Global Hawk Ground Segment (GHGS) consists of a Mission Control Element (MCE), a Launch and Recovery Element (LRE) and associated ground communications equipment, for operation and control of the Northrop Grumman RQ-4A Global Hawk high-altitude

Raytheon AN/MSQ-131 GHGS

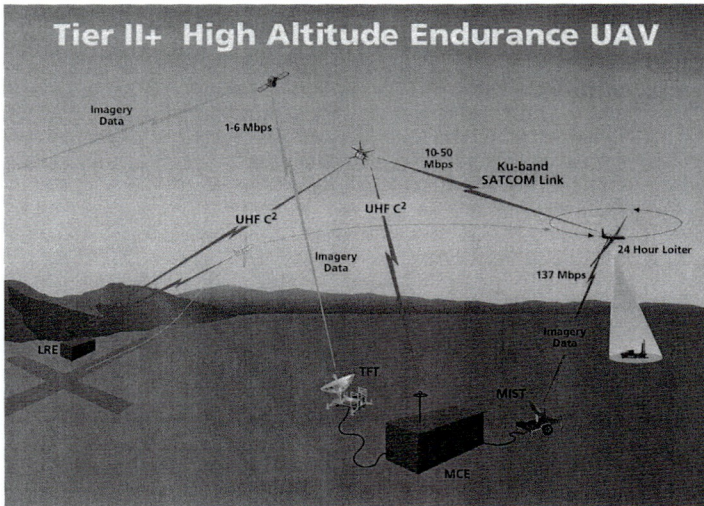

Global Hawk control and communication diagram 0518117

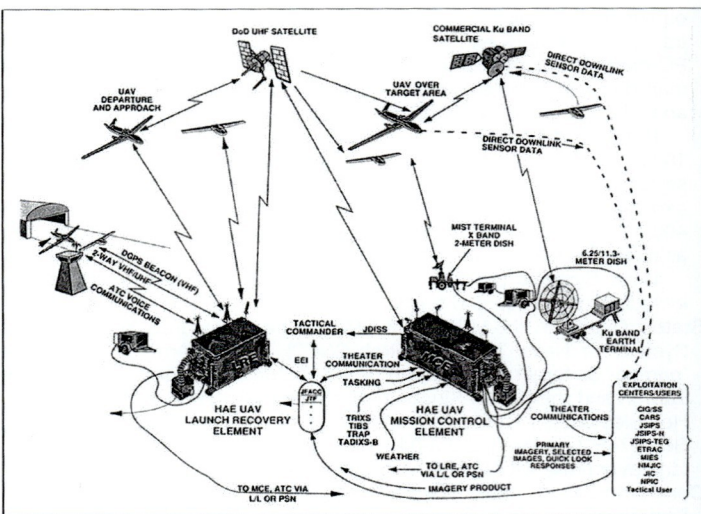

How the Common Ground Segment operates 0518323

Workstations inside the shelter-mounted LRE 0518299

endurance UAV (which see). Development of the GHGS was begun by the former Raytheon E-Systems' Falls Church Division.

Description: The GHGS was originally developed as an integral part of the DARPA/DARO HAE UAV programme and is designed to provide theatre commanders with a continuous all-weather surveillance capability and intelligence. As such, it is broken down into separate elements to permit operation in geographically dispersed areas. For example, the LRE could be deployed to the UAV operating location while the MCE remains with the command's primary exploitation site or main operating base. Through C^2 links, both elements remain functionally connected at all times. They are housed in mobile, air-conditioned buildings.

The GHGS is designed to be dismantled within 24 hours of receiving movement orders, and reassembled in less than 24 hours after arrival at the bed-down site. The complete infrastructure required to support 30 days of automatic operation (GHGS, aircraft support equipment, support personnel and the maintenance support kit) can be transported in three C-141 StarLifters.

The **MCE** (USAF designation **RD-2A**) contains industry-standard, high-performance hardware and software to satisfy all image processing objectives while minimising cost and risk. It is responsible for key elements of the mission plan including flight, communications, sensor and dissemination planning, sensor processing, and aircraft and mission payload control. Control of the UAV is handed over to the MCE by the LRE once the aircraft is airborne and the MCE can command and control up to three air vehicles simultaneously.

Sensor data obtained from each UAV are transmitted to the MCE via wideband RF line of sight or satellite datalink (see diagram). Data are then disseminated in near-realtime to existing command and control systems, or directly to suitably equipped tactical commanders or exploitation centres. Voice reporting of quick-look reports and dissemination of selected images are supported by means of standard commercial and military communications.

The **LRE** (USAF designation **RD-2B**) prepares, launches and recovers the UAV. It verifies the health and status of the various subsystems aboard the air vehicle, receives the mission plan from the MCE and loads it into the UAV. During launch and recovery, the LRE is responsible for air vehicle control; co-ordination with local and en-route traffic control facilities; and air vehicle handoff, once airborne, to the MCE. As described above, it can be located separately from the MCE but remains functionally connected to it at all times.

Use of common hardware and software, both within and across the two elements, provides inherent functional redundancy and a high degree of reliability in support of flight safety and operational coverage. Raytheon has stated that 85 per cent by value of GHGS hardware will be commercial-off-the-shelf equipment; software re-use rate will be 84 per cent.

Specifications

Dimensions
TFT antenna diameter..6.25 m (20 ft 6.0 in)
Volume
MCE 1 shelter...43.5 m3 (1,536 cu ft)
MCE 1 ECU pallet...17.4 m3 (616 cu ft)
MCE 1 support pallet..11.6 m3 (410 cu ft)
LRE shelter...19.63 m3 (693 cu ft)

MCE workstations inside the GHGS shelter 0518324

Specifications

Weights	
MCE 1 shelter	7,484 kg (16,500 lb)
ECU pallet	1,950 kg (4,300 lb)
MCE 1 support pallet	1,724 kg (3,800 lb)
100 kW generator	4,536 kg (10,000 lb)
MIST	2,849 kg (6,280 lb)
TFT	6,327.5 kg (13,950 lb)
MCE airlift weight	29,095 kg (57,530 lb)
LRE weight	4,586 kg (10,111 lb)
LRE airlift weight	11,657 kg (25,700 lb)

Status: Over time, the AN/MSQ-131 GHGS is known to have been deployed in support of the RQ-4 UAV.

Contractor: Raytheon Company, Intelligence & Information Systems Falls Church, Virginia.

Raytheon TCS

Type: Tactical control system.

Development: The TCS programme was initiated by the UAV Joint Project Office (JPO) in September 1997 in order to create a single common ground control system that would (a) be interoperable with the whole present and future family of US land-based and ship-based tactical UAVs; (b) be able to process and disseminate payload imagery and target data to more than 20 service C4I systems; and (c) receive and disseminate payload information from the RQ-4 Global Hawk endurance UAV. It is also being designed to disseminate sensor and surveillance data to other US and Allied C4I systems. Contracting agency is the US Naval Air Systems Command.

The TCS is planned to be scalable through five levels of UAV integration, as follows:

Level 1: Indirect receipt and retransmission of secondary imagery and/or data.
Level 2: Level 1 plus direct receipt of imagery and/or data.
Level 3: Level 2 plus control of the UAV payload.
Level 4: Level 3 plus control of the air vehicle, except for take-off and landing.
Level 5: Level 4 plus control of UAV including the take-off and landing phases.

These capabilities allow a battlefield commander to receive and, in some cases, direct the collection of imagery from any tactical UAV asset in the area of operations.

Description: The TCS is an open architecture system that aims to provide an interoperable and scaleable command, control, communications and data dissemination system for the family of US and Allied tactical and medium endurance UAVs. This approach, to which software is the key, promotes commonality, interoperability, affordability and easy introduction of new air vehicles when required.

The concept is to design the fundamental architecture and system interfaces between the various air vehicles and their control systems to use common and transportable software. By having core software that supports the elements common to all UAVs, and software at subsystem level established and controlled by joint interoperability interfaces, the software can be ported to any number of TCS hardware configurations. The initial core of software will provide varying levels of interaction; software and software-related hardware will be scaleable to meet individual user's needs. Maximum use will be made of COTS and Government Off-The-Shelf (GOTS) hardware and software wherever possible.

TCS will work across the spectrum of UAVs, including full-sized corps and division control systems; US Navy shipboard, submarine, airborne and Marine Expeditionary Force (MEF) control systems; and the downsized GCSs used by the MEF and US Army brigades and battalions. It will be capable of being hosted on a variety of computers. The TCS will support direct connectivity to standard DoD tactical radios (VHF, UHF and HF), mobile subscriber equipment and military and commercial satellite communications. Software interfaces will permit information exchange between the TCS and numerous C4I systems such as Joint STARS, Guardrail, Trojan Spirit II and JSIPS.

Status: Five prototype systems have been produced and fielded: a ship-based (USS *Mount Whitney*) one, two HMMWV-mounted shelterised systems, and two land-based, non-mobile shelterised centres (one for the UK and one for Canada). Early operational demonstrations were undertaken using Gnat 750 (Predator surrogate), RQ-5A Hunter (TUAV surrogate), RQ-6A Outrider, RQ-1 Predator and other UAVs. One system was delivered to Joint Forces Command in December 2000 to demonstrate Level 5 Predator launch and recovery capability; another was developed for testing with the US Army's RQ-7A Shadow 200 TUAV in 2001, and the first EMD TCS systems were tested with the US Navy RQ-8A Fire Scout in 2002.

In March 2001, the TCS demonstrated Level 5 flight control of a Predator during testing at Fort Huachuca, Arizona. This was the first time that a UAV had been controlled at Level 5 by a non-OEM (original equipment manufacturer) control system. In November 2001, the TCS concluded Level 5 flight demonstration of an AAI Shadow 600 while simultaneously interfacing with airborne US Army Shadow 200 UAVs at Level 2. Following flight testing, TCS demonstrated a capability to interface directly with every major C4I system currently in use by the US military. TCS has also been proposed as a replacement for the existing Pioneer UAV GCS and over time, the architecture has continued to be promoted. Most recently, Raytheon reports TCS as being designed 'to operate with all branches of the US Armed Services' and as having been 'utilised' by customers in Canada and the United Kingdom. In terms of usage, 19 November 2009 saw Raytheon announce that the TCS had 'recently' been deployed aboard the USS *McInerney* in support of an MQ-8B counter-narcotics mission in Central America. At the time of the announcement, an accompanying release described the deployment as being a 'huge step' for the TCS programme and as 'solidifying' the architecture's position as the 'future ground control system for the US Navy'.

Contractor: Raytheon Company, Intelligence & Information Systems Falls Church, Virginia. (See text for other supporting organisations.)

Rockwell Collins Athena 111m

Type: Miniaturised flight control system.

Description: The Rockwell Collins Athena 111m (formerly the Athena Technologies GS-111m) is described as being a miniaturised flight control system that has been specifically designed for use aboard small, high performance, fixed-wing, ducted-fan or helicopter UAV applications. As such, it integrates a 'complete' INS/GPS navigation solution with multi-I/O laws and includes all the I/O data hardware required to interface with servo actuators, datalink communications and payloads. Rockwell Collins' patented Feedback LTI'zation design process and flight control algorithms are claimed to offer 'high tolerance' to variations in UAV aerodynamics and configurations. Other system elements include a range of 'high' reliability solid-state components that include accelerometers, rate gyros, a magnetometer, air data/pressure/AoA/sideslip sensors and a differential-ready, WAAS-enabled GPS receiver. Control modes include airspeed, altitude, climb rate, AoA, sideslip, 3-D automatic navigation and automatic landing and take-off. Athena 111m is housed in an aluminium enclosure; offers onboard data logging and analogue and digital interfaces.

Specifications

Dimensions	
Length (including mountings)	99 mm (3.90 in)
Width (including mountings)	40 mm (1.56 in)
Depth (including mountings)	66 mm (2.60 in)
Weights	
Unit	0.2 kg (0.5 lb)
Performance	
IMU	
Update rate	50 Hz (100 Hz optional)
Maximum angular rate	±200°/s
Maximum g range	±7 g
Analogue/digital sampling resolution	24-bits
Accuracy	
Heading/pitch/roll	0.1° (one sigma)
Maximum airspeed	160 kt (296 km/h; 184 mph) (up to 650 kt (1,204 km/h; 748 mph) option)

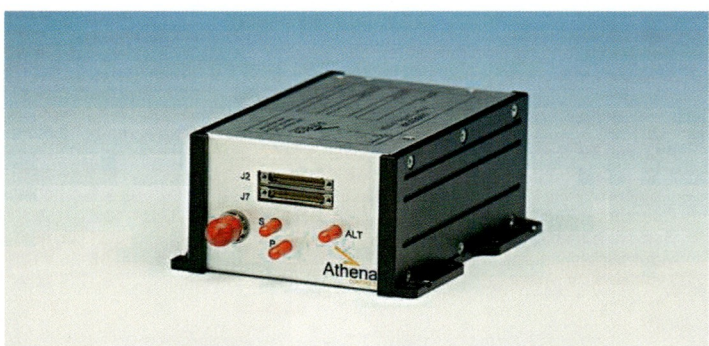

The Athena 111m flight control system *(Rockwell Collins)* 1325172

The Athena 111m flight control system *(Rockwell Collins)* 1325171

Specifications

Altitude (at S/L)	13.7 m (45 ft)
Altitude (at 12,200 m (40,000 ft))	<60.9 m (<200 ft)
Maximum power (at 9-18 V or 18-36 V)	4.5 W
Temperature	
Non-operating	–54 to +85°C (–65 to +158°F)
Operating	–40 to +70°C (–40 to +185°F)
Relative humidity (non-condensing)	95%
Vibration/shock	MIL-STD-810 compliant

Status: Over time, Athena 111m has continued to be promoted and is noted as having been installed aboard/mandated for installation aboard the Aurora GoldenEye-50/-80 and the EMT Luna UAVs. Rockwell Collins' 0.8 kg (1.7 lb), 140 × 127 × 64 mm (5.5 × 5.0 × 2.5 in) Athena 111 flight control system is reported as being used aboard the BQM-167A air target vehicle.

Contractor: Rockwell Collins
Warrenton, Virginia.

Rockwell Collins Athena 311

Type: Flight control system.

Description: Rockwell Collins (formerly Athena Technologies) describes the Athena 311 (formerly the GS-311) as being a "high performance, integrated and affordable" INS/GPS/attitude and heading reference/flight controls system that is designed for UAV and military applications. As such, it provides a "tactical grade" strapdown navigation solution that provides attitude and heading information with "static and dynamic accuracies [that are] superior to those of traditional spinning-mass gyros". Again, the device is noted as offering a "full" range of analogue and digital input-output interfaces and incorporates dedicated computers for flight systems management and payload control. Autopilot and mission management functions include autonomous landing and route navigation, with the autopilot design making use of patented, "highly-automated" control-synthesis techniques that "guarantee" stability and performance robustness throughout the flight envelope. Rockwell Collins also notes that it provides an "integrated development environment" for customer-developed software used in the Athena xxx range.

Specifications

Dimensions	
Length	152 mm (6.0 in)
Width	127 mm (5.0 in)
Depth	102 mm (4.0 in)
Weights	
Unit	2.2 kg (4.8 lb)
Performance	
Heading: Range	±180°
One sigma accuracy	<1°
Update rate	50 Hz
Start-up time: Initial attitude	10 s
INS/GPS	1 minutes
Maximum angular rate	±200°/s
Maximum g rate	±7 g
Full data accuracy	15 minutes
Pitch	±90°
Roll	±180°
One sigma attitude accuracy	0.1°
Airspeed: Maximum	165 kt (306 km/h; 190 mph) (up to 650 kt (1,204 km/h; 748 mph) option)
Accuracy	<1.0 kt (1.9 km/h; 1.2 mph) at 60 kt (111 km/h; 69 mph)
Altitude: Typical	Up to 15,240 m (50,000 ft)
Accuracy (at S/L)	13.7 m (45 ft)
Accuracy (at 12,192 m (40,000 ft))	<60.9 m (<200 ft)
Operating temperature	–40 to +70°C (–40 to +158°F)
Relative humidity (non-condensing)	95%
Input supply voltage	18-36 V DC
Typical input power	12 W

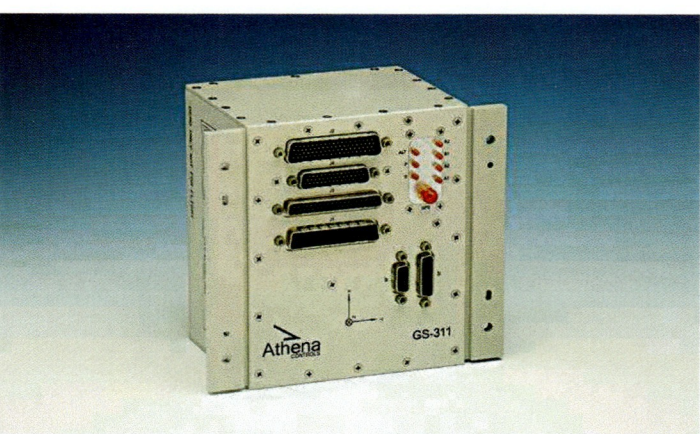

The Athena 311 flight control system *(Rockwell Collins)* 1325173

Status: Over time, Athena 311 has continued to be promoted and is noted as having been installed aboard Alenia's Sky-X UAV.

Contractor: Rockwell Collins
Warrenton, Virginia.

Rockwell Collins Athena 411

Type: Inertial navigation, global positioning and Air Data Attitude Heading Reference (ADAHR) system.

Development: The Rockwell Collins Athena 411 has been selected for installation aboard the U-TacS WK450 UAV.

Description: The Rockwell Collins (formerly Athena Technologies) Athena 411 inertial navigation, global positioning, ADAHR and flight control system is designed for military and UAV applications and is described as integrating solid-state gyros and accelerometers, a magnetometer, a GPS receiver and air data pressure transducers into a single sealed enclosure. Rockwell Collins characterises Athena 411 as being a 'high reliability strap-down system' and claims that it provides attitude and heading measurements that have 'static and dynamic [accuracies that are] superior to traditional spinning mass vertical and directional gyros'. Other system features include:
- the ability to maintain accurate roll, pitch and heading in 'dynamic environments and sustained turns'
- use of a WAAS-enabled, differential ready GPS application
- dual airspeed and altitude air data measurement
- use of Kalman filter algorithms to provide a 'full' INS and attitude solution.

Specifications

Dimensions	
Unit: Length	114 mm (4.5 in)
Width	109 mm (4.3 in)
Height	125 mm (4.9 in)
Weights	
Unit	1 kg (2.2 lb)
Performance	
INS/GPS	
Update rate	100 Hz (50 Hz option)
Start-up time	10 s (initial altitude); 1 minute (full alignment)
Heading range	±180°
Attitude: Pitch and roll	±90° (pitch); ±180° (roll)
Altitude: range	–305 to +15,240 m (–1,000 ft to +50,000 ft)
accuracy	13.7 m (45 ft, at sea level); <61 m (<200 ft, at 12,192 m (40,000 ft))
GPS range	Worldwide
Lat/Long position accuracy	GPS C/A code, differential ready and WAAS-enabled
GPS antenna power	5 V
Airspeed accuracy	<1.08 kt (2.00 km/h; 1.24 mph) at 60 kt (111 km/h; 69 mph); <2.16 kt (4 km/h; 2.49 mph) at 40 kt (74 km/h; 46 mph)
Maximum airspeed	200 kt (371 km/h; 230 mph - standard); up to 650 kt (1,204 km/h; 748 mph - option)
AoA range	±8° (at maximum airspeed)
Angle-of-sideslip range	±8° (at maximum airspeed)
IMU	
Angular rates	±200°/s (maximum)
g range	±7 g (maximum)
1 sigma attitude heading accuracy with GPS	
Pitch/roll accuracy	0.1°
Heading accuracy	0.1° (external magnetometer performance dependent)
Extended GPS outage	
Pitch/roll accuracy	0.5° (external magnetometer performance dependent)
Heading accuracy	1° (external magnetometer performance dependent)
General	
Enclosure	Aluminium
Power	6 W (nominal, at 18-36 V DC); 12 W (peak)
Serial I/O	RS-232 (× 2); RS-422 (× 2 plus 1 × RS-422 DGPS and 1 × RS-422 service port)
Temperature	–40°C to +70°C (–40°F to +158°F, operating)
Vibration	MIL-STD-810 compliant (operating)
Humidity	95% (non-condensing, sealed enclosure)

Status: Over time, the Athena 411 inertial navigation, global positioning and ADAHR system has continued to be promoted and has been selected for installation aboard the U-TacS WK450 and AeroViroment Global Observer UAVs.

Contractor: Rockwell Collins
Warrenton, Virginia.

Rockwell Collins Athena 511

Type: Integrated INS/GPS/air data sensor and flight control system.

Description: The Athena 511 (formerly the GS-511) integrated INS/GPS/air data sensor and flight control system is designed for UAV applications and can (if required) incorporate a flight control module. As applied to the US Army's Grey Eagle UAV, other system features include:
- use of a tactical grade IMU
- an integral, tri-axial magnetometer (with automatic hard iron compensation)
- use of a differential-ready GPS receiver
- integral air data system pressure transducers (including static, airspeed and flow angle measurements)
- use of a Kalman filter INS to facilitate 'high' accuracy navigation and control state vector
- an electrical and mechanical design that meets military temperature, shock, vibration, electromagnetic interference and humidity requirements
- use of a software module that provides Rockwell Collins' 'full' suite of 'flight-proven' inertial reference and navigation functions.

Specifications

Dimensions	
Unit	156 × 156 × 156 mm (6.1 × 6.1 × 6.1 in) or smaller
Weights	
Unit	<2.7 kg (<6.0 lb)
Performance	
INS/GPS	
Update rate	100 Hz (50 Hz option)
Start-up time	10 s (initial altitude); 1 minute (full alignment)
Heading range	±180°
Attitude: Pitch and roll	±90° (pitch); ±180° (roll)
Altitude: Range	−305 to +15,240 m (−1,000 ft to +50,000 ft)
Baroinertial accuracy	13.7 m (45 ft, at sea level); <61 m (<200 ft, at 12,192 m (40,000 ft))
GPS range	Worldwide
Lat/Long position accuracy	GPS receiver accuracy
GPS antenna power	5 V
Airspeed: Accuracy	<1.0 kt (1.9 km/h; 1.2 mph) at 60 kt (111 km/h; 69 mph); <2.0 kt (3.7 km/h; 2.3 mph) at 40 kt (74 km/h; 46 mph)
Maximum	165 kt (306 km/h; 190 mph - standard); up to 650 kt (1,204 km/h; 748 mph - option)
AoA range	±8° (at maximum airspeed)
Angle-of-sideslip range	±8° (at maximum airspeed)
IMU	
Angular rates	±1,000°/s (maximum)
Rate gyro bias	1°, 2° or 3°/h
g range	±37 g (maximum)
1 sigma attitude heading accuracy with GPS	
Pitch/roll accuracy	0.3°
Heading accuracy	0.1°
General	
Enclosure	Aluminium
Power	17 W (nominal, at 18-36 V DC)
Signal I/O	Outside air temperature (input); RS-232 (× 1); RS-422 (× 4)
Temperature	−40°C to +70°C (−40°F to +158°F, operating)

Status: Over time, Athena 511 has continued to be promoted and is noted as having been selected for installation aboard the US Army's Grey Eagle UAV and as having been installed aboard the prototype Excalibur VTOL UAV.

Contractor: Rockwell Collins
Warrenton, Virginia.

Rockwell Collins GuS™

Type: INS/GPS/air data/flight control unit.

Description: Described as being the same size as a "small cell phone", the Rockwell Collins (formerly Athena Technologies) GuS™ navigation unit is billed as offering a 'complete' navigation solution with the option of hosting flight control software if required. System features include:
- use of solid-state accelerometers, rate gyros, a triaxial magnetometer and air data pressure sensors
- integral automatic magnetometer calibration
- use of a differential-ready, wide area augmentation system-enabled, C/A code GPS receiver
- the incorporation of 'all' the analogue/digital I/O interface hardware needed to accommodate a 'variety' of host vehicle configurations and payloads
- use of a single board architecture
- provision of a 'full' INS/GPS/air data attitude heading reference system Kalman-filter solution
- use of a high-speed micro controller
- a flight data recording capability
- an optional flight control package that includes 3-D waypoint navigation; joystick attitude (pitch/roll/heading), altitude/airspeed/heading hold and climb rate control; ground station altitude/airspeed/heading hold and climb rate control; onboard data logging; stall protection and speed/attitude/load limiting and tolerance of GPS outage and datalink loss.

Specifications

Dimensions	
Length	115.0 mm (4.52 in)
Width	51.0 mm (2.01 in)
Depth	25.5 mm (1.00 in)
Volume	131 cm3 (8.0 cu in)
Weights	
Unit	113 g (0.25 lb)
Performance	
IMU	
Update rate	50 Hz (100 Hz optional)
Maximum angular rate	±200°/s
Maximum g range	±7 g (±18 g option)
Analogue/digital sampling resolution	24-bits
Accuracy	
Heading (one sigma)	0.6°
Pitch/roll (one sigma)	0.2°
Maximum airspeed	200 kt (371 km/h; 230 mph) (up to 650 kt (1,204 km/h; 748 mph) option)
Altitude (at S/L)	13.7 m (45.0 ft)
Altitude (at 12,192 m (40,000 ft))	<60.9 n (<200 ft)
Temperature	
Non-operating	−54/+85°C (−65/+185°F)
Operating	−40/+70°C (−40/+158°F)
Relative humidity (non-condensing)	95%
Vibration/shock	MIL-STD-810 compliant
Maximum power (at 9-18 V)	3.5 W

Status: Over time, GuS™ has continued to be promoted for UAV applications.

Contractor: Rockwell Collins
Warrenton, Virginia.

UAV Factory portable GCS

Type: Portable GCS.

Development: UAV Factory announced the launch of its second generation portable GCS on 8 January 2012.

Description: UAV Factory describes its second generation portable GCS as being an off-the-shelf 'universal' solution that is designed to control

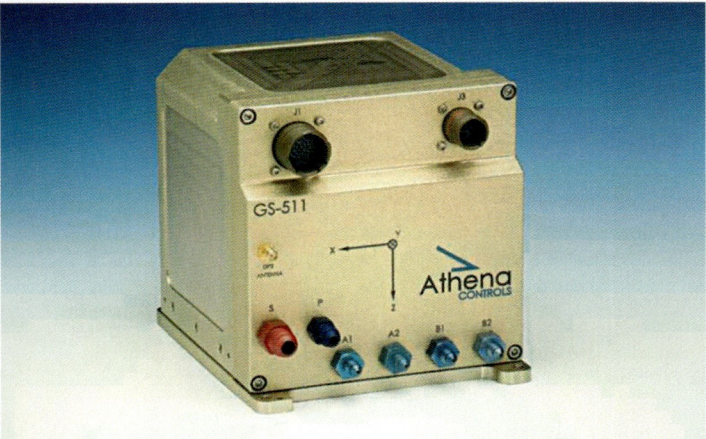

The Athena 511 integrated INS/GPS/air data sensor and flight control system *(Rockwell Collins)* 1325174

A vehicle-mounted application of the UAV Factory's portable GCS *(UAV Factory)* 1509459

both unmanned vehicles and their payloads. As such, it features a modular 320 × 270 × 80 mm (12.6 × 10.6 × 3.1 in) electronics compartment to facilitate configuration for 'many different applications' and which can accommodate a variety of application specific hardware including RF autopilot modems, video receivers, datalinks and data storage/recording devices. Other system features include:
- docking for a ruggedised Panasonic CF-31 Toughbook computer (to be procured separately from the GCS)
- a 43 cm (17 in) touch screen display
- provision for two hot swapable Lithium-Ion batteries (providing an up to two hour operating time)
- a system status monitor
- an integrated power distribution system (with two 12 V DC/50 W outputs)
- plug and play design
- 10-32 V DC input range
- the ability to operate within a –20°C to +60°C (–4°F to +140°F) temperature range
- an all-up weight of 18.9 kg (41.7 lb - excluding the CF-31 computer)
- over-voltage/over-current/reverse polarity/battery over-charge protection.

UAV Factory's portable GCS is housed in 1,000 (L) × 420 (W) × 170 (D) mm (39.4 × 16.5 × 6.7 in) 'military grade' container that can be carried by a single individual (using grab handles, a carrying strap or a wheeled chassis) and can be mounted on an optional foldable stand if required. System accessories include a joystick, a mouse and antennas (including an external GPS element). In addition to the control of UAVs, ground robots, bomb disposal robots, remotely operated vehicles and other robotic devices, the UAV Factory's second generation portable GCS is further noted as being able to be configured to control and monitor measurement and sensing equipment.

Status: Most recently, UAV Factory has continued to promote its second generation portable GCS.

Contractor: UAV Factory USA LLC,
Irvington, New York.UAV Factory Ltd Europe,
Jelgava, Latvia.

LAUNCH AND RECOVERY SYSTEMS

Canada

MicroPilot MP CAT

Type: Foldable pneumatic catapult launcher.

Development: MP CAT was launched at the August 2009 AUVSI North American Convention and Trade Show.

Description: MicroPilot describes the MP CAT pneumatic catapult launcher as being a 'simple, inexpensive [then year USD15,000], carry anywhere' device that is designed for use with mini-UAVs with maximum take-off weights of up to 20 kg (44.1 lb). As such, the launcher is foldable and comprises two transport cases (one for the launcher and a second for a li-ion battery-powered (14 × 4 V batteries) compressor). The MP CAT compressor can also draw power from a 12 V car/automobile battery via the vehicle's cigarette lighter socket.

Specifications

Dimensions	
Launcher length: Folded	1.4 m (4.6 ft)
Rail deployed	4.0 m (13.1 ft)
Weights	
System	25 kg (55.1 lb)
Performance	
Launch angle	11°
Launch velocity: 7 kg (15.4 lb) UAV	15.9 m/s (52.2 ft/s, 3.0 bar pressure); 17.7 m/s (58.1 ft/s, 3.5 bar pressure); 19.4 m/s (63.7 ft/s, 4.0 bar pressure); 20.9 m/s (65.6 ft/s, 4.5 bar pressure); 22.5 m/s (73.8 ft/s, 5.0 bar pressure)
10 kg (22.1 lb) UAV	13.3 m/s (43.6 ft/s, 3.0 bar pressure); 14.8 m/s (48.6 ft/s, 3.5 bar pressure); 16.2 m/s (53.2 ft/s, 4.0 bar pressure); 17.5 m/s (57.4 ft/s, 4.5 bar pressure); 18.8 m/s (61.7 ft/s, 5.0 bar pressure)
15 kg (33.1 lb) UAV	10.9 m/s (35.8 ft/s, 3.0 bar pressure); 12.1 m/s (39.7 ft/s, 3.5 bar pressure); 13.2 m/s (43.3 ft/s, 4.0 bar pressure); 14.3 m/s (46.9 ft/s, 4.5 bar pressure); 15.4 m/s (50.5 ft/s, 5.0 bar pressure)
20 kg (44.1 lb) UAV	9.4 m/s (30.8 ft/s, 3.0 bar pressure); 10.5 m/s (34.5 ft/s, 3.5 bar pressure); 11.5 m/s (37.7 ft/s, 4.0 bar pressure); 12.4 m/s (40.7 ft/s, 4.5 bar pressure); 13.3 m/s (43.6 ft/s, 5.0 bar pressure); 14.1 m/s (46.3 ft/s, 5.5 bar pressure)

Status: Most recently, MicroPilot has continued to promote the MP CAT foldable pneumatic catapult launcher.

Contractor: MicroPilot
Stony Mountain, Manitoba.

Finland

Robonic MC series

Type: Pneumatic catapults.

Development: Robonic (owned by Sagem, part of the Safran Group) operates in Tampere, Finland. It has developed and produced pneumatic catapult launchers for UAVs and aerial targets since 1983. In addition, the company operates a UAV/UAS test flight centre (designated as the Robonic Arctic Test UAV Flight Centre) just above the Arctic Circle at Kemijärvi in Finland. Robonic also notes its ability to provide launch-related consultancy and launcher rental/lease. Historically, Robonic's first launcher design (a towed equipment) was delivered to the Finnish Defence Forces during 1985, with the first MC series system appearing during 1995. The MC series equipment provided 'much improved' mobility and carried the designation suffix 'H' for 'heavy duty', 'L' for 'light' or 'LLR' for 'light and long ramp rail'. Additional development has produced a 'small' UAV launcher and 'L' series launchers capable of accommodating wide mass and speed envelopes with a greater degree of transportability by road, air and sea. The launchers described in this entry represent the company's third-generation designs.

Over time, Robonic catapults have been used to launch a wide range of UAVs and targets including the AT4, Banshee (MC91 and MC2555LLR launchers), Do-DT25 (MC2055/2555L/2555LLR), Do-DT35 (MC2055/2555L/2555LLR), Do-DT55 (MC2555L/2555LLR), Falco (MC2555LLR), Falconet, KDR, Hermes 450 (MC2055/2555L/2555LLR), Irisjet (MC2555LLR), Ranger (MC2555LLR), Sperwer A/A MkII/B (MC2555LLR) and Voodoo (MC2555LLR).

Description: Robonic launchers operate on high pressure compressed air and are designed for rapid 'reload' (by means of a hydraulic mechanism). As such, the launchers are designed for all-terrain use and can be mounted on various types of on- or off-road vehicles and trailers. Equally, they can be set up at permanent ground sites. Within the various applications, the launcher frames are hinged so that the catapults can be

The Elbit Hermes 450 tactical UAV on a Robonic MC2055L launcher *(Robonic)* 1046301

Sperwer A on MC2555LLR at Kemijärvi *(Robonic)* 1346508

A Robonic MC2555L launcher with an EADS Dornier Do-DT25 target *(Robonic)* 1046302

An EADS 3Sigma Irisjet ready for launch on a Robonic MC2555LLR launcher at the Vidsel range in Sweden *(Robonic)* 1311295

Launch and recovery systems > Finland > Robonic MC series

An EADS Dornier Do-DT55 target being launched from a Robonic MC2555L launcher (Robonic) 1120713

The Robonic MC2555LLR launcher packaged in a 6 m (20 ft) ISO container (Robonic) 1120775

transported, stored in small spaces, and where required, 'easily' disconnected from host vehicles (and stored on their own feet) if the particular truck is needed for other duties. No special tools are needed to set up ramps for launch or transportation and Robonic launchers can be erected in a 'short' time by a single individual. Robonic launchers are designed to accommodate a wide variety of UASs, with each UAS type having a dedicated 'adapter' to act as the interface between the launcher and the air vehicle.

Variants: MC0315L: MC0315L is a 'fully' pneumatic device that is designed to launch 'small' UAVs and target drones, offers a maximum pressure of up to 10 bar (in order to support the 'widest' range of small AVs without modification) and is designed for field deployment under 'wide temperature extremes' with 'minimal' maintenance requirements. The MC0315L's small size allows it to be easily repositioned to optimise launch with the prevailing wind direction and the equipment incorporates its own compressor in order to be self-sustaining. If required, MC0315L can accommodate nitrogen-filled and high pressure air vessels in place of its integral compressor. The device incorporates integral safety features and procedures (for 'all' operating conditions) and can be set-up in no more than 15 minutes. The MC0315L is further noted as being ready for repeat launches in a 'matter of minutes'.

MC2055L: The MC2055L is described as being a highly mobile pneumatic catapult launcher that features adjustable launching pressure and hydraulic recovery. Use of a closed pneumatic system allows pressures to recover quickly, thereby enabling the equipment to be ready for a UAV's launch preparations within a 'matter of minutes'. Logistics advantages are claimed to include 'full and easy' transportability by land, sea or air. MC2055L can be mounted on a towed trailer or be truck-mounted. It can be stored or transported in a 20 ft (6.1 m) ISO container.

MC2555L and MC2555LLR: MC2555L and MC2555LLR are described as being 'more compact' mobile pneumatic launchers that incorporate adjustable launching pressure and hydraulic recovery. The remarks given for the MC2055L also apply to the MC2555L.

(A: MC0315L, B: MC2055L, C: MC2555L, D: MC2555LLR):

DIMENSIONS

Length for transport: A	3.495 m (11 ft 5.6 in)
B	7.10 m (23 ft 3.5 in)
C	5.20 to 6.35 m (17 ft 0.7 in to 20 ft 10.0 in)
D	5.80 to 7.15 m (19 ft 0.0 in to 23 ft 6.0 in)
Length deployed: A	3.495 m (11 ft 5.6 in)
B	18.00 m (59 ft 0.7 in)
C	15.20 m (49 ft 10.4 in)
D	16.80 m (55 ft 1.2 in)
Width: A	0.92 m (3ft 0.2 in)
B	2.40 m (7 ft 10.5 in)
C	2.27 m (7 ft 4.8 in)
D	2.27 m (7 ft 4.8 in)
Height: A (0 to 15° elevation)	0.75-1.40 m (2 ft 5.5 in - 4 ft 7.1 in)
B (for transport)	2.10-2.35 m (6 ft 10.7 in - 7 ft 8.5 in)
B (8 to 15° elevation)	4.00-5.50 m (13 ft 1.5 in - 18 ft 0.5 in)
C (for transport, incl trailer)	2.10 m (6 ft 10.7 in)
C (8 to 15° elevation)	3.50-4.80 m (11 ft 5.8 in - 15 ft 9.0 in)
D (front section, 8 to 15° elevation)	3.70-5.00 m (12 ft 1.7 in - 16 ft 0.5 in)
D (for transport)	2.10-2.21 m (6 ft 10.7 in - 7 ft 0.3 in)

WEIGHTS

Launcher weight: A	255 kg (562 lb)
B	4,200 kg (9,259 lb)
C	4,500 kg (9.921 lb)
D	4,800 kg (10,582 lb)
Power pack weight: A	25 kg (55.1 lb)
AV maximum weight: A	40 kg (88.2 lb)
B, C	30-500 kg (66.1-1,102 lb)

PERFORMANCE

Road transport speed: B, C, D	43 kt (80 km/h; 50 mph)
Launch angle: A	5-15°
B, C, D	8-15°
Maximum AV launch velocity: A	15 m/s (49 ft/s) (30 kg (66.1 lb) AV)
B, C	20-55 m/s (65 -180 ft/s)
Max acceleration: A	<12 g (up to 15 m/s (49 ft/s))
Set-up time: A, C, D	<15 min
B	<20 min
Recovery/reload time: A, B, C	<5 min
D	<4 min
Operating temperature range:	
A	–20 to +50°C (–4 to +122°F)
B, C, D	–30 to +50°C (–22 to +122°F)
Max available power: B	1.25 MW
C, D	1.5 MW
Launch recovery time: D	< 5 min

Status: Robonic launchers have been supplied to the Finnish Defence Forces, EADS Deutschland, Sagem (Safran Group), Meggitt Defence Systems and the Canadian Centre for Unmanned Vehicle Systems (CCUVS) and the US Army Targets Management Office. Robonic launchers have seen service with the Royal Netherlands Army and Robonic has worked with EADS on the UK CATS (Common Aerial Target System) programme; with Flight Refuelling Ltd on the UK Watchkeeper programme and with Galileo Avionica on the Falco and Mirach 100/5 programmes. In March 2006, Robonic also noted it was working with BAE Systems (to 'explore issues' associated with launching the HERTI air vehicle from the MC2555LLR launcher) and was pursuing several other programmes internationally. Over time, Robonic has continued to promote the MC0315L and MC2555LLR and has established the Robonic Arctic Test UAV Flight Centre (RATUFC) at Kemijärvi in Finland. Here, the range covered an area of 11,000 km^2 (9,471 sq miles) and was noted as being capable of handling both catapult and runway launched UAVs. A total of four types of air vehicles are known to have been test flown at this location, namely Sagem's Sperwer A and B, Patroller and Selex Galileo's Falco.

Contractor: Robonic Oy Ltd
Tampere.

Israel

HEC HEC-BRK-150/250

Type: Braking systems.

Description: Hydromechanical Engineering was founded in 1984 and is primarily concerned with test and ground support equipment. The latter includes test benches and dynamometers for testing UAV power plants as well as a run-in and fuel system diagnostic trolley for UAVs. The company has now extended its UAV product line to produce the HEC-BRK braking system.

The HEC-BRKs are cable braking systems to assist UAVs upon landing. The generic system consists of two stainless steel and aluminium braking drums and a 50 m (64 ft) steel cable 50 mm (2 in) above the landing surface (optionally 40 or 45 mm; 1.6 to 1.8 in) with the former anchored to the sides of the landing area and linked by the cable. Of the two cited sub-variants, the BRK-150 configuration's drums weigh 45 kg (99.2 lb).

The generic capability meets MIL-STD-810D requirements with brake force adjustable to a wide range of UAVs in terms of landing weights and speeds up to 75 kt (140 km/h; 87 mph). Roll-back after landing takes a maximum of 30 s and the cycle time between landings is a maximum of 40 s. The specification data given refers to the BRK-150.

Specifications

Dimensions	
Length	670 mm (26.4 in)
Width	670 mm (26.4 in)
Height	270 mm (10.6 in)
Weights	
System weight	90 kg (198.5 lb)
UAV landing weight	300–1,500 kg (661.5–3,307.5 lb)

Status: Over time, HEC has continued to promote its cable braking systems for UAVs.

Contractor: Hydromechanical Engineering Company Ltd
Rishon-Le-Zion.

The HEC hydraulic UAV launcher *(HEC)*

HEC hydraulic launcher

Type: Hydraulic launcher.

Description: Capable of launching UAVs of up to 500 kg (1,102 lb) in weight, the launch velocity and acceleration of HEC's hydraulic launcher can be varied to suit specific types of air vehicle. The system is mounted on either an all-terrain trailer or the flatbed of a truck, and its launch ramp comprises up to four sections which fold up for transportation or storage. HEC's hydraulic launcher is self-contained, requires two operators and facilitates pre-flight testing and maintenance with the UAV mounted on it.

Specifications

Weights	
Max air vehicle weight	500 kg (1,102 lb)

Status: Over time, HEC has continued to promote its hydraulic UAV launcher capability.

Contractor: Hydromechanical Engineering Company Ltd
Rishon-Le-Zion.

Italy

Selex Galileo launch and recovery stations

Type: Truck-mounted launch and recovery systems.

Development: Developed for use with the Meteor Mirach family of UAVs.

Description: *Launch station:* Mounted on a six-wheeled Fiat truck, the launch station comprises a hoist crane for air vehicle lifting; launch ramp; service console; launch console; firing console; and a 20 kW, 220 V AC (50 Hz) power generator. It enables loading the air vehicle on to the launching ramp; connection of the power supply to the air vehicle on the ramp; performing a preflight functional check of the air vehicle and loading of the mission plan; take-off booster ignition and air vehicle launch. The preflight test sequence is computer-controlled and fully automatic, using a step by step procedure and go/no-go testing.

Recovery station: For retrieval of the air vehicle after landing, the recovery station comprises a hoist crane for air vehicle lifting; a parachute bag; an air vehicle support saddle; and communications equipment. It, too, is mounted on a six-wheeled Fiat truck.

Status: Over time, Selex Galileo launch and recovery systems have been procured by the Italian Army.

Contractor: Selex Galileo
Ronchi dei Legionari.

Mirach 26 on Galileo launch station *(Selex)*

Galileo recovery station and Mirach 150 *(Selex)*

Norway

OKT Norge 10/150

Type: Pneumatic launcher.

Development: OKT Norge has developed, constructed and produced a series of hydraulic and pneumatic launchers. In the beginning the aim was to launch Northrop target drones for the Norwegian Navy. However, since 1989 the company has mainly produced pneumatic launchers.

Description: The standard launcher, the 10/150, has the capacity to launch air vehicles of up to 115 kg, with a launching speed of 150 km/h (42 m/s). It is designed for stable and reliable functioning in variable conditions and temperatures. Structurally it is very strong, and operation is straight-forward. A 6.7 kW (9 hp) four-stroke petrol engine provides power for the two-cylinder air compressor. The pneumatic power transmission provides a uniform and predictable acceleration, with a high launching speed and little strain on the air vehicle, and can be recharged in 15 minutes. The unit is virtually maintenance free. Other launchers can be produced to meet individual customers' requirements.

Specifications

Dimensions
- Length: overall (extended) .... 9.00 m (29 ft 6.3 in)
- rail .... 7.50 m (24 ft 7.3 in)
- overall (for transportation) .... 5.30 m (17 ft 4.7 in)
- Max width .... 2.10 m (6 ft 10.7 in)

Weights
- Launcher weight .... 1,400 kg (3,086 lb)
- UAV launch weight .... Up to 115 kg (253 lb)

Performance
- UAV launch velocity .... 42 m (138 ft)/s

Status: Over time, the 10/150 has been procured by the Norwegian armed forces and the French arm of EADS. OKT Norge also produced a pneumatic launcher for INTA of Spain, able to launch a 230 kg (507 lb) air vehicle at 75 kt (140 km/h; 87 mph). Over time, OKT Norge has continued to promote 'custom-made' pneumatic UAV launchers, with specific configurations being created in a 'close relationship' with the particular customer.

Contractor: OKT Norge AS
Stavanger.

OKT Norge 10/150 with ramp extended to launch a Fox TS3 target *(OKT Norge)* 0505129

OKT Norge 10/150 launcher in transportation configuration *(OKT Norge)* 0505128

South Africa

ATE Vulture launch and recovery systems

Type: Dedicated launch and recovery systems.

Development: These systems were developed specifically for use with the ATE Vulture tactical UAV system for the South African Army.

Description: Launch of the Vulture UAV is fully automated and in unprepared terrain, using a vehicle-mounted (S100 flatbed) atmospheric catapult launcher based upon a vacuum principle. The catapult is activated by a pump, drawing a vacuum into the cylinder; a drive piston, cable and pulley combination. The advantages of this system are:
- constant acceleration of the air vehicle along the launch tube, due to constant pressure differential between the vacuum and atmospheric air
- safe launch, due to high atmospheric pressure being on the outside of the tube
- low life-cycle cost
- 'guaranteed' performance due to being able to measure vacuum before launch
- a launch 'box' that measures no more than 100 × 100 × 12 m (330 × 330 × 40 ft)
- no pyrotechnic requirements
- high above ground launch
- the ability to be turned into the wind.

Recovery of the Vulture UAV is also fully automated in unprepared terrain, using a ground-based system which features automated approach of the air vehicle using a laser tracker. The aircraft is then 'captured' and decelerated in an energy absorption device (strap configuration with friction brakes, mounted on an S100 flatbed carrier

Vulture recovery system, showing energy absorption device, airbag and laser tracker *(ATE)* 1030826

Recovery sequence, showing Vulture captured by energy-absorbing straps, decelerating and dropping on to airbag *(ATE)* 1030819

Vulture recovery sequence 3 *(ATE)* 1030821

A general view of the Vulture S100 flatbed vehicle-based atmospheric catapult launcher *(ATE)* 1414590

Vulture recovery sequence 2 *(ATE)* 1030820

vehicle), from which it descends on to a large inflatable airbag. Recovery 'box' is of the same dimensions as for the launch box. Other advantages of this method include:
- the ability to be turned into the wind
- 'pin-point' recovery accuracy
- 'soft' AV deceleration
- damage-free recovery
- 'easy access' to the air vehicle after recovery.

Status: Vulture UAV launch and recovery systems have been procured (starting in 2006) by the South African Army's Artillery arm. Over time, ATE has continued to promote such systems.

Contractor: Advanced Technologies & Engineering Co (Pty) Ltd
Halfway House.

Spain

Aries UAV launchers

Type: Portfolio of UAV launch systems.

Description: Most recently, Aries Ingeniería y Sistemas's portfolio of UAV launch systems has comprised the ALPPUL LP-02, ATLAS ME-01, BULL EL-01, HERCULES AH-01 and LAE equipments, the known details of which are as follows:

ALPPUL LP-02: ALPPUL LP-02 is specifically designed to launch UAVs and targets at 'elevated' speeds and masses using low-pressure pneumatic energy conversion. A launch pressure of less than 10 bar facilitates the use of 'standard' components and is claimed to reduce life-cycle costs, maximise safety throughout operation of the equipment and to produce an 'easy-to-use' system that is 'quick' to set up and assemble.

ATLAS ME-01: The ATLAS ME-01 launcher is designed for use with medium-weight UAVs and targets and is claimed to offer: a 'very high' performance; an 'autonomous driving force'; the ability to be customised; a 'high' degree of mobility (foldable and trailer-transportable); and 'fully ensured' repeatability in 'all' environments and operational scenarios. ATLAS ME-01 is also noted as being 'quick' to set up and assemble.

BULL EL-01: BULL EL-01 is designed for use with light UAVs and makes use of the transformation of elastic energy accumulated in rubber cords into kinetic energy as its launch medium. Other system features include disassembly into two main components, 'easy' manual assembly and transportability by a single individual with the necessary equipment being carried in a backpack.

HERCULES AH-01: Standing for the High-Energy Rail Catapult UAV Launcher Evolved System, Aries describes the HERCULES AH-01 the evolutionary successor to the company's RO-1 equipment. As such, the system is noted as including the 'latest innovations' developed by the contractor and as being suitable for launching UAVs and targets in 'all' types of environments and conditions.

HERCULES AH-01 is billed as providing 'safe and point-based' launch for 'any' kind of TUAV and as being able to accommodate a 'wide' range

The RO-01 launcher in truck-mounted configuration *(Aries)* 1120777

of vehicle weights, geometries, configurations and launch speed requirements.

LAE: The LAE device is described as being a pneumatic high-power device that is capable of launching UAVs of up to 400 kg (881.8 lb) in weight at speeds of up to 70 m/s (300 ft/s). LAE is further noted as being of 'novel' design and as employing a combined acceleration and braking system.

Supplemental equipment: Alongside the described UAV launchers, Aries Ingeniería y Sistemas also produces a 'wide' range of ground and test equipment for UASs, with the spread including: a 'versatile' mobile engine test cell; UAV starters; braking systems and payloads; and UAV functional automatic test equipment. Of these, the engine test cell can be used with both jet and reciprocating power plants and can measure parameters such as engine thrust, fuel consumption, pressure and temperature ranges, as well as electrical power data. The equipment's control and data modules are fully automatic and are operated remotely from an independent mobile unit. Aries further notes that the cell can be customised to work with specific engines and auxiliary/ground power units with only 'minor adjustments'. Again, the engine test cell is reported as having been used by Spain's *Instituto Nacional de Ténica Aerospacial* (INTA, National Institute of Aerospace Technology) organisation to test Microturbo jet engines. Elsewhere, Aries also characterises itself as being a provider of a 'full' range of UAS services including specialised consultancy, system engineering, ISR/UAS training, 'full' UAS supply, maintenance and in-theatre operational services.

In terms of historical UAV launcher activity, Aries is known to have produced systems designated as the RO-1 and the RO-2. Of these, the RO-2 is described as having made use of rubber cored, nylon covered bungee cords as the launch medium, with an electric winch being used to tension the cords. Computation of the architecture's launch cradle position was based on AV weight and required launch speed and was calculated automatically. The release position of the launch cradle was adjusted by means of a pneumatic clutch and an energy absorber was

The Aries RO-01 at Spain's CEDEA range with a SIVA UAV *(IHS/Kenneth Munson)* 0558656

used to stop the launch cradle's forward motion post-launch. Again, RO-2 could take the form of a stand-alone or a trolley-mounted structure and had a launch velocity, an acceleration distance and a ramp angle of up to 25 m/s (82 ft/s), >8 m (26.2 ft) and up to 30° respectively.

Specifications
Dimensions
Ramp length: HERCULES AH-01..14.4 m (47.25 m)
    LAE..12 m (39.37 m)
Weights
Payload: ALPPUL LP-02..up to 300 kg (661.4 lb)
    ATLAS ME-01..up to 150 kg (330.7 lb)
    BULL EL-01..up to 14 kg (30.9 lb)
    HERCULES AH-01..up to 500 kg (1,102.3 lb)
    LAE..up to 400 kg (881.8 lb)
Overall: ALPPUL LP-02..7,000 kg (15,432.3 lb)
    ATLAS ME-01..2,000 kg (4,409.2 lb)
    BULL EL-01..10 kg (22.1 lb)
    HERCULES AH-01..6,000 kg (13,227.7 lb)
    LAE..14,000 kg (30,864.7 lb)
Performance
Launch velocity: ALPPUL LP-02................................up to 34 m/s (112 ft/s)
    ATLAS ME-01..up to 38 m/s (125 ft/s)
    BULL EL-01..up to 15 m/s (49 ft/s)
    HERCULES AH-01..up to 65 m/s (213 ft/s)
    LAE..up to 70 m/s (230 ft/s)
Continuous acceleration: HERCULES AH-01..........<8 g
Acceleration distance: HERCULES AH-01..............11 m (36.1 ft)
Ramp angle: HERCULES AH-01............................up to 20°
    LAE..5°

Status: Aries's RO-01 launcher (the HERCULES AH-01's predecessor) was developed to meet the Spanish SIVA (Aerial Surveillance Integrated System) requirement and as of early 2010, was being reported as having been in service with the Spanish Armed Forces (in support of the SIVA project). Most recently, Aries has continued to promote its ALPPUL LP-02, ATLAS ME-01, BULL EL-01, HERCULES AH-01 and LAE launchers and the no longer promoted RO-01 system is known to have been selected by South African contractor Denel for use with its Seeker UAV..
Contractor: Aries Ingeniería y Sistemas SA
Madrid.

Sweden

Bülow BTW 832

Type: Tow target winch.
Development: Bülow's first ram air turbine winch was the MBV-2S, of which 10 entered Swedish Air Force service in 1965 and 10 others with the Swiss Air Force. The BTW 832 is a computer-controlled development of the MBV-2S.
Description: The pod-mounted BTW 832 is a universal winch capable of reeling and docking all kinds of tow targets: wings, darts, sleeves and banners. In the case of sleeve targets, only the nosecone or MDI is docked, the cloth sleeve itself being jettisoned.

The pod is equipped with standard NATO 356 mm (14 in) bomb hooks for mounting to military aircraft; a special mounting plate is available for use from civil aircraft. A long, downward-tiltable launching arm permits a lowered towing/docking point, free of interference from the tow aircraft.

Reel-in/reel-out phases are computer controlled automatically, permitting the winch to be operated by the pilot if the tow aircraft is a single-seater. In fact, the radio control box enables him to operate two winches and two targets simultaneously if required. The winch pod is equipped with a built-in generator for electrical current self-supply.

Specifications
Dimensions
Length overall..2.55 m (8 ft 4.4 in)
Max width..0.45 m (1 ft 5.7 in)
Max depth (excl connectors)....................................0.40 m (1 ft 3.75 in)
Towline capacity (standard)
2.65 mm diameter...3,000 m (9,840 ft)
1.85 mm diameter...6,000 m (19,680 ft)
Weights
Weight empty, without drum......................................160 kg (353 lb)
Total weight, incl towline, target and MDI
    transmitter...310 kg (683 lb)

Performance
Max speed of towing aircraft....................................>M1.0
Aircraft speed for reel-in/reel-out......150–300 kt (278–556 km/h; 173–345 mph) IAS
Max reel-in/reel-out speed..12 m (39.4 ft)/s
Operating ceiling for reel-in/reel-out........................7,620 m (25,000 ft)
g limit..+5

Status: Over time, the BTW 832 has continued to be promoted and is understood to have been procured.
Contractor: Bülow Air Target Systems AB
Vätö.

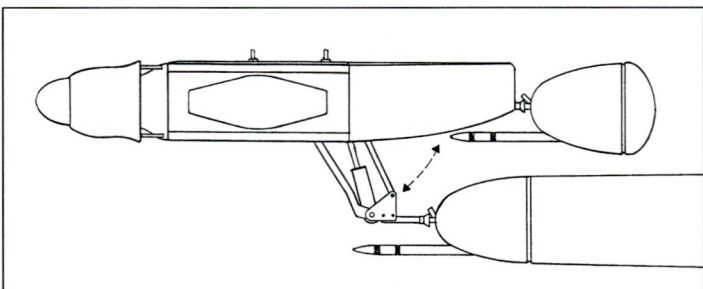

Diagram illustrating the lowerable launching arm of the BTW 832   0054197

The ram-air-driven BTW 832 under the wing of a Learjet   0054196

Bülow HTW 550 and TTW 500/550

Type: Target-towing winches for helicopter (HTW 550) and fixed-wing aircraft use (TTW 500/550).
Development: The HTW 550 helicopter towing winch was designed for Swedish Navy use to tow large sea-skimming targets such as the HRL 110

The Bülow TTW 550 *(Bülow)*   0054200

Bülow TTW 500 internal winch *(Bülow)*   0054199

HTW 550 helicopter winch towing a sea-skimming HRL 110 target *(Bülow)* 0054201

sleeve target. The TTW 550 is a modified version for installation on fixed-wing aircraft; the TTW 500 is similar except for its power source.

Description: In the **HTW 550**, the winch unit is encased in a streamline housing of GFRP and suspended by a thick, 12 m (39.4 ft) cable from the helicopter's cargo hook. It is guided and monitored from the cockpit by a control/display box, to which it is connected by a thin wire paid out through the cabin door. The petrol engine power plant drives the line drum through a hydraulic motor. For parking before and after a mission, the winch unit is clamped in a sturdy frame furnished with skid gear. Although designed specially for towing sea-skimming targets, the HTW 550 is a multipurpose reeling machine and can tow any kind of air target or radar calibration spheres. Options include a target altitude display in the helicopter and an MDI receiving station.

The **TTW 550** is a modified version, designed for fixed-wing aircraft use by mounting on racks attached to the aircraft's seat rails. It consists of three major parts: the winch unit, the hydraulic unit and the launching/pick-up unit. The first two of these are normally located near the CG, with the launch/pick-up arm on seat rails just inside the door. The arm can be rotated hydraulically between down for towing, half-way up for take-off and landing, and fully up for changing the MDI and sleeve. Cable speed adjustment is controlled, using a hand-held control box, by an electrical proportional flow valve.

The **TTW 500** is essentially similar to the TTW 550 except that the hydraulic system is electrically powered. Current consumption is 190 to 240 A (300 A peak) at 24 to 30 V.

Power plant: HTW 550 and TTW 550 are each powered by a 13.4 kW (18 hp) four-stroke air-cooled petrol engine; TTW 500 is powered by tow aircraft electrical system or separate battery pack.

Specifications

Dimensions
Towline capacity (2.3 mm diameter)
HTW 550 .... >3,000 m (9,840 ft)
TTW 500/550 .... 3,400 m (11,155 ft)

Weights
Total weight with empty line drum
HTW 550 .... 315 kg (694 lb)
TTW 500 .... 210 kg (463 lb)
TTW 550 .... 250 kg (551 lb)

Performance
Reel-in/reel-out speed: HTW 550 .... 5 m (16.4 ft)/s
TTW 500/550 .... 5–8 m (16.4–26.2 ft)/s

Status: Over time, the TTW 500 target towing winch has continued to be promoted. Again over time, the Swedish Navy and Jet Air (on behalf of the Danish armed forces) are understood to have procured examples of the HTW 550 and TTW 500 (two units) equipments respectively.

Contractor: Bülow Air Target Systems AB
Vätö.

Switzerland

RUAG launchers

Type: Family of hydraulic catapult launchers.

Description: RUAG Aviation (originally F+W) has developed a family of hydraulic catapult launchers for use with UAVs and targets, with identified configurations comprising a unit mounted on an all-terrain vehicle (marked 'A' within the specification data) and the trailer-mounted, air transportable (A400M, C-27, C-130 and C.160) Archer system ('B'). Both equipments can be erected by a two-man crew and comprise four portable track sections that are held together by clamps and pins and connected to a vehicle- or trailer-based forward section. No specialist tools are required for the assembly process and the track sections frames have a triangular profiled and are made of welded stainless steel tubes. RUAG also notes that even at full length, the fully assembled launch ramp can be 'easily' manoeuvred into an appropriate launch orientation.

Air vehicle launch is by means of a low-inertia, roller-mounted carriage that is accelerated by means of a synthetic strap that provides a traction force of up to 25 kN. A strap winder (driven by two hydraulic motors) is located close beneath the point of maximum launch velocity. When the carriage reaches the point above the strap winder, it causes the winch to reverse its rotation, thus decelerating and releasing the aircraft.

A general view of the trailer-based, air transportable variant of RUAG's hydraulic catapult UAV launcher *(RUAG)* 1329023

A Ranger UAV on a RUAG hydraulic catapult launcher *(RUAG)* 0079343

The empty carriage is 'quickly' stopped by the reversely driven motors, now acting as hydraulic pumps against the high gas pressure. Launch energy is supplied by a piston accumulator pressurised by nitrogen gas. The temperature-dependent gas pressure (maximum 350 bar; 5,076 lb/sq in) determines the catapult power, with the gas being permanently contained in a closed system. The piston accumulator is recharged in about 12 minutes by a small pump which is powered by either the vehicle battery or an external source. A microprocessor-based control is used to check different parameters such as oil pressure and volume, to actuate the electro-valves and control operational safety. Necessary electrical power is supplied from a vehicle battery or a 230 V line.

Specifications

Dimensions
- Catapult: Assembled length: A/B .... 18.50 m (60.69 ft)
- Width: A/B .... 1.96 m (6.43 ft)
- Catapult: Unassembled length: A .... 5.93 m (19.45 ft)
- Length: B .... 4.00 m (13.12 ft)
- Vehicle width: A .... 1.96 m (6.43 ft)
- B .... 2.10 m (6.89 ft)
- Height: A .... 3.32 m (10.89 ft)
- B .... 2.60 m (8.53 ft)
- Optional height: B .... 2.25 m (7.38 ft)

Weights
- Total weight: A .... 1,700 kg (3,748 lb - approx, excl vehicle)
- B .... 3,500 kg (7,716 lb - approx)

Payload
- A .... Up to 350 kg (772 lb)
- B .... Up to 320 kg (705 lb)

Performance
- Max power: A/B .... 600 kW (805 hp)
- Max lift-off velocity
  - A .... 33 m (108.3 ft/s)
  - B .... 63–66 kt (117–122 km/h; 73–76 mph)
- Max acceleration: A/B .... 7–8 g
- Max acceleration rate: A/B .... 100 g/s
- Acceleration time: A/B .... 0.7 s
- Acceleration distance: A/B .... 14 m (46 ft)
- Recharge time: A/B .... 12 min (approx)

Status: Over time, RUAG has continued to promote its UAV launch systems, with examples being understood to have been procured by the Swiss Air Force for use with the Ranger UAV.

Contractor: RUAG Aviation
Emmen.

Ukraine

SIS UAV launchers

Type: Family of UAV launchers.

Development: The Design Bureau "VZLET" (a division of SIS since 2003) has been designing UAVs and associated equipment since 1997.

Description: Over time, SIS has produced a range of launchers to support its UAVs, the known details of which are as follows:

CGM-6C: Designed to launch UAVs of up to 50 kg (110.2 lb) in weight, the CGM-6C is described as being a modular equipment that incorporates a 6 m (19.7 ft) long launch rail and can be 'speedily' set up on unprepared ground. Disassembled, CGM-6C can be carried in a 'standard military off-road car'.

CLLGW-4C: Designed to handle UAVs of up to 20 kg (44.1 lb) in weight, the CLLGW-4C is understood to comprise an electric winch mechanism and either an AV cradle or launch rail (as required). CLLGW-4C is understood to weigh 98 kg (216.1 lb), have a length of 2 m (6.6 ft) and to be suitable for use at a range of operating temperatures and altitudes. Like CGM-6C, CLLGW-4C requires no ground preparation prior to set-up.

CLLTS-3/-3C: Designed for use with UAVs with a launch weight of up to 30 kg (66.1 b), the CLLTS-3/-3C equipment are understood to make use of an inclined, 3 m (9.8 ft) long launch rail, with launch impetus being provided by a 12 m (39.3 ft) rubber bungee. CLLTS-3/-3C equipment can be deployed by a single individual. As with SIS's other catapult launchers, the CLLTS-3/-3C equipment require no ground preparation prior to set-up.

CTP-3: The CTP-3 pneumatic launcher is designed for use with UAVs weighing of up to 15 kg (33.1 lb) and can be carried in a light military off-road vehicle. When disassembled, CTP-3 comprises two modules, the largest of which is understood to measure 1.5 m (4.9 ft).

Specifications

Dimensions
- Length: CGM-6C launch rail .... 6 m (19.7 ft)
- CLLGW-4C system .... 2 m (6.6 ft)
- CLLTS-3/-3C bungee .... 12 m (39.3 ft)
- CLLTS-3/-3C launch rail .... 3 m (9.8 ft)
- CTP-3 launch rail .... 3 m (9.8 ft)

Weights
- UAV: CGM-6C .... up to 50 kg (110.2 lb)
- CLLGW-4C .... up to 20 kg (44.1 lb)
- CLLTS-3/-3C .... up to 30 kg (66.1 lb)
- CTP-3 .... up to 15 kg (33.1 lb)

Performance
No data supplied

Status: Most recently, SIS has continued to promote its range of UAV launchers.

Contractor: Scientifically Industrial Systems Ltd
Kharkov.

United Arab Emirates

Adcom pneumatic launcher

Type: Trailer-mounted pneumatic launch system.

Description: The Adcom pneumatic launcher is designed for use with ground-launched aerial vehicles and can be deployed on 'confined, restricted or unimproved' sites. Functionally, compressed air is used to pressurise accumulator tanks which, in turn, propel a launch cradle to take-off velocity.

Specifications

Dimensions
- Ramp length .... 9.5 m (31.2 ft)
- Acceleration distance .... 4.5 m (14.8 ft)
- Overall width .... 2.5 m (8.2 ft)

Weights
- Payload .... Up to 150 kg (331 lb)

Performance
- Velocity .... 30–45 m/s (98-148 ft/s)
- Maximum pressure capacity .... 172.4 bar (2,500 psi)
- Launch preparation time: First .... 5 minutes
- Subsequent .... 30 minutes

Status: Over time, the Adcom pneumatic launcher has been promoted.

Contractor: Adcom Systems
Abu Dhabi.

United Kingdom

Airborne Systems retarder/recovery parachutes

Type: Retarder and recovery parachutes.

Development: Over time, Airborne Systems Europe (formerly Irvin-GQ) has developed a number of recovery systems for RPVs and UAVs and has worked closely with the air vehicle designers ('from the outset') to ensure a balanced solution to the requirement. The company adopts a 'system' approach which covers all related areas of technology such as mechanisms, flight dynamics, three-dimensional dynamic modelling and the associated engineering disciplines. Parachutes, harnesses and airbags have been developed and tested over more than 90 years to demanding specifications.

Description: Solutions to customer requirements range from basic stand-alone parachutes of a conventional round design to a complete self-contained package incorporating high-drag and stable cruciform retarders, either singly or in clusters. Recovery systems range from hand-packed parachutes to high-density, hydraulically packed cassettes which are returned to the manufacturer for re-assembly. Conflicting requirements of mass, volume, rate of descent and operability are resolved as part of the overall air vehicle configuration.

Over time, a comprehensive range of recovery systems has been produced, with features including safe, easy and reliable operation, repairability, fail-safe operation and damage tolerance. Most of the contractor's recovery systems have been designed to meet specific customer requirements, have been proven in service and are able to provide off-the-shelf starting points for new systems, with only 'occasional re-packaging' being required. Options include: modular

An Airborne Systems UAV recovery parachute in action *(Airborne Systems)*

construction; single- or multi-shot operation; full spares support; packing and repair; packing, training and documentation; commercial and full ISO9001 standard compliance; a bespoke design/development service and build to print.

Status: Over time, Airborne Systems Europe retarder/recovery parachutes have been installed aboard a range of UAVs and targets including the Banshee, CL-289, Crecerelle, MQM-107 Streaker, Phoenix, and SAGEM Sperwer.

Contractor: Airborne Systems Europe
Bridgend, Glamorgan.

Tasuma launch systems

Type: Aerial target launcher.

Description: Tasuma has developed a family of UAV/target launchers, the known detail of which are as follows:

A3 Observer launcher: As its designation implies, the A3 Observer launcher has been specifically designed for use with the Cranfield A3 Observer UAV. As such, it offers an adjustable elevation angle of between 5 and 20°, is actuated by an electro-hydraulic ram and features four corner jacks with which to level its trailer/launcher assembly for operations on uneven ground.

LTL 1: The LTL 1 launcher is designed for use with 'large' targets (similar in design to the Vindicator and Banshee types) and has been supplied to Pakistan.

TML 2: The TML 2 launcher is designed to permit the safe launching of small target aircraft weighing up to 7.5 kg (16.5 lb). As such, it fits into a 'compact' transit and storage container and can be carried and operated by one person. other features include triple safety devices, power band tensioning by means of a 'small' geared hand winch unit and use of a remote manual trigger.

TML 3: TML 3 is designed for the launch of surveillance UAVs such as Tasuma's CSV 20 and can be operated by a single person (a second individual being required for system transportation). TML 3 features three safety devices, power band tensioning by means of a 'small' geared hand winch and vehicle release by means of a remote manual trigger.

TML 3 Ultima: TML 3 Ultima is a TML 3 derivative that is designed for use with the Belgian Army's Ultima surveillance UAV. As such, it is built from stainless steel (as an anti-corrosion measure) and features a custom carriage unit that is tailored to the Ultima airframe.

TML 4: The TML 4 launcher is a lightweight, portable launcher that is designed for use with 'light' target and surveillance UAVs that weigh between 18 and 35 kg (40 and 77 lb).

Specifications

A: A3 Observer B: LTL 1 C: TML 2 D: TML 3 E: TML 3 Ultima F: TML 4

Dimensions
Launcher length: B..6.0 m (19.7 ft)
C..1.8 m (5.9 ft)
D..4.6 m (15.1 ft)
E..5.0 m (16.4 ft)
F..4.5 m (14.8 ft)
Operating length: A..7.5 m (24.6 ft)
Folded length: A..4.0 m (13.1 ft)
Trailer length: A..5.5 m (18.1 ft)
Transit length: B..4.0 m (13.1 ft)
Transport length: D..2.0 m (6.6 ft)
E..2.1 m (6.9 ft)
F..3.2 m (10.5 ft)

Weights
Maximum launch weight: A..50 kg (110 lb)
B..88 kg (194 lb)
C..7.5 kg (16.5 lb)
D..15 kg (33 lb)
E..20 kg (44 lb)
F..35 kg (77 lb)
Minimum launch weight: A..20 kg (44 lb)
B..50 kg (110 lb)
D/E..6 kg (13 lb)
F..18 kg (40 lb)
Launcher weight: C..7.2 kg (16 lb)
D..32 kg (71 lb)
E..52 kg (115 lb)
F..135 kg (320 lb)

Performance
Launch speed: A..20-35 m/s (66-112 ft/s)
B..100 km/h (62 mph)
C..40 km/h (25 mph)
D/E..60 km/h (37 mph)
F..75 km/h (47 mph)

Status: Most recently, Tasuma (UK) has continued to promote its range of UAV launch systems.

Contractor: Tasuma (UK) Ltd
Blandford Forum, Dorset.

UTSL AVL

Type: Bungee-powered UAV (Target and UAS) launch system.

Description: UTSL describes its AVL (Air Vehicle Launcher) as being an 'exceptionally powerful' bungee-powered UAV launch system that incorporates a propeller guard (to ensure) target starter safety and lanyard operated safety and shuttle releases. The equipment is 'completely' weather proof and can be supplied with optional covers and heating systems for cold weather functionality. UTSL's AVL can be configured for different AV weights and launch speeds (thereby making 'each system uniquely versatile') and can be readied for target launch within 15 minutes of arrival at an operating site. The UTSL AVL is typically mounted on a mobile platform (such as UTSL's own ground support trailer that can carry three air vehicles and associated operating systems, a hydraulic launcher stowage system and a logistical/support package that includes first-line spares, aerial target fuel (and an associated fuelling system), a target starter system and batteries), with permanent installation on a single-use vehicle as an alternative. Here, the two described applications can handle 50 to 100 kg (110.2 to 220.5 lb) AUW AVs, with a 'mini' configuration handling 10 to 50 kg (22.1 to 110.2 lb) AUW vehicles. Other system features include a removable shuttle, a launch speed indicator and a safe lanyard release.

A ship-portable UTSL AVL standard configuration *(UTSL)*

Launch and recovery systems > United Kingdom – United States > UTSL AVL – Butler UAV recovery systems

A standard portable UTSL AVL with MSAT-500 NG mounted *(UTSL)* 1435443

UTSL's LWL can accommodate AVs with all-up weights of between 6 and 22 kg *(UTSL)* 1435444

Specifications

Dimensions
Launcher: Length unfolded..13 m (42.7 ft)
Length folded...5 m (16.4 ft)
Weights
Typical AV.. 90 kg (198.4 lb)
AV range... 55-130 kg (121-286 lb)
Performance
AV release speed.........................>35 m/s (>115 ft/s), depending on AV mass

Status: UTSL reports its AVL as being 'in service with various customers around the world' and continues to promote the equipment. The UTSL Light Weight Launcher (LWL - see separate entry) caters for air vehicles in the 6-22 kg range.

Contractor: Universal Target Systems Ltd
Challock, Kent.

UTSL LWL

Type: Lightweight UAV/target launcher.

Development: *IHS Jane's All The World's Aircraft:Unmanned* first became aware of the UTSL LWL UAV launcher during 2010.

Description: UTSL describes its LightWeight Launcher (LWL) as being an 'exceptionally light yet powerful' device that has recently been upgraded with new gas struts and an improved arresting system and which, when compared with previous iterations, offers reductions in overall length and weight. The LWL can accommodate a wide range of operational temperatures and can be configured with reduced or increased power in order to make it suitable for use with 'varying' UAV types. The LWL architecture is 'completely' weather proof and is supplied with a ruggedised housing case that unfolds to form the launcher's base. An LWL can be readied for UAV/target launch within 10 minutes of on-site arrival, can be set-up as a stand-alone (non-use of the housing case as a base) device when installed on uneven surfaces and is typically packaged (together with a tactical control station, spare AVs and other operating equipment) aboard a 4 × 4 vehicle such as a Land Rover or a Supacat. Other system features include a ruggedised DC electric winch, an 'easily removable' shuttle, a launch speed indicator, integral safety systems, safe lanyard release and a 'minimal' maintenance requirement.

Specifications

Dimensions
Length...2.1 m (6.9 ft), unfolded
Weights
Launcher.. 20 kg (44.1 lb)
Typical AV.. 6-22 kg (13.2-48.5 lb)
Performance
Air vehicle release speed.........................>16 m/s (>52.5 ft/s, depending on AV mass)
Operational temperature range......................–25°C to +60°C (–13°F to +140°F)

Status: Most recently, UTSL has continued to promote the LWL UAV/target launcher.

Contractor: Universal Target Systems Ltd
Challock, Kent.

United States

Butler UAV recovery systems

Type: Parachute recovery systems.

Development: The Butler Parachute Systems Group (BPSG) Inc specialises in the design, engineering, manufacture and testing of parachutes, recovery systems and related items, with a particular emphasis on the rapid development of systems for 'unusual' applications. The company also provides recovery systems design, consulting, manufacturing and testing services to government agencies, aerospace firms and other parachute companies. Again, BPSG does not undertake build-to-print for government organisations and its Butler Unmanned Parachute Systems (BUPS) LLC subsidiary specialises in the development of recovery systems for UAVs and weapons delivery. BPSG's Flight Test Support Services LLC provides a parachute and recovery system test capability for equipments ranging in size from sub-munitions to approximately 1,814 kg (4,000 lb) and at speeds of up to 850 kt (1,574 km/h; 978 mph).

A third affiliate (Butler Aerospace Technology (BAT) Inc) has developed the BAT Sombrero Slider (US Patent 5,890,678, issued during 1999) that provides self-modulating, continuous control of the inflation process of any axis-symmetric parachute canopy. The BAT Sombrero Slider also prevents the occurrence of inversion-type malfunctions, thereby improving reliability by 'several' orders of magnitude. Butler's H-X series of canopies are equipped with this device and several versions of the HX family are approved by the US Federal Aviation Administration (FAA) under TSO C-23d. All BPSG division have utilised the BAT Sombrero for all new products developed since 1999.

Description: UAV-related recovery systems that have (over time) been developed by BUPS include the following:

Mirach 100/5: During 1994, Butler was awarded a 'substantial' contract to design, develop and test a complete recovery system for the Mirach 100/5 target drone. The resultant architecture is a two-stage system, with a 2.44 m (8 ft) reefed drogue parachute and an 11.28 m (37 ft) diameter main canopy. As such, it can recover up to 300 kg (661 lb) at M0.9 at altitudes of up to 4,575 m (15,000 ft). The system has been tested and qualified at up to 315 kg (695 lb) and 468 kt (867 km/h; 538 mph). Again, the system includes completely reusable drogue and main riser releases (with a common body) and employs redundant, mechanically actuated, time-delay pyrotechnics for drogue release. The main chute is deployed using a single, electrically activated pyrotechnic.

Predator: During 1994 (and within 120 days), Butler Parachute Systems (BPSG's predecessor) designed, built and tested a recovery system for the Predator UAV. Capable of accommodating a maximum weight of 862 kg (1,900 lb), the architecture utilised a rocket extraction system, with a pilot chute providing for positive main chute deployment until such time as full line stretch was reached.

The main chute used had a nominal diameter of 21.49 m (70.5 ft) and featured an extended skirt, tri-conical canopy with drive vents to provide a 'modest' glide ratio with minimum oscillation. A completely new test vehicle (together with aircraft and ground handling equipment) was designed and built by Butler specifically for this programme. Moving forward, 2002 saw BUPS working with General Atomics Aeronautical Systems to develop a new modular recovery system for the Predator. Here, a new General Atomics-sourced carbon fibre shell (with BUPS provided flaps and 'other components') was used and the whole is designed to be easily installed/removed and swapped with a fuel bladder that is installed during operational missions (the described recovery system being employed during training sorties).

A new rocket mount complements the modular nature of the revised recovery system and an HX variant of the original main canopy is used. Here, a BAT Sombrero Slider application allows an expansion of the architecture's performance envelope to 998 kg (2,200 lb) and speeds of up to 160 kt (296 km/h; 184 mph).

Shadow 200T: Designed and built for AAI Corporation. The main canopy, a 4.88 m (16 ft) diameter HX-200, is housed in a BPS-designed and manufactured container with a vacuum-formed shell, stainless steel stiffener and nylon flaps. The system weighs less than 3.2 kg (7lb) and can be installed or removed in 20 minutes.

During 2008 (and after the Shadow UAVs weight had increased to 168 kg (370 lb), BUPS designed, tested and began manufacture of an updated version of the Shadow recovery system. Here, the new configuration incorporated the HX-400/16 (certified under FAA TSO C-23d) which provides twice the drag area and reduces the rate of descent to less than 70 per cent of that of the original system.

The new architecture also utilises a BPSG-designed and manufactured container (of a slightly different shape to that of the original) with a vacuum-formed shell, stainless steel stiffeners and nylon flaps. The new container weighs approximately 5 kg (11 lb) and can be installed or removed in 20 minutes.

Status: Most recently, BUPS is noted as 'continuing to design, develop and manufacture UAV recovery systems'. Again, the BPSG is reported as continuing to design, develop and manufacture parachute systems for a 'wide range of other applications'.

Contractor: Butler Unmanned Parachute Systems LLC
Roanoke, Virginia.

ESCO UAV launch systems

Type: Hydraulic launchers.

Development: The Engineered Arresting Systems Corporation (ESCO) has more than 40 years' experience in the design, development and manufacture of equipment that will launch and/or recover fighter and civil aircraft, the Space Shuttle, and numerous UAVs.

Description: ESCO notes that key features of all its launch and recovery systems are flexibility and mobility and claims to be able to build customised systems to 'nearly every set of requirements'. The company offers launchers for both fixed and mobile applications, making use of a variety of power sources, launch rail options and manual or automatic functionality. ESCO launchers can accommodate many different UAV configurations, with their design being based on performance and cycle times. Using its 40 year's worth of experience in the field, ESCO has developed a computer simulation model that predicts launch performance, using data from design parameters such as UAV weight, cylinder size and reeve ratio to ensure that performance criteria are maximised.

Predicted launch performance results (such as acceleration (*g*-loading), velocity and time) can be plotted and reviewed to ensure that requirements are being met in the most efficient manner. ESCO launchers utilise dry nitrogen (maintained in a closed system and pressurised in a hydraulic accumulator) as their launch medium and the contractor claims that its hydraulic-pneumatic technology provides 'reliable, repeatable launch dynamics' under all 'required operational environments'. ESCO claims that experience has shown that its closed-loop hydraulic-pneumatic approach is able to provide requisite performance within the −40°C to +66°C (−40°F to +150°F) temperature range, at altitudes of up to 3,658 m (12,000 ft) above sea level.

HP-2002: HP-2002 is designed for use with UAVs with launch weights of between 45 and 113 kg (100 and 250 lb) and is trailer-mounted.

HP-3003: HP-3003 is described as being a high mobility launcher that can handle UAVs that have a weight of between 68 and 240 kg (150 and 530 lb) and a launch speed of up to 75 kt (139 km/h; 86 mph). The system is fully integrated on a towable trailer and folds down to create a smaller footprint.

HP-3402: HP-3402 is a truck-mounted launcher that can accommodate UAVs with weights of up to 204 kg (450 lb) and launch them at speeds of up to 70 kt (130 km/h; 81 mph).

HP-3407: HP-3407 is a truck-mounted launcher that makes use of the M-811 military vehicle and is specifically designed for use with R4E SkyEye surveillance UAV. Like the HP-3401 launcher, HP-3407 features a hinged launch rail for ease of transport.

HP-3502: The HP-3502 launcher is noted as being able to handle UAVs with launch weights of up to 556 kg (1,225 lb) and launch velocities of up to 80 kt (148 km/h; 92 mph). The HP-3502 launcher is mounted on an M-814 or M-900 truck.

Specifications

Weights	
UAV launch weight: HP-2002	45-113 kg (100-250 lb)
HP-3003	68-240 kg (150-530 lb)
HP-3402	up to 204 kg (450 lb)
HP-3407	up to 340 kg (750 lb)
HP-3502	up to 556 kg (1,225 lb)
Performance	
UAV launch velocity:	
HP-2002/HP-3407	up to 65 kt (120 km/h; 75 mph)
HP-3003	up to 75 kt (139 km/h; 86 mph)
HP-3402	up to 70 kt (130 km/h; 81 mph)
HP-3502	up to 80 kt (148 km/h; 92 mph)
Launch acceleration:	
HP-2002/HP-3502	up to 10 g
HP-3003	up to 12 g
HP-3402/HP-3407	up to 8 g

Status: ESCO's UAV launchers of the type described are understood to have been procured for customers including Flight Refuelling in the UK and Developmental Sciences in the US. The HP-3407 is noted as having seen service with 'various' offshore users. ESCO continues to market the HP-2002, HP-3003, HP-3402, HP-3407 and HP-3502 equipments.

Contractor: Engineered Arresting Systems Corporation (ESCO) (a division of Zodiac Aerospace)
Aston, Pennsylvania.

ESCO UAV recovery systems

Type: Parachute and net retrieval systems.

Description: Over time, ESCO has developed a range of UAV recovery systems, the known details of which are as follows:

Mid-Air Retrieval System (MARS): MARS was a mid-air recovery system for parachute-borne UAVs that was adaptable to a 'wide' range of helicopters and fixed-wing aircraft and made use of ESCO's Model 80H tension-controlled, energy-absorbing winch.

Mobilenet® 2000: As its designation suggests, Mobilenet® 2000 is a mobile UAV recovery system that incorporates a nylon capture net and ESCO's Water Twister energy absorbing technology. Functionally, the UAV is flown into the net, with the system's energy absorbing sub-system killing its forward speed. As such, the application is billed as being 'simple' to set-up (requiring no special tools or operator skills) and other system features include 'permanent' installation on two trailers (with an integral Water Twister energy absorber on each trailer), retractable wheels (for 'rapid' deployment), soil installation stakes, a spur-gear hand winch (for manual stanchion raising and lowering), internally threaded anchors (for use on runways) and an options package that includes a hydraulic power unit, a hydraulic jackhammer (to assist with emplacement) and custom net lengths.

Shipboard Pioneer ARresting System (SPARS): SPARS was a net recovery system that was specifically developed for use with US Navy Pioneer UAVs aboard surface ships.

STARS: STARS is described as being a fixed, land-based UAV net recovery system.

Specifications

Dimensions	
Standard net: Length	60.0 m (196.9 ft)
Centreline height	1.5 m (4.9 ft)
Mobile mode system: Length	5.4 m (17.7 ft)
Width	1.6 m (5.3 ft)
Height	1.3 m (4.3 ft)
Weights	
System	3,629 kg (8,000 lb, nominal, excluding spares)
UAV	500-650 kg (1,102-1,433 lb)
Performance	
Deceleration force	<2 g
Engagement speed	150 kt (278 km/h, 173 mph)
Runout	183 m (600.4 ft)
Deployment area	60 m (196.9 ft)

Status: Looking at the cited systems individually, MARS is understood to have been first fielded during the 1960s and is reported to have effected in excess of 20,000 UAV retrievals during its service life. MARS is credited as having successfully recovered UAVs with weights of between 91 and 1,814 kg (200 to 4,000 lb) on an operational basis under battlefield conditions. Turning to ESCO's net systems, a combined vertical/horizontal net application is known to have been successfully demonstrated during UAV-CR trials conducted at the US Army's Yuma Proving Grounds in the early 1990s, while the SPARS system was used operationally during the 1991 Gulf War. ESCO has continued to market its UAV recovery systems with an emphasis on the Mobilenet 2000® application.

Contractor: Engineered Arresting Systems Corporation (ESCO) (a division of Zodiac Aerospace)
Aston, Pennsylvania.

Meggitt RM series

Type: Reeling machines.

Description: The RM series are pilot-operated, two-way, reeling systems that permit the recovery and re-use of tow targets. As such, they incorporate automatic reeling functions ('out', 'stop' and 'in'); are fitted with dual/redundant towline cutters; have a fail-safe braking system and are available with centreline or side-mounted target launchers. Over time, *IHS Jane's* sources have identified the following RM series variants:

RM-12A: RM-12A is described as being an 'improved' version of the earlier RM-12 equipment whose small size and lightweight construction permit its use on small trainer-type aircraft such as Swiss Air Force PC-9s. A single, 'compact' control panel in cockpit allows the pilot to monitor the system without interfering with her/his normal workload.

RM-24: RM-24 is described as being a 'mid-size' system that is still 'small enough' for use on trainer-type aircraft. Single-stage turbine and planetary geared transmission integrate with an electric fail-safe brake and the baseline system is sized to carry and deploy sleeve-type targets. This said, RM-24 is noted as being configurable for 'any' type of tow body.

RM-30A: RM-30A is described as being a 'high-performance' equipment whose cockpit control panel provides selectable digital display of towline length and tension. Other system features include a

An RM-30A1 reeling machine mounted on one of the wing stations of an Apache Aviation Hawker Hunter `aggressor' aircraft *(Apache Aviation)* 1299072

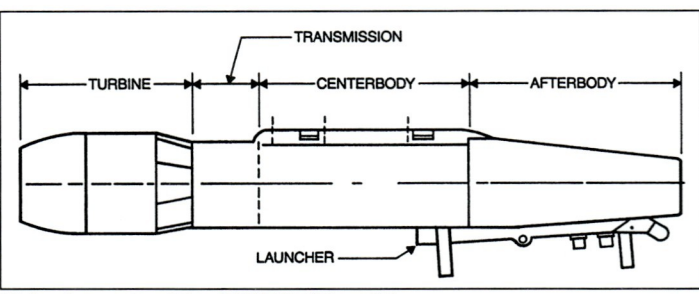

Main elements of the RMK-35 *(Meggitt)* 0518304

fixed-blade ram air turbine (with chain drive transmission); a fail-safe brake system (mounted directly on the towline spool) and a Microprocessor Logic Control Module (MLCM) that provides user-defined system control with functional presets, self-test and a non-volatile memory. The RM-30A and RM-30A1 (see following) systems have been certified for use with Alpha Jet, F-4, F-5, F-15, F-16, Falcon 20, Gulfstream G-100, Hunter, Il-28, L-59 and Learjet 35/36 aircraft.

RM-30A1: RM-30A1 is described as being an 'upgraded' version of the RM-30A that features an 'advanced' ram air turbine that produces 50 per cent more horsepower than that of the RM-30A, thereby enabling faster reeling speeds. Again, RM-30A1 eliminates all pneumatic and hydraulic technology (RM-30A1 is 'all electric') and is compatible with those AVs specified for the RM-30A.

RM-30B: RM-30B is derived from Meggitt's RM-30A and RMK-35 equipments and offers the RMK-35's load capacity combined with the towline capacity and reeling speed of the RM-30A (whose turbine it also shares). RM-30B has been developed for use aboard 'modern' jet fighters and utilises a single reeling machine configuration to support both short gunnery and long tow missile target missions. Again, RM-30B is compatible with the RADOPS real-time scoring system and its cockpit digital display includes machine status and bullet scores. RM-30B is compatible with the A-4, A-7, F-4, F-5, F-15, F-16 and F/A-18 fast jets.

RM-62: RM-62 has been derived from the RM-24, RM-30B and RMK-35 equipments (which see) and has been developed to operate from 'modern' jet fighters in support of short towline length air-to-air gunnery and long towline length AAM and SAM target missions. The equipment's control/display panel includes switches for pre-selection of the length of towline to be reeled out. RM-62 is compatible with the TDU-25, TDU-34/A, TDU-37B and Blazer 2 targets and the RADOPS real-time scoring system.

RMK-35: RMK-35 (also known as the AGTS-36) is derived from the RM-30A system and is designed for use with TDK-39 tow target (which see). System features include an automatically programmed control unit (with recovery and lock-in of a target to the launcher being executed via 'simple discrete commands') and an afterbody that contains a control module and scoring system components. RMK-35 is compatible with the F-4, F-15 and F-16 fast jets.

Power plant: Single-stage ram air turbine with bidirectional rotation and nominal speed of 4,000 rpm. Power rating at 250 kt (463 km/h; 288 mph) CAS at S/L is 8.9 kW (12 hp) for RM-12A, 17.9 kW (24 hp) for RM-24, 22.4 kW (30 hp) for RM-30A/30B and RMK-35, 33.6 kW (45 hp) for RM-30A1 and 46.2 kW (62 hp) for RM-62.

Specifications

Dimensions
- Length: RM-12A ... 1.43 m (4 ft 8.4 in)
- RM-24 ... 2.34 m (7 ft 8.0 in)
- RM-30A/30A1 ... 2.84 m (9 ft 4.0 in)
- RM-30B ... 3.09 m (10 ft 1.5 in)
- RMK-35 ... 2.91 m (9 ft 6.5 in)
- RM-62 ... 2.83 m (9 ft 3.4 in)
- Diameter: RM-12A ... 0.25 m (10.0 in)
- RM-24 ... 0.31 m (1 ft 0.3 in)
- RM-30A/30A1/30B, RMK-35 ... 0.44 m (1 ft 5.3 in)
- RM-62 ... 0.46 m (1 ft 6.3 in)
- Height: RM-12A ... 0.28 m (11.1 in)
- RM-24 ... 0.38 m (1 ft 2.8 in)
- RM-30A/30A1 ... 0.48 m (1 ft 6.8 in)
- RM-30B, RMK-35 ... 0.53 m (1 ft 8.9 in)
- RM-62 ... 0.57 m (1 ft 10.3 in)
- Spool volume: RM-12A ... 4.95 dm3 (302 cu in)
- RM-24 ... 17.91 dm3 (1,092 cu in)
- RM-30A/30A1/30B ... 41.00 dm3 (2,500 cu in)
- RMK-35 ... 25.75 dm3 (1,570 cu in)
- RM-62 ... 77.90 dm3 (4,750 cu in)
- Towline length: RM-12A ... 1,524 m (5,000 ft)
- RM-24 ... 3,962 m (13,000 ft)
- RM-30A/30A1/30B ... 9,754 m (32,000 ft)
- RMK-35 ... 610 m (2,000 ft)
- RM-62 ... 13,106 m (43,000 ft)

Weights
- Weight excl towline: RM-12A ... 63.5 kg (140 lb)
- RM-24 ... 90.7 kg (200 lb)
- RM-30A/30A1/30B ... 199.6 kg (440 lb)
- RMK-35 ... 208.7 kg (460 lb)
- RM-62 ... 249.5 kg (550 lb)
- Weight incl towline: RM-12A ... 84.8 kg (187 lb)
- Load rating: RM-12A, RM-24 ... 1,701 kg (3,750 lb)
- RM-30A/30A1/30B, RMK-35, RM-62 ... 2,721 kg (6,000 lb)
- Towline load limit: RM-12A, RM-24 ... 1,134 kg (2,500 lb)
- RM-30A/30A1/30B, RMK-35, RM-62 ... 1,814 kg (4,000 lb)
- Reel-in load limit: RM-12A ... 227 kg (500 lb)
- RM-24 ... 363 kg (800 lb)
- RM-30A/30A1/30B, RMK-35, RM-62 ... 615 kg (1,355 lb)

Performance
- Tow aircraft max speed (towing)
  - except RM-62 ... 600 kt (1,111 km/h; 690 mph) CAS
  - RM-62 ... 500 kt (926 km/h; 575 mph) CAS
- Tow aircraft max speed (reeling)
  - all ... 250 kt (463 km/h; 288 mph) CAS
- Max altitude (towing): RM-24, RM-62 ... 12,200 m (40,000 ft)
- Max altitude (reeling): RM-24, RM-62 ... 7,620 m (25,000 ft)
- Max acceleration (towing): all ... 6 g
- Average reel-in speed: RM-12A ... 229 m (750 ft)/min
- RM-24 ... 457 m (1,500 ft)/min
- RM-30A, RM-30B ... 610 m (2,000 ft)/min
- RM-30A1 ... 792 m (2,600 ft)/min
- RMK-35 ... 244 m (800 ft)/min
- RM-62 ... 1,067 m (3,500 ft)/min
- Operating voltage: all ... 28 V DC
- Operating current (reeling)
  - except RM-24 ... 3 A
  - RM-24 ... 4 A

Status: Over time, RM series users are known to have included the Swiss Air Force (RM-12A) and the armed forces of Japan, South Korea, the UK and the US (all RM-30A/-30A1). Most recently, Meggitt Defense Systems has continued to promote the RM-30A1, RM-30B and RM-62 reeling machines.

Contractor: Meggitt Defense Systems Inc
Irvine, California.

Sierra Nevada PLS

Type: Precision landing system.

Description: The Sierra Nevada PLS is designed to provide precision or automatic take-off and landing guidance for 'all' types of UAVs whether fixed-based, deployed or shipboard. As such, it is scalable, employs DGPS and a tracking radar (which can be employed independently or in combination and can provide independent or blended navigation solutions) and combines a DGPS ground system with the ground tracking and airborne subsystems from the company's Tactical Automatic Landing System (TALS). Here, the DGPS ground system includes GPS, datalink and processor/maintenance subsystems. Other features of the architecture include standard VHF or customer specified data-links and omnidirectional coverage that facilitates custom design of approach/departure paths.

Specifications

Performance
- DGPS: Coverage ... Landing surface to 130 n miles (241 km; 150 miles) - standard VHF datalink
- Navigation sensor accuracy (H × V)
  - Land-based application ... 0.6 × 0.9 m (2.0 × 3.0 ft)
  - Shipboard application ... 0.5 × 0.6 m (1.5 × 2.0 ft)
- Tracking radar: Frequency ... 35 GHz
- Coverage ... Landing surface to ±12° of runway centreline out to a range of 8.0 n miles (14.8 km; 9.2 miles)

The PLS's DGPS ground unit *(Sierra Nevada)* 1112227

Specifications

Navigation sensor accuracy	
Range bias	±0.3 m (1.0 ft)
Azimuth bias	< 0.2°
Elevation bias	< 0.1°

Status: Over time, the Sierra Nevada PLS has formed part of Raytheon's Global Hawk Launch and Recovery Element (LRE).
Contractor: Sierra Nevada Corporation
Sparks, Nevada.

Sierra Nevada TALS and UCARS-V2

Type: Automatic UAV recovery/tactical automatic landing systems.
Description: *TALS:*

Sierra Nevada's Tactical Automatic Landing System (TALS) provides a day/night, all-weather, automatic landing and take-off capability for UAVs operating in tactical or fixed-base land environments. As such, it is capable of interfacing with virtually any fixed- or rotary-wing UAV and offers automatic landing capabilities ranging in scope from a simple position sensor for air-derived guidance systems to a fully integrated, ground-derived air vehicle controller. TALS consists of a ground-based radar Track Subsystem **(TS)** and an air vehicle-mounted Airborne Transponder Subsystem **(ATS)**. Looking at the system in more detail, the TALS TS comprises a Track Control Unit (TCU), a pedestal, an antenna/radome assembly and an interrogator unit and precisely

The TALS/ATOLS land-based UAV automatic take-off and landing system *(Sierra Nevada)* 1112228

measures the UAV's position relative to the touchdown point. The TS is capable of tracking a UAV augmented with an AS at a maximum distance of approximately 8.0 n miles (14.8 km; 9.2 miles). Of the described elements, the TCU's primary function is to provide power and control for the other TS hardware line replaceable units. The TCU also includes the computing resources for the system's recovery software and is the interface for bi-directional RS-422 data transfers between the TS and a GCS. The equipment's pedestal is a two-axis gimbal system with a high gain, parabolic, dish-tracking antenna mounted to its gimbal. When power is removed, elevation and azimuth stow pins are deployed to prevent pedestal movement and it has lifting handles that allow a two-man team to carry it. The pedestal's legs contain adjustable bases which allow for precise leveling of the base to within ± 0.5° using an integral bubble level. The TS's antenna/radome assembly comprises a 35 GHz, high-gain, 46 cm (18 in) parabolic dish tracking antenna; a four-part waveguide antenna feed (with support arms) and a low composites radome. This assembly is driven by elevation and azimuth motors. The equipment's interrogator unit is the TS's radio frequency transceiver and provides the signal processing necessary to track the AS. As such, it is frequency-locked to 70 MHz below the approximately 35 GHz AS transponder frequency and incorporates a top-mounted solar shield to reduce solar loading.

The TALS AS provides a unique point of reference on the UAV (enabling the TS to detect and precisely track the vehicle) and comprises transponder and antenna units. Of these, the transponder is installed within the UAV and provides a point source beacon signal for the TS to acquire and track. Upon the establishment of TS interrogation, the AS enters its transponder mode, where it listens for TS interrogations and responds to them (thereby providing TS tracking of the latest position). The equipment's antenna unit is mounted externally and a circular protrusion on its top surface accommodates a 360° omnidirectional element. The front face of the assembly incorporates a higher gain directional horn antenna, which is used during initial TS acquisition of the UAV. When the UAV is 305 m (1,000 ft) from the touchdown point, a logic signal is transmitted which switches antennas to the omni unit for final approach, touchdown and roll-out.

The TALS concept of operations allows for rapid set-up at either a remote tactical site or a fixed airfield location and is normally positioned abeam the UAV touchdown point 12–18 m (40–60 ft) to the left or right of landing zone centreline. Prior to landing, the UAV is automatically or manually flown into a defined window of airspace and an auto recovery sequence is initiated. A TALS automatic landing is conducted via a recovery sequence that begins with activation of both the system's TS and AS. The UAV is then flown manually or automatically into the volume of airspace designated as the TALS Acquisition Window (AW). Once the UAV is in the AW, TALS is manually or automatically commanded into the acquisition mode where it acquires the air vehicle. As the UAV is acquired, TALS automatically transitions to the track mode and flies the UAV into the Recovery Initiation Window (RIW). Having met the parameters of the RIW, the TALS is commanded into auto-recovery mode and begins to land the UAV. The TALS flies the UAV down the glideslope, initiating a flare prior to touchdown, and provides steering, braking and engine cut after touchdown. The glideslope for landing is typically 3.5 to 6° (minimum 3°, maximum 15°). The recovery sequence for rotary-wing UAVs is modified to take advantage of the unique rotary wing flight characteristics. An 'international' variant of TALS is designated as the Automatic Take-Off and Landing System (ATOLS).

UCARS-V2:

Directly descended from Sierra Nevada's AN/UPN-51(V) UAV Common Automatic Recovery System (UCARS - deployed by the US Marine Corps in support of its Pioneer UAVs), UCARS Version 2 (UCARS-V2) provides an all-weather, day/night, automatic landing and take-off capability for UAVs operating from ships and/or fixed-base land sites. As such, UCARS-V2 is capable of interfacing with 'virtually any' fixed- or rotary-wing UAV and provides an automatic landing capability that ranges in form from a simple position sensor for air-derived guidance systems to a fully integrated, ground-derived air vehicle controller. Functionally, UCARS-V2 provides accurate position sensing for automatic recovery guidance and control for the UAV and consists of Airborne (resident in the UAV) and Track Sub-systems (**AS** and **TS**). Here, the TS locates and precisely tracks the AS relative to the desired TouchDown Point (TDP). A video camera is used to check the pointing position in relation to the TS antenna position and a TS Alignment Set (TSAS) is used for both post-installation/post-maintenance validation of TS functionality and for daily readiness tests before/between missions.

Looking at the architecture's subsystems in more detail, the TS measures UAV motion and position relative to the TDP and outputs the stabilised air vehicle position and motion data required for automatic precise positioning. Physically, the TS comprises a Track Control Unit (TCU), a pedestal, a base unit, an Inertial Measurement Unit (IMU), an interrogator, a tracking antenna, a camera and four interconnect cables. Of these, the TCU is described as being an all-weather, convection-cooled equipment whose primary function is the provision of power and control for the TS's other hardware weapons replaceable assemblies. The TCU also includes the computing resources for hosting the architecture's recovery software and is the interface for bi-directional RS-422 data transfers between the TS and GCS. For its part, the TS pedestal is a two-axis gimbal system with a high gain, parabolic, dish tracking antenna mounted on its gimbal. So mounted, the antenna can move through ± 125° in azimuth and −35 to +70° in elevation, with antenna drive being provided by elevation and azimuth motors. The antenna's large vertical movement range facilitates its stabilisation in high sea states and electronically operated circuits drive the antenna gimbals to the 0° positions (home) and then apply brakes to prevent high wind/sea state gimbal damage when power is removed. The 35 GHz high-gain tracking antenna used is provided with an integral, weather resistant, high performance radome, with the complete assembly comprising a 457 mm (18 in) parabolic reflector dish, a four-part waveguide antenna feed (with support arms) and a low composites radome.

The TS interrogator unit is the equipment's radio frequency transceiver and provides the signal processing necessary to track the AS. As such, it is frequency-locked to 70 MHz below the approximately 35 GHz AS transponder frequency and a solar shield is mounted on top of the unit to reduce solar loading. The TS camera is a CCD color video unit that features a fixed (set for infinity) focal length and is attached to the TS pedestal antenna. It allows the operator to qualitatively evaluate acquisition and track performance. The system's IMU is used to sense ship motion and helps stabilise TS measurements on a moving ship. The IMU typically requires 6 minutes to initialise.

The UCARS-V2 AS comprises a transponder, an antenna, and two radio frequency cables. Of these, the transponder is mounted within the UAV and provides a point source beacon signal for the TS to acquire and track. Upon determination that the TS is interrogating, the AS enters transponder mode, where it listens for TS interrogations. The AS replies allow the TS to track the UAV's latest position. The AS antenna unit is mounted on the outside of the host UAV and receives/transmits radio frequency signals to the TS interrogator. A circular protrusion on top of this antenna contains an omni-directional element that provides 360° coverage. The front of the antenna radome incorporates a higher gain directional horn that is used initially when the UAV is acquired by the TS. When the UAV is 305 m (1,000 ft) from touchdown, a logic signal is transmitted by the TS to the transponder, which in turn automatically switches antennas to the omnidirectional unit for final approach and touchdown.

Within its concept of operations, UCARS-V2 is designed for semi-permanent installation aboard a ship or rapid set-up at remote tactical sites or fixed airfields, in which locations it provides an all-weather UAV landing capability. Aboard ship, the system is normally positioned forward and to the side of the intended landing point/net, while on land, the architecture is normally located abeam of the UAV touchdown point, some 12 to 18 m (40 to 60 ft) to the left or right of landing zone centreline. Prior to landing, the UAV is automatically or manually flown into a defined window of airspace and an auto recovery sequence is initiated. To begin a recovery sequence both the UCARS-V2 TS and AS are activated and the UAV is flown manually or automatically into a volume of airspace designated as the UCARS acquisition window. Once the UAV is so located, the UCARS is manually or automatically commanded into the acquisition mode where it acquires the UAV. As the UAV is acquired, UCARS automatically transitions to the track mode and flies the UAV into a Recovery Initiation Window (RIW). Having met the parameters of the RIW, the UCARS is commanded into auto-recovery and begins to land the UAV. UCARS flies the UAV down the glideslope into a net for shipboard operations or, when used on land, down the glideslope to a flare prior to touchdown and then provides steering, braking, and engine cut after touchdown. The glideslope for landing is typically 3.5 to 6° (3° minimum, 15° maximum). The architecture's recovery sequence for rotary-wing UAVs is modified to take advantage of the unique rotary wing flight characteristics. An 'international' variant of UCARS-V2 is designated as the Automatic Take-Off and Landing System - Shipboard (ATOLS-S).

Specifications

Weights
- ATS: System (incl antenna)..1.32 kg (2.9 lb)
- TS: Interrogator/solar shield..3.2 kg (7.1 lb)
- TCU..22.7 kg (50.1 lb)
- TCU pedestal cables: W1 and W2..2.8 kg (6.2 lb)
- W3..0.59 kg (1.3 lb)
- Pedestal..35.6 kg (78.5 lb)
- Antenna/radome assembly..3.6 kg (7.9 lb)

Performance
- ATS: Operating modes......................................Off; BIT; beacon; transponder; suspend
- Frequency......................................35 GHz (±150 MHz−70 MHz transmitter frequency offset)
- Transmitter power......................................Not greater than 200 MW
- Bandwidth......................................Up to 20 MHz
- Operating altitude (min)......................................Up to 4,575 m (15,000 ft)
- Antenna polarisation......................................Vertical
- Power requirements at 25 W max (provided by UAV)......................................22 to 32 V DC
- Antenna gain (typical)
  - horn......................................14.8 dBi
  - omnidirectional......................................4 dBi
- BIT......................................Monitors/detects all AS circuits for malfunctions; puts AS into suspend mode if one is detected
- Operating temperature......................................−32/+50°C (−26/+122°F)
- TS: Operating modes......................................Off; self-test; standby; acquisition; track; autoland; suspend
- Frequency: receiver......................................35 GHz (±150 MHz, AS dependent)
- transmitter......................................Offset 70 MHz below receiver frequency
- Track PRF......................................1,000 Hz (±5%)
- Detection/tracking range
  - min, omnidirectional antenna......................................6.0 m (20 ft)
  - min, directional antenna......................................15.0 m (50 ft)
  - max......................................8.0 n miles (14.8 km; 9.2 miles); 2.0 n miles (3.7 km; 2.3 miles) - directional antenna, 6 mm/h (0.2 in/h) rain rate
- Angular coverage: azimuth......................................±130°
- elevation......................................−10/+70°
- Power requirements......................................115 V AC (60 Hz, single phase, 3 wire, 200 VA typical, 700 VA peak)
- Antenna output power......................................<33 dBm
- Antenna gain......................................Approx 41 dBi on each of four lobes
- Operating temperature......................................−32/+50°C (-26/+122°F)
- Density altitude (min)......................................Up to 2,745 m (9,000 ft)
- UAV acquisition capacity......................................1 air vehicle or 1 target
- UAV track capacity......................................1 AS at a time
- Acquisition window scan time......................................<20 s (20° azimuth × 5° elevation window; greater for larger window)
- UAV track speed
  - min......................................Zero - stationary with respect to TS
  - max......................................150 kt (278 km/h; 173 mph) - UAV angular rate relative to TS pedestal <200°/s in azimuth, <40°/s in elevation; angular acceleration<400°/s2 in azimuth, <300°/s2 in elevation
- Track glideslope angle......................................6° (nominal)
- Touchdown accuracy......................................UAV/wind conditions dependent

Specifications

Weights
- AS: System (incl antenna, transponder, bulkhead, feedthroughs and cables)......................................Approx 1.6 kg (3.5 lb)
- TS: Camera......................................1.7 kg (3.8 lb)
- Interrogator/solar shield......................................3.0 kg (6.6 lb)
- TSAS electronics enclosure......................................6.0 kg (13.2 lb)
- TCU......................................23.0 kg (50.7 lb)
- Pedestal/base (incl IMU)......................................41.8 kg (92.2 lb)
- Interconnect cables......................................3.6 kg (7.9 lb)
- Antenna......................................2.9 kg (6.4 lb)

Performance
- AS: Operating modes......................................Off; BIT; beacon; transponder; suspend
- Frequency......................................35 GHz (±85 MHz −70 MHz transmitter frequency offset)
- Detection range......................................10 m (33 ft) to 2.5 n miles (4.6 km; 2.9 miles); 10 m (33 ft) to 0.8 n mile (1.5 km; 0.9 mile) in 25 mm/h (0.98 in/h) rain
- Transmitter power......................................Not greater than 200 MW
- Bandwidth......................................Up to 20 MHz
- Operating altitude......................................Up to 4,575 m (15,000 ft)
- Antenna polarisation......................................Vertical
- Power requirements at 25 W max (provided by UAV)......................................28 V DC (24 V DC min to 29 V DC max)
- Typical antenna gain: horn......................................14.8 dBi
- omnidirectional......................................4 dBi
- BIT......................................Monitors/detects all AS circuits for malfunctions; puts AS into suspend mode if one is detected
- Operating temperature......................................−32/+50°C (-26/+122°F)

The UCARS-V2/ATOLS-S shipboard UAV automatic take-off and landing system *(Sierra Nevada)* 1112229

Specifications

TS: Operating modes	Off; self-test; standby; acquisition; track, autoland; suspend
Frequency: receiver	35 GHz (±85 MHz, AS dependent)
transmitter	Offset 70 MHz below receiver frequency
Track PRF	1,000–2,000 Hz
Track data accuracy	
horizontal	0.5 m (1.5 ft) max when UAV is within 61 m (200 ft) of TS
vertical	0.3 m (1.0 ft) max when UAV is >2.1 m (7.0 ft) above TDP and within 61 m (200 ft) of TS
Detection tracking range: min	10 m (33 ft)
max	2.5 n miles (4.6 km; 2.9 miles); 0.8 n mile (1.5 km; 0.9 mile) in a rain rate up to 25 mm/h (0.98 in/h)
Angular coverage: azimuth	±125°
elevation	–35/+70°
Power requirements	115 V AC (60 Hz, single phase, 3 wire, 350 VA continuous, 1,000 VA surge)
Antenna output power	<2 W (peak)
Antenna gain	41 dBi on each of four lobes
Operating temperature	–32/+50°C (–26/+122°F)
Operating altitude	Up to 3,050 m (10,000 ft)
UAV acquisition capacity	1 air vehicle or 1 target
UAV track capacity	1 AS at a time
Update rate	25 Hz
Acquisition window scan time	<20 s (typical window)
UAV acquisition speed: min	Zero
max	120 kt (222 km/h; 138 mph) - UAV angular rate relative to TS pedestal <1°/s; angular acceleration <1°/s2
UAV track speed	
min	Zero
max	150 kt (278 km/h; 173 mph) - UAV angular rate relative to TS pedestal <200°/s in azimuth, <40°/s in elevation; angular acceleration <400°/s2 in azimuth, <300°/s2 in elevation
UAV acquisition speed	
min	Zero
max	150 kt (278 km/h; 173 mph) - UAV angular rate relative to TS pedestal <200°/s in azimuth, <40°/s in elevation; angular acceleration <400°/s2 in azimuth, <300°/s2 in elevation

Status: TALS is reported as having been procured by the US Army (for use with the service's RQ-7 Shadow TUAV), while UCARS-V2 is noted as having been a 'system of choice' for both the US Navy and the US Coast Guard. UAVs that are known to have been integrated with UCARS include the CL-227/-327 (TALS/UCARS-V2), Eagle Eye (TALS/UCARS-V2), FireScout (TALS/UCARS-V2), Hunter (TALS/UCARS-V2), Outrider (TALS/UCARS-V2), Pioneer (TALS/UCARS-V2), Predator (TALS), Ranger (UCARS-V2), Sentry (TALS/UCARS-V2) and Shadow 200/400 (TALS/UCARS-V2) types. TALS-type technology is also incorporated in Sierra Nevada's Dual-Thread Automatic Take-off and Landing (DT-ATLS) architecture and (over time) the contractor has continued to promote both the TALS and UCARS-V2 systems.

Contractor: Sierra Nevada Corporation
Sparks, Nevada.

A general view of a line-up of three UAV Factory pneumatic catapult launchers *(UAV Factory)* 1509458

A head-on view of a Penguin B mini UAV becoming airborne after launch from a UAV Factory pneumatic catapult launcher *(UAV Factory)* 1509456

UAV Factory launch systems

Type: A portfolio of UAV launch systems.

Description: Over time, UAV Factory has developed a portfolio of UAV launch systems for use (in the first instance) with the contactor's Penguin B UAV, the known details of which are as follows:

Car-top launcher: Designed to be mounted on any car/automobile that can accept a Thule Rapid Aero™ roof rack, UAV factory characterises its car-top launcher as being a 'convenient, reliable, low-cost' option that provides a runway-independent means of launching its Penguin B UAV. As such, the 9.7 kg (including a removable electric UAV engine starter and its associated Lithium Ion battery) device can be fitted or removed in 'less than two minutes' and can be used as a carrier as well as a launcher for the Penguin B. Here, the UAV being transported has its wing tips removed and secured with safety pins. To achieve launch, the host vehicle has to achieve a velocity of 70 km/h (44 mph). Again, the starter assembly incorporates a starter handle and is removed once the UAV's engine is running. The UAV is mounted on the launcher by means of a three point cradle that mates with the AV's centre rear fuselage.

Pneumatic catapult launcher: The UAV Factory's pneumatic catapult launcher is a man-portable, 4,685 (L) × 1,100 (H) × 1,279 (W) mm (184.5 × 43.3 × 50.4 in) device that is capable of accommodating UAVs with launch weights of up to 35 kg (77 lb). As such, the architecture's launch rail is mounted on tripod legs (with the front pair folding for transport), with the launcher incorporating an integrated compressor unit and a remote control box in addition to the rail. A four arm cradle is used to attach the UAV to the launcher. Other system features include:

- a 110 kg (242.5 lb - packed weight), ruggedised, 1,313 (L) × 740 (W) × 543 (D) mm (51.7 × 27.7 × 21.4 in) transportation case
- compressor reverse polarity protection and thermal shutdown
- remote control safety features that include an audible malfunction alarm, voltage and pressure displays and a permanent launch counter
- the ability to be customised to accommodate a client specified UAV that meets the system's launch weight criteria
- an options/accessory package that includes operator and maintenance training, a maintenance kit (supporting the device's 100 launch maintenance cycle) and a Penguin B UAV training dummy.

The UAV Factory pneumatic catapult launcher can operate within the –20°C to +50°C (–4°F to +122°F) temperature range and has maximum launch energy, maximum launch speed and maximum launch pressure values of 6,000 J, 23 m/s (75.5 ft/s) and 10 bar (145 psi) respectively.

Status: Most recently, UAV Factory has continued to promote its car-top and pneumatic catapult UAV launchers.

Contractor: UAV Factory USA LLC,
Irvington, New York. UAV Factory Ltd Europe,
Jelgava, Latvia.

Contractors

Argentina

Argentine Army
Ejército Argentino
Azopardo 250, 4to piso, 1328, Buenos Aires, Argentina
Tel: (+54 11) 43 46 61 00
(+54 11) 43 46 11 16
(+54 11) 43 46 11 17
e-mail: centroinfo@ejercito.mil.ar
Web: www.ejercito.mil.ar

Armenia

Ministry of Defence
Bagrevand 5, 0044, Yerevan, Armenia
Tel: (+374 10) 29 46 99
e-mail: modpress@mil.am
press@mil.am
Web: www.mil.am

Australia

Aerosonde Pty Ltd
(a subsidiary of AAI Corporation, US)
Unit 1 585 Blackburn Road, Notting Hill, Victoria, 3168, Australia
Tel: (+61 3) 95 18 73 00
Fax: (+61 3) 95 18 73 41
e-mail: salesenquiries@aerosonde.com
Web: www.aerosonde.com

Air Affairs Australia Pty Ltd (AAA)
PO Box 233 Pyrmont, New South Wales, 2009, Australia
Unit 113, 243 Pyrmont Street, Sydney, New South Wales, 2009, Australia
Tel: (+61 2) 44 23 67 55
Fax: (+61 2) 44 23 12 33
e-mail: sales@airaffairs.com.au
Web: www.airaffairs.com.au

BAE Systems Australia
40 River Boulevard, Richmond, Victoria, 3121, Australia
Tel: (+61 3) 99 18 40 57
(+61 3) 99 18 40 00
Fax: (+61 3) 99 18 49 00
Web: www.baesystems.com

Cyber Technology (WA) Pty Ltd
1C Ambitious Link, Bibra Lake, Western Australia, 6163, Australia
Tel: (+61 8) 94 18 92 00
Fax: (+61 8) 94 18 92 01
Web: www.cybertechuav.com.au

V-TOL Aerospace Pty Ltd
18/ 1645 Ipswich Road, Rocklea, Queensland, 4106, Australia
Tel: (+61 7) 32 75 28 11
Fax: (+61 7) 32 75 25 52
e-mail: info@v-tol.com
Web: www.v-tol.com

Austria

BRP Powertrain GmbH & Co KG
Welser Strasse 32, A-4623, Gunskirchen, Austria
Tel: (+43 7246) 60 10
Fax: (+43 7246) 63 70
e-mail: officepowertrain@brp.com
Web: www.rotax-aircraft-engines.com

Schiebel Elektronische Geraete GmbH
(a subsidiary of Schiebel Industries AG, Austria)
Margaretenstrasse 112, A-1050, Wien, Austria
Tel: (+43 1) 54 62 60
Fax: (+43 1) 545 23 39
e-mail: pr@schiebel.net
info@schiebel.net
Web: www.schiebel.net

Schiebel Industries AG
Margaretenstrasse 112, A-1050, Wien, Austria
Tel: (+43 1) 54 62 60
Fax: (+43 1) 545 23 39
Web: www.schiebel.net

Belarus

Independent Development Laboratory (INDELA)
11 Petra Glebki Street, 220104, Minsk, Belarus
Tel: (+375 17) 253 85 10
e-mail: mail@indelagroup.com
Web: www.indelauav.com

Belgium

Flemish Institute for Technological Research (VITO)
Boeretang 200, B-2400, Mol, Belgium
Tel: (+32 14) 33 55 11
Fax: (+32 14) 33 55 99
e-mail: vito@vito.be
Web: www.vito.be

Brazil

Aeromot Indústria Mecânico-Metalúrgica Ltda
(a subsidiary of Aeromot Aeronaves e Motores SA, Brazil)
Avenue Industries, 1,290th, 90200-290, Porto Alegre, Rio Grande do Sul, Brazil
Tel: (+55 51) 33 57 85 50
Fax: (+55 51) 33 71 16 55
e-mail: industria@aeromot.com.br
Web: www.epicos.com
www.aeromot.com.br

Avibras Indústria Aeroespacial SA
Rodovia dos Tamoios km 14, Jacareí, PO Box 278 12315-020, São Paulo, Brazil
Tel: (+55 12) 39 55 60 00
Fax: (+55 12) 39 51 62 77
e-mail: gspd@avibras.com.br
Web: www.avibras.com.br

Flight Technologies
Parque Tecnológico, Rodovia Presidente Dutra, km 138, Eugênio de Melo, 12247-004, São José dos Campos, São Paulo, Brazil
Tel: (+55 12) 32 01 70 12
e-mail: imprensa@flighttech.com.br
Web: www.flighttech.com.br

Canada

Bristol Aerospace Ltd (BAL)
(a subsidiary of Magellan Aerospace Corporation, Canada)
PO Box 874 Winnipeg, Manitoba, R3C 2S4, Canada
660 Berry Street, Winnipeg, Manitoba, R3H 0S5, Canada
Tel: (+1 204) 775 83 31
Fax: (+1 204) 783 20 91
(+1 204) 774 01 95
Web: www.bristol.ca
www.magellan.aero

CDL Systems Ltd
Harvest Hills Office Park, Building 5000, Suite 5301, 333 - 96th Avenue North East, Calgary, Alberta, T3K 0S3, Canada
Tel: (+1 403) 289 17 33
Fax: (+1 403) 289 39 67
e-mail: sales@cdlsystems.com
Web: www.cdlsystems.com

Draganfly Innovations Inc
2108 St George Ave, Saskatoon, Saskatchewan, S7M 0K7, Canada
Tel: (+1 306) 955 99 07
Fax: (+1 306) 955 99 06
e-mail: info@draganfly.com
Web: www.draganfly.com

General Dynamics Ordnance and Tactical Systems - Canada Inc (GD OTS)
Head Office
5 Montée des Arsenaux, Le Gardeur, Québec, J5Z 2P4, Canada
Tel: (+1 450) 581 30 80
Fax: (+1 450) 585 73 02 (reception)
e-mail: info@can.gd-ots.com
Web: www.gd-otscanada.com
www.snctec.com

Meggitt Training Systems Canada (MTSC)
(a subsidiary of Meggitt plc, UK)
3-1735 Brier Park Road North West, Medicine Hat, Alberta, T1C 1V5, Canada
Tel: (+1 403) 528 87 82
Fax: (+1 403) 529 26 29
e-mail: mtscanada@meggitt.com
Web: www.meggitttrainingsystems.com
www.meggitt.com

MicroPilot
72067 Road 8E, Sturgeon Road, PO Box 720 Stony Mountain, Manitoba, R0C 3A0, Canada
Tel: (+1 204) 344 55 58
Fax: (+1 204) 344 57 06
e-mail: info@micropilot.com
Web: www.micropilot.com

Mist Mobility Integrated Systems Technology Inc (MMIST)
3 Iber Road, Ottawa, Ontario, K2S 1E6, Canada
Tel: (+1 613) 723 04 03
Fax: (+1 613) 723 89 25
e-mail: info@mmist.ca
Web: www.mmist.ca

Pratt & Whitney Canada (P&WC)
Head Office
(a subsidiary of Pratt & Whitney, US)
1000 Marie-Victorin Boulevard,
Mail Code 01CQ5 Longueuil, Québec, J4G 1A1, Canada
Tel: (+1 450) 677 94 11
Freephone: (+1 800) 268 80 00
Fax: (+1 450) 647 36 20
Web: www.pwc.ca

China

Beijing University of Aeronautics & Astronautics (BUAA)
UAV Research Institute
37 Xue Yuan Road, Haidian District, Beijing, China
Tel: (+86 10) 62 01 72 51
(+86 10) 82 31 76 58
Fax: (+86 10) 62 02 83 56
Web: www.buaa.edu.cn

China Aerospace Science and Industry Corporation (CASIC)
Building 8A, Fucheng Road, Haidian District, 100048, Beijing, China
Tel: (+86 10) 68 37 35 22
(+86 10) 68 37 36 22
Fax: (+86 10) 68 37 36 26
e-mail: ht@casic.com.cn
Web: www.casic.com.cn

Contractors

China

China Aerospace Science and Technology Corporation (CASC)
Fuchenglu 16, 100048, Haidian, Beijing, China
Tel: (+86 10) 68 76 74 92
Fax: (+86 10) 68 37 22 91
e-mail: casc@spacechina.com
Web: www.spacechina.com

China Aviation Industry Corp (AVIC)
China Aviation Industrial Building, 128 Jianguo Road, PO Box 2399 100022, Chaoyang, Beijing, China
Tel: (+86 10) 58 35 69 84
  (+86 10) 84 38 05 66
Fax: (+86 10) 65 66 65 18
e-mail: avic_website@avic.com.cn
Web: www.avic.com.cn

China National Aero-Technology Import & Export Corporation (CATIC)
Head Office
CATIC Plaza, 18 Beichen Dong Street, 100101, Beijing, Chaoyang, China
e-mail: public@catic.cn
Web: www.catic.cn

Xian ASN Technical Group
No. 34 Fenghui Road, 710065, Xian, Shaanxi, China
Tel: (+86 29) 88 45 10 20
Fax: (+86 29) 88 45 10 32
e-mail: asngroup@163.com
Web: www.asngroup.com.cn

Czech Republic

Lom Praha s p o z Vtulapvo
944 Mladoboleslavská,
PO Box 18 CZ-197 06, Praha, 9, Czech Republic
Tel: (+420 255) 28 21 11
Fax: (+420 255) 81 50 86
e-mail: vtulapvo@vtusp.cz
Web: www.vtul.cz

Prvni Brnenska Strojirna Velka Bites as
Vlkovska 279, CZ-595 12, Velká Bítes, Czech Republic
Tel: (+420 566) 82 23 04
Fax: (+420 566) 82 23 72
e-mail: sales@pbsvb.cz
Web: www.pbsvb.cz

Finland

Patria Aviation Oy
Naulakatu 3, FI-33100, Tampere, Finland
Tel: (+358 20) 46 91
Fax: (+358 20) 469 26 97
e-mail: systems@patria.fi
Web: www.patria.fi

Robonic Oy Ltd
Pinninkatu 53 C, FI-33100, Tampere, Finland
Tel: (+358 3) 273 05 86
Fax: (+358 3) 273 05 88
Web: www.robonic.fi

France

Alcore Technologies SA
BP 7111 F-95054, Cergy-Pontoise, France
Tel: (+33 1) 30 37 42 21
Fax: (+33 1) 34 64 41 50
e-mail: info@alcore-tech.com
Web: www.alcore-tech.com

Aviation Design
38 ZI Le Chenet, Milly La Foret, F-91490, France
Tel: (+33 1) 64 98 93 93
Fax: (+33 1) 64 98 93 88
e-mail: aviation.design@wanadoo.fr
Web: www.adjets.com

Bertin Technologies
Parc d'activities du Pas du lac, 10 Avenue Ampére, F-78180, Montigny-le-Bretonneux, France
Tel: (+33 1) 39 30 60 00
Fax: (+33 1) 39 30 09 50
e-mail: communication@bertin.fr
Web: www.bertin.fr

Cassidian
Headquarters
(a subsidiary of EADS Deutschland GmbH, Germany)
ZAC de la Clef Saint-Pierre, 1 boulevard Jean Moulin, CS 40001, F-78996, Elancourt, France
Tel: (+33 1) 61 38 50 00
Fax: (+33 1) 61 38 70 70
Web: www.cassidian.com

Cose Sarl
32 rue Amelin, F-93440, Dugny, France
Tel: (+33 1) 48 37 42 53
Fax: (+33 1) 48 37 63 84
e-mail: cose@cose.fr
Web: www.cose.fr

Dassault Aviation
Head Office
(jointly owned by Groupe Industriel Marcel Dassault, France (50.55 per cent), EADS France (46.32 per cent), Public (3.13 per cent))
78 quai Marcel Dassault, F-92552, Saint Cloud, Cédex 300, France
Tel: (+33 1) 47 11 40 00
Fax: (+33 1) 47 11 49 01
Web: www.dassault-aviation.com

Helipse
16 Route de la Grande Riviere, F-16400, Crown, France
Tel: (+33 05) 45 65 49 92
Fax: (+33 05) 45 65 49 45
e-mail: helipse@club-internet.fr
Web: www.helipse.com

Infotron
17 Rue Ampere, F-91300, Massy, France
Tel: (+33 1) 60 13 66 00
Fax: (+33 1) 60 13 66 10
Web: www.infotron.fr

MBDA
Headquarters
(a joint venture between BAE Systems, UK (37.5 per cent), EADS NV, Netherlands (37.5 per cent) and Finmeccanica, Italy (25 per cent))
1 Avenue Réaumur, F-92358, Le Plessis-Robinson, Cedex, France
Tel: (+33 1) 71 54 10 00
Fax: (+33 1) 71 54 00 01
Web: www.mbda-systems.com

Microturbo
Headquarters, Management and Central Services
(a joint venture between Safran, France and Turbomeca, France)
8 Chemin du Pont de Rupé,
BP 62089 F-31019, Toulouse, Cedex 2, France
Tel: (+33 5) 61 37 55 00
Fax: (+33 5) 61 70 74 45
e-mail: microturbo_dc@csi.com
Web: www.microturbo.com

Roxel SAS
(a subsidiary of MBDA, UK)
La Boursidiere, Immeuble Jura, F-92357, Le Plessis-Robinson, Cedex, France
Tel: (+33 1) 41 07 82 95
Fax: (+33 1) 46 30 22 37
e-mail: info@roxelgroup.com
Web: www.roxelgroup.com

SAS Fly-n-Sense
Aéroparc Bordeaux-TechnoWest, 25 rue Marcel Issartier, F-33702, Merignac Cedex, France
Tel: (+33 5) 35 54 00 97
Fax: (+33 5) 56 47 31 48
e-mail: info@fly-n-sense.com
Web: www.fly-n-sense.com

Sagem
Sagem Défense Sécurité
(a subsidiary of Safran Group, France)
Le Ponant de Paris, 27 rue Leblanc, F-75512, Paris, Cedex 15, France
Tel: (+33 1) 58 11 78 00
  (+33 1) 53 23 20 16
Fax: (+33 1) 07 96 62 94
  (+33 1) 58 11 78 20
  (+33 1) 58 11 78 50
Web: www.sagem-ds.com

Secapem SA
Headquarters
11 rue Henri Beaudelet - Zone Industriel, F-77330, Ozoir-la-Ferrière, France
Tel: (+33 1) 64 40 00 57
Fax: (+33 1) 64 40 21 55
e-mail: info@secapem.com
Web: www.secapem.com

Survey Copter
Route de l'aérodrome, F-26700, Pierrelatte, France
Tel: (+33 4) 75 00 09 96
Fax: (+33 4) 75 49 92 85
Web: www.survey-copter.com

Thales France
160 Boulevard de Valmy, F-92704, Colombes, Cedex, France
Tel: (+33 1) 41 30 30 00
Fax: (+33 1) 41 64 57 57
Web: www.thalesgroup.com

WorkFly
16 rue de la Fertile Plaine, F-93330, Nelly-sur-Marne, France
Tel: (+33 1) 43 00 31 54
e-mail: contact@workfly.fr
Web: www.workfly.net

Zodiac Data Systems SAS
3 avenue du Canada, Z A de Courtaboeuf 2, LP 880, F-91966, Courtaboeuf, Cedex, France
Tel: (+33 1) 64 86 34 00
Fax: (+33 1) 64 86 34 12
e-mail: isite@zodiacaerospace.com
Web: www.zds-fr.com

Germany

AirRobot GmbH and Co KG
Head Office
Werler Straße 4, D-59755, Arnsberg, Germany
Tel: (+49 2932) 54 77 40
Fax: (+49 2932) 54 77 45
e-mail: service@airrobot.de
Web: www.airrobot.com

Cassidian Optronics GmbH
(a subsidiary of Carl Zeiss AG, Germany (24.9 per cent) and Cassidian, Germany (75.1 per cent))
Carl-Zeiss-Straße 22, D-73447, Oberkochen, Germany
Tel: (+49 7364) 20 65 30
Fax: (+49 7364) 20 36 97
e-mail: optronics@optronics.zeiss.com
Web: www.zeiss.de/optronics

EADS Deutschland GmbH
Headquarters
(a subsidiary of EADS NV, Netherlands)
D-81663, München, Germany
Willy-Messerschmitt-Straße, D-85521, Ottobrunn, Germany
Tel: (+49 89) 60 70
Fax: (+49 89) 60 72 64 81
e-mail: eadsweb@eads.net
Web: www.eads.com

EADS Deutschland GmbH
Cassidian Electronics
Wörthstraße 85, D-89007, Ulm, Germany
Tel: (+49 731) 392 49 43
Fax: (+49 731) 392 72 33
Web: www.eads.com

Germany

EMT Ingenieurgesellschaft
Grube 29, D-82377, Penzberg, Germany
Tel: (+49 8856) 922 50
Fax: (+49 8856) 20 55
e-mail: sales@emt-penzberg.de
vertrieb@emt-penzberg.de
Web: www.emt-penzberg.de

Göbler Hirthmotoren KG
Max Eyth Strasse 10,
PO Box 62 D-71726, Benningen, Neckar, Germany
Tel: (+49 7144) 855 10
Fax: (+49 7144) 54 15
e-mail: info@hirth-engines.de
Web: www.hirth-engines.de
www.hirth-uavengines.de

Limbach Flugmotoren GmbH & Co KG
Kotthausenerstrasse 5, D-53639, Königswinter, Germany
Tel: (+49 2244) 920 10
Fax: (+49 2244) 92 01 30
e-mail: sales@limflug.de
Web: www.limflug.de

Microdrones GmbH
Gutenbergstrasse 86, D-57078, Siegen, Germany
Tel: (+49 271) 770 03 80
Fax: (+49 271) 77 00 38 11
e-mail: info@microdrones.com
Web: www.microdrones.com

Rheinmetall Defence Electronics GmbH
Headquarters
(a subsidiary of Rheinmetall AG, Germany)
Brueggeweg 54, D-28309, Bremen, Germany
Tel: (+49 421) 457 01
Fax: (+49 421) 457 29 00
e-mail: info@rheinmetall-de.com
Web: www.rheinmetall-defence.de

Greece

EADS 3 Sigma SA
Head Office
Industrial Park,
OT 23 GR-73200, Chania Crete, Greece
Tel: (+30 28210) 801 33
(+30 28210) 801 34
(+30 28210) 801 68
(+30 28210) 801 65
Fax: (+30 28210) 800 32
e-mail: admin@eads-3sigma.gr
Web: www.eads-3sigma.gr

India

Aerial Delivery Research and Development Establishment (ADRDE)
Station Road,
Post Box 51 Agra, 282001, India
Tel: (+91 040) 25 89 32 74
Fax: (+91 020) 25 89 31 02
e-mail: director@adrde.drdo.in
Web: www.drdo.org

Aeronautical Development Establishment (ADE)
DRDO, Ministry of Defence, Suranjan Das Road, CV Raman Nagar, Bangalore, 560 093, India
Ministry of Defence, R&D Organisation, New Thippasandra, Bangalore, 560 075, India
Tel: (+91 80) 25 05 70 37
Fax: (+91 80) 25 28 31 88
e-mail: director@ade.drdo.in
Web: www.drdo.com

Aurora Integrated Systems Pvt Ltd
60/4 Srirampura Cross, Opp JNC ASR, Jakkur Post, Bangalore, Karnataka, 560064, India
Tel: (+91 80) 65 83 65 08
(+91 80) 65 46 52 37
Fax: (+91 80) 30 72 36 03
e-mail: sales@aurora-is.com
Web: www.aurora-is.com

Kadet Defence Systems
243, West Chowbagha, Kolkata (Calcutta), West Bengal, 700 039, India
Tel: (+91 33) 23 45 55 94
Fax: (+91 33) 23 45 55 93
e-mail: info@kadetmodels.com
Web: www.kadet-uav.com

MKU Pvt Limited
D-20, Ist Floor Defence Colony, New Delhi, 110 024, India
Tel: (+91 11) 46 54 35 12
Fax: (+91 11) 46 54 35 14
Web: www.mku.com

Swallow Systems Pvt Ltd
6, Pratapgunj, Near Natraj Talkies, Vadodara, Gujarat, 389350, India
Tel: (+91 265) 279 24 71
e-mail: info@swallowuav.com
Web: www.swallowuav.com

Israel

Aeronautics Ltd
PO Box 169 IL-81101, Yavne, Israel
Nahal Snir 10 Street, Yavne, Israel
Tel: (+972 8) 943 36 00
Fax: (+972 8) 932 89 12
e-mail: info@aeronautics-sys.com
Web: www.aeronautics-sys.com

Bental Industries Ltd
(a subsidiary of TAT Technologies Ltd, Israel)
IL-12436, Kibbutz Merom Golan, Israel
Tel: (+972 6) 696 01 99
Fax: (+972 6) 696 01 87
e-mail: bental_marketing@bental.co.il
Web: www.bental.co.il

BlueBird Aero Systems Ltd
Hamatechet Street, Industrial Park,
PO Box 5041 IL-60920, Kadima, Israel
Tel: (+972 9) 899 93 35
Fax: (+972 9) 899 93 45
e-mail: info@bluebird-uav.com
Web: www.bluebird-uav.com

Controp Precision Technologies Ltd
PO Box 611 IL-45105, Hod Hasharon, Israel
5 Hanagar Street, IL-45240, Hod Hasharon, Israel
Tel: (+972 9) 744 06 61
Fax: (+972 9) 744 06 62
e-mail: info@controp.co.il
sales@controp.co.il
Web: www.controp.com

ELTA Systems Ltd (IAI/ELTA)
(a subsidiary of Israel Aerospace Industries Ltd, Israel)
100 Yitzhak, Hanasi Boulevard,
PO Box 330 IL-77102, Ashdod, Israel
Tel: (+972 8) 857 23 33
e-mail: market@elta.co.il
Web: www.iai.co.il
www.elta-iai.com

Elbit Systems Electro-Optics Elop Ltd
(a subsidiary of Elbit Systems Ltd, Israel)
Advanced Technology Park,
PO Box 1165 IL-76111, Rehovot, Israel
Tel: (+972 8) 938 62 11
Fax: (+972 8) 938 62 37
Web: www.el-op.com
www.elbitsystems.com

Elbit Systems Ltd
Advanced Technology Center,
PO Box 539 IL-31053, Haifa, Israel
Tel: (+972 4) 831 64 04
Fax: (+972 4) 831 69 44
Web: www.elbitsystems.com

Elbit Systems SAR and Data Links - Elisra Ltd
29 Hamerkava Street,
PO Box 150 IL-58101, Holon, Israel
Tel: (+972 3) 557 31 02
Fax: (+972 3) 557 75 79
e-mail: info@tadspec.com
Web: www.tadspec.com

Elisra
Corporate Headquarters
(partly owned by Elbit Systems Ltd, Israel (70 per cent) and IAI ELTA Electronics Industries Ltd, Israel (30 per cent))
48 Mivtza Kadesh Street, IL-51203, Bene Baraq, Israel
Tel: (+972 3) 617 51 11
Fax: (+972 3) 617 58 50
e-mail: marketing@elisra.com
Web: www.elisra.com

Hydromechanical Engineering Ltd
18 Rozansky Street, IL-75706, Rishon le Zion, Israel
Tel: (+972 3) 951 04 85
Fax: (+972 3) 951 04 76
e-mail: hec@zahav.net.il
Web: www.hec-eng.com

Innocon Ltd
Company Headquarters
32 Abanai Street, IL-58856, Holon, Israel
Tel: (+972 3) 558 63 30
Fax: (+972 3) 558 38 82
e-mail: innocon@innoconltd.com
Web: www.innoconltd.com

Israel Aerospace Industries Ltd (IAI)
Ben Gurion International Airport, IL-70100, Tel-Aviv, Israel
Tel: (+972 3) 935 85 09
(+972 3) 935 31 11
(+972 3) 935 81 11
(+972 3) 935 85 14
Fax: (+972 3) 935 33 96
(+972 3) 935 85 16
Web: www.iai.co.il

Israel Aerospace Industries Ltd (IAI)
MBT Systems Missiles and Space Group
Yehud Industrial Zone,
PO Box 105 IL-56000, Beer Yacov, Israel
e-mail: marketing@mbt.iai.co.il
Web: www.iai.co.il

Israel Aerospace Industries Ltd (IAI)
Tamam
PO Box 75 IL-56100, Yehud, Israel
Tel: (+972 3) 531 52 05
e-mail: infotmm@iai.co.il
Web: www.iai.co.il/tamam

Israel Military Industries (IMI)
PO Box 1044 IL-47100, Ramat Hasharon, Israel
Tel: (+972 3) 548 52 22
Fax: (+972 3) 548 61 25
e-mail: imimrktg@imi-israel.com
Web: www.imi-israel.com

Phantom Technologies Ltd
68 Amal Street, IL-49513, Petah Tikvah, Israel
Tel: (+972 3) 921 57 20
Fax: (+972 3) 921 54 34
e-mail: sales@phantom.co.il
Web: www.phantom.co.il

Rafael Advanced Defense Systems Ltd
Corporate Headquarters
Rafael Manor
PO Box 2250 IL-31021, Haifa, Israel
Tel: (+972 4) 879 47 17
Fax: (+972 4) 879 46 57
e-mail: customersupport@rafael.co.il
intl-mkt@rafael.co.il
Web: www.rafael.co.il

Contractors

Steadicopter Ltd
3D Aerial Robotics,
PO Box 567 IL-20692, Yokneam Illit, Israel
Tel: (+972 4) 959 29 59
Fax: (+972 4) 959 76 06
Web: www.steadicopter.com

Top I Vision
25 Hathiya Street, IL-58402, Holon, Israel
Tel: (+972 3) 933 54 69
    (+972 3) 933 54 66
Fax: (+972 3) 933 93 27
e-mail: info@topivision.com
Web: www.topivision.com

UVision Air Ltd
4 Bazelet Street, Sapir Industrial Park, IL-44862, Zur Igal, Israel
Tel: (+972 9) 749 68 22
Fax: (+972 9) 749 68 23
Web: www.uvisionuav.com

Urban Aeronautics Ltd
10 Nahal Snir Street,
PO Box 13137 IL-81224, Yavne, Israel
Tel: (+972 8) 943 36 40
Fax: (+972 8) 943 36 44
Web: www.urbanaero.com

Vision Map Ltd
13 Mozes Street, IL-67442, Tel-Aviv, Israel
Tel: (+972 3) 609 10 42
Fax: (+972 3) 609 10 43
Web: www.visionmap.com

Italy

Alenia Aeronautica SpA
Headquarters
(a subsidiary of Finmeccanica, Italy)
Via Campania 45, I-00187, Roma, Italy
Tel: (+39 6) 42 08 81
Fax: (+39 6) 42 82 45 28
e-mail: communication@alenia-aeronautica.it
        international-sales@aeronautica.it
        press-office@alenia-aeronautica.it
Web: www.alenia-aeronautica.it
     www.aiad.it

Selex Galileo
UAVs and Simulators
Via Mario Stoppani 21, I-34077, Ronchi dei Legionari, Gorizia, Italy
Tel: (+39 0481) 47 81 11
Fax: (+39 0481) 47 83 13
e-mail: sales.marketing@selexgalileo.com
Web: www.selexgalileo.com
     www.galileoavionica.it

Selex Galileo
Via dei Castelli Romani 2, I-00040, Pomezia, Roma, Italy
Tel: (+39 06) 91 19 61
Fax: (+39 06) 912 15 90
e-mail: sales.marketing@selexgalileo.com
Web: www.selexgalileo.com

Zanzottera Technologies srl
Zone Industrial Via Italia, I-22010, Grandola ed Uniti (Como), Italy
Tel: (+39 03) 443 26 86
Fax: (+39 03) 443 28 25
Web: www.zanzotteraengines.com

Japan

Fuji Heavy Industries Ltd (FHI)
Aerospace
1-1-11 Yonan, Utsunomiya-shi, Tochigi, 320-8564, Japan
Tel: (+81 28) 684 77 77
Web: www.fhi.co.jp

Gen Corp
5652-83 Sasaga, Matsumoto-shi, Nagano-Ken, 399-0033, Japan
Tel: (+81 263) 26 07 37
Fax: (+81 263) 26 07 57
e-mail: aviation@gen-corp.jp
Web: www.gen-corp.jp

Mitsubishi Heavy Industries Ltd (MHI)
Nagoya Guidance and Propulsion Systems Works
1200 O-aza Higashi-Tanaka, Komaki, Aichi, 485-8561, Japan
Tel: (+81 568) 79 21 13
Web: www.mhi.co.jp

Yamaha Motor Co Ltd
2500 Shinngai, Iwatashi, Shizuoka-Ken, 438-8501, Japan
Tel: (+81 538) 32 11 70
Fax: (+81 538) 37 42 59
Web: www.yahama-motor.co.jp

Jordan

Jordan Aerospace Industries (JAI)
Head Office
Ali Saydo Al Kurdi str., Building No. 31,
PO Box 815570 11180, Amman, Sweifieh, Jordan
Tel: (+962 6) 593 55 50
Fax: (+962 6) 593 75 15
e-mail: sales@sama-aircraft.com
Web: www.sama-aircraft.com

Korea, South

Korea Aerospace Industries Ltd (KAI)
Head Office
802- Yucheon-Ri- Sanam-Myeon, Sacheon, 100-737, Korea, South
Tel: (+82 55) 851 10 00
Fax: (+82 55) 851 10 04
Web: www.koreaaero.com

Korea Aerospace Research Institute (KUVSA)
Aeronautics Programme Office, 45 Eoeun-Dong, Yuseong-Gu,
PO Box 113 Daejeon, 305333, Korea, South
Tel: (+82 42) 860 21 64
Fax: (+82 42) 860 20 15
e-mail: garden@kari.re.kr
Web: www.korea-uvs.org

Korean Air Lines Co Ltd (KAL)
Aerospace
KAL Bldg 10th floor, 117 Seosomun-ro, Chung-gu, Seoul, 100 110, Korea, South
Tel: (+82 2) 751 73 11
Fax: (+82 2) 751 73 47
e-mail: bep@koreanair.co.kr
Web: techcenter.koreanair.com
     www.kal-asd.co.kr

Uconsystem Company Ltd
1319 Gwanpyong-Dong, Yusung-gu, Daejeon, 305-509, Korea, South
Tel: (+82 42) 936 22 51
    (+82 42) 936 22 54
Fax: (+82 42) 936 22 56
e-mail: paulkang@foosung.com
Web: www.uconsystem.com

Malaysia

CTRM Aviation Sdn Bhd
Composites Technology City, Batu Berendam, 75450, Melaka, Malaysia
Tel: (+60 3) 317 41 05
Fax: (+60 3) 317 72 13
Web: www.ctrm.com.my

Netherlands

Dutch Space BV
PO Box 32070 NL-2303 DB, Leiden, Netherlands
Newtonweg 1, NL-2303 DB, Leiden, Netherlands
Tel: (+31 71) 524 50 00
Fax: (+31 71) 524 59 99
e-mail: info@dutchspace.nl
Web: www.dutchspace.nl

FlyCam Lemond
Kortenhoevenseweg 36a, NL-4128 CR, Lexmond, Netherlands
Tel: (+31 347) 34 23 82
Fax: (+31 347) 34 23 16
e-mail: info@flycam.nl
        flycam@worldonline.nl
Web: www.flycam.nl
     www.helicopter.nl

Geocopter BV
Olivier van noortweg 7, LX Venlo, NL-5928, Netherlands
Tel: (+31 77) 396 77 59
e-mail: info@geocopter.nl
Web: www.geocopter.nl

TNO Defence, Security and Safety
PO Box 96864 NL-2509 JG, 's-Gravenhage, Netherlands
Oude Waalsdorperweg 63, NL-2597, 's-Gravenhage, Netherlands
Tel: (+31 88) 866 10 00
    (+31 70) 328 09 61
e-mail: infodesk@tno.nl
Web: www.tno.nl

Norway

OKT Norge AS
N-4354, Voll, Norway
Tel: (+47 51) 42 01 62
Fax: (+47 51) 42 22 80
e-mail: info@oktnorge.no
Web: www.oktnorge.no

Pakistan

Advanced Computing & Engineering Solutions (Pvt) Ltd (ACES)
House No 156, Street No 5 F11/1, Islamabad, Pakistan
Tel: (+92 51) 222 44 53
    (+92 51) 222 44 51
Fax: (+92 51) 222 44 54
e-mail: info@aces.com.pk
Web: www.aces.com.pk

East West Infiniti Pvt Ltd
146 Industrial Triangle, Kahuta Road, 44000, Islamabad, Pakistan
Tel: (+92 51) 449 20 41
Fax: (+92 51) 449 18 45
e-mail: info@eastwestin.com
Web: www.eastwestin.com

Institute of Industrial Control Systems (IICS)
(a subsidiary of Global Industrial & Defence Solution (GIDS), Pakistan)
Dhoke Nusah, Dakhli Gangal, Near Chatri Chowk,
PO Box 1398 46000, Rawalpindi, Pakistan
Tel: (+92 51) 447 00 70
Fax: (+92 51) 447 00 76
e-mail: info@iics.com.pk
Web: www.iics.com.pk

Integrated Dynamics
250 Main Korangi Creek Road, 75190, Karachi, Sindh, Pakistan
e-mail: info@idaerospace.com
        sales@idaerospace.com
Web: www.idaerospace.com

Pakistan Aeronautical Complex (PAC)
Headquarters
(a subsidiary of Defence Export Promotion Organisation, Pakistan)
Rebuild Factory Kamra, District Attock, Pakistan
Tel: (+92 51) 90 99 22 34
Fax: (+92 51) 922 55 00
e-mail: pacit@pac.org.pk
Web: www.pac.org.pk

Pakistan – South Africa

Pakistan Aeronautical Complex (PAC)
Aircraft Manufacturing Factory (AMF)
Pac Kamra, District Attock, Pakistan
Tel: (+92 51) 909 90 52 36
Fax: (+92 51) 922 55 13
e-mail: amf@pac.org.pk
Web: www.pac.org.pk

Satuma
Union Council Road, Sihala, Islamabad, Pakistan
Tel: (+92 51) 448 58 61
    (+92 51) 448 58 62
Fax: (+92 51) 448 58 63
e-mail: support@satuma.com.pk
Web: www.satuma.com.pk

Poland

Air Force Institute of Technology
Instytut Techniczny Wojsk Lotniczych (ITWL)
Ksiecia Bolestawa 6,
PO Box 96 PL-01-494, Warszawa, Poland
Tel: (+48 22) 685 10 13
Fax: (+48 22) 836 44 71
e-mail: poczta@itwl.pl
Web: www.itwl.pl

WB Electronics SA (WBE)
ul Poznanska 129/133, Ozarow, PL-05-850, Mazowiecki, Poland
Tel: (+48 22) 731 25 00
Fax: (+48 22) 731 25 01
e-mail: info@wb.com.pl
Web: www.wb.com.pl

Romania

SC Electromecanica Ploiesti SA
Industrial Group of the Army
Ploiesti-Targoviste Road, Km 8, Ploiesti, Prahova, Romania
Tel: (+40 2) 44 59 07 80
Fax: (+40 2) 44 51 33 01
e-mail: elmec@elmecph.ro
Web: www.elmec.ro

Russian Federation

A S Yakovlev Design Bureau JSC (YAK)
(a subsidiary of Joint Stock Company United Aircraft Corporation, Russia)
68 Leningradsky Prospect, 125315, Moskva, Russian Federation
Tel: (+7 495) 158 34 32 (public relations and marketing)
    (+7 495) 787 31 57 (directorate)
    (+7 495) 158 36 61 (exhibition and advertising)
    (+7 495) 151 15 63
    (+7 495) 152 06 80 (scientific and technical information)
Fax: (+7 495) 787 28 44
e-mail: okb@yak.ru
Web: www.yak.ru

Augur Aeronautical Centre
4 ul Stepana Shutova, Korpus 1, 109380, Moskva, Russian Federation
Tel: (+7 495) 359 10 01
Fax: (+7 495) 359 10 65
e-mail: augur@augurballoons.ru
Web: www.augur.pbo.ru

ENICS JSC
ENICS Joint Stock Company
120 Korolenko Street,
PO Box 8 420127, Kazan, Russian Federation
Tel: (+7 843) 570 76 07
Fax: (+7 843) 570 95 41
e-mail: uav@enics.ru
Web: www.enics.ru

Klimov JSC
(a subsidiary of Russian Aircraft Corporation MIG, Russia)
11 Kantemirovskaya street, 194100, St Petersburg, Russian Federation
Tel: (+7 812) 301 90 50
Fax: (+7 812) 301 90 42
e-mail: klimov@klimov.ru
       service@klimov.ru
Web: www.klimov.ru

Kumertau Aviation Production Enterprise (KUMAPE)
15a Novozarinskaya Street, 453350, Kumertau, Bashkortostan, Russian Federation
Tel: (+7 34761) 423 00
    (+7 34761) 422 53
Fax: (+7 34761) 476 24
    (+7 34761) 439 13
e-mail: kumape@bashnet.ru
Web: www.kamov.ru

Lutch Design Bureau JSC
25 Pobedy Boulevard, 152920, Rybinsk, Yaroslavl Region, Russian Federation
Tel: (+7 4855) 28 58 22
Fax: (+7 4855) 28 58 35
e-mail: kb@kb-lutch.ru
Web: www.kb-lutch.ru

MIL Moscow Helicopter Plant
MIL Helicopters
(a subsidiary of Oboronprom United Industrial Corporation, Russia)
Garshina Street, 26 / 1, Tomilino, 140070, Moskva, Lyubertsy, Russian Federation
Tel: (+7 495) 669 70 54
    (+7 495) 669 21 75 (HR)
    (+7 495) 669 72 13 (commerce service)
e-mail: mvz@mi-helicopter.ru
Web: www.mi-helicopter.ru

Production Association Urals
Optical Mechanical Plant
33B Vostochnay Street, 620100, Ekaterinburg, Russian Federation
Tel: (+7) 229 81 09
e-mail: infouomz@uomz.com
Web: www.uomz.ru

Russian Aircraft Corporation MIG (RAC MIG)
(a subsidiary of Joint Stock Company United Aircraft Corporation, Russia)
Boulevard 7, 1st Botkinsky Drive, 125284, Moskva, Russian Federation
Tel: (+7 499) 795 80 10
Fax: (+7 495) 250 19 48
e-mail: mig@migavia.ru
Web: www.migavia.ru

Sokol Design Bureau
Chaadaev Street, 603035, Nizhnyi Novgorod, Russian Federation
Tel: (+7 831) 229 85 01
Fax: (+7 831) 276 46 04
e-mail: okb@sokolnn.ru
       info@sokolnn.ru
Web: www.sokolplant.ru

Strela Orenburg Production Association
ulitsa Shevchenko 26, 460005, Orenburg, Russian Federation
Tel: (+7 3532) 75 71 00
Fax: (+7 3532) 75 54 60
e-mail: po_strela@mail.ru
       strela_market@mail.ru
Web: www.pa-strela.com

Transas Ltd
54-4 Maly Prospect V O, St Petersburg, Russian Federation
Tel: (+7 812) 325 31 31
Fax: (+7 812) 325 31 32
e-mail: sea@transas.com
Web: www.transas.ru

Tupolev PSC
(a subsidiary of Joint Stock Company United Aircraft Corporation, Russia)
Academician Tupolev Embankment 17, 105005, Moskva, Russian Federation
Tel: (+7 495) 267 25 33
Fax: (+7 495) 267 27 33
    (+7 495) 261 08 68
    (+7 495) 261 71 41
e-mail: tu@tupolev.ru
Web: www.tupolev.ru

Zala Aero
PO Box 1424 426011, Izhevsk, Russian Federation
Tel: (+7 3412) 43 05 05
Fax: (+7 3412) 22 15 33
e-mail: info@zala.aero
       nz@zala.aero (press)
Web: www.zala.aero

Singapore

Singapore Technologies Aerospace Ltd (ST Aerospace)
Headquarters
(a subsidiary of Singapore Technologies Engineering Ltd, Singapore)
540 Airport Road, Paya Lebar, 539938, Singapore
Tel: (+65) 62 80 71 11
Fax: (+65) 62 80 97 13
    (+65) 62 80 82 13
e-mail: mktg.aero@stengg.com
       comms.aero@stengg.com
Web: www.staero.aero

South Africa

Advanced Technologies and Engineering Co (Pty) Ltd (ATE)
998, 16th Road, 1685, Halfway House, South Africa
Tel: (+27 11) 266 76 00
    (+27 11) 266 76 46 (marketing investigation)
Fax: (+27 11) 314 17 76
e-mail: info@ate-southafrica.com
Web: www.ate-southafrica.com

Cassidian Optronics (Pty) Ltd
(a subsidiary of Cassidian Optronics GmbH, Germany and Denel (Pty) Ltd, South Africa)
Nellmapius Drive, 0157, Irene, South Africa
Tel: (+27 12) 674 02 15
Fax: (+27 12) 674 01 98
e-mail: optronics@optronics.zeiss.com
Web: www.zeiss.com

Denel Dynamics
PO Box 7412 0046, Centurion, South Africa
Tel: (+27 12) 671 19 11 (switchboard)
    (+27 12) 671 10 01 (marketing)
Fax: (+27 12) 671 17 79
e-mail: market@deneldynamics.co.za
Web: www.deneldynamics.co.za

Saab Avitronics
Recording and Monitoring Systems
(a subsidiary of Saab Grintek, South Africa and Saab Avitronics, Sweden)
PO Box 8492 0046, Centurion, South Africa
185 Witch-Hazel Avenue, Highveld Technopark, Centurion, South Africa
Tel: (+27 12) 672 60 00
    (+27 12) 674 35 00
Fax: (+27 12) 674 35 40
    (+27 12) 672 62 22
Web: www.saabgroup.com

Tellumat Pty Ltd
PO Box 30451 7966, Tokai, South Africa
64-74 White Road, 7945, Tokai, South Africa
Tel: (+27 21) 710 29 11
Fax: (+27 21) 712 12 78
e-mail: helpdesk@tellumat.com
       avionics@tellumat.com
Web: www.tellumat.com

Spain

Aerovision Vehículos Aereos SL
Mikeletegi Pasealekua 2, Parque Technológico, E-20009, San Sebastián, Spain
Tel: (+34 943) 24 72 78
Fax: (+34 943) 62 77 27
e-mail: info@aerovision-uav.com
Web: www.aerovision-uav.com

Alpha Unmanned Systems
C/ La Granja 74, Alcodendas, E-28108, Madrid, Spain
Tel: (+34 91) 657 27 23
e-mail: info@alphaunmannedsystems.com
Web: www.alphaunmannedsystems.com
www.uavnavigation.com

Aries Ingeniería y Sistemas SA
Defense and Security / UAS (ASD)
Paseo de la Castellana, 130, E-28046, Madrid, Spain
Tel: (+34 915) 70 27 37
e-mail: info@aries.com.es
Web: www.aries.com.es

Artes Jet
Venus 10 No.3, E-08012, Barcelona, Spain
Tel: (+34 93) 795 07 19
e-mail: artesjet@gmail.com
Web: www.artesjet.com

Indra Sistemas SA
Carretera de Loeches 9, E-28850, Torrejón de Ardoz, Madrid, Spain
Tel: (+34 91) 627 10 00
Web: www.indracompany.com

Instituto Nacional de Tecnica Aeroespacial (INTA)
Carretera de Ajalvir Km 4, E-28850, Torrejón de Ardoz, Madrid, Spain
Tel: (+34 91) 520 12 00
(+34 91) 520 64 33
Fax: (+34 91) 520 15 86
(+34 91) 520 10 74
e-mail: info@inta.es
relaciones.institucionales@inta.es
Web: www.inta.es
www.cab.inta.es

Sistemas de Control Remoto (SCR)
C/ Isla de la Palma 36 10, Poligono Industrial Norte, San Sebastian, E-28703, Madrid, Spain
Tel: (+34 91) 651 82 27
Fax: (+34 91) 653 91 25
Web: www.scrtargets.com

Tekplus Aerospace
Calle D, Polígono Industrial A Granxa, Parcela 87, E-36400, Porrina, Madrid, Spain
Tel: (+34 986) 29 56 70
e-mail: info@grupotekplus.com
Web: www.grupotekplus.com

UAV Navigation
Avienda de los Pirineos 7, San Sebastian de los Reyes, E-28703, Madrid, Spain
Tel: (+34 91) 153 25 90
e-mail: info@uavnavigation.com
Web: www.uavnavigation.com

Sweden

Air Target Sweden AB (ATS)
Österögatan 1, SE-164 40, Kista, Sweden
Tel: (+46 8) 730 22 33
Fax: (+46 8) 730 34 24
e-mail: info@airtarget.se
Web: www.airtarget.com

Bülow Air Target Systems AB
Våtö Prästgård, SE-760 21, Våtö, Sweden
Tel: (+46 176) 582 10
Fax: (+46 176) 582 12
e-mail: info@bulowairtarget.com
Web: www.bulowairtarget.com
www.bulow.se

CybAero AB
PO Box 1271 SE-581 12, Linköping, Sweden
Tel: (+46 13) 465 29 00
Fax: (+46 13) 991 30 32
e-mail: info@cybaero.se
Web: www.cybaero.se

Forsvarets Materielverk (FMV)
Swedish Defence Materiel Administration
Administration, SE-SE-115 88, Stockholm, Sweden
Tel: (+46 8) 782 40 00
Fax: (+46 8) 667 57 99
e-mail: registrator@fmv.se
info@fmv.se
Web: www.fmv.se

Saab Aerosystems
Bröderna Ugglas gata, SE-581 88, Linköping, Sweden
Tel: (+46 13) 18 00 00
Fax: (+46 13) 18 00 11
(+46 13) 18 24 11
e-mail: infoaerosystems@saab.se
Web: www.saabgroup.com

Saab Microwave Systems
Solhusgatan 10, Kallebaecks Teknikpark, SE-412 89, Göteborg, Sweden
Tel: (+46 31) 794 90 00
Fax: (+46 31) 794 90 02
Web: www.saabgroup.com

Switzerland

Ruag Aviation
Aircraft and Defence Systems
Seetalstrasse 175,
PO Box 301 CH-6032, Emmen, Switzerland
Tel: (+41 412) 68 41 11
Fax: (+41 412) 68 39 88
(+41 412) 68 25 88
e-mail: info.aviation@ruag.com
Web: www.ruag.com

Swiss UAV
Bachmatten 2a, CH-4435, Niederdorf BL, Switzerland
Tel: (+41 61) 508 20 11
Fax: (+41 61) 951 26 70
e-mail: info@swiss-uav.com
Web: www.swiss-uav.com

Unmanned Systems AG
Baarermattstrasse 10, CH-6300, Zug, Switzerland
Tel: (+41 41) 741 33 33
Fax: (+41 41) 741 33 32
Web: wwww.uasystems.ch

Taiwan

Carbon Based Technology Inc (CBT)
3F, No.30, Keya Road, Daya District, 42881, Taichung City, Taiwan
Tel: (+886 4) 25 65 85 58
Fax: (+886 4) 25 65 85 59
e-mail: service99@uaver.com
Web: www.uaver.com

Chung Shan Institute of Science and Technology (CSIST)
No 15, Shi Qi Zi, Gaoping Village, Longtan Township, Taoyuan, Taiwan
Tel: (+886 3) 411 21 17
Fax: (+886 3) 411 54 60
e-mail: csist@csistdup.org.tw
Web: www.csistdup.org.tw

Turkey

Turkish Aerospace Industries Inc (TAI)
P O Box 18 TR-06692, Kavaklidere, Ankara, Turkey
Tel: (+90 312) 811 18 00
Fax: (+90 312) 811 14 25
(+90 312) 811 14 08
e-mail: marketing@tai.com.tr
Web: www.tai.com.tr

Ukraine

Kharkov State Aircraft Manufacturing Company (KSAMC)
134 Sumskaya Street, 61023, Kharkov, Ukraine
Tel: (+380 57) 707 07 81
(+380 57) 700 34 39
Fax: (+380 57) 707 08 34
e-mail: info@ksamc.com
Web: www.ksamc.com

Scientifically Industrial Systems Ltd
Design Vzlet
PO Box 9233 61004, Kharkov, 4, Ukraine
Tel: (+380 57) 724 11 58
Fax: (+380 57) 225 14 91
e-mail: kbvzlet@mail.ru
Web: www.kbvzlet.com

United Arab Emirates

Adcom Systems
Industrial City of Abu Dhabi,
PO Box 25298 Abu Dhabi, United Arab Emirates
Tel: (+971 2) 550 06 30
Fax: (+971 2) 550 06 31
e-mail: info@adcom-systems.com
Web: www.adcomsystems.ae

United Kingdom

Aesir Ltd
3 Saint David's Square, Peterborough, Cambridgeshire, PE1 5QA, United Kingdom
Tel: (+44 1733) 55 25 45
(+44 1733) 77 31 63
e-mail: enquiries@aesir-uas.com
Web: www.aesir-uas.com

Airborne Systems Ltd
Bettws Road, Llangeinor, Bridgend, CF32 8PL, United Kingdom
Tel: (+44 1656) 72 70 21
Fax: (+44 1656) 72 11 00
e-mail: sales@airborne-sys.com
Web: www.airborne-sys.com

Allsopp Helikites Ltd
South End Farm, Damerham, Fordingbridge, Hampshire, SP6 3HW, United Kingdom
Tel: (+44 1725) 51 87 50
Fax: (+44 1725) 51 87 86
e-mail: helikites@yahoo.com
Web: www.helikites.com

BAE Systems Military Air & Information
Headquarters
(a subsidiary of BAE Systems plc, UK)
Warton Aerodrome, Preston, Lancashire, PR4 1AX, United Kingdom
Tel: (+44 1772) 63 33 33
Fax: (+44 1772) 63 47 24
Web: www.baesystems.com

Blue Bear Systems Research Ltd
Building 32, Twinwoods Business Park, Thurleigh Road, Milton Ernest, Bedford, Bedfordshire, MK41 6JE, United Kingdom
Tel: (+44 1234) 21 20 01
Fax: (+44 1234) 21 57 95
e-mail: enquiries@bbsr.co.uk
Web: www.bbsr.co.uk

Cosworth Ltd
Engineering and Manufacturing
The Octagon, St James Mill Road, Northampton, Northamptonshire, NN5 5RA, United Kingdom
Tel: (+44 1604) 59 83 00
Fax: (+44 1604) 59 83 01
Web: www.cosworth.com

United Kingdom – United States   **Contractors** 421

Cubewano Ltd
Unit 8b, Reddicap Trading Estate, Sutton Coldfield, Birmingham, B75 7BU, United Kingdom
Tel: (+44 121) 378 17 65
e-mail: information@cubewano.com
Web: www.cubewano.com

Cyberflight Ltd
The Machine House, Unit 1, Newfields, Moira, Derbyshire, DE12 6EG, United Kingdom
Tel: (+44 1283) 22 23 36
Fax: (+44 1283) 22 21 36
e-mail: info@cyberflightuavs.com
tc@cyberflightuavs.com
Web: www.cyberflightuavs.com

ESL Defence Limited
(a subsidiary of AAI Corporation, US)
16 Compass Point, Ensign Way, Hamble, Southampton, SO31 4RA, United Kingdom
Tel: (+44 23) 80 45 51 10
Fax: (+44 23) 80 74 42 00
e-mail: esl@esldefence.co.uk
Web: www.esldefence.co.uk

Fan Wing Ltd
Second Floor, 43-45 Dorset Street, London, W1U 7NA, United Kingdom
Tel: (+44 7855) 37 40 06
Web: www.fanwing.com

Meggitt Defence Systems Ltd UK (MDS)
The Boulevard, Orbital Park, Ashford, Kent, TN24 0GA, United Kingdom
Tel: (+44 1233) 50 53 00
Fax: (+44 1233) 50 37 07
Web: www.meggitt-defence.co.uk

Nitrohawk Ltd
Purland Chase, Ross-on-Wye, Hereford and Worcestershire, HR9 5RR, United Kingdom
Tel: (+44 1989) 56 31 88
Web: www.nitrohawk.co.uk

QinetiQ
Company Headquarters
Cody Technology Park, Ively Road, Farnborough, Hampshire, GU14 0LX, United Kingdom
Tel: (+44 1252) 39 20 00
Fax: (+44 1252) 39 33 99
Web: www.qinetiq.com

RCV Engines Limited
4 Haviland Road, Ferndown Industrial Estate, Wimborne, Dorset, BH21 7RF, United Kingdom
Tel: (+44 1202) 87 70 44
Fax: (+44 1202) 87 18 36
e-mail: sales@rcvengines.com
Web: www.rcvengines.com

Rubicon Systems Design Ltd
15 Dowding Road, Biggin Hill, Westerham, Kent, TN16 3BE, United Kingdom
Tel: (+44 1959) 57 49 92
Fax: (+44 1959) 57 07 86
e-mail: info@rubiconsd.com
Web: www.rubiconsd.com

Selex Galileo
Head Office
(a subsidiary of Finmeccanica SpA, Italy)
Christopher Martin Road, Basildon, Essex, SS14 3EL, United Kingdom
Tel: (+44 1268) 52 28 22
Web: www.selexgalileo.com

Selex Galileo
(Selex S&AS)
2 Crewe Road North, Edinburgh, EH5 2XS, United Kingdom
Tel: (+44 131) 332 24 11
Fax: (+44 131) 343 40 11
e-mail: sales.marketing@selexgalileo.com
Web: www.selexgalileo.com

Selex Galileo
Unit 7 ParcAberporth, Blaenannerch, Cardigan, Ceredigion, SA43 2AD, United Kingdom
Tel: (+44 1239) 81 41 57
Fax: (+44 1239) 81 49 75
e-mail: sales.marketing@selexgalileo.com
Web: www.selexgalileo.com

SkyshipsRemote
Leighs Lodge, Willows Green, Chelmsford, Essex, CM3 1QJ, United Kingdom
Tel: (+44 1245) 36 29 80
Fax: (+44 1245) 36 29 81
e-mail: info@skyships.co.uk
Web: www.skyships.co.uk

Tasuma UK Ltd
Blandford Heights Industrial Estate, Unit 11 Uplands Way, Blandford Forum, Dorset, DT11 7UZ, United Kingdom
Tel: (+44 1258) 48 88 33
Fax: (+44 1258) 48 88 66
e-mail: tasuma@btconnect.com
Web: www.tasuma-uk.com

Thales UK
Manor Royal, Crawley, West Sussex, RH10 9HA, United Kingdom
Tel: (+44 1293) 58 00 00
Web: www.thalesgroup.com/uk

Thales UK
Vicon House, Western Way, Bury St Edmunds, Suffolk, IP33 3SP, United Kingdom
Tel: (+44 1284) 75 05 99
Fax: (+44 1284) 75 05 98
e-mail: uk.enquiries@thalesgroup.com
Web: www.thalesgroup.com

Thermoteknix Systems Ltd
Teknix House, 2 Pembroke Avenue, Waterbeach, Cambridge, CB25 9QR, United Kingdom
Tel: (+44 1223) 20 40 00
Fax: (+44 1223) 20 40 10
e-mail: sales@thermoteknix.com
Web: www.thermoteknix.com

UAV Engines Ltd (UEL)
Lynn Lane, Shenstone, Lichfield, Staffordshire, WS14 0DT, United Kingdom
Tel: (+44 1543) 48 18 19
Fax: (+44 1543) 48 13 93
e-mail: uav@uavenginesltd.co.uk
Web: www.uavenginesltd.co.uk

Universal Target Systems Ltd (UTSL)
Prestige House, Landews Meadow, Green Lane, Challock, Kent, TN25 4BL, United Kingdom
Tel: (+44 1233) 74 00 55
Fax: (+44 1233) 74 00 66
e-mail: sales@utsl.co.uk
info@utsl.co.uk
Web: www.utsl.co.uk

Warrior (Aero-Marine) Ltd
The Big House, Lower Quay Road, Hook, Haverfordwest, Pembrokeshire, SA62 4LR, United Kingdom
Tel: (+44 14) 37 89 98 83
Web: www.warrioraero.com

Wilksch Airmotive Ltd
Unit SE12A, Gloucestershire Airport Ltd, Cheltenham, Staverton, Gloucestershire, GL51 6SP, United Kingdom
Tel: (+44 870) 170 96 70
Fax: (+44 870) 170 96 78
e-mail: sales@wilksch.com
Web: www.wilksch.net

United States

AAI Corporation
124 Industry Lane, Hunt Valley, Maryland, 21030-0126, United States
Tel: (+1 410) 666 14 00
(+1 978) 657 20 20 (Textron systems, public relations)
Fax: (+1 410) 628 32 15
e-mail: aaireg@aai.textron.com
tscpublicrelations@systems.textron.com
Web: www.aaicorp.com

AME Unmanned Air Systems
133 West Park Loop, Huntsville, Alabama, 35806, United States
Tel: (+1 256) 319 72 84
Fax: (+1 256) 722 01 44
Web: www.chandlermay.com

AV Inc
Headquarters
181 West Huntington Drive, Suite 202, Monrovia, California, 91016, United States
Tel: (+1 626) 357 99 83
Fax: (+1 626) 359 96 28
e-mail: info@avinc.com
Web: www.aerovironment.com
www.avsuav.com

Acuity Technologies
Headquarters
3475 Edison Way, Building P, Menlo Park, California, 94025, United States
Tel: (+1 650) 369 67 83
Fax: (+1 650) 249 35 29
e-mail: info@acuitytx.com
Web: www.acuitytx.com

Aerovel Corporation
83 Oak Ridge Road, White Salmon, Washington, 98672-8114, United States
Tel: (+1 541) 490 41 03
Web: www.aerovelco.com

Alliant Techsystems Inc
Corporate Headquarters
7480 Flying Cloud Drive, Minneapolis, Minnesota, 55344, United States
Tel: (+1 952) 351 30 00
Freephone: (+1 800) 345 85 94
Fax: (+1 952) 351 30 09
e-mail: atk.corporate@atk.com
Web: www.atk.com
www.atkethics.com

Alliant Techsystems Inc
Propulsion and Controls
PO Box 241 Elkton, Maryland, 21922, United States
Tel: (+1 410) 392 10 00
Web: www.atk.com

American Dynamics Flight Systems
8264 Preston Court, Suite A, Jessup, Maryland, 20794, United States
Tel: (+1 301) 358 07 47
Fax: (+1 301) 483 03 03
e-mail: info@adflightsystems.com
Web: www.americandynamics.us

Arcturus UAV
539 Martin Avenue, Suite 110, Rohnert Park, California, 94928, United States
Tel: (+1 707) 206 93 72
Fax: (+1 707) 206 93 93
Web: www.arcturus-uav.com

Aurora Flight Sciences Corporation (AFS)
9950 Wakeman Drive, Manassas, Virginia, 20110, United States
Tel: (+1 703) 369 36 33
Fax: (+1 703) 369 45 14
Web: www.aurora.aero

BAE Systems
Unmanned Aircraft Programs
3292 East Hemisphere Loop, Tucson, Arizona, 85706, United States
Tel: (+1 520) 573 63 00
Fax: (+1 520) 573 20 57
Web: www.baesystems.com

BAE Systems Electronic Systems
(E&IS)
450 Pulaski Road, Greenlawn, New York, 11740-1609, United States
Tel: (+1 631) 261 70 00
Web: www.baesystems.com/eis

BAE Systems Electronic Systems
(E&IS)
12-18 Hartwell Avenue, Lexington, Massachusetts, 02421-7306, United States
Tel: (+1 781) 863 48 21
Web: www.baesystems.com

Contractors

BAE Systems Electronic Systems (E&IS)
(a subsidiary of BAE Systems Inc, US)
600 Main Street, Johnson City, New York, 13790, United States
Tel: (+1 607) 770 20 00
Fax: (+1 607) 770 35 24
Web: www.baesystems.com

Brandebury Tool Company Inc (BTC)
7901 A Cessna Avenue, Gaithersburg, Maryland, 20879, United States
Tel: (+1 301) 519 20 34
Web: www.microuav.com

Brock Technologies Inc
3774 East, 43rd Place, Tucson, Arizona, 85713, United States
Tel: (+1 520) 790 54 84
e-mail: info@brocktechnologies.com
sales@brocktechnologies.com
Web: www.brocktechnologies.com

Butler Parachute Systems Inc
1820 Loudon Avenue North West,
PO Box 6098 Roanoke, Virginia, 24017, United States
Tel: (+1 540) 342 25 01
Fax: (+1 540) 342 40 37
Web: www.butlerparachutes.com

Composite Engineering Inc
5381 Raley Boulevard, Sacramento, California, 95838, United States
Tel: (+1 916) 991 19 90
Web: www.compositeeng.com
www.cei.to

Cyber Aerospace LLC
5147 South Harvard Avenue, Suite 138, Tulsa, Oklahoma, 74135, United States
Tel: (+1 504) 722 74 02
Fax: (+1 918) 493 62 34
e-mail: info@cyberaerospace.com
Web: www.cyberaerospace.com

DRS Defense Solutions
Headquarters
(a subsidiary of DRS Technologies Ltd, US)
530 Gaither Road, Suite 900, Rockville, Maryland, 20850, United States
Tel: (+1 240) 238 39 00
Fax: (+1 240) 238 39 77
e-mail: marketing@drs-ds.com
Web: www.drs-ds.com

Desert Aircraft Inc
1815 South Research Loop, Tucson, Arizona, 85710, United States
Tel: (+1 520) 722 06 07
Fax: (+1 520) 722 56 22
e-mail: info@desertaircraft.com
Web: www.desertaircraft.com

Dragonfly Pictures Inc
600 West End of Second Street, Essington, Pennsylvania, 19029-0202, United States
Tel: (+1 610) 521 61 15
Fax: (+1 610) 521 30 74
Web: www.dragonflypictures.com

Engineered Arresting Systems Corporation
Zodiac Aerospace
(a subsidiary of Zodiac Aerospace, France)
2550 Market Street, Aston, Pennsylvania, 19014-3426, United States
Tel: (+1 610) 494 80 00
Fax: (+1 610) 494 89 89
e-mail: sales.aston@zodiacaerospace.com
Web: www.esco.zodiacaerospace.com

FLIR Commercial Systems Inc
Santa Barbara
70 Castilian Drive, Goleta, California, 93117, United States
Freephone: (+1 888) 747 35 47
Fax: (+1 805) 964 97 97 (international)
Web: www.flir.com

FLIR Systems Inc
Corporate Headquarters
27700 South West Parkway Avenue, Wilsonville, Oregon, 97070, United States
Freephone: (+1 800) 727 35 47
Web: www.flir.com

Flir Systems Inc
25 Esquire Road, North Billerica, Massachusetts, 01862, United States
Tel: (+1 978) 901 80 00
Freephone: (+1 800) 464 63 72
e-mail: service@flir.com
Web: www.flir.com

General Atomics Aeronautical Systems Inc (GA-ASI)
Reconnaissance Systems
13322 Evening Creek Drive, San Diego, North, California, 92128, United States
Tel: (+1 858) 964 67 00
(+1 858) 455 30 00
Fax: (+1 858) 964 69 52
(+1 858) 455 36 21
e-mail: pr-asi@ga-asi.com
Web: www.ga-asi.com
www.uav.com
www.ga.com

General Atomics Aeronautical Systems Inc (GA-ASI)
Aircraft Systems Group
(a subsidiary of General Atomics, US)
14200 Kirkham Way, Poway, California, 92064, United States
Tel: (+1 858) 312 28 10
Fax: (+1 858) 312 42 47
e-mail: pr-asi@ga-asi.com
Web: www.ga-asi.com

Goodrich ISR Systems
3490 Route 1, Building 12, Princeton, New Jersey, 08540-5914, United States
Tel: (+1 609) 520 06 10
Fax: (+1 609) 520 16 63
e-mail: sui_info@goodrich.com
Web: www.sensorsinc.com
www.goodrich.com

Griffon Aerospace
106 Commerce Circle, Madison, Alabama, 35758, United States
Tel: (+1 256) 258 00 35
Fax: (+1 256) 258 00 39
Web: www.griffon-aerospace.com

Hamilton Sundstrand
Power Systems
Auxiliary Power Engine and Control Systems
4400 Ruffin Road,
PO Box 85757 San Diego, California, 92186 5757, United States
Tel: (+1 858) 627 65 65
(+1 858) 627 65 27
Fax: (+1 860) 660 41 85
e-mail: businessdev@hs.utc.com
Web: www.hs-powersystems.com

Herley Industries Inc (HRLY)
3061 Industry Drive, Lancaster, Pennsylvania, 17603-4025, United States
Tel: (+1 717) 397 27 77
Fax: (+1 717) 735 81 23
Web: www.herley.com

Honeywell Aerospace
Defense and Space Electronic Systems
9201 San Mateo Boulevard North East, Albuquerque, New Mexico, 87113-2227, United States
Web: www.honeywell.com

ISL Bosch
Aeronautical Systems Operations (ASO)
205 Lawler Drive, Brownsboro, Alabama, 35741-9455, United States
Tel: (+1 256) 852 50 33
Fax: (+1 256) 852 58 98
Web: www.islinc.com

ITT Electronic Systems Reconnaissance & Surveillance Systems (ITT RSS)
(a subsidiary of ITT Corporation, US)
18705 Madrone Parkway, Morgan Hill, California, 95037, United States
Tel: (+1 408) 210 67 18
e-mail: rss.marketing@itt.com
Web: www.ittrss.com
www.defense.itt.com
www.exelisinc.com

ITT Exelis
1650 Tysons Boulevard, Suite 1700, Mclean, Virginia, 22102, United States
Tel: (+1 703) 790 63 00
Fax: (+1 703) 790 63 60
Web: www.exelisinc.com

Imsar LLC
940 South 2000 West 40, Springville, Utah, 84663, United States
Tel: (+1 801) 798 84 40
Fax: (+1 801) 798 28 14
e-mail: sales@imsar.com
Web: www.imsar.com

Insitu Group Inc
(a subsidiary of The Boeing Company, US)
118 East Columbia River Way, Bingen, Washington, 98605, United States
Tel: (+1 509) 493 86 00
Fax: (+1 509) 493 86 01
Web: www.insitu.com

Irvine Sensors Corporation
3001 Red Hill Avenue, Building 4, Suite 108, Costa Mesa, California, 92626-4526, United States
Tel: (+1 714) 549 82 11
Freephone: (+1 800) 468 46 12
Fax: (+1 714) 444 87 73
e-mail: investorrelations@irvine-sensors.com
productinfo@irvine-sensors.com
Web: www.irvine-sensors.com

Jet Propulsion Laboratory
NASA (JPL)
(a subsidiary of National Aeronautics & Space Administration, US)
4800 Oak Grove Drive, Pasadena, California, 91109, United States
Tel: (+1 818) 354 01 12
Fax: (+1 818) 393 46 41
Web: www.jpl.nasa.gov

Kaman Aerospace Corporation (KAC)
(a subsidiary of Kaman Corp, US)
Old Windsor Road,
PO Box 2 Bloomfield, Connecticut, 06002-0002, United States
Tel: (+1 860) 242 44 61
Web: www.kamanaero.com

L-3 Communications Systems - East
1 Federal Street, Camden, New Jersey, 08103, United States
Tel: (+1 856) 338 30 00
Fax: (+1 856) 338 33 45
Web: www.l-3com.com

L-3 Communications Systems - West
640 North 2200 West,
PO Box 16850 Salt Lake City, Utah, 84116-0850, United States
Tel: (+1 801) 594 20 00
Freephone: (+1 800) 543 47 05
Fax: (+1 801) 594 35 72
Web: www.l-3com.com

L-3 Unmanned Systems
Production Facility
(a subsidiary of L-3 Communications, US)
9040 Glebe Park Drive, Easton, Maryland, 21601, United States
Tel: (+1 410) 820 75 00
Fax: (+1 410) 820 85 00
e-mail: bai.sales@l-3com.com
us-e.sales@l-3com.com
Web: www.l-3com.com

United States

L-3 Unmanned Systems
Headquarters
(a subsidiary of L-3 Communications, US)
4240 International Parkway, Suite 100, Carrollton, Texas, 75007, United States
Tel: (+1 469) 568 23 76
Fax: (+1 469) 568 21 00
e-mail: us-d.info@l-3com.com
Web: www.l-3com.com/uas
www.l-3com.com

Lockheed Martin Aeronautics
(LM Aero)
Advanced Development Programs
Department 002J, Building 608, MZ, 1011 Lockheed Way, Palmdale, California, 93599-3740, United States
Tel: (+1 661) 572 62 63
Web: www.lockheedmartin.com/aeronautics

Lockheed Martin Information Systems and Global Services
2339 Route 70 West, Cherry Hill, New Jersey, 08358-0001, United States
Tel: (+1 859) 486 52 39
Fax: (+1 859) 486 52 07
Web: www.lockheedmartin.com

Lockheed Martin Maritime Systems and Sensors
(a subsidiary of Lockheed Martin Mission Systems & Sensors, US)
1210 Massillon Road, Akron, Ohio, 44315-0001, United States
Web: www.lockheedmartin.com/ms2

Lockheed Martin Maritime Systems and Sensors (MS2)
300 M Street SE, Suite 700, Washington, DC 20003-3442, United States
Tel: (+1 202) 863 34 59
Web: www.lockheedmartin.com

Lockheed Martin Missiles and Fire Control
Headquarters
(a subsidiary of Lockheed Martin Electronic Systems, US)
650003 PT-42 Dallas, Texas, 75265-0003, United States
Tel: (+1 972) 603 10 00
Fax: (+1 407) 603 10 09
(+1 407) 356 03 08
Web: www.lockheedmartin.com

Lockheed Martin Missiles and Fire Control
Dallas
1902 West Freeway, Grand Prairie, Texas, 75051, United States
Tel: (+1 972) 603 10 00
Web: www.lockheedmartin.com

Lockheed Martin Mission Systems & Sensors (MS2)
Headquarters
55 Charles Lindbergh Boulevard, Mitchel Field, New York, 11553-3682, United States
Tel: (+1 516) 228 20 00
Web: www.lockheedmartin.com/ms2

Lockheed Martin Rail Systems
55 Charles Lindbergh Boulevard, Mitchel Field, New York, 11553-3682, United States
Tel: (+1 516) 228 20 00
Web: www.lockheedmartin.com/ms2

MAV6 LLC
1619 Walnut Street, Vicksburg, Mississippi, 39180, United States
Tel: (+1 703) 340 13 04
e-mail: info@mav6.com
Web: www.mav6.com

MBDA Inc
Headquarters
(a subsidiary of MBDA, UK)
1300 Wilson Boulevard, Suite 550, Arlington, Virginia, 22209, United States
Tel: (+1 703) 387 71 20
Fax: (+1 703) 875 91 04
Web: www.mbdainc.com

MLB Company
3335 Kifer Road, Santa Clara, California, 95051-0719, United States
Tel: (+1 650) 966 10 22
(+1 408) 738 10 22
e-mail: info@mlbuav.com
Web: www.spyplanes.com

Meggitt Defense Systems Inc
(a subsidiary of Meggitt plc, UK)
9801 Muirlands Boulevard, Irvine, California, 92618-2521, United States
Tel: (+1 949) 465 77 00
Fax: (+1 949) 465 95 60
(+1 949) 660 87 46
Web: www.meggitt.com

Micro Systems Inc (MSI)
35 Hill Avenue, Fort Walton Beach, Florida, 32548-3858, United States
Tel: (+1 850) 244 23 32
Fax: (+1 850) 243 13 78
e-mail: marketing@gomicrosystems.com
Web: www.gomicrosystems.com
www.herley-msi.com

Mission Technologies Inc
(Mi-Tex)
14785 Omicron Drive, Suite 102, San Antonio, Texas, 78245-3222, United States
Tel: (+1 210) 677 06 02
(+1 210) 698 12 71
Fax: (+1 210) 698 09 84
(+1 210) 677 06 85
e-mail: mitex@mitex-sa.com
Web: www.mitex-sa.com

Mitsubishi Electric and Electronics US Inc (MEUS)
5665 Plaza Drive,
PO Box 6007 Cypress, California, 90630-0007, United States
Tel: (+1 714) 220 25 00
e-mail: customercare@meus.mea.com
Web: www.mitsubishielectric-usa.com

National Aeronautics & Space Administration (NASA)
Ames Research Center
NASA Ames Research
Moffett Field, California, 94035-1000, United States
Tel: (+1 650) 604 50 00
e-mail: nssc-contactcenter@nasa.gov
Web: www.nasa.gov

Naval Research Laboratory (NRL)
4555 Overlook Avenue South West, Washington, District of Columbia, 20375, United States
Web: www.nrl.navy.mil

Naval Surface Warfare Center (NSWC)
6149 Welsh Road, Suite 203, Dahlgren, Virginia, 22448, United States
Tel: (+1 540) 653 82 91
(+1 540) 249 82 91
Freephone: (+1 877) 845 56 56
Web: www.navsea.navy.mil

Naval Surface Warfare Center (NSWC)
IHDIV Public Affairs, 3767 Strauss Avenue, Building 20, Suite 113, Indian Head, Maryland, 20640-5150, United States
Tel: (+1 540) 744 65 05
e-mail: ihdiv.nswc.pao@navy.mil
Web: www.navsea.navy.mil

Northrop Grumman Electronic Systems
Headquarters
(a subsidiary of Northrop Grumman Corporation, US)
Mail Stop A255,
PO Box 17319 Baltimore, Maryland, 21203-7319, United States
1580-A West Nursery Road, Linthicum, Maryland, 21090, United States
Tel: (+1 410) 765 10 00
(+1 410) 993 68 48 (media relations)

Freephone: (+1 800) 443 92 19
Web: www.es.northropgrumman.com
www.northropgrumman.com

Northrop Grumman Electronic Systems
Defensive Systems
600 Hicks Road,
Mail Stop H6097 Rolling Meadows, Illinois, 60009-1098, United States
Tel: (+1 847) 259 96 00
Fax: (+1 847) 870 57 05
Web: www.es.northropgrumman.com
www.northropgrumman.com

Northrop Grumman Electronic Systems
Navigation Systems
(a subsidiary of Northrop Grumman Electronic Systems, US)
21240 Burbank Boulevard, Woodland Hills, California, 91367-6675, United States
Tel: (+1 866) 646 28 79
e-mail: customerservice.nsd@northropgrumman.com
Web: www.es.northropgrumman.com
www.northropgrumman.com

Orbital Sciences Corporation
Missile Defense Systems
3380 South Price Road, Chandler, Arizona, 85248, United States
Tel: (+1 602) 899 60 00
Fax: (+1 602) 899 88 00
Web: www.orbital.com

Piasecki Aircraft Corporation
(PiAC)
519 West Second Street,
PO Box 360 Essington, Pennsylvania, 19029-0360, United States
Tel: (+1 610) 521 57 00
Fax: (+1 610) 521 59 35
e-mail: piasecki_jw@piasecki.com
info@piasecki.com
Web: www.piasecki.com

Prioria Robotics Inc
104 North Main Street, Suite 300, Gainsville, Florida, 32601, United States
Tel: (+1 352) 505 21 88
Fax: (+1 352) 505 21 89
e-mail: info@prioria.com
Web: www.prioria.com

Proxy Aviation Systems Inc
7940 Airpark Road, Ste C, Gaithersburg, Maryland, 20879-4198, United States
Tel: (+1 301) 216 28 51
Fax: (+1 301) 216 28 52
e-mail: info@proxyaviation.com
Web: www.proxyaviation.com

Quadra-Aerrow International
911 New London Turnpike, Glastonbury, 06033, United States
Tel: (+1 905) 440 44 35
Web: www.quadraengines.com

Raytheon Company
Electronic Warfare Systems
6380 Hollister Avenue, Goleta, California, 93117-3114, United States
Tel: (+1 805) 967 55 11
(+1 805) 879 21 67
Fax: (+1 805) 879 25 79
e-mail: saspr@raytheon.com
Web: www.raytheon.com

Raytheon Missile Systems
(RMS)
Headquarters
(a subsidiary of Raytheon Company, US)
1151 East Hermans Road, Tucson, Arizona, 85756, United States
Tel: (+1 520) 794 88 61
(+1 520) 794 30 00
Web: www.raytheon.com

Raytheon Network Centric Systems (NCS)
Headquarters
(a subsidiary of Raytheon Company, US)
2501 West University Drive, Mc Kinney, Texas, 75071, United States
Tel: (+1 972) 952 48 77
e-mail: ncspr@raytheon.com
Web: www.raytheon.com

Raytheon Space and Airborne Systems (SAS)
Headquarters
(a subsidiary of Raytheon Company, US)
2000 East El Segundo Boulevard, El Segundo, California, 90245, United States
Tel: (+1 310) 647 10 00
e-mail: saspr@raytheon.com
Web: www.raytheon.com

Rockwell Collins Inc
Control Technologies
3721 Macintosh Drive, Vint Hill Tech Park, Warrenton, Virginia, 20187, United States
Tel: (+1 540) 428 33 00
Fax: (+1 540) 428 33 01
Web: www.rockwellcollins.com

Rolls-Royce Corporation
(a subsidiary of Rolls-Royce plc, UK)
PO Box 420 Indianapolis, Indiana, 46206-0420, United States
Tel: (+1 317) 230 20 00
Freephone: (+1 888) 255 47 66
Fax: (+1 317) 230 67 63
Web: www.rolls-royce.com

Scaled Composites LLC
Headquarters
Hangar 78 Airport, 1624 Flight Line, Mojave, California, 93501, United States
Tel: (+1 661) 824 45 41
Fax: (+1 661) 824 41 74
e-mail: info@scaled.com
Web: www.scaled.com

Schiebel Technology Inc (STI)
(a subsidiary of Schiebel Industries AG, Austria)
70 Main Street, Suite 11, Warrenton, Virginia, 20186, United States
Tel: (+1 540) 351 17 31
Fax: (+1 540) 351 17 36
e-mail: info@schiebel.net
Web: www.schiebel.net

Science Applications International Corporation (SAIC)
(a part of Science Applications International Corporation, US)
4001 North Fairfax Drive, 4th FLoor, Arlington, Virginia, 22203, United States
Web: www.saic.com

Sierra Nevada Corporation (SNC)
Headquarters
444 Salomon Circle, Sparks, Nevada, 89434, United States
Tel: (+1 775) 331 02 22
(+1 720) 407 32 23
Fax: (+1 775) 331 03 70
e-mail: generalinfo@sncorp.com
ssg@sncorp.com
Web: www.sncorp.com

Swift Engineering Inc
1141-A Via Callejon, San Clemente, California, 92673, United States
Tel: (+1 949) 492 66 08
Fax: (+1 949) 366 82 49
e-mail: info@swiftengineering.com
Web: www.swiftengineering.com

TCOM LP
7115 Thomas Edison Drive, Columbia, Maryland, 21046-2113, United States
Tel: (+1 410) 312 23 06
Fax: (+1 410) 312 24 55
e-mail: aerostat@tcomlp.com
Web: www.tcomlp.com

The Boeing Company
Defense, Space and Security
(a subsidiary of The Boeing Company, US)
PO Box 516 St Louis, Missouri, 63166, United States
Tel: (+1 314) 232 02 32
(+1 562) 797 20 20
Web: www.boeing.com/bds

The Center for Interdisciplinary Remotely Piloted Aircraft Studies (CIRPAS)
Hangar 507, 3200 Imjin Road, Marina, California, 93933, United States
Tel: (+1 831) 384 27 76
Fax: (+1 831) 384 32 77
Web: www.cirpas.org

UAV Factory Ltd
50 South Buckhout Street, Irvington, New York, 10533, United States
Tel: (+1 914) 591 30 70
Fax: (+1 914) 591 37 15
e-mail: info@uavfactory.com
Web: www.uavfactory.com

Williams International
2280 East West Maple Road,
PO Box 200 Walled Lake, Michigan, 48390-0200, United States
Tel: (+1 248) 624 52 00
Fax: (+1 248) 624 53 45
e-mail: publicrelations@williams-int.com
Web: www.williams-int.com

INDEXES

Alphabetical Index

10/150 *launch and recovery* (Norway) 402
37M *UAVs* (France) 50
420K *UAVs* (United States) 249

A

A3GT *aerial targets* (France) 307
A3 Observer launcher *launch and recovery* (United Kingdom) 407
A-3 Remez *UAVs* (Ukraine) 171
A-4 Albatros *UAVs* (Ukraine) 172
A-5 *UAVs* (Ukraine) 173
A-10 *UAVs* (Ukraine) 173
A-11 *aerial targets* (Ukraine) 341
A-12 *UAVs* (Ukraine) 173
A160 *UAVs* (United States) 213
A-160 *UAVs* (Ukraine) 174
A160T *UAVs* (United States) 213
Ababeel *aerial targets* (Pakistan) 324
Ababil *UAVs* (Iran) 73
ABJT *aerial targets* (Pakistan) 322
Accipiter *UAVs* (Taiwan) 169
ACE *UAVs* (United States) 244
AD-150 *UAVs* (United States) 201
ADCOM-3D *control and communications* (United Arab Emirates) 384
Adcom Systems Smart Eye 1 *UAVs* (United Arab Emirates) 174
ADM-141A TALD *UAVs* (Israel) 100
ADM-141C ITALD *UAVs* (Israel) 100
ADM-160B *UAVs* (United States) 283
ADM-160C *UAVs* (United States) 283
ADNAV GCS *control and communications* (United Arab Emirates) 384
ADS 95 *UAVs* (Switzerland) 162
Advanced Cockpit GCS *control and communications* (United States) 388
Aelius *UAVs* (France) 36
Aerolight *UAVs* (Israel) 77
Aerosky *UAVs* (Israel) 77
Aerosonde
UAVs (Australia) 3
UAVs (United States) 197
Aerosonde Mark 4.7 *UAVs* (United States) 196
Aerostar *UAVs* (Israel) 77
AGCS *control and communications* (Israel) 378
AirMule *UAVs* (Israel) 106
Aist *UAVs* (Russian Federation) 144
AL-4 *UAVs* (Taiwan) 166
AL-20 *UAVs* (Taiwan) 166
AL-40 *UAVs* (Taiwan) 167
AL-120 *UAVs* (Taiwan) 167
Aladin *UAVs* (Germany) 56
Alamak *control and communications* (Italy) 379
ALBA *aerial targets* (Spain) 335
Albhatros *UAVs* (Spain) 156
Alkyon *aerial targets* (Greece) 312
ALPPUL LP-02 *launch and recovery* (Spain) 403
Al Saber *UAVs* (Austria) 8
Altair *UAVs* (United States) 231
Altius Mk II *UAVs* (India) 66
Aludra *UAVs* (Malaysia) 121
AM 03089 *aerial targets* (Brazil) 294
A/MH-6X *UAVs* (United States) 216
Anka *UAVs* (Turkey) 170
Anka Block A *UAVs* (Turkey) 170
Anka Block B *UAVs* (Turkey) 170
AN/MSQ-131 GHGS *control and communications* (United States) 393
AP04 *control and communications* (Spain) 381
AP04OEM2 *control and communications* (Spain) 381
AP04R *control and communications* (Spain) 381
AP04SDK *control and communications* (Spain) 381
APID 55 *UAVs* (Sweden) 160
AR 70 *UAVs* (Germany) 53
AR 100 *UAVs* (Germany) 53
AR 150 *UAVs* (Germany) 53
Archer *launch and recovery* (Switzerland) 405
Archimede *UAVs* (Italy) 109
AS-113/12U *control and communications* (Sweden) 383
AS-133/12U *control and communications* (Sweden) 383
AS-135/12U *control and communications* (Sweden) 383
Asio *UAVs* (United Kingdom) 191
ASN-7 *aerial targets* (China) 301
ASN-9 *aerial targets* (China) 302
ASN-12 *aerial targets* (China) 303
ASN-15 *UAVs* (China) 27
ASN-104 *UAVs* (China) 27
ASN-104 GCS *control and communications* (China) 370
ASN-105B *UAVs* (China) 27
ASN-105B GCS *control and communications* (China) 370
ASN-206 *UAVs* (China) 28
ASN-207 *UAVs* (China) 29
ASN-209 *UAVs* (China) 29
ASN-213 *UAVs* (China) 30
ASN-216 *UAVs* (China) 30
ASN-217 *UAVs* (China) 31
ASN-229A *UAVs* (China) 31
ASR-4 *aerial targets* (United Kingdom) 346
AT-3 *UAVs* (United States) 199
AT-10 *UAVs* (United States) 200
Athena 111 *control and communications* (United States) 395
Athena 111m *control and communications* (United States) 395
Athena 311 *control and communications* (United States) 396
Athena 411 *control and communications* (United States) 396
Athena 511 *control and communications* (United States) 397
Atlantic *UAVs* (Spain) 155
ATLAS ME-01 *launch and recovery* (Spain) 403
ATT-01 *aerial targets* (Romania) 327
Au-23 *aerial targets* (Russian Federation) 329
Au-29 *aerial targets* (Russian Federation) 329
AutoCopter *UAVs* (United States) 275
AV8-R *UAVs* (United States) 220
AvantVision *UAVs* (Brazil) 12
Avenger *UAVs* (United States) 239
Aves *UAVs* (France) 36
Avian *UAVs* (Taiwan) 169
Avibras Falcao *UAVs* (Brazil) 12
AVIC TL-8 *aerial targets* (China) 300
AVL *launch and recovery* (United Kingdom) 407
AV Switchblade *UAVs* (United States) 209

B

BA-131 *UAVs* (United States) 244
Baaz *aerial targets* (Pakistan) 324
Banshee *aerial targets* (United Kingdom) 346
Barracuda *UAVs* (Germany) 53
Bat *UAVs* (United States) 259, 260
Bayonet *UAVs* (United States) 228
Berta *UAVs* (Russian Federation) 137
B-Hunter *UAVs* (International) 70
Bicopt CH *UAVs* (France) 49
Biodrone *UAVs* (France) 37
Bird Eye 400 *UAVs* (Israel) 90
BJ9906 *aerial targets* (China) 304
BLA-06 *UAVs* (Russian Federation) 144
Black Brant *aerial targets* (Canada) 296
BlackLynx *UAVs* (Italy) 107
Black Sparrow *aerial targets* (Israel) 319
Blimp 37M *UAVs* (France) 50
Block 50 Advanced Cockpit *control and communications* (United States) 388
Blue Bear Blackstart *UAVs* (United Kingdom) 185
Blue Bear iStart *UAVs* (United Kingdom) 186
Blue Horizon *UAVs* (Israel) 89
Blue Magpie *UAVs* (Taiwan) 167
Blueye *UAVs* (Israel) 81
Boomerang *UAVs* (Israel) 82
Border Eagle *UAVs* (Pakistan) 126
BQM-34 *aerial targets* (United States) 361
BQM-74 *aerial targets* (United States) 363
BQM-74F *aerial targets* (United States) 363
BQM-167 *aerial targets* (United States) 354
BroadSword *UAVs* (United States) 240
BroadSword XL *UAVs* (United States) 240
Brumby Mk 1 *UAVs* (Australia) 4
Brumby Mk 2 *UAVs* (Australia) 4
BT20 *UAVs* (United States) 220
BTT-1 Imp *aerial targets* (United Kingdom) 345
BTT-3 *aerial targets* (United Kingdom) 346
Btumby Mk 3 *UAVs* (Australia) 4
BTW 832 *launch and recovery* (Sweden) 404
BULL EL-01 *launch and recovery* (Spain) 403
Bullseye *aerial targets* (India) 317
Busard *UAVs* (France) 46
Buster *UAVs* (United States) 258
BZK-005 *UAVs* (China) 21

C

C2 systems *control and communications* (United States) 389
C 18 *UAVs* (Spain) 159
C 30 *UAVs* (Spain) 159
C 50 *UAVs* (Spain) 159
Cabure *UAVs* (Argentina) 2
Calineczka *UAVs* (Israel) 104, 105
Camcopter 5.1 *UAVs* (Austria) 7
Camcopter S-100 *UAVs* (Austria) 8
Cardinal *UAVs* (Taiwan) 168
Carine *aerial targets* (France) 304
Carolo *UAVs* (Germany) 59
CASPA *control and communications* (United Kingdom) 385
Casper 200 *UAVs* (Israel) 104
Casper 250 *UAVs* (Israel) 104
Casper 350 *UAVs* (Israel) 105
CB 350 *UAVs* (France) 44
CB 750 *UAVs* (France) 44
CBACS *control and communications* (South Africa) 380
CEI SSAT *aerial targets* (United States) 357
Centauro *UAVs* (Spain) 159
CGM-6C *launch and recovery* (Ukraine) 406
CH-3 *UAVs* (China) 23
CH-91 *UAVs* (China) 19
CH-92 *UAVs* (China) 19
CH-901 *UAVs* (China) 19
Chang Hong *UAVs* (China) 21
Chang Kong 1 *aerial targets* (China) 300
Chukar *aerial targets* (United States) 363
Chung Shyang II *UAVs* (Taiwan) 168
CK1 *aerial targets* (China) 300
CLLGW-4C *launch and recovery* (Ukraine) 406
CLLTS-3/-3C *launch and recovery* (Ukraine) 406
CMDL *control and communications* (United States) 390
Cobra *UAVs* (United States) 282
Commando *UAVs* (Spain) 156
Condor *UAVs* (United States) 229
Copter 1B *UAVs* (France) 50
Copter City *UAVs* (France) 50
Corax *UAVs* (United Kingdom) 184

Coyote
aerial targets (United States) 365
UAVs (United States) 211
CQ-10A Snow Goose *UAVs* (Canada) 16
CQ-10B SnowGoose *UAVs* (Canada) 16
Crecerelle *aerial targets* (United Kingdom) 346
CropCam *UAVs* (Canada) 15
CSV 40 *UAVs* (United Kingdom) 193
CSV 50 *UAVs* (United Kingdom) 193
CSV 65 *UAVs* (United Kingdom) 193
CSV X0 *UAVs* (United Kingdom) 193
CTP-3 *launch and recovery* (Ukraine) 406
CU-161 *UAVs* (France) 47
Cutlass
UAVs (Israel) 92
UAVs (United States) 247
CyberBug *UAVs* (United States) 224
CyberEye I *UAVs* (United Kingdom) 186
CyberEye II *UAVs* (Australia) 5
CyberQuad *UAVs* (Australia) 6
CyberScout *UAVs* (United States) 224
CyberWraith *UAVs* (Australia) 6
CyBird *aerial targets* (Australia) 293

D

Damselfly *UAVs* (United Kingdom) 192
Dan *aerial targets* (Russian Federation) 331
Danem *aerial targets* (Russian Federation) 331
Delilah *UAVs* (Israel) 101
Demon *UAVs* (United Kingdom) 180
Demonstrator Firebird *UAVs* (United States) 261
Desert Hawk *UAVs* (United States) 250
Diana *aerial targets* (Spain) 336
DLV-52 *control and communications* (Israel) 378
DM-65 *UAVs* (United Kingdom) 187
Do-DT25 Mosquito *aerial targets* (Germany) 309
Do-DT35 Hornet *aerial targets* (Germany) 309
Do-DT55 *aerial targets* (Germany) 309
Dominator II *UAVs* (Israel) 79
Dominator XP *UAVs* (Israel) 79
Dorna *UAVs* (Iran) 75
Do-SK6 *aerial targets* (Germany) 310
Do-SK10 *aerial targets* (Germany) 311
Dozor *UAVs* (Russian Federation) 145
DP-6 *UAVs* (United States) 227
DP-11 *UAVs* (United States) 228
DP series *UAVs* (United States) 226
Draganflyer X4 *UAVs* (Canada) 14
Draganflyer X4/E4 *UAVs* (Canada) 14
Draganflyer X4-ES *UAVs* (Canada) 14
Draganflyer X4-P *UAVs* (Canada) 14
Draganflyer X6-ES *UAVs* (Canada) 14
Draganflyer X8 *UAVs* (Canada) 14
Draganflyer X series *UAVs* (Canada) 14
Dragon Eye *UAVs* (United States) 278
Dragonfly 2000 *UAVs* (Israel) 88
Dragonflyer X6 *UAVs* (Canada) 14
DVF-2000 *UAVs* (France) 42
DVF 2000 *UAVs* (France) 51

E

E1 *UAVs* (Mexico) 122
E08 *UAVs* (Russian Federation) 137
E26T *UAVs* (Russian Federation) 138
E95 *aerial targets* (Russian Federation) 330
Eagle *UAVs* (France) 40
Eagle Eye P1T *UAVs* (Pakistan) 124
Eagle Eye PI *UAVs* (Pakistan) 124
Eagle Eye PII *UAVs* (Pakistan) 124
Easycopter *UAVs* (France) 38
Eclipse *aerial targets* (France) 305
Eel *UAVs* (United States) 220
Ehécatl *UAVs* (Mexico) 123
Eitan *UAVs* (Israel) 93, 95
Eleron *UAVs* (Russian Federation) 137
EL/K-1861 GDT *control and communications* (Israel) 375
EL/K-1862 CDT *control and communications* (Israel) 376
EL/K-1865 ADT *control and communications* (Israel) 376
EL/K-1871 DGDT
control and communications (Israel) 377
EL/K-1891 SCS *control and communications* (Israel) 378
Embler *UAVs* (United Kingdom) 178
Erasmus *UAVs* (India) 68
E-Swift Eye *UAVs* (United Kingdom) 188
ETOP *UAVs* (Israel) 91
Euro Hawk *UAVs* (United States) 269
Evolution *UAVs* (United States) 247
Excalibur *UAVs* (United States) 204
Excalibur 1b *aerial targets* (Canada) 297
Expedition *UAVs* (United States) 275
Explorer
UAVs (Pakistan) 127
UAVs (United States) 275
Express *UAVs* (United States) 275
EyesFly *UAVs* (France) 52

F

Falco *UAVs* (Italy) 111
Falco GCS *control and communications* (Italy) 379
Falconet II *aerial targets* (United Kingdom) 343
FanCopter *UAVs* (Germany) 57
FanWing UAVs *UAVs* (United Kingdom) 189
Featherlite *UAVs* (France) 36
FINDER *UAVs* (United States) 277
FireBee *UAVs* (India) 67
Firebee I *aerial targets* (United States) 361
Firejet *aerial targets* (United States) 355
Fire Scout *UAVs* (United States) 263, 264
Fire Shadow *UAVs* (United Kingdom) 190
Fire-X *UAVs* (United States) 263
Flamingo *UAVs* (Pakistan) 130
Flexrotor *UAVs* (United States) 203
FlyEye *UAVs* (Poland) 132
FNS 900 *UAVs* (France) 44
FoxCar *UAVs* (United States) 222
Fox TS1 *aerial targets* (France) 305
Fox TS3 *aerial targets* (France) 306
FPASS *UAVs* (United States) 250
FR50 *UAVs* (France) 42
FR101 *UAVs* (France) 42
FS-01 *UAVs* (Brazil) 13
FS-02 *UAVs* (Brazil) 12
Fulmar *UAVs* (Spain) 155
Fury *UAVs* (United States) 201
Futura *UAVs* (France) 38

G

GA22 *UAVs* (United Kingdom) 181
GAIC Harrier Hawk *UAVs* (China) 25
Gavilán *UAVs* (Mexico) 122
GC-201 *UAVs* (Netherlands) 124
GCS *control and communications* (Israel) 372
GCS03 *control and communications* (Spain) 382
Global Hawk *UAVs* (United States) 269
Global Hawk ICS
control and communications (United States) 390
Global Observer *UAVs* (United States) 207
Gnat *UAVs* (United States) 232
Gnat-XP *UAVs* (United States) 232
GoldenEye *UAVs* (United States) 205
GQM-163A *aerial targets* (United States) 365
GQM-173 *aerial targets* (United States) 353
GSAT-200 NG *aerial targets* (United Kingdom) 351
GT-400 *aerial targets* (United States) 358
Guardian *UAVs* (Argentina) 1
GuS™ *control and communications* (United States) 397

H

H-4 *UAVs* (Japan) 113
H250 *UAVs* (France) 36
HA304 *UAVs* (United States) 263
HAA *UAVs* (United States) 252
HADA *UAVs* (Spain) 158
HALE UAV *UAVs* (China) 22
Harop *UAVs* (Israel) 92
Harpy *UAVs* (Israel) 92
Havoc *UAVs* (United States) 221
Hawk *UAVs* (Pakistan) 127
Hawkeye
aerial targets (United Kingdom) 346
UAVs (United Kingdom) 194
Hawkeye III *UAVs* (United Kingdom) 194
HE-190 *UAVs* (France) 44
HEC-BRK-150 *launch and recovery* (Israel) 401
HEC-BRK-250 *launch and recovery* (Israel) 401
Helivision *UAVs* (Israel) 103
HERCULES AH-01 *launch and recovery* (Spain) 403
Hermes 90 *UAVs* (Israel) 83
Hermes 450 *UAVs* (Israel) 84
Hermes 900 *UAVs* (Israel) 85
Hermes 1500 *UAVs* (Israel) 86
Heron 1 *UAVs* (Israel) 93
Heron TP *UAVs* (Israel) 95
Heros *UAVs* (Czech Republic) 32
HERTI *UAVs* (United Kingdom) 182
HIDL *control and communications* (International) 371
High Altitude Airship *UAVs* (United States) 252
Higinn X1 ground station
control and communications (Denmark) 370
Hoder *UAVs* (United Kingdom) 178
Hodhod *UAVs* (Iran) 75
Horus 100 *UAVs* (Brazil) 12
Horus 200 *UAVs* (Brazil) 13
HoverEye *UAVs* (France) 39
HP-2002 *launch and recovery* (United States) 409
HP-3003 *launch and recovery* (United States) 409
HP-3402 *launch and recovery* (United States) 409
HP-3407 *launch and recovery* (United States) 409
HP-3502 *launch and recovery* (United States) 409
HST *UAVs* (Pakistan) 130
HTW 550 *launch and recovery* (Sweden) 404
Hud Hud *UAVs* (Pakistan) 125
Huginn X1 *UAVs* (Denmark) 35
Huma-1 *UAVs* (Pakistan) 126
Hummingbird *UAVs* (United States) 213
Hunter *UAVs* (International) 70
Hurricane *UAVs* (Ukraine) 173
Hydraulic launcher *launch and recovery* (Israel) 401

I

IAI Ghost *UAVs* (Israel) 91
Igla *UAVs* (Russian Federation) 138
I-Gnat *UAVs* (United States) 232
I-Gnat ER *UAVs* (United States) 232
IMMC *control and communications* (United States) 388
Imperial Eagle *UAVs* (India) 63
In-Tact *control and communications* (Israel) 373
Integrator *UAVs* (United States) 243
Ion Tiger *UAVs* (United States) 278
Iris Jet *aerial targets* (Greece) 313
Iris Prop *aerial targets* (Greece) 314
Irkut-1A *UAVs* (Russian Federation) 138
Irkut-2F *UAVs* (Russian Federation) 139
Irkut-2M *UAVs* (Russian Federation) 139
Irkut-2T *UAVs* (Russian Federation) 139
Irkut-3 *UAVs* (Russian Federation) 140
Irkut-10 *UAVs* (Russian Federation) 140
Irkut-20 *UAVs* (Russian Federation) 140
Irkut-200 *UAVs* (Russian Federation) 141
Irkut-850 *UAVs* (Russian Federation) 141

ISIS *UAVs* (United States)....253
IT 180-5 *UAVs* (France)....45

J

J/AQM-1 *aerial targets* (Japan)....320
Jasoos *UAVs* (Pakistan)....130
Javelin X *aerial targets* (India)....316
Jordan Arrow *aerial targets* (Jordan)....321
Jordan Falcon *UAVs* (Jordan)....115
JX-2 *aerial targets* (India)....317

K

Kalkara *aerial targets* (United States)....355
Karrar *UAVs* (Iran)....74
Kestrel *UAVs* (United Kingdom)....184
KillerBee *UAVs* (United States)....282, 286
Kingfisher *UAVs* (Australia)....4
Kingfisher Mk 1 *UAVs* (Australia)....4
Kingfisher Mk 2 *UAVs* (Australia)....4
Kiwit *UAVs* (South Africa)....150
K-MAX *UAVs* (United States)....245
Koax *UAVs* (Switzerland)....164
Koliber *UAVs* (Poland)....131
Komar *aerial targets* (Poland)....326
K-UCAV *UAVs* (Korea, South)....116
KUS-7 *UAVs* (Korea, South)....118
KUS-9 *UAVs* (Korea, South)....119
KuSDL *control and communications* (United States)....391
KZO *UAVs* (Germany)....61

L

La-17 *aerial targets* (Russian Federation)....332
LAE *launch and recovery* (Spain)....403
Lakshya *aerial targets* (India)....315
Lakshya-2 *aerial targets* (India)....315
Launch and recovery stations *launch and recovery* (Italy)....401
Launch systems *launch and recovery* (United States)....413
LEMV *UAVs* (United States)....263
Lipan M3 *UAVs* (Argentina)....1
Little Bird *UAVs* (United States)....216
LN-100G *control and communications* (United States)....393
LN-200 *control and communications* (United States)....393
LN-251 *control and communications* (United States)....393
LOCATS *aerial targets* (South Africa)....333
Locusta *aerial targets* (Italy)....319
LRS *control and communications* (Israel)....372
LTL 1 *launch and recovery* (United Kingdom)....407
LT series *UAVs* (China)....23
LUNA *UAVs* (Germany)....57
LWL *launch and recovery* (United Kingdom)....408

M

M2600 *UAVs* (United States)....286
Machatz *UAVs* (Israel)....93
MAGIC ATOLS *control and communications* (France)....371
MALD *UAVs* (United States)....283
MALD-J *UAVs* (United States)....283
MALD-V *UAVs* (United States)....283
Mamok *UAVs* (Czech Republic)....32
Manta
UAVs (Czech Republic)....32
UAVs (United States)....212
MANTIS *UAVs* (United Kingdom)....183
Mantis *UAVs* (Spain)....157
Mariner *UAVs* (United States)....236
Mark 5 *UAVs* (United States)....197
Marque *control and communications* (Sweden)....382
Marque Scoring Station *control and communications* (Sweden)....383
MARS *launch and recovery* (United States)....409
MASS *UAVs* (Finland)....35
MAV *UAVs* (United States)....241
MAV6 M1400-I Blue Devil II *UAVs* (United States)....256
Maveric *UAVs* (United States)....279
Maya *UAVs* (France)....38
MBDA TiGER *UAVs* (United States)....257
MC91 *launch and recovery* (Finland)....399
MC0315L *launch and recovery* (Finland)....399
MC2055L *launch and recovery* (Finland)....399
MC2555L *launch and recovery* (Finland)....399
MC2555LLR *launch and recovery* (Finland)....399
MCALS *UAVs* (United States)....283
MD4-200 *UAVs* (Germany)....60
MD4-1000 *UAVs* (Germany)....60
MDI 154 *aerial targets* (France)....307
MDI189 *control and communications* (France)....370
Mercator *UAVs* (Belgium)....11
Mercury *UAVs* (United States)....207
Mi-34BP *UAVs* (Russian Federation)....144
MicroB *UAVs* (Israel)....82
MicroFalcon I *UAVs* (Israel)....102
Midge *UAVs* (United Kingdom)....188
Milano *UAVs* (Spain)....158
MiniFalcon *UAVs* (Israel)....103
Mini Marque *control and communications* (Sweden)....382
Mini Marque Scoring Station *control and communications* (Sweden)....383
Mini Panther *UAVs* (Israel)....98
Mirach 100/5 *aerial targets* (Italy)....319
Mk II *aerial targets* (Pakistan)....323
MLB Super Bat *UAVs* (United States)....259
Mobilenet® 2000 *launch and recovery* (United States)....409
Mobius *UAVs* (United States)....248
model 15M *UAVs* (United States)....287
model17M® *UAVs* (United States)....287
model 22+™ *UAVs* (United States)....287
model 22M™ *UAVs* (United States)....287
model 25M *UAVs* (United States)....287
model 28M *UAVs* (United States)....287
model 31M *UAVs* (United States)....287
model 32M® *UAVs* (United States)....287
model 38M® *UAVs* (United States)....287
model 53M/System 250 *UAVs* (United States)....287
model 67M/System 365/System 365H *UAVs* (United States)....287
model 71M® *UAVs* (United States)....287
model 74M™ *UAVs* (United States)....287
model 76M *UAVs* (United States)....287
Model 355 *UAVs* (United States)....261
MoD Krunk-25 *UAVs* (Armenia)....3
Modular Airborne Sensor System *UAVs* (Finland)....35
Mohadjer *UAVs* (Iran)....75
Molynx *UAVs* (Italy)....107
MONTAGE *control and communications* (United States)....392
Mosquito *UAVs* (Israel)....96
MP1028g *control and communications* (Canada)....369
MP2028g *control and communications* (Canada)....369
MP2028XP *control and communications* (Canada)....369
MP2128g *control and communications* (Canada)....369
MP2128HELI *control and communications* (Canada)....369
MP2128LRC *control and communications* (Canada)....369
MP21283x *control and communications* (Canada)....369
MP CAT *launch and recovery* (Canada)....399
MP-Vision *UAVs* (Canada)....15
MQ-1 *UAVs* (United States)....233
MQ-1C *UAVs* (United States)....235
MQ-5 *UAVs* (International)....70
MQ-8 *UAVs* (United States)....264
MQ-8B *UAVs* (United States)....264
MQ-8C *UAVs* (United States)....263
MQ-9 *UAVs* (United States)....236
MQ-9 Reaper *UAVs* (United States)....236
MQM-74 *aerial targets* (United States)....363
MQM-107 Streaker *aerial targets* (United States)....355
MQM-170A *aerial targets* (United States)....357
MQM-171A *UAVs* (United States)....240
MQ-X *UAVs* (United States)....255
MRS-2000 *control and communications* (Israel)....373
MRS-2000M *control and communications* (Israel)....373
MRTT *aerial targets* (United Kingdom)....344
MSAT-500 *aerial targets* (United Kingdom)....352
MSAT-500 NG *aerial targets* (United Kingdom)....352
MSST *aerial targets* (United States)....353
MUDL *control and communications* (United States)....391
Mukhbar *UAVs* (Pakistan)....130
Museco *UAVs* (Germany)....59
MV5B1 *UAVs* (France)....52

N

Nearchos *UAVs* (Greece)....62
Neo S-300 *UAVs* (Switzerland)....164
Neptune *UAVs* (United States)....228
Neuron *UAVs* (International)....70
NI 100 *UAVs* (Korea, South)....117
NI 300 *UAVs* (Korea, South)....117
Nietoperz-3L *UAVs* (Poland)....132
Night Eagle *UAVs* (China)....20
Night Intruder 100N *UAVs* (Korea, South)....117
Night Intruder 300 *UAVs* (Korea, South)....117
Nishan Mk II *aerial targets* (Pakistan)....323
Nishant *UAVs* (India)....63
Nishan TJ-1000 *aerial targets* (Pakistan)....323
NRUAV *UAVs* (Israel)....97
NSU *control and communications* (United States)....388
NT 150 *UAVs* (Switzerland)....163

O

OCPJ-07x *aerial targets* (Poland)....326
Odin *UAVs* (United Kingdom)....179
Odysseus *UAVs* (United States)....225
OPATS *control and communications* (Switzerland)....384
Optoelektron 1 *UAVs* (Czech Republic)....33
Orbiter *UAVs* (Israel)....79
Orion *UAVs* (Israel)....95
Orion HALL *UAVs* (United States)....206
OSGCS *control and communications* (United States)....386
OSPGCS *control and communications* (United States)....387
OSRVT *control and communications* (United States)....387
Otus *UAVs* (United Kingdom)....193
Outlaw *aerial targets* (United States)....357
Owl *UAVs* (United States)....199

P

P.1HH HammerHead *UAVs* (Italy)....111
P330 *UAVs* (Germany)....59
Panther *UAVs* (Israel)....98
Parwaz *UAVs* (Pakistan)....130
Patroller *UAVs* (France)....46
Pchela/Shmel *UAVs* (Russian Federation)....147
Pelican *UAVs* (United States)....223
Pelicano *UAVs* (Spain)....157
Penguin B *UAVs* (United States)....290
Perseas *aerial targets* (Greece)....314
Petrel *aerial targets* (United Kingdom)....348
Phantom Eye *UAVs* (United States)....215
Phantom Ray *UAVs* (United States)....216
Phoenix *UAVs* (Ukraine)....173
Phoenix Jet *aerial targets* (Australia)....293
Picador *UAVs* (Israel)....80
Pico *UAVs* (Italy)....110

Pilotless Target Aircraft *aerial targets* (India) 315
Pitagora *UAVs* (Italy) 110
PLS *launch and recovery* (United States) 410
PM-8 *aerial targets* (Sweden) 338
Pneumatic launcher *launch and recovery* (United Arab Emirates) 406
Pop-up Helicopter *aerial targets* (Canada) 298
portable GCS *control and communications* (United States) 397
PR53 *aerial targets* (France) 308
PR55 *aerial targets* (France) 308
Predator *UAVs* (United States) 233
Predator B *UAVs* (United States) 236
Predator C *UAVs* (United States) 239
Production-Ready Firebird *UAVs* (United States) 261
Project 360 *UAVs* (France) 43
Prospector *UAVs* (Germany) 61
Proteus *UAVs* (United States) 284
Puma AE *UAVs* (United States) 208

Q

QF-4 *aerial targets* (United States) 353
QF-16 *aerial targets* (United States) 354

R

R-50 *UAVs* (Japan) 113
R-300 *UAVs* (Turkey) 170
Raffaello *UAVs* (Italy) 110
Rainbow-3 *UAVs* (China) 23
Ranger *UAVs* (Switzerland) 162
RATUFC *launch and recovery* (Finland) 399
Raven
UAVs (United Kingdom) 184
UAVs (United States) 208
RD series *aerial targets* (Russian Federation) 329
RemoEye 002 *UAVs* (Korea, South) 120
RemoEye 006 *UAVs* (Korea, South) 120
RemoH-C100 *UAVs* (Korea, South) 121
Responder *UAVs* (United States) 200
Retarder/recovery parachutes *launch and recovery* (United Kingdom) 406
RM-12A *launch and recovery* (United States) 409
RM-24 *launch and recovery* (United States) 409
RM-30A *launch and recovery* (United States) 409
RM-30A1 *launch and recovery* (United States) 409
RM-30B *launch and recovery* (United States) 409
RM-62 *launch and recovery* (United States) 409
RMAX *UAVs* (Japan) 113
RMK-35 *launch and recovery* (United States) 409
RO-1 *launch and recovery* (Spain) 403
RO-2 *launch and recovery* (Spain) 403
Roadrunner *UAVs* (South Africa) 150
Rover *UAVs* (Pakistan) 128
ROVER 1 *control and communications* (United States) 391
ROVER 3 *control and communications* (United States) 391
ROVER 4 *control and communications* (United States) 391
ROVER 5 *control and communications* (United States) 391
ROVER 6 *control and communications* (United States) 391
RPCS *control and communications* (Israel) 372
RQ-1 *UAVs* (United States) 233
RQ-4 *UAVs* (United States) 269
RQ-5 *UAVs* (International) 70
RQ-7 *UAVs* (United States) 197
RQ-8A *UAVs* (United States) 264
RQ-11 *UAVs* (United States) 208
RQ-11A *UAVs* (United States) 208
RQ-11B *UAVs* (United States) 208
RQ-15A *UAVs* (United States) 228
RQ-21A *UAVs* (United States) 243
RQ-170 *UAVs* (United States) 255
RT-3 *aerial targets* (Romania) 327
RT-11D *aerial targets* (Romania) 328
Rufus *UAVs* (Israel) 105
Rustom *UAVs* (India) 64
RVM01 *aerial targets* (Slovenia) 333
RVT
control and communications (Israel) 372
control and communications (United States) 389

S

S4 *UAVs* (Mexico) 123
S-100 *UAVs* (Austria) 8
S-280 OSGCS configuration *control and communications* (United States) 386
S-300 *UAVs* (Switzerland) 164
S-788 OSGC configuration *control and communications* (United States) 386
Saeghe *aerial targets* (Iran) 318
SASS LITE *UAVs* (United States) 245
Scancopter CB 350 *UAVs* (France) 44
Scancopter CB 750 *UAVs* (France) 44
ScanEagle *UAVs* (United States) 217
scoring systems *control and communications* (Sweden) 383
Scout™ *UAVs* (Canada) 14
Scrab I *aerial targets* (Spain) 337
Scrab II *aerial targets* (Spain) 337
Scythe *UAVs* (United States) 222
Sea ALL *UAVs* (United States) 278
Sea Eagle *UAVs* (Ukraine) 173
Seagnos 80 *UAVs* (France) 37
Searcher *UAVs* (Israel) 98
Seeker *UAVs* (South Africa) 153
Seeker FNS 900 *UAVs* (France) 44
Sentinel
UAVs (South Africa) 151
UAVs (United States) 255
SETA-3 S1 *aerial targets* (Germany) 312
SGT-20 *aerial targets* (United States) 358
Shadow *UAVs* (Pakistan) 128
Shadow 200 *UAVs* (United States) 197
Shadow 300 *UAVs* (United States) 197
Shadow 400 *UAVs* (United States) 198
Shadow 600 *UAVs* (United States) 198
Shahbaz *aerial targets* (Pakistan) 322
Shahzore *aerial targets* (Pakistan) 323
Sherpa *UAVs* (Canada) 18
Sherpa™ Provider 600 *UAVs* (Canada) 18
Sherpa™ Provider 1200 *UAVs* (Canada) 18
Sherpa™ Provider 2200 *UAVs* (Canada) 18
Sherpa™ Provider 10,000 *UAVs* (Canada) 18
Sherpa™ Ranger *UAVs* (Canada) 18
Shooting Star *aerial targets* (Pakistan) 325
Shot Stalker *UAVs* (United States) 255
Shoval *UAVs* (Israel) 93
Silver Fox *UAVs* (United States) 212
Simsek *aerial targets* (Turkey) 340
SIVA *UAVs* (Spain) 158
Sivrisinek *UAVs* (Turkey) 170
Skat *UAVs* (Russian Federation) 143
Skate *UAVs* (United States) 206
Skeeter *aerial targets* (United States) 354
Skeldar *UAVs* (Sweden) 161
Skimmer *UAVs* (India) 69
Skua *aerial targets* (South Africa) 334
Skyblade II *UAVs* (Singapore) 148
Skyblade III *UAVs* (Singapore) 148
Skyblade IV *UAVs* (Singapore) 149
Sky Dot *UAVs* (India) 66
Skyhook *UAVs* (United Kingdom) 179
Skylark I *UAVs* (Israel) 87
Skylark II *UAVs* (Israel) 88
SkyRaider *UAVs* (United States) 280
Sky Warrior *UAVs* (United States) 235
SkyWatcher *UAVs* (United States) 281
Sky-X *UAVs* (Italy) 108
Sky-Y *UAVs* (Italy) 109
Skyzer 100 *UAVs* (Israel) 81
Sleeve targets *aerial targets* (Sweden) 338
SL-RAMBO *control and communications* (Israel) 374
Slybird *UAVs* (India) 63
SM3B *aerial targets* (Sweden) 339
SM6 *aerial targets* (Sweden) 339
SM9A *aerial targets* (Sweden) 340
Smart *UAVs* (Korea, South) 119
Smart Eye *UAVs* (United Arab Emirates) 177
Smart Eye 2 *UAVs* (United Arab Emirates) 174
SMCP-JU *aerial targets* (Poland) 326
SMCP-WU *aerial targets* (Poland) 326
Snipe *aerial targets* (United Kingdom) 349
Sniper *UAVs* (Spain) 156
SnowGoose™ *UAVs* (Canada) 16
Soar Dragon *UAVs* (China) 25
Sofar *UAVs* (Israel) 104
Sojka III *UAVs* (Czech Republic) 33
Sokol *UAVs* (Bulgaria) 13
SPADES *UAVs* (Netherlands) 123
Sparrow *UAVs* (Israel) 89
SPARS *launch and recovery* (United States) 409
Spear *UAVs* (United States) 221
Spectre *aerial targets* (United Kingdom) 346
Sperwer A *UAVs* (France) 47
Sperwer B *UAVs* (France) 47, 49
Spotter *UAVs* (United Kingdom) 196
Spy Arrow *UAVs* (France) 51
SpyLite *UAVs* (Israel) 83
SRST *control and communications* (Israel) 374
Stalker *UAVs* (United States) 255
StarLink *control and communications* (Israel) 375
STARS *launch and recovery* (United States) 409
Sting *UAVs* (Israel) 89
Stingray *UAVs* (Pakistan) 131
Super Swiper *UAVs* (United States) 230
Swift *aerial targets* (Ukraine) 341
Swiper *UAVs* (United States) 230
Szerszen *aerial targets* (Poland) 326

T

T-15 *UAVs* (United States) 203
T-16 *UAVs* (United States) 203
T-20 *UAVs* (United States) 204
T200 *UAVs* (Germany) 59
Talarion *UAVs* (Germany) 54
Talash *UAVs* (Iran) 76
TALS *launch and recovery* (United States) 411
Taranis *UAVs* (United Kingdom) 185
Taxan *aerial targets* (France) 308
TCDL *control and communications* (United States) 388
TCS *control and communications* (United States) 395
TCTS *control and communications* (United Kingdom) 385
TDDL *control and communications* (Israel) 375
TDK-39 *aerial targets* (United States) 359
Téléporteur *UAVs* (France) 42
TERP *UAVs* (India) 68
tethered aerostats *UAVs* (United States) 287
Tethered aerostats and blimps *UAVs* (United States) 202
TGT *aerial targets* (United States) 359
TGX *aerial targets* (United States) 359
Tianyi *UAVs* (China) 22
TigerShark *UAVs* (United States) 222
Tipchak *UAVs* (Russian Federation) 143
Tip-Jet DragonFly *UAVs* (Switzerland) 163
TIX *aerial targets* (United States) 359
TJ-1000 *aerial targets* (Pakistan) 323
TLX-1 *aerial targets* (United States) 359
TML 2 *launch and recovery* (United Kingdom) 407

TML 3 *launch and recovery* (United Kingdom) 407
TML 3 Ultima *launch and recovery* (United Kingdom) 407
TML 4 *launch and recovery* (United Kingdom) 407
Tornado *aerial targets* (Pakistan) 323
tow targets *aerial targets* (United States) 359
TPDM-01 *aerial targets* (Romania) 328
TPT *aerial targets* (United States) 359
Tracker *UAVs* (France) 42
TRAP *aerial targets* (Canada) 297
Tårnfalken *UAVs* (France) 47
Trogon *UAVs* (India) 68
trueHWIL *control and communications* (Canada) 369
TRX *aerial targets* (United States) 359
TTW 500/550 *launch and recovery* (Sweden) 404
Tu-143 Reis *UAVs* (Russian Federation) 145
Tu-243 Reis *UAVs* (Russian Federation) 145
Tu-300 Korshun *UAVs* (Russian Federation) 145
Tunder *aerial targets* (Pakistan) 325
Turais *UAVs* (United States) 279
Turna *aerial targets* (Turkey) 340

U

U8E *UAVs* (China) 24
UAV recovery systems *launch and recovery* (United States) 408
UAV surveillance system *UAVs* (Poland) 132
UCARS-V2 *launch and recovery* (United States) 411
UGCS
  control and communications (Israel) 372
  control and communications (United States) 386
Ugglan *UAVs* (France) 47
ULB *UAVs* (United States) 216
Ultima *aerial targets* (Belgium) 293
UMAS™ *control and communications* (Israel) 372
Unmanned Systems CT-450 Discoverer I *UAVs* (Switzerland) 165
Unmanned Systems Orca *UAVs* (Switzerland) 165

Uqab *UAVs* (Pakistan) 124
Urban View *UAVs* (India) 67
UVision Air Wasp *UAVs* (Israel) 106

V

V-150 *UAVs* (Sweden) 161
Vector *UAVs* (Pakistan) 129
Vidar *UAVs* (United Kingdom) 179
Vigil *UAVs* (South Africa) 151
Vigilant *UAVs* (United Kingdom) 196
Viking 400 *UAVs* (United States) 248
Vindicator II *aerial targets* (Canada) 299
Vision *UAVs* (Pakistan) 128
VISIONAIR *control and communications* (Spain) 382
Voodoo *aerial targets* (United Kingdom) 350
V-TVR *control and communications* (Israel) 375
Vulture *UAVs* (South Africa) 151
Vulture GCS *control and communications* (South Africa) 380
Vulture launch and recovery systems *launch and recovery* (South Africa) 402
Vulture programme *UAVs* (United States) 225

W

Warrigal *UAVs* (Australia) 7
Warrior Alpha *UAVs* (United States) 232
Wasp III *UAVs* (United States) 210
Watchdog *UAVs* (Brazil) 13
Watchkeeper *UAVs* (United Kingdom) 194
WBBL-UAV *UAVs* (United States) 279
Whirlwind Scout *UAVs* (China) 20
Whisper *UAVs* (United States) 227
Wing Loong *UAVs* (China) 23
Wizard *control and communications* (United Kingdom) 385
WJ-600 *UAVs* (China) 24
WK 450 *UAVs* (Israel) 84
WZ-2000 *UAVs* (China) 26

X

X-47 *UAVs* (United States) 274
X-48B *UAVs* (United States) 219
X-48C *UAVs* (United States) 219

X-240 Mk II *UAVs* (Switzerland) 164
Xian ASN-106 *aerial targets* (China) 303
Xian ASN-211 *UAVs* (China) 30
Xianglong *UAVs* (China) 25
XPA-1B *UAVs* (United Kingdom) 182

Y

Yabhon-GRN 1 *aerial targets* (United Arab Emirates) 341
Yabhon-GRN 2 *aerial targets* (United Arab Emirates) 341
Yabhon-HMD *aerial targets* (United Arab Emirates) 342
Yabhon-M *UAVs* (United Arab Emirates) 175
Yabhon N *aerial targets* (United Arab Emirates) 342
Yabhon-R *UAVs* (United Arab Emirates) 176
Yabhon-RX *UAVs* (United Arab Emirates) 176
YACS *control and communications* (Japan) 380
Yarara *UAVs* (Argentina) 2
Yastreb-2 *aerial targets* (Bulgaria) 295
Yastreb-2S *UAVs* (Bulgaria) 14
Yilong *UAVs* (China) 23
YMQ-18A *UAVs* (United States) 213

Z

ZALA 421-02 *UAVs* (Russian Federation) 133
ZALA 421-04 *UAVs* (Russian Federation) 134
ZALA 421-05H *UAVs* (Russian Federation) 133
ZALA 421-06 *UAVs* (Russian Federation) 134
ZALA 421-08 *UAVs* (Russian Federation) 135
ZALA 421-09 *UAVs* (Russian Federation) 135
ZALA 421-11 *UAVs* (Russian Federation) 135
ZALA 421-12 *UAVs* (Russian Federation) 136
ZALA 421-16 *UAVs* (Russian Federation) 136
ZALA 421-21 *UAVs* (Russian Federation) 137
Zephyr *UAVs* (United Kingdom) 190
Zygo *UAVs* (United Kingdom) 188

Manufacturers' Index

A

AAI Corporation
Aerosonde *UAVs* (United States) 197
Aerosonde Mark 4.7 *UAVs* (United States) 196
Mark 5 *UAVs* (United States) 197
OSGCS *control and communications* (United States) 386
OSPGCS *control and communications* (United States) 387
OSRVT *control and communications* (United States) 387
RQ-7 *UAVs* (United States) 197
S-280 OSGCS configuration *control and communications*
 (United States) 386
S-788 OSGC configuration *control and communications*
 (United States) 386
Shadow 200 *UAVs* (United States) 197
Shadow 300 *UAVs* (United States) 197
Shadow 400 *UAVs* (United States) 198
Shadow 600 *UAVs* (United States) 198
UGCS *control and communications* (United States) 386

Acuity Technologies Inc
AT-3 *UAVs* (United States) 199
AT-10 *UAVs* (United States) 200
Owl *UAVs* (United States) 199
Responder *UAVs* (United States) 200

Adcom Systems
ADCOM-3D *control and communications* (United Arab Emirates) 384
Adcom Systems Smart Eye 1 *UAVs* (United Arab Emirates) 174
ADNAV GCS *control and communications* (United Arab Emirates) 384
Pneumatic launcher *launch and recovery* (United Arab Emirates) 406
Smart Eye *UAVs* (United Arab Emirates) 177
Smart Eye 2 *UAVs* (United Arab Emirates) 174
Yabhon-GRN 1 *aerial targets* (United Arab Emirates) 341
Yabhon-GRN 2 *aerial targets* (United Arab Emirates) 341
Yabhon-HMD *aerial targets* (United Arab Emirates) 342
Yabhon-M *UAVs* (United Arab Emirates) 175
Yabhon N *aerial targets* (United Arab Emirates) 342
Yabhon-R *UAVs* (United Arab Emirates) 176
Yabhon-RX *UAVs* (United Arab Emirates) 176

Advanced Computing & Engineering Solutions (Pvt) Ltd
Eagle Eye P1T *UAVs* (Pakistan) 124
Eagle Eye PI *UAVs* (Pakistan) 124
Eagle Eye PII *UAVs* (Pakistan) 124
Uqab *UAVs* (Pakistan) 124

Advanced Technologies & Engineering Co (Pty) Ltd
Kiwit *UAVs* (South Africa) 150
Roadrunner *UAVs* (South Africa) 150
Sentinel *UAVs* (South Africa) 151
Vigil *UAVs* (South Africa) 151
Vulture *UAVs* (South Africa) 151
Vulture GCS *control and communications* (South Africa) 380
Vulture launch and recovery systems *launch and recovery*
 (South Africa) 402

Aeroart SAS
Aelius *UAVs* (France) 36
Aves *UAVs* (France) 36
Featherlite *UAVs* (France) 36
H250 *UAVs* (France) 36
Seagnos 80 *UAVs* (France) 37

Aeroland UAV Inc
AL-4 *UAVs* (Taiwan) 166
AL-20 *UAVs* (Taiwan) 166
AL-40 *UAVs* (Taiwan) 167
AL-120 *UAVs* (Taiwan) 167

AeroMech Engineering Inc
Fury *UAVs* (United States) 201

Aeronautical Development Establishment
Lakshya *aerial targets* (India) 315
Lakshya-2 *aerial targets* (India) 315
Nishant *UAVs* (India) 63
Pilotless Target Aircraft *aerial targets* (India) 315
Rustom *UAVs* (India) 64

Aeronautics Defense Systems Ltd
Aerolight *UAVs* (Israel) 77
Aerosky *UAVs* (Israel) 77
Aerostar *UAVs* (Israel) 77
Dominator II *UAVs* (Israel) 79
Dominator XP *UAVs* (Israel) 79
GCS *control and communications* (Israel) 372
LRS *control and communications* (Israel) 372
Orbiter *UAVs* (Israel) 79
Picador *UAVs* (Israel) 80
RPCS *control and communications* (Israel) 372
RVT *control and communications* (Israel) 372
UMAS™ *control and communications* (Israel) 372

Aeronaves e Motores SA (Aeromot)
AM 03089 *aerial targets* (Brazil) 294

Aerosonde Pty Ltd
Aerosonde *UAVs* (Australia) 3

Aerospace Long-March International Trade Company Ltd,
CH-901 *UAVs* (China) 19

Aerospace Long-March International Trade Company Ltd
CH-91 *UAVs* (China) 19
CH-92 *UAVs* (China) 19

Aerostar International Inc,
Tethered aerostats and blimps *UAVs* (United States) 202

AeroTactiX Ltd
Skyzer 100 *UAVs* (Israel) 81

Aerovision Vehículos Aereos SL
Fulmar *UAVs* (Spain) 155

Aeryon Laboratories Inc,
Scout™ *UAVs* (Canada) 14

Aesir Unmanned Autonomous Systems
Embler *UAVs* (United Kingdom) 178
Hoder *UAVs* (United Kingdom) 178
Odin *UAVs* (United Kingdom) 179
Vidar *UAVs* (United Kingdom) 179

Air Affairs Australia
Phoenix Jet *aerial targets* (Australia) 293

Airborne Systems Europe
Retarder/recovery parachutes *launch and recovery* (United Kingdom) 406

Air Force Research Institute
Mamok *UAVs* (Czech Republic) 32
Manta *UAVs* (Czech Republic) 32
Optoelektron 1 *UAVs* (Czech Republic) 33
Sojka III *UAVs* (Czech Republic) 33

AirRobot GmbH & Co KG
AR 70 *UAVs* (Germany) 53
AR 100 *UAVs* (Germany) 53
AR 150 *UAVs* (Germany) 53

Air Target Sweden AB
AS-113/12U *control and communications* (Sweden) 383
AS-133/12U *control and communications* (Sweden) 383
AS-135/12U *control and communications* (Sweden) 383
Marque *control and communications* (Sweden) 382
Marque Scoring Station *control and communications* (Sweden) 383
Mini Marque *control and communications* (Sweden) 382
Mini Marque Scoring Station *control and communications* (Sweden) 383
scoring systems *control and communications* (Sweden) 383

Albadeey Technologies
ABJT *aerial targets* (Pakistan) 322
Hud Hud *UAVs* (Pakistan) 125

Shahbaz *aerial targets* (Pakistan) 322
Shahzore *aerial targets* (Pakistan) 323

Alcore Technologies SA
Biodrone *UAVs* (France) 37
Easycopter *UAVs* (France) 38
Futura *UAVs* (France) 38
Maya *UAVs* (France) 38

Alenia Aeronautica SpA
BlackLynx *UAVs* (Italy) 107
Molynx *UAVs* (Italy) 107
Sky-X *UAVs* (Italy) 108
Sky-Y *UAVs* (Italy) 109

A-Level Aerosystems
ZALA 421-02 *UAVs* (Russian Federation) 133
ZALA 421-04 *UAVs* (Russian Federation) 134
ZALA 421-05H *UAVs* (Russian Federation) 133
ZALA 421-06 *UAVs* (Russian Federation) 134
ZALA 421-08 *UAVs* (Russian Federation) 135
ZALA 421-09 *UAVs* (Russian Federation) 135
ZALA 421-11 *UAVs* (Russian Federation) 135
ZALA 421-12 *UAVs* (Russian Federation) 136
ZALA 421-16 *UAVs* (Russian Federation) 136
ZALA 421-21 *UAVs* (Russian Federation) 137

Alliant Techsystems Inc
GQM-173 *aerial targets* (United States) 353
MSST *aerial targets* (United States) 353

Alpha Unmanned Systems
Atlantic *UAVs* (Spain) 155
Commando *UAVs* (Spain) 156
Sniper *UAVs* (Spain) 156

American Dynamics Flight Systems
AD-150 *UAVs* (United States) 201

Arcturus UAV
T-15 *UAVs* (United States) 203
T-16 *UAVs* (United States) 203
T-20 *UAVs* (United States) 204

Aries Ingeniería y Sistemas SA
ALPPUL LP-02 *launch and recovery* (Spain) 403
ATLAS ME-01 *launch and recovery* (Spain) 403
BULL EL-01 *launch and recovery* (Spain) 403
HERCULES AH-01 *launch and recovery* (Spain) 403
LAE *launch and recovery* (Spain) 403
RO-1 *launch and recovery* (Spain) 403
RO-2 *launch and recovery* (Spain) 403

Armenian Ministry of Defence
MoD Krunk-25 *UAVs* (Armenia) 3

A S Yakovlev OKB
Pchela/Shmel *UAVs* (Russian Federation) 147

Augur Aeronautical Centre Inc
Au-23 *aerial targets* (Russian Federation) 329
Au-29 *aerial targets* (Russian Federation) 329

Aurora Flight Sciences Corporation
Excalibur *UAVs* (United States) 204
GoldenEye *UAVs* (United States) 205
Orion HALL *UAVs* (United States) 206
Skate *UAVs* (United States) 206

Aviation Design
Carine *aerial targets* (France) 304

Avibras Indústria Aeroespacial SA
Avibras Falcao *UAVs* (Brazil) 12

AVIC Aviation Techniques Co Ltd
AVIC TL-8 *aerial targets* (China) 300

AV Inc
AV Switchblade *UAVs* (United States) 209
Global Observer *UAVs* (United States) 207
Mercury *UAVs* (United States) 207
Puma AE *UAVs* (United States) 208
Raven *UAVs* (United States) 208

RQ-11 *UAVs* (United States) 208
RQ-11A *UAVs* (United States) 208
RQ-11B *UAVs* (United States) 208
Wasp III *UAVs* (United States) 210

Aviotech Ltd
RVM01 *aerial targets* (Slovenia) 333

Aviotechnica S.p. Ltd
Sokol *UAVs* (Bulgaria) 13
Yastreb-2 *aerial targets* (Bulgaria) 295
Yastreb-2S *UAVs* (Bulgaria) 14

B

BAE Systems Air Systems
Corax *UAVs* (United Kingdom) 184
Demon *UAVs* (United Kingdom) 180
GA22 *UAVs* (United Kingdom) 181
HERTI *UAVs* (United Kingdom) 182
Kestrel *UAVs* (United Kingdom) 184
MANTIS *UAVs* (United Kingdom) 183
Raven *UAVs* (United Kingdom) 184
Taranis *UAVs* (United Kingdom) 185
XPA-1B *UAVs* (United Kingdom) 182

BAE Systems Australia
Brumby Mk 1 *UAVs* (Australia) 4
Brumby Mk 2 *UAVs* (Australia) 4
Btumby Mk 3 *UAVs* (Australia) 4
Kingfisher *UAVs* (Australia) 4
Kingfisher Mk 1 *UAVs* (Australia) 4
Kingfisher Mk 2 *UAVs* (Australia) 4

BAE Systems Electronics, Intelligence and Support
NSU *control and communications* (United States) 388

BAE Systems Flight Systems
Kalkara *aerial targets* (United States) 355
MQM-107 Streaker *aerial targets* (United States) 355
QF-4 *aerial targets* (United States) 353

BAE Systems, Land Systems Dynamics
LOCATS *aerial targets* (South Africa) 333

BAE Systems Unmanned Aircraft Programs
Coyote *UAVs* (United States) 211
Manta *UAVs* (United States) 212
Silver Fox *UAVs* (United States) 212

Beijing University of Aeronautics and Astronautics
BZK-005 *UAVs* (China) 21
Chang Hong *UAVs* (China) 21

Belgian Defence
Ultima *aerial targets* (Belgium) 293

Bertin Technologies
HoverEye *UAVs* (France) 39

Bülow Air Target Systems AB
BTW 832 *launch and recovery* (Sweden) 404
HTW 550 *launch and recovery* (Sweden) 404
PM-8 *aerial targets* (Sweden) 338
TTW 500/550 *launch and recovery* (Sweden) 404

Blue Bear Systems Research
Blue Bear Blackstart *UAVs* (United Kingdom) 185
Blue Bear iStart *UAVs* (United Kingdom) 186

BlueBird Aero Systems Ltd
Blueye *UAVs* (Israel) 81
Boomerang *UAVs* (Israel) 82
MicroB *UAVs* (Israel) 82
SpyLite *UAVs* (Israel) 83

Boeing Defense, Space and Security
A160 *UAVs* (United States) 213
A160T *UAVs* (United States) 213
Hummingbird *UAVs* (United States) 213
Phantom Ray *UAVs* (United States) 216
QF-16 *aerial targets* (United States) 354
YMQ-18A *UAVs* (United States) 213

Brandebury Tool Company Inc
Scythe *UAVs* (United States) 222

Bristol Aerospace Ltd
Black Brant *aerial targets* (Canada)..................................296
Excalibur 1b *aerial targets* (Canada) 297

Brock Technologies Inc
AV8-R *UAVs* (United States)..220
BT20 *UAVs* (United States)..220
Eel *UAVs* (United States)..220
Havoc *UAVs* (United States)..221
Spear *UAVs* (United States)...221

Butler Unmanned Parachute Systems LLC
UAV recovery systems *launch and recovery* (United States).....................408

C

Carbon-Based Technology Inc, Uaver Division
Accipiter *UAVs* (Taiwan) 169
Avian *UAVs* (Taiwan) 169

Cassidian (an EADS company)
Barracuda *UAVs* (Germany)..53
Do-DT25 Mosquito *aerial targets* (Germany) 309
Do-DT35 Hornet *aerial targets* (Germany) 309
Do-DT55 *aerial targets* (Germany)...................................309
Do-SK6 *aerial targets* (Germany) 310
Do-SK10 *aerial targets* (Germany) 311
DVF-2000 *UAVs* (France)...42
Eagle *UAVs* (France)..40
Eclipse *aerial targets* (France) 305
Fox TS1 *aerial targets* (France) 305
Fox TS3 *aerial targets* (France) 306
SETA-3 S1 *aerial targets* (Germany)................................312
Talarion *UAVs* (Germany) 54
Tracker *UAVs* (France)...42

Changzhou Aircraft Factory
Chang Kong 1 *aerial targets* (China)..............................300
CK1 *aerial targets* (China)..300

Chengdu Aircraft Industry (Group) Company
HALE UAV *UAVs* (China)...22
Tianyi *UAVs* (China)...22
Wing Loong *UAVs* (China)..23
Yilong *UAVs* (China)...23

China Aerospace Science and Industry Corporation
LT series *UAVs* (China)...23

China Aerospace Science and Technology Corporation
CH-3 *UAVs* (China)..23
Rainbow-3 *UAVs* (China)...23

China National Aero-Technology Import & Export Corporation
U8E *UAVs* (China)..24

Chung Shan Institute of Science and Technology
Blue Magpie *UAVs* (Taiwan)...167
Cardinal *UAVs* (Taiwan)...168
Chung Shyang II *UAVs* (Taiwan) 168

Comando de Aviación Naval Argentina
Guardian *UAVs* (Argentina)...1

Composite Engineering Inc
BQM-167 *aerial targets* (United States)........................354
CEI SSAT *aerial targets* (United States).........................357
Firejet *aerial targets* (United States).............................355
Kalkara *aerial targets* (United States)...........................355
MQM-107 Streaker *aerial targets* (United States).....355
Skeeter *aerial targets* (United States)..........................354

CTRM Aviation Sdn Bhd
Aludra *UAVs* (Malaysia)...121

Cubic Defense Applications
HIDL *control and communications* (International) 371
TCDL *control and communications* (United States) 388

Curtiss-Wright Controls Embedded Computing
IMMC *control and communications* (United States)......388

CybAero AB
APID 55 *UAVs* (Sweden) 160

Cyber Aerospace
CyberBug *UAVs* (United States) 224
CyberScout *UAVs* (United States)..................................224

Cyberflight Ltd
CyberEye I *UAVs* (United Kingdom)...............................186
DM-65 *UAVs* (United Kingdom)......................................187
E-Swift Eye *UAVs* (United Kingdom).............................188
Zygo *UAVs* (United Kingdom)..188

Cyber Technology (WA) Pty Ltd
CyberEye II *UAVs* (Australia)..5
CyberQuad *UAVs* (Australia)..6
CyberWraith *UAVs* (Australia) 6
CyBird *aerial targets* (Australia)......................................293

D

Dassault Aviation
Neuron *UAVs* (International) 70

Defense Advanced Research Projects Agency
Odysseus *UAVs* (United States) 225
Vulture programme *UAVs* (United States)..................225

Denel Aerospace Systems
Seeker *UAVs* (South Africa)..153
Skua *aerial targets* (South Africa)..................................334

Draganfly Innovations Inc
Draganflyer X4 *UAVs* (Canada) 14
Draganflyer X4/E4 *UAVs* (Canada) 14
Draganflyer X4-ES *UAVs* (Canada)...................................14
Draganflyer X4-P *UAVs* (Canada).....................................14
Draganflyer X6-ES *UAVs* (Canada)...................................14
Draganflyer X8 *UAVs* (Canada) 14
Draganflyer X series *UAVs* (Canada)...............................14
Dragonflyer X6 *UAVs* (Canada)...14

Dragonfly Pictures Inc
Bayonet *UAVs* (United States) 228
DP-6 *UAVs* (United States)...227
DP-11 *UAVs* (United States)...228
DP series *UAVs* (United States)......................................226
Whisper *UAVs* (United States)...227

DRS Defense Solutions
Neptune *UAVs* (United States) 228
RQ-15A *UAVs* (United States) 228

E

EADS 3 Sigma SA
Alkyon *aerial targets* (Greece)..312
Iris Jet *aerial targets* (Greece)...313
Iris Prop *aerial targets* (Greece) 314
Nearchos *UAVs* (Greece)...62
Perseas *aerial targets* (Greece).......................................314

Ejército Argentino
Lipan M3 *UAVs* (Argentina) 1

Elbit Systems Ltd
Hermes 90 *UAVs* (Israel) 83
Hermes 450 *UAVs* (Israel)...84
Hermes 900 *UAVs* (Israel)...85
Hermes 1500 *UAVs* (Israel)...86
Skylark I *UAVs* (Israel)..87
Skylark II *UAVs* (Israel)...88
UGCS *control and communications* (Israel).....................372
WK 450 *UAVs* (Israel)..84

Electromecanica Ploiesti (Elmec)
ATT-01 *aerial targets* (Romania) 327
RT-3 *aerial targets* (Romania) 327

RT-11D *aerial targets* (Romania) 328
TPDM-01 *aerial targets* (Romania) 328

Elisra Electronic Systems Ltd
In-Tact *control and communications* (Israel) 373
MRS-2000 *control and communications* (Israel) 373
MRS-2000M *control and communications* (Israel) 373
SL-RAMBO *control and communications* (Israel) 374
SRST *control and communications* (Israel) 374
StarLink *control and communications* (Israel) 375
TDDL *control and communications* (Israel) 375
V-TVR *control and communications* (Israel) 375

Elta Systems Ltd
EL/K-1861 GDT *control and communications* (Israel) 375
EL/K-1862 CDT *control and communications* (Israel) 376
EL/K-1865 ADT *control and communications* (Israel) 376
EL/K-1871 DGDT *control and communications* (Israel) 377
EL/K-1891 SCS *control and communications* (Israel) 378

EMIT Aviation Consult Ltd
Blue Horizon *UAVs* (Israel) 89
Dragonfly 2000 *UAVs* (Israel) 88
Sparrow *UAVs* (Israel) 89
Sting *UAVs* (Israel) 89

Emmen Aerospace Inc
Condor *UAVs* (United States) 229
Super Swiper *UAVs* (United States) 230
Swiper *UAVs* (United States) 230

EMT Ingenieurgesellschaft
Aladin *UAVs* (Germany) 56
FanCopter *UAVs* (Germany) 57
LUNA *UAVs* (Germany) 57
Museco *UAVs* (Germany) 59

Enator Miltest AB
SM3B *aerial targets* (Sweden) 339
SM6 *aerial targets* (Sweden) 339
SM9A *aerial targets* (Sweden) 340

Engineered Arresting Systems Corporation (ESCO)
HP-2002 *launch and recovery* (United States) 409
HP-3003 *launch and recovery* (United States) 409
HP-3402 *launch and recovery* (United States) 409
HP-3407 *launch and recovery* (United States) 409
HP-3502 *launch and recovery* (United States) 409
MARS *launch and recovery* (United States) 409
Mobilenet® 2000 *launch and recovery* (United States) 409
SPARS *launch and recovery* (United States) 409
STARS *launch and recovery* (United States) 409

ENICS JSC
Berta *UAVs* (Russian Federation) 137
E08 *UAVs* (Russian Federation) 137
E26T *UAVs* (Russian Federation) 138
E95 *aerial targets* (Russian Federation) 330
Eleron *UAVs* (Russian Federation) 137
Igla *UAVs* (Russian Federation) 138

F

FanWing Ltd
FanWing UAVs *UAVs* (United Kingdom) 189

Flight Refuelling Ltd
Falconet II *aerial targets* (United Kingdom) 343
MRTT *aerial targets* (United Kingdom) 344

Flight Technologies
AvantVision *UAVs* (Brazil) 12
FS-01 *UAVs* (Brazil) 13
FS-02 *UAVs* (Brazil) 12
Horus 100 *UAVs* (Brazil) 12
Horus 200 *UAVs* (Brazil) 13
Watchdog *UAVs* (Brazil) 13

Flying Robots SA
FR50 *UAVs* (France) 42
FR101 *UAVs* (France) 42
Téléporteur *UAVs* (France) 42

Flytronic
FlyEye *UAVs* (Poland) 132

Fuji Heavy Industries Ltd, Aerospace Company
J/AQM-1 *aerial targets* (Japan) 320

G

Gen Corporation
H-4 *UAVs* (Japan) 113

General Atomics Aeronautical Systems Inc
Advanced Cockpit GCS *control and communications* (United States) 388
Altair *UAVs* (United States) 231
Avenger *UAVs* (United States) 239
Block 50 Advanced Cockpit *control and communications* (United States) 388
Gnat *UAVs* (United States) 232
Gnat-XP *UAVs* (United States) 232
I-Gnat *UAVs* (United States) 232
I-Gnat ER *UAVs* (United States) 232
Mariner *UAVs* (United States) 236
MQ-1 *UAVs* (United States) 233
MQ-1C *UAVs* (United States) 235
MQ-9 *UAVs* (United States) 236
MQ-9 Reaper *UAVs* (United States) 236
Predator *UAVs* (United States) 233
Predator B *UAVs* (United States) 236
Predator C *UAVs* (United States) 239
RQ-1 *UAVs* (United States) 233
RVT *control and communications* (United States) 389
Sky Warrior *UAVs* (United States) 235
Warrior Alpha *UAVs* (United States) 232

General Dynamics Ordnance and Tactical Systems — Canada Inc
TRAP *aerial targets* (Canada) 297

Geocopter BV
GC-201 *UAVs* (Netherlands) 124

Griffon Aerospace
BroadSword *UAVs* (United States) 240
BroadSword XL *UAVs* (United States) 240
MQM-170A *aerial targets* (United States) 357
MQM-171A *UAVs* (United States) 240
Outlaw *aerial targets* (United States) 357

Guizhou Aviation Industry Group
GAIC Harrier Hawk *UAVs* (China) 25
Soar Dragon *UAVs* (China) 25
WZ-2000 *UAVs* (China) 26
Xianglong *UAVs* (China) 25

H

HELIPSE
HE-190 *UAVs* (France) 44

Herley Industries Inc
C2 systems *control and communications* (United States) 389

Honeywell Defense and Space Electronics Systems
MAV *UAVs* (United States) 241

Hydra Technologies de Mexico
E1 *UAVs* (Mexico) 122
Ehécatl *UAVs* (Mexico) 123
Gavilán *UAVs* (Mexico) 122
S4 *UAVs* (Mexico) 123

Hydromechanical Engineering Company Ltd
HEC-BRK-150 *launch and recovery* (Israel) 401
HEC-BRK-250 *launch and recovery* (Israel) 401
Hydraulic launcher *launch and recovery* (Israel) 401

I

Indra Sistemas SA
Albhatros *UAVs* (Spain) 156
Mantis *UAVs* (Spain) 157
Pelicano *UAVs* (Spain) 157

Infotron
IT 180-5 *UAVs* (France) 45

Innocon Ltd
MicroFalcon I *UAVs* (Israel) 102
MiniFalcon *UAVs* (Israel) 103

Insitu Group Inc
Integrator *UAVs* (United States) 243
RQ-21A *UAVs* (United States) 243
ScanEagle *UAVs* (United States) 217

Institut Kulon NII OAO
Aist *UAVs* (Russian Federation) 144
BLA-06 *UAVs* (Russian Federation) 144

Instituto Nacional de Técnica Aeroespacial
ALBA *aerial targets* (Spain) 335
Diana *aerial targets* (Spain) 336
HADA *UAVs* (Spain) 158
Milano *UAVs* (Spain) 158
SIVA *UAVs* (Spain) 158

Instytut Techniczny Wojsk Lotniczych
Komar *aerial targets* (Poland) 326
SMCP-JU *aerial targets* (Poland) 326
SMCP-WU *aerial targets* (Poland) 326
Szerszen *aerial targets* (Poland) 326

Integrated Defence Systems
Huma-1 *UAVs* (Pakistan) 126

Integrated Dynamics
Border Eagle *UAVs* (Pakistan) 126
Explorer *UAVs* (Pakistan) 127
Hawk *UAVs* (Pakistan) 127
Mk II *aerial targets* (Pakistan) 323
Nishan Mk II *aerial targets* (Pakistan) 323
Nishan TJ-1000 *aerial targets* (Pakistan) 323
Rover *UAVs* (Pakistan) 128
Shadow *UAVs* (Pakistan) 128
TJ-1000 *aerial targets* (Pakistan) 323
Tornado *aerial targets* (Pakistan) 323
Vision *UAVs* (Pakistan) 128

International Aviation Supply SRL
Archimede *UAVs* (Italy) 109
Pico *UAVs* (Italy) 110
Pitagora *UAVs* (Italy) 110
Raffaello *UAVs* (Italy) 110

Iran Aircraft Manufacturing Industries
Ababil *UAVs* (Iran) 73
Karrar *UAVs* (Iran) 74

ISL Aeronautical Systems
ACE *UAVs* (United States) 244
BA-131 *UAVs* (United States) 244
SASS LITE *UAVs* (United States) 245

Israel Aerospace Industries Ltd, Malat Division
AGCS *control and communications* (Israel) 378
Bird Eye 400 *UAVs* (Israel) 90
Eitan *UAVs* (Israel) 93, 95
ETOP *UAVs* (Israel) 91
Heron 1 *UAVs* (Israel) 93
Heron TP *UAVs* (Israel) 95
IAI Ghost *UAVs* (Israel) 91
Machatz *UAVs* (Israel) 93
Mini Panther *UAVs* (Israel) 98
Mosquito *UAVs* (Israel) 96
NRUAV *UAVs* (Israel) 97
Orion *UAVs* (Israel) 95
Panther *UAVs* (Israel) 98
Searcher *UAVs* (Israel) 98
Shoval *UAVs* (Israel) 93

Israel Aerospace Industries, MBT Systems Missiles and Space Group
Cutlass *UAVs* (Israel) 92
Harop *UAVs* (Israel) 92
Harpy *UAVs* (Israel) 92

Israel Military Industries Ltd
ADM-141A TALD *UAVs* (Israel) 100
ADM-141C ITALD *UAVs* (Israel) 100
Delilah *UAVs* (Israel) 101

J

Jordan Advanced Remote Systems
Jordan Arrow *aerial targets* (Jordan) 321
Jordan Falcon *UAVs* (Jordan) 115

K

Kadet Defence Systems
FireBee *UAVs* (India) 67
Javelin X *aerial targets* (India) 316
JX-2 *aerial targets* (India) 317
Trogon *UAVs* (India) 68

Kaman Aerospace Corporation,
K-MAX *UAVs* (United States) 245

Korea Aerospace Industries Ltd
K-UCAV *UAVs* (Korea, South) 116
NI 100 *UAVs* (Korea, South) 117
Night Intruder 100N *UAVs* (Korea, South) 117
Night Intruder 300 *UAVs* (Korea, South) 117

Korea Aerospace Research Institute
Smart *UAVs* (Korea, South) 119

Korean Air Lines
KUS-7 *UAVs* (Korea, South) 118
KUS-9 *UAVs* (Korea, South) 119

L

L-3 Communications Unmanned Systems
FoxCar *UAVs* (United States) 222
TigerShark *UAVs* (United States) 222

L-3 Communication Systems-West
CMDL *control and communications* (United States) 390
Global Hawk ICS *control and communications* (United States) 390
KuSDL *control and communications* (United States) 391
MUDL *control and communications* (United States) 391
ROVER 1 *control and communications* (United States) 391
ROVER 3 *control and communications* (United States) 391
ROVER 4 *control and communications* (United States) 391
ROVER 5 *control and communications* (United States) 391
ROVER 6 *control and communications* (United States) 391

L-3 Unmanned Systems Inc
Evolution *UAVs* (United States) 247
Mobius *UAVs* (United States) 248
Viking 400 *UAVs* (United States) 248

Lockheed Martin Advanced Development Programs
ISIS *UAVs* (United States) 253

Lockheed Martin Aeronautics Company
Desert Hawk *UAVs* (United States) 250
FPASS *UAVs* (United States) 250
MQ-X *UAVs* (United States) 255
RQ-170 *UAVs* (United States) 255
Sentinel *UAVs* (United States) 255
Shot Stalker *UAVs* (United States) 255
Stalker *UAVs* (United States) 255

Lockheed Martin Mission Systems and Sensors
HAA *UAVs* (United States) 252
High Altitude Airship *UAVs* (United States) 252
ISIS *UAVs* (United States) 253

Lockheed Martin Mission Systems and Sensors,
K-MAX *UAVs* (United States) 245

Lutch Design Bureau JSC
Tipchak *UAVs* (Russian Federation) 143

M

MAV6 LLC
MAV6 M1400-I Blue Devil II *UAVs* (United States) 256

Mavionics GmbH
Carolo *UAVs* (Germany) 59
P330 *UAVs* (Germany) 59
T200 *UAVs* (Germany) 59

MBDA
Fire Shadow *UAVs* (United Kingdom) 190
MBDA TiGER *UAVs* (United States) 257

Meggitt Defence Systems Ltd UK
ASR-4 *aerial targets* (United Kingdom) 346
Banshee *aerial targets* (United Kingdom) 346
BTT-1 Imp *aerial targets* (United Kingdom) 345
BTT-3 *aerial targets* (United Kingdom) 346
CASPA *control and communications* (United Kingdom) 385
Crecerelle *aerial targets* (United Kingdom) 346
Hawkeye *aerial targets* (United Kingdom) 346
Petrel *aerial targets* (United Kingdom) 348
Snipe *aerial targets* (United Kingdom) 349
Spectre *aerial targets* (United Kingdom) 346
Voodoo *aerial targets* (United Kingdom) 350
Wizard *control and communications* (United Kingdom) 385

Meggitt Defense Systems Inc
GT-400 *aerial targets* (United States) 358
RM-12A *launch and recovery* (United States) 409
RM-24 *launch and recovery* (United States) 409
RM-30A *launch and recovery* (United States) 409
RM-30A1 *launch and recovery* (United States) 409
RM-30B *launch and recovery* (United States) 409
RM-62 *launch and recovery* (United States) 409
RMK-35 *launch and recovery* (United States) 409
SGT-20 *aerial targets* (United States) 358
TDK-39 *aerial targets* (United States) 359
TGT *aerial targets* (United States) 359
TGX *aerial targets* (United States) 359
TIX *aerial targets* (United States) 359
TLX-1 *aerial targets* (United States) 359
tow targets *aerial targets* (United States) 359
TPT *aerial targets* (United States) 359
TRX *aerial targets* (United States) 359

Meggitt Training Systems Canada
Pop-up Helicopter *aerial targets* (Canada) 298
Vindicator II *aerial targets* (Canada) 299

Microdrones GmbH
MD4-200 *UAVs* (Germany) 60
MD4-1000 *UAVs* (Germany) 60

MicroPilot
CropCam *UAVs* (Canada) 15
MP1028g *control and communications* (Canada) 369
MP2028g *control and communications* (Canada) 369
MP2028XP *control and communications* (Canada) 369
MP2128g *control and communications* (Canada) 369
MP2128HELI *control and communications* (Canada) 369
MP2128LRC *control and communications* (Canada) 369
MP21283x *control and communications* (Canada) 369
MP CAT *launch and recovery* (Canada) 399
MP-Vision *UAVs* (Canada) 15
trueHWIL *control and communications* (Canada) 369

Micro Systems Inc
MONTAGE *control and communications* (United States) 392

Mil Helicopters
Mi-34BP *UAVs* (Russian Federation) 144

Mission Technologies Inc
Buster *UAVs* (United States) 258

Mist Mobility Integrated Systems Technology Inc
CQ-10A Snow Goose *UAVs* (Canada) 16
CQ-10B SnowGoose *UAVs* (Canada) 16
Sherpa *UAVs* (Canada) 18
Sherpa™ Provider 600 *UAVs* (Canada) 18
Sherpa™ Provider 1200 *UAVs* (Canada) 18
Sherpa™ Provider 2200 *UAVs* (Canada) 18
Sherpa™ Provider 10,000 *UAVs* (Canada) 18
Sherpa™ Ranger *UAVs* (Canada) 18
SnowGoose™ *UAVs* (Canada) 16

MKU Pvt Ltd
Bullseye *aerial targets* (India) 317
Erasmus *UAVs* (India) 68
TERP *UAVs* (India) 68

MLB Company
Bat *UAVs* (United States) 259
MLB Super Bat *UAVs* (United States) 259

N

National Development Complex
Vector *UAVs* (Pakistan) 129

Naval Research Laboratory
Dragon Eye *UAVs* (United States) 278
FINDER *UAVs* (United States) 277
Ion Tiger *UAVs* (United States) 278
Sea ALL *UAVs* (United States) 278

Navmar Applied Sciences Corporation
FoxCar *UAVs* (United States) 222
TigerShark *UAVs* (United States) 222

Neural Robotics Inc
AutoCopter *UAVs* (United States) 275
Expedition *UAVs* (United States) 275
Explorer *UAVs* (United States) 275
Express *UAVs* (United States) 275

NO ORGANISATION DATA
Flexrotor *UAVs* (United States) 203
R-50 *UAVs* (Japan) 113
RMAX *UAVs* (Japan) 113

Northrop Grumman
Fire Scout *UAVs* (United States) 263
Fire-X *UAVs* (United States) 263
MQ-8C *UAVs* (United States) 263

Northrop Grumman Aerospace Systems
Demonstrator Firebird *UAVs* (United States) 261
HA304 *UAVs* (United States) 263
LEMV *UAVs* (United States) 263
Model 355 *UAVs* (United States) 261
Production-Ready Firebird *UAVs* (United States) 261

Northrop Grumman Corporation, Navigation Systems Division
LN-100G *control and communications* (United States) 393
LN-200 *control and communications* (United States) 393
LN-251 *control and communications* (United States) 393

Northrop Grumman Integrated Systems
X-47 *UAVs* (United States) 274

Northrop Grumman Unmanned Systems
Bat *UAVs* (United States) 260
B-Hunter *UAVs* (International) 70
BQM-34 *aerial targets* (United States) 361
BQM-74 *aerial targets* (United States) 363
BQM-74F *aerial targets* (United States) 363
Chukar *aerial targets* (United States) 363
Euro Hawk *UAVs* (United States) 269
Firebee I *aerial targets* (United States) 361
Fire Scout *UAVs* (United States) 264
Global Hawk *UAVs* (United States) 269
Hunter *UAVs* (International) 70
MQ-5 *UAVs* (International) 70
MQ-8 *UAVs* (United States) 264
MQ-8B *UAVs* (United States) 264
MQM-74 *aerial targets* (United States) 363
RQ-4 *UAVs* (United States) 269
RQ-5 *UAVs* (International) 70
RQ-8A *UAVs* (United States) 264

Nostromo Defensa SA
Cabure *UAVs* (Argentina) 2
Yarara *UAVs* (Argentina) 2

O

OKT Norge AS
10/150 *launch and recovery* (Norway) 402

Orbital Sciences Corporation, Launch Systems Group
Coyote *aerial targets* (United States) 365
GQM-163A *aerial targets* (United States) 365

P

Pakistan Aeronautical Complex, Aircraft Manufacturing Factory
Ababeel *aerial targets* (Pakistan) 324
Baaz *aerial targets* (Pakistan) 324

Patria Aerostructures Oy
MASS *UAVs* (Finland) 35
Modular Airborne Sensor System *UAVs* (Finland) 35

Piaggio Aero Industries SpA,
P.1HH HammerHead *UAVs* (Italy) 111

Piasecki Aircraft Corporation
Turais *UAVs* (United States) 279
WBBL-UAV *UAVs* (United States) 279

Prioria Robotics Inc
Maveric *UAVs* (United States) 279

Proxy Aviation Systems Inc
SkyRaider *UAVs* (United States) 280
SkyWatcher *UAVs* (United States) 281

Q

QinetiQ
Zephyr *UAVs* (United Kingdom) 190

Qods Aviation Industries
Dorna *UAVs* (Iran) 75
Hodhod *UAVs* (Iran) 75
Mohadjer *UAVs* (Iran) 75
Saeghe *aerial targets* (Iran) 318
Talash *UAVs* (Iran) 76

R

Rafael Advanced Defense Systems Ltd
DLV-52 *control and communications* (Israel) 378

Rafael Armament Development Authority Ltd
Black Sparrow *aerial targets* (Israel) 319

Raytheon Company, Intelligence & Information Systems
AN/MSQ-131 GHGS *control and communications* (United States) 393
TCS *control and communications* (United States) 395

Raytheon Missile Systems
ADM-160B *UAVs* (United States) 283
ADM-160C *UAVs* (United States) 283
Cobra *UAVs* (United States) 282
KillerBee *UAVs* (United States) 282
MALD *UAVs* (United States) 283
MALD-V *UAVs* (United States) 283
MCALS *UAVs* (United States) 283

Raytheon Space and Airborne Systems
ISIS *UAVs* (United States) 253

Rheinmetall Defence Electronics GmbH
KZO *UAVs* (Germany) 61
Prospector *UAVs* (Germany) 61

Robonic Oy Ltd
MC91 *launch and recovery* (Finland) 399
MC0315L *launch and recovery* (Finland) 399
MC2055L *launch and recovery* (Finland) 399
MC2555L *launch and recovery* (Finland) 399
MC2555LLR *launch and recovery* (Finland) 399
RATUFC *launch and recovery* (Finland) 399

Rockwell Collins
Athena 111 *control and communications* (United States) 395
Athena 111m *control and communications* (United States) 395
Athena 311 *control and communications* (United States) 396
Athena 411 *control and communications* (United States) 396
Athena 511 *control and communications* (United States) 397
GuS™ *control and communications* (United States) 397

RUAG Aviation
ADS 95 *UAVs* (Switzerland) 162
Archer *launch and recovery* (Switzerland) 405
OPATS *control and communications* (Switzerland) 384
Ranger *UAVs* (Switzerland) 162

Russian Aircraft Corporation (RSK) MiG
Skat *UAVs* (Russian Federation) 143

S

Saab Aerosystems AB
Skeldar *UAVs* (Sweden) 161
V-150 *UAVs* (Sweden) 161

Sagem Défense Sécurité
Busard *UAVs* (France) 46
CU-161 *UAVs* (France) 47
Patroller *UAVs* (France) 46
Sperwer A *UAVs* (France) 47
Sperwer B *UAVs* (France) 47, 49
Tårnfalken *UAVs* (France) 47
Ugglan *UAVs* (France) 47

SAS Fly-n-Sense
CB 350 *UAVs* (France) 44
CB 750 *UAVs* (France) 44
FNS 900 *UAVs* (France) 44
Scancopter CB 350 *UAVs* (France) 44
Scancopter CB 750 *UAVs* (France) 44
Seeker FNS 900 *UAVs* (France) 44

Scaled Composites LLC
Proteus *UAVs* (United States) 284

Schiebel Elektronische Geräte GmbH
Al Saber *UAVs* (Austria) 8
Camcopter 5.1 *UAVs* (Austria) 7
Camcopter S-100 *UAVs* (Austria) 8
S-100 *UAVs* (Austria) 8

Scientifically Industrial Systems Ltd
A-3 Remez *UAVs* (Ukraine) 171
A-4 Albatros *UAVs* (Ukraine) 172
A-5 *UAVs* (Ukraine) 173
A-10 *UAVs* (Ukraine) 173
A-11 *aerial targets* (Ukraine) 341
A-12 *UAVs* (Ukraine) 173
A-160 *UAVs* (Ukraine) 174
CGM-6C *launch and recovery* (Ukraine) 406
CLLGW-4C *launch and recovery* (Ukraine) 406
CLLTS-3/-3C *launch and recovery* (Ukraine) 406
CTP-3 *launch and recovery* (Ukraine) 406
Hurricane *UAVs* (Ukraine) 173
Phoenix *UAVs* (Ukraine) 173
Sea Eagle *UAVs* (Ukraine) 173
Swift *aerial targets* (Ukraine) 341

Scientific Production Corporation, Irkut JSC
Irkut-1A *UAVs* (Russian Federation) 138
Irkut-2F *UAVs* (Russian Federation) 139
Irkut-2M *UAVs* (Russian Federation) 139
Irkut-2T *UAVs* (Russian Federation) 139
Irkut-3 *UAVs* (Russian Federation) 140
Irkut-10 *UAVs* (Russian Federation) 140
Irkut-20 *UAVs* (Russian Federation) 140
Irkut-200 *UAVs* (Russian Federation) 141
Irkut-850 *UAVs* (Russian Federation) 141

Secapem SA
A3GT *aerial targets* (France) 307
MDI 154 *aerial targets* (France) 307
MDI189 *control and communications* (France) 370

S-T > Manufacturers' Index

PR53 *aerial targets* (France) 308
PR55 *aerial targets* (France) 308
Taxan *aerial targets* (France) 308

Selex Galileo
Alamak *control and communications* (Italy) 379
Asio *UAVs* (United Kingdom) 191
Damselfly *UAVs* (United Kingdom) 192
Falco *UAVs* (Italy) 111
Falco GCS *control and communications* (Italy) 379
Locusta *aerial targets* (Italy) 319
Mirach 100/5 *aerial targets* (Italy) 319
Otus *UAVs* (United Kingdom) 193

Sierra Nevada Corporation
PLS *launch and recovery* (United States) 410
TALS *launch and recovery* (United States) 411
UCARS-V2 *launch and recovery* (United States) 411

Singapore Technologies Aerospace Ltd
Skyblade IV *UAVs* (Singapore) 149

Sistemas de Control Remoto
Scrab I *aerial targets* (Spain) 337
Scrab II *aerial targets* (Spain) 337

Sky-Watch
Higinn X1 ground station *control and communications* (Denmark) 370
Huginn X1 *UAVs* (Denmark) 35

Sokol Experimental Design Bureau
Dan *aerial targets* (Russian Federation) 331
Danem *aerial targets* (Russian Federation) 331

ST Aerospace Ltd
Skyblade II *UAVs* (Singapore) 148
Skyblade III *UAVs* (Singapore) 148

Steadicopter Ltd
Helivision *UAVs* (Israel) 103

Strela Orenburg Production Association (PO Strela)
La-17 *aerial targets* (Russian Federation) 332

Surveillance and Target Unmanned Aircraft
Flamingo *UAVs* (Pakistan) 130
HST *UAVs* (Pakistan) 130
Jasoos *UAVs* (Pakistan) 130
Mukhbar *UAVs* (Pakistan) 130
Parwaz *UAVs* (Pakistan) 130
Shooting Star *aerial targets* (Pakistan) 325
Stingray *UAVs* (Pakistan) 131
Tunder *aerial targets* (Pakistan) 325

Survey-Copter
37M *UAVs* (France) 50
Bicopt CH *UAVs* (France) 49
Blimp 37M *UAVs* (France) 50
Copter 1B *UAVs* (France) 50
Copter City *UAVs* (France) 50
DVF 2000 *UAVs* (France) 51

Swift Engineering Inc
KillerBee *UAVs* (United States) 286

SwissCopter AG/UASystems Ltd
NT 150 *UAVs* (Switzerland) 163
Tip-Jet DragonFly *UAVs* (Switzerland) 163

Swiss UAV
Koax *UAVs* (Switzerland) 164
Neo S-300 *UAVs* (Switzerland) 164
S-300 *UAVs* (Switzerland) 164
X-240 Mk II *UAVs* (Switzerland) 164

T

Tactical Aerospace Group
M2600 *UAVs* (United States) 286

Tasuma (UK) Ltd
A3 Observer launcher *launch and recovery* (United Kingdom) 407

CSV 40 *UAVs* (United Kingdom) 193
CSV 50 *UAVs* (United Kingdom) 193
CSV 65 *UAVs* (United Kingdom) 193
CSV X0 *UAVs* (United Kingdom) 193
Hawkeye *UAVs* (United Kingdom) 194
Hawkeye III *UAVs* (United Kingdom) 194
LTL 1 *launch and recovery* (United Kingdom) 407
TML 2 *launch and recovery* (United Kingdom) 407
TML 3 *launch and recovery* (United Kingdom) 407
TML 3 Ultima *launch and recovery* (United Kingdom) 407
TML 4 *launch and recovery* (United Kingdom) 407

TCOM LP
model 15M *UAVs* (United States) 287
model 17M® *UAVs* (United States) 287
model 22+™ *UAVs* (United States) 287
model 22M™ *UAVs* (United States) 287
model 25M *UAVs* (United States) 287
model 28M *UAVs* (United States) 287
model 31M *UAVs* (United States) 287
model 32M® *UAVs* (United States) 287
model 38M® *UAVs* (United States) 287
model 53M/System 250 *UAVs* (United States) 287
model 67M/System 365/System 365H *UAVs* (United States) 287
model 71M® *UAVs* (United States) 287
model 74M™ *UAVs* (United States) 287
model 76M *UAVs* (United States) 287
tethered aerostats *UAVs* (United States) 287

Tekplus Aerospace
C 18 *UAVs* (Spain) 159
C 30 *UAVs* (Spain) 159
C 50 *UAVs* (Spain) 159

Tellumat (Pty) Ltd, Defence Division
CBACS *control and communications* (South Africa) 380

Thales France
MAGIC ATOLS *control and communications* (France) 371
Spy Arrow *UAVs* (France) 51

The Boeing Company, Boeing Defense, Space and Security
A/MH-6X *UAVs* (United States) 216
Little Bird *UAVs* (United States) 216
Phantom Eye *UAVs* (United States) 215
ScanEagle *UAVs* (United States) 217
ULB *UAVs* (United States) 216
X-48B *UAVs* (United States) 219
X-48C *UAVs* (United States) 219

The Center for Interdisciplinary Remotely Piloted Aircraft Studies
Pelican *UAVs* (United States) 223

Top I Vision
Calineczka *UAVs* (Israel) 104, 105
Casper 200 *UAVs* (Israel) 104
Casper 250 *UAVs* (Israel) 104
Casper 350 *UAVs* (Israel) 105
Rufus *UAVs* (Israel) 105
Sofar *UAVs* (Israel) 104

Track System a.s.
Heros *UAVs* (Czech Republic) 32

Transas Avia
Dozor *UAVs* (Russian Federation) 145

Tupolev PSC
Tu-143 Reis *UAVs* (Russian Federation) 145
Tu-243 Reis *UAVs* (Russian Federation) 145
Tu-300 Korshun *UAVs* (Russian Federation) 145

Turkish Aerospace Industries Inc,
Turna *aerial targets* (Turkey) 340

Turkish Aerospace Industries Inc
Anka *UAVs* (Turkey) 170
Anka Block A *UAVs* (Turkey) 170
Anka Block B *UAVs* (Turkey) 170
R-300 *UAVs* (Turkey) 170
Simsek *aerial targets* (Turkey) 340
Sivrisinek *UAVs* (Turkey) 170

U

UAV Factory Ltd Europe
Penguin B *UAVs* (United States) 290

UAV Factory Ltd Europe,
Launch systems *launch and recovery* (United States) 413
portable GCS *control and communications* (United States) 397

UAV Factory USA LLC,
Launch systems *launch and recovery* (United States) 413
Penguin B *UAVs* (United States) 290
portable GCS *control and communications* (United States) 397

UAV Navigation
AP04 *control and communications* (Spain) 381
AP04OEM2 *control and communications* (Spain) 381
AP04R *control and communications* (Spain) 381
AP04SDK *control and communications* (Spain) 381
GCS03 *control and communications* (Spain) 382
VISIONAIR *control and communications* (Spain) 382

Uconsystem Company Ltd
RemoEye 002 *UAVs* (Korea, South) 120
RemoEye 006 *UAVs* (Korea, South) 120
RemoH-C100 *UAVs* (Korea, South) 121

Ultra Electronics
HIDL *control and communications* (International) 371

Universal Target Systems Ltd
AVL *launch and recovery* (United Kingdom) 407
GSAT-200 NG *aerial targets* (United Kingdom) 351
LWL *launch and recovery* (United Kingdom) 408
MSAT-500 *aerial targets* (United Kingdom) 352
MSAT-500 NG *aerial targets* (United Kingdom) 352
Spotter *UAVs* (United Kingdom) 196
TCTS *control and communications* (United Kingdom) 385
Vigilant *UAVs* (United Kingdom) 196

Unmanned Systems AG
Unmanned Systems CT-450 Discoverer I *UAVs* (Switzerland) 165
Unmanned Systems Orca *UAVs* (Switzerland) 165

Urban Aeronautics Ltd
AirMule *UAVs* (Israel) 106

U-TacS (UAV Tactical Systems Ltd), Thales UK and Elbit Systems joint company,
Watchkeeper *UAVs* (United Kingdom) 194

UVision Air Ltd
UVision Air Wasp *UAVs* (Israel) 106

V

VITO NV
Mercator *UAVs* (Belgium) 11

V-TOL Aerospace Pty Ltd
Warrigal *UAVs* (Australia) 7

W

WorkFly
EyesFly *UAVs* (France) 52
MV5B1 *UAVs* (France) 52

X

Xian ASN Technical Group
ASN-7 *aerial targets* (China) 301
ASN-9 *aerial targets* (China) 302
ASN-12 *aerial targets* (China) 303
ASN-15 *UAVs* (China) 27
ASN-104 *UAVs* (China) 27
ASN-104 GCS *control and communications* (China) 370
ASN-105B *UAVs* (China) 27
ASN-105B GCS *control and communications* (China) 370
ASN-206 *UAVs* (China) 28
ASN-207 *UAVs* (China) 29
ASN-209 *UAVs* (China) 29
Xian ASN-106 *aerial targets* (China) 303
Xian ASN-211 *UAVs* (China) 30

Y

Yamaha Motor Co Ltd
YACS *control and communications* (Japan) 380

Yuhe Group Company Ltd
BJ9906 *aerial targets* (China) 304